Palaeoclimatology

## The Geological Time Scale for the Phanerozoic Eon and the Neoproterozoic Era

| Era | Period | Epoch | Base Age (Ma) |
|---|---|---|---|
| Cenozoic[a] | Quaternary | Holocene | 0.0117 |
| | | Pleistocene | 2.583 |
| | Neogene | Pliocene | 5.3 |
| | | Miocene | 23.0 |
| | Palaeogene | Oligocene | 33.9 |
| | | Eocene | 56.0 |
| | | Palaeocene | 66.0 |
| Mesozoic | Cretaceous | Upper | 100.5 |
| | | Lower | 145.0 |
| | Jurassic | Upper | 163.5 |
| | | Middle | 174.1 |
| | | Lower | 201.3 |
| | Triassic | Upper | 237.0 |
| | | Middle | 247.2 |
| | | Lower | 251.9 |
| Palaeozoic | Permian | Upper | 259.1 |
| | | Middle | 273.0 |
| | | Lower | 299.0 |
| | Carboniferous | Upper | 323.2 |
| | | Lower | 359.0 |
| | Devonian | Upper | 382.7 |
| | | Middle | 393.3 |
| | | Lower | 419.2 |
| | Silurian | | 443.8 |
| | Ordovician | | 485.4 |
| | Cambrian | | 541.0 |
| Proterozoic | Neoproterozoic | | 1000.0 |

[a]During the nineteenth century, geological time was divided into Primary, Secondary, Tertiary and Quaternary Eras. Mesozoic and Palaeozoic strata were regarded as belonging to the Secondary Era. The Tertiary was equivalent to the Cenozoic Era without the Quaternary. Those older designations were done away with in the latter part of the twentieth century, although *Tertiary Era* is often misused for *Cenozoic Era*. Of the older terms, Quaternary has managed to hang on in the form of a geological Period.

From Cohen, K.M., Finney, S.C., Gibbard, P.L., and Fan, J.-X. (2013; updated). The ICS International chronostratigraphic chart. *Episodes* 36: 199–204. (http://www.stratigraphy.org/ICSchart/ChronostratChart2018–08.pdf).

# Palaeoclimatology

From Snowball Earth to the Anthropocene

*Colin P. Summerhayes*

*Scott Polar Research Institute*
*Cambridge University*
*United Kingdom*

*Registered Offices*
John Wiley & Sons, Inc., 111 River Street, Hoboken, NJ 07030, USA
John Wiley & Sons Ltd, The Atrium, Southern Gate, Chichester, West Sussex, PO19 8SQ, UK

*Editorial Office*
9600 Garsington Road, Oxford, OX4 2DQ, UK

For details of our global editorial offices, customer services, and more information about Wiley products visit us at www.wiley.com.

Wiley also publishes its books in a variety of electronic formats and by print-on-demand. Some content that appears in standard print versions of this book may not be available in other formats.

*Library of Congress Cataloging-in-Publication Data*

Names: Summerhayes, C. P., author.
Title: Palaeoclimatology : from snowball earth to the anthropocene / Colin
    P. Summerhayes, Scott Polar Research Institute, Cambridge University.
Description: First edition. | Hoboken, NJ : Wiley-Blackwell, 2020. |
    Includes bibliographical references and index.
Identifiers: LCCN 2020008857 (print) | LCCN 2020008858 (ebook) | ISBN
    9781119591382 (paperback) | ISBN 9781119591474 (adobe pdf) | ISBN
    9781119591504 (epub)
Subjects: LCSH: Paleoclimatology.
Classification: LCC QC884 .S917 2020 (print) | LCC QC884 (ebook) | DDC
    551.609–dc23
LC record available at https://lccn.loc.gov/2020008857
LC ebook record available at https://lccn.loc.gov/2020008858

Cover Design: Wiley
Cover Image: A view of some the numerous small crevassed glaciers typical of the Antarctic Peninsula, which are seen here cutting across the Mid-Jurassic to Lower Cretaceous volcanic rocks of the exposed magmatic core of the ancient island arc underlying the Peninsula, on the 700 m high east side of the northern entrance to the Lemaire Channel (author photo, January 2017).

Set in 9.5/12.5pt STIXTwoText by SPi Global, Pondicherry, India
Printed and bound in Singapore by Markono Print Media Pte Ltd

10   9   8   7   6   5   4   3   2   1

*To my grandchildren, Reid, Torrin, and Jove Cockrell, and Zoe, Jake, Phoebe and Alexa Summerhayes, in the hope that you can work towards freeing the future from the negative aspects of anthropogenic climate change.*

# Contents

# Author Biography

Colin Summerhayes is an Emeritus Associate of the Scott Polar Research Institute of Cambridge University. He has carried out research on aspects of past climate change in both academia and industry: at the New Zealand Oceanographic Institute; Imperial College London; the University of Cape Town; the Woods Hole Oceanographic Institution; the UK's Institute of Oceanographic Sciences Deacon Laboratory (IOSDL); the UK's Southampton (now National) Oceanography Centre; the Exxon Production Research Company; and the BP Research Company. He has managed research programmes on climate change for the UK's Natural Environment Research Council, the Intergovernmental Oceanographic Commission (IOC) of UNESCO, and the Scientific Committee on Antarctic Research (SCAR) of the International Council for Science. He is a former director of both the IOSDL and SCAR, and of the IOC's Global Ocean Observing System Project (a component of the Global Climate Observing System). He has written or co-edited several books on aspects of past or modern climate, including North Atlantic Palaeoceanography (1986), Upwelling Systems: Evolution Since the Early Miocene (1992), Upwelling in the Oceans (1995), Oceanography: an Illustrated Guide (1996), Understanding the Oceans (2001), Oceans 2020: Science, Trends and the Challenge of Sustainability (2002), Antarctic Climate Change and the Environment (2009), Understanding Earth's Polar Challenges: International Polar Year 2007–2008 (2011), Earth's Climate Evolution (2015), and The Anthropocene as a Geological Time Unit (2019).

# Acknowledgement

Research for an all-embracing book like this is impossible without the help of many people. In particular, I thank Julian Dowdeswell for facilitating my research by making me an Emeritus Associate of Cambridge University's Scott Polar Research Institute. Many other individuals donated time and effort to helping me with advice or materials or discussions along the way. They included Jane Francis (Leeds, then BAS), Alan Haywood (Leeds), Joe Cann (Leeds), Anthony Cohen (Open University), Rob Larter (BAS), Eric Wolff (BAS, then Cambridge U), John Lowe (Royal Holloway), Nick McCave (Cambridge), Paul Pearson (Cardiff), Paul Valdes (Bristol), Richard Alley (Penn State), Keith Alverson (then UNESCO, Paris), David Archer (Chicago), Mike Arthur (Penn. State), Peter Barrett (Wellington, NZ), Andre Bérger (Louvain, Belgium), Bob Berner (Yale, now deceased), Bob Binschadler (then NASA), David Bottjer (Los Angeles), Tom Bracegirdle (Cambridge), Diana Clements (London), Michel Crucifix (Louvain), Peter Dexter (then Melbourne), Marie Edmonds (Cambridge), Ian Fairchild (Birmingham), Richard Fortey (British Museum of Natural History), Terrence Gerlach (US Geological Survey), John Gould (Southampton), Joanna Haigh (London), Vicki Hammond (Edinburgh), Brian Hoskins (Imperial, London), Ian Jamieson (Woking, now deceased), Jim Kennett (Santa Barbara), Jeremy Leggett (London), Tim Lenton (Exeter), Cherry Lewis (Bristol), Peter Liss (Norwich), Alan Lord (London), Bryan Lovell (Cambridge), Cornelia Lüdecke (Munich), Valérie Masson-Delmotte (Gif-sur-Yvette), Paul Mayewski (Maine), Malcolm Newell (IMareST), Jerry North (Texas A & M), Judy Parrish (Idaho), Raymond Pierrehumbert (Oxford), Clement Poirier (Caen), Chris Rapley (UCL), Ralph Rayner (IMarEST, London), Dana Royer (Wesleyan University), Ed Sarukhanian (then WMO, Geneva), Chris Scotese (Arlington, Texas), Emily Shuckburgh (Oxford and Cambridge), Martin Siegert (Bristol, then London), Mike Sparrow (Cambridge and WMO Geneva), Will Steffen (Canberra), Eric Steig (Washington State), Iain Stewart (Plymouth), Bryan Storey (Christchurch, NZ), Jai Syvitski (Colorado), Pieter Tans (NOAA), Jörn Thiede (Kiel), John Turner (BAS), Heinz Wanner (Bern), Colin Waters (Leicester), Andy Watson (then Norwich), Mark Williams (Leicester), Phil Woodworth (Liverpool), Jan Zalasiewicz (Leicester), and others who I have failed to remember.

Dialogue is important for comprehensive understanding, and in that context I have enjoyed discussions and e-mails over the years with somewhat sceptical acquaintances particularly Bob Carter (now deceased), Howard Dewhirst, Roger Dewhurst, Henry Dodwell, Don Easterbrook, Ashley Francis, David Gee, Bob Heath, Lydia Holden, David Jenkins, Malcolm McClure, Euan Mearns, Ross Muir, and Mike Ridd amongst others. My search for meaning has led me to read widely in the literature sceptical of one or other aspect of man-made global warming. Besides that, I have attended London lectures by sceptical scientists including Richard Lindzen (then at MIT), and Matt Ridley, participated in 2013 in a public debate with sceptics Bob Carter and Vincent Courtillot on the website of the Global Warming Policy Foundation (GWPF), and had a piece published in 2016 on Euan Mearns' web site *Energy Matters*, besides contributing comments on that and other blog sites. I have corresponded with climate change sceptics from the geological world through the letters column of *Geoscientist*, the monthly magazine of the Geological Society of London, and with them and others through well over 1000 e-mails since the publication of the Geological Society's *Statement on Climate Change* in 2010 and its update in 2013. These interactions have helped me to tighten up my arguments and to alert me to scientific papers that I might not have come across otherwise.

I apologize if I have inadvertently left anyone out. Any errors in the text are my own. But I should also thank the staff of WILEY/Blackwell for their assistance in producing a high quality product from my efforts.

Last but not least, I would like to thank my long-suffering wife Diana for putting up with my mental absences on Planet Climate. Yes, that's where I was in those moments when my eyes were glazed and I didn't hear the question.

Readers familiar with my earlier WILEY book *Earth's Climate Evolution* (2015), will notice some similarities, not least that the basic structure of the new book remains the same as its predecessor. However, the new book is not only much more detailed and more extensively illustrated than its predecessor, but also extends the range of that earlier book back in time from 450 million to 800 million years ago, and forward in time into the proposed Anthropocene Epoch. In addition it now includes learning targets appropriate for a textbook, and can be readily adapted as the basis for a university course in Palaeoclimatology.

# 1

# Introduction

**LEARNING OUTCOMES**

- Understand that palaeoclimatology is concerned with the development of Earth's climate as measured by using proxies for environmental variables.
- Know about some aspects of the history of palaeoclimatology and how it emerged as a field.
- Understand some of the questions that underpin studies of palaeo climate.
- Understand the relevance of palaeoclimate studies to an understanding of modern and possible future climate states.

## 1.1  What Is Palaeoclimatology?

Information about the climate of the past is referred to as palaeoclimate data (American spelling drops the second 'a'). It relies on the use of proxies of climate variables to estimate the past behaviour of Earth's ocean–atmosphere-ice system. Whilst proxy measures have their drawbacks, the fact that they mutually support one another provides a measure of confidence in palaeoclimatology.

The study of past climates used to be the exclusive province of geologists. They would interpret past climate from the character of rocks – coals represented humid climates; polished three-sided pebbles and cross-bedded red-stained sands represented deserts; grooved rocks indicated the passage of glaciers; corals indicated tropical conditions, and so on. Since the 1950s we have come to rely as well on geochemists using oxygen isotopes and the ratios of elements like magnesium to calcium (Mg/Ca) to tell us about past ocean temperatures. And in recent years we have come to realize that cores of ice contain detailed records of past climate change as well as bubbles of fossil air. Glaciologists have joined the ranks. Nowadays the study of our climate system is the province of Earth System Science [1], a topic that treats the Earth's surface as parts of an integrated system in which everything is connected, much as Alexander von Humboldt first suggested back in the early 1800s.

Geologists are fond of saying 'the climate is always changing'. They are correct, but that ignores the all-important question – why? What we really need to know is - how has the climate been changing, at what rates, with what regional variability, and in response to what driving forces? With the answers to those questions we can establish with reasonable certainty what the natural variability of Earth's climate is, and determine how it is most likely to evolve as we pump more greenhouse gases into the atmosphere. This book attempts to answer those questions in a way that should be readily understandable to anyone with a basic scientific education. It describes a voyage of discovery by scientists obsessed with exposing the deepest secrets of our changing climate through time. I hope that readers will find the tale as fascinating as I found the research that went into it.

The drive to understand climate change is an integral part of the basic human urge to understand our surroundings. As in all fields of science, the necessary knowledge to underpin that understanding accumulates gradually. At first we see dimly, but eventually the subject matter becomes clearer. The process is a journey through time in which each generation makes a contribution. Imagination and creativity play their parts. The road is punctuated by intellectual leaps. Exciting discoveries change its course from time to time. No one person could have discovered in his or her lifetime what we now know about the workings of the climate system. Thousands of scientists have added their pieces to the puzzle. Developing our present picture of how the climate system has worked through time has required contributions from an extraordinary range of different scientific disciplines from astronomy to zoology. The breadth of topics that must be understood for us to have a complete picture has made the journey slow, and still makes full understanding of climate change and global warming difficult to grasp for those not committed to serious investigation of a very wide ranging literature. The pace of advance is relentless, and for many it is difficult to keep up. And yet, as with most fields of scientific enquiry, there is still much to learn – mostly, these days, about progressively finer levels of detail. Uncertainties remain. We will never know everything. But we do know enough to make reasonably confident statements about what is happening now and what is likely to happen next. Looking back at the progress that has been made is like watching a time-lapse film of the opening of a flower. Knowledge of the climate system unfolds through time until we find ourselves at the doorstep of the present day and looking at the future.

This book is the story of climate change as revealed by the geological record of the past 800 million years (800 Ma). It is a story of curiosity about how the world works, and ingenuity in tackling the almost unimaginably large challenge of understanding climate change. The task is complicated by the erratic nature of the geological record. Geology is like a book whose pages recount tales of Earth's history. Each copy has some pages missing. Fortunately the American, African, Asian, Australasian, and European editions all miss different pages. Combining them lets us assemble a good picture of how Earth's climate has changed through time. Year-by-year the picture becomes clearer as researchers develop new methods to probe its secrets.

## 1.2 What Can Palaeoclimatology Tell Us About Future Climate Change?

Whilst the story of Earth's climate evolution has a great deal to teach us, it is largely ignored in the ongoing debate on global warming. And yet the examination of the past to tell us what the future may hold is not a new idea. It was first articulated in 1795 by one of the *fathers of geology*, James Hutton. But it is not something the general public hears much about when it comes to understanding global warming. This book is a wake-up call, introducing the reader to what the geological record tells us.

Increasing scrutiny of the palaeoclimate record over the past few decades helps us to explain why our present climate is the way it is. Most of the fluctuation from warm to cold climates and back through time takes place because of changes in the balance of Earth's interior processes. Changes over millions of years involve periods of excessive volcanic activity associated with the break up and drift of continents, which fills the air with $CO_2$ and keeps the climate warm, while periods of continental collision build

mountains and encourage the chemical weathering of exposed terrain that sucks $CO_2$ out of the atmosphere, to keep the climate cool. Continental drift moves continents through climatic zones, sometimes leaving them in the tropics, sometimes at the poles. It also changes the locations of the ocean currents that transport heat and salt around the globe. Individual volcanic eruptions large enough to eject dust into the stratosphere provide short-term change from time to time, while the equally erratic but more persistent volcanic activity of large igneous provinces involving the eruption of millions of cubic metres of lava over a period of a million years or so can change the climate for longer periods and did so enough at times to cause substantial biological extinctions.

External changes are important too. The Sun is the climate's main source of energy. Orbital variations in the Earth's path around the sun, combined with regular changes in the tilt of the Earth's axis, superimpose additional change on these millions of years-long changes, through cycles lasting 400 000 to 20 000 years (400 to 20 Ka). Variations in the Sun's output superimpose yet another series of changes, with variability at millennial, centennial, and decadal scales, although their impact is surprisingly small. Examples of these include the 11-year sunspot cycle and its occasional failure. Best known of those failures was the Maunder Minimum between 1645 and 1715 AD at the heart of the Little Ice Age. Large but rare meteorite impacts had similar, albeit temporary, and fortunately extremely rare, effects.

Internal oscillations within the ocean–atmosphere system, like El Niño events and the North Atlantic Oscillation, cause further changes at high frequencies but low amplitudes and are usually regional in scope. Whatever the climate may be at any one time, it is modified by internal processes like those oscillations, and by the behaviour of the atmosphere in redistributing heat and moisture rapidly, by the ocean in redistributing heat and salt slowly, and by the biota – for example through the 'biological pump' in which plankton take $CO_2$ out of surface water and transfer it to deep water and eventually to sediments when they die. These processes can make attribution of climate change difficult, as can the smearing of the annual record in deep water sediments by burrowing organisms.

In spite of the potential for considerable variation in our climate, close inspection shows that at any one time the climate is constrained within a well-defined natural envelope of variability. Excursions beyond that natural envelope demand specific explanation. As we shall see, one such excursion is the warming of our climate since late in the last century.

Other processes are important too, notably weathering. In almost every churchyard you'll find gravestones so old that their inscriptions have disappeared. Over the years drop after drop of a mild acid has eaten away the stone from which many old gravestones were carved, obliterating the names of those long gone. We know that mild acid as rainwater, formed by the condensation of water vapour containing traces of atmospheric gases like carbon dioxide ($CO_2$) and sulphur dioxide ($SO_2$). It's the gases that make it acid. Rain eats rock by weathering.

Weathering is fundamental to climate change. Over time it moves mountains. Freezing and thawing cracks new mountain rocks apart. Roots penetrate cracks as plants grow. Rainwater penetrates surfaces, dissolving as it goes. In due course, scientists of a chemical bent realized not only that $CO_2$ contributed to weathering, but also that the $CO_2$ in the dissolved products of weathering reached the sea to form food for plankton, eventually reaching the seabed in the remains of dead organisms that would form the limestones and hydrocarbons of the future. One day, volcanoes would spew that $CO_2$ back into the atmosphere for the cycle to begin all over again.

The greenhouse gases are also important modifiers of climate, especially carbon dioxide ($CO_2$), which connects the climate to life and is part of the carbon cycle. The carbon cycle includes the actions of land plants, which extract $CO_2$ from the air by photosynthesis. Furthermore, as David Beerling reminds us, '*roots and their symbiotic fungal associates secrete organic acids that attack mineral particles in soils to liberate nutrients needed for growth. Roots also anchor soils, slowing erosion and giving the minerals more time to be dissolved by rainwater. Meanwhile, above ground organic debris accumulates in a litter layer to form a continuous moist acidic environment that helps break down soil minerals*' [2] ... Plants and their fungal partners dissolve rocks five times faster than normal, irrespective of their environment [2].

When plants die, they rot, returning their $CO_2$ to the air. Some are buried, preserving their carbon from that same fate, until heat from the Earth's interior turns them back into $CO_2$ and returns it to the air. This natural cycle has been in balance for millions of years. We have disturbed it by burning fossil carbon in the form of coal, oil, or gas.

There is a certain paradox to the story of plants. As they evolved, developing longer stems and broader leaves and more extensive roots, they extracted more of that $CO_2$ plant food from the air, causing its concentration to fall, and so cooling the climate. Disaster was averted as Earth's natural thermostat kicked in, with cooler conditions slowing the rates of chemical weathering and the supply of rain (a product of the concentration of water vapour, which diminishes in a cooling world). Ongoing volcanic activity built the $CO_2$ levels of the atmosphere back up again for the cycle to repeat. In time, new varieties of plant (mostly grasses, including such food crops as maize and sugar cane) evolved capable of extracting carbon dioxide more efficiently from the atmosphere in our increasingly $CO_2$-poor world. Our climate is the result of this delicate dance of plants, $CO_2$, weathering, and climate [2].

In this book we focus on the past 800 million years (Ma) of Earth's climate, starting with the period when our climate was rocked by a succession of massive glaciations during which Earth may have looked like a gigantic snowball. This period overlapped with the development of fungi, and there is some speculation that plants may have played a key role in the development of those glaciations. Land plants are thought to have played a significant role in modulating Earth's climate over the past 450 Ma, through capturing carbon from the air. For most of the past 450 Ma our planet has been a lot warmer than it is now. Our climate is usually of the greenhouse variety, with abundant $CO_2$ warming the part of the atmosphere in which we live - the troposphere. This long history of warmth is not widely recognized, because in the past 50 Ma Earth's atmosphere lost much of its $CO_2$ and moved into an icehouse climate characterized by cool conditions and polar ice. That cooling intensified to the point where over the past 2.6 Ma Earth developed large ice sheets in both polar regions, a period that has earned a popular title – the Ice Age. We are living in a geologically brief warm interlude within that Ice Age. Forty to fifty million years ago, before an ice sheet formed on Antarctica, global temperatures were very much warmer than they are today. The Sun did not cause our climate to cool into the Ice Age; its luminosity has been steadily but slowly increasing over time, like all main sequence stars of its type. Warming and cooling over long time scales have much more to do with the amounts of the greenhouse gases $CO_2$ and water vapour in the atmosphere. Warming and cooling over short time scales, in contrast, have more to do with orbital variability and its control on incoming solar radiation (insolation).

In this book we will look at the behaviour of these several variables in the past and how they are expected to vary in the future, to see what that tells us about what comes next. Will we stay in the icehouse, or move back into the greenhouse? Today, both Greenland and Antarctica are losing land ice at significant rates, and the Arctic is losing its sea ice; we appear to be moving away from the icehouse and towards the greenhouse. Where will that trajectory take us? Already geologists speak of our entering a new geological epoch – the Anthropocene – in which our human activities have grown so extensive that they impact all parts of the Earth System: atmosphere, ocean, ice, and land surface [3]. We have become a geological force rivalling nature. Where will our activities take the climate?

## 1.3 Using Numerical Models to Aid Understanding

Climate modellers have contributed to our understanding of the past variability of our climate system. Since the 1950s our ability to use computers has advanced apace. We now use them not only to process palaeoclimate data and find correlations, but also to run numerical models of past climate systems, testing the results against data from the rock record. Applying numerical models to past climates that were much colder or much warmer than today's has an additional benefit. It helps climate modellers to test the robustness of the models they use to analyse today's climate and to project change into the future. Research into past climates by both of these research streams – the practical and the theoretical - adds to our confidence in understanding the workings of Earth's climate system and in predicting its likely future.

It is a common fallacy that the climate cannot be modelled because it is, like the weather, chaotic. In fact, the natural conservation of energy, heat, and momentum does make it possible to predict the general evolution of a chaotic system like the climate in statistical terms. An important feature of chaotic systems like the climate system is that although they are unpredictable in the short term, they evolve towards a stable statistical spread of outcomes after a long period. These outcomes commonly have very stable and predictable average behaviours. Indeed, we know that Earth's climate variability takes place within a range narrow enough to have prevented the ocean from boiling away or the planet becoming permanently covered with ice (although that did happen temporarily in the Snowball Earth episodes of the late Proterozoic).

## 1.4 The Structure of this Book

This book looks at these various processes and puts them into perspective in their proper historical context. Chapter 2 follows the evolution of thinking about climate change by natural scientists, philosophers, and early geologists from the late 1700s on. It touches on the debates of the early 1800s on the virtues of gradual change versus sudden change, and highlights the growing realization that the world cooled towards an Ice Age in geologically recent times. Chapter 3 takes us into the minds of nineteenth century students of the Ice Age, and examines the astonishing discovery that its climate cycles were probably controlled by metronomic variations in the behaviour of the Earth's orbit as it responded to the gravitational influences of the great gas planets, Venus, and Jupiter.

The arrival of new technologies on the scene, often from different disciplines, changes the way in which science works – think of the effect of the telescope on Galileo's perception of astronomy. Geology is no exception. In Chapter 4, we explore the extraordinary mid-nineteenth century experimental discovery of the absorptive properties of what we now know as the greenhouse gases like water vapour, carbon dioxide and methane, which changed the way we view past climates. At the end of that century a Swedish chemist, Svante Arrhenius, made the first calculations of what emissions of $CO_2$ would do to the climate. Few people realize that he did so at the urging of a geological colleague, to try to see if variations in atmospheric $CO_2$ might explain the fluctuations in temperature of the Ice Age. Back in 1899 an American geologist, Thomas Chamberlin, used Arrhenius's findings to construct an elegant hypothesis as to how $CO_2$ controlled climate, but it was soon forgotten for lack of data. Much of what he had to say on the subject has since been proved correct.

In Chapter 5 we examine the evolution of ideas in the early part of the twentieth century about the way in which the continents move relative to one another through continental drift, which geophysicists realized in the late 1960s was driven by the newly discovered process of plate tectonics. Once again, new technologies played a key role – in this case the echo-sounder and the magnetometer. Knowing the past positions of the continents provides us with the maps of past geography – the palaeogeographical base maps – needed to determine the past locations of sedimentary deposits that are sensitive to climate, like coal swamps and salt pans. Along the way we see how studies of past climates benefited from novel access to the accurate dating of rocks, minerals, and fossils at the smallest possible intervals of time. Once again, a new technology was key – radiometric dating by the use of natural radioactivity.

Chapter 6 describes how the new science of palaeoclimatology developed, with Earth scientists plotting their indicators of past climates on maps, using yet another new technology – oxygen isotopes to determine the temperature of past seawater. Geologists investigated the origins of sedimentary cycles, coming up with hypotheses explaining the evolution of climate from the Carboniferous glaciation roughly 300 Ma ago to the end of the Cretaceous at 65 Ma ago. Yet another new technology changed the picture again, this time in the shape of numerical models of the climate system, which capitalized on the rapid development of the computer. We see early attempts to use numerical models to find out why the Cretaceous period was so warm, and note that until the mid-1980s the analysis of palaeoclimates virtually ignored $CO_2$.

Chapter 7 takes us into the Cenozoic Era, which includes what used to be known as the Tertiary between 65 Ma ago and 2.6 Ma ago, and the Quaternary, which lasted from 2.6 Ma ago to the present. Here we follow the cooling of our climate from the warmth of the Cretaceous seas that flooded western Europe and central North America 60–100 Ma ago to the current Ice Age that characterizes the Pleistocene period (2.6 Ma to 11 700 years ago) and the Holocene period (starting 11.7 Ka ago). We look at how climate changed, and at how our knowledge of climate change was dramatically expanded by drilling into the largely undisturbed sediments of the deep ocean floor, starting in 1968. As we saw in Chapter 6, many of the theories to explain the changes in climate of the Cenozoic Era prior to the 1980s developed in the absence of substantial knowledge about the past composition of the air.

A clear understanding of the roles of greenhouse gases in the climate system demanded an ability to measure those gases and examine their properties, capabilities that were limited until the mid 1950s, and which then took another 30 years to penetrate the world of geological thought. Chapter 8 explores the massive strides made over the past 50 years in enhancing that knowledge base, and in formulating theories to explain how greenhouse gases behave within the air and ocean. Along with that understanding came the realization that to understand the climate problem we must see our planet holistically, as a whole, not in a reductionist way. Alexander von Humboldt was right. Everything is connected. One key consequence was the development of a new field of scientific endeavour, biogeochemistry, which has proved especially important for understanding how the carbon cycle works. Answering questions about the evolution of the climate system also came to involve a more international approach in which national scientists increasingly worked with each other across borders on major scientific issues like climate change that were not susceptible to resolution by individual investigators or even individual nations. Amongst other matters, Chapter 8 introduces the work of the Intergovernmental Panel on Climate Change

Chapter 9 reminds us of the amazing discovery that ice cores contain bubbles of fossil air containing pristine samples of $CO_2$ and other greenhouse gases. We also see how palaeoclimatologists eventually learned how to measure amounts of $CO_2$ in the atmosphere in the ages before the oldest ice cores (which span the past 800 Ka), by using fossil leaves, tree rings, planktonic remains, soils, corals, and cave deposits. These data are being used to check numerical models of past climates and to test the theory that warm periods of the past occurred when $CO_2$ was most abundant.

Our planet's climate has experienced large cycles through time. Chapter 10 explores how those cycles relate to changes in plate tectonic processes, sea level, emissions of $CO_2$, and the weathering of emerging mountain chains as continents collided. It investigates the evidence for changes to our climate, and the creation of major biological extinctions, caused by occasional meteorite impacts and/or massive eruptions of plateau basalts. The chapter concludes with an analysis of the Snowball Earth conditions some 800–600 Ma ago.

In Chapter 11 we examine the evidence for how $CO_2$ and climate changed together through the Mesozoic and Cenozoic Eras, and explore three case histories. One is from the Palaeocene-Eocene boundary 56 Ma ago, when a massive injection of carbon into the air caused dramatic warming, at the same time making the seas more acid. It took the Earth 100 000 years to recover – now there's a lesson from the past! The second is from the mid-Miocene, some 17–14 Ma ago, when $CO_2$ levels rose to about 500 ppm And the third is from the mid-Pliocene, about 3 Ma ago, when $CO_2$ rose to levels much like today's, but when temperatures were warmer and sea level was higher – another lesson from the past. These periods are not precise analogues for today, because the world was configured slightly differently then. But they can teach us something about what is happening now and may happen in the future.

Chapter 12 begins our exploration of the Ice Age of the past 2.6 Ma, noting how much of what we know comes from cores of sediment extracted with great difficulty from the ocean bed. It was a big surprise in 1976 when it emerged that marine sediment cores display signs of change in the Earth's orbit and the tilt of the Earth's axis through time. These cores also display unexpected millennial signals.

Our exploration of Ice Age climate continues in Chapter 13, where we examine the contribution made by ice cores collected in recent decades. We see what the records tell us from Greenland and from Antarctica, and explore the linkages between the poles. The latest research shows that during the warming from the Last Glacial Maximum, $CO_2$ in the Antarctic region rose synchronously with temperature, and not after it as had been thought. The chapter ends with a survey of plausible explanations for the fluctuations of the Ice Age, concluding that $CO_2$ played a crucial, though not primary, role in the changes from glacial to interglacial and back over the past 800 Ka.

In Chapter 14 we focus on the changes that took place over the past 11.7 Ka forming the latest interglacial, the Holocene. Insolation – the amount of heat received due to the motions of the Earth's orbit and the tilt of the Earth's axis – was greatest in the Northern Hemisphere at the beginning of the Holocene, but the great North American and Scandinavian ice sheets kept the Northern Hemisphere cool until they had completely melted by the middle Holocene. All that while Northern Hemisphere insolation was in decline, moving Earth's climate towards a neoglacial period the peak of which we reached in the Little Ice Age of the past few hundred years. $CO_2$ played a small part in this cooling.

Chapter 15 focuses on the end of the Holocene – the past 2000 years and up to the present, including the Anthropocene. It reviews cyclical changes in solar output. It explores the concept of the Mediaeval Warm Period centred on 1100 AD, and the subsequent Little Ice Age. Multiple sources of palaeoclimatic data now make it abundantly clear that the years since 1970 were the warmest of the past 2000. Yet astronomical calculations show that despite variations in the sun's output, our climate should still be like that of the Little Ice Age. Only by adding our emissions of greenhouse gases like $CO_2$ to palaeoclimate models can we recreate the climate that we see today. We can't blame the Sun, which has experienced declining output since 1990.

The final Chapter 16, provides an overview of Earth's climate evolution, concluding that from the evidence of previous chapters we should expect to see sea level rises of up to 6–9 m as temperatures rise 2–3 °C above the 'preindustrial' levels typical of the years before the industrial revolution. Those conditions were typical of recent interglacials that were warmer than our own. We will not see such rises in sea level this century, because it takes a long time for the Earth system to arrive at an equilibrium in which the ocean is heated as fully as it can be for a given level of atmospheric $CO_2$, and no more ice will melt.

As in any other field of science, this 200-year history of past climate studies was punctuated with arguments and disagreements, but the influence of $CO_2$ on climate eventually emerged as highly significant. The exciting developments documented in this book revolutionized as much as plate tectonics the way in which Earth science is done. The demands of climate science now require sedimentologists and palaeontologists to become familiar with the host of related disciplines that deal with processes taking place on and above the Earth, and to take a holistic approach to interpreting their data. Due to the rapid evolution of these topics and techniques, including the use of computers to model palaeoclimate behaviour, much of what we now know is quite recent, little publicized except in scientific journals, hence little known by the wider public.

One thing you will need to consider carefully is context. In this book you will see evidence that $CO_2$ does correlate with temperature. Correlation is not causation, but that is a trite observation that ignores context. When you know that $CO_2$ is a greenhouse gas that both absorbs and re-emits radiation, you should expect a rise in temperature from an emission of $CO_2$ – this is simple causation, or

cause and effect. Equally, you should expect a rise in temperature to cause the ocean to emit $CO_2$; another prediction that is easy to test. What then becomes interesting are the instances when the two do NOT correlate, for which we have to find alternative hypotheses, which usually involve the operation of feedbacks from elsewhere in the climate system – like the fact that ice-covered regions, having high albedo, reflect the Sun's energy so maintain cool conditions locally, and the realization that much of the finer scale variability in the Earth System is driven by extraterrestrial orbital change, which causes temperature to change, which in turn caused $CO_2$ to change. We have to think! What are the connections? What are the feedbacks? What is the role of life? Climate is not a simple matter of just the physical science linkage between $CO_2$ and temperature – plants and of ocean chemistry have key roles to play, too. Even so, considering all the many possibilities, nobody has managed yet to explain plausibly what, if not our emissions of greenhouse gases and related feedbacks, has caused the global warming since 1970. Keep in front of you the notion that the acid test of research findings like those laid out in this book is to see whether or not they make accurate predictions in the real world. Our ability to predict well is a measure of our scientific progress.

## 1.5 Why Is This History Not More Widely Known?

Few of the results of the growing body of research on palaeoclimates reach the general public. That is largely because most of the results of modern science are hidden behind a pay-wall erected by the publishers of scientific journals. If you are a working scientist in academia or a government research laboratory, your institution will subscribe to the main scientific journals you need to support your research, which you can therefore access free via your institution's library. If you are not in such an environment, you will have to subscribe to those journals yourself (a daunting task when you need to access many different journals). Alternatively you can pay a fee to access a single article from a single journal (if you know it's there). That pay-wall limits access to information. It explains why even many professional Earth scientists I meet, especially from industry, know little of what the most up-to-date Earth science studies tell us about climate change and global warming. For the most part they have specialized in other aspects of the Earth sciences that were relevant to their careers. This unavoidable disconnect between the academic scientific community and the wider community helps to fuel dissension.

# References

**1** Lenton, T. (2016). *Earth System Science: A Very Short Introduction*, 153 pp. Oxford University Press.

**2** Beerling, D. (2007). *The Emerald Planet: How Plants Changed Earth's History*, 288 pp. Oxford University Press.

**3** Zalasiewicz, J., Waters, C.N., Williams, M., and Summerhayes, C.P. (eds.) (2019). *The Anthropocene as a Geological Time Unit: A Guide to the Scientific Evidence and Current Debate*, 361 pp. Cambridge University Press.

# 2

# The Great Cooling

**LEARNING OUTCOMES FROM THE EIGHTEENTH AND NINETEENTH CENTURIES**

- Understand and explain the steps by which the geologists of the late eighteenth and nineteenth centuries came to realise that the climate of the past 50 million years had cooled.
- Understand and explain the nineteenth century controversy that raged over the reality and nature of the Ice Age, and the arguments for deposition of glacial material from either an ice sheet or floating icebergs.
- Explain the thinking behind James Hutton's proposals that '*the present is the key to the past*', and also that '*the past is the key to the future*'. How can palaeoclimatology help us understand the continuing evolution of our climate?
- Understand and explain Alexander von Humboldt's concept of the holistic view of nature, and the climatic significance of his discovery of isothermal lines.
- Understand how Charles Lyell changed Victorian thought with his *Principles of Geology*.
- Explain the contribution that the Geikie brothers, Archibald and James, made to our understanding of the origin and extent of the Ice Age.
- Understand the significance of Nansen's observations on what was happening beneath the Greenland ice sheet.

## 2.1 The Founding Fathers

Geologists have known for over 200 years that climate is one of the main controls on the accumulations of minerals and organic remains that end up as sedimentary rocks and fossils. As early as 1686, Robert Hooke, a fellow of London's Royal Society living in Freshwater on the Isle of Wight, deduced from fossils discovered on the Isle of Portland, Dorset, that the climate there had once been tropical [1]. His perceptive observation lay unremarked upon until the Keeper of France's Royal Botanical Gardens – the *Jardin du Roi* (the post-revolutionary *Jardin des Plantes*) – in Paris realized that differences in climate might explain the differences between living and fossil organisms found at the same place. This was the naturalist Georges-Louis Leclerc, the Comte de Buffon (1707–1788) [2], friend to Voltaire, and a member of both the French Academy of Sciences and the literary Académie Française (Figure 2.1). Louis XV appointed Buffon to head not only the *Jardin*, but also the *Cabinet du Roi*, which was to become Paris's Natural History Museum.

Buffon planned to take his place in history with a vast 50-volume encyclopaedia: the *Histoire Naturelle, Générale et Particuliére*. The 36 volumes that he actually produced were among the most widely read publications of the time.

His reconstruction of geological history appeared in 1788 in *Les Époques de la Nature* (*The Epochs of Nature*), the supplement to volume 5 [3]. It is arguably '*the first scientific, evidence-based history of the Earth, from beginning to end*' [4], and is often overlooked because only fragments have previously been translated into English. Buffon realised that each geographical region had its own distinctive plants, animals, and climate – a basic principle of what we now call biogeography. Finding in Siberia and Europe the fossil remains of animals that now inhabit the tropics, he deduced that the climate there must have been warmer in the past. Buffon also realised that ancient sedimentary rocks contained the fossils of animals no longer living, like the 'horns of Ammon' (which we now call ammonites), and belemnites, thus in effect putting forward the notion of biological extinctions a generation before Baron Cuvier, who is more generally associated with establishing this concept in geology [4]. Buffon also correctly proposed that coal seams were the compressed and carbonized remains of prehistoric swamp forests, much like those found in modern Guyana.

Buffon thought that the temperature of the air reflected the temperature of the Earth, rather than the heat from the Sun, and interpreted animal remains to show that the Earth was cooling from its original molten state [2]. Sir

**Figure 2.1** Georges Leclerc, Comte de Buffon. Oil painting by François-Hubert Drouais, Musée Buffon, Montbard.

**Figure 2.2** James Hutton.

Humphry Davy, FRS, discoverer of sodium and potassium and inventor of the coal miners' safety lamp, was another who shared this popular notion, penning it in 1829 in his *Consolations in Travel, or the Last Days of a Philosopher*, shortly before he died. Measurements of the temperature of the Earth and its atmosphere by French scientist Joseph Fourier had knocked this idea on its head in 1824, however – an advance that Davy overlooked.

Buffon and other savants of the late eighteenth century considered that Earth's past history must be explained with reference to what is happening now [5]. Among them was James Hutton (1726–1797) (Figure 2.2), a Scot who profoundly influenced geological thought [6, 7]. Born in Edinburgh, Hutton studied medicine and chemistry there and in Paris and Leyden. Taking up farming, first in Norfolk, then on his paternal acres in Berwickshire, he developed an interest in geology, and exploited his chemical knowledge to become partner in a profitable sal ammoniac business. By 1768 he was established in Edinburgh, pursuing his geological interests. In March and April 1785 he presented to meetings of the Royal Society of Edinburgh his innovative *Theory of the Earth*, which was published in the first volume of the *Transactions of the Royal Society of Edinburgh* in 1788 [8]. Encouraged to seek observations to support his theory, he found several telling examples enabling him in 1795 to expand his ideas into a two-volume book *Theory of the Earth with Proofs and Illustrations* [9]. His 1788 paper, with slight modifications, forms the first chapter of his 1795 book. His friend John Playfair brought Hutton's ideas to a wider audience in *Illustrations of the Huttonian View of the World*, published in 1802.

Hutton popularized the notion that '*the present is the key to the past*' [8, 9]. As he put it '*In examining things present, we have data from which to reason with regard to what has been*'.

In following that approach, Hutton echoed Isaac Newton's dictum that '*We are to admit no more causes of natural things than such as are both true and sufficient to explain their appearances*' [9]. But that's not all. Hutton went on to say, '*and, from what has actually been, we have data for concluding with regard to that which is to happen hereafter*' [8, 9]. Here was an extraordinary notion both for its time and now – that examples from the past preserved in the geological record could provide examples of what might happen on Earth in future if the same guiding conditions were repeated.

Observing the processes at work on his farm and in the surroundings, Hutton saw that – far from being everlasting – today's hills and mountains were sculpted by the slow forces of erosion. The eroded materials were transported by rivers and dumped in the ocean, where they accumulated to great thickness before being raised into mountains. The process then began again, in yet another great geological cycle. Hutton's idea that the ruins of earlier worlds lay beneath our feet was demonstrated by younger and undisturbed strata resting unconformably on older folded and eroded beds, notably at Siccar Point on Scotland's east coast. To some degree Hutton's concept repeats ideas proposed originally in the notebooks of Leonardo Da Vinci in about 1500 [10]. Perhaps this is an early example of the convergent evolution of ideas.

The slowness of geological processes led Hutton to conclude that Earth's history was unimaginably long. Indeed, in dramatic contrast with biblical scholars, he found that Earth's history showed '*no vestige of a beginning, – no prospect of an end*' [8, 9]. Hutton's contemporary, Buffon, had calculated that the Earth was as much as 75 000 years old [3], although his private notebooks suggested that he thought the true answer was closer to three million years

[4]. To avoid an argument with the church, since his conclusions were at variance with biblical analyses, Buffon described his suggested date as purely theoretical, and got on with the science [4]. Did Buffon know about Hutton's 1785 reading of his *Theory of the Earth*? Or did both men reach much the same conclusion – that the Earth was exceedingly old – through convergent evolution of ideas, both of their articles being published in 1788?

Hutton was wrong in one respect – not all operations of nature are equable and steady. Earthquakes and volcanic eruptions are sudden, as are meteorite strikes of the kind that cratered the surface of the moon. Even so, he realised that earthquakes and volcanic eruptions, although discontinuous, are recurrent and integral parts of ongoing nature. Neither he nor many other savants of his time knew about the kind of catastrophic meteorite impact that we now believe led to the extinction of the dinosaurs 65 Ma ago, nor about the vast eruptions of plateau basalts in Large Igneous Provinces that may have contributed to climate change and biological extinctions at various times in the past.

When Buffon died, Georges Cuvier (1769–1832) (Figure 2.3) took his place as France's leading natural historian. Cuvier was a key figure in establishing the scientific fields of comparative anatomy and palaeontology. He was elected a member of the Academy of Sciences in 1795, Professor of Natural History at College de France in 1799, and Professor at the *Jardin des Plantes* in 1802. He also became a foreign member of the Royal Society of London in 1806, and was ennobled Baron Cuvier in 1819.

Cuvier used his knowledge of anatomy to identify fossil species and their likely interrelationships. While Buffon thought that Siberian fossils of woolly rhinoceros and elephant were the remains of animals still living, Cuvier showed that they were extinct, and identified the elephant remains as mammoths [5]. Both men knew that these animals were found frozen into the tundra '*with their skin, their fur and their flesh*' [11, 12]. Unlike Hutton, Cuvier was keen on the moulding of geological history by catastrophic events. So, he attributed this freezing to an environmental catastrophe: '*this event took place instantaneously, without any gradation... [and]... rendered their country glacial*' [5, 13]. Here we have the first inkling of the idea of the Ice Age.

Cuvier's senior colleague Jean-Baptiste de Monet, Chevalier de la Marck, commonly known as Lamarck (1744–1829), challenged his call for catastrophic change. Studying the sequence of fossil molluscs from the region around Paris, he concluded in 1802 that many of them belonged to genera that are now tropical, and that they represented a slow change of climate with time [5].

At about the same time, in the late 1790s, William Smith (1769–1839), a land surveyor engaged in building the network of canals that now cross the English countryside, began using distinctive fossils to identify and map the occurrence of particular strata. This led him to publish in 1801 a 'prospectus' for the production of a geological map of England [5], something he achieved in 1815. He had invented the science of 'stratigraphy' – the use of fossil remains to establish the succession of strata – which now underpins our appreciation of changes in climate through geological time.

French geoscientists were quick to seize upon this new approach to geohistory. In 1802, Alexandre Brongniart (1770–1847), the newly appointed young director of the porcelain factory at Sèvres, near Paris, visited England to find out more about the mass production of ceramics by the Wedgwood factory [5]. In London, he dined with fellows of the Royal Society, where he is likely to have become aware of the novel ideas and unpublished maps of William Smith. Searching for new deposits of clay, and working with Cuvier to identify fossils, Brongniart began a systematic survey of the Parisian region. Brongniart and Cuvier used fossils à la Smith to determine the order of the layers of sedimentary rock of the Paris Basin and map the outcrops of the strata. They concluded that the area had been submerged at times by the sea and at times by freshwater – a first indication that environmental conditions could change with time in a relatively small area, and something that went beyond anything attempted by Smith in its high level of detail. In 1808, they delivered a preliminary report of their paper on the Paris Basin, with an accompanying draft geological map that was eventually published in 1811 [5].

Brongniart and Cuvier were not the first to map the sedimentary divisions of the Paris Basin. The famous French chemist, Antoine-Laurent Lavoisier (1743–1794), who had discovered oxygen and hydrogen, and was guillotined

**Figure 2.3** Georges Cuvier.

during the French Revolution, beat them to it. Lavoisier's 1789 memoir on the topic [14] was brought to light by sedimentologist Albert Carozzi of the University of Illinois, in 1965 [15]. Lavoisier saw in the alternating deep and shallow water (littoral) deposits of the Paris Basin evidence for a succession of transgressions (floodings) and regressions (retreats) of the sea. His vision of how these packages of sediment were built up through time by the alternating rising and falling of sea level is like the modern understanding of the origin of sedimentary cycles. It involved '*a very slow oscillatory movement of the sea ... [each oscillation] requiring several hundreds of thousand years for completion*' [14, 15]. Lavoisier's cross-sections of the Basin provide an outline for the correct classification of its Tertiary deposits. He was a man far ahead of his time.

These parallel French and British efforts were major developments in the evolution of palaeontology and geology. They provided an essential platform for the development of palaeoclimatic studies, and influenced the thinking of those who followed.

Aside from Cuvier, one of the most influential scientific men in Europe in the early1800s was the German naturalist Baron Alexander von Humboldt (1769–1859) (Figure 2.4, Box 2.1) [16–23]. By 1797 Humboldt was planning an overseas expedition, learning to use a wide range of scientific and navigational instruments, and visiting experts in Vienna and Paris. While in Paris Humboldt met botanist Aimé Bonpland (1774–1854), who was to be his travelling companion. Humboldt focused on physical geography,

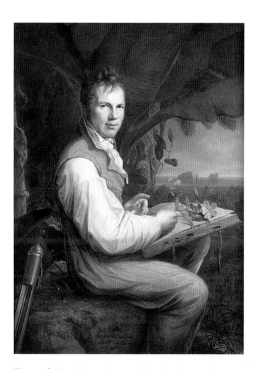

**Figure 2.4**  Alexander von Humboldt working on his botanical specimens.

geology, geomorphology, and climatology, while Bonpland focused on flora and fauna. Visiting Madrid, they obtained royal assent to scientifically examine Spain's American territories as a contribution to understanding the physical make-up of the world. They sailed from La Coruña on 5 June 1799, visiting South and Central America, the West Indies, and the United States, returning to France in August 1804. The major scientific outcomes were Humboldt's seminal *Essay on the Geography of Plants* [17], published in 1805, and his treatise on *Isotherms and the Distribution of Heat over the Earth's Surface* [18], published in German in 1816, which Humboldt regarded as the foundation for comparative climatology. Wider recognition followed publication of his more general works: *Views of Nature*, in 1808 [19], and a more general travelogue: *Personal Narrative of a Journey to the Equinoctial Regions of the New Continent* [20], in three volumes in 1814, 1819, and 1825. Further travels, to Russia and Siberia, in 1829, led to publication of *Fragments of the Geology and Climatology of Asia* in 1831, followed by *Aspects of Nature*, in 1849 [21]. These works laid the foundations for the study of physical geography, biogeography, meteorology, and climatology.

As we see from Romanowski's translation of *Humboldt's Essay on the Geography of Plants* [17], Humboldt's view of nature was holistic, reflecting the need expressed by Enlightenment philosophers for a unifying system for the sciences. Humboldt was seeking unity in diversity, a fundamental view of nature as a whole, in which he was going beyond what botanists had previously been concerned with, namely classification. He saw that nature's parts were intimately related and were only understandable with reference to the whole, with plants growing where they did in response to relationships between biology (plants, animals, soils), meteorology (temperature, winds, humidity, and cloudiness), geography (altitude, latitude, and distance from coast) and geology. Everything, in his mind, was connected. As Andrea Wulf reminds us in her 2015 book *The Invention of Nature*, Humboldt '*saw vegetation through the lens of climate and location: a radically new idea that still shapes our understanding of ecosystems today*' [22]. Indeed, from reading his *Views of Nature*, we can consider him as the father of the concept of the 'web-of-life' and of ecology, although ecology as the science of the relationships of an organism with its environment was not named as such until Ernst Haeckel (1834–1919) used the term in his 1866 work *Generelle Morphologie der Organismen*.

Humboldt envisaged these relationships by means of a 'Tableau Physique' [17], a fold-out plate showing volcanic Mount Chimborazo (6236 m), Ecuador's highest mountain, with different ecosystems and plants at different altitudes between the sea and the snow line (Figure 2.5). It also represented – translated into a horizontal sense – the succession of ecosystems between the hot equator and the snowy

**Box 2.1  Baron Alexander von Humboldt**

Humboldt was born in Tegel, now the location of Berlin's major airport. At the age of 19, he developed a lifelong interest in botany, which led him to investigate the laws that govern not only the diversity of plant life, but also everything that impinged on the environment. Entering the University of Göttingen to study natural sciences in 1789, he travelled to Mainz to meet Georg Forster, the naturalist from Captain Cook's second voyage. Forster encouraged Humboldt to study the basalts of the Rhine, a topic that Humboldt wrote up in his first book, in 1790. Next year the two men travelled to England together, visiting Sir Joseph Banks, who had been the naturalist on Cook's first expedition, and Captain William Bligh, who had been on Cook's third. These encounters gave Humboldt a desire to travel and study regions not yet explored scientifically, and – like Forster – to combine science and travel writing. In England he met the physicist Henry Cavendish, who introduced him to the work of Antoine Lavoisier. Humboldt's study of Lavoisier convinced him of the importance of measurement and experimentation and of the value of scientific cooperation and the exchange of ideas. Scientific networking is not new. In June 1791, Humboldt joined Freiburg's School of Mining, run by one of the great men of geological science, Abraham Gottlob Werner (1750–1817) [24]. Werner led the so-called Neptunists, who thought that all rocks were once precipitates in the ocean. Humboldt initially followed Werner on this, for example in his work on the Rhine basalts, but eventually joined the so-called Vulcanist or Plutonist school led by James Hutton, who showed that granites were created from molten rock. Despite his mining studies, Humboldt found time to continue research on plant life, winning the Saxon gold medal for his work. In 1792, aged 22, he joined the Prussian Mining Service, rising to become inspector of mines. During his early twenties Humboldt dreamt of writing a *Physique du Monde*, a total description of the physics of the world. His dream would come to fruition in his five-volume work *Cosmos: A Sketch for a Physical Description of the Universe* [23], starting with a first volume in 1845. In recognition of his outstanding contributions, many geographical features are named after him, including the Humboldt Current off the coast of Peru, as well as numerous towns, forests, streets, parks, universities, colleges, and schools, a lunar crater, and several plants and animals. Humboldt was awarded the Copley Medal by London's Royal Society in 1852.

**Figure 2.5** The centrepiece of Humboldt's *Tableau Physique* showing the vertical succession of plant life in relation to the flanks of Mt. Chimborazo [17]. Side panels listing the species characteristic of each altitude flank the original tableau [17].

poles, reflecting Humboldt's realization that vegetation followed climatic zones that could be defined vertically or horizontally by temperature [18]. It would be fair to say that Humboldt was the first to understand climate as the result of interactions between the atmosphere, the oceans, and the landmasses (not least the ability of forests to regulate moisture). Writing in *Cosmos*, he referred to the '*perpetual interrelationship*' between air, winds, ocean currents, elevation, and plant cover [23].

Humboldt's search for meaning stimulated Darwin, who took Humboldt's narratives and *Views of Nature* [19] on the *Beagle*, enabling Darwin in due course to find an all-encompassing biological framework through evolution. Perhaps that is not entirely surprising, since Humboldt had published what we would now regard as Darwinian ideas, writing about '*the gradual transformation of species*' [17].

## 2.2  Charles Lyell, 'Father of Palaeoclimatology'

The ideas of Buffon, Hutton, Humboldt, Cuvier, and Brongniart had a considerable influence on a young Scottish geologist, Charles Lyell (1797–1875) (Figure 2.6, Box 2.2). Lyell was famed for turning Hutton's big idea into a fundamental geological principle that has stood the test of time, though with certain modifications. He was destined to become the greatest geologist of his time [25, 26].

It seems oddly fitting that he was born in 1797, the year that Hutton died.

The key to Lyell's understanding of the Earth lay in the subtitle to his *Principles of Geology* [25, 26, 27], namely '*An attempt to explain the former changes of the Earth's surface by causes now in operation*', which demonstrates the influence of Hutton on his thinking. According to palaeontologist and essayist Stephen J. Gould, Lyell's *Principles* was perhaps the most important scientific textbook ever

**Figure 2.6**   Charles Lyell.

written [28]. Lyell adopted the concept of gradualism, believing that the same natural laws and processes that operate in the universe now have always operated, and apply everywhere, with no tendency to increase or decrease in general intensity. Lyell's gradualistic and steady state approach was later named 'uniformitarianism' by William Whewell (1794–1866), a natural philosopher and Master of Trinity College, Cambridge. In effect, Lyell took Hutton's ideas and magnified them a hundred-fold, showing how they applied to the many different aspects of geology from fossil life to volcanoes. In doing so he was labouring to overcome the catastrophist theories of scientists like Cuvier, who held that geological change occurred in rare episodes of global paroxysm. Lyell believed that what appeared from the geological record to be the results of catastrophic events could instead have arisen by the slow and steady action of processes observable today. Like Hutton, he called for immense periods of time to wear down the land and deposit the sediments eventually represented by different uplifted strata. This would not endear him to strict interpreters of Genesis.

Gould reminds us that Lyell argued that only his gradualist approach provided the basis for a proper methodical study of the geological record [28]. If that record was one of

---

**Box 2.2   Sir Charles Lyell [25, 26]**

Born at Kinnordy, near Dundee, Scotland, Lyell was brought up at Bartley Lodge in England's New Forest. Son of a wealthy naturalist after whom the plant *Lyellia* was named, he was fascinated by natural history. Studying classics at Oxford between 1816 and 1819, he also attended lectures on geology given by William Buckland. Deciding to become a lawyer, he entered Lincoln's Inn in London in 1820, but his interests drew him into the emerging science of geology. Lyell rose to fame with the publication of his *Principles of Geology* in three volumes between 1830 and 1833. This was the first comprehensive geological textbook, its 12th edition emerging in 1875 just after his death. His reputation was further enhanced by publication in 1838 of a companion volume *Elements of Geology*. Originally intended as a supplement to the *Principles*, it formed an independent practical guide to the new science of geology. Together the two books put the study of geology on a firm footing. Lyell's influence was further assured from his naming a number of geological periods – the Recent (now the Holocene), the Pleistocene, the Pliocene, the Miocene and the Eocene. From 1831 to 1833 Lyell was the first Professor of Geology at London's fledgling King's College, but he later earned his living

as a geological writer. His influence stretched far and wide through publications, lectures, and his association with the Geological Society of London. Having been elected a Fellow of the Society in 1819, and published his first paper there in 1823, he became one of its joint secretaries from 1823 to 1826, its Foreign Secretary from 1829 to 1835, a Vice President for 20 sessions, and its President in both 1835–1837 and 1849–1851. His talent was recognized early. He was elected a Fellow of the Royal Society in 1826, and got its Copley Medal in 1858. The Geological Society awarded him its prestigious Wollaston Medal in 1866. Recognizing his huge contribution to understanding the Earth, he was knighted by Queen Victoria in 1848 at the age of 51, and made a Baronet in 1864. The year he died, the Geological Society inaugurated the prestigious Lyell Medal. Lyell has a crater on the Moon and a crater on Mars named after him, along with an Antarctic glacier and several mountains. He was buried in Westminster Abbey, an honour reserved for few scientists. His burial memorial reports that '*For upwards of half a century he has exercised a most important influence on the progress of geological science, and for the last twenty-five years he has been the most prominent geologist in the world*'.

paroxysms, then how could we use modern processes – the only mechanism subject to direct observations and experiment – to resolve the past? Gould explained that as a famous statement of advocacy (and let's not forget that Lyell was trained as a lawyer) '*Lyell condemn[ed] catastrophism as a doctrine of despair, while labelling his uniformitarian reform as the path to scientific salvation*' [28]. Yet, clearly, volcanoes like Vesuvius, which Lyell had visited, did behave in paroxysmal fashion. Lyell's answer to that conundrum was to point out that volcanic eruptions usually impose only a fleeting influence upon history; for the most part volcanoes are quiescent, hence the substantial population on Vesuvius's lower slopes. A record of a lava flow interrupting a sedimentary sequence did not mean that geology moved in jumps. It could mislead an interpreter into thinking that the temporary action represented by the flow was a more important reflection of geological processes than the ongoing gradual accumulation of the sediments above and below that brief event [28].

Lyell drew heavily on contemporary geological literature to produce the *Principles* [25–27]. Particularly influential was *Conchiologia fossile subapennina* (*The Fossil Seashells of the SubAppenines*) published in 1814 by the Italian geologist Giovanni Battista Brocchi (1772–1826), curator of the museum of natural history in Milan [5, 27]. Lyell was fascinated by Italy. Besides honeymooning there, he studied the geology with local guides, read the Italian literature, and met local specialists. He may have read Brocchi's work in Italian, or the English translation made in 1816 from a copy given to William Buckland during a visit to Milan in 1816 [29].

An expert on the fossil seashells of the Apennines, Brocchi used the change in the percentage of living forms in fossil assemblages with time as a means of the relative dating of their encasing formations. Using that approach he produced a definitive study of the historical geology of Italy, an advance comparable to that made by Smith in England and Brongniart and Cuvier in France. Comparing modern and ancient molluscs, he noticed that the recent species of older Tertiary strata now inhabit warmer climates, suggesting, much as Lamarck had seen in the Parisian region, that the world was cooling. Lyell took note both of the approach and its conclusion.

Young Lyell hoped to meet Cuvier during his first visit to the continent on a tour with his family in June 1818. Cuvier was away, so Lyell peeked into his office, looked at some of his fossil specimens, and read his paper on the 'Geology of the Country Around Paris'. Lyell went on to climb the glaciers around Chamonix and the Grindelwald glacier in Switzerland, giving him a first inkling of the power of ice. This was the first of many visits to all parts of the United Kingdom, much of Europe, and to North America, which

would make him the best-travelled of the geologists of his generation. Seeing the most rocks is one route to becoming an excellent geologist, and Lyell saw plenty. Equally important is becoming fully submerged in the world of ideas about the subject, which Lyell managed by meeting and corresponding with all of the major geological figures of his time in Europe and North America.

Lyell eventually met Cuvier, along with Brongniart and Brongniart's former student Constant Prévost (1787–1856), when visiting Paris in 1823 to improve his French [25]. He was impressed to find that young Prévost, unlike Cuvier, thought that the changes in strata in the Paris Basin came about gradually, not as the result of a series of catastrophic events. Others, like Karl von Hoff (1771–1837) in Germany, also concluded that given enough time ordinary agencies could effect major changes. Over the years Lyell and Prévost worked closely together, recognizing strong similarities between the Mesozoic strata of Normandy and southern England.

Lyell became expert at identifying fossil molluscs. By 1828, following Brocchi, he had used the percentages of modern molluscs in each epoch, and the relations of strata to one another, to subdivide the Tertiary Period into several geological Epochs. This statistical approach was a novelty at the time. Perhaps Lyell was following Humboldt's dictum that all science should be based on numbers. The following year he met Gérard Deshayes (1795–1875), a French palaeontologist with an even larger collection of fossil molluscs, who had arrived at similar views. Lyell persuaded Deshayes to expand on that work and combine it with his own, publishing the results in the *Principles* where he named the four periods of the Tertiary as Eocene ('dawn of the recent', with 3.5% modern species), Miocene (or 'less recent') with 17% modern species, and Pliocene (or 'more recent'), divided into Early Pliocene (with 35–50% modern species) and Late Pliocene (with 90–95% modern species).

Later Lyell worked closely with Danish palaeontologist Henrick Beck (1799–1863) to extract yet more information from fossil molluscs, finding that Europe's Eocene had a tropical climate, while the Pliocene climate was more like today's, with the Miocene lying in between. In Chapter 10 of volume 2 of the *Principles* he established that there was '*a great body of evidence, derived from independent sources, that the general temperature has been cooling down during the epochs which immediately preceded our own.*' Later palaeobotanical work confirmed that. Large pointed leaves with many stomata and thin cuticles typical of warm humid climates characterized Europe's early Tertiary, and the tropical rainforest flora of the Eocene London Clay [30].

Lyell was much influenced by Humboldt, who he met in Paris in 1823, and again in Potsdam in1850. He was particularly taken with Humboldt's holistic view of nature and his observations of the way in which the distribution of plants reflected both geography and climate. Lyell was also among the first to appreciate the geological significance of Humboldt's 'isothermal lines' – lines of equal temperature that could be used to divide the world into climatic zones [18]. Observing that the positions and sizes of continents and the development of mountain ranges distorted those lines and climatic zones, he made a crucial intellectual leap. Recognizing that that many of Europe's older rocks had been deposited in much warmer climates than today's, he deduced that if the Earth's climate zones had not changed, then the land must have moved – the geography must have changed with time (Figure 2.7). Writing to Gideon Mantell (1790–1852) in February 1830, and swearing him to secrecy, he said '*I will give you a receipt* [i.e. recipe] *for growing tree ferns at the pole, or if it suits me, pines at the equator; walruses under the line [the Equator], and crocodiles in the arctic circle*' [31]. This exciting new idea profoundly changed the way people thought about the distant past.

Lyell acknowledged his debt to Humboldt in a letter to his geological friend George Poulett Scrope (1797–1876): '*Give Humboldt due credit for his beautiful essay on isothermal lines: the geological application of it is mine, and the coincidence of time 'twixt geographical and zoological changes is mine, right or wrong*' [31]. Would his theory hold the test of time? In the same letter Lyell confessed: '*That all my theory of temperature will hold, I am not so sanguine as to dream. It is new, bran new*' [31] (at that time the term bran-new was interchangeable with brand-new).

I focus on Lyell because he was the first scientist to concentrate intently on the geological record of past climates, and it would be fair to call him the 'father of palaeoclimatology'. He devoted three chapters of Volume I of the *Principles* to showing how the climates of past times could be recognized from the types and distributions of sedimentary rocks and their enclosed fossils – especially the seashells he so enjoyed studying. Not only that, but he also incorporated seven chapters on 'aqueous causes', under which he listed rivers, torrents, springs, currents, tides, and

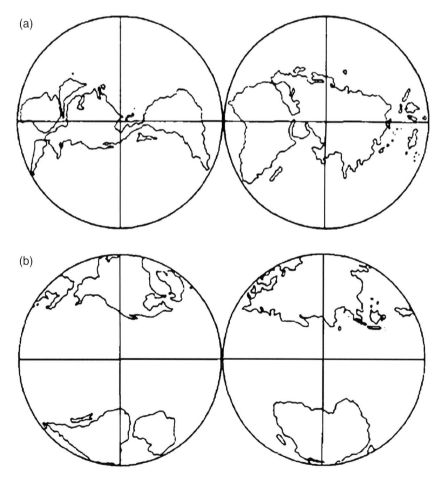

**Figure 2.7** Lyell's attempt to show how changes in the positions of the continents through time might contribute to extremes of (a) heat, or (b) cold.

icebergs as agents of change in the inorganic world, all of them likely to change as climate did. Lyell began his climate chapters by recapping the approach he took in a scientific paper in 1825–1826, where he deduced the likely conditions of deposition of fossil freshwater limestones [32]. His chapters on climate rehearsed the standard arguments for climate change from fossil evidence. He agreed with Buffon, Cuvier, and Brocchi that fossil evidence showed that Europe's climate was much warmer in former times. Unlike Cuvier, he found no need for some catastrophe to explain the cooling, and, like Fourier, he thought Buffon wrong to suggest that this was due to the solid Earth having been hotter in former times.

Lyell's thinking on the geographical control of climate matured as he gathered more data – especially from his visits to North America in the 1840s. For example in the 12th edition of his *Principles* published in 1875, anticipating later notions of the break-up of formerly continuous continents, he observed that '*If we go back ... to the Eocene period... we find such a mixture of forms now having their nearest living allies in the most distant parts of the globe, that we cannot doubt that the distribution of land and sea bore scarcely any resemblance to that now established*' [33]. Along the same lines he noted that '*In the case of the great Ohio or Appalachian coal-field, ... it seems clear that the uplands drained by one or more great rivers were chiefly to the eastward, or occupied a space now covered by the Atlantic Ocean*' [33]. Nothing he had discovered in 45 years of publishing the *Principles* detracted from his conclusion that '*Continents therefore, although permanent for whole geological epochs, shift their positions entirely in the course of ages*' [33]. He was well aware of how geography – manifested as the positions of continents, their coasts and their topography – modified climatic zones, observing that '*... on these geographical conditions the temperature of the atmosphere and of the ocean in any given region and at any given period must mainly depend ...*' [33].

Moving the continents around was dramatic stuff at the time, though unfortunately for Lyell he had no other information than his climate theory to back him up. That would be left to others as more data arrived. In due course Lyell's uniformitarian assumption that the Earth's average temperature had remained more or less constant through time would be proved wrong, but many of his other assumptions about climate remain valid, including the notion that the continents had changed position through time.

Lyell broke with tradition in abandoning the theory in vogue in the early nineteenth century, and embraced by – among others – his old Oxford tutor, William Buckland (1784–1856) (Figure 2.8), that the erratic blocks of rock littering the British landscape were the relics of Noah's flood. Like Hutton, Lyell was keen to take the scripture out of

**Figure 2.8**   William Buckland.

**Figure 2.9**   Charles Darwin.

geology and the geology out of scripture. Buckland's attempts to relate geology to scripture are hardly surprising, given that he was an Anglican Clergyman. An influential man, he was twice elected President of the Geological Society, from 1824 to 1826 and again from 1839 to 1841. But he had an open mind, and eventually abandoned the 'deluge' hypothesis.

Just as Humboldt had influenced Lyell, so Lyell too influenced a younger man with a big future, Charles Darwin (1809–1882) (Figure 2.9, Box 2.3).

Lyell's notion of significant change arising from slow processes operating steadily over the eons of geological time provided Darwin with the long periods through which tiny natural variations, which we now understand as genetic mutations, could accumulate and give rise to different species through natural selection. The two men met in

**Box 2.3  Charles Darwin [34]**

Charles Darwin is probably the world's most famous scientist. He obtained his BA degree from the University of Cambridge in 1831, after having spent a year (not very diligently) studying medicine at the University of Edinburgh, as his medical doctor father Robert had done. Medicine turned Charles's off, but he did learn a great deal of natural history from one of his lecturers, Robert Edmond Grant, an expert on sponges, and he began to study sea creatures intently, even going to sea with local fishermen and inspecting their catches. He also learned the basics of taxidermy from a former African slave, John Edmonstone, and the basics of geology from lectures by Robert Jameson. Both Grant and Jameson introduced him to the evolutionary ideas of the French savant Lamarck. Darwin was no stranger to the idea of evolution, a concept that his grandfather Erasmus had long espoused and written about in the previous century. Recognizing that Charles was never going to make a doctor, his father decided that he should plan to go instead into the Church and become a respectable parson. In Cambridge he became an avid collector of beetles, rising to be one of the country's most promising young entomologists even as an undergraduate. Natural history was exciting, and something he could pursue as a hobby while carrying out his not very onerous duties as a parson, should he eventually decide to become one. In his last year in Cambridge he developed a close friendship with an influential botanist, the Reverend John Stevens Henslow, who introduced him to the world of plants. Henslow persuaded Darwin to read Alexander von Humboldt's tales of his botanical travels in South America. After his graduation, seized with the idea of following in Humboldt's footsteps, and with Henslow's help, Darwin began planning a scientific expedition to Tenerife in the Canary Islands. On the island he would need not only to collect plants, beetles, and birds, but also to make geological measurements. To improve his understanding of geology and to learn its tricks of measurement he took advice from Cambridge's leading geological light, the Reverend Adam Sedgwick, Woodwardian Professor of Geology. Together they undertook a field trip to the wilds of North Wales. For Darwin it was a concentrated short course in the subject. He was now an accomplished young natural historian with skills in botany, zoology, entomology, and geology, a perfect background for his selection as the person to replace Henslow as the shipboard scientist on the proposed voyage around the world of HMS *Beagle*. Sailing on the ship in late 1831, Darwin took with him the first volume of Lyell's *Principles*, fresh from its publication in late 1830. On his return he would become friends with Lyell, who had him made a fellow of the Geological Society of London in 1836. Like Lyell he served the Society as an officer, being a member of Council from 1837 to 1850, one of the two joint secretaries from 1838 to 1841, and a Vice President from 1844 to 1845. Also like Lyell, he was made a fellow of the Royal Society, was awarded the Geological Society's Wollaston Medal (1859), was awarded the Royal Society's Copley Medal (1864), and was buried in Westminster Abbey.

October 1836 shortly after Darwin's return, and became friends. Lyell nominated Darwin to the Council of the Geological Society, and later helped to ensure that on 2 July 1858 Darwin's paper on natural selection was read at the Linnaean Society in London alongside that of Alfred Russell Wallace, who had reached the idea independently. Humboldt, too, influenced Darwin, the two having met in London in 1842. As Richard Holmes pointed out in 2008 in *The Age of Wonder*, '*Science is truly a relay race with each discovery handed on to the next generation ... and the world of modern science begins to rush towards us*' [35].

Lyell's *Principles* considerably influenced thought in Victorian times. As James Secord points out in his introduction to the Penguin Classic version in 1997, the *Principles* was '*a manifesto for fundamental change in the organisation of intellectual life*', capping the campaign by Lyell and others '*to sever all links of geology to a theology based in scripture*' [36]. After Lyell, geologists no longer accepted the biblical flood as having any worth in analysing Earth's history. For a modern view of Lyell's merits we may turn to *Time's Arrow, Time's Cycle*, the 1987 work by palaeontologist Stephen Jay Gould, for whom Lyell '*doth bestride my world of work like a colossus*' [37].

Nevertheless, the *Principles* had its flaws. For example, the first edition published in 1830–1832 did not mention the Ice Age. Like James Hutton, Lyell knew that mountain glaciers transport rock debris that is dumped in piles called moraines, where the glaciers melt, and that rivers sweep the glacial sand and mud away to the sea. But he also knew that polar mariners had seen drifting icebergs transporting large amounts of rock (Figure 2.10). That observation led him to speculate that melting icebergs would dump their loads on the seabed to '*offer perplexing problems to future geologists*' [33]. In Volume 3 of the first edition of the *Principles* he speculated that the huge erratic blocks of rock littering the landscape in the Alps

(a)

(b)

**Figure 2.10** Ice Transporting Rocks at Sea. Photos from the author: (a) Stranded iceberg carrying a load of rocks in the Fridtjof Channel, Antarctic Peninsula; (b) ice floe carrying a load of rocks in the Erebus and Terror Gulf, Antarctic Peninsula.

and the Jura had been transported by floating ice, not by ice sheets sliding over land as some Swiss geologists thought at the time. And he did not connect the erratic blocks littering the Swiss landscape with those of the United Kingdom.

## 2.3 Agassiz Discovers the Ice Age

One great mind is never enough – it requires many for us to get to grips with how the world works. And so it was left to another geological genius, Jean Louis Rodolphe Agassiz

(1807–1873) (Figure 2.11, Box 2.4), to point out how important ice may have been in the geological history of Europe and North America.

Like Lyell, Agassiz upset the comfortable world of established thought. In the summer of 1837, aged just 30, he turned the world of geological ideas upside-down with a proposal made at the annual meeting of the Swiss Society of Natural Sciences, of which he was the new President. A vast ice sheet must have carried erratic blocks across Europe in a recent Ice Age. Living in Switzerland, he knew that far from the snouts of glaciers the rock surfaces were scratched and smoothed, and strewn with boulders and

**Figure 2.11**   Louis Agassiz drawing radiates, 1872.

rubble like that still being carried and deposited by the ice. His new idea countered the suggestion by the first scientific explorer of the Alps, Horace Bénédicte de Saussure, in the late 1700s, that fast rushing streams deposited these boulders in catastrophic events. Hutton too disagreed with Saussure, proposing in his *Theory of the Earth* [8, 9] that a former extension of Swiss glaciers accounted for the distribution of boulders of Mont Blanc granite that de Saussure attributed to a deluge. Hutton wrote '*There would then have been immense valleys of ice sliding down in all directions towards the lower country, and carrying large blocks of granite to great distance, where they would be variously deposited and many of them remain an object of admiration to after ages, conjecturing from whence or how they came*'[9].

Later Swiss observers agreed with Hutton. In 1818, Jean Pierre Perraudin, a guide and chamois hunter, interpreted the gouges in hard unweathered rock as indicating the former widespread extent of Alpine glaciers. His remarks came to the attention of Ignace Venetz, chief engineer for the Swiss Canton du Valais, who in 1821 deduced from the positions of old terminal moraines down slope from present glacier terminations that the climate had warmed and the glaciers shrunk. He told that to the annual meeting of the Swiss Society of Natural Sciences in 1829, suggesting that glaciers had once extended over the Jura and into the European plain. Jean de Charpentier (1786–1855), Director of Mines of the Canton de Vaud, applied Venetz's observations more widely, proposing to the annual meeting of the same Society in 1834 that widespread erratic blocks and moraines had been deposited by ice; Swiss glaciers had formerly been much more extensive.

Agassiz, who attended Charpentier's lecture, had been one of his students. A trip into the field with Charpentier in 1836 to study the evidence for glacial transport convinced him. We'll let Elizabeth Agassiz tell us about her

---

**Box 2.4   Louis Agassiz**

Agassiz was born in Switzerland in 1807. He studied medicine and natural sciences at Zurich, Heidelberg, and Munich before moving to Paris where he studied with Humboldt and Cuvier. In 1832 he was appointed Professor of Natural History at Neuchatel in the Swiss Jura, and published 5 volumes of research on fossil fish between 1833 and 1843. In 1846 he visited the USA and was invited to stay there, becoming Head of the Lowell Scientific School of Harvard University in 1847. Harvard made him a Professor of Zoology and of Geology, and he founded the Museum of Comparative Zoology there in 1859, serving as its Director until he died. He was awarded the Geological Society of London's Wollaston Medal in 1836 for his work on fossil fish, and was elected one of the Society's Foreign Fellows in 1841. He thought highly of Charles Lyell, and named an ancient jawless fish after him – *Cephalaspis lyelli* – which lived in Scottish lakes in Old Red Standstone times in the Silurian and Late Devonian periods between 420 and 360 Ma ago. Like Lyell, Humboldt, and Darwin, Agassiz became one of the world's best-known scientists of the 1800s. Mountains, glaciers, and a Martian crater are named after him, as are several fish, beetles, a fly and a tortoise. A fossil glacial lake is named after him, as is the Agassiz Glacier in the USA's Glacier National Park.

husband's ensuing lecture to the same Society in 1837: '*In this address he announced his conviction that a great ice-period, due to a temporary oscillation of the temperature of the globe, had covered the surface of the earth with a sheet of ice, extending at least from the north pole to Central Europe and Asia .... "Siberian winter", he says, "established itself for a time over a world previously covered with a rich vegetation and peopled with large mammalia, similar to those now inhabiting the warm regions of India and Africa. Death enveloped all nature in a shroud, and the cold, having reached its highest degree, gave to this mass of ice, at the maximum of tension, the greatest possible hardness". In this novel presentation the distribution of erratic boulders, instead of being classed among local phenomena, was considered "as one of the accidents accompanying the vast change occasioned by the fall of the temperature of our globe before the commencement of our epoch" .... This was, indeed, throwing the gauntlet down to the old expounders of erratic phenomena upon the principle of floods, freshets, and floating ice'* [38].

Much astonishment and not a little ridicule greeted Agassiz's proposal that an ice sheet like that now covering Greenland formerly covered much of northwest Europe as far south as the Mediterranean. The great German geologist Leopold von Buch (1774–1853) attended the meeting and could hardly conceal his indignation and contempt for this young upstart [38]. It was von Buch who had first identified the erratic blocks littering the north German plain as having come from Scandinavia, though by some unknown means. Even Humboldt, who knew Agassiz from the time they had spent together in Paris, counselled his young friend to abandon '*these general considerations (a little icy besides) on the revolutions of the primitive world – considerations which, as you well know, convince only those who give them birth*' [38].

Some of the reluctance to accept Agassiz's idea stemmed from the fact that very few European scientists knew anything about the extent of ice sheets. The vast extent of the Antarctic ice sheet would not be fully appreciated until after the visits to the Ross Sea and Ross Ice Shelf with H.M.S. *Erebus* and H.M.S. *Terror* in 1841 and 1842 by James Clark Ross (1800–1862), whose book on his expedition was not published until 1847. The icy mass of East Antarctica had been only glimpsed before, by Von Bellingshausen in 1820, and by Dumont d'Urville and by Charles Wilkes in 1840. It was not even known (by Europeans) that a vast continuous ice sheet covered Greenland. Even so, Agassiz had leapt ahead of himself by claiming that ice extended as far as the Mediterranean, when glacial erratic blocks were actually confined to the Alps and northernmost Europe.

Undeterred, Agassiz wrote up his ideas for an English-speaking audience in a short paper published in 1838 [39].

In it he set out the evidence for glacial activity, and noted that grooved and polished rocks beneath Swiss glaciers are usually overlain by fine sand followed by rounded pebbles and then by angular blocks – the opposite of the sequence expected from transport by currents. The fine sand came from the disintegration of rock fragments and most likely caused the polishing. He called for research to see if this same relationship applied in the polar regions.

Agassiz put it all together in his 1840 book *Etudes sur les Glaciers* [40], using the distribution of boulders that could only have been transported by ice, and the gouges in the smoothed surfaces of rocks over which glaciers had passed, along with other features, to divine the former existence of ice sheets and glaciers where none now existed, and even to propose that ice sheets could move erratic blocks up hill, for example to the top of the Jura Mountains near his university, where the 3000 ton *Pierre-a-Bot* (toadstone) occurred. Charpentier published his own *Essay on the Glaciers and the Erratic Terrain of the Rhone Basin* in 1841. His Swiss-scale vista was eclipsed by Agassiz's Europe-wide vision, which became global after his arrival in North America in 1846, where he discovered further evidence for former ice sheets. Agassiz started a major research programme on Alpine glaciers, spending a decade working at an alpine research station and climbing all over the Alps with fellow researchers and students, literally starting the study of glaciology and publishing the results in 1847 in *Système Glaciare*.

Agassiz was keen to convert William Buckland, the leading proponent of the biblical deluge hypothesis for explaining erratic blocks. Buckland was intrigued enough by Agassiz's theory to visit Switzerland and see the evidence. Becoming convinced that Agassiz might be right, he invited Agassiz to visit Britain to see if evidence of a past ice sheet could be found there too. Agassiz duly arrived on this mission in August 1840, visiting Scotland and lecturing on his new theory at the meeting of the British Association for the Advancement of Science in Glasgow. Lyell attended the meeting, but remained unconvinced.

Touring Scotland and other parts of Britain, Agassiz and Buckland found the evidence they were looking for – moraines, erratic blocks and polished and grooved rocks showed that great sheets of ice like that covering Greenland must formerly have covered the mountainous areas of Great Britain. Buckland even managed to convince Lyell that the piles of rocks near Lyell's Scottish home were moraines deposited at the edge of this former ice sheet [41]. In November and December 1840, Agassiz, Buckland and Lyell gave lectures at the meetings of the Geological Society of London on their discoveries of evidence for former British ice sheets [42, 43]. The initial reaction was hostile [31]. Buckland concluded the 1840 meeting in high spirits by condemning to '*the pains of eternal itch without the privilege of scratching*' anyone who challenged the evidence supporting the ice age theory [43]. But although the papers were read at the Society's meetings, and précis were published by the Society's Secretaries to convey the main points to the readers of the Society's Proceedings, the full papers were never published [41, 42]. Part of the problem was that another Scottish lion of the British geological scene, Roderick Murchison (1792–1871), who had visited Scotland with Agassiz and Buckland, was unconvinced. Murchison had been President of the Geological Society in 1831–1833, and was again in 1841–1843. During his Presidential Address to the Society in the latter term, he chose to attack the ice age theory. He did not back away from that stance until 1862, when he finally recanted in an address to the Geological Society of London. Sending a copy of his 1862 paper to Agassiz he wrote: '*I have the sincerest pleasure in avowing that I was wrong in opposing as I did your grand and original idea of my native mountains. Yes! I am now convinced that glaciers did descend from the mountains to the plains as they do now in Greenland*' [44]. The evidence had mounted up.

In his 1840 paper [42] Agassiz explained that rivers draining the massive ice sheet that had brought the erratic boulders to the plains of northern Europe had also given rise to widespread outwash gravels, for which there was no other explanation. The existence of this ice sheet indicated that a period of intense cold – an Ice Age – had intervened between the warm conditions of the Tertiary period and those of today. Modern mountain glaciers were the remnants of that former ice sheet. Having found polished rocks, erratic blocks, and outwash gravels across much of Scotland, Ireland, and the north of England, along with rounded hillocks of ice-cut rock named 'roches moutonnés', he deduced that an ice sheet had covered these areas too. The distribution of erratic blocks suggested that they had moved in all directions away from 'centres of dispersion', which would not be expected for deposition from floating ice. The main centres of dispersion in the British Isles were the mountains of Ben Nevis, the Grampians, Ayrshire, the English Lake District, Wales, Antrim, Wicklow, and the west of Ireland. Floating ice from Scandinavia explained the origin of erratic blocks on the east coast of England.

I was somewhat surprised to learn from a recent paper by Geir Hestmark [45] that the Ice Age had been discovered before Agassiz, in Norway, by Jens Esmark, who became in 1815 the first Professor of Mining Sciences at the newly established Norwegian University, and who was elected to the Royal Swedish Academy of Science in 1825. Inspired by Horace de Saussure's studies of the Alps, Esmark made extensive studies in Norway's mountains, realising from

landforms and deposits close to sea level that glaciers had formerly extended from the mountains to the coast [46, 47]. He deduced that long ramparts crossing valleys and made of unsorted sand, gravel, and boulders were abandoned end moraines. One such ridge – Vassrygg, near the coast at Lysefjorden – is now named the Esmark Moraine. In front of it is a sandy plain – a typical outwash deposit – and in the valley behind it lies Lake Haukalivatn. Esmark deduced that great masses of ice had filled and hollowed out the valley. Later, visiting an active glacier (Rauddalsbreen) 1020 m up in the hills he found a modern end moraine similar to the Vassrygg, and now named Otto Tank's Moraine in memory of Esmark's travelling companion and student. As at Vassrygg, the moraine is accompanied by a sandy outwash plain. Esmark noted that the glacier had actually pushed boulders across a valley and 100 m up the opposite slope, a clear sign of the transporting power of ice. Esmark deduced that enormous glaciers once covered Norway, reaching the sea at Stavanger. He deduced that at some point in relatively recent times Earth had been shrouded in ice and snow, and went on to claim that the mysterious boulders scattered across northern Europe had been transported there by glaciers. The boulders atop the Jura had to be explained likewise, he wrote, well before Agassiz had appeared on the scene. Here, then, we have the causal explanation for the Norwegian fjords, and the solution to the missing masses of rock from the glaciated valleys. Much of that rock must lie beneath the adjacent ocean.

## 2.4 Lyell Defends Icebergs

Lyell met Charpentier in 1832 in Switzerland, while on his honeymoon, and so had been exposed to Charpentier's ideas [25]. He also met Agassiz several times during the 1830s, and they worked together on fossil fish for a while [25]. They met again when Agassiz visited Buckland in 1840, and presented their papers on glaciers together to the Geological Society meetings late that year. But Lyell was a hard man to convince, and the extent to which he accepted the notion that sheets of ice had transported boulders was distinctly limited. Having seen moraines in the Alps, he was not going to deny the role of mountain glaciers in transporting erratic blocks. But – what happened beyond the mountains? Lyell thought icebergs had done the work.

Lyell used the word 'till', a Scottish farmers' term, to describe the widespread unstratified jumbled mass of erratic blocks, pebbles, and clay that we now call 'boulder clay' covering parts of the British Isles. He lumped 'till', together with other deposits from the glacial era (like Agassiz's outwash fans of gravel), into what he called the 'glacial drift', a term chosen on the one hand to support his iceberg theory, and on the other hand to replace the former term 'diluvium', which came from the biblically inspired 'flood' hypothesis formerly used to explain the distribution of this recent debris [25].

Agassiz was unable to explain in detail the origin of the till, but assumed that it was derived from the ice sheet in some way, not least because the boulders in the till were typically striated and gouged like ice-transported rocks. He imagined that boulders now found as erratic blocks might have slid down the Alpine slopes and out across his proposed European ice sheet in some catastrophic fashion, to be left behind when the ice beneath them melted. Lyell, in contrast offered a non-catastrophic 'steady-state' mechanism, the supply of rocks, pebbles, and rock flour from floating icebergs. We now know more than both Lyell and Agassiz. The fine-grained clayey element of 'boulder clay' is the rock 'flour' or powdered rock derived by rock fragments interacting with each other and with the surrounding country rock as ice sheets move over the ground.

In his 1840 paper [42, 44], Lyell suggested that '*the assumed glacial epoch*' had arisen as Scottish glaciers first advanced to the sea as they did in South Georgia, then remained stationary while the intervening hollows filled with snow and ice on which boulders could slide to their present positions – much as Agassiz was suggesting for the erratic blocks of western Europe. The ice then retreated, leaving moraines and debris behind. To explain the origin of boulder clay or till away from mountainous areas where glaciers could provide a means of transport, Lyell called, as he had in his *Principles*, on the transport of rocks and sediment by floating ice [42]. He rejected Agassiz's idea that some catastrophic event had caused boulders to fall off the Alps and slide out over the European ice sheet, because he believed that the Alps rose gradually, consistent with his uniformitarian principles.

Although he was exposed back in 1832 to Charpentier's observation that Swiss glaciers moved boulders, Lyell was equally impressed during his visit to Sweden in 1834, where he saw along the coast granite boulders that appeared to have been carried by floating ice [13]. He was also impressed by accounts from mariners of boulders carried on icebergs. Not long after his excursion with Agassiz he was told of similar observations made by Joseph Hooker on the James Clark Ross expedition to Antarctica in 1839–1841, and Darwin had reported rocks being carried out to sea by icebergs broken off from glaciers in southern Chile, as Lyell reported in his *Elements of Geology*. Lyell just missed observing this phenomenon for himself when he was crossing the Atlantic on his way to and from America in the mid to late 1840s [48]. Writing to his sister Carry from the Steamship Britannia in June 1846, he reported

'*We passed fifty icebergs or more in daylight ... One iceberg ... which came close to us when I was below, had a large rock twelve feet square on the top and as much gravel and dark sand on its side*' [31]. While this confirms my own observation that such occurrences are rare, it was enough to make Lyell stick to icebergs as accounting for the distribution of erratic blocks away from mountainous glaciated regions like Scotland and Scandinavia, no matter what Agassiz said.

Like Agassiz, Lyell also saw glacial erratics in North America. During his visit to the USA in 1853, his host James Hall took him to see trains of erratic boulders in the Berkshire Hills of western Massachusetts. Lyell's biographer explains [49]: '*The boulders were distributed in long parallel rows, extending in nearly straight lines across ridges and valleys from their starting points on the Canaan Ridge. Their direction was nearly at right angles to the lines of the ridges and bore no relation to the direction of the streams and rivers. The boulders were rounded like the glacial boulders called in Switzerland roches moutonnés .... one of the larger boulders ... [near the meeting house in Richmond] was fifty-two feet long, forty feet wide, and, although partially buried, fifteen feet high [50]. The boulders rested on a deposit resembling the European "northern drift." Where the underlying rock was exposed, its surface was polished, striated, and furrowed, with the furrows running in the same direction as the trains of boulders. Lyell thought that the trains of boulders must have been transported by floating ice at a time when the Berkshire hills stood at a much lower level, with only their highest ridges protruding above the sea. He thought their transport could not be explained by glaciation, because if glaciers had transported the boulders, the trains of boulders should have been distributed down the valleys instead of across them. In fact, the boulders had been transported by glaciers, but by continental glaciers rather than by mountain glaciers, the only ones with which Lyell was familiar*' [49]. Nowadays we would say 'transported by continental ice sheets' rather than by continental glaciers.

The contrast between Lyell and Agassiz was one of vision. Lyell stuck to what he knew to be true – glaciers occupied valleys and carried boulders down them, and, where they met the ocean, icebergs carrying boulders might break off and carry their burden of rocks out to sea. Agassiz could envision a merging of mountain glaciers into great sheets of ice covering entire landscapes, ploughing across and shaping the land and dumping clay and boulders en route. 'God's Great Plough' he called the ice sheet.

Although in later years he would back away from Lyell's adherence to transport by icebergs, Charles Darwin initially followed Lyell's line closely in a paper on the glaciers of Caernarvonshire, in Wales [51]. Investigating the moraines near Lakes Ogwyn and Idwell in the Welsh mountains, he deduced that the glaciers from the valleys in which those lakes now sat had formerly united and plunged down the valley of Nant-Francon towards Bethesda, where they had dumped in the sea a whitish earth full of rounded and angular boulders that were deeply scored like the rocks *in situ* over which a glacier had passed. Following Lyell's line he assumed that the boulders had been dropped into this mud from floating icebergs, and that the land had since been uplifted. '*By this means*' he said '*we may suppose that the great angular blocks of Welch [sic] rocks scattered over the central counties of England were transported*' [51]. He concluded: '*that the whole of this part of England was, at the period of the floating ice, deeply submerged. I do not doubt that at this same period the central parts of Scotland stood at least 1300 feet beneath the present level, and that its emergence has since been very slow. The mountains at this period must have formed islands, separated from each other by rivers of ice, and surrounded by the sea*' [51]. Lyell would have approved.

Like Lyell, Darwin accepted that there must also have been vast thicknesses of land ice locally as a source for floating icebergs. His letter to W.H. Fitton [52] is a reminder that one may often not be able to 'see' what is under one's nose. On a field trip to Capel Curig in North Wales he wrote: '*the valley about here, & the Inn, at which I am now writing, must once have been covered by at least 800 or 1000 ft in thickness of solid Ice! – Eleven years ago, I spent a whole day in the valley, where yesterday every thing but the Ice of the Glacier was palpably clear to me, and then I saw nothing but plain water, and bare Rock. These glaciers have been grand agencies.*' But he then went on to extol the virtues of the power of drifting icebergs to distribute erratic blocks. '*I am the more pleased with what I have seen in N. Wales, as it convinces me that my views, on the distribution of the boulders on the S. American plains having been effected by floating Ice, are correct*' [52]. It would take a lot for Darwin to withdraw support from Lyell.

Lyell stuck to the iceberg theory more or less unchanged throughout his life. In the second edition of his *Elements of Geology*, published in 1841, he admitted that small glaciers might once have existed in Scotland, but dismissed the theory that the widespread British deposits of 'glacial till' comprising mixed boulders and clay had been deposited beneath an ice sheet, preferring still to think of them as deposited from floating icebergs. By the time he published *Antiquity of Man,* in 1863, he had accepted the refrigeration of the climate in the post-Tertiary Pleistocene that Agassiz had postulated, and that this had led to large areas of Britain and northwestern Europe becoming covered by 'glacial drift'. Lyell's hypothesis that boulder clay was deposited from floating icebergs required that much of England north of a line joining the estuaries of the Thames

in the east and the Severn in the west, as well as much of the northwest European plain, must have been submerged. He explained away the grooves carved into exposed rocks on hillsides as having been made by stones embedded in the bottoms of icebergs, rather than – as Agassiz would have it – by stones embedded in a moving ice sheet.

Lyell accepted that the ice originated in glacial dispersion centres on highlands in Scandinavia, Scotland, Wales, and the English Lake District. But he thought that those centres were limited in extent and discharged their ice into a surrounding ocean rather than into a surrounding ice sheet like that of Greenland. Lyell also agreed with Agassiz's suggestion that within those distribution centres, glacial lakes dammed by ice were locally important, the beaches of different lake levels explaining the terraces or 'parallel roads' around Scotland's Glen Roy. His interpretation of the terraces around Glen Roy was not original – it had first been proposed by Scottish geologist John MacCulloch in 1817, when he was President of the Geological Society of London.

Where Lyell and Agassiz differed profoundly was in explaining the origin of the glacial drift. In *Antiquity of Man* Lyell expanded on his marine glacial theory, suggesting that during the Ice Age much of England and northwest Europe must have been submerged to depths of more than 600 ft, Scotland to depths of as much as 2000 ft and Wales to a depth of 1350 ft. By 1875, in the 12th edition of *Principles*, these figures had changed to 'perhaps' 500 ft in Scotland and 2000 ft in Wales [33]. As is clear from that edition, much of his argument for submergence rested upon the occurrence of seashells at high altitudes among the boulder clay. For someone who denied any role for catastrophism in geology, Lyell was sailing perilously close to the wind in invoking unexplained forces that could periodically lift the UK and Europe above the sea and then submerge it, during the small amount of geological time represented by the Ice Age.

By 1848, Darwin began to realise that sticking to the Lyellian view required some contorted thinking [53]. Trying to answer a common criticism of the time – that floating ice could not carry erratic blocks from a lower to a higher level – he suggested that with repeated subsidence of the land, floating ice could gradually deposit boulders at progressively higher levels. Special pleading indeed! The subsidence would have to have been significant and more or less immediate, something for which there was no apparent mechanism, and would have to have been continually repeated. A certain Mr. Nicol objected: '*that when the parent rock was once submerged, no further supply of boulders could be derived from it*' [53]. Darwin confessed '*this appears to me an objection of some force*' [53]. Well he might! Not to be deterred, he argued that the piling up of ice by storms along a shore would raise boulders above their original level. Rather a weak response considering that erratic boulders of immense size occurred 900 to 1000 ft above the strata from which they had been carved. He would recant, as we see in Chapter 3.

Enter Archibald Geikie (1835–1924), a young Scottish geologist, who roundly criticized Lyell's iceberg transport theory early in the 1860s (Figure 2.12, Box 2.5).

From his detailed examination of the Ice Age geology of Scotland, Geikie concluded that the land must have been shaped by the actions of a giant ice sheet, the remains of which mantled its surface as 'drift' deposits [54]. Geikie called for the iceberg theory to be abandoned forthwith. He went on to suggest that he hoped he might have convinced Lyell of the correctness of his conclusions. Lyell did read Geikie's book, writing to his wife in May 1863 that '*Geikie's book on the Glacial Period in Scotland is well done …*' [31]. Nevertheless, that same year, 1863, Lyell published

*Antiquity of Man* with its illustrations showing Great Britain drowned beneath an iceberg-flooded sea! In the 12th edition of his *Principles* Lyell continued with his ice-flooded sea, but conceded a little ground to Geikie, noting that in Scotland '*some examples of this … striation may have been due to the friction of icebergs on the bed of the sea during a period of submergence; others to a second advance of land glaciers over moraines of older date*' [33].

Lyell did reverse his conclusion about seaborne transport in one case. In the 12th edition of *Principles*, he reported that on a visit to Switzerland in 1857 the local geologists had convinced him that an ice sheet had filled the Valley of Switzerland between the Alps and the Jura and transported down into the valley and up the other side the erratic blocks now found 50 miles away from the Alps atop the Jura Mountains [33]. Writing to his father-in-law, Leonard Horner, from Zurich in 1857, he said '*If the hypothesis now adopted here to account for the drift and erratics of Switzerland, the Jura, and the Alps be not all a dream, we must apply the same to Scotland, or to the parts of it that I know best. All that I said in May 1841 on the old glaciers of Forfarshire … I must reaffirm*' [31]. In a letter to J.W. Dawson in February 1858 he went further to call for glaciers (not icebergs) to transport erratics and drift onto the plains of the River Po in northern Italy [31].

By the 12th (1875) edition of his *Principles*, Lyell's conversion to the Ice Age cause was more or less complete. In it he recalls seeing that many of the rocky surfaces exposed in Switzerland are '*… smoothed and polished, and scored with parallel furrows, or with lines and scratches produced by hard minerals. …. The discovery of such markings at heights far above the surface of the existing glaciers, and for miles beyond their present terminations,*' he said '*… affords geological evidence of the former extension of the ice beyond its present limits in Switzerland and other countries*' [33]. Although this meant that Agassiz had been right all along about the Swiss erratics, Lyell could not accept that Agassiz's theory could be extended beyond the Swiss region, where icebergs must

**Figure 2.12** Archibald Geikie.

---

**Box 2.5  Archibald Geikie**

Archibald Geikie was born in Edinburgh, and educated at the university there. He became an assistant for the British Geological Survey in 1855, worked extensively on the geology of Scotland, was elected a Fellow of the Royal Society in 1865, and was appointed Director of the Geological Survey of Scotland when it was formed in 1867. While in that post he became the first Murchison Professor of Geology and Mineralogy at the University of Edinburgh in 1871, and held those two posts together until 1881, when he was appointed to be Director-General of the Geological Survey of the UK and Director of the Museum of Practical Geology in London. Geikie was President of the Geological Society of London 1891–1892, was awarded the Murchison Medal by that Society in 1895, received the Royal Medal from the Royal Society in 1896, and became its President in 1909. He was knighted in 1891.

have ruled, except in other mountainous places like Scotland (and presumably Scandinavia).

Next into the lists was yet another Scottish geologist, James Geikie (1839–1915) (Figure 2.13, Box 2.6), younger brother of the more famous Sir Archibald.

Following in his illustrious brother's footsteps, James amassed a vast storehouse of knowledge of the geology of the glacial and interglacial periods of the Ice Age from all over the world, publishing his tome *The Great Ice Age* in 1874 [55]. The comprehensive 3rd edition published in 1894 included a chapter on the glaciations of North America, by the great American geologist T.C. Chamberlin. James's most telling fact came from Nansen's observation that ground moraines beneath the Greenland ice sheet were visible in arches and tunnels under the ice front, where one could see a bluish clay charged with blunted and scratched boulders – evidently the precursor to boulder clay or till. Geikie confirmed '*Most ... icebergs [are] free from inclusions [meaning rocks, pebbles or clay] of any kind. ... Nothing has been observed to lead us to believe that*

*parallel striations and markings, like those produced by glaciers, are ever the result of iceberg action*' [55].

Having examined glaciers in the field, he went on to note that '*The finer material – the "flour of rocks" resulting from this action [glaciers with rock debris embedded in their bases and grinding away at the rocks beneath]... renders the glacial rivers turbid and milky*' [55]. If there were no river to wash the 'flour' out, it would accumulate beneath the ice. A further argument against the iceberg hypothesis was that it should lead to deposits that were sorted, the coarser material being deposited first, the finer having settled last, not at all like the higgledy-piggledy nature of boulder clay.

A final telling point in his mind was that the topography itself was witness to the passage of an ice sheet, with humpbacked features like roches moutonnés carved from underlying rock, drumlins – mounds made of squeezed up boulder clay, kames – collapsed piles of gravel, eskers – long winding ridges of sand and gravel deposited by streams beneath the ice, and kettle holes – hollows created when buried blocks of ice melt away and now filled with water – witness the many glacial lakes of Canada and Finland. From the difference between river-cut and ice-carved topography, James Geikie deduced that the British ice sheet had covered the northern Pennine hills, the backbone of England, but not the southern part, though we now know that in the Anglian glacial advance, 480 000 years ago, the Pleistocene ice sheet reached as far south as Finchley and Hornchurch on the northern outskirts of London (Figure 2.14) [55].

Onto the scene in 1875 strode yet another Scottish geologist, James Croll, at the time about to join the Geikies and Lyell as a fellow of the Royal Society. We'll see more of him in Chapter 3. Like the Geikies, Croll could not find any evidence that icebergs produced striations when ploughing through the seabed, but found ample evidence for ice sheets having done so when passing over land [57]. Re-examining the travellers' tales that Lyell had used to support his theory that icebergs transported stones, Croll found that those kinds of icebergs were rare. Besides there were no reports of icebergs carrying clay. My own observations support his conclusion. Only one of the many icebergs I have seen around Greenland or in the Antarctic carried rocks, as did one ice floe among a myriad others

**Figure 2.13** James Geikie.

---

**Box 2.6   James Geikie**

James Geikie was born in Edinburgh and educated at the university there. He served on the Geological Survey from 1862 to 1882, when he succeeded his brother Archibald as Murchison Professor of Geology and Mineralogy at Edinburgh University. He was elected a Fellow of the Royal Society in 1875, and awarded the Geological Society of London's Murchison Medal in 1889. He published a standard textbook *Outlines of Geology* in 1886. An Alaskan glacier is named after him.

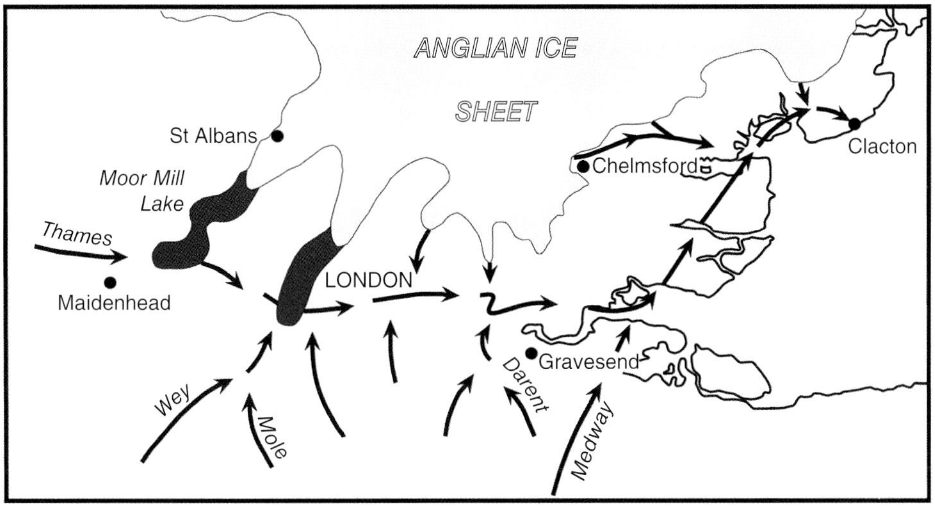

**Figure 2.14** The extent of the Anglian ice sheet, 480 000 years ago, near London. *Source:* From figure 7 in Ref. [56].

that I sailed through in Erebus and Terror Gulf in the western Weddell Sea (Figure 2.8). Even so, Ice Age iceberg drift does explain the occurrence of blocks of granite and related continental rocks dredged from the Mid-Atlantic Ridge in the northern North Atlantic [58], and the African continental shelf south of Cape Town [59].

Croll agreed with the Geikies that the clay of boulder clay is eroded from the ground by the action of ice sheets on land, and left behind as the ice melted. Where the boulder clay contained marine shells, for example in Caithness in Scotland, Croll showed that they were as fragmented and striated as the adjacent boulders. He inferred that the Scandinavian ice sheet had extended across the North Sea, and had ripped marine clays and shells from the bed of the sea en route, depositing them on Scotland.

Modern geologists confirm that the boulder clay of Europe's glacial 'drift' is the 'ground moraine' or 'bottom moraine' of the thick ice sheets that covered much of Europe and North America. It is quite different from 'end moraines' – the piles of debris dumped at the terminations of glaciers or ice sheets. They also agree that ice sheets scouring the bed of the North Sea en route from Scandinavia, or scouring the Irish Sea en route to Wales, incorporated marine shells from the seabed and subsequently plastered them on Britain as part of the boulder clay when the ice melted. This obviated the need for Britain and northwest Europe to have oscillated up and down to accommodate Lyell's glacial theory. As in Switzerland, ice sheets can move both down and up. The gods, they say, have feet of clay – in Lyell's case it was boulder clay.

Ironically, we are left with the word 'drift' as a descriptor of glacial deposits deposited by moving ice sheets, despite its original application to explain deposits from drifting icebergs. Lyell's influence lingers long! But he was not alone.

Other influential figures, like Roderick Impey Murchison, one time President of the Geological Society, agreed with Lyell, for example in his Presidential addresses to the Society in 1842 and 1843, which helps to explain why we had to wait until the immaculate fieldwork and compilations of the brothers Geikie in the 1860s and 1870s exposed the fallacy of believing that the 'drift' of Europe and north America originated from icebergs.

Initially, neither Lyell nor Agassiz seem to have realised that Agassiz's theory, calling for the formation of a vast ice sheet on land, implied a large drop in sea level, something pointed out as early as 1842 by Charles MacLaren [48, 60]. Eventually, Lyell did recognise the link between the formation of an ice sheet and a drop in sea level, and drew attention to it in the 10th and later editions of *Principles*.

Could parts of the British Isles have subsided, by several hundred feet, as Lyell imagined to explain raised beaches and terraces and raised deposits of unbroken marine shells? Opposing Lyell, James Geikie [55] favoured the explanation provided by a Mr. Jamieson [61] that the immense weight of the ice sheet had depressed the land surface, which then rebounded slowly as the ice melted. This process, called isostatic adjustment, which Lyell knew little about, would have raised coastal deposits and terraces. James seems to have forgotten that his own brother Archibald had already suggested more or less the same process as Jamieson in 1863 to explain the elevation of evidently marine deposits [54]. We now know that when ice sheets are abundant sea level is low, and the land is depressed beneath the weight of ice. As the ice melts, relatively rapidly, the sea rises, quickly flooding the margins of the depressed landmass, which itself rises more slowly allowing time for shell beds to form on the flooded margins. When all the ice sheets have melted, sea level stabilizes but the submerged lands continue to rise, eventually raising the shell beds above sea level. This process might have helped to delude Lyell into thinking that the land had been submerged beneath an iceberg-flooded sea rather than pushed down by an ice sheet. Scotland and Scandinavia are still slowly rising, as Lyell knew, having visited Sweden in 1834 and delivered the Bakerian Lecture on this topic to the Royal Society in the winter of that year [25]. In contrast, the south of England, which had bulged up south of the front of the ice sheet as the north of Britain sank under its load of ice, is now slowly sinking while Scotland rises.

James Geikie's monumental tome shows that the great ice sheets of the glacial period extended over most of the north European plain, the southern boundary being a roughly east–west line starting in the west between the Severn and the Thames estuaries at around 51° 30'N. In North America the main Laurentide ice sheet centred over Hudson's Bay covered most of Canada and the northern USA east of the Rockies, extending down to 37°35'N close to the junction of the Ohio and Mississippi Rivers in Illinois.

Continuing his study of what he called the 'glacial epoch' and what we know as the Ice Age, Lyell reported in his 1863 volume, *Antiquity of Man*, that Swiss geologists considered that there had been at least two phases of glacial action in the Alps, the first carrying erratics to the Jura mountains before the glaciers retreated, and the second filling Lake Geneva with ice but not reaching the Jura before retreating again. Lyell saw some parallels with the geology of the UK, associating the first retreat of the Alpine glaciers with the period of his supposed submergence of England and the deposition of boulder clay from icebergs. Being a canny Scot he realized that such chronological

comparisons were 'very conjectural'. He was confident, however, that when the ice advanced most the sea would have been at its coldest, and that this would be reflected in the species of seashells, a prescient observation.

James Geikie's study of the Ice Age greatly extended that of Lyell, identifying up to six glacial periods with intervening interglacial periods since the end of Pliocene time [55]. Chamberlin had also identified several glacial periods in North America. Geikie included the Alps, where by the end of the nineteenth century, geologists agreed that there had been at least four periods of advance of the ice – the Günz, Mindel, Riss, and Würm glaciations, separated by warm interglacial episodes when the climate may have been slightly warmer than it is today [62]. The difficulty in establishing a chronology for the successive events of the Ice Age lay partly in the fact that each succeeding advance of the ice sheet tended to obliterate the evidence deposited by its predecessors – ice sheets were indeed, as Agassiz put it, 'God's Great Plough'.

The realisation that there were several pulses of ice advance during the Ice Age leads us on to consider what the drivers or such change might be, in Chapter 3.

Note added in proof. In 1865, Archibald Geikie and his brother James set out for Norway to examine glaciers in the field, to provide them with evidence bearing on the origin of glacial remains in Scotland. They visited glaciers in Hølandsfjord south of Tromso, and Øksfjord north of Tromsø. This was a test in the field of Agassiz's and Geikie's land-ice theory, and it confirmed in Geikie's mind that the former glaciation of Scotland was the result of land-ice processes, not icebergs, although – not yet aware of isostatic uplift – he still called on submergence to explain tills with marine faunas [63, 64].

## References

**1** Lamb, H.H. (1966). Britain's climate in the past. In: *The Changing Climate – Selected Papers*, 170–195. London: Methuen.

**2** Georges-Louis Leclerc, Comte de Buffon (1749-1788) *Histoire Naturelle, Générale et Particuliére*, in 36 volumes, see https//www.buffon.cnrs.fr/ice_book detail-in-text-buffon-buffon hn-34-7.

**3** de Buffon, C. (2018). *The Epochs of Nature*. (trans. and compiled by J. Zalasiewicz, A.-S. Milon, and M. Zalasiewicz, with an Introduction by J. Zalasiewicz, S. Sörlin, L. Robin and J, Grinevald). Chicago University Press.

**4** Zalasiewicz, J., Milon, A.-S., and Zalasiewicz, M. (2018). Buffon the geologist. *Geoscientist* 28 (3): 17–19.

**5** Rudwick, M.J.S. (2005). *Bursting the Limits of Time: the Reconstruction of Geohistory in the Age of Reason*, 708 pp. Chicago, London: University of Chicago Press.

**6** Geikie, A. (1897). *Founders of Geology*. Macmillan, London: https://archive.org/index.php.

**7** Baxter, S. (2004). *Revolutions in the Earth: James Hutton and the True Age of the World*, 245 pp. Phoenix: Orion Books, London.

**8** Hutton, J. (1788). Theory of the Earth; or an investigation of the laws observable in the composition, dissolution, and restoration of land upon the Globe. *Transactions of the Royal Society of Edinburgh, I, Part II*: 209–304.

**9** Hutton, J. (1795). *Theory of the Earth, with Proofs and Illustrations*. Edinburgh.

10 Capra, F. (2013). *Learning from Leonardo: Decoding the Notebooks of a Genius*. San Francisco: Berrett-Koehler. ISBN 978-1-60994-989-1, 380 pp.

11 Cuvier, G. (1812) Discourse sur les révolutions de la surface du globe et sur les changements qu'elles ont produits dans la règne animal, in *Recherches sur les Ossememnts Fossiles*. 1st ed in 4 volumes, Deterville, Paris. New edn Christian Bourgeois, Paris, 1985, in French.

12 Cuvier, G. (1818). *Essay on the Theory of the Earth*, 5e. New York: Kirk & Mercein. T. Cadell, London, 1827.

13 Bard, E. (2004). Greenhouse effect and ice ages: historical perspective. *Comptes Rendus Geoscience* 336: 603–638.

14 Lavoisier, A. (1789). Observations generales sur les couches horizontales, qui ont ete deposees par la mer, et sur les consequences qu'on peut tirer de leurs dispositions, relativement a l'anciennete du globe terrestre. *Memoires de l'Académie des Sciences*: 351–371.

15 Carozzi, A.V. (1965). Lavoisier's fundamental contribution to stratigraphy. *The Ohio Journal of Science* 65 (2): 71–85.

16 McCrory, D. (2010). *Natures Interpreter – The Life and Times of Alexander von Humboldt*, 242 pp. Cambridge: Lutterworth.

17 von Humboldt, A. and Bonpland, A. (1805) *Essai sur la Géographie des Plantes*, Levrault Schoell et Cie, Strasbourg, 154 pp. [English version 2009, edited by S.T. Jackson, S.T., and translated by S. Romanowski, University of Chicago Press, Chicago IL, 274pp].

18 von Humboldt, A. (1817) Von den isothermen linien und der verteilung der wärme auf dem erdkörper, *Mémoires de Physique et de Chimie de la Société d'Arceuil* 3, 462–602; reprinted in: von Humboldt, A. (1989) Studienausgabe, Sieben Bände, hrsg. von Hanno Beck. Wissenschaftliche Buchgesellschaft, Darmstad,Band VI, 18–97; English translation (1820-21) "On isothermal lines, and the distribution of heat over the globe". *Edinburgh Philosophical Journal* **III** (5), 2-20; **III** (6), 256-274; **IV** (7), 23-37; **IV** (8), 262-280; and **V** (9), 28-38.

19 von Humboldt, A. (2014). *Views of Nature; Or Contemplations on the Sublime Phenomena of Creation*, 344 pp (trans. by M.W. Person from the German version published in 1808, and ed. by S.T. Jackson and L.D. Walls). University of Chicago Press.

20 von Humboldt, A. (1907). *Personal Narrative of a Journey to the Equinoctial Regions of the New Continent During the Years 1799–1804*, in 3 vols (trans. T. Ross). London: George Bell & Sons.

21 von Humboldt, A. (1849). *Aspects of Nature in Different Lands and Different Climates, with Scientific Elucidations* (trans. M. Sabine). Philadelphia, PA: Lea and Blanchard, 475pp (first published as *Ansichten der Natur*, in 1808).

22 Wulf, A. (2015). *The Invention of Nature: The Adventures of Alexander von Humboldt – The Lost Hero of Science*, 473 pp. London: John Murray.

23 von Humboldt, A. (1845-1861). *Cosmos: A Sketch of a Physical Description of the Universe*, in 5 vols, (trans. E.C. Otte). London: Henry Bohn.

24 http://www.humboldt-foundation.de/web/kosmos-view-onto-germany-93-1.html.

25 Wilson, L.G. (1972). *Charles Lyell; the Years to 1841: The Revolution in Geology*, 553 pp. New Haven, CT and London: Yale University Press.

26 Bonney, T.G. (1895). *Charles Lyell and Modern Geology*, 222 pp. Macmillan New York.

27 Rudwick, M.J.S. (1998). Lyell and the principles of geology, in *Lyell: The Past is the Key to the Present* (eds D.J. Blundell and A.C. Scott). *Geological Society of London Special Publication* 143: 3–15.

28 Gould, S.J. (1999). Lyell's pillars of wisdom. *Natural History* 108 (4), reprinted in Gould, S.J., 2000, *The Lying Stones of Marrakech*. Harmony Books, New York, 147-168.

29 Vai, G.B. (2009). Light and shadow: the status of Italian Geology around 1807. In: *The Making of the Geological Society of London* (eds C.L.E. Lewis and S.J. Knell). *Geological Society of London Special Publication* 317: 179–202.

30 Kraüsel, R. (1961). Palaeobotanical evidence of climate. In: *Descriptive Palaeoclimatology* (ed. A.E.M. Nairn), 227–254. New York: Interscience Publishers.

31 Lyell, K.M. (1881). *Life, Letters and Journals of Sir Charles Lyell, Bart*, vol. 1 and 2. London: John Murray, 475 pp. and 489 pp.

32 Lyell, C. (1826). On a recent formation of freshwater limestone in Forfarshire, and on some recent deposits of freshwater marl; with a comparison of recent with ancient freshwater formations; and an Appendix on the gyrogonite or seed-vessel of the chara. *Transactions of the Geological Society of London* 2: 73–96.

33 Lyell, C. (1875). *Principles of Geology*, 12e. London: John Murray.

34 Desmond, A. and Moore, J. (1991). *Darwin*, 808 pp. London: Michael Joseph.

35 Holmes, R. (2008). *The Age of Wonder: The Romantic Generation and the Discovery of the Beauty and Terror of Science*. London: Harper Press.

36 Secord, J. (1997). Introduction. In: *Principles of Geology* (ed. C. Lyell), ix–xliii. London: Penguin Classics.

37 Gould, S.J. (1987). *Time's Arrow, Time's Cycle: Myth and Metaphor in the Discovery of Geological Time*. Harvard, CT: Harvard University Press.

38 Agassiz, E.C. (ed.) (1885). *Louis Agassiz, his Life and Correspondence*. Boston and New York: Houghton Mifflin.

**39** Agassiz, L. (1838). On the erratic blocks of the Jura. *Edinburgh New Philosophical Journal* 24: 176–179.

**40** Agassiz, L. (1840). *Etudes sur les Glaciers*. Neuchâtel: Jent and Gassman. See also Agassiz, A. (1967) *Studies on Glaciers. Preceded by the Discourse of Neuchatel* (trans. and ed, A. Carozzi). Hafner, New York.

**41** Davies, G.H. (2007). *Whatever Is Under The Earth*, 356 pp. Geological Society of London.

**42** Agassiz, L. (1840). On the evidence of the former existence of glaciers in Scotland, Ireland and England. *Proceedings of the Geological Society of London* III (II, 72): 327–332. See also Buckland, pp 332-337 and 345-348; and Lyell, pp 337-345 in the same issue. Lyell applied to the Society on 5 May 1841 to have his complete paper withdrawn, and Buckland did the same in June 1842.

**43** Woodward, H.B. (1908). *The History of the Geological Society of London*. London: Longmans, Green & Co.

**44** McPhee, J. (1998). *Annals of the Former World*, 696 pp. New York: Farrar, Straus, and Giroux.

**45** Hestmark, G. (2017). Jens Esmark's mountain glacier traverse 1823 – the key to his discovery of Ice Ages. *Boreas* https://doi.org/10.1111/bor.12260. ISSN 0300-9483.

**46** Esmark, J. (1824) Bidrag til vor Jordklodes Historie. *Magazin for Naturvidenskaberne*. Anden Aargangs første Bind, Første Hefte, 28–49.

**47** Esmark, J. (1826). Remarks tending to explain the Geological History of the Earth. *The Edinburgh New Philosophical Journal* 2: 107–121.

**48** Dott, R.H. (1998). Charles Lyell's debt to North America: his lectures and travels from 1841 to 1853. In: *Lyell: The Past is the Key to the Present* (eds D.J. Blundell and A.C. Scott). *Geological Society of London Special Publication* 143: 53–69.

**49** Wilson, L.G. (1998). *Lyell in America – Transatlantic Geology 1841–1853*, 429 pp. Baltimore, MD: Johns Hopkins University Press.

**50** Lyell, C. (1854). On certain trains of erratic blocks on the western borders of Massachusetts, United States. *Notices of the Proceedings of the Royal Institution of Great Britain* 2 (1854–1858): 86–97.

**51** Darwin, C.R. (1842). Notes on the effects produced by the ancient glaciers of Caernarvonshire, and on the boulders transported by floating ice. *London, Edinburgh, Dublin Philosophical Magazine* 21: 180–188.

**52** Darwin, C.R. (1842) Letter to W.H. Fitton, 28 June 1842. *Letter 632, Darwin Correspondence Project*. Available from: http://www.darwinproject.ac.uk/letter/entry-632.

**53** Darwin, C.R. (1848). On the transport of erratic boulders from a lower to a higher level. *Quarterly Journal of the Geological Society of London* 4: 315–323.

**54** Geikie, A. (1863) On the phenomena of the glacial drift of Scotland, *Transactions of the Geological Society of Glasgow* I (II). Available from http://digital.nls.uk/early-gaelic-book-collections/pageturner.cfm?id=77409534.

**55** Geikie, J. (1874). *The Great Ice Age, and its Relation to the Antiquity of Man*, 3e. Stanford: London.

**56** Clements, D. (2012) The Geology of London. Geologists Association Guide no 68.

**57** Croll, J. (1875). *Climate and Time in their Geological Relations: a Theory of Secular Changes of the Earth's Climate*. New York: Appleton and Co.

**58** Huggett, Q.J. and Kidd, R.B. (1983/84). Identification of ice-rafted and other exotic material in deep-sea dredge hauls. *Geo-Marine Letters* 3: 23–29.

**59** Needham, H.D. (1962). Ice-rafted rocks from the Atlantic Ocean of the coast of the Cape of Good Hope. *Deep Sea Research* 9 (11–12): 475–486.

**60** MacLaren, C. (1842). Review of the Glacial theory of Prof. Agassiz. *American Journal of Science* 42: 346–365.

**61** Jamieson, J. (1882) *Geological Magazine* p.400. Cited in Geikie, J. (1894) *The Great Ice Age*. University of Chicago Press, Chicago, IL.

**62** Penck, A. and Bruckner, E. (1901-1909) *Die Alpen im Eiszeitalter*, 3 vols. Chr. Herm. Tauchnitz Verleger, Leipzig; summarized by Sollas, W.J. (1911) *Ancient Hunters*, Macmillan, London, esp. pp. 18–28.

**63** Geikie, A. (1866). Notes for a comparison of the glaciation of the west of Scotland with that of Arctic Norway. *Proceedings of the Royal Society of Edinburgh* 5: 530–556.

**64** Worsley, P. (2019). Archibald Geikie as a glacial geologist. In: *Aspects of the Life and Works of Archibald Geikie*, vol. 480 (eds. J. Betterton, J. Craig, J.R. Mendum, et al.). Geological Society, London, Special Publications.

# 3

# Ice Age Cycles

**LEARNING OUTCOMES**

- Understand and explain the astronomical theory of climate change
- Explain how James Croll modified our understanding of the impact of orbital change on Earth's climate.
- Explain Croll's theory for sea level change with time and the evidence he found for the changing of sea level with time.
- Explain what Croll's research signified as far as future climate prediction was concerned.
- Explain the theory of climate change that Lyell stressed in preference to Croll's theory, and his reasoning.
- Explain the evidence Croll found for the existence of interglacial periods.

## 3.1 The Astronomical Theory of Climate Change

Back in 1830, when Lyell was writing volume 1 of his *Principles*, he knew that glaciers had advanced and retreated. That led him to wonder if significant changes in our climate through time might be driven by some regular astronomical control on the amount of sunlight the Earth received [1]. To address this question, Lyell turned to his friend the English astronomer John Herschel (1792–1871), another Fellow of the Royal Society, who was knighted in 1831. Herschel explained that the amount of sunlight falling anywhere on the Earth's surface varies with regular changes in the 'eccentricity' of the Earth's orbit around the Sun, in the 'precession of the equinoxes', and in the tilt (or 'obliquity') of the Earth's axis [2].

Starting with eccentricity, it seems self-evident that if Earth were the only planet orbiting the Sun it would follow a circular orbit. But it is not and does not. Earth's orbit is influenced by the gravitational pull of its giant sister planets, which converts our orbit into an ellipse with the Sun off-centre, a state called 'eccentric'. The eccentricity of the Earth's orbit slowly changes from more or less circular, or 'centric' (with the Sun close to the centre) to 'eccentric' and back following cycles of about 2.4 Ma, 400 Ka and 100 Ka [3, 4]. That affects climate, because when the orbit is at its most elliptical, the amount of radiation Earth gets from the Sun at 'perihelion' (the point on the orbit closest to the Sun) is 23% more than it gets at 'aphelion' (the point on the orbit furthest from the Sun). Nevertheless the total energy received by the Earth over a year is maximal for the most elliptical orbit and minimal for the circular orbit [4]. At present, eccentricity is small, and in any case its variation leads to quite small changes in insolation.

The way in which that radiation is distributed over the planet depends on the fact that the position of the Earth migrates slowly around the 360° arc of its orbit from perihelion to aphelion and back in a cycle that takes about 23 Ka and 19 Ka [4, 5]. The position of the Earth on that arc can be described in terms of its longitude in relation to the Sun. Perihelion is currently reached close to the northern winter solstice (December), while the northern summer solstice (June) then lies 180° of longitude away at aphelion. The northern spring and autumn equinoxes lie at longitudes halfway between these two positions [6]. Earth's migration around the arc explains why during the 23 Ka cycle every point on the planet receives an annual maximum and minimum of solar energy. As Michel Crucifix points out *'Depending on whether the perihelion is reached in March, June, September or December, the actual amount of energy received in any of those months will be higher or lower. In order to model this phenomenon, we need to define the angle formed by the Sun, the point reached by the Earth at the vernal equinox – 21st March, by convention – and the perihelion. This angle is longitude of the perihelion'* [5]. The migration of the perihelion and its effects form the 'climate precession', often loosely referred to as the 'precession of the equinoxes' (for more detail see References [4, 5]). This migration has little effect on the climate when the Earth's orbit is more or less circular, but a large effect when the orbit is more elliptical (more eccentric). For instance, currently, the Earth is closest to the Sun (at perihelion) in midwinter, making Northern Hemisphere winters warm, and is furthest from the Sun (at aphelion) in mid-summer, making Northern Hemisphere summers cool. In about 11.5 Ka, these positions will be reversed for the Northern Hemisphere, making winters cooler and summers warmer. Given an elliptical orbit, the season will be longer at aphelion than at perihelion, whether in summer on in winter. The Earth's orbit is currently close to circular, so these differences have minimal effect at present, although Northern Hemisphere summer is presently 7.5 days longer than winter. The effect will increase as the orbit becomes more eccentric with time, because the eccentricity determines the amplitude of the precession cycle [4]. The effects of precession are greatest in the tropics and least in the polar regions.

Most people know that the Earth's spin axis is tilted at an angle of around 23° to the plane of the Earth's orbit, which accounts for the seasons. If the axis were upright we would still have climatic zones with more heat at the equator and less at the poles, but no seasons. The tilt of the axis fluctuates from around 21.5° to around 24.5° and back on cycles of 41 and 54 Ka, with a modulation at 1.2 Ma [3, 4]. The higher the angle, the more the seasonal difference, with summers receiving more energy from the Sun, and winters less. The effects of the tilt cycle are greatest in temperate and polar latitudes, where snow accumulates when tilt is small and winters are cold. However, in tropical regions the intensity of the monsoons is enhanced when tilt (obliquity) is large, because that strengthens the meridional thermal gradient between the Equator and the poles. Monsoons are also stronger when summer insolation is highest, which is when perihelion is reached in June (for northern summer monsoons) [5].

Interaction between these three cycles controls the amount of **in**coming **sol**ar radi**ation – 'insolation'** – received at any point on the Earth's surface through time. These cycles are so regular that astronomers can use them to calculate the amount of insolation anywhere on the Earth's surface over periods of millions of years. The Sun being the major driver of Earth's climate, these calculations provide geologists with a first order means of estimating climate change. Obliquity is particularly important for controlling polar ice through its control on the seasonal cycle, which leads to the melting of ice in polar summers when tilt is greatest [5]. Obliquity plays the same role in both hemispheres during the same local season, while precession has an opposite effect [4].

Herschel presented his ideas on the astronomical theory of climate change to the Geological Society in December 1830. He speculated that extreme variations in the eccentricity of the Earth's orbit might exaggerate the difference between summer and winter temperature, and that these might combine with the effect of precession of the equinoxes to produce '*periodical fluctuations in the quantity of solar heat received by the earth, every such fluctuation being of course accompanied with a corresponding alteration of climates; and therefore, if sufficiently extensive and continued, giving room for variation in the animal and vegetable productions of the same region at different and widely remote epochs*' [2]. Herschel thought that the variation of the tilt of the Earth's axis was insufficient to affect the climate. He regarded '*the excentricity [sic] as the only element whose variation can possibly have any effect of the kind in view*' [2]. He went on to say '*by reason of the precession of the equinoxes combined with the motion of the apogee of the earth's orbit, the two hemispheres would alternately be placed in climates of a very opposite nature, the one approaching a perpetual spring, the other to extreme vicissitudes of a burning summer and a rigorous winter*' [2]. Obscure prose indeed, but I'm sure you get the drift. Observing that the Earth's orbit was becoming more centric, he thought that this might indicate a cooling. We have to remember that these were early days and that Herschel was arguing without the benefit of detailed calculations of orbital changes, though they could have been determined from the work of the great French astronomer, Laplace.

Based on his discussions with Herschel, Lyell observed in volume 1 of *Principles* that it is '*of importance to the geologist to bear in mind that in consequence of the precession of the equinoxes, the two hemispheres receive alternately, each for a period of upwards of 10,000 years, a greater share of solar light and heat. This cause may sometimes tend to counterbalance inequalities resulting from other circumstances of a far more influential nature; but, on the other hand, it must sometimes tend to increase the extremes of deviation, which certain combinations of causes produce at distant epochs*' [1]. This dense prose hid the fact that Lyell suspected that these astronomical changes might affect climate enough to be detectable in the geological record, but had no idea if that were the case.

At the time, in 1830, Agassiz had not yet 'discovered' the Ice Age. The honour of being the first to conclude that astronomical forces controlled what happened during the Ice Age goes to French mathematician Joseph Alphonse Adhémar (1797–1862). He had the brilliant intuition that Agassiz's glaciations must be periodic and thus controlled by celestial mechanics, publishing his ideas in 1842 [7]. He calculated that periods of cooling and warming would correspond to the 21 Ka precession cycle, and that they would alternate between the two hemispheres.

We now know that precession does not affect the mean insolation, only the way in which insolation is distributed across the different months of the year [5]. In contrast, obliquity controls the mean annual energy at any point, as well as the meridional gradient of energy and the poleward flux of moisture needed to feed ice sheets as they form [5]. Whilst precession acts on northern and southern ice sheets asynchronously (as noted by Adhémar), obliquity affects them synchronously [5]. Precession is influenced by eccentricity and has its greatest effects in low latitudes, whilst obliquity has its greatest effects at high latitudes, as pointed out long ago by André Berger [8], who also provides an elegant review of the astronomical theory of climate change [9].

## 3.2 James Croll Develops the Theory

The first scientist to investigate the astronomical theory of climate change in detail was Scottish physicist-cum-geologist James Croll (1821–1890) (Figure 3.1, Box 3.1), the man who disagreed with Lyell's ideas about boulder clay.

Thinking about the Ice Age, Croll felt sure that *'The recurrence of colder and warmer periods evidently points to some great, fixed, and continuously operating cosmic law'* [10]. Immersing himself in the studies of celestial mechanics by Frenchmen Pierre Simon de Laplace and Urbain Leverrier, discoverer of the planet Neptune, he worked out a more complex theory to explain the effect of seasonal contrasts of insolation, publishing his results in 1864 [10]. His paper refined predictions of the timing of the glacial epochs of the Ice Age, and showed how changes in the Earth's orbit provided a periodic extraterrestrial mechanism for initiating 'multiple' glacial epochs. These would occur every 22 Ka, and when there was a glacial period in the north there would be an interglacial in the south, as suggested by Adhémar. Unlike Adhémar, Croll thought

**Figure 3.1**   James Croll.

that eccentricity was important, because the accumulation of snow would be encouraged by decreased sunlight and longer winters when the orbit was most elliptical and the Earth was furthest from the sun (at aphelion).

One of the most important and overlooked aspects of Croll's analysis of the astronomical theory of climate change was his realization that the change in heat received by the Earth due to orbital changes was *'not enough by itself'* to cause glaciation. He concluded correctly that *'glacial cycles may not arise directly from cosmical causes, they may do so indirectly!'* [11]. Feedbacks were needed. The effect of decreased insolation was amplified by increasing accumulation of snow and ice, which increased the reflectiveness of the Earth's surface – its albedo. This increased the development of mists and fogs, reflecting yet more solar energy.

Croll also considered how changes in insolation would affect winds and ocean currents [10]. Cooling of the poles in glacial periods would steepen the thermal gradient in

---

**Box 3.1   James Croll**

Croll's scientific career began late in life. Born on a farm near Wolfhill in Perthshire, Scotland in 1821, he was largely self-educated in mathematics and astronomy. Leaving school at age 13, he worked as a millwright, a carpenter, a tea merchant, the keeper of a temperance hotel, a life insurance salesman, and writer for a temperance newspaper, before becoming caretaker of the museum at Anderson College in Glasgow in 1859 at the age of 38. That gave him access to a fine scientific library, exposing him to the work of Herschel and Adhémar and enabling him to develop his own ideas about one of the 'hot' topics of the era – the origin of the Ice Age. His findings led to him being invited to work at the Geological Survey of Scotland, which he did from 1867 to 1881. For his Ice Age research he was elected a Fellow of the Royal Society in 1876 and awarded an Honorary Degree by the University of St Andrews. He retired because of ill health in 1880 and died in 1890. The Quaternary Research Association now awards the James Croll Medal.

the air between pole and equator, making winds stronger. Glaciation in the Northern Hemisphere would weaken the Gulf Stream, which carried warm water from the equator to the Arctic. It would also strengthen the N.E. Trade Winds, which would force the Equatorial Current south, limiting the supply of warm water to the Gulf Stream from the South Atlantic; '*The Gulf-stream would consequently be greatly diminished, if not altogether stopped*' [10]. If that were so, he speculated that the climate of northern Europe would resemble that of Greenland [12]. This is more or less what occurred during the Ice Age.

Lyell was intrigued, and began corresponding with Croll. That interaction led to Croll being appointed in 1867 to a clerical position as keeper of maps and correspondence in the Geological Survey of Scotland, where the Director, Sir Archibald Geikie, encouraged his research. In 1875, Croll summarized his research findings in an influential book *Climate and Time* [12] (Figure 3.2).

Given the importance he attached to the decrease of heat from the Gulf Stream as one of the positive feedbacks enhancing the glaciation of the north, Croll was keen to find out more about the nature of ocean circulation, then a topic of much speculation [13]. He closely monitored the work of physiologist William Carpenter (1813–1885), a scientist keen to test the notion of Edward Forbes (1815–1854) that there was no life in the deep ocean – it was 'azoic'. As

registrar of the University of London and Vice-President of the Royal Society, Carpenter used his influence to access a Royal Navy ship, HMS *Porcupine,* which, in 1870, dredged numerous creatures from the deep sea and so killed off Forbes's azoic theory.

Carpenter's work with the *Porcupine* showed most of the deep North Atlantic to be extremely cold. He thought this meant that a deep current originating in the Arctic carried cold water south into the interior of the Atlantic, and '*embarked on the development of his "magnificent generalization" that the cold temperatures were part of a large-scale general ocean circulation*' [13]. In his conceptual model, this deep current replaces the warmer surface water that flows from the equator towards the poles. He saw '*this flow of water toward the equator and its eventual return towards the poles [as] just as much a physical necessity as that interchange of air which has so large a part in the production of winds*' [13]. Density was an important driver in Carpenter's model of the general circulation, with lighter, warmer water at the surface moving north connected to denser colder water at depth moving south and eventually returning to the surface near the Equator. These were important insights into ocean circulation, which helps to regulate Earth's climate, although we now know that the return to surface takes place around Antarctica, not at the equator.

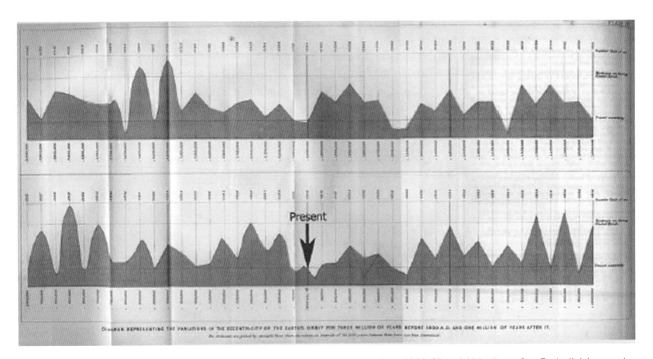

**Figure 3.2** Croll's Orbital Variations. Variations in the Earth's orbit for 3 Ma before 1800 CE and 1 Ma thereafter. Each division on the horizontal axis represents 50 000 years. Croll thought that when the orbit was most eccentric (i.e. elliptical, with high values along the vertical axis) there would be longer winters allowing more snow to accumulate. In contrast, low values of eccentricity would equate with warm climates. The arrow indicating 'present' conditions denotes the position of 1800 CE, when eccentricity was close to its lowest value.

Croll disagreed with Carpenter's model, setting out his own model of how ocean currents contributed to glaciations [12, 13]. There just wasn't enough oceanographic information to enable them to resolve their differences [13]. The controversy did have one happy outcome in providing a raison d'être for the world-encircling oceanographic expedition of HMS *Challenger* in 1872–1876, which would create a much clearer picture of ocean circulation than was available to either man.

Although Croll focused his attention on precession and eccentricity, he suspected that changes in the tilt of the Earth's axis might also affect the climate, especially at the poles, where the longer summers at times of maximum tilt would melt more snow and ice than at other times [12]. He surmised that particular combinations of eccentricity, precession, and axial tilt would lead to periods of ice melt that would raise sea level, and that oscillations of sea level should be associated with changes from interglacial to glacial conditions – another prescient conclusion. Lacking calculations of the changes in tilt through time, Croll could do little more than speculate about its effects. We now know that changes in axial tilt do have a strong effect in the polar regions. Croll was on the right track, and that got geologists thinking in the right direction. Noting that tilt was at a maximum 11.7 Ka ago, he speculated that this might have led to a rise in sea level that would explain the occurrence of raised beaches from about that period in Scotland and Scandinavia. This was a perceptive observation, although it ignored the effect of the isostatic upward adjustment of the land in response to the removal of the last ice sheet.

Croll was also one of the first to note the occurrence beneath the Scottish boulder clay of buried river channels, the depths of whose beds showed that they must have been cut when sea level was much lower. The cutting of deep channels implied that the sea level had dropped, so steepening the gradient of the rivers' beds. Whilst Croll thought the channels might have been cut during warm periods, it seems more likely that they were cut when ice sheets were extensive and sea level was low, both as the climate was cooling and as it was warming. Channel cutting would cease when the ice sheets advanced to the edge of the continental shelf, although even then ice streams might gouge their way into the surface of the continental shelf beneath an ice sheet, much as happened on the continental shelf west of the Antarctic Peninsula during low stands of sea level.

In one particularly perceptive leap of the imagination Croll deduced that '*If the glacial epoch resulted from a high condition of eccentricity, we have not only a means of determining the positive date of that epoch, but we also have a means of determining geological time in absolute measure*'

[12]. This turned out to be true, although not in precisely the way Croll imagined, as we shall see later. Following that leap to its logical conclusion, he calculated eccentricity not only for 3 Ma into the past, but also for 1 Ma into the future (Figure 3.2), making him the first to take a mathematical approach to estimating future climate change.

## 3.3 Lyell Responds

Having read Croll's 1864 paper [10], and corresponded with him, Lyell modified the last three editions of his *Principles of Geology* (numbers 10 in 1866, 11 in 1872, and 12 in 1875), to introduce a new Chapter 13, on '*Vicissitudes in Climate – How Far Influenced by Astronomical Changes*' [14–16]. It explained how changes in the eccentricity of the Earth's orbit in combination with the precession of the equinoxes would cause alternations of climate with a period of around 21 Ka, repeated thousands of times throughout the geological past, which might explain '*some of the indications of widely different climates in former times*' [16]. Lyell attributed to Croll the observation that winters would be at their coldest when the Earth was at aphelion (farthest from the sun) and the orbit was at its most eccentric. The development of large amounts of ice at those times, Lyell went on to note, '*must have given rise at certain periods to some differences in the ocean's level*' [16]. This is a rather late acknowledgment of the notion that ice ages might be times of lowered sea level, which Charles Maclaren had proposed as early as 1842 [17].

Despite the attractive features of the astronomical theory of climate change, Lyell thought that Croll had given insufficient credence to Lyell's own principle that '*abnormal geographical conditions*' – meaning the existence of land near or over the poles – were '*far the most influential in the production of great cold*' [16]. He reminded his readers '*The simple fact that totally different climates exist now in the same hemisphere and under the same latitude would alone suffice to prove that their occurrence cannot be exclusively due to astronomical influence*' [16]. For example '*the climates of South Georgia and Tierra del Fuego are at present so different that the former might be supposed to belong to a glacial period, while the latter, by its flowers and hummingbirds in the winter, and the genera of marine molluscs in the adjoining sea, might indicate to the traveller, as well as to some future geologist, such a temperature as has been spoken of as perpetual spring. This contrast is due to geographical causes*' [16]. The confusion in Lyell's mind arose because the movements of continents, which could indeed cause changes in climate, took place on a time scale of millions of years, while the variations of the astronomical theory were measured in tens or hundreds of thousands of years.

Although Lyell considered '*the former changes of climate and the quantity of ice now stored up in polar latitudes to have been governed chiefly by geographical conditions*' [16], he accepted that the combination of a large excess of polar land with maximum eccentricity of the Earth's orbit '*would produce an exaggeration of cold in both hemispheres*' [16]. To see when maximum eccentricity occurred, he had colleagues draw up a table showing the eccentricity of the Earth's orbit over the last 1 Ma. The table, on page 285 of the 12th edition of *Principles*, shows that major eccentricity occurred at intervals of about 100 Ka, more or less in agreement with modern calculations, with the greatest eccentricity within relatively recent times occurring some 200–210 Ka ago [16].

Changes in the tilt of the Earth's axis, Lyell agreed, might also have some effect on the climate, greater tilt causing colder winters at the poles. If that condition were combined with maximum eccentricity and abundant polar land, '*this would favour a glacial epoch*' he thought [16]. In fact, greater tilt would bring more solar energy into the polar regions both on average and especially in summer, making a glaciation less likely.

Despite their differences, Croll and Lyell agreed in one key respect – as Croll noted '*the geological agents are chiefly the ordinary climatic agents. Consequently, the main principles of geology must be the laws of the climatic agents, or some logical deductions from them. It therefore follows that, in order to [pursue] a purely scientific geology,* **the grand problem must be one of geological climate** [my emphasis]. It is through geological climate that we can hope to arrive ultimately at principles which will afford a rational explanation of the multifarious facts which have been accumulating during the past century*' [12]. Where the two men differed was in emphasis, Lyell stressing the pre-eminence of geography, and Croll that the existence of warm interglacial periods goes '*to prove that the long epoch known as the Glacial was not one of continuous cold, but consisted of a succession of cold and warm periods [which] is utterly inexplicable on every issue of the cause of the glacial epoch which has hitherto been advanced*' [12].

Lyell was concerned enough about Croll's astronomical challenge to his geographical theory of climate change to write to Herschel and the Astronomer Royal, Sir George Biddell Airey. The correspondence led him to tentatively accept Croll's theory as a minor cause of climate change [11]. He must have felt he had an edge over Croll, in that Croll had concluded that, following his theory, ice ages should have recurred through time. It did not disturb Croll that no evidence for them had been found in the warmer Tertiary; after all, the inadequacy of the geological record might explain their absence [12]. Lyell disagreed, considering that eccentricity alone could not be the cause of the

post-Pliocene ice ages, because there was no evidence for ice ages in the Tertiary or the Cretaceous formations, or indeed back to the Carboniferous [16]. '*This absence of recurrent periods of cold is perfectly explicable*', he went on '*if I am right in concluding that they can only be brought about by an abnormal quantity of land in high latitudes*' [16]. As we shall see later, he was wrong about the Carboniferous, where there was a major glaciation, but then he was limited by the span of knowledge at the time.

Whilst we might think this a stubborn adherence to what might be becoming an outdated idea, Lyell regarded the geographical principle of climate change as being one of his major contributions to the science of geology, and it is hard to let go of your favourite ideas, especially when compelling evidence for the competing theory is weak or absent. Besides we now know that Croll and Lyell may both have been right – land in the polar regions does help to build up substantial accumulations of ice in accordance with Lyell's view, and Earth's orbital changes do modify the climate in accordance with Croll's view, though not to the extent of forming glaciations in warm periods like the early Tertiary and Cretaceous, as Croll thought. That is where greenhouse gases come in, as we shall see later. And neither man had considered them.

Croll's work influenced many eminent scientists including Lyell, Darwin, and James Geikie [11, 18]. Darwin was more forthcoming than Lyell in his praise for Croll's theory, '*in part because it provided a valuable mechanism for speciation*' [11]. He wrote to Croll on 24 November 1868 that '*I have never, I think, in my life, been so deeply interested by any geological discussion*', agreeing with Croll that the advocates of the iceberg theory (such as Lyell) had formed '*too extravagant notions regarding the potency of floating ice as a striating agent*' [11] and that '*scored rocks throughout the more level parts of the United States result from true glacier action*' [19].

## 3.4 Croll Defends His Position

Croll realized that one barrier to getting his astronomical theory accepted was that scientists tend to find what they are looking for [12]. Before publication of his astronomical theory in 1864 there was no impetus to seek evidence for interglacial periods. That the evidence existed he was sure, recalling reading some years before a paper describing fossiliferous sediments found between deposits of glacial till and containing '*rootlets and stems of trees, nuts, and other remains showing that it had evidently been an old interglacial land surface*' [12]. After 1864, evidence for interglacial deposits began to emerge [18], including evidence that warm interglacial conditions had extended as far north as

75° 32'N in the Arctic [12]. While Croll took this data aboard [12], Lyell largely ignored in his last edition of 1875 [16] the detailed findings about glacial and interglacial geology that James Geike began publishing in 1874 [18]. By then Lyell was at the end of his life. From 1875 on, Croll and Geikie carried the day, their major opponent having vanished from the scene.

Unfortunately for the astronomical theorists, while orbital variations could be calculated fairly accurately, nineteenth century geologists could only date rocks crudely. It was impossible to precisely relate sedimentary sequences to astronomical variables, or to test Croll's notion that glaciations alternated between the hemispheres. All that Croll could do was use rates of erosion and deposition to suggest that the glacial epoch ended around 80 Ka ago [12]. The right answer was around 20 Ka ago. But his was a good best guess for those times.

James Geikie supported Croll's theory with the caveat '*it must be confessed that a complete solution of the [Pleistocene Ice Age] problem has not yet been found*' [18]. He realized that Lyell's requirement for land in the Polar Regions to explain glaciations did not address the real issue: the complex alternation of cold and warm epochs. It was unreasonable to suppose that land moved in and out of the polar regions sufficiently rapidly to explain the origin of glacial to interglacial cycles. Lyell had also called on increases in elevation of land to explain the origin of cold periods. Geikie took issue with that too, considering it unlikely that highlands had popped up and down fast enough to account for the observed cycles [18].

Croll's work was widely discussed, but generally disregarded, not least because of geologists' continuing inability to date variations at a fine enough scale. By the time Croll died, geologists realized that glacial conditions had persisted much later than his proposed peak glaciation 80 Ka ago, and many thought his theory must be wrong. James Geikie hoped that '*some modification of his views will eventually clear up the mystery. But for the present we must be content to work and wait*' [18]. Advances in the theory and in dating rocks were needed to revive Croll's theory.

## 3.5  Even More Ancient Ice Ages

Agassiz's discovery changed the way field geologists thought about the rocks they saw. One of the first to respond, in 1855, was Professor Andrew Constable Ramsay (1814–1891), another of those influential Scottish geological fellows of the illustrious Royal Society of London. He was President of the Geological Society of London from 1862 to 1864, and in 1872 became Director-General of the Geological Survey of Britain. He was awarded the Wollaston Medal of the Geological Society in 1871, and the Royal Medal of the Royal Society in 1880, and was knighted in 1881.

Ramsay had a special interest in the effects produced by ice, and applied it to interpreting the origin of the roughly 290 Ma old Permian breccias skirting the English and Welsh coal fields. The breccias comprise many large, polished, and striated angular fragments of rock stuck in a 'marly paste' [20]. He deduced '*that they are chiefly formed of the moraine matter of glaciers, drifted and scattered in the Permian sea by the agency of icebergs*' [20]. The Permian strata overlie the older Carboniferous Coal Measures, in which the coal was assumed to be the remains of swamps and forests growing in '*a moist, equable, and temperate climate, possibly such as that of New Zealand*' [20]. Croll was delighted to see evidence for a major glaciation far earlier in time than the relatively recent Ice Age, because it supported his notion that regular changes in the Earth's orbit should have led to glaciations in the distant past [10], and meant that Earth's climate could not be explained simply in terms of a gradual cooling.

Ramsay's interest in glaciations led him to influence Archibald Geikie. As a 23-year old, young Geikie was on a geological ramble in Fife when Ramsay led him '*to enquire more narrowly into the received theories respecting the cause of the dressed rock surfaces, and the accumulations of boulder clay*' [21]. At the time, Geikie, like practically every geologist in Britain, was used to following Lyell in '*regarding the striations on the hills [as] due to the abrading force of icebergs, by whose operations also the boulder-clay was held to have been transported over the surface of the submerged land*' [21]. Following careful observation, by the summer of 1861 Geikie '*finally abandoned the attempt to explain the origin of the rock-dressing and the boulder-clay by the action of icebergs*' [21]. Instead, '*These phenomena seemed only explicable on the supposition that the whole of this country [Scotland] was covered with ice and snow, like large tracts of Greenland at the present day; and that, by the constant downward and seaward movement of this ice covering, the rocks were ground down, and the boulder-clay was produced*' [21]. Evidently Ramsay, who had been studying the evidence for ice action in Britain, Europe, and Canada, had come to much the same conclusion, but it was young Geikie's comprehensive research that clinched the matter. Whilst Lyell accepted Geikie's evidence for the glaciation of Scotland's mountains, he drew the line at terrestrial extensions of glacial ice into lowlands, preferring to see evidence for ice in such places as evidence for his iceberg theory.

By the time Croll wrote his extensive work on climate and time [12], he knew that the great coal formations of the Carboniferous Period had been deposited in cycles, with thin coal beds made from the remains of forests formed on land alternating with thin beds of marine clay, suggesting a succession of geologically rapid changes from a terrestrial to a marine environment and back. He interpreted this succession as possibly representing repeated changes from warm interglacial periods when the sea level was low and coal forests grew, to cold glacial periods when the sea level was high and marine clays were deposited. Although alternations between warm and cold conditions were consistent with his astronomical theory, he missed the point that during cold glacial conditions the sea level should have been low rather than high, with more water being trapped in ice at those times. We will explore the origin of cyclical deposits of the Carboniferous later.

## 3.6 Not Everyone Agrees

By the end of the nineteenth century a lot more was known about how the modern climate system worked than when Lyell first set down his *Principles*. Other geologists had begun to write treatises about the climates of the geological past. One was Dutch palaeo-anthropologist and geologist Marie Eugène François Thomas Dubois (1858–1940) (Box 3.2), famed for discovering 'Java Man' (*Pithecanthropus erectus*, later named *Homo erectus*).

Aside from his fascination with the link between man and apes, Dubois was intrigued by the climatic changes of the past, publishing an expanded essay on this topic in 1895 [22]. His article brought to an English-speaking audience a wide range of references to the growing understanding of past climate change, especially from Germany. Notable amongst these was the work of Melchior Neumayer (1845–1890) [23], who had shown the biogeographical link between Brazil and Africa across the South Atlantic, and Swiss geologist and naturalist Oswald Heer (1809–1883) [24–26], who Dubois considered to be the 'father of paleoclimatology', noting that: '*To him we owe nearly all the information which we possess about ancient climates*' [22]. Heer was Professor of Botany at the University of Zurich.

For his services to science the Geological Society of London awarded him its Wollaston Medal in 1874.

Dubois drew attention to the growing evidence from fossil plants and animals that the climate of the Arctic had declined from almost tropical in the early Tertiary to the cold conditions of today. There was ample evidence from early Tertiary and former times – especially the Jurassic and the Cretaceous – that whilst they had been warmer than today, their heat had been distributed in climate zones between the equator and the pole as they presently exist. No displacement of the poles seemed necessary to explain the distribution of these zones. That seemed to knock on the head, at least temporarily, Lyell's theory of continental displacement as a cause of climate change. For earlier times, like the Permo–Carboniferous, where there was evidence for the action of ice far from the present poles, Dubois preferred to explain their occurrence as caused by increased precipitation rather than by excessive cold, citing as an example the occurrence of glaciers in temperate New Zealand today.

How then could the Cenozoic cooling be explained? Dubois believed that the answer must lie with the energy emitted by the Sun, which, he thought, must have declined with time, but he also recognized that ocean currents play an important role in transporting heat away from the tropics. He thought that periodic changes in the Sun's output, like those causing the sunspot cycle, might explain the warm interglacial periods of the Ice Age. Whilst the astronomical variations called upon by James Croll might explain some of that variation in Pleistocene times, they were inadequate to explain the general cooling of the Earth in Cenozoic times. '*It now seems to be completely proved*' said Dubois '*that no other source of heat than the sun can ever have exercised an appreciable influence either on the meteorological condition of the Earth, or on the climates. To any very considerable changes of the solar heat we must, therefore, look for the cause of the geological changes of climate*' [22].

Dubois can perhaps be taken of typical of the geologists of the time. He was unaware of the effects of greenhouse

---

**Box 3.2  Marie Eugène François Thomas Dubois**

Raised in Eijsden, in Limburg, The Netherlands, Dubois was fascinated by natural history. He studied medicine at the University of Amsterdam, obtaining his degree in 1844. Specializing in comparative anatomy, he developed an interest in human evolution and the link between man and apes. In 1887 he joined the Dutch army to get himself posted to the Dutch East Indies – now Indonesia – because he felt sure that the 'missing link' between apes and man lay in the tropics. There he searched caves on Sumatra and Java, finding 'Java Man' in 1891. He returned to Europe in 1895, was awarded an honorary doctorate by the University of Amsterdam in 1897, and became Professor of Geology there in 1899. He also served from 1897 to 1928 as keeper of palaeontology, geology, and mineralogy at Teyler's Museum in Haarlem.

gases in modifying the Sun's heat, and neglected the possibility that the continents may have changed their positions with relation to the poles and to each other. Breakthroughs in the sciences would be needed before the prevailing paradigm could change. We turn in the next chapter to the efforts of nineteenth century scientists to explore the possibility that changes to Earth's climate resulted from changes to the composition of the atmosphere.

## References

**1** Lyell, C. (1830). *Principles of Geology*, vol. 1. London: John Murray, followed by Volume 2 in 1832 and Volume 3 in 1833.

**2** Herschel, J. (1830). On the astronomical causes which may influence geological phenomena, read before the Geological Society, December 15th 1830. *Transactions of the Geological Society* III: 293, second series 244.

**3** Laskar, J., Robutel, P., Joutel, F. et al. (2004). A long-term numerical solution for the insolation quantities of the Earth. *Astronomy & Astrophysics* 428: 261–285.

**4** Berger, A. and Loutre, M.-F. (1994). Long term variations of the astronomical season. In: *Topics in Atmospheric and Interstellar Physics and Chemistry* (ed. C. Boutron), 33–61. Les Ulis, France: Les Éditions de Physique.

**5** Crucifix, M. (2019). Pleistocene glaciations. Chapter 3. In: *Climate Changes in the Holocene, Impacts and Human Adaptation* (ed. E. Chiotis), 77–106. CRC Press publishers, Taylor and Francis, ISBN 97808153938.

**6** Yin, Q.Z. and Berger, A.L. (2012). Individual contribution of insolation and $CO_2$ to the interglacialclimates of the past 800,000 years. *Climate Dynamics* 38: 709–724.

**7** Adhémar, J. (1842). *Révolutions de la Mer, Déluges Périodique*, 2e, 184 pp. Paris: Carilian-Goeury et V. Dalmont, Lacroix-Comon, Paris, 1860, 258 pp; see review in Bard, E. (2004) Greenhouse effect and ice ages: historical perspective. *ComptesRendus Geoscience* **336**, 603-638.

**8** Berger, A. (1978). Long-term variations of caloric insolation resulting from the Earth's orbital elements. *Quaternary Research* 9 (2): 139–167. https://doi.org/10.1016/0033-5894(78)90064-9.

**9** Berger, A. (2012). A brief history of the astronomical theories of paleoclimates. In: *Climate Change, Inferences from Paleoclimate and Regional Aspects* (eds. A. Berger, D. Mesinger and D. Sijacki), 107–129. Springer.

**10** Croll, J. (1864). On the physical cause of the change of climate during geological epochs. *London, Edinburgh and Dublin Philosophical Magazine and Journal of Science* XXVIII, 4th series, August: 121–136.

**11** Fleming, J.R. (2006). James Croll in context: the encounter between climate dynamics and geology in the second half of the nineteenth century. *History of Meteorology* 3: 43–53.

**12** Croll, J. (1875). *Climate and Time in Their Geological Relations: A Theory of Secular Changes of the Earth's Climate*. New York: Appleton and Co.

**13** Mills, E.L. (2009). *The Fluid Envelope of Our Planet – How the Study of Ocean Currents Became a Science*, 434. Toronto: University Press.

**14** Lyell, C. (1866). *Principles of Geology*, 10e. London: John Murray.

**15** Lyell, C. (1872). *Principles of Geology*, 11e. London: John Murray.

**16** Lyell, C. (1875). *Principles of Geology*, 12e. London: John Murray.

**17** MacLaren, C. (1842). Review of the glacial theory of prof. Agassiz. *American Journal of Science* 42: 346–365.

**18** Geikie, J. (1894). *The Great Ice Age, and Its Relation to the Antiquity of Man*, 3e. Stanford: London.

**19** Fleming, J.R. (1998). Charles Lyell and Climatic change: speculation and certainty. *Geological Society Special Publication* 143: 161–169.

**20** Ramsay, A.C. (1855). On the occurrence of angular, subangular, polished and striated fragments and boulders in the Permian breccia of Shropshire, Worcestershire, etc; and on the probable existence of glaciers and icebergs in the Permian Epoch. *Proceedings of the Geological Society of London* 11: 185–205.

**21** Geikie, A. (1863). *On the Phenomena of the Glacial Drift of Scotland, Transactions of the Geological Society of Glasgow* I (II). Glasgow: John Gray http://digital.nls.uk/early-gaelic-book-collections/pageturner.cfm?id=77409534.

**22** Dubois, E. (1895). *The Climates of the Geological Past and Their Relation to the Evolution of the Sun*, 167 pp. Sonnenschein, London. This was a translation of an essay published in German in 1893 as *Die Klimate der geologischen Vergangenheit und ihre Beziehung zur Entwickelungsgeschichte der Sonne*, H. C. A. Thieme, Nijmegen, which in turn was an expansion of a paper initially published in Dutch in 1891 as De klimaten der voorwereld en de geschiedenis der zon, *Natuurkundig Tijdschrift voor Nederlandsch-Indie, Batavia* **51**, 37-92.

**23** Neumayr, M. (1887) *Erdgeschichte*. Vol. II, Leipzig; and his 1889 lecture: *Die klimatischen Verhaltnisse der Vorzeit*, Schriften des Vereins zur Verbreitung naturwiss.

Kenntnisse. Wien. A translation of this lecture entitled *The Climates of Past Ages* appeared in 1890 in *Nature* **42**, pp 148 and 175.

**24** Heer, O. (1864). *Die Urwelt der Schweiz.* [Lieferungen 7–11], 289–496. Zürich: Friedrich Schulthess.

**25** Heer, O. (1879). *Die Urwelt der Schweiz.* Zweite, umgearbeitete und vermehrte Auflage. Zürich: Friedrich Schulthess, xix + 713 pp.

**26** Heer, O. (1868–1882). *Flora Fossilis Arctica – Die fossile Flora der Polarländer*, Zweiter Band enthaltend: 1. *Fossile Flora der Bären-Insel.* 2. *Flora fossilis Alaskana.* 3. *Die Miocene Flora und Fauna Spitzbergens.* 4. *Contributions to the Fossil Flora of North-Greenland*, Mit 59 Tafeln, Winterthur. Verlag von Wurster & Comp., **2**, 307 pp, hdl:10013/epic.35335.d001.

# 4

# Trace Gases Warm the Planet

---

**LEARNING OUTCOMES**

- Know and understand the observational, theoretical and experimental contributions of eighteenth and nineteenth century scientists to our understanding of how the climate system works, in particular:
    Saussure's Alpine observations and his experimental hot box;
    Fourier's theoretical contribution;
    Tyndall's experiments with gases;
    Arrhenius's numerical modelling.
- Know and understand the contributions of Tyndall, Arrhenius, and Chamberlin to our understanding of the nature and origin of glacial-interglacial cycles.
- Know and understand the contributions of Tyndall, Chamberlin and Suess to our understanding of how Earth's climate operates over long periods of time.

---

## 4.1 De Saussure's Hot Box

In the late eighteenth and early nineteenth century physicists were puzzling over what it was that kept Earth's atmosphere warm. Amongst them was Horace Bénédicte de Saussure (1740–1799) (Figure 4.1, Box 4.1), a Genevan aristocrat, brilliant scientist, and author of the widely known *Voyages dans les Alpes* [1].

De Saussure led the second expedition to the summit of Mt. Blanc, having stimulated an associate to make the first expedition in 1786 (Figure 4.2). Proximity to the Alps enabled him to investigate the behaviour of the atmosphere. Carrying barometers and thermometers to the summits of Alpine peaks, he measured the temperature and relative humidity of the air and the strength of solar radiation at different heights. In 1767, to test the idea that the air is colder on peaks because the sunlight is weaker there, he invented the 'heliothermometer', a solar energy collector or solar oven comprising a black-lined, well-insulated box with a lid made up of three layers of glass with air between them (Figure 4.3). As the temperatures in his 'hot box' reached 110 °C regardless of altitude or outside air temperature, he concluded that sunlight was constant everywhere within a given locality, and that some other physical process must cause the air to cool upwards. Radiation was involved, but what kind, and from where?

## 4.2 William Herschel's Accidental Discovery

The explanation starts with visible light. When the sun shines, tiny rainbows flit about the rooms in my house. They come from crystals hanging in the windows. Physicists had long puzzled about the origin of the colours emerging from crystals and seen in rainbows, until, in 1665, young Isaac Newton used a prism to show that white light, or visible radiation, is made of a spectrum of colours mixed together. It was 135 years before another young genius extended Newton's work by discovering infra-red radiation. He was the German-born musician and astronomer Frederick William Herschel (1738–1822) (Box 4.2), father of Sir John Herschel (see Chapter 3).

In 1800, the inquisitive Herschel thought it would be interesting to measure the temperatures of the different colours of the spectrum emerging from a prism placed in a beam of white light. Much to his amazement he found the highest temperatures just beyond the visible red end of the spectrum. It was an accident: he had placed one thermometer just outside the visible spectrum emerging from his prism, to represent the ambient temperature of the environment. He realized that he had discovered an invisible form of light beyond the visible spectrum. Because it vibrates at a longer (i.e. lower) wavelength than does visible red light, we call it **infra**red light.

*Palaeoclimatology: From Snowball Earth to the Anthropocene*, First Edition. Colin P. Summerhayes.
© 2020 John Wiley & Sons Ltd. Published 2020 by John Wiley & Sons Ltd.

All warm objects emit heat as invisible infrared radiation. If you heat them enough they give off visible radiation – we speak of things becoming red hot, and at extreme temperatures, white hot. You can use an infrared light detector to see people and objects at night, because they emit heat. Engineers use this principle to find heat leaks from buildings. You use infrared rays to operate your TV by remote control. Warmed by the visible and infrared radiation from the sun, the surface of the Earth and the ocean emit infrared radiation that we can't see. The atmosphere is heated from the bottom by the infrared radiation from the Sun-heated surface of the planet.

## 4.3  Discovering Carbon Dioxide

The man who introduced the term 'gas' to chemistry, Flemish chemist Jan Baptist van Helmont (1579–1644), was the first to combine carbon with air to produce a gas.

Burning charcoal in a closed vessel, he found that the weight of the ash was less than the original weight of charcoal. Part of the charcoal was transmuted into an invisible substance that he named 'gas sylvestre'. He thought it was the same as the gas produced by fermenting fruit juice, which rendered the air of caves unbreathable.

One hundred years later, Scottish chemist Joseph Black (1728–1799), a friend of James Hutton, discovered that treating limestone with acids or heat yielded a gas he called

**Figure 4.1**  Horace Bénédicte de Saussure.

**Figure 4.2**  Statue of Saussure and his Alpine guide Jaques Balmat, who had made the first ascent of Mt. Blanc in 1786, in Chamonix.

---

**Box 4.1  Horace Bénédicte de Saussure**

Born in Conches, near Geneva, Saussure was a candidate for a professorship of mathematics by the age of 20, and at the age of 22 obtained one in philosophy. Fascinated by the Alps, in 1787 he made the second ever ascent of Mt. Blanc, and his scientific measurements there captured the imagination. Rector of Geneva University from 1774 to 1776, he was made a Fellow of the Royal Society of London and a foreign member of the Academies of Sciences of both France and Sweden. Well connected, he visited Buffon in Paris, and discussed electricity with Benjamin Franklin in London. A genus of high alpine plants – the *Saussurea* – is named after him, as is the mineral Saussurite, and he appeared on the Swiss 20 Franc note (1979–1995). Saussure was the first to break the abhorrence of the English-speaking public for the dangers of the Alps, and to foster the rise of mountaineering. He also popularized the use of the term 'geology' [2]. As a geologically minded natural philosopher he gave Hutton ideas about the former extent of glaciers. His *Voyages dans les Alpes* supplied Hutton with many of the illustrations of geological processes on which he based his Theory of the Earth (see Chapter 2).

**Figure 4.3** Saussure's hot box. The heliothermometer, comprising several boxes encased one inside another, each of whose sides are glazed. Each case is isolated thermally from its neighbour by cork, and its bottom is painted black to minimize heat losses by reflection. Mercury thermometers, placed on the glass windows, make it possible to read the temperatures inside the various encased boxes.

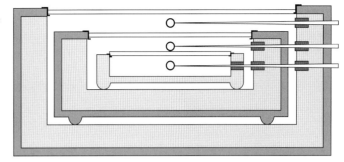

---

**Box 4.2   William Herschel**

William Herschel was born in Hanover and moved to England aged 19 as a musician and composer. Links between Britain and Germany were strong at the time, Britain having adopted a Hanoverian family to become its rulers. William turned his attention to astronomy, building his own telescopes and grinding his own lenses. He discovered the planet Uranus in 1781, and two of its major moons in 1787, as well as two of the moons of Saturn in 1789, spending much of his life close to his observatory, in Slough, in what was then Buckinghamshire, where I went to school. One of our school's houses was named after him. In 1781, he was elected a fellow of the Royal Society and received their Copley Medal. In 1782 he was appointed the King's Astronomer, not to be confused with the Astronomer Royal. He was knighted in 1816, and became the first President of the Royal Astronomical Society when it was founded in 1820. His son John followed in his footsteps as an astronomer, also winning the Royal Society's Copley Medal, being made a fellow of the Society and obtaining a knighthood.

---

'fixed air'. Fixed air was denser than air, did not burn and did not support animal life. When bubbled through limewater it precipitated calcium carbonate. He proved Helmont right – the gas was produced by fermentation – and he showed that it was also respired by animals. Black also discovered latent heat. Asking the question 'Why doesn't ice melt on a sunny day?' he found that when ice is heated its temperature increases to the freezing point and stays there until all the ice has melted. Similarly, if you boil water its temperature stays the same until it has all evaporated. A specific quantity of heat is needed to make these changes. This 'lost' or 'hidden' heat he called 'latent heat'. It is fundamental to the science of thermodynamics [3].

## 4.4   Fourier, the 'Newton of Heat', Discovers the 'Greenhouse Effect'

Enter Jean Baptiste Joseph Fourier (1768–1830) (Figure 4.4), who we met in Chapter 2 concluding that the Earth's internal heat did not affect the temperature of the atmosphere. He originated Fourier's Law regarding the rate of transfer of heat through a material, and the Fourier Transform, commonly used to transform a mathematical function of time into a new function such as frequency. Fourier's contributions to the study of heat earned him fellowship in the Royal Society of London in 1823. The Université Joseph Fourier in Grenoble, in the Department of Isère – of which he was made Prefect by Napoleon – is named after him.

Fourier made planetary temperature a proper object of study in physics, and set the stage for the development of that field over much of the nineteenth century [4]. He had been studying heat for some 20 years before summarizing his work in his 1822 magnum opus *Theorie Analytique de Chaleur* [5]. His ideas on the atmosphere were presented to the *Academie Royale des Sciences* in 1824, published that same year in the *Annales de Chimie et de Physique* (as well as later in 1827) [6], and translated into English in the *American Journal of Science* in 1837 [7]. The oft-quoted 1827 paper is reproduced in translation in *The Warming Papers* [4]. Science historian J.R. Fleming tells us that so few people seem to have read Fourier's various papers that *'We may … safely say that … [Fourier's contribution to the history of the greenhouse effect] … is not well known [even] inside the atmospheric sciences'* [5].

Thinking about Saussure's 'hot box', Fourier knew that glass was transparent to sunlight and opaque to infrared radiation, which might contribute to heating the oven. But knowing that convection redistributes heat, he reasoned that the glass lid of Saussure's 'hot box' would also contribute to warming the oven by preventing the heat from blowing away [4, 6, 7]. He pointed out that infra-red radiation (which he called non-luminous heat) is the only means by which a planet loses heat, and that whilst the air is largely transparent to sunlight it is relatively opaque to infrared

**Figure 4.4** Joseph Fourier.

**Figure 4.5** John Tyndall.

radiation from the Earth's surface, which keeps the atmosphere warmer than it would be if it were transparent to infrared radiation [4, 6, 7]. The same applied to the ocean as to the atmosphere.

In effect, Fourier, who regarded himself as 'the Newton of heat' [5], had discovered what we now call the 'Greenhouse Effect'. He did not call it that, and he had no idea precisely what agency absorbed the infrared radiation to warm the atmosphere and the ocean. Using the term 'Greenhouse Effect' to describe what greenhouse gases do to the atmosphere is actually a misnomer. Greenhouses trap heat firstly because glass does not conduct infrared radiation (you can see through glass into the interior of your oven, but the glass remains cool), *and* secondly by preventing heat from dissipating by convection (as Fourier had observed – see earlier).

## 4.5  Tyndall Shows How the 'Greenhouse Effect' Works

Fourier's 'non-luminous' or 'radiant' heat became the focus of attention of another scientific genius of the nineteenth century, John Tyndall (1822–1893) (Figure 4.5, Box 4.3), the man who discovered why the sky is blue.

In 1859, having become aware of the work of Saussure and Fourier, Tyndall began experimenting on the absorption and radiation of heat by gases. A meticulous experimenter, he used a spectrophotometer of his own design, whose tube could be filled with different mixtures of gases at variable pressures. The work required an almost superhuman effort in which he spent many hours improving his apparatus. Following a brief synopsis of his results that appeared in his book *The Glaciers of the Alps in 1860* [8], he presented a seminal paper on radiation and gases to the Royal Society in their Bakerian Lecture in February 1861 [9]. In it he explained how he measured the relative powers of air, hydrogen, oxygen, nitrogen, water vapour, ozone, carbon dioxide (then called carbonic acid), and other compounds, to absorb infrared radiation (radiant heat). His apparatus is also described in his 1863 book *Heat and Mode of Motion* [10].

Tyndall was thunderstruck by his findings. Of the components in air, he found that water vapour absorbed the most radiant heat, and oxygen and nitrogen almost none, with ozone and carbon dioxide absorbing moderate amounts. Using a second apparatus, equipped with thermometers, he found that gases that absorb infrared radiation re-emit it as heat. Here was the cause of the trapping of heat in the atmosphere that Fourier had deduced as the basis for what we now call the greenhouse effect.

Using an improved apparatus, Tyndall confirmed his previous findings, reporting his results to the Royal Society in January1862 [11]. This time he included methane (marsh gas), which he found to be around 4.5 times more absorbent than $CO_2$, and human breath. He continued to improve his apparatus (Figure 4.6) [12], summarizing his results in the Rede Lecture 'On Radiation', in May 1865 [13], where he confirmed that water vapour absorbs a great deal of infrared radiation. Recognizing that the Sun's energy warmed the Earth's surface, causing it to emit heat, Tyndall deduced that even though water vapour comprises a mere 0.5% of the atmosphere, its enormous powers of absorption of radiant heat meant that it has a huge influence on the temperature

**Box 4.3  John Tyndall**

Born in County Carlow, Ireland, Tyndall moved to work in England as a land surveyor at the age of 22 before moving to Germany in 1848 to study mathematics at the University of Marburg, where the great German chemist Robert Bunsen (1811–1899) was one of his teachers. Following the award of his PhD in 1850, he returned to London, where he gained the support of Michael Faraday (1791–1867). His work on magnetism led in 1853 to his being appointed Professor of Natural Philosophy (Physics) at the Royal Institution in Albemarle Street off Piccadilly in London. As an experimental physicist he became well known for his public lectures at the Institution, and for writing popular books on heat, sound, and light. A fascination with the effects of pressure on ice led him to study Alpine glaciers. Spending several summers in the Alps, he became an expert mountaineer, twice ascending the highest and second highest peaks, Mont Blanc and Monte Rosa, and becoming the first to ascend the Weisshorn, in 1861. Elected a fellow of the Royal Society in 1852, he was awarded its Royal Medal in 1853 for his work on magnetism but declined the honour. He accepted the Royal Society's Rumford Medal in 1864 for his work on the absorption and radiation of heat by gases. Like Lyell, he has a Tasmanian mountain named after him. Tyndall met an untimely death when his wife accidentally administered to him an overdose of the chloral hydrate he took to treat his insomnia.

of the atmosphere, '*protecting its [Earth's] surface from the deadly chill which it would otherwise sustain.... In consequence of this differential action upon solar and terrestrial heat, the mean temperature of our planet is higher than is due to its distance from the sun*' [13].

It was clear to Tyndall that water vapour absorbs much of the infrared radiation emitted by the Earth's solid or watery surface, re-emitting it in all directions to keep the atmosphere warm. '*Similar remarks*' he wrote '*would apply to the carbonic acid [$CO_2$] diffused through the air*' [9]. Every variation of these minor constituents, he thought, '*must produce a change of climate*', and could have produced '*all the mutations of climate which the researches of geologists reveal ... they constitute true causes, the extent alone of the operation remaining doubtful*' [9].

As Tyndall explained, water vapour has an important effect on atmospheric temperatures. Its absence from the dry air of deserts makes them extremely cold at night. I experienced that cold at first hand on a camping trip to the Big Bend National Park in the Chihuahua Desert in west Texas at Thanksgiving in November 1979. It was so cold at night that my girlfriend and I had to don every article of clothing we had brought with us and squeeze into our double sleeping bag for warmth, and when we woke up the following morning the dishes were solidly frozen into the water in the washing up bowl outside the tent.

Not everyone agreed with Tyndall about the overwhelming power of water vapour as a major forcing agent in warming the atmosphere. Beginning with Arrhenius (see Section 4.6), climate scientists realized that the

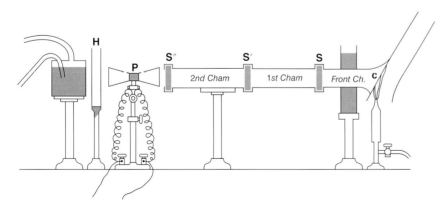

**Figure 4.6**  Tyndall's revised apparatus 1863–1864 [12]. The new apparatus was a 49.4-in. long tube with three chambers. A heat source at one end played on a copper plate. Heat passed through a first chamber full of dry air, then through a plate of rock salt into two chambers filled with the gas or vapour to be measured and separated by a second plate of rock salt, then through a final plate of rock salt to a detector comprising a thermoelectric pile in which the heating of different metal strips created a current that could be read by a galvanometer. Beyond the pile was a compensating cube to neutralize the radiation coming from the heat source, the cube being separated from the detector by an adjusting screen that helped to adjust for the compensation. *Source:* From Ref. [12].

abundance of water vapour in the atmosphere is limited by the hydrological cycle: if the air gets too humid, it rains. In effect water vapour acts as a secondary agent in global warming, by amplifying the warming caused by increases in $CO_2$, which has no such limit. Increasing $CO_2$ warms the atmosphere, causing more water vapour to evaporate from oceans and lakes, so further increasing the temperature by positive feedback. $CO_2$ does its work even in the stratosphere, where there is almost no water vapour.

Tyndall showed that whilst carbon dioxide ($CO_2$) is 'in general' a weak absorber of radiant heat, it absorbs practically all of the radiant heat from a carbonic oxide flame, suggesting what we now know, which is that because of its particular chemical structure $CO_2$ preferentially absorbs heat in specific parts of the electromagnetic spectrum. He was unable to measure the true strength of the absorption of infrared radiation by the various gases, which turns out to be dependent on the exact wavelength of the radiation [4], as that example suggests.

We now know that the $CO_2$ absorption spectrum comprises a collection of very narrow peaks at set wavelengths (for more on that topic see Chapter 8), which broaden and coalesce with increasing pressure [4] (for more on that topic see Chapter 9). This helps in part to explain why Tyndall's results showed that when a gas like $CO_2$ is very dilute or at a low enough pressure, the absorption increases linearly with the concentration of the gas, but that when the concentration of gas (its partial pressure) is sufficiently high, it absorbs all or most of the radiant heat – a phenomenon known today as 'band saturation'. As a result, '*even today, a detailed calculation of the absorption and emission of IR [infrared radiation] by a column of atmosphere is not trivial; it can be done by computer models known as line-by-line codes, based on megabytes of detailed spectral information for the various greenhouse gases, but these calculations are too computationally expensive, that is to say slow, to be done in the full climate models that are used to predict things such as, say, the climate sensitivity or global warming forecasts*' [4].

Comparing the lack of absorption of infrared radiation by single element gases like hydrogen ($H_2$), oxygen ($O_2$) and nitrogen ($N_2$) with its abundant absorption by compound gases like carbon dioxide ($CO_2$) and water ($H_2O$), Tyndall concluded that something about the chemical bonds in the compound gases encouraged the absorption of radiant heat. We now know that the structure of these compound gases encourages vibration enabling them to capture and re-emit infrared radiation. Thanks to his work we now use infrared radiation detectors and spectrometers to measure the abundance of $CO_2$ in the air and in the atmosphere of other planets. Hospitals still use Tyndall's system for measuring the carbon dioxide in human breath to monitor the health of patients under anaesthetics during surgical operations.

Although Tyndall gets the credit for having demonstrated the role of $CO_2$ in absorbing and re-emitting radiation, Edouard Bard reminds us that '*Jacques Joseph Ebelmen (1814–1852), professor at the "École des mines" in Paris and Director of the Royal Works of porcelains in Sèvres, was the first to suggest that past changes in the carbon cycle could have changed the atmospheric concentration of "carbonic acid" and, as a direct consequence, the climate of the Earth*' [14]. Ebelmen was an eminent French scientist in his day. His is one of the 72 names of famous scientists inscribed on the Eiffel Tower at the time of its construction, and the Ebelmen Award of the International Association of Geochemistry is named in his honour. Bob Berner and Kirk Maasch [15] tell us that that before his untimely death at the age of 37, Ebelmen had developed essentially all of the fundamental concepts not only of the geochemical carbon and sulphur cycles and their effects on atmospheric $CO_2$ and $O_2$, but also on the whole process of chemical weathering: for example with volcanic emissions supplying $CO_2$ to the atmosphere and the weathering of magnesium silicates removing it. He realized that a close balance was needed between $CO_2$ uptake by silicate weathering and $CO_2$ release by volcanism, to prevent a build-up of its concentration in the atmosphere. He also recognized that changes in the rates of its uptake and release could change $CO_2$ levels in the atmosphere over geological time. Ebelmen wrote in 1845 (translation by Berner and Maasch [15]) that '*many circumstances … tend to prove that in ancient geologic epochs the atmosphere was denser and richer in carbonic acid and perhaps oxygen, than at present. To a greater weight of the gaseous envelope should correspond a stronger condensation of solar heat and some atmospheric phenomena of a greater intensity*' [16]. Ebelmen's theoretical work was brought to the attention of the wider community by T.S. Hunt in 1880 [17], but largely ignored [15]. Given the vibrancy of the intellectual exchange networks of the nineteenth century, even in the absence of the Internet, it seems surprising that apparently neither Lyell nor Tyndall knew anything of Ebelmen's work (or, if they did, they did not acknowledge it). But it seems quite likely that by Tyndall's time Ebelmen's ideas may have been in the scientific air and may have helped to trigger his research.

Tyndall was not quite the first to experiment with the effects of $CO_2$ on atmospheric temperature. A Mrs. Eunice Foote had carried out rather simple experiment along the same lines three years before him in 1856, with the results being reported by Professor Joseph Henry of the Smithsonian Institution at the 10th annual meeting of the American Association for the Advancement of Science (AAAS) in Albany, New York, in 1856 [18–20]. Foote had used tubes of gases left in the sunshine to demonstrate that temperatures rose more in moist than in dry air, and more in carbonic acid gas ($CO_2$) than in normal air, indicating the power of both water vapour and $CO_2$ to warm the atmosphere. Science reporter David Wells recorded that '*An atmosphere of that gas [$CO_2$] would give to our earth a much higher temperature; and if there once was, as some suppose, a larger proportion of that gas in the air, an increased temperature must have accompanied it, both from the nature of the gas and the increased density of the atmosphere*' [18, 19]. Mrs. Foote's two-page note containing more or less that same quote was published in the *American Journal of Science and Arts* for 1856 [20]. Perhaps its findings travelled through the scientific air across the Atlantic to Albemarle Street.

Essentially, the natural greenhouse effect works like this: some of the outgoing infrared radiation from the Earth's surface escapes directly to space, but much is absorbed and re-emitted in all directions in the atmosphere by trace gases in the form of complex molecules like ozone, water vapour, $CO_2$, and methane ($CH_4$). The absorption of infrared energy makes these molecules vibrate and thus warm the air. This natural warming helps to raise the average temperature of the atmosphere at the surface of the Earth by some 33°C above Earth's 'black body' temperature. Without it we would be shivering at an unpleasant −18°C average instead of a comfortable +15°C (more on that in Chapter 9). As Bill Hay points out, if the oven in your kitchen has a light in it and a glass pane in its door you will be familiar with the fact that glass allows visible radiation to pass through, but blocks the oven's heat radiating at longer (infrared) wavelengths [21]. Greenhouses and blankets work in much the same way as the glass, though both also work by preventing convection. Convection works to ensure that $CO_2$ and heat are widely mixed throughout the atmosphere.

Convection also moves water vapour around, but as Gilbert Plass pointed out [22], the distribution of water vapour is subject to the limits imposed by temperature. Very cold air contains no moisture so water vapour tends to be concentrated within the troposphere whilst $CO_2$ extends into the stratosphere; there is also little water vapour at the poles. Water also rains out rapidly from the atmosphere, whilst $CO_2$ can continue to increase. Water vapour's main effect is to reinforce temperature change by positive feedback. $CO_2$ increases temperature, hence evaporation, hence water vapour, which further increases temperature, which warms surface waters so that they release more $CO_2$. $CO_2$, in contrast, tends to be present everywhere and at all altitudes. Whilst one-for-one water is a less effective greenhouse gas than $CO_2$, with a global warming potential of 0.28 to $CO_2$'s 1.0, the greater abundance of water vapour in the atmosphere means that its potency is more (4.0 as against 1.42 for $CO_2$) [21].

In contrast, the main atmospheric gases – oxygen and nitrogen – have no effect on infrared radiation, as they lack the absorptive properties of the more complex trace gases. Under these circumstances, adding or subtracting $CO_2$ would be expected to raise or lower the temperature of the atmosphere, as we see next, not least by helping to increase the amount of water vapour in it.

Is there evidence for water vapour increasing as the world warms? Whilst we do need better data on the distribution of water vapour through the atmosphere, Section 2.4.4.3 of Chapter 2 of the report on the Physical Science Basis for the 5th Assessment Report (AR5) of the Intergovernmental Panel on Climate Change (IPCC) tells us '*The interannual variability and longer-term trends in column-integrated water vapour over oceans are closely tied to changes in SST [sea surface temperature] at the global scale, and interannual anomalies show remarkable agreement with low level specific humidity anomalies.... The rate of moistening at large spatial scales over oceans is close to that expected from the Clausius-Clapeyron relation (about 7% per °C) with invariant relative humidity.... Satellite measurements also indicate that the globally-averaged upper tropospheric relative humidity has changed little over the period 1979–2010 while the troposphere has warmed, implying an increase in the mean water vapour mass in the upper troposphere*' [23]. One result is that the hydrological cycle is changing: dry areas, like the Mediterranean basin, are getting drier, whilst other areas are getting wetter and experiencing exceptional flooding [23].

## 4.6 Arrhenius Calculates How $CO_2$ Affects Air Temperature

Someone had to take Tyndall's findings further, and calculate by how much $CO_2$ might change the temperature of the atmosphere. That person was Swedish chemist Svante August Arrhenius (1859–1927) (Figure 4.7, Box 4.4).

Stimulated by the work of an eminent geologist colleague, Arvid Gustaf Högbom, who had worked on the geochemistry of carbon, and being aware of Tyndall's observations on the radiative properties of gases, Arrhenius became intrigued by the question of what caused the onset of glacial and interglacial periods. In 1895, he presented to the Swedish Physical Society a paper published the following year suggesting that changes of the order of 40% in $CO_2$, a minor atmospheric constituent, might trigger feedback phenomena that could account for glacial advances and retreats [24]. '*A simple calculation shows that the temperature in the Arctic regions would rise about 8° to 9°C, if the carbonic acid [$CO_2$] increased to 2.5 to 3 times its present value. In order to get the temperature of the ice age between*

**Figure 4.7** Svante Arrhenius.

*the 40th and 50th parallels, the carbonic acid in the air should sink to 0.65–0.55 of its present value (lowering of temperature 4°–5°C)'* [24].

To get that result, Arrhenius had to invent the field of climate modelling [14]. As explained by David Archer and Raymond Pierrehumbert [4], his energy budget model took into account the radiative energy exchanges between the atmosphere and space, between the atmosphere and the ground, and between the ground and space owing to the transmission of infrared radiation through the atmosphere. It also took into account the increase in water vapour in the atmosphere as the atmosphere warms, and the change in albedo (reflection of solar energy) as snow and ice melt with increased warming. As in modern climate models, his calculations assumed a four-dimensional space with a grid based on average temperatures for every 10th degree of latitude, extending vertically through the atmosphere, and repeated for every season. To keep the calculations manageable, he considered the entire atmosphere as a single layer whose radiative properties were

characterized by a single vertically averaged temperature. To further simplify his calculations, he assumed that the transport of heat by winds and ocean currents remained unchanged with time, as did the extent of cloud cover, which allowed him to focus on the variation of the temperature with the transparency of the air. Even so, without the aid of a computer, this was a long and tedious process to say the least – but perhaps helped by those long Swedish nights and the absence of distractions like television.

Arrhenius did the best he could, given the limitations of the data at the time. But the infrared spectral data that he had at his disposal only went up to wavelengths of $13\,\mu m$ ($\mu$) – not far enough for him to pick up the major $CO_2$ absorption band near $15\,\mu$, which is a major feature for global warming [4, 21](see also Chapter 8). Moreover his absorption data were largely at intervals of $0.5$–$1.0\,\mu$, whereas individual absorption spectra for both water vapour and $CO_2$ are now known to occur in multiple spectral lines each around $1/100\,\mu$ wide or less [25] (for mere details see Chapter 8). This was unknown to Arrhenius and his contemporaries and would remain so until the 1950s. Arrhenius knew of the limitations on water vapour caused by rainfall at 100% humidity. His model required that whatever the temperature, the relative humidity should stay at around 80%.

Arrhenius predicted that doubling the $CO_2$ content of the atmosphere from its present value, for which he used 300 ppm, would raise the temperature of the surface of the Earth by about 6 °C. Most modern estimates put the likely warming for a doubling of $CO_2$ at 2.5–4 °C. He got lucky, because his model contained two significant sources of error – first in not having the full spectrum of absorption of $CO_2$, which biassed the result on the high side, and second in using just one layer for the entire atmosphere, which biassed the result on the low side. These two errors cancelled one another [4]. As Archer and Pierrehumbert point out [4], *'the genius in the work of Arrhenius is that he turned Fourier's rather amorphous and unquantified notion of planetary temperature into exactly the correct conceptual framework, even going so far as to get the notion of water vapour feedback right. Most importantly he correctly*

---

**Box 4.4   Svante August Arrhenius**

Arrhenius was born near Uppsala, Sweden. He studied at the university of Uppsala and at the Institute of Physics of the Swedish Academy of Sciences in Stockholm. In 1884, he published a dissertation on the chemical theory of electrolytes, and worked on electrochemistry in several continental laboratories before returning to a lectureship in physics at the Stockholm Hogskola in 1891. There he was promoted to Professor and served as Rector before becoming the first Director of the Nobel Institute for Physical Chemistry. He was elected to the Swedish Academy in 1901, awarded the Nobel Prize for Chemistry in 1903, and elected a foreign member of London's Royal Society in 1911.

*identified the importance of satisfying the energy balance both at the top of the atmosphere and at the surface ... [C]orrect spectroscopy was not brought together with a correct conceptual framework [like that of Arrhenius] in a multilevel model until the seminal work of Manabe in the early 1960s'* [4]. The importance of Arrhenius's contribution is that *'While we can now compute the effects of $CO_2$ on climate at a level of detail and confidence that Arrhenius could hardly have dreamed of, we are basically doing the same energy book-keeping as Arrhenius ... [did, but] in vastly elaborated detail with vastly better fundamental spectroscopic data'* [4].

Arrhenius predicted that to get the temperature drop of 4–5 °C thought likely for the last glaciation, the $CO_2$ concentration of the atmosphere must have fallen to between 186 and 165 ppm We now know it was about 180 ppm, so he was not far off. Nevertheless, he was unable to provide a mechanism for the changes in $CO_2$ with time between glacial and interglacial periods. His theory implied that the whole Earth should have undergone about the same variations of temperature at the same time, the atmosphere being well mixed, meaning that glacial periods should have been simultaneous in both hemispheres, contrary to Croll's notion that glacial and interglacial periods alternated between the hemispheres. Arrhenius disputed the notion that orbital variations drove variations in the climate as Croll required, but offered no alternative in its place. As we shall see later, the fluctuations of the Ice Age require a combination of the mechanisms of Croll, Arrhenius, and Lyell [14].

Arrhenius was well aware that during Tertiary times, Arctic vegetation had been like that of the mid latitudes, implying that temperatures must have been 8–9 °C above what they are there today. This, he calculated, would have required an increase in $CO_2$ of 2.5–3.0 times more than the current value of 300 ppm

To underscore the relevance of his model to geology, Arrhenius reprinted in his paper a lengthy summary of the views of his geological colleague, Högbom [24]. Högbom calculated that the burning of coal would supply about one thousandth part of the carbonic acid ($CO_2$) in the atmosphere, *'completely compensating the quantity of carbonic acid that is consumed in the formation of limestone ...'* and thus not building up in the atmosphere [24]. At the time, Arrhenius went along with that, showing no concern about man's influence on the climate.

Even so, just after the turn of the century, in 1907, the date of the preface to his 1908 book *Worlds in the Making* [26]. Arrhenius did look into the question of how changes in $CO_2$ might affect our present climate. By then, he was referring to Tyndall's theory as the 'hot-house' (i.e. green-house) theory. He first calculated by how much the temperature of the Earth's atmosphere would fall if deprived of

its $CO_2$, and what the added effect would be of depriving it of its water vapour, showing that *'comparatively unimportant variations in the composition of the air have a very great influence'* [26]. His new calculations showed that *'doubling of the percentage of carbon dioxide in the air [from its value then] would raise the temperature of the Earth's surface by 4°'* [6]. Recognizing that we humans were now burning significant amounts of coal, and assuming that it continued to be burned at the current rate (i.e. with linear increase), he suggested: *'the slight percentage of carbonic acid in the atmosphere may by the advances of industry be changed to a noticeable degree in the course of a few centuries'* [26]. But, being an astute fellow, he realized that this change in climate would be significantly more rapid if the rate of consumption of fossil fuels increased – as indeed it did.

Knowing that $CO_2$ was good for plants, he suggested that increasing the $CO_2$ content of the atmosphere by burning fossil fuels might be beneficial, making the Earth's climates warmer and more equable, stimulating plant growth, providing more food for a larger population, and even preventing the recurrence of another glacial period [26]; these were prescient observations, as we shall see. Arrhenius also knew that in the normal course of events the content of $CO_2$ in the atmosphere did not continually increase as a result of volcanic activity, because it was taken out of the atmosphere by weathering of minerals and assimilation by plants. This is what we may call the 'slow cycle' of carbon on geological timescales, in contrast with the burning of fossil fuel, which is happening much more rapidly in what we may call the 'fast cycle' of carbon.

## 4.7 Chamberlin's Theory of Gases and Ice Ages

It seems odd in retrospect that neither Lyell nor Croll, both publishing their key works in 1875, seem to have picked up on the significance of Tyndall's discoveries, reported in the 1860s and early 1870s, of the importance of water vapour and $CO_2$ as greenhouse gases capable of influencing the climate. Lyell only mentions $CO_2$, in the 12th edition of *Principles*, to dismiss the ideas of some geologists that it was an abundance of $CO_2$ in the air that encouraged the flourishing of vegetation that led to the coal deposits of the Carboniferous. That he knew about Tyndall's discoveries is evident from a letter to Joseph Hooker dated October 1866, in which he said *'I suspect that the vapour to which you allude, and on which Tyndall has written so much, may equalize the heat and cold caused by greater proximity to, and distance from, the sun'* [27]. Fleming reminds us that in his 1866 edition of *Principles* Lyell had taken some of Tyndall's discoveries into account in noting that plants could be saved from extinction during an Ice Age '*by the heat of the earth's surface ... being prevented from radiating off freely into space by a blanket of aqueous vapour caused by the melting of snow and ice*' [28]. But, as Fleming points out Lyell '*was grasping at straws, aware of new theoretical problems, but taking from them only the aspects that reinforced his own preconceptions*' [28]. One might have thought that as both Croll and Lyell were fellows of the Royal Society they might have been more aware of the implications of Tyndall's findings, he being also a fellow. Perhaps this lack of connection between the frontiers of atmospheric physics and geology is a reflection of the increasing specialization of the sciences at the time.

The geologist who did pick up Tyndall's findings and run with them, albeit indirectly, was Thomas Chrowder Chamberlin (1843–1928) (Figure 4.8, Box 4.5).

Rather than following Tyndall directly, Chamberlin was following Arrhenius, who had been stimulated by Tyndall's work. Like Lyell, Tyndall, Croll, and Arrhenius, he too seems to have been unaware of the groundbreaking work of Ebelmen.

In 1899, having seen Arrhenius's seminal paper in 1896, and attracted by the idea that $CO_2$ may have helped to control the climates of former times, Chamberlin set about developing a working hypothesis to explain what its role might have been [29, 30]. He surmised that $CO_2$ in the air combines with water in rain to form a weak acid that decomposes the silicate minerals of shales, sandstones, and volcanic rocks, thus supplying mineral salts including carbonate ions ($CO_3^{2-}$) to the ocean and leaving behind residual silicates and oxides [29, 30]. This chemical 'weathering' sucks $CO_2$ out of the atmosphere slowly over long periods of time. When the dissolved carbonate ions arrive in the ocean they are used by marine creatures to form shells of calcium and magnesium carbonate that fall to the seabed to accumulate over time and be compressed into limestones and dolomites. $CO_2$ is also lost

**Figure 4.8** Thomas Chrowder Chamberlin.

from the atmosphere through photosynthesis by plants, then trapped as carbonaceous compounds in the organic matter disseminated widely in small amounts in marine sediments, and in organic-rich deposits that ultimately form coal, bitumen, oil, and natural gas. This loss cools the atmosphere. Compensating for this loss over the aeons of time is the supply of $CO_2$ to the atmosphere from volcanoes, and also from the decomposition of organic matter. Chamberlin concluded, '*the state of the atmosphere at any time is dependent upon the relative rates of loss and gain. [The] agencies of permanent supply [volcanoes and the decomposition of organic matter] and of permanent loss [weathering and the deposition of carbonates] are ... rather slow in action, and ... on the whole mutually compensatory ... but these relations are believed to be subject to sufficient fluctuation to give a basis for pronounced*

---

**Box 4.5  Thomas Chrowder Chamberlin**

Chamberlin was born in Mattoon, Illinois, and educated at Beloit College, Wisconsin, and the University of Michigan. On the faculty at Beloit he mapped the glacial deposits of Wisconsin, and devised the terminology used to describe the North American Pleistocene. He was the first to demonstrate that there had been multiple Pleistocene glaciations in North America. He joined the Geological Survey of Wisconsin in 1873, and became its Chief Geologist in 1876. He was appointed Head of the Glaciological Division of the US Geological Survey in 1881, president of the University of Wisconsin in 1887, and professor of geology at the University of

Chicago in 1892. He held positions as president of the academies of science of Wisconsin (1885–1886), Chicago (1897–1915), and Illinois (1907) and of the AAAS (1908–1909). In 1893 he launched the *Journal of Geology*, and in 1904 a college textbook – *Geology*, probably the most influential textbook on geology in the United States prior to the Second World War. He was elected to the National Academy of Sciences in 1903, and won the Penrose Gold Medal of the Society of Economic Geologists in 1924, and the Penrose Medal of the Geological Society of America in 1927. He has craters named after him on the Moon and Mars.

*climatic change*' [29, 30]. His hypothesis reads remarkably like Ebelmen's in several respects, probably due to the convergent evolution of ideas, which can lead to independent generation of the same concept – as Darwin and Wallace found out in 1858.

Chamberlin was the first geologist to fully appreciate the role of the ocean in controlling the amount of $CO_2$ in the atmosphere. By his time, chemists knew that $CO_2$ dissolves in the ocean, and that the $CO_2$ dissolved in the ocean should be in equilibrium with that diffused in the air [29, 30]. Reducing the concentration or partial pressure of the $CO_2$ in the air would force some of the $CO_2$ in the ocean to diffuse into the air to restore the equilibrium. Chemists also knew that warm water contains less gas than cold water. So Chamberlin appreciated that a warming ocean would release more $CO_2$ to the air, whilst a cooling ocean would take more $CO_2$ from it. If $CO_2$ were lost from the air by weathering faster than it could be replaced by diffusion from the ocean, the atmosphere and the ocean would cool, and the cooler ocean would hold more $CO_2$, preventing its escape to the air. Bearing these various new facts in mind, Chamberlin was the first to suggest that '*the ocean, during a glacial episode instead of resupplying the atmosphere ... would withhold its carbon dioxide to a certain extent ... [and], when the temperature is rising after a glacial episode ... the ocean gives forth its carbon dioxide at an increased rate, and thereby assists in the amelioration of climate*' [29, 30].

In a nutshell, Chamberlin's theory of $CO_2$ and climate change runs like this: when land is raised and extended by mountain building, the rate of weathering and dissolution of silicate and carbonate rocks increases, consuming $CO_2$ from the atmosphere. The dissolved constituents are carried to the sea to form limestones, which keep $CO_2$ out of the atmosphere. The extension of the land reduces the area available for marine life (which Chamberlin assumes is more productive than terrestrial life), which in turn reduces the resupply of $CO_2$ to the atmosphere from decomposing marine organic remains. The lessened $CO_2$ concentration in the atmosphere cools the planet. The cooling ocean absorbs $CO_2$, further reducing its supply to the atmosphere. Over time, erosion reduces the land level, and weathering is reduced, permitting a rise in the $CO_2$ concentration in the atmosphere, supplied by the ongoing emission of volcanic gases. Warming ensues. The warming ocean floods the margins of the continents, further reducing weathering and enhancing $CO_2$ in the atmosphere. The warming evaporates water vapour from the ocean, warming the air yet further, which releases yet more $CO_2$ from the ocean. Whilst increasingly abundant marine life takes $CO_2$ out of the atmosphere by photosynthesis, Chamberlin assumes that this is balanced by decomposition that puts $CO_2$ back into the atmosphere. '*As a result ... geological history has*

*been accentuated by an alternation of climatic episodes embracing, on the one hand, periods of mild, equable, moist climate nearly uniform for the whole globe, and on the other, periods when there were extremes of aridity and precipitation, and of heat and cold, these last denoted ... occasionally by glaciation*' [29, 30].

In many ways, Chamberlin's theory is much like that in vogue today. Applying it to the Ice Age of the past 2 Ma or so [31], he surmised that as $CO_2$ began to decrease in the atmosphere, starting the cooling process, ice sheets and snowfields expanded. These reflected more solar energy, accentuating cooling – the positive feedback that Croll had identified. Areas of frozen ground – permafrost – expanded too. Combined, these effects limited the area exposed to chemical weathering, which in any case would have been reduced by the increasingly cold conditions, thus limiting the consumption of $CO_2$. Bearing in mind that the concentration of $CO_2$ in the atmosphere reflected a balance between supply and consumption, there would be a tendency for the continued supply of $CO_2$ from volcanic gases and by other means to restore the initial equilibrium by building up the $CO_2$ concentration in the atmosphere and thus warming the ocean, which would then release more $CO_2$ thus causing further warming, which would evaporate water vapour, causing yet further warming, leading eventually to the retreat of the ice sheets. Chemical weathering would then be renewed, starting the cycle over again. Chamberlin thought that these patterns would be influenced and modified, but not controlled, by variations in solar heating caused by orbital changes of the kind suggested by Croll.

In his later work Chamberlin considered that '*a vital factor in initiating a glacial period is a reversal of the deep-sea circulation*' [5]. In interglacials like the present, he surmised, deep ocean currents transfer heat to high latitudes keeping the poles relatively warm. Reversal of this circulation cools the polar oceans leading them to take up $CO_2$ at the expense of the atmosphere. Cooling puts less water vapour into the atmosphere, which accentuates the fall in temperature. Eventually cold water occupies the ocean floor. In due course, this cold water reaches the surface in the tropics, warms, and starts to release more $CO_2$ than the polar waters are absorbing, so ending the glacial cycle. The oceanographic concepts behind his thinking are rather like those of Carpenter and Croll, which we explored in Chapter 3, and which probably stimulated Chamberlin's thinking in this direction, but we no longer think that cold deep water from the poles rises in the tropics. Nevertheless, the notion of the reversal of deep ocean circulation is now integral to our understanding of the glacial to interglacial transition. What Chamberlin lacked was the trigger for it.

Chamberlin's work was eventually largely forgotten, but as Fleming tells us [5], he gave his contemporaries an almost modern understanding of multiple glaciations, of the role of the atmosphere as a geological agent, and of the role of deep-ocean circulation in the ice age. His work '*was filled with fundamentally sound insights and represents a surprisingly modern voice from the past*' [5].

With time, Chamberlin regretted having been quite so willing to accept Arrhenius's results. He thought that the role of $CO_2$ in the atmosphere had been overemphasized, and that not enough attention had been given to the role of the ocean, which he considered his distinctive contribution to the subject. We now know that Chamberlin's dilemma should not have been a question of choice between atmospheric $CO_2$ and deep ocean circulation. Both play key roles in the transitions from glacial to interglacial period and back. He simply lacked the data to support his $CO_2$ theory, a state of affairs that has long since been remedied.

Like all geologists in the nineteenth century, one of Chamberlin's greatest problems was the absence of any ready means of accurately dating the boundaries between glacial and interglacial deposits. Taking a conservative view, he estimated that the last glaciation might have ended some 20 Ka ago – '*the time since the last ice retired from the site of Niagara River*', and that interglacial intervals lasted 20–30 Ka [31]. Not a bad guess, as we shall see in later chapters.

In Chamberlin's final paper on the causes of ice ages he addressed the enigma of the glaciation in the Permo-Carboniferous period 300 Ma ago [32]. Evidence for the occurrence of extensive glaciation in India, Brazil, Australia, and South Africa had grown throughout the last half of the nineteenth century. A Dr. Blandford discovered a late Carboniferous boulder bed of glacial origin at Talchir, India, in 1856. Tillites (essentially beds of boulder clay) of about the same age were identified in South Australia in 1859; in South Africa in 1870; and in Brazil in 1888 [33]. Establishing the nature and cause of this earlier glaciation proved difficult in the face of inexact dates and the absence of information about contemporaneous conditions in other parts of the world. The distribution of the tillites was extraordinary, the chief areas being within the tropics. Another confusing factor was the association of glacial remains with coal deposits, which seemed to imply tropical conditions and high concentrations of $CO_2$ in the atmosphere. Lyell had noted in the 12th edition of his *Principles* '*That the air was charged with an excess of carbonic acid in the Coal period has long been a favourite theory with many geologists, who have attributed partly to that cause an exuberant growth of plants*' [34]. He thought this inference '*most questionable*', because although $CO_2$ was supplied to the atmosphere from volcanoes and other sources, there were ample causes in action to prevent it building up in the atmosphere, notably, in his view, the burial of plant

remains – a good point, as it turns out. Chamberlin was reduced to suggesting, on what were thought to be flimsy grounds at the time, that the widespread deposition of carbonates and coals in the Carboniferous may have depleted the air of $CO_2$, so encouraging development of a glaciation [32]. He was not far from the truth, as it happens. The various glacial deposits were judged to belong to the Gondwana Series – chiefly land and freshwater deposits whose distribution around the Indian Ocean implied unusual connections of some kind between the different continents.

Chamberlin was not alone in his speculations about past geography and climate. In the second half of the nineteenth century geologists began to realize from the distributions of fossil plants and animals that there must have been close links between different continents. One of the first to point this out for the southern continents was Eduard Suess (1831–1914) (Figure 4.9, Box 4.6).

Noticing that fossils of the extinct order of seed ferns known as *Glossopteris* were common to southern Africa, India, and Australia, Suess postulated in volume II of his book *The Face of the Earth* [35], that they had formerly been connected in a supercontinent that he named 'Gondwanaland' (i.e. 'Land of the Gonds') after an ancient people of India. Suess thought that the present separation of the fragments of Gondwanaland was due to flooding of its low parts by the ocean, not to the fragments moving apart from one another. He was the first to recognize that an ocean, which he named Tethys, must have connected the Atlantic and Pacific Oceans more or less along the line of the Mediterranean Sea and Persian Gulf in the distant

**Figure 4.9** Eduard Suess in 1869.

**Box 4.6 Eduard Suess**

Born in London to a German merchant family, Eduard Suess moved to Vienna with his family as a teenager and spent much of his career as professor of geology at the university there. In his magnum opus *Das Antlitz der Erde (The Face of the Earth)*, Suess pointed out that organic life was limited to a narrow zone at the surface of the Earth's crust or lithosphere, which he named the 'biosphere'. Highly respected, and considered as one of the fathers of the science of ecology, he was elected a member of the Royal Swedish Academy of Sciences in 1895, awarded the Copley Medal by London's Royal Society in 1903, and the Wollaston Medal of the Geological Society of London in 1896, and had craters named after him on the Moon and Mars.

geological past, a significant geological development that must have affected the regional climate. Suess also believed that the rises and falls of the sea (transgressions and regressions) must be connected from continent to continent and should be correlatable across the Earth. We will revisit that important concept in later chapters.

When setting down his ideas between 1897 and 1899, Chamberlin might not have known about Suess's concept of the unity of the southern continents in Gondwanaland, because the three volumes of Suess's book were not translated into French (*La Face de la Terre*, in four volumes) until between 1897 and 1918, or into English (in five volumes) between 1904 and 1924.

Suess's idea that the continents had been connected by former 'land bridges' now sunk into the ocean became popular amongst palaeontologists seeking to explain the distribution of similar species on widely separated land masses. We may laugh at it now, with the benefit of hindsight, but in those days little was known about the composition of the deep ocean floor, so it did not seem unreasonable to assume that former land bridges simply sank to form ocean basins. So influential was Suess that as an undergraduate in the early 1960s I still had to write essays on the possible existence of 'land bridges'. But I was never taught about Chamberlin or about $CO_2$.

The tree-fern *Glossopteris* (or 'tongue fern', in Greek), which would help in coming years to establish the reality of a southern supercontinent, was named by botanist Adolphe Brongniart (1801–1876), the son of Alexandre Brongniart, whom we met as a friend of Lyell's in Chapter 2. Adolphe's first memoir on fossil plants appeared in 1822, and his *Histoire des Végétaux Fossiles* was published between 1828 and 1837, earning him the title 'father of

palaeobotany' [36]. The boundary between the Gondwanan flora, in which *Glossopteris* plays such an important role – especially in the Permian – and that of the northern continents has been described as '*the most profound floristic boundary in the history of land vegetation*' [37]. Captain Scott of Antarctic fame would unwittingly identify Antarctica as a former part of Gondwana, by collecting from Mt. Buckley on the Beardmore Glacier, on his tragic return from the South Pole in February 2012, a slab of rock containing fossil *Glossopteris* leaves. The eminent palaeobotanist Albert Charles Seward (1863–1841), who later became President of the Geological Society of London (1922–1924), and won the Society's Wollaston Medal in 1930, analysed Scott's leafy rock sample in 1914 [38]. Suess would have been pleased to see the results, though not with the human cost.

As following chapters show, we would have to wait until the late 1950s and beyond before science advanced far enough to test the ideas put forward by nineteenth century pioneers like Tyndall, Arrhenius, and Chamberlin concerning $CO_2$ in the atmosphere and its role in the climate system. Atmospheric physicists and chemists would have to understand much more about how the greenhouse effect worked, which would require the development of more powerful analytical techniques, along with advanced computers for the processing of data. And palaeoclimatologists would have to follow Chamberlin and learn much more about atmospheric and ocean chemistry and circulation, as well as the role of the greenhouse gases, in order to appreciate fully how the Earth's climate system worked in the past.

As Lyell had suspected, the migration of continents is fundamental to understanding the change in their climates through time, so it is to continental drift that we turn next.

## References

**1** De Saussure, H.B. (1779, 1786, 1796). *Voyages Dans les Alpes, Précédés d'un Essai sur l'Histoire Naturelle des Environs de Genève*, in 3 volumes. Neuchatel: Fauche Barde, Manget, et Fauche-Borel.

**2** Geikie, A. (1897). *Founders of Geology*. London: Macmillan and Co. www.archive.org.

**3** Uglow, J. (2002). *The Lunar Men*. London: Faber & Faber, 588 pp.

**4** Archer, D. and Pierrehumbert, R. (2011). *The Warming Papers*, 3–6. Chichester: Wiley Blackwell.

**5** Fleming, J.R. (1998). *Historical Perspectives on Climate Change*. Oxford University Press, 194 pp.

**6** Fourier, J.B.J. (1824). Remarques Générales Sur Les Températures Du Globe Terrestre Et Des Espaces Planétaires. *Annales de Chimie et de Physique* 27: 136–167.

**7** Burgess, E. (1837). General remarks on the temperature of the terrestrial globe and the planetary spaces, by Baron Fourier. *American Journal of Science* 32: 1–20.

**8** Tyndall, J. (1860) *The Glaciers of the Alps* – available on the Internet as a Project Gutenberg e-Book http://www.gutenberg.org/ebooks/34192.

**9** Tyndall, J. (1861) on the absorption and radiation of heat by gases and vapours, and on the physical connexion of radiation, absorption and conduction, *London, Edinburgh and Dublin Philosophical Magazine and Journal of Science*. Reproduced in Tyndall, J. (1868) *Contributions to Molecular Physics in the Domain of Radiant Heat*, (the 3rd edition, of 1872, is downloadable from www.archive.org. The paper is also reproduced with comments in Archer, D., and Pierrehumbert, R., (2011) *The Warming Papers*, Wiley-Blackwell, Chichester, 21–44.

**10** Tyndall, J. (1863). *Heat and Mode of Motion*, 1e. New York: D. Appleton, 480 pp. (7th ed., Longmans, Green, London, 1887, 591 pp).

**11** Tyndall, J. (1868) Further researches on the absorption and radiation of heat by gaseous material" section II of *Contributions to Molecular Physics in the Domain of Radiant Heat*", (the 3rd edition, of 1872, can be downloaded from www.archive.org).

**12** Tyndall, J. (1864). The absorption and radiation of heat by gaseous and liquid matter. *London, Edinburgh and Dublin Philosophical Magazine and Journal of Science* XXVIII: 81–106, Figure 2 (also in Tyndall, J., 1872, *Contributions to Molecular Physics in the Domain of Radiant Heat – a Series of Memoirs Published in the 'Philosophical Transactions' and 'Philosophical Magazine', with Additions*. Longmans, Green and Co., London 477 pp, where it appears in Part V as Figure 14, p.173).

**13** Tyndall, J. (1868) *On Radiation*, the 1865 Rede Lecture, D Appleton and Co of New York. Also included in his 1871 book "*Fragments of Science*", available on the Internet as a Project Gutenberg e-Book.

**14** Bard, E. (2004). Greenhouse effect and ice ages: historical perspective. *Comptes Rendus Geoscience* 336: 603–638.

**15** Berner, R.A. and Maasch, K.A. (1996). Chemical weathering and controls on atmospheric $O_2$ and $CO_2$: fundamental principles were enunciated by J.J. Ebelmen in 1845. *Geochimica Cosmochimica Acta* 60 (9): 1633–1637.

**16** Ebelmen, J.J. (1845). Sur les produits de la décomposition des espèces minéraux de la famille des silicates. *Annales des Mines* 7: 3–66.

**17** Hunt, T.S. (1880). The chemical and geological relations of the atmosphere. *American Journal of Science* 19: 349–363.

**18** Wells, D.A. (1857) Annual of scientific discovery: or, year-book of facts in science and art, for 1857.

**19** Sorensen, R.P., 2011, Eunice Foote's Pioneering Research On CO2 And Climate Warming. *Search and Discovery Article #70092*, American Association of Petroleum Geology.

**20** Foote, E. (1856). Circumstances affecting the heat of the Sun's rays. *American Journal of Science and Arts* XXII, 2nd Series, No.66, Article XXXI: 382–383.

**21** Hay, W.W. (2013). *Experimenting on a Small Planet: A Scholarly Entertainment*. New York: Springer, 983 pp.

**22** Plass, G.N. (1961). The influence of infrared absorptive molecules on the climate. *New York Academy of Science* 95: 61–71.

**23** IPCC (2013), *Climate Change 2013: The Physical Science Basis*. Working Group I Contribution to the IPCC Fifth Assessment Report. Intergovernmental Panel on Climate Change, WMO Headquarters, Geneva.

**24** Arrhenius, S. (1896) On the influence of carbonic acid in the air upon the temperature of the ground, *London, Edinburgh, and Dublin Philosophical Magazine and Journal of Science*, Series 5, 41 (251), 39 pp.

**25** Pierrehumbert, R.T. (2010). *Principles of Planetary Climate*. Cambridge University Press. (Fig. 4.7, p.221).

**26** Arrhenius, G. (1908). *Worlds in the Making: The Evolution of the Universe*. New York: Harper and Bros.

**27** Lyell, K.M. (1881). *Life, Letters and Journals of Sir Charles Lyell, Bart*, vol. 1 and 2, 475 pp. and 489 pp. London: John Murray.

**28** Fleming, J.R. (1998). Charles Lyell and Climatic change: speculation and certainty. *Geological Society Special Publication* 143: 161–169.

**29** Chamberlin, T.C. (1897). A group of hypotheses bearing on climate changes. *Journal of Geology* 5, Oct-Nov issue: 653–683.

**30** Chamberlin, T.C. (1899). An attempt to frame a working hypothesis of cause of glacial periods on an atmospheric basis. *Journal of Geology* 7, Jul-Dec issue: 545–584.

**31** Chamberlin, T.C. (1899). An attempt to frame a working hypothesis of cause of glacial periods on an atmospheric basis. Part II, special application of the hypothesis to the known glacial periods. *Journal of Geology* 7, Jul-Dec issue: 667–685.

**32** Chamberlin, T.C. (1899). An attempt to frame a working hypothesis of cause of glacial periods on an atmospheric basis. Part III, localisation of glaciation. *Journal of Geology* 7, Jul-Dec issue: 751–787.

**33** Holmes, A. (1965). *Principles of Physical Geology*, 727–728. London: Nelson.

**34** Lyell, C. (1875). *Principles of Geology*. London: John Murray.

**35** Suess, E. (1888). *Das Antlitz der Erde*, vol. II, 704 pp. Vienna: F. Tempsky (Wikipedia gives the date wrongly as 1861 for the naming of Gondwanaland).

**36** Seward, A.C. (1892). *Fossil Plants as Tests of Climate; Being the Sedgwick Prize Essay for the Year 1892*. London: C.J. Clay & Sons, 151 pp.

**37** Schopf, J.M. (1968) Distribution of the Glossopteris-Gangamopteris flora". *Geological Society of America Annual Meeting, Mexico City*, p 267.

**38** Seward, A.C. (1914). Antarctic Fossil Plants. In *British Antarctic ("Terra Nova") Expedition, 1910. Natural History Report, British Museum (Natural History). Geology* 1 (1): 1–49.

# 5

# Changing Geography Through Time

---

**LEARNING OUTCOMES**

- Understand and explain the contribution of Alfred Wegener to the theory of continental drift.
- Understand and explain the development and underpinning of sea floor spreading and the theory of plate tectonics in relation to the contributions of Wegener, Holmes, Du Toit, Runcorn, Hess, Dietz, Vine, Matthews, Le Pichon, McKenzie, and Wilson.
- Understand and explain the development and application of radiometric techniques to refining the dating of rocks and hence of continental movements.
- Know about and be able to explain in broad terms the development and use of palaeogeographic maps in palaeoclimatology.
- Be able to describe in broad details the coming together of continental fragments to form the Pangaea supercontinent, and its subsequent break up and the dispersal of its fragments, in relation to geological time.
- Be able to describe and explain changes in sea level through Phanerozoic time.

---

## 5.1 The Continents Drift

Climate is intimately linked to geography because the amount of energy received at any point on the Earth's surface varies with latitude. As we saw in Chapter 2, it was Humboldt who realized through his 'vegetation zones' and 'isothermal lines' that the changes in climate with latitude create climatic zones, and Charles Lyell who realized from geological evidence that the continents must have moved across those climate zones with time [1]. Some later geologists thought that the continents might have been fixed, with climate shifts reflecting displacement of the Earth's pole and axis, a phenomenon known as 'polar wandering'. The problem in applying the polar wandering concept to the southern continents was that evidence for glaciation in the Carboniferous and Permian around 300 Ma ago, as we saw in Chapter 4, was spread across the widely separate southern continents and India. Where might the pole have been then?

The matter was complicated by Eduard Suess's notion that the Earth was shrinking, making the continental crust collapse in places to form the ocean basins [2]. He thought the world's mountain belts arose where the edges of floating pieces of crust crept up over sunken pieces. Suess's notion of continental creep was taken up and modified by American geologist Frank Bursley Taylor (1860–1938), who lectured on it to a Geological Society of America meeting in Baltimore in December 1908, and published his ideas in 1910 [3]. Taylor saw the Tertiary mountain ranges of Europe and Asia '*as the product of southward creep of the entire crustal sheet of Eurasia*' [3], which extended into Alaska. To account for the southward bend of the mountain ranges of South East Asia around the northern edge of India, Suess and Taylor required Africa and India to remain still while southward moving Eurasia moulded itself around them to form the Alps, the Himalayas and the South East Asian ranges. Taylor also saw North America as creeping towards the west or southwest, and away from the fault-bounded margins of Greenland.

Combining these notions with the observation that America's Appalachian mountains looked as if they had once been continuous with Europe's Caledonide mountains running from Norway to Ireland, Taylor wrote that '*it seems probable that a considerable part of the present [North Atlantic] oceanic interval is due to Tertiary and perhaps to older crustal movements which divided the original chain near Greenland and carried the parts away on divergent lines – to the northeast and the southwest*' [3]. Noting the remarkable size and shape of the Mid-Atlantic ridge and its position mid-way between America and Africa, he suggested: '*The ridge is a submerged mountain range of a different type and origin from any other on the earth … It is apparently a sort of horst ridge – a residual ridge along a line of parting or rifting – the earth-crust having moved away from it on both sides … It is probably much nearer the truth to*

*Palaeoclimatology: From Snowball Earth to the Anthropocene*, First Edition. Colin P. Summerhayes.
© 2020 John Wiley & Sons Ltd. Published 2020 by John Wiley & Sons Ltd.

*suppose that the mid-Atlantic ridge has remained unmoved, while the two continents on opposite sides of it have crept away in nearly parallel and opposite directions'* [3]. Taylor thought Africa moved east away from the Mid-Atlantic ridge before the Mesozoic era, whilst South America moved away to the west in the Tertiary. Elsewhere in the Southern Hemisphere, Antarctica appeared to have held fast, whilst Australia moved north. Observing that *'both polar areas were areas of crustal dispersion or spreading, and the continental sheets, excepting Africa, all crept toward the equatorial zone'*, Taylor declined *'to attempt any discussion of the ultimate cause of the Tertiary mountain making'* [3].

Here, then we have the suggestion that the continents may move apart, forming mountain ranges at their leading edges where they meet other continental fragments (as between Europe and Africa, or Asia and India) or are thrust over the ocean basins (as on the western coast of North America), and leaving some new kind of seabed behind them in newly opened ocean basins: a prescient observation, though it caused barely a ripple at the time.

The suggestion that the Americas had moved away from Africa and Europe was not new. Noticing the nice 'fit' between west Africa and eastern South America, several eminent thinkers and scientists – starting with Abraham Ortelius in 1596 – suggested that these continents had been torn apart. By 1805, Humboldt had realized from the similarities of their coastal plants that there had been an ancient connection between Africa and South America [4]. French geographer Antonio Snider-Pellegrini (1802–1885) argued in 1859 that this separation was sudden and had something to do with Noah's flood [5]. Geological evidence played no part in his speculations. As we saw in Chapter 2, Charles Lyell had also realized that there must once have been land where we now

have the (North) Atlantic Ocean, and that continents must *'shift their positions entirely in the course of ages'* [1].

Examining the fit between South America and Africa, a 32-year old German meteorologist, Alfred Wegener (1880–1930) (Figure 5.1, Box 5.1), was led in 1912 to propose his own controversial notion of 'continental drift'.

Wegener's musings were triggered by an Atlas [6]. In her 1960 biography of Alfred, his wife Else quoted from a letter he wrote in January 1911: *'My roommate ... got Andree's Handatlas (Scobel 1910) as a Christmas gift. For hours we admired the magnificent maps. A thought occurred to me: ... Does not the east coast of South America fit the west coast of Africa as though they had been contiguous in the past? Even better is the fit seen on the bathymetric map of the Atlantic*

**Figure 5.1**   Alfred Wegener.

---

**Box 5.1   Alfred Wegener**

Born in Berlin, Wegener studied natural sciences at the University there, receiving a doctorate in astronomy in 1904. He then pursued a career in meteorology, focusing on the upper atmosphere. He pioneered the use of balloons to study currents of air. He and his brother, Kurt, broke the world endurance record for hot air balloons, staying aloft for 52 hours in 1906. He was the meteorologist on the 1906 Danish expedition to Greenland, before taking up a post at the University of Marburg, where he lectured on meteorology, astronomy, and navigation for explorers. In 1911 his collected meteorology lectures were published as *The Thermodynamics of the Atmosphere*. Wegener also worked at the German Naval Observatory and the University of Hamburg. He was the first to explain the formation of raindrops and of the haloes around the sun formed by ice crystals in Arctic air. He was back in Greenland in 1912, becoming the first explorer to overwinter on the ice cap. In due course, he became a world expert on polar meteorology and glaciology, and in 1924 was appointed Professor of Meteorology and Geophysics at the University of Graz, in Austria. He died on the ice cap in Greenland in November 1930. The Alfred Wegener Institute for Polar and Marine Research, in Bremerhaven, Germany, is named after him, as are craters on the Moon and Mars. The Institute awards a Wegener Medal, and the European Union of Geosciences awards an Alfred Wegener Medal.

*when comparing ... the break-off into the deep sea [the edge of the continental shelf]. I must follow this up'* [7]. So we can date the spark of Wegener's revolutionary idea to Christmas 1910, just as Frank Taylor's idea was being published in the USA, in yet another convergence of ideas.

Wegener scoured the geological literature for evidence to support his idea [8]. He linked continents together by matching the shapes of opposing continental shelves, by matching geological structures and fossil groups across continental boundaries, and by bringing together continental fragments having similar palaeoclimatic information, such as corals, coals, desert dunes and glacial deposits. This formed the basis for his claim that about 300 Ma ago the continents had formed a single mass that he called the 'Urkontinent', and that this had subsequently split, with the individual fragments moving gradually to their present positions. His theory explained the hitherto seemingly odd distribution of Cretaceous corals and Carboniferous coals. And by bringing the Southern Hemisphere continents together, he showed that the Carboniferous–Permian glaciation 300 Ma ago was centred about a South Pole located near Durban (Figure 5.2).

Wegener startled his audience with this sweeping proposal at a meeting of the Geological Association in Frankfurt am Main on 6 January 1912, and repeated it four days later at a meeting of the Society for the Advancement of Natural Science in Marberg. He titled his Frankfurt talk *The Geophysical Basis of the Evolution of the Large-Scale Features of the Earth's Crust (Continents and Oceans)*, and it

was published that same year, in German, as *The Origins of Continents* [8].

Like others, Wegener saw that the Earth's topography has a bimodal distribution, with the continents standing consistently higher than the ocean floors. He realized from newly acquired gravity data that the ocean floors differed geologically from the continents in being made of dense rocks like basalt, full of heavy iron and magnesium silicate minerals, on which less dense continental rocks like granite, full of relatively light minerals such as quartz and feldspar, would float – rather like icebergs in an ocean – following the principle of 'isostasy', or hydrostatic equilibrium. The dense oceanic rocks he called 'sima', using Suess's term for the dense global subcrust of basaltic rocks rich in silica and magnesium: the light ones he called 'sial', modified from Suess's term 'sal' for the less dense granitic rocks of the continents that were rich in silica and aluminium. As an example of isostasy, Wegener cited the northern polar regions, where the land was now rising owing to the removal of the weight of the great ice sheets. Clearly, given enough time, the Earth's solid crust could flow.

Wegener thought hard about what might drive continental drift. In 1912 he wrote that the Mid-Atlantic Ridge represented a zone in which the floor of the Atlantic, as it keeps widening, is continuously tearing open and making space for fresh, relatively fluid and hot basaltic material (sima) rising from depth, a prescient observation, but one that, sadly, he did not follow up [8]. Later, influenced by his

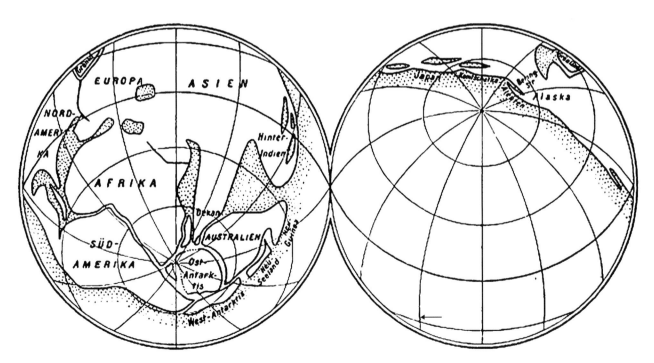

**Figure 5.2** Wegener's continental palaeoreconstruction for the Carboniferous.

Arctic experience, he dropped this early rendition of what we now call 'sea-floor spreading', replacing it with the idea that the lighter rocks of the continents ploughed through the denser rocks of the ocean basins like icebergs through the ocean, under the influence of '*accidental currents in the globe*' – which later came to be called convection currents.

Wegener's aim of writing something more substantial to explain his ideas was delayed by participating in an expedition to Greenland in 1912. There he made the longest traverse of the ice cap ever undertaken, a 750 mile crossing at the widest point, and was the first scientist to overwinter on the ice cap. A further delay came with the advent of the First World War, in which he served as an army officer. Convalescing from a wound, he expanded on his theory to write one of the most influential and controversial books in the history of science: *The Origin of Continents and Oceans*, published in German in 1915, and revised in 1920, 1922, and 1929 [9]. Due to the war, the book was not appreciated widely beyond Germany until the third German edition (of 1922) was translated into English in 1924 [10].

In his 1920 text, Wegener provided three reconstructions of past continental positions: one for the Carboniferous (300 Ma ago), with the South Pole estimated to lie just off the coast of Durban in South Africa; one for the Eocene (50 Ma ago) and one for the Ice Age (Quaternary) [9]. By 1924 he had expanded his suite of maps to nine geological periods, an example of which appears in Figure 5.3 [11]. They showed the assembly of continental fragments in Palaeozoic time and their subsequent disintegration and their dispersion in Mesozoic and Cenozoic times.

Wegener's 'Urkontinent' included Suess's 'Gondwanaland' (the southern continents plus India) at its southern end, with its separate continental pieces migrated back to their 'correct' initial positions instead of being connected by sunken land bridges. He was still referring to 'Gondwanaland' in his 1920 book [9], but by 1922 he had started to use the term 'Pangaea' (meaning 'all the Earth' in Greek) for his 'Urkontinent' and had dropped the term 'Gondwanaland' entirely [10].

Wegener gave us a first crude way of looking at Earth's palaeogeography: the locations of the continental fragments on which climate-sensitive sediments were deposited through time. His maps were a boon to palaeoclimatologists seeking to interpret past climate change from the sedimentary record. Wegener himself used the climate record of the rocks to estimate probable past polar positions and improve the 'polar wandering curve': the line joining the successive estimated positions of past poles from the Carboniferous to today [12]. Unfortunately, he lacked the magnetic data on past polar positions that would help later investigators improve on his reconstructions of past continental positions. Even so, his maps were not vastly different from modern reconstructions. One major difference is that while he showed India in its more or less correct southern location in the Carboniferous between Africa, Antarctica, and Australia,

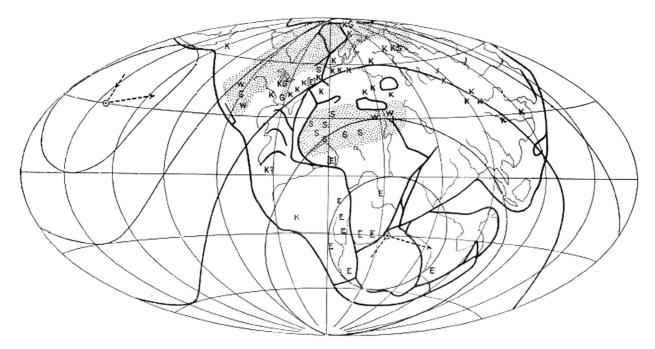

**Figure 5.3** Köppen and Wegener's Palaeogeographic Map for the Carboniferous showing regions of ice, marshes, and deserts. E = signs of ice; K = coal; S = salt; G = gypsum; W = desert sands; dotted area = dry regions.

he created an imaginary land bridge between India and Asia across what we now know was the ocean that Suess named 'Tethys' (Figure 5.3).

Wegener can claim several 'firsts': first use of the expression 'continental drift' to describe the movement of the continents: first reconstruction of continental positions based on the edges of the continental shelf rather than the coastline; first amassing of large volumes of geological and paleontological evidence to support his reconstructions; and first statement of the proposal that his supercontinent broke apart due to rifting like that forming today's East African Rift Valley. By considering the changes in flora and fauna with time on the different continents, he was even able to suggest when they broke apart: Australia and Africa in the Jurassic; South America and Africa in the Lower to Middle Cretaceous; and India and Madagascar at the end of the Cretaceous [10, 11]. There was no longer any need for Suess's land bridges.

Geologists are a conservative bunch, and many could not accept the radical idea that the continents may formerly have been joined together, not least because its advocate was a meteorologist [12]. Following Suess, 'everyone knew' that land bridges explained how similar plants and animals came to be found on the different continents. And anyway, Wegener could not explain precisely what moved the continents. '*Utter, damned rot!*' said the President of the American Philosophical Society [12]. But not everyone was so dismissive. Some recalled that even Humboldt had noted in his 5-volume *Cosmos* that there were striking similarities between the rocks on either side of the South Atlantic, and that he had invoked '*secular currents in the Earth's interior*' (which we might now see as convection currents) as a means of homogenizing the Earth's internal density [13]. But the lack of a plausible mechanism for continental drift would consign Wegener's theory to the dustbin in the minds of his many critics. He would eventually be proved right and they wrong, but not until long after he was dead.

Was criticism of Wegener's background justified? In fact he was far from being 'just a meteorologist'. He had trained in the natural sciences, including geology, held a PhD in astronomy, practiced as a meteorologist and had considerable experience of glaciology from expeditions to the Greenland ice sheet, all of which led him to adopt a highly interdisciplinary approach to scientific questions that was unusual at the time. His observations of the behaviour of ice floes and icebergs around Greenland probably helped him come to his drift theory.

A major problem in getting his theory accepted lay in the absence of a plausible mechanism for the movement of the continents: geologists just did not know about the processes taking place deep beneath the Earth's crust. As he

**Figure 5.4** Arthur Holmes.

put it '*Relative continental movements have been determined by purely empirical means …. [but] the Newton of continental drift has not yet* appeared *[to explain how they took place]*' [7]. He was adamant that '*Continental drift, divergence and convergence, earthquakes, volcanism, transgressions and polar wander are interconnected in a grand causative scheme*' [6]. How right he was.

Inspired by Wegener's start, a plausible scheme to explain the drifting of the continents came in 1928 from a young English geologist, Arthur Holmes (1890–1965) (Figure 5.4, Box 5.2).

Holmes suggested that the continents were split apart by rifting induced by convection currents within the warm, viscous rocks of the Earth's mantle, which stretches 35–3000 km deep beneath the lithosphere, Earth's rigid outer crust. As the convection currents transported continental fragments away from one another, new deep ocean floor formed between them by the crystallization of basalt magma welling up from below beneath the mid-ocean ridge [14]. Elaborating on this idea in the first edition of his *Principles of Physical Geology*, Holmes explained that '*the basaltic layer becomes a kind of endless travelling belt on the top of which a continent can be carried along*' [15] (Figure 5.5). Where the fronts of advancing continents met old ocean floor, it sank beneath them into the Earth's interior along ocean deeps like the Peru–Chile Trench. The down-going slabs of basaltic seafloor would eventually be heated, compressed, and metamorphosed to form an unusually dense rock that geologists name 'eclogite', which '*merged into and became part of the descending [convection] current, so gradually sinking out of the way, and providing room for the crust on either side to be drawn inwards by the horizontal currents beneath them*' [15]. The eclogite, Holmes thought, would be heated up by its surroundings, eventually forming pockets of magma that rose and formed the source of the basaltic intrusions and lavas that fed the swell forming the mid-ocean ridge [15].

---

**Box 5.2    Arthur Holmes**

Arthur Holmes, from Gateshead, Northumberland, had a significant influence on the development of geology. A precocious pioneer in geochronology, he carried out the first uranium–lead dating of rocks whilst still an undergraduate at the Royal School of Mines in London, publishing *The Age of the Earth* in 1913, when just 23 years old. He taught geology at Durham University before becoming Head of the geology department at the University of Edinburgh. He was widely known for his 1944 textbook *Principles of Physical Geology*, in whose title we see an attempt to emulate Lyell. I learned a lot from the massive 1288-page second edition of his book (1965). Holmes was much honoured by his peers. He won the Geological Society's Murchison Medal in 1940 and its Wollaston Medal in 1956. He became a fellow of the Royal Society in 1942 and won its Copley Medal, as well as the Geological Society of America's Penrose Medal. He has a crater on Mars named after him. The European Geosciences Union awards the Holmes Medal.

---

Holmes was not clear in his own mind about the process, because in the figure illustrating continental drift in his 1928 paper, and re-used in the second edition of his *Principles of Physical Geology* in 1965, he showed the mid-ocean ridge as a relict continental fragment, apparently because he thought of Iceland as such a fragment [14]. In contrast, in the first edition of his book, in 1944, he showed the oceanic crust between the continents, including a mid-ocean island or swell, as basaltic [15]. Nevertheless, close inspection of his 1944 figure illustrating continental drift (Figure 5.5) shows that buried within that new basaltic crust were fragments of former continental crust [15]. The true nature of the oceanic crust (including its lack of such continental remnants) would require data that had not emerged by 1964 when Holmes penned the preface to his second edition.

Like Wegener, Holmes recognized that these ideas were '*purely speculative*' and merited testing with independent evidence [15]. Despite widespread resistance, especially from American geologists, Wegener's and Holmes' ideas slowly began to attract support, as Naomi Oreskes showed [16]. Indeed, even back in 1926 the great American geologist Reginald Daly recognized that although many geologists found the idea of continental drift '*bizarre, shocking*', nevertheless '*an increasing number of specialists in the problem are already convinced that it must be seriously entertained as the true basis for a sound theory of mountain building [and that] every educated person cannot fail to be interested by this revolutionary concept*' [17].

South African Alexander Du Toit (1878–1949, Box 5.3) was an early supporter of continental drift, presenting detailed geological evidence for it in 1937 in his influential book *Our Wandering Continents* [18], which he dedicated to Wegener.

Pointing out that Wegener's '*illuminating hypothesis ... can be tested on the basis of prediction*' Du Toit went on to

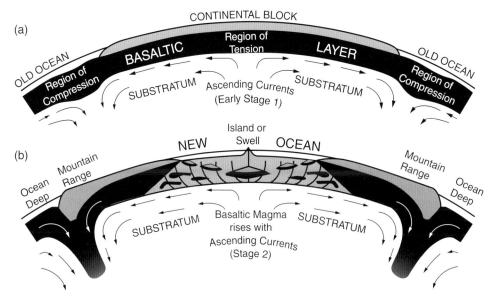

**Figure 5.5**  Arthur Holmes' sketch of the mechanism for continental drift [15]. New basaltic crust (shown as incorporating fragments of old continental (grey) and basaltic (black) lithosphere) lies under the new ocean.

| Box 5.3   Alexander Du Toit |
| --- |

Du Toit was born in Cape Town and educated there and at the Royal Technical College in Glasgow. Joining the Geological Commission of the Cape of Good Hope led him to map much of South Africa and, in 1926, to write a book on its geology. In 1923, the Carnegie Institute of Washington gave him a grant to study the geology of Argentina, Paraguay, and Brazil, to see how their geology matched that of the west coast of Africa, as a test of continental drift. He ended his career as a consulting geologist for the mining house De Beers, retiring in 1941. The Geological Society of London awarded him its Murchison Medal in 1933, and he was elected a fellow of the Royal Society in 1943. Like Wegener, he has a crater on Mars named after him.

say that he had made predictions on the basis of the theory and '*verified [them] by field work*' [18]. His detailed studies, funded by the Carnegie Foundation, led him to propose that 'Pangaea' was an amalgamation of the southern continents grouped together as 'Gondwanaland', and of the northern continents grouped together as 'Laurasia', the two being separated at times by Suess's Tethys Ocean, which disappeared as the southern continents collided with the northern ones in Cenozoic times (during the past 50 Ma). He favoured Holmes's concept of continents drifting under the influence of convection currents in the Earth's subcrust.

Even so, as Oreskes reminds us, '*Wegener's work contradicted the edifice and rhetoric of practice that Americans had laboriously constructed and articulated over the course of nearly a century. Wegener violated the norms of American geological practice*' in being theoretical and not '*putting facts first*' [16]. American geologists did not trust general deductive theories like Wegener's. Indeed, one of them, Charles Shuchert, reminded Wegener that facts are facts, and it is from facts that we make our generalizations [16]. In effect, Shuchert rejected continental drift because he interpreted it to be incompatible with Lyell's uniformitarianism; he believed that the Earth today is more or less as it always has been [16]. And in making his rejection, he was ignoring the geological facts that Wegener had supplied.

## 5.2   The Sea Floor Spreads

In the absence of geophysical evidence for the movement of the continents or the opening of the ocean basins, geologists organized themselves into different camps. The 'drifters' followed Wegener and Du Toit; the 'fixists' thought the ocean basins permanent; the 'contracters' thought the Earth was shrinking, with mountains forced up as the space for rocks diminished (á la Suess); and the 'expanders' thought the continents were being forced apart as the Earth expanded. This diversity of opinion provided me with ample meat for arm-waving undergraduate essays in the early 1960s. It's amazing what theories one can dream up in the absence of hard data.

Growing evidence for movement of the continents through time came from the striking discovery by Achille Delesse in 1849 that magnetic minerals like magnetite became aligned parallel to the Earth's magnetic field in solidifying volcanic lavas. But due to the lack of a suitably sensitive magnetometer, it was 100 years before the study of the magnetic field trapped in ancient rocks could begin in earnest. Experimental physicist Patrick Maynard Stuart Blackett (1897–1974), of Imperial College London, invented the appropriate device. Blackett won the Nobel Prize in Physics in 1948 for his work on cosmic rays, became a fellow of the Royal Society, of which he was made President in 1965, and was made a life peer, Baron Blackett of Chelsea, in 1969. He also has a crater on the moon named after him.

Blackett's magnetometer triggered many sophisticated palaeomagnetic studies in the late 1950s by, amongst others, British scientists Stanley Keith Runcorn (1922–1995) [19], and Edward (Ted) Irving (1927–2014) [20]. Like Blackett, both men received several honours for their contributions including being made fellows of the Royal Society. Their studies pointed to what appeared to be successive changes in the geographic position of the magnetic North and South Pole through time. In due course they reasoned that it was not the magnetic poles that had moved, at least not much, but rather the continents on which the sampled rocks were found. A sequence of measurements of magnetic polar position from rocks of different ages on each continent could define the 'polar wandering curve' for that continent. In effect this showed what the palaeo-**latitude** of the measured rocks was in relation to the magnetic pole through time. From the difference between the positions of the polar wandering paths in rocks of the same age on different continents, the **relative** palaeo-**longitude** of those rocks could be obtained. The differences between the polar wandering paths of Europe and North America disappeared when the Atlantic was closed, proving that the continents had drifted apart. At last scientists could determine with some confidence where the continents had been in relation to one another through time. Blackett was not to be left out, publishing his own comparisons of palaeolatitudes with the evidence of ancient climates from geological data [21]. Thus, by the end of the 1950s, palaeomagnetic data had vindicated Wegener: the continents had moved, although we still did not know how.

This revolution in understanding the way the Earth works picked up pace with a landmark paper widely circulated in 1960 by Harry Hammond Hess (1906–1969) (Figure 5.6, Box 5.4), Professor of Geology at Princeton, and eventually published in 1962 as *History of Ocean Basins* [22].

Hess's knowledge of the shape of the ocean basins came from his vast collection of echo-soundings. Before the Second World War, few research vessels had an echo-sounder, and none had continuous depth recorders printing out the depth as a line on a rolling strip of paper. Widespread use of echo-sounders and line printers by some US Navy vessels during the war and by research ships after it ended provided abundant continuous depth records, or bathymetric profiles, enabling the shape of the ocean floor to be mapped in much more detail than could be achieved by lead-line soundings. In 1957, Bruce Heezen (1924–1977), a graduate student under Maurice Ewing (1906–1974) at the Lamont Geological Observatory of Columbia University, working with geologist and cartographer Marie Tharp (1920–2006), used these new data to make the first physiographic map of the North Atlantic. It confirmed that there was a deep V-shaped rift running down the axis of the Mid-Atlantic Ridge. By 1977, Heezen and Tharp had expanded their map to cover all the ocean basins. Heezen was honoured by having a US Navy survey ship named after him – the USNS *Bruce C Heezen* – in 1999.

Maurice Ewing, meanwhile, had begun using seismic techniques to build up a picture of the Earth's crust beneath the ocean basins, showing it to be just a few kilometres thick. He had also dredged the Mid-Atlantic Ridge, in 1947, finding it to be made of basalt and other igneous rocks [23]. For his contributions to marine science Ewing was awarded fellowship of the US National Academy of Science, the US National Medal of Science in 1973, the Vetlesen Prize in 1960, the Wollaston Medal of the Geological Society of London in 1969, and fellowship in the Royal Society in 1972.

**Figure 5.6**   Harry Hammond Hess.

Mapping the focal points of earthquakes showed the deep ones to be concentrated in the deep trenches around the Pacific margin. Heezen showed that the shallow ones tended to be concentrated along the axis of the mid-ocean ridge. Sir Edward Bullard (1907–1980), of Cambridge University, a fellow of the Royal Society and the Geological Society of London's Wollaston Medallist for 1967, found that heat flow was also high along those axes. The geophysical evidence that Wegener and Holmes lacked was beginning to accumulate.

Thinking about the significance of this growing body of information, and building on Wegener's and Holmes's ideas, Hess surmised that hot mantle material oozes up from the Earth's interior beneath the world-encircling belt of mid-ocean ridges, creating new seafloor that spreads away on either side from the active zone at the ridge crest and eventually cools and sinks to form the deep ocean basins [22]. While there were many similarities between his ideas and those of Holmes, there were significant differences. Holmes imagined that basaltic magma welled up beneath the mid-ocean ridges, creating a basaltic crust that moved away on either side [15], while Hess imagined that the rising material was hydrated mantle material in the form of a dense, coarse-grained rock rich in iron and

---

**Box 5.4   Harry Hess**

Born in New York City, Hess taught for a year at Rutgers University (1932–1933) and spent a year at the Geophysical Laboratory in Washington, DC, before joining Princeton in 1934, where he remained, becoming chairman of the geology department from 1950 to 1966. As a naval officer in the Second World War, Hess, who ended up as a Rear Admiral in the US Naval Reserve, was made Captain of the USS *Cape Johnson*, an attack transport ship equipped with a relatively new tool in oceanography and navigation – the echo-sounder. Hess used its soundings to map the deep ocean floor in more detail than had been achieved before, which deepened his understanding of sea-floor processes and led him to articulate novel ideas about the origin of the ocean basins. He was honoured with the establishment of the Harry H. Hess medal by the American Geophysical Union.

magnesium that geologists refer to as 'peridotite'. Where peridotite crops out on the seabed its constituent minerals are commonly partly or entirely converted to 'serpentine', a green filamentous magnesium–iron silicate belonging to the family of minerals that includes asbestos. Geologists refer to these rocks as 'serpentinized peridotites' [22].

Recalling the fate of Wegener's ideas about continental drift, Hess was cautious, describing his ideas as '*an essay in geopoetry*' [22]. In a paper in the journal *Nature* in 1961, Robert (Bob) Sinclair Dietz (1914–1995), a scientist with the US Coast and Geodetic Survey who had read the widely circulated draft of Hess's paper, named Hess's process the 'sea-floor spreading hypothesis' [24]. Ironically, as Oreskes points out [16], the process Dietz described was actually far more like that described by Holmes – that oceanic crust is generated by sub-oceanic intrusion and submarine eruption of basaltic lava [15] (Figure 5.5). The sea-floor spreading mechanism provided the missing cause for Wegener's continental drift, and followed the sea-floor spreading model that Wegener had originally proposed, though later abandoned. The sea-floor spreading hypothesis, which still needed testing, was the trigger for the astonishing development of the theory of plate tectonics in the late 1960s. Dietz was honoured for his achievements in marine science with the Penrose Medal of the Geological Society of America in 1988.

The first acid test of Hess's geopoetry came from the world of magnetics. Back in 1906, Bernhard Brunhes (1867–1910) discovered that some rocks are magnetized in opposition to the Earth's present magnetic field: the compass needle points south when it should be pointing north, or vice versa. By the early 1960s, we knew that Earth's magnetic field had reversed polarity at least 170 times in the past 80 Ma, probably due to changes in the Earth's molten core. The distinctive pattern of these reversals back through time provided a magnetic polarity timescale that could be used to date rocks in the absence of other means.

In 1962, surveying the Carlsberg Ridge, part of the mid-ocean ridge in the Indian Ocean, from HMS *Owen*, graduate student Fred Vine (1939–) and his supervisor Drummond Matthews (1931–1997), both from Cambridge University, found the ridge's crest to be paralleled on either side by puzzling stripes of different magnetic character. Making a profound intellectual leap, they surmised in a paper in *Nature* in 1963 that if Hess and Dietz were right about sea-floor spreading, younger rocks along the mid-ocean ridge crest should have today's magnetic polarity, with the magnetic and geographic north poles being in the same direction, whilst the older rocks further away from the ridge crest should have the opposite magnetic polarity, having formed at the ridge crest when the magnetic field was reversed [25]. Given the multiple

changes in magnetic polarity with time, the sideways moving seafloor should show a pattern of ridge-parallel magnetic stripes increasing in age away from the ridge, with the pattern on one side of the ridge being an exact replica of the pattern on the opposite side. By matching the magnetic patterns on the seafloor to the magnetic polarity time scale, one could in theory date the sea floor far from the ridge crest. Vine and Matthews were right, and for that insight both were made fellows of the Royal Society. In due course, the Geological Society rewarded Matthews with the Bigsby Medal, in 1975, and the Wollaston Medal, in 1989, and awarded Vine the Bigsby Medal, in 1971, and the Prestwich Medal, in 2007.

Many scientific discoveries are 'ideas waiting to happen', and so it proved with the Vine and Matthews hypothesis. Their idea had been hit upon independently by Canadian geophysicist Lawrence Whittaker Morley (1920–2013), director of the geophysics branch of the Canadian Geological Survey in Ottawa (1950–1969). He too had submitted a paper to *Nature* in 1963 proposing much the same theory. As reported in Morley's obituary in the *Canadian Globe and Mail*, a reviewer of the paper gave his opinion that the paper might be interesting for '*talk at cocktail parties, but it is not the sort of thing that ought to be published under serious scientific aegis*' [26]. Another geologist later described Morley's paper as '*probably the most significant paper in the Earth sciences ever to be denied publication*' [26]. *Nature* published the Vine and Matthews paper a few months after Morley's failed submission. But Morley was not forgotten: eventually, the seafloor-spreading hypothesis, confirming the reality of continental drift, became known as the Vine–Matthews–Morley hypothesis. Morley was duly honoured with the Royal Canadian Geographical Society's Gold Medal in 1995, and made an Officer of the Order of Canada in 1999.

By 1965, Fred Vine, working with Canadian geologist John Tuzo Wilson (1908–1993) (Figure 5.7, Box 5.5) from the University of Toronto, showed that these patterns were widespread in the ocean basins and could be predicted to occur where they were not yet found. Proof of the correctness of their theory came from dating rocks drilled by the Deep-Sea Drilling Project's *Glomar Challenger* from differently magnetized zones on either side of the mid-ocean ridge, starting in 1968 (more about that in Chapter 6). The model had a further advantage: if the ages of the magnetic stripes were known, then geologists could determine the timing of the separation of the continents and the opening of the ocean basins. This would be a boon to palaeoclimatologists.

Capitalizing on the new information about the age of the sea floor, Sir Edward Bullard worked with colleagues at Cambridge, including young Alan Gilbert Smith

**Figure 5.7** Tuzo Wilson.

(1937–2017), to show in 1965 that, by applying the principles of sea-floor spreading they could fit together the edges of the continental shelves, which were the true margins of the continents, to close the Atlantic tightly – demonstrating the validity of Wegener's theory. Schuchert had demanded facts. Here they were.

By late 1967 to early 1968 geophysicists like Xavier Le Pichon (1937–) at Lamont, Dan McKenzie (1942–) at Cambridge, and Jason Morgan (1935–) at Princeton had realized that the Earth's crust consists of a set of thin but rigid giant plates apparently being moved about by underlying convection in the hot mantle [27]. These 'tectonic plates' (Figure 5.8) are thinnest beneath the oceans and thickest beneath the continents. Some, like the Pacific Plate, consist entirely of oceanic crust with a thin surface veneer of relatively young sediments. Others, like the North American Plate, comprise a mixture of continental parts (North America) and oceanic parts (the North Atlantic west of the Mid-Atlantic Ridge). Where the edges of plates moving in opposite directions meet, one sinks beneath the other in a process known as 'subduction',

which is accompanied by deep earthquakes (Figure 5.9). Great ocean trenches like those around the edges of the Pacific form as the down-going or subducting plate buckles. The solid, subducting plate melts deep in the Earth's hot and semi-molten mantle beneath the thin and rigid crust, and the freshly melted material rises as magma, which spews out as lava through volcanoes in island arcs, such as the Aleutians, next to deep-ocean trenches. Much of the Pacific Ocean comprises an oceanic plate that is being subducted beneath the surrounding continents, with the resulting ring of volcanoes around its margin forming the so-called 'Pacific ring of fire'. Volcanoes have an important influence on climate, as we shall see in later chapters. It is through their $CO_2$ emissions and the associated movement of continents that plate tectonics influences climate.

Sometimes whole mountain ranges rise up over a subducting oceanic plate, as in the Andes, where the Pacific and related plates dive under South America (Figure 5.10). Where continents meet, the over-riding one is thrust up as mountain ranges, like those of the Alps, between the European and African plates, or the Himalayas, between the Indian and Asian plates. Within the ocean basins, as McKenzie predicted and John Sclater confirmed, heat flow declines as the seabed moves away from the narrow zone of crustal formation at the ridge crest. As a result the ocean crust cools, becomes denser, and sinks, which explains why the deepest parts of the oceans basins contain the oldest (hence coldest) ocean crust. Convection of some kind was assumed to 'close the circle' between magma rising beneath ridges, plates moving sideways, and ocean crust descending into the Earth's mantle. It is accepted that there has to be some kind of connection between the rising material and the sinking material, although it is no longer the case that mid-ocean ridges are expected to lie over the ascending branches of convection currents (Figure 5.10).

As you might imagine, mapping convection currents is impractical without being able to drill into the hot and ductile asthenosphere – the upper hot and mobile part of the Earth's mantle beneath the rigid crust. What we do know of what goes on down there is somewhat fuzzy, and based on crude geophysical imaging of the subcrust, earthquake focal depths, the occasional rocks that rise from extreme

---

**Box 5.5  Tuzo Wilson**

Wilson was born in Ottawa, graduated in geophysics from the University of Toronto in 1930, and obtained a PhD from Princeton in 1936. After serving in the Canadian army in the Second World War, he retired with the rank of colonel. Honoured for his novel discoveries in plate tectonics, he was elected a fellow of the Royal Societies of London, Canada and Edinburgh, and won several medals including the EUG's Alfred Wegener Medal and, in 1978, the Geological Society of London's Wollaston Medal. The Canadian Geophysical Union awards the John Tuzo Wilson Medal to honour his achievements.

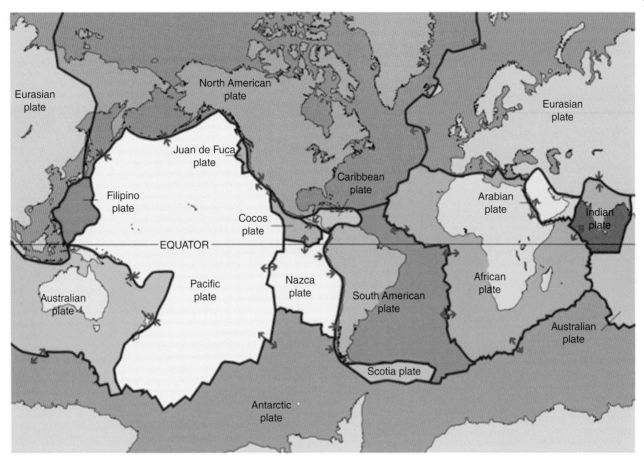

**Figure 5.8** Tectonic plates. *Source:* Courtesy of Geological Society of London at http://www.geolsoc.org.uk/Plate-Tectonics/ Chap2-What-is-a-Plate.

depths – like Kimberlite pipes containing diamonds – and subtle differences in the geochemical makeup of igneous (volcanic and plutonic) rocks in island arcs, on volcanic islands and at mid-ocean ridges. We need not concern ourselves here with that level of detail, because we are concerned with what happens on the surface in relation to or under the control of climate. Nevertheless, it is obvious that the rising or erosion of mountain chains as a result of plate tectonic activity must affect climate both locally and regionally.

Plate tectonics revolutionized the Earth Sciences by explaining how and why everything was connected – the locations of the continents, mountain ranges, ocean trenches, water depths, deep, and shallow earthquakes, regions of high heat flow, magnetic patterns, and volcanoes (Figures 5.9 and 5.10). Plate tectonics also allowed geoscientists to reconstruct past geographies, one of the key controls of climate, and enabled geophysicists to use the dated magnetic stripes on the seafloor, along with the past positions of the poles as determined by palaeomagnetic analysis of continental rocks, to reconstruct the motions of the plates – and hence the continents – through time with a

fair amount of precision. Amongst other things, it became possible to calculate how rapidly the continents were moving apart: the North Atlantic is opening at a rate of around 2.5 cm/year – the rate a fingernail grows.

Plate tectonics provided us not just with past geographies, but also with mechanisms for supplying $CO_2$ through volcanic activity, and for removing $CO_2$ by chemical weathering in newly uplifted mountains, as we shall see in subsequent chapters.

We now use magnetic lineations on the seabed and palaeomagnetic data from land to reconstruct continental positions since the breakup of Pangaea began 180 Ma ago. Making reconstructions prior to that is more difficult, because there is no older seafloor anywhere in the ocean basins. To position the continents in pre-Jurassic time we use palaeomagnetic data. As we see in Chapter 6, given the 'palaeogeography' of past continental fragments, the original geographic position of climatically sensitive sediments on those fragments can now be established with a high degree of certainty, confirming Lyell's notion that climatic zones much like those of today have prevailed through time, although modified by geographic changes like the

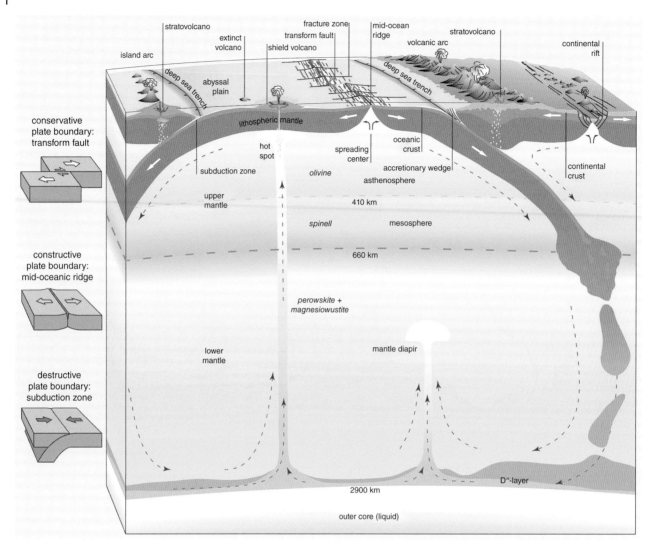

**Figure 5.9** Schematic overview of plate tectonic processes at the surface and in the Earth's crust and mantle [28]. *Source:* From figure 2 in Ref. [28].

development of seaways and mountain belts, as he suspected. Plate tectonic theory fully vindicated Wegener, whose ideas have now been brought in from the cold, as beautifully illustrated by Robin Muir Wood [29].

At this point it is worth pausing to reflect why it took so long for geologists to accept continental drift. Oreskes argues convincingly that slow but growing acceptance reflected a trend amongst geologists '*to move their discipline away from observational field studies and an inductive epistemic stance toward instrumental and laboratory methods and a more deductive stance... a move from the field to the laboratory*' [16]. Equally, as Oreskes points out, by the middle of the twentieth century, geologists found themselves in a world increasingly inhabited by physicists, whose instruments began providing valuable new 'geological' insights: '*New methods and machines provide rare*

*opportunities for conceptual flexibility*' [16]. As I show in this book, chemistry and physics would play a far more important role in the interpretation of palaeoclimatic phenomena in the post Second World War world.

Attitudes to science and the philosophy of science were changing, too, as philosophers like Karl Popper and, later, Thomas Kuhn began arguing that science begins with an idea, a hypothesis, to test which data is collected [16]. In that sense, uniformitarianism was a very useful concept in geology, but it proved to be wrong when it was taken to imply that monumental changes could never occur [16].

In a sense these changes are embellishments to the scientific revolution that began at the beginning of the nineteenth century when geologists like Lyell strove to take theology out of geology (see Chapter 2). As we shall see in later chapters, geologists are now having to take on board

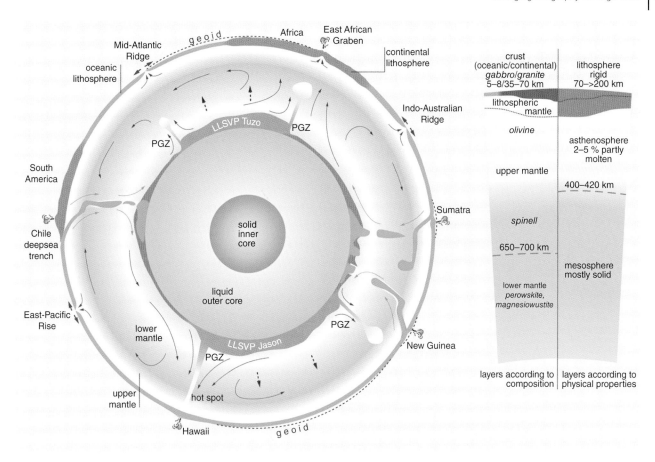

**Figure 5.10** Convection currents in the Earth's interior in relation to the locations of rift valleys (e.g. East Africa), mid-ocean ridges, deep-sea trenches, island arcs and associated volcanoes (e.g. Sumatra, New Guinea), as well as zones where plumes of hot mantle material emerge from plume generation zones (PGZ) at the core–mantle boundary, giving rise to hot spots at the Earth's surface, like that forming the Hawaiian island chain. Note that local accumulation of relatively 'light' low velocity material at the core mantle boundary forms two 'large low shear-wave velocity provinces' (LLSVPs) labelled Jason and Tuzo, whose low-density presence distorts the geoid at the surface. Mantle plumes seem to be generated preferentially around the edges of the LLSVPs [28]. *Source:* From figure 1 in Ref. [28].

Humboldt's notion that everything is connected, as we move towards Earth System Science with its emphasis on the roles of biogeochemical cycles and greenhouse gases.

In geology, as elsewhere, resistance to change is not uncommon. When I worked for Exxon Production Research Company (EPRCo) in Houston in the mid to late 1970s, Exxon's exploration geologists working in Texas told me that this new-fangled plate tectonics was a waste of time – it wasn't going to help them find oil in places like Texas and Oklahoma. And they were not alone: in 1972, the *Bulletin of the American Association of Petroleum Geologists* published a paper by the Meyerhoffs, father and son, denying the reality of plate tectonics and, by implication, continental drift [30]. One of their key criteria was the dredging of granitic and dolomitic rocks from the mid-ocean ridge in the northern North Atlantic. What they had failed to recognize, in assuming that the ridge must therefore be continental in origin, was that this part of the ridge

lay directly beneath the path of eastwardly drifting icebergs during times of glacial maxima. Those rocks were glacial erratics. That fact killed their theory.

By 1972, Keith Runcorn and Donald Harvey Tarling (1935–) could report at a NATO Advanced Study Institute at the University of Newcastle-upon-Tyne that '*These developments have … essentially ended the long debate about whether or not the classical lines of geological evidence, palaeoclimatic, palaeontological distributions, global tectonic patterns and lithological relationships, support or refute drift*' [31]. At that meeting, Kenneth Creer, Professor of Magnetism at the same university from 1966 to 1972, used palaeomagnetic data to compile a suite of maps of past continental positions for the main geological periods: Cambrian, Ordovician, Silurian to Early Devonian, Devonian, lower to middle Carboniferous, upper Carboniferous to Permian, Triassic, Jurassic, and Early Cretaceous [32]. Others demonstrated how plate tectonics explained the opening of the Atlantic and Indian Oceans [33–35].

These were exciting times for the Earth sciences, and for me, as, shortly after leaving Oxford University at the end of 1964 to join the New Zealand Oceanographic Institute as a marine geologist, I began using the new sea floor spreading concept to analyse the origins of seafloor ridges, trenches, and plateaux around New Zealand. Operating from the New Zealand Navy's HMNZS *Endeavour* in the Southern Ocean south of the country, I surveyed the submerged continental crust of the Campbell Plateau, the eastern margin of which had broken off from Antarctica's Marie Byrd Land and the Ross Sea area some 80 Ma ago. My research on the adjacent Macquarie Ridge, running south from New Zealand's Fjordland showed in 1966 that it was an island arc with an associated trench [36, 37]. Following Tuzo Wilson's 1965 proposal that plate tectonics required a new class of fault – the 'transform fault' to connect the oppositely directed ends of island arcs, I deduced that the famous Alpine Fault cutting across New Zealand had to be a transform fault connecting the west-facing Macquarie–Fjordland island arc in the south to the east-facing Tonga–Kermadec island arc in the north [38]. Wilson was pleased to see that result when he visited our research institute over the lollipop factory on Thorndon Quay in Wellington in 1967. My New Zealand colleagues were not all as kind to me as Wilson – many of the older kiwi geologists resented a young upstart from the UK revising the history of their great Alpine Fault. In those days their notions of geology stopped at the coast.

I was also lucky enough to be assigned as the New Zealand marine geology representative to the Scripps Institution of Oceanography's *Nova Expedition* to investigate the submarine geology of the Southwestern Pacific in 1967 aboard the research vessel *Argo*. The expedition was led by Henry William (Bill) Menard (1920–1986) (Figure 5.11), author of one of my bibles of the time, the 1964 pre-plate tectonics text *Marine Geology of the Pacific* [39]. Menard later became the 10th Director of the US Geological Survey (1978–1981), a member of the US National Academy of Sciences and a winner of the Geological Society of America's Penrose Medal.

In the pre-plate tectonics era, Menard had thought long and hard about mid-ocean ridges, proposing, rather like Holmes and Hess before him, that they lay above rising convection currents that pulled the overlying crust apart [39]. He had correctly assumed that the process would be accompanied by rifting at the ridge crest, and that heat transported upwards by convection would partially melt the upper mantle, leading to intrusions into the fractured crust accompanied by high heat flow. Like both Holmes and Hess, he saw the great oceanic trenches as formed by

**Figure 5.11** Bill Menard.

down-bowing of the crust caused by the drag of the downward plunging arms of convection cells. Nevertheless, writing his book in 1963 he just missed the new interpretation of the pattern of magnetic anomalies on the seafloor [25] and hence the beginning of the plate tectonic revolution. In that sense his book was out of date by the time it was published in 1964.

Menard was a great teacher, especially for a novice marine geologist like me, and I had some fascinating company aboard ship to bounce ideas off, including Dan Karig, Jean Francheteau, and Sean Solomon, students who would later become leaders in their own right in the fast evolving field of plate tectonics. I learned a great deal about the new plate tectonics from them and from seeing its major features as we surveyed the southwest Pacific from the *Argo*. Sailing from Noumea in New Caledonia to Auckland, New Zealand, via Brisbane in Australia, we examined several of the features expected from the seafloor spreading hypothesis: first the spreading seafloor of the Tasman Sea; next the Lord Howe Rise and Norfolk Ridge, submerged thinned continental fragments pushed away from Australia by the opening of the Tasman Sea; then an active island arc and trench – the Tonga–Kermadec Ridge and Kermadec Trench; and, finally, the South Fiji Basin, which was currently being created by seafloor spreading driven by a subordinate convection cell tucked in behind the Tonga–Kermadec Ridge to form what is now called a 'back arc basin' [40].

Later, in October 1990, I was able to study a mid-ocean ridge at first hand during a research cruise to the Reykjanes Ridge, from which Iceland emerges in the North Atlantic [41]. Lindsay Parson of the UK's Institute of Oceanographic

Sciences Deacon Laboratory, of which I was then director, led the cruise, and I shared a cabin with Sir Anthony Laughton, FRS, my predecessor and an expert on mid-ocean ridges, who had been awarded the Geological Society of London's Murchison Medal in 1989. We sailed from Bergen to New York aboard the MV *Maurice Ewing,* of the Lamont–Doherty Earth Observatory, on a storm-tossed sea. The occasional equinoctial gale had our captain steaming for gentler seas from time to time, which rather interrupted our seabed studies. Losing scarce research time to storms is all part of the trials and tribulations of open ocean research.

My involvement in the early days of plate tectonics would stand me in good stead as the topic developed in future years to underpin the palaeogeographic and palaeoclimatic studies in which I became increasingly involved in the 1970s and 1980s.

## 5.3 The Dating Game

Palaeoclimatologists need to know precisely when climatic changes occurred. For that, they needed a means of dating rocks much more precisely than by using fossils. The breakthrough in dating rocks came about from the discovery of radioactivity in 1896 by Antoine Henri Berquerel (1852–1908). Ernest Rutherford (1871–1937), a New Zealander working at McGill University in Montreal, proved that radioactivity involved an alchemist's dream – the transmutation of one chemical element into another – and that it did so at a set rate, the 'half-life', which was unique to each element. The half-life is the time by which half of what you started with, say uranium, has decayed into its 'daughter' product, which in the case of uranium is the element lead (Pb). Rutherford, who won the Nobel Prize in Chemistry in 1908 for his work on radioactive decay, realized that the decay of uranium into lead could be used to date rock samples. In 1907, Bertram Boltwood (1870–1927) applied that understanding to date some rocks. His samples appeared to range in age from 92 to 570 Ma old, but improvements to his technique changed their age range to 250–1300 Ma. In 1911, 21-year old Arthur Holmes further developed the uranium–lead method, dating a rock from Ceylon to 1.6 billion years [42]. Because each radioactive element has its own half-life, geologists can use them as clocks to date past events.

The following year Frederick Soddy (1877–1956), found that elements could occur in more than one form, with identical properties except for their atomic weight. He coined the word 'isotope' (meaning one place) for these identical entities of different mass. We now know that chemical elements contain nuclei made up of positively charged protons along with particles of the same size but no charge: neutrons. Individual chemical elements are characterized by their numbers of protons. They usually contain the same number of neutrons. Isotopes occur where there are more or fewer neutrons than protons. Within any one element – say potassium – there may be both stable and radioactive isotopes. For example, potassium has three naturally occurring isotopes, two of which are stable ($^{39}$K and $^{41}$K), and one of which ($^{40}$K) is not. The superscript numbers refer to the sum of protons and neutrons in the nucleus. Radioactive $^{40}$K, with a half-life of 1.3 billion years, decays into the stable argon isotope $^{40}$Ar. Similarly, the radioactive rubidium isotope $^{87}$Rb, with a half-life of 49 billion years, decays into the stable strontium isotope $^{87}$Sr. I used a mass spectrometer to measure rubidium and strontium isotopes when I was a student at Oxford in 1963–1964, in order to date the age of the igneous intrusion of Garabal Hill on the edge of Loch Fyne in the Scottish Highlands. It was 392 Ma old.

Radiometric dating, as it is called, has pinned down the finer details of the geological timescale, commonly by dating volcanic lavas and ash embedded within fossiliferous sediments. The minerals in igneous rocks like lavas and granites contain the dates of their formation, unlike the minerals in sedimentary rocks, which contain the dates of formation of the rocks from which they were weathered.

We have now entered the world of 'geochemistry'. Chemists had been examining geological materials as chemical objects in their own right for perhaps 200 years or more before it was realized that the tools and principles of chemistry could be used to understand and explain geological processes and the origins of the Earth's atmosphere, ocean, and crust. This new way of doing things spawned a new subfield of geology. Stimulated by the development of new analytical techniques in the Second Word War, and the arrival on the scene of new analytical instruments able to analyse progressively smaller amounts of material with progressively greater accuracy, this new geological subdivision grew rapidly, attracting its own journal, *Geochimica et Cosmochimica Acta,* by 1950, and its own society, *The Geochemical Society,* by 1955. The geochemist's bible, *Geochemistry,* emerged in 1954 from the pen of one of the founders of this new subject, Victor Moritz Goldschmitt (1888–1947), regarded by some as the 'father of modern geochemistry'. My interest in this new field led to my earning a PhD in geochemistry at Imperial College London in 1970. As we follow the development of the science of past climate change we will see an increasingly prominent role played by geochemists.

By 1949, Willard Libby (1908–1980) of the University of Chicago had discovered that a naturally occurring radioactive isotope of carbon could be used to date organic remains up to about 60 Ka old. The abundant naturally occurring stable isotopes of carbon are $^{12}$C, with six protons and six neutrons, and $^{13}$C, with six protons and seven neutrons. Radioactive $^{14}$C (or carbon-14), discovered in 1940, has eight neutrons, forms naturally in the atmosphere, and decays with a half-life of 5730 years to the nitrogen isotope $^{14}$N. Both fossil plants and calcareous skeletons made from calcium carbonate contain carbon and so can be dated by radiocarbon dating provided they are young enough. Libby won the Nobel Prize in Chemistry in 1960 for his contributions. Amongst his colleagues was a young Austrian chemist, Hans Suess (1909–1993), a grandson of the famous Eduard Suess. In 1950, Hans migrated to the USA to work at the University of Chicago. Libby showed him how to use $^{14}$C to date Earth materials [43]. Suess soon established a radiocarbon laboratory at the nearby US Geological Survey office, publishing his first dates, on fossil wood, in 1954. We will meet him again.

New techniques and refinements to old ones continue to provide measurements of progressively smaller intervals of time. Just to cite one example, a variant of the potassium–argon (K–Ar) dating method was developed in the 1970s to ascertain the ratio between two argon isotopes $^{40}$Ar and $^{39}$Ar in volcanic rocks. It allowed ages as young as 2000 years (2 Ka) to be determined.

With these various astonishing developments geologists could accurately date the ages of magnetic reversals and fossil zones, enabling palaeoclimatologists to pin down with increasing confidence precisely when climate changes had occurred. Dating also helped to refine plate tectonic reconstructions of past continental positions, another crucial aid in palaeoclimatology.

## 5.4 Base Maps for Palaeoclimatology

By the early 1970s geologists were producing suites of continental reconstruction maps. Amongst them was a much-admired set compiled in 1973 by Alan Smith and colleagues [44]. Smith's maps led to his being awarded the Geological Society's Bigsby Medal in 1981 and the Lyell Medal in 2008.

Reconstructing past continental positions is of critical interest not only to palaeoclimatologists, but also to petroleum companies wanting to reduce the risk involved in the increasingly expensive drilling required to find commercial quantities of oil and gas. Plotting information about past geography and environments on maps of past continental positions from different geological periods provides exploration geologists with an accurate regional perspective with which to determine oil and gas prospectivity ahead of the drill in frontier areas. It tells the explorers where to expect the organic-rich sediments that will eventually form

petroleum source rocks, the porous sandy deposits that could form reservoirs for oil and gas, and the clay minerals that could clog those pores. Source rocks, reservoir sands and pore-clogging clays result in many different ways from the climate of their depositional environment.

Recognizing the potential value of palaeogeographic mapping, in 1981 the Exploration Division of BP Research Company (BPRCo) at Sunbury-on-Thames, under the leadership of Peter Llewellyn, set up Extramural Research Awards with Smith's group in the Earth Sciences department at Cambridge, to supply palaeoreconstruction maps of past continental positions, and with Professor Brian Funnell's group in the School of Environmental Sciences at the University of East Anglia (UEA), to provide information on past mountainous areas, shorelines, depositional environments and climate-sensitive sediment types that could be superimposed on Smith's maps [45] to provide BP's explorers with a technical edge in their worldwide search for oil and gas. Richard Tyson would do much of the legwork under Funnell's supervision at UEA.

BP's Global Palaeoreconstruction Group, located in BPRCo's Palaeontology Branch (later renamed Stratigraphy Branch), managed the project. I led that group from September 1982 to 1985, following Ian Hoskin, the leader from 1981 to 1982, and being followed by David Smith, who led the project to its completion in 1987. We planned to produce maps for each of the 50 geological stages of the past 245 Ma. In some cases there was not enough data to plot a single stage, so it was combined with another; we thus created 31 maps through time, each representing a period averaging about 8 Ma long. In 1994, Alan Smith, David Smith and Brian Funnell published the reconstructed continental positions and shorelines as *Atlas of Mesozoic and Cenozoic Coastlines* [46] (Figure 5.12). The oldest reconstruction was of the Early Triassic (Scythian stage) at 245 Ma. Tyson and Funnell published the palaeogeographic environmental data in a number of research papers [47, 48]. This was '... *the first known attempt to draw a continuous series of global maps with such a fine resolution*' [46].

A major difference between Wegener's and later reconstructions is that later groups knew that Pangaea took the shape of the letter C, with its east side open to the Tethys Ocean separating eastern Laurasia in the north from eastern Gondwanaland in the south (Figure 5.12). Wegener also thought that Greenland had remained connected to Europe until the Quaternary, but its separation actually began in the Eocene.

The main uncertainties in today's reconstruction maps are the positions of continental fragments that were compressed into mountain ranges at times between those for which each map was made [46]. This applies not only to pieces involved in building the Alps and the Himalayas, but

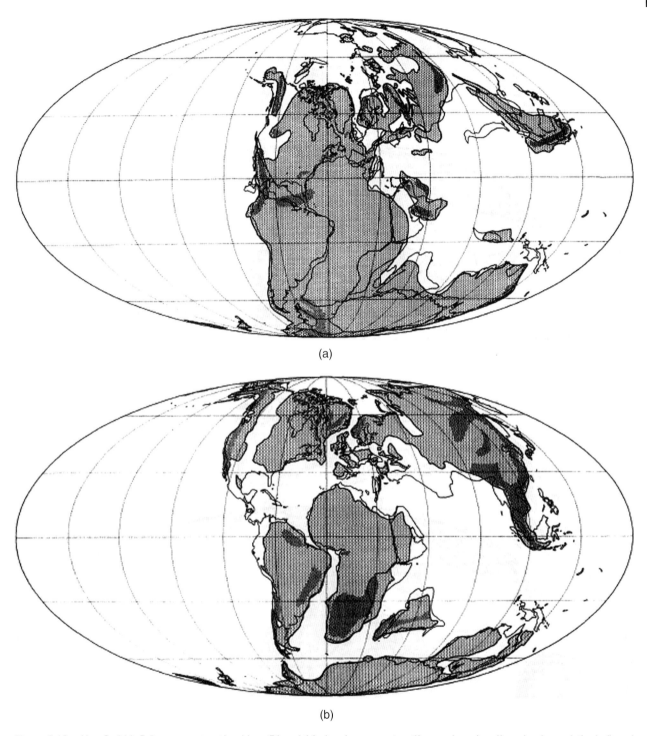

(a)

(b)

**Figure 5.12** Alan Smith's Palaeoreconstruction Maps 31 and 16, showing current outlines, palaeoshorelines, land area (stippled), and highland (shaded). From Maps 31 and 16 in Ref. [46]; (a) Early Triassic 245 Ma palaeocoastline map; (b) mid Cretaceous (Albian) 105 Ma palaeocoastline map. *Source:* From Ref. [46].

also to Pacific continental margins, which accumulated small continental fragments that had migrated considerable distances along major transcurrent faults. Palaeomagnetic data will continue to help in determining the original positions and pathways of those fragments, as well as unravel-

ling the geology of complex areas like South East Asia. Fortunately, these uncertainties involve quite small areas and do not affect the main elements of the reconstructions. These days, new reconstruction maps refine rather than greatly change the reconstructions made in the early 1970s.

To check on the results of this in-house project, I persuaded BPRCo to join a multi-oil company consortium funding the work of the Palaeogeographic Atlas Project, led by Alfred M Ziegler (1938–) at the University of Chicago, which began in 1975 [49]. Ziegler's proposal for this project was written '*in the expectation of stimulating the interest and support of the petroleum industry.... The time has come to apply the findings of plate tectonics to the field of paleogeography*' [49]. Work began in 1975 with the help of seed money from the Shell Development Company. The US National Science Foundation provided co-funding from 1977 [50], and the project continued with funding from some seven additional companies into the 2000s, with the latest publication listed on the Internet as 2007 [51].

I was pleased to be working with Ziegler, having been a fellow student of his in the department of geology and mineralogy at Oxford in 1963–1964. A palaeontologist, Ziegler was working closely with geologist Chris Scotese (1953–), Chicago's equivalent to Alan Smith at Cambridge. As part of his PhD (1976–1983) Scotese produced seven palaeoreconstruction maps for the Palaeozoic [52], and seven for the Mesozoic and Cenozoic [53], showing shallow seas, highlands, and lowlands. Ziegler's team would plot onto refined versions of those maps palaeogeographic and palaeoclimatic data derived from the literature [54, 55]. Ultimately they hoped to produce one map for every 16 Ma during the Mesozoic and Cenozoic [53].

The differences between the reconstructions of Smith from Cambridge and Scotese from Chicago were small. The main difference lay in time interval. Cambridge produced maps at 8 Ma intervals for Mesozoic and Cenozoic times from the Triassic to the present, but no maps for the Palaeozoic (pre-Triassic), whilst Chicago produced maps at 16 Ma intervals, for Palaeozoic, Mesozoic, and Cenozoic times. To ensure a close match between the approaches of the Cambridge and Chicago schools, BP Research Co invited Scotese to spend a sabbatical at Sunbury-on-Thames in 1985, working with me on applications of palaeoreconstructions – more on that in Chapter 6. Nowadays, Chris Scotese supplies palaeoreconstruction maps through his PALEOMAP Project (www.scotese.com).

## 5.5  The Evolution of the Modern World

Modern geological and geophysical methods confirm that during the Late Palaeozoic the various scattered fragments of continental crust were in the process of being swept together to form a single supercontinent – 'Pangaea', much as suggested by Wegener. But Pangaea was not the first such supercontinent. Geological evidence suggests that the modern style of plate tectonics, in which 90% of the force required to drive the process comes from the sinking of subducting lithosphere, began around one billion years ago in the Late Proterozoic, leading to the formation of an earlier supercontinent – 'Rodinia' [56]. Rodinia assembled 1.1–0.9 billion years ago [57], and broke up between 750 and 633 Ma ago.

By Cambrian times, the continental fragments that form Gondwana (I drop the 'land' because 'wana' means land) had come together [57]. They were separated by the Iapetus Ocean from Laurentia, comprising North America and Greenland; from Baltica, comprising Scandinavia and Russia, by the Tornquist Sea; and from separate fragments comprising Siberia, and North China. By the Mid Silurian (430 Ma ago), Baltica and Laurentia had merged, closing the Iapetus Ocean to form Laurussia, separated from Gondwana by the Rheic Ocean. By early Carboniferous times, around 360 Ma ago, the ocean between Gondwana in the south and Laurussia in the north had shrunk to a narrow seaway. A similarly narrow seaway separated the North American and European block from Siberia and Chinese fragments. By late Carboniferous-Early Permian times, around 300 Ma ago, these seaways had closed, making the supercontinent a reality. Closure took place in the central upright part of the C-shaped supercontinent, which still enclosed a broad Tethyan Ocean between its two eastern-facing arms (Figure 5.12a).

This configuration would not last [57]. In the Middle Jurassic (200–180 Ma ago), Pangaea began to break apart. Linear domes grew over rising magma. Huge cracks developed along the crests of the domes, then widened to form gigantic linear rift valleys like that in East African today. The core of each rift valley, or 'graben' (after the German for trench), was a down-dropped block of the Earth's crust bordered on either side by steep faults commonly associated with volcanoes, like Mount Kilimanjaro in East Africa. Lakes like East Africa's Lake Nyasa (now Lake Malawi) grew in the down-dropped blocks. Rifting was accompanied by the intrusion of thin vertical dykes of basalt. Jurassic basaltic dykes and sills – horizontal sheets tens to hundreds of metres thick and extending over thousands of square kilometres – are widespread as precursors of eventual continental separation, for example in Antarctica, South Africa, and Tasmania.

In most places rifting proceeded to the point where the continent split and basalt magma welled up from below to form new ocean floor between the separating continental parts. This second stage is just beginning in today's Red Sea, where the continental rocks on either side of the original rift valley have thinned and sunk, letting in the sea, and new basaltic ocean floor is just beginning to emerge along the axis. The Gulf of California is an example of the third stage: the continental sides of the rift have moved apart, the deep basin between them is floored by new fresh basaltic deep ocean floor, and sea floor spreading has begun. If the East African rift were to progress over a few million years to stage three, we would see Tanzania, Kenya, and Somalia moving east to form a large new island in the Indian Ocean. Where rifting aborted before reaching stage two, we are left with failed rifts like the one beneath the North Sea, holding the North Sea's oil and gas fields, and the Rhine Graben, through which the Rhine flows on its way to the sea.

Pangaea's first major break, at about 160 Ma ago, split the South America/Africa block from the North America block, opening a seaway comprising the central North Atlantic and a proto Gulf of Mexico and Caribbean region [57]. That break connected the Tethys Ocean in the east with the Pacific Ocean in the west. To the north lay Laurasia, comprising North America, Europe, and Asia. To the south lay Gondwana, containing the rest of today's continental bits and pieces.

Laurasia began to break up in the Late Jurassic, with rift basins forming between Europe and Greenland and between Greenland and North America [57]. The aborted North Sea rift formed at that time. Figure 5.12b provides a representative view. Starting in the Late Cretaceous, around 90 Ma ago, the North Atlantic Ocean began to open like a zipper, from south to north. The process was extremely slow, with new ocean floor not forming between Britain and Greenland until some 60 Ma ago, in Palaeocene times [58]. Widespread volcanic activity including outpourings of plateau basalts and the injection of swarms of basaltic dykes, accompanied the opening of this northern section between 60 and 50 Ma in Britain and Greenland. This activity formed the columnar basaltic wonders of the Giant's Causeway in Northern Ireland, and of Fingal's Cave on Scotland's island of Staffa. Following a visit in 1829, the sound of the waves in the cave stimulated Felix Mendelssohn to produce Hebrides Overture Opus 26, known as the Fingal's Cave Overture.

Extensive volcanic activity continues in Iceland, over a plume of molten material from deep within the Earth's mantle, which started life with the opening of the Norwegian–Greenland Sea 60 Ma ago. The slowly opening zipper finally created a deep-water connection between the North Atlantic and the Arctic through Fram Strait between Greenland and Svalbard only 12 Ma ago, in Miocene times.

Meanwhile, in the south, Gondwana began slowly disintegrating. About 180 Ma ago it split into two [57]. The Africa and South America block slid sedately north, whilst the other half (Madagascar, India, Antarctica, Australia, and New Guinea) stood still. This latter block started moving south about 170 Ma ago, then began to break apart in the lower Cretaceous, between 140 and 130 Ma ago (Figure 5.12b). At that time Antarctica and Australia continued moving south away from Madagascar and India.

At about the same time South America began to move west away from Africa. Rifting between Antarctica and Australia began around 110 Ma ago, in the Mid Cretaceous, as Antarctic moved south and Australia stayed still. About 85 Ma ago, New Zealand split away from Antarctica, and India split away from Madagascar. India then moved rapidly north to collide with Asia around 50 Ma ago, creating the Himalayas and the Tibetan Plateau. Australia started to move slowly north at about the same time, and the Australia-New Guinea block collided with the Indonesian island arc around 30 Ma ago. Antarctica has been where it is now for about 90 Ma.

The northward movement of Africa, Arabia, India, and Australia to collide with the southern margin of Laurasia swallowed up the Tethys Ocean between Gondwana and Laurasia [57]. Remnants of Tethys exist today as the Mediterranean, Black, and Caspian Seas. With the opening of the North Atlantic, the Tethys did for a while extend west from the Indian Ocean through the Mediterranean to the Pacific (Figure 5.12b). But that connection eventually closed at both ends, first with the collision of Africa with Iberia beginning in the Late Cretaceous, and second with the rising of the central American isthmus to close the central American seaway and create the Caribbean Sea in the Late Pliocene between 4 and 2.8–2.5 Ma ago. Today's Indian Ocean contains nothing of Tethys – it formed as a new ocean basin in the wake of the northward movement of India. The opening and closing of these various oceanic gateways radically changed ocean circulation and hence climate.

Plate tectonics also influenced climate in yet another significant way by causing major fluctuations in sea level. The creation of 2000-km wide and 2-km high ridges of hot rock in the middle of the ocean basins inevitably displaced seawater onto the edges of the adjacent continental margins. Sea levels rose steadily from about 200 Ma ago to a peak around 90 Ma ago as sea-floor spreading expanded and caused ridges to grow, then declined as sea-floor spreading declined and caused ridges to cool and subside. Investigating the ways in which different strata lapped onto continental margins, several of my colleagues at Exxon Production Research Co (EPRCo) in Houston, Texas, led by Peter Vail (1930–), defined a curve of eustatic sea level through time (Figure 5.13) [59]. As a colleague of Vail's, I was much influenced by his thinking when I worked for EPRCo between 1976 and 1982. Another close colleague of mine, Bilal Haq (1942–), later worked with Vail to refine his efforts, concluding that Cretaceous sea-level stood 100–250 m above today's level [60]. Their results, derived from seismic data, paralleled those derived by Walt Pitman (1931–) of Lamont from rates of sea floor spreading and resultant changes in mid-ocean ridge volume [61]. Similar

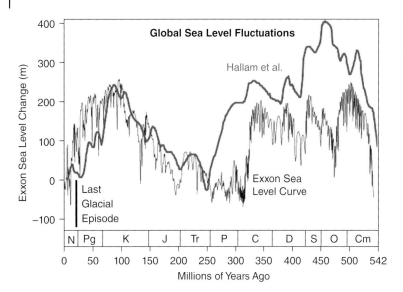

**Figure 5.13** Sea level curves through time. *Source:* From Vail (Exxon) [59, 60] and Hallam (University of Birmingham) [62].

results were obtained by Tony Hallam (1933–) of Birmingham University, a winner of the Geological Society's Lyell Medal in 1990 [62] (Figure 5.13). Later, Ken Miller and colleagues from Rutgers University derived a new global sea level curve based on a combination of oxygen isotope data, stratigraphic data and the subsidence history of the New Jersey continental margin, concluding that Late Cretaceous sea-levels were only $100 \pm 50$ m above today's levels [63].

Dietmar Müller, from Sydney University, and colleagues used a comprehensive analysis of the volumes of mid-ocean ridges through time, along with a re-analysis of Miller's data from the New Jersey margin, to conclude that Mesozoic sea levels lay mid-way between the high estimates of Haq and the low estimates of Miller [64]. These high sea levels help to explain the widespread distribution of Cretaceous chalk over western Europe, for example (Figure 5.12b). These various studies, including geochemical analyses by Sean McCauley and Don DePaulo of the Center for Isotope Geochemistry of the Lawrence Berkeley National Laboratory [65], suggest that the rate of production of mid-ocean ridges declined by about 50% over the past 100 Ma.

Studies of seamounts in the southwestern Pacific produced further evidence for an uplift of the ocean floor of the kind that would have made sea level rise during the Cretaceous. From his wartime naval echo-soundings, Harry Hess discovered that all over the Pacific Ocean were submerged flat-topped seamounts that he named 'guyots' after the nineteenth century Swiss–American geographer Arnold Henry Guyot (1807–1884) [66]. Guyot, a friend of Louis Agassiz, emigrated to America at Agassiz's urging, in 1848, ending up as professor of physical geography and

geology at Princeton; hence the connection to Hess, who was a professor there. Hess concluded that the guyots' flat tops were eroded at sea level before sinking to their present depths of 1–2 km. His conclusion echoed that of Charles Darwin, who had deduced that atolls sat atop sunken volcanoes, and had reasoned from the preponderance of atolls in the region that a vast area of the southwestern Pacific must have subsided [67]. Mapping the distribution of atolls and guyots, Bill Menard realized that the guyots occupied an elongate region with its NW–SE trending axis about halfway between New Zealand and Hawaii. As the amounts of subsidence indicated by the guyots increased towards the axis, Menard deduced that the rise must have arisen as a broad upward bulge in the Earth's mantle – a kind of super-swell – which he named the Darwin Rise [39]. While it does not appear to have been a mid-ocean ridge associated with sea floor spreading, it does seem to have been a centre for mid-plate or intra-plate volcanism between 100 and 60 Ma ago. Atolls and guyots are abundant there, so this must have been a major source for some of the $CO_2$ that filled the Cretaceous atmosphere. More recently, the chains of island and guyots that define the Darwin Rise have come to be seen as the result of the Pacific plate moving northwest over multiple fixed plumes or 'plumelets' of vertically rising magma [68, 69], much in the same way that the Hawaiian island chain originated (Figure 5.10).

In 2002 and 2008, David Rowley, of the University of Chicago, calculated that the average rate of ridge production evident from the visible distribution of ocean crust had been more or less constant at around 3.4 km$^2$/year [70, 71]. That led him to question Müller's hypothesis that rates of ridge production and hence the size of ridges were greater in the Cretaceous than since [64]. Rowley was not arguing that there was no possibility that rates of spreading may have been higher in the past, he was criticizing what he regarded as unjustifiable assumptions back beyond the period for which we had accurate seafloor data. He made no attempt to evaluate the evidence for rates of seafloor spreading derived from the flooding of continental margins (à la Vail or Haq or Pitman or Hallam [59–62]), nor evidence derived from the geochemical record (à la Miller or McCauley or DePaulo [63, 65]). Why does this matter? Because rates of sea-floor spreading affect rates of production of $CO_2$, hence climate. In geology, we have to take all

factors into consideration. Going solely with Rowley is to ignore multiple alternative strands of evidence; more on that in Chapter 10.

With these various advances, and despite the fact that we still do not know precisely how plate tectonics works [72], we now have a much better understanding of it [28, 56, 57], which provides us with a sound palaeogeographic framework for studying climate change. In Chapter 6 we explore what the distribution of climate-sensitive strata through time on the different continents has to tell us about the changes in Earth's climate with time.

## References

**1** Lyell, C. (1875). *Principles of Geology*, 12e. London: John Murray.

**2** Suess, E. (1885-1908). *Das Antlitz der Erde*, vol. 1–3. Vienna: F. Tempsky (trans. H.B.C. Sollas, "The Face of the Earth", Clarendon Press, Oxford, 1904-1924).

**3** Taylor, F.B. (1910). Bearing of the tertiary mountain belt in the origin of the Earth's plan. *Bulletin of the Geological Society of America* 21: 179–227.

**4** von Humboldt, A. and Bonpland, A. (1805). *Essai sur la Géographie des Plantes*, 154 pp. Strasbourg: Levrault Schoell et Cie. [English version 2009, ed. S.T. Jackson, and trans. S. Romanowski, University of Chicago Press, Chicago IL, 274pp].

**5** Snider-Pellegrini, A. (1859). *La Creation et ses Mystères Dévoiles*, 487 pp. Paris: Librarie A. Frank (see Figures 9 (before) and 10 (after) between his pages 314 and 315.

**6** Jacoby, W. (2012). Alfred Wegener – 100 years of mobilism. *Geoscientist* 22 (9): 12–17.

**7** Wegener, E. (1960). *Alfred Wegener: Tagebücher, Briefe, Erinnerungen*. Wiesbaden: F.A. Brockhaus.

**8** Wegener, A. (1912). Die Enstehung der Kontinente, *Peterm. Mitt.*, 185–195, 253–256, and 305–309; and (somewhat abbreviated) in Geologische Rundschau 3 (4), 276–292.

**9** Wegener, A. (1920). *Die Entstehung der Kontinente und Ozeane*. 2nd ed., Die Wissenschaft, Band 66. Vieweg und Sohn, Braunschweig, 135 pp.

**10** Wegener, A. (1924). *The Origins of Continents and Oceans* (translation of the 3rd (1922) edition), 212 pp. London: Methuen.

**11** Köppen, W. and Wegener, A. (1924). *Die Klimate der Geologischen Vorzeit*, 255 pp. Berlin: Borntraeger. Note that in Google one sees this commonly referred to as published in English as "Climates of the Geological Past" by Van Nostrad, London, 1863, which is an erroneous reference to a book with a similar title "Climates of the Past: An Introduction to Paleoclimatology" by M

Schwarzbach (Van Nostrad, London, 1963) translated from the German 2nd edition by R Muir. The original Köppen and Wegener book was reissued in German and English in 2015.

**12** Demhardt, I.J. (2006). Alfred Wegener's hypothesis on continental drift and its discussion in Petermanns Geographische Mitteilungen. *Polarforschung* 75 (1): 29–35.

**13** von Humboldt, A. (1845-1861). *Cosmos: A Sketch of a Physical Description of the Universe*, 5 vols (trans. E.C. Otte). London: Henry Bohn.

**14** Holmes, A. (1928). Radioactivity and Earth movements. *Transactions of the Geological Society of Glasgow* 18: 559–606.

**15** Holmes, A. (1944). *Principles of Physical Geology*. London: Thomas Nelson & Sons.

**16** Oreskes, N. (1999). *The Rejection of Continental Drift*, 420pp. Oxford University Press.

**17** Daly, R. (1926). *Our Mobile Earth*. New York: Charles Scribner's Sons.

**18** Du Toit, A. (1937). *Our Wandering Continents: An Hypothesis of Continental Drifting*. Edinburgh & London: Oliver & Boyd.

**19** Runcorn, S.K. (1959). Rock magnetism. *Science* 129: 1002–1011.

**20** Irving, E. (1959). Paleomagnetic pole positions. *Royal Astronomical Society Geophysics Journal* 2: 51–77.

**21** Blackett, P.M.S. (1961). Comparison of ancient climates with the ancient latitudes deduced from rock magnetic measurements. *Proceedings of the Royal Society of London. Series A, Mathematical and Physical Sciences* 263 (1312): 1–30.

**22** Hess, H.H. (1962). History of ocean basins. In: *Petrologic Studies: A Volume to Honor A.F. Buddington* (eds. A.E.J. Engel, H.L. James and B.F. Leonard), 599–620. Geological Society of America.

**23** Ewing, M. (1948). Exploring the mid Atlantic ridge. *National Geographic Magazine* XCIV (3): 275–294.

24 Dietz, R.S. (1961). Continent and ocean basin evolution by spreading of the sea floor. *Nature* 190: 854–857.

25 Vine, F.J. and Matthews, D.H. (1963). Magnetic anomalies over oceanic ridges. *Nature* 199: 947–949.

26 Csillag, R. (2013). Pioneering geophysicist Lawrence Morley broke new ground. *The Globe and Mail*, Tuesday 14 May, available from: http://www.theglobeandmail.com/news/national/pioneering-geophysicist-lawrence-morley-broke-new-ground/article11925771 (last accessed 29 January 2015).

27 Le Pichon, X. (1968). Sea-floor spreading and continental drift. *Journal of Geophysical Research* 73 (12): 3661–3687.

28 Meschede, M. (2016). Plate tectonics. In: *Encyclopedia of Marine Geosciences* (eds. J. Harff, M. Meschede, S. Petersen and J. Thiede), 676–680. Dordrecht: Springer.

29 Wood, R.M. (1985). *The Dark Side of the Earth*, 246 pp. London: George Allan and Unwin.

30 Meyerhoff, A.A. and Meyerhoff, H.A. (1972). The new global tectonics: major inconsistencies. *American Association of Petroleum Geologists Bulletin* 56: 269–336.

31 Tarling, D.H. and Runcorn, S.K. (eds.) (1973). *Implications of Continental Drift to the Earth Sciences*, vol. 1, 622 pp. London: Academic Press (see Introduction p. ix–x).

32 Creer, K.M. (1973). A discussion of the arrangement of palaeomagnetic poles on the map of Pangaea for epochs in the Phanerozoic. In: *Implications of Continental Drift to the Earth Sciences*, vol. 1 (eds. D.H. Tarling and S.K. Runcorn), 47–76. London: Academic Press.

33 Heirtzler, J.R. (1973). The evolution of the North Atlantic Ocean. In: *Implications of Continental Drift to the Earth Sciences*, vol. 1 (eds. D.H. Tarling and S.K. Runcorn), 191–196. London: Academic Press.

34 Francheteau, J. (1973). Plate tectonic model of the opening of the Atlantic Ocean south of the Azores. In: *Implications of Continental Drift to the Earth Sciences*, vol. 1 (eds. D.H. Tarling and S.K. Runcorn), 197–202. London: Academic Press.

35 Laughton, A.S., Sclater, J.G., and McKenzie, D.P. (1973). The structure and evolution of the Indian Ocean. In: *Implications of Continental Drift to the Earth Sciences*, vol. 1 (eds. D.H. Tarling and S.K. Runcorn), 203–212. London: Academic Press.

36 Summerhayes, C.P. (1966). Marine geology of the New Zealand Subantarctic seafloor. MSc thesis, Victoria University, Wellington, NZ.

37 Summerhayes, C.P. (1966). Structure and geology of the seafloor south of New Zealand. *New Zealand Marine Sciences Society Newsletter* 8: 36–37, (Abstract only).

38 Summerhayes, C.P. (1967). New Zealand region volcanism and structure. *Nature* 215: 610–611.

39 Menard, H.W. (1964). *Marine Geology of the Pacific*, 271 pp. New York: McGraw-Hill.

40 Menard, H.W. (1969). *Anatomy of an Expedition*, 255 pp. New York: McGraw Hill.

41 Parson, L.M., Murton, B.J., Searle, R.C. et al. (1993). En echelon volcanic ridges at the Reykjanes Ridge: a life cycle of volcanism and tectonics. *Earth and Planetary Science Letters* 117: 73–87.

42 Holmes, A. (1911). The association of lead with uranium in rock-minerals and its application to the measurement of geological time. *Proceedings of the Royal Society of London. Series A, Mathematical and Physical Sciences* 85: 248–256. Also see the biography of Holmes by Lewis, C. (2002) The Dating Game*: One Man's Search for the Age of the Earth*, Cambridge Univ. Press. 272 pp.

43 Waencke, H. and Arnold, J.R. (2005). Hans E. Suess, 1909-1993, a biographical memoir. In: *Biographical Memoirs*, vol. 87, 20 pp. Washington D.C.: National Academy of Sciences.

44 Smith, A.G., Briden, J.C., and Drewry, G.E. (1973). Phanerozoic world maps. In: *Organisms and Continents Through Time* (ed. N.F. Hughes), *Special Papers on Palaeontology*, No. **12**, 1–42. London: Palaeontological Association.

45 Smith, A.G., Hurley, A.M., and Briden, J.C. (1981). *Phanerozoic Palaeocontinental World Maps*, 102 pp. Cambridge University Press.

46 Smith, D.G., Smith, A.G., and Funnell, B.M. (1994). *Atlas of Mesozoic and Cenozoic Coastlines*, 99 pp. Cambridge University Press.

47 Tyson, R.V. and Funnell, B.M. (1987). European Cretaceous shorelines, stage by stage. *Palaeogeography Paleoclimatology Paleoecology* 59: 69–91.

48 Funnell, B.M. (1990). Global and European shorelines, stage by stage. In: *Cretaceous Resources, Events and Rhythms* (eds. R.N. Ginsburg and B. Beaudoin), 221–235. Netherlands: Kluwer Acadamic Publication.

49 Ziegler, A.M. (1975). *A Proposal to Produce an Atlas of Paleogeographic Maps*. Department of Geophysical Sciences, University of Chicago. Unpub. Ms., 16 pp.

50 Ziegler, A.M. (1982). *The University of Chicago Paleogeographic Atlas Project: Background – Current Status – Future Plans*. Department of Geophysical Sciences, University of Chicago. Unpub. Ms, 19 pp.

51 *The Paleogeographic Atlas Project in* 2002. Available from https://www.uchicago.edu/research/center/paleogeographic_atlas_project/ (last accessed 29 January 2015).

52 Scotese, C.R., Bambach, R.K., Barton, C. et al. (1979). Paleozoic base maps. *Journal of Geology* 87: 217–277.

53 Ziegler, A.M., Scotese, C.R., and Barrett, S.F. (1983). Mesozoic and Cenozoic paleogeographic maps. In: *Tidal Friction and the Earth's Rotation II* (eds. P. Brosche and J. Sundermann), 240–252. Berlin: Springer-Verlag.

**54** Scotese, C.R., Gahagan, L.M., and Larson, R.L. (1988). Plate tectonic reconstructions of the Cretaceous and Cenozoic ocean basins. *Tectonophysics* 155: 27–48.

**55** Scotese, C.R., Boucot, A.J., and McKerrow, W.S. (1999). Gondwanan palaeogeography and palaeoclimatology. *Journal of African Earth Sciences* 28 (1): 99–114.

**56** Stern, R.J. (2005). Evidence from ophiolites, blueschists, and ultrahigh-pressure metamorphic terranes that the modern episode of subduction tectonics began in Neoproterozoic time. *Geology* 33 (7): 557–560.

**57** Torsvik, T.H. and Cocks, L.R.M. (2017). *Earth History and Palaeogeography*, 317 pp. Cambridge University Press.

**58** Bott, M.H.P. (1973). The evolution of the Atlantic North of the Faeroe Islands. In: *Implications of Continental Drift to the Earth Sciences*, vol. 1 (eds. D.H. Tarling and S.K. Runcorn, S.K), 175–189. London: Academic Press.

**59** Vail, P.R. and Mitchum, R.M. (1977). Seismic stratigraphy and global changes of sea level. *American Association of Petroleum Geologists Memoir* 26: 49–212.

**60** Haq, B.U., Hardenbol, J., and Vail, P.R. (1988). Mesozoic and Cenozoic chronostratigraphy and cycles of sea-level change. *Society of Economic Paleontologists and Mineralogists Special Publication* 42: 71–108.

**61** Pitman, W.C. (1978). Relationship between eustasy and stratigraphic sequences of passive margins. *Geological Society of America Bulletin* 89: 1389–1403.

**62** Hallam, A. (1984). Pre-Quaternary Sea level changes. *Annual Review of Earth and Planetary Sciences* 12: 205–243.

**63** Miller, K.G., Kominz, J.V., Browning, J.V. et al. (2005). The Phanerozoic record of global sea-level change. *Science* 312: 1293–1298.

**64** Müller, R.D., Sdrolias, M., Gaina, C. et al. (2008). Long-term sea-level fluctuations driven by ocean basin dynamics. *Science* 319: 1357–1362.

**65** McCauley, S.E. and DePaolo, D.J. (1997). The marine $^{87}$Sr/$^{86}$Sr and, $\delta\,^{18}$O records, Himalayan alkalinity fluxes, and Cenozoic climate models. In: *Tectonic Uplift and Climate Change* (ed. W.F. Ruddiman), 427–467. New York: Plenum.

**66** Hess, H.H. (1946). Drowned ancient islands of the Pacific basin. *American Journal of Science* 244: 772–791.

**67** Darwin, C. (1896). *The Structure and Distribution of Coral Reefs*, 3e, 344 pp. London: Smith and Elder.

**68** Janney, P.E. and Castillo, P.R. (1999). Isotope geochemistry of the Darwin rise seamounts and mantle dynamics beneath the South Central Pacific. *Journal of Geophysical Research* 104 (B5): 10571–10590.

**69** Foulger, G.R. (2010). *Plates Vs Plumes: A Geological Controversy*, 220 pp. Chichester: Wiley.

**70** Rowley, D.B. (2002). Rate of plate creation and destruction: 180 Ma to present. *Bulletin of the Geological Society of America* 114 (8): 927–933.

**71** Rowley, D.B. (2008). Extrapolating oceanic age distributions: lessons from the Pacific region. *Journal of Geology* 116: 587–598.

**72** Ananthaswamy, A. (2012). Earthly Powers, *New Scientist* **2878**, August 18th, 38–41.

# 6

# Mapping Past Climates

---

**LEARNING OUTCOMES**

- Know what climate-sensitive sediments/plants/animals are typically found in which climate regions on land or in the ocean, or at the poles and how their modern distributions in relation to latitude differ from ancient ones.
- Know and understand the derivation of the world's major climate zones, and explain how Köppen and Wegener used them to ascertain where the continents had been in the past (before the advent of palaeomagnetism).
- Know and understand how scientists modelled past climates conceptually and numerically, and what the limitations of numerical models of climate are.
- Know and understand how oxygen isotopes may be used to derive palaeotemperatures.
- Understand and be able to explain the origins of cyclical sedimentation.
- Know and understand the effects on the accumulation of marine and terrestrial climate-sensitive sediments of the plate tectonic changes taking place during the break-up of Pangaea, especially with regard to the distribution of warmth and sea level in Cretaceous times.
- Ignoring the role of greenhouse gases (which we address in detail later), understand and be able explain why it took so long to develop a comprehensive understanding of past climate change in the twentieth century, and what technologies and new ideas were needed for success.

---

## 6.1 Climate Indicators

Because many climatically sensitive deposits have well defined latitudinal ranges, and climate zones are reasonably well defined, we can make crude forecasts about what kinds of sediments we might expect in different geographical locations [1–4]. Coral reefs tend to be concentrated between the 30th parallels. Thick accumulations of sediments rich in the calcium carbonate remains of marine organisms tend to accumulate in warm seas – as did the Cretaceous chalk of the White Cliffs of Dover – although muds from tropical rivers may locally mask that tendency [5]. Glacial deposits or glacial striae (scratches made on rocks by glaciers carrying boulders) tend to occur at high latitudes and/or at high elevations, along with boulder clays or tillites, moraines, and glacial landforms like drumlins, kames, and eskers [6]. Salt (halite) and gypsum formed by evaporation normally occur in mid- to low-latitude arid areas with other 'evaporite' deposits [7]. They may be associated with dunes and other indications of deserts including 'dreikanters': pyramid-shaped pebbles facetted by the wind [8]. Sand grains blow up the long fore-slope and

avalanche down the steep lee-slope of dunes, making them advance downwind and providing an internal structure known as dune bedding or 'cross bedding' at an angle of ~30° to the desert surface. We use dune bedding to estimate past wind directions [9]. Iron oxide is common in hot deserts, making them reddish; their pebbles show a haematite (iron oxide) glaze, or 'desert varnish'.

Of course, not all deserts are hot, and not all dunes form in deserts. As Alan Eben Mackenzie Nairn (1927–2007), reminded us in his book *Descriptive Palaeoclimatology*, in 1961, '*the association of one or more criteria, such as evaporite deposits representing hot dessicating conditions with the dune bedded sandstone, may remove the ambiguity*' [10]. A Scottish palaeomagnetist and stratigrapher, Nairn was as a fellow of the University of Durham and King's College, Newcastle-upon-Tyne. In 1965, he was a co-founder of the journal *Palaeogeography, Palaeoclimatology, and Palaeoecology*. By 1991, his book was being described as '*the first modern book on palaeoclimatology*' [11]. As a sign of the times, Nairn confessed he was not fond of the astronomical theory of climate change, writing: '*the effect of the varying distance between the earth and the sun from*

82 *Palaeoclimatology: From Snowball Earth to the Anthropocene*

*perihelion to aphelion, the basis of Croll's theory of the origin of ice ages, is not now thought to be a significant factor in climate*' [10]. By 1976, he would be shown to be completely wrong, as we shall see in Chapter 12.

Nairn's book contains articles on ancient deserts, evaporites, red beds, cold climates, and fossils as climatic indicators. Arid conditions can be recognized from mud cracks. Peats tend to form in mid- to high-latitude bogs. Organic-rich deposits also form by the accumulation of terrestrial plant remains in humid tropical settings. Both may give rise to coals. Marine sediments rich in the organic remains of plankton may occur along coasts where winds blow parallel to the shore, especially at mid-latitudes off desert coasts, often in association with phosphate-rich phosphorite rock. Laterites are red soils rich in iron and aluminium that form in hot wet areas. Extreme tropical weathering converts them to bauxite, a mixture of iron and aluminium oxides and hydroxides. In arid and semi-arid regions soils may acquire a hard crust rich in calcium, forming caliche – also known as hardpan or calcrete.

Types of animals and plants also provide climate signals; think of crocodiles and penguins for instance, or fir trees and banana plants. Even such lowly creatures as the marine plankton may signal oceanic climates, some species preferring warm and others cold seas. Siliceous oozes made of the remains of radiolaria may be common beneath the tropical ocean, while those made of the remains of diatoms may be common beneath high latitude seas [1–3]. Clay minerals give away their climate zones: kaolinite comes from tropical weathering, and chlorite from mechanical weathering in Polar Regions. Illite tends to predominate in between [1–3].

Seasonal variation is signalled by the growth rings in trees and corals, and by 'varves' (alternating light and dark layers representing the change from summer to winter deposition) in lakes and closed marine basins. We can also use a variety of chemical indicators to simulate past temperatures. The range of indicators has grown in recent years [12–14]; they help to establish the likely palaeolatitude of the environment of deposition at the time a rock was formed. Ultimately, the truest analysis of past climates comes from combining many different lines of evidence [4, 10].

## 6.2 Palaeoclimatologists Get to Work

The genius of Wegener was to leap beyond Lyell in determining where and when the continents had moved. Wegener's concept of drifting continents provided a testable means of predicting where past climate zones were. Although he didn't put the locations of climatic indicators onto past continental positions in his 1920 maps [15], he did use them to support his theory and to indicate where lines of latitude probably lay. Most of the coal deposits of the Carboniferous lay along what he thought was the palaeo-equator, extending from modern Texas through Germany to China (Figure 5.3). Salt and gypsum deposits typical of arid climates in low- to mid-latitudes lay just north and south of this equatorial zone. He thought that the *Glossopteris* ferns from the Southern Hemisphere were deposited in sub-polar peat bogs, because they surrounded a region carrying evidence for glaciation (Figure 6.1) [16]. Based on his biogeographic analysis and the distribution of

**Figure 6.1** Floral distribution in the Carboniferous and Permian [16].

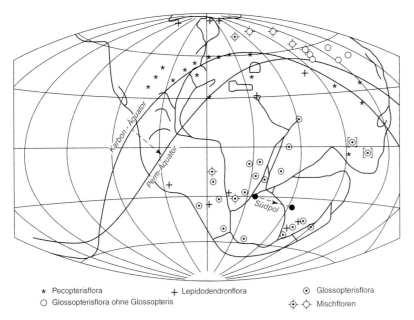

★ Pecopterisflora          + Lepidodendronflora          ⊙ Glossopterisflora
○ Glossopterisflora ohne Glossopteris                    ◈ ◇ Mischfloren

**Figure 6.2**  Wladimir Köppen in 1875.

samples indicating glaciation, he concluded that the South Pole lay near Durban, South Africa. He plotted some of these biogeographical features on his first paleoclimate map in 1922 [17].

Wegener expanded his understanding of past climates by collaborating with Wladimir Köppen (1846–1940) (Figure 6.2, Box 6.1), a Russian climatologist of German extraction who was one of the founders of modern meteorology and climatology.

Köppen saw that the combination of dryness and temperature driven by the distribution of radiation and precipitation creates largely latitudinal climatic zones characterized by their vegetation. His system divides the land surface up on the basis of annual and monthly temperatures and precipitation and the seasonality of precipitation, in such a way

as to coincide as much as possible with the world's patterns of vegetation and soils (Figure 6.3) [23]. It comprises six major climate types designated by capital letters (Figure 6.4, Table 6.1).

Köppen was initially unconvinced of Wegener's theory of continental drift, but later changed his mind [18]. Indeed, he liked his son-in-law's theory enough to publish a paper in support of it in 1921 [25]. A correction seems in order here – it has been suggested that the polar wandering positions that Wegener gave to Köppen for his 1921 paper were based on palaeomagnetic data [26], but the use of palaeomagnetic data to establish past pole positions was not developed until the 1950s, as we saw in Chapter 5.

In 1924 the two men published *Die Klimate der geologischen Vorzeit* (published in English in 2015 as *The Climates Of the Geological Past*) [16]. Its central feature was an innovative suite of crude paleoclimatic maps for selected time periods between the Devonian and the Pleistocene (e.g. Figure 5.3). These were the first comprehensive global palaeoclimatic maps. They featured Wegener's continental reconstructions, the distributions of climate sensitive indicators, and selected geographic features: the positions of the North and South Poles, the Equator and the 30° and 60° lines of latitude. Other maps showed the flora of the Carboniferous and Permian (Figure 6.1), the flooded areas of the continents in the Jurassic, and the corals of the Cretaceous.

The two compared their data (Figure 5.3) with Köppen's model of the climate system (Figure 6.4 and Table 6.1). The salt and gypsum deposits occurred where such evaporites are found today, in the arid belts north and south of the Equator (Figures 6.4 and 6.5). Cretaceous corals occurred in the equatorial zone between the 30th parallels more or less

---

**Box 6.1  Wladimir Köppen**

Wladimir Köppen was born into the family of a well-respected member of the Imperial Russian Academy of Science of German descent residing in St Petersburg [18]. He studied in St Petersburg, Heidelberg, and Leipzig, where he submitted his doctoral thesis in 1872 before returning to St Petersburg to join the Central Physical Observatory as a meteorologist. In 1875, he was appointed to the newly formed Deutsche Seewarte (the German Marine Observatory) in Hamburg. Much as Matthew Fountaine Maury (1806–1873) had begun doing in the United States in the 1850s and 1860s, Köppen began using ships' reports to map the winds over the ocean, contributing to the Seewarte's sailing handbooks for the Atlantic, Pacific, and Indian Oceans [19]. Continuing to develop his ideas on climatology, in 1884

he published the first comprehensive map of global climate zones. It formed the basis for the Köppen climate classification system that appeared in 1901 [20], was later expanded [21], and is still in use today [22]. Alfred Wegener's association with Köppen was close. He married Köppen's daughter Else in 1913. When Köppen retired in 1919, Alfred Wegener replaced him at the Seewarte. When Wegener took up a new position at the University of Graz, Austria, in 1924, their combined households moved to Graz, where Köppen continued research until his death in 1940. He is remembered through St Petersburg's Köppen Laboratory of Geochronology, by having a street named after him in Hamburg, and through the Köppen Prize awarded annually by Hamburg's Centre for Earth System Research.

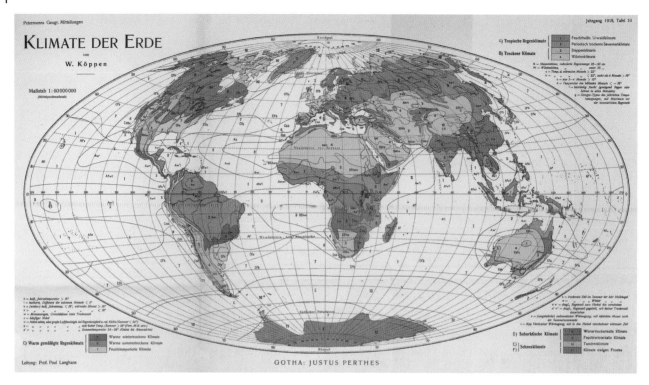

**Figure 6.3** Map of Earth's climate zones [23] (also in Thiede [18] figure 11).

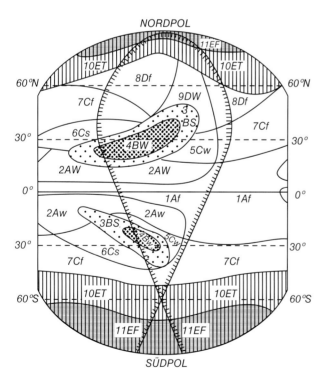

**Figure 6.4** Köppen's climate classification system. The heavy serrated line denotes a large northern and a small southern continent. See Table 6.1 for explanation of symbols; f = constantly humid; s = dry summers; w = dry winters. *Source:* From figure 7 in Ref. [24].

**Table 6.1** Köppen's classification system [24].

A) Moist tropical climates with high temperature and rainfall; average temperature of the coldest months >18 °C;

B) Dry climates with little rain and a large daily temperature range; this category is divided into S = semi-arid or steppe, and W = arid or desert;

C) Humid mid-latitude climates with warm, dry summers and cool, wet winters;

D) Continental climates in the interiors of large land masses, with low overall precipitation and a wide range of seasonal temperature; snow and frost with warmest month >10 °C and coldest month <−3 °C;

E) Cold climates, where permanent ice and tundra are present and temperatures are below freezing for most of the year; warmest month <10 °C.

F) Polar with warmest month <0 °C; T = tundra

These are divided into subgroups designated by lower case letters (Figure 6.4). For example: Af = tropical rainforest; Aw = savanna; Bs = grassland; Cf = deciduous forest; Dfc = Boreal forest (taiga). An additional localized climate type is H = cold Alpine climate, which is important in mid-latitudes for water storage (snow in winter) and release (spring thaw).

like today. Glacial indications occurred around the poles. Coals formed under temperate humid conditions north of the arid belts as well as in the humid tropics. These findings vindicated Lyell's suggestion that a shifting of the continents through time might explain the global distribution

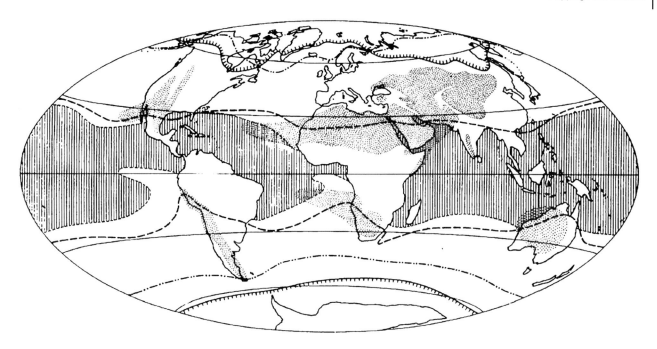

**Figure 6.5** Arid regions and isotherms today. Hatched line = annual mean temperature – 2 °C; dash/dot line = mean temperature of warmest month at sea level = 10 °C; dashed line = mean temperature of coldest month at sea level = 18 °C. hatched area = temperature of sea surface in coldest month at least 22 °C; dotted area = arid regions including arid highlands. *Source:* From figure 1 in Ref. [16].

of fossils and the location of past climate-sensitive deposits. Even so, much like Lyell, Köppen, and Wegener were limited in their ability to produce maps of different time slices by the inadequacies of rock dating, which limited their maps for the Tertiary to the early Tertiary, the Miocene, and the transition from the Pliocene to the Quaternary [16]. They were also limited by a lack of data in producing palaeoclimatic maps for Palaeozoic times. Indeed, they were reduced to plotting Devonian data on their assumed map of Pangaea, the fragments of which, we now know, had not yet assembled. They concluded that the equator in the Devonian was much further south than it was in Carboniferous and Permian times, when the southern end of Gondwana had been glaciated. Similarly rough conclusions applied to the preceding Silurian and Cambrian periods and the Late Proterozoic, although it was obvious to them that there had been a prominent glaciation in the latter period [16].

By 1937, the South African geologist Alexander Du Toit [27] was advising his readers to consult Köppen and Wegener's maps, while providing additional supporting data of his own. He was repaying the compliment: they had used his published comparisons of the geology of Africa and South America to support their continental reconstructions. In turn, Du Toit influenced English-born, New Zealand-trained geologist Lester King (1907–1989), who became professor of geology at the University of Natal, in Durban in 1935. King deduced that the Gondwana glaciation passed from west to east through time, starting with early Carboniferous deposits in western Argentina, then moving through upper Carboniferous deposits in South Africa to Early Permian ones in India, and finishing with Mid Permian tillites in Australia [28], presumably reflecting migration of Gondwana across the pole. Within the glacial deposits, King found evidence for multiple advances and retreats, like those of the Quaternary Ice Age. Consistent with the notion that Gondwana travelled '*through a succession of climatic girdles*', King found that the main phase of coal formation in Gondwana ranged from late Carboniferous in Brazil, through Early Permian in Africa and India, to Late Permian in eastern Australia [28].

Wegener, King, and Du Toit, but not Köppen, were the authors I was exposed to, when as an undergraduate student in geology at University College London in 1960–1963, I learned about palaeoclimates from our head of department, Professor Sydney Hollingworth, winner of the Geological Society's Murchison Medal in 1959. Hollingworth's presidential address to the Geological Society of London in 1961 dwelt upon *The Climate Factor in the Geological Record*. Under his tutelage, and with urging from the sedimentology lecturer Alec Smith and geology lecturer Eric Robinson, I became fascinated by the prospect of divining past climates from the geological record. Following in the footsteps of Lyell, Wegener, Du Toit and King, we students learned how climate-sensitive sediments and fossils occurred in distinct climatic zones. What we needed to know was the palaeogeography: where had the continental fragments on which those sediments were deposited been located through time? Thanks to the pioneers of continental drift we had some idea, but much of what they had to say was dismissed by the geological community. Ahead of the Vine and Matthews era, we were reduced to writing arm-waving essays like *Continental Drift – Pros and Cons.*

## 6.3   Refining Palaeolatitudes

Forty years after the publication of Köppen and Wegener's book, another seminal palaeoclimatic publication appeared, stimulated by the tremendous advances in palaeomagnetic studies of continental rocks that had been made in the late 1950s and the very early 1960s, and were outlined in Chapter 5. Edited by A.E.M. Nairn, it contained the proceedings of a NATO-funded conference at the University of Newcastle-upon-Tyne in January 1963, which brought together an eclectic mix of palaeomagnetists, palaeontologists, and palaeoclimatologists [29]. The NATO meeting was in many respects a follow-up to Nairn's 1961 book on *Descriptive Palaeoclimatology.*

It is worth bearing in mind that a conference held in January 1963 would predominantly review research results from earlier times – mostly no later than the middle of the preceding year, 1962 – so it is not surprising that only one of the 54 papers at that meeting, by Australian geologist Rhodes Fairbridge (1914–2006), referred to Hess's 1962 'geopoetry' paper on sea-floor spreading. Indeed, the conference preceded by nine months the proof of the sea-floor spreading concept by Vine and Mathews. What a difference that nine months would make! Nairn's 1964 volume [29] contained almost no reconstructions of past continental positions. Its successor – the report of the 1972 NATO Advanced Study Institute at Newcastle University, published almost a decade later, in 1973 – contained several.

The paper from the 1963 conference that is most remembered in palaeoclimate circles is the classic by Jim Briden and Ed Irving [30], which posed the question: '*with reference to palaeoclimatology, has the balance of rainfall and temperature and their gradients been the same in the past as they are today?*' In other words, did Lyell's uniformitarian views hold water when the details of past climates were examined? This question could be addressed by recognizing '*some feature, which may be called a palaeoclimatic indicator, and which may reasonably be assumed to indicate the occurrence of a particular climatic condition, say heavy rainfall or low temperature, at the time it was formed … [and by] … the use of some model of past climatic zonation of the*

*Earth, so that the indicator can be placed in its correct palaeoclimatic zone*' [30]. This is what Köppen and Wegener had done [16], but as Briden and Irving pointed out there was a drawback to using modern analogues to determine past climates. For example, while the spread of modern corals was limited by the 18 °C isotherm, past corals may not have had the same limit. Equally, the position of the 18 °C isotherm may have varied through time with respect to the Equator. '*Palaeomagnetism*', they affirmed, '*affords the means of estimating numerically the palaeolatitude spectra of palaeoclimatic indicators in a manner which is not subject to these fluctuations in the climatic model, being based on an entirely different type of observation and analysis*' [30].

Their palaeomagnetic data enabled Briden and Irving to plot accurately on North America, Eurasia, and Australia for the first time the palaeolatitudes for the main geological periods of the past 540 Ma (see Frontispiece), superimposing on each map (more or less as had Köppen and Wegener) the distribution of selected palaeoclimatic indicators: red beds, desert sandstones, evaporites, glacial beds, and coal. Unlike Köppen and Wegener, they did not reconstruct past continental positions, nor did they show any data from Africa or South America.

Their other novel contribution was to determine the past latitudinal distributions for their various palaeoclimatic indicators. Most carbonates clustered between the 40th parallels, with the bulk between the 30th parallels, as they do today. Most fossil coral reefs also occurred between the 30th parallels, like modern reefs. Red beds indicative of arid environments occupied similar latitudes. Dune-bedded sandstones occur today between latitudes 18° and 40°; in the past they occurred between 20°N and 30°S, while in the Permian they occurred within 10° of the equator – '*much lower latitudes than is common at present*' [30]. Most fossil evaporites (primarily salt and gypsum) also occurred within 30° of the palaeo-equator, whereas modern terrestrial evaporites show maxima at 25°S and 40°N; the discrepancy may be explained by some of the fossil evaporites being marine rather than terrestrial. Fossil coals were bimodal. Most occurred in tropical and temperate humid zones; very few occurred at palaeolatitudes between 15° and 30°, indicating the presence of an arid zone there. Figure 6.6 is a recent update of the palaeolatitudinal zonation of climate sensitive deposits [31].

Briden and Irving found that, while ancient marine carbonates and coral reefs displayed strikingly similar distributions to their modern counterparts, this was less true of the indicators of arid conditions, which were concentrated much closer to the equator than at present, especially in Palaeozoic times, suggesting some disturbance to latitudinal climate zoning by the distribution of land. There was

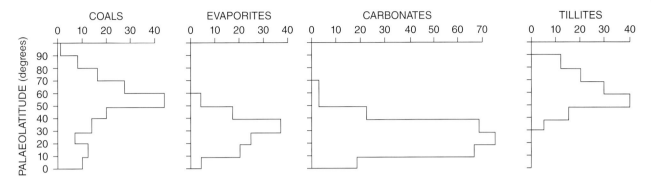

**Figure 6.6** Palaeolatitudinal zonation of climate sensitive deposits. Frequency in number of deposits, against palaeolatitude [31].

also a rather abrupt change from low latitude coals in the Carboniferous to temperate latitude coals in later times [30]. Briden and Irving speculated that these patterns came about because of the creation of a large land area (Wegener's Pangaea) in low latitudes, encouraging the initial development of equatorial coals and later development of dry conditions all across the supercontinent's interior. In due course, the Geological Society of London rewarded the pair for their efforts with the Murchison Medal for Briden in 1984, and the Wollaston Medal for Irving in 2005. Irving was also elected a Fellow of the Royal Societies of London and Canada, and a member of the US National Academy of Sciences, and was honoured with several other medals.

Another attendee at the meeting, Walter Bucher, of New York's Columbia University, was not impressed. '*The main conflict*' he pointed out '*arises from the implicit assumption that the width of the latitudinal climatic belts has always been essentially the same as at present. Yet, during the Cenozoic it has certainly undergone drastic changes in width.... Within the first third of that time, warm temperate floras grew on Ellesmere Island, Greenland, Iceland and Spitsbergen ..., and conditions were favourable for limestone deposition [there]... The change to present conditions started slowly, speeded up during the Middle Miocene, and led at an accelerating rate to the glacial conditions in the shadow of which we still live ... Why should the present width of climatic zones and the conditions it implies for temperature, wind direction and rainfall be applied to the fossil record?*' [32].

Reinforcing Bucher's message, another palaeontologist, Erling Dorf, of Princeton University, showed that the Arctic and temperate forests of the Cenozoic migrated from around 50–65°N in the Eocene to 35–45°N in the late Pliocene. He reminded his audience that tropical floras characterized the Eocene London Clay, affirming that '*the present epoch in geological history is rather abnormal in many ways, but especially in its climatic characteristics*' because it is just an interglacial within the recent Ice Age [33]. Yet another participant, Curt Teichert of the US Geological Survey, observed in that same report that

'*relationships between climate and coral-reef growth are not very straightforward, because coral evolution seems to have been influenced more by intrinsic biologic factors than by climate*' [29].

The Bucher–Dorf–Teichert message from the palaeontologists was that one could not apply rigidly the Huttonian–Lyellian dictum that the present is the key to the past. Nevertheless there had to be something to it, or the likes of Wegener, Köppen, Du Toit, Briden, and Irving would not have been able to confirm that most climatic indicators were broadly where one would expect to find them if past climate zones resembled present ones (e.g. Figure 6.6). Briden and Irving did recognize exceptions: notably the abundant development of tropical coals in the Carboniferous, but not later, and the widespread development of arid deposits in the interior of Pangaea. The solution to the riddle would not come until paleoclimate data were plotted on accurate reconstructions of continental positions at rather fine geological intervals, and until accurate means were found to determine past temperatures.

## 6.4 Oxygen Isotopes to the Rescue

That last requirement was in the process of being met. The man who won a Nobel Prize in 1934 for discovering deuterium, American chemist Harold Urey (1893–1981), discovered at the University of Chicago in the late 1940s that the isotopes of oxygen measured in seashells are related to the temperature of the seawater in which they grew [34–36]. It works like this: while oxygen carries 8 protons in its nucleus, the number of neutrons varies between 8 and 10, thus giving rise to stable isotopes known as oxygen-16 (or $^{16}O$), with 8 protons and 8 neutrons, and oxygen-18 (or $^{18}O$), with 8 protons and 10 neutrons. As the ocean warms, water molecules carrying the light isotope, $^{16}O$, evaporate preferentially, enriching warm surface waters in the heavier one, $^{18}O$. Planktonic organisms such as foraminifera, growing in the water, use that oxygen to construct their skeletons, which thus reflect the isotopic composition, and hence the temperature, of the water. Urey published his oxygen isotopic temperature scale in 1951, opening a magnificent new vista for studies of the changes in climate with time. Of this discovery it has been said, '*The measurement of the paleotemperatures of the ancient oceans stands as one of the great developments of the earth sciences; a truly remarkable scientific and intellectual achievement*' [37]. Urey was showered with honours during his career, among them election to the Royal Society of London in 1947 and the US National Medal of Science in 1964. He also has a lunar crater and an asteroid named after him.

In practice, the widespread use of stable isotopes in palaeoclimate studies awaited the development of the isotope ratio mass spectrometer to provide the necessary accuracy and precision, something that was achieved around 1950 [38, 39]. In due course, the relationship between oxygen isotopes and temperature turned out to be not as simple as first supposed, because ice volume also affects this ratio, although only at times when there were large volumes of ice on Earth – as we see in later chapters.

Analyses of oxygen isotopes in fossil shells, together with studies of climate-sensitive fossil plants and animals, confirmed Lyell's observation that global temperatures fluctuated through time. They were relatively warm between 540 and 340 Ma ago, cold during the Permo-Carboniferous glaciation between 340 and 260 Ma ago, warm again between 260 and 40 Ma ago, and cold from 40 Ma ago to the present. To some extent these patterns reflected the influence on climate of the changing positions of the continents. But other factors also affected temperature, like the concentrations of greenhouse gases in the air, as we see later.

Heinz Lowenstam (1912–1993) (Box 6.2), who had been part of Urey's Chicago University group, but had moved to Caltech, presented oxygen isotope data from the Permian and the Cretaceous to the 1963 NATO conference in Newcastle [40].

---

**Box 6.2   Heinz Lowenstam**

Lowenstam was born in Germany. He started out studying palaeontology at the universities of Frankfurt and Munich. Unfortunately, he fell foul in 1936 of a new Nazi law prohibiting the awarding of doctorates to Jews, which was passed the week before his PhD thesis defence. He and his wife Ilse migrated to the United States in 1937. There his prior work was accepted by the University of Chicago, which gave him a doctorate in 1939.

---

Within the Cretaceous, Lowenstam found that while temperatures were similar to those found today in the tropics, they declined less rapidly towards the North Pole than they do today [40]. Evidently the Cretaceous ocean was warmer than today's ocean, but the amount of warming varied with time. While the 18 °C isotherm was shifted north from about 32°N (today) to 60°N in the Santonian (86–84 Ma ago), it was shifted rather less far north in the Albian (112–100 Ma ago), the Cenomanian (100–94 Ma ago) and the Maastrichtian (70–65 Ma ago). Estimated average temperatures for polar waters of 10 °C for the Cenomanian, 15 °C for the Albian, and 16–17 °C for the Santonian are in sharp contrast to those of today – around 0 °C, which '*points towards a considerably more uniform temperature distribution of the oceanic surface waters during the Cretaceous periods as compared with today*' [40]. Lowenstam estimated Cretaceous deep water temperatures as being around 10 °C in the Cenomanian, 15 °C in the Albian and 16–17 °C in the Santonian, implying that the Cretaceous oceans were considerably more uniform than they are today, where there is a large difference between warm surface waters and bottom waters that average between 1.5 °C and 4 °C. These various findings underscore the limits to applying modern climate zones to ancient environments.

Isotopic evidence for both cool and warm temperatures in the Permian of Australia, and for significant temperature variations between the different ages of the Cretaceous within specific areas like Europe, led Lowenstam to stress '*that palaeobiogeographic studies must be limited to short time-stratigraphic intervals to serve as a meaningful palaeo-climatological tool*' [40].

Another member of Urey's international team was Italian geologist Cesare Emiliani (1922–1995), who we shall come across again in later chapters. In 1961, Emiliani analysed the ratios of $^{16}O$ to $^{18}O$ in the benthic (bottom-dwelling) foraminifera collected from cores of deep-sea sediment from the early Cenozoic, and found that the bottom waters in which those creatures grew were significantly warmer than they are today – further proof that climate and ocean circulation had changed profoundly [41].

## 6.5   Cycles and Astronomy

As we saw in Chapter 3, James Croll thought that periodic changes in the Earth's orbit might have caused not only the fluctuations of the Ice Age, but also cycles earlier in Earth's history. Following up on Croll's ideas, in 1895 the prominent American geologist Grove Karl Gilbert (1843–1918) (Box 6.3) thought that variations in the Earth's orbital behaviour might also explain oscillations in the carbonate content of Cretaceous marls in Colorado. Later, Wilmot Hyde (Bill) Bradley (1899–1979) (Box 6.3) of the US Geological Survey, suggested in 1929 that cycles in the oil shales of the Eocene Green River Formation in Wyoming might also have been caused by variations in orbital precession [42].

These pioneering approaches seem to have been largely ignored or forgotten when an explanation was sought in the first half of the twentieth century for the so-called 'cyclothems' of coal-rich Carboniferous strata like those of the British Coal Measures. These are repeated sedimentary cycles several metres thick, comprising coal formed in a swamp forest, then shallow marine shales, followed by lagoonal deposits and deltaic sands, and capped with mudstone and clay containing rootlets from the next coal seam. How might they have originated? The authors of several of my undergraduate textbooks, written in the late 1950s, invoked unexplained tectonic processes to alternately lift and lower the land, enabling the sea to flood the coastal plain and then retreat, so giving rise to these cycles [43, 44]. Having the land surface raise and lower hundreds of times might seem realistic in a tectonically active setting, but not on the stable continental margins where most Carboniferous cyclothems were found. Nevertheless, these geologists were in illustrious company, since – as we saw in Chapter 2 – Lyell too had called upon large and unexplained changes in land level to enable his icebergs to drop glacial erratics on British highlands, and even Darwin had followed him down that same illusive path.

In due course sedimentologists realized that these cyclothems were deposited on flat plains in slowly subsiding basins that eventually accumulated thick piles of sediment [45]. Wet and swampy environments on the plains encouraged the accumulation of organic matter away from the oxidizing conditions that would otherwise have encouraged decomposition. Across these plains, rivers and their associated deltas migrated and lakes formed from time to time, leading to a geological record of alternating coal seams and mudstones [46]. The sea invaded the subsiding basins at times, leading to the deposition of marine clays. What we see is thus the end result of an interplay between the tectonic processes of Earth movement, causing basins to subside, not necessarily uniformly; sedimentary processes causing the lateral migration of river channels and deltas to shut off coal formation temporarily at one site and move it laterally to another; and eustatic changes in sea level reflecting changes in the volume of water in the oceans caused either by glacial-interglacial fluctuations in some distant polar region, or by alternately warming (i.e. expanding) and cooling (i.e. shrinking) of the ocean's mass [47]. Croll knew the basins were subsiding, but he did not cater for the effects of river systems swinging back and forth across the flat and swampy plains through time. His insight that some of the cycles between coal and mud were due to elevations or depressions in sea level caused by glacial–interglacial changes driven by variations in the Earth's orbit was well ahead of its time, even though he got the association round the wrong way around (see Chapter 3). By 1977, the idea that cyclic deposition of sedimentary sequences at all scales was probably controlled by eustatic rather than tectonic changes in sea level, was being widely promoted by EPRCo researchers led by Pete Vail [48]. Nowadays it is accepted that cyclothems are millennial-scale sedimentary cycles controlled by the rhythms of Earth's orbit and their effects on climate and sea level [49], although local tectonics may influence the pattern.

Support for this leap in the imagination required a significant advance on the work of James Croll that we read about in Chapter 3. Ludwig Pilgrim, a German mathematician whose efforts have long been overlooked, kicked off the necessary work in 1904 [50]. He calculated in minute detail the changes through time expected in the eccentricity of the Earth's orbit, the precession of the equinoxes, and the tilt of the Earth's axis, and linked them to the probable chronology of the ice ages.

Next on the scene was the man who would 'solve' the mystery of the ice ages, Serbian engineer Milutin Milankovitch (1879–1958) (Figure 6.7, Box 6.4).

Milankovitch realized that Croll lacked the detail needed to solve the problem, while Pilgrim lacked the understanding of the operation of the climate system. But he was happy enough with Pilgrim's work to use the German's figures to make his own calculations. Before World War I, he published several papers documenting the emerging results of his theory, which he refined during the war (see Box 6.4). His theory demonstrated how astronomical changes altering the amount of solar radiation could account for the glaciations of the Ice Age, as we shall see later.

**Figure 6.7** Milutin Milankovitch.

Wladimir Köppen was struck by the similarity between Milankovitch's curves and the sequence of glaciations established for Europe by geographers Albrecht Penck (1858–1945), from Germany, and Eduard Bruckner (1862–1927), from Austria [56], which seemed to confirm Milankovitch's theory. He was so impressed by Milankovitch's conclusions that he invited him to contribute to *Climates of the Geological Past*, the book that he and Wegener were writing about past climates [16]. Milankovitch was much influenced by Köppen, the experienced climatologist, who told him that it was long periods of low summer temperature that produced glaciation by preserving winter's snow cover; that contradicted Croll, who thought that long winters caused glaciations by extending snow cover. In fact Köppen's idea came from Joseph John Murphy, who had deduced in 1869 that the cause of glaciations is the occurrence of Northern Hemisphere summer at aphelion (i.e. when summers were too cool to melt the snows of the previous winter) [57].

Milankovitch recognized that '*Köppen, with his ingenious insight, was the first to discover the connection between the secular march of insolation explored mathematically and the proved historical climates of the Earth*' [51–55]. At Köppen's urging, he produced for the Köppen and Wegener book a set of graphs showing the variation in summer radiation with time at middle latitudes between 55°N and 65°N over the past 600 Ka [16] (Figure 6.8). These showed four cold periods that Köppen recognized as the four glacial periods of the Penck–Bruckner scheme (Günz, Mindel, Riss, and Würm), identified from studies of gravels in river terraces north of the Alps [56]. Milankovitch's graph was a great leap forward. It provided a time calibration for glacial events, as well as explaining their occurrence [55]. At last we had an Ice Age calendar with which to date glacial epochs. Milankovitch's contribution to the Köppen and Wegener book drew his own work to the attention of a wider audience.

---

**Box 6.4   Milutin Milankovitch**

Milankovitch was born into an affluent family owning extensive farms and vineyards in Serbia. Being more inclined towards science and engineering than to managing the family estates, he went off to attend the University of Vienna, where he earned a PhD in engineering in 1904. After some years as a civil engineer, building bridges and dams in Vienna, he returned to his native land to take up a post at the University of Belgrade, where he lectured on mechanics, theoretical physics and astronomy. But, like all young men, he needed a challenge – a way to make his mark on the world. Starting in 1911, he chose climate, deciding to develop a mathematical theory enabling him to determine the temperature of the Earth at different times and places, as well as that of the other planets in the solar system. An ambitious goal! Milankovitch was a reserve army officer, and when war broke out in 1914 he was interned for a while. At the urging of a Hungarian university professor who was familiar with his work, he was eventually paroled and allowed to work in Budapest, where he could access the library of the Hungarian Academy of Sciences. He spent the war years refining his theory for predicting the world's climates through time and describing the climates of Mars and Venus, publishing his work in 1920 as *Mathematical Theory of Heat Phenomena Produced by Solar Radiation* [51]. In 1941, Milankovitch synthesized all of his results into a magnum opus known as *The Canon* [52–54]. Published first in German [52], it was translated into Serbian in 1977, then into English in 1969 [53] and again in 1998 [54]. Aleksander Grubic of Belgrade University published the key elements of Milankovitch's theory, from a study of the 1998 version of *The Canon*, in 2006 [55]. For his applications of celestial mechanics to climatology, Milankovitch is often regarded as the founder of cosmic climatology. His efforts were rewarded in the naming of craters on the Moon and Mars, and in the establishment of the Milutin Milankovitch Medal, for climatological investigations, by the European Geophysical Union, in 1993, among other accolades.

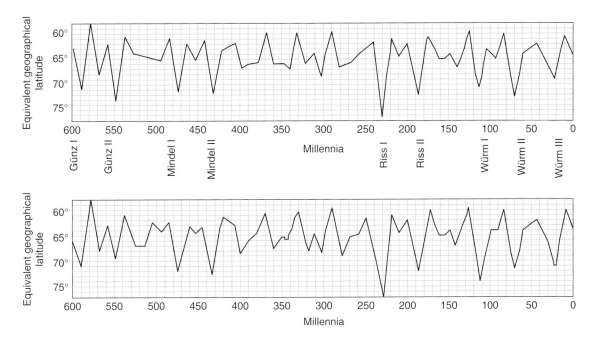

**Figure 6.8** Variation in insolation at 65°N. Amplitudes of the secular variations of the summer radiation at the northern latitude of 65°. Top, the old graph; bottom, the new graph. The vertical axes represent the equivalent geographic latitude for the variation in radiation at 65°N prior to 1800 over the past 600 000 years. Where the radiation curve moved downward (i.e. towards latitude 70°N) conditions were colder than normal, while when the curve moved up (i.e. towards latitude 60°N), conditions were warmer than normal. The glacial advances recorded in the field are labelled from left to right as Günz I and II, Mindel I and II, Riss I and II, and Würm I, II, and III.

Climatologists now appreciate that long cool summers could be critical for the inception of glaciation [10]. They might also help to explain cyclic sedimentation in periods when there were no major ice sheets, by affecting the thermal expansion of seawater, and hence eustatic change in sea level.

With the arrival of computers on the scene, it proved possible after the Second World War to refine Milankovitch's theory. Leading the process in the 1970s was Belgian meteorologist André Berger (1942–), followed by French astronomer Jacques Laskar (1955–). In 2004, Laskar presented a new solution for the astronomical calculation of Earth's insolation due to changing orbital properties over the past 250 Ma. He and his team expect it to be useful for calibrating palaeoclimate data back to 50 or even 65 Ma ago [58]. Their solutions were improved in 2011 [59].

Fascinating though this topic is, it will not detain us further until we get to the Ice Age in later chapters. This is partly because less of the geological record is preserved in older strata, and age control declines the further back in time we go, so we cannot be sure precisely what stratigraphic cycles mean in older strata [60]. For more information about the role of Milankovitch cycles in driving climate change a good source is the chapter on Orbital Forcing in Andrew Miall's book *The Geology of Stratigraphic Sequences* [61].

## 6.6 Pangaean Palaeoclimates (Carboniferous, Permian, Triassic)

In April 1972, experts at a NATO Advanced Study Institute at the University of Newcastle-upon-Tyne reviewed evidence for the relationship between continental positions and past climates [62]. Among them was Pamela Lamplugh Robinson (1919–1994) (Figure 6.9, Box 6.5) of the Zoology Department of University College London.

To support her studies of fossil vertebrates, Robinson created world maps of palaeogeography for upper Permian and upper Triassic times, following Köppen and Wegener in showing on her maps the distribution of climate indicators [64]. For the NATO meeting [65], Robinson used Köppen's conceptual model of climate zones (Figure 6.4) to demonstrate the likely distribution of climatic regions on a hypothetical continent of low and uniform relief (Figure 6.10). She then used an idealized diagram of world wind and pressure systems to show the likely distribution of annual precipitation on such a continent, which could represent modern Africa or ancient Pangaea [65]. She realized that both the climate zones and the precipitation zones integrated the operations of the atmosphere and ocean and were essentially controlled by latitude. This meant that shifting the hypothetical continent north or south would change the

**Figure 6.9**  Pamela Lamplugh Robinson.

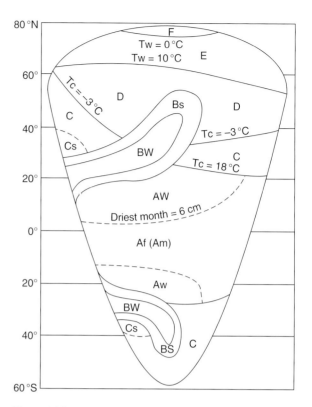

**Figure 6.10**  Robinson's conceptual climate model [65]. Distribution of climatic regions on a hypothetical continent of low and uniform relief, after Köppen (compare Figure 6.4).

location of the climate and precipitation zones on its surface: they would stay fixed, while it moved. Her conceptual model ignored the effects of topography and of ocean currents like the Gulf Stream. Nevertheless, her approach helped to demonstrate how meridional movement of a continent to north or south would lead to changes in the sequences of climate-sensitive sediments at any one location along that journey. Uplift to form mountains would complicate the picture by inviting rainfall on the windward side and aridity in the rainshadow on the leeward side.

Robinson's climate zone model (Figure 6.10) reminds us that while aridity is common at around latitudes 20°–30° on western coasts (e.g. the Sahara in the Northern Hemisphere), its latitudinal position rises poleward as one progresses inland

---

**Box 6.5  Pamela Robinson**

Pamela Robinson was a vertebrate palaeontologist, and an expert in the fossil vertebrates of Gondwana [63]. She had a somewhat unusual career. Her university studies as a pre-med student at the University of Hamburg in 1938 were interrupted by the threat of war and she returned to England to work in a munitions factory until 1945, finally registering as a geology undergraduate at University College London in 1947. Graduating in 1951, she moved to the zoology department to study a giant Triassic lizard for her PhD, ending up her career in the same department as Reader in Palaeozoology in 1982. In 1957 she began the first of many visits to India, where she helped to establish the Geological Studies Unit of the Indian Statistical Institute in Calcutta, and where she initiated a research programme in vertebrate palaeontology and Gondwana stratigraphy. She is well known for her benchmark review *The Indian Gondwana Formations*, published in the *First*

*Symposium on Gondwana Stratigraphy* (1967). She was Alexander Agassiz visiting professor at Harvard University in autumn 1972, and in 1973 was awarded the Wollaston Fund of the Geological Society of London for her work in India. Her biography describes her as '*an excellent, if demanding, teacher, with an immense breadth and depth of knowledge of biology and geology. She could be patient, helpful, charming, and thoroughly entertaining, but also intimidating, imperious, and quite terrifying*' [63]. I can vouch for the accuracy of that description, having been taught by Pamela during my undergraduate days in the Geology Department at UCL. Pamela smoked, and one of her colleagues, Tom Barnard, the Professor of Micropalaeontology, hated smoking. Alan Lord, a former UCL colleague, told me '*They would stand in the lab until she finished her cigarette, Tom would then invite her to his office whereupon she would light a new cigarette just to annoy him*.' She was quite a character.

eastwards to around 45° (e.g. Mongolia in the Northern Hemisphere); thus one can have the same kind of aridity under two quite different temperature regimes. These patterns explain what led Köppen to stipple certain areas to denote aridity on the maps that he had produced with Wegener 50 years earlier. Later on we'll examine the validity of the Robinson–Köppen assumption about the location of arid zones.

Robinson applied her conceptual climate modelling approach to a suite of continental reconstruction maps [65], which were rather like those of Alan Smith (Figure 5.12a). On each map she plotted the likely positions of high pressure maxima, winds, and the Inter-Tropical Convergence Zone (ITCZ) for the northern summer (July) and winter (January) seasons, for the Late Triassic (235–200 Ma ago), and Late Permian (250–260 Ma ago). Applying first principles, she then deduced which regions were likely to have been dry year round, which had sharply seasonal (monsoonal) rainfall, and which were likely to have been humid at high latitude. To test her predictions she compared her estimates with the distribution of climate-sensitive sedimentary rocks.

Starting with the Triassic, she suggested that during the northern summer (July) the warming of the landmass would have led to a major centre of low pressure developing over northeastern Pangaea, which would have deflected the ITCZ northward over the coast of eastern Laurasia. As the ITCZ is the boundary between the northeast Trades and the southeast Trades, this displacement would have sucked in wet air from the south over the Tethyan Ocean, causing summer monsoonal rains to fall over the northern coasts of Tethys, much as happens in southern Asia today. In the Southern Hemisphere, the winter cooling of southern Pangaea (Gondwana) would have formed a high-pressure maximum there, creating a dry winter season. The winds blowing from that centre across land towards western Laurasia (North America) would have led to dry summers in the latter region. In January, these conditions would have reversed, with the ITCZ being pushed far to the south over eastern Gondwana, bringing monsoonal rains to the southern margins of Tethys, in what is now Arabia and northern India. Robinson thought that smaller high-pressure cells would have developed over both poles. Today's polar high pressure cells are surrounded by low pressure zones, which, if they occurred in the same way in the past, would have brought seasonal rains to places like Alaska and Japan in the northern summer, and to coastal Australia, Antarctica and southern South America in the southern summer.

Robinson considered that conditions would have been slightly different in Permian times, because – compared with the younger Triassic period – the Equator lay some 10° further north, the North Pole lay 10° north of the coast of Laurasia, and the South Pole still lay in Antarctica and close to Africa. This meant that there would have been less divergence between the northern and southern extremes of the ITCZ. The arid conditions of the interior would have shifted north, covering most of North America and Greenland; monsoon rains would still have characterized the northern and southern coasts of Tethys; and the humid temperate conditions at the southern end of Gondwana would have extended further into the continent. At that time, she thought, the more central position of the Equator within Tethys would have encouraged development of a warm ocean current flowing east along the coasts of India and Australia at the northern margin of Gondwana, increasing the chances of heavy rains along those coasts.

Did her model work? She found evaporites where her model predicted that climates were dry year round, and coals where the climate was humid, so 'On the whole, agreement between the model and the pattern of distribution of the four types of "climate-sensitive" rocks is a good one', although she accepted that 'there are some anomalies' [65]. Why were there no equatorial coals in the Permian and Triassic like there were in the Carboniferous? Robinson reminded us that in the Carboniferous the northern and southern components of Pangaea were still in the process of coming together, and so were separated by an equatorial seaway on either side of which monsoonal conditions would have provided the rainfall necessary to sustain extensive coal swamps. That Carboniferous seaway, which Köppen and Wegener had not included in their own maps (Figure 5.3), had disappeared by Permian and Triassic times (Figure 5.12a), and the monsoon rains could not penetrate far enough into the arid hinterland to support any longer the vegetation necessary for equatorial coal deposits to form where Carboniferous ones had done along the palaeo-equator.

The 1972 conference clarified other aspects of the climatic history of Palaeozoic times. For instance, much of the discussion about past climate change prior to the conference was rather confused because many geologists thought that coal must have formed in a tropical climate. Coal deposits first became widespread during the late Carboniferous. They contain beautifully preserved structures of a wide variety of terrestrial plants that once formed parts of a swamp community. The lack of herbaceous plants and the abundance of tree ferns or lianas with giant leaf fans, along with the remains of trees with smooth cortex and little bark, show that they formed in rainforests that may have been tropical or subtropical [66]. At the 1972 conference, palaeobotanist Bill Chaloner (1928–), from Royal Holloway College, near London, who was to be elected a Fellow of the Royal Society in 1979, and was awarded the Geological Society's Lyell Medal in 1994, showed that trees that grew at temperate latitudes differed considerably from those in tropical locations. Temperate trees carried rings representing seasonal change, while tropical ones did not

[67]. Most of the Carboniferous and Permian coals of North America and Europe, near Köppen and Wegener's palaeo-equator, lacked tree rings; they were tropical. Those of the same age from southern Gondwana, including Antarctica, and formed near Köppen and Wegener's South Pole, carried tree rings; they were temperate [67]. Problem solved. Tying coals lacking trees with rings to the palaeo-equator removed the necessity for Carboniferous coals to signify global warmth. Coals could just as well have formed in cool humid environments, which would be signalled by trees with rings. We no longer had to think of the Carboniferous as a period that was especially warm *globally*. Zonal conditions ruled, much as Lyell had suspected.

Like Humboldt and Köppen, Chaloner saw that '*climate is the overriding influence controlling the distribution of plant communities*' [68]. Hence, palaeoclimatic information could be extracted from fossil plant remains by observing what climate zones their nearest living relatives inhabited, the shape of their leaves, or the character of their wood – especially the presence or absence of rings representing seasonality [68]. Seeing that Ian Woodward, then at Cambridge University, had discovered in 1987 that the frequency of stomata (pores) on leaves was proportional to the abundance of $CO_2$ in the atmosphere [69], he noted that this '*offers promise for direct palaeobotanical evidence for past changes in the level of this climatically significant atmospheric constituent*' [68], something we follow up on later.

Robinson concurred with Wegener and Du Toit that, during the Carboniferous and Permian, today's southern continents were clustered over a South Pole near South Africa, where we find the extensive Dwyka Tillite. Glacial conditions covered South Africa, Antarctica, India, and much of southern Australia and South America. Lyell would probably have been pleased to see the association of cooling with high latitude land (see Figure 2.7).

We now know more about the Permo-Carboniferous glacial period. Emerging evidence suggests that global ice volume reached a peak at the Carboniferous–Permian boundary, causing a significant global fall of sea level at about 300 Ma ago. In due course, that was followed by a rise in sea level manifest as a global transgression during the following Sakmarian stage (295–290 Ma ago), signifying the beginning of the major deglaciation of Gondwana [70].

Gondwana continued to warm through the Permian and into the Triassic, as the supercontinent moved north away from the South Pole. During this time, while Pangaea's maritime margins were humid, much of its immense interior was arid and desert-like; imagine a gigantic version of modern Australia. Where evaporation exceeded precipitation, vast deposits of salt accumulated in the Permian of western Europe. By Late Permian (Zechstein) times

(250–270 Ma ago), salts were being deposited in a basin extending from west central Poland to northeastern England and from Denmark to southern Germany [7]. Deposition began in the Early Permian, around 280 Ma ago, and extended up into Triassic time, diminishing towards its end 200 Ma ago. Laminations within the deposits suggest climate cycles of more or less aridity. The salt may have been deposited during particularly arid times, rather than continuously. Most past evaporites were deposited in warm arid regions between 45°N and 40°S, with a peak in the desert regions centred on about latitude 32° [7].

Knowing how the continents were distributed through time was a boon to palaeontologists, who could now begin understand, rather than just guessing, why the fossils of animals and plants, were distributed in the way that they were across today's continents [71]. It was simple: the break-up of Pangaea disrupted former land links. Knowing the timing of the different breaks, one could understand the divergence of fossil lineages from one another on today's different continents.

As we saw in Chapter 5, one of those palaeontologists, Fred Ziegler of the University of Chicago, realized that it would benefit the wider community to construct an accurate series of palaeogeographic maps to show how fossil plants and animals and climate-sensitive deposits had been distributed through time, which led to the inception of the Paleogeographic Atlas Project. In 1979, Ziegler and his colleagues publish a suite of seven continental reconstruction maps for the Palaeozoic onto which were plotted the locations of climate-sensitive sediments [72]. They found that '*The distribution of climatically sensitive sediments shown on our reconstructions for the Paleozoic is in good agreement with expectations based on the model of the Earth's present atmospheric and oceanic circulation patterns*' [72]. They went on to explain: '*We do not mean to imply that climate has been constant through time. The proportion of land, and its latitudinal array, must have been very important in controlling world temperature and precipitation. The heat derived from solar radiation is absorbed and redistributed in the oceans, and by contrast, lost over land areas during the nights and the winters. From this, one would expect that the world climate of periods like the Recent, the Permo-Carboniferous and the late PreCambrian, with much land in high latitudes, would be generally cool and this is confirmed by glaciations of these times. At the other extreme were times like the early Paleozoic and the late Mesozoic with large expanses of shelf seas associated with relatively low latitude continents. The occurrence during such times of carbonates in higher latitudes than present may be evidence of more uniform temperature conditions*' [72]. Lyell would have been pleased to see the emphasis on latitude as a controlling factor in climate.

**Figure 6.11** Judith Totman Parrish.

One of the co-authors of the 1979 paper on *Paleozoic Paleogeography* was Ziegler's former PhD supervisor from Oxford, W.S. (Stuart) McKerrow (1922–2004),

a palaeo-ecologist and the Geological Society of London's Lyell Medallist for 1981. McKerrow went on with Chris Scotese to write about palaeogeography and palaeoclimatology, notably in Africa, making ample use of the usual climatic indicators [73]. Among their cold climate indicators were glendonites: carbonate pseudomorphs after crystals of ikaite, a calcium carbonate hexahydrate (CaCO$_3$.6H$_2$O) that forms in organic-rich marine or brackish sediments at near freezing temperatures and decomposes when the temperature rises above 5 °C.

In 1977–1978, the Paleogeographic Atlas Project was expanded to apply to the continental reconstruction maps the likely circulation patterns of the atmosphere and ocean (à la Robinson). Judith (Judy) Totman Parrish (Figure 6.11) was invited to supervise that part of the programme [74]. In due course, Parrish would rise through the ranks to become President of the Geological Society of America in 2008–2009. She built upon and expanded Robinson's conceptual approach to palaeoclimatology in a set of landmark papers published in 1982/1983. Basically, she superimposed conceptual distributions of likely past air pressure on continental reconstruction maps for different time slices (Figure 6.12), and used the pressure maps to determine likely palaeo-wind directions and areas of high or low rainfall, to compare with palaeoclimate data [75].

**Figure 6.12** Past distribution of atmospheric pressure for the earliest Triassic, with northern winter above and northern summer below. Heavy solid lines are isobars; arrows represent wind directions; H = high-pressure centre; L = low-pressure centre. *Source:* From figure 6 in Ref. [75].

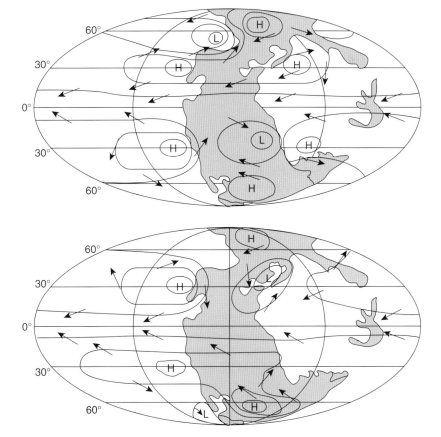

Where the winds blow parallel to the coast, the surface waters move offshore. Nutrient-rich subsurface waters well up to replace them, stimulating high productivity [76]. Under appropriate conditions this can lead to the deposition of the organic-rich rocks that are the source rocks for oil, and to rocks rich in phosphorus: phosphorites, which can be mined for fertilizer. Parrish produced a set of papers predicting where upwelling might have occurred in the Palaeozoic [77, 78], and the Mesozoic and Cenozoic [79], and where rainfall might have been high, perhaps leading to coal deposition, in Mesozoic and Cenozoic times [80]. As we shall see later, her predictions correlated well with palaeoclimatic data from the field.

Compared with what Robinson had to offer, Parrish benefitted by having improved reconstruction maps, more time slices, and more data on the distribution of climate-sensitive deposits. She found that the present day rainy zones around 55°N and 55°S and at the Equator were present in the past: '*from this, it can be concluded that atmospheric circulation has not been radically different from its present configuration, despite some apparently great differences in some climatic parameters such as the equator-to-pole temperature gradient*' [80]. She agreed with Robinson that the Triassic world was generally dry, with seasonal rainfall on eastern coasts. Sea level was low at the time.

Applying Parrish's conceptual palaeoclimatic model helped to refine the Chicago group's analysis of climate change in the Carboniferous [81]. Her maps showed that the collision between Gondwana and Laurasia in the Late Carboniferous created a supercontinent that stretched from pole-to-pole (e.g. Figure 6.12), changing the climate from mainly zonal to strongly monsoonal and causing increasing asymmetry of climate patterns from east to west [82]. That dried out the equatorial region by the Permian, leading to the demise of formerly flourishing coal swamps along an equatorial seaway, as mentioned earlier, and increased seasonality. Formation of a single large north–south oriented land mass also dried the interior and deflected to both north and south the former through-flowing warm equatorial currents, which then carried heat to high latitudes along the east coasts of Pangaea. The mountain belt created along the suture line between Gondwana and Laurasia would have interacted with the monsoonal circulation much as the Tibetan Plateau does today, accentuating the pressure difference between north and south [82]. As noted by Briden and Irving [30], coals forming after the suture were abundant at high latitudes not at the Equator.

According to Parrish, '*monsoonal circulation attained its maximum strength, in effect creating a megamonsoon, in the Triassic*', when the supercontinent was divided in two by the Equator [82]. She attributed the megamonsoon to global warming causing more evaporation from the ocean following the Carboniferous–Permian glaciation, leading to stronger and more widespread seasonal rainfall alternating with increasing aridity. Mid-Pangean aridity helps to explain the great expansion of evaporites in the Triassic, although this coincided with extensive rainfall along the western margin. By the Early Jurassic, as Pangaea began to break apart, circulation began to revert to more zonal rather than meridional as the monsoon pattern weakened with the rifting of Laurasia from Gondwana [82].

Neither Robinson nor Parrish had much to say about the likely range of temperature from the coast to the interior of Pangaea. Numerical modelling experiments suggest that mean monthly summer temperatures there exceeded 35 °C, or more than 6 °C above today's maximum temperatures for the interior [11]. Indeed, daytime highs may have approached 45–50 °C. Numerical climate models agree that most of the interior would have been dry, especially between 40°N and 40°S [11].

Certain caveats must be applied in palaeoclimate studies. Bruce Sellwood and Brian Price remind us that despite enthusiasm for using sedimentary facies (or types) as indicators of past climate, the data have to be interpreted with care [83]. The most climatically informative sediments are tills, laterites, evaporites, and aeolianites (e.g. dunes). Other criteria provide supplementary evidence of climate. But, because sedimentary rocks are subject to post-depositional change (diagenesis), they seldom faithfully record subtle climate signals. They are imperfect receivers of the climate signal, although certain settings (e.g. deserts and ice caps) preserve climate signals better than others [72].

Among the best preservers of climate signals are fossil plants. As Ziegler and his team pointed out, they occupy realms with pronounced climate signals, are sedentary (no seasonal migrations), are not subject to diagenetic alteration (unlike isotopes), represent ground truth (unlike model outputs), and are abundantly preserved in many places [84]. The team assigned the fossil vegetation of Eurasian floras from the Triassic and Jurassic periods to one or other of 10 biological zones (biomes). Most plants fell into the dry subtropical, warm temperate and cool temperate biomes. There was a general absence of tropical rainforests. Tropical coals swamps disappeared in the Early Triassic, except locally in the Asian monsoon region, and the equatorial belt became arid in the Triassic. Coal swamps emerged at mid- to high-latitudes during the late Mid Triassic. Warm temperate floras reached above 70°N in the Triassic and up to 70°N in the Jurassic, and there was no hint of the cold temperate, or Arctic or glacial climates of today. Triassic warmth contrasted with the glacial conditions of the southern continents in the Carboniferous and Permian, reflecting the drift of Gondwana north away from the South Pole.

During the Triassic, generally arid conditions prevailed over North America and Europe within 5° and 50° north of the equator. They were interrupted in the Late Triassic Carnian period (216–228 Ma ago) by a warm, wet monsoonal phase [85]. Substantial changes occurred within the marine invertebrate fauna at the end of the Early Carnian, and there was a major change in the terrestrial biota at the end of the Carnian. Michael Sims of Trinity College, Dublin, and Alastair Ruffell of University College London, interpreted these developments to suggest that the final coalescence of Gondwana and Laurasia to form Pangaea was followed in the mid-Carnian by rifting preceding the break-up of the supercontinent. This rifting would have been associated with volcanism and the emission of $CO_2$, which might have led to sufficient warming to have caused the development of the monsoonal conditions [85], along the lines suggested by Australian geologist John Veevers [86].

Among those reviewing the relationship between continental positions and past climates was Lawrence A Frakes (1930–, Box 6.6), who produced a series of papers on Palaeozoic glaciation in Gondwana, starting in 1969 [87].

As Frakes pointed out in 1981 [88], the idea that variations in the age and distribution of late Palaeozoic glacial

deposits on Gondwana resulted from the drift of the supercontinent over the pole was first elaborated in 1937 by Du Toit [27], then in 1961 by Lester King [28], and in 1970 by Crowell and Frakes [87]. Palaeomagnetic studies had established by 1981 that South America and South Africa were the first parts of Gondwanaland to cross the pole, and Australia was the last [89]. It was not entirely obvious to Frakes why glaciation ceased by the early Late Permian, as Gondwana remained at fairly high latitudes then, as did its southernmost fragments during the continental break-up that followed in the early Mesozoic. One possibility was that more of Gondwana now lay at or closer to latitude 65°S, where conditions were warm enough to melt ice and prevent its further accumulation. Global warming of unspecified cause – and an associated decrease in albedo – might account for these changes, along with a decrease in the precipitation required to build an ice sheet [88]. A decrease in the requisite precipitation might have resulted from the gradual shift of the continents or from shifts in the locations of warm ocean currents. We have to remember that at the time $CO_2$ had only just been discovered in ice cores, and nothing much was known about its past distribution.

## 6.7 Post-Break Up Palaeoclimates (Jurassic, Cretaceous)

With sea floor spreading taking place in all of the new seaways as well as in the pre-existing Pacific Ocean, the rate of production of new ocean crust increased significantly, forming several new mid-ocean ridges during Late Jurassic and Cretaceous times [90, 91]. These massive new upstanding ridges displaced ocean water, thus raising sea level (Figure 5.13) and drowning low standing parts of the former fragments of Pangaea, creating warm shallow seas in North America and Europe (Figure 5.12b) [92]. Following Peter Vail's lead (Figure 5.13), Parrish thought that Cretaceous sea levels stood on average 170 m higher than today [12], while Dietmar Müller and colleagues put it at about 150 m, rising to about 170 m in the Cenozoic between 75 and 85 Ma ago [90]. Sea level began to fall from these high levels when the Izanagi Plate (East of Japan) with its associated mid-ocean ridge in the northeast Pacific was subducted beneath East Asia around 60 Ma ago [90].

The new seaways caused by continental displacements changed the pattern of ocean currents, introducing a new element into the story of climate change. For example, creation of a north–south passage by the opening of the Atlantic Ocean (Figure 5.12b) increased the opportunity for oceanic transport of heat from the tropics to the poles,

---

**Box 6.6    Lawrence A. Frakes**

Larry Frakes was born in the USA, and started his career with John Crowell at the University of California, Los Angeles (1964–1971), studying late Palaeozoic glaciations on Gondwana fragments. Later, at Florida State University, he worked with Elizabeth Kemp on global reconstructions of Eocene–Oligocene palaeotemperatures, making an early contribution to climate modelling, and publishing key findings in the journal *Nature* in 1972. Later, working with Jane Francis and Neville Alley at Adelaide University (1987–1999), where he was appointed the Foundation Douglas Mawson Professor of Geology and Geophysics (1985), his research overturned the concept of a uniformly warm Cretaceous through discoveries of evidence for glacial activity. He and Jane Francis found evidence for glaciation in most periods of the Phanerozoic (e.g. through the occurrence of dropstones and related criteria), culminating in a paper to *Nature* in 1988. Frakes was awarded the Antarctic Service Medal by the US National Science Foundation (NSF), and has a mountain named after him in Marie Byrd Land, Antarctica.

which might tend to make polar glaciations less likely, other conditions being equal.

The climate of the Jurassic has been described as 'equable', in the sense of warm but not very variable. Warm it was, compared with today, but there were strong seasonal contrasts in continental interiors where, during the Early Jurassic (195 Ma ago) the annual range of temperature was up to 40 °C in Eurasia, at about 60°N, and more than 45 °C in Gondwana at about 60°S [11]. Hardly 'equable'!

Tony Hallam is a fount of knowledge about Jurassic climate [93, 94]. He tells us that there were no significant polar ice caps then, but dropstones indicate the presence of seasonal ice in the Mid Jurassic of Siberia, where winter temperatures probably hovered close to 0°C. Most of Africa, Madagascar, India, South America, North America south of the Canadian border, western Europe and western Asia would have been dry. Monsoons would have made the margins of these Pangean fragments seasonally wet. Year-round humidity characterized high latitudes, South East Asia and southern-most South America. Coral reefs were confined to a tropical belt mostly between the 30th parallels. This was also a time of warm shallow seas in which carbonate minerals could precipitate to form 'oolites', accumulations of 'ooliths' (from the Greek for 'egg-stone'), which are ovoid grains made up from layers of carbonate precipitate. Oolitic limestones are a common feature of the Jurassic across Britain. These golden coloured rocks were used to build many of Oxford's colleges. Portland Stone, from the Isle of Portland in the English Channel, is a pale grey version of these limestones in which ooliths are dispersed through a matrix of fine calcitic mud, or 'micrite'. It was used to build many of England's prominent buildings including St Paul's Cathedral.

Parrish's palaeoclimate models and Ziegler's data showed that monsoonal circulation with extended wet and dry periods allowed evaporites to form seasonally in equatorial regions in the Mesozoic [80], as Briden and Irving also found [30]. North Africa and northern South America became wetter with time as the North Atlantic opened.

In a seminal study of the climate of the past 540 Ma, the Phanerozoic Eon, Larry Frakes and his colleagues from the University of Adelaide in South Australia, Jane Francis and Joseph Sytkus, reported in 1992 that sea surface temperatures in the low latitudes of the Middle to Late Jurassic were 26–28 °C, while bottom waters were about 17 °C [95]. Their oxygen isotope data showed that water temperatures cooled towards the lower Cretaceous. The Frakes team provided evidence for transport by ice at high latitudes in the Late Jurassic and Early Cretaceous; chiefly the occurrence of boulders and dropstones of exotic rock types embedded in fine-grained mudstones, harking back to Lyell's ice-rafting (see Section 2.4 and Figure 2.9). But, in the absence of glacial deposits like tillites it seemed likely that the north polar environment was periglacial, with seasonal winter ice forming on rivers and shorelines and incorporating exotic materials from the banks and bases of rivers and from cliffs. Seaward transport of floating ice explained the occurrence of dropstones offshore [95]. Lyell's theory that cool conditions would result from the polar locations of continents did not work all the time. Something operated against it.

While evidence for episodes of cooler conditions in the Early Cretaceous has been proposed, a recent detailed study of palaeotemperatures from the lower Cretaceous (Berriasian–Barremian between 145 and 125 Ma ago) showed that sea surface temperatures were much warmer than today, averaging 26 °C at 53°S and 32 °C at 15–20°N [96]. It seems that the climate was warm and stable with a weaker meridional temperature gradient (0.2–0.3 °C/° of latitude) than we have today (0.5 °C per degree). The temperatures appear to be no different from those of the Late Cretaceous Cenomanian and Turonian periods (100–88 Ma ago). If there were cool or cold periods in the Early Cretaceous, as was formerly supposed, they may have been just short cold snaps or seasonal extremes [96]. One such Early Cretaceous cold snap gave rise to glacial tillites in the Flinders Range of Australia [97].

Parrish's conceptual model implied that Cretaceous opening of the entire Atlantic (Figures 5.12b and 6.13) brought rainfall to the formerly dry east coasts of both North and South America, while the interiors of Asia and Africa remained dry. Much the same applied in the Late Cretaceous (Maastrichtian, 70–65 Ma ago), but the widening North Atlantic would have encouraged the development of westerlies, bringing rain to western Europe. An equatorial Tethyan current separated the northern and southern fragments of Pangaea (Figures 5.12b and 6.13), and there was probably a proto Gulf Stream in the North Atlantic. When the break-up was well underway, sea level was at a maximum (Figure 5.13), with flooded continental margins (Figure 5.12b and 6.13) [80, 84].

Parrish interpreted her climate model to suggest where winds blew parallel to the coast, generating upwelling currents, for example in the Mid Cretaceous (Figure 6.13) [79]. These are locations where one might expect organic-rich rocks to form [76, 98, 99].

Based on the studies of phosphorite that I carried out off the coast of northwest Africa for my PhD (1967–1970), and off southwest Africa while at the University of Cape Town (1970–1972), I had independently developed a similar approach to Parrish's for predicting the likely occurrence of organic-rich rocks. Whereas she focused more on the winds, I focused more on what was happening within

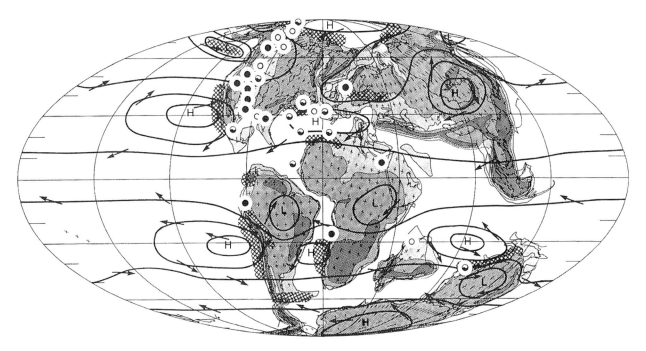

**Figure 6.13** Upwelling predictions from qualitative circulation models – Cenomanian (99.6–93.6 Ma ago). The map shows highland in dark shading, lowland in medium shading, and flooded continental edges in light shading. Lines represent isobars, with H = high pressure and L = low pressure. Upwelling indicated by cross-hatching along continental margins. Dots = locations of samples of organic rich rocks. *Source:* From figure 3, in Ref. [79].

the body of the ocean – especially on the depletion of oxygen in the oxygen minimum zone. This zone arises because sinking and decomposing remains of dead plankton consume oxygen at intermediate depths at rates faster than it can be replenished by mixing from well-oxygenated surface waters and bottom waters, making the ocean into an oxygen sandwich. Where oxygen depletion is extreme, conditions may become anoxic (zero oxygen), and sediments reaching the seabed there may be well preserved leading to organic enrichment, especially on continental margins – as off Peru, California, and Namibia today. I applied this model extensively in my research for EPRCo in Houston (1976–1982), with the object of predicting where explorers might find oil-rich source rocks in ancient basins. Parrish and I presented papers on our complementary approaches at a NATO meeting organized by Jörn Thiede and Erwin Suess, in September 1981, in Villamoura, Portugal, on the topic of *Coastal Upwelling – Its Sediment Record* [78, 98].

When Chris Scotese from Ziegler's group joined me for a sabbatical at BPRCo, in the mid-1980s, we devised a method of quantifying the Robinson–Parrish approach to climate modelling. The end result was a set of palaeogeographic maps, complete with isobars, that I used to show where upwelling currents may have formed organic-rich deposits in past times [99]. While our results did not differ much from Parrish's, we felt that quantifying the principles gave

us more credible results. The theoretical underpinning for the role of upwelling in generating organic rich sediments is available elsewhere [76].

Knowing that plants are strongly related to climate [84], Bob Spicer, of Oxford University, and colleagues, used Cretaceous fossil plant remains to show that cool temperate rain forests in polar coastal areas were conifer-dominated and deciduous [100]. At high latitudes and in continental interiors, winter temperatures likely fell below freezing, but some plants retained leaves year round, with reduced leaf size and thick cuticles. At mid-latitudes, conifers, ferns, and cycads dominated open canopy woodlands and forests, giving way in the Late Cretaceous to broad-leaved angiosperms including shrubs and small trees. Forests were patchy at low latitudes.

Working with Parrish, Spicer suggested that Late Cretaceous-Early Cenozoic floras from high palaeolatitudes (75–85°N) experienced a similar light regime to that at present. Their plant data suggested that mean annual air temperatures at sea level there were 10 °C in the Cenomanian (100–94 Ma ago), rose to 13 °C in the Coniacian (88–86 Ma ago), dropped to 5 °C in the Maastrichtian (71–65 Ma ago), and were 6–7 °C in the Palaeocene (65–55 Ma ago) [101]. They thought that polar winter temperatures were freezing in the Maastrichtian, and that '*Permanent ice was likely above 1700 m at 75°N in the Cenomanian, and above 1000 m at 85°N in the Maastrichtian*' [101].

Jane Francis (1956–) (Figure 6.14) (Box 6.7), an eminent palaeobotanist, specialized in using the fossil plants of the polar regions as indicators of past climates.

Francis likes to point out 'the Antarctic paradox', which is that '*despite the continent being the most inhospitable … on Earth with its freezing climate and a 4-km thick ice [sheet], some of the most common fossils preserved in its rock record are those of ancient plants. These fossils testify to a different world of warm and ice-free climates, where dense*

**Figure 6.14**  Dame Jane Francis.

---

**Box 6.7  Dame Jane Francis**

Jane Francis was a palaeobotanist with the British Antarctic Survey from 1984 to 1986, before spending five years as a post-doctoral researcher with Larry Frakes at the University of Adelaide in Australia. Returning to the UK in 1991, she joined Leeds University where she rose to become professor of palaeoclimatology in the school of earth and environment, director of the Centre for Polar Science in 2004 and dean of the faculty of the environment in 2008. For her polar research she was awarded the US Antarctic Service Medal, the US Navy Antarctic Medal and, in 2002, the UK's Polar Medal – only the fourth woman to receive that honour. She received the President's Award of the international Paleontological Society in the 1980s, became President of the UK's Palaeontological Association for 2010–2012, was appointed to head the British Antarctic Survey in October 2013, was awarded the Coke Medal of the Geological Society of London in 2014, was made a Dame Commander of the Order of St Michael and St George in 2017, and was appointed Chancellor of the University of Leeds in 2018. Francis led the UK's involvement in ANDRILL the international Antarctica drilling programme.

---

*vegetation was able to survive very close to the poles. The fossil plants are an important source of information about terrestrial climates in high latitudes, the regions on Earth most sensitive to climate change*' [102]. Francis and her team found that fossil plants from the mid-Cretaceous are abundant on Alexander Island on the west side of the Antarctic Peninsula at around 70°S. Conifers, tree ferns and ginkgos were abundant there, with shrubs, mosses, and plants known as liverworts in the rich undergrowth. Ginkgo trees were common in the distant past, but are rare today. You may have come across one the Maidenhair Tree, *Ginkgo biloba*, used in traditional Asian medicine. Evidence from the plants and their associated soils showed that the climate was warm and humid; probably dry in summer and wet in winter. This was around the time when Antarctica reached the South Pole 90 Ma ago. Dinosaurs roamed the woods.

Younger Cretaceous strata are preserved on the opposite, eastern, side of the Antarctic Peninsula, on James Ross Island and Seymour Island, at about 64°S. Flowering plants (*angiosperms*) were abundant. Their modern equivalents live in warm temperate or subtropical conditions including wet tropical mountain rainforests or cool temperate rainforests. Analysing the shapes of the margins of leaves, which are related to temperature, told Francis that mean annual temperatures in this part of Antarctica, then some 2000 km from the pole, averaged around 17–19 °C. Winter temperatures must have been above freezing, and rainfall ranged from around 600 to 2400 mm/year, with peaks in the growing season [102]. Evidence from plants from various sites on the Antarctic Peninsula also told Francis that temperatures there had declined from an average of around 20 °C in Albian times (c. 100 Ma ago) to c. 7.5 °C by the end of the Eocene 34 Ma ago (Figure 6.15) [103].

Most of the perceptions of climate change that we have examined so far come from sediments and fossils found on land. But ocean sediments also have something to tell us. Here we benefit from the application of oil company technology to the solution of fundamental science questions. Many of the advances in our understanding of the evolution of our climate following the break-up of Pangaea come from using a floating drill rig to sample the sediments deep beneath the 72% of the planet's surface covered by the ocean. Drill cores obtained through the Deep Sea Drilling Project (DSDP) and its successors (Figure 6.16, Box 6.8) extend as far back as the Early Jurassic in a few places: the age of the oldest known deep marine sediments formed since the break-up of Pangaea. We have many more Cretaceous deep ocean drill cores, and yet more from the Cenozoic as we'll see in Chapter 7. The marine microfossils from those cores tell us a great deal about past climates [104].

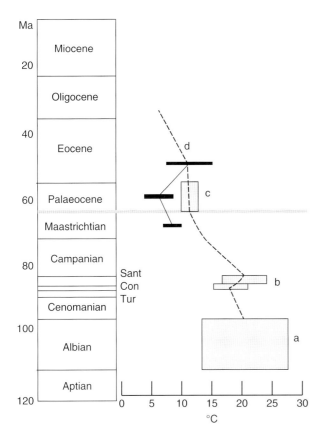

Figure 6.15 Estimated mean annual temperatures from fossil plants on the Antarctic Peninsula. *Source:* From figure 5A in Ref. [103].

**Figure 6.16** Deep Ocean Drilling Vessel RV JOIDES Resolution (1989–).

Integrating data from marine and terrestrial sources, Larry Frakes and his team tell us '*the period from the mid-Cretaceous (mid-Albian) to the mid-Early Eocene (...105–55 Ma) was one of the warmest times in the late Phanerozoic*'. [95]. The average global temperature then was probably at least 6 °C higher than today, the poles were most likely free of permanent ice, and there was no evidence for seasonal ice rafting. As the high latitude oceans were warm, the equator-to-pole temperature gradient was low, resulting in relatively weak atmospheric and oceanic circulation. '*Temperate climates extended right up to the poles during the*

---

**Box 6.8 Deep Ocean Drilling**

The DSDP on the 120 m-long drilling vessel *Glomar Challenger* was run by the US NSF and collected samples from the summer of 1968 through 1972. It was followed from 1975 to 1985 by the International Phase of Ocean Drilling (IPOD), on the same ship, funded by the NSF along with Germany, France, the UK, Japan, and the Soviet Union. A new phase, the Ocean Drilling Program (ODP), began with the advent of a larger drill ship, the DV *JOIDES Resolution*, and ran from 1985 to 2003, then was replaced by the Integrated Ocean Drilling Program (IODP). From 2012 this became the International Ocean Discovery Program (also IODP), which involves the United States, 17 European countries, India, China, and South Korea. It employs two ships, the *JOIDES Resolution* and Japan's *Chikyu*, which is equipped with a 'riser' – a device for preventing blowouts – thus enabling drilling into deep sediment sections on continental margins where natural gas may be a potential hazard. As Bill Hay tells it [105], the development of the DSDP benefitted from the efforts of Cesare Emiliani to obtain long

cores to study the history of the ocean. In 1963 he submitted his LOCO (Long Cores) proposal to the NSF, and the drill ship *Submarex* duly collected some test drill cores of late Tertiary and Quaternary age from the Nicaragua Rise late that year. In 1964, the major US oceanographic institutions Lamont, Woods Hole and Scripps (of which, more in Chapter 7) formed the Joint Oceanographic Institutions for Deep Earth Sampling (JOIDES), and used the drill ship *Caldrill* for a drilling campaign on the Blake Plateau off Florida in 1965. The successes of *Submarex* and *Caldrill* led JOIDES to propose to NSF in 1966 that there be an 18-month programme of ocean drilling – the DSDP – and a uniquely outfitted research drill ship, the *Glomar Challenger*, was commissioned. Complete with a dynamic positioning system, it was named after the first major oceanographic survey ship HMS *Challenger*. In recent years a comparable programme to the DSDP has developed for the continents: the International Continental Scientific Drilling Programme (ICDP).

*Cretaceous and Early Tertiary, allowing the growth of forest vegetation at high latitudes. The plants were able to tolerate the rather extreme light regime that they would have experienced...'* [95].

That analysis seems to neglect Parrish and Spicer's conclusion that the high latitudes of the Maastrichtian (70–65 Ma ago) were rather cold [101]. More recent data from deep-sea sites confirms that bottom waters were about 12 °C during the Mid Cretaceous, reached 20 °C during the latest Cenomanian (100–94 Ma ago) and Turonian (94–88 Ma ago), and cooled to 9 °C by the Maastrichtian at the end of the Cretaceous [106]. Kenneth MacLeod of the University of Missouri, and colleagues, confirmed in 2013 that Turonian seas were particularly warm, with surface water temperatures of 30–35 °C, and bottom temperatures of 18–25 °C [107].

Fossil leaves from Alaska during the Albian (112–100 Ma ago) and Cenomanian (100–94 Ma ago) suggest temperatures of around 10 °C, warming in the Coniacian (89–86 Ma ago) to about 13 °C, then cooling during the Campanian-Maastrichtian (84–65 Ma ago) to around 2–8 °C [101]. Winter temperatures may have declined below freezing there. Under these conditions dinosaurs thrived in Arctic deltas among mild to cold temperate forests of deciduous conifers and broad-leaved trees. At the other end of the world there were rainforests on the Antarctic Peninsula and in Tierra del Fuego. The climate there was like that of New Zealand and Tasmania today [101]. The continental interiors, like central Asia, remained very dry [95].

Myriads of tiny planktonic plants, the *coccolithophoridae*, flourished in the warm shallow seas that flooded western Europe during the Cretaceous period between 145 and 65 Ma ago. The remains of their calcium carbonate skeletons sank to the shallow seabed to form white ooze, now consolidated and uplifted as the chalk of the White Cliffs of Dover and the French coast. In effect that makes the chalk a biological rock. When you rub a piece of chalk between your fingers the dust that comes off is made of the miniscule platelets, or coccoliths, that covered these tiny creatures. Strange to think that while watching my science teacher scribble on the blackboard I was seeing fossil coccoliths scrawled on slate. And I doubt my mother ever knew that she was dusting her face with fossils when powdering her nose. Incidentally, chalk is implicit in the very name of the Cretaceous, *creta* being the Latin word for chalk.

As with the Jurassic, in the past geologists thought the Cretaceous had a warm, equable climate with a lack of seasonal extremes. Nowadays, Hallam tells us that the mid-latitude Cretaceous climate was most probably seasonal, and the concept of an equable climate belongs on the scrap heap [93]. Besides that, Frakes and his team found that

while the Mid to Late Cretaceous was generally warm, abundant evidence of cyclic sedimentation shows that the warmth was interrupted by cool periods lasting from a few thousand to two million years, which might be related to variations in the Earth's orbit of the kind identified by Croll and Milankovitch [95]. Mid-Cretaceous sea surface temperatures in Israel, then located at about 10°N, were between 29 °C and 31 °C, but may have dropped to 21 °C in the late Campanian (84–71 Ma ago). Equatorial bottom waters were around 10 °C cooler than surface waters, but in restricted basins like the South Atlantic they were as warm as 22 °C [95]. The high latitude ocean was cooler, with Antarctic shelf waters ranging from 9 to 16 °C [108].

The tropical ocean was significantly warmer than it is today during parts of the Cretaceous, notably during the Turonian (94–89 Ma ago), when sea surface temperatures in the equatorial Atlantic reached 33–42 °C [105]. But, as Frakes and his team noted, conditions were not permanently warm [95]. There were periodic coolings of 1–3 °C [109]. In addition, the fluctuations in sea level identified from seismic records, for example by Pete Vail and colleagues from EPRCo [48, 110, 111], strongly suggest the fluctuating presence of at least small ice caps in the polar regions during Jurassic, Cretaceous, and Early Eocene times, despite their warm greenhouse climates [112]. In later chapters we will address the question of what caused the cooling of the Late Cretaceous – changes in oceanic heat transport or declining concentrations of atmospheric $CO_2$?

Cretaceous chalk is widespread across Europe, and similar calcareous deposits are common in other parts of the world, for example across the central United States in a belt that extends from Texas to Alberta, where there was a wide shallow seaway. The explanation is simple. The creation of multiple new mid-ocean ridges between the separating fragments of Pangaea pushed sea level up, flooding the margins of the moving continental fragments with shallow water in which carbonates could readily precipitate or coccolithophorids could flourish under the warm conditions. The latest data from Bil Haq in 2013 suggests that *'average sea levels throughout the Cretaceous remained higher than the present day mean sea level (75–250 m above PDMSL [present day mean sea level]). Sea level reached a trough in mid Valanginian [c.133.9–140.2 Ma ago] (~ 75 m above PDMSL), followed by two high points, the first in early Barremian [c.125–130 Ma ago] (~ 160–170 m above PDMSL) and the second, the highest peak of the Cretaceous, in earliest Turonian [c.88.6–93.6 Ma ago] (~ 240–250 m above PDMSL). The curve also displays two ~ 20 Myr-long periods of relatively high and stable sea levels (Aptian through early Albian [c.100–125 Ma ago] and Coniacian through Campanian [c.71–88 Ma ago]). The short-term curve identifies 57 third-order eustatic events in the Cretaceous, most have been documented in several basins, while a smaller number are included provisionally as eustatic, awaiting confirmation. The amplitude of sea-level falls varies from a minimum of ~ 20 m to a maximum of just over 100 m and the duration varies between 0.5 and 3 [Ma]. The causes for these relatively rapid, and at times large amplitude, sea-level falls in the Cretaceous remain unresolved, although based mainly on oxygen-isotopic data, the presence of transient ice cover on Antarctica as the driver remains in vogue as an explanation. This idea has, however, suffered a recent setback following the discovery of pristine foraminiferal tests in the Turonian of Tanzania whose oxygen- isotopic values show little variation, implying absence of glacioeustasy at least in the Turonian [88.6–93.6 Ma ago]. The prevalence of 4th-order (~ 400 [Ka]) cyclicity through most of the Cretaceous (and elsewhere in the Paleozoic and Cenozoic) strongly implies that the periodicity on this time scale, presumably driven by long-term orbital eccentricity, may be a fundamental feature of depositional sequences throughout the Phanerozoic' [113].*

More evidence has emerged from a study of the remains of the dinoflagellate cyst *Impletosphaeridium clavus*, peaks in whose abundance occur in the muds of the Maastrichtian (latest Cretaceous) and Danian (earliest Palaeocene) of Seymour Island, Antarctica [114]. Such peaks suggest the existence of blooms of this species, which typically occur at high latitudes in association with the melting of winter sea ice. If winter sea ice was forming along the eastern side of the Antarctic Peninsula at 65°S, 70–60 Ma ago, then conditions may well have been suitable for glaciers to form on the highlands of the continental interior. And if they were

present, and periodically melted under the influence of orbital variations in insolation, then we have a mechanism for the periodic changing of sea level.

Planet Earth was quite different in warm Cretaceous times from the way it is now. It was a world largely without polar ice. The difference in temperature between pole and equator was about 20 °C, while today it is about 33 °C. The weaker Equator–pole thermal gradient would have weakened westerly winds like the jet stream. Polar climates would have become much more seasonal than they now are. As Bill Hay points out, '*If there were no perennial ice in the Polar Regions, the temperatures there could alternate between cold in winter and warm in summer, and that means that the polar atmospheric pressure systems would change between summer and winter*' [105]. Hay supposed that these changes meant that the Hadley cell that governs the positions of the westerlies (Figure 6.17) could have expanded poleward, and that the westerly and easterly winds at high latitudes would have become seasonal and disorganized. This would have had a knock-on effect on the circulation of the surface ocean, which is driven by the winds. While the Trade Winds and east–west flowing ocean currents beneath them would still have existed in the tropics, Hay suspected that at higher latitudes there would have been '*a chaotic pattern of giant eddies generated by storms*' [105]. Without the steady westerly winds of the middle latitudes, the vertical structure of the ocean that we are familiar with would have broken down: '*no great surface gyres, no subtropical and polar frontal systems, no clear separation between surface and deep waters, no "Great Conveyor"*' [105]. Hay went on to suppose that '*upwelling would have*

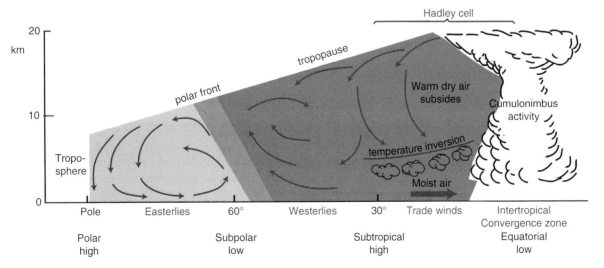

**Figure 6.17** Schematic view of the latitudinal atmospheric circulation below 20 km altitude, from the equator to the pole, showing the Inter-Tropical Convergence Zone near the equator, where strong upward motions take place, tropical clouds form and intense thunderstorm activity occurs, the return flow along the tropopause and the slow air subsidence in the subtropics. The polar front and the polar circulation are also shown. The Hadley Cell extends from the Equator to about latitude 30°; the Ferrel Cell extends from latitudes 30–60°, and the Polar Cell extends north of the Polar Front. *Source:* From figure 2.46a in Ref. [115].

*depended on the development of cyclonic eddies, which pump water upward'.* That situation would have persisted until the cooling towards the modern ocean that took place at the end of the Eocene, of which more in Chapter 7. Hay's vision begs the question: by how much might the jet streams have moved position? Crowley and North used numerical models to suggest that such displacements were probably minor [11].

Surprising though it may seem, Hay's apocalyptic vision of ocean and climate change did not occur to Robinson, or Parrish, or Scotese and me when we constructed our maps of past climate, since we basically used annual pressure models derived from the modern climate to derive our palaeoclimatic maps. There was a danger in that approach because the Cretaceous world was so different from today's. As ever, concepts evolve with time. Proof of the pudding would lie not in mental concepts, which might be based on unsound premises, but in the development of numerical models of the atmosphere and ocean based on sound physical principles.

## 6.8 Numerical Models Make Their Appearance

By the very early 1980s climatologists were using a brand new tool – the numerical general circulation model (or GCM) – to simulate the behaviour of the present climate system. These models can also be used to explore the relationships between climate and geology in the past [116]. Some global warming contrarians like to portray these numerical models as computer games. Games are designed for you to pit your wits against a series of known obstacles to win. GCMs are different. Climate scientists use them to find out how the climate system works, and to discover what hidden properties of the system emerge when models are run for long periods, such as whether the climate might tip from one stable state to another as warming continues.

I became familiar with the operation of numerical models when I joined the UK's Institute of Oceanographic Sciences Deacon Laboratory as its director in 1988, and found myself among other things responsible for oversight of a Southern Ocean modelling project FRAM (the Fine Resolution Antarctic Model). This was the first high-resolution, ocean-scale numerical model capable of simulating typical oceanic eddies no more than about 100 km across. There was no way at the time that we could gain a comprehensive understanding of how the Southern Ocean worked from just the 100 years' worth of scattered ocean data points we possessed. In that remote region they were far too sparsely distributed in time and space. But, given those data points, and certain other starting conditions, FRAM

could apply natural laws, such as the First Law of Thermodynamics, and Newton's Three Laws of Motion, at closely spaced points on a 27 km grid, and at several levels down through the ocean, to show precisely how the Southern Ocean worked at all levels through time. It was as if the static school atlas of ocean currents had suddenly come alive. This was really exciting. We could see in real time the sinuous motions of currents and the spinning of eddies [117]. Comparing the output to sea surface temperatures as seen from an ocean-observing satellite [118] showed that the model results were very close to the real world. FRAM really did show how the Southern Ocean worked. It was a breakthrough.

Such models, of the ocean, or the atmosphere, or of both combined, provide us with a unique and verifiable means of connecting widely scattered data points, and of understanding why the data are distributed the way they are. More than that, they tell us where to go to test ideas about how the ocean circulates or how the climate system works. They are vital aids. For instance, trying to sample every square metre of the ocean so as to understand its circulation is simply impossible. It can only be done from expensive research ships or through a massive and costly collection of autonomous floats and data buoys, and we have to remember that the ocean covers 72% of the surface of the planet! Satellites alone will not do the trick, because they cannot see below the ocean's surface.

Michel Crucifix of the Institut d'Astronomie et de Géophysique G. Lemaître, of the Université Catholique de Louvain, explains *'The aim of climate modeling is to understand past changes in climate that are currently unexplained and to be able to predict successfully the future evolution of climate'* [119]. It is a myth that the system is too chaotic for us to do that. As John Barrow explains: *'The standard folklore about chaotic systems is that they are unpredictable … [but in fact] classical … chaotic systems are not in any sense intrinsically random or unpredictable … An important feature of chaotic systems is that, although they become unpredictable when you try to determine the future from a particular uncertain starting value, there may be a particular stable statistical spread of outcomes after a long time, regardless of how you started out. The most important thing to appreciate about these stable statistical distributions of events is that they often have very stable and predictable average behaviours'* [120]. For an example, look at Boyle's Law, $PV/T$ = a constant, where P (= pressure), V (= volume), and T (= temperature). These are the average properties of a confined gas comprising a number of molecules whose chaotic interactions are unpredictable. *'The lesson of this simple example is that chaotic systems can have stable, predictable, long-term, average behaviours'* [120].

Crucifix explained how we use that understanding in modelling climate: '*The ocean–atmosphere–cryosphere–biosphere system is a complex system in the sense that it is made of different components that may interact with each other on a very wide range of time-scales ... These interactions are generally nonlinear, that is the response is not proportionate to the amplitude of the excitation. A physical system with at least three components interacting nonlinearly with each other may be chaotic .... In other words, its evolution cannot be predicted accurately beyond a certain time horizon because any error on the initial conditions grows exponentially with time. The atmosphere is chaotic. This is the reason we cannot forecast weather much beyond about 6 days. Yet, we can predict global warming. Indeed, conservation of energy, heat, and momentum makes it possible to predict the general evolution of a chaotic system in statistical terms. This statistical description of weather is nothing but the definition of climate*' [119]. Actually, strictly speaking '*the climate system has no steady state because oceanic and atmospheric currents vary constantly. This is why theoreticians prefer to use the notion of "attractor." The attractor is, loosely speaking, a closed trajectory that the climate system follows more or less closely*' [119] (Figure 6.18).

Crucifix reminds us that there is also the possibility to consider that Earth's climate can exist in any one of a number of states between which it can flip. That has led to the concept of 'tipping', where too large a perturbation may kick a currently stable state into a new stable state [121]. Figure 6.19a shows an example in which a slow change in environmental conditions can induce a rapid transition towards a new state once a 'threshold' (a tipping point) is crossed, from which it will take considerable effort to return to the original state. '*However, this is only one of a very rich set of possibilities*' [121]. For example '*A glacial maximum is inherently unstable, and $CO_2$ outgassing ejects the system towards an interglacial. A background glaciation process then brings the system back to a glacial state, from where it is ejected again. In this model, the glacial-interglacial process no longer requires an externally forced tipping: it is the manifestation of a self-sustained oscillation, also known as a limit cycle ... Instead of tipping, this is ping pong*' (Figure 6.19b) [121]. As Crucifix explains, a problem with these possibilities is that they are 'deterministic' (with trajectories determined by original conditions and forcing, but the climate system varies on all timescales. Instead we have to take a stochastic approach that describes variables in terms of probability distributions. Conceptual models like those in Figure 6.19 can be extremely useful in enhancing our understanding of underlying processes. But to understand complex environmental systems we also have to rely '*on complex numerical models, which allow us to infer emergent constraints on the basis of physical laws of ice, atmospheric and oceanic motion, and such models tend to include hundreds of parameters*' [121]. We need to use models of different levels of complexity including models of intermediate complexity (EMICs) and GCMs.

For many geologists, this is a new world, brought to us courtesy of the massive increase in computing power since the early 1980s. As John Barrow reminds us [102] '*The advent of small, inexpensive, powerful computers with good interactive graphics has enabled large, complex, and disordered situations to be studied observationally – by looking at a computer monitor. Experimental mathematics is a new tool. A computer can be programmed to simulate the evolution of complicated systems, and their long-term behaviour*'

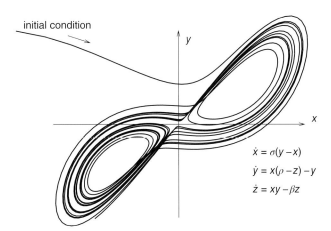

$$\dot{x} = \sigma(y - x)$$
$$\dot{y} = x(\rho - z) - y$$
$$\dot{z} = xy - \beta z$$

**Figure 6.18** A classic example of a chaotic system with a strange attractor. The particle leaves its initial condition and is attracted to a certain region of the space called the attractor. The 'strange attractor' featured here displays chaotic behaviour. Its shape can be statistically defined after a long integration time (i.e. it is ultimately predictable, like climate), but individual trajectories of the particle (like weather) cannot be predicted beyond a certain limited time horizon. *Source:* From figure 4.1 in Ref. [119].

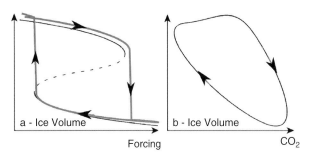

**Figure 6.19** Two examples of conceptual models for paleoclimate dynamics. (a) The Budyko–Sellers model envisions two stable states for a range of forcing. Large forcing deviations from the resting state may precipitate a transition; (b) In a limit cycle, glacial interglacial stages succeed each other as a result of internal dynamics, without the need for forcing. In this case, insolation forcing is but a pacemaker that controls the timing of transitions. *Source:* From figure 1 in Ref. [121].

*observed, studied, modified and replayed. By these means, the study of chaos and complexity has become a multidisciplinary subculture within science. The study of the traditional, exactly soluble problems of science has been augmented by a growing appreciation of the vast complexity expected in situations where many competing influences are at work'* [120]; for example in the climate system. Mathematics is essential to understanding the complexities of the climate system.

Mathematical models of the climate system are not reality. Nor are they perfect. But they are useful. Uncertainties arise for several reasons. First, they encompass the interaction between components of very different time-scales, ranging from cloud formation and precipitation, on the scale of hours, to long-lived ice sheets. Second, to be addressed efficiently, the operations of the different elements of the climate system must be simplified. Third, the horizontal and vertical spacing of the points on the global grid, dictated by the capacity of the computer, restricts the resolution of the outputs. Early GCMs were also limited

because computer power was too small to simulate the circulation of both the ocean and the atmosphere.

Figure 6.20 is an example of the inputs to and interactions between the components of an Earth System Model of Intermediate Complexity (EMIC) – the Climate and Biosphere model (CLIMBER). CLIMBER and its fellow EMICs are powerful tools for investigating Earth System dynamics, emphasising the interaction among all components of the Earth System. CLIMBER consists of fully coupled components that simulate atmospheric dynamics, vegetation cover and structure dynamics, terrestrial carbon cycle, inland ice dynamics and ocean biogeochemistry and dynamics. There is a caveat: because of the computational demands of the model its individual components have to be described in a less comprehensive manner than, for example, in models of the general circulation of the atmosphere and the ocean. Nevertheless, the model has successfully simulated the long-term natural changes of the past 9000 years, capturing, for example, the greening of the Sahara in the early to mid-Holocene, and simulating the

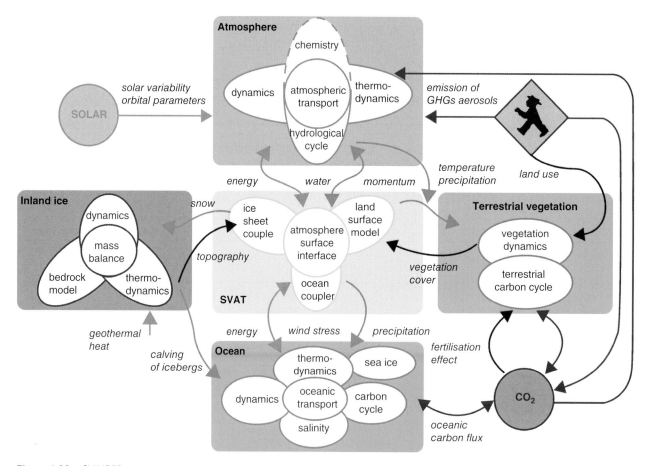

**Figure 6.20** CLIMBER, a comprehensive Earth System Model of Intermediate Complexity for investigating Earth system interactions. These models can be used for example to simulate Earth System behaviour where change is highly non-linear (e.g. the pattern of Heinrich Events in the North Atlantic). *Source:* From figure 6.20 in Ref. [115].

changes in carbon fluxes between dynamically interacting atmosphere, vegetation, and ocean [115].

Given these limitations, model outputs should be seen as aids to understanding how past climate systems worked [116]. They are tools grounded in the application of fundamental laws. We can verify their outputs by comparing them to climatic indicators from the sedimentary record. Where there are discrepancies between record and output, the challenge is to work out which is wrong. For example, a disagreement between model output and oxygen isotope data may come about because the isotope data come from species that live deep in the water column, while the model represents sea surface temperatures.

GCMs have an advantage over the conceptual models of Robinson or Parrish, in that numerical modellers can tweak the controlling parameters of their model to see what effect each has. From that, they can evaluate the sensitivity of the climate to different controls, providing insights into the operation of the climate system through time. Without coupled ocean–atmosphere GCMs, it would be impossible for geologists to relate knowledge about meteorological and oceanographic processes to geological information about past climates. In the remainder of this chapter we will ignore the effects on climate of $CO_2$, which we will explore in detail later.

Apparently the first to use GCMs in analysing Cretaceous climate was Eric James Barron (1951–) (Box 6.9) a colleague and former student of palaeoclimatologist William (Bill) Hay (1934–) (Box 6.10).

As is obvious from the titles of his papers in 1981 (*Ice-free Cretaceous Climate? Results from Model Simulations* [122]), and 1983 (*A Warm Equable Cretaceous: the Nature of the Problem* [123]), Barron was fascinated with the question of why the Cretaceous was warm and ice-free. Could it have to do with the distribution of land-masses, with the opening of oceanic gateways allowing warm water to penetrate poleward or with the growth of mountain chains as continents continued to split apart? In 1984, together with Warren Washington, Barron used an atmospheric GCM, combined with a simple energy balance model, to suggest that Cretaceous geography alone could have warmed global surface temperature by 4.8 °C [124]. The actual temperatures were warmer, so geography was not the sole forcing factor. The results might have been slightly different had he coupled his atmospheric model to an ocean model, but that required more computing power than was available at the time.

Investigating the effect of the opening mid-Cretaceous seaways, in 1986 Barron used a slightly more advanced GCM to show that a well-developed zonal tropical rain belt, located 10° south of the Equator in January and 20° north of it in July, developed when the ITCZ was located over the zonal Tethyan Ocean. '*Clearly*' he said, '*the zonal subtropical ocean has influenced the general circulation pattern through the importance of latent heating in driving the circulation*' [116]. The creation of seaways, like the zonal equatorial ocean evident in Figures 5.12b and 6.13, provided a source for rain that then fell on Pangaea's formerly dry interiors. His model showed that there would have been a mid-latitude rain belt at around 40° north and south of the equator, and that it would have extended to 50° in both hemispheres in their respective summers. Palaeoclimatic evidence in the form of widespread laterites, bauxite, and kaolin-rich clays confirms an increase in precipitation in mid-latitudes at the time, as do the widespread coals of that age in North America [116].

Evaluating the effects of topography on atmospheric circulation and precipitation patterns, Barron found that '*The greater the starting temperature and relative humidity, the more topography will influence temperature and evaporation-precipitation patterns ... If the topographic expression is high enough, evaporites can be formed in the lee of a mountain range at* any *latitude. During warmer climatic periods, the role of topography may be accentuated. The greatest potential for extensive evaporites is in the lee of a mountain range in the tropics where temperatures and relative humidities are high*' [116].

Like Parrish and myself, Barron tried to see if his numerical models could predict the likely location of upwelling conditions, hence of organic rich rocks (the potential source rocks for petroleum). Indeed they could. Even so, Barron accepted that his GCM had several limitations for that purpose, including crude coastlines, lack of seasons, poor representation of ocean circulation, and lack of bathymetry. Of course, the same limitations applied to the outputs of the Parrish model [79] and the Scotese and Summerhayes model [99].

Barron was also keen to use a GCM to test Pamela Robinson's notion (Figure 6.10) that Köppen's arid zones would increase in latitude from west to east, just as happens today in Asia (Figures 6.3 and 6.4) [125]. Carrying out a number of experiments, together with Bill Hay, he found that mountainous areas or high plateaus distort zonal climate boundaries. In none of their experiments did zonal climate boundaries slope steeply polewards from east to west '*suggesting that the "standard climatic pattern" attributed to Köppen may be an artefact of the topography and disposition of contemporary continents, and more related to the Tibetan-Himalayan and North American uplifts and to the configuration of South America than to the general atmospheric circulation*' [125]. Without the Tibetan Plateau, '*Climates may have been more zonal in the past than they are today*' [125]. Figures 6.3, 6.4 and 6.10 were thus unlikely to be completely practical guides to the past distributions of atmospheric pressure, wind, and rain. Caveat emptor.

This tends to confirm what we know from the distribution of palaeoclimatic indicators (Figures 5.3 and 6.6) [30]. What of Bill Hay's suggestion, then, that the Hadley Cell (Figure 6.17) could have expanded its range northwards when the equator-to-pole thermal gradient was much lower in the warm climates of the mid-Cretaceous [105]? To test that idea, H. Hasegawa of the University of Tokyo, and colleagues, mapped the distribution of palaeo-desert deposits and palaeo-wind directions for that period. Where air descending from high altitude in the northern limb of the Hadley Cell reaches the surface, it diverges to the north, forming the mid-latitude westerlies, and to the south, forming the easterly Trade Winds. Hasegawa's maps showed that the zone of divergence shifted polewards during the Early and Late Cretaceous, suggesting poleward expansion of the Hadley Cell [126], as Hay suggested. But it shifted equatorwards during the hot, Mid Cretaceous 'super-greenhouse' period (of which more later), suggesting shrinkage of the Hadley Cell at that time, contrary to Hay's expectation.

Fred Ziegler also realized the importance of including such factors as topography into a GCM. Along with co-author John Kutzbach, he presented a paper on this topic at a Royal Society discussion meeting in London in 1993 [127]. Kutzbach hails from the Center for Climatic Research at the Gaylord Nelson Institute for Environmental Studies in Madison, Wisconsin. His contributions were recognized by election to the US National Academy of Sciences in 2006, the award of the American Geophysical Union's Revelle Medal in 2006, and of the European Geophysical Society's Milankovitch Medal in 2001.

Using a GCM to model the climate of the Late Permian, Kutzbach, and Ziegler found that adding mountains, plateaus, inland seas, and large lakes to palaeogeographic base maps produced outcomes different from model simulations lacking such features [127]. Mountains and plateaus became focal points for enhanced precipitation, intensifying monsoonal circulation. Inland seas and lakes damped the seasonal range of mid-continental temperature. Agreement between Kutzbach's model results and Ziegler's palaeoclimate data was much better for the model with lakes and inland seas than for the one without. Failure to include lakes and inland seas may explain the extreme seasonality of continental interiors suggested by models lacking those constraints.

Anticipating that topography was likely to have modified climate in the past, much as it does today, a team from Chevron experimented with different topographies in their palaeoclimate model runs in 1992 [128]. Their most convincing results emerged from the model containing mountain ranges with variable heights up to 3 km. Inputting more simplified (lower) topography produced more simplified global circulation patterns, much as Ziegler and Kutzbach had found. Evidently, palaeotopography provides an important boundary condition in applying numerical models.

Nevertheless, uncertainties in modelling remain, not least, for example, in establishing past palaeogeography and palaeotopography accurately. Equally, GCMs suffer from having a high spatial scale (e.g. a 300 km grid), which makes them incapable of resolving regional climatic patterns accurately. Although GCMs have evolved to the point of being able to consider feedbacks from both land ice and vegetation, making them more like Earth System Models and allowing more realistic coupling between the physical climate system and the biosphere, assumptions based on present-day vegetation are not likely to have been applicable to times prior to the evolution of angiosperms (130 Ma ago) and the expansion of grasses (34 Ma ago) [33].

That finding underscores Barron's observation that we should see models as 'thought experiments'. As palaeoclimate modeller Paul Valdes of the University of Bristol points out, even the most comprehensive numerical model will not include every detail of the ocean and atmosphere that may be important for the climate, especially at regional and local scales [129]. Then, too, the models may not include all likely forcing factors (such as atmospheric composition) appropriately. Finally, because the geological record itself is more gap than record, it may be difficult to find the field data required to test (or, in model speak, to 'verify') the results supplied.

The shortcomings of palaeoclimate modelling have lessened with time as the modelers' skills have improved. The topic was reviewed in 2011 by a committee of the US National Research Council, led by Isabel Montañez of the University of California at Davis and Richard Norris of Scripps Institution of Oceanography [112]. Remarking on the abundant evidence for anomalous polar warmth during past greenhouse periods (e.g. the Middle Cretaceous to Eocene, and the Pliocene), they pointed out that '*To date, climate models have not been able to simulate this warmth without invoking greenhouse gas concentrations that are notably higher than proxy estimates ...[prompting] modeling efforts to explain high latitude warmth through vegetation ..., clouds..., intensified heat transport by the ocean ...., and increased tropical cyclone activity ... The ability to successfully model a reduced latitudinal temperature gradient state, including anomalous polar warmth, presents a first-order check on the efficacy of climate models as the basis for predicting future greenhouse conditions*' [112]. Mismatches between model outputs, modern observations, and palaeoclimate records may suggest important deficiencies in scientific knowledge of climate and the construction of climate models [112]. Past analyses of warm worlds might help to resolve these disparities.

The Montañez–Norris committee of the National Research Council of the US National Academy of Sciences was concerned about tropical climate stability and the answer to the question: were the tropics warmer in the past than now, or do they have some natural 'thermostat' that keeps them to a limit of about 30–32 °C, like the tropical Pacific temperatures of today [112]? They found that tropical ocean temperatures during past greenhouse times were much warmer than modern tropical maxima – possibly as high as 42 °C, and also that the temperatures of tropical continental interiors were anomalously high (30–34 °C). A tropical thermostat seemed unlikely under the circumstances. Clearly, we need to know more about the tropical climates of these past warm worlds in order not only to better understand Earth's climate system, but also what to expect of a warmer world in the future.

Climate models have advanced a good deal since Barron's day, as has the power of computers. Nowadays climate models incorporate a very wide range of parameters that may have some effect on the climate system, including soil dynamics, vegetation dynamics, ocean biogeochemistry, ice-sheet dynamics and river hydrology [119]. Although the coarse scale of global grid points in GCMs (300 km) prevents them resolving regional details, the outputs of a global model can be used to drive regional dynamical models that resolve much smaller length scales (30–50 km), enabling details of the climate of specific regions to be obtained.

To assist in understanding past climate change, and also to test the ability of climate models to match climate data, various climate models have been applied to past time slices for which palaeoclimate data were both reliable and abundant. The international modelling community has pooled its efforts through the Palaeoclimate Modeling Intercomparison Project (PMIP), which simulated the climates of the mid-Holocene (6 Ka ago) and the Last Glacial Maximum (21 Ka ago) with different atmospheric GCMs, and organized systematic comparisons between model outputs and palaeoclimate data [130]. PMIP showed '*that climate models correctly reproduce a number of observed features of the mid-Holocene climate*' [119].

There is now an abundant literature on the newly emergent discipline of climate modelling and its application to past climates, which concerns us most here. In effect, it forms a new sub-field of science: theoretical palaeoclimatology. Thomas Crowley and Gerald North, then resident at Texas A & M University, explained in 1991 that, '*For the first time, physicists and atmosphere and ocean scientists are applying quantitative climate models to interpret many fascinating observations uncovered by geologists. In some cases we have a good understanding of the observed changes. In other cases there is a considerable gap between models and data*'

[11]. It is the task of the palaeoclimatologist to resolve such discrepancies. '*Responsible use of climate models can help in the formulation of theories of climate change and in some cases such models can lead the geologist to collect and analyze data in new ways*' [11]. Far from being just 'computer games', palaeoclimate models provide us with valuable physical insights into how past climates may have operated.

Despite that, palaeoclimate models do have one interesting problem, as Paul Valdes reminds us: '*If anything, the models are underestimating change, compared with the geological record. According to evidence from the past, the Earth's climate is sensitive to small changes, whereas the climate models seem to require a much bigger disturbance to produce abrupt change*' [131]. This means that: '*Simulations of the coming century with the current generation of complex models may be giving us a false sense of security*' [131]. Caveat emptor.

## 6.9 From Wegener to Barron

Why did it take so long to reach agreement on how the climate had changed with time? Geologists were not prepared to agree that Köppen and Wegener might have been right until the palaeomagnetists entered the arena in the late 1950s and early 1960s, with geophysicists like Briden and Irving confirming that the climate sensitive deposits of the past did indeed fall into climate zones much like those of today (Figure 6.6). The advent of the plate tectonics and sea-floor spreading paradigm in the late 1960s radically changed peoples' view of what was possible – and, indeed, probable. That helps to explain why it took almost 50 years until the landmark NATO meeting in 1972, before Robinson broke new ground by using conceptual palaeoclimate models to show how continental displacement affected climate-sensitive deposits on land, basically confirming the validity of the Köppen and Wegener model and applying it to 'modern' plate tectonic reconstructions for the Permian and Triassic. Her work implied that in order for geologists to understand past climates, they first had to learn and apply the principles of meteorology. As we shall see later, they also had to learn and apply the principles of oceanography.

New techniques, like oxygen isotope palaeothermometry, added another new dimension to palaeoclimate studies from the 1950s on. Our understanding of orbital controls on climate grew with the work of Milankovitch. We began amassing palaeoclimatic data from land, with Ziegler's 'Paleogeographic Atlas Project', from the mid-1970s on, and from the ocean with the advent of the DSDP and its successors, in late 1968. By the mid-1980s, 60 years after Köppen and Wegener, the geography of past continental fragments and the original geographic position of climatically sensitive sediments could be established with a fair degree of certainty. The advent of numerical modelling of the climate system introduced a thoroughly modern understanding of how the climate system works, further improving our appreciation of how past climates evolved. Nevertheless, there is more to learn about how those deposits formed, as we shall see.

Much of the work reviewed in this chapter dates from before the mid-1980s, when there was little or no discussion in the geological community of the possible role of $CO_2$ as a modifier of Earth's climate through time. Influenced by Lyell's thinking, most geologists simply attributed the glaciation of the Permo-Carboniferous and the warmth and high sea levels of the Mid Cretaceous to the changing positions of the continents. More land over the pole led to cooling; more land over the Equator led to warming. To many of them, the results of increasingly detailed palaeoclimatic studies summarized in this chapter confirmed Lyell's notion that climatic zones much like those of today must have prevailed through time. Much effort was devoted to considering how those climatic zones might have been modified by geographic changes like the development of seaways and mountain belts, with their attendant effects on winds and ocean currents. Little if any thought was given to the possible role of the changing composition of the atmosphere. The questions of why was the Cretaceous so warm remained unanswered.

Having examined the evidence for climate change from the Carboniferous to the Cretaceous, it is now time to turn our attention to the Cenozoic – the past 65 Ma. During much of this time the Earth cooled towards the Ice Age of the Pleistocene. In later chapters we review that change.

## References

**1** Lisitzin, A.P. (1972). Sedimentation in the world ocean. *Society of Economic Paleontologists and Mineralogists Special Publication* 17: 218.

**2** Kennett, J.P. (1982). *Marine Geology*, 813 pp. New Jersey: Prentice-Hall.

**3** Parrish, J.T. and Barron, E.J. (1986). Paleoclimates and economic geology. *Society of Economic Paleontologists and Mineralogists Short Course* 18: 162.

**4** Nairn, A.E.M. (1961). The scope of palaeoclimatology. In: *Descriptive Palaeoclimatology* (ed. A.E.M. Nairn), 1–7. New York: Interscience Publishers.

**5** Fairbridge, R.W. (1964) The importance of limestone and its Ca/Mg content to palaeoclimatology, in *Problems in Palaeoclimatology* (ed A.E.M. Nairn), Proceedings of the NATO Palaeoclimates Conference, University of Newcastle-upon-Tyne, January 1963, Intersci. Publishers, London, 431–477.

6 Flint, R.F. (1961). Geological evidence of cold climate. In: *Descriptive Palaeoclimatology* (ed. A.E.M. Nairn), 140–155. New York: Interscience Publishers.

7 Green, R. (1961). Palaeoclimatic significance of evaporites. In: *Descriptive Palaeoclimatology* (ed. A.E.M. Nairn), 61–88. New York: Interscience Publishers.

8 Opdyke, N.D. (1961). The palaeoclimatological significance of desert sandstone. In: *Descriptive Palaeoclimatology* (ed. A.E.M. Nairn), 45–60. New York: Interscience Publishers.

9 Runcorn, S.K. (1964) Paleowind directions and palaeomagnetic latitudes, in *Problems in Palaeoclimatology* (ed A.E.M. Nairn), Proceedings of the NATO Palaeoclimates Conference, University of Newcastle-upon-Tyne, January 1963, Intersci. Publishers, London, 409–419.

10 Nairn, A.E.M. (ed.) (1961). *Descriptive Palaeoclimatology*. London, New York: Interscience Publishers.

11 Crowley, T.J. and North, G.R. (1991). *Paleoclimatology*, 339 pp, Oxford Monographs on Geology and Geophysics 18. Oxford University Press.

12 Parrish, J.T. (1998). *Interpreting Pre-Quaternary Climate from the Geologic Record*, 338 pp. New York: Columbia University Press.

13 Vaughan, A.P.M. (2007). Climate and geology – a Phanerozoic perspective. In: *Deep-Time Perspectives on Climate Change: Marrying the Signal from Computer Models and Biological Proxies* (eds. M. Williams, A.M. Haywood, F.J. Gregory and D.N. Schmidt), 5–59. London: Micropalaeontological Society Special Publication, The Geological Society.

14 Markwick, P.J. (2007). The palaeogeographic and palaeoclimatic significance of climate proxies for data-model comparisons. In: *Deep-Time Perspectives on Climate Change: Marrying the Signal from Computer Models and Biological Proxies* (eds. M. Williams, A.M. Haywood, F.J. Gregory and D.N. Schmidt), 251–312. London: Micropalaeontological Society Special Publication, The Geological Society.

15 Wegener, A. (1920). *Die Entstehung der Kontinente und Ozeane*, Die Wissenschaft, Band 66, 2e, 135 pp. Braunschweig: Vieweg und Sohn.

16 Köppen, W. and Wegener, A. (1924). *Die Klimate der geologischen Vorzeit*, 1–255. Berlin: Borntraeger. Translated as the *Climates of the Geological Past* (eds. Thiede, J., Lochte, K., and Dummermuth, A.(2015), Borntraeger, Stuttgart. Note that in Google one sees this commonly referred to as published in English as *Climates of the Geological Past* by Van Nostrad, London, 1863, which is an erroneous reference to a book with a similar title *Climates of the Past: An Introduction to Paleoclimatology* by M Schwarzbach (Van Nostrad, London, 1963).

17 Wegener, A. (1922). published in English in 1924 as *The Origin of Continents and Oceans*, 212 pp. London: Methuen and Co.

18 Thiede, J. (2018). Wladimir Köppen's scholarly legacy: a rambler between close and distant worlds and times. Preface to. In: *Wladimir Köppen – Scholar for Life, and Wladimir Köppen -Ein Gelehrtenleben für die Meteologie* (ed. J. Thiede) for the English version and Else Wegener-Köppen for the German version, 13–23. Stuttgart: Borntraeger Science Publishers.

19 Lewis, J.M. (1996). Winds over the world sea: Maury and Köppen. *Bulletin of the American Meteorological Society* 77 (5): 935–952.

20 Complete Dictionary of Scientific Biography (2008). *Köppen*. Wladimir Peter www.encyclopedia.com.

21 Köppen, W.P. (1923). *Klimate der Erde*. Berlin und Leipzig: Walter de Gruyter and Co.; revised second edition 1931, *Grundriss der Klimakunde*, Walter de Gruyter and Co, Berlin und Leipzig.

22 Peel, M.C., Finlayson, B.L., and McMahon, T.A. (2007). Updated world map of the Koppen-Geiger climate classification. *Hydrology and Earth System Sciences* 11: 1633–1644. (with free download of the current map of climate zones).

23 Köppen, W. (1918). Klassifikation der Klimate nach Temperatur, Niederschlag und Jahreslauf. *Petermann's Geographische Mitteilungen* 64 (9.12): 243–248.

24 Köppen, W. (1931). *Grundriss der Klimakunde*. Berlin and Leipzig: Walter de Gruyter & Co.

25 Köppen, W. (1921) Polwanderungen, verschiebungen der kontinente und klimageschichte. *Dr. A. Petermanns Mitteilungen aus Justus Perthes' Geographischer Anstalt* **67** (1–8), 57–63 (Polar wandering, movement of the continents and climate history).

26 Demhardt, I.J. (2006). Alfred Wegener's hypothesis on continental drift and its discussion in Petermanns Geographische Mitteilungen. *Polarforschung* 75 (1): 29–35.

27 Du Toit, A.L. (1937). *Our Wandering Continents: An Hypothesis of Continental Drifting*, 366 pp. London: Oliver and Boyd.

28 King, L.C. (1961). The palaeoclimatology of gondwanaland during the Palaeozoic and Mesozoic eras. In: *Descriptive Palaeoclimatology* (ed. A.E.M. Nairn), 307–331. New York: Interscience Publishers.

29 Nairn, A.E.M., (ed.) (1964) *Problems in Palaeoclimatology*. Proc. NATO Palaeoclimates Conf., University of Newcastle-upon-Tyne, January 1963, Intersci. Publishers, London, 705 pp.

30 Briden, J.C., and Irving, E. (1964) Palaeolatitude spectra of sedimentary palaeoclimatic indicators, in *Problems in Palaeoclimatology* (ed A.E.M. Nairn), Proc. NATO Palaeoclimates Conf., University of Newcastle-upon-Tyne, January 1963, Intersci. Publishers, London, 199–224.

31 Summerhayes, C.P. (1990). Palaeoclimates. *Journal of Geological Society of London* 147: 315–320.

32 Bucher, W.H. (1964) The third confrontation, in *Problems in Palaeoclimatology* (ed A.E.M. Nairn), Proc. NATO

Palaeoclimates Conf., University of Newcastle-upon-Tyne, January 1963, Intersci. Publishers, London, 3–9.

33  Dorf, E. (1964). The use of fossil plants in palaeoclimatic intepretation, in *Problems in Palaeoclimatology* (ed A.E.M. Nairn), Proc. NATO Palaeoclimates Conf., University of Newcastle-upon-Tyne, January 1963, Intersci. Publishers, London, 13–31.

34  Urey, H.C. (1947). The thermodynamic properties of isotopic substances. *Journal of the Chemical Society* 61: 562–581.

35  Urey, H.C. (1948). Oxygen isotopes in nature and in the laboratory. *Science* 108: 489–496.

36  Urey, H.C., Lowenstam, H.A., Epstein, S., and McKinney, C.R. (1951). Measurement of paleotemperatures and temperatures of the Upper Cretaceous of England, Denmark, and the Southeastern United States. *Bulletin of the Geological Society of America* 62 (4): 399–416.

37  Arnold, J.R., Bigeleisen, J., and Hutchison, C.A. (1995). *Harold Clayton Urey 1893–1981*, Biographical Memoir, 363–411. Washington, DC: National Academies Press.

38  Nier, A.O. (1947). A mass spectrometer for isotope and gas analysis. *Review of Scientific Instruments* 19 (6): 398–411.

39  McKinney, C.R., McCrea, J.M., Epstein, S. et al. (1950). Improvements in mass spectrometers for the measurement of small differences in isotope abundance ratios. *Review of Scientific Instruments* 21 (8): 724–230.

40  Lowenstam, H.A. (1964). Palaeotemperatures of the Permian and Cretaceous Periods, in *Problems in Palaeoclimatology* (ed A.E.M. Nairn),. Proc. NATO Palaeoclimates Conf., University of Newcastle-upon-Tyne, January 1963, Intersci. Publishers, London, London, 227–248.

41  Emiliani, C. (1961). The temperature decrease of surface sea-water in high latitudes and of abyssal-hadal water in open oceanic basins during the past 75 million years. *Deep Sea Research* 8: 144–147.

42  Bradley, W.H. (1929). The varves and climate of the Green River Epoch. *U.S. Geological Survey Professional Paper* 158: 87–110.

43  Wells, A.K. and Kirkaldy, J.F. (1959). *Outline of Historical Geology*, 398 pp. London: Thomas Murby and Co.

44  Krumbein, W.C. and Sloss, L.S. (1958). *Stratigraphy and Sedimentation*, 497 pp. San Francisco: W.H. Freeman and Co.

45  Klein, G.d.V. and Willard, D. (1989). Origin of the Pennsylvanian coal-bearing cyclothems of North America. *Geology* 17 (2): 152–155.

46  Leeder, M.R. and Strudwick, A.E. (1987). Delta-marine interactions: a discussion of sedimentary models for Yoredal-type cyclicity in the Dinantian of Northern England. In: *European Dinantian Environments* (eds. J. Miller, A.E. Adams and V.P. Wright), 115–130. Chichester: Wiley.

47  Wanless, H.R. and Shepard, F.P. (1936). Sea level and climatic changes related to late Paleozoic cycles. *Bulletin of the Geological Society of America* 47: 1177–1206.

48  Vail, P.R. and Mitchum, R.M. (1977). Seismic stratigraphy and global changes of sea level. *American Association of Petroleum Geologists Memoir* 26: 49–212.

49  Tucker, M.E., Gallagher, J., and Leng, M.J. (2009). Are beds in shelf carbonates millennial-scale cycles? An example from the mid-carboniferous of northern England. *Sedimentary Geology* 214: 19–34.

50  Pilgrim, L. (1904). Versuch einer rechnerischen behandlung des eiszeitenproblems. *Jahreschefte fur Vaterlandische Naturkunde in Wurtemberg* 60.

51  Milankovitch, M. (1920). *Theorie Mathématique des Phénoménes Thermiques Produits par la Radiation Solaire*. Paris: Gauthier-Villars.

52  Milankovitch, M. (1941). *Kanon der erdbestrahlung und seine anwendung auf das eiszeitenproblem*, Special Publications **132**, Section of Mathematics and Natural Sciences **33**. Belgrade: Königliche Serbische Akademie.

53  Milankovitch, M. (1969). *Canon of Insolation and the Ice Age Problem*. Jerusalem: Israel Program for Scientific Translations.

54  Milanković, M. (1998). *Canon of Insolation and the Ice-Age Problem*. Belgrade: Zavod za udzbenike i nastavna sredstva.

55  Grubic, A. (2006). The astronomic theory of climatic changes of Milutin Milankovich. *Episodes* 29 (3): 197–203.

56  Penck, A. and Bruckner, E. (1909-11). *Die Alpen im Eiszeitalter (The Alps in the Ice Age)*, vol. 1–3. Leipzig: Tauchnitz.

57  Murphy, J.J. (1869). On the nature and cause of the glacial climate. *Quarterly Journal of the Geological Society of London* 25: 350–356.

58  Laskar, J., Robutel, P., Joutel, F. et al. (2004). A long-term numerical solution for the insolation quantities of the Earth. *Astronomy and Astrophysics* 428: 261–285.

59  Laskar, J., Fienga, A., Gastineau, M., and Manche, H. (2010). A new orbital solution for the long-term motion of the Earth. *Astronomy and Astrophysics* 532 (A 89): 15.

60  Smith, D.G., Bailey, R.J., Burgess, P.M., and Fraser, A.J. (2015). Strata and time: probing the gaps in our understanding. In Smith, D.G., Bailey, R.J., Burgess, P.M., and Fraser, A.J. (eds.), *Strata and Time: Probing the Gaps in our Understanding. Geological Society of London Special Publication* 404: 1–10.

61  Miall, A.D. (2010). *The Geology of Stratigraphic Sequences*, 2e, 522 pp. New York: Springer.

62  Tarling, D.H. and Runcorn, S.K. (eds.) (1973). *Implications of Continental Drift to the Earth Sciences*, vol. 1, 622 pp. London: Academic Press.

63  Milner, A.C. (2004). *Robinson, Pamela Lamplugh (1919–1994)*. Oxford Dictionary of National Biography, Oxford University Press http://www.oxforddnb.com/view/article/63228.

64  Robinson, P.L. (1971). A problem of faunal replacement on Permo-Triassic continents. *Palaeontology* 14 (1): 131–153.

65  Robinson, P.L. (1973). Palaeoclimatology and continental drift. In: *Implications of Continental Drift to the Earth Sciences*, vol. 1 (eds. D.H. Tarling and S.K. Runcorn), 451–476. London: Academic Press.

66  Kraüsel, R. (1961). Palaeobotanical evidence of climate. In: *Descriptive Palaeoclimatology* (ed. A.E.M. Nairn), 227–254. New York: Interscience Publishers.

67  Chaloner, W.G. and Creber, G.T. (1973). Growth rings in fossil woods as evidence of past climates. In: *Implications of Continental Drift to the Earth Sciences*, vol. 1 (eds. D.H. Tarling and S.K. Runcorn), 425–437. London: Academic Press.

68  Chaloner, W.G. and Creber, G.T. (1990). Do fossil plants give a climate signal? *Journal of the Geological Society of London* 147: 343–350.

69  Woodward, F.I. (1987). Stomatal numbers are sensitive to increases in $CO_2$ from pre-industrial levels. *Nature* 327: 617–618.

70  Koch, J.T. and Frank, T.D. (2011). The Pennsylvanian-Permian transition in the low-latitude carbonate record and the onset of major Gondwanan glaciation. *Palaeogeoraphy, Palaeoclimatology, Palaeoecology* 308: 362–372.

71  Cox, C.B. (1974). Vertebrate palaeodistributional patterns and continental drift. *Journal of Biogeography* 1: 75–94.

72  Ziegler, A.M., Scotese, C.R., McKerrow, W.S. et al. (1979). Paleozoic paleogeography. *Annual Reviews of Earth and Planetray Science* 7: 473–503.

73  Scotese, C.R., Boucot, A.J., and McKerrow, W.S. (1999). Gondwanan palaeogeogaphy and palaeoclimatology. *Journal of African Earth Sciences* 28 (1): 99–114.

74  Ziegler, A.M. (1982). *The University of Chicago Paleogeographic Atlas Project: Background – Current Status – Future Plans*. Department of Geophysical Sciences, University of Chicago. Unpub. ms. 19 pp.

75  Parrish, J.T., Bradshaw, M.T., Brakel, A.T. et al. (1996). Palaeoclimatology of Australia during the Pangean interval. *Palaeoclimates* 1: 241–281.

76  Summerhayes, C.P. (2016). Upwelling. In: *Encyclopedia of Marine Geosciences* (eds. J. Harff, M. Meschede, S. Petersen and J. Thiede), 900–912. Dordrecht: Springer.

77  Parrish, J.T. (1982). Upwelling and petroleum source beds, with reference to the Palaeozoic. *Amercan Association of Petroleum Geologists Bulletin* 66: 750–754.

78  Parrish, J.T., Ziegler, A.M., and Humphreville, R.G. (1983). Upwelling in the Paleozoic Era. In: *Coastal Upwelling – Its Sediment Record, Part B, Sedimentary Records of Ancient Coastal Upwelling*, NATO Conference Series IV, Marine Sciences (eds. J. Thiede and E. Suess), 553–578. New York and London: Plenum.

79  Parrish, J.T. and Curtis, R.L. (1982). Atmospheric circulation, upwelling, and organic-rich rocks in the Mesozoic and Cenozoic. *Palaeogeography, Palaeocllimatology, Palaeoecology* 40: 31–66.

80  Parrish, J.T., Ziegler, A.M., and Scotese, C.R. (1982). Rainfall patterns and the distribution of coals and evaporates in the Mesozoic and Cenozoic. *Palaeogeoraphy, Palaeocllimatology, Palaeoecology* 40: 67–101.

81  Rowley, D.B., Raymond, A.R., Parrish, J.T. et al. (1985). Carboniferous paleogeographic, phytogeographic and paleoclimatic reconstructions. *International Journal of Coal Geology* 5: 7–42.

82  Parrish, J.T. (1993). Climate of the supercontinent Pangea. *The Journal of Geology* 101: 215–133.

83  Sellwood, B.W. and Price, G.D. (1994). Sedimentary facies as indicators of Mesozoic palaeoclimate. In: *Palaeoclimates and their Modelling – with Special Reference to the Mesozoic Era* (eds. J.R.L. Allen, B.J. Hoskins, B.W. Sellwood, et al.), 17–25. London: Chapman & Hall [Originally published as *Philosophical Transactions of the Royal Society Series B*, **342**, 1993, 203-343; based on a discussion mtg held on Feb 24-25 1993].

84  Ziegler, A.M., Parrish, J.M., Jiping, Y. et al. (1994). Early Mesozoic phytogeography and climate. In: *Palaeoclimates and their Modelling – with Special Reference to the Mesozoic Era* (eds. J.R.L. Allen, B.J. Hoskins, B.W. Sellwood, et al.), 89–97. London: Chapman & Hall; [Originally published as *Philosophical Transactions of the Royal Society Series B*, **342**, 1993, 203-343; based on a discussion meeting held on Feb 24-25 1993].

85  Sims, M.J. and Ruffell, A.H. (1990). Climatic and biotic change in the late Triassic. *Journal of the Geological Society of London* 147: 321–327.

86  Veevers, J.J. (1989). Middle/late Triassic (230±5 Ma) singularity in the stratigraphic and magmatic history of the Pangaean heat anomaly. *Geology* 17: 784–787.

87  Frakes, L.A. and Crowell, J.C. (1969). Late Paleozoic glaciation, I. South America. *Bulletin of the Geological Society of America* 80: 1007–1042.

88  Frakes, L.A. (1981). Late Paleozoic paleoclimatology. In: *Paleoreconstruction of the Continents, Geodynamics Series 2* (eds. M.W. McElhinny and D.A. Valencio), 39–44. American Geophysical Union.

89  Crowell, J.C. and Frakes, L.A. (1970). Phanerozoic glaciation and the causes of ice ages. *American Journal of Science* 2688: 193–224.

90 Müller, R.D., Sdrolias, M., Gaina, C. et al. (2008). Long-term sea-level fluctuations driven by ocean basin dynamics. *Science* 319: 1357–1362.

91 Pitman, W. (1978). Relationship between eustacy and stratigraphic sequences of passive margins. *Bulletin of the Geological Society of America* 89: 1389–1403.

92 Smith, D.G., Smith, A.G., and Funnell, B.M. (1994). *Atlas of Mesozoic and Cenozoic Coastlines*, 99 pp. Cambridge University Press.

93 Hallam, A. (1985). A review of Mesozoic climates. *Journal of the Geological Society of London* 142: 433–445.

94 Hallam, A. (1994). Jurassic climates as inferred from the sedimentary and fossil record. In: *Palaeoclimates and Their Modelling – with Special Reference to the Mesozoic Era* (eds. J.R.L. Allen, B.J. Hoskins, B.W. Sellwood, et al.), 79–88. London: Chapman & Hall; [Originally published as *Philosophical Transactions of the Royal Society Series B*, **342**, 1993, 203-343; based on a discussion meeting held on Feb 24-25 1993].

95 Frakes, L.A., Francis, J.E., and Syktus, J.I. (1992). *Climate Modes of the Phanerozoic – The History of the Earth's Climate over the Past 600 Million Years*, 274 pp. Cambridge University Press.

96 Littler, K., Robinson, S.A., Brown, P.R. et al. (2011). High sea surface temperatures during the early Cretaceous Epoch. *Nature Geoscience* 4: 169–172.

97 Alley, N.F. and Frakes, L.A. (2003). First known Cretaceous glaciation: Livingston Tillite member of the Cadna-owie formation, South Australia. *Australian Journal of Earth Science* 50: 139–144.

98 Summerhayes, C.P. (1983). Sedimentation of organic matter in upwelling regimes. In: *Coastal Upwelling – Its Sediment Record, Part B, Sedimentary Records of Ancient Coastal Upwelling*, NATO Conference Series IV, Marine Sciences (eds. J. Thiede and E. Suess), 29–72. New York and London: Plenum.

99 Scotese, C.R. and Summerhayes, C.P. (1986). Computer model of paleoclimate predicts coastal upwelling in the Mesozoic and Cenozoic. *Geobyte* 1 (3): 28–44 and 94.

100 Spicer, R.A., Rees, P.M., and Chapman, J.L. (1994). Cretaceous phytogeography and climate signals. In: *Palaeoclimates and their Modelling – with Special Reference to the Mesozoic Era* (eds. J.R.L. Allen, B.J. Hoskins, B.W. Sellwood, et al.), 69–78. London: Chapman & Hall; [Originally published as *Philosophical Transactions of the Royal Society Series B*, **342**, 1993, 203-343; based on a discussion meeting held on Feb 24-25 1993].

101 Spicer, R.A. and Parrish, J.T. (1990). Late Cretaceous-early tertiary palaeoclimates of northern high latitudes: a quantitative view. *Journal of the Geological Society of London* 147: 329–341.

102 Francis, J.E., Ashworth, A., Cantrill, D.J., et al. (2008) 100 million years of Antarctic climate evolution: evidence from fossil plants in *Antarctica: A Keystone in a Changing World* (eds. A.K. Cooper, P.J. Barrett, H. Stagg, B. Storey, E. Stump, and W. Wise), Proc. 10th Int. Sympos. Antarctic Earth Sci., Santa Barbara, Cal., 26 August – 1 September 2007, 19–27 (http://www.nap.edu/catalog/12168.html).

103 Francis, J.E. and Poole, I. (2002). Cretaceous and early tertiary climates of Antarctica: evidence from fossil wood. *Palaeogeography, Palaeoclimatology, Palaeoecology* 182: 47–64.

104 Kennett, J.P. (1982). *Marine Geology*, 813 pp. New Jersey: Prentice-Hall, Englewood Cliffs.

105 Hay, W.W. (2013). *Experimenting on a Small Planet: A Scholarly Entertainment*, 983 pp. New York: Springer.

106 Huber, B.T., Norris, R.D., and MacLeod, K.G. (2002). Deep-sea paleotemperature record of extreme warmth during the Cretaceous. *Geology* 30 (2): 123–126.

107 MacLeod, K.G., Huber, B.T., Berrocoso, A.J., and Wendler, I. (2013). A stable and hot Turonian without glacial $\delta^{18}$O excursions is indicated by exquisitely preserved Tanzanian foraminifera. *Geology* 41 (10): 1083–1086.

108 Cramer, B.S., Miller, K.G., Barrett, P.J., and Wright, J.D. (2011). Late Cretaceous-Neogene trends in deep ocean temperature and continental ice volume: reconciling records of benthic foraminiferal geochemistry ($\delta^{18}$O and Mg/Ca) with sea level history. *Journal of Geophysical Research* 116: C12023.

109 Forster, A., Schouten, S., Baas, M., and Damste, J.S.S. (2007). Mid-Cretaceous (Albian – Santonian) sea surface temperature record of the tropical Atlantic Ocean. *Geology* 35: 919–922.

110 Vail, P.R., Mitchum, R.W., and Thompson, S. (1977). Seismic stratigraphy and global changes of sea level 4. Global cycles of relative changes of sea level. *American Association of Petroleum Geologists Memoir* 26: 82–97.

111 Haq, B.U., Hardenbol, J., and Vail, P.R. (1987). Chronology of fluctuating sea levels since the Triassic. *Science* 235: 1156–1167.

112 Montañez, I.P., Norris, R.D., Algeo, T. et al. (2011). *Understanding Earth's Deep Past: Lessons for Our Climate Future*, 194 pp. Washington, DC: The National Academies Press.

113 Haq, B.U. (2013). Cretaceous Eustasy revisited. *Global and Planetary Change* 113 https://doi.org/10.1016/j.gloplacha.2013.12.007.

114 Bowman, V.C., Francis, J.E., and Riding, J.B. (2013). Late Cretaceous winter sea ice in Antarctica? *Geology* 41: 1227–1230.

115 Steffen, W., Sanderson, A., Tyson, P.D. et al. (2004). *Global Change and the Earth System: A Planet Under*

*Pressure*, 336 pp, Global Change – the IGBP Series. Berlin: Springer.

116 Barron, E.J. (1986). Mathematical climate models: insights into the relationship between climate and economic sedimentary deposits. In: *Paleoclimates and Economic Geology*, *Lecture Notes for Short Course* **18** (eds. J.T. Parrish and E.J. Barron), 31–83. Society of Economic Paleontologists and Mineralogists.

117 Webb, D.J., Killworth, P.D., Coward, A.C., and Thompson, S.R. (1991). *The FRAM Atlas of the Southern Ocean*, 67 pp. Swindon: Natural Environment Research Council.

118 Sparrow, M.D., Heywood, K.J., and Brown, J. (1996). Current structure of the South Indian Ocean. *Journal of Geophysical Research, Oceans* 101 (C3): 6377–6391.

119 Crucifix, M. (2009). Modeling the climate of the Holocene. In: *Natural Climate Variability and Global Warming: A Holocene Perspective* (eds. R.W. Battarbee and H.A. Binney), 98–122. Chichester and Oxford: Wiley Blackwell.

120 Barrow, J.D. (2010). Simple really: from simplicity to complexity – and Back again. In: *Seeing Further: The Story of Science and the Royal Society* (ed. B. Bryson), 360–383. London: Harper Press.

121 Crucifix, M. (2016). Tipping Ice Ages. *PAGES Magazine* 24 (1): 6–7.

122 Barron, E.J., Thompson, S.L., and Schneider, S.H. (1981). Ice-free Cretaceous climate? Results from model simulations. *Science* 212: 501–508.

123 Barron, E.J. (1983). A warm, equable Cretaceous: the nature of the problem. *Earth Science Reviews* 19: 305–338.

124 Barron, E.J. and Washington, W.M. (1984). The role of geographic variables in explaining paleoclimates: results from Cretaceous climate model sensitivity studies. *Journal of Geophysical Research* 89: 1267–1279.

125 Hay, W.W., Barron, E.J., and Washington, S.L. (1990). Results of global atmospheric circulation experiments on an Earth with a meridional pole-to-pole continent. *Journal of the Geological Society of London* 147: 385–392.

126 Hasegawa, H., Tada, R., Jiang, X. et al. (2012). Drastic shrinking of the Hadley circulation during the mid-Cretaceous supergreenhouse. *Climate of the Past* 8: 1323–1337.

127 Kutzbach, J.E. and Ziegler, A.M. (1994). Simulation of late Permian climate and biomes with an atmosphere-ocean model: comparisons with observations. In: *Palaeoclimates and Their Modelling – with Special Reference to the Mesozoic Era* (eds. J.R.L. Allen, B.J. Hoskins, B.W. Sellwood, et al.), 119–132. London: Chapman and Hall [Originally published as *Philosophical Transactions of the Royal Society, Series B*, **342**, 1993, 203-343; based on a discussion mtg held on Feb 24-25 1993].

128 Moore, G.T., Sloan, L.C., Hayashida, D.N., and Umrigar, N.P. (1992). Paleoclimate of the Kimmeridgian/Tithonian (late Jurassic) world: II. Sensitivity tests comparing three different paleotopographic settings. *Palaeogeography, Palaeoclimatology, Palaeoecology* 95: 229–252.

129 Valdes, P.J. (1994). Atmospheric general circulation models of the Jurassic. In: *Palaeoclimates and Their Modelling – with Special Reference to the Mesozoic Era* (eds. J.R.L. Allen, B.J. Hoskins, B.W. Sellwood, et al.), 109–118. London: Chapman and Hall [Originally published as *Philosophical Transactions of the Royal Society Series B*, **342**, 1993, 203-343; based on a discussion mtg held on Feb 24-25 1993].

130 Joussaume, S., and Taylor, K.E. (2007) Status of the Paleoclimate Modeling Intercomparison Project (PMIP). PMIP web site at http://pmip.lsce.ipsl.fr.

131 Valdes, P. (2011). Built for stability. *Nature Geoscience* 4: 414–416.

# 7

# Into the Icehouse

LEARNING OUTCOMES

- Understand and be able to explain the contribution made to the study of palaeoclimates by new developments in oceanography including sampling technologies and the evolving methods of chemical stratigraphy.
- Understand and explain the development of thinking about the origins of the Antarctic ice sheet including the evolution of the Antarctic Circum-polar Current, drawing on the concept of ocean gateways.
- Understand and explain the origins of Arctic climate through the Cenozoic, including the development of high latitude forests in the Eocene, and eventual glaciation, drawing on the concept of ocean gateways.
- Be able to compare and contrast the development of climate at high latitudes in the southern and northern hemispheres.
- Understand and explain the evolution of the vertical structure of ocean circulation (surface, intermediate, and bottom waters) through the Cenozoic.
- Understand and be able to explain the difference between ice volume and temperature through time.

## 7.1 Climate Clues from the Deep Ocean

Much of what we now know about climate change through the Cenozoic comes from studying the climate record in deep ocean sediments. That's not entirely surprising, because the oceans are an important driver of climate change. They cover 72% of the Earth's surface. They store huge amounts of solar energy, with the top 3.5 m of the ocean containing as much heat as the entire atmosphere. And they move this heat around the planet in giant ocean currents that girdle the Earth and transport significant amounts of heat from the Equator to the poles. Europe is much warmer than Labrador, despite both lying at the same latitude, because of heat transported north by the Gulf Stream. In effect, the ocean is the flywheel of the climate system, storing solar energy and moving it around much more slowly than would be the case if the Earth's surface were all land. Deep-ocean sediments carry the record of that oceanic behaviour, hence of climate change.

We call the students of the ocean 'oceanographers'. The science of oceanography came of age with publication in 1942 of a magnificent reference work – *The Oceans* – by scientists at Scripps Institution of Oceanography: Harald Ulrik Sverdrup (1888–1957), an import from Norway, and

Martin Johnson and Richard Fleming. While *The Oceans* helped to explain how the oceans worked, it also showed that even then we didn't know much more about the distribution of the sediments of the deep-ocean floor than the scientists on the round-the-world expedition of HMS *Challenger* (1872–1876) had known 70 years before. Much the same applied at the end of 1964, when I began my career as a marine geologist with the New Zealand Oceanographic Institute. That lack of knowledge largely reflected the costs and difficulties of sampling the ocean floor. On average it is about 4 km deep, and all that water gets in the way. A breakthrough in our understanding of the role of the ocean in past climate change called for new technologies to sample the deep-sea floor.

Before World War II, weighted tubes known as gravity corers were dropped from ships to retrieve cores of deep-sea sediment of up to about 3 m long. Inspecting cores collected from the South Atlantic by the German *Meteor* Expedition of 1925–1927, Wolfgang Schott (1905–1989) identified three distinct layers characterized by different species of planktonic foraminifera [1]: the upper layer contained abundant *Globorotalia menardii*, a species typical of warm surface waters; the second layer contained cold-water types; and the deepest layer contained *menardii* again. Schott deduced that the middle layer represented the

*Palaeoclimatology: From Snowball Earth to the Anthropocene*, First Edition. Colin P. Summerhayes.
© 2020 John Wiley & Sons Ltd. Published 2020 by John Wiley & Sons Ltd.

last glaciation, when the Atlantic was cooler, and that the other two layers represented warm interglacials, the last one being the Holocene. Changes in planktonic foraminifera could serve as a sort of thermometer for past times.

A simple technological breakthrough 20 years later enabled us to expand on Schott's findings, when the Swedish geologist Börje Kullenberg (1906–1991) had the bright idea of adding an internal piston to avoid the compression of sedimentary layers under the weight of gravity corers. This allowed much longer cores to be collected without the layers being compressed. The piston corer became the tool of choice for studying past climates in ocean sediments. First deployed in Sweden's Gullmar Fjord in 1945, it obtained an undisturbed core of sediment 20.3 m long. Following successful trials in the Mediterranean, the device was first widely used aboard the *Albatross* on the round-the-world Swedish Deep Sea Expedition of 1947–1948, which obtained 57 cores, totalling 500 m in length.

Widespread use of the Kullenberg piston corer began to expose the secrets of climate change hidden in deep-ocean sediments. Many cores were collected by the worldwide research expeditions of the ships of three great American oceanographic institutions. The Scripps Institution of Oceanography, based in the village of La Jolla near San Diego, California, began life as the Marine Biological Association of San Diego, in 1903, before becoming Scripps in 1925. It is referred to as Scripps. The Woods Hole Oceanographic Institution (WHOI), on Cape Cod, Massachusetts, where I worked in the 1970s, formed in 1930. The Lamont Geological Observatory of New York's Columbia University, known as Lamont, was established in 1949 and renamed the Lamont–Doherty Earth Observatory in 1993. The comparable major research institutions of the UK, Germany, and France also began to collect piston cores across the world's oceans. The UK's National Institute of Oceanography, at Wormley, in Surrey, was created in 1948, morphing with time into the Institute of Oceanographic Sciences Deacon Laboratory, before moving to Southampton in 1995 to become the heart of the Southampton (now National) Oceanography Centre. In France, the Institute Français de Recherche pour l'Exploitation de la Mer (IFREMER) was formed in 1984 and the Institut Polaire Français Paul Emile Victor (IPEV) in 1992. In Germany, the Alfred Wegener Institute for Marine and Polar Research (AWI) was formed in 1980.

Piston corers themselves evolved. In the early 1970s, when I was an Assistant Scientist at WHOI, my colleague Charlie Hollister (1936–1999) developed the giant piston corer. This monster could easily collect large-diameter cores up to around 20 m long, some 8 m longer than the average piston corer, providing much larger samples for analysis [2]. My last marine research expedition, between the Azores and Marseilles in July 1995, was aboard the brand new French research vessel *Marion DuFresne*, where we used the latest French giant piston corer to collect cores as long as 39 m off the coast of Portugal. Before we boarded the ship in the Azores, the ship's coring crew had collected cores up to 52 m long on the Bermuda Rise. The mind boggles!

Opening up the next dimension, rather like a bird-watcher graduating from binoculars to a telescope, called for deep-ocean drilling, profiled in Chapter 6. Whilst piston corers can be launched from medium-sized to large research ships, deep-sea drilling requires a specialized vessel, of which there is usually only one available to the global research community at any one time. The international Deep Sea Drilling Project (DSDP) and its successors have supplied such drilling platforms since late 1968, enabling us to penetrate the older geological record that is buried too deep to be reached by piston corers. Even there, piston coring has enhanced recovery, through operation of the hydraulic piston corer, developed in 1979. This allows the recovery of undisturbed cores 10–100 m long in the soft uppermost sediments of the seafloor, which are poorly sampled by the rotary drilling process.

Piston coring and deep sea drilling underpin the new science of palaeoceanography, which has unlocked many of the secret files of past climate change. I learned all about piston coring early on in my career, when I was lucky enough to join the Scripps research vessel *Argo* in the southwest Pacific in 1967, as mentioned in Chapter 5 [3]. We collected a piston core every day. Lessons learned on that cruise would come in handy when I began collecting my own piston cores for my PhD at Imperial College London, while on expeditions down the Moroccan coast. The *Argo* cruise also exposed me to some of the hazards of ocean research, when the cylindrical 1 ton weight that drove the piston corer into the sediment broke free of its lashings on the after deck in a big storm and began to roll from one side of the ship to the other, smashing everything in its path. Stopping it was rather like trying to lasso a raging bull, and Bill Menard pulled a tendon trying to leap over the rolling monster when it unexpectedly headed his way. We headed to Brisbane for repairs. (As an unexpected bonus, that visit introduced me to Australian aboriginal art.) Later, when I worked for EPRCo in Houston, I was much involved with the DSDP, analysing drill cores for their content of organic matter and methane, and serving as an advisor on the DSDP's Organic Geochemistry Panel.

## 7.2 Palaeoceanography

Palaeoceanography is the study of how the ocean behaved in the past in response to changes in the positions of the continents, the patterns of the winds, the presence or absence of polar ice and changes in solar radiation determined by variations in the Earth's orbit [4]. It can tell us about the distribution of the water masses, the average positions of major currents and oceanic fronts, the regional patterns of sea surface temperatures, changes in global ice volume and extent, changes in the vertical thermal structure of the ocean, changes in bottom currents and bottom water masses, changes in the positions and intensity of the Trade Winds, and past changes in the biogeography (distribution) of different ocean-dwelling organisms. Changes with time reflect changes in controlling processes, enabling us to decipher the history of ocean and atmospheric circulation and ice and to understand their effects on global climate.

The earliest use of the word 'palaeoceanography' that I could find was by William (Bill) R. Riedel (1927–), an Australian specialist in studies of the radiolaria, a form of siliceous plankton, who joined Scripps in 1951. Riedel used it in a paper submitted in August 1960, and published in 1963 [5]. New Zealander Jim Kennett (1940–) (Figure 7.1, Box 7.1) used the term in his 1971 paper on the Cenozoic palaeoglacial history of Antarctica, referring to it as 'paleo-oceanography' [6], and again in the report of DSDP's Leg 29 (exploring the climate history of the Southern Ocean south of New Zealand), which he co-led with Bob Houtz of Lamont in March and April 1973 [7]. 'Paleo-oceanography' was soon to lose the hyphen. Kennett was one of the initial developers of this new field in the early 1970s, although the title of 'father of palaeoceanography' is generally bestowed on Italian isotope geochemist Cesare Emiliani for his pioneering work in the 1950s.

**Figure 7.1** Jim Kennett. *Source:* Photo courtesy of Jim Kennett.

Like all new research fields, this one led to the formation of a new scientific journal: *Paleoceanography* (note the American spelling). Jim Kennett helped establish the new journal in 1986. It also led to a new scientific conference series: I attended the first International Conference on Paleoceanography in Zurich in 1983, and the series continues.

Amongst the pioneers of palaeoceanography was Swedish scientist Gustaf Arrhenius (1922–), grandson of Svante Arrhenius of $CO_2$-fame. Arrhenius is a member of the Royal Swedish Academy of Sciences and the 1998

---

**Box 7.1   Jim Kennett**

Kennett was one of the primary developers of palaeoceanography. As a youngster he spent much of his spare time studying geology at New Zealand's National Museum and Geological Survey. He was particularly taken with the usefulness of foraminifera as recorders of ocean history, and completed a PhD in Wellington in 1965 on Neogene foraminifera across the New Zealand region. The two of us shared an office at the New Zealand Oceanographic Institute for a while, and I was much influenced by his growing awareness of the as yet un-named field of past ocean studies. With the advent of the DSDP, many of those involved in interpreting the deep-sea sediment record began to see obvious connections between the oceans, the ice, the atmosphere and the biosphere. Interpreting these connections required a broader, more integrative Earth Systems approach to geology than was taught in most geology schools. Kennett was amongst the first to take such an approach, which permeates his textbook *Marine Geology* published in 1982 [4]. This was the first comprehensive textbook on ocean history and processes following the plate tectonics revolution, and it remains an authoritative source book. I was thrilled to see it appear, and I envied the new generation of marine geologists who would kick off their studies and careers with such a comprehensive text to guide them. In 1987 Kennett was appointed director of the Marine Science Institute of the University of California at Santa Barbara. There he used deep-ocean drilling to provide a 160 000-year high-resolution record of climate change in the Santa Barbara Basin, showing that it was synchronous with that in Greenland ice cores. He is well known as an imaginative thinker, having proposed that Quaternary climate shifts (deglaciations) may have been accentuated by the destabilization of seafloor methane hydrates and that the abrupt cooling at the onset of the cold Younger Dryas episode 13 000 years ago was caused by a meteorite impact. He was awarded the Shephard Medal of the Society of Economic Palaeontologists and Mineralogists in 2002.

recipient of its Hans Pettersson Gold Medal. In 1952, after sailing on the *Albatross* cruise and completing his PhD at the University of Stockholm, he joined Scripps. Amongst his *Albatross* cores were several from the eastern equatorial Pacific, in which he found that productivity, measured by the accumulation of planktonic skeletal remains on the deep sea floor, was high during what appeared to be the glacial stages of the Ice Age. This increase in productivity, he proposed, came about when the increased temperature gradient between the Equator and the poles during glacial periods strengthened the Trade Winds, thus enhancing the upwelling currents that bring nutrients to the surface at the Equator [8, 9]. This was a striking advance in understanding the workings of the atmosphere–ocean system in an Ice Age world.

Another pioneer was Cesare Emiliani (1922–1995) (Figure 7.2, Box 7.2), whom we met in Chapter 6, working with Harold Urey at the University of Chicago.

Emiliani realized early on, from studies of deep-sea benthic foraminifera, that the bottom waters of the Miocene and Oligocene ocean were warmer than those of today, although they fell from 10 °C in the mid Oligocene to 2 °C in the late Pliocene in the eastern equatorial Pacific, a sign of cooling driven by sinking of cold water in high latitudes [10, 11]. In contrast, his studies of pelagic foraminifera showed that ocean surface temperatures in the equatorial Atlantic were rather uniform through Oligocene and Miocene times [11]. Later research confirmed that picture. In addition, he made substantial contributions to the study of the record of the Ice Age in deep-ocean sediments, as we'll see later.

---

**Box 7.2   Cesare Emiliani**

Emiliani was born in Bologna, Italy, and gained his PhD in geology there in 1945, specializing in micropalaeontology, before moving to Chicago and gaining a second PhD, specializing in isotopic palaeoclimatology, in 1950. He worked in Urey's geochemistry lab until he moved in 1957 to the Institute of Marine Sciences of the University of Miami, which gave him ready access to the ships he needed in order to sample deep-sea sediments for their foraminifera. In 1967, he became director of the Institute's Geology and Geophysics Division. In due course, the Institute became the Rosenstiel School of Marine and Atmospheric Science. Emiliani was honoured by having the coccolithophorid genus *Emiliana* named after him, and was awarded the Swedish Vega Medal, and the Alexander Agassiz Medal of the US National Academy of Sciences.

---

Emiliani was also one of the first to apply oxygen isotopes to the study of the temperatures of past surface and bottom waters. By 1965, British physicist Nicholas J. Shackleton (1937–2006) (Figure 7.3, Box 7.3), from the Godwin Laboratory of Cambridge University, had radically improved the technology of oxygen isotopic analysis. Realizing that substantial advances could be made if many more samples were available, he rectified inadequacies in the mass spectrometric instruments used to analyse the proportions of different oxygen isotopes, paving the way for multiple oxygen isotopic analyses on very small samples [12]. Shackleton became a major force in palaeoceanography over the next 40 years.

I first met Shackleton at WHOI when I was working there in the early 1970s. We subsequently discussed the results emerging from the DSDP on many occasions. As the DSDP drew to a close, we co-organized a Geological Society

**Figure 7.2**   Cesar Emiliani. Photo from the early 1950s when he was doing his pioneering research at the University of Chicago.

**Figure 7.3**   Nicholas Shackleton [13].

### Box 7.3   Nicholas Shackleton

Nick Shackleton was the leading British oxygen isotope analyst and one of the stars of the new and growing field of palaeoceanography. A distant relative of polar explorer Sir Ernest Shackleton, in due course he became famous in his own right. He was also owner of the largest collection of antique clarinets in the world, as well as being a talented clarinettist. His renown in scientific circles came from his innovative and pioneering research. As his Cambridge colleagues Nick McCave and Harry Elderfield said: '*his lifetime achievements define the emergence of our understanding of the operation of Earth's natural climate system*' [13]. Amongst many honours for his services to science, the Geological Society of London awarded him its Lyell Medal in 1987 and its prestigious Wollaston Medal in 1996. He was also made a fellow of the Royal Society in 1985, received its Royal Medal in 2003, and was knighted in 1998 – the ultimate British accolade.

of London conference on the Palaeoceanography of the North Atlantic, to review the advances that had come from deep-sea drilling. The meeting, in November 1984, showed how climate had influenced ocean circulation to create the Atlantic sedimentary record over the past 180 Ma [14]. It helped to ensure the success of the bid to replace the DSDP with the next phase of ocean drilling.

Whilst I was working for BPRCo, in the mid-1980s, I persuaded Shackleton to come and talk to us about isotope stratigraphy. At the time I was responsible for the micropalaeontologists who provided BP explorers with the ages of the fossils in drill cuttings and cores, along with information about what the fossils told us about the depositional environments and climates in which they had lived, in order to help the explorers evaluate the petroleum potential of their sedimentary basins. I thought we could use Nick's isotopic stratigraphy techniques to tell us yet more about past environments through 'chemical stratigraphy'. It was amusing to see him turning heads in the senior dining room at BP's research centre. The senior managers dining there in their pinstriped suits were not used to seeing hippie-looking men with shoulder-length hair, open-necked shirts and sandals in their carpeted and waitress-serviced inner sanctum. Only once did I see Nick wearing a tie – at a conference at the Royal Society in London. The horizontal black and white stripes spaced irregularly down this neckpiece turned out to be the magnetic polarity time scale used to interpret the ages of the magnetic stripes on the deep-sea floor on either side of mid-ocean ridges. Typical!

In 1988 I was invited to organize the Geological Society's 3rd Lyell meeting, on Palaeoclimates, which took place in February 1989 [15]. The papers presented there examined in depth the several different ways we then had of looking at past climates: the biostratigraphic approach, using the fossils of marine plankton and land plants; the geochemical approach, using oxygen isotopes and alkenones (more on that below); high resolution studies based on the astronomical approach; and a new kid on the block, numerical modelling (reviewed at the end of Chapter 6). The conference concluded that deep-ocean drilling into ancient marine sediments had an increasingly important role to play in showing us how climate had changed with time and in providing the data needed to test global circulation models. Papers from the meeting were published in volume 147 of the *Journal of the Geological Society of London* in 1990.

Looking for a chemical method independent of oxygen isotopes for estimating the temperatures of past surface waters, a Bristol University group led by chemist Geoff Eglinton (1927–2016) discovered in 1985 that the relative proportions of organic chemical compounds known as alkenones, derived from marine algae, correlated with the temperature of ocean surface waters [16]. This was one of the most important discoveries in organic geochemistry in the last 50 years. For his work in applying the tools of organic chemistry to the understanding of geological problems, Eglinton was made a fellow of the Royal Society, got the Society's Royal Medal in 1997, and the Geological Society's Wollaston Medal in 2004. I worked closely with Geoff when we both served on the DSDP's Organic Geochemistry Panel.

Alkenones are lipids. They are long-chained organic compounds belonging to a class known as ketones, and come from cell membranes in a few species of haptophyte algae, such as the widespread marine coccolithophorid *Emiliana huxleyi* [17]. In effect they are fossil organic molecules. They are distinguished from one another by the numbers of carbon atoms in the chain ($C_{37}$, $C_{38}$, $C_{39}$), their degree of unsaturation (meaning the number of double bonds they contain), and the structure of the group at the end of the chain (methyl or ethyl). The ones of interest for palaeothermometry are the $C_{37}$ methyl ketones containing two or three double bonds [18]. The temperature of the surface waters in which the algae grew is now determined from the $U^{k'}_{37}$ index, measured as the ratio of $C_{37.2}/(C_{37.2} + C_{37.3})$, where $C_{37.2}$ represents the amount of the di-unsaturated ketone and $C_{37.3}$ the amount of the tri-unsaturated form [19]. The beauty of this technique is that it provides results even where the skeletons of the organisms have been dissolved.

Another new tool in palaeothermometry capitalized on the discovery that the ratio of magnesium to calcium

(Mg/Ca) in the calcium carbonate skeletons of planktonic foraminifera was directly related to temperature [20]. First to observe a tantalizing relationship between the Mg content of biogenic carbonates and their growth temperature was the American geochemist Frank Wigglesworth Clarke (1847–1931), after whom the Geochemical Society's F.W. Clarke Award is named [21]. This relationship only became geologically useful 70 years later, in the mid 1990s, through the application of powerful modern analytical techniques like the electron microprobe, which allowed the trace element concentration of a single foraminiferal chamber to be measured to high accuracy [22, 23]. Progress has been rapid since then [21].

One of the scientists developing Mg/Ca palaeothermometry was my former fellow student and bridge-playing opponent at Imperial College, Henry (Harry) Elderfield (1943–2016) (Figure 7.4). Harry and I went to sea together in 1975 on WHOI's *RV Chain* cruise 119 to collect long sediment cores from the Nile Fan off the coast of Egypt. We celebrated the end of the cruise by climbing to the top of the Great Pyramid of Cheops, along with Al Driscoll, WHOI's chief coring technician. For his many outstanding geochemical contributions, Elderfield was made a fellow of the Royal Society in 2001, was awarded the Geological Society's Prestwich Medal in 1993 and its Lyell Medal in 2003, and received the Urey Medal of the European Association of Geochemistry in 2007.

Using Mg/Ca ratios to detect past temperature change is especially helpful in the tropics, where glacial–interglacial changes are small (<5 °C) and thus more difficult to detect by other means. The Mg/Ca method works well there because its relationship to temperature is exponential, steepening towards higher temperatures. It indicates tropical cooling of around 3 °C in glacial times, as do data from alkenones [21]. The Mg/Ca ratio changes only slightly at

temperatures typical of glacial–interglacial bottom waters, so is not so useful at demonstrating change there.

Carrie Lear, one of Elderfield's students, worked with him to apply the Mg/Ca technique to pre-Quaternary marine sediments. In 2000, they used Mg/Ca ratios in benthic foraminifera to produce a deep-sea bottom water temperature record for the past 50 Ma [24]. This closely replicated the oxygen isotope $\partial^{18}O$ curve (which represents combined temperature and ice volume), defined a cooling of about 12 °C from Mid Eocene to present, and confirmed that ice first formed on Antarctica 34 Ma ago. Although the event was not accompanied by an abrupt decrease in deep-sea temperature, it was accompanied by a fairly steep cooling of 3.5 °C from 38 to 30 Ma that crossed the Eocene–Oligocene boundary. How did that relate to the formation of the ice sheet? Combining the $\partial^{18}O$ and Mg/Ca data enabled extraction of the history of the $\partial^{18}O$ signature of seawater, which equates to ice volume. The lack of direct association between the ice volume curve and the deep-sea bottom water curve indicated that ice built up rapidly on Antarctica without a concomitant decrease in the temperature of polar surface waters [24]. The lack of immediate cooling in the bottom waters suggested to Lear et al. '*that a mechanism other than temperature decline must have promoted the initiation of Antarctic ice*' [24]. This may be explained by assuming that the development of the seaway between Antarctic and Australia reduced the previous aridity of this polar region by providing a major new source of moisture to facilitate the development of an ice sheet [24]. Nevertheless, the 3.5 °C decline in temperature across the boundary strongly suggested that surface waters were cooling throughout this time, forming sea ice, and hence creating dense cold Antarctic Bottom Water. The calculated ice volume showed that the ice sheet then decayed towards the Mid Miocene Climatic Optimum at about 17 Ma ago, during which time bottom water temperatures remained fairly stable, before growing again after the Mid Miocene Climatic Optimum, from 14 Ma ago, at which point bottom waters cooled by a further 2 °C. Following a renewed shrinkage of the ice sheet in the Pliocene, the bottom waters fell by a further 3 °C in the late Pliocene and Pleistocene, as the northern Hemisphere ice sheets grew [24]. This interpretation naturally relied heavily on the validity of the Mg/Ca ratio as a signal of bottom water temperatures. We will revisit these changes in more detail in Chapter 11.

Testing questions about that validity, Elderfield found in 2010 that the carbonate ion concentration in oceanic bottom water affected the Mg/Ca ratio of epifaunal benthic foraminifera (those living on the seabed) but not that of the benthic infauna (those living within the sediment) [25].

Another of Elderfield's students, Aradhina Tripati, used the Mg/Ca palaeothermometry technique to investigate changes in tropical sea surface temperatures in the Late

**Figure 7.4**   Harry Elderfield.

Palaeocene through Middle Eocene (59–40 Ma ago) from an Ocean Drilling Program (ODP) site on Allison Guyot in the western central Pacific [26]. They observed long-term warming into the Early Eocene – reaching a peak between 51 and 48 Ma ago, with values slightly warmer than those in the present tropical Pacific – followed by cooling of 4 °C over the next 10 Ma. The Mg/Ca temperatures were warmer than those calculated from oxygen isotopes, which had suggested a cool tropics for that period. Tripati concluded that '*absolute SST [sea surface temperature] values inferred from the $\partial^{18}O$ of mixed-layer planktonic foraminifera in deep-sea sediments are biased towards colder values due to (1) secondary calcification, and (2) assumptions about surface water $\partial^{18}O$*' [26]. They went on to estimate that there had been '*reduced equator-to-pole thermal gradients during the early Paleogene, with minimum gradients coincident with peak tropical warmth during the early Eocene. Gradients increase during the middle Eocene, with tropical SST cooling by 3–4 °C and high-latitude SST cooling by several degrees. This finding is consistent with a weaker ocean circulation and heat transport during the early Paleogene, relative to modern*' [26].

As we will see later in this chapter, additional evidence confirms that the deep ocean cooled gradually from the Mid Eocene onwards, whilst ice growth occurred rapidly at certain periods, starting in the earliest Oligocene [27].

The apparent lack of warming of the Eocene tropics mentioned above and derived from oxygen isotope data led to what Eric Barron called the 'cool tropics paradox' [28]. Puzzling over this paradox, scientists eventually came to realize that re-crystallization on the deep-sea floor distorted the original isotopic signal from tropical surface waters, biassing the record towards cold temperatures [29, 30]. Science continually advances, and as Tripati and colleagues found, the latest techniques exposed the problem and made the paradox disappear.

In 1986, a new organic molecular proxy for sea surface temperature was discovered by researchers on the Dutch Island of Texel, in the Netherlands. They named it TEX$_{86}$, shorthand for the Tetraether index based on the 86-tetraether-carbon chain in lipids derived from the cell membranes of the marine picoplankton known as the Archaea (picoplankton are just 0.2–2.0 mμ in diameter, and Archaea are single-celled micro-organisms lacking a nucleus; they are similar in size to bacteria). The membrane lipids are glycerol dialkyl glycerol tetraethers, known by the shorthand GDGT. TEX$_{86}$ data confirmed that Eocene tropical temperatures were 5–10 °C warmer than previous reconstructions based on $\partial^{18}O$ [29–31]. This tells us that in the warmer worlds of the past, both the tropics and the poles were warm and thus the Equator to pole temperature gradient was less than in past cool periods. Under those conditions wind

strengths would be weaker than today, weakening oceanic circulation patterns.

By the early 2000s, we thus had four independent geochemical tools with which to map past climate and ocean temperature in deep sea sediment cores: oxygen isotope ratios ($\partial^{18}O$), which also provided a means of documenting changes in ice volume; alkenones; Mg/Ca ratios; and TEX$_{86}$. These were supplemented by calculations of sea surface temperature based on the assemblages of planktonic microfossils, a technique developed by John Imbrie that we explore in Chapter 12. As time went by we were also able to use carbon isotope ratios ($\partial^{13}C$), and the ratios of cadmium to calcium (Cd/Ca) as tracers to document the changing characteristics of deep waters with time, by applying developments by Ed Boyle at MIT [32]. Robert Anderson of Lamont showed how an isotope of thorium ($^{230}Th$) could be used to measure the flux of sediment particles to the seabed as a guide to past ocean productivity [33]. The striking advances represented by these new technologies enabled us to investigate climate variability in marine sediment cores from the Cenozoic in unprecedented detail for the first time.

When new techniques are added to the toolkit, they have to be tested in a wide range of environments before they are widely accepted. TEX$_{86}$ is no exception. Concerned that proxy-based estimates of sea surface temperature for the southwest Pacific in the Eocene were too high, a team led by Christopher Hollis of the department of palaeontology of GNS Science in Lower Hutt, New Zealand, recently tested the TEX$_{86}$, $\partial^{18}O$ and Mg/Ca proxies for sea surface temperature against one another and against numerical model results [34]. They were worried that proxies based on lipid organic material (for example the GDGTs underpinning the TEX$_{86}$ palaeothermometer), might respond differently to the inorganic palaeothermometers like $\partial^{18}O$ and Mg/Ca, which were based on chemical analyses of fossil shells. They found that the high temperatures estimated for surface and bottom waters could be reconciled with numerical model outputs if the proxies were biassed towards summer seasonal maxima or influenced by warm surface currents. Evidently, multiple approaches to evaluating palaeotemperatures are advisable.

Application of these different tools demonstrates that the cycles of orbital insolation occur throughout the Cenozoic. Indeed, they form the 'heartbeat' of Oligocene time [35]. It appears that particular combinations of insolation minima in both eccentricity and obliquity led to sustained cooling accompanied by the build up of ice at the Oligocene–Miocene boundary (23 Ma ago), and may also be responsible for the occurrence of other such 'thresholds' in climate change through time [35]. More on that later.

## 7.3 The World's Freezer

It was not just the remote floor of the oceans that we needed to study to find out more about the history of our climate. We also needed much more information from another remote spot, the coldest place on Earth, Antarctica. The world's freezer, Antarctica sucks in heat from the tropics, and pours out cold. The world's biggest ice cube, at 30 million km$^3$, it holds 70% of the world's fresh water in frozen form. How cold can it get there? Satellite data collected in 2010 showed that up on the peak of the 3000 m high Polar Plateau temperatures were as low as −93.2 °C in winter (data from the National Snow and Ice Data Centre reported at an American Geophysical Union meeting, December 2013). More recently, in 2018, they have recorded a winter low of −98 °C on the Polar Plateau.

Rather like the deep ocean, Antarctica has only been open to detailed and systematic scientific investigation since the Second World War. This expansion in scientific knowledge began with the International Geophysical Year (IGY) of 1957–1958, when a dozen nations put research stations on the continent. Many of them are still in operation, and they have since been joined by the research stations of several other nations. Much of the research carried out in the Antarctic today focuses on climate change.

Like the deep ocean, Antarctica is hard to get to, and a dangerous place to be. In the ocean you can fall off a ship and drown. In Antarctica you can fall into a crevasse and die. In both places you are subject to fierce storms. Living conditions for researchers are basic, both on the ice and at sea. And the speed of research on the ground around the pole is about the same as it is in the deep ocean: caterpillar-tracked trucks and one-man skidoos move at about the speed you can pedal a bicycle, some 13 mph, just like research ships. Antarctica is probably a worse environment, in that for six months of the year it is dark, and the sea surface is covered with sea ice, both of which conditions limit climate research.

To see how Antarctica fits into the climate story we are going to have to become better acquainted with it (Figure 7.5). It helps to think of the continent as two distinct pieces. The largest bit, East Antarctica, lies mostly east of the Greenwich

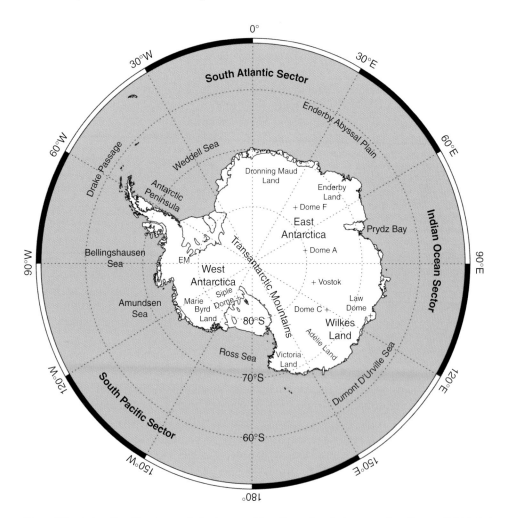

**Figure 7.5** Antarctica. Showing main features and locations of main ice core sites. *Source:* British Antarctic Survey.

Meridian and south of Africa and Australia. It forms 75% of the continent and is rimmed by mountains. Buried beneath its ice are the Gamburtsev Mountains, with peaks up to 3500 m above sea level. More than 98% of East Antarctica, and all of the interior, is covered with the largest mass of ice on the planet: up to 4776 m thick, averaging around 1800 m. If this 24 million km$^3$ of ice melted, it would raise sea level 52 m. The weight of ice has depressed the land beneath by about 600 m. The base of the ice is mostly above sea level, but locally it can be as much as 2 km below sea level.

The smaller bit, West Antarctica, lies south of the Pacific Ocean and South America. Although it forms 25% of the continent, it has only 10% of the ice: around 3 million km$^3$, equivalent to 5 m of sea level rise. West Antarctica has a mean elevation of just 850 m. It is really an archipelago of large islands. Most of its ice sheet is grounded below sea level, by over 2 km in places, making it inherently vulnerable to melting. The main peak, Mount Vinson in the Ellsworth Mountains, rises to 4892 m.

The top of the East Antarctic Ice Sheet is the windswept Polar Plateau, the almost featureless white plain that Scott, Amundsen, and Shackleton crossed on their way towards the South Pole early in the twentieth century. They slogged it on foot or by sled and ski. Nowadays you can fly there in three hours from McMurdo Station, the US base on Ross Island in the Ross Sea, on a 4-engined, ski-equipped Lockheed C-130 Hercules. How times change!

The Polar Plateau rises slightly towards a central crest with three broad peaks: Dome Argus, or Dome A, at 4093 m; Dome Fuji, or Dome F, at 3810 m; and Dome Charlie, or Dome C, at 3233 m. The Chinese recently built a base at Dome A, and the French and the Italians have a base at Dome C. The Americans built their Amundsen–Scott base at the South Pole itself, which lies on one edge of the plateau, at 2835 m. It is not the geographic centre of the continent, which lies at the Pole of Inaccessibility at 85° 50′S, 65° 47′E. The Soviet Union built its Mirny base there, but the major Soviet, and now Russian, base is Vostok, now famous for supplying a 400 Ka record of climate change and for the lake beneath the ice.

The height of the continent keeps it cool, because cooling increases with height by roughly 6.5 °C per 1000 m in humid conditions, and 10 °C per 1000 m if it's dry. East Antarctica is 3000 m high on average, with an average annual temperature of −50 °C. It is so cold that the air is extremely dry. Precipitation is low enough for it to be classified as a desert. With its mean elevation of just 850 m, West Antarctica is warmer and wetter, with an average temperature of only −20 °C, and snowfall averaging 50–100 cm/year, mostly on the coast.

The East Antarctic Ice Sheet is bounded on the west by the Transantarctic Mountains, which cross the continent from the Ross Sea to the Weddell Sea. It lies atop an ancient

piece of continental crust that migrated to its present position during the breakup of Gondwana, as we saw in Chapter 5. Marked by the formation of new ocean floor, the break-up started with the separation from Africa in the early Cretaceous around 140 Ma ago, continued with that from India around 125 Ma ago, and ended with that from Australia around 40 Ma ago and that from South America about 20 Ma ago. The leading edge of Antarctica reached the South Pole about 100 Ma ago. Since then, the continent moved slowly back and forth across the pole, which has been within about 2000 km of its present position for the last 90 Ma [36].

East Antarctica's ancient crustal core was bordered on the side facing the Pacific Ocean by small continental fragments merged into one of the volcanic 'island arcs' that form when ocean crust is 'subducted' beneath the advancing edge of a moving continental fragment, a process described in Chapter 5. Antarctica's island arc was a southward continuation of the Andes of South America, now interrupted by the Drake Passage between the Antarctic Peninsula and Tierra del Fuego in Patagonia at the southern tip of Argentina (Figure 7.5). Drake Passage formed about 23 Ma ago. It was named after its discoverer, the sixteenth century English freebooter Sir Francis Drake, who found it when blown off course in 1578, whilst trying to reach the west coast of North America and to steal gold from Spanish galleons sailing out of Lima.

The islands beneath West Antarctica are separated from the Transantarctic Mountains of East Antarctica by a 1000 m deep oceanic trough, the opposite ends of which are the Weddell Sea and the Ross Sea. This fault-bounded depression is a back-arc basin, a sea-floor spreading structure typically found tucked in behind an island arc: we came across a back-arc basin, the South Fiji Basin, forming behind the Tonga–Kermadec island arc, in Chapter 5. The island arc running from the Peninsula through West Antarctica originally continued into New Zealand, which migrated away around 80 Ma ago.

Examining the geological record for evidence of climate change in and around Antarctica is extremely difficult and costly, not least because only 46 000 km$^2$ of rock, or about 0.33% of the continent, is exposed to the geologist's hammer or drill. Captain Scott's team collected numerous geological samples from Antarctica, including coal from the cliffs of the Beardmore Glacier. As we saw in Chapter 4, amongst the plant remains from the glacier was the first recorded occurrence from the continent of the fossil Gondwanan seed-fern *Glossopteris*. It was the recovery of geological specimens and scientific data that ultimately distinguished Scott's expedition from Amundsen's.

During and after the IGY of 1957–1958, palaeontologists scoured the few peaks protruding through the ice for fossils that could tell us about the icy continent's past climate. All that these rocks provided were a few snapshots of climate change through time. Reviewing the status of Antarctic glacial geology in 1964, geologist R.L. Nichols was moved to state: '*this writer believes that as yet there is no good evidence for Tertiary Antarctic glaciation*' – it was all thought to be Quaternary, as in the Northern Hemisphere [37]. Alas for him, in the same published collection of papers Cam Craddock (1930–2006), then at the University of Minnesota, published radiometric potassium-argon (K–Ar) dates showing that a glaciated surface overlain by basalt in the Jones Mountains of West Antarctica was older than 10 Ma [38]. Not long after that, small basaltic cones overlain and underlain by evidence of glaciation yielded ages ranging between 2.8 and 3.6 Ma in Taylor Valley west of McMurdo Sound. Evidently Antarctic ice had a significant pre-Quaternary history [39, 40].

Through his studies of the foraminifera of the New Zealand region, Jim Kennett recognized a significant cooling in the late Miocene that corresponded to a sea level fall. Knowing about Craddock's 1964 paper [38], he surmised that the cooling resulted from expansion of the Antarctic ice sheet [41]. Kennett broadened his studies when he emigrated to the USA in 1966 to work with Stan Margolis at Florida State University on piston cores from the Southern Ocean. By 1970, their work had pushed back the inception of Antarctic ice rafting to the Early and Middle Eocene [42]. The following year, their core studies led them to conclude that the Southern Ocean had been cool through much of the Cenozoic, with periodic major cooling during the Early Eocene, late Middle Eocene, and Oligocene, causing '*considerable ice-rafting of continental sediments to present-day Subantarctic regions*' [6]. They argued that there had been a warming and a reduction of ice-rafted sands in the lower and Middle Miocene, followed by a cooling trend in the Late Miocene [6]. The discovery that the Southern Ocean had cooled over the past 40 Ma tied in nicely to Emiliani's finding that deep waters had cooled at low latitudes over the same period. A connection via the deep ocean circulation was implied.

Back in the late 1960s – early 1970s, Kennett was intrigued by the revolution in plate tectonics that showed how New Zealand had migrated away from Antarctica. He teamed up with palaeomagnetist Norman Watkins, from the University of Rhode Island, to examine the spreading of NZ and Australia away from Antarctica. Comparing Kennett's palaeontological data with Watkins' palaeomagnetic data, they realized in 1971 that the Antarctic Circumpolar Current (ACC) only developed after Australia separated from Antarctica some time in the early Cenozoic [43].

The problem shared by geologists keen to understand the climatic history of Antarctica was that the ice was in the way. The solution was obvious: they had to drill through it into the rocks beneath. That would be very expensive as well as logistically challenging. Then came the Eureka moment. A clever way around the problem would be to sample the sediments that glaciers and ice streams had dumped in the ocean, and which had been spread out over the adjacent ocean floor by currents. Sampling the ocean sediments close to the continent would have the added benefit of finding microfossils that would tell us about the behaviour of the Southern Ocean through time, another key part of the climate story. Sampling could be carried out either by drilling through the ice shelves adjacent to the land, or by deep-ocean drilling further offshore. Land drilling in ice-free areas was also possible, for example in the McMurdo Dry Valleys discovered by Captain Scott in 1903.

## 7.4 The Drill Bit Turns

Drilling through strata on land in Antarctica began in 1971, in the international Dry Valley Drilling Project (DVDP) [44], and was extended seawards by drilling through the 'fast ice' close to shore. Drilling further offshore, from the DSDP's *Glomar Challenger*, began shortly afterwards, on DSDP Leg 28, between December 1972 and February 1973 [45]. Although the onshore and fast ice drilling took place in parallel with the DSDP drilling deep offshore, for convenience in telling the story I'll first review the results of the drilling on land and from the fast ice.

The ice-free McMurdo Dry Valleys were explored and mapped in 1957–1958 by Victoria University of Wellington students Barrie McKelvey and Peter Webb as part of New Zealand's programme for the IGY. Initiated by the US Office of Polar Programs, the DVDP project was led by Lyle McGinnis, a geophysicist at the University of Northern Illinois, and valley mapper Peter Webb, by then Chairman of Geology at Northern Illinois. The Project was eventually sponsored and staffed by New Zealand, the USA and Japan, continuing until 1976 [46]. In November 1973, the drill was tested on the volcanic rocks at McMurdo Station on Ross Island, reaching a depth of over 300 m. It was then taken to the Antarctic mainland, near the mouth of Taylor Dry Valley, where it cored through 300 m of glacial sediment, recovering material as old as 6 Ma. By now Peter Barrett (1940–) (Figure 7.6, Box 7.4), a young palaeoclimatologist from Victoria University in Wellington, was involved in the project.

Having been to sea on DSDP Leg 28 (more on that below), Barrett and Webb were convinced that to reach further back in time than the 6 Ma sampled by the DVDP, they would have to use the sea ice as a drilling platform. Drilling from the fast ice (sea ice frozen to the land, as opposed to drifting pack ice) would enable them to penetrate hundreds of metres of sediment beneath the sea floor off the

**Figure 7.6** Peter Barrett.

and the Cape Roberts Project in 1997–1999. The latter provided a continuous core, with a remarkable 98% recovery, through 1600 m of strata beneath the sea ice, and a record of coastal tundra and oscillating warm ice sheets from 33 to 17 Ma ago. Unfortunately, the oldest glacial strata lay more or less directly on 330 Ma-old Beacon Sandstone, and strata had been eroded off the top, so another drilling project was needed to fill in both the older and younger ice and climate stories [47, 48].

By now, the scientific value of Antarctic offshore drilling had been established. Following the creation of a new international drilling consortium (ANDRILL), Tim Naish of Victoria University of Wellington, and Ross Powell of the University of Northern Illinois, led the drilling of the first ANDRILL hole, AND-1B, in 2006–2007. Penetrating the 85 m-thick McMurdo Ice Shelf and the underlying 870 m of ocean, it cored 1284 m into the sediments beneath, getting the geologists back to 13 Ma ago [49]. The top 80 m was made up of cycles, in which sediments with alternately strongly glacial and weakly glacial character matched the eight major climate cycles of the last 800 Ka. Further back in time, only a few cycles were preserved, most having been eroded. But those few provided useful snapshots of Antarctic climate, with weaker glacial–interglacial cycles at intervals of 40 Ka. There was also one big surprise. Between 2 and 5 Ma ago, in the Pliocene, warm conditions melted the Ross Ice Shelf, replacing it by open water full of planktonic siliceous diatoms [50]. Might this tell us something about our modern global warming? We shall see as we go on.

Later, Dave Harwood of the University of Nebraska, and Fabio Florindo of Rome's Institute of Geophysics, drilled a

coast. This would be a better option than waiting several years until the DSDP returned to the Ross Sea. Selecting a site 12 km seaward of the mainland, they set up their drill rig on 2 m of sea ice over 120 m of water in November 1975. The 50 m of core recovered just scratched the surface, but it proved the concept and led Barrett, Webb and colleagues to create further multinational drilling projects in the area, including: the McMurdo Sediment and Tectonic Studies (MSSTS) programme in 1976, the Cenozoic Investigation in the Western Ross Sea (CIROS) programme in 1984–1986,

---

**Box 7.4  Peter Barrett**

Barrett's interest in geology came from exploring and mapping caves in his high school years in the Waitomo district of New Zealand. He went to Ohio State University in the early 1960s to help to map the strata of the Transantarctic Mountains, and discovered the first four-legged vertebrate dinosaur fossil in Antarctica, a 200 Ma-old amphibian. That helped to confirm that Antarctica had once been part of Gondwana. After completing a PhD on the geological history of the central Transantarctic Mountains, he joined Victoria University of Wellington. There he helped to establish the stratigraphy of Antarctica's Gondwana sequence, comprising Devonian quartz sandstone, Carboniferous–early Permian glacial deposits, and Permian–Triassic coal measures. In the early 1970s, Barrett became much involved with drilling, both through the ice adjacent to the Antarctic coast, and through the DSDP, leading several projects that revealed

much about Antarctica's climate history. Through the Scientific Committee on Antarctic Research (SCAR), he promoted studies of the history of the Antarctic ice sheet, and was a leader of SCAR's major research programme on Antarctic Climate Evolution from 2004 to 2010. SCAR has provided the forum for planning international drilling activities in the Ross Sea region since the early 1980s, with Barrett providing leadership and advice. Professor of geology at Victoria University, and founding director of the Antarctic Research Centre there (1972–2007), Barrett was instrumental in establishing the NZ Climate Change Research Institute at Victoria University in 2008. He was awarded fellowship in the Royal Society of New Zealand in 1993, the Marsden Medal by the NZ Association of Scientists in 2004, the first SCAR President's Medal in 2006, the NZ Antarctic Medal in 2010, and honorary fellowship of the Geological Society of London in 2011.

second hole, AND-2B, from the sea-ice seaward of the Dry Valleys, reaching a depth of 1139 m below the sea floor and finding evidence for a warm, pulsating East Antarctic ice sheet 15–20 Ma ago [51].

Four decades of research in mapping and dating the landscape of the McMurdo Dry Valleys by George Denton of the University of Maine, and David Sugden of the University of Edinburgh, and their students shows that the region was cold and dry for the last 14 Ma, refrigerated by the huge East Antarctic ice sheet [47, 48]. The last vestiges of shrubby vegetation were found there recently by Adam Lewis, a student at Boston University working with Dave Marchant, a student of Denton and Sugden in the 1990s. The remains occurred in thin, scattered moraines at 1700 m above sea level, dated from volcanic ash layers at 14.0 and 13.6 Ma respectively. They marked the transition from warm to cold glaciers. The evidence for a cold stable ice sheet on East Antarctica co-existing with the warm unstable ice sheet on West Antarctica might give us clues about what may happen as global warming continues.

At least drilling from the 'fast ice' and from the ice shelf eliminated any possibility of the drill rig being damaged by icebergs. Drilling offshore might be a gamble in contrast, given that Antarctica's continually moving ice streams, glaciers, and ice shelves hive off icebergs that drift along in Antarctica's coastal currents. Would the DSDP's *Glomar Challenger* be able to operate at high latitudes, where the weather was severe, and icebergs were common? The acid test took place between December 1972 and February 1973 on DSDP Leg 28 [45], led by Dennis Hayes of Lamont and Larry Frakes (who we met in Chapter 6) then at Florida State University and soon to move to Monash University in Australia. Also aboard were Barrett and Webb.

Coring from south of Fremantle to the Antarctic margin, they established that Australia had been moving away from Antarctica for the last 55 Ma. Coring the Ross continental shelf as far south as 78°S, they found evidence of grounded ice on the continent as far back as 25 Ma ago. Recovering some 1405 m of core from various sites, they deduced from the distribution of ice-rafted fragments of rock and the fossils of micro-organisms that major continental glaciation had begun at least by the late Oligocene, some 25 Ma ago. The climate began deteriorating then, and cool waters pushed north until the end of the Miocene, around 5 Ma ago, when there was a rapid build-up and subsequent retreat of the ice sheet.

Between March–April 1973, *Glomar Challenger* was back in Antarctic waters on DSDP Leg 29, led by Jim Kennett, then at the Graduate School of Oceanography in Rhode Island, and Bob Houtz of Lamont [7]. Earlier, Kennett had participated as a micropalaeontologist on DSDP Leg 21 to the Southwest Pacific (November 1971–January 1972), where he and his colleagues had found a vast regional unconformity that had developed in deep water during the Oligocene [52]. They deduced that this represented massive erosion caused by the development of fast-moving bottom waters in the Ross Sea area in response to extensive glaciation. They also thought that the movement apart of Australia and Antarctica had allowed the inception of the ACC in the Late Oligocene, and that there must be a connection of some kind between that current and the deep waters that caused the erosion [52].

The Leg 29 team showed that the ACC must have developed in the middle to late Oligocene close to 30 Ma ago [53]. Although the slow separation of Australia from Antarctica began much earlier, around 80–90 Ma ago, the Tasman Rise, a ridge extending south from Tasmania, blocked development of a full ocean connection between the Indian Ocean and Pacific Ocean south of Australia until 30 Ma ago. The team confirmed that the onset of substantial glaciation near the Eocene–Oligocene boundary dramatically increased the volume of ocean bottom water, causing massive worldwide erosion in the deep sea. '*The separation of Australia from Antarctica*' they considered '*led to a fundamental change in the world's oceanic circulation and its climate that marks the onset of the modern climate regime*' [53]. Subsequent deep-ocean drilling around Antarctica refined that picture.

Appreciating the value of the growing field of oxygen isotope stratigraphy, Kennett persuaded Shackleton to analyse the oxygen isotopes in benthic calcareous microfossils from Leg 29. In 1975, they concluded that sea surface temperatures on the Campbell Plateau south of New Zealand had declined from 19 °C in the Early Eocene to 11 °C by the Late Eocene and 7 °C in the Oligocene, with a dramatic drop of 1–2 °C at the Eocene–Oligocene boundary, some 34 Ma ago [54]. In deeper water nearby, on the Macquarie Ridge, Early Oligocene bottom water temperatures were as low as those of the present day, implying that the mean annual temperature in high southern latitudes must have been near freezing by the beginning of the Oligocene, and that deep ocean bottom waters were being created then around Antarctica. From that time, glaciers descended to sea level and sea ice was abundant. The isotopic evidence suggested that a small ice sheet formed during the Oligocene, and a large one in the middle Miocene at around 14 Ma ago [54].

Kennett integrated micropalaeontological and oxygen isotope results from deep sea drilling around the continent on Legs 28, 29, 35, and 36 to infer in 1977 that the Antarctic ice sheet first formed in the earliest Oligocene [55]. His summary of Antarctic climate history, in which continental shifts changed ocean currents, which changed climate and even influenced the evolutionary pathways of life in the ocean, was highly influential. Cores from the ODP in the 1980s from the Prydz Bay Shelf and the Kerguelen Plateau confirmed the timing of the first continental ice sheet at 34 Ma ago [47], a key contributor being Jim Zachos of the University of California, Santa Cruz [56].

Using oxygen isotopes from surface dwelling and benthic foraminifera in deep sea drill samples from the Southern Ocean, Zachos reported in 1994 that sea surface temperatures in high southern latitudes rose from around 10–12 °C in the Late Palaeocene to as high as 15 °C in the Early Eocene, declined back to 10–12 °C in the early Middle Eocene, then fell to 6 °C in the Late Eocene and dropped to 4 °C in the Early Oligocene [57]. Deep-water temperatures followed sea surface temperatures but were around 2 °C cooler. Tropical and subtropical sea surface temperatures averaged around 24–25 °C in the early Palaeogene, much like today, while deep-water temperatures then were around 3 °C warmer there than around Antarctica. A global oxygen isotope profile for the Cenozoic, based on Zachos's work [58], appears in Figure 7.7 [59]. Nevertheless, we must bear in mind that these various interpretations relied on oxygen isotopes, while the later work of Elderfield and his students also incorporated temperature analyses from Mg/Ca rations, which, as we saw earlier [24], did *not* support the idea that cold surface waters (hence cold bottom waters) formed around Antarctica immediately on the formation of the Antarctic ice sheet 34 Ma ago. Oxygen isotopes carry the mixed signals from ice volume on the one hand and water temperature on the other hand.

Fossil plants confirmed the picture emerging from drilling and from oxygen isotopes. Jane Francis found botanical evidence from Seymour Island showing that after the peak warmth of the mid to late Cretaceous, Antarctica began to cool (compare Figure 6.15 with Figure 7.7) [59, 60]. Warmth returned during the Palaeocene and Early Eocene between 65 and 50 Ma ago, when there were moist, cool temperate rainforests like those of southern Chile today, before cooling began again in the Mid to Late Eocene. According to Francis, '*Richly fossiliferous Eocene sediments have yielded the only fossils of land mammals in the whole Antarctic continent ... along with fossil wood, fossil leaves, a rare flower, plus .... giant penguins*' [60]. Analyses of leaf margin shapes suggested that between the Late Palaeocene and the Middle Eocene, average Antarctic temperatures fell from around 13–11 °C, ranging from around 1–2 °C in winter to 24–25 °C in summer. Independent confirmation came from oxygen isotopes in fossil Antarctic molluscs, which suggested that annual average temperatures fell to around 10 °C by the Late Eocene. Cooling in the Late Eocene was confirmed by the growing abundance of relatives of today's Monkey Puzzle conifer trees, along with widespread growth of the Southern Beech, *Nothofagus*,

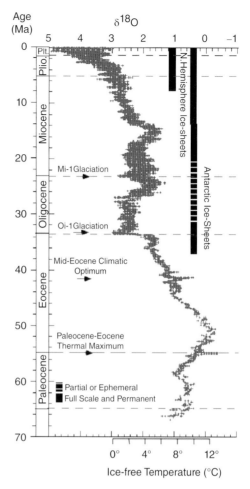

**Figure 7.7** Global compilation of oxygen isotope records for benthic foraminifera through the Cenozoic. Solid bars span intervals of ice-sheet activity in the Antarctic and Northern Hemisphere. Pre-ice sheet values represent mainly bottom-water temperatures (see scale below the diagram); later values represent a mixture of bottom-water and ice volume effects. *Source:* From figure 8.12 in Ref. [59], and based on Ref. [58].

which became progressively more dominant as floral diversity declined in the Middle Eocene around 45 Ma ago.

Declining diversity is yet another sign of cooling. *Nothofagus* is now common in the hills behind Ushuaia in Tierra del Fuego, where it shades the path of the hike up from the city into the snow clad hills directly behind (Figure 7.8 left panel). As one approaches the snowfield these trees reduce to ground-hugging dwarfs (Figure 7.8 top right panel). The centre bottom panel is a close-up showing the typical character of Southern Beech leaves, and at bottom right is an example of a fossil leaf of the same species from Antarctica, where this plant was once widespread (Figure 7.8).

Confirmation of Antarctica's Early Eocene warmth came in 2012 from a core drilled off the Wilkes Land coast at a palaeolatitude of 70°S, where scientists from Integrated

**Figure 7.8** Southern Beech in the hills behind Ushuaia (author photos), and fossil *Nothofagus* leaf from the Antarctic Peninsula.

Ocean Drilling Programme (IODP) expedition 318 analysed fossil pollen and spores and used organic geochemical proxies to show that the coastal lowlands of the time supported a highly diverse, near-tropical forest including palms [61]. Winters were warmer than 10 °C and essentially frost-free despite polar darkness.

The pattern of cooling suggested that ice may have been present on the continent, at least during winter, by the Middle Eocene, 47–44 Ma ago, though not as a major ice sheet like that of today [59, 60]. Ice-rafted debris suggests the presence of valley glaciers during the Eocene, and '*by the end of the Eocene it is possible that an ice sheet extended over much of the peninsula, although average Seymour Island temperatures did not reach below zero*' [60].

After three decades of surveying the Antarctic continental shelf, John B. Anderson of Rice University developed the SHADRIL project to take a series of short (~100 m) cores from gently dipping marine strata of Eocene to Oligocene age beneath the continental shelf on the east side of the Antarctic Peninsula. He and his colleagues found that the onset of full glacial conditions from a beginning near the Eocene–Oligocene boundary was a long, slow process that took several million years from the development of mountain glaciers to a full-bodied ice sheet [62]. This slow development, he pointed out, would

tend to favour the Kennett hypothesis for the gradual development of ice on the continent. Was he right? We will see further on.

The first big Antarctic ice sheets were similar in extent to modern Antarctic ice sheets, but warmer, thinner, and highly dynamic [63]. Independent confirmation of the changing thermal conditions on the continent came from the ages of glacial tillites on King George Island west of the Peninsula, dated at 45–41 Ma. It seems likely that Middle Eocene ice caps would then have been present on the continent above elevations of 1000 m, with perhaps a few valley glaciers reaching sea level [59, 64].

In 1999 Peter Barrett summarized the results of earlier work, noting that '*The earliest evidence of glaciers forming on the Antarctic continent since the Gondwana glaciation of Carboniferous-early Permian times comes as sand grains in fine-grained uppermost lower Eocene and younger deep-sea sediments from the South Pacific, with isolated sand grains interpreted to record ice-rafting events centred on 51, 48 and 41 Ma. Cores from Maud Rise and Kerguelen Plateau have similar sand grains at around 15 Ma .... Lonestones in cores of deep-water late Eocene to early Oligocene mudstones in the CIROS-1 drill hole indicate that glaciers (but not necessarily ice sheets) were discharging ice at sea level through those times .... The eustatic sea-level curve ... indicates that over this period there were several short periods of substantial sea level change (nominally ~100 m, but more likely a few tens, at 58, 55, 49, 42 and 39 Ma), suggesting the growth and collapse of ice sheets thousands of km across ... [although] ... isotopic data provide no support for such variations in ice volume beyond around 34 Ma'* [65].

The earliest Oligocene, when the first major ice sheet formed, was marked by an ice rafting event on the Kerguelen Plateau and a change in the mineralogy of clays from smectite to illite and chlorite on the Kerguelen Plateau and Maud Rise, representing a shift from chemical to physical weathering of the Antarctic continent as a consequence of the progressive cooling on land [65]. In Prydz Bay, on the Antarctic coast south of India, *'grounded ice first deposited debris beyond the coast from earliest Oligocene (or possibly late Eocene) times ... The ice sheet margin extended to the edge of the continental shelf, implying that the ice sheet of that time was of similar size and extent to the present ice sheet in its expanded form'*. Things were different in the Ross Sea, where *'close to the Transantarctic Mountains glaciers were calving at sea level in Eocene times'* [65]. Drilling in 1999 found evidence for glaciers grounding at sea level in the Ross Sea area from 33 Ma ago onwards. Ice cover increased substantially around 15 Ma ago, since when the East Antarctic Ice Sheet has been a more or less permanent feature with temperatures close to present values.

Despite Antarctica's icing up, it took a while for terrestrial plant life to be wiped out. As Jane Francis says, *'Even though many warmth-loving plant taxa disappeared during the mid-Eocene, floras dominated by Nothofagus [the southern beech] remained for many millions of years'* [59, 60]. One of the most remarkable Antarctic floras, comprising twigs of *Nothofagus* along with moss cushions, liverworts and the fruits, stems, and seeds of several other plants, is preserved near the head of the Beardmore Glacier at 85°S in the Sirius Group of sediments at Oliver Bluffs in the Transantarctic Mountains. This suggests a tundra environment with a mean annual temperature of around −12 °C, summers up to +5 °C and winters below freezing. This 'interglacial' environment existed at a time when glaciers temporarily retreated from the Bluffs, allowing dwarf shrubs to colonize a tundra surface only about 500 km from the pole. These deposits are regarded as frozen remnants from the time of the warmer ice sheet that pulsated on the continent between 34 and 14 Ma ago. The plants and insect remains from Oliver Bluffs are like the 14 Ma-old Miocene remnants discovered by Adam Lewis in the McMurdo Dry Valleys [66].

One of the fascinating problems in reconstructing Cenozoic climate is that the global oxygen isotope data sets and data on sea-level suggest that there must have been more ice around that was predicted by general circulation models (GCMs) for the Eocene–Oligocene boundary 34 Ma ago. Could that ice have been located in the northern hemisphere? The then current model took it for granted that the area of West Antarctica was relatively small at the time. Was that correct? Doug Wilson and Bruce Luyendyk of the Marine Science Unit of the University of California at Santa Barbara thought not. From the mass of Oligocene sediment found in nearby deep-sea drilling cores they realized that a significant highland must have existed in the vicinity of West Antarctica. This led them to question the geophysical models for rifting and subsidence of the Ross Sea and uplift of the Transantarctic Mountains, which led in turn to the realization that, 34 Ma ago, West Antarctica was part of a mountain belt of which the Transantarctic Mountains are a remnant. There was thus a much larger land area than had been supposed available for the growth of ice on a proto-West Antarctica 34 Ma ago. Having a larger land area available provided the basis for a larger ice sheet, which matched the requirements based on the sea level and isotope data [67]. A Northern Hemisphere ice sheet was not necessary after all.

Does the evidence accumulated since the mid 1970s bear out Kennett's theory that the first ice sheets formed about 34 Ma ago in response to the opening of the Drake Passage and Tasman Rise and the formation of an ACC that isolated Antarctica thermally by preventing warm currents from impinging on the continent? This was still the prevailing view in the early 1980s [4, 68]. While John Anderson's data from the Peninsula region suggested that ice had built up gradually during the mid Cenozoic [62], Jim Zachos's isotope data (Figure 7.7) [58, 59] showed that there must have been a major cooling event [56], most likely associated with a sudden increase in ice volume at the Eocene–Oligocene boundary. Ken Miller, of Rutgers University and his colleagues documented a global sea-level fall of some 55 m at that boundary, consistent with the idea of a rapid growth of ice [69]. But, the evidence from Elderfield's group suggests that even with all that new land ice, there was no accompanying sudden formation of cold Antarctic bottom water [24], which in turn implies that there was little formation of sea ice. That begs the question – what precisely did the formation of Antarctic ice have to do with global cooling?

## 7.5 Global Cooling

So much for Antarctica, which contains the record of Earth's most extreme cooling during the Cenozoic Era. What about the rest of the world? In the nineteenth century, geologists like Charles Lyell, Eugène Dubois, and T.C. Chamberlin were well aware that the world had cooled towards the Ice Age throughout the Cenozoic, just as we saw happening at the regional scale in and around Antarctica. Close inspection shows that the picture was not quite that simple. The Palaeocene, which intervened between the Cretaceous and the Eocene from 65 to 55 Ma ago, had been about as warm as the late Cretaceous, but the Eocene had been warmer.

Back in April 1972, at the NATO Palaeoclimate meeting in Newcastle-upon-Tyne, the warming through the Palaeocene towards the Eocene had intrigued Larry Frakes, then at Florida State University, and his colleague Elizabeth Kemp. They were keen to explain '*the poleward expansion of a zone of warm and humid climate during the early part of the Tertiary*' and '*the gradual decrease in temperature which followed*' [70]. This warming was manifest in the Eocene temperate forests at high latitudes in Greenland, Spitzbergen, the high Canadian Arctic, and Alaska, as recognized in the nineteenth century by palaeobotanist Oswald Heer (see Chapter 3). The occurrence of such fossil floras at 81°N in Grinnell Land on Canada's Ellesmere Island, for example, '*points to the absence of a polar ice cap in the Early Tertiary*', said Frakes and Kemp [70], echoing for the Arctic what Margolis and Kennet had discovered in the Southern Ocean in 1971 [6]. Jan Zalasiewicz and Mark Williams of the University of Leicester later reminded us that the ancient forest buried on Ellesmere Island was discovered during the First International Polar Year of 1882–1883 by '*the handsomely whiskered First Lieutenant Adolphus Washington Greely (1844–1935) of the United States Army*' [71]. Greely's experiences were horrific. Stuck in the ice, his party survived by eating their boots and their dead companions. A relief expedition two years later found just six men left alive, including Greely: a reminder that in the early days, new geological information came at a sometimes terrible price.

The Norwegians followed the Americans, led by Otto Sverdrup (1854–1930), a pal of fellow Arctic explorer Fridtjof Nansen (1861–1930). Sverdrup had commanded Nansen's *Fram* on its celebrated drift across the Arctic Ocean in 1893–1896. In 1898, whilst trying to circumnavigate Greenland in the *Fram*, he was caught by the ice and had to winter-over on Ellesmere Island, something he did again three times between 1899 and 1902. The expedition's geologist, Per Schei, sampled the Eocene forest. His specimens, including trees like the 60 m tall redwood *Metasequoia*, showed the forest to have been lush and the climate humid. Summer temperatures reached 15 °C, falling to 0 °C in winter, no more than about 700 km from the pole [71].

Plant fossils confirmed that warm, humid conditions typical of the tropics were widespread at around 60°N in the Eocene, and there was growing evidence that such conditions also prevailed in the Southern Hemisphere including the Antarctic Peninsula [70]. Eocene laterites indicating tropical warmth and high rainfall extended from 55°N to 45°S, while today they are confined to Köppen's 'Type A' climate zones (wet and tropical) between the 30th parallels. By the Mid Eocene, coals were forming both in equatorial regions and at high latitudes, making Eocene coals more widespread than Cretaceous ones [64]. Evaporites formed at mid-latitudes. Rainfall continued to be high along the high-latitude western margins of the Pacific [64].

In contrast, plant fossils indicated widespread cooling in the Oligocene, especially in North America between 34 and 31 Ma ago, although there was disagreement about how sharp the decline was globally [70]. Laterites indicating warm, humid conditions were less geographically extensive in the Oligocene. Sparse oxygen isotope data confirmed this picture, with shallow water invertebrate shells from Australia and New Zealand showing a drop from 17 to 22 °C in the Eocene to 12 °C in the Oligocene.

Frakes and Kemp wondered if continental displacements might have something to do with these patterns of warming and cooling. To answer that question they reconstructed continental positions for the Eocene and Oligocene [70]. This was most likely done in 1971 for the April 1972 NATO conference in Newcastle, so predated the definitive suite of reconstruction maps by Smith, Briden and, Drewry that became available in 1973 [72]. As there was no significant latitudinal difference of the Arctic between the two time periods, changing proximity to the pole did not account for the cooling.

By the mid-1980s we knew much more about the distribution of plants during the Cenozoic than did Heer in the late 1800s or Frakes and Kemp in 1972. Jack Albert Wolfe (1936–2005) of the US Geological Survey office in Denver, Colorado, argued that if the boundary between tropical and paratropical rain forests bore the same relationship to temperature in the past that it does today in East Asia, where the boundary today lies at 50°N, then that boundary must have lain at 70°N in the Early Eocene. Broad-leaved evergreens would then have extended north of that to the pole [73]. As more data arrived, it seemed that climate zones could shift through time; they were not as static as might be supposed from the earlier work of Briden and Irving (e.g. see Figure 6.6), which was based on a limited suite of samples [74].

Wolfe pointed out that the great poleward expansion of broad-leaved evergreen vegetation in the warm Palaeocene and Eocene must have produced a different carbon cycle from that operating today [73]. Yet another version of the carbon cycle would have operated during the Oligocene and Miocene, when broad-leaved deciduous forests covered large parts of the Northern Hemisphere at middle latitudes, evergreen forests were more restricted than in earlier times and there were no extensive grasslands or deserts. The grasslands and deserts that we see today are late developments in the Cenozoic, something we return to in Chapter 11.

Large leaves with continuous (i.e. smooth) margins are typical of tropical plants, whilst small leaves with serrated edges indicate cooler climes. Based on those modern relationships, one can infer past temperatures from leaf shape. Accepting that assumption, Wolfe plotted the likely zonal distribution of vegetation on the reconstruction maps of Alan Smith for the Late Palaeocene–Early Eocene (50–55 Ma ago), for warm intervals of the Middle Eocene, for cool intervals of the Middle to Late Eocene (46–50 Ma and 37–41 Ma ago), and for the Early Miocene (18–22 Ma ago). The maps showed the gradual origin and spread of vegetation types of low biomass including savanna (incorporating woodland and scrub), steppe, taiga, tundra and true desert, all of which seemed to be Miocene or younger [73]. Tundra vegetation consists of dwarf shrubs, sedges and grasses, mosses, and lichens, locally with scattered and stunted trees. Taiga (coniferous forest) covers much of Canada, Russia, and Siberia south of the tundra. These low-biomass types of vegetation originated and spread in response to prolonged cooling and the steepening of the Equator-to-pole thermal gradient. The steeper thermal gradient intensified the subtropical high-pressure systems, and the associated summer drought along the western sides of the continents caused deserts to develop there. Savanna spread at low latitudes, where temperature increased and precipitation decreased. As a result, the mass of carbon in land plants decreased during the Cenozoic [75]. Later we will explore the relation of these developments to atmospheric $CO_2$.

Marine plankton also changed in response to the Cenozoic cooling [64]. The four plankton provinces characterizing the Eocene ocean expanded to six by the Oligocene, as two cold Polar Regions became established, signalling the development of a more differentiated latitudinal temperature gradient. Globigerinid planktonic foraminifera became restricted to mid- and low-latitude sites, and nannofossils moved equatorward.

To see if the Eocene to Oligocene cooling might be related to the separation of the continents, Frakes, and Kemp pondered the implications of the five main changes that took place in continental positions during that time [70]. First, India continued moving north, shrinking the Tethyan Ocean between India and Tibet and expanding the Indian Ocean in its wake. Warming began to decline after reaching a peak in the Early Eocene 50 Ma ago, when the advancing edge of the Indian fragment of Gondwana,

much of it now subducted beneath Tibet, collided with the southern edge of Laurasia, much of it now crumpled up to form the Himalayas. That collision closed the east–west Tethyan Ocean connecting the Atlantic and Pacific through the tropics. Removal of the Tethyan seaway between Laurasia and Gondwana dried the climate along the Mediterranean–Caspian corridor and its surrounding lands. Second, the Australia–New Guinea–New Zealand complex separated from Antarctica, and the Drake Passage opened between South America and Antarctica, allowing the creation of the ACC between 50 and 60°S. As Judy Parrish noted [76], the new wind and ocean current regime brought moisture to West Australia. Equatorial easterlies made east Africa wet, but there would not yet have been a substantial monsoonal regime over Asia because that would require massive (later) uplift to form the Tibetan Plateau. Third, the northward movement of Australia–New Guinea narrowed the space for the Pacific Equatorial Current between Australia and S.E. Asia, diminishing the supply of warm water to the widening Indian Ocean. Fourth, continued westward migration of the Americas shrank the Pacific. Fifth, a slight widening of the narrow gap between Greenland and Scandinavia offered the prospect of more oceanic exchange between the Arctic and Atlantic. These opening or closing 'oceanic gateways' ought to have influenced the climate.

Knowing that today's easterly Trade Winds blow warm water west to accumulate in the Pacific Warm Pool just north and east of New Guinea, Frakes and Kemp suggested that during the Eocene, when the Pacific Ocean was wide, the warm pool accumulated more heat than during the Oligocene, when the ocean was narrower [70]. The north- and south-moving branches of the Eocene equatorial current would have moved that warm water towards high latitudes along the coasts of Asia and Australasia, thus warming the polar regions more than would have been possible during the Oligocene. The Earth would also have been slightly warmer than it is now because there was no substantial polar ice to reflect the sun's radiation.

Analysing the rather limited and scattered oxygen isotope data from the North Pacific, Australia, and New Zealand, the pair went on to suggest that in the Oligocene *the extensive cooling indicated by the oxygen isotope curves could have led to the development of the first extensive glaciers on Antarctica'* [70], where evidence for glaciation was beginning to appear. *'We favour a model which allows limited or regional ice formation in the Eocene, construction of the bulk of the Antarctic ice sheet in the Oligocene, and culmination of this process in the Late Miocene by the development of extensive ice shelves .... Ice formed during the Oligocene probably exceeded the amount built up during the Eocene because the principal source, the new seaway south of*

*Australia ... was much closer to the interior of the old polar continent*' [70].

This begs the question, what caused the extensive cooling from the Eocene into the Oligocene? Was it, as Frakes and Kemp suggested [70], the narrowing of the Pacific Ocean, which weakened the supply of warm water to the polar regions? Their analysis ignored two potentially important controls on global heat exchange. One was a change in the global content of $CO_2$ in the atmosphere, which we examine in Chapter 9. The other was the potential of the newly formed ACC to isolate Antarctica thermally from the rest of the world as Kennett had suggested. We'll look at the implications of that next.

By 1975, as more and more cores arrived from the DSDP, which was then only seven years old, the analysts of oxygen isotopes began to derive a comprehensive picture of global climate change for the past 70 Ma. One of the leading analysts was Sam Savin, from Case-Western Reserve University in Cleveland, Ohio [77]. A striking feature of his new $\partial^{18}O$ record was its fall from the end of the Palaeocene to the present in a series of major steps indicative of cooling. These steps were far more apparent in the polar ocean samples analysed by Shackleton and Kennett [54], than in the tropical ocean samples analysed by Savin and his team [77]. One of these cooling steps occurred at the boundary between the Eocene and the Oligocene. As we saw earlier, in the mid 1970s when these data started to become available, that cooling was thought to represent the change in ocean circulation caused by the separation of Australia and South America from Antarctica. That opening would have allowed the formation of an ACC and associated strong westerly winds that isolated Antarctica from the warmer climes to the north, encouraging the formation of a major ice sheet, which in turn led to the formation of cold, dense bottom water capable of changing the global circulation of the world ocean [54].

Refinements to the curve of changing oxygen isotopes with time continued, for example with one published in 1987 by Ken Miller (1956–) from Lamont, whom I remembered as the fast-talking, cigar-smoking PhD student I had played poker with when I was on the staff at WHOI in the 1970s. Miller and his team compiled oxygen isotope data from planktonic foraminifera and benthic foraminifera collected by deep ocean drilling to suggest that ice-free conditions characterized the Palaeocene and Eocene, 65–34 Ma ago, and that continental ice emerged later [78]. Evidently, the positive isotopic shifts seen in earliest Oligocene and middle Miocene sediments of the Southern Ocean by Shackleton and Kennett in 1975 [54] were indeed global.

As touched on earlier, one significant refinement to the interpretation of the curve of changing $\partial^{18}O$ with time came from Shackleton's realization that only about half of the oxygen isotope signal from the base of the Oligocene onwards represented the cooling of the ocean, while the rest represented the trapping of water on land as ice [79, 80]. We now know that the $\partial^{18}O$ trend in benthic foraminifera is a proxy for two properties of the climate: first, deep ocean temperatures, which represent the temperatures of surface waters at the high latitudes where those waters formed; and second, the seawater $\partial^{18}O$ value, which reflects how much water is tied up in ice on land [27].

This picture was further refined by Jim Zachos and colleagues, who confirmed in 2001 that temperatures had risen from warm conditions in the Late Cretaceous towards a climatic optimum in the Early Eocene between 50 and 54 Ma ago, when deep-ocean temperatures reached 10–12 °C, as suggested in Figure 7.7 [58, 59]. If oxygen isotope data are to be believed, those temperatures then fell to around 5–6 °C at the Eocene Oligocene boundary 34 Ma ago, when the first major ice sheet formed on Antarctica (Figure 7.7). However, as Lear et al. showed [24], by using Mg/Ca ratios to help to ascertain which part of the oxygen isotope curve represents the temperature of bottom water and which represents ice volume, it appears that the deep-water temperature fall across the Eocene–Oligocene boundary was *less* than expected from the oxygen isotope data (Figure 7.7).

Measurements of oxygen isotope data from benthic foraminifera of late Middle Eocene to Early Oligocene age (38–28 Ma ago) by Miriam Katz from Rensselaer Polytechnic University and her team, also showed the jump at the Eocene–Oligocene boundary [81] representing the cooling that accompanied formation of the Antarctic ice sheet. Katz's carbon isotopic data showed that a large $\partial^{13}C$ offset developed between mid-depth waters (about 600 m deep) and waters deeper than 1000 m in the western North Atlantic in the Early Oligocene, indicating the development of a low $\partial^{13}C$ zone at intermediate water depths in the Early Oligocene at around 31–30 Ma ago [81]. Low (i.e. negative) $\partial^{13}C$ means abundant $^{12}C$ and associated nutrients derived from the decomposition of marine planktonic remains. In the modern ocean, 1000 m is about the depth of Antarctic Intermediate Water (AAIW), which sinks at the Polar Front. Katz deduced that '*ventilation by AAIW leads to a relatively low $O_2$ and low $\partial^{13}C$ layer (~700-1000 m) [deep]... At the same time, the ocean's coldest waters became restricted to south of the ACC [Antarctic Circumpolar Current], probably forming a bottom-ocean layer, as in the modern ocean*' [81]. In effect, the existence of AAIW suggests the existence of the Polar Front (the source of AAIW today), which is a feature of the ACC, telling us that the ACC should have been well established by the early Oligocene

Katz deduced that the modern four-layer ocean structure of surface, intermediate, deep, and bottom waters probably developed in the Oligocene as a consequence of the development of the ACC and its associated frontal structures. In effect, '*intermediate water circulation today is a consequence of the ACC, which blocks warm surface waters entrained in subtropical gyres from reaching Antarctica; this thermally isolates the continent and the surrounding ocean, allowing large-scale ice sheets to persist*' [81]. As an aside, that helps to explain what keeps Antarctica cold today, whereas the Arctic lacks this thermal isolation, receiving warm subsurface water from the North Atlantic via the Norwegian–Greenland Sea.

Katz and her team thought it possible that the ACC began to develop in the Middle Eocene, with shallow through-flow via the Drake Passage, strengthening, and deepening with the opening of the Tasman gateway between Australia and Antarctica in the Late Eocene to Early Oligocene and the deepening of the Drake Passage in the Oligocene. Whilst it was widely thought that the ACC did not develop until the Late Oligocene, Katz thought that even a shallow ACC might have thermally isolated Antarctica from the rest of the ocean from the Late Eocene onwards [81]. Echoing Frakes and Kemp, their results suggested that '*changes in tectonic gateways affected the middle Eocene to early Miocene ocean circulation, which in turn affected global climate*' [81]. However, Katz's team's data from the Atlantic continental slope, and their comparison of one ocean basin with another, suggested to them that the thermal structure and circulation of the ocean were most strongly affected more or less right at the Eocene–Oligocene boundary rather than before. We might therefore conclude that this boundary represented a key stage of the opening of the Drake Passage.

How do Katz's conclusions about the opening of the Drake Passage square with other interpretations? Some other data do not entirely support development of the ACC as the sole cause of the build-up of ice on the continent [82]. By 2009, several pieces of evidence would change the picture [59]. It seems clear that narrow (and probably relatively shallow) gaps opened between Antarctica and Tasmania and Antarctica and South America and Antarctica in the Mid to Late Eocene before the first ice sheets developed at the Eocene–Oligocene boundary. However, the dinoflagellate phytoplankton population around the Antarctic coast, especially in the narrow gaps between Tasmania and Antarctica and South America and Antarctica, proved to be highly endemic (i.e. local), precluding the operation of a major circumpolar current connecting these environments in the Late Eocene (Figure 7.9). Moreover, as we saw earlier, Lear and Elderfield's Mg/Ca palaeothermometer record argues against the sudden development of cold Antarctic Bottom Water at the Eocene–Oligocene boundary [24]. And they

do not require full development of the ACC to explain the development of the ice sheet, calling instead on the formation of a substantial water mass between Australia and Antarctica (like that depicted in Figure 7.9) to serve as a source of moisture to feed a growing ice sheet [24]. Lear's Mg/Ca thermometry [24] does suggest that cooling bottom water was a feature of the circulation near Antarctica in the Late Eocene, as we might expect if sea ice began to form along the coast.

Numerical modelling of the behaviour of the Eocene Southern Ocean suggests that during the Eocene it comprised two major clockwise gyres: one on the South Pacific side (proto-Ross Gyre) and one on the South Atlantic and Indian Ocean side (proto-Weddell Gyre) of the continent, with no major connection between them (Figure 7.9). In addition, Peter Barker and Ellen Thomas found that the inception of siliceous biogenic sedimentation near 34 Ma ago did not necessarily indicate deep-water connections and circum-Antarctic through-flow [82]. The gyres run clockwise, being driven at the Antarctic coast by the coastal easterlies, and further offshore by the strong westerlies of the West Wind Drift.

Clearly, the geographical isolation of Antarctica by an eventual ACC driven by powerful westerly winds would, as Kennett first suggested, have contributed to Antarctica's thermal isolation. Nevertheless, as we shall see in Chapter 9, the icing up of Antarctica may also have been driven at least in part by cooling forced by the gradual decline in the $CO_2$ content of the air. Moreover, recent geophysical and geological investigations of the Scotia Sea, which floors the Drake Passage between South America and Antarctica, showed Ian Dalziel and colleagues that '*a now submerged [Neogene] volcanic arc may have formed a barrier to deep eastward ocean circulation [even with a deep open Drake Passage] until after the mid Miocene Climatic Optimum. Inception and development of a full deep Antarctic Circumpolar Current may therefore have been important, not in the drop in global temperatures at the Eocene–Oligocene boundary as long surmised, but in the subsequent late Miocene global cooling and intensification of Antarctic glaciation*' [83]. That would leave us with a shallow ACC in the Drake Passage and a deep ACC in the Tasman Gateway, both driven by the strong winds of the West Wind Drift, together with the Weddell and Ross gyres mentioned in the previous paragraph. Dalziel agreed that '*more work is necessary to understand the tectonic development of the North Scotia Ridge and the tortuous, enigmatic path of the ACC in space and time*' [84].

Zachos's and Lear's $\partial^{18}O$ data showed that short warm periods occurred in the Late Oligocene and Middle Miocene (Figure 7.7), when the ice sheet was in a decaying phase [24]. In 2013, Mawbey and Lear showed from analyses of

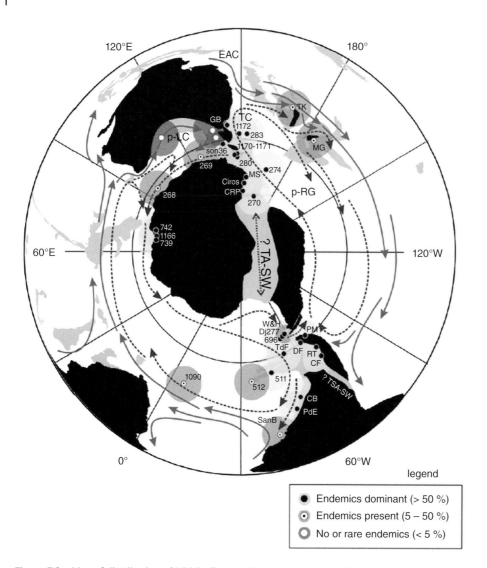

**Figure 7.9** Map of distribution of Middle Eocene dinocysts, overlain with ocean circulation pattern inferred from global circulation models, showing where endemic (i.e. local) species predominate. Numbered and lettered locations are deep ocean drill sites. TA-SW = hypothetical Trans-Antarctic Seaway; TSA-SW = hypothetical Trans-South American Seaway; EAC = East Australian Current; p-LC = proto Leeuwin Current; p-RG = proto Ross Gyre; TC = Tasman Current. *Source:* From figure 8.11 in Ref. 59.

benthic infauna from deep-sea cores in the equatorial Atlantic that bottom water temperatures varied cyclically across the Oligocene-Miocene boundary near 23 Ma ago (where there is a prominent cooling event representing a brief glaciation labelled Mi-1), '*with the main cooling and warming steps followed by ice growth and decay respectively*' [85]. These cycles appeared to be driven by Earth's eccentricity cycles [85]. Two cooling steps led into the glaciation, which was followed by two warming steps associated with ice sheet decay. The overall cooling appeared related to '*an orbital configuration that led to a relatively long interval of reduced seasonality*' [85]. This brief glaciation is associated with a sea-level fall off −50 m, and the recovery was associated with a rise in bottom water temperature of >2 °C [85].

There is ample evidence for a decline in the volume of the Antarctic ice sheet as we progress from the Oligocene–Miocene boundary to the Mid Miocene Climatic Optimum, with the loss of ice calculated as ranging between 30 and 80% [86]. During the Optimum, the Ross Sea experienced warm conditions with summer temperatures of around 10 °C, and the ice sheet margins retreated inland, allowing tundra vegetation to grow on ice-free terrain [86]. In Chapter 11 we explore the role of changing $CO_2$ and orbital forcing as an explanation for such changes.

An oxygen isotopic shift heralded the formation of a major ice sheet on East Antarctica in the Mid Miocene 14 Ma ago [24, 58]. In the mid-1980s, this shift was thought to have initiated the modern ocean–atmosphere system with its cold polar regions, strong Equator-pole temperature gradient and well-developed thermocline (warm waters sitting over cold) [4, 87]. A corollary of that concept was that the southern Subtropical Convergence separating subtropical waters from the sub-Antarctic waters of the ACC developed then, further isolating Antarctica from tropical warmth [64]. The subsequent work of Katz's team suggests that whilst those changes may have been accentuated in the Mid Miocene, they had been initiated in the Oligocene [81].

High latitude bottom waters were as cool as about 5 °C in the Mid to Late Miocene, but had dropped to 2.5 °C by the Pleistocene [24], reflecting the further development of the world's cold deep water. The circulation of North Atlantic Deep Water gradually strengthened through that period, as the world cooled, intensifying around 2.75 Ma ago [64]. We will explore the details of Arctic change in the next section.

By the Mid Miocene, the rising of the Tibetan Plateau caused by the collision of India with Asia would have strengthened monsoon circulation [76]. Sea level was dropping, and east Africa had become increasingly arid.

There was a further cooling in the latest Miocene 5–6 Ma ago. Coinciding with the effects of the ongoing collision between Africa (Morocco) and Europe (Spain), which was gradually closing the equatorial Tethyan Ocean at the Straits of Gibraltar, that cooling created sufficient land ice to lower sea level enough to cut the connection between the Atlantic and the Mediterranean. The Mediterranean Sea almost completely dried out, leading to the deposition on its bed of thick accumulations of salt in a number of discrete basins [88, 89]. It is unclear whether the floor of the Mediterranean comprised one large brackish lake or a series of smaller ones in which salts might precipitate in relatively deep water. In the eastern Mediterranean, the river Nile subsequently dumped vast masses of sediment on top of these Miocene salt deposits to form a giant submarine fan, which mantles the continental slope seawards of the Nile Delta. Its weight has squeezed the salt into vertical finger-like intrusions (diapirs) up to a kilometre or so across, like those beneath the continental slope of the Gulf of Mexico [90]. As Harry Elderfield and I found during our *RV Chain* research cruise to the eastern Mediterranean in 1975, the salt is percolating up through the Nile Fan [91].

Besides these major changes, the oxygen isotope record (Figure 7.7) reveals much low-level, short-term, variability, most of it reflecting changes in the amount of insolation received from the Sun due to regular changes in the tilt of the Earth's axis [58]. These changes have a period of around 40 Ka and persisted through much of the Neogene and Oligocene.

Can we use the new techniques of palaeoceanography to separate the influence of temperature and ice volume in the Cenozoic? The answer is 'yes'. For example, comparing Mg/Ca-based temperatures with measured $\partial^{18}O$ across the Eocene–Oligocene boundary confirms that the $\partial^{18}O$ shifts were dominated by changes in global ice volume [21, 24]. In 2011, a team led by Benjamin Cramer of Rutgers University used Mg/Ca ratios and $\partial^{18}O$ records from benthic foraminifera, along with a record of sea level change as a proxy for ice volume, to evaluate trends in ice volume and deep ocean temperature for the past 108 million years (Figure 7.10) [27]. They showed that *ice volume and temperature did not change proportionately through time.* Cenozoic deep ocean cooling occurred in two prolonged intervals: a gradual decline in temperature from about 14 °C in the Early Eocene to about 6 °C in the Late Eocene and another gradual decline from about 6–0 °C from the Late Miocene to the Pliocene. In contrast, the Cenozoic increase in ice volume (and decrease in sea level) took place in three steps: one at the Eocene–Oligocene boundary 34 Ma ago, one in the Middle Miocene 14 Ma ago, and one in the Plio-Pleistocene from 2.6 Ma ago. According to Cramer, '*These differences are consistent with climate models that imply that temperatures ... should change only gradually on timescales > 2 [Ma], but growth of continental ice sheets may be rapid in response to climate thresholds due to feedbacks that are not yet fully understood*' [27].

Can numerical models help us to understand the cooling? In 2000, a team led by Karen Bice of WHOI used atmosphere and ocean GCMs with fixed atmospheric $CO_2$ to see if changes in palaeogeography could affect heat transport and so explain the high-latitude and deep-water warmth of the early Cenozoic and its subsequent decline [92]. Using an atmospheric GCM connected to a slab ocean, they found little difference between different time slices (Early Eocene, Middle Eocene, Early Oligocene, Early Miocene, and Middle Miocene), suggesting that variations in geography were not enough by themselves to explain the global Cenozoic cooling trend. In contrast, a full three-dimensional ocean model forced by the output of the atmospheric GCM produced substantial differences in the magnitude of heat transport from the equator to the poles between the different time periods.

Given the tropical conditions typical of the Eocene in the Arctic, Bice's model outputs are quite instructive. At that time, there was much higher poleward heat transport in the Southern Hemisphere than in the Northern. This is because warm equatorial water could flow south to Antarctica in the absence of an ACC, the Drake Passage being closed. In contrast, heat transport north of the equator was dominated by a strong east–west current taking warm equatorial waters from the Pacific through the narrowing Tethyan Ocean between India and Asia and across the Atlantic and back into the Pacific via the Central

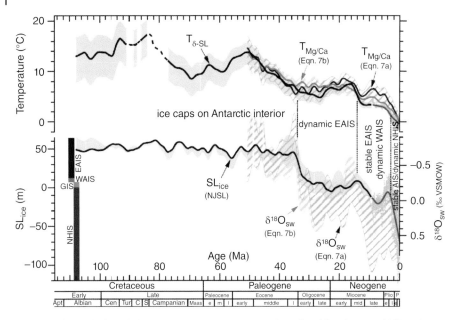

**Figure 7.10** Variations in bottom water temperature and sea level for the past 108 Ma. Data smoothed to show only variations on >5 Myr timescales. Vertical bars indicate the approximate cumulative range of sea level variations due to ice sheets. EAIS = east Antarctic ice sheet, WAIS = west Antarctic ice sheet, GIS = Greenland ice sheet, NHIS = northern hemisphere ice sheets. NJSL = New Jersey sea level. Shading indicates confidence levels; cross-hatching indicates uncertainties in either temperature (top) or seawater $\partial^{18}O$ (bottom) estimates. The latter do not necessarily reflect realistic uncertainty in ice volume calculations. *Source:* From figure 9, in Ref. [27].

American Seaway. Links to the Arctic were weak, due to the slow opening of the northernmost Atlantic. So what made Ellesmere Island more or less tropical in the Eocene? The answer seems to be the atmosphere. Sea levels were high, warm conditions caused more evaporation, and the atmosphere transported significant latent heat poleward in humid air, to be released in the Arctic. All that would be required to achieve this would be higher than modern equatorial temperatures.

Over time, the Equator-to-pole heat transport in the Southern Hemisphere reduced with the gradual opening of the Southern Ocean's gateways, as the distance widened between Antarctica and both South America and Australia. In the Northern Hemisphere, the gradual closing of the Tethyan gateway by the collision of India and Asia, along with the closing off of the Mediterranean and the narrowing of the central American and Indonesian–Pacific gateways, blocked east–west equatorial transport and diverted warm currents towards the Arctic. Over time, warm salty water reaching the Norwegian–Greenland Sea eventually cooled and became dense enough to sink and move south at depth as North Atlantic Deep Water, so strengthening the Atlantic Meridional Overturning Circulation (AMOC) between 55 and 14 Ma ago, as the northern North Atlantic basins opened. Northern Hemisphere ocean overturning and heat transport increased further as the central American seaway closed and the Indonesian–Pacific seaway narrowed further towards modern times. *Clearly geography plays a key*

*role in controlling heat transport by the ocean* [92]. But as we shall see later, changes in $CO_2$ may be equally important, if not more so.

## 7.6 Arctic Glaciation

In contrast to the major development of ice on Antarctica starting 34 Ma ago, it had long been thought that the ice sheets of the Northern Hemisphere did not develop until the Pleistocene, beginning around 2.6 Ma ago. But by the early 1990s, reports of so-called 'dropstones' in sediments from Svalbard suggested that there might have been some seasonal winter ice there in Eocene times, whilst ice-rafted debris had been recovered in sediments of Late Miocene age (5.5 Ma ago) from the central Arctic Ocean [64]. Alaskan tillites suggested intermittent glaciation in Alaska's Wrangel Mountains over the past 10 Ma [64]. Pack ice may have been present intermittently in the Arctic basin since 13–10 Ma, and there were signs of ice rafting in the Russian sector of the Pacific from the Late Oligocene onwards [64]. This and related evidence convinced Ana Christina Ravelo of the University of California, Santa Cruz, and her colleagues in 2007 that '*small ice sheets were present in the Miocene and earliest Pliocene, the marked increase in global ice volume, or the onset of significant Northern Hemisphere glaciation, occurred gradually from 3.5–2.5 Ma*' [93]. By 2.6 Ma ago, vast regions of the Northern

Hemisphere were in the grip of huge ice sheets that covered large areas of the Eurasian Arctic, northeast Asia and Alaska, and by 2.5 Ma, North America [63] (more on that in Chapter 13).

Coming at the question of Arctic glaciation from a different perspective, Aradhina Tripati and her colleagues thought that the changes in the deep-sea oxygen isotope ($\partial^{18}O$) record during the Cenozoic were larger than could be explained by the formation of ice on land just in Antarctica, and concluded, '*there was ice stored on Antarctica and in the Northern Hemisphere at 44.5 Ma, 42 Ma, 38 Ma, and after 34 Ma*' [94]. If true, perhaps this meant that the transition from greenhouse to icehouse was not as sudden as had been initially suggested, though the transition from mountain ice caps to ice sheets on Antarctica may well have taken place over a short period 34 Ma ago. Was Tripati right? No. As we saw earlier, by 2009 Wilson and Luyendyk had deduced that the missing ice had not been in the Northern Hemisphere as Tripati thought, but on West Antarctica [67].

The only way to test ideas about the timing of Arctic glaciation was to drill close to the North Pole. This was accomplished in 2004, on the Arctic Ocean Coring Expedition (ACEX), as Leg 302 of the IODP, under the leadership of Jan Backman of Stockholm University and Kathryn Moran of the University of Rhode Island [95]. The expedition drilled close to the North Pole on the Lomonosov Ridge, using an ice-strengthened drill ship, the *Vidar Viking*, protected by icebreakers: the *Sovetskyi Soyuz* from Russia and the *Oden* from Sweden.

The results were stunning, as Jan Backman and Kathryn Moran explain [96]. During the Eocene, the Arctic was an isolated deep-water basin containing brackish water with a salinity of 21–25‰. From time to time its surface became fresh enough to support blooms of the freshwater fern *Azolla*, which formed vast organic mats. The main *Azolla* event lasted for about 1 Ma between the late Early Eocene and the early Middle Eocene. Isolation from the adjacent Pacific and Atlantic Oceans ended about 45 Ma ago, but there was no substantial deep-water connection until the Miocene (see below). Starting about 46 Ma ago, conditions cooled to the point where sea ice formed a seasonal cover, transporting ice-rafted debris, mainly from the Eurasian margin. The rate of supply of ice-rafted sediment increased with time into the Miocene, sea ice becoming permanent about 13 Ma ago, shortly after the Mid Miocene warm climatic optimum 15 Ma ago. At that time Arctic Intermediate Water began to form on the Eurasian shelf, presumably in response to increased density caused by the excretion of salt into subsurface waters as sea ice formed at the surface.

The full-scale connection of the Arctic Ocean to the North Atlantic was delayed because, although sea-floor

**Figure 7.11** Jörn Thiede.

spreading began in the Norwegian–Greenland Sea in the Eocene around 54 Ma ago, the Fram Strait between Svalbard and Greenland did not open fully until some time between the Early Miocene (20–25 Ma ago) and the Late Miocene (9.8 Ma ago) [97]. The ACEX results show that the ventilation of the Arctic by deep Atlantic waters began at 17.5 Ma ago at about the same time as Arctic outflow to the Atlantic began [96].

In 2010, Jörn Thiede (1941–) (Figure 7.11, Box 7.5) and colleagues brought together all of the evidence on Arctic ice rafting to paint a picture of the history of Northern Hemisphere glaciation for the Cenozoic [98].

Thiede's analysis shows that we cannot say anything about the sea ice cover of the Arctic Ocean for the Late Eocene to the Early Miocene (44.4–18.2 Ma ago), or for much of the Miocene (11.6–9.4 Ma ago), because sediment is missing in the cores from the Lomonosov Ridge. But it is plain that there was abundant sea ice in the Arctic even in the Late Eocene and that it was a continuous feature of the Arctic winter from possibly 47.5 Ma ago onwards [98].

Examining the evidence from deep-ocean drilling for ice-rafting, Thiede's group concluded that Greenland must have had an ice sheet much earlier than was formerly supposed. Ice rafting in these areas was most likely from icebergs rather than, as in the Arctic Ocean, from sea ice. The data '*suggest the more or less continuous existence of the Greenland ice sheet since 18 Ma [ago], maybe much longer*' [98]. That is consistent with evidence cited by Backman and Moran for some glaciation on Greenland as early as 38 Ma ago [96].

What might explain the development of conditions cold enough for sea ice to form in the Arctic Ocean 47 Ma ago? Thiede and his team '*speculate that this may be related to a reorganization of Siberian drainage patterns as a result of the tectonic changes subsequent to the collision of the plate carrying the Indian subcontinent northwards, once it*

**Box 7.5    Jörn Thiede**

Born in Berlin in April 1944, Thiede studied at the universities of Kiel (Germany), Buenos Aires (Argentina) and Vienna (Austria), receiving his PhD from the University of Kiel in 1971. His research has focused on attempting to understand the history of the ocean's shape, water, and life. He has led many expeditions into high latitude northern waters on Germany's research icebreaker *Polarstern* and on the drill-ship *JOIDES Resolution*, notably within the ODP's drilling programme on North Atlantic Arctic Gateways. He has held research and teaching posts at the Universities of Aarhus (Denmark), Bergen and Oslo (Norway), Oregon State (Corvallis, USA), Kiel (Germany) and Longyearbin (Svalbard). He helped to set up and became the first director of the GEOMAR Research Centre for Marine Sciences at Kiel University (1987–1995) and was the director of Germany's AWI for Polar and Marine Research (Bremerhaven, Germany) from 1997 to 2007. He currently holds positions with the Academy of Sciences and Literature of Mainz, in Germany, and the St Petersburg State University in Russia. Thiede has been chairman of the European Polar Board (1999–2002) and president of the SCAR (2002–2006), helping to reorganize both institutions. He helped to prepare for the International Polar Year 2007–2009, and initiated a novel, though ultimately unsuccessful, EU project to develop an international icebreaker (Aurora Borealis) for the Arctic. His current research as chief scientist of the Köppen Laboratory of the State University in St Petersburg involves late Cenozoic palaeoceanography, especially in the Arctic. For his contributions to the study of climate change, and especially Arctic climate change, he has received several awards, amongst them Germany's prestigious Leibniz Prize in 1988, and the Murchison Medal of the Geological Society of London in 1994. He was appointed a foreign member of the Russian Academy of Sciences in 2003. In 2016 he and two colleagues produced the first Encyclopaedia of Marine Geosciences.

*collided with the southern margins of the Eurasian plate'* [98]. That collision began about 50 Ma ago and may have caused Siberia to tilt down to the north, away from Tibet, so increasing the runoff of freshwater from central Asia to the Arctic, making conditions more suitable for the formation of sea ice there [98].

These new data paint a different picture from the old one. We now have evidence for ice at both poles dating back into the Eocene. Ice was evidently beginning to be present in the Arctic around 46 Ma ago, and, as Barrett told us, there is evidence for ice rafting events in the South Pacific at 51, 48, and 41 Ma [65]. Once you have ice, it increases the Earth's albedo, thus providing a positive feedback to whatever caused the Late Eocene cooling [96].

What sort of temperatures were we dealing with? The global isotopic and related data sets and their translation into global temperature records (Figures 7.7, and 7.10) [27, 58] are matched by palaeoclimate data from the Arctic. A multinational team led by Gifford Miller found in 2010 that Late Cretaceous Arctic temperatures averaged 15 °C, and were still at about 10 °C in the Eocene 50 Ma ago, when *Azolla* dominated the vegetation of the relatively fresh Arctic surface waters [99]. Temperatures rose by 3 °C in the later Eocene, when forests of *Metasequoia* dominated an Arctic landscape of organic-rich floodplains and wetlands. Despite the cooling and floral turnover at the Eocene–Oligocene transition 34 Ma ago, warm conditions persisted into the Early Miocene 23–16 Ma ago, when the central Canadian islands were covered in mixed conifer-hardwood forests like those of southern maritime Canada and New England today. *Metasequoia* was still present, although less abundant than in the Eocene. This was the period of winter sea ice, according to the ACEX team [96].

Late Miocene to Pliocene riverside forests of pine, birch, and spruce populated the Canadian Arctic islands after a pronounced Mid Miocene cooling [99]. Miller's Palaeoclimate data from Ellesmere Island suggest mean annual temperatures there were 14 °C warmer than today as recently as the mid Pliocene warm period. Given that the Arctic Ocean was rather fresh in the Eocene, we cannot call on poleward heat transport by the ocean to keep conditions warm at that time. Miller and his team pointed out that '*Taken very broadly, the Arctic changes parallel global changes during the Cenozoic, except that the changes in the Arctic were larger than those globally averaged … In general, global and Arctic temperature trends parallel changing atmospheric CO$_2$ concentrations, which is the likely cause for most of the temperature changes'* [99]. We will look at the role of CO$_2$ in later chapters.

Once again, we see the slow accumulation of fragments of knowledge filling out the spaces in the jigsaw puzzle. Some pieces don't fit and have to be discarded; others fit well. The Cenozoic palaeoclimate data confirm that as in the Cretaceous (Chapter 6), despite the fact that there was land in the polar regions, the world did not become cool as a consequence of Antarctica's arrival at the South Pole 90 Ma ago, contrary to what would be implied by Lyell's theory. It took another 45 Ma before a major ice sheet developed on Antarctica. The data also confirm that, whilst there was considerable evidence for the persistence of climate zones more or less unchanged through time, they expanded during periods of extreme warmth such as the

Mid to Late Cretaceous and the Eocene, and began to contract with the cooling that began in the late Eocene. The warm world of the Cretaceous and early Cenozoic gave way after 50 Ma ago to a cool mid to late Cenozoic world in which sea ice formed at both poles and large ice sheets then formed apparently first on Antarctica and then in the Arctic, though the jury is still out on the dating of the first Greenland glaciation. Much the same level of global cooling affected the late Cenozoic as had occurred in the Carboniferous and early Permian.

The cooling from the Middle Eocene (50 Ma) onwards was a puzzle. It was reasonable to explain the small changes from warm to cold that were superimposed on the broad envelope of Cenozoic cooling as being the product of Milankovitch's orbital variations on timescales of 10–100 Ka, but those did not explain the underlying cooling trend. Could the Sun be at fault? There was no evidence to suggest that the trend was the product of variations in solar output. If we only had the Sun's output to consider, then – as Jim Hansen and Makiko Sato pointed out [100] – the planet should have warmed, because solar luminosity increases as the Sun continues to burn hydrogen to form helium by nuclear fusion, and so is slowly getting brighter, like other 'main sequence' stars. Solar physics models suggest that during this time the Sun should have brightened by about 0.4%, equivalent to a forcing of about $1\,\mathrm{W\,m^{-2}}$. We should have seen modest warming throughout the Cenozoic Era.

Most geologists prior to the mid-1980s looked for physical explanations for the cooling, like changes in continental shape and position. Remember that Frakes and Kemp thought the world cooled because the Pacific shrank, diminishing the supply of heat to the Polar Regions. Shackleton and Kennett thought that the separation of Australia and Antarctica created the ACC, causing ice to form on Antarctica and cold deep water to cool the ocean.

What was really going on? Why were there warm forests in the early Cenozoic of high latitudes? Why did sea ice begin forming at both poles in the Late Eocene? Could we blame changes in ocean circulation? Or did $CO_2$ play an unexpected role? How had atmospheric chemistry changed? Was that what geologists had missed from their thinking? As we shall see in later chapters, to fully understand climate change geologists would now have to break away from their conservative bastion and let in light being shed by discoveries in other disciplines, especially oceanography, meteorology, glaciology, atmospheric and ocean chemistry, and geochemistry. Above all they would have to consider the behaviour of the carbon cycle. Having established how palaeoclimatology developed as a discipline, we now turn to the discoveries being made outside geology about the nature and influence on climate of the greenhouse gases.

## References

**1** Schott, W. (1935). Wissensch. *Ergebn. Deutschen Atlantischen Expedition Vermess. Forschungsschiff "Meteor" 1925–1927* 3 (3): 43–134.

**2** Hollister, C.D., Silva, A.J., and Driscoll, A. (1973). A giant piston corer. *Ocean Engineering* 2: 159–168.

**3** Menard, H.W. (1967). *Anatomy of an Expedition*, 255 pp. New York: McGraw Gill.

**4** Kennett, J.P. (1982). *Marine Geology*, 813 pp. New York: Prentice-Hall.

**5** Riedel, W.R. (1963). The present record: paleontology of pelagic sediments. In: *The Sea: Ideas and Observations on Progress in the Study of the Seas, Vol. 3, the Earth Beneath the Sea – History* (ed. M.N. Hill), 866–887. London: Wiley.

**6** Margolis, S. and Kennett, J.P. (1971). Cenozoic paleoglacial history of Antarctica recorded in subantarctic deep-sea cores. *American Journal of Science* 271: 1–36.

**7** Kennett, J.P., Houtz, R.E., Andrews, P.B., et al. (1975). *Initial Reports of the Deep Sea Drilling Project* 29, US Government Printing Office Washington D.C. 1187 pp.

**8** Arrhenius, G. (1952). Sediment cores from the East Pacific: properties of the sediment, *Reports of the Swedish Deep-Sea Expedition 1947-48* 1, 1–288.

**9** Arrhenius, G. (1966). Sedimentary record of long-period phenomena. In: *Advances in Earth Science, Contributions to the International Conference on the Earth Sciences, MIT, 1964* (ed. P.M. Hurley), 155–174. MIT Press.

**10** Emiliani, C. (1954). Temperatures of Pacific bottom waters and polar superficial wàters during the tertiary. *Science* 119: 853–855.

**11** Emiliani, C. (1956). Oligocene and Miocene temperature of the equatorial and subtropical Atlantic Ocean. *Journal of Geology* 64: 281–288.

**12** Shackleton, N.J. (1965). The high-precision isotopic analysis of oxygen and carbon in carbon dioxide. *Journal of Scientific Instruments* 42: 689–692.

**13** McCave, I.N., and Elderfield, H. (2011). Sir Nicholas John Shackleton, *Biographic Memoirs of Fellows of the Royal Society*. published online 29 June 2011.

**14** Summerhayes, C.P., and Shackleton, N.J., (eds.) (1986). *North Atlantic Palaeoceanography*, Special Publication **21**, Geological Society of London and Blackwells, 473 pp.

**15** Summerhayes, C.P. (1990). Palaeoclimates. *Journal of the Geological Society of London* 147: 315–320.

**16** Brassell, S.C., Eglinton, G., Marlowe, I.T. et al. (1986). Molecular stratigraphy: a new tool for climatic assessment. *Nature* 320: 129–133.

**17** Lawrence, K.T., Herbert, T.D., Dekens, P.S., and Ravelo, A.C. (2007). The application of the alkenone organic proxy to the study of Plio-Pleistocene climate. In: *Deep-Time Perspectives on Climate Change: Marrying the Signal from Computer Models and Biological Proxies* (eds. M. Williams, A.M. Haywood, F.J. Gregory and D.N. Schmidt), 539–562. Geological Society, London: Micropalaeontological Society Special Publication.

**18** Herbert, T.D. (2003). Alkenone paleotemperature determinations. In: *The Oceans and Marine Geochemistry*, Treatise on Geochemistry (eds H.D. Holland and K.K. Turekian), vol. 6 (ed. H. Elderfield), 391–432. Oxford: Elsevier-Pergamon.

**19** Prahl, F.G. and Wakeham, S.G. (1987). Calibration of unsaturation patterns in long-chain ketone compositions for paleotemperature assessment. *Nature* 330: 367–369.

**20** Elderfield, H. and Ganssen, G. (2000). Past temperature and $\partial^{18}$o of surface ocean waters inferred from foraminiferal Mg/Ca ratios. *Nature* 405: 442–445.

**21** Lea, D.W. (2003). Elemental and isotopic proxies of past ocean temperatures. In: *The Oceans and Marine Geochemistry*, Treatise on Geochemistry (eds. H.D. Holland and K.K. Turekian), vol. 6 (ed. H. Elderfield), 365–390. Oxford: Elsevier-Pergamon.

**22** Nürnberg, D., Bijma, J., and Hemleben, C. (1996). Assessing the reliability of magnesium in foraminiferal calcite as a proxy for water mass temperatures. *Geochimica Cosmochimica Acta* 60 (5): 803–814.

**23** Rosenthal, Y., Boyle, E.A., and Slowey, N. (1997). Temperature control on the incorporation of Mg, Sr, Fe, and Cd into benthic foraminiferal shells from the little Bahama Bank: prospects for thermocline paleoceanography. *Geochimica Cosmochimica Acta* 61 (17): 3622–3643.

**24** Lear, C.H., Elderfield, H., and Wilson, P.A. (2000). Cenozoic deep-sea temperatures and global ice volumes from Mg/Ca in benthic foraminiferal calcite. *Science* 287: 269–272.

**25** Elderfield, H., Greaves, M., Barker, S. et al. (2010). A record of bottom water temperature and seawater $\partial^{18}$O for the Southern Ocean over the past 440 kyr based on Mg/Ca of benthic foraminiferal *Uvigerina* spp. *Quaternary Science Reviews* 29: 160–169.

**26** Tripati, A.K., Delaney, M.L., Zachos, J.C. et al. (2003). Tropical Sea-surface temperature reconstruction of the early Paleogene using Mg/Ca ratios of planktonic foraminifera. *Paleoceanography* 18 (4): 1101. https://doi.org/10.1029/2003PA000937.

**27** Cramer, B.S., Miller, K.G., Barrett, P.J., and Wright, J.D. (2011). Late Cretaceous–Neogene trends in deep ocean temperature and continental ice volume: reconciling records of benthic foraminiferal geochemistry ($\partial^{18}$O and Mg/Ca) with sea level history. *Journal of Geophysical Research* 116: C12023, 23 pp.

**28** Barron, E.J. (1987). Eocene equator-to-pole surface ocean temperatures: a significant climate problem? *Paleoceanography* 2: 729–739.

**29** Pearson, P.N., Ditchfield, P.W., Singano, J. et al. (2001). Warm tropical sea surface temperatures in the late Cretaceous and Eocene epochs. *Nature* 413: 481–487.

**30** Pearson, P.N., van Dongen, B.E., Nicholas, C.J. et al. (2007). Stable warm tropical climate through the Eocene epoch. *Geology* 35: 211–214.

**31** Montañez, I.P., Norris, R.D., Algeo, T. et al. (2011). *Understanding Earth's Deep Past: Lessons for Our Climate Future*, 194 pp. Washington, D.C.: National Academies Press.

**32** Boyle, E.A. and Keigwin, L.D. (1982). Deep circulation of the North Atlantic over the last 200,000 years: geochemical evidence. *Science* 218 (4574): 784–787.

**33** Anderson, R.F. (2003). Chemical tracers of particle transport. In: *The Oceans and Marine Geochemistry*, *Treatise on Geochemistry* (ed. H.D. Holland and K.K. Turekian), vol. 6 (ed. H. Elderfield), 247–291. Oxford: Elsevier-Pergamon.

**34** Hollis, C.J., Taylor, K.W.R., Handley, L. et al. (2012). Early Paleocene temperature history of Southwest Pacific Ocean: reconciling proxies and models. *Earth and Planetary Science Letters* 349-350: 53–66.

**35** Cronin, T.M. (2010). *Paleoclimates: Understanding Climate Change Past and Present*, 441 pp. New York: Columbia University Press.

**36** Di Venere, V.J., Kent, D.V., and Dalziel, I.W.D. (1994). Mid-Cretaceous paleomagnetic results from Marie Byrd Land, West Antarctica: a test of post-100 ma relative motion between East and West Antarctica. *Journal of Geophysical Research* 99 (B8), 15, 115-15, 139.

**37** Nichols, R.L. (1964). Present status of Antarctic glacial geology. In: *Antarctic Geology* (ed. R.J. Adie), 123–137. Amsterdam: North-Holland.

**38** Craddock, C., Bastien, T.W., and Rutford, R.H. (1964). Geology of the Jones Mountains area. In: *Antarctic Geology* (ed. R.J. Adie), 171–187. Amsterdam: North-Holland.

**39** Armstrong, R.L., Hamilton, W., and Denton, G.H. (1968). Glaciation in Taylor Valley, Antarctica, older than 2.7 million years. *Science* 159: 187–189.

**40** Denton, G.H., Armstrong, R.L., and Stuiver, M. (1970). Late Cenozoic glaciation in Antarctica. The record in the McMurdo sound region. *Antarctic Journal of the United States* 5: 15–21.

**41** Kennett, J.P. (1967). Recognition and correlation of the Kapitean Stage (upper Miocene, New Zealand). *New Zealand Journal of Geology and Geophysics* 10: 1051–1063.

42 Kennett, J.P. and Margolis, S.V. (1970). Antarctic glaciation during the tertiary recorded in sub-Antarctic deep-sea cores. *Science* 170: 1085–1087.

43 Watkins, N.D. and Kennett, J.P. (1971). Antarctic bottom water: major change in velocity during the late Cenozoic between Australia and Antarctica. *Science* 173: 813–818.

44 Torii, T. (1981). A Review of the Dry Valley Drilling Project, 1971–76. *Polar Record* 20 (129): 533–541.

45 Hayes, D.E., Frakes, L.A., Barrett, P.J. et al. (1973). *Initial Reports of the Deep Sea Drilling Project, Vol. 28, Freemantle, Australia to Christchurch, New Zealand, December 1972–February 1973*. Washington DC: US Government Printing Office.

46 McGinnis, L.D. (ed.) (1981). *Dry Valley Drilling Project*, Antarctic Research Series **81**. Washington D.C: American Geophysical Union.

47 Barrett, P.J. (2007). Cenozoic climate and sea level history from Glacimarine strata off the Victoria Land Coast, Cape Roberts Project, Antarctica, in glacial processes and products (ed. M. J. Hambrey, P. Christoffersen, N. F. Glasser, and B. Hubbart). *International Association of Sedimentologists Special Publicaiton* 39: 259–287.

48 Barrett, P.J. (2009). A history of Antarctic Cenozoic glaciation – view from the margin. In: *Antarctic Climate Evolution, Developments in Earth and Environmental Sciences* **8** (eds. F. Florindo and M. Siegert), 33–83. Amsterdam: Elsevier.

49 Naish T.R., Powell, R.D., Barrett, P.J., et al. (2008). Late Cenozoic climate history of the Ross Embayment from the AND-1B drill hole: culmination of three decades of Antarctic margin drilling, in *Antarctica: A Keystone in a Changing World* (eds A.K. Cooper, P.J. Barrett, H. Stagg, B. Storey, E. Stump, and W. Wise), Proc. 10th International Symposium on Antarctic Earth Sciences, Santa Barbara, California, 26 August – 1 September 2007, 71–82 (http://www.nap.edu/catalog/12168.html).

50 Naish, T.R. et al. (2009). Obliquity-paced Pliocene West Antarctic Ice Sheet oscillations. *Nature* 458: 322–329.

51 Harwood, D., Florindo, F., Talarico, F., and Levy, R.H. (2008-2009). Studies from the ANDRILL Southern McMurdo Sound project, Antarctica – initial science report on AND-2A. *Terra Antarctica* 15 (1).

52 Kennett, J.P., Burns, R.E., Andrews, J.E. et al. (1972). Australian – Antarctic continental drift, palaeocirculation changes and Oligocene deep-sea erosion. *Nature* 239 (91): 51–55.

53 Kennett, J.P., Houtz, R.E., Andrews, P.B., et al. (1975). Cenozoic paleoceanography in the Southwest Pacific Ocean, Antarctic glaciation, and the development of the Circum-Antarctic Current, *Initial Reports DSDP* **29**, US Government Printing Office. Washington D.C. 1155–1169.

54 Shackleton, N.J., and Kennett, J.P. 1975. Paleotemperature history of the Cenozoic and the initiation of Antarctic glaciation: oxygen and carbon isotope analyses in DSDP sites 277, 279, and 280, *Initial Reports DSDP* 29, US Government Printing Office. Washington D.C. 743–755.

55 Kennett, J.P. (1977). Cenozoic evolution of Antarctic glaciation. *Journal of Geophysical Research* 82: 3843–3860.

56 Zachos, J.C., Breza, J., and Wise, S.W. (1992). Early Oligocene ice sheet expansion on Antarctica, sedimentological and isotopic evidence from the Kerguelen Plateau. *Geology* 20: 569–573.

57 Zachos, J.C., Stott, L.D., and Lohmann, K.C. (1994). Evolution of early Cenozoic marine temperatures. *Paleoceanography* 9 (2): 353–387.

58 Zachos, J., Pagani, M., Sloan, L. et al. (2001). Trends, rhythms and aberrations in global climate 65 Ma to present. *Science* 292: 686–693.

59 Francis, J.E., Marenssi, S., Levy, R. et al. (2009). From greenhouse to icehouse – the Eocene/Oligocene in Antarctica. In: *Antarctic Climate Evolution, Developments in Earth and Environmental Sciences* **8** (eds. F. Florindo and M. Siegert), 309–368. Amsterdam: Elsevier.

60 Francis, J.E., Ashworth, A., Cantrill, D.J., et al. (2008). 100 million years of Antarctic climate evolution: evidence from fossil plants, in in *Antarctica: A Keystone in a Changing World* (eds A.K. Cooper, P.J. Barrett, H. Stagg, B. Storey, E. Stump, and W. Wise), Proc. 10th International Symposium on Antarctic Earth Sciences, Santa Barbara, California, 26 August – 1 September 2007,19–27 (http://www.nap.edu/catalog/12168.html)

61 Pross, J., Contreras, L., Bijl, P.K. et al. (2012). Persistent near-tropical warmth on the Antarctic continent during the early Eocene epoch. *Nature* 488 https://doi.org/10.1038/nature11300.

62 Anderson, J.B., quoted in Schultz, C. (2013). Tectonic, climate and cryospheric evolution of the Antarctic Peninsula. *EOS* **94** (23), 210; [A review of the book of the same title published by Anderson, J.B., and Wellner, J.S. (2011) as *AGU Special Publication* **63**, 218 pp].

63 Bertler, N.A.N. and Barrett, P.J. (2010). Vanishing polar ice sheets. In: *Changing Climates, Earth Systems and Society, International Year of Planet Earth* (ed. J. Dodson), 49–83. Berlin: Springer.

64 Frakes, L.A., Francis, J.E., and Syktus, J.I. (1992). *Climate Modes of the Phanerozoic – The History of the Earth's Climate Over the Past 600 Million Years*, 274 pp. Cambridge University Press.

65 Barrett, P. (1999). Antarctic climate history over the last 100 million years. Proceedings of the Workshop on Geological Records of Global and Planetary Changes, *Terra Antarctica Reports* 3, 53–72.

66 Barrett, P.J. (2013). Resolving views on Antarctic Neogene glacial history – the Sirius debate. *Transactions of the Royal Society of Edinburgh* 104 (1): 31–53.

**67** Wilson, D.S. and Luyendyk, B.P. (2009). West Antarctic paleotopography estimated at the Eocene-Oligocene climate transition. *Geophysical Research Letters* 36: L16302. https://doi.org/10.1029/2009GL039297.

**68** Kennett, J.P. (1980). Paleoceanographic and biogeographic evolution of the Southern Ocean During the Cenozoic, and Cenozoic microfossil datums. *Palaeogeography, Palaeoclimatology, Palaeoecology* 31: 123–152.

**69** Miller, K.G., Browning, J.V., Aubry, M.-P. et al. (2008). Eocene-Oligocene global climate and sea-level changes: St. Stephens Quarry, Alabama. *Bulletin of the Geological Society of America* 120 (1–2): 34–53.

**70** Frakes, L.A. and Kemp, M.E. (1973). Palaeogene continental position and evolution of climate. In: *Implications of Continental Drift to the Earth Sciences*, vol. 1 (eds. D.H. Tarling and S.K. Runcorn), 539–558. London: Academic Press.

**71** Zalasiewicz, J. and Williams, M. (2012). *The Goldilocks Planet – The Four Billion Year Story of Earth's Climate*, 303 pp. Oxford University Press.

**72** Smith, A.G., Briden, J.C., and Drewry, G.E. (1973). Phanerozoic World Maps. In: *Organisms And Continents Through Time*, *Special Paper on Palaeontology*, vol. 12 (ed. N.F. Hughes), 1–43. London: Palaeontological Association.

**73** Wolfe, J.A. (1985). Distribution of major vegetational types during the tertiary. In: *The Carbon Cycle and Atmospheric $CO_2$: Natural Variations Archean to Present*, *Geophysical Monograph*, vol. 32 (eds. E.T. Sundquist and W.S. Broecker), 357–375. Washington D.C.: American Geophysical Union.

**74** Briden, J.C., and Irving, E. (1964). Palaeolatitude spectra of sedimentary palaeoclimatic indicators, in *Problems in Palaeoclimatology* (ed A.E.M. Nairn), Proc. NATO Palaeoclimates Conference, University of Newcastle-upon-Tyne, January 1963, Intersci. Publishers, London, 199–224.

**75** Olsen, J.S. (1985). Cenozoic fluctuations in biotice parts of the global carbon cycle. In: *The Carbon Cycle and Atmospheric $CO_2$: Natural Variations Archean to Present*, *Geophysical Monograph*, vol. 32 (eds. E.T. Sundquist and W.S. Broecker), 377–396. Washington D.C.: American Geophysical Union.

**76** Parrish, J.T., Ziegler, A.M., and Scotese, C.R. (1982). Rainfall patterns and the distribution of coals and evaporites in the Mesozoic and Cenozoic. *Palaeogeography, Palaeoclimatology, Palaeoecology* 40: 67–101.

**77** Savin, S.M., Douglas, R.G., and Stehli, F.G. (1975). Tertiary marine paleoemperatures. *Bulletin pf the Geological Society of America* 86: 1499–1510.

**78** Miller, K.G., Fairbanks, R.G., and Mountain, G.S. (1987). Tertiary oxygen isotope synthesis, sea level history, and continental margin erosion. *Paleoceanography* 2: 1–19.

**79** Shackleton, N.J. and Opdyke, N.D. (1973). Oxygen isotope and palaeomagnetic stratigraphy of equatorial Pacific core V28-238: oxygen isotope temperatures and ice volumes on a $10^5$ year and a $10^6$ year scale. *Quaternary Research* 3: 39–55.

**80** Shackleton, N.J. (1987). Oxygen isotopes, ice volume and sea level. *Quaternary Science Reviews* 6: 183–190.

**81** Katz, M.E., Cramer, B.S., Toggweiler, J.R. et al. (2011). Impact of Antarctic Circumpolar Current development on late Paleogene ocean structure. *Science* 332: 1076–1078.

**82** Barker, P.F. and Thomas, E. (2004). Origin, signature and paleoclimatic influence of the Antarctic Circumpolar Current. *Earth Science Reviews* 66: 143–166.

**83** Dalziel, I.W.D., Lawver, L.A., Pearce, J.A. et al. (2013). A potential barrier to deep Antarctic circumpolar flow until the late Miocene. *Geology* 41 (9): 947–950.

**84** Dalziel, I.W.D. (2014). Drake passage and the Scotia arc: a tortuous space-time gatewayfor the Antarctic Circumpolar Current. *Geology* 42 (4): 367–368.

**85** Mawbey, E.M. and Lear, C.H. (2013). Carbon cycle feedbacks during the Oligocene-Miocene transient glaciation. *Geology* https://doi.org/10.1130/G34422.1.

**86** Ganssen, E., DeConto, R.M., Pollard, D., and Levy, R.H. (2016). Dynamic Antarctic ice sheet during the early to mid-Miocene. *Proceedings of the National Academy of Sciences of the United States of America* 113 (13): 3459–3464.

**87** Vincent, E. and Berger, W.H. (1985). Carbon dioxide and polar cooling in the Miocene: the Monterey hypothesis. In: *The Carbon Cycle and Atmospheric $CO_2$: Natural Variations Archean to Present*, *Geophysical Monograph* **32** (eds. E.T. Sundquist and W.S. Broecker), 455–468. Washington D.C.: American Geophysical Union.

**88** Hodell, D.A., Elmstrom, K.M., and Kennett, J.P. (1986). Late Miocene benthic $\delta^{18}O$ changes, global ice volume, sea level and the 'Messinian Salinity Crisis'. *Nature* 320: 411–414.

**89** Cita, M.B. and McKenzie, J.A. (1986). The terminal Miocene event. In: *Mesozoic and Cenozoic Oceans*, *Geodynamics Series*, vol. 15 (ed. K.J. Hsu), 123–140. Washington D.C.: American Geophysical Union.

**90** Ross, D.A., Uchupi, E., Summerhayes, C.P. et al. (1978). Sedimentation and structure of the Nile Cone and Levant Platform area. In: *Sedimentation in Submarine Canyons, Fans, and Trenches* (eds. D.J. Stanley and G. Kelling), 261–275. Dowden, Hutchinson and Ross.

**91** Elderfield, H. and Summerhayes, C.P. (1978). Salt diffusion in eastern Mediterranean Sea sediments. *Deep Sea Research* 25: 837–841.

**92** Bice, K.L., Scotese, C.R., Seidov, D., and Barron, E.J. (2000). Quantifying the role of geographic change in

Cenozoic ocean heat transport using uncoupled atmosphere and ocean models. *Palaeogeography, Palaeoclimatology, Palaeoecology* 161: 295–310.

**93** Ravelo, A.C., Billups, K., Dekens, P.S. et al. (2007). Onto the ice ages: proxy evidence for the onset of Northern Hemisphere glaciation. In: *Deep-Time Perspectives on Climate Change: Marrying the Signal from Computer Models and Biological Proxies, Micropalaeontological Society Special Publication* (eds. M. Williams, A.M. Haywood, F.J. Gregory and D.N. Schmidt), 563–573. London: The Geological Society.

**94** Tripati, A.K., Dawber, C.F., Ferretti, P., et al. (2007). Evidence for synchronous glaciation of Antarctica and the Northern Hemisphere during the Eocene and Oligocene: insights from Pacific records of the oxygen isotopic composition of seawater. 10^th International Symposium on Earth Science, *Extended Abstract* **186**, U.S. Geological Survey and The National Academies, USGS OF-2007-1047.

**95** Moore, T.C., et al. (2006). Sedimentation and subsidence history of the Lomonosov Ridge, in *Proc. IODP, 302: Edinburgh* (eds J. Backman, K. Moran, D.B. McInroy, L.A. Mayer, and the Expedition 302 Scientists), Integrated Ocean Drilling Program Management International, Inc., doi:https://doi.org/10.2204/iodp.proc.302.105.2006 (http://publications.iodp.org/proceedings/302/105/105_.htm).

**96** Backman, J. and Moran, K. (2008). Introduction to special section on Cenozoic paleoceanography of the Central Arctic Ocean. *Paleoceanography* 23 https://doi.org/10.1029/2007/PA001516.

**97** Engen, O., Faleide, J.I., and Dyreng, T.K. (2008). Opening of the Fram Strait gateway: a review of plate tectonic constraints. *Tectonophysics* 450: 51–69.

**98** Thiede, J., Jessen, C., Knutz, P. et al. (2010). Millions of years of Greenland ice sheet history recorded in ocean sediments. *Polarforschung* 80 (3): 141–159.

**99** Miller, G.H., Brigham-Grette, J., Alley, R.B. et al. (2010). Temperature and precipitation history of the Arctic. *Quaternary Science Reviews* 29: 1679–1715.

**100** Hansen, J.E. and Sato, M. (2012). Paleoclimate implications for human-made climate change. In: *Climate Change: Inferences from Paleoclimate and Regional Aspects* (eds. A. Berger, F. Mesinger and D. Šijački), 21–48. Berlin: Springer https://doi.org/10.1007/978-3-7091-0973-1_2.

# 8

# Greenhouse Gas Theory Matures

---

**LEARNING OUTCOMES**

- Understand and be able to describe the infrared spectrum and explain its significance in terms of greenhouse gas (radiative transfer) theory (i.e. the relationship between $CO_2$ and water vapour and infrared absorption).
- Be able to describe the array of technical advances that led to our current understanding of the role of greenhouse gases in the climate system (e.g. advances in spectrometry, satellite spectrometry, computing, ocean and atmospheric measurement of $CO_2$, numerical modelling of the effect of $CO_2$).
- Know how to convert concentrations of carbon to those of $CO_2$.
- Understand and be able to describe the pattern of increasing $CO_2$ with time (seasonal variability, hemispheric differences, isotopic signature ($\partial^{13}C$), the concept of 'background air', the related decrease of atmospheric oxygen).
- Understand how air-borne $CO_2$ exchanges with that in the ocean, and know how to counter the misconception that $CO_2$ does not last long in the air.
- Understand and be able to describe 'polar amplification' and its implications, and why Antarctica is cooler than the Arctic.
- Understand and explain the meaning of Earth System Science and biogeochemistry.
- Understand and be able to describe the slow carbon cycle.
- Understand and be able to describe the roles of the solubility pump and the biological pump in the fast carbon cycle.

---

Having reviewed climate change through time in previous chapters, we are left with many unanswered questions that can only be adequately addressed by exploring the role of greenhouse gases, especially $CO_2$, through time. In keeping with Lyell's dictum that the present is the key to the past, before palaeoclimatologists could hope to understand the role of $CO_2$ in the climates of the past, they first needed to know what its role is in the modern climate system, and until the 1950s even meteorologists did not know much about that. Progress had to wait until our ability to characterize and measure $CO_2$ in the ocean and atmosphere evolved. In this chapter we look at the developing understanding of the role of $CO_2$ in today's climate, before assessing in Chapter 9 what its role was in the climates of the past. As for other topics in this book, this story is an unfolding narrative painstakingly pieced together by many individuals working with progressively better technology over several decades, starting well over 100 years ago as we saw in Chapter 4. Much of what we now know, and which has been addressed in reports from 1990 onwards by the UN's Intergovernmental Panel on Climate Change (IPCC), is based on the comprehensive scientific understanding developed in the pre-IPCC years

between 1950 and 1990. The IPCC's reports, dating from 1990 onwards, widen and deepen that understanding, but do not fundamentally change it.

## 8.1 $CO_2$ in the Atmosphere and Ocean (1930–1955)

In the late 1930s, steam engineer, inventor, and amateur meteorologist Guy Callendar (1898–1964) started thinking about how $CO_2$ might have affected climate [1, 2]. His pronouncements fell on deaf ears. Besides, his calculations contained some errors. Like Arrhenius, he assumed wrongly that the burning of fossil fuels and emission of $CO_2$ would increase linearly, whilst in fact they increased exponentially. And he failed to take into account the effect on the climate of feedback from water vapour [3].

What held $CO_2$ science back? Firstly, spectroscopists couldn't measure variation across the infrared radiation spectrum with the necessary refinement. They didn't know where in the spectrum $CO_2$ had the most effect. Secondly, they lacked computers to calculate the flux of infrared

*Palaeoclimatology: From Snowball Earth to the Anthropocene*, First Edition. Colin P. Summerhayes.
© 2020 John Wiley & Sons Ltd. Published 2020 by John Wiley & Sons Ltd.

radiation at different wavelengths. Three communities pushed for improvement. Astronomers wanted better measures of the absorption and emission spectra of gases in the Earth's atmosphere, to correct for their effect on the infrared spectra of light from other planets. Meteorologists wanted to know how the absorption and emission of infrared radiation affected the vertical temperature structure of the atmosphere, to improve weather forecasting. But the biggest driver for change after the Second World War was military, to detect heat sources represented by enemy fighter jet engines or air-to-air or ground-to-air missiles. It was important to be able to determine where in the spectrum infrared radiation was blocked by absorption by gases like $CO_2$, and where the spectrum was transparent, letting that radiation pass.

By the early 1950s, spectroscopists knew that the spectrum of infrared light emitted by Earth's Sun-warmed surface ranged from wavelengths of 5–100 mμ, with about 50% peaking between 8.5 and 28 mμ, and that $CO_2$ preferentially absorbed infrared radiation at wavelengths of 13–18 mμ, right in the middle of that peak. Hence absorption of infrared light by $CO_2$ must significantly impede the transmission of infrared radiation through the atmosphere, causing the air to warm (Figure 8.1) [4].

Military funding was needed to build spectrometers of very high resolution, so as to establish the spectrum of $CO_2$ precisely as the basis for accurate calculations of its absorptive effects. In 1950, the US Office of Naval Research (ONR)

funded the design of an ultra-high-resolution spectrometer at the Laboratory of Astrophysics and Physical Meteorology of Johns Hopkins University in Baltimore, where scientists like Robert P. Madden did the research [5]. By then the absorption spectrum of $CO_2$ between 13 and 18 mμ was known to contain clusters of hundreds and perhaps thousands of individual absorption lines, between which the spectrum was transparent allowing light from a source to pass without being absorbed (Figure 8.2). An ultra-high-resolution instrument was needed to separate the individual lines so that the amount of light absorbed by each of them could be calculated accurately [6]. High resolution demanded a long light path: 14 m long in the case of Madden's instrument. Later, ONR funded development of a more advanced spectrometer with a light path of 182 m [7]. Data began to emerge as early as 1952 [8], although Madden's report was not declassified and published until 1961 [9]. Here, then, we have a practical application of greenhouse gas theory [10].

The spectroscopic data from these various military-funded experiments were collected by the US Air Force at their Cambridge (Massachusetts) Research Laboratories in the late 1960s, and are now available from the **Hi**gh-Resolution **Tran**smission Molecular Absorption (HITRAN) Database, being further developed at the Atomic and Molecular Physics Division of the Harvard-Smithsonian Centre for Astrophysics [11]. NASA uses HITRANS data to study the changing properties of the atmosphere, whilst numerical modellers use the data to simulate and predict the effects of $CO_2$ and other greenhouse gases in the atmosphere. This is another of those examples of a spinoff from military research adding value in the civilian science sphere.

Calculation was exceptionally tedious before computers arrived on the scene. The marriage of high-resolution spectroscopy and the computer constituted a major

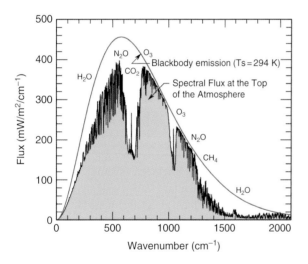

**Figure 8.1** The effect of different greenhouse gases ($CO_2$, methane [$CH_4$], nitrous oxide [$N_2O$], ozone [$O_3$] and water vapour [$H_2O$]) on thermal radiation emitted from the Earth's surface in the infrared region of the spectrum, as observed from a satellite. The red line is the calculated black body emission of infrared radiation from the Earth's surface, whilst the black line is the measured outgoing infrared radiation at the top of the atmosphere. Where the black and red lines diverge, these gases are trapping outgoing radiation, thus warming the atmosphere. The areas where the two curves match are 'windows' through which infrared radiation escapes without entrapment en route. The units of radiance (y-axis) are milliwatts per square metre per wavelength. The x-axis is wave number, which is inversely proportional to wavelength. *Source:* From Ref. [4].

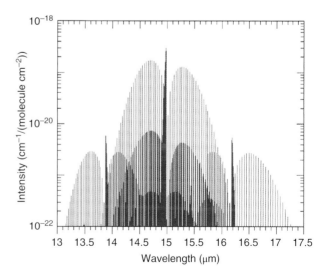

**Figure 8.2** Details of the spectrum of $CO_2$. The solid red areas represent hundreds of closely spaced absorption lines. *Source:* From http://ozonedepletiontheory.info/what-is-radiation.html.

technical breakthrough, rather like Galileo using his new-fangled telescope to discover the four moons of Jupiter, which ultimately led him to prove Copernicus right, to the dismay of the conservative church.

Amongst those working on how to measure infrared radiation in relation to the temperature of the atmosphere in the 1950s and 1960s was John (later Sir John) Houghton (1931–2020), then a research scientist at Oxford, later to become director of the UK's Meteorological Office, and leader of the scientific panel (Working Group 1) of the IPCC. A much respected scientist and fellow of the UK's Royal Society, Houghton was the recipient in 2006 of the Japan Prize, the equivalent in the environmental sciences of the Nobel Prize. In the mid 1950s, he designed the first airborne infrared spectrometers, later refining them so that they could be deployed on NASA's Nimbus 4 satellite in 1970. That enabled us to measure Earth's infrared radiation from space for the first time, a breakthrough reliant on the development of three new technologies: spectrometers to make the measurements, computers to process the data, and satellites to carry the payload. The tale is told in Houghton's 2013 autobiography *In the Eye of the Storm* [12]. His work on infrared radiation is ably summarized in his 1966 book with Des Smith *Infra-red Physics*, a must-read for anyone interested in $CO_2$ and infrared radiation [13]. Summarizing decades of experimental work they remind us that '*a large proportion of [the] infra-red radiation emitted by the Earth's surface ... [is] ... absorbed and in turn re-emitted by different layers of the atmosphere, in amount dependent on the concentration of the absorber, the temperature, and the absorption coefficient, which itself depends on pressure and temperature*' [13]. This is not theory, but measurement.

## 8.2 $CO_2$ in the Atmosphere and Ocean (1955–1979)

One of the key people working on the absorption and emission of infrared radiation with the new data emerging from Johns Hopkins was Canadian physicist Gilbert Norman Plass (1920–2004) (Figure 8.3). In 1956 he capitalized on improvements in spectroscopy and computing to recalculate the radiation flux due to $CO_2$ in the atmosphere [14–18]. He knew that each of the many thousand closely spaced absorption lines in the $CO_2$ spectrum (Figure 8.2) broadened as the concentration of $CO_2$ increased, eventually eliminating the transparent gaps through which infrared energy could pass between the lines. Furthermore, when the $CO_2$ concentration was sufficiently high, even its weaker absorption bands became effective, enabling a greater amount of radiation to be absorbed, as discussed later.

In 1956, Plass calculated that doubling the amount of $CO_2$ in the atmosphere would cause a rise in the average

**Figure 8.3** Gilbert Plass.

temperature of the atmosphere at the Earth's surface to 3.6 °C to restore equilibrium in the balance between solar radiation received and radiation emitted back into space from the top of the atmosphere [14]. He qualified that estimate in 1961 by pointing out that '*The actual temperature changes for the earth may even be somewhat larger than these values because of the action of water vapour in reinforcing temperature changes caused by other factors*' [18]. Like Callendar and Arrhenius, he made some erroneous assumptions [3], but his findings woke people to the potential effects of an exponential increase of $CO_2$ in the atmosphere.

Looking to the future, Plass calculated that '*If no other factors change, man's activities are increasing the average temperature by 1.1°C per century*' [14]. Considering the effect of the absorption of $CO_2$ by the ocean, and taking account of its slow circulation, Plass calculated that '*it would probably take at least 10,000 years for the atmosphere–ocean system to come to equilibrium after a change in the atmospheric $CO_2$ amount*' [14]. He warned, '*the influence of the extra $CO_2$ on the climate will become increasingly important in the near future as continuously greater amounts of $CO_2$ are released into the atmosphere by man's activities*' [14]. This was in 1956: it's not something new.

The oceanographers of the mid 1950s were just as interested as Plass in how $CO_2$ was distributed. They knew that about 90 billion tonnes (1 billion tonnes = 1 Gigatonne, or Gt) of carbon were routinely exchanged per year as part of the chemical equilibrium between the ocean (with a carbon reservoir of 38 000 Gt) and the atmosphere (with a carbon reservoir of 780 Gt).[1] One oceanographer,

---

[1] To convert from carbon (C) to carbon dioxide ($CO_2$) multiply by the ratio of the molecular weight of **carbon dioxide** to that of **carbon** (44/12) [= 3.67]. 1 GtC = 3.67 $GtCO_2$.

**Figure 8.4**   Roger Revelle in August 1952.

Roger Revelle (1909–1991) (Figure 8.4; Box 8.1), the director of Scripps, thought that studying the radioactive isotope of carbon, $^{14}C$, in the atmosphere and ocean might help him to understand the behaviour of the marine part of the global carbon cycle. He needed to hire an expert on $^{14}C$. The key person at the leading edge of applying $^{14}C$ to the study of natural materials was Hans Suess (Figure 8.5), whom we met in Chapter 5. Suess had already deduced that $^{14}C$ in the atmosphere was being diluted by the $^{12}C$ derived from the burning of fossil fuels [19]. In 1955 Revelle hired Suess to work with him.

The two men knew that fossil fuel combustion produced $CO_2$ at a rate two orders of magnitude greater than its usual rate of production from volcanoes. Echoing Plass, they concluded in a landmark paper in 1957 that '*human beings are now carrying out a large-scale geophysical experiment of a kind that could not have happened in the past nor be reproduced in the future. Within a few centuries we are returning to the atmosphere and oceans the concentrated organic carbon stored in sedimentary rocks over hundreds of millions of years. This experiment, if adequately documented, may yield*

---

**Box 8.1   Roger Revelle**

Born in Seattle, Washington State, on 7 March 1909, Roger Revelle moved to California with his family in 1917, graduated from Pomona College in 1929, and attended the University of California as a geology graduate student in 1930, moving to the University's Scripps Institution of Oceanography in 1931. Scripps was tiny at the time, and oceanography itself was a young science. Revelle began by studying carbonate sediments from the deep ocean, moving on to study carbon dioxide in seawater and its associated buffer mechanism, which limits ocean acidity, co-authoring a paper on the *Buffer Mechanism of Sea Water* in the Scripps Bulletin in 1934. That was the beginning of his interest in the carbon cycle and eventually in measuring carbon dioxide in the atmosphere. Obtaining his PhD in 1936, Revelle spent a year at Norway's Geophysical Institute, meeting many of the prominent European oceanographers of the day. At the same time he received a commission in the US Naval Reserve. Returning to teach at Scripps in 1937, he then fulfilled military research roles with the Navy during the Second World War. He planned and initiated the organization of the Oceanographic Section of the US Navy Hydrographic Office, and helped to plan the ONR, being appointed head of its Geophysics Branch in 1946. At the same time he worked closely with Scripps director, Harald Sverdrup, to reorganize and

develop Scripps as a premier oceanographic institution. Revelle returned to Scripps as associate director in 1948 and became its director in 1950. His activities from then on are too numerous to mention, including providing advice at the highest level of the US government and to international agencies like UNESCO. In 1957, for the International Geophysical Year (IGY) he organized the Special (now Scientific) Committee on Oceanic Research (SCOR) on behalf of the International Council of Scientific Unions (ICSU), and became SCOR's first president. That same year he was elected a member of the US National Academy of Sciences. He was instrumental in planning UNESCO's oceanographic office, now the Intergovernmental Oceanographic Commission (IOC). During the IGY, Scripps was designated as a participant in the Atmospheric Carbon Dioxide Program, and Revelle hired Charles David Keeling in 1956 to head the program. Revelle did a great deal to inform the public and policy makers about the role of $CO_2$ in global warming, and to ensure that it was measured routinely. He served government in and out of Washington in a myriad ways. He was awarded the prestigious Tyler Ecology Energy Prize in 1984, the Balzan Foundation Prize in 1986, and the US National Medal of Science in 1990. In 1995, Scripps launched the research vessel Roger Revelle in his honour.

**Figure 8.5** Hans Suess in 1972.

**Figure 8.6** Charles Keeling (https://library.ucsd.edu/dc/object/bb3857862s).

*a far-reaching insight into the processes determining weather and climate. It therefore becomes of prime importance to attempt to determine the way in which carbon dioxide is partitioned between the atmosphere, the oceans, the biosphere and the lithosphere'* [20].

Echoing Plass's calls for more data, Revelle hinted at the need for funds: *'An opportunity exists during the International Geophysical Year [the IGY of 1957-58] to obtain much of the necessary information'* [20]. His wish was granted. During the IGY he got funds to employ Charles David Keeling (1928–2005) (Figure 8.6) to set up a carbon dioxide measuring station 3000 m up on the Mauna Loa volcano in Hawaii. Keeling began measuring $CO_2$ in well-mixed oceanic air in 1958 [21], using a system he developed whilst a post-doctoral fellow in geochemistry at the California Institute of Technology. His measurements, which continue at Mauna Loa today, established the steady rise in $CO_2$ with time, as well as its seasonal ups and downs (Figure 8.7). They are also now made at numerous other sites, including one at the South Pole (Figure 8.7). Why Mauna Loa? Keeling knew that the content of $CO_2$ in the atmosphere near ground level varies widely from place to place, depending on proximity to major human sources like cities or factories. He could only obtain the well-mixed 'background' concentration in remote areas. The coasts of open ocean areas with onshore breezes were ideal. Wouldn't the Mauna Loa volcano emit $CO_2$? Yes, but the recording station was upwind of any volcanic emissions. Measurements at multiple sites confirm that the Mauna Loa record is not corrupted.

Apart from the trend, two things are clear from Figure 8.7: first, the $CO_2$ record from the South Pole has lower levels of $CO_2$ (because it is from the Southern Hemisphere and the gradient of $CO_2$ increases towards the north, the main source of anthropogenic $CO_2$); second, the South Pole record also shows much less interannual variability (reflecting the relative dearth of land plants in the Southern Hemisphere compared with the Northern, where most of the land is found). This latter observation will be important when we compare records of $CO_2$ from land plants in the north with records of $CO_2$ from ice in the south. Greenland ice is not used for past atmospheric $CO_2$ measurements because its dust contains dolomite, a carbonate rock that reacts with the acids derived from precipitation and from the oxidation of fine-grained organic material to produce $CO_2$ as a by-product [22].

After the development of mass spectrometric analytical methods in the 1950s, geochemists began measuring the distribution of isotopes of various kinds in rocks, sediments, biological materials, the oceans, and the atmosphere [23]. Keeling measured the carbon isotopes in atmospheric $CO_2$ as a way of characterizing this molecule. Carbon has two stable isotopes: 98.9% is $^{12}C$ (with six protons and six neutrons), and 1.1% is $^{13}C$ (with one extra neutron). During photosynthesis, the enzyme *Rubisco* discriminates against the heavier of the two isotopes, giving land plants a higher proportion of $^{12}C$ and hence a more negative $^{13}C/^{12}C$ (or $\partial^{13}C$) ratio than the air. Back in 1960 the air from Mauna Loa and the South Pole contained $\partial^{13}C$ ratios typical of land plants (Figure 8.8) [21]. The ratio decreased in the summer (i.e. values were more negative, with less $^{13}C$), along with $CO_2$, when photosynthetic activity was greatest, and increased in the winter (i.e. had more $^{13}C$), along with $CO_2$,

**Mauna Loa Observatory, Hawaii and South Pole, Antarctica**
**Monthly Average Carbon Dioxide Concentration**
Data from Scripps $CO_2$ Program Last updated July 2019

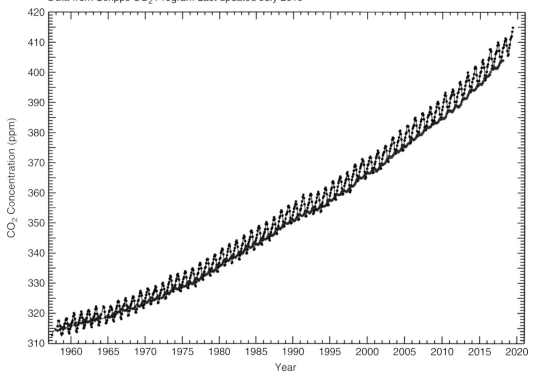

**Figure 8.7** The rise in global $CO_2$ since 1957. *Source:* from Scripps Institution of Oceanography, showing the 'global' record from Mauna Loa compared with the record from South Pole (in red). https://scrippsco2.ucsd.edu/data/atmospheric_co2/primary_mlo_co2_record.

when plants were no longer active. The ratio is opposite in the Northern Hemisphere compared with the Southern (Figure 8.8) because of the opposition of the seasons, but the limited area of land plant production in the south means that the northern seasonal signal dominates the global signal.

Keeling also found that atmospheric $CO_2$ was more abundant in the Northern Hemisphere, where most industrial sources were located, than in the Southern Hemisphere [24]. The amount of $CO_2$ released to the atmosphere annually from the burning of fossil fuels can be calculated from records of fuel production compiled internationally, for example by the US Oak Ridge National Laboratory in Tennessee [25]. The estimated addition closely matches the observed amount. The $\partial^{13}C$ ratio decreased (became more negative, with less $^{13}C$) in the same direction, north, confirming the source for this pattern as the burning of fossil fuel, which is depleted in $^{13}C$ because it originates from dead $^{12}C$-rich plant carbon. Thus, while short term variations in $\partial^{13}C$ reflect annual changes in the biosphere, the long term underlying trend of a decline in atmospheric $\partial^{13}C$ of around 1.5 parts per thousand (‰) over the past 100 years represents the contribution from burning fossil

fuels [25, 26] (Figure 8.8). Knowing the carbon isotopic composition of the burned fuels, one can estimate their likely effect on the $\partial^{13}C$ ratio of the $CO_2$ in the atmosphere. It closely matches the observed values [27]. The $^{14}C$ content of atmospheric $CO_2$ shows a similar effect, decreasing at the same time as the content of $CO_2$ has increased, which is to be expected if the $^{14}C$ is being diluted by $CO_2$ from fossil fuels in which all the original $^{14}C$ has decayed away.

The combustion of fuel removes oxygen from the atmosphere, as is shown by the inverse relation between growing $CO_2$ and declining $O_2$ [28] (Figure 8.9). Keeling's son Ralph (1959–), the present director of the Scripps $CO_2$ Program, continues his father's work. $CO_2$ science has become a family business!

Keeling's research stimulated a global growth in measurements of $CO_2$ in background air away from industrial sources. For example, from 1969 to 1981 the Canadian government ran a research programme to monitor atmospheric $CO_2$ at Ocean Weather Station P, in the N.E. Pacific, in collaboration with Scripps, later expanding the programme to incorporate sites at Sable Island off Nova Scotia, and Alert in the Canadian Arctic. By 1981 $CO_2$ in background air was being measured at some 40 sites around the world.

**Mauna Loa Observatory, Hawaii and South Pole, Antarctica**
**Monthly Average $\delta^{13}$C Ratio**
Data from Scripps CO$_2$ Program Last updated April 2019

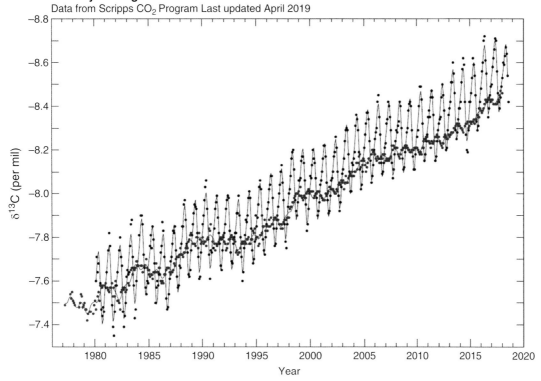

**Figure 8.8** The change in $\partial^{13}$C with time. *Source:* from Scripps Institution of Oceanography https://scrippsco2.ucsd.edu/graphics_gallery/isotopic_data/mauna_loa_and_south_pole_isotopic_c13_ratio_yaxis_inverted.html. $\partial^{13}$C is in per mil. The scale of $\partial^{13}$C is inverted so that seasonal patterns of CO$_2$ concentration (Figure 8.7) and $\partial^{13}$C appear with the same phasing. Black dots: monthly average reduced isotopic ratio of atmospheric carbon dioxide, $\partial^{13}$C, versus time at Mauna Loa Observatory, Hawaii. Red dots: monthly average atmospheric carbon dioxide reduced isotopic ratio, $\partial^{13}$C, versus time at the South Pole. Note that Northern Hemisphere and Southern Hemisphere signals have opposite phase.

Callendar was still going strong in 1957. For comparison with the measurements that would be made during the IGY, he reviewed the published analyses of CO$_2$ in the air from around the world [29]. Like Keeling, he noticed that high values were reported near towns. His equivalent of Keeling's 'background air' was what he called 'free air'. To obtain free air CO$_2$ he devised a system for rejecting samples collected in or near towns and within buildings. He also concluded that measurements made before 1870 were unreliable. The resulting data showed that free air in the northeast Atlantic region contained 290 parts per million (ppm) CO$_2$ in 1900, which rose to 326 ppm over Scandinavia in the mid 1950s. Those figures compare favourably with the global average of 315 ppm reported by Keeling in 1958. The rising CO$_2$ closely followed the rise in the consumption of fossil fuels [29].

Was that CO$_2$ rise real? As we shall see, it matches well with other sources of data. But in 2007 it was questioned by Ernst-Georg Beck (1948–2010), a biology teacher at the Merian technical grammar school in Freiburg, and co-founder of the European Institute for Climate and Energy (EIKE). Beck compiled the results of some 90 000 CO$_2$

analyses made since 1812, and used them in an attempt to demonstrate that the CO$_2$ content of air had fluctuated considerably since then [30]. Unfortunately, he ignored the simple and straightforward Keeling/Callendar sampling protocol, and the reasons for it. Just to cite one example, he accepted for his global compilation of atmospheric data the CO$_2$ measurements made at Giessen, in Germany, where CO$_2$ values ranged from lows of around 300 ppm to highs of up to 550 ppm There is no doubt that the Giessen measurements were made well, but that's not the point. Giessen lies about 140 km southeast of the industrial centres of the Ruhr Valley, in an area of prevailing westerly winds, and just 50 km north of another major industrial centre: Frankfurt. It is not an ideal site at which to determine the atmospheric background level of CO$_2$. Values recorded there most likely represent contamination by down-wind transport from industrial sources and population centres. Beck's misleading analysis was roundly criticized [31, 32]. But it still has adherents on the Internet.

Although Revelle and Suess's landmark 1957 paper [20] is much quoted, it contains two statements in apparent

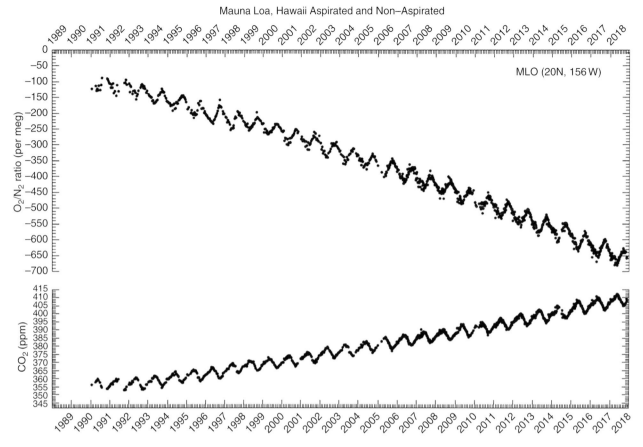

**Figure 8.9** The inverse relation between $CO_2$ rise and declining $O_2$. $CO_2$ values are from samples from Mauna Loa and Baring Head (New Zealand); Oxygen values (in $O_2/N_2$ ratios) are from samples in Alert (Canada) and Cape Grim (Australia). *Source:* From Scripps Institution of Oceanography http://scrippso2.ucsd.edu/assets/pdfs/plots/daily_avg_plots/mlo.pdf.

contradiction. Their calculations showed that the average lifetime of a $CO_2$ molecule in the atmosphere before it is dissolved into the sea is about 10 years, from which one might conclude that any new emission of $CO_2$ would be quickly taken up by the ocean rather than staying in the atmosphere. But, as Spencer Weart points out, the paper also contains a much overlooked implication: '*While it was true that most of the $CO_2$ molecules added to the atmosphere would wind up in the oceans within a few years, most of these molecules (or others already in the oceans) would promptly be evaporated out*' by the normal processes of ocean–atmosphere exchange [33]. Bearing that ongoing exchange in mind, Revelle and Suess stated: '*In contemplating the probably large increase in $CO_2$ production by fossil fuel combustion in coming decades we conclude that a total increase of 20 to 40% in atmospheric $CO_2$ can be anticipated*' [20]. In other words, much of the extra $CO_2$ pumped into the atmosphere by our emissions would stay there, with obvious implications for Earth's climate [3].

The first part of that paragraph reminds me that there is a popular misconception that $CO_2$ does not last long in the atmosphere. The misconception is based on evidence from the behaviour of $CO_2$ containing the radioactive isotope [14]C, which was created by the nuclear bomb tests of the 1950s and 1960s, and which was found to have a relatively short residence time of 5–10 years in the air. Whilst that confirms that any individual molecule of $CO_2$ is likely to have a short residence time in the air, it has no bearing whatsoever on what the bulk of the $CO_2$ is doing over time. That is because, as implied in the second part of the above paragraph, $CO_2$ is constantly being exchanged between the tiny reservoir of carbon in the atmosphere (720 Gt) and the vast reservoir of carbon in the ocean (38 400 Gt), both of which are dominated by the stable isotopes of carbon: mainly [12]C with a little [13]C. Every year some 90 Gt of carbon dominated by the [12]C isotope goes from the atmosphere into the ocean, and more or less the same amount comes out in the other direction. The tiny fraction of airborne $CO_2$ containing [14]C is lost in the noise of this massive exchange. Most of it is mixed down into the ocean interior above the thermocline, and some disappears further down through the sinking of dense cold surface waters into the ocean depths, where the $^{14}C$ decays. In effect it is massively diluted out of the exchange equation by the vast amounts of $^{12}C$ and its little sister $^{13}C$, and so its abundance and brief residence time have no bearing on the overall and very long residence time of total $CO_2$ in the atmosphere. Exchanges to and fro across the air–sea interface, and between surface and deep waters are the key factors ignored by those who favour the short residence time (like Harde) [34], as explained by Peter Köhler and colleagues [35]. Köhler points out, first, that Harde confuses the residence time of an individual molecule (years) with the adjustment time for an enhancement of atmospheric $CO_2$ (centuries); second, that you can't model the evolution of atmospheric $CO_2$ with a single equation – you have to solve at least two simultaneously to consider the roles of the two main carbon reservoirs (atmosphere and surface ocean); and, third, you have to consider the Revelle factor, which limits how much of our carbon emissions can be taken up by the oceans (about 20–30% of our emissions will remain in the atmosphere for thousands of years) [35]. And there's still the proportion of carbon emissions exchanged with the land surface to consider. Harde also erred in assuming that the change in temperature from glacial to interglacial was about 8 °C, when in fact that was the difference in Antarctica (the global difference was more like 4 °C), and in other inaccurate assumptions about the adequacy of $CO_2$ data from ice cores [35].

Swedish meteorologists, Bert Bolin (1925–2007) and Erik Eriksson of the International Meteorological Institute in Stockholm would spell these points out brilliantly in 1958, in yet another landmark paper [36]. Their work took into account the slow mixing time of the ocean, the impact of the ocean's carbonate buffer chemistry, and the accelerating emission of $CO_2$ with time [3, 36]. In summary: we put excess $CO_2$ into the atmosphere; it constantly exchanges with the $CO_2$ in the ocean; some stays in the ocean because physical and biological processes prevent it getting back into the atmosphere; some is trapped in land plants and soils; but the constant exchange keeps the atmospheric concentration high; and the excess $CO_2$ causes the atmospheric concentration to rise.

We still needed to know more about the ocean's role in the $CO_2$ story, and Roger Revelle led the charge. Back in 1960, on the occasion of the founding of UNESCO's Intergovernmental Oceanographic Commission (IOC), based in Paris, Revelle highlighted '*the present attempt to determine the total carbon dioxide content in the atmosphere and the change in this content with time as a result of the input from fossil fuel combustion and the loss to the ocean and biosphere. One of the questions we are asking is: Where is the carbon dioxide absorbed by the ocean? Does it remain in the surface layers or does it extend throughout the ocean volume?*' [37]. We will explore the answers to those questions below.

Meanwhile, the growing understanding of the role of $CO_2$ in the air and ocean outlined above by the likes of Callendar, Plass, Bolin, Keeling, and Revelle began to trigger public warnings, despite the fact that the climate had cooled slightly since the mid-1940s. In 1965, one of President Lyndon B. Johnson's Science Advisory Committees warned that burning fossil fuels would lead to a rise in $CO_2$ of 25% by the year 2000, which might be expected '*to produce measurable and perhaps marked changes in climate ... [that could be] ... deleterious from the point of view of human beings*' [38]. The actual rise has been close to 40%. Revelle wrote the $CO_2$ section of the report, aided by marine geochemist Wallace Smith (Wally) Broecker (1931–2019), of the Lamont-Doherty Geological Observatory of New York's Columbia University (Figure 8.10, Box 8.2), along with others including Charles Keeling.

Broecker was to play a key role in improving understanding of the role of the ocean in controlling atmospheric levels of $CO_2$. Part of the answer to Revelle's questions posed before the IOC in 1960 (above) would come from improved measurements of the global distribution of carbon in the ocean made in the Geochemical Ocean Sections Study (GEOSECS) by a team led by Broecker and others, who launched GEOSECS as part of the International Decade of Ocean Exploration (IODE), in 1970. The programme ran from 1971 to 1978, making some 6000 measurements of dissolved organic carbon and total ocean alkalinity. However, still more accurate measurements would be required before the anthropogenic contribution to oceanic $CO_2$ could be established with confidence [39].

**Figure 8.10** Wally Broecker.

---

**Box 8.2    Wally Broecker**

Wally Broecker, sometimes known as the 'Grandfather (or the Dean) of Climate Science' was a long-standing member of the Lamont-Doherty Geological (now Earth) Observatory of Columbia University, in New York, as well as being Columbia's Newberry Professor of Earth and Environmental Sciences. He was one of the world's really smart guys, one of the world's greatest geoscientists. His major interest was the ocean's role in climate change, and he was amongst the pioneers of radiocarbon and isotope dating, the key agents for creating maps of past climates. Wally was the first to recognize the Ocean Conveyor Belt (see Chapter 12), arguably the most important discovery in the history of oceanography. He was a leader in studies of the biogeochemical cycle of carbon and how it influenced climate change. He has been awarded a slew of prizes including the Vetlesen Prize in 1987, the Geological Society's Wollaston Medal in 1990, the US National Medal of Science, in 1996, the Tyler Prize for Environmental Achievement in 2002, and the Swedish Crafoord Prize in 2006 – considered the equivalent of a Nobel Prize in the geosciences. In 2007 he was made a fellow of the Royal Society of London.

---

By 1970, Bert Bolin was explaining the mysteries of the carbon cycle and the effect of $CO_2$ on the climate to the readers of *Scientific American* [40], reminding them that at current rates of increase there would be between 375 and 400 ppm in the atmosphere by the year 2000. Clearly that present exponential growth could not continue. Tampering with the planet's biological and geochemical balances might well prove injurious to humankind; we must understand these balances much better, and may have to leave them close to their natural state at the beginning of the Industrial Revolution [40]. '*Out of a simple realization of this necessity may come a new industrial revolution*' he suggested.

In the same issue, Fred Singer agreed that by releasing the energy stored in fossil fuels, man was accelerating the slow cycles of nature [41]. The waste products of power generation, especially $CO_2$, were interfering with the fast cycles of the biosphere, not least because our emissions were relatively rapid compared with the rates of preceding natural processes. Whilst it seemed highly likely that this would lead to global warming, Singer advised caution [41], reminding readers that calculating the precise effect of an increase in $CO_2$ was difficult, not least because we did not yet understand enough about factors like the reflective effects of cloud cover or changing atmospheric turbidity

(e.g. from industrial or volcanic aerosols). Besides, some of the emitted $CO_2$ would enhance plant growth [41].

The tenor of Revelle's report and Bolin's article was reflected in the Club of Rome's report *Limits to Growth*, in 1972, which forecast a rise of $CO_2$ to 380 ppm by the year 2000 [42] (not far off – it was actually 370 ppm according the NASA-GISS web site). But, neither the Club of Rome nor the Conference on Environment and Development organized by UNEP, which took place in Stockholm in 1972, paid a great deal of attention to $CO_2$. Most likely that reflects the fact that it was not until 1970 that temperatures began to recover from the cool conditions of the 1950s and 1960s. The Conference recommended setting up systems for the collection and analysis of $CO_2$, particulates, and so on, to learn more about the relationship between $CO_2$, aerosols, and weather and climate.

Meanwhile, by 1967, Syukuro Manabe and Richard Wetherald of the Geophysical Fluid Dynamics Laboratory in Washington DC had already provided the first fully sound estimate of the warming that would arise from a doubling of $CO_2$ [43]. Their primary object was to incorporate the radiative properties of atmospheric gases (like water vapour, $CO_2$, and ozone) into a general circulation model (GCM) of the atmosphere. Their model showed that as $CO_2$ increased in abundance, the temperature of the troposphere warmed whilst the temperature of the stratosphere cooled, a prediction that observations soon showed was correct. They were well aware of the effects of water vapour, calculating that doubling $CO_2$ would increase global temperatures by 1.3 °C if the water vapour content of the atmosphere remained constant, but by 2.4 °C if water vapour increased to maintain the same relative humidity.

Their calculations would be further refined with time, as we can see from Raymond Pierrehumbert's weighty 2010 tome *Principles of Planetary Climate* [44]. Pierrehumbert summarized much of this understanding into a short paper published in *Physics Today* in 2011, underscoring the fact that our understanding of planetary temperature anywhere relies on the theory of infrared radiative transfer, '*one of the most productive physical theories of the past century … [which] has unlocked myriad secrets of the universe including that of … the connection between global warming and greenhouse gases*' [45]. Pierrehumbert reminds us that some confusion arose in 1900, when Anders Angström argued, in opposition to Arrhenius, that infrared absorption by $CO_2$ was saturated, but, as Pierrehumbert pointed out, Angström was 'doubly wrong'. '*First, modern spectroscopy shows that $CO_2$ is nowhere near being saturated [as Angström had assumed]; [second, while Angström assumed] that $CO_2$ could have no influence on radiation balance because water vapour already absorbs all the [infrared radiation] that $CO_2$ would absorb … [in fact] radiation in the* *portion of the spectrum affected by $CO_2$ escapes to space from the cold, dry upper portions of the atmosphere, not from the warm, moist lower portions [considered by Angström] ... [Furthermore] the individual water vapour and $CO_2$ spectral lines interleave but do not totally overlap. That structure limits the competition between $CO_2$ and water vapour'* [45]. Ramanathan also provides a good summary of the theory of infrared radiative forcing [46]. His and Pierrehumbert's papers should be required reading for all climate studies, as this theory underpins all of the atmospheric studies referred to here from Plass's and beyond.

By 1975, Manabe and Wetherald had developed a GCM for *'solving the three-dimensional fluid dynamical and thermodynamical equations governing transfer of heat, moisture and momentum around the world'*, a step that taxed the powers of the biggest computers of the time [47]. Their model confirmed that $CO_2$ causes the troposphere to warm and the stratosphere to cool. A doubling of $CO_2$ in the atmosphere would lead to a 2–3 °C rise in temperature at middle latitudes, increasing by three to four times more in high latitudes (especially the Arctic), due to a corresponding decrease in albedo caused by the loss of snow and ice. This is the 'polar amplification' predicted by Arrhenius. Their model confirmed that land warms faster than the ocean, and that global precipitation increases as the world warms, because warming evaporates water vapour from the ocean, increasing the intensity of the hydrological cycle [3, 47]. The model could be improved upon, for instance by changing cloudiness, adding ocean circulation (rather than a swamp ocean), and including seasonal change.

By 1977 Revelle was chairing a panel that reviewed Energy and Climate for the US National Research Council (NRC). In bold fashion he opened the report by noting that *'Worldwide industrial civilization may face a major decision over the next few decades – whether to continue reliance on fossil fuels as principal sources of energy or to invest the research and engineering effort, and the capital, that will make it possible to substitute other energy sources for fossil fuels within the next 50 years. The second alternative presents many difficulties, but the possible climatic consequences of reliance on fossil fuels for another one or two centuries may be so severe as to leave no other choice'* [48]. These choices still face us today. The primary cause for concern was absorption of infrared radiation (heat) by $CO_2$, warming the lower atmosphere and potentially shifting currently stable patterns of agriculture and food production. The report warned that *'it may be necessary to reverse the trend in consumption of fossil fuels. Alternatively, carbon dioxide emissions will somehow have to be controlled or compensated for. In the face of so much uncertainty regarding climatic change, it might be argued that the wisest attitude would be* laissez-faire. *Unfortunately, it will take a millennium for the effects of a*

*century of use of fossil fuels to dissipate if the decision is postponed until the impact of man-made climate changes'* [48]. Although large-scale effects on climate were not yet readily apparent, the authors realized that growing population and energy use would increase emissions, ensuring that such effects would emerge in due course. It seemed likely to them that by 2150–2200 peak atmospheric concentrations of $CO_2$ might be 4–8 times the pre-industrial level (i.e. roughly 1100–2200 ppm), and that *'concentrations much higher than today's may persist for many centuries'* [48]. Such a rise in $CO_2$, they thought, might lead to a global temperature increase of more than 6 °C, a level last seen in the Mesozoic climate of 70–100 Ma ago. Warming of the atmosphere and ocean would also lead to a rise in sea level, and melting of Arctic sea ice – potentially opening up the Northeast and Northwest Passages to shipping. These findings provided *'the stimulus for greater efforts at conservation and a more rapid transition to alternate energy sources than is justified by economic considerations alone'* [48]. The Panel recommended a comprehensive global research programme be established, with the assistance of the World Meteorological Organization (WMO), the IOC, and ICSU, to improve understanding of the carbon cycle, of how $CO_2$ behaves in the environment, and of how the climate system works. A global monitoring system was needed of the distribution in time and space of key physical properties and processes in the ocean and atmosphere, not least to provide early warnings of climate change *'before their consequences become unpleasantly obvious'* [48]. These conclusions, amongst others, helped to establish the IPCC in 1988 (see later).

Independently of these studies of $CO_2$, meteorologists were compiling meteorological data to see if the world was really warming. Amongst them was Hubert Horace Lamb (1913–1997) (Figure 8.11, Box 8.3), an English climatologist.

One of Lamb's tasks at the UK Met Office was to answer the question: 'is the climate changing?' Temperatures had risen from 1900 to 1940, followed by a decline into the 1960s. Considering, in 1963, the greater frequency of severe winter weather in the UK in recent decades, Lamb asked whether this was: *'(i) Only a temporary lapse from the warm climate attained in the early 20$^{th}$ century, possibly to be followed by a renewed trend to still greater warmth? (ii) A return to conditions normal in the past century or two? Or (iii) the beginning of a climatic decline to still harsher conditions?'* (i.e. towards the next Ice Age) [49]. At the time he thought (i) unlikely, and (iii) more likely, concluding: *'it seems prudent to assume that the longer-term temperature trend is at present on balance downward and likely to remain so'* [49]. That was also the view of palaeoclimatologist George Kukla, who observed in 1972 that the rise in incoming solar radiation (insolation) that had pushed the Earth into the interglacial state of the Holocene (the last

**Figure 8.11** Hubert Lamb.

---

**Box 8.3    Hubert Horace Lamb**

Lamb started his working life as a meteorologist for the UK's Met Office, where he was responsible for long-range weather forecasting and studies of climate change. In 1972, he founded the Climate Research Unit (CRU) of the University of East Anglia's school of environmental sciences, based in Norwich. In August 2006, the CRU building was renamed the Hubert Lamb Building in his honour. That same year in a report listing the top 100 world-changing discoveries, innovations, and research projects to come out of the UK universities in the last 50 years, Lamb was hailed as instrumental in establishing the study of climate change as a serious research subject. For his contributions to climate science, Lamb won several awards, amongst them the Symons Gold Medal of the Royal Meteorological Society and the Murchison Medal of the Royal Geographical Society.

---

11 700 years) had recently gone into a decline that would last for 8000 years and cause our climate to cool by 1 °C in the next 100 years [50].

Although some other scientists also agreed that the world might be cooling towards the next Ice Age, it is a myth that there was some kind of scientific consensus on the subject [51]. In fact only around 10% of scientific papers on climate change published between 1965 and 1979 predicted global cooling; the others all predicted warming [51]. Amongst the 'cooling' papers was one by Steven Schneider, then a young climate modeller at NASA's Goddard Institute for Space Studies, who had used what turned out to be a flawed climate model to suggest in 1971 that atmospheric dust particles injected by industrial processes and the burning of fossil fuels could counter the effect of atmospheric $CO_2$, cooling the climate by as much as 3.5 °C if human outputs of aerosols increase six- to eight-fold over the next 50 years [52]. By 1973, Schneider, by then at the National Centre for Atmospheric Research (NCAR), in Boulder, had used a much more advanced numerical model to correct his forecast from a cooling to a warming of between 1.5 and 3.5 °C (above the values for 1900) for a doubling of $CO_2$ [53]. The media, as usual looking for scary things to lead with, had backed the wrong horse and banged the cooling drum.

Lamb eventually accepted that $CO_2$ played a key role in governing climate change [54]. Recognizing that the orbital geometry calculations of André Berger suggested that over the long term we were heading towards a colder climate, he concluded that our future climate would reflect the balance between orbital effects and human emissions, modulated by shorter term phenomena like El Niño events, and volcanic eruptions like that of Mt. Pinatubo. '*Man may be obliged in the future either to seek to avert, or slow down, the onset of a new ice age by deliberately increasing the $CO_2$ in the atmosphere*' he wrote [54] – an intriguing geo-engineering notion. He did not think that urban heat islands significantly distorted global temperature records, because that would not explain warming in remote locations, nor the melting of glaciers.

Another prominent climate scientist to enter the ring in the late 1960s was the Soviet meteorologist Mikhail Ivanovich Budyko (1920–2001) (Figure 8.12, Box 8.4).

**Figure 8.12**    Mikhail Budyko.

## Box 8.4 Mikhail Budyko

A brilliant and original thinker, Mikhail Budyko was one of the founders of modern climatology, well-known for his calculations of the planetary 'albedo' – the amount of solar radiation reflected back into space by light coloured surfaces like ice, snow, and deserts. He recognized that more ice and snow would reflect more energy, cooling the Earth, which would lead to more snow and ice and so on: a positive feedback. Conversely, warming would lead to less snow and ice, and more exposure of the darker ocean, meaning less reflected energy and more energy absorbed by the ocean: another positive feedback. Budyko was head of the Division for Climate Change Research of the State Hydrological Institute in Leningrad (now St. Petersburg), a member of the Russian Academy of Sciences, and a 1994 recipient of the Robert E. Horton Medal of the American Geophysical Union.

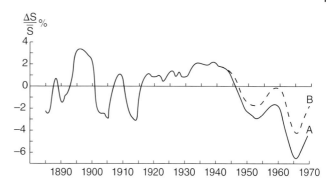

**Figure 8.14** Budyko's measurements show effects of growing aerosol concentration. A = actual measurement; B = adjustment for cloudless sky. Five-year running means of the anomalies of direct radiation expressed in per cent of its normal amount, calculated from the data observed at a number of stations. Attenuation is equal to about 6%. *Source:* From figure 5 in Ref. [57].

Much like Lamb, Budyko found that between 1881 and 1960 the annual temperature of the Northern Hemisphere had risen by 0.6 °C from 1900 to 1940, then fell by 0.2 °C (see Figure 8.13 [55]). Puzzling over what caused the decline, he found that it correlated with a decrease in direct solar radiation measured under a cloudless sky. Thinking that this was due to rising industrial output, he wrote: '*the decrease in radiation after 1940 could ... depend on the increase in dust in the atmosphere due to man's activity*' [56]. By 1977 he calculated that direct solar radiation had decreased by 6% between the late 1950s and the late 1960s due to our growing emission of fine particles (aerosols), which could have reduced the global temperature by around 0.5 °C (Figure 8.14) [57].

Being aware that $CO_2$ was a greenhouse gas, in 1971 Budyko agreed that adding $CO_2$ emissions to the atmosphere would be expected to warm the climate [58]. By 1977, he calculated that a continued rise of $CO_2$ would cause temperatures to rise by 0.6–0.7 °C by the end of the century [57] – a prophetic estimate. By 1979, he concluded that: '*man's influence on the atmosphere has a considerable significance ... As a result of the combustion of steadily increasing amounts of coal, oil, and other forms of fossil fuel as well as the reduction in the amount of carbon in living organisms (from the felling of forests) and a reduction in the amount of carbon in soil humus, the $CO_2$ level in the atmosphere has recently been rising. It is assumed that this mass might increase by a factor of 6-8 within 100-200 years. The important point is that in such a case the $CO_2$ level would attain the values typical of the average for the Phanerozoic. Such a change in the atmospheric composition would rule out any*

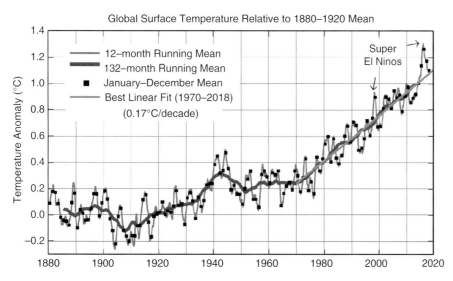

**Figure 8.13** Global surface temperatures relative to 1880–1920. *Source:* From figure 1 in Ref. [55].

*further cooling of the climate and will probably result in the return to warmer climatic conditions typical of the Pre-Quaternary period'* [59]. In bringing in the geological dimension he had been influenced by his co-author, geologist Alexander Borisovitch Ronov (1913–1996) of the V.I. Vernadsky Institute of the USSR Academy of Sciences.

Budyko realized that warming due to increased $CO_2$, plus the increase in $CO_2$ itself, might provide more favourable conditions for the plant growth needed to feed the growing human population [60]. Besides, warming would expand the area of northern Russia and Siberia capable of growing wheat. It would also reduce Arctic sea ice, which could be a boon to shipping [57]. The USSR could benefit from global warming. There would be winners and losers.

Like Schneider, Budyko understood that two anthropogenic effects modify air temperature: emissions of $CO_2$ and of aerosols. His Soviet data on aerosols (atmospheric opacity) (Figure 8.14) help to explaining the hiatus in global warming in the 1950s and 1960s [57]. The effects of aerosols then declined as nations introduced legislation to limit air pollution from particulates, spurred by such events as the smogs in London in 1952 and New York in 1953. Smith and colleagues agreed, noting that industrial emissions of sulphur dioxide increased to 1970, then decreased as clear air legislation took effect [61]. The rise in the emission of sulphur aerosols after the Second World War paralleled the rise in $CO_2$ until 1970 and is likely to have prevented the warming signal due to $CO_2$ from becoming manifest between 1945 and 1970, much as implied by Budyko's data. In effect the decline in aerosols after 1970 is what allowed the warming signal due to $CO_2$ to emerge.

Oceanographers like Revelle and climatologists like Budyko were not alone in agreeing with Plass. By 1972, the UK's Meteorological Office had become convinced by Keeling's data of the dangers inherent in continued emissions of $CO_2$ [62]. By 1975, as we saw above, Schneider had joined those forecasting a rise in temperature due to increasing atmospheric $CO_2$ [53]. He took pains to remind his audience that the surface temperature response to increasing $CO_2$ was logarithmic, not linear: its rate of rise declined as the $CO_2$ increased (but see Ramanathan's 1977 results, below).

$CO_2$ received only a passing mention in the 1975 report by the US National Academy of Sciences' Panel on Climate Variation (NAS, 1975), the main focus of which was to recommend national and international research programmes for the period 1980–2000 to increase understanding of climatic variation [63]. Even so, one member of that Panel, Wally Broecker, had a great deal to say about the $CO_2$ question, which seemed to put him somewhat at odds with the Panel's conclusions. That same year he published a paper with the provocative subtitle: '*Are we on the brink of*

*a pronounced global warming?'* [64]. That paper, in which he famously predicted that the recent cooling trend would soon give way to a rise in temperature as $CO_2$ continued to rise, popularized the use of the term 'global warming'. In a memorable phrase from the 1990s, Broecker later described the world's climate as an 'angry beast' that we are continuing to poke through growing emissions of $CO_2$, without being sure how it may respond [65]. In 1978, a similar warning came from John Mercer of Ohio University's Institute of Polar Studies, concerned that polar warming could destabilize the West Antarctic Ice Sheet and lead to a global rise in sea level [66].

Despite Broecker's provocation, by 1976 the meteorological community was not in complete agreement that $CO_2$ was contributing or might contribute to climate change. At that time, climatologist Steve Schneider, referring to the 0.5 °C rise in global temperature over the previous 75 years, wrote '*climatic theory is still too primitive to prove with much certainty whether the relatively small increases in $CO_2$ and aerosols up to 1975 were responsible for this climate change'* [67, 68]. So we should not be too surprised that many geologists of the 1970s and before did not consider the carbon dioxide theory of climate change worth a second thought.

Before the $CO_2$ model of climate change could be accepted, it was necessary to quantify the errors and uncertainties built into the radiative-convective models being used to calculate climate change. Tommy Augustsson, of the NASA-Langley Research Centre of George Washington University, and Veerabhadran (Ram) Ramanathan of the NCAR at Boulder in Colorado, carried out that task in 1977 [69]. They calculated that while most of the absorption of infrared radiation by $CO_2$ occurred in the main spectral bands attributed to $CO_2$ vibrations centred on about 15 mμ, about 30% occurred in the so-called weak spectral bands attributed to $CO_2$ vibrations between 12–18 mμ, 9–10 mμ, and 7–8 mμ (see Figure 8.1 for the location of those bands, which are not marked as affected by $CO_2$). They found that while the increase in temperature tailed off logarithmically as $CO_2$ increased in the 15 mμ band, it was almost linear with increasing $CO_2$ in the weak bands. Hence '*the warming effect of $CO_2$ on the global surface temperature may never saturate out even for large increases in $CO_2$ concentrations'* [69]. They estimated that doubling $CO_2$ would raise temperature 1.98–3.2 °C.

Broecker was keen to see if the dissolution of deep ocean carbonate sediments might ultimately neutralize the $CO_2$ generated by burning fossil fuels, a project he worked on with Taro Takahashi in 1978 [70]. They calculated that the process would take 1500 years or longer. James Walker of the University of Michigan and James Kasting of Penn State University agreed, concluding that '*dissolution of pelagic carbonates is not likely to influence the levels of atmospheric carbon dioxide before several hundred years have elapsed'* [71].

## 8.3 CO$_2$ in the Atmosphere and Ocean (1979–1983)

By 1979 interest in the CO$_2$ question was beginning to move out of national laboratories and into the international arena. In February 1979, the first World Climate Conference took place in Geneva, Switzerland, organized by the UN's WMO, which is headquartered there [72]. Other UN agencies provided assistance, amongst them UNESCO, the UN Environment Programme (UNEP), the UN's Food and Agricultural Organisation (FAO) and the UN's World Health Organization (WHO). Robert M. White (1923–2015), who had been head of the newly formed National Oceanic and Atmospheric Administration (NOAA) from 1970 to 1977, chaired the Conference. White was a member of the US Academy of Sciences, and would become President of the National Academy of Engineering (1983–1995) and, in 1980, of the American Meteorological Society.

CO$_2$ was not a primary focus for the meeting. The preceding decade had seen one of the strongest El Niño events of the twentieth century (in 1972–1973), a five-year Sahelian drought, failure of the Indian monsoon in 1974, damage to the Brazilian coffee crop from cold waves in 1975, drought in Europe, and cold waves in the United States. Recognizing the importance of variations in climate, the WMO had decided to initiate a World Climate Programme (WCP), which would include a World Climate Research Programme (WCRP). The conference was designed to provide a platform for the launch of these initiatives, by assessing the state of man's knowledge of our climate, considering the effects of climate variability and change on human society, and addressing the issue of human influences on climate. There are echoes here of the recommendations of the 1975 NAS report on climate variation [63].

Kicking off the meeting, White reminded his audience *'If natural disasters had not been enough to motivate governments and the scientific community to action, the ominous possibilities for man-induced climate changes would have triggered our presence here .... In recent years, we have come to appreciate that the activities of humanity can and do affect climate'.* As to what could be done about that, he went on *'It is essential that we join together to consider what we can do collectively and individually about climatic issues in the interests of all ...The possibility that actions by individual nations may influence the climates of others may demand new types of international action .... We thus see emerging a need for some mechanism to develop global environmental impact assessments that will be accepted by all*

nations ...Let us hope that this conference marks the commencement of a new level of collaboration for the protection and productive use of climatic resources'* [73]. Note the call for a 'mechanism', and use of the word 'assessments'.

Amongst the experts making presentations was the great German climatologist Hermann Flohn, of the University of Bonn (1912–1997, Box 8.5), who had been discussing this topic at a Dahlem Workshop in Berlin in 1976 [74].

Flohn thought that the probability of human-induced future global warming was high and would increase with time, such that *'soon after the turn of the century, a level may possibly be reached that exceeds all warm periods of the last 1000-1200 years ... This risk must be avoided even at very high costs'* [75]. He reminded the conference that global changes are larger by a factor of about three in the polar regions, hence *'The most fascinating, and also the most controversial problem of the future evolution of our climate is the possibility of a complete disappearance of the drifting ice of the Arctic Ocean ... an increase of the atmospheric CO$_2$ content might lead rather rapidly to an ice-free Arctic Ocean'* [75]. Budyko, another participant at the Conference, had reached that same conclusion in 1977 [57], as had Revelle [48]. The disappearance of ice, Flohn thought, would lead to *'an increase of cold season snowfall along the northern coasts of the continents and arctic islands ... [along with] reduction and northward displacement of the winter-rain belts in the Mediterranean, Near East, and southwestern North America, together with frequent summer droughts in the belt 45°-50°N and an extension of the subtropical dry areas towards north'* [75]. These are things we have now begun to see with increasing frequency. The news was not all bad: *'The possibility of a significant melting of the polar ice caps is ... small'*, he thought [75].

Not everyone agreed with Flohn's diagnosis. F.K. Hare of the University of Toronto, thought that the climate system was highly variable, and we did not yet know enough about it to be able to predict with confidence whether temperatures would rise or fall [76]. R.E. Munn of the University of

---

**Box 8.5 Hermann Flohn**

Hermann Flohn produced the first north–south cross section through the atmosphere, with widespread implications for understanding atmospheric processes at the global scale. His analysis of the atmospheric circulation of the Southern Hemisphere contributed to his becoming one of the world's most famous climatologists after the Second World War. He won the WMO International Meteorological Association Prize in 1986, and was a member of the Leopoldina (the German Academy of Sciences) and the Royal Belgium Academy.

Toronto and L. Machta of NOAA agreed, noting that as well as predictions of warming. '*There are predictions for naturally occurring cooling trends during the next few decades*' [77], as suggested for example by Kukla [50]. If the climate cooled, '*the $CO_2$ induced global warming would then be welcome*' [77]. They recognized that '*few predictions call for significant climatic effects before 2000 AD*' and, further, that '*Few, if any, scientists believe the $CO_2$ problem in itself justifies a curb, today, in the use of fossil fuels or deforestation*' [77]. But, given the exponential rise in the consumption of fossil fuels and emissions of $CO_2$, they concluded '*There is ... a sense of urgency in determining whether there is likely to be any real environmental or socioeconomic threat from growing atmospheric $CO_2$*' [77]. We needed more information: '*Studies of the climate impacts of an increase in the concentrations of greenhouse gases, and of the resulting impacts on society, should be pursued internationally with great vigour*' [77].

Summing up the results of this ground-breaking meeting, the Secretary-General of the WMO, David Arthur Davies (1913–1990), of the UK, said in the Foreword to the 1979 report '*This publication may safely be considered as the most profound and comprehensive review of climate and of climate in relation to mankind yet published*' [78]; in effect it could be seen as the first climate change textbook. On the basis of their deliberations, the experts adopted 'The Declaration of the World Climate Conference', which agreed '*it appears plausible that an increased amount of carbon dioxide in the atmosphere can contribute to a gradual warming of the lower atmosphere, especially at high latitudes ... [which] may be detectable before the end of this century and become significant before the middle of the next century*' [72]. As a result, the declaration went on more controversially, to avoid serious environmental problems it might become necessary to '*redirect ... many aspects of the world economy*' [72]. We might have to *do* something about this warming.

Concern about the potential effects on climate of the doubling of $CO_2$ in the atmosphere led the US Congress and the Executive to ask the National Academy of Sciences to consider the topic afresh, and in the summer of 1979 a team of experts led by meteorologist Jule Gregory Charney (1917–1981), of the Massachusetts Institute of Technology (MIT), undertook an evaluation of the models being used to estimate likely effects of $CO_2$ on climate [79]. His team found that doubling atmospheric $CO_2$ would increase global temperature by about 3°C. Charney, the father of modern dynamical meteorology, was awarded the Carl-Gustaf Rossby Research Medal of the American Meteorological Society for contributions to atmospheric science.

Accepting the World Climate Conference's observations about the potential threat to future climate posed by increasing amounts of atmospheric $CO_2$, the 8th WMO Congress

meeting in Geneva in May 1979 created a Project on Research and Monitoring of Atmospheric $CO_2$, which became part of the WCRP. The 31st session of the WMO Executive Committee, meeting in Geneva in June 1979, agreed that a conference on the $CO_2$ question should be held in 1981. As a first step towards that, WMO, UNEP, and ICSU convened a Meeting of Experts on the Assessment of the Role of $CO_2$ on Climate Variations and their Impact, which took place in November 1980, in Villach, Austria. Attendees agreed that they should convene the first Scientific Conference on Analysis and Interpretation of Atmospheric $CO_2$ Data, which took place in Bern, Switzerland, in September 1981, under the Chairmanship of Professor Hans Oeschger and with the assistance of Charles Keeling. At the meeting, William Kellogg of NCAR suggested investigating the roles of positive and negative feedback loops in determining the exchanges of $CO_2$ between atmosphere, ocean and biosphere sources and sinks [80].

Meanwhile, Ramanathan was evaluating the latitudinal and seasonal changes likely from doubling the $CO_2$ content of the atmosphere. In 1979 he showed that there were two reasons why the Antarctic should be colder than the Arctic [81]: first, Antarctica is land surrounded by ocean, whilst the Arctic is ocean surrounded by land; and second, because the atmosphere in the Southern Hemisphere is thinner than that in the North, the surface pressure is 20% lower at 80°S than at 80°N, and as the opacity of the atmosphere to $CO_2$ at infrared wavelengths is proportional to atmospheric pressure, the atmosphere is less opaque in the south at comparable latitudes, and so warms less. Differences between the tropics and the poles arise because of the distribution of water vapour. Because the $CO_2$ absorption bands at 12–18 mμ overlap $H_2O$ absorption bands, where both occur together in the tropics enhanced emission by $CO_2$ is partly absorbed by $H_2O$, which enhances tropospheric heating. At high latitudes, where there is little $H_2O$, $CO_2$ emissions warm the surface. Polar warming is also amplified by decreases in the albedo resulting from melting sea ice.

Ramanathan found that '*At 75°N the $CO_2$-induced enhancement in surface temperature is roughly 2 times as great in summer as compared with winter, while at 85°N this factor is greater than 3*' [81]. He noted that '*This effect is not, however, due to seasonal variability of the $CO_2$ radiative heating, but is solely the result of ice albedo feedback*' [81]. This led him to observe that '*The high-latitude summer enhancement in surface temperature could be of considerable importance with regard to the stability of arctic sea ice*' [81] (implying that the sea ice would melt). He calculated that with $1.33 \times CO_2$ (which was then at 372 ppm) the mean increase in surface temperature for the Northern Hemisphere would be 1.45°C, whereas at 85°N in June (mid-summer) it would be 6.5°C [81]. He concluded that a significant influence of $CO_2$ on the climate would first appear during Arctic summers, not Antarctic summers, because heating rates would be 50% lower in the south. His $CO_2$ signal did indeed appear, in the year he published his paper, in the shape of the beginning of a remarkable decline in summer Arctic sea ice. Clever fellow! **From the palaeoclimate perspective his observations about the differences between the two poles are important to our evaluation of changes in ice cover with time**.

Amongst those concerned about the possible effects of $CO_2$ and other greenhouse gases was John Houghton (see Section 8.1). Houghton, an expert in radiation physics, drew attention in 1979 to the effects on temperature not only of $CO_2$ but also of the release of chlorofluoromethanes (what we now refer to as chlorofluorocarbons, or CFCs) and aerosols from human-made sources [82].

Whilst these activities were going on, Revelle was overseeing a two-year collaboration between the US Department of Energy (USDOE) and the American Association for the Advancement of Science to produce a research agenda addressing the 'Environmental and Societal Consequences of a Possible $CO_2$-Induced Climate Change' as a contribution to the Department's 'Carbon Dioxide Effects Research and Assessment Program'. The prophetic final report was published in December 1980 [83]. It recognized that the problem is global, it is beyond human experience (in that the amounts eventually emitted will be larger than seen over the past 10 000 years), and that it is long-range (with $CO_2$ continuing to rise over the next century and a half, and requiring several hundred or more years before the added $CO_2$ can be sequestered naturally in deep ocean waters). Revelle realized, first, that Society has limited experience of responding to slow pervasive environmental changes of this kind, which will be imperceptible to most people from year-to-year; second, that much of the increase in emissions of $CO_2$ will come from increasing use by the developing world rather than from the currently industrialized countries; and, third, that international agreements to mitigate the problem will be difficult given that each nation will tend to operate in its own interests. Alleviation of the problem could come from: a rapid expansion of nuclear power; breakthroughs in solar power; greater use of natural gas (which produces less $CO_2$ than burning coal or oil); development of fuels from biomass; increased energy conservation; constraints on the use of coal; and reversal of the destruction of forests. His report recognized that whilst increased atmospheric $CO_2$ could stimulate plant growth, increasing temperature could put plant growth at risk. Amongst the consequences of inaction were such possibilities as loss of the West Antarctic ice sheet; loss of Arctic sea ice; melting of Arctic permafrost (releasing large volumes of methane); rising sea level; the effects on coral and

plankton or rising ocean acidity; the effects of more intense heat waves on human health; the effects of warming on agriculture (especially through changing supplies of water); and possible increases in human migration away from zones of increasing aridity. Revelle concluded '*the study of $CO_2$-induced impacts on human affairs is just beginning*' [83]. Monitoring and modelling the global climate system in much more detail was absolutely essential to understand better how the system worked and the effects on it of increasing greenhouse gases in the atmosphere. This was a multidisciplinary challenge involving physics, biology, chemistry, geology, hydrology, glaciology, agriculture, fisheries, forestry, energy, medicine, and the social sciences including economics. The first IPCC report was 10 years away.

Meanwhile, Manabe was using yet more advanced climate models to assess the sensitivity of the climate in an increase in $CO_2$. In October of 1980, he and Ronald Stouffer showed that a quadrupling of $CO_2$ would warm the Arctic more than the rest of the globe, with larger warming in winter and in summer, leading to the removal of Arctic sea ice in summer months and reduced snow cover in the Northern Hemisphere [84]. Surface air temperature would warm by about 4 °C. Because warm air holds more moisture, more moisture would be taken poleward, increasing rates of precipitation and runoff.

Enter NASA, the US's National Aeronautics and Space Administration. NASA's atmospheric physicists, located at its Institute for Space Studies at the Goddard Space Flight Centre in New York, had been studying the atmospheres of other planets in the solar system. Knowing that the greenhouse effect explained the atmospheric temperatures of Mars (low $H_2O$ and low $CO_2$ = cold) and Venus (high $H_2O$ and high $CO_2$ = hot), they realized that it ought also to be able to explain the atmosphere of the home planet, Earth. Stimulated by Charney's 1979 paper [79], and led by Jim Hansen, they published in 1981 a compelling analysis of the effects of increasing $CO_2$ on Earth's climate [85]. Their model included feedbacks from clouds and from changes in the surface albedo caused by changes in the cover of reflective snow and ice, and suggested a rise of 2.8 °C for a doubling of atmospheric $CO_2$. That rise is known as the 'climate sensitivity', defined as the equilibrium change in global mean surface temperature for a doubling of atmospheric $CO_2$ above pre-industrial values (more on that in Chapter 11). Hansen showed that exchange of $CO_2$ with the ocean would slow the full impact of warming, and that warming could be masked for short periods (one to two years) by the ejection of reflective materials into the stratosphere by large volcanoes. Human emissions of fine particles – aerosols – could also have an effect, though whilst some reflected radiation and so cooled the air, others absorbed radiation and so warmed it. Trace gases such as ozone ($O_3$) could also warm the planet, and were increasing. Warming was broadly consistent with the human output of $CO_2$, especially in the Southern Hemisphere [85].

Other factors affected the warming pattern in the Northern Hemisphere, where temperatures remained low between 1940 and 1970 (Figure 8.13). Taking into account the rise in $CO_2$ plus volcanic activity plus solar variability, Hansen was able to recreate the global temperature curve from 1880 to 1980, giving some confidence to their methods [85]. Projecting the amount of likely energy use to 2100, he suggested that the signal of $CO_2$ warming should be detectable (reaching values more than one standard deviation from the mean trend) above the noise level of natural climate variability by the mid to late 1980s, and significantly detectable (reaching values more than two standard deviations from the mean trend) by the mid to late 1990s. The prediction was accurate, so it is wrong to say, as doubters do, that climate models have not been tested. By now the predictions of these various early modelling efforts have been tested many times, and are still holding up well.

Much like Arrhenius, Hansen's team calculated that the climate would warm further and faster at the poles than elsewhere. With a 2 °C rise in global temperature, Antarctic temperatures would rise by up to 5 °C, which could trigger the decay of the West Antarctic Ice Sheet and cause a rise in sea level of 5 m, possibly over a century or less [85], much as Mercer suggested in 1978 [66]. The 5–10 °C Arctic warming expected for doubled $CO_2$ in the twenty-first century should eventually melt all the sea ice in summer, opening the Northwest and Northeast Passages to navigation. Predicted temperatures would likely exceed those of the last interglacial.

Echoing Plass and Revelle, Hansen concluded that '*The climate change induced by anthropogenic release of $CO_2$ is likely to be the most fascinating global geophysical experiment that man will ever conduct*', requiring challenging efforts in global observation and climate analysis [85]. And the implications were becoming more stark. Given past evidence that it takes several decades to complete a major change in fuel use, Hansen thought that substantial climate change was almost inevitable, depending on the rate of growth of energy use and the mix of fuels. He thought that full exploitation of coal resources might become undesirable and it would be wise to conserve energy and develop alternative energy sources.

In October 1982, with Taro Takahashi, Hansen convened the 4th Maurice Ewing Symposium, at Lamont, to discuss 'Climate Processes and Climate Sensitivity' [86]. The meeting was supported with a grant from the EXXON Research and Engineering Company, whose President, E.E. David,

provided the first paper in the published volume of the proceedings. '*Few people doubt*' he wrote, '*that the world has entered an energy transition away from dependence upon fossil fuels and toward some mix of renewable resources that will not pose problems of $CO_2$ accumulation. The question is how do we get from here to there while preserving the health of our political, economic and environmental support systems*' [87]. Mr. David was '*upbeat about the chances of coming through this most adventurous of human experiments with the ecosystem*' [87].

Much as in his 1981 paper, Hansen's contribution to the Ewing Symposium explored the feedbacks within the system that conspired to boost temperature increases beyond those attributable to $CO_2$ alone, such as that from water vapour [88]. But following in the footsteps of Arrhenius and Plass, he also investigated the likely change in climate due to the lesser amounts of $CO_2$ in the atmosphere of the last glacial maximum, 20 Ka ago. We shall explore his findings on that topic in Chapter 13.

Wally Broecker contributed two papers to the conference, exploring possible causes for the rise in atmospheric $CO_2$ at the end of the last glaciation – and especially the ocean's role – concluding that $CO_2$ acts as a feedback amplifier [89, 90]. The ocean's precise role in controlling atmospheric $CO_2$ remained somewhat obscure at the time. That same year as the conference, 1982, Broecker had pointed out in his classic *Tracers in the Sea*, co-written with Tsung-Hung Peng, that our current emissions '*will alter the ocean's chemistry even on geologic time scales*' [91]. How can that be? Consider this: the surface mixed layer equilibrates with the atmosphere in about a year. The waters of the main ocean thermocline equilibrate with the air on time scales of several tens of years. The waters of the deep sea equilibrate with the air on time scales of many hundreds of years. The calcite in marine sediments will equilibrate with the air on time scales of several thousands of years. Finally, the excess calcium dissolved in the sea as the result of the calcite-$CO_2$ reaction will be removed on the time-scale of many tens of thousands of years [91]. Geologists take note.

Drawing on Hansen's findings, Revelle took his message about $CO_2$ to the public in 1982, reminding readers (much as he had in 1957) that by burning fossil fuels to release $CO_2$ '*Man was inadvertently conducting a great geophysical experiment*' on the planet [92]. In the short- to medium-term much of the $CO_2$ emitted by burning fossil fuel would be retained in the atmosphere rather than in the ocean because of the 'buffer factor' which meant that the more $CO_2$ the surface ocean contained, the less it could absorb from the atmosphere. Not all of the rise in $CO_2$ came from burning fossil fuels; a goodly percentage came from the massive land clearances around the globe after about 1860, when forests were chopped down to create agricultural land and pastures, a process that was limited ultimately by the amount of useable land. Up to then, some 40–50% of the $CO_2$ produced by humans had remained in the air. The agreement between numerical models and observed trends in temperature, Revelle noted, strongly suggests that $CO_2$ and volcanic aerosols are responsible for much of the global variation in temperature of the past 100 years, although '*so far the warming trend has not [yet] risen above the "noise level" of yearly fluctuations in temperature … an unmistakeable warming trend should appear in the 1990s*' [92]. Recapping on Manabe's modelling experiments [84], Revelle thought that increases in temperature would be least in the tropics and most at the poles, although he recognized that the models were as yet imperfect, and so the results contained many uncertainties. Improvements were needed. Nevertheless, there was much to be concerned about, since '*For a doubling of [$CO_2$] the climatic changes predicted by the model are larger than any since the end of the last ice age about 12,000 years ago*' [92]. Of particular concern were the effects on the distribution of the world's water resources, and the possible melting of West Antarctica, which could raise sea level by 5–6 m. Despite the uncertainties, '*steps should be taken to obtain more evidence and to consider the consequences of a continuing increase in atmospheric carbon dioxide*', not least because '*the planning and construction of water-resource developments in major river basins can take several decades*' [92]. As he had done in 1980 [83], he warned that although we were facing slow pervasive environmental shifts that would be imperceptible to most people, it would be prudent to consider now how we might avoid unfavourable effects.

In April 1980, the US Senate Committee on Energy and Natural Resources convened a hearing on $CO_2$ and climate, one outcome of which was the Energy Security Act of 1980, which called on the Director of the Office of Science and Technology to work with the National Academy of Sciences to carry out a comprehensive study of the projected impact on the level of $CO_2$ in the atmosphere in relation to fossil fuel use. In response the Academy formed a Carbon Dioxide Assessment Committee under the chairmanship of William Nierenberg, Director of the Scripps Institution of Oceanography, with Revelle as one of the members. The Committee's report was published in 1983 [93].

Much of what the Nierenberg report had to say had already been said by the Revelle reports of 1977 [48] and 1980 [83]. This is an extremely complex area of research crossing multiple discipline boundaries, Nierenberg noted, and '*no individual may be considered an expert on the entire problem*' [93]; it has to be addressed by teams of competent experts from each relevant discipline. The report's stance was conservative, noting '*we believe there is reason for caution, not panic*' [93]. Nevertheless, there was no doubt in Nierenberg's mind that the continued growth in $CO_2$ emissions resulting in increased warming was not welcome. Warming could prove '*to be exceedingly bad news for particular parts of the world*' [93], the report noted, citing Bangladesh as one example. Corrective action would be necessary and could not wait upon definitive proof. Indeed, there may yet be unanticipated surprises. It makes sense, said Nierenberg, '*to anticipate changing climates*' and to work towards '*developing … nonfossil sources of energy*' [93].

$CO_2$ was then at around 340 ppm, and was projected to rise to about 400 ppm by 2000, to 425 ppm by 2025 and to 600 ppm by 2065. By 2000 it was actually 370 ppm, reaching 400 in 2015 (Figure 8.7). Doubling $CO_2$ from 340 ppm could raise the global average surface temperature by 3 °C, and Arctic temperatures by 5–10 °C [93]. If $CO_2$ reached 400 ppm in 2000, temperature could reach 1 °C over 1900 values (which it did by 2014 – Figure 8.13). Volcanic eruptions large enough to put dust (and $SO_2$ gas that would form sulphuric acid droplets) into the stratosphere to veil the Earth could cause local cooling of about 0.2 °C, as happened in 1964 after the 1963 eruption of Mt Agung on the island of Bali. Global temperatures had risen by since then, the report noted, and were about 0.5 °C warmer in the mid 1970s than in the 1880s, but it was not yet possible at that stage to confirm that this rise could be attributed to rising $CO_2$ [93]. Indeed, it was still not entirely clear what the level of $CO_2$ in the air had been in pre-industrial times; calculations suggested it may have been around 265 ppm in 1850, but '*this [was] likely to be a lower limit*' [93]. Knowledge of the behaviour of $CO_2$ in the ocean was growing, and Nierenberg was increasingly confident that the ocean took up about 40% of anthropogenic $CO_2$ emissions. As a result, it seemed likely that ocean pH had reduced on average by close to 1 pH unit. As the ocean warmed in future $CO_2$, would become less soluble, and more oceanic $CO_2$ would be expelled into the air (or, conversely, the ocean would take up less $CO_2$ from the air). Regarding sea level, the authors pointed out that disappearance of the West Antarctic ice sheet would likely take centuries rather than decades [93]. In the meantime, sea level was rising due to the thermal expansion of seawater, and might reach 60–70 cm if $CO_2$ doubled over the twenty-first century. Adding an estimated 41 cm from the melting of land ice would bring the total rise by 2100 to just over 1 m [93].

To progress understanding of $CO_2$ in the ocean, the IOC and SCOR, the Scientific Committee on Oceanic Research of the International Council for Science (ICSU), appointed Revelle to chair their new Committee on Climate Change and the Ocean. By 1984, that committee had established a $CO_2$ Advisory Panel under Revelle's chairmanship. His panel recommended an observation programme and a sampling strategy to determine the global oceanic inventory of

$CO_2$ to an accuracy of 10–20 Petagrams of carbon (PgC) [39] (one Petagram is equivalent to one Gigaton [1 billion metric tons]). Later we'll see where that recommendation led.

## 8.4 Biogeochemistry: The Merging of Physics and Biology

By 1983, it was widely recognized that Earth's living and inanimate systems are inextricably intertwined, much as Humboldt told us in the early1800s (Chapter 2). To address these interactions in a more effective way, a whole new area of science sprang into being: biogeochemistry – the study of the chemical, physical, geological, and biological processes and reactions that govern the composition of the natural environment. It focuses on chemical cycles that are driven by or have an impact on biological activity, with a particular emphasis on the cycles of carbon, nitrogen, sulphur, and phosphorus, which play key roles in biological systems and their interactions with the Earth [94].

Whilst some of the roots of biogeochemistry can be traced back to Humboldt, others can be traced back to Eduard Suess, who invented the term 'biosphere', and James Hutton, who maintained that geological and biological systems were linked. In his 1926 book *The Biosphere*, Soviet mineralogist and geochemist Vladimir Ivanovich Vernadsky (1863–1945), founder of the Ukrainian Academy of Sciences, now the National Academy of Sciences of the Ukraine, popularized the use of Suess's term by arguing that life is the geological force that shapes the Earth. He is often credited with first use of the term biogeochemistry, but its use grew only gradually. Some of the impetus behind the development of this 'new' field came from surprising discoveries in the oceans: for example that 'chemotrophic' bacteria lived off hydrogen sulphide emitted from hot vents on mid-ocean ridges, and that microbes were altering sediment composition down to depths of 800–1000 m beneath the deep-sea floor, as found in cores collected by the DSDP [95].

How should the community proceed? One could not apply biogeochemical principles by keeping the physical and biological sciences separate, as had formerly been the trend; the IGY of 1957–1958, for example, had been just that: geophysics with no biology. At the urging of Swedish meteorologist Bert Bolin and Dutch chemist Paul Crutzen, the International Council of Scientific Unions (ICSU), whose name (but not its acronym) changed in 1998 to the International Council for Science, and in 2018 to the International Science Council, decided that something must be done. A proposal was tabled for developing an International Geosphere-Biosphere Programme (IGBP) linking geophysics, chemistry, biology, and geology in a more 'holistic' framework than

before, to address cross-disciplinary topics like acid rain, ozone depletion, the build-up of $CO_2$ and methane in the atmosphere, and biogeochemical cycles. ICSU agreed to review this proposal during its 20th session, in Ottawa, Canada in September 1984. Meanwhile, the US NRC had considered the proposal at a meeting in 1983 [96]. The topic was reviewed within the ICSU meeting, at a one-day symposium on global change that would be informed by the US NRC report and by a collection of papers prepared for the meeting and eventually published in 1985 [97]. The symposium recommended that ICSU develop an interdisciplinary programme to study interrelations between the geosphere and the biosphere, so as to understand global change over long time-scales, and the influence of humans on the environment. ICSU launched the IGBP in 1987.

This development put efforts to understand the functioning of the Earth as an integrated system into effect on a grand scale for the first time [98]. An important underlying context was the increasing scale and significance of the role of humans as agents of global change, and the need to use scientific knowledge to underpin management of our global life support system, to enhance biological productivity, and to respond to the needs of a growing population [99]. The creation of the IGBP recognized the urgency of improving '*understanding of the pathways and rates of exchange for the primary constituents of living organisms (carbon, nitrogen, phosphorus, sulfur, hydrogen, and oxygen) and their relation to the other great domains of planet Earth*' [100]. As Bill Fyfe of the University of Western Ontario put it in his concluding remarks to the Ottawa symposium, '*The Earth is changing and man is contributing to the change at an accelerating rate. We must acquire the fundamental data on geosphere interactions necessary to understand the impact of change on the biosphere*' [99]. At the urging of Thompson Webb, John Kutzbach and Alayne Street-Perrott [101], one of the core programmes of the IGBP would be PAGES, the Past Global Changes programme dealing with past climate change.

Biogeochemistry lay at the heart of the IGBP. Recognizing the birth of this new field of science, Springer launched the journal *Biogeochemistry* in 1984/1985, and the American Geophysical Union launched the journal *Global Biogeochemical Cycles* in 1987. By the mid 1990s, biogeochemistry was considered to be one of the core elements of the field of geochemistry, featuring in 2005 as Volume 8 of the 9-Volume *Treatise on Geochemistry* [102]. W.H. Schlesinger, the editor of Volume 8, explained that we rely on biogeochemists '*to understand fully the full biogeochemical cycles of water and the various chemical elements and the human impacts on each of them ... Biogeochemistry must emerge as the critical scientific discipline that informs planetary stewardship through the rest of this century'* [102]. Discoveries in biogeochemistry have significantly advanced our understanding of Earth's climate evolution.

In 2004, Will Steffen (Executive Director of the IGBP) and colleagues published a comprehensive review of the outcomes of the IGBP [103], highlighting how Earth operates as a single integrated system, the **Earth System,** comprising a number of interlocking cycles, like the hydrological cycle and the carbon cycle, which cannot be disentangled from one another, and which are ultimately forced by the Sun from outside and by plate tectonic and volcanic activity from within. Each cycle affects the others through positive or negative feedbacks that produce emergent properties. The Earth System has to be studied as a complex integrated whole – the opposite of the Cartesian approach. Integral to 'Earth System Science' is the growing role of humans, who are both affected by and influence the system. By the early 1980s it was clear that the magnitude and rates of human-driven changes to the global environment were unprecedented, and that there was no previous analogue for the current operation of the Earth System [103, 104]. That realization led, in turn, to the recognition that we are now living in a new geological epoch, the Anthropocene [105] (see Chapter 15).

## 8.5  The Carbon Cycle

What we are concerned with mostly in this book is the carbon cycle, which lies at the heart of biogeochemistry. There are in fact two interconnected carbon cycles, one fast and one slow. Many texts describe these two. Here I draw on descriptions from the *Treatise on Geochemistry* [106, 107]. In the fast cycle, photosynthesis chemically reduces inorganic carbon ($CO_2$) to form organic matter plus oxygen:

$$CO_2 + H_2O \leftrightarrow CH_2O + O_2$$

When organic matter decomposes, the equation is reversed, and oxygen is used to produce $CO_2$ and water. This cycle operates on time-scales of days to millennia.

Carbon as $CO_2$ is exchanged between a number of 'reservoirs' of different sizes in Gigatons (Gt), including the atmosphere (780 Gt), vegetation (550 Gt), litter (300 Gt), soil (1200 Gt), the surface ocean (dissolved inorganic carbon 700 Gt, and dissolved organic carbon 25 Gt), surface biota (3 Gt), and the deep ocean (dissolved inorganic carbon 36 300 Gt; dissolved organic carbon 975 Gt) (Figure 8.15) [108]. As mentioned earlier, about 90 Gt are exchanged annually between the ocean and the air. Very little organic matter reaches the sea floor, where it may escape decay and become incorporated into sediment and

thus eventually into the lithosphere, whence it is transferred to the slow carbon cycle. Sediment accumulation of both organic and inorganic carbon in the sea is around 0.2 Gt/year, a tiny amount given the vast size of the oceanic reservoir [107].

The slow carbon cycle (Figure 8.16) determines the abundance of $CO_2$ in the Earth's air and oceans on timescales of tens to hundreds of millions of years [109]. It is controlled largely by tectonics, volcanism, and chemical weathering, but also by biology. $CO_2$ is released from the mantle to the air and oceans via volcanic activity and seafloor spreading, and removed from the air mainly by reaction with silicate minerals through chemical weathering in mountains [104], but also by land plants [110]. $CO_2$ both dissolves in the ocean directly, and arrives there as a product of chemical weathering on land. Plankton use dissolved carbon to form their skeletons, which supply carbonate minerals that accumulate in deep-sea sediments and eventually become part of the lithosphere [109]. Most of the carbonates are eventually subducted into the Earth's mantle, where they are heated, releasing their carbon as $CO_2$ to the air and ocean, beginning the cycle again. The chemistry of the slow carbon cycle would also operate whether or not there was life on the planet. But it was enhanced by the activity of plants at times when plant life was abundant, as in the past 540 Ma that geologists call Phanerozoic time [110].

Volcanoes are an important natural source of $CO_2$. Terry Gerlach of the US Geological Survey showed in 2011 that present day volcanoes emit rather modest amounts of $CO_2$ globally: about as much annually as all human activities in Florida [111]. Their $CO_2$ comes from the degassing of magma. In the absence of humans, volcanoes provide the main means of restoring $CO_2$ lost from the atmosphere and ocean by silicate weathering and the deposition and burial of sedimentary carbonates and organic carbon. Gerlach estimated that subaerial and submarine volcanoes emit between 0.18 and 0.44 Gt $CO_2$/year, the average being 0.26 Gt $CO_2$/year [111]. The human-induced emission of 35 Gt $CO_2$/year estimated for 2010 was 135 times greater than the 0.26 Gt $CO_2$/year estimated from volcanoes of all sorts, a ratio that Gerlach called the anthropogenic $CO_2$ multiplier (ACM): an index of the dominance of anthropogenic over volcanic $CO_2$ emissions. The ACM rose gradually from around 18 in 1900 to 38 by 1950, then rapidly to its 2010 level of 135, due the exponential rise in documented $CO_2$ emissions from the burning of fossil fuels [111].

Volcanoes do not erupt steadily, releasing a continuous stream of gas. They erupt periodically and unpredictably. Large, infrequent eruptions cause significant divergence from the global estimate of 0.26 Gt $CO_2$/year. But, as

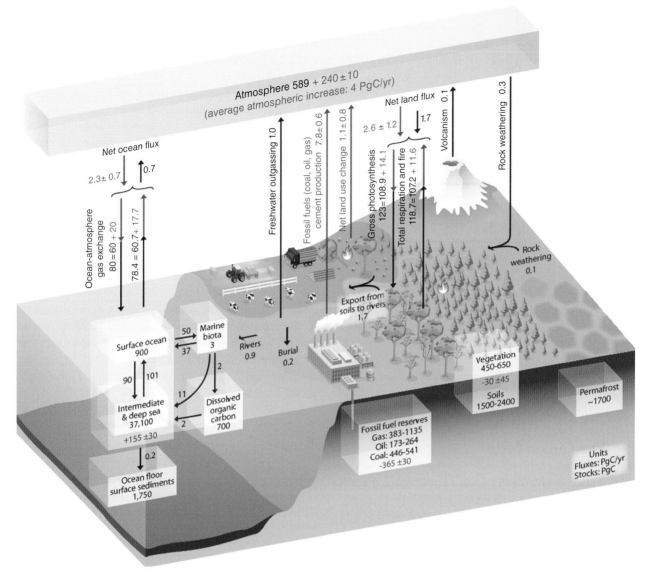

**Figure 8.15** Simplified schematic of the global carbon cycle. Numbers represent reservoir mass, also called 'carbon stocks' in PgC (1 PgC = $10^{15}$ gC, or 1 GtC) and annual carbon exchange fluxes (in PgC/year). Black numbers and arrows indicate reservoir mass and exchange fluxes estimated for the time prior to the Industrial Era (c. 1750). The sediment storage is a sum of 150 PgC of the organic carbon in the mixed layer and 1600 PgC of the deep-sea $CaCO_3$ sediments available to neutralize fossil fuel $CO_2$. Red arrows and numbers indicate annual 'anthropogenic' fluxes averaged over 2000–2009. These fluxes are a perturbation of the carbon cycle post 1750. These fluxes (red arrows) are: fossil fuel and cement emissions of $CO_2$, net land use change, and the average atmospheric increase of $CO_2$ in the atmosphere also called '$CO_2$ growth rate'. The uptake of anthropogenic $CO_2$ by the ocean and by terrestrial ecosystems, often called 'carbon sinks', are the red arrow components of net land flux (top right) and net ocean flux (top left). Red numbers in the reservoirs denote cumulative changes of anthropogenic carbon over the Industrial Period 1750–2011. A positive cumulative change means that a reservoir has gained carbon since 1750. The cumulative change of anthropogenic carbon in the terrestrial reservoir is the sum of carbon cumulatively lost through land use change and carbon accumulated since 1750 in other ecosystems. Note that the mass balance of the two ocean carbon stocks (surface ocean and intermediate and deep ocean) includes a yearly accumulation of anthropogenic carbon (not shown). Uncertainties are reported as 90% confidence intervals. The change of gross terrestrial fluxes (red arrows of Gross Photosynthesis and Total Respiration and Fires) has been estimated from model results. The change in air–sea exchange fluxes (red arrows of ocean atmosphere gas exchange) have been estimated from the difference in atmospheric partial pressure of $CO_2$ since 1750. Individual gross fluxes and their changes since the beginning of the Industrial Era have typical uncertainties of more than 20%, whilst their differences (net land flux and net ocean flux in the figure) are determined from independent measurements with a much higher accuracy. To achieve an overall balance, the values of the more uncertain gross fluxes have been adjusted so that their difference matches the net land flux and net ocean flux estimates. Fluxes from volcanic eruptions, rock weathering (silicates and carbonates weathering reactions resulting in a small uptake of atmospheric $CO_2$), export of carbon from soils to rivers, burial of carbon in freshwater lakes and reservoirs and transport of carbon by rivers to the ocean are all assumed to be pre-industrial fluxes, that are unchanged during 1750–2011. The atmospheric inventories have been calculated using a conversion factor of 2.12 PgC/ppm. *Source:* From IPCC AR5 Ref. [108].

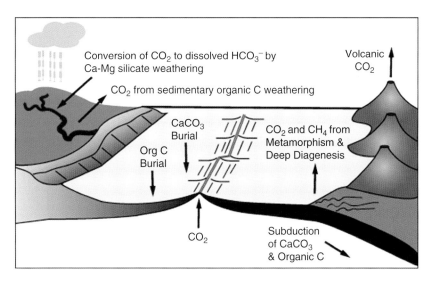

**Figure 8.16** The elements of the long-term carbon cycle. *Source:* From figure 1 in Ref. [109].

Gerlach noted, modern volcanic paroxysms are unlikely to have breached the upper limit of 0.44 Gt $CO_2$/year. When Mt St Helens blew up in May 1980, it contributed a mere 0.01 Gt $CO_2$. Even Mt Pinatubo, erupting in June 1991, only released around 0.05 Gt $CO_2$. Although the actual rate of $CO_2$ emission per hour during a big eruption may be about the same as what we humans emit over the same period, eruptions don't last long, so big eruptions have small effects averaged over the long term. As Gerlach noted, '*humanity's ceaseless emissions release an amount of $CO_2$ comparable to the 0.01 gigaton of the 1980 Mount St Helens paroxysm every 2.5 hours and the 0.05 gigaton of the 1991 Mount Pinatubo paroxysm every 12.5 hours*' [111].

Exploring further, Gerlach calculated that if the global volcanic output equalled or exceeded 35 Gt $CO_2$/year, the annual mass of volcanic $CO_2$ emissions would be more than three times the known annual mass of erupted magma. Such a supply would suggest that $CO_2$ makes up more than 30 wt % of global magma supply, which, if true, would make all volcanoes explosive. In reality $CO_2$ concentrations are more like 1.5 wt % in magmas from ocean ridges, plumes, and subduction zones. It is geologically implausible that magma production of the amount required to supply $CO_2$ to the atmosphere at present rates (more than 40 times the rate of current mid-ocean ridge magma supply) comes from hitherto undiscovered magma sources on the seabed. Indeed, Gerlach pointed out, the release of such vast amounts of volcanic $CO_2$ into the ocean would quickly turn it acid [111]. To get the volumes of $CO_2$ required would call for magma output rates higher than known outputs of past massive continental plateau basalt eruptions, and over much shorter times than they lasted.

Independent analyses by other volcanological experts confirm Gerlach's findings, and were taken into account in reaching his conclusions. For example, Marty and Tolstikhin calculated that the flux of $CO_2$ into the atmosphere and ocean from submarine volcanic sources (mid-ocean ridges, island arcs and plumes) is about 0.264 Gt/year [112], whilst the flux from subaerial volcanoes is about $0.065 \pm 0.046$ Gt/year [113]. Some $CO_2$ is emitted from hydrothermal vents on mid ocean ridges. As Chris German pointed out, whilst the first hydrothermal vent sites discovered on the deep sea floor contained less than twice the $CO_2$ present in seawater, we now know that few vent sites have $CO_2$ levels less than or equal to those in seawater, whilst many sites have an order of magnitude more $CO_2$ than seawater, and a few have two orders of magnitude more [114]. However, these contributions are rapidly diluted away from immediate vent sources. Independent confirmation of the weakness of the submarine volcanic signal comes from the limited abundance in the ocean of the $^3$He isotope, which is an expression of submarine volcanic activity. Helium-3 is modestly abundant at oceanic depths between about 1500 and 3000 m, decreasing rapidly away from mid ocean ridge crests. John Lupton of NOAA's Pacific Marine Environmental Laboratory (PMEL) in Washington State has found $CO_2$-rich fluid locally venting onto the seabed on the flanks of a submarine volcano in an island arc setting [115], but this is thought to be a highly localized phenomenon. Similar venting has also been found in a sedimentary setting in the Okinawa Trough.

An independent study by Giuseppe Etiope and Nils-Axel Mörner agreed that natural emissions of $CO_2$ lie far below those from human activities, but pointed out that geological seepage of light hydrocarbons from deep within the Earth, including methane, ethane, and propane, could be a neglected source of greenhouse gases. These might prove to be more important than volcanic emissions of $CO_2$ as natural sources of greenhouse gas. Even so, amounts emitted are trivial compared with emissions of greenhouse gases from human activities [116].

Updating the information base for volcanic emissions, in 2013, Michael Burton and colleagues from the Italian National Institute of Geophysics and Vulcanology, in Pisa, concluded that subaerial volcanic emission amounted to some 540 Mt/year $CO_2$ (Mt = million tonnes), whilst the flux from mid-ocean ridges reached 97 Mt/year $CO_2$, for a grand total of 637 Mt/year $CO_2$ [117]. Adding in possible contributions of 300 Mt/year $CO_2$ released by metamorphism [116], gave a possible total of 937 Mt/year $CO_2$ [117]. These numbers match within an order of magnitude to the rates of $CO_2$ removal estimated for chemical weathering [117]. Whilst the overall flux estimate is higher than previously thought, it remains '*insignificant relative to anthropogenic emissions, which are two orders of magnitude greater at 35,000 Mt/yr*' [117]. Uncertainties remain, especially for emissions from mid-ocean ridges, not least because as $CO_2$ forms in those environments it reacts with hot rock in a process that sequesters $CO_2$ as carbonate minerals; the amount of $CO_2$ sequestered by oceanic crustal reactions is estimate to be c. 150 Mt/year, which is of similar magnitude to the $CO_2$ flux of $97 \pm 40$ Mt/year, and it is '*probable that reactions in the oceanic crust absorb more $CO_2$ than is emitted from [mid ocean ridges]*' [117].

There is no reason to suppose that the very large number of seabed volcanic structures (seamounts and guyots), identified from satellite gravity overpasses and known to be widespread especially across the floor of the Pacific Ocean, is an unidentified source of $CO_2$ emissions. The vast majority of these structures are extremely ancient and inactive, and oceanographic surveys confirm that they are not associated with the $^{3}$He emissions indicative of submarine volcanism, which are confined to the volcanically active areas around mid-ocean ridges.

Aside from these calculations of the amounts of $CO_2$ emitted, we can use the carbon isotopic composition of $CO_2$ to indicate its likely source. The $CO_2$ emitted from magmatic sources and uncontaminated by air usually has a carbon isotopic signature of around $-2\ \partial^{13}C$ [118], which is quite different from that of air or petroleum. But the volume emitted is too small to have an effect on the composition of the air on human time scales, unlike the change from burning fossil fuels over the last century.

In summary, the data show that global emissions of $CO_2$ from within the crust by volcanic and related processes on land and beneath the ocean amount to less than 1% of the annual anthropogenic emissions of $CO_2$. There is no justification in the rock record for the unbelievably high volumes of magma production or unbelievably high concentrations of magmatic $CO_2$ required to support the notion that volcanoes have supplied the increases in atmospheric $CO_2$ since the start of the Industrial Revolution. Volcanic $CO_2$ is dwarfed by anthropogenic outputs.

## 8.6 Oceanic Carbon

On the millennial time scale it is the vast amount of dissolved inorganic carbon in the ocean that determines the equilibrium amount of $CO_2$ in the atmosphere. $CO_2$ dissolves in the ocean to form carbonic acid ($H_2CO_3$), which dissociates to form bicarbonate ions ($HCO_3^-$) and carbonate ions ($CO_3^{2-}$), which are by far the main forms of dissolved inorganic carbon in seawater. Bicarbonate ions in turn dissociate to form hydrogen ions ($H^+$) and carbonate ions ($CO_3^{2-}$). The concentration of hydrogen ions dictates the ocean's acidity, or pH. As $CO_2$ dissolves in the ocean it initially alters ocean carbonate chemistry, causing the ocean to become more acid and reducing its pH. That process cannot continue indefinitely, because the carbonate system in the ocean is buffered to the extent that the more acid the ocean becomes, the less $CO_2$ it is able to absorb from the atmosphere. The buffering effect keeps the ocean's pH to a narrow range between pH 7.5–8.5 [119]. Whilst that may appear to be a rather small range, the pH scale is logarithmic, and the pH decrease of 0.1 unit between 1751 and 1994 is equivalent to an increase of 30% in acidity [107]. Because buffering increases as dissolved $CO_2$ increases, when we put more $CO_2$ into the air, the ocean becomes increasingly resistant to taking it up, and more of it stays in the air.

This long-term equilibrium is modified in the short-term by three key factors. One is ocean mixing, which moves $CO_2$ absorbed from the air by the ocean deeper in the water column with time, enabling more $CO_2$ to be adsorbed at the surface. An example is the sinking of relatively dense sub-Antarctic surface water from the surface of the Southern Ocean to form Antarctic Intermediate Water at an average depth of around 1000 m in the South Atlantic. The other two factors are 'pumps'. The 'solubility pump' operates because $CO_2$ is twice as soluble in cold water as it is in warm water. Warm the water and $CO_2$ comes out; cool it and $CO_2$ goes in. Thanks to this pump, the sinking of cold Arctic and Antarctic water takes $CO_2$ from the air down into the deep ocean, making the atmospheric concentration of $CO_2$ lower than it would be otherwise [107]. The 'biological pump' transfers $CO_2$ from surface waters to the deep ocean through the decomposition at depth of sinking dead organic matter that grew by using $CO_2$ in photosynthesis in the sunlit zone at the ocean's surface [120, 121] (Figure 8.17) [103]. This process also enriches deep waters in dissolved carbon, thus lowering the $CO_2$ content of

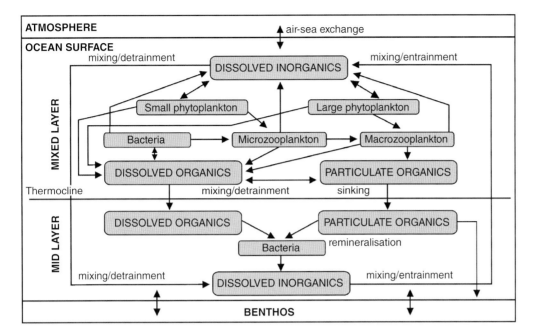

**Figure 8.17** The operation of the biological pump and the marine carbon cycle (a conceptual diagram from the Joint Global Ocean Flux Study, JGOFS). Marine organisms assimilate $CO_2$ in surface waters. It is primarily recycled through a complex food web of small organisms in the upper water column, which generates dissolved carbon that will mix with deeper waters plus particulates that will sink into the deep ocean. Some 25% of carbon fixed by phytoplankton in the upper layer sinks to the interior where it is oxidized and recycled. Only about 1–2% of surface production reaches the seabed, where it is consumed and recycled by benthic organisms, whilst the rest is buried in ocean sediment. Dissolved $CO_2$ and nutrients can return to the surface through wind-driven upwelling (e.g. along continental margins, along the equator, or in the Antarctic Circum-polar Current), or during deep winter mixing (e.g. in the northern North Atlantic). *Source:* From figure 2.29 in Ref. [103].

surface waters and hence making less $CO_2$ available to the air. If these pumps were switched off, the preindustrial concentration of $CO_2$ in the atmosphere would likely have been 720 ppm rather than its measured 280 ppm [107].

Most primary production by ocean plankton is recycled as part of the fast carbon cycle, with the release of dissolved inorganic carbon to the upper ocean within the sunlit top 100 m of the water column [120]. This recycling process keeps most carbon out of the biological pump. Rates of primary production (in gC/m²/year) are highest (about 420) in coastal regions where upwelling currents bring nutrients to the surface, somewhat less (about 250) in other coastal areas, and only about 130 in the open ocean. But because 90% of the ocean is open, 80% of carbon fixation occurs there rather than in the more productive continental margins [120]. About 75% of production in upwelling and coastal regions is siliceous diatoms, which make up only 25% of the open ocean productivity where calcareous phytoplankton predominate away from the poles, where siliceous diatoms hold sway.

Organic matter commonly sinks at rates of around 100 m/day in the form of dense aggregates or clumps known as 'marine snow', some of which comprises the faecal pellets of zooplankton. The rates of decomposition en route are so

high that the rate of accumulation of organic carbon in marine sediments is only 0.5% of the rate of carbon fixation at the ocean surface. Mapping organic matter on the seabed shows that around 94% of the sedimentary organic carbon preserved there is buried on continental shelves and slopes, with only 6% accumulating in the open ocean. As the open ocean plays host to 80% of primary production, the accumulation of only 6% of organic carbon there suggests that this area has an overall preservation efficiency of 0.02%. On the continental shelves and slopes, the preservation efficiency is more like 1.4% [120]. To put it another way, only 10% of primary productivity escapes from the production zone in the upper 100 m; only 10% of that reaches the seabed at 4000 m; 90% of that is degraded at the seabed, leaving only 0.1% of the original production locked in sediments under the open ocean [122]. As implied by the efficiency numbers, significantly more sinking organic matter is trapped on continental margins. Some of the organic material deposited on the continental shelf and upper continental slope is resuspended by bottom currents, and drifts slowly down slope in turbid bottom waters. Some moves to the deep ocean much faster through turbidity currents. These lateral processes may mix organic matter from land with that from marine production in surface waters, on the deep sea floor.

Because of the difficulty of measuring either changes in the ocean's inventory of carbon or the exchange of carbon across the air-sea interface, the uptake of anthropogenic carbon by the oceans used to be calculated with models that simulated the chemistry of carbon in seawater, the air-sea exchanges of $CO_2$, and oceanic circulation [107]. However, with more than 4 million observations of the partial pressure of $CO_2$ in the oceans made in recent years, we can now calculate the uptake of $CO_2$ more directly [123]. Models and data are in good agreement.

By the mid-1980s, oceanographers recognized that we still knew far too little about how the ocean worked, not least because it was grossly under-sampled. Yet the ocean plays a key role in the climate system by storing masses of heat and moving it slowly around the globe. As much heat is stored in the top 3.5 m of the ocean as in the entire atmosphere. To reduce uncertainties about global ocean dynamics and the rates of ventilation and mixing in the deep ocean, and to improve understanding of the role of the ocean in the global climate system, the WCRP launched a major international global experiment – the World Ocean Circulation Experiment (WOCE) – conceived in the mid-1980s and having a field phase between 1990 and 1998, with a modelling phase extending to 2002 [124].

Today, the oceans contain about 50 times as much $CO_2$ as the atmosphere, so small changes in the ocean carbon cycle can have large atmospheric consequences. Realizing that it was essential to increase focus on the role of the ocean in the $CO_2$ story, the US scientific community began developing a strategy for large-scale, coordinated studies of oceanic biogeochemical processes. By 1987, several countries were following that lead, and the SCOR was persuaded to combine these efforts in the international JGOFS [39], which aimed to understand how physical, chemical, and biological processes influence the exchange of $CO_2$ between the ocean and the air, and to elucidate the role of biological feedbacks in amplifying the chemical and physical effects. Its planners followed the recommendations of Revelle's $CO_2$ Advisory Panel in designing an observation programme to determine the global oceanic inventory of $CO_2$. In 1989, JGOFS would become a core project of the fledgling IGBP. Together, WOCE and JGOFS would improve our understanding of the workings of the ocean and its role in the carbon cycle. As director of the UK's major deep-sea oceanographic laboratory at Wormley between 1988 and 1995, I helped to oversee the development of the UK's substantial contribution to WOCE and my institute's contribution to JGOFS.

Reviewing Revelle's legacy, Chris Sabine from NOAA, Hugh Ducklow from WHOI, and Maria Hood from the IOC had this to say (with my apologies for the acronym soup!): '*The first global survey of ocean $CO_2$ was carried out under the joint sponsorship of IOC and the Scientific Committee on Oceanic Research (SCOR) in the Joint Global Ocean Flux Study (JGOFS) and the World Ocean Circulation Experiment (WOCE) in the 1990s. With these programs and underway $pCO_2$ measuring systems on research vessels and ships of opportunity, ocean carbon data grew exponentially, reaching about a million total measurements by 2002 when Taro Takahashi (Lamont-Doherty Earth Observatory) and others provided the first robust mapping of surface ocean $CO_2$. Using a new approach developed by Nicolas Gruber (ETH Zurich) and colleagues with JGOFS-WOCE and other synthesized data sets, one of this article's authors (Sabine) with a host of coauthors estimated [125] that the total accumulation of anthropogenic $CO_2$ between 1800 and 1994 was $118 \pm 19$ Pg C, just within the uncertainty goals set by JGOFS and IOC prior to the global survey. Today, ocean carbon activities are coordinated through the International Ocean Carbon Coordination Project (IOCCP). Ocean carbon measurements now accumulate at a rate of over a million measurements per year—matching the total number achieved over the first three decades of ocean carbon studies. IOCCP is actively working to combine these data into uniform data sets that the community can use to better understand ocean carbon uptake and storage*' [39]. Through the IOCCP, which became a standing project of SCOR and the IOC in 2005, the '*IOC is achieving the international cooperation in ocean carbon observations that Roger Revelle was promoting back in 1960*' [39].

Observations confirm that the ocean is accumulating anthropogenic $CO_2$ in much the same way that Keeling showed was happening in the air. As of 1980 the oceans had absorbed around 40% of the emissions made to that time [107]. On a time-scale of several thousand years, Dave Archer and colleagues calculated that around 90% of anthropogenic $CO_2$ emissions will end up in the oceans [126].

Sabine's study showed that our (anthropogenic) emission of $CO_2$ is not distributed uniformly through the oceans [125]. Most is found in the North Atlantic, where cold $CO_2$-rich surface water sinks into the depths to form North Atlantic Deep Water. There are substantial, though lesser, amounts in a belt along the north side of the Antarctic Circumpolar Current, where cold $CO_2$-rich surface waters sink to form Antarctic Intermediate Water and Subantarctic Mode Water. These various sources make the Atlantic Ocean much richer in anthropogenic $CO_2$ than the Pacific Ocean. The surface waters least rich in anthropogenic $CO_2$ are the upwelling waters of the Southern Ocean south of the Polar Front, which bring old naturally $CO_2$-rich water to the surface and supply it to the air. Thus, the Southern Ocean is a source for old $CO_2$ south of the front, and a sink for new anthropogenic $CO_2$ north of it.

## 8.7 A Growing International Emphasis

For the purposes of my tale, the next big development after ICSU's 1984 Ottawa meeting was a conference convened by the UNEP, the WMO, and ICSU, which took place in the fall of 1985 in Villach, Austria, on the 'Assessment of the Role of Carbon Dioxide and of Other Greenhouse Gases in Climate Variations and Associated Impacts' [127]. The meeting participants concurred that '*we must place the $CO_2$ question in the context of a global picture which involves understanding the linkages of carbon to the biogeochemical cycles of nutrients, and treat these cycles in the context of climate feedbacks rather than as isolated, independent systems*' [128]. Much remained to be done, including: clarifying the pre-industrial concentration of $CO_2$, quantifying the oceanic sink for $CO_2$, studying the impact of El Niño events on $CO_2$, and studying the natural variability of $CO_2$ between glacial and interglacial cycles (which we examine in Chapters 12 and 13) and during climatic events like the Little Ice Age and Medieval Optimum (which we examine in Chapters 14 and 15). Evidently, geologists too would have to contribute to the growing understanding of the behaviour and effects of $CO_2$ within the climate system.

The Villach meeting concluded that '*As a result of the increasing concentrations of greenhouse gases, it is now believed that in the first half of the [21st century] a rise of global mean temperature could occur which is greater than any in man's history*' [127]. They considered that the effect of doubling $CO_2$ on temperature lay in the range 1.5–4.5 °C,

and on sea level in the range 0.2–1.4 m. To address this problem, '*governments and funding agencies should increase research support and focus effort on crucial unsolved problems related to greenhouse gases and climate change*' [127]. Participants also suggested that governments should support periodic assessments of the state of scientific understanding and its practical implications. To see how that suggestion might be taken forward, ICSU, UNEP, and the WMO set up, in 1985, a seven-member Advisory Group on Greenhouse Gases to provide recommendations based on current research. However, this small group of scientists, which last reported in 1992, lacked the resources to cover the increasingly complex interdisciplinary nature of climate science. Something more was needed. Further advice came from a review of the scientific literature in 1986 by SCOPE (the Scientific Committee on the Problems of the Environment), a body of ICSU (then the International Council for Scientific Unions), edited by Bert Bolin (Box 8.6) and others [129]. These were key steps towards formation of the IPCC (Box 8.7).

The reports of IPCC's Working Group 1 on the science of climate change are referred to as a 'consensus', meaning the broad agreement of the group of scientists who contributed to it. But one has to bear in mind that, following the IPCC's Terms of Reference (see (i) in Box 8.7), the report is not a comprehensive study of all aspects of climate science. Indeed, parts of its reports may be inaccurate. For instance, in my Antarctic science community, we were unhappy with the climate modelling for Antarctica in the 2007 IPCC

---

| **Box 8.6   Bert Bolin** |
|---|

Bert Rickard Johannes Bolin (1925–2007) was a Swedish meteorologist who served as the first chairman of the IPCC. Graduating from Uppsala University in 1946, Bolin earned Master's and PhD degrees in Meteorology at Stockholm University, spending 1950 at the Institute for Advanced Study in Princeton, New Jersey, where he worked with Jule Charney, John von Neumann and others on the first computerized weather forecast, using ENIAC, the first electronic computer. He was Professor of Meteorology at Stockholm University from 1961 to 1990, and much involved in international climate research, including organizing use of the new satellite tools for climate research, which led to the formation of the ICSU Committee on Atmospheric Sciences (CAS) in 1964, which he chaired. The Committee started the Global Atmospheric Research Programme (GARP) in 1967, which Bolin also chaired, and GARP became the WCRP in 1980. Bolin served on ICSU's Advisory Group on Greenhouse Gases from 1985 to 1992, and in 1987 contributed to the 500-page

Brundtland Report, which (like other reports of the time) recommended the setting up of the IPCC. Under his chairmanship (from 1988 to 1997), the IPCC produced its First and Second Assessment Reports (1990 and 1995). The first report led to the United Nations Framework Convention on Climate Change (UNFCCC), and the second to the Kyoto Protocol. Bolin has also been scientific director of the European Space Research Organisation (now the European Space Agency). Awards for his work in climate research include the International Meteorological Organization Prize (1981), the Carl-Gustaf Rossby Research Medal (1984), the Tyler Prize for Environmental Achievement (1988), the Körber European Science Prize (1990), the Milutin Milankovic Medal (1993), and the Blue Planet Prize (1995), Bolin was a member of the Swedish, Norwegian, and Russian Academies of Sciences. In November 2007, he published the partly autobiographical *A History of the Science and Politics of Climate Change: The Role of the Intergovernmental Panel on Climate Change.*

---

**Box 8.7    The IPCC**

By 1987, the 10th Congress of the WMO saw the need for an objective, balanced, and internationally coordinated scientific assessment of the effects of increasing concentrations of greenhouse gases on the Earth's climate and on ways in which these changes might impact socio-economic patterns, and in 1988 the WMO and the UNEP, under the leadership of G.O.P. Obasi (for WMO) and M.K. Tolba (for UNEP), jointly established the IPCC as the mechanism for that assessment. It took the form of groups of experts, which first reported in 1990, repeating their assessments at intervals of about five years. The IPCC's two Terms of Reference, spelled out in the Preface to its first assessment report included:

i) assess the scientific information that is related to the various components of the climate change issue, such as emissions of major greenhouse gases and modification of the Earth's radiation balance resulting therefrom, and that [is] needed to enable the environmental and socio-economic consequences of climate change to be evaluated,

ii) formulate realistic response strategies for the management of the climate change issue.

To achieve those (limited) tasks the IPCC established three working groups to:

a) assess available scientific information on climate change.

b) assess environmental and socio-economic impacts of climate change.

c) formulate response strategies.

The activities of Working Group 1 (chaired by Dr. John Houghton, Director of the UK's Meteorological Office), dealing with the science of climate change and the first Term of Reference, were informed by reviews of the scientific literature, by the work of the Advisory Group on Greenhouse Gases, and by the 1986 report by ICSU's SCOPE [129]. Working Group 1 does not do original research. It reviews pertinent published reports and publishes its findings every five years or so in reports of about 1000 pages that can be downloaded free from the IPCC web site (www.ipcc.ch). Although these seem like textbooks on the evolving state of climate science, they contain only limited reviews of past climate change. The latest complete one – the 5th Assessment Report – concluded that '*It is extremely likely that human influence has been the dominant cause of the observed warming since the mid-20th century*' [130]. In other words, it agreed that the forecasts of Plass (1956), Revelle (1965, for President Johnson's Science Advisory Committee), Manabe (1967), Broecker (1975), Charney (1979), Flohn (1979), Ramanathan (1979), and Hansen (1981 and 1982) were coming to pass.

The IPCC reports inform the deliberations of the annual Conference of the Parties to the UNFCCC, which was established at the UN's Earth Summit in Rio de Janeiro in 1992, and which entered into force in March 1994. In December 2015 the Conference of the Parties (i.e. the national governments) agreed that efforts should be made to prevent global warming from rising more than 2.0 °C above the 1900 level, and in 2018 the IPCC produced a special report on how to avoid exceeding 1.5 °C [131].

---

report, because the global climate models used for the Antarctic did not properly represent the Antarctic climate system [132]. Similarly, glaciologists took exception to the sea level projections in the 2007 report because they did not include results from dynamic models for the mechanical decay of ice sheets (admittedly because those data were not yet in the public domain). Even so, at the time the IPCC reports are published they are comprehensive and useful reviews of the state of the art of climate science, especially in regard to the role of greenhouse gases, and useful guides as to what research needs to be done next. They do not, however, provide much detail on palaeoclimates, which can tell us about the extremes of climate change addressed in this book.

The international Global Carbon Project now monitors $CO_2$ emissions and atmospheric and oceanic concentrations in relation to land use, fossil fuel use, and carbon uptake by land and ocean, and produces an annual Global Carbon Budget, the latest version of which was published in December 2018 [133]. Most of it relates to 2017, when total emissions of $CO_2$ amounted to $11.3 \pm 0.9$ GtC/year (that's C, not $CO_2$), and averaged 405 ppm $CO_2$ [133]. Cumulative $CO_2$ emissions for 1870–2017 were $615 \pm 80$ GtC, and partitioned between the atmosphere ($250 \pm 5$ GtC), the ocean ($150 \pm 20$ GtC, and the land ($190 \pm$ GtC). Evidently, uncertainty mainly derives from estimates of land use change.

## 8.8    Reflection on Developments

Reflecting on developments since Plass's pronouncements in the late 1950s, a good beginning has been made towards understanding the operation of the planetary climate system and of the role of $CO_2$ within it. These advances owe a great deal to an extraordinary expansion in the numbers and kinds of satellites, and a comparable advance in the number of research ships, floats, buoys, and other means of observing the ocean. For example, at the time of writing there were some 3500 *Argo* floats measuring the temperature and salinity of the upper ocean, and there were some 3000 satellites operating in space, many observing Earth's environment.

Following the launch of the first satellite, Sputnik, on 4 October 1957, the use of satellites to look at the Earth has mushroomed. TIROS-I was the first Earth observing satellite, launched in 1960 to observe the weather; now there are dozens doing the same. SEASAT was the first ocean observing satellite, launched in 1978: there are now many more. Ships, planes, and now people can know where they are by using the Global Positioning System (GPS), which has been operational since 1978 and generally available since 1994. The GPS is a boon to oceanographers trying to unravel the secrets of the motion of the oceans. And these days we all talk to each other and send messages via communication satellites, the first of which, Telstar, was launched in 1962. Autonomous instruments like *Argo* floats use communications satellites to send their data back to base; their positions are based on GPS.

Just as Earth-observing operations, both in situ and remotely from satellites, have grown dramatically over the years, so too has the research to understand what the operational data mean, and into the processes acting within the climate system. Research continues through the WCRP, the IGBP (which ended in 2015, but whose activities continue in the Future Earth project) and groups like the International Ocean Carbon Coordination Project. One might justifiably say that the 1st World Climate Conference in 1979 had momentous outcomes. Climate science, like medical science, has become high-tech, and most of the developments in both fields have taken place since the 1950s.

So, by the late 1970s, some 20 years after publication of his key papers in 1956, one of Plass's key wishes had come true. $CO_2$ was being measured in the atmosphere and ocean. We now have abundant and ongoing measurements of $CO_2$ and other radiatively active gases in the air, and a comprehensive theoretical understanding of how the climate system works and of the roles of $CO_2$ and water vapour in it. Not everything is known with absolute certainty yet. Ramanathan reminded us back in 1987 that we did not know enough about the role of clouds, the tops of which could reflect radiation, whilst their moisture could trap it [134], and this is still an area of some difficulty (although not a show-stopper) [107]. But even before the IPCC began deliberating in 1988, much was known about $CO_2$ and climate and what the future might hold. And at about that time development had begun on a global observing system

capable of making an annual health check on the climate of the planet (the Global Climate Observing System), and JGOFS had set out to make a quantum leap in measurements of $CO_2$ in the ocean. Momentous advances were made as we moved towards the end of the twentieth century and into the twentyfirst. We still do not yet know everything we need to. Some uncertainties remain, but they are steadily getting fewer and smaller.

Are climate projections for the end of the century trustworthy? There is every reason to suppose that the theory of infrared radiative transfer (e.g. as spelled out by Pierrehumbert [45] and Ramanathan [46]) is correct, not least because it explains the present trends of climate and helps to explain past climatic behaviour. Nevertheless, as Anderson and colleagues pointed out in 2016, it has the shortcoming that it does not address feedbacks, for example like the storage of immense amounts of heat by the ocean and the gradual mixing of that heat throughout the entire water column, which will take centuries [135]. That means that today's temperature lags behind what will eventually be possible once the ocean and the atmosphere reach a state of radiative equilibrium. A similar argument applies to the role of ice, which takes a great deal of heat to melt, thus representing a further aspect of the present disequilibrium. To address the role of feedbacks we have to go beyond radiative transfer equations and use Earth System Models, which include complex GCMs of the climate system that address a myriad of processes and interactions between land, ocean, ice, and atmosphere [135]. Projections of 'baseline' warming by Earth System Models are robust with few uncertainties, because they rely on well-understood radiative transfer equations that address the greenhouse effect. But Earth System Models must also address the added impact of feedbacks, where there is considerable uncertainty about the magnitude of change, but not about the sign of the temperature response, which for the majority of feedbacks (e.g. except reflective aerosols) is positive [135]. These feedbacks include the facts that warming is accompanied by increasing water vapour (a greenhouse gas), melting of snow and ice is increasing (reducing albedo), uptake of $CO_2$ by plants decreases (e.g. as warming dries out soil), and production of $CO_2$-consuming plankton decreases (as warming oceans become more stratified and feed fewer nutrients to the ocean surface) [135]. The models must also address the time-dependency of the processes involved – like the time required for warming to reach full ocean depth, and to melt ice [135]. The core physics used by early investigators like Arrhenius and Callendar does explain much of the observed variation in global temperature due to greenhouse gases, and the addition of the effects of feedbacks adds further understanding. Whilst Earth System Models help to explain warming at the global scale now and in palaeoclimate projections, modelling at the regional scale is required to provide high geographical resolution [135]. Palaeoclimate Earth System Models, whilst generally robust, tend to run cool – i.e. to project temperatures cooler than suggested by proxy measurements. There is more research to be done.

One of the key elements that has a bearing on our understanding of how $CO_2$ may have behaved in the climate systems of the past is the question of duration: how long will $CO_2$, once emitted, stay in the atmosphere? In 2007, a team led by Alvaro Montenegro of the University of British Columbia assessed the climate response on millennial time-scales and calculated '*that 75% of the anthropogenic $CO_2$ has an average perturbation lifetime of ~1,800 years with the remaining 25% having average lifetime much longer than 5,000 years*' [136]. They showed that if all of the available fossil fuels (then estimated as ~5000 GtC) were burned over, say, the next 200 years, global temperatures would rise by 6–8 °C and remain at least 5 °C higher than preindustrial levels for more than 5000 years. Later, in 2009, a subset of those same researchers refined their calculations to show that '*For emissions up to about 1000 [Gt], 50% of the $CO_2$ anomaly is taken up within 100 yr and another 30% is absorbed within 1000 yr, which is similar to IPCC estimates … Above 1000 [Gt], the time to absorb 50% of the emissions increases dramatically, and more than 2000 yr are needed to absorb half of a 5000-[Gt] perturbation*' [136, 137]. In addition, due to the logarithmic relationship between $CO_2$ and its radiative forcing, '*Temperature anomalies may last much longer*' [137, 138]: up to 10000 years or more. These studies agree with those of Susan Solomon of the US NOAA, and colleagues, who demonstrated that '*the climate change that takes place due to increases in carbon dioxide concentration is largely irreversible for 1,000 years after emissions stop … Following cessation of emissions, removal of atmospheric carbon dioxide decreases radiative forcing, but is largely compensated by slower loss of heat to the ocean, so that atmospheric temperatures do not drop significantly for at least 1,000 years*' [139].

In addition, one of the key things we have learned, which will have a bearing on interpretations of past climate change, is that water vapour is the dominant contributor to warming (~50% of the effect), followed by clouds (~25%), and $CO_2$ (~20%). According to Gavin Schmidt of NASA's Goddard Institute for Space Studies, these proportions remain about the same even with a doubling of $CO_2$ in the atmosphere [140]. This is likely to be typical of past climates too.

Following the ideas of James Hutton (Chapter 2), the palaeoclimate data that we evaluate in following chapters will lead us to a view on future change. That view will not be based solely on the outputs of the IPCC, useful though they may be. It will reflect all that we have learned in these pages, starting with the work of Ebelmen in 1845 and Tyndall in 1859 (see Chapter 4), about $CO_2$ in the atmosphere and its role in the climate system.

In summary, then, we have learned progressively more about the likely effects of rising $CO_2$ on our climate since scientists started taking a close look at the matter in the mid to late 1950s, 35 years before the IPCC produced its first report. Indeed, it was in response to that rapidly growing understanding of the operation of the greenhouse effect that the IPCC was formed to replace many ad hoc national assessments with the regular international assessments needed to address what was obviously a global problem. Since its first report in 1990, the IPCC has embellished but not substantially changed the picture of what had been discovered by the climate science community by that time.

How would geologists apply this new knowledge to their interpretation of past climate change? And would they be able to grant Plass's second wish – '*to be able to determine atmospheric $CO_2$ concentration for past geological periods*' [14]? We shall explore those questions in later chapters.

## References

**1** Callender, G.R. (1938). The artificial production of carbon dioxide and its influence on temperature. *Quarterly Journal Royal Meteorological Society* 64: 223–237.

**2** Callendar, G.R. (1949). Can carbon dioxide influence climate? *Weather* 4: 310–314.

**3** Archer, D. and Pierrehumbert, R.T. (2011). *The Warming Papers – The Scientific Foundation for the Climate Change Forecast*. Chichester and Oxford: Wiley Blackwell.

**4** Schmidt, G. (2010) Taking the measure of the Greenhouse Effect. *Science Briefs. Goddard Institute for Space Studies*, NASA (https://www.giss.nasa.gov/research/briefs/schmidt_05).

**5** Madden, R.P. (1957) *Study Of $CO_2$ Absorption Spectra Between 15 And 18 Microns*, Progress Report to Office of Naval Research, Contract 248(01), declassified. Available from Armed Forces Technical Information Agency, Arlington, Virginia, 50pp plus references, tables and figures.

**6** Czerny, M. and Turner, A.F. (1930). Über den astigmatismus bei spiegelspektrometern. *Zeitschrift für Physik* 61 (11–12): 792–797.

**7** Strong, J. (1962). *Infrared Spectroscopy and Infrared Properties of Atmospheric Gases*. Fort Belvoir: Defense Technical Information Centre.

**8** Cloud, W.H. (1952) *The 15 Micron Band of CO₂ Broadened by Nitrogen and Helium*, ONR Progress Report, Johns Hopkins University, Baltimore.

**9** Madden, R.P. (1961). A high-resolution study of $CO_2$ absorption spectra between 15 and 18 microns. *Journal of Chemical Physics* 35: 2083. http://dx.doi.org/10.1063/1.1732212) 15 pp.

**10** Anding, D. (1967) *Band-Model Methods For Computing Atmospheric Slant-Path Molecular Absorption*, IRIA Technical Report, 181 pp (available from NTIS – National Technical Information Service, Department of Commerce, Washington DC).

**11** http://www.cfa.harvard.edu/hitran (December 2013) The HITRAN Database.

**12** Houghton, J. (2013). *In the Eye of the Storm: The Autobiography of Sir John Houghton*. Oxford: Lion Books.

**13** Houghton, J. and Smith, D. (1966). *Infra-red Physics*. Oxford: Oxford University Press.

**14** Plass, G.N. (1956). The carbon dioxide theory of climate change. *Tellus* VIII: 140–154.

**15** Plass, G.N. (1956). The influence of the 15-micron carbon dioxide band on the atmospheric infra-red cooling rate. *Quarterly Journal Royal Meteorological Society* 82: 310–329.

**16** Plass, G.N. (1956). Effect of carbon dioxide variations on climate. *American Journal of Physics* 24 (5): 376–387.

**17** Plass, G.N. (1959). Carbon dioxide and climate. *Scientific American* 201 (1): 41–47.

**18** Plass, G.N. (1961). The influence of infrared absorptive molecules on the climate. *New York Academy of Sciences* 95: 61–71.

**19** Keeling, C.D. (1980). The Suess effect: $^{13}$Carbon – $^{14}$Carbon interrelations. *Environment International* 2 (4–6): 229–300.

**20** Revelle, R. and Suess, H.E. (1957). Carbon dioxide exchange between atmosphere and ocean and the question of an increase of atmospheric co₂ during the past decades. *Tellus* IX (1): 18–27.

**21** Keeling, C.D. (1960). The concentration and isotopic abundance of carbon dioxide in the atmosphere. *Tellus* XII (2): 200–203.

**22** Ahn, J., Brook, E.J., Mitchell, L. et al. (2012). Atmospheric $CO_2$ over the last 1000 years: a high-resolution record from the West Antarctic Ice Sheet (WAIS) divide ice core. *Global Biogeochemical Cycles* 26 https://doi.org/10.1029/2011GB004247.

**23** Yakir, D. (2006). The stable isotopic composition of atmospheric $CO_2$. In: *The Atmosphere, Treatise on Geochemistry* (eds H.D. Holland and K.K. Turekian), vol. 4 (ed. R.F. Keeling), 175–213. Oxford: Elsevier-Pergamon.

**24** Keeling, C.D., Piper, S.C., Bacastow, R.B., et al. (2001) Exchanges of atmospheric $CO_2$ and $^{13}CO_2$ with the terrestrial biosphere and oceans from 1978 to 2000.

I. global aspects, *SIO Reference Series*, **01-06** (Revised from SIO Reference Series, No. 00-21), Scripps Institution of Oceanography, San Diego, 91pp.

**25** Marland, G., Andres, R.J., Boden, T.A., and Johnston, C. (1998) *Global, Regional and National CO₂ Emissions Estimates from Fossil Fuel Burning, Cement Production and Gas Flaring: 1751–1995* (Revised 1998), Report ORNL/CDIAC NDP-030/Rb, http://cdiac.esd.ornl.gov/ndps/ndp030.html.

**26** Francey, R.J., Allison, C.E., Etheridge, D.M. et al. (1999). A 1000-year high precision record of $\partial^{13}C$ in atmospheric $CO_2$. *Tellus Series B.: Chemistry, Physics and Meteorology* 51 (2): 170–193.

**27** Andres, R.J., Marland, G., Boden, T., and Bischoff, S. (2000). Carbon dioxide emissions from fossil fuel consumption, 1751–1991, and an estimate of their isotopic composition and latitudinal distribution. In: *The Carbon Cycle* (eds. T.M.L. Wigley and D. Schimel), 53–62. Cambridge University Press.

**28** Manning, A.C. and Keeling, R.F. (2006). Global oceanic and land biotic carbon sinks from the Scripps atmospheric oxygen flask sampling network. *Tellus* 58B: 95–116.

**29** Callendar, G.S. (1958). On the amount of carbon dioxide in the atmosphere. *Tellus* 10: 243–248.

**30** Beck, E.-G. (2007). 180 years of atmospheric $CO_2$ gas analysis by chemical methods. *Energy and Environment* 17 (2): 259–282.

**31** Meijer, H.A.J. (2007). Comment on "180 years of atmospheric $CO_2$ gas analysis by chemical methods" by Ernst-Georg Beck. *Energy and Environment* 18 (5): 635–636.

**32** Keeling, R.F. (2007). Comment on "180 years of atmospheric $CO_2$ gas analysis by chemical methods" by Ernst-Georg Beck. *Energy and Environment* 18 (5): 637–639.

**33** Weart, S. (2008) *The Discovery of Global Warming* (available from http://Amazon.com); 2011 expanded and updated edition available on-line courtesy of the American Institute of Physics, at http://www.aip.org/history/climate/index.htm.

**34** Harde, H. (2017). Scrutinizing the carbon cycle and $CO_2$ residence time in the atmosphere. *Global and Planetary Change* 152: 19–26. https://doi.org/10.1016/j.gloplacha.2017.02.009.

**35** Köhler, P., Hauck, J., Völker, C. et al. (2017). Comment on "scrutinizing the carbon cycle and $CO_2$ residence time in the atmosphere" by H. Harde. *Global and Planetary Change* 152 https://doi.org/10.1016/j.gloplacha.2017.02.009.

**36** Bolin, B. and Eriksson, E. (1958). Changes in the carbon dioxide content of the atmosphere and sea due to fossil fuel consumption. In: *The Atmosphere and the Sea in Motion: Scientific Contributions to the Rossby Memorial*

*Volume* (ed. B. Bolin), 130–142. New York: Rockefeller Institute Press.

**37** Revelle, R. (1960) *Summary of statement on international cooperation in oceanography in Scripps Institution of Oceanography Archives*, Proceedings of Preparatory Meeting of the Intergovernmental Conference on Oceanographic Research, Paris, March 21–29, 1960, NS/2503/620, UNESCO, Paris, 2 pp.

**38** Revelle, R., Broecker, W., Craig, H. et al. (1965). Atmospheric carbon dioxide. In: *Restoring the Quality of Our Environment: Report of the Environment Pollution Panel*, 126–127. Washington DC: President's Science Advisory Committee, White House.

**39** Sabine, C.L., Ducklow, H., and Hood, M. (2010). International carbon coordination: Roger Revelle's legacy in the intergovernmental oceanographic commission. *Oceanography* 23 (3): 48–61. https://doi.org/10.5670/oceanog.2010.23.

**40** Bolin, B. (1970). The carbon cycle. *Scientific American* 223 (3): 125–132.

**41** Singer, F. (1970). Human energy production as a process in the biosphere. *Scientific American* 223 (3): 175–190.

**42** Meadows, D.H., Meadows, D.L., Randers, J., and Behrens, W.W. (1972). *The Limits to Growth: A Report for the Club of Rome's Project on the Predicament of Mankind*. New York: Universe Books.

**43** Manabe, S. and Wetherald, R.T. (1967). Thermal equilibrium of the atmosphere with a given distribution of relative humidity. *Journal of Atmospheric Sciences* 24 (3): 241–259.

**44** Pierrehumbert, R.T. (2010). *Principles of Planetary Climate*, 652 pp. Cambridge University Press.

**45** Pierrehumberty, R. (2011). Infrared radiation and planetary temperature. *Physics Today*, S-0031-9228-1101-010-6: 33–38.

**46** Ramanathan, V. (1998). Trace-gas greenhouse effect and global warming: underlying physics and outstanding issues. *Ambio* 27 (3): 187–197.

**47** Manabe, S. and Wetherald, R.T. (1975). The effects of doubling $CO_2$ concentration on the climate of a general circulation model. *Journal of Atmospheric Sciences* 32 (1): 3–15.

**48** National Research Council (1977). *Energy and Climate: Studies in Geophysics*, 158 pp. Washington DC: The National Academies Press.

**49** Lamb, H.H. (1963). What can we find out about the trend of our climate? *Weather* 18: 194–216; reprinted in Lamb, H.H. (1966) *The Changing Climate*, Methuen, London, 196–214.

**50** Kukla, G. (1972). Insolation and glacials. *Boreas* 1 (1): 63–96.

**51** Peterson, T., Connolley, W., and Fleck, J. (2008). The myth of the 1970s global cooling scientific consensus. *Bulletin of the American Meteorological Society* 89 (9): 1325–1337.

**52** Rasool, S.I. and Schneider, S.H. (1972). Atmospheric carbon dioxide and aerosols: effects of large increases on global climate. *Science* 173: 138–141.

**53** Schneider, S.H. (1975). On the carbon dioxide – climate confusion. *Journal of Atmospheric Sciences* 32: 2060–2066.

**54** Lamb, H.H. (1995). *Climate History and the Modern World*, 433 pp, 2e. London: Routledge.

**55** Hansen, J., Sato, M., Ruedy, R., Schmidt, G.A., and Lo, K (2019) Global temperature in 2018 and beyond. http://www.columbia.edu/~mhs119/Temperature.

**56** Budyko, M.I. (1969). The effect of solar radiation variations on the climate of the Earth. *Tellus* 21: 611–619.

**57** Budyko, M.I. (1977). On present-day climatic changes. *Tellus* 29: 193–204.

**58** Budyko, M.I. (1971) *Klimat I Zhizn*, Gidrometeoizdat, Leningrad, 472pp, translated as Budyko, M.I., (1974), *Climate and Life* (ed D.H. Miller), Academic Press, New York and London, 508pp.

**59** Budyko, M.I. and Ronov, A.B. (1979). Chemical evolution of the atmosphere in the Phanerozoic. *Geochemistry International* 16 (3): 1–9; Translated from *Geokhimiya* **5**, 1979, 643-653.

**60** Budyko, M.I. (1996). Past changes in climate and societal adaptations. In: *Adapting to Climate Change* (eds. J.B. Smith, N. Bhatti, G.V. Menzhuin, et al.), 16–26. New York: Springer.

**61** Smith, S.K., Van Aardenne, J., Klimont, Z. et al. (2011). Anthropogenic sulfur dioxide emissions: 1850–2005. *Atmospheric Chemistry and Physics* 11: 1101–1116.

**62** Sawyer, J.S. (1972). Man-made carbon dioxide and the "greenhouse" effect. *Nature* 239: 23–26.

**63** NAS (1975). *Understanding Climate Change: A Program for Action*. Washington DC: National Research Council of the National Academy of Sciences ISBN 0-309-02323-8.

**64** Broecker, W.S. (1975). Climatic change: are we on the brink of a pronounced global warming? *Science* 188: 460–463.

**65** Cross, T. P. (2012) Wallace Broecker '53 battles the angry climate beast. *Columbia College Today*, summer 2012; http://www.college.columbia.edu/cct/summer12/features4.

**66** Mercer, J.H. (1978). West Antarctic ice sheet and $CO_2$ greenhouse effect: a threat of disaster. *Nature* 271: 321–325.

**67** Schneider, S.H. and Mesirow, L. (1976). *The Genesis Strategy: Climate and Global Survival*. New York: Plenum.

**68** Schneider, S.H. (1996). *Laboratory Earth, the Planetary Gamble We Cannot Afford to Lose*, 184 pp. London: Weidenfeld and Nicolson.

**69** Augustsson, T. and Ramanathan, V. (1977). A radiative-convective model study of the CO2-climate problem. *Journal of Atmospheric Science* 34: 448–451.

**70** Broecker, W.S. and Takahashi, T. (1978). Neutralization of fossil fuel $CO_2$ by marine calcium carbonate. In: *The Fate of Fossil Fuel $CO_2$ in the Oceans* (eds. N.R. Andersen and A. Malahoff), 213–248. New York: Plenum Press.

**71** Walker, J.C.G. and Kasting, J.F. (1992). Effects of fuel and forest conservation on future levels of atmospheric carbon dioxide. *Global and Planetary Change* 5 (3): 151–189.

**72** WMO (1979) *First World Climate Conference*, February 1979, WMO, Geneva.

**73** White, R.M. (1979) *Climate at the Millennium*, Proceedings Report of First World Climate Conference, February 1979, WMO, Geneva.

**74** Flohn, H. (1977). Man-induced changes in heat budget and possible effects on climate. In: *Global Chemical Cycles and their Alterations by Man*, 207–224. Berlin: Dahlem Workshop.

**75** Flohn, H. (1979) *A Scenario of Possible Future Climates – Natural and Man-Made*, Proceedings Report of First World Climate Conference, February 1979, WMO, Geneva.

**76** Hare, F.K. (1979) *Climatic Variation and Variability*, Proceedings Report of First World Climate Conference, February 1979, WMO, Geneva.

**77** Munn, R.E., and Machta, L. (1979) *Human Activities that Affect Climate*, Proceedings Report of First World Climate Conference, February 1979, WMO, Geneva.

**78** Davies, D.A. (1979) *Foreword*, Report of First World Climate Conference, February 1979, WMO, Geneva.

**79** Charney, J.G., Arakawa, A., Baker, J.D. et al. (1979). *Carbon Dioxide and Climate: A Scientific Assessment*, 22 pp. Washington DC: National Academy of Sciences.

**80** Kellogg, W.W. (1981) Feedback Mechanisms in the Climate System affecting Future Levels of Carbon Dioxide. In: Papers presented at the *Scientific Conference on Analysis and Interpretation of Atmospheric $CO_2$ Data, Bern, September 1981*. World Climate Report 14, WMO, Geneva, 243–251

**81** Ramanathan, V., Lian, M.S., and Cess, R.D. (1979). Increased atmospheric $CO_2$: zonal and seasonal estimates of the effect on the radiation energy balance and surface temperature. *Journal of Geophysical Research* 84 (C8): 4949–4958.

**82** Houghton, J.T. (1979). Greenhouse effects of some atmospheric constituents. *Philosophical Transactions of the Royal Society of London. Series A* 290: 515–521.

**83** USDOE (1980) *Environmental and Societal Consequences of a Possible $CO_2$-Induced Climate Change – a Research Agenda*. Technical Report DOE/EV/10019–01, vol. 1 of 2, US Department of Energy, US Department of Commerce, Springfield VA

**84** Manabe, S. and Stouffer, R.J. (1980). Sensitivity of the global climate to an increase of $CO_2$ concentration in the atmosphere. *Journal of Geophysical Research, Oceans* 85 (C10): 5529–5554.

**85** Hansen, J., Johnson, D., Lacis, A. et al. (1981). Climate impact of increasing atmospheric carbon dioxide. *Science* 213: 957–966.

**86** Hansen, J.E. and Takahashi, T. (1984). *Climate Processes and Climate Sensitivity*, 368 pp, *Geophysical Monographs* **29**. Washington D.C.: American Geophysical Union.

**87** Davies, E.E. (1984). Inventing the future: energy and the $CO_2$ "greenhouse" effect. In: *Climate Processes and Climate Sensitivity*, *Geophysical Monographs* **29** (eds. J.E. Hansen and T. Takahashi), 1–5. Washington DC: American Geophysical Union.

**88** Hansen, J.E., Lacis, A., Rind, D. et al. (1984). Climate sensitivity: analysis of feedback mechanisms. In: *Climate Processes and Climate Sensitivity*, *Geophysical Monographs*, vol. 29 (eds. J.E. Hansen and T. Takahashi), 130–163. Washington DC: American Geophysical Union.

**89** Broecker, W.S. and Takahashi, T. (1984). Is there a tie between atmospheric $CO_2$ content and ocean circulation? In: *Climate Processes and Climate Sensitivity*, *Geophysical Monographs* **29** (eds. J.E. Hansen and T. Takahashi), 314–326. Washington DC: American Geophysical Union.

**90** Broecker, W.S. and Peng, T.H. (1984). The climate-chemistry connection? In: *Climate Processes and Climate Sensitivity*, *Geophysical Monographs* **29** (eds. J.E. Hansen and T. Takahashi), 327–336. Washington DC: American Geophysical Union.

**91** Broecker, W.S. and Peng, T.H. (1982). *Tracers in the Sea*, 690 pp. New York: Lamont-Doherty Geological Observatory.

**92** Revelle, R. (1982). Carbon dioxide and world climate. *Scientific American* 247 (2): 35–43.

**93** National Academy of Sciences (1983) *Changing Climate*. Report of the Carbon Dioxide Assessment Committee. National Academies Press, Washington DC, 496pp

**94** Schlesinger, W.H. (1997). *Biogeochemistry: An Analysis of Global Change*, 2e. San Diego, Calif: Academic Press ISBN 0-12-625155-X.

**95** Parkes, R.J., Cragg, B.A., and Wellsbury, P. (2000). Recent studies on bacterial populations and processes in subseafloor sediments: a review. *Hydrogeology Journal* 8: 11–28.

**96** National Research Council (1983). *Toward an International Geosphere-Biosphere Program – A Study of Global Change*, 81 pp. Washington DC: National Academies Press.

**97** Malone, T.C. and Roederer, J.G. (eds.) (1985). *Global Change*, 512 pp. Cambridge University Press.

**98** Mooney, H.A., Duraiappah, A., and Larigauderie, A. (2013) Evolution of Natural and Social Science Interactions in Global Change Research Programs. *Proceedings of the National Academy of. Sciences,* Early Edition. http://www.pnas.org/cgi/doi/10.1073/pnas.1107484110.

99 Fyfe, W.S. (1985). The international geosphere-biosphere program: global change. In: *Global Change* (eds. T.S. Malone and J.G. Roederer), 499–508. Cambridge University Press.

100 Malone, T.F. (1985). Preface. In: *Global Change* (eds. T.S. Malone and J.G. Roederer), xi–xxi. Cambridge University Press.

101 Webb, T., Kutzbach, J., and Street-Perrott, A. (1985). 20,000 years of global climatic change: paleoclimate research plan. In: *Global Change* (eds. T.S. Malone and J.G. Roederer), 182–218. Cambridge University Press.

102 Schlesinger, W.H. (2005). Volume Editor's introduction. In: *Biogeochemistry*, *Treatise on Geochemistry* (eds H.D. Holland and K.K. Turekian), vol. 8 (ed. W.H. Schlesinger), xv–xvii. Oxford: Elsevier-Pergamon.

103 Steffen, W., Sanderson, A., Tyson, P.D. et al. (2004). *Global Change and the Earth System: A Planet Under Pressure*, 336 pp, Global Change – the IGBP Series. Berlin: Springer.

104 Kump, L.R., Kasting, J.F., and Crane, R.G. (2016). *The Earth System*, 462 pp, 3e. Chennai, India: Pearson Education Ltd.

105 Waters, C.N., Zalasiewicz, J., Summerhayes, C. et al. (2016). The Anthropocene is functionally and stratigraphically distinct from the Holocene. *Science* 351 (6269): 10. https://doi.org/10.1126/science.aad2622.

106 Falkowski, P.G. (2005). Biogeochemistry of primary production in the sea. In: *Biogeochemistry*, *Treatise on Geochemistry* (eds H.D. Holland and K.K. Turekian), vol. 8 (ed. W.H. Schlesinger), 185–213. Oxford: Elsevier-Pergamon.

107 Houghton, R.A. (2005). The contemporary carbon cycle. In: *Biogeochemistry*, *Treatise on Geochemistry* (eds H.D. Holland and K.K. Turekian), vol. 8 (ed. W.H. Schlesinger), 473–513. Oxford: Elsevier-Pergamon.

108 IPCC (2013) *Chapter 6: Carbon and Other Biogeochemical Cycles* – Final Draft Underlying Scientific-Technical Assessment. Working Group I Contribution to the IPCC Fifth Assessment Report, Climate Change 2013, *The Physical Science Basis*, IPCC, 167 pp.

109 Berner, R.A. (1999). A new look at the long-term carbon cycle. *GSA Today* 11 (9): 1–6.

110 Beerling, D. (2007). *The Emerald Planet*, 288 pp. Oxford University Press.

111 Gerlach, T. (2011). Volcanic versus anthropogenic carbon dioxide. *EOS* 92 (24): 201–202.

112 Marty, B. and Tolstikhin, I.N. (1998). $CO_2$ fluxes from mid-ocean ridges, arcs and plumes. *Chemical Geology* 145: 233–248.

113 Williams, S.N., Schaeffer, S.J., Calvache, M.L., and Lopez, D. (1992). Global carbon dioxide emissions to the atmosphere by volcanoes. *Geochimica Cosmochimics Acta* 56: 1765–1770.

114 German, C.R. and Von Damm, K.L. (2003). Hydrothermal processes. In: *The Oceans and Marine Geochemistry*, *Treatise on Geochemistry* (eds H.D. Holland and K.K. Turekian), vol. 6 (ed. H. Elderfield), 181–222. Oxford: Elsevier-Pergamon.

115 Lupton, J.E., Butterfield, D.A., Lilley, M. et al. (2006). Submarine venting of liquid carbon dioxide on a Mariana Arc volcano. *Geochemistry, Geophysics, and Geosystems* https://doi.org/10.1029/2005GC001152, 002006.

116 Mörner, N.-A. and Etiope, G. (2002). Carbon degassing from the lithosphere. *Global and Planetary Change* 33 (1–2): 185–203.

117 Burton, M.R., Sawyer, G.M., and Granieri, D. (2013). Deep carbon emissions from volcanoes. *Reviews in Mineralogy and Geochemistry* 75: 323–354.

118 Wardell, L.J., Kyle, P.R., and Campbell, A.R. (2003). Carbon dioxide emissions from fumarolic ice towers, Mount Erebus Volcano, Antarctica. *Geological Society of London Special Publicaiton* 213: 231–246.

119 Varney, M. (1996). The marine carbonate system. In: *Oceanography, an Illustrated Guide* (eds. C.P. Summerhayes and S.A. Thorpe), 182–194. London: Manson.

120 De La Rocha, C.L. (2003). The biological pump. In: *The Oceans and Marine Geochemistry*, *Treatise on Geochemistry* (eds H.D. Holland and K.K. Turekian), vol. 6 (ed. H. Elderfield), 83–111. Oxford: Elsevier-Pergamon.

121 Sigman, D.M. and Haug, G.H. (2003). Biological pump in the past. In: *The Oceans and Marine Geochemistry*, *Treatise on Geochemistry* (eds H.D. Holland and K.K. Turekian), vol. 6 (ed. H. Elderfield), 491–528. Oxford: Elsevier-Pergamon.

122 Eglinton, T.I. and Repeta, D.J. (2003). Organic matter in the contemporary ocean. In: *The Oceans and Marine Geochemistry*, *Treatise on Geochemistry* (eds H.D. Holland and K.K. Turekian), vol. 6 (ed. H. Elderfield), 145–180. Oxford: Elsevier-Pergamon.

123 Takahashi, T., Sutherland, S.C., and Kozyr. A. (2011) *Global Ocean Surface Water Partial Pressure of $CO_2$ Database: Measurements Performed During 1957–2010 (Version 2010)*. Report ORNL/CDIAC-159, NDP-088(V2010), Carbon Dioxide Information Analysis Center, Oak Ridge National Laboratory, U.S. Department of Energy, Oak Ridge, Tennessee, http://dx.doi.org/10.3334/CDIAC/otg.ndp088(V2010).

124 Morel, P. (1989) The Global Climate System: Current knowledge and Uncertainties, in *Carbon Dioxide and Other Greenhouse Gases: Climatic and Associated Impacts* (eds R. Fantechi and A. Ghazi), Proc. European Commission Symposium, Brussels, 3-5 November 1986, Kluwer Acad. Press, Dordrecht, 5-10.

**125** Sabine, C.L., Feely, R.A., Gruber, N. et al. (2004). The oceanic sink for anthropogenic $CO_2$. *Science* 305 (5682): 367–371.

**126** Archer, D.E., Kheshgi, H., and Maier-Reimer, E. (1998). Dynamics of fossil fuel $CO_2$ neutralization by marine $CaCO_3$. *Global Biogeochemical Cycles* 12: 259–276.

**127** WMO (1986) Report of the International Conference of the Assessment of the Role of Carbon Dioxide and of Other Greenhouse Gases in Climate Variations and Associated Impacts, 9–15 October 1985, Villach, Austria. WMO Report 661, 78 pp.

**128** Oeschger, H., Degens, E., Meszaros, E., et al., 1986, *Cycling of Carbon and Other Radiatively Active Constituents*, Proceedings of Villach meeting, October 1985, WMO Report 661, 48–55.

**129** Bolin, B., Doos, B., Jager, J., and Warrick, R.A., (eds.), (1986) The Greenhouse Effect, Climate Change and Ecosystems. SCOPE Report 29, John Wiley and Sons, Chichester

**130** IPCC (2013) Summary for Policy Makers. Working Group I Contribution to the IPCC Fifth Assessment Report – *Climate Change* 2013, The Physical Science Basis, IPCC, 36 pp.

**131** IPCC (2018) *Global Warming of 1.5°C*. An IPCC special report on the impacts of global warming of 1.5°C above pre-industrial levels and related global greenhouse gas emission pathways, in the context of strengthening the global response to the threat of climate change, sustainable development, and efforts to eradicate poverty. Special Report on Global Warming of 1.5°C. IPCC, available at https://www.ipcc.ch/sr15.

**132** Turner, J., Bindschadler, R., Convey, P. et al. (2009). *Antarctic Climate Change and the Environment*, 526 pp. Cambridge: SCAR.

**133** Le Quéré, C., Andrew, R.M., Friedlingstein, P. et al. (2018). Global carbon budget 2018. *Earth System Science Data* 10: 2141–2194.

**134** Ramanathan, V., Callis, L., Cess, R. et al. (1987). Climate-chemical interactions and effects of changing atmospheric trace gases. *Reviews of Geophysics* 25 (7): 1441–1482.

**135** Anderson, T.R., Hawkins, E., and Jones, P.D. (2016). $CO_2$, the greenhouse effect and global warming: from the pioneering work of Arrhenius and Callendar to today's earth system models. *Endeavour* 40 (3): 178–187.

**136** Montenegro, A., Brovkin, V., Eby, M. et al. (2007). Long term fate of anthropogenic carbon. *Geophysical Research Letters* 34: L19707. https://doi.org/10.1029/2007GL030905.

**137** Eby, M., Zickfield, K., Montenegro, A. et al. (2009). Lifetime of anthropogenic climate change: millennial time scales of potential $CO_2$ and surface temperature perturbations. *Journal of Climate* 22: 2501–2511.

**138** Archer, D. and Brovkin, V. (2008). The millennial atmospheric lifetime of anthropogenic $CO_2$. *Climate Change* 90: 283–297.

**139** Solomon, S., Plattner, G.-K., Knutti, R., and Friedlingstein, P. (2009). Irreversible climate change due to carbon dioxide emissions. *Proceedings of the National Academy of Sciences* 106 (6): 1704–1709.

**140** Schmidt, G.A., Ruedy, R.A., Miller, R.L., and Lacis, A.A. (2010). Attribution of the present-day total greenhouse effect. *Journal of Geophysical Research* 115: D20106. https://doi.org/10.1029/2010JD014287.

# 9

# Measuring and Modelling CO$_2$ Back Through Time

---

**LEARNING OUTCOMES**

- Understand and explain the significance of the carbonate compensation depth (CCD) to studies of the abundance of CO$_2$ in the atmosphere.
- Understand and be able to explain what ocean acidification is, and why it is called 'the other CO$_2$ problem'.
- Know and be able to reproduce the four Urey and Garrels equations explaining silicate weathering, the development of carbonate sediments, and the processes that return CO$_2$ to the atmosphere.
- Understand and be able to explain the relationships between (i) plate tectonic processes (sea-floor spreading and subduction) and (ii) the slow carbon cycle, in terms of negative and positive feedbacks, with particular reference to the warmth of the Mid Cretaceous and the cold of the late Cenozoic.
- Understand and be able to explain the 'faint young Sun paradox' in relation to the greenhouse gas concentration in the atmosphere.
- Understand and be able to explain the role of plants in controlling atmospheric CO$_2$, with particular reference to the development of the Carboniferous glaciation.
- Understand and be able to explain the role of tectonic uplift in controlling climate, with reference to the changing climate of the Cenozoic.
- Describe the different methods for establishing past atmospheric concentrations of CO$_2$.
- Be able to explain the origins of the Ordovician glaciation.

---

## 9.1 CO$_2$ – The Palaeoclimate Perspective

By the late 1970s, then, the atmospheric and ocean sciences had provided a comprehensive theoretical understanding of the role of the greenhouse gases in the climate system. Later work would strengthen it. How, and when, would geologists apply this new knowledge to their interpretations of change in past climates?

Early in the last century a few geologists knew about the potential for global warming created by the emissions of CO$_2$ from industry. For instance in the Foreword to R.L. Sherlock's 1922 book *Man as a Geological Agent,* geologist A.S. Woodward pointed to '*the question whether man, by his prodigious combustion of coal and other carbonaceous substances, is producing more carbonic acid than can be eliminated by ordinary processes ... [which may eventually lead to] an unwelcome change in his atmospheric surroundings*' [1]. Although both Woodward and Sherlock supported the ideas of Arrhenius and Chamberlin (reviewed in Chapter 4), nobody in the geological world

seems to have volunteered prior to the 1950s to investigate the role of CO$_2$ in past climate change. That is not entirely surprising, since the questions would be difficult to answer in the absence of the theoretical and practical underpinnings to greenhouse gas science that developed from the 1950s onwards.

As Gilbert Plass pointed out in 1956, the first step to applying the new knowledge was '*to be able to determine atmospheric CO$_2$ concentration for past geological periods*' [2, 3]. In the absence of those determinations, we had calculations to go by. Plass calculated that if chemical weathering halved the CO$_2$ content of the atmosphere, it would lower average global temperature by 3.8 °C and could precipitate an Ice Age [4]. That convinced him (in agreement with Tyndall and Arrhenius) that CO$_2$ had played a role in past climate change. And he knew that at least one geologist, Thomas Chamberlin, agreed that more CO$_2$ led to warm periods and less led to cool ones, thus helping to explain the fluctuations between glacial and interglacial periods of the Ice Age [2], as we saw in Chapter 4. Plass

*Palaeoclimatology: From Snowball Earth to the Anthropocene*, First Edition. Colin P. Summerhayes.
© 2020 John Wiley & Sons Ltd. Published 2020 by John Wiley & Sons Ltd.

knew that Chamberlin had proposed that the chemical weathering of silicate minerals in mountain belts removed $CO_2$ from the atmosphere that was added back by volcanic emissions – an idea that Nobel chemist Harold Urey had recently set out in his 1952 book *The Planets* [5]; Urey's chemical equations are shown below in Section 9.2.

From the fossil evidence for warmer times during much of the geological record, Plass deduced that the air then must have contained much more $CO_2$ than it does now [6], a prediction that subsequent data and models show to have been the case, as we shall see. He accepted Urey's argument that losses of $CO_2$ from the air by chemical weathering would eventually be balanced by the addition of $CO_2$ emitted from volcanoes, which turns out to be reasonable on long geological time scales. And, agreeing with Chamberlin, he surmised that the burial of masses of organic matter to form coal in the early Carboniferous reduced the level of $CO_2$ in the air sufficiently to generate the late Carboniferous to Early Permian glaciation [4], an idea now widely accepted. In contrast with that acceptance, the notion that changes in the supply of $CO_2$ by volcanoes and its extraction by chemical weathering might explain the warming and cooling of the Pleistocene Ice Age had been played down since Arrhenius's time, although Chamberlin was close to explaining the mechanism by invoking the ocean's role in expelling $CO_2$ when warm and retaining it when cool (Chapter 4).

Plass was, like Wegener, far ahead of his time, and it would be years before the geological community caught up with him. $CO_2$, the greenhouse effect, and the work of Tyndall, Arrhenius, Chamberlin, and Plass rated not a single mention in the textbooks that I was recommended to read as a geology undergraduate at University College London in the early 1960s. In his 1961 presidential address to the Geological Society of London on 'The Climate Factor in the Geological Record', my then head of department, Prof. Sydney Hollingworth, who we met in Chapter 6, mentioned $CO_2$ just twice, but not as a major control on past climate [7]. That can be taken as typical of the geological thinking of the time.

Not all geologists were quite so oblivious. Arthur Holmes had read Plass's 1959 paper in *Scientific American* [6], and in his 1965 landmark textbook *Principles of Physical Geology* [8] he reminded his readers about the greenhouse effect and the implications of Tyndall's work of a century before, whilst saying little about the variation of $CO_2$ in past atmospheres [8]. But he went on to note that although sea ice had been melting and glaciers retreating in the early part of the twentieth century, '*During the last decade a few glaciers have ceased to melt away and have begun to advance again*' [8]. He interpreted that correctly to mean that '*other factors besides the $CO_2$ content of the atmosphere are concerned with climatic changes, but what their relative effects may be still remains problematical*' [8]. Although Holmes knew about the theory, he also knew it wasn't working in the 1960s quite the way it was supposed to. $CO_2$ was increasing, but the world was not warming, as we saw in Chapter 8. Nevertheless, he was concerned about what burning more fossil fuels would do to the climate, noting – '*The $CO_2$ effect has been mentioned here only to illustrate the remarkable consequences of burning fuel hundreds of times as fast as it took to accumulate*' [8].

Holmes's neglect of the possible role of $CO_2$ in affecting past climates was not surprising, because at that time there was no direct evidence for atmospheric $CO_2$ having been abundant in the past. Despite the contributions of Tyndall, Arrhenius and Chamberlin, neither Köppen, nor Wegener, nor Du Toit had considered the possible role of $CO_2$ in the climate story. Most geologists were in the dark as far as greenhouse gases were concerned, as were most meteorologists, only Callendar, the engineer and amateur meteorologist, was on the same wavelength as Plass (Chapter 8).

By the early 1970s, things had begun to change. In 1971 (with an English translation in 1974), Russian meteorologist Mikhael Budyko, whom we met in Chapter 8, postulated that the warmth of the Mesozoic and early Tertiary compared with later periods may have been caused by a decrease in '*the atmospheric transparency for long-wave radiation ... [caused by] a high carbon dioxide content in the atmosphere*' [9]. He and his geologist colleague Alexander Ronov tested this idea by using the known volumes of volcanic rocks, carbonate sediments, and the organic carbon content of sedimentary rocks to calculate the likely amounts of oxygen and $CO_2$ in the atmosphere over the past 550 Ma. Despite their crude approach, the results, published in detail first in 1979 [10], and then in a book in 1987 [11], confirmed that it was highly likely that changing atmospheric $CO_2$ had affected past climates.

Budyko and Ronov estimated that volcanism and $CO_2$ were abundant in the Devonian (400 Ma ago) and the Permian (275 Ma ago), and low during the Permo–Carboniferous glaciation (300 Ma ago) and at the Permo–Triassic boundary (250 Ma ago). They rose to intermediate levels from the Mid Triassic through the Jurassic and the Cretaceous, before declining to the low values of the Pliocene. Budyko and Ronov assumed – as Urey had suggested – that volcanism had a primary role in supplying $CO_2$ to the air, and that the fall in volcanism and $CO_2$ during the Cenozoic cooled the Earth and produced the Quaternary glaciations [10, 11]. They were puzzled by the fact that the amount of $CO_2$ that they had calculated for the Carboniferous atmosphere had not fallen as low as during the recent Ice Age. We now know that they overestimated the abundance of $CO_2$ for the Carboniferous.

Despite their estimates, we still lacked actual measurements of past $CO_2$. However, we were about to acquire an indirect measure of past atmospheric $CO_2$, from fluctuations in the CCD. This is the boundary at the deep-sea floor between sediment rich in the calcium carbonate skeletons of foraminifera and coccolithophorids, and deeper sediment made of non-fossiliferous red clay (Figure 9.1) [12]. It is the point at which the rate of dissolution of the mineral calcite exceeds the rate of supply. Red clay is always present as a tiny fraction of deep-sea carbonate oozes, but below the CCD it is all that is left after the carbonate has gone. Major factors controlling the position of the CCD are the effects of increasing hydrostatic pressure with depth on the solubility of carbonate minerals (calcite and aragonite), and the increasing concentration of $CO_2$ in old bottom waters due to the decomposition of sinking organic matter [13, 14].

It was not until the DSDP began collecting core samples on transects across the CCD that geologists realized that depth of this boundary changed with time. Because it was initially thought that the CCD was a pressure solution horizon (reflecting ocean depth), it was believed that fluctuations in the CCD must represent past vertical movements of the seabed. The first person to realize that this was not the case was the University of Miami's Bill Hay, on Leg 4 of the DSDP (in the western Atlantic and Caribbean in early

1969). At the end of the Leg 4 report, in 1970 [15], Hay noted that in effect there were two CCDs, one being the lower limit of occurrence of pelagic foraminifera, and the other being the slightly deeper lower limit of the occurrence of the remains of calcareous nannofossils (coccolithophorids). Their vertical changes through time, which were of the order of 1 km, he said '*are more easily explained by assuming that the only vertical motion of points on the ocean floor has been downward as they moved away from the crest of the Mid-Atlantic Ridge, but that the zone of calcium carbonate compensation has fluctuated through considerable distances in the water*' [15]. In other words the changes in the CCD represented major changes in ocean chemistry, leading Hay to point out that '*We were becoming aware of just how much we did not know*' [14]. Unfortunately, as Hay said, tongue in cheek, about his published discovery of the fluctuating CCD – '*Nobody read it*' [14].

In fact the CCD represents the effect of changing atmospheric $CO_2$ on ocean chemistry and thus on the stability of the main calcium carbonate minerals, calcite, and aragonite. Their solubility depends not only on depth – representing pressure – but also on the acidity of subsurface water. As mentioned in Chapter 8, when $CO_2$ dissolves in the ocean it reacts with water to produce either inorganic bicarbonate ions ($HCO_3^-$) or carbonate ions ($CO_3^{2-}$). In the process, hydrogen ions ($H^+$) are also

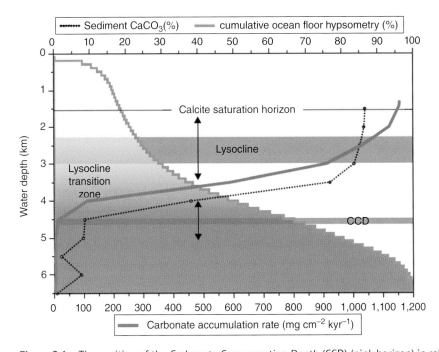

**Figure 9.1** The position of the Carbonate Compensation Depth (CCD) (pink horizon) in relation to deep-ocean topography (orange line). The CCD occurs where the downward flux of carbonate organic remains is balanced by dissolution. The dotted line shows the decline in percentage of CaCO₃ in the sediments, whilst the blue line shows the carbonate accumulation rate declining to zero with depth. The lysocline is the horizon where dissolution first becomes noticeable. It lies below the calcite saturation horizon, where there is no dissolution of calcium carbonate. *Source:* From figure 1 in Ref. [12].

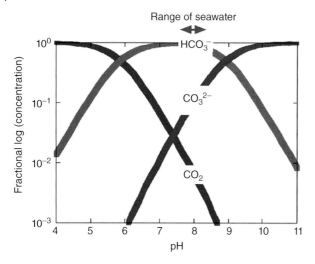

**Figure 9.2** Relative proportions of the three inorganic forms of $CO_2$ dissolved in seawater (note log scale at left). Increasing $CO_2$ (blue) from right to left between pH 8.7 and 7.7 causes an initial slight increase in bicarbonate ions ($HCO_3^-$) (in red) and a significant decrease in carbonate ions ($CO_3^{2-}$) (in purple), as well as the decrease in pH (bottom scale), the decreasing pH representing an increase in H+ ions. *Source:* From figure 2 in Ref. [16].

produced: the greater the solution of $CO_2$, the lower the concentration of carbonate ions and the greater the concentration of hydrogen ions, hence the lower the pH (Figure 9.2) [16]. The corresponding rise in acidity dissolves $CaCO_3$, causing the CCD to shallow, decreasing the area covered by carbonate sediment and increasing the area covered by red clay. Today the concentrations of dissolved $CO_2$ and carbonate ions ($CO_3^{2-}$) are about the same in the surface waters of both the Atlantic and Pacific, but there is more dissolved $CO_2$ and less $CO_3^{2-}$ in the old deep Pacific water than in the younger deep Atlantic. Hence, the CCD in the Atlantic is deeper than that in the Pacific; calcite particles sinking in the Atlantic dissolve below 5000 m, whilst in the Pacific they dissolve below 4200–4500 m. Aragonite dissolves at shallower depths than calcite, so it is calcite dissolution that defines the CCD. The process of dissolution is limited by the fact that when $CaCO_3$ dissolves, it uses up H+ ions, thus buffering the ocean against further acidification. That buffering process is not immediate, so that if the rate of supply of $CO_2$ to the ocean is too rapid, conditions can become more acid than expected, and remain that way for a time until equilibrium is restored.

Fluctuations in the CCD on a stable ocean floor that was slowly subsiding as it aged and moved away from the mid-ocean ridge meant that ocean chemists were wrong to assume that the chemistry of the ocean had remained more or less constant for millions of years. Another deep-sea drilling geologist like Hay who had come across the CCD in deep-ocean cores was Arthur Fischer, of Princeton, who in the mid-1970s began working with his PhD student, Mike Arthur, on the carbonate-rich, deep marine, open ocean (or pelagic) Cretaceous sediments from Gubbio in Italy's Apennine Mountains. In 1977, they assumed that the sedimentary cycles they observed there represented '*significant variation in the general nature of the ocean as a physical-chemical-biological system*' [17]. During warm times carbonate sedimentation had declined in deep waters probably due, amongst other things, to an increased concentration of $CO_2$ in deep water causing a rise in the CCD. Rather than connecting the warming to an increase in atmospheric $CO_2$, they attributed it in part to atmospheric warming caused by an increase in atmospheric water vapour (a greenhouse gas), perhaps driven by higher sea levels in response to plate tectonic activity [17]. But by 1981, Fischer had changed his mind, attributing the warming to an increase in atmospheric $CO_2$ [18] (more on that in Chapter 10).

Theoretically, as explained above, we should be able to use changes in the CCD through time to tell us about the relationship between past ocean chemistry and past atmospheric $CO_2$, and Figure 9.3 confirms that we can, with the CCD (red in panel 9.3a) tracking both long-term ocean cooling (blue in 9.3b) and changes in atmospheric $CO_2$ (d in Figure 9.3) in an assembly of deep-ocean drilling data for the Equatorial Pacific in the Cenozoic [12]. Note the sharp deepening of the CCD, indicating an abrupt increase in bottom water circulation, when the Antarctic ice sheet formed 34 Ma ago, represented also in the $\partial^{18}O$ (temperature) signal from benthic foraminifera; this also coincides with a marked fall in sea level (not shown).

The irregular nature of the CCD between 46 and 34 Ma ago is attributed to the incidence of carbonate accumulation events (CAEs in Figure 9.3a) lasting between 250 000 and one million years and causing the CCD to shoal. These are attributed to temporary increases in the productivity of surface waters that overwhelm the underlying potential of bottom waters to cause the shoaling of the CCD [12]. The CAE-3 productivity event (temporary deepening of the CCD) immediately precedes the Mid-Eocene Climatic Optimum (MECO), a warm event where the CCD shoals (Figure 9.3a). Note also the pronounced CCD shoaling of the Mid Miocene (18.5–16 Ma ago), which coincides with the start of the warmth of the Mid Miocene climatic optimum (seen in the $\partial^{18}O$ signal, Figure 9.3b) and an increase in atmospheric $CO_2$ (Figure 9.3d). It is possible that these events represent oscillations responding to instabilities in the ocean–atmosphere system as the climate declined from Middle Eocene times through to the end of the Eocene 34 Ma ago.

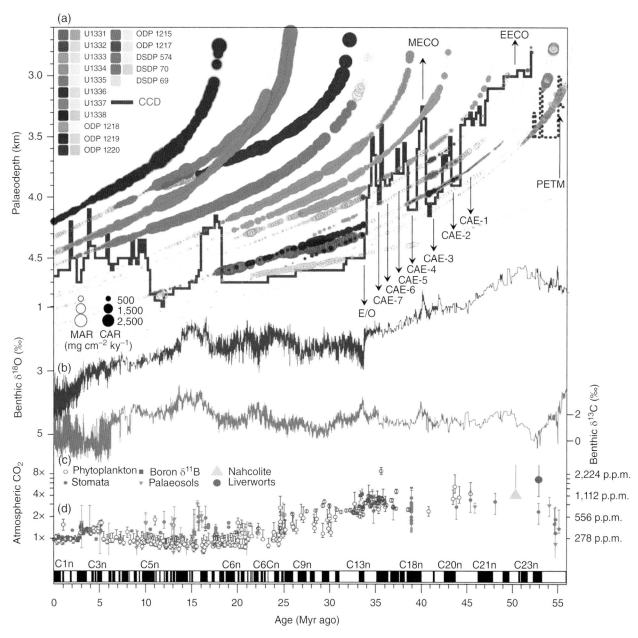

**Figure 9.3** (a) The decline in the CCD through time in the Equatorial Pacific (red line) as derived from the 'back-tracking' of deep-sea cores numbered at top left. The 'backtracked' depths for each core are colour coded and decline from right to left. Circles are scaled by either carbonate accumulation rate (open circles) or mass accumulation rates (coloured). All sites are within 3.5° of the Equator, except for the red dashed line. Black and white bars along the x-axis represent the palaeomagnetic time scale; (b) benthic oxygen isotopic values (blue); (c) benthic carbonate isotopic values (green); (d) atmospheric $CO_2$ compilation (log scale relative to pre-industrial level of 278 ppm) based on Refs. [19 and 20]; MECO = Mid Eocene Climatic Optimum; EECO = Early Eocene Climatic Optimum; PETM = Palaeocene–Eocene Thermal Maximum; CAE = Carbonate Accumulation Event. *Source:* From figure 2 in Ref. [12].

In the modern ocean we see this same basic chemical process – enhancement in oceanic $CO_2$ caused by enhancement in atmospheric $CO_2$ – taking place at the ocean's surface to create 'ocean acidification' (often referred to as 'the other $CO_2$ problem'), as the surface ocean absorbs $CO_2$ from the atmosphere (Figure 9.4, Box 9.1).

## 9.2 Modelling CO₂ Back Through Time

Enter the geochemical modellers. Along with the application of more powerful geochemical tools, the 1950s and 1960s saw a significant increase in theoretical geochemistry and the application of chemical models to understanding geological problems. Amongst the leading practitioners in

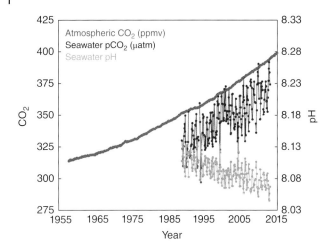

**Figure 9.4** Time series of $CO_2$ and ocean pH at Mauna Loa, Hawaii, from NOAA Ocean Acidification Program (December 2018), Ref. [21].

this new field was American geochemist Robert Minard Garrels (1916–1988) (Box 9.2).

Building upon the work of Harold Urey [5], Garrels documented the equations explaining how $CO_2$ in rainwater created the acidity needed to dissolve the silicate rocks of mountain belts, pointing out that the $CO_2$ extracted from the air in this way would eventually be restored, for instance by reactions in the oceans and sediments, not to mention via eruption from volcanoes [25]. First, the mix of $CO_2$ and $H_2O$ makes rainwater into a weak acid:

$$CO_2 + H_2O \rightarrow H_2CO_3 \qquad (9.1)$$

**Box 9.2 Bob Garrels**

Garrels revolutionized the field of aqueous geochemistry with the publication of his 1965 book *Solutions, Minerals and Equilibria*, based on applying analytical physical chemistry to geological and geochemical problems. He was fascinated by the cycles of carbon and other elements, and developed several numerical models to explain in quantitative terms how these cycles work to produce what we see in sedimentary rocks. During his career he worked at various times for the US Geological Survey, Northwestern University, Harvard, Scripps, the University of Hawaii, and the University of South Florida. Garrels' many efforts were rewarded with election to the US National Academy of Sciences in 1961, presidency of the Geochemical Society in 1962, and several medals, amongst them the Penrose Medal of the Geological Society of America in 1978 and the Wollaston Medal of the Geological Society of London in 1981.

That weak acid reacts with magnesium silicate, in the form of the mineral olivine, to provide magnesium and bicarbonate ions plus silicic acid in solution [26, 27]:

$$Mg_2SiO_4 + 4CO_2 + 4H_2O \leftrightarrow 2Mg^{2+} + 4HCO_3^- + H_4SiO_4. \qquad (9.2)$$

If the magnesium silicate is replaced by calcium silicate, $CaSiO_3$, the reaction produces calcium carbonate and silica, $CaCO_3 + SiO_2$ [26]. These equations for the dissolution of silicates by carbonic acid are commonly

**Box 9.1 Ocean Acidification – The Other $CO_2$ Problem**

At times when the atmosphere contains more $CO_2$, the dissolution of $CO_2$ in the surface ocean produces more $H^+$ ions and fewer $CO_3^{2-}$ ions (Figure 9.2). As a result the surface ocean becomes slightly more acid and less alkaline (its pH increases). This process, which is going on today, is commonly known as 'ocean acidification'. (We measure acidity by the pH scale, a measure of the free hydrogen ($H^+$) ions in the system. The scale runs from 1 to 14. Solutions are described as acid when their pH is less than 7). Today's slightly alkaline surface ocean, with a pH of around 8.05 units, is lower by about 0.2 pH units than it was in 1751 [21]. Because the pH scale is logarithmic (for readers with a technical bent, pH is the logarithm of the reciprocal of the hydrogen ion activity in a solution), this apparently tiny change is roughly equivalent to a 150% increase in hydrogen ion concentration, referred to either as an increase in ocean acidity (hence 'ocean acidification') or a decrease in ocean alkalinity (which is the same thing). As we put more $CO_2$ into the surface ocean it will move more towards the acid end of the pH scale (Figure 9.2). Fortunately, as we saw in Chapter 8, seawater is chemically buffered against moving into the acid half of the scale where pH is <7. The 'antacid' is the $CaCO_3$ sediment of the ocean floor. Unfortunately, the present rate of $CO_2$ rise is so rapid that it is faster than the normal process of neutralization (buffering) by the dissolution of $CaCO_3$ [22]. As a result, ocean acidification is beginning to erode the shells of living creatures in the surface waters that make their skeletons from $CaCO_3$, such as corals and planktonic organisms at the base of the food chain [23, 24].

called the 'Urey Reactions', following their description by Harold Urey [5].

When the bicarbonate ions reach the ocean via rivers, they combine with calcium ions to produce calcium carbonate, water, and carbon dioxide, thus:

$$4HCO_3^- + 2Ca^{2+} \rightarrow 2CaCO_3 + 2H_2O + 2CO_2 \quad (9.3)$$

Marine organisms use the $CaCO_3$ to build skeletons made of calcite or aragonite, which ends up as carbonate ooze before hardening to limestone. Half of the original $4CO_2$ is thus prevented from returning rapidly to the atmosphere.

The rates of these reactions depend on temperature, doubling with a $10\,°C$ rise [14]. When $CO_2$ increases in the air, so does the temperature, increasing the rates of reaction so that more $CO_2$ is consumed by weathering. This keeps the $CO_2$ from expanding out of hand. Here we have part of Earth's natural thermostat. $CO_2$ is also removed from the air by photosynthesis, as we saw in Chapter 8. A proportion of the organic matter created in this way is trapped in sediments, providing yet another path for the removal of $CO_2$ from the atmosphere.

$CO_2$ is returned to the atmosphere when organic matter decomposes or is burned. It is also returned to the atmosphere when limestone ($CaCO_3$) is dissolved by weak carbonic acid (rainwater), but with no net addition of $CO_2$ to the atmosphere, producing calcium and bicarbonate ions, thus:

$$H_2CO_3 + CaCO_3 \rightarrow Ca^{2+} + 2HCO_3^- \quad (9.4)$$

In the ocean, organisms cause the calcium and bicarbonate ions to recombine to form carbonate minerals, water, and $CO_2$, thus:

$$Ca^{2+} + 2HCO_3^- \rightarrow CaCO_3 + H_2O + CO_2 \quad (9.5)$$

Note that there is no net change in the amount of $CO_2$ in the atmosphere: one molecule is consumed in these reactions, and one is returned. But the processes take a long time. There is a lag between the use and return of $CO_2$ molecules.

These reactions conceal the voyage of a typical molecule of $CO_2$ from its emission in volcanic gas to its solution in a bead of rainwater, to its impact on an exposed mountainside and its subsequent transfer through rivers to the sea, where it gets absorbed by phytoplankton to form organic matter that is eaten by zooplankton, incorporated into their $CaCO_3$ skeletons, and ends up as calcareous ooze on the ocean bed awaiting transformation to limestone and the repeat of the cycle depicted in Figure 8.16.

By 1974, Garrels was able to show that the geochemical cycles of carbon and sulphur help to control the amount of oxygen in the atmosphere, the oxidation state of minerals in rocks, and life processes in general [28]. And by 1976,

working at Northwestern University with Abraham Lerman and Fred Mackenzie, he demonstrated from the pattern of change in sedimentary rocks with time that changes in climate reflect changes in the abundance of $CO_2$ in the atmosphere that fluctuate within rather narrow limits, indicating the operation of natural feedbacks within the climate system [29]. To put it another way, Earth's surface is not in a perfectly balanced steady state. Whilst it tends to a balance over time, the *'flux rates and reservoir sizes of major components of Phanerozoic ... cycles tend ... to fluctuate markedly'* [29]. The size of the $CO_2$ reservoir in the atmosphere depends on the rates at which $CO_2$ is supplied by sources (volcanoes and metamorphism) or deposited in sinks as carbonate or organic rich sediment. *'The $CaCO_3$ [sedimentary] reservoir and its fluxes can be regarded as the atmospheric $CO_2$ buffering system'* [29], much as described in the equations presented above. *'Higher atmospheric $CO_2$ would result in more dissolved calcium carbonate'* [29]. Amongst other things, the authors noted that *'The response times for ... perturbations are of the order of millions of years'* during which *'negative feedbacks ...[operate] to restore the system to new steady states that are not drastically different from the original state'* [29]. Garrels realized that many of the Earth's geochemical cycles are intimately linked, demonstrating in 1984 that *'Reduction of carbonate carbon to organic carbon is ... mirrored in a roughly comparable oxidation of reduced sulfur to oxidized sulfur over time periods of tens of millions of years ... This relationship is a consequence of a remarkable coupling among the sedimentary reservoirs in a world that cannot afford wild fluctuations in the oxygen and carbon dioxide contents of the atmosphere without eliminating life'* [30].

Essentially, Garrels found that the world of the past 550 Ma, the Phanerozoic, was not markedly different from the world of today in terms of rates of deposition and erosion. He had found restraints on the bounds of physical conditions on the planet in former times, which acted like a thermostat to keep Earth's climate within a fairly small natural envelope – a sort of 'Goldilocks zone'.

In 1981, Garrels' co-author, Fred Mackenzie, continued to develop this theme, bringing in the role of plate tectonics. Working with J.D. Pigott, he found a strong correlation between the distribution of preserved masses of sediment, the cyclicity of sea level, and the distribution of carbon and sulphur amongst their major reservoirs, which he related to the two main tectonic modes of Phanerozoic time: oscillatory and submergent [31]. The *'submergent mode of active plate convergence...subduction of sediments, large ridge volume [i.e. abundant sea-floor spreading and associated volcanism], and high sea-level gave rise to low erosion and sedimentation rates, less restricted environments of carbonate deposition, and relatively high atmospheric $CO_2$ levels* *(high temperatures) resulting from an increased rate of production of $CO_2$ from diagenetic and metamorphic reactions at subduction zones*' [31]. Continental cratons tended to be submerged around their edges by the high sea levels of those times. All of these processes were weakened in the 'oscillatory' or 'emergent' mode, characterized by generally elevated continental interiors with less flooded margins. Decreased atmospheric $CO_2$ levels led to lower global mean temperatures. There were fewer mid-ocean ridges, and more abundant evaporites due to the aridity of the emergent continents [31]. Oscillations between the emergent and submergent modes could well lead to atmospheric $CO_2$ varying through time, causing times of global warming (submergent periods) or cooling (emergent periods) [31].

Examination of Phanerozoic formations of oolite rocks (composed of ooids – ovoid pellets made from layers of minerals grown in situ) showed that ooids from prior to the Carboniferous tended to be calcitic, whilst those from later oolites tended to be aragonitic, suggesting that '*atmospheric $CO_2$ levels were higher prior to Carboniferous time*', favouring the deposition of calcite, whilst $CO_2$ levels fell after the Carboniferous, favouring the deposition of aragonite and calcite high in Mg [31].

That same year, 1981, James Callan Gray Walker, from the Space Physics Research Laboratory of the University of Michigan, Ann Arbor, who was to win the F.W. Clarke Award of Geochemical Society for his contributions to geochemistry, published with colleagues a landmark paper suggesting that a negative feedback involving surface temperature, $CO_2$, and chemical weathering controlled the evolution of the Earth's climate through time by modulating the strength of the greenhouse effect [32]. Walker was stimulated to undertake this study in an attempt to explain how the Earth managed to have a watery surface back in the Archaean some 3.8 billion years ago, given that stellar evolution implied that a 'main sequence' star like the Sun would have produced 25% less luminosity back then [33]. His team set out to follow the suggestion of Owen and others that the greenhouse effect of $CO_2$ and water vapour together could be great enough to solve the cool Sun problem provided that there had been a large enough amount of $CO_2$ in the atmosphere of the early Earth [33]. Ramanathan, whom we met investigating the greenhouse effect in Chapter 8, was one of Owen's co-authors. In addition, Walker was well aware of the ideas put forward by Budyko and Broecker, which we examined in Chapter 8.

The papers by Mackenzie and Pigott [31] and Walker et al. [32] influenced a young Yale geochemist, Robert Arbuckle (Bob) Berner (1935–2015) (Figure 9.5, Box 9.3) to expand his studies of the geochemical cycles of carbon and sulphur to examine how they might contribute to the $CO_2$ levels of the atmosphere and hence to changes in climate through time.

**Figure 9.5**   Robert (Bob) Arbuckle Berner.

---

**Box 9.3   Bob Berner**

Geochemist Bob Berner received his PhD from Harvard in 1962 [34]. He was elected to the National Academy of Sciences and received numerous awards, amongst them the V.M. Goldschmitt Medal of the Geochemical Society in 1995, the Murchison Medal of the Geological Society of London in 1996, and the Vernadsky Medal of the International Association of Geochemistry in 2012. Berner and Garrels knew one another reasonably well, Berner having taken courses from Garrels while studying for his PhD.

---

Walker's, Mackenzie's and also Berner's papers [31, 32, 35] got Garrels thinking, and while he was at the University of South Florida, in 1983, he got together with Berner, and with Antonio Lasaga from Penn State, to produce a classic paper on 'The carbonate-silicate geochemical cycle and its effect on atmospheric carbon dioxide over the past 100 million years' [36]. Their quantitative model became known as the BLAG model, after the authors' initials. The BLAG team built on Walker's and Mackenzie's notions by proposing that climate evolution was controlled by a negative feedback loop balancing the supply of $CO_2$ to the air from the Earth's interior, via volcanoes, with the consumption of $CO_2$ from the air by chemical weathering of silicate rocks [36]. Supply and consumption broadly matched over long periods of time, preventing either a build-up of $CO_2$ leading to a runaway greenhouse effect like that on Venus, or a depletion to near zero and an icehouse effect like that on Mars.

In the BLAG model, high rates of sea-floor spreading and volcanic activity in the Cretaceous – estimated from sea level curves that were thought to reflect rates of sea-floor spreading, hence volcanic activity – supplied CO$_2$ to strengthen the greenhouse effect and increase temperatures. In the inelegant jargon of the field, the supply of CO$_2$ from the Earth's sub-crust (the mantle) to the air, via volcanic activity associated with sea-floor spreading, was referred to as 'mantle degassing'. Warmth encouraged evaporation and increased rainfall. An increasingly warm moist climate rich in CO$_2$ increased the rate of chemical weathering, which acted as a negative feedback preventing further build-up of CO$_2$ in the atmosphere, and, over time, reducing the atmospheric concentration of CO$_2$. Rates of supply of CO$_2$ slowed when sea-floor spreading slowed in the Cenozoic. The resulting cooling also slowed chemical weathering. The model showed Earth's thermostat at work. It made a valuable contribution, not only in making geologists think more deeply about the carbon cycle and the greenhouse effect, but also in providing testable predictions. The model also showed a small CO$_2$ peak in the mid Eocene, which was not discussed at any length in the BLAG paper, but which does correlate with the rise in bottom water temperature deduced from oxygen isotopes, which we will explore later.

At the Chapman Conference in Tarpon Springs, in January 1984, the authors refined their model to take into account also the natural reservoirs of carbon, pyrite (iron sulphide), gypsum (calcium sulphate), and atmospheric oxygen, as well as such factors as the formation and oxidation of pyrite, the reduction of sulphate at mid-ocean ridges, and the weathering and burial of organic carbon [37]. While this modified the original results, removing some of the model's problems, it did not change the prediction that the Cretaceous atmosphere 100 Ma ago had been much richer than today in CO$_2$. It did, however, reduce the predicted amount of CO$_2$ in that atmosphere by a factor of 2, to 13 times as much as today. This inevitably reduced the predicted Cretaceous global temperature at that time from 25 to 23 °C, more in keeping with the evidence from oxygen isotopes. As in the 1983 version of the model, CO$_2$ and temperature changed together, the temperature falling to a low point (around 15 °C) by 55 Ma ago, rising to a peak (18.5 °C) around 43 Ma ago, falling to a new low (14.5 °C) around 18 Ma ago, rising to a new peak (15.5 °C) 3–5 Ma ago, before finally falling to Holocene pre-industrial levels (on average about 14 °C).

One of the implications of these various studies is that land plants affect the levels of atmospheric CO$_2$ through a stabilizing feedback loop [38]. The evolution of plants with strong roots enhances the chemical weathering of rocks, thus helping to drive down the concentration of CO$_2$ in the air [39]. The sequestering of CO$_2$ as skeletal remains in limestones, or as organic matter in coals, adds to this effect. Hence biological activity is one of the key drivers of CO$_2$

trends. For example, the rise of ectomycorrhizal fungi during the Cenozoic may have contributed to enhancing weathering rates [40].

## 9.3 The Critics Gather

All scientists are trained to be sceptical, and the BLAG model was not exempt from critiques. Some of them came to the fore in 1984 at the Chapman Conference in Tarpon Springs, papers from which were published as Monograph 32 of the American Geophysical Union in 1985.

Based on studies of the $\partial^{13}$C carbon isotope ratios in deep ocean sediments, Nick Shackleton suggested that much more organic matter was trapped in Cenozoic sediments prior to the middle Miocene (around 14 Ma ago) than since [41]. He attributed the post-mid Miocene reduction in trapping of organic matter to the increasing oxygen content of bottom waters that came about as polar waters cooled from 10 to 2 °C. Polar cooling created more sea ice, which excreted salt and made surface waters dense enough to sink, taking oxygen with them. Shackleton concluded that dissolved CO$_2$ in the Palaeocene ocean was about the same as it is today, placing a constraint on the predictions of the BLAG model. Clever though he was, he was wrong in this case, as is evident from Figure 9.3.

Somewhat more controversial was Shackleton's claim that CO$_2$ was not more than twice its present level in the Cretaceous – nothing like the factor of 13 claimed by the BLAG modellers. Behind his calculations lay the assumption that the carbon input to the ocean had been constant through time [41]. If it had not, then his criticism would not be justified. As we see later, it was not. Besides that, more recently it was found that using $\partial^{13}$C ratios from marine sediments as a proxy for atmospheric CO$_2$ could be suspect because of carbon isotopic fractionation by phytoplankton in so-called 'vital effects', such as changing growth rate [42].

One of those questioning Berner's assumptions about the Cretaceous was Michael Allen Arthur (Figure 9.6, Box 9.4) of the University of Rhode Island (whom we met above as a student of Al Fischer), and his colleagues [43].

The BLAG model assumed that over the past 100 Ma the carbon cycle had been in balance, and the rate of burial of organic carbon had been small compared with rates of fixing of carbon by weathering, or of supply of carbon by outgassing from the Earth as a by-product of metamorphism, volcanism, and geothermal activity. Arthur and his team presented '*evidence that during the middle to late Cretaceous (110 to 70 m.y. ago), high rates of production and burial of Corg [=organic carbon], and intense and widespread volcanism ....may have had profound effects on the global CO$_2$ cycle that were not accounted for by the [BLAG] model*' [43].

**Figure 9.6**   Michael Allen Arthur in the field.

---

**Box 9.4   Mike Arthur**

Mike Arthur, professor of geosciences at Penn State University since 1991, was a student of Al Fischer (See Chapter 10) at Princeton, where he earned his PhD in 1979. Like his mentor, he is a very broad geological thinker. He was one of the first geoscientists to apply stable isotopic techniques to palaeoceanographic and palaeoclimatic problems. As a rigorous practitioner of the multi-proxy approach to palaeoenvironmental analysis, he has made significant contributions to our understanding of the controls on organic carbon burial and the global carbon cycle, the expression of orbital forcing of climate in sedimentary systems, the biogeochemical cycles of sulphur, iron, nitrogen, and phosphorus, and their relationship to the oxidation state of the oceans and atmosphere [44]. He was honoured with the F.P. Shepard Medal of the Society for Sedimentary Geology in 1996 and the Laurence L Sloss Award of the Geological Society of America in 2007.

---

Both Arthur and Berner accepted the observation of Harvard's Heinrich Holland (1927–2012) that, prior to the Industrial Revolution, out-gassing by volcanoes was the major source for the replacement of atmospheric $CO_2$ drawn down by weathering [45]. As we saw in Chapter 5, the creation of numerous hot, young mid-ocean ridges starting 180 Ma ago with the break-up of Pangaea [46], undoubtedly pumped large quantities of $CO_2$ into the atmosphere during Jurassic and Cretaceous times. To understand how this might have affected the production of $CO_2$, it was important to know how much the rate of ridge production had changed through time. Initially, it was thought that it had changed a lot, with periods of rapid sea-floor spreading and mid-ocean ridge production causing the flooding of the continents in the Jurassic and Cretaceous (Figures 5.12b and 6.13) [47, 48]. As we saw in Chapter 5,

David Rowley called that long-held notion into question [49, 50], but Dietmar Müller and his team disagreed. Reconstituting the pattern of Mesozoic mid-ocean ridges from the magnetic anomaly patterns in the Pacific they concluded that the highest rates of production of new crust occurred 140 Ma ago and declined towards the end of the Cretaceous [46], in keeping with what was understood from patterns of the change in sea level with time (Figure 5.13) as documented by Pete Vail [47], Tony Hallam [48], and Bilal Haq [51]. I take Vail's and Haq's sea level curves as independent evidence for the likely correctness of Müller's reconstructions of past ridge volumes [46], and hence for likely changes in the rates of production of $CO_2$ by mantle degassing through time. Besides, as we shall see later, Müller's data agree with independent changes in ocean chemistry, whilst Rowley's concept does not.

Arthur argued not only that there was more mid-ocean ridge activity supplying $CO_2$ to the atmosphere during the Cretaceous, but also that there was widespread mid-plate volcanic activity right across the Pacific basin, contributing to what Bill Menard called the Darwin Rise (see Chapter 5) [52]. Arthur's team went further [43], suggesting that volcanic activity in the middle of ocean crustal areas away from mid-ocean ridge crests may also have been equally abundant on the Kula, Farallon, and Phoenix Plates, which formerly occupied much of the eastern Pacific seabed but had since been subducted beneath the surrounding continents. This is 'mid-plate volcanism'. In his Chapman Conference paper [43], Arthur was anticipating Müller's conclusions [46]. If Arthur was right, then the supply of $CO_2$ to the atmosphere by volcanic activity in the Cretaceous might have been as much or even more than supposed in Berner's models, which had not taken into account the Darwin Rise or contributions from long-subducted mid-ocean ridge crests.

Large amounts of the organic matter supplied to the ocean basins during Cretaceous times were buried during what became called Oceanic Anoxic Events (more on those in Chapter 10). While the increased atmospheric $CO_2$ of the time may have been balanced to some extent by the increased rates of burial of organic carbon in those events, it was nevertheless entirely feasible that some of the volcanogenic $CO_2$ had accumulated in the air as suggested by Berner's models. Reviewing evidence that might constrain the rise in atmospheric $CO_2$ in the Cretaceous, Arthur concluded that the $CO_2$ concentration in the air then was less than eight times present values [43]. Rates of burial of organic carbon were also unusually high during Carboniferous and Permian times 300–250 Ma ago, reflecting the rise in vascular land plants and the abundance of flooded and swampy lowlands [53]. In that instance there was enough carbon burial to draw down atmospheric $CO_2$

to very low levels, leading to a glaciation [53]. That was not the case during the Cretaceous, although there may have been limited amounts of polar ice during Cretaceous times.

By 1990, Berner had simplified his approach to modelling the carbon cycle [26]. His revised model showed that natural long-term processes, involving carbon stored in rocks, brought about changes in atmospheric $CO_2$ with time that were gradual but large enough to have caused the climate to change through the operation of $CO_2$ as a greenhouse gas. These changes involved the slow carbon cycle in which the long lag between the chemical weathering of silicate minerals in the 'Urey Reactions' and the thermal de-carbonation of the carbonate products of that weathering over time by metamorphism, would lead from time to time to a large imbalance in $CO_2$ fluxes that affected the Earth's climate à la Chamberlin. Evidently, the Urey Reactions must control the role of $CO_2$ in the climate system on long geological timescales.

Like the BLAG model, the new one (a later version of which came to be named GEOCARB) reproduced the low $CO_2$ concentrations of the Carboniferous and the late Cenozoic, leading Berner to conclude that climates should have been warm in the early Palaeozoic and the Mesozoic, and cool in the late Palaeozoic and late Cenozoic, in agreement with palaeoclimatic data. He appreciated that the model was no better than the data and assumptions that went into it, all of which needed refinement. The semi-quantitative trends emerging from the model were not hard facts. They showed how one could '*use the geochemical carbon cycle as a means of exploring the various factors affecting atmospheric CO₂ over long geological times*' [26]. In effect, Berner had managed to convert into a quantitative model both the conceptual model of $CO_2$ and climate change proposed by T.C. Chamberlin (Chapter 4) and the qualitative model of $CO_2$ and climate change proposed by Budyko and Ronov (Chapter 8).

By 2001, Berner had revised his GEOCARB model for Phanerozoic time to incorporate the behaviour of the sulphur cycle in GEOCARB-III [54]. And by 2006 it had been modified again to account for changes in both oxygen and $CO_2$ [55]. This was a significant advance on the original model. It showed a fall in $CO_2$ at around 440–450 Ma ago in the GEOCARBSULF-volc model close to the boundary between the Ordovician and Silurian, where there was a major but short-lived glaciation (of which more at the end of this chapter (Figure 9.7)). Berner calculated the proportion of volcanic weathering from the oceanic record of strontium isotopes ($^{87}Sr/^{86}Sr$), and used it to modify the equations for calculating atmospheric $CO_2$ [55]. The key remaining questions were: to what extent did the model represent reality? And what proxies could we use to estimate the atmospheric $CO_2$ of past times?

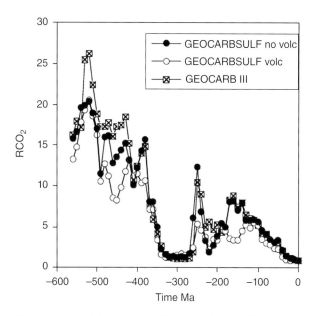

**Figure 9.7** Modelled plot of $CO_2$ versus time for Phanerozoic time (the past 542 Ma). $RCO_2$ refers to the ratio of the mass of $CO_2$ in the atmosphere at time 't' to that in the present atmosphere (at 1990). The plots show $RCO_2$ versus time from the GEOCARB and GEOCARBSULF models. *Source:* From figure 18 in Ref. [55].

**Figure 9.8** Maureen Raymo.

By 1988, Berner's model was under fire for not paying enough attention to mountain building. The criticism came from Maureen Raymo (Figure 9.8, Box 9.5) and her former professor Bill Ruddiman (1943–) (Figure 9.9, Box 9.6), who thought that the uplift and chemical weathering of highlands changed global climate [56, 57].

Together with Flip Froelich, Raymo and Ruddiman thought that '*if the concept of a global temperature-weathering feedback was correct, then global chemical weathering rates should have dropped through the Cenozoic in concert with falling mantle degassing rates and surface temperature*' [56]. To determine those weathering rates, they suggested using the ocean-wide ratios of strontium isotopes $^{87}Sr/^{86}Sr$

---

**Box 9.5    Maureen Raymo**

Maureen Raymo, a marine geologist and palaeoceanographer, studied at Brown University before receiving her PhD at New York's Columbia University in 1989. She has been a research professor at Lamont Doherty Earth Observatory and a member of Columbia's Earth Institute since 2011, following periods at the University of California, Berkeley, MIT, Woods Hole Oceanographic Institution, and Boston University. She has also been a visiting investigator with the National Institute of Oceanography in Goa, India, and at the University of Melbourne, Australia. She specializes in studies of how ice ages wax and wane and how sea levels change with time. Over the years, that has led her to become much involved with deep-ocean drilling, and to chair the Science Advisory Structure Executive Committee for the Integrated Ocean Drilling Program (IODP) (2010–2011). She has received several awards for her work, including the Cody Award in Ocean Sciences 2002, the Wollaston Medal of the Geological Society of London in 2014 (the first woman in its 183-year history to receive that award), the Milutin Milankovitch Medal of the European Geosciences Union, also in 2014, and was elected to the US National Academy of Sciences in 2016.

Figure 9.9    Bill Ruddiman.

---

**Box 9.6    Bill Ruddiman**

Trained as a marine geologist, much of Ruddiman's career has been spent examining one or other aspect of past climate change, a topic on which he has written many research papers, edited several volumes and published three books. Ruddiman gained a PhD from Columbia University. After working for the US Naval Oceanographic Office he returned to New York to join what was then Lamont–Doherty Observatory of Columbia University, where he spent most of his career until moving to the University of Virginia in 1991 as professor in the department of environmental sciences, of which he was chairman from 1993 to 1996. Having retired in 2001 he holds emeritus status with that university and continues publishing. During his career he was director of the CLIMAP program (see Chapter 12 for details) (1982–1983) and was on the Executive Committee of the COHMAP project (see Chapter 14 for details) (1982–1990). In 2010 he won the Geological Society of London's Lyell Medal. Amongst other accolades he has a Distinguished Career Award from the American Quaternary Association (2012).

---

in carbonate sediments. This ratio is high in continental granites and low in oceanic basalts. They inferred that the increase in this ratio in carbonate sediments over the past 40 Ma came from increased chemical weathering of continental rocks. High values of that ratio represented increased chemical weathering in the rapidly uplifting Himalaya–Tibet region, which caused atmospheric $CO_2$ values to fall following the collision of India with Tibet [57, 58].

By 1997, Ruddiman had put together a book on *Tectonic Uplift and Climate Change* [59]. In it he pointed out that tectonic uplift may control climate in one of two ways [60]. The **direct effect** works as follows: through the planetary moist lapse rate of around 6.5 °C/1000 m, uplift cools surfaces at high altitudes. The accumulation of snow and ice there further cools those areas by reflecting solar energy. The growth of frost and ice accentuates mechanical erosion. Mountains also encourage rainfall on their windward sides, which leads to rain shadows on their lee sides, and they can create and intensify the seasonal monsoonal circulation, as happens over Tibet, which has a mean elevation of 5 km over an area of more than 4 million km². Heating of the Tibetan Plateau in summer leads to hot rising air, which is compensated for by the inflow of moist air from the ocean at sea level. The ensuing heavy rains enhance erosion on the windward mountain slopes of

northern India. In winter, cold dry air sinks over the Plateau as it cools and becomes snow covered, causing a reversal of the monsoon winds over the surrounding lowlands and the Indian Ocean. This has been going on for the past 40 Ma or so since the Tibetan Plateau began to rise with the collision between India and Asia 50 Ma ago. Large mountain and plateau areas can also cause and amplify meanders in the jet stream, affecting local climate. By changing the circulation of the winds, mountains can also change ocean circulation, as shown today by the action of monsoonal winds in the Indian Ocean. The **indirect effect** of tectonic uplift can be seen in the action of rain on windward slopes, which chemically weathers silicate rocks, thus drawing down atmospheric CO$_2$ and lowering temperature.

Ruddiman suggested that the Himalaya and the Tibetan Plateau were uplifted gradually from about 40 Ma ago, and that there must have been significant relief and erosion 20 Ma ago to account for the massive accumulation of sediment in the Bay of Bengal [59]. The evidence suggested that '*the Tibetan Plateau reached a threshold size sufficient to induce the Asian monsoon about 8 million years ago*', he wrote [59]. Uplift led to the drying of the Asian interior through a rain-shadow effect north of the Himalaya and the Tibetan Plateau. The counter-clockwise circulation around the low pressure cell over Tibet in summer brought hot dry air from central Asia south over Arabia, the mideast and northeast Africa, shifting vegetation to dry-adapted types.

Other uplifts had more local effects. African uplift, giving rise to a relatively high plateau in the east and south over what seemed to be mantle hot spots (Figure 5.10), began in the Miocene, probably about 8 Ma ago. Both that rise and the accelerated uplift of the Himalaya and Tibetan Plateau at that time may have been caused by global reorganization of the world's tectonic plates. Uplift of the Andes is also relatively recent, beginning in the Eocene and accelerating with plate reorganization in the Oligocene around 26 Ma ago. And both the Alps and the North American interior plateaus are Cenozoic features related to subduction. All of these uplifts changed regional climate. But only the Tibetan uplift seems to have been extensive enough to have a global effect.

Ruddiman, Raymo, and their colleagues thought that the pervasive cooling of the late Cenozoic may have been enhanced by its own feedbacks. They suggested that decreasing CO$_2$ caused cooling, leading to formation of ice sheets and mountain glaciers, which accentuated mechanical erosion, exposing yet more silicate minerals to chemical weathering and thus further lowering CO$_2$ [61]. They found no need to call upon changes in heat transport towards the Arctic by the ocean to explain the high latitude late Cenozoic changes in vegetation that we explored in Chapter 7. The tectonic opening or closing of oceanic gateways was discontinuous through that period, with each having discrete beginnings and/or ends, and all occurring in scattered locations. Rather than being an ongoing source of global climate change, like the Himalayan–Tibetan uplift, Ruddiman's team thought that those gateways were more likely to have had important effects on the formation of deep and bottom water, and on the creation of regional perturbations to longer-term trends.

Raymo and Ruddiman's uplift-weathering hypothesis elevated the importance of chemical weathering to a major factor in reducing CO$_2$ levels and thus leading to global cooling. It assumed that increased chemical weathering in mountains must be a key factor controlling CO$_2$ and climate, and so challenged Berner's BLAG thermostat model of the climate system, in which mountains played little part. One of the implications of their hypothesis was that even steady-state plate motions could lead to non-steady state effects on climate, through variations in continental relief with time, since continental collisions resulting in the uplift of plateaux the size of Tibet were rare and episodic: '*Despite the continuous presence throughout geological history of high mountain terrain along the convergent margins of the world, it may be the rarer occurrence of plateaux that can drive climate away from steady state and decouple rates of horizontal and vertical tectonic movement*' [62].

Geophysicists Peter Molnar and Phil England of MIT reversed that hypothesis, proposing that weathering drove uplift by reducing mountain mass, so allowing the Himalayas to rise through isostatic adjustment [63, 64]. It wasn't the uplift that drove the climate change, but the climate change that drove the uplift! Bearing that proposal in mind, Cronin concluded in 2010 that '*The jury is still out on the hypothesis that Neogene climatic cooling was forced by Himalayan uplift, changes in the rate or intensity (or both) of continental erosion, and CO$_2$ drawdown, or whether climate changes drove plateau erosion*' [65]. Even so, elsewhere evidence was accumulating that ties uplift (in Africa and the Andes) to Neogene changes in climate. What we need to resolve this issue are improvements in proxies for past elevation.

Could strontium isotopes be used to suggest the pattern of uplift in the way Raymo supposed? Joel Blum, of Dartmouth College, New Hampshire, thought not, noting that '*the Sr isotopic composition released by weathering of a single rock type changes dramatically with the age of the soil. Thus, the marine $^{87}$Sr/$^{86}$Sr ratio cannot be considered a direct proxy for silicate weathering rates*' [66]. Others disagreed. John Edmond (1943–2001), a hard-drinking, red-bearded, Scottish geochemist and a fellow of the Royal Society from MIT, who was one of my seagoing expedition companions at Christmas 1973 en route from Dakar to Cape Town, sided with Raymo. He argued that: '*tectonically active mountain belts are the loci for accelerated drawdown of atmospheric CO$_2$; hence their initiation and evolution have a direct influence on global climate*' [67]. Edmond concluded that given the lack of any increase in the rates of sea-floor spreading and the volcanic activity associated with mid-ocean ridges over the past 50 Ma, the intensive mountain building of that period was sufficient not only to have prevented any runaway greenhouse effect, but also to have accounted for the observed climatic deterioration. '*The resulting $CO_2$ drawdown*' he said '*is now close to complete, such that the Earth has entered into what will likely be a prolonged glacial epoch with the atmospheric $CO_2$ held at a kinetic minimum by the resulting exposure of aluminosilicate rock*' [67]. In other words we are in for a re-run of the lengthy Carboniferous–Permian glaciation.

Berner fought back. In his own paper in Ruddiman's 1997 book, he concluded that Edmond's argument about strontium isotope ratios was flawed by its reliance on analyses of river chemistry that paid insufficient attention to the presence of carbonate minerals that would distort the strontium isotope data [68]. He accepted that mountain uplift is an important control on rock weathering and atmospheric $CO_2$, agreeing that '*to have appreciable silicate weathering it is necessary to have sufficient relief so as to enable erosive removal of any protective clay overburden*' [68]. But, *given sufficient relief* (his italics), he insisted that climate also plays an important role in weathering through its control on temperature, rainfall, and vegetation [68]. Agreeing that topography is important, he incorporated a special term for topographic relief into the successor to the BLAG model The outputs from his new GEOCARB model of the variation of $CO_2$ through time matched independent estimates of past levels of atmospheric $CO_2$ obtained from fossil soils (palaeosols), for example in showing that a major fall in $CO_2$ preceded the Carboniferous–Permian glaciation, and was sustained throughout that glaciation. '*Focussing only on mountain uplift,*' he concluded, '*with neglect of … other factors, is unnecessarily narrow. Earth system science requires that all factors be evaluated and first-order attempts at quantification at least be attempted*' [68]. Humboldt and Lyell would have agreed.

Together with Lee Kump of Penn State University, Mike Arthur examined the assumptions about rates of sea-floor spreading and $CO_2$ emission for the Cenozoic. They knew that atmospheric $CO_2$ declined from about 20 Ma ago, consistent with the uplift history of the Himalaya. They assumed that from the end of the Cretaceous '*the rate of volcanism (and release of $CO_2$) has in general decreased, rather than increased, throughout the Cenozoic*' [69]. That assumption led them to '*reject the hypothesis of increased global weathering rates as the cause of Cenozoic cooling and of the Sr isotope trend toward more radiogenic values*' [69]. Instead Kump and Arthur thought that whilst rates of chemical weathering and erosion probably increased as the Himalaya rose, mass balance considerations meant that chemical erosion rates must have decreased elsewhere, or else atmospheric $CO_2$ would have shown a steep decline as the mountains went up. They explained the increase with time in the marine $^{87}Sr/^{86}Sr$ ratio as reflecting the changing

chemistry of the rocks exposed with time as the Himalaya were eroded and deeper, more crystalline basement rocks with higher $^{87}Sr/^{86}Sr$ ratios were gradually exposed. The ratio was therefore not a reliable index of the amount or rate of chemical weathering. Rather than having us consider rates of weathering, they thought we should be thinking about 'weatherability'.

For example, in 2000, Kump and Arthur pointed out that as well as there being an increase in the rate of chemical weathering in the region of the Himalaya and the Tibetan Plateau, there would have been an increase in chemical erosion in areas of continental glaciation, especially Antarctica, where mechanical erosion exposed fresh rocks to chemical attack, making the rocks more weatherable [70]. They also highlighted the key role played by plants, whose root systems penetrate the soil allowing acid-carrying fluids and $CO_2$ to permeate it. Plants also bind fine particles to the soil, thus increasing the surface area exposed to soil pore waters. Tree roots and associated fungi accelerate weathering by acidifying the weathering environment, hence affecting the concentration of $CO_2$ in the atmosphere, and thus global temperature [70]. Decomposition of soil organic matter also plays a role in controlling atmospheric $CO_2$ levels. Whilst warming tends to increase the rate of decomposition, cooler temperatures will slow it. The rise of land plants 450 Ma ago increased the weatherability of silicates by about sevenfold, and, as a result, atmospheric $CO_2$ dropped from about 12 times pre-industrial levels to about 5 times [71]. At the same time, the increasing burial of organic remains helped to draw down formerly high levels of $CO_2$. In developing the GEOCARB model, Berner considered both weatherability and the sequestration of organic matter in sediments to have been important.

Kump and Arthur concluded that a rise in $CO_2$ brought the Permian glaciation to an end. The widespread aridity of Permian Pangaea decreased the rate of chemical weathering and reduced the supply of nutrients to the ocean, hence reducing ocean productivity. Less weathering and less productivity meant less extraction of $CO_2$ from the air, allowing a long-term build-up of volcanic $CO_2$, which eventually led to global warming and a more active hydrological cycle, higher rates of weathering, more nutrient run-off, and increased productivity [70].

In 2003, Ravizza and Zachos pointed out that if the curve of $^{87}Sr/^{86}Sr$ reflected the uplift and erosion of Himalaya, one would expect changes in its slope to be related to events in Himalayan uplift, but they are not [72]. It may be no coincidence that the rapid rise in $^{87}Sr/^{86}Sr$ ratios in the Cenozoic began around 35 Ma ago, close to the initiation of Antarctic glaciation. Glaciation exposes large areas of fresh rock to weathering, hence raising the $^{87}Sr/^{86}Sr$ ratio [72], much as

Kump and Arthur suggested [69, 70]. In summary, the $^{87}Sr/^{86}Sr$ curve doesn't tell us about the drawdown of $CO_2$.

Does mountain building lead to cooling? Consideration of the likely balance between rates of $CO_2$ drawdown by chemical weathering throughout the Cenozoic, and the rates of supply of $CO_2$ to the atmosphere from volcanic and metamorphic sources tended, for a while, to suggest not [72–74]. Then again, cooling and ice growth during the Neogene are not associated with systematic decreases in atmospheric $CO_2$ [75–77]. Ravizza and Zachos suggested that steady state levels of $CO_2$ could be reduced without changes in global weathering flux, by changes in 'weatherability' [72] as shown by Kump and Arthur [69, 70]. Tectonic and climatic factors could act to allow weathering fluxes to remain high enough to balance $CO_2$ input in spite of overall cooling. This concept provides a defence for the causative link between Himalayan uplift and global cooling, since physical weathering can be argued to increase 'weatherability'. It provides the basis for arguing that Cenozoic cooling might have been enhanced by a limited decoupling of average global temperatures and weathering rates. Glaciation itself may have contributed by greatly increasing the area of fresh mineral surfaces available to chemical weathering. As we shall see soon, the 'orogenesis (mountain building) leads to cooling' theory is not dead yet, although it does not provide a single, unified explanation for Cenozoic climatic variation.

Bill Hay added another astute observation to the mix [14]. The rising of the Himalayas did not simply expose old rocks to chemical weathering that would tend to lower $CO_2$, it also led to massive deposition of eroded sediments in the enormous submarine fans of the Indus in the Arabian Sea and the Ganges in the Bay of Bengal. Even though individual layers of sediment in the fans contain rather small amounts of organic matter, the fans are so large that the vast amounts of organic matter locked up in them must have removed significant amounts of $CO_2$ from the climate system.

## 9.4  Fossil CO₂

A substantial advance in geological thinking about the climatic effects of changes in atmospheric $CO_2$ had to wait for a dramatic breakthrough: the discovery of fossil atmospheric $CO_2$ in 1978 by a pioneering Swiss team led by Hans Oeschger (1927–1998) (Figure 9.10), head of the division of Climate and Environmental Physics at the University of Bern [78]. Analysing fossil air from bubbles in the Camp Century Ice Core from Greenland, Oeschger's team found low values for atmospheric $CO_2$ in Ice Age ice, consistent with Callendar's estimated late nineteenth century value

**Figure 9.10**  Hans Oeschger.

of 290 ppm. The idea of being able to analyse fossil air grabbed the imagination of climate scientists and geologists. We could now test Plass's ideas with data.

Oeschger's team published more results in January 1984 at a Chapman Conference on 'Natural Variations in Carbon Dioxide and the Carbon Cycle', convened by Eric Sundquist and Wally Broecker in Tarpon Springs, Florida [79]. Oeschger showed that for the last 1000 years atmospheric $CO_2$ had hovered around 260–270 ppm, with little fluctuation, and that by the late nineteenth century it had reached around 290 ppm, in agreement with Callendar's data [80]. It looked as if an association of $CO_2$ and temperature extended back through time, although the origin of the $CO_2$ signal was unclear and the details of the $CO_2$-temperature relationship had yet to be worked out.

Corroboration of Callendar's and Oeschger's independent findings came from yet another source: tree rings, whose chemical composition is related to the air in which the trees lived. The relationship between the carbon isotopic composition of the atmosphere and its $CO_2$ content led to efforts to use the $\partial^{13}C$ signal in annual tree rings to determine past levels of atmospheric $CO_2$. First to establish that link, in 1978, was geochemist Minze Stuiver from the University of Washington [81]. Building on Stuiver's $\partial^{13}C$ results, Tsung-Hung Peng of the Oak Ridge National Laboratory in Tennessee concluded in 1985, that the pre-1850 atmosphere contained about 266 ppm of $CO_2$, consistent with the data from ice cores [82]. Plass would have been pleased to see these data emerge from the dim and distant past.

It was time to revisit Callendar's results. The arrival on the scene of the exciting new measurements of fossil air from ice cores triggered a meeting of experts in Boulder, Colorado, in June 1983, under the aegis of the World Climate Research Programme. The results of the meeting were published as an obscure WMO Technical Report [83], and summarized in the scientific newsletter *EOS* by W. Elliott [84]. The experts reviewed a wider range of techniques than were available to Callendar for measuring the content of $CO_2$ in modern and fossil air. Much as Callendar had done (see Chapter 8), they rejected many of the pre-existing measurements of $CO_2$ as unrepresentative of background air. In addition, they analysed solar spectral data collected from 1902 to 1956 by the Smithsonian Solar Constant Program, and $CO_2$ values derived by analysing the carbon isotopic composition of tree rings and the carbon chemistry of old ocean water. They concluded from this multi-dimensional research programme that the air in the mid nineteenth century most probably contained between 260 and 280 ppm $CO_2$, like the levels found in ice cores [83, 84].

Oeschger's team continued refining their measurements, and in November 1986 they showed that the $CO_2$ content of bubbles in ice from the Siple Dome core in the Ross Sea region increased exponentially from about 280 ppm in 1760 towards the values of the present day [85]. The measurements from the youngest ice mapped on to the measurements of $CO_2$ in air made since 1957 by Keeling and his team [85]. The ice record and the air record formed an uninterrupted time series of $CO_2$ increase that has since been verified by several groups of scientists in ice cores from both Antarctica and Greenland (Figure 9.11) [86]. Not only that, but the growth of $CO_2$ in the air matched both the growth in consumption of fossil

fuels, and the change in the $\partial^{13}C$ composition of the $CO_2$ in the air bubbles in the Siple Dome core and in the air from Mauna Loa [85]. As more data from ice cores became available, and as analytical techniques improved, the analytically determined pre-industrial level of $CO_2$ converged on 280 ppm.

The rise in $CO_2$ since 1760 is not surprising. By then, steam engines were being used increasingly to power the industrial revolution. James Watt developed his new version of the steam engine between 1773 and 1775, patenting his design in 1769 and going into full production in 1776 – a momentous year in more ways than one. His new engine set off a demand for fossil fuel that continued to rise; for example from 5.1 Gt/year in 2000 to 8.4 Gt/year in 2011 [91].

Oeschger's team went one step further. They devised a numerical model of the carbon cycle to replicate their observations of atmospheric $CO_2$ and its carbon isotopic composition [85]. As to what controlled the natural atmospheric concentration of $CO_2$ through time, the answer seemed to lie in the ocean, where – because the ocean contains about 60 times the amount of carbon in the atmosphere (Figure 8.15) – small changes in productivity or circulation cause large changes in atmospheric $CO_2$.

$CO_2$ was not the only greenhouse gas showing signs of change with time. By 1986, it was clear from studies of fossil air in ice cores that the methane ($CH_4$) content of the air was also increasing, partly due to an increase in rice growing and cattle production. $CH_4$ in ice core air showed little change for the past 1000 years before an exponential rise began in the late 1700s, doubling its abundance in the atmosphere [92]. The rising values for the past 1000 years mapped on to the air measurements of methane made since 1978 [93], much as shown for $CO_2$ in Figure 9.11.

As time went by, an important caveat emerged on the usefulness of $CO_2$ extracted from ice cores. High $CO_2$ concentrations may result from chemical reactions between rock dust rich in carbonate, and acidic materials derived from volcanic eruptions. Oxidation of organic matter trapped in ice may also produce $CO_2$. Those caveats apply mainly to Greenland, where the dust of glacial intervals is most abundant and rich in carbonate. It does not apply to Antarctica, where there is little dust and virtually none of it is carbonate or organic.

As Sundquist points out, ice core records are most compelling where they agree despite different locations, accumulation conditions, and analytical procedures [86]. We now rely on Antarctic ice for changes in $CO_2$ through Ice Age times, but on both Greenland and Antarctica for changes in $CH_4$ over the same period, because methane is not subject to distortions arising from the chemistry of dust.

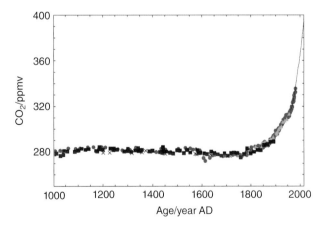

**Figure 9.11** Concentrations of $CO_2$ in the atmosphere from bubbles in ice cores over the past 1000 years (symbols) and from atmospheric measurements at Mauna Loa between 1957 and 2013 (solid line). Ice core data sourced from Law Dome [87, 88], Siple Dome [89], EPICA Dronning Maud Land [90] and South Pole [90].

## 9.5   Measuring CO$_2$ Back Through Time

So far, so good, and as we'll see in detail in Chapter 13, we can now measure CO$_2$ in bubbles of fossil air trapped in Antarctic ice cores back to 800 Ka ago. The key new question for geochemists was: how do we obtain hard data on the likely abundance of CO$_2$ in the atmosphere through time in the eras **before** fossil air became trapped in ice cores? As Crowley and North wrote in 1999 '*Documenting the magnitude of past CO$_2$ fluctuations is an important problem in paleoclimatology*' [94]. Geochemists and palynologists have now cracked that nut.

Nowadays, we can use several independent techniques to estimate the abundance of CO$_2$ in past atmospheres. We've already looked at the CCD for example, as an indirect measure (Section 9.1 and Figure 9.3). We can also use fossil leaves. Examining a leaf under a microscope you will find its surface dotted with tiny pores, or stomata. They let in the CO$_2$ that plants use to build their bodies, and prevent loss of water. When CO$_2$ increases in the air, the number of pores per unit area of leaf (the stomatal density) decreases, as does the percentage of leaf cells that are pores (the stomatal index) [95]. The stomatal index provides a reliable yardstick for estimating CO$_2$ abundance back to the Ordovician (c. 450 Ma ago), when land plants evolved [39, 96]. And, as we shall see in more detail below, we can also estimate past levels of CO$_2$ from the alkenones extracted from marine plankton; from the $\partial^{13}$C ratios in carbonate sediments and in foraminifera; from boron isotopes ($\partial^{11}$B) in planktonic foraminifera; from the occurrence of the sodium bicarbonate mineral nahcolite (NaHCO$_3$); and from carbonate in soils. These various independent data sources support one another to provide a convincing view of the variation of atmospheric CO$_2$ through time that is consistent with separately obtained geological data, palaeontological data and the backward predictions (or hindcasts, as we might call them) of Berner's GEOCARB model.

A leading expert in pulling all this diverse data together to show how atmospheric CO$_2$ varied through time is palaeobotanist Dana Royer (Figure 9.12, Box 9.7), who specializes in determining past levels of CO$_2$ from studies of fossil leaves [97–100].

In 2004, Royer and colleagues used a variety of proxy measurements (the $\partial^{13}$C of soil carbonates and phytoplankton, the stomatal characteristics of the leaves of C$_3$ plants, and the $\partial^{11}$B of planktonic foraminifera) to estimate how much CO$_2$ there was in the air at 10 million year intervals back to 450 Ma ago (Figure 9.13) [101], comparing the proxy data with the output of Berner's GEOCARB-III model of the variation on CO$_2$ through the Phanerozoic [54].

**Figure 9.12**   Dana Royer.

---

**Box 9.7   Dana Royer**

Royer earned a BA in geology and environmental studies at the University of Pennsylvania in 1998, and a PhD at Yale 2002, under the supervision of paleobotanist Leo Hickey and low-temperature geochemist Bob Berner. After spending a year with Dave Beerling at the University of Sheffield and a three-year post-doc at Penn State, he joined the department of earth and environmental sciences at Wesleyan University in 2005. He explores how fossil plants can be used to quantify the climate and ecology of ancient ecosystems, as well as the physiological mechanisms that underpin these relationships. He also compiles Phanerozoic CO$_2$ records and investigates CO$_2$-temperature coupling over multimillion year timescales.

---

By 2006, Royer had refined his analysis, in some cases to time steps of as little as one million years, as well as adding further proxies (the $\partial^{13}$C of alkenones in marine plankton, and of liverworts), finding a slightly closer correspondence between the proxy data and the model outputs of GEOCARB-III (Figure 9.14) [102].

Royer's research showed that '*For periods with sufficient CO$_2$ coverage, all cool events are associated with CO$_2$ levels below 1000 ppm. A CO$_2$ threshold of below c. 500 ppm is suggested for the initiation of widespread, continental glaciations, although this threshold was likely higher during the Paleozoic due to a lower solar luminosity at that time ... A pervasive, tight correlation between CO$_2$ and temperature*

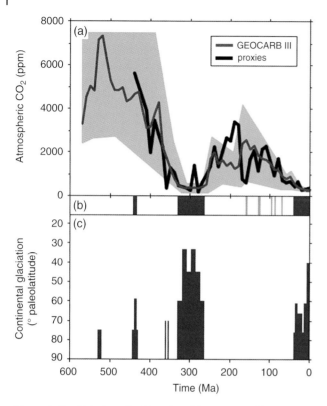

**Figure 9.13** Relation of cold and cool intervals to $CO_2$ through time for the past 600 Ma. (a) Comparison of model predictions from GEOCARB III (red line, with range shown by pink shading) with proxy reconstructions of $CO_2$ (black line) at 10 Ma time-steps. (b) Intervals of glacial (dark blue) or cool climates (light blue shading). (c) Latitudinal distribution of direct glacial evidence (tillites, striated bedrock, etc.) throughout the Phanerozoic. *Source:* From figure 2 in Ref. [101].

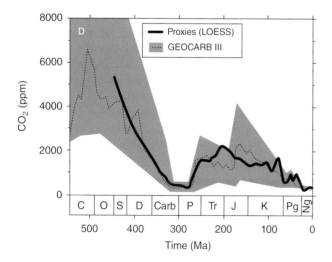

**Figure 9.14** Atmospheric $CO_2$ through the Phanerozoic, comparing the results of GEOCARB-III from reference [54] with the latest proxy measurements of $CO_2$ smoothed to locally weighted regression (LOESS). *Source:* From figure 1D in Ref. [102].

is found both at coarse [10 million year timescales] and fine resolutions up to the temporal limits of the data set (million-year timescales), indicating that $CO_2$, operating in combination with many other factors such as solar luminosity and paleogeography, has imparted strong control over global temperatures for much of the Phanerozoic' [102]. Royer summarized his 2006 results as follows: 'the overarching pattern is of high $CO_2$ (4000+ ppm) during the early Paleozoic, a decline to present-day levels by the Pennsylvanian [Upper Carboniferous, 318-299 Ma ago], a rise to high values (1000–3000 ppm) during the Mesozoic, then a decline to the present-day' [102] (Figure 9.14). Although $CO_2$ was higher during the early Palaeozoic prior to 420 Ma ago than during the Jurassic and Cretaceous, it affected temperature less because the Sun's output was 6% less during the early Palaeozoic [102].

Royer's proxy measures of $CO_2$ confirm that the Carboniferous–Permian glaciation 326–267 Ma ago coincided with low levels of atmospheric $CO_2$ (Figure 9.14) [102]. At the time (2006) he thought that the event occurred in two phases, with no evidence for ice from 310 to 300 Ma, when there was a pulse of high $CO_2$. But earlier, in 2003, Isbell and colleagues had already demonstrated that this late Palaeozoic ice age could be divided into three distinct periods of glaciation (two in the Carboniferous and one spanning the latest Carboniferous–Permian boundary) [103]. Later, in 2008, from field work in Australia, Fielding and colleagues identified eight discrete glacial events (four in the Carboniferous and four in the Permian), each lasting 1–8 Ma, which were separated by non-glacial intervals of equal duration [104]. A range of data suggested that the acme of the Carboniferous–Permian glaciation lay in the Early Permian [105].

As Jesse Koch and Tracy Frank pointed out in 2011 'An Early Permian glacial expansion is also supported by geochemical proxy records of ice volume, pCO₂ (atmospheric carbon dioxide), and temperature. Late Pennsylvanian–Early Permian stable isotopic records from around the world are consistent with a drop in atmospheric pCO₂ in the earliest Permian, as indicated by carbon isotope records from marine carbonates, soil-forming minerals, and sedimentary organic matter' [106]. Nevertheless, in their stratigraphic review of the Carboniferous–Permian boundary around the world, Koch and Frank found widespread evidence for a major fall in sea level then. They attributed the fall to a combination of tectonic effects (collision of Gondwana with the Europe-America block), eustatic fluctuations in sea level caused by Milankovitch-type cycles in ice volume, and the beginning of increased ice volumes across much of Gondwana [106]. The common cause for the fall in sea level was the drawdown in ocean volume caused by growth in the ice sheet, which was large enough to create the largest drop in sea level seen in the Palaeozoic [106]. Transgressive events associated with a subsequent rise in sea level '*suggest an end to major Gondwana glaciation by the mid-Sakmarian [c.290 Ma ago, although]....the persistence of small ice centers through the middle Permian suggests an asynchronous deglaciation of Gondwana after [that]*' [106].

As we know from Chapter 6, the glaciation occurred at a time when Antarctica and Australia drifted across the South Pole [107]. The fall of atmospheric $CO_2$ from high levels in the early Palaeozoic to low values in the Carboniferous (Figures 9.13 and 9.14) paralleled the rise of land plants, which increased extraction of $CO_2$ from the air by photosynthesis [108]. Mosses and liverworts emerged in the Ordovician some 470 Ma ago, and accelerated the chemical weathering of rock and soil, which drew $CO_2$ out of the air [109]. As $CO_2$ fell, a brief glaciation followed near the Ordovician–Silurian boundary about 440 Ma ago (explored below in Section 9.6). Although this glaciation did not seem to correlate with any change in $CO_2$ in the GEOCARB data (Figures 9.13 and 9.14), it did correlate with Berner's GEOCARBSULF-volc volcanic weathering model [55], which showed a fall in $CO_2$ values at around 440–450 Ma ago (Figure 9.7). The vascular land plants dominating today's land vegetation evolved in the Devonian, about 400 Ma ago. They absorbed $CO_2$ through their leaves, and their roots altered the chemistry of rocks and soils 10 times more effectively than did mosses and liverworts, making the substrate even more susceptible to chemical weathering that would extract $CO_2$ from the air [108]. That drove atmospheric $CO_2$ down, and the entrapment of trees in swamps as Laurasia and Gondwana met (ultimately creating massive coal deposits) operated as a further $CO_2$ sink, bringing $CO_2$ levels low enough to lead to the Carboniferous glaciation.

Royer's analyses (Figure 9.13) and the results of Berner's GEOCARBSULF-volc model (Figure 9.7), which agree quite well with each other (except for the Ordovician case discussed above), indicate a rise in $CO_2$ through the Permian towards the Triassic. We can test the model results by examining the data on the ground from an area of critical climate importance – the Arctic – using 'inverse modelling'. What that means in jargon-free terms is attempting '*to determine for a given era just which values of CO₂ reconcile climate model output with some subset of geological data, and then compare this CO₂ prediction with geochemical or other estimates*' [110]. Hyde and colleagues used Siberia in the Carboniferous–Permian between 300 and 360 Ma ago as an example. At the time Siberia was a small continental fragment whose northern tip drifted from latitude 52°N at 360 Ma ago to latitude 72°N by 260 Ma ago. Summer temperatures would have decreased during that time, and it is cool summers that invite glaciations

to develop. The absence of glaciation under such conditions can be used to infer likely levels of $CO_2$, whilst bearing in mind the increase of the solar constant by 1% over this period. The model required $CO_2$ levels to increase with time to keep Siberia ice-free, and geological evidence suggests that Siberia was indeed largely ice-free (or at least lacked an ice sheet) at its northernmost position 260 Ma ago. Hyde and his team found reasonable agreement with Berner's GEOCARBSULF model, suggesting that $CO_2$ was indeed abundant before the end Permian event (Figure 9.7) [110].

As implied in the publications of Dana Royer [101, 102], changes in $CO_2$ through time can be estimated from changes in the carbon isotopic ($\partial^{13}C$) composition of bulk carbonate sediments and foraminifera. Because planktonic organic matter selectively absorbs $^{12}C$, the atmosphere and the ocean become relatively enriched in $^{13}C$ when that $^{12}C$-rich organic matter is trapped in marine sediments or coals. When we find an increase in the $\partial^{13}C$ signal in carbonate sediments, we know that more organic matter rich in $^{12}C$ was trapped in sediments, thereby reducing the abundance of $CO_2$ in the air. We can do this thanks to Jan Veizer, of Ottawa–Carlson University in Canada, winner of the 1995 Logan Medal of the Geological Association of Canada, who compiled with his colleagues a global record of changes in $\partial^{13}C$ ratio in carbonate sediments [111]. There were substantial rises in $\partial^{13}C$, suggesting a lowering of $CO_2$ in the air, at the Silurian–Devonian boundary (around 416 Ma ago) and in the Carboniferous and Permian (around 300–250 Ma ago), the last two being times of significant glaciation (Figures 9.7, 9.13, and 9.14). $\partial^{13}C$ levels in subsequent time periods were modest, like those in the Devonian (~400–370 Ma ago), although there were small peaks in the Late Jurassic (c. 150 Ma ago), Mid Cretaceous (c. 100 Ma ago) (associated with the formation of organic-rich black shales that would have trapped $^{12}C$) and in the mid Cenozoic (c. 30 Ma ago).

Nick Shackleton, whom we met in Chapter 7, developed a clever twist on $\partial^{13}C$ analyses. He recognized that during photosynthesis the phytoplankton, known as the 'grass of the sea', would preferentially extract from seawater the lighter of the two carbon isotopes, $^{12}C$, leaving surface waters relatively enriched in the heavier isotope $^{13}C$. When they sank and decomposed, those phytoplankton would release their $^{12}C$-rich carbon load to deep waters that fed the benthic organisms living on the deep sea floor. As a result, while the carbonate skeletons of the planktonic foraminifera that live near the ocean surface and eat phytoplankton, are slightly enriched in $^{13}C$, the benthic foraminifera living on the deep sea floor are slightly enriched in $^{12}C$. In 1983, he used the gradient in $\partial^{13}C$ between planktonic and benthic species of foraminifera from ancient sediments at the same site to infer changes in the $CO_2$ content of the air for the past 100 Ka [112]. His estimates were close to the $CO_2$ values observed in the Vostok ice core from Antarctica, which we will examine in Chapter 13.

In the marine realm, we can also estimate atmospheric $CO_2$ from $\partial^{13}C$ ratios in complex sedimentary organic molecules known as alkenones. As we saw in Chapter 7, they are long-chained organic compounds produced by a few haptophyte algae (phytoplankton) like *Emiliana huxleyi* [113]. Their carbon isotopic character can be used to estimate atmospheric $CO_2$ back to the Cretaceous [75]. Using this technique, Mark Pagani of Yale University, together with Jim Zachos of the University of Santa Cruz, whom we met in Chapter 7, and colleagues, found in 2005 that the mid to late Eocene air contained 1000–1500 ppm $CO_2$, which declined to modern levels through the Oligocene [75]. Pagani and Zachos used oxygen isotopes to document the cooling of the planet after the middle Eocene [114]. Much the same relationship between cooling and $CO_2$ for the Cenozoic is evident from Figure 9.3.

One of the problems with Pagani's alkenone-based reconstruction of past $CO_2$ levels was that it combined records from different localities to produce a single 'stacked' record. Recognizing that the process of stacking may have introduced regional biases in $CO_2$ trends and magnitudes, Pagani worked with Yi Ge Zhang from Yale, and colleagues, to derive a continuous alkenone-based $CO_2$ record at one deep-sea drilling site in the western equatorial Pacific. Their new $CO_2$ record spans the past 40 Ma (Figure 9.15) [115]. Whilst their results were broadly similar to other alkenone-based $CO_2$ reconstructions, they found $CO_2$ reaching 400–500 ppm in the Middle Miocene. Those values were higher than previous records from alkenones, but

close to those estimated from multiple proxies (Figures 9.3 and 9.16). They suggested that $CO_2$ levels were high during the Middle Miocene Climatic Optimum 17–14 Ma ago, a period of global warmth after which there was a decline in both $CO_2$ and temperature (compare Figures 9.3, 9.15, and 9.16). There is more on variability in the Miocene climate in Chapter 11.

Still in the ocean realm, we can also use boron isotopes to estimate past oceanic acidity (pH) and, from that, atmospheric $CO_2$. As Ravizza and Zachos explain [72], boron has two stable isotopes, $^{10}B$ and $^{11}B$, of which $^{10}B$ is the most abundant (80%). Variations in the two are calculated as the ratio $\partial^{11}B$, which varies with the pH of seawater. Boron is incorporated into calcite, so we can estimate pH from the $\partial^{11}B$ of marine carbonates, something first attempted in 1993 and now more common [76, 77]. Between around 60 and 20 Ma ago, oceanic pH increased from 7.4 to modern values (near 8.2), as atmospheric $CO_2$ fell from about 1000 ppm to 200–400 ppm This is similar to the levels calculated independently from alkenone data (see Figure 9.15). These are early days in the use of $\partial^{11}B$; further work will refine its usefulness.

Carbon isotopes in soils provide an additional control on atmospheric $CO_2$, which enters soils from the air and from the decomposition of soil organic matter, leading to the precipitation of soil carbonates. The $\partial^{13}C$ ratio in soil carbonates depends on temperature and factors such as depth, porosity, diffusion, and rates of respiration. Where these factors are well known and the processes are well understood, the $\partial^{13}C$ ratio should reflect large-scale changes in atmospheric $CO_2$ [116, 117]. $CO_2$ values from analyses of soil carbonates are referred to as 'pedogenic'. The data tend to show a rather broad spread at any one-time interval, and can be unreliable, but they may help where other data are unavailable (see Figure 9.3).

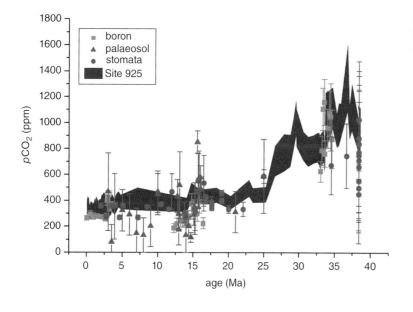

**Figure 9.15** Comparison between different proxy-based $pCO_2$ estimates from the past 40 Ma. *Source:* From figure 4 in Ref. [115].

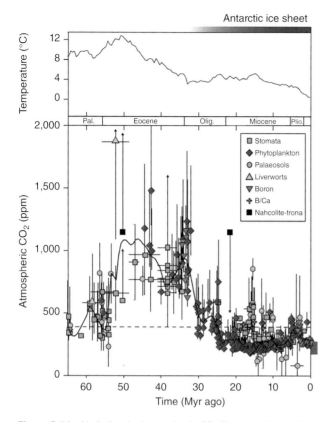

**Figure 9.16** Variation in Atmospheric CO₂ Through the past 65 Ma. Deep-sea temperatures (upper panel) generally track the estimates of atmospheric CO₂ (lower panel) reconstructed from terrestrial and marine proxies and showing error bars. Symbols with arrows indicate either upper or lower limits. Vertical grey bar at right indicates glacial–interglacial CO₂ range from ice cores. The top dark bar represents development of the Antarctic ice sheet. Horizontal dashed line indicates present-day atmospheric CO₂ concentration in 2011 (390 ppm). *Source:* From figure 1 in Ref. [19].

Mineralogy in arid environments can also help to fix past CO₂ levels. The sodium bicarbonate mineral nahcolite (NaHCO₃), which is abundant in the fossil muds of the 'Piceance Creek Member' of the Eocene 'Green River Formation' of the western United States, precipitates only under elevated levels of CO₂. Tim Lowenstein and Robert Demicco remind us that its co-precipitation there with common salt in the form of halite (NaCl) anchors minimum levels of CO₂ at >1125 ppm in the early Eocene atmosphere between 49 and 51 Ma ago [118]. They consider that the nahcolite and halite in the Piceance Creek Member probably precipitated at temperatures of between 25 to 35 °C, as happens today in the Dead Sea, in Jordan. This suggests that atmospheric CO₂ ranged from 1125 to 2985 ppm, averaging about 2100 ppm at that time. In similar but younger deposits, the main mineral is trona (NaHCO₃.Na₂CO₃.2H₂O), which crystallizes at temperatures above 25 °C in modern alkaline saline lakes. Trona

precipitates at lower levels of CO₂ than nahcolite, and is common at today's levels.

Proxy measurements do have their downsides. Alkenone and leaf stomatal proxies for past levels of atmospheric CO₂ have two key deficiencies [119]. First, the sensitivity of both methods declines above 1000 ppm, due to CO₂ saturation. Second, they use calibrations based on living taxa or their nearest living relatives, which may not have precisely the same relationship to their environment as their fossil equivalents did. Independent mineral-based methods like the $\partial^{11}B$ technique may be more reliable. Methods based on soil carbonate are less reliable because of their high sensitivity to soil CO₂, which varies with soil moisture and productivity [117]. The nahcolite–trona transition has only been found in two parts of the world to date: the southwestern USA and Henan Province in China. Still, it provides a well-established, independent boundary for palaeo-CO₂ levels for the Eocene.

The latest twist in the tale comes from a comprehensive re-evaluation in 2014 of leaf–gas exchange, carbon isotopic composition and the stomatal characteristics of fossil leaves, by a team including Dana Royer and David Beerling [120]. Their analysis shows that atmospheric CO₂ was constrained within an envelope ranging from 1000 to 200 ppm, and that this condition arose after the development of forests in the Devonian 390 Ma ago (Figure 9.17). Up until that time atmospheric CO₂ exceeded 1000 ppm, forests

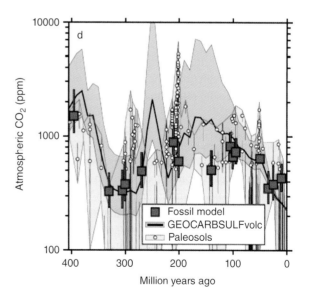

**Figure 9.17** Atmospheric CO₂ concentration for the past 400 Ma, modelled using fossil information for multiple species from published studies (red squares), compared with estimates from the GEOCARBSULFvolc model (black line, with 10–90 percentile error shaded in blue, and palaeosol carbonate values (open circles) with their error envelope in yellow (for palaeosol CO₂ estimates under 500 ppm, only the error ranges are plotted, because this method loses precision at low CO₂). *Source:* From figure 4d in Ref. [120].

then captured and sequestered vast amounts of $CO_2$, and, through their interactions with rocks, enhanced chemical weathering. Nevertheless there were likely to have been short periods when $CO_2$ excursions overcame the climate system's ability to balance them with weathering, for example at the Triassic–Jurassic boundary, where $CO_2$ temporarily exceeded 1000 ppm

In evaluating the significance of palaeo-$CO_2$ data, we need to bear in mind that in some cases, increased burial of $CO_2$ in organic rich deposits may have caused atmospheric $CO_2$ to decline, for example at times when ocean bottom waters became oxygen-free, or anoxic, so preserving substantial accumulations of organic matter on the deep sea floor. These unusual circumstances are designated 'oceanic anoxic events', and we will examine them more closely in Chapters 10 and 11. Accumulation of organic matter at those times transferred $CO_2$ from the air to a sink in sediments, so temporarily reducing the concentration of $CO_2$ in the air.

## 9.6 $CO_2$, Temperature, Solar Luminosity, and the Ordovician Glaciation

Royer realized in 2004 [101] that his data conflicted with the interpretation of past oxygen isotope data by Jan Veizer, who suggested in the year 2000 that global temperature and $CO_2$ were decoupled through time [121]. Why the discrepancy? Royer concluded that in estimating palaeotemperatures from $\partial^{18}O$ ratios, Veizer had not corrected his temperature proxy – the $\partial^{18}O$ ratio of ancient shallow water carbonates – for the changing acidity of seawater – the pH – through time. The dependence of $\partial^{18}O$ ratios on pH, and hence the rationale for such a correction, had been recently spelled out by Richard Zeebe [122, 123]. Once that correction was applied, the temperature and $CO_2$ data agreed (compare Figures 9.7 and 9.18) [101]. This correction disposed of the idea proposed by Shaviv and Veizer in 2003 that the cosmic ray flux rather than the flux of $CO_2$ through time controlled global temperature [124]. Indeed, what Royer's recalculated temperature data show is that there was a fall in temperature close to the Ordovician–Silurian Boundary between about 440–450 Ma ago (Figure 9.18), which ties in rather nicely in time with the fall in $CO_2$ determined from the GEOCARBSULF-volc model (Figure 9.7), suggesting a plausible link between a lessening of $CO_2$ and the known glaciation of that geological boundary (Figure 9.7).

The Shaviv and Veizer temperature reconstruction of 2003 [124] was strongly criticized in 2004 by a group led by Stefan Rahmstorf of the Potsdam Institute for Climate Impact Research, who found that there was no significant correlation between Shaviv's original cosmic ray flux

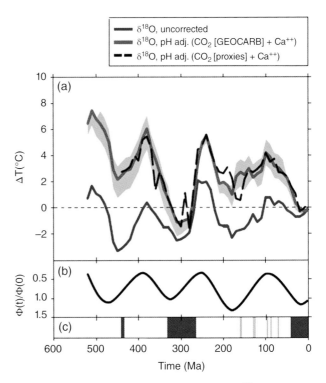

**Figure 9.18** pH correction for shallow-marine $\partial^{18}O$ carbonate curve. (a) blue curve = temperature deviations relative to today calculated from Ref. [124]; red and black dashed curves are adjustments to the blue curve for pH effects; orange band = sensitivity levels for the adjustments. (b) black line = cosmic ray flux relative to present day. (c) intervals of glacial and cool climate. *Source:* From figure 4 in Ref. [101].

(adjusted to give the black curve of Figure 9.18) and Veizer's temperature analysis (blue curve of Figure 9.18) [125]. Rahmstorf explained that '*The explained variance claimed by Shaviv and Veizer [2003] is the maximum achievable by optimal smoothing of the temperature data, and by making several arbitrary adjustments to the cosmic ray data (within their large uncertainty) to line up their peaks with the temperature curve*' [125].

Fifteen years later, in 2019, Scott Wing and Brian Huber from the Smithsonian's National Museum of Natural History (NMNH), along with other palaeotemperature experts, produced an update of the picture of changing global temperature with time through the Phanerozoic (Figure 9.19) [126, 127]. This is an interim product. It shows no significant correlation with Shaviv's suggested cosmic ray flux (black curve in Figure 9.18). Recent analyses of oxygen isotopes by clumped oxygen isotope thermometry, a new technique, confirm that Palaeozoic ocean temperatures prior to 350 Ma ago were warmer that the averages for the Mesozoic or Cenozoic [128], consistent with the observation that the Palaeozoic atmosphere contained more $CO_2$. The clumped isotope technique showed that tropical temperatures during the icehouse climate of

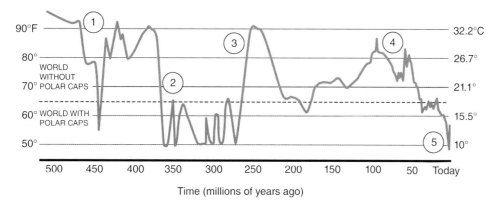

**Figure 9.19** A preliminary global temperature curve showing that marine life diversified in extreme heat (1) before land-based plants absorbed CO$_2$ and polar ice caps formed (2). Volcanoes and erosion swung CO$_2$ levels up and down (3), but mammals evolved in a warm period (4). Now, humans are rapidly warming the climate again (5). From Smithsonian Institution National Museum of National History [128], produced by the 2019 PhanTASTIC Collaboration Team comprising Richard Alley, Kristin Bergmann, Gavin Foster, Ethan L. Grossman, Gregory A. Henkes, Bärbel Hönisch, Brian T. Huber, Linda C. Ivany, Lorraine E. Lisiecki, Dan Lunt, Kenneth G. Macleod, Isabel P. Montanez, Christopher J. Poulsen, Dana Royer, Gavin A. Schmidt, Kathryn E. Snell, Christopher Scotese, Jessica Tierney, Paul J. Valdes, Scott L. Wing and James C. Zachos.

the Carboniferous were like those of today (25–30 °C) [128], as suggested by Figure 9.19. As we shall see later, CO$_2$ was probably in the range 1000–2000 ppm even in the Palaeozoic, not at the high values originally calculated by the GEOCARB model (Figure 9.7).

This latest interpretation of Phanerozoic palaeotemperatures improves upon the one inferred by Veizer and Prokoph in 2015, based on their compilation of oxygen isotopes through time, in which they estimated that temperatures of Cambrian time were in the vicinity of 30–60 °C [129], whereas Figure 9.19 suggests that they were about 32–35 °C. The solution to this conundrum may lie in a change through time of some 6% in the oxygen isotopic composition of seawater, which would keep tropical surface temperatures in the 20–32 °C range suitable for marine life [129].

Various lines of geological and palaeontological evidence confirm, as we saw in Chapter 6, that the Earth was much warmer than today during most of Jurassic and Cretaceous time, when CO$_2$ levels were high (Figures 9.7 and 9.13). Nevertheless, there is increasing evidence at those times for cool pulses 1–3 million years long. Several of them correlate with periods of low CO$_2$. Royer noted that '*scrutiny of the paleotemperature record reveals that there were multiple discrete cool events from at least the early Jurassic to late Cretaceous ... each only lasting <3my [Ma] in length. Thus the most plausible climate history for the Jurassic and Cretaceous is one dominated by warm climates but punctuated by brief (<3my [Ma]; typically <1 my [Ma]) cool pulses*' [102]. We will investigate that level of climate variability later.

Independent confirmation of Royer's results comes from many sources. For example, a team led by Gregory Retallack from the University of Oregon used stomatal data from fossil leaves and $\partial^{13}$C analyses to show that there were significant

fluctuations in the levels of atmospheric CO$_2$ in the Late Permian and Early Triassic of Australia [130]. They found that '*Successive atmospheric CO$_2$ greenhouse crises coincided with unusually warm and wet paleoclimates for a paleolatitude of 61°S ... [which] punctuated long-term, cool, dry, and low-CO$_2$ conditions, and may account for the persistence of low diversity and small size in Early Triassic plants and animals*' [130]. The short-lived pulses of high CO$_2$ may have been caused by the intrusion of coals by flood basalts in Asia, they thought. Analyses of stomatal frequencies in the leaves of different plant groups from locations in Greenland and Ireland across the Triassic–Jurassic boundary suggested a rise of CO$_2$ from around 1000 ppm to 2000–2500 ppm, agreeing with published $\partial^{13}$C records [131]. In addition, CO$_2$ values between 853 and 1033 ppm were calculated from stomatal data from leaves from the Messel Formation in Germany, which was deposited during the warm Eocene period [132].

A similar compilation of results from carbon isotopic analyses of fossil plant remains (liverworts from five continents) and from models was presented in 2007 for the Mesozoic and early Cenozoic by a team from the University of Sheffield led by Benjamin Fletcher, and including David Beerling and Bob Berner [133]. Their results show that atmospheric CO$_2$ rose from ~420 ppm in the Triassic (200 Ma ago) to a peak of ~1130 ppm in the Mid-Cretaceous (100 Ma ago), then declined to ~680 ppm by 60 Ma ago, coincident with Mesozoic climate change as determined from oxygen isotopes from marine carbonate fossils. Evidently climate and CO$_2$ were not decoupled during this interval, contrary to Veizer's suggestion [121]. The team found that '*These reconstructed atmospheric CO$_2$ concentrations drop below the simulated threshold for the initiation of* *glaciations on several occasions and therefore help explain the occurrence of cold intervals in a "greenhouse world"* [133]. In view of their close correlation to the $\partial^{18}O$ record, fossil bryophytes (non-vascular plants, such as mosses) provided a more accurate representation of past levels of $CO_2$ than did leaf stomatal indices or palaeosols, the latter being the least accurate. Given this association, they said, we have a plausible *'explanation for Jurassic and Cretaceous climates without the need to invoke the influence of cosmic rays on cloud cover and planetary albedo'* [133].

It is obvious from Figure 9.3 that there was a fall in $CO_2$ (represented by proxies) at the same time as a drop in the temperature of bottom water ($\partial^{18}O$) 34 Ma ago as the first great Antarctic ice sheet formed at the end of the Eocene. This went along with a rise in $\partial^{13}C$ in benthic foraminifera and a deepening of the CCD, both indicative of a lessening of atmospheric $CO_2$ (Figure 9.3) [12]. That is an important observation because, as Mark Pagani and colleagues pointed out in 2011, the *'onset of Antarctic glaciation reflects a critical tipping point for Earth's climate and provides a framework for investigating the role of carbon dioxide ($CO_2$) during major climate change'* [20]. Prior to Pagani's 2011 study, the widely spaced and rather sparse data across that boundary in deep-sea sediments conflicted with one another. To resolve this problem, Pagani and colleagues carried out detailed $\partial^{13}C$ analyses of alkenones across the Eocene–Oligocene boundary from a number of deep-sea cores from a variety of environments. They found that *'low-latitude records show a persistent $CO_2$ decline beginning about two million years before the onset of rapid cooling 33.7 million years ago …that continues just beyond the climate event'* [20]; furthermore *'the partial pressure of atmospheric $CO_2$ fell from 1200 to 1000 ppm to 700 to 600 ppm …consistent with model estimates for a threshold $CO_2$ level required for rapid Antarctic glaciation'*. They concluded *'that the available evidence supports a fall in $CO_2$ as a critical condition for global cooling and cryosphere evolution ~34 million years ago …[and that the data] implicate the role of silicate weathering rates over the influence of short-term organic-carbon burial rates as the primary cause for long-term change in atmospheric carbon dioxide'* [20].

Further evidence has emerged to support the geochemical theory. In 2008 Richard Zeebe of the University of Hawaii, and Ken Caldeira of the Carnegie Institution in Stanford California, set out to test the notion that *'atmospheric $CO_2$ concentrations over millions of years are controlled by a $CO_2$-driven weathering feedback that maintains a mass balance between the $CO_2$ input to the atmosphere from volcanism, metamorphism and net organic matter oxidation, and its removal by silicate rock weathering and subsequent carbonate mineral burial'* [134]. Using the $CO_2$ concentrations from ice cores they found no more than a 1–2% imbalance between the supply and uptake of $CO_2$ during the past 610 Ka, providing *'support for a weathering feedback driven by atmospheric $CO_2$ concentration that maintains the observed fine mass balance'* [134]. If this process did not take place, the release of $CO_2$ from volcanism and metamorphism would double the amount of $CO_2$ in the atmosphere within less than 600 Ka. Those emissions are balanced by weathering acting as a stabilizing feedback controlled by the concentration of atmospheric $CO_2$, much as Chamberlin concluded in the 1890s.

New data and ideas are constantly arriving. In 2014, Flip Froehlich and Sambuddha Misra discovered that the isotope of Lithium, $^7Li$, in deep-ocean sediments could be used as an indicator of the weathering of continental rocks [135]. They found that the $^7Li$ chemistry of the Palaeocene–Eocene ocean *'indicates that continental relief during this period of the Early Cenozoic was one of peneplained (flat) continents characterized by high chemical weathering intensity and slow physical and chemical weathering rates, yielding low river fluxes of suspended solids, dissolved cations, and clays delivered to the sea'* [135]. Only when mountain building was reinitiated in the Oligocene–Miocene, following the collision of India with Asia, *'did continental weathering take on modern characteristics of rivers with high suspended loads… with much of the cations released during weathering being sequestered into secondary clay minerals'* [135]. Hence *'The early Cenozoic climatic optimum was a result of increased supply of carbon dioxide to the ocean–atmosphere system as well as diminished removal of carbon dioxide from the atmosphere through the weathering of silicate rocks'* [135]. What we are seeing is the absence of the negative feedback mechanism provided by weathering, rather than some steady increase in the supply of $CO_2$. The absence of a $CO_2$ weathering sink explains how the hot Eocene climate persisted for millions of years. This hypothesis also *'provides another piece of circumstantial evidence in support of the Late Cenozoic Uplift-Weathering Hypothesis'* [135].

Nevertheless some concern was expressed that the chemical weathering associated with the acceleration of mountain building over the past 15 Ma should have lowered atmospheric $CO_2$ by more than observed. Li and Elderfield [136], and Torres, West and Li [137], concluded recently that the loss of $CO_2$ by interaction with silicates would have been balanced to some extent by the weathering of iron sulphide (pyrite). Oxidation of sulphides produces sulphuric acid, which then attacks carbonate minerals ($CaCO_3$) to release $CO_2$. The balance between these two forms of chemical weathering explains better the observed patterns of atmospheric $CO_2$ over the past 15 Ma [136, 137].

Bill Hay recently brought up an additional point that has a bearing on the long-term cycling of $CO_2$ [14].

Coccolithophores and planktonic foraminifera that build their skeletons out of calcium carbonate evolved in *shallow* seas during the Jurassic and spread throughout the open ocean during the Cretaceous. As a result, whereas previously carbonate sediments were primarily deposited in shallow seas, they are now deposited right across the *deep ocean*, as carbonate oozes. Eventually these deep-ocean carbonates will be subducted at ocean trenches, and their trapped CO$_2$ will return to the atmosphere via volcanic eruptions that tap the sedimentary loads of down-going subducted slabs of ocean crust. But that is tens to hundreds of millions of years hence; for instance, today there is virtually no subduction on the margins of the Atlantic, where most deep-sea carbonates are currently being deposited. The processes of deposition and destruction have not yet reached equilibrium. That may partly help to explain the slow decline in atmospheric CO$_2$ from the Cretaceous into the Cenozoic.

Before concluding, we need to consider the role of solar luminosity in Earth's climate history. The topic was recently considered at length by Gavin Foster of Southampton's National Oceanography Centre, along with Dana Royer from Wesleyan University and Dan Lunt of Bristol University [138]. First they used a large new compilation of proxy data to document the change in CO$_2$ through time over the past 450 Ma (Figure 9.20).

Then they considered the effects of changing solar luminosity. Back in 1972, Carl Sagan and George Mullen had deduced that because the oldest dated sediments around 3.8 billion years old indicated the presence of running water, the early Earth's climate must have been warm – possibly as warm as or warmer than today [139]. However, given that the Earth's black body temperature was close to 255 K, and that the Sun was a typical 'main sequence' star in which solar luminosity (basically the Sun's output of energy) increased with time, the Earth's surface should have been frozen at least until about 2 billion years ago. This dilemma is known as the 'faint young Sun paradox'. The only way to warm the early Earth up enough for running water to exist (they thought) was to have an atmosphere containing abundant greenhouse gases, and they calculated that the most likely combination in the early Earth's atmosphere was ammonia (NH$_3$) and water vapour [139].

By 1979, Tobias Owen and Robert Cess of the State University of New York's Laboratory for Planetary Atmospheres Research, and Ram Ramanathan of the National Center for Atmospheric Research, at Boulder, had deduced that the early atmosphere was much more likely to have contained an important mix of CO$_2$ and water vapour, rather than ammonia [33]. They calculated how much CO$_2$ would be required in that early atmosphere to keep global temperatures at a level appropriate for running water over the past 4.5 billion years, given that solar luminosity was calculated to have increased by 25% over that period (Table 9.1).

In 2012, Feulner reviewed the various data in support of the faint young Sun concept and its impact on climate [140]. The latest evidence suggested that the Sun's luminosity had increased by 30% over the past 4.5 billion years, and that maximum ocean temperatures had been much the same for the past 3.5 billion years as they were now. This indicates the persistence of a temperate climate through time, much as suggested by Owen and colleagues [33]. '*The solution to the faint young Sun problem seems to lie in ... the concentration of greenhouse gases in the early atmosphere*', resolving the inconsistency between low solar luminosity (implying freezing temperatures) and signs of liquid water [140]. Calculations suggest that pCO$_2$ reached

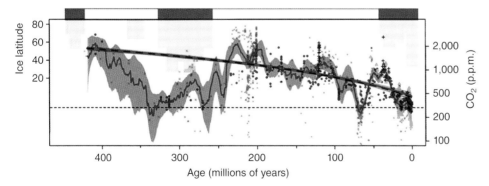

**Figure 9.20** Temporal evolution of climate and atmospheric CO$_2$. CO$_2$ is depicted as blue circles, from leaf stomata; pink crosses, from $\partial^{13}$C in soil carbonate; green triangles, from boron isotopes in foraminifera; dark blue dots, from liverwort $\partial^{13}$C; dark blue crosses, from $\partial^{13}$C of alkenones. Blue wiggly line = most likely LOESS fit through the data, with 68 and 95% confidence intervals shown as dark and light grey bands. Red line is a linear best fit (curved due to log-scale for y-axis). Black line is least squares fit through the LOESS best fit (blue line). Dashed horizontal line is pre-industrial CO$_2$ (278 ppm). Icehouse time intervals are indicated by a black band and greenhouse intervals by a white band at top. Blue vertical bars = latitudinal extent of continental ice deposits. *Source:* From figure 1 in Ref. [138].

**Table 9.1** Predicted global surface temperature, $T_s$, throughout the evolution of the Earth's atmosphere. S = the solar constant (incoming radiation at surface).

| Time (10⁹ year BP) | S (W/m²) | PCO₂ (bar) | T_s (°K) |
|---|---|---|---|
| 4.25 | 1.039 | 0.31 | 310 |
| 3.5 | 1.096 | 0.070 | 296 |
| 3.0 | 1.133 | 0.33 | 293 |
| 2.5 | 1.171 | 0.018 | 292 |
| 2.0 | 1.209 | 0.0086 | 290 |
| 1.5 | 1.247 | 0.0029 | 288 |
| 1.0 | 1.284 | 0.00065 | 286 |
| 0.5 | 1.322 | 0.00032 | 287 |
| 0.0 | 1.360 | 0.00032 | 290 |

*Source:* From table 1 in Ref. [33].

c. 0.06 bar (or about 200 times preindustrial levels) for the Early Archaean (3.5 billion years ago), and c. 0.01 bar (roughly 30 times preindustrial levels) for the Late Archaean (2.5 billion years ago) [140], much like the estimates of Owen (Table 9.1). Nevertheless, factors such as the faster rotation of the early Earth and the smaller area of continent still need to be taken into consideration as possible factors [140]. An early atmosphere richer in CO₂ would also be heavier than our present one, and its increased mass would have led to slight increases in temperature through increases in the density of the air (cramming more gas molecules together heats the air), as pointed out by Chemke et al. in 2016 [141]. Increasing the density of the air also leads to broadening of the lines in the CO₂ spectrum, which further prevents the escape of infra-red radiation (heat) to outer space, hence further warming the atmosphere [140, 141]. Chemke's data suggested that having running water on the early Earth required either a much increased concentration of greenhouse gases, or a much increased atmospheric mass (for which there is no supporting evidence).

Foster and his team converted the CO₂ values from Figure 9.20 into climate forcing values, and added the solar forcing for the past 450 Ma (Figure 9.21a) [138]. Realizing that the actual climate would be a blend of these two forcing functions, they merged the two to calculate the combined forcing due to changes in both solar luminosity and CO₂ (Figure 9.21b).

Foster and his team noted that *'The marked "double-hump" pattern of the CO₂ reconstruction that is also common to other compilations [Figure 9.20]... is likely caused by changes in the inputs and outputs of CO₂ in response to the supercontinent cycle. For example, the low CO₂ during the Carboniferous [c.300 Ma ago] and during the later parts of the Cenozoic [last 65 Ma] was likely a result of a reduced*

*volcanic flux and/or enhanced silicate weathering, at least in part due to higher continental relief during [a] supercontinent construction phase. During the intervening greenhouse intervals the reverse was likely true (high volcanic flux and/or low silicate weathering due to low relief). An additional factor in the decline in CO₂ was the overall enhancement of silicate weathering due to the expansion of the terrestrial biosphere over the last 400 [Ma] ... It is probably this expansion, along with the silicate weathering-negative feedback (facilitated by sufficiently active plate tectonic regime), that was the key in keeping [the combined CO₂ and solar forcing] relatively constant over the long term ....[the] new CO₂ record largely agrees with the GEOCARB carbon cycle model that incorporates these long-term changes in silicate weathering ... This serves to further underscore the importance of the rise of the terrestrial biosphere, along with plate tectonics and silicate weathering, in ensuring climate stability through time and provides additional support for our broad understanding of the Earth's long-term carbon cycle as encapsulated by the GEOCARB model'* [138].

With all of the above background in mind we can now turn to the enigma of the late Ordovician glaciation around 440 Ma ago, mentioned above in passing. I use the word enigma because this glaciation occurred before the arrival of the vascular plants that extracted CO₂ from the air via stomata, and had roots that would accentuate chemical weathering, which explains why the CO₂ record of Foster et al. does not begin until after the Ordovician glaciation (Figure 9.20). Using his geochemical models Berner related this glaciation to a fall in CO₂ caused by the weathering of basaltic volcanic rocks (Figure 9.7) [55]. Nardin and colleagues concluded from a study of oxygen isotopes that there had been a large, long cooling trend during the Ordovician, followed by an abrupt cooling during the late Ordovician glaciation [142]. They used a combined climate and biogeochemical models to suggest that the concentration of CO₂ in the atmosphere had halved during that period, most probably due (35%) to the weathering of fresh volcanic rocks on the continents, combined with the movement of the continent through the inter-tropical convergence zone (ITCZ). Their data suggested that global temperatures likely fell to around 13.5 °C during the glacial interval [142].

Various strands of evidence suggest that the (hypothetical) fresh volcanic rocks were flood basalts created within a Large Igneous Province (LIP) [143], which may have been centred on Argentina, where volcanic outpourings appear to have covered an area at present as large as 27 000 km³ [144]. However, Courtillot reported in 2007 that the postulated plateau basalt was *'as yet unrecognized'* [145], and the basalts identified in Argentina were largely pillow basalts [144] that (in my view) may simply have been parts of the

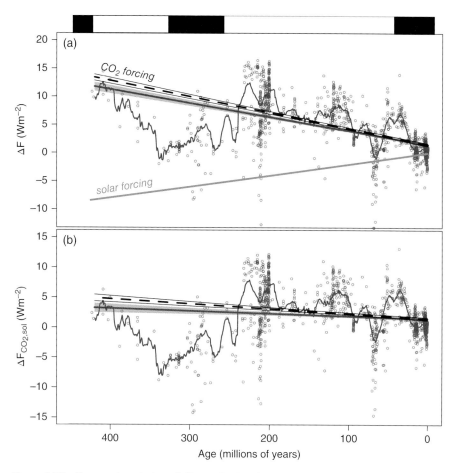

**Figure 9.21** Temporal evolution of climate forcing for the past 450 Ma: (a) $CO_2$ data (blue circles) and LOESS best fit (blue line) converted to $CO_2$ forcing in Watts/m². Red line is a linear best fit and 95% confidence interval for least squares regression through the $CO_2$ forcing data. Orange line is solar forcing due to changing luminosity. (b) combined forcing in W/m² from $CO_2$ and solar luminosity, with LOESS best fit (blue line), least squares fit through the LOESS best fit data (black line), and red line as linear best fit and 95% confidence interval for least squares regression through the combined forcing data. Icehouse time intervals = black band, and greenhouse intervals = white band. *Source:* From figure 2 in Ref. [138].

original deep sea floor accreted onto South America during subduction along the Gondwanan margin. Moreover the area of basaltic rock in Argentina is much smaller than in other LIPs. Nevertheless, indirect evidence for a volcanic trigger for the Late Ordovician glaciation has been found in the form of enrichments in mercury (Hg) in marine strata of late Ordovician age [146]. Volcanism is the major natural source of Hg, which can be used as a proxy for ancient volcanic events, and enrichments in this element are common in marine strata deposited during other Phanerozoic mass extinction events associated with known LIPs [146]. We discuss LIPs and extinctions in Chapter 10.

Figure 9.22, from Lenton and others in 2012, illustrates some of the features of the climate changes leading up to the Ordovician glaciation, including the aforementioned decline in oxygen isotopes ($\partial^{18}O$) [109]. Lenton suggested that the fall in $CO_2$ that led to the cooling was at least in part caused by the expansion of the first land plants [109].

Land plants are able to break down silicate minerals by chemical weathering to release essential nutrients, but the plants of most of the Ordovician were probably non-vascular, like modern mosses. Could they weather silicates? Lenton's experiments showed that they could, and that this could have reduced temperatures by 1.4°–7.2 °C and sustained a cold interval some 10 Ma long, aided by the origination and expansion, during the Hirnantian, of the first vascular plants (Figure 9.22), whose roots would help attack silicate minerals [109]. Lenton agreed that the fall in the strontium isotope ratios in seawater ($^{87}Sr/^{86}Sr$) suggested an increase in the abundance and weathering of volcanic rock (Figure 9.23), as previously noted by Berner [55]. While that chemical weathering of silicates acted as a sink for $CO_2$, Lenton's team argued that it would also have released phosphorus, which, on transfer to the ocean, would have stimulated the growth of marine plants and their subsequent rapid burial, helping to further draw

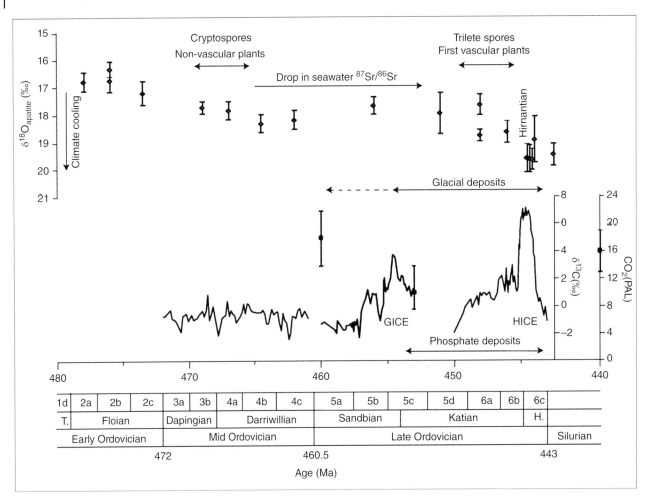

**Figure 9.22** Global changes during the Ordovician period. (i) appearances of non-vascular and vascular plants; (ii) timeline of climate change as recorded by conodont $\partial^{18}O$ values (diamonds, left-hand scale); (iii) the appearance of non-vascular plants is followed by an increase in the abundance and weathering of volcanic rock, as recorded by seawater $^{87}Sr/^{86}Sr$; (iv) the weathering seems to coincide with a drop in atmospheric $CO_2$ concentrations (solid squares, and outer right-hand scale), though proxy-based estimates are scarce and highly uncertain; (v) glacial deposits appear early in the Late Ordovician and increase in frequency as it progresses; (vi) phosphate deposits appear then in marine settings reflecting the increase in supply of phosphorus to the ocean by weathering; (vii) the carbon isotope profiles (wiggly lines) show that two prominent positive $\partial^{13}C$ excursions mark the Late Ordovician (GICE and HICE – see text for details). T = Tremadocian age; H = Hirnantian age. *Source:* From figure 1 in Ref. [109].

down atmospheric levels of $CO_2$. These burials were episodic and recorded by the locking up of the $^{12}C$ loved by marine organisms, leaving the ocean and its carbonates relatively enriched in $^{13}C$, as shown by the positive $\partial^{13}C$ excursions in the late Ordovician and labelled either GICE (Gutenberg isotopic carbon excursion) or HICE (Hirnantian isotopic carbon excursion) (Figure 9.22). At the same time, the decomposition of the growing amounts of marine plankton would have tended to make subsurface waters anoxic. Indirectly, therefore, the activities of land plants in supplying nutrients to the ocean may have stimulated the processes that eventually killed off their marine cousins [109]. Solar luminosity would have been lower at the time (Figure 9.21), which means that a smaller fall in $CO_2$ would have been required to create a glaciation than would be the

case later on when solar luminosity was higher. Modelling these various interactions, Lenton's team predicted the development of climatic conditions shown in Figure 9.23.

These findings help to confirm Berner's notion that the chemical weathering of mid to late Ordovician basalts helps to explain the origins of the Hirnantian Glaciation at the end of the Ordovician, c. 445 Ma ago, as suggested by his GEOARBSULF-volc model (Figure 9.7) [55]. Pogge von Strandmann and colleagues agree, based on a new method for identifying Earth's 'weathering thermostat' [147]. In 2017 they used lithium isotopes ($\partial^7Li$) to assess the impact of silicate weathering through the late Ordovician into the early Silurian, finding a positive excursion in their $\partial^7Li$ measurements that indicated a decline in silicate weathering as the world cooled. They deduced that the glaciation

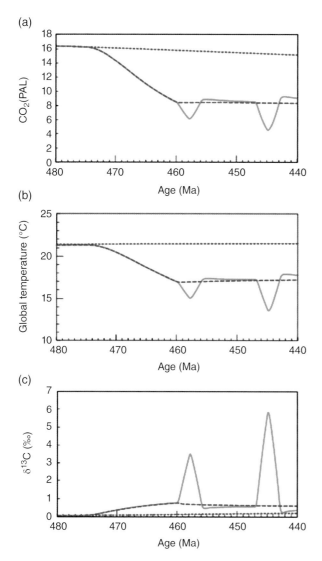

**Figure 9.23** Model results. Predicted Ordovician variations in (a), atmospheric $CO_2$ relative to PAL (present-day atmospheric levels); (b), global temperature; and (c), $\delta^{13}C$ of marine carbonates; for: baseline model with Ordovician geological forcing and changing solar luminosity only (red dotted line), including enhancement of silicate weathering by non-vascular plants (blue dashed line), and adding transient enhancement of phosphorus weathering by early plants (green solid line). *Source:* From figure 3 in Ref. [109].

was initiated by a lessening of mid-Oligocene basaltic volcanic activity and associated outgassing of $CO_2$, '*which triggered abrupt global cooling, and much lower weathering rates*' [147]. The reader needs to beware of confusing weathering flux (which one might think of as driven by humidity and temperature) with weatherability (which depends on the area of exposed weatherable silicate minerals and rocks – as per the earlier references to Kump and Arthur).

The glaciation was associated with the second largest mass extinction in the geological record (loss of 85% of marine species [148]), and caused an ice sheet to grow over

parts of Gondwana (Figure 9.24) [149]. At that time the south pole lay near the coast of Senegal, just inland from Dakar, and the ice sheet extended from Arabia to Amazonia, accompanied by a small ice cap over the mountains of southernmost South Africa. In the Early Ordovician (480 Ma ago), the pole was located just off the coast of Gondwana, near what is now the position of Tunis; presumably the higher abundance of $CO_2$ in the atmosphere at that time (Figure 9.7) kept that edge of Gondwana free of the formation of an ice sheet that would form later as the climate cooled. Analyses of Palaeozoic planktonic fossils (chitinozoans) from the Hirnantian allows us to reconstruct the position of the polar front during glacial times (Figure 9.25) [150].

According to Pogge von Strandmann, the glaciation is likely to have proceeded rapidly: '*once global temperature dropped to a critical threshold, northern high latitude sea-ice expanded abruptly, causing a further decrease in global temperatures and a rapid expansion of an ice sheet on the southern polar land surfaces ... these ice-albedo and heat transport feedbacks operate far faster than the long term carbon cycle*' [147] The cooling reduced chemical weathering, eventually allowing $CO_2$ to gradually build up in the atmosphere, warming it enough to melt the ice, and reverse the $\partial^7 Li$ trend. These findings are broadly consistent with those of an earlier review, which found that there were two main cycles of glacial advance during the Hirnantian, each associated with an extinction caused by anoxic conditions in bottom waters [151]. Once again, we need to remember that the reduction in weathering could be driven by either a fall in temperature and humidity, or a decrease in the weatherability (extent) of exposed silicate rock.

Pogge von Strandmann's lithium isotope research is significant in providing the first specific evidence for the existence of chemical weathering as a control on $CO_2$ and hence on temperature. In his Ordovician example (above) he found evidence for the slowing of weathering as the planet cooled [147], whilst in research on Ocean Anoxic Event 2 (OAE2), at about 93.5 Ma ago, when the climate was warm, he found increased levels of weathering associated with $CO_2$ removal [152]. This may mean that we can use lithium isotopes to investigate Earth's thermostat back through time. As a caveat, we have to bear in mind the possibility that the abundance of lithium may also depend on the amount of rock exposed at the surface, and there was a significant amount of mountain building during the Ordovician. The weathering of those newly exposed mountains would also tend to bring down $CO_2$ levels.

Hammerlund and colleagues agreed that the major positive $\partial^{13}C$ excursion in the middle Hirnantian most likely represented the burial under reducing (anoxic) conditions of large amounts of $^{12}C$-rich marine carbon, which would

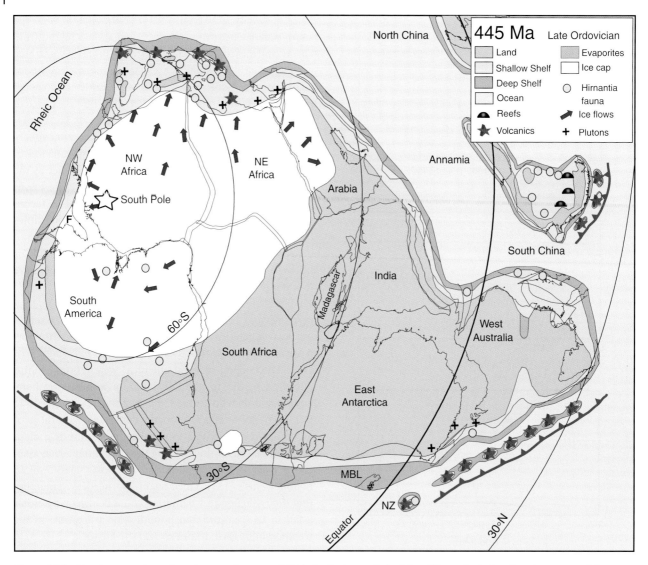

**Figure 9.24** Gondwana and nearby palaeocontinents at 445 Ma in the latest Ordovician (Hirnantian), showing the extent of the glacial icecap, and also the distribution of the Hirnantian brachiopod Fauna (yellow dots). MBL = Marie Byrd Land, Antarctica; NZ = New Zealand. The strings of island arcs shown are diagrammatic. Solid blue lines are subduction zones, with teeth on their downward sides. *Source:* From figure 6.9 in Ref. [149].

have drawn down atmospheric $CO_2$ [148], consistent with the findings of Lenton [109] and Pogge von Strandmann [147]. Hammerlund argued that decomposing marine organic remains intensified reducing conditions in the ocean, leading to anoxia, and that a post-glacial rise in sea level would have brought oxygen depleted water onto the continental shelf, where carbonate sedimentation was focused, leading to the massive extinction of marine organisms [148].

More recently, in 2018, Shen and colleagues used nitrogen isotopes in bulk sediments, along with the degradation products of chlorophyll from the Vinni Creek Formation of Nevada to show that eukaryotic algae dominated marine export production leading up to the Hirnantian glaciation

[153]. Shen thought that the productivity increased due to an increase in the supply of phosphorus to the ocean, stimulated by increased terrestrial weathering by early land plants, and that an increase in algal cell size probably increased the rate of sinking of dead organic matter through the biological pump. The coincidence of this marine community shift with an extended marine transgression increased the burial of organic matter in the late Katian (just prior to the Hirnantian), drawing down atmospheric $CO_2$. The ensuing Hirnantian glaciation reduced shallow water habitats, leading to a loss of 26% of families and 49% of genera of marine biota, making this one of the 'Big Five' biological extinctions [153]. Global temperature probably fell by about 4 °C from the late Katian to the late Hirnantian (446–444 Ma) [153].

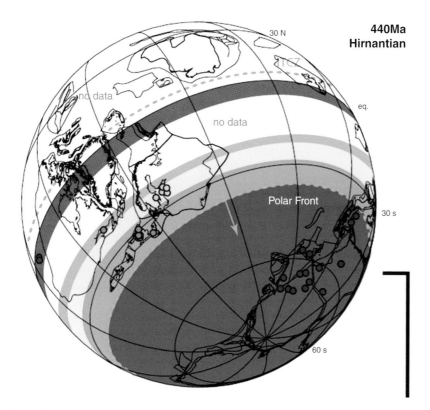

**440Ma
Hirnantian**

**Figure 9.25** Compilation globe showing the inferred Hirnantian climate belts as inferred from chitinozoan distribution data. Dark blue = polar; light blue = subpolar; yellow = transitional; orange = subtropical; red = tropical; dashed lines = intertropical convergence zone and polar front. *Source:* From figure 24.5 in Ref. [150].

This discussion begs the question, when was the land surface colonized by plants? The latest research suggested, in 2018, that embryophytes originated in the mid-Cambrian, whilst tracheophytes evolved in the Late Ordovician to Silurian [154]. Embryophytes are complex multicellular eukaryotes with specialized reproductive organs, which are thought to have evolved from a group of complex green algae. They include liverworts, bryophytes, mosses, and hornworts. Tracheophytes are vascular plants containing lignified tissues (the xylem) for conducting water and minerals throughout the plant. Vascular plants include the club-mosses, horsetails, ferns, gymnosperms (including conifers), and angiosperms (flowering plants).

Following the Hirnantian (latest Ordovician) glaciation, conditions stayed cool during the early Silurian Llandovery Epoch, and the early retreat of Gondwana's south polar ice sheet was punctuated from the Hirnantian into the Llandoverian by several episodes of ice re-advance accompanied by changes in sea level caused by fluctuations in ice volume [155]. These were part of what Page and others call the Early Palaeozoic Icehouse, which contained seven glacial maxima, lasted for some 30 Ma, and extended from the Ordovician (c. 450 Ma ago) through to Wenlock times in the Silurian (c. 420 Ma ago) [156].

How cold did it get? Seth Finnegan and his team use a new technique, 'clumped isotope palaeothermometry' to constrain ocean temperatures, and thereby estimate ice volumes through the Late Ordovician–Early Silurian glaciation [157]. Tropical near-surface ocean temperatures averaged 33°–37°C '*except for short-lived cooling by 5°C during the final Ordovician stage*' [157].

It seems likely, given these constraints, that temperatures were significantly lower during the Hirnantian glaciation than suggested by Royer in Figure 9.20. However, it seems equally likely that the CO₂ values at the time were above those associated with later glaciations (Figure 9.7), because solar luminosity was lower at the time (Figures 9.20 and 9.21). Consistent with that assumption, palaeobiogeographic analyses of the distribution of planktonic graptolites suggests that despite the relatively high levels of atmospheric CO₂ the climate of the late Ordovician was like that of the present day [158]. Indeed it seems highly likely that the combination of high CO₂ and low solar luminosity prevented even the Early to Middle Ordovician from being in a super-greenhouse state [158].

Development of a glaciation implies significant changes to ocean circulation of the kind addressed by Pat Wilde and William Berry of the University of California, Berkeley, in the mid-1980s [159]. They remind us that during warm periods (which we may think of as Fischer's greenhouse climates), deep ocean circulation tends to be driven by the sinking of dense, warm, salty water from mid-latitude and tropical continental shelves; these waters, being warm are relatively poor in oxygen, which may encourage development of an extensive oxygen-poor oxygen minimum zone and even anoxic deep waters. In contrast, during glacial periods (Fischer's icehouse climates) deep ocean circulation is characterized by the sinking at high latitudes of dense, cold, surface water rich in oxygen, which fills the deeper parts of the ocean and can push towards the surface the deep anoxic and nutrient-rich bottom water representative of the previous warm climate. The changing climate also causes sea level to rise in warm periods and fall in cold periods, creating the potential for oxygen poor subsurface waters from cold periods to transgress onto the continental shelves during warm periods, there causing extinctions. Much the same process, with oxygenated deep waters during glacial periods, was identified by Page and others [156]. The sea level rises associated with transgressions of anoxic water onto the continental shelves help to explain the extinctions of the Late Ordovician [156, 159, 160]. A general fall in sea level towards the Hirnantian glacial period is well documented by Haq and Schutter, who indicate the presence there of some large falls and rises [161], which were most likely associated with the extinctions.

In 2016, Lenton and colleagues also demonstrated that the development of terrestrial life towards the end of the Ordovician and into the following Silurian was likely to have been responsible for the development of a relatively oxygen-rich atmosphere [162]. They argue that the evolution of land plants, which colonized the land surface from ~470 Ma onward, caused global terrestrial net primary productivity to reach some 30% of today's level by ~445 Ma ago, leading to a significant increase in chemical weathering and a corresponding increase in the burial of organic matter, causing a corresponding rise in $O_2$ to present atmospheric levels by ~400 Ma ago. The rise in carbon burial led to a sustained increase of +2‰ in the carbon isotope ($\partial^{13}C$) record, indicative of the entrapment of substantial amounts of $^{12}C$-rich organic matter in marine sediments. At more or less the same time, between the Early Silurian and Middle Devonian (435–392 Ma ago), as indicated by the isotopic record of molybdenum (Mo) in marine sediments, the state of the deep ocean shifted from widespread anoxia to widespread oxygenation, consistent with an increase in the level of atmospheric oxygen [162]. '*The evolution of land plants is the obvious candidate, with the first nonvascular plants (ancestors of extant mosses, liverworts, and hornworts) colonizing the land in the Middle to Late Ordovician (<470–445 Ma ago), followed by the first vascular plants in the Silurian*

(<445–420 Ma ago) and Early Devonian (<420–390 Ma ago)' [162]. The presence of substantial fluctuations in the $\partial^{13}C$ levels between 455 and 420 Ma ago could be interpreted to suggest that during the transition from the less well oxidized oceans of the Ordovician to the well oxygenated oceans of the Silurian the system was unstable, oscillating between more and less well oxygenated states. Lenton's research implies (i) that early land plants were much more productive and globally extensive than has been previously assumed; (ii) that the development of an oxygen-rich atmosphere instigated new fire-mediated negative feedbacks that have helped to stabilize atmospheric $O_2$ concentration up to the present day; and (iii) that this atmosphere enabled '*the subsequent evolution of large, mobile, intelligent animals with a high respiratory oxygen demand, including ourselves*' [162]; a case of the evolution of plants influencing not only climate but also the rise of more complex life.

A surprising feature of the Late Ordovician glaciation is its association with an increase in $\partial^{13}C$, the opposite of what occurred in association with the glaciations of the Neoproterozoic, as we will see in Chapter 10. Conceivably this reflects the fact that the Neoproterozoic atmosphere contained much less oxygen, primitive land plants were more widespread in the Phanerozoic than in the Neoproterozoic (which also lacked shelled marine organisms), and the Neoproterozoic was likely to have had a completely anoxic water column at least in waters below wave-base.

This seems an appropriate point to remind readers that there was a vast difference in oceanic life between Palaeozoic and later Mesozoic and Cenozoic times. As Eichenseer and colleagues pointed out in 2019 '*In the mid-Mesozoic, the Earth–life system was revolutionized by the rise of calcifying plankton. Before the widespread occurrence of planktonic calcifiers, $CaCO_3$ precipitation was largely confined to the continental shelves and linked to the success of benthic calcifiers, such as corals and brachiopods. The [subsequent] evolutionary success of calcifying plankton, especially of coccolithophores, shifted the carbonate factory from the shelves to the open ocean. Calcareous tests sank to the ocean floor and either dissolved or accumulated, depending on the local $CaCO_3$ saturation state. Since the proliferation of planktonic calcifiers, changes in the atmospheric and oceanic $CO_2$ content have been compensated on geologically short timescales by increased $CaCO_3$ deposition or dissolution in the deep sea*' [163]. Evidently, then, Ordovician oceanic conditions and environmental processes differed significantly, along with the rest of the Palaeozoic ocean, from what came later. Things changed. Coccolith flux to the deep sea increased by two orders of magnitude in the Mid Jurassic, and the deep Tethyan ocean of the Late

Jurassic contained abundant coccolith ooze. Eichenseer also observed that the sinking of relatively heavy foraminifera in Mesozoic and Cenozoic times took decaying organic matter into deep water, which had the effect of oxygenating the shallow seas of the continental shelves, which had formerly been the locus of periodic anoxia when sea levels rose bringing deep oxygen-depleted water onto the shelves. Well-oxygenated shelves in the post-Palaeozoic helped to stabilize the carbon cycle, thus decreasing the potential for catastrophic environmental change. Deep sea carbonate sedimentation was a key consequence of the development of calcifying marine organisms in post-Palaeozoic times [163].

## 9.7  Some Summary Remarks

These various studies show that although we lack '*a paleo-CO₂ "gold standard" analogous to the ice-cores of the Late Quaternary period*' [86], we do have considerable evidence, from a number of methods, for global trends in the abundance of $CO_2$ in the air through Phanerozoic time. Supporting evidence for the cycle of organic carbon production and consumption through time comes from the nature of the rock record – for example from periods when carbonate sediments, or coals or organically enriched black shales accumulated.

We also have adequate representation of the changes in global temperature over time. Just as we need to be cautious in evaluating proxy data for $CO_2$, we also need to be alert to the difficulties of determining palaeotemperatures accurately, especially in older geological intervals (Jurassic and earlier), where the carbonates used for $\partial^{18}O$ measurements may have been more subjected to chemical change (diagenesis) involving post-depositional precipitation of carbonate, and thus overprinting of the original seawater signal. These difficulties have spurred the development of independent measurements of palaeotemperature, for example via alkenones and the Mg/Ca ratio, as explained in Chapter 7.

With both the model and the database having been refined, things now hang together well. Convergence between the very different approaches provided by geochemical modelling on the one hand and analyses of proxies for $CO_2$ on the other hand gives us confidence that the science is on the right track and that $CO_2$ has in the past played a significant role in controlling past climate. But $CO_2$ will not have acted alone. More water evaporates from the ocean in a warming world, and water vapour itself is a powerful greenhouse gas. There may also have been fluctuations in methane ($CH_4$) another powerful greenhouse gas that is released from wetlands, which were probably more abundant in warmer, wetter worlds. Unfortunately,

we lack a way to determine past levels of methane back beyond the range of ice cores, as it tends to be rapidly oxidized to $CO_2$ in the atmosphere.

Back in 1991, a review of the state of the art of paleoclimatology concluded that whilst it was reasonable to assume that $CO_2$ provided an important control on climate, '*A major problem with the CO₂ scenario is that insufficient proxy evidence has been gathered to rigorously support the CO₂ model*' [94]. But, in the decades since then much of the missing proxy data to support the model has now been gathered, thanks to the efforts of Dana Royer, David Beerling, Mark Pagani and Paul Pearson.

I end this chapter with some quotes from a review led by Isabel Montañez whom we met in Chapter 6. They give a useful summary of the net result of all these efforts over the years in a 2011 report to the US National Academy of Sciences, which concluded: '*The carbon fluxes in and out of the surface and sedimentary reservoirs over geological time-scales are finely balanced, providing a planetary thermostat that regulates Earth's surface temperature. Initially, newly released CO₂ (e.g., from the combustion of hydrocarbons [or from volcanoes]) interacts and equilibrates with Earth's surface reservoirs of carbon on human timescales (decades to centuries). However, natural "sinks" for anthropogenic CO₂ [such as weathering] exist only on much longer timescales, and it is therefore possible to perturb climate for tens to hundreds of thousands of years … Transient (annual to century-scale) uptake by the terrestrial biosphere (including soils) is easily saturated within decades of the CO₂ increase, and therefore this component can switch from a sink to a source of atmospheric CO₂ … Most (60 to 80 percent) CO₂ is ultimately absorbed by the surface ocean, because of its efficiency as a sweeper of atmospheric CO₂, and is neutralized by reactions with calcium carbonate in the deep sea at timescales of oceanic mixing (1,000 to 1,500 years). The ocean's ability to sequester CO₂ decreases as it is acidified and the oceanic carbon buffer is depleted. The remaining CO₂ in the atmosphere is sufficient to impact climate for thousands of years longer while awaiting sweeping by the "ultimate" CO₂ sink of the rock weathering cycle at timescales of tens to hundreds of thousands of years…Lessons from past hyperthermals [periods of extreme warmth, discussed in Chapter 11] suggest that the removal of greenhouse gases by weathering may be intensified in a warmer world but will still take more than 100,000 years to return to background values for an event the size of the Paleocene-Eocene Thermal Maximum*' [119]. We'll explore the Palaeocene–Eocene Thermal Maximum in detail in Chapter 11.

I have spent some time on the seemingly convoluted development of thinking about the levels of $CO_2$ in past atmospheres, not least because it displays how science works. It is not an authoritarian business. Propositions are continually being tested. Arguments rage. Analytical methods improve. Databases grow. Models improve. Knowledge and understanding evolve. The frontier is always being pushed forward. Old ideas are discarded as new ones take their place. There is always more research to do! And it can take several decades to assemble the necessary knowledge base (think of the 60 years between Gilbert Plass's early work [2] and the findings of Foster et al. [138]). But as time goes by we converge on robust answers that stand the test of time. There should be no doubt now about the positive relationship between $CO_2$ and climate over long periods of geological time. We should think of the change in $CO_2$ with time as a key forcing factor in the climate system (Figures 9.20 and 9.21). Over the course of the Cenozoic,

atmospheric $CO_2$ fell from around 1000 ppm to as low as 170 ppm, equivalent to a negative climate forcing of $-10 W/m^2$ [65]. No wonder the climate cooled! There was no comparable change in any other known forcing factor.

What emerges from these various studies of $CO_2$ is that throughout the 550 Ma of the Phanerozoic, warm periods with abundant atmospheric $CO_2$ alternated with cool periods with much less $CO_2$. On the long timescales of change considered thus far we cannot identify any alternative source of atmospheric heating than the changes in the abundance of greenhouse gases observed from proxies and calculated from models. We now turn to look in more detail at the history of those fluctuations – the 'pulse of the Earth' – in Chapter 10.

# References

1 Woodward, A.S. (1922). Foreword. In: Man as a Geological Agent (ed. R.L. Sherlock). London: H.F. & G. Witherby.

2 Plass, G.N. (1956a). The carbon dioxide theory of climate change. *Tellus* VIII: 140–154.

3 Plass, G.N. (1956b). The influence of the 15 micron carbon dioxide band on the atmospheric infra-red cooling rate. *Quarterly Journal Royal Meteorological Society* 82: 310–329.

4 Plass, G.N. (1961). The influence of infrared absorptive molecules on the climate. *New York Academy of Sciences* 95: 61–71.

5 Urey, H.C. (1952). The Planets: Their Origin and Development. Yale University Press.

6 Plass, G.N. (1959). Carbon dioxide and climate. *Scientific American* 201 (1): 41–47.

7 Hollingworth, S.E. (1962). The climate factor in the geological record. *Quarterly Journal of the Geological Society of London* 118: 1–21.

8 Holmes, A. (1965). Principles of Physical Geology, 465–467. London: Nelson and Sons.

9 Budyko, M.I. (1974). Climate and Life, 508pp. New York and London: Academic Press [Translation of a book published in Russian in 1971].

10 Budyko, M.I. and Ronov, A.B. (1979). Chemical evolution of the atmosphere in the Phanerozoic. *Geochemistry International* 16 (3): 1–9; translated from *Geokhimiya* **5**, 1979, 643–653.

11 Budyko, M.I., Ronov, A.B., and Yanshin, A.L. (1987). History of the Earth's Atmosphere, 139 pp. New York: Springer-Verlag.

12 Pälike, H., Lyle, M.W., Nishi, H. et al. (2012). A Cenozoic record of the equatorial Pacific carbonate compensation depth. *Nature* 488: 609–614.

13 Sverdrup, H.U., Johnson, M.W., and Fleming, R.H. (1942). The Oceans, Their Physics, Chemistry and General Biology, 1087 pp. New York: Prentice Hall.

14 Hay, W.W. (2013). Experimenting on a Small Planet: A Scholarly Entertainment, 983 pp. New York: Springer.

15 Benson, W.E., Gerard, R.D., and Hay, W.W. (1970). Summary and Conclusions, *Initial Reports DSDP* **4**, US Government Printing Office. Washington, D.C., 659–673.

16 Raven, J. et al. (2005). Ocean Acidification Due to Increasing Atmospheric Carbon Dioxide, 60 pp. *Policy document* **12**/05. London: the Royal Society.

17 Fischer, A.G. and Arthur, M.A. (1977). Secular variations in the pelagic realm. In: *Deep Water Carbonate Environments* (ed. H.E. Cook and P. Enos). *Society of Economic Palaeontologists and Mineralogists Special Publication* 25: 19–50.

18 Fischer, A.G. (1981). Climatic oscillations in the biosphere. In: Biotic Crises in Ecological and Evolutionary Time (ed. M.H. Nitecki), 103–131. New York: Academic Press.

19 Beerling, D.J. and Royer, D.L. (2011). Convergent Cenozoic $CO_2$ history. *Nature Geoscience* 4: 418–420.

20 Pagani, M. et al. (2011). The role of carbon dioxide during the onset of Antarctic glaciation. *Science* 334: 1261–1264.

21 https://oceanacidification.noaa.gov.

22 Caldeira, K. and Wickett, M.E. (2003). Anthropogenic carbon and ocean pH. *Nature* 425: 394–395.

23 Riebesell, U., Zondervan, I., Rost, B. et al. (2000). Reduced calcification of marine plankton in response to increased atmospheric $CO_2$. *Nature* 407: 364–367.

24 Doney, S.C., Ruckelshaus, M., Duffy, J.E. et al. (2012). Climate change impacts on marine ecosystems. *Annual Reviews of Marine Science* 4: 11–37.

25 Garrels, R.M. and Mackenzie, F.T. (1971). Evolution of Sedimentary Rocks, 397 pp. New York: Norton.

26 Berner, R.A. (1990). Atmospheric carbon dioxide levels over Phanerozoic time. *Science* 249: 1382–1386.

27 http://en.wikipedia.org/wiki/Weathering (December 2013).

28 Garrels, R.M. and Perry, E.A. (1974). Cycling of carbon, sulfur and oxygen through geologic time. In: The Sea, vol. 5 (ed. E. Goldberg), 303–336. Chichester: Wiley.

29 Garrels, R.M., Lerman, A., and Mackenzie, F.T. (1976). Controls of atmospheric O$_2$ and CO$_2$: past, present and future. *American Scientist* 64: 306–317.

30 Garrels, R.M. and Lerman, A. (1984). Coupling of the sedimentary sulfur and carbon cycles – an improved model. *American Journal of Science* 284: 989–1007.

31 Mackenzie, F.T. and Pigott, J.D. (1981). Tectonic controls of Phanerozoic sedimentary rock cycling. *Journal of the Geological Society of London* 138: 183–196.

32 Walker, J.C.G., Hays, P.B., and Kasting, J.F. (1981). A negative feedback mechanism for the long term stabilization of earth's surface temperature. *Journal of Geophyscal Research* 86: 9976–9982.

33 Owen, T., Cess, R.D., and Ramanathan, V. (1979). Enhanced CO$_2$ greenhouse to compensate for reduced solar luminosity on early Earth. *Nature* 277: 640–642.

34 Berner, R.A. (2013). From black mud to earth system science: a scientific autobiography. *American Journal of Science* 313: 1–60.

35 Berner, R.A. (1982). Burial of organic carbon and pyrite sulfur in the modern ocean: its geochemical and environmental significance. *American Journal of Science* 282: 451–173.

36 Berner, R.A., Lasaga, A.C., and Garrels, R.M. (1983). The carbonate-silicate geochemical cycle and its effect on atmospheric carbon dioxide over the past 100 million years. *American Journal of Science* 283: 641–683.

37 Lasaga, A.C., Berner, R.A., and Garrels, R.M. (1985). An improved geochemical model of atmospheric CO$_2$ fluctuations over the past 100 million years. In: The Carbon Cycle and Atmospheric CO$_2$: Natural Variations Archean to Present, *Geophysical Monograph* 32 (eds. E.T. Sundquist and W.S. Broecker), 397–411. Washington D.C.: American Geophysical Union.

38 Cowie, J. (2013). Climate Change: Biological and Human Aspects, 558 pp. Cambridge University Press.

39 Beerling, D. (2007). The Emerald Planet. 288 pp. Oxford University Press.

40 Taylor, L.L., Leake, J.R., Quirk, J. et al. (2009). Biological weathering and the long-term carbon cycle: integrating Mycorrhizal evolution and function into the current paradigm. *Geobiology* 7 (2): 171–191.

41 Shackleton, N.J. (1985). Oceanic carbon isotope constraints on oxygen and carbon dioxide in the Cenozoic atmosphere. In: The Carbon Cycle and Atmospheric CO$_2$: Natural Variations Archean to Present, *Geophysical Monograph* 32 (eds. E.T. Sundquist and W.S. Broecker), 412–417. Washington D.C.: American Geophysical Union.

42 Laws, E.A., Popp, B.N., Bidigare, R.B. et al. (1995). Dependence of phytoplankton carbon isotopic composition on growth rate and [CO$_2$]$_{aq}$: theoretical considerations and experimental results. *Geochimica Cosmochimica Acta* 59 (6): 1131–1138.

43 Arthur, M.A., Dean, W.E., and Schlanger, S.O. (1985). Variations in the global carbon cycle during the Cretaceous related to climate, volcanism, and changes in atmospheric CO$_2$. In: The Carbon Cycle and Atmospheric CO$_2$: Natural Variations Archean to Present, *Geophysical Monograph* 32 (eds. E.T. Sundquist and W.S. Broecker), 504–529. Washington D.C.: American Geophysical Union.

44 http://www.geosociety.org/awards/07speeches/sloss.htm (December 2013).

45 Holland, H.D. (1978). The Chemistry of the Atmosphere, 351 pp. New York: Wiley.

46 Müller, R.D., Sdrolias, M., Gaina, C. et al. (2008). Long-term sea-level fluctuations driven by ocean basin dynamics. *Science* 319: 1357–1362.

47 Vail, P.R., Mitchum, R.M. Jr., Todd, R.G. et al. (1977). Seismic stratigraphy and global changes of sea level. In: *Seismic Stratigraphy – Applications to Hydrocarbon Exploration* (ed. C.E. Payton). *American Association of Petroleum Geologists Memoir* 26: 49–212.

48 Hallam, A. (1984). Pre-quaternary sea level changes. *Annual Reviews of Earth and Planetary Science* 12: 205–243.

49 Rowley, D.B. (2002). Rate of plate creation and destruction: 180 Ma to present. *Buletin of the Geological Society of America* 114 (8): 927–933.

50 Rowley, D.B. (2008). Extrapolating oceanic age distributions: lessons from the Pacific region. *Journal of Geology* 116: 587–598.

51 Haq, B.U., Hardenbol, J., and Vail, P.R. (1988). Mesozoic and Cenozoic chronostratigraphy and cycles of sea-level change. *Society of Economic Palaeontologists and Mineralogists Special Publication* 42: 71–108.

52 Menard, H.W. (1964). Marine Geology of the Pacific, 271. New York: McGraw-Hill.

53 Berner, R.A. (1989). Biogeochemical cycles of carbon and sulfur and their effect on atmospheric oxygen over Phanerozoic time. *Palaeoeography Palaeoclimatology Palaeoecology* 75: 97–122.

54 Berner, R.A. and Kothavala, Z. (2001). GEOCARB III: a revised model of atmospheric CO$_2$ over Phanerozoic time. *American Journal of Science* 301: 182–204.

55 Berner, R.A. (2006). GEOCARBSULF: a combined model for Phanerozoic atmospheric O$_2$ and CO$_2$. *Geochimica Cosmochimica Acta* 70: 5653–5664.

56 Raymo, M.E., Ruddiman, W.F., and Froelich, P.N. (1988). Influence of late Cenozoic mountain building on ocean geochemical cycles. *Geology* 16: 649–653.

57 Raymo, M.E. and Ruddiman, W.F. (1992). Tectonic forcing of late Cenozoic climate. *Nature* 359: 117–122.

58 Raymo, M.E. (1997). Carbon cycle models: how strong are the constraints? In: Tectonic Uplift and Climate Change (ed. W.F. Ruddiman), 367–381. New York and London: Plenum.

59 Ruddiman, W.F. (ed.) (1997). Tectonic Uplift and Climate Change, 535 pp. New York and London: Plenum.

60 Ruddiman, W.F. and Prell, W.L. (1997). Introduction to the uplift-climate connection. In: Tectonic Uplift and Climate Change (ed. W.F. Ruddiman), 3–15. New York and London: Plenum.

61 Ruddiman, W.F., Raymo, M.E., Prell, W.E., and Kutzbach, J.E. (1997). The uplift-climate connection: a synthesis. In: Tectonic Uplift and Climate Change (ed. W.F. Ruddiman), 471–515. New York and London: Plenum.

62 Raymo, M.E. and Ruddiman, W.F. (1992). Tectonic forcing of late Cenozoic climate. *Nature* 359: 117–122.

63 Molnar, P. and England, P. (1990). Late Cenozoic uplift of mountain ranges and global climate change: chicken or egg? *Nature* 346: 29–34.

64 England, P. and Molnar, P. (1990). Surface uplift, uplift of rocks, and exhumation of rocks. *Geology* 18: 1173–1177.

65 Cronin, T.M. (2010). Paleoclimates: Understanding Climate Change Past and Present, 441 pp. New York: Columbia University Press.

66 Blum, J.D. (1997). The effect of late Cenozoic glaciation and tectonic uplift on silicate weathering rates and the marine $^{87}Sr/^{86}Sr$ record. In: Tectonic Uplift and Climate Change (ed. W.F. Ruddiman), 259–288. New York and London: Plenum.

67 Edmond, J.M. and Huh, Y. (1997). Chemical weathering yields from basement and orogenic terrains in hot and cold climates. In: Tectonic Uplift and Climate Change (ed. W.F. Ruddiman), 329–351. New York and London: Plenum.

68 Berner, R.A. and Berner, E.K. (1997). Silicate weathering and climate. In: Tectonic Uplift and Climate Change (ed. W.F. Ruddiman), 353–365. New York and London: Plenum.

69 Kump, L.E. and Arthur, M.A. (1997). Global chemical erosion during the Cenozoic: Weatherability balances the budgets. In: Tectonic Uplift and Climate Change (ed. W.F. Ruddiman), 399–426. New York and London: Plenum.

70 Kump, L.R., Brantley, S.L., and Arthur, M.A. (2000). Chemical weathering, atmospheric $CO_2$, and climate. *Annual Reviews of Earth and Planetary Science* 28: 611–667.

71 Doughty, C.E., Taylor, L.L., Girardin, C.A.J. et al. (2014). Montane forest root growth and soil organic layer depth as potential factors stabilizing Cenozoic global change. *Geophysical Research Letters* 41: 983–990.

72 Ravizza, G.E. and Zachos, J.C. (2003). Records of Cenozoic ocean chemistry. In: The Oceans and Marine Geochemistry, vol. 6, *Treatise on Geochemistry* (eds H.D. Holland and K.K. Turekian) (ed. H. Elderfield), 551–581. Oxford: Elsevier-Pergamon.

73 Berner, R.A. and Caldeira, K. (1997). The need for mass balance and feedback in the geochemical carbon cycle. *Geology* 25 (10): 955–956.

74 Broecker, W.S. and Sanyal, A. (1998). Does atmospheric $CO_2$ police the rate of chemical weathering? *Global Biogeochemical Cycles* 12 (3): 403–408.

75 Pagani, M., Zachos, J.C., Freeman, K.H. et al. (2005). Marked decline in atmospheric carbon dioxide concentrations during the Paleogene. *Science* 309: 600–603.

76 Pearson, P.N. and Palmer, M.R. (1999). Middle Eocene seawater pH and atmospheric carbon dioxide concentrations. *Science* 284 (5421): 1824–1826.

77 Pearson, P.N. and Palmer, M.R. (2000). Atmospheric carbon dioxide concentrations over the past 60 million years. *Nature* 406: 695–699.

78 Berner, W., Stauffer, B., and Oeschger, H. (1978). Past atmospheric composition and climate, gas parameters measured in ice cores. *Nature* 276: 53–55.

79 Sundquist, E.T. and Broecker, W.S. (1985). The Carbon Cycle and Atmospheric $CO_2$: Natural Variations Archean to Present, *Geophysical Monograph* **32**, 132–142. Washington D.C.: American Geophysical Union.

80 Oeschger, H., Stauffer, B., Finkel, R., and Langway, C.C. (1985). Variations of the $CO_2$ concentration of occluded air and of anions and dust in polar ice cores. In: The Carbon Cycle and Atmospheric CO2: Natural Variations Archean to Present, *Geophysical Monograph* **32** (eds. E.T. Sundquist and W.S. Broecker), 132–142. Washington D.C.: American Geophysical Union.

81 Stuiver, M. (1978). Atmospheric $CO_2$ increases related to carbon reservoir changes. *Science* 199: 253–258.

82 Peng, T.-H. (1985). Atmospheric CO2 variations based on the tree-ring $^{13}C$ record. In: The Carbon Cycle and Atmospheric CO2: Natural Variations Archean to Present, *Geophysical Monograph* **32** (eds. E.T. Sundquist and W.S. Broecker), 123–131. Washington D.C.: American Geophysical Union.

83 World Climate Research Program (1983). Report of the WMO (CAS) Meeting of Experts on the CO2 Concentrations from Pre-Industrial Times to IGY, *WCP-53, Report* **10**, WMO, Geneva.

84 Elliott, W., Bojkov, D., Brewer, P. et al. (1984). The pre-1958 atmospheric concentration of carbon dioxide. *Eos* 65 (26): 416–417.

85 Oeschger, H. (1989). Information on the history of atmospheric $co_2$ and the carbon cycle from ice cores, in

*Carbon Dioxide and Other Greenhouse Gases: Climatic and Associated Impacts* (eds R. Fantechi and A. Ghazi), Proc. European Commission Symposium, Brussels, 3–5 Nov. 1986, Kluwer Acad. Press, Dordrecht, 40–54.

86  Sundquist, E.T. and Visser, K. (2005). The geologic history of the carbon cycle. In: Biogeochemistry, *Treatise on Geochemistry* (eds H.D. Holland and K.K. Turekian), vol. 8 (ed. W.H. Schlesinger), 425–513. Oxford: Elsevier-Pergamon.

87  Etheridge, D.M., Steele, L.P., Langenfelds, R.L. et al. (1996). Natural and anthropogenic changes in atmospheric CO$_2$ over the last 1000 years from air in Antarctic ice and firn. *Journal of Geophysical Research* 101: 4115–4118.

88  MacFarling Meure, C., Etheridge, D., Trudinger, C. et al. (2006). Law dome CO$_2$, CH$_4$ and N$_2$O ice core records extended to 2,000 years BP. *Geophysical Research Letters* 33 (14): L14810. https://doi.org/10.1029/2006GL026152.

89  Friedli, H., Lötscher, H., Oeschger, H. et al. (1986). Ice core record of the $^{13}$C/$^{12}$C ratio of atmospheric CO$_2$ in the past two centuries. *Nature* 324: 237–238.

90  Siegenthaler, U., Monnin, E., Kawamura, K. et al. (2005). Supporting evidence from the EPICA Dronning Maud Land ice core for atmospheric CO$_2$ changes during the past millennium. *Tellus Series B: Chemistry, Physics, Meteorology* 57: 51–57.

91  EIA (2012). *International Energy Statistics*, US Energy Information Administration; (http://www.eia.gov/cfapps/ipdbproject/iedindex3.cfm?tid=1&pid=7&aid=1&cid=regions&syid=2000&eyid=2012&unit=TST).

92  Khalil, M.A.K., and Rasmussen, R.A. (1989). Trends of atmospheric methane: past, present and future, in *Carbon Dioxide and Other Greenhouse Gases: Climatic and Associated Impacts* (eds R. Fantechi and A. Ghazi), Proceedings European Commission Symposium, Brussels, 3-5 Nov. 1986, Kluwer Acad. Press, Dordrecht, 91-Retallack

93  Etheridge, D.M., Steele, L.P., Francey, R.J., and Langenfelds, R.L. (1998). Atmospheric methane between 1000 AD and present: evidence of anthropogenic emissions and climatic variability. *Journal of Geophysical Research, Atmosphere* 103: 15,979–15,993.

94  Crowley, T.J. and North, G.R. (1991). Paleoclimatology, Oxford Monographs on Geology and Geophysics **18**, 339 pp. Oxford University Press.

95  Woodward, F.I. (1987). Stomatal numbers are sensitive to increases in CO$_2$ from pre-industrial levels. *Nature* 327: 617–618.

96  Beerling, D.J. (1999). Stomatal density and index: theory and application. In: Fossil Plants and Spores: Modern Techniques (eds T.P. Jones and N.P. Rowe), 251–256. Geological Society London.

97  Royer, D.L. (2001). Stomatal density and stomatal index as indicators of atmospheric CO$_2$ concentration. *Review of Palaeobotany and Palynology* 114: 1–28.

98  Royer, D.L., Berner, R.A., and Beerling, D.J. (2001). Phanerozoic atmospheric CO$_2$ change: evaluating geochemical and paleobiological approaches. *Earth-Science Reviews* 54: 349–392.

99  Royer, D.L., Wing, S.C., Beerling, D.J. et al. (2001). Palaeobotanical evidence for near present-day levels of atmospheric CO$_2$ during part of the tertiary. *Science* 292: 2310–2313.

100  Beerling, D.J. and Royer, D.L. (2002). Fossil plants as indicators of the Phanerozoic global carbon cycle. *Annual Reviews of Earth and Planetary Science* 30: 527–556.

101  Royer, D.L., Berner, R.A., Montañez, I.P. et al. (2004). CO$_2$ as a primary driver of Phanerozoic climate. *GSA Today* 14 (3): 4–10. https://doi.org/10.1130/1052-5173(2004)014<4:CAAPDO>2.0.CO;2.

102  Royer, D.L. (2006). CO$_2$ – forced climate thresholds during the Phanerozoic. *Geochimica Cosmochimic Acta* 70: 5665–5675.

103  Isbell, J.L., Miller, M.F., Wolfe, K.L., and Lenaker, P.A. (2003). Timing of late Paleozoic glaciation in Gondwana: was glaciation responsible for the development of Northern Hemisphere cyclothems? In: *Extreme Depositional Environments: Mega End Members in Geologic Time* (Ed. M.A. Chan and A.A. Archer). *Geological Society of America Special Paper* 370: 5–24.

104  Fielding, C.R., Frank, T.D., Birgenheier, L.P. et al. (2008a). Alternating glacial and non-glacial intervals characterize the late Paleozoic Ice Age: stratigraphic evidence from eastern Australia. *Journal of the Geological Society of London* **165**: 129–140.

105  Fielding, C.R., Frank, T.D., and Isbell, J.L. (eds.) (2008b). Resolving the Late Paleozoic Ice Age in Time and Space, 354, vol. 441. *Geological Society of America Special Paper*.

106  Koch, J.T. and Frank, T.D. (2011). The Pennsylvanian–Permian transition in the low-latitude carbonate record and the onset of major Gondwanan glaciation. *Palaeogeography, Palaeoclimatology, Palaeoecology* 308: 362–372.

107  Frakes, L.A., Francis, J.E., and Syktus, J.I. (1992). Climate Modes of the Phanerozoic – The History of the Earth's Climate Over the Past 600 Million Years. 274. Cambridge University Press.

108  Lenton, T.M. and Watson, A. (2011). Revolutions That Made the Earth, 423. Oxford University Press.

109  Lenton, T.M., Crouch, M., Johnson, M. et al. (2012). First plants cooled the Ordovician. *Nature Geoscience* 5: 86–89.

110  Hyde, W.H., Grossman, E.L., Crowley, T.J. et al. (2006). Siberian glaciation as a constraint on Permian–carboniferous CO$_2$ levels. *Geology* 34 (6): 421–424.

111 Veizer, J., Alab, D., Azmy, K. et al. (1999). $^{87}Sr/^{86}Sr$, $^{13}C$ and $^{18}O$ evolution of Phanerozoic seawater. *Chemical Geology* 161: 59–88.

112 Shackleton, N.J., Hall, M.A., Line, J., and Cang, S. (1983). Carbon isotope data in core V19-30 confirm reduced carbon dioxide concentration in the ice age atmosphere. *Nature* 306: 319–322.

113 Lawrence, K.T., Herbert, T.D., Dekens, P.S., and Ravelo, A.C. (2007). The application of the Alkenone organic proxy to the study of Plio-Pleistocene climate. In: Deep-Time Perspectives on Climate Change: Marrying the Signal from Computer Models and Biological Proxies, *Micropalaeontological Society Special Publication* (eds. M. Williams, A.M. Haywood, F.J. Gregory and D.N. Schmidt), 539–562. London: The Geological Society.

114 Zachos, J., Pagani, M., Sloan, L. et al. (2001). Trends, rhythms and aberrations in global climate 65 Ma to present. *Science* 292: 686–693.

115 Zhang, Y.G., Pagani, M., Lu, A. et al. (2013). A 40-million –year history of atmospheric $CO_2$. *Philosophical Transactions of the Royal Society A: Mathematical, Physical and Engineering Sciences* 371: 20130096, 20 pp.

116 Ekart, D.D., Cerling, T.E., Montañez, I.P., and Tabor, N.J. (1999). A 400 million year carbon isotope record of pedogenic carbonate: implications for paleoatmospheric carbon dioxide. *American Journal of Science* 299: 805–827.

117 Cerling, T.E. (1991). Carbon dioxide in the atmosphere: evidence from Cenozoic and Mesozoic paleosols. *American Journal of Scienc* 291: 377–400.

118 Lowenstein, T.K. and Demicco, R.V. (2006). Elevated Eocene atmospheric $CO_2$ and its subsequent decline. *Science* 323: p1928.

119 Montañez, I.P., Norris, R.D., Algeo, T. et al. (2011). Understanding Earth's Deep Past: Lessons for Our Climate Future, 194 pp. Washington, D.C.: National Academies Press.

120 Franks, P.J., Royer, D.L., Beerling, D.J. et al. (2014). New constraints on atmospheric $CO_2$ concentration for the Phanerozoic. *Geophysical Research Letters* 41 (13): 4685–4694. https://doi.org/10.1002/2014GL060457.

121 Veizer, J., Godderis, Y., and François, L.M. (2000). Evidence for decoupling of atmospheric $CO_2$ and global climate during the Phanerozoic Eon. *Nature* 408: 698–701.

122 Zeebe, R.E. (1999). An explanation of the effect of seawater carbonate concentration on foraminiferal oxygen isotopes. *Geochimica et Cosmochimica Acta* 63: 2001–2007.

123 Zeebe, R.E. (2001). Seawater pH and isotopic paleotemperatures of Cretaceous oceans. *Palaeogeography, Palaeoclimatology, Palaeoecology* 170: 49–57.

124 Shaviv, N.J. and Veizer, J. (2003). Celestial driver of Phanerozoic climate? *GSA Today* 13 (7): 4–10.

125 Rahmstorf, S., Archer, D., Ebel, D.S. et al. (2004). Cosmic rays, carbon dioxide, and climate. *EOS* 85 (4): 38–41.

126 Wing, S.L., Huber, B.T., Moerman, J.W., and the PhanTASTIC team (2018). Towards A Phanerozoic Technique-Averaged Surface Temperature Integrated Curve. American Geophysical Union, Fall Meeting 2018, abstract #PP11F-1323, Bibcode: 2018AGUFMPP11F1323W.

127 Voosen, P. (2019). A 500-million-year survey of Earth's climate reveals dire warning for Humanity. *Science* https://doi.org/10.1126/science.aay1323.

128 Henkes, G.A., Passey, B.H., Grossman, E.L. et al. (2018). Temperature evolution and the oxygen isotope composition of Phanerozoic oceans from carbonate clumped isotope thermometry. *Earth and Planetary Science Letters* 490: 40–50.

129 Veizer, J. and Prokoph, A. (2015). Temperatures and oxygen isotopic composition of Phanerozoic oceans. *Earth-Science Reviews* 146: 92–104.

130 Retallack, G.R., Sheldon, N.D., Carr, P.F. et al. (2011). Multiple early Triassic greenhouse crises impeded recovery from late Permian mass extinction. *Palaeogeography, Palaeoclimatology, Palaeoecology* 308: 233–251.

131 Steinthorsdottir, M., Jeram, A.J., and McElwain, J.C. (2011). Extremely elevated $CO_2$ concentrations at the Triassic/Jurassic boundary. *Palaeogeography, Palaeoclimatology, Palaeoecology* 308: 418–432.

132 Grein, M., Konrad, W., Wilde, V. et al. (2011). Reconstruction of atmospheric $CO_2$ during the early middle Eocene by application of a gas exchange model to fossil plants from the Messel formation, Germany. *Palaeogeography, Palaeoclimatology, Palaeoecology* 309: 383–391.

133 Fletcher, B.J., Brentnall, S.J., Anderson, C.W. et al. (2007). Atmospheric carbon dioxide linked with Mesozoic and early Cenozoic climate change. *Nature Geoscience* 1: 43–48.

134 Zeebe, R.E. and Calderia, K. (2008). Close mass balance of long-term carbon fluxes from ice-core $CO_2$ and ocean chemistry records. *Nature Geoscience* 1: 312–315.

135 Froehlich, P. and Misra, S. (2014). Was the late Paleocene-early Eocene hot because Earth was flat? *Oceanography* 27 (1): 36–49.

136 Li, G. and Elderfield, H. (2013). Evolution of carbon cycle over the past 100 million years. *Geochimica Cosmochimica Acta* 103: 11–25.

137 Torres, M.A., West, J.A., and Li, G. (2014). Sulphide oxidation and carbonate dissolution as a source of CO2 over geological timescales. *Nature* 507: 346–349.

**138** Foster, G.L., Royer, D.L., and Lunt, D.J. (2017). Future climate forcing potentially without precedent in the last 420 million years. *Nature Communications* 8 https://doi.org/10.1038/ncomms14845.

**139** Sagan, C. and Mullen, G. (1972). Earth and Mars: evolution of atmospheres and surface temperatures. *Science* 177 (4043): 52–56.

**140** Feulner, G. (2012). The faint young sun problem. *Reviews of Geophysics* 50: RG2006. https://doi.org/10.1029/2011RG000375.

**141** Chemke, R., Kaspi, Y., and Halevy, I. (2016). The thermodynamic effect of atmospheric mass on early Earth's temperature. *Geophysics Research Letters* 43 https://doi.org/10.1002/2016GL071279.

**142** Nardin, E., Goddéris, Y., Donnadieu, Y. et al. (2011). Modeling the early Paleozoic long-term climatic trend. *GSA Bulletin* https://doi.org/10.1130/B30364.1.

**143** Lefebvre, V., Servais, T., François, L., and Averbuch, O. (2010). Did a Katian large igneous province trigger the Late Ordovician glaciation? A hypothesis tested with a carbon cycle model. *Palaeogeography, Palaeoclimatology, Palaeoecology* 296: 310–319.

**144** Retallack, G.R. (2015). Late Ordovician basalts of Sierra del Tigre, Argentine Precordillera, and the Hirnantian mass extinction. *Large Igneous Provinces Commission*, http://www.largeigneousprovinces.org/15nov.

**145** Courtillot, V. and Olson, P. (2007). Mantle plumes link magnetic superchrons to Phanerozoic mass depletion events. *Earth and Planetary Science Letters* 260: 495–504.

**146** Jones, D.S., Martini, A.M., Fike, D.A., and Kaiho, K. (2017). A volcanic trigger for the Late Ordovician mass extinction? Mercury data from South China and Laurentia. *Geology* 45 (7): 631–634.

**147** Pogge von Strandmann, P.A.E., Desrochers, A., Murphy, M.J. et al. (2017). Global climate stabilisation by chemical weathering during the Hirnantian glaciation. *Geochemical Perspectives Letters* 3: 230–237. https://doi.org/10.7185/geochemlet.1726.

**148** Hammarlund, E.U., Dahl, T.W., Harper, D.A.T. et al. (2012). A sulfidic driver for the end-Ordovician mass extinction. *Earth and Planetary Science Letters* 331–332: 128–139.

**149** Torsvik, T.H. and Cocks, L.R.M. (2017). Earth History and Palaeogeography, 317 pp. Cambridge University Press.

**150** Vandenbroucke, T.R.A., Armstrong, H.A., Williams, M. et al. (2013). Chapter 24 Late Ordovician zooplankton maps and the climate of the Early Palaeozoic Icehouse. *Geological Society of London, Memoirs* 38: 399–405.

**151** Melchin, M.J., Mitchell, C.E., Holmden, C., and Storch, P. (2013). Environmental changes in the late Ordovician – early Silurian: review and new insights from black shales and nitrogen isotopes. *GSA Bulletin* 125 (11/12): 1635–1670.

**152** Pogge von Strandmann, P.A.E., Jenkyns, H.C., and Woodfine, R.G. (2013). Lithium isotope evidence for enhanced weathering during Oceanic Anoxic Event 2. *Nature Geoscience* 6: 668–672.

**153** Shen, J., Pearson, A., Henkes, G.A. et al. (2018). Improved efficiency of the biological pump as a trigger for the Late Ordovician glaciation. *Nature Geoscience* 11: 510–514.

**154** Morris, J.L., Puttick, M.N., Clark, J.W. et al. (2018). The timescale of early land plant evolution. *Proceedings of the National Academy of Sciences of the United States of America* 115 (10): E2274–E2283.

**155** Davies, J.R., Waters, R.A., Molyneux, S.G. et al. (2016). Gauging the impact of glacioeustasy on a mid-latitude early Silurian basin margin, mid Wales, UK. *Earth-Science Reviews* 156: 82–107.

**156** Page, A.A., Zalasiewicz, J.A., Williams, M., and Popov, L.E. (2007). Were transgressive black shales a negative feedback modulating glacioeustasy in the Early Palaeozoic Icehouse? In: Deep-Time Perspectives on Climate Change: Marrying the Signal from Computer Models and Biological Proxies, The Micropalaeontological Society, Special Publications (eds. M. Williams, A.M. Haywood, F.J. Gregory and D.N. Schmidt), 123–156. London: The Geological Society.

**157** Finnegan, S., Bergmann, K., Eiler, J.M. et al. (2011). The magnitude and duration of Late Ordovician–Early Silurian glaciation. *Science Express* 27 January 2011, https://doi.org/10.1126/science.1200803 4 pp.

**158** Vandenbroucke, T.R.A., Armstrong, H.A., Williams, M. et al. (2009). Ground-truthing Late Ordovician climate models using the paleobiogeography of graptolites. *Paleoceanography* 24: , PA4202. https://doi.org/10.1029/2008PA001720.

**159** Wilde, P. and Berry, W.B.N. (1984). Destabilization of the oceanic density structure and its significance to marine "extinction" events. *Palaeogeography, Palaeoclimatology, Palaeoecology* 48: 143–162.

**160** Wilde, P. (1991). Oceanography in the Ordovician. In: Advances in Ordovician Geology, vol. 90-9 (eds. C.R. Barnes and S.H. Williams), 283–298. Geological Survey of Canada Paper.

**161** Haq, B.U. and Schutter, S.R. (2008). A chronology of Paleozoic sea-level changes. *Science* 322: 64–68.

**162** Lenton, T.M., Dahl, T.W., Daines, S.J. et al. (2016). Earliest land plants created modern levels of atmospheric oxygen. *Proceedings of the National Academy of Sciences of the United States of America* 113 (35): 9704–9709.

**163** Eichenseer, K., Balthasar, U., Smart, C.W. et al. (2019). Jurassic shift from abiotic to biotic control on marine ecological success. *Nature Geoscience* 12: 638–642.

# 10

# The Pulse of the Earth

---

**LEARNING OUTCOMES**

- Understand and be able to explain how geological cycles emerged in the 1980s as a key concept underpinning the relationship between greenhouse gases and climate, and the concept of greenhouse and icehouse 'states'.
- Understand and be able to explain Goddéris's concept of palaeogeographic controls on climate, with reference to the Carboniferous glaciation, Carboniferous coals, and the weathering history of Pangaea.
- Understand and be able to explain the role of mountain building and weathering in the cooling of the Cenozoic over the past 50 Ma.
- Understand and be able to explain for how long – during Phanerozoic time the past 540 Ma the Earth has been in an icehouse state, in a greenhouse state, or an ice-free hothouse state? And how much of the time spent in an icehouse state (like the Pleistocene) has been spent in warm interglacial periods like the Holocene?
- Understand and be able to explain the interaction of seawater with the ocean crust as a mechanism for modifying $CO_2$ abundance and the carbonate compensation depth (CCD), in relation to changes in the Mg/Ca ratio.
- Describe the operation and origin of the 'calcite-aragonite metronome'.
- Understand and explain the long-term relationship between $CO_2$, ice sheets, and sea level, and contrast that with the roles of mantle upwelling in raising continents and of mid-ocean ridge growth.
- Describe Earth System Science, and explain its relation to Gaia theory.
- Understand and be able to explain the postulated link between flood basalts, climate and biological extinctions.
- Explain the development of the Snowball Earth hypothesis.

---

## 10.1 Climate Cycles and Tectonic Forces

Rather like human history, Earth's history did not run smoothly. Patches where nothing much seemed to be happening apart from business as usual were punctuated by episodes of relatively rapid change. Dutch geologist, Johannes (Jan) Herman Frederick Umbgrove (1899–1954, Box 10.1) dubbed this periodicity *The Pulse of the Earth*, the title for his classic textbook first published in 1942 [1], which I was still poring over as an undergraduate in 1960. Like Alexander Humboldt and Eduard Suess, he thought of the Earth as a single integrated dynamic system.

Umbgrove ascribed the alternating periods of Earth history to '*deep-seated forces [that are] the paramount source of all subcrustal energy, which manifests itself with a [~250 Ma] periodicity observed in a whole series of phenomena in the earth's crust and on its surface … [including] the magmatic cycles and rhythmic cadence of world-wide transgressions and regressions [of the sea], the pulsation of climate, and – lastly – the pulse of life*' [1]. Writing before plate tectonic theory, and while Wegener's continental drift concept was still discredited, he could not ascribe a mechanism to his earthly pulsations. He favoured Suess's land bridges rather than Wegener's moving continents, and discounted the notions of Arrhenius and Chamberlin, concluding that things other than $CO_2$ primarily controlled Earth's climate. With the benefit of hindsight it's easy to be critical, but like others of his generation he lacked the data to say anything much different. Even so, his concept of the Earth's pulse had certain attractions and lingered on.

The next key figure to pick up the baton in our story, in the late 1970s, was Princeton stratigrapher Alfred (Al) George Fischer (1921–2017) (Figure 10.1, Box 10.2), whom we met in passing in Chapter 9.

As mentioned in Chapter 9, In 1977, Fischer and his PhD student, Mike Arthur (see Box 9.4), drew wide-ranging conclusions from their study of the carbonate-rich, deep marine, open ocean (or pelagic) Cretaceous sediments from Gubbio in Italy's Apennine Mountains [2]. Bolstering their conclusions were analyses of similar pelagic sections

*Palaeoclimatology: From Snowball Earth to the Anthropocene*, First Edition. Colin P. Summerhayes.
© 2020 John Wiley & Sons Ltd. Published 2020 by John Wiley & Sons Ltd.

---

**Box 10.1 Johannes Herman Frederick Umbgrove**

A product of Leiden University, Umbgrove spent three years as a palaeontologist in what is now Indonesia with the geological survey of the Dutch East Indies, before returning to Leiden. In 1930, he became professor of stratigraphy and palaeontology at the University of Delft. He was an 'all-rounder', taking an interest in many different branches of geology, and publishing on the palaeogeography of the Dutch East Indies, the palaeontology of corals and coral reefs, volcanology, and the geology of the Netherlands. Amongst other honours, he was made an Honorary Fellow of the Royal Society of Edinburgh and of the New York Academy of Sciences, as well as being a fellow of the Royal Dutch Academy of Sciences.

**Figure 10.1** Al Fischer.

found on land or in deep ocean drill cores, some collected by Fischer as stratigrapher and sedimentologist on the first Deep Sea Drilling Project (DSDP) expedition in 1968 [3]. Stimulated by the new concept of plate tectonics, they reasoned that '.... *the episodic development of great glacial ages suggests that patterns of energy distribution and perhaps the energy budget as a whole have undergone fluctuations ... One might expect, then, that variations in sedimentation through time reflect not only locally generated changes but also carry an overprint produced by shifts in the state of the oceans, of the biosphere, and perhaps of the earth as a whole – changes of a sort not considered by Hutton, Lyell, and other classical uniformitarianists*' [2]. Given that this 'overprint' was subtle, it had to be sought in sediments deposited in open ocean pelagic realms far from the influence of land and tectonic 'noise' [2].

Their examination of pelagic sections convinced them that these sediments carried the signal of a cycle with a period of about 32 Ma [2]. It represented an oscillation between two oceanic states that they termed 'oligotaxic' and 'polytaxic'. Like all scientists, geologists are not immune from the curse of jargon. Oligotaxic times were periods of low global biodiversity, reduced complexity of biological communities, and widespread extinction of free-swimming organisms including plankton. They were also typically cool, like the Pleistocene and present seas, and associated with lowered sea level marked by regression of the shoreline. In contrast, polytaxic times were periods of high rates of speciation and high biodiversity. They were times of warm seas and high sea level marked by transgressions of the shoreline. They were also associated with weaker latitudinal temperature gradients, less vigorous ocean circulation, an intensified and expanded oxygen

---

**Box 10.2 Al Fischer**

Born in Germany, Al Fischer emigrated to the United States at the age of 15, obtaining bachelors and masters degrees from the university of Wisconsin. He worked first for the Virginia Polytechnic Institute, then for Standard Oil, and finally for the university of Kansas, before obtaining a PhD from Columbia University in 1950. He then spent five years as a petroleum geologist for Esso, in Peru, before joining the staff at Princeton University in 1956. He moved to the University of Southern California in Los Angeles in 1984, becoming professor emeritus there in 1991. As a 'biogeohistorical visionary' he would become one of the world's best-known stratigraphers. He became an expert on cyclic sedimentation in Mesozoic and Cenozoic sequences, and thus one of the pioneers of 'cyclostratigraphy'. His work in this field culminated in publication by the Society for

Sedimentary Geology, in 2004, of *Cyclostratigraphy: Approaches and Case Histories*, containing several papers that he wrote or co-authored. His lasting contribution lay in linking variations in biodiversity, through changes in climate and the chemistry of oceans and atmospheres, to variations in rates of seafloor spreading, changes in sea level, and fluctuations in continental igneous activity. He was a great integrator and synthesizer of multifarious facts. Amongst his accolades, Fischer was awarded the Twenhofel medal of the Society for Sedimentary Geology in 1982, the Penrose Medal of the Geological Society of America in 1993, the Geological Society of London's Lyell Medal in 1992, and the Mary Clark Thompson Medal of the National Academy of Sciences in 2009. He was elected to US National Academy of Sciences in 1994.

minimum zone, widespread deposition of organic-rich sediment, and a rise in the CCD.

Cool (oligotaxic) episodes were centred on the Permo–Triassic Boundary (250 Ma ago), the Triassic–Jurassic Boundary (200 Ma ago), the Bathonian–Callovian Boundary (165 Ma ago), the early Neocomian (~140 Ma ago), the Cenomanian (95 Ma ago), the early Palaeocene (62 Ma ago), the mid Oligocene (30 Ma ago), and the Pleistocene–Holocene (the last 2.6 Ma). Warm (polytaxic) episodes favourable to the accumulation of petroleum source beds were centred on the Early Jurassic (190 Ma ago), the Late Jurassic (155 Ma ago), the Mid Cretaceous (110 Ma ago), the Late Cretaceous (85 Ma ago), the Eocene (50 Ma ago), and the Miocene (15 Ma ago). These cycles lay within a broader climatic cycle of 200–300 Ma duration that tended to emphasize one state over the other, such as pronounced warm (polytaxic) episodes during Jurassic-Cretaceous time and cool (oligotaxic) periods during the Cenozoic [2].

The cycles seemed to be unrelated to plate tectonics or magnetic reversals, but were related to cycles in terrestrial biodiversity, in sea level, and in the $\partial^{13}C$ ratio – probably due to periodic burial of isotopically light ($^{12}C$-rich) organic matter, rather than to variation in productivity. Fischer and Arthur were unsure about what made the polytaxic climates warm, suggesting that: '*processes within the earth's interior influence sea levels by changing the earth's surface configuration, and may simultaneously affect the atmosphere and therefore climates through vulcanism*' [2]. On reflection it seems surprising that they did not mention fluctuations in $CO_2$ as providing any control on climate, not least because as we saw in Chapter 9, Garrels, Lerman, and Mackenzie had published their thoughts on that topic in the May–June issue of *American Scientist* the previous year (1976). It is possible that Garrels' paper appeared after Fischer had submitted his work for publication.

By 1981, thinking deeply about these cycles, and considering the many advances being made deep ocean drilling, Fischer realized that new developments in geology were dragging historical geology away from the uniformitarian view of Hutton and Lyell in which the present state and functioning of the Earth were taken as the 'norm' [4, 5]. The palaeontological record, for instance, he observed, is neither uniform nor gradual, but rather a record of sharp discontinuities. The discovery of ancient glaciations implied intervals of major climatic deterioration. Former warm periods might be explained by an increase in the abundance of greenhouses gases (like $CO_2$) in the atmosphere, as suggested by Arrhenius and Chamberlin. Orogenic (mountain-building) events seemed to be periodic, as Umbgrove had suggested, as did changes in sea level. There were regular changes in the thermal structure

and behaviour of the ocean. Clearly, the Earth had not persisted in an invariant state of which the present was representative, as Hutton and Lyell had inferred.

Fischer wrote that '*While most of this change has come about slowly – at what might be thought of as a uniformitarian pace – the role of catastrophe cannot be dismissed as it was by Lyell*' [5]. Unlike Lyell, he knew (as we see in more detail later) that an asteroid had hit the Earth at the end of the Cretaceous, killing off the dinosaurs, and that some regions had experienced long periods of massive volcanic eruptions of highly liquid lava that emerged through long fissures and spread out to create floods of basalt that covered vast areas and no doubt filled the air with noxious fumes (such as the Deccan Traps of India, the flood basalts of Siberia, and the Columbia River basalts of the northwest United States). Flood basalts were periodic, each eruption forming a new layer. The stacking up of hundreds of these layers today makes the outcrops of these deposits look like stairs – hence the name 'trap', from the Dutch word for stair.

Fischer thought that '*We may now view earth history as a matter of evolution in which some changes are unidirectional (at least, in net effect), others are oscillatory or cyclic, and still others are random fluctuations, while the whole is punctuated by smaller or greater catastrophes. The prime tasks of modern historical geology are to separate the local signals from the global ones, to plot the relationships of global patterns both to time and to each other, and to search for the forces that drive these varied processes*' [5]. There was a role here for a blend of Lyell's uniformitarianism and Cuvier's catastrophism. Fischer's conclusion was echoed in 1993 by Derek V. Ager, one-time professor of geology at Imperial College, London, former head of the department of geology and oceanography of the University College of Swansea, in Wales, and a former president of the United Kingdom's Geologist's Association [6]. Adapting Napoleon's aphorism, Ager concluded that '*the history of any one part of the earth, like the life of a soldier, consists of long periods of boredom and short periods of terror*' [6]. And so it is that today we find geological thought combining the uniformitarian and the catastrophic approaches. The on-going mundaneness of the day-to-day can be interrupted by the special event. The two camps have merged.

Fischer identified two great tectonic–climatic cycles with a periodicity of around 300 Ma, which he proposed were driven by cycles in mantle convection leading to cyclic changes in the abundance of $CO_2$ in the air [5]. These cycles began with the Ice Age of the late Proterozoic (around 650 Ma ago) (later referred to as the time of 'Snowball Earth', of which more later), continued with the inferred high $CO_2$ greenhouse warm conditions of the

Early–middle Palaeozoic, the low $CO_2$ conditions of the Ice Age of the late Palaeozoic (late Carboniferous to early Permian), and the high $CO_2$ conditions of the warm Mesozoic greenhouse state, and ended in the low $CO_2$ icehouse state of the late Cenozoic to the present. Volcanism was abundant and sea level was high in what he called the 'greenhouse states', whilst both were low in what he called the 'icehouse states'. This was most probably because the continents were dispersed and mid-ocean ridges were abundant and spreading fastest during the greenhouse states (with growing mid-ocean ridges displacing the sea onto continental margins), while continents tended to be aggregating and mid-ocean ridges were less active during the icehouse states (the shrinking mid-ocean ridges leading to regression of the sea on continental margins). He considered that the continents tend to be thicker and to have less freeboard when aggregated in icehouse states, and that the post-Eocene drop in sea level might mark the start of the next phase of continental aggregation, which began with the collisions of India and Tibet, Australia and South East Asia, and Africa and Europe [4, 5].

Fischer thought that the primary source for the increase in atmospheric $CO_2$ in the greenhouse states was abundant volcanism when mid-ocean ridges were spreading fastest and granites were being emplaced beneath volcanoes at the leading edges of the moving continents. Large mid-ocean ridges made sea level high, flooding the continents and reducing the area susceptible to the weathering that would draw down the $CO_2$ content of the air. When spreading and volcanism declined and sea level was low, weathering would catch up, reduce $CO_2$, and lower the temperature, eventually leading to conditions suitable for glaciation in the icehouse states. He credited Budyko and Ronov (whose ideas we investigated in Chapter 9) with drawing attention to the association between volcanism, chemical weathering, sedimentation, and $CO_2$. And, citing Heinrich Holland's work on the carbon cycle [7], of which more later, he argued that '*carbon dioxide in the oceanic/atmospheric reservoir has a residence time of about 500,000 years … [which is] not long in terms of the timescales here considered: an imbalance in input versus loss that continued over tens of millions of years could effect considerable changes*' [5]. That much was also evident to Chamberlin at the end of the nineteenth century, as we saw in Chapter 4.

Fischer was a great integrator. What we see in his work is the seepage into mainstream geology of ideas from the worlds of geochemistry (Walker, Holland, Garrels, and Mackenzie) atmospheric chemistry and physics (Keeling and Plass), and ocean chemistry (Revelle, Suess, and Broecker), discussed in Chapters 8 and 9. And as we saw in Chapter 9, his conclusions are borne out by the later palynological data of Royer and the geochemical models of Berner.

In 1985, Thomas Worsley of Ohio University published a refinement of Fischer's ideas [8]. Worsley concluded that the build-up of heat beneath supercontinents like Pangaea might account for their uplift and eventual rupture. Eventually, the passive margins of the newly expanding oceans would become sufficiently old and dense to spontaneously self-subduct, a process that seemed likely to him to occur within 200 Ma of initial rifting. In due course, Worsley argued, the Atlantic and the Indian Ocean should close to form a new supercontinent. Mad idea? Not when you realize that that the Atlantic Ocean has opened and closed before, more or less along the same lines. The Iapetus Ocean separated Europe from North America in the early Cambrian. Its closure thrust up the Caledonian mountain chain that runs from Norway through Scotland and Ireland and continues south in the Appalachians of North America. In the late Cambrian–early Ordovician another ancient ocean, the Rheic Ocean, separated Gondwana from Laurussia (carrying North America and Baltica – which included Scandinavia and parts of Europe) (Figure 9.24). Its closure created the Variscan orogeny, which formed the Pyrenees–Alleghenny–Ouachita mountain chain. The present North Atlantic represents a new break, not quite along the old line. In Worsley's conceptual model, sea level was low on supercontinents and high during their break-up, matching Vail's curves of sea level through time (Figure 5.13) [9]. The distributions of stable isotopes of carbon, sulphur, and strontium through time matched the model, supporting the idea that episodic plate tectonic processes drove biogeochemical cycles, including the slow carbon cycle and, by inference, $CO_2$. Fischer's and Worsley's cycles followed the cycle of ocean basin evolution proposed in 1966 by Tuzo Wilson, whom we met in Chapter 5 [10, 11].

Despite the realization emerging in the geological community of the early 1980s that fluctuations in $CO_2$ had most likely played an important role in determining the changes in Earth's climate with time, many of the papers published on past global change in that era did so without mentioning $CO_2$ as a driver – such as a 1984 paper by Bil Haq on 200 Ma of Earth history [12]. We should not be surprised. New concepts take time to work their way through the system. The BLAG model was first published in 1983, Fischer's key paper on the topic did not emerge until 1984. As we saw in Chapter 8, even within the climate science community it was only in the very late 1970s and early 1980s that convincing studies of the role of $CO_2$ in the climate system began to emerge in papers by the likes of Charney, Ramanathan, Hansen, and Broecker. If you got your geological education before the mid-1980s you might not have been exposed to these developments.

Following Fischer's lead, the link between $CO_2$ and climate was now 'in the geological air'. Writing in 1986 and 1987, Bob Sheridan of the University of Delaware used the links between tectonics, $CO_2$, and climate to propose a theory of 'pulsation tectonics' [13, 14]. This involved plumes of hot magma periodically erupting from the boundary between the Earth's core and mantle, speeding sea floor spreading and causing mid-ocean ridges to grow, which increased sea level and displaced seawater onto the continents. The increased volcanic activity associated with seafloor spreading added $CO_2$ to the atmosphere, which warmed it and, in concert with the expanded ocean area, increased evaporation. Adding water vapour to the atmosphere further increased warming, leading to a warm wet climate. Lessening of plume activity lessened all the other factors, including the output of $CO_2$ and evaporation of seawater, leading to a cooler climate as weathering extracted $CO_2$ from the atmosphere.

By 1992, Larry Frakes, whom we met in Chapter 6, had adopted Fischer's cycles concept as a key element in understanding palaeoclimates. With his colleagues Jane Francis and Josef Syktus, Frakes '*divided climate history into Warm Modes and Cool Modes, in a way not unlike Fischer's ....*"*Greenhouse*" *and* "*Icehouse*" *states, but our Modes are of shorter duration*' [15]. Frakes and his team '*questioned the theory that the Mesozoic climates were uniformly warm and ice free and instead propose[d] a Cool Mode in the Middle Mesozoic*' [15], listing the resulting modes as follows:

**Cool Mode 5: Early Eocene to present, 55–0 Ma ago**
*Warm Mode 4: Early Cretaceous to Early Eocene 105–55 Ma ago*
**Cool Mode 4: Late Jurassic to Early Cretaceous 167–105 Ma ago**
*Warm Mode 3: latest Permian to Middle Jurassic 253–167 Ma ago*
**Cool Mode 3: Early Carboniferous to Late Permian 333–253 Ma ago**
*Warm Mode 2: Early Silurian to Early Carboniferous 436–333 Ma ago*
**Cool Mode 2: Late Ordovician to Early Silurian 445–436 Ma ago**
*Warm Mode 1: earliest Cambrian to Late Ordovician 540–445 Ma ago*
**Cool Mode 1: latest Precambrian to earliest Cambrian 615–540 Ma ago**

The evidence for these modes comprised climate-sensitive fossil animals and plants, along with sedimentary rock types and characteristics (tillites, carbonates, evaporates, aeolian sandstones, calcrete, kaolinite, coal, and so on), as well as oxygen isotopes and the relative heights of sea level. Warm Mode 2 included a brief glaciation recognized in South America, and there was evidence for some cool periods within Warm Mode 4, and for some warm periods during Cool Mode 4, which was identified as cool from growing evidence of ice-rafted debris of that age at high latitudes. Like Fischer, the Frakes team accepted that $CO_2$ had some part to play as a driving force in changing climate, with more $CO_2$ being supplied when sea floor spreading was rapid than when sea floor spreading slowed and mountains were built, although they did not go out of their way to provide details.

The cool modes were associated with low sea level and seemed to require a considerable extent of land at high latitudes, with long intervals of cooling leading to glaciation. They tended to be associated with a marked increase in $\partial^{13}C$ rising to a peak during extreme cooling, and with a marked decrease in the abundance of evaporites. Whilst land at high latitudes seemed to be a necessary condition for the development of ice sheets, it was not sufficient. Global cooling, and especially the development of cool summers to prevent snowmelt were required too. This independently imposed cooling implied a decrease in atmospheric $CO_2$, possibly reflecting in turn the sequestration of large amounts of $^{12}C$-rich organic carbon as coal or peat on land, or as organic-rich black shales in the ocean, which would have increased $\partial^{13}C$ in the atmosphere and upper ocean as cooling and the trapping of $^{12}C$-rich $CO_2$ in sediments progressed.

The warm modes tended to appear quite suddenly, but to end gradually. Their beginnings were usually preceded by an abrupt decrease in $\partial^{13}C$ (hence increase in $CO_2$), and their ends saw a gradual increase in $\partial^{13}C$ (hence decrease in $CO_2$). They were characterized by high sea level, a rise in volcanic activity, and an increase in the accumulation of organic carbon with time, suggesting tectonic control in the form of an upsurge of seafloor spreading and volcanic emissions of $CO_2$ associated with sea level rise and warming, followed by the pulling down of $CO_2$ by the accumulation of organic carbon in sediments, leading eventually to cooling.

Clearly, the Frakes team was convinced of the strong link between plate tectonic activity and climate that had been alluded to back in 1979 by Budyko and Ronov [16]. They noted that some 70% of above-average volcanism in tectonic regimes occurred in warm modes, '*consistent with the hypothesis that global climates are influenced, and perhaps forced, by volcanic outgassing*' [15]. This activity was usually most enhanced towards the middle of a warm mode. Whilst orbital changes were obviously important controls on climate, especially within ice ages, the Frakes team considered that they did not contribute to the much longer-lasting warm and cool modes, nor did they initiate the late Cenozoic glaciation.

Looking towards the future, Frakes's team observed '*The great variability of Phanerozoic climates has not seen a clear trend towards overall warming or cooling in the last 570 m.y., but rather can be characterized as alternating cool and warm intervals of long period*' [15]. Most ancient climates were relatively warm. With the increase in geological information of recent decades '*it has come to be recognized that climates have varied more often than previously accepted ... Further work may also reveal greater variability on both short and long wavelengths ...*' [15]. We will see plenty of evidence of that in later chapters.

As in all the sciences, nothing stands still forever in geology. These early ideas have become more refined, for example in an analysis by Alan Vaughan of the British Antarctic Survey in 2007 (Figure 10.2) [17]. Vaughan capitalized on Royer's 2004 curve of palaeotemperature and listing of multiple short-lived cool periods through time [18], to

show that Frakes's Warm 1 and Warm 2 periods tended to be interrupted by short-lived cool periods, whilst his Cool 2 and Cool 3 periods were interrupted by short-lived warm periods [17]. Vaughan saw Frakes's Warm 3, Cool 4 and Warm 4 periods as one long warm period containing numerous short-lived cool intervals, some of which tended to cluster close to the Cool 4 period but by no means mapped onto it. Vaughan also shortened Frakes's Cool 5 period by starting it towards the end rather than the start of the Eocene (Figure 10.2).

Summarizing his view of Phanerozoic climate, Vaughan concluded that '*through the Phanerozoic, two overlapping stable climate regimes appear to have dominated: a high-$CO_2$ (>1000 ppmv), largely warm climate regime, punctuated by many short-lived episodes of glaciation; and a low-$CO_2$ (<1000 ppmv), largely cool regime, marked by protracted episodes of superglaciation*' [17]. Vaughan's first high-$CO_2$

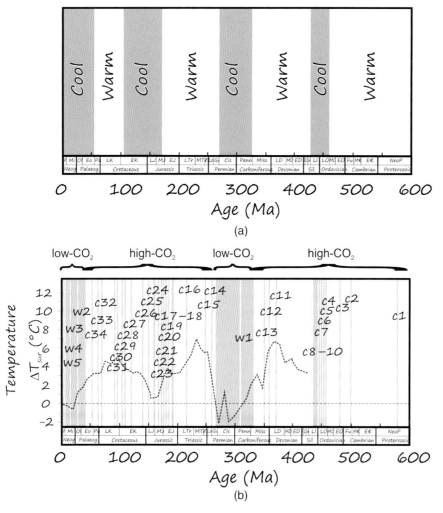

**Figure 10.2** Vaughan's view of alternating cool and warm periods through time. *Source:* from figure 1 in Ref. [17]. (a) Late Neoproterozoic and Phanerozoic climate modes of Frakes and colleagues [15]. (b) Dark grey = glacial or cool conditions; pale grey to white = warm conditions, through Late Neoproterozoic and Palaeozoic to recent time. Cool intervals are labelled (e.g. c1–c19), as are warm intervals (e.g. w1–w5). Modified from Royer et al. [18], and including the palaeotemperature curve from that source (dashed line). Brackets above (b) show durations of high and low $CO_2$ modes for Phanerozoic climate. From figure 1 in Ref. [17].

phase ran from the start of the Cambrian to the start of the Permo–Carboniferous glaciation, the first of his low-$CO_2$ climates (Figure 10.2). The second high $CO_2$ phase ran from the Late Permian to the Late Eocene, after which Antarctic glaciation heralded the start of the next low-$CO_2$ climate phase. We know from Royer's work that Vaughan's short-lived episodes of glaciation during the warm climate regime were associated with low $CO_2$ [18, 19].

By Vaughan's time, geologists knew that there were areas scattered around the world characterized by the eruption of flood basalts with volumes of >$100\,000\,km^3$ of lava, which must have been major sources of $CO_2$ from the Earth's mantle, as we shall see later. Fischer knew they existed [5], but had not built them into his conceptual model as sources of $CO_2$. Vaughan's high-$CO_2$ climate modes tended to coincide with times of high rates of emplacement of flood basalts and of high rates of continental dispersal in the supercontinent cycle [17]. These were also periods of high rates of magmatic activity, enhanced hydrothermal activity at mid-ocean ridges, a high rate of supply of $CO_2$, and relatively low rates of continental weathering.

Vaughan's low-$CO_2$ modes, which are generally of much shorter duration, coincided with low rates of emplacement of flood basalts, and times of amalgamation in the supercontinent cycles, marked by large-scale mountain building and high rates of crustal exhumation (deep rocks being brought to the surface as mountains are eroded). These are periods of low rates of magmatic activity, low rates of hydrothermal activity at mid-ocean ridges, relatively low fluxes of $CO_2$ and high rates of continental weathering. He thought that the short-lived cool episodes during the warm high-$CO_2$ modes were probably glacial, being associated with rapid drops in sea level. They tended to be associated with the deposition of organic rich sediments, and with increases in $\partial^{13}C$ (due to the trapping of $^{12}C$ rich organic matter in sediments), and in $\partial^{18}O$ (which is abundant when evaporated water rich in $^{16}O$ is trapped in ice leaving the surface ocean enriched in $^{18}O$) [17].

On long time-scales, plate tectonics continued to provide an important background control on $CO_2$ and climate. Let us consider the Carboniferous glaciation for example, not least because '*its demise constitutes the only recorded turnover to a greenhouse state*' [20]. It has long been thought that the evolution and expansion of land plants caused $CO_2$ to decline rapidly in the late Palaeozoic, not just because plants take in $CO_2$, but also because their root systems enhance the weathering of silicate rocks, which also absorbs $CO_2$ and increases the storage of carbon on land. However, as Yves Goddéris of Géosciences Environnement Toulouse observed in 2017, '*plant expansion and carbon uptake substantially predate [the Carboniferous] glaciation*' [21]. Goddéris and his team found evidence to suggest that

the uplift of the Variscan (or Hercynian) mountains between Euramerica (Larussia) and Gondwana during the formation of the Pangaean supercontinent in the Late Carboniferous[1] led to an increase in physical weathering that removed a thick pre-existing soil cover and so enhanced the chemical weathering that drew $CO_2$ down below the levels required to initiate glaciation. Chemical weathering then gradually declined as erosion lowered the mountains and re-established thick soils. Meanwhile '*the assembly of Pangaea promoted arid conditions in continental interiors, [which] were unfavourable for chemical weathering. These changes allowed $CO_2$ concentrations to rise to levels sufficient to terminate the glacial event*' [21]. Nevertheless, drawdown of $CO_2$ was also a side effect of the highest rates of global organic carbon burial in the past half billion years during the Carboniferous–Permian (330–260 Ma ago), in large part due to the accumulation and burial of peat in vast tropical lowland basins, where thick, regionally extensive coals formed under humid climates [22]. This not only led to anomalously low $pCO_2$, but also anomalously high $pO_2$ [22], which encouraged the development of insect flight and the development of gigantic insects (e.g. dragonflies with 1 m wingspan) [22].

Goddéris's theory is that palaeogeography is a key factor controlling the long-term evolution of atmospheric $CO_2$ by modulating the efficiency of silicate weathering [23]. He argues that palaeogeography strongly controls $CO_2$ consumption by silicate rock weathering through modulating runoff and air temperature above the continental surface. Continents in low latitudes with more heat and humidity would experience more chemical weathering of silicates, promoting the consumption of $CO_2$, whilst supercontinent assembly coincides with high $pCO_2$ values due to the development of arid conditions, which weaken the efficiency of silicate weathering. The polar distribution of continents has a similar effect due to cooling and albedo. Continental positions favoured the development of high atmospheric $CO_2$ in the early Palaeozoic, whilst the development of Gondwana favoured low $CO_2$ and the development of the Carboniferous glaciation. Since the warm Mid Cretaceous, palaeogeography has promoted the decline of $CO_2$ and the onset of cold climatic conditions. Goddéris considers that the role of palaeogeography has been largely underestimated in climate models, and developed the GEOCLIM model to correct that [23].

In one sense Goddéris is reintroducing the concept of 'weatherability' proposed by Kump and others, which we discussed in Chapter 9. Chemical weathering is indeed important, but only if there are exposed rocks on which it can act. More rock means more weathering even if the

---

[1] These mountains include Europe's Pyrenees and North America's Ouachitas and Appalachians.

intensity of weathering is low. A similar argument was put forward by Rugenstein and colleagues in 2019, although they used the term 'surface reactivity' to represent the availability of fresh mineral surface area or an increase in the supply of reactive minerals [24]. In other words it is not the intensity of weathering (what one might call the 'weathering flux') that is the key. Both weathering flux and weatherability (or surface reactivity) are required – which is implicit in Goddéris's palaeogeographic argument. Rugenstein argued that the long-term cooling of the Neogene was the result of the draw-down of $CO_2$ due not to an increased weathering flux, but rather to an increase in the availability of weatherable surfaces caused by an increase in mountain building [24].

As Montañez and Poulsen point out, the late Palaeozoic glaciation (broadly between 335 and 260 Ma ago) was dynamic '*characterized by discrete periods of glaciation separated by periods of ice contraction during intermittent warmings, moderate-size ice sheets emanating from multiple ice centers throughout southern Gondwana, possible glaciation of the Northern Hemisphere, and atmospheric $CO_2$ as a primary driver of both ice sheet and climate variability*' [20]. Individual glaciations may have lasted between <1 Ma to ~8 Ma, and may have been asynchronous, emanating from numerous ice centres [20]. Glaciation may have begun in the form of mountain glaciers during the latest Devonian, but the main Gondwanan ice sheet appears to have developed south of latitude 60°S about 347 Ma ago, and to have expanded to about 30°S between c.330 and 279 Ma ago, before shrinking back south of 60°S and gradually decreasing in area between 279 and 260 Ma ago [20]. Ice sheets were gradually replaced by mountain glaciers. Many shorter-term glacial advances are attributed to long-period modulation of obliquity. Sea levels probably fell by between 20 and 70 m at times of maximal ice volume. Climate model simulations suggest that $CO_2$ was the primary driver for the building up and breaking down of glaciation [20].

Clearly, one-dimensional models focused on single elements (like the effect of land plants) are not adequate to explain our climate history. One must consider the plate tectonic environment, mountain building, aridity, physical and chemical weathering, and soil development as well as the evolution and spread of plants.

Ocean gateways are also important. For instance, Korte and colleagues used a high-resolution oxygen isotope data set to show '*an especially abrupt earliest Middle Jurassic (c.174 Ma ago) mid-latitude cooling of seawater by as much as 10 °C in the north–south Laurasian Seaway, a marine passage that connected the equatorial Tethys Ocean to the Boreal Sea*' [25]. This coincided with large-scale regional lithospheric doming of the North Sea region, which may have blocked the northward transport of warm water

through the seaway, triggering cooling from northern boreal sources [25].

Worsley continued working on global cycles at the grand scale, this time with his Ohio University colleague David Kidder. Finding a correlation in time between prolonged episodes of mountain building and icehouse climate, they saw this as '*prima facie evidence for orogenically driven $CO_2$ drawdown and carbon burial*' [26, 27], much as Raymo and Ruddiman had suggested (see Chapter 9).

Kidder and Worsley divided Earth's climates into icehouse, greenhouse and a kind of super-greenhouse that they described as 'hothouse' (Figure 10.3) [27]. As they pointed out '*Hothouse climate has been approached or achieved more than a dozen times in Phanerozoic history. Geologically rapid onset of hothouses in $10^4$–$10^5$ yr occurs as HEATT (haline euxinic acidic thermal transgression) episodes, which generally persist for less than 1 million years. Greenhouse climate preconditions conducive to hothouse development allowed large igneous provinces (LIPs) [with their emissions of $CO_2$], combined with positive feedback amplifiers, to force the Earth to the hothouse climate state. The two most significant Cenozoic LIPs (Columbia River Basalts and much larger Early Oligocene Ethiopian Highlands) failed to trigger a hothouse climate from icehouse preconditions, suggesting that such [icehouse] preconditions can limit the impact of $CO_2$ emissions at the levels and rates of those LIPs*' [27].

In the haline euxinic acidic thermal transgression (HEATT) model of Earth's climate (Figure 10.3) [27], the average global surface ocean temperatures in the icehouse state ranged from roughly 15 °C, with 280 ppm $CO_2$ (expressed as $1 \times CO_2$) to 21 °C (with $2 \times CO_2$). Surface ocean temperatures in the cool greenhouse state ranged from 21 to 24 °C (with $4 \times CO_2$), whilst those in the warm greenhouse state ranged from 21 to 30 °C (with $16 \times CO_2$), and those in the hothouse state reached almost 33 °C (with $32 \times CO_2$). We can take these conditions to represent the natural envelope of the climate system in the Phanerozoic. The 280 ppm $CO_2$ chosen by Kidder and Worsley for the low end of their icehouse scale represents the present preindustrial interglacial Earth with 280 ppm $CO_2$ in the air. At peak icehouse conditions, as we shall see in Chapters 12 and 13, $CO_2$ fell to 180 ppm, and global average temperatures fell by a further 4–5 °C.

Bill Hay, whom we met in Chapter 6, was much taken with Kidder and Worsley's identification of this new hothouse state for climate, and with their division of greenhouse states into cool (with small polar ice caps and Alpine glaciers, but no ice sheets capable of calving to produce icebergs) and warm (where the only ice may be seasonal sea ice at the poles). Mapping out the alternations between icehouse, cool greenhouse, warm greenhouse,

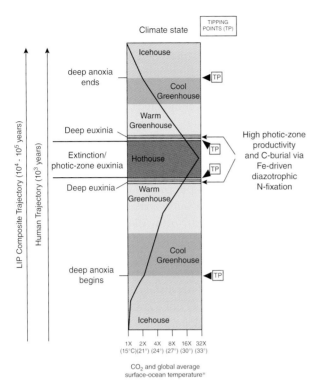

**Figure 10.3** Progression of developments during a HEATT episode [26, 27]. $CO_2$ thresholds needed for achieving cool greenhouse, warm greenhouse, and hothouse planetary states are suggested (1X through 32X $CO_2$) using the 280 ppm pre-industrial level as 1X $CO_2$ and today's solar constant. High productivity of nitrogen-fixing bacteria and green-algal phytoplankton, coupled with increased carbon-burial rate and efficiency as anoxia expands, hampers achievement of the HEATT peak unless warming factors and feedbacks can overcome that obstacle. Similar carbon-burial rates after the HEATT peak accelerate cooling from hothouse to warm greenhouse. *Source:* From figure 3 in Ref. [27].

and hothouse states through time over the past 750 Ma [28], Hay calculated that Earth had been in an icehouse state (with substantial ice at one or both poles) for just 25% of Phanerozoic time (the past 540 Ma) and in a greenhouse state (with little or no ice at either pole) for the remaining 75% (with 4% of the total spent in the hothouse state). How about interglacials like the one we live in? Hay calculated that these represented only about 10% of the time spent in the icehouse state, i.e. 2.5% of Phanerozoic time – a trifling amount. Evidently, we live in unusual times under geologically rare conditions. What might it take to tip our climate back into the greenhouse state typical of 75% of Phanerozoic time? That is one of the prime questions for this book.

Evidence for several kinds of geological and biological events following regular cycles of similar lengths through Phanerozoic time continues to accumulate, with marine organisms showing cycles of roughly 62 and 140 Ma, with stratigraphic sequences showing a cycle of around 56 Ma,

and with a 59 Ma cycle in the strontium isotope record and in the atmospheric $CO_2$ record [29]. By 2013, Michael Rampino of New York University and Andreas Prokoph of Carleton University in Ottawa were ready to pose the question – if these 60 and 140 Ma cycles are real, is there an underlying cause in large scale Earth processes [29]? Do these cycles have something to do with mantle convection or plume activity? Mantle plumes are now thought to be responsible for the eruption of flood basalts, as we shall see below. However, by 2015, Rampino was considering the possibility that dominant ~26 to 30 Ma periodicity might reflect an extraterrestrial forcing, possibly cosmic showers, related to a ~30–42 Ma half-period of the solar system's oscillation about the plane of the Milky Way galaxy [30].

Dieter Müller and Adriana Dutkiewicz of Sydney University took a different approach, demonstrating that the '*periodicities of 26 to 30 [Ma] [which] occur in diverse geological phenomena including [atmospheric $CO_2$], mass extinctions, flood basalt volcanism, ocean anoxic events, deposition of massive evaporites, sequence boundaries, and orogenic events [are in fact controlled by fluctuations in] the vast oceanic carbon crustal reservoir*' driven largely by the dynamics of migration of subduction zones – '*a previously overlooked mechanism that connects plate tectonic pulsing with fluctuations in atmospheric carbon and surface environments*' [31]. To arrive at this perspective the authors carried out a spectral analysis of the latest $CO_2$ data, represented in Figure 9.20, which revealed that they contained previously unrecognized 26–32 Ma cycles akin to the 26–30 Ma periodicity previously identified in Mesozoic–Cenozoic geological events including extinctions [31]. Tectonic processes that might explain these cycles include '*degassing of $CO_2$ along mid-ocean ridges, storage of $CO_2$ in the ocean crust due to hydrothermal alteration of basalt, storage of $CO_2$ in the oceanic mantle due to serpentinization of mantle peridotites* [see Chapter 5] *along the trench-parallel outer rise where the subducting plate bends and is faulted, and venting of arc volcanoes where a portion of subducted $CO_2$ reenters the atmosphere*' [31]. The authors recognized that oceanic crust is a significant sink for carbon, which accumulates at an estimated rate of $3.4 \times 10^{12}$ mol C/year due to the long-term alteration of the basaltic oceanic crust by low-temperature (hydrothermal) fluids (essentially circulating seawater), which precipitate calcite in veins and voids. Most alteration occurs within 20 Ma of basalt emplacement, with the rest occurring in crust aged between 20 and 50 Ma on the ridge flanks [31]. Precipitation increases with bottom water temperature, which explains why the $CO_2$ content of Cretaceous ocean crust is higher than the ocean crust of the late Cenozoic. Müller's tectonic model implies a doubling of ridge lengths from 200 Ma ago as Pangaea began splitting apart, to a maximum during the early Cretaceous (145–100 Ma ago), followed by a decrease. His analysis of seafloor spreading and subduction rates (in turn related to volcanic degassing along island arcs) shows a dominant period of 26 Ma, matching that observed in atmospheric $CO_2$ suggesting *'a compelling connection between tectonic processes, atmospheric $CO_2$ cycles, and surface environmental conditions'*, as reflected in the geological record [31].

These various analyses suggest a strong link should exist between tectonic processes and ocean chemistry, which we turn to next.

## 10.2    Ocean Chemistry

Mike Arthur figured out back in 1980 that the elemental and stable isotopic composition of marine sediments represented the chemical history of seawater [32]. Given the vast amount of $CO_2$ in the ocean and the small amount in the atmosphere, he realized that there had to be a link between ocean and atmospheric $CO_2$ that was discoverable from the ocean's chemical history. Back then, it was commonly assumed that the ocean's composition had remained constant through time, but Arthur – an integrator like his mentor Al Fischer – recognized that ocean composition reflects, on the one hand, the interplay between the composition of the atmosphere, the climate, weathering, and the passage of dissolved chemical species to the ocean via rivers, and, on the other hand, the rates of ocean circulation, inputs of hydrothermal fluids associated with sea-floor spreading, and changes in biological and non-biological processes affecting the extraction and storage of materials (e.g. plankton extracting calcium [$Ca^{2+}$] ions and $CO_2$ and using them to make $CaCO_3$ skeletons that get stored in bottom sediment). The result of this interplay was that geochemists could use the sedimentary chemical record to infer changes in climate.

Arthur also reminded us that all the $CO_2$ in the atmosphere is probably cycled through plants once in every 10 years or less, which is why in order to understand climate, we have to understand the carbon cycle that controls atmospheric $CO_2$. That means we have to have a comprehensive understanding of the operation of the biosphere [32]; we have to know about carbon sources and sinks, the availability of nutrients, productivity, the burial of organic matter, the accumulation and dissolution of carbonates, and the rates and reactions involved in the weathering of silicate and carbonate rocks. It turns out that evaporites are an important part of the equation. The massive formation of evaporitic salt deposits in isolated sedimentary basins in arid environments absorbs a great deal of calcium ($Ca^{2+}$) ions, for example in gypsum (calcium sulphate), thus transferring Ca from the carbonate to

the evaporite reservoir. As a result there is less Ca available for $CaCO_3$ to form, which results in a net transfer of $CO_2$ from the ocean to the atmosphere, accompanied by a rise in the CCD. This scenario typifies the Early Cretaceous of the narrow and slowly opening South Atlantic, where 2–3 km of evaporites accumulated through 2 Ma in the isolated Angola and Brazil Basins, preceding an abrupt rise in the CCD in the Aptian (125–112 Ma ago). Arthur went on to point out that the accumulation of 1.5 million km$^3$ of salt in the Mediterranean in a period of 1 Ma in the Messinian stage of the upper Miocene (7.2–5.3 Ma ago) made the Atlantic less salty. That made it easier to form sea ice in the northern North Atlantic, which may have increased albedo there sufficiently to contributed to the progressive cooling that eventually led to the formation of the northern hemisphere ice sheets [32].

Support for idea that the history of plate tectonics is represented in the chemistry of seawater comes from James Walker. Studying the global geochemical cycles of carbon, sulphur, and oxygen, he concluded that there was *'a significant flux of hydrothermal sulfide to the deep sea, at least during the Cretaceous'* [33]. Assuming it came from hydrothermal vents on spreading ridges, this supports Müller's theory that there was more seafloor spreading in the Mesozoic than since [34].

By 1983 further evidence had accumulated to show that plate tectonics influenced atmospheric $CO_2$ and seawater chemistry, which in turn influenced marine organisms. Based on studies of the primary mineralogy of oolites and early marine carbonate cement, Sandberg had divided Phanerozoic time into three intervals of 'aragonitic seas' (Aragonite I = late Precambrian to early Cambrian; Aragonite II = late Carboniferous to early Jurassic; and Aragonite III = early/middle Cenozoic to present) and two intervals of 'calcitic seas' (Calcite I = Cambrian to late Carboniferous; and Calcite II = early Jurassic to early/middle Cenozoic) [35]. According to Stanley and Hardie, Sandberg, following the work of Mackenzie and Pigott that we reviewed in Chapter 9, *'speculated that the observed mineralogical oscillations were mainly the result of tectonically induced changes in atmospheric $pCO_2$. The rationale for this explanation was that first-order sea level changes driven by sea-floor spreading correlate with Sandberg's calcite sea–aragonite sea oscillations and with the "icehouse–greenhouse" cycles of Fischer: when [mid-ocean ridge] activity was high, global sea level and atmospheric $pCO_2$ would have been high, and "greenhouse" conditions would have prevailed'* [36]. By 1996, Hardie had shown that Sandberg's oscillations, together with synchronous oscillations in the mineralogy of marine potash evaporites, could be explained by secular shifts in the Mg/Ca ratio of seawater driven by changes in spreading rates along mid-ocean ridges, and that aragonite would also be associated with the deposition of high-Mg calcite [37].

How had climate change influenced marine biology? Reviewing progress in 1998, and referring to Sandberg's numbered aragonitic or calcitic periods, Stanley and Hardie provided details: '*In early and middle Paleozoic calcite seas (Calcite I), reefs were dominated by calcitic ... rugose corals and ... stomatoporoids ... [while] during the period of late Palaeozoic – early Mesozoic aragonite seas (Aragonite II), aragonitic groups of sponges, scleractinian corals, and phylloid algae, as well as high-Mg calcite red algae, were principal reef builders. During Late Cretaceous time, at the acme of Calcite II, massive rudists displaced aragonitic hermatypic corals .... During Calcite II, calcitic nannoplankton formed massive coccolith chalks in [the] warm shallow seas of the Late Cretaceous, after the Mg/Ca ratio of seawater had reached a very low value and calcium concentration, a very high value. As the Mg/Ca ratio of seawater rose and calcium concentration fell during the Cenozoic Era, individual coccoliths, on average, became less massive and encrusted cells less thickly. By Pliocene time, during Aragonite III, the prominent genus Discoaster secreted only narrow-rayed coccoliths that covered less than 25% of the cell surface ... [and] the aragonitic green alga Halimeda emerged as the dominant skeletal sediment producer in reef tracts ... In today's aragonite seas (Aragonite III) scleractinian corals are again dominant reef builders, along with high-Mg calcite coralline algae*' [36]. The two scientists observed that '*the Mg/Ca ratio of seawater [exerts] strong control over the success of individual reef-building taxa .... [indicating] that throughout Phanerozoic time a chain of causation has extended from mid-ocean ridge processes, via seawater chemistry, to the mineralogical and biological composition of reef communities and bioclastic carbonate deposits*' [36]. These changes are broadly consistent with the changes in atmospheric $CO_2$ deduced by Vaughan [17] (Figure 10.2), although he does not show the low $CO_2$ in the Proterozoic that we might expect from the presence of Aragonite I there [35, 36], but we do now know more about the Neoproterozoic, which might explain that. These broad conclusions have been more firmly cemented with time, as we see below.

These various findings on the past chemistry of the atmosphere and ocean were brought together into a coherent picture of the behaviour of Earth's climate by Heinrich (Dick) Holland (1927–2012, Box 10.3).

Holland saw a close correspondence between carbonate mineralogy and sea level [38]. Comparing the distribution of aragonite-rich versus calcite-rich carbonates with the distribution of sea level through time, as mapped by Pete Vail [9], Bil Haq [39], and Tony Hallam [40], he recognized that aragonite tended to be associated with times of low sea level (early Cambrian, Carboniferous through Early

---

**Box 10.3   Heinrich Holland**

Holland was born to Jewish parents in Germany, and was sent to England as a child, to escape the Nazis. He ended up in the United States with his parents in 1940. Graduating from Princeton with a degree in chemistry, and acquiring a PhD in geology from Columbia University in New York in 1952, he subsequently served on the staffs of both Princeton and Harvard. A brilliant scholar, he was made a member of the US National Academy of Sciences, and received the V.M. Goldschmidt Award of the Geochemical Society in 1994, the Penrose Gold Medal of the Society of Economic Geologists in 1995, and the Leopold von Busch Medal of the Deutsche Geologische Gesellschaft in 1998.

---

Jurassic, and Neogene), and calcite with times of high sea level. He deduced that this correlation represented changes with time in the amount of hot hydrothermal fluid exhaled from mid-ocean ridges. This fluid was more abundant when rates of production of basaltic ocean crust and mid-ocean ridges were high, which in turn raised sea level. Seawater circulating through these fractured rocks lost much of its dissolved magnesium (Mg) and sulphate ions [41]. The exhalation of this Mg-depleted seawater as hydrothermal fluid at ridge crests changed ocean chemistry, affecting the calcium carbonate mineralogy of skeletons formed by marine creatures [38] as noted by Hardie [36, 37]. When rates of ridge production were high, the depletion of seawater in Mg favoured the deposition of calcite. When rates of ridge production were low, the enrichment of seawater in Mg favoured the deposition of aragonite. Aragonitic carbonate deposits are thus more common when continents collide and oceanic crust is being destroyed, rather than being created. Jan Zalasiewicz and Mark Williams from Leicester University call the oscillation between the two minerals the 'calcite metronome' [42]. Cool climates are associated with aragonitic carbonate deposits, warm ones with calcitic deposits [17], confirming Arthur's thesis [32]. We would expect these mineralogical consequences to reflect the changes from periods of high to low $CO_2$ depicted in Figure 10.2.

Additional geochemical evidence soon arrived to support Holland's observations. In 2010, Rosalind Coggon of Imperial College London and colleagues analysed Mg/Ca and Sr/Ca ratios in carbonate veins that precipitated in the basalts on the flanks of mid-ocean ridges from circulating fluids derived from seawater [43]. Before the Neogene, and back to 170 Ma ago, the ratios of these elements were lower than they are in the modern ocean, presumably because the rate of sea-floor spreading, and hence the production of hydrothermal fluids at mid-ocean ridge crests, has declined since the Cretaceous, increasing the Mg content and Mg/Ca ratio of seawater [41]. The change in the Mg/Ca ratio occurs because the submarine chemical weathering of basalt on mid-ocean ridges by hot percolating seawater is a source of calcium (Ca) and a sink for magnesium (Mg) through the breakdown of minerals such as feldspar and olivine during the alteration of pillow basalts, sheeted dikes, and gabbros, to form clay minerals (hydrous aluminium silicates), such as smectite, saponite, and chlorite [41, 44].

In 2013, Dietmar Müller and colleagues from Sydney University demonstrated that the Mg/Ca ratio in seawater is controlled not just by crustal accretion at mid-ocean ridge crests and the associated production there of high temperature hydrothermal fluids, but also by the gradual growth and destruction of mid-ocean ridges and their relatively cool flanks, the latter being the sites of extensive off-ridge, low-temperature hydrothermal fluid flow derived from heated percolating seawater [44]. Their main thesis is pictured in Figure 10.4. They base their approach on the fact that only ~30% of the total global hydrothermal fluid flux occurs at the ridge crest. Most heat loss occurs by passive convection on ridge flanks and involves low temperature hydrothermal circulation in crust as old as 65 Ma, where chemical reactions (the chemical weathering of sea floor basalt by percolating seawater) take place at temperatures as low as 25 °C [44]. Hence ridge flank contributions to the global geochemical budget are likely to be substantial. The aragonite seas of the Jurassic were associated with a mid-ocean ridge system only half as long as the modern system. The change from aragonite seas to calcite seas between 200 and 140 Ma (Figure 10.4) coincided with the break-up of Pangaea, which increased ridge length by 25% by 140 Ma, thus increasing the volume of seabed basalt available for alteration to clay minerals. The ridge volume subject to the most alteration (and represented by the increase in fluid flow through the ridge and its flanks) increased substantially in the Cretaceous period, peaking around 120 Ma ago (Figure 10.4). Evidently fluid flow through the crust, altering the chemistry of both the oceanic crust and the ocean, was at a maximum during the Cretaceous, decreasing the Mg/Ca ratio and favouring the precipitation of abundant low-Mg calcite and the accumulation of calcareous fossils in deposits like the Chalk. At those same times volcanic activity associated with sea-floor spreading made $CO_2$ more abundant in the atmosphere (Figure 9.20), which made bottom waters and their associated percolating hydrothermal fluids more corrosive, increasing their submarine chemical weathering potential and raising the CCD. In contrast subduction dominated during the late Cenozoic, reducing the volume of that

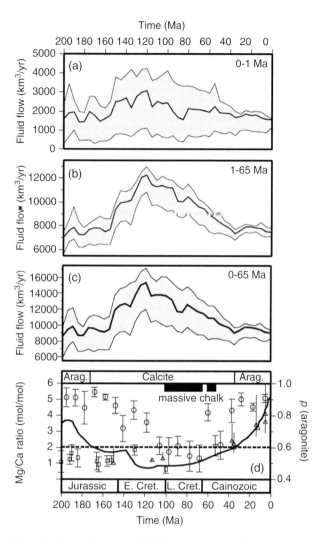

**Figure 10.4** Hydrothermal fluid flux and marine carbonate geochemistry. (a–c) Global mean fluid flux in ocean crust (uncertainties in grey) (a) aged 0–1 Ma; (b) aged 1–65 Ma; (c) aged 0–65 Ma. (d) Calculated oceanic Mg/Ca ratio through time; horizontal dashed line at Mg/Ca ratio of 2 represents boundary between low-Mg calcite (Mg/Ca < 2) and aragonite (Arag.) + high-Mg calcite (Mg/Ca > 2). Circles with error bars = proportional abundance (p) of aragonitic reef builder fossils through time. Mg/Ca ratio of seawater is estimated from (i) fluid inclusions (triangles) or (ii) echinoderms (squares). Top line shows domination of aragonite or calcite in sediments. Black bar = age range of massive chalk deposits. Cret. = Cretaceous; E. = early; L. = late. *Source:* From figure 3 in Ref. [44].

hydrothermal activity and its associated chemical weathering (Figure 10.4). At the same time the production of atmospheric $CO_2$ fell, making bottom water less acidic, lowering the CCD (see below and Figure 9.3) and driving the return to aragonite seas (Figure 10.4). The calcite metronome and cycles in the Mg/Ca ratio and in marine organisms [36, 37], provide further geochemical, mineralogical, and palaeontological evidence, like Walker's sulphur cycle [33], that Müller was right about rates of sea floor spreading through time [34], and Rowley was wrong [45] (see Chapter 5). In summary, we can see a strong link between the rate of production of $CO_2$ and the rate of chemical weathering not only on land (as Berner's work suggested – see Figure 8.16), but also beneath the ocean through the impact of relatively acidic bottom waters on submarine basalts at ridge crests and on ridge flanks (Figure 10.4). Müller's main conclusions remain robust [46].

As we saw in Chapter 9, the CCD also changes with $CO_2$ output, making the CCD a chemical signal of climate change. In warm climates it is relatively shallow because dissolution of $CO_2$ makes the ocean slightly more acid, and it is deeper in cool periods (e.g. Figure 9.3). Heiko Pälike of the UK's National Oceanography Centre, Southampton, and colleagues, thought that the position of the CCD through time should tell us something about the changes in the balance through time between the supply of $CO_2$ from volcanic and metamorphic outgassing and its removal by the weathering of silicate and carbon-bearing rocks [47]. As we saw in Figure 9.3, the CCD deepened significantly at the Eocene–Oligocene boundary. This took place along with growth of the Antarctic ice sheet, a fall in sea level, and a shift of carbonate deposition from continental shelves to the deep sea, consistent with a pronounced fall in atmospheric $CO_2$ (Figure 9.3) something we explore more in Chapter 11.

It took roughly 25–30 years to get from Plass's papers in 1956 to Arthur's 1980 geochemical paper (published in 1982) [32], Fischer's papers on global climate change and stratigraphy in 1981 [4] and 1984 [5], and Walker's sulphur chemistry paper in 1986 [33]. In parallel with these, we benefitted from Holland's *Chemistry of the Atmosphere and Oceans*, in 1978 and *Chemical Evolution of the Ocean and Atmosphere*, in 1984. These stunning advances meant that from the early 1980s onwards we were in an Earth System world, where it was clear that everything was connected, and the entire globe – including the 72% covered by ocean – was becoming well enough sampled to evaluate processes operating at the global scale. As we saw in Chapter 9, in that same period geochemists like Garrels and Mackenzie and Berner and oxygen isotope specialists like Shackleton were changing our understanding of the operations of the carbon cycle and its relation to $CO_2$ and climate, and it would not be long before palynologists like Royer and Beerling were demonstrating the important role of plants in the evolution of $CO_2$ in the atmosphere. Between them, geochemists and palynologists were changing the game of palaeoclimatology. The paradigm was beginning to shift.

## 10.3 Black Shales

At times in Earth history, sediments rich in organic matter formed widespread black shales in the deep sea and on continental margins. These deposits are of interest not only as the potential source rocks for oil, but also because of their possible climatic significance. They are abundant in the Cretaceous of the deep Atlantic, where they were first drilled in 1968 on DSDP Leg 1 by Al Fischer and colleagues, who reported finding grey–black, bituminous, laminated Albian to Cenomanian radiolarian mudstones at Site 5A, just east of the Bahamas [3]. As ocean drilling progressed, organic-rich deep-sea black shales stuffed with $^{12}C$ were found to be common in the Aptian (125–112 Ma ago), at the Cenomanian–Turonian boundary (94 Ma ago), and in the Toarcian (Early Jurassic, 183–176 Ma ago). Study of these fascinating deposits suggested that at those times the oceans may have been largely devoid of oxygen (i.e. anoxic), allowing the preservation of organic remains that would otherwise have been degraded by bacteria. Occurrences were labelled 'Oceanic Anoxic Events' [48–50]. Depositional conditions seem to have been warm, with abundant $CO_2$.

Sheridan thought that these deposits formed during that part of his pulsation cycle when atmospheric $CO_2$, temperature, sea level and the CCD were rising [13, 14]. At those times the ocean would have been more thermally stratified, with poorly oxygenated bottom waters. Vegetation would have been lush on land, providing abundant fine-grained organic matter to the ocean, to be preserved as black shales where oxygen levels were low. He painted a convincing picture.

Portuguese geologist João Trabucho-Alexandre and his team noted that some of these black shales formed in lakes associated with the rift phase or early stages of opening of ocean basins [51]. Others formed as the sea flooded subsiding rift valleys, or in shallow shelf seas as continental margins subsided or sea level rose. More formed on the margins and in the deeps of the opening ocean basins, when surface waters were highly productive and bottom waters poorly oxygenated. Only a few were found in the deeps of the fully mature ocean basins, which tend to be well oxygenated [51].

In 1987, I looked into the nature and origin of the deep-water black shales of the Atlantic. I agreed with Sheridan that their formation most likely reflected internal conditions within the ocean [52]. What might those have been during the Cretaceous? In 1982, Garret Brass of the Rosensthiel School of Marine and Atmospheric Science of the University of Miami had suggested that the deeps of the Cretaceous North Atlantic would have been filled with warm, salty, dense, and oxygen-poor water derived from tropical regions, where evaporation in shallow continental margin seas made the warm surface waters salty and dense enough to sink into the deep ocean. That warm water would have contained much less oxygen than today's cold bottom water, thus making the development of anoxia and the accumulation of organic matter more likely [53]. Incidentally, being warm it would also have been less able to dissolve $CO_2$, thus ensuring that the atmosphere contained more $CO_2$ than it would under cooler conditions like those of today, which helped to keep the air warm.

Today's Mediterranean Sea provides an example of Brass's 'haline' circulation to contrast with the 'thermohaline' circulation of today's global ocean. Atlantic water makes its way at the surface through the Straits of Gibraltar to the Egyptian coast. There, evaporation in the Levantine Sea makes the surface waters sufficiently dense that they sink, returning to the Atlantic as subsurface water passing over the Gibraltar sill. As they warm at the surface in the Levantine Sea they lose dissolved oxygen to the air, warm water holding less gas than cold. Imagine that process characterizing the whole ocean. Combined with the sinking and decomposition of dead organic matter, it probably created a vastly expanded oxygen minimum zone, encouraging the accumulation of organic matter on the seabed.

Oliver Friedrich of Germany's Bundesanstalt für Geowissenschaften und Rohstoffe, in Hannover, tested Brass's model by using $\partial^{18}O$ and Mg/Ca ratios from benthic foraminifera to reconstruct the intermediate water characteristics of the tropical proto-Atlantic Ocean between 92 and 95 Ma ago [54]. The temperatures ranged from 20 to 25 °C, the warmest ever found for depths between 500 and 1000 m. Friedrich and colleagues found evidence for highly saline conditions, confirming an influx of water from surrounding epicontinental seas. The existence of these warm waters accentuated the stratification of the Atlantic basin, preconditioning it for prolonged periods of oxygen depletion.

Much the same loss of oxygen happens today in the subsurface waters of the Red Sea. The exit of its oxygen-depleted deep water contributes to the stratification and oxygen depletion of the adjacent Arabian Sea at the northeastern end of the Indian Ocean. In 1987 I suggested that in much the same way, a subsurface current of intermediate depth water poor in oxygen and rich in nutrients had most likely entered the Atlantic from the Pacific beneath the westward moving Atlantic surface water, thus contributing to stratification and oxygen depletion in the Cretaceous deep Atlantic [52]. During the Cretaceous there was no central American isthmus to block the westward moving surface water or the eastward moving subsurface current (Figures 5.12b and 6.13), whose patterns would have mimicked the circulation of modern Pacific equatorial waters.

Two other factors accounted for the abundant accumulation of organic matter in the Cretaceous sediments of the

Atlantic. One was runoff from the surrounding land, which carried terrestrial organic material derived from lush, warm, tropical and subtropical forests. The other was wind-driven upwelling along certain continental margins. There, the upwelling of nutrient-rich subsurface waters stimulated high productivity that enhanced oxygen depletion in bottom waters, reinforcing the already strong oxygen minimum imported from the Pacific [52].

The palaeoclimate map for the Cenomanian shown in Figure 6.13 suggests that wind-driven upwelling should have been well developed then along the margin of northwest Africa and in the narrow gap between West Africa and Guyana, where the richest deposits of marine organic matter are found [52]. More sophisticated numerical modelling by Robin Topper of Utrecht University in 2011 confirmed that these were the areas most likely to be subject to upwelling currents in the Cenomanian [55]. Their model confirmed my 1987 prediction that a subsurface current brought intermediate water into the North Atlantic basin [52]. It was focused along the southern margin of the basin where upwelling was best developed [55].

I thought that the unusual enrichment of Cenomanian sediments in organic matter between West Africa and Guyana might also reflect the breaking apart of Africa and South America. That would have led to an oceanic connection between the North and South Atlantic allowing highly saline, oxygen-depleted and nutrient-rich waters from the south to enter the North Atlantic, much as Red Sea water enters and influences the Arabian Sea. Such an influx would have accentuated both productivity, through upwelling of the nutrient-rich subsurface water, and preservation of organic matter at depth in the saline, oxygen-depleted deep water [52]. The influx of this 'new' deep water would have caused significant accumulation of organic-rich sediments until the reserve of nutrients was exhausted or until continued widening of the connection to the south diminished the influx of highly saline water [52]. That nutrient limit could account for the organic enrichment in the Cenomanian and its subsequent decline.

More recently, David Kidder and Tom Worsley suggested that Brass's evaporation-driven haline circulation model may have typified ocean circulation during the extremely warm 'hothouse' intervals of climate that developed in response to the massive volcanic eruptions that produced flood basalts [26, 27]. Their 'hothouse' climate state is like an extreme version of the greenhouse climate state, driven ultimately by the addition of masses of $CO_2$ to the atmosphere from flood basalt eruptions. The *hothouse model explains the systemic interplay among factors including warmth, rapid sea level rise, widespread ocean anoxia, ocean euxinia [oxygen depletion] that reaches the photic zone, ocean acidification, nutrient crises, latitudinal expansion of desert belts, intensification and latitudinal expansion of cyclonic storms, and more*' [27]. In their model, sinking warm salty tropical waters would have permeated the deep ocean, eventually making their way to the surface at the poles, where they would have warmed the polar regions, eliminated polar ice and so helped to reduce the Equator-to-pole thermal gradient, thus reducing the strength of major global wind systems.

Bill Hay linked these developments to ocean productivity, pointing out that in a world of weaker winds the supply of dust to the atmosphere would have been severely curtailed compared to what it is now, thus limiting the supply of iron (a limiting nutrient), resulting in a 'nutrient crash' [28]. At the same time the warming of the ocean depleted it of oxygen. Waters poor in oxygen, if not actually anoxic, would have filled the subsurface waters of the ocean basins. Loss of land ice and thermal expansion of water volume would have flooded the continental margins with oxygen-depleted water. There would have been many more tropical storms, extending to higher latitudes and to deeper depths than they do today, which would have helped to maintain warm conditions in polar regions, not least by promoting the development of a warming cover of clouds. This cycle came to an end eventually as warm humid conditions on land encouraged the chemical weathering of silicates, which brought down the $CO_2$ content of the atmosphere, making conditions cooler [28].

The Frakes team noticed that oceanic anoxic events tended to be associated with large $\partial^{13}C$ peaks, for example during Early Cretaceous Aptian times (125–112 Ma ago) [15]. Did this indicate cooling, or was it due to the tying up of lots of $^{12}C$-rich organic matter in oceanic anoxic events? At high latitudes in the Early Cretaceous, there were some indicators of cold climate in the form of dropstones, which may have derived from river or shore ice. This suggests that the Early Cretaceous climate could have been cooler than was previously thought, and mountain glaciers (though not ice sheets) may have been present near the poles at that time [15].

My research back in 1984 showed that '*Deposition of sediment rich in organic matter in the Gulf [of Mexico] was not confined to a Barremian-Aptian "oceanic anoxic event" but continued at high rates throughout the Early Cretaceous, possibly because the North Atlantic (and its offshoot, the Gulf of Mexico) were separated from the rest of the world's oceans by sills*' [56]. Organic matter tends to accumulate in silled basins where the oxygen content is low, as in today's Cariaco Trench on the continental shelf off Venzuela and in today's Black Sea. This tells us that individual deep basins within the Atlantic province may preserve the history of both global events (oceanic anoxic events) and local conditions (e.g. isolation of the deep Gulf of Mexico).

As Fischer pointed out, one must take care to distinguish between global and local (or regional) effects when constructing the narrative of Earth's climate history, a lesson we will return to in later chapters.

Whilst much of the organic matter in deep North Atlantic black shales originated from the remains of marine plankton, especially along the upwelling margins of northwest Africa and Guyana, much of it elsewhere in the basin was terrestrial in origin [4, 52]. This land-derived material most likely got to the deep ocean in dense, rapidly moving currents of water stuffed with suspended sediment – the so-called 'turbidity currents' – which dumped their loads in 'turbidites' (graded deposits with distinctive layers of basal sand and overlying mud). More dilute suspensions of turbid water flowed slowly down the continental margin and across the basin floors to form 'hemipelagic muds': mixtures of pelagic planktonic remains and land-derived 'terrigenous' mud supplied via rivers or winds [52]. Independent confirmation for the proposal that many of the laminated black shales from the Cretaceous of the deep Atlantic were in fact thin turbidites comes from detailed sedimentological studies in both the North Atlantic [57] and the South Atlantic [58]. Some terrestrial components would also have arrived as wind-blown desert dust, along with minor inputs of charcoal (fusain) blown in from forest fires.

The abundance of terrestrial plant remains, especially in the western North Atlantic and off Portugal, attests to high productivity on land at the time [59] and a humid temperate coastal climate [60]. Terrestrial organic matter was preferentially pumped into the deep Atlantic basin at times of lowered sea level, when rivers would have discharged their loads close to the continental slope, making the sediments on the slope unstable and liable to slump, so generating turbidity currents [56].

Accepting the latest results of Trabucho-Alexandre [57] and Stow [58], much of the Mid Cretaceous deep sea floor comprises stacled layers of very thinly bedded, darkly coloured, organic-rich turbidites, which were deposited rapidly and give the appearance of laminated deposits, sandwiched between layers of bioturbated, oxygenated, light-coloured, organic-poor hemipelagic sediments, which were deposited slowly. This pattern suggests that the Cretaceous deep-sea floor was oxygenated, with deposition interrupted from time to time by the arrival of organic-rich turbidity currents originating on the nearby continental margin, where there must have been a strong oxygen minimum zone. The thinly bedded organic-rich turbidite layers of the deep sea are likely to tell us more about conditions on the continental margins – the source of the sediments – than about conditions on the basin floor, beneath what was most probably a rather unproductive open ocean

[61]. It was once thought that the fine-scale laminations of the Mid Cretaceous Atlantic black shales represented oscillations in the oxygen saturation of bottom waters, but it now seems more likely that the arrival of organic-rich material via bottom currents may have caused the poorly oxygenated bottom waters to have become anoxic at or near the sediment water interface, preventing bioturbation by benthic organisms and so preserving thinly bedded turbidites as laminated structures.

There is another possible interpretation for the Aptian $\partial^{13}C$ peak. A team of Swiss scientists, led by Christina Keller, argued in 2011 that the prominent negative carbon isotope excursion that preceded the Aptian oceanic anoxic event (OAE-1) was caused by major volcanic activity on the Ontong Java Plateau in the western Pacific, which drove an increase in atmospheric $CO_2$ [62]. Examining floral changes in Italy, they noted that at the beginning of the isotope event the climate was warm-temperate. The temperature rose through the duration of the isotope event, with the highest temperatures coinciding with arid conditions. This may have reflected a northward shift in the hot-arid northern Gondwana floral province in response to the increase in atmospheric $CO_2$. '*Over 200 ka after the onset of OAE-1, reduced volcanic activity and/or increased black shale deposition allowed for a drawdown of most of the excess $CO_2$ and a southward shift of floral belts*', they found [62].

Careful study of one of these events, Oceanic Anoxic Event 2 (OAE-2), formed about 93.5 Ma ago at the Cenomanian–Turonian boundary, shows that it was marked by a high concentration of $CO_2$, rapid global warming, and a low-oxygen marine environment, all of which seem to be connected to high levels of silicate weathering [63]. From analyses of lithium isotopes ($\partial^7Li$), together with calcium, osmium, and strontium tracers, Pogge von Strandmann and his team suggest that '*the eruption of a large igneous province led to high atmospheric $CO_2$ concentrations and rapid global warming, which initiated OAE-2. The ... warming was accompanied by a roughly 200,000-year pulse of accelerated weathering of mafic silicate rocks, which [eventually] removed $CO_2$ from the atmosphere [cooling things down] ... The weathering also delivered nutrients to the oceans that stimulated primary productivity [which led to rapid accumulation of organic matter and the oxygen-poor conditions under which it could accumulate]*' [63]. These processes created the warming and eventual cooling that limited the bounds of this temporary event. Evidence from carbon and lead isotopes suggests that there was massive subaerial and/or shallow marine volcanism during OAE-2 linked to LIPs in the Caribbean or Madagascar or underwater on the Ontong Java Plateau in the Pacific. The data suggest that weathering of ~0.6–2.0 million km$^3$ of basaltic material under an intensified hydrological cycle

began ~30 Ka before the onset of OAE-2, which then lasted for 445 Ka [63]. Silicate weathering is calculated to have reached its peak 200–300 Ka after the start of volcanism, and then declined over a period of 100–300 Ka, allowing $\partial^7Li$ levels to be restored to former levels and the oceans to recover more oxic conditions [63].

Isabel Montañez and Richard Norris agreed. The recent discovery of large magnitude but short-lived $\partial^{13}C$ excursions at the onset of several Mesozoic oceanic anoxic events, they said '*is compelling evidence for greenhouse gas forcing of these abrupt climate events, possibly by methane hydrate release from seafloor gas hydrates ... methane release by magmatic intrusion into organic-rich sediments ... or other greenhouse gas sources such as volcanism*' [64].

Finally, several Cretaceous black shale deposits show cyclicity reminiscent of that imposed by orbital variations in insolation (discussed in Chapter 6), demonstrating '*the sensitivity of oceanic conditions to perturbation of atmospheric circulation and continental weathering brought on by global warming*' [64].

## 10.4 Sea Level

These various studies suggest a close relationship between sea level and $CO_2$ through time. Does it exist? Gavin Foster and Eelco Rohling, then of the National Oceanography Centre, Southampton, thought so. They found a well-defined relationship between $CO_2$ and sea level extending over the past 40 Ma (Figure 10.5) that '*strongly supports the dominant role of $CO_2$ in determining Earth's climate on these time scales and suggests that other variables that influence long-term global climate (e.g., topography, ocean circulation) play a secondary role*' [65]. They started from the premise that sea level largely represents ice volume, and the observation that in ice cores covering the past 800 Ka, fluctuations in the level of atmospheric $CO_2$ closely match changes in sea level (more on that in Chapter 12). This is because $CO_2$ is the principal greenhouse gas that amplifies orbital forcing and so determines to a large extent the thermal state of the Earth system across glacial–interglacial cycles and thus the amount of ice stored on land. The relationship is strong despite small leads and lags. For the past 800 Ka, $CO_2$ did not rise above 300 ppm, not much different from the pre-industrial values identified by Oeschger and Callendar. Over the past ~6000 years a lack of change in sea level shows that the ice sheets were stable, hence the threshold for major ice retreat must be higher than 300 ppm $CO_2$, which is close to the pre-industrial maximum of 280 ppm.

Examining the relationship between $CO_2$ and sea level for periods in the Cenozoic when $CO_2$ was quite abundant (Figure 10.5) [65], Foster and Rohling found that for $CO_2$

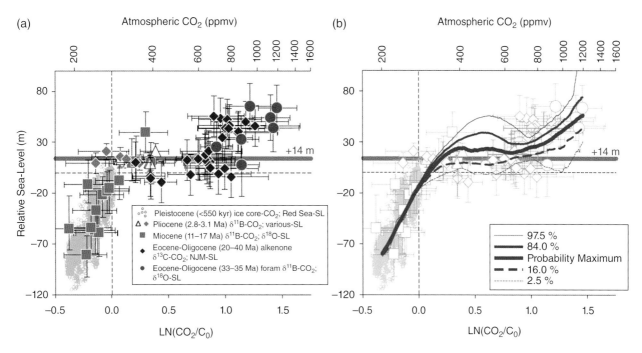

**Figure 10.5** Cross-plot of estimates of atmospheric $CO_2$ and coinciding sea level. (a) Data are split according to time period and technique used. (b) Results from a probabilistic analysis of the data that fully accounts for uncertainty in both X and Y parameters. Vertical and horizontal dashed lines denote the pre-industrial conditions of 0 m and 280 ppm $CO_2$. The horizontal orange line shows +14 m, which is the sea level rise associated with the total melting of the West Antarctic and Greenland Ice Sheets. *Source:* From figure 3 in Ref. [65].

levels between 200 and 400 ppm in Pliocene and Miocene times, the relationship between $CO_2$ and sea level was more or less the same as that for the past few hundred thousand years, with sea level rising in proportion to the logarithm of atmospheric $CO_2$. In contrast, for $CO_2$ levels between 400 and 650 ppm in Pliocene, Miocene, and Oligocene times, sea level estimates remained on a plateau of about +22 m (±12 m) compared to today's level. For $CO_2$ levels above 650 ppm, in the Eocene, rises in $CO_2$ were associated with rising sea level. Peter Barrett (pers. comm.) reminded me that there is a further plateau in sea level that is a consequence of the 64 m upper limit for ice-driven sea level rise once all the ice sheets have gone, no matter how high $CO_2$ rises.

Foster and Rohling explained the sigmoidal nature of this relationship (Figure 10.5) as follows. During the Eocene, when $CO_2$ was above 1000 ppm, sea levels were 60–70 m higher than today, as there were no major ice sheets [65]. $CO_2$ declined from 1000 to 650 ppm towards the Oligocene, and sea level fell as the East Antarctic Ice Sheet (EAIS) grew. $CO_2$ continued falling, from 650 to 400 ppm, but sea level did not respond, probably because, as the oxygen isotope data suggest, very little continental ice grew or retreated during that time. That is probably because most of the land ice was on Antarctica; the Northern Hemisphere ice sheets had not yet begun to form,

although mountain glaciers and sea ice may have developed in the Arctic. The average sea level of +22 m suggests that there was neither a Greenland Ice Sheet (GrIS) nor a West Antarctic Ice Sheet (WAIS), those being equivalent together to about +14 m, and that the EAIS may have been smaller than today by the equivalent of about 10 m of sea level. According to Foster and Rohling, '*Presumably $CO_2$ was too high, hence the climate too warm to grow more continental ice after the "carrying capacity" of the EAIS had been reached*' [65].

Evidently, then, sea level does not exhibit a simple response to changing $CO_2$. It is modulated by the behaviour of large ice sheets especially at levels of $CO_2$ between 400 and 650 ppm. The data '*suggests that 300–400 ppm is the approximate threshold $CO_2$ value for retreat and growth, respectively, of WAIS [West Antarctic Ice Sheet] and GrIS [Greenland Ice Sheet] (and possibly a more mobile portion of EAIS [East Antarctic Ice Sheet])*' [65]. This implies, further, that sea levels of 20–30 m above present at times during the Pliocene and Miocene, when $CO_2$ reached 280–400 ppm, mainly represent melting of the ice sheets of Greenland and West Antarctica, with a possible contribution from East Antarctica [65]. Sea level fell below present levels only after $CO_2$ dropped below 280 ppm some 2.6–2.8 Ma ago, when the Laurentide and Fennoscandian ice sheets began to grow in the Northern Hemisphere.

Roderik Van de Wal of the University of Utrecht, and colleagues, took a slightly different approach to test the notion that the gradual cooling of the climate through the Cenozoic could be attributed to a decrease in $CO_2$ (Figure 10.6) [66]. Collecting data on Northern Hemisphere temperature and $CO_2$ for the past 20 Ma, they found that whilst the relationship between the two was positive and clear for some data sets, it was weak for others, notably for some of the $CO_2$ estimates derived from alkenones and from boron isotopes. Relying on the assumption that there was a relationship between temperature and $CO_2$, as demonstrated by ice core data, they excluded the apparently wayward alkenone and boron-based $CO_2$ data sets, retaining the $CO_2$ estimates based on B/Ca ratios, combined alkenones and boron isotopes, and stomatal data that were consistent with the ice core data.

Using the temperature-$CO_2$ relationship emerging from their data sets, Van de Wal's team found a gradual decline of 225 ppm $CO_2$ from about 450 ppm in the Mid Miocene warm period around 15 Ma ago to a mean level of 225 ppm during the last 1 Ma, coinciding with a fall in temperature of about 10 °C (Figure 10.6). The inception of Northern Hemisphere ice around 2.7 Ma ago took place once the long-term average concentration of $CO_2$ had dropped below $265 \pm 20$ ppm.

There is another important factor in changing sea level, and that is a change in the base level of the land caused by tectonic processes leading to uplift or subsidence [67, 68]. For example, subduction leads to uplift where deep ocean sediments are scraped off the down-going plate and piled up in wedges on the edges of continents, whereas basins subside due to sediment loading creates space for sediments and potentially lowers sea level if the sediment supply dries up. A further change may come about from the location of continents in relation to the Large Low Shear-Wave Velocity Provinces (LLSVPs) at the core–mantle boundary, mentioned in Chapter 5. As Torsvik and Cocks point out, the location of Pangaea over the mantle upwelling above the Tuzo LLSVP 250 Ma ago pushed Gondwana up, significantly lowering sea level without any contribution from ice sheets [69]. Sea level reached an all-time low near the Permo–Triassic boundary, when the supercontinent was in effect high and dry. Subsequent dispersal of the continents moved fragments away from Tuzo and towards regions of negative dynamic topography, allowing global sea level to rise by 50–100 m after Early Jurassic times [69]. The global sea level curve will thus reflect long slow tectonic changes of two kinds including, first, the migration of Pangaea over Tuzo, and, second, long slow changes in mid-ocean ridge volume causing changes in $CO_2$ output and associated warming, complicated by the incidence of glaciation (storing water on land as ice) and by regional tectonic change.

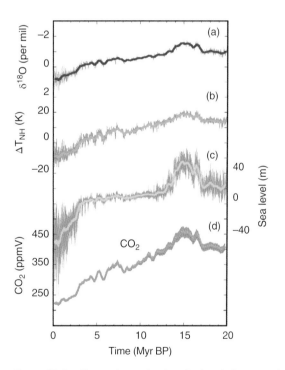

**Figure 10.6** Change in sea level and related climate variables over past 20 Ma. (a) Analysis of the smoothed $\partial^{18}O$ record of Zachos leads to (b) Northern Hemisphere temperature change with respect to pre-industrial conditions, and (c) sea level (related to ice volume). (d) The reconstructed $CO_2$ record comes from inverting the relation between Northern Hemisphere temperatures and $CO_2$ data. Thick lines represent 400 Ka running mean. Grey error bars indicate standard deviation of model input and output. *Source:* From figure 1 in Ref. [66].

## 10.5   Biogeochemical Cycles, Gaia and Cybertectonic Earth

So far, Chapters 9 and 10 have shown that geological thinking about the evolution of Earth's climate began to change from about 1980 onwards thanks to growing appreciation of the operation of the slow carbon cycle involving volcanic emission, weathering, and sedimentation. This 'revolution' in thinking about the role of the carbon cycle in Earth's climate evolution owes much to the rapid expansion of geochemistry from the mid-1950s, to the comprehensive sampling of deep ocean sediments by the DSDP and its successors beginning in 1968, to the discovery of gas bubbles in ice cores in 1978, to the modelling of the carbon cycle from around 1980 onwards, which was stimulated by a growing ease of access to fast numerical computers, and to continuing efforts to improve or find new proxies for past atmospheric levels of $CO_2$ in the 2000s.

The rise of biogeochemistry and its holistic approach to science (described in Chapter 8) also played a part in the evolution of Earth System Science, which '*seeks to integrate various fields of academic study to understand the Earth as a system. It considers interaction between the atmosphere, hydrosphere, lithosphere (geosphere), biosphere, and heliosphere*' [70, 71]. Some journals have even changed their names to keep up with this trend, such as the *Proceedings of the Indian Academy of Sciences (Earth and Planetary Sciences)*, which in 2005 became the *Journal of Earth System Science*. Reflecting these various developments, in the past 20 years we have seen many a geology department change its name to 'department of Earth Sciences'. In December 2013, the American Geophysical Union (AGU) launched a new journal focused on Earth System Science, titled *Earth's Future*. AGU's rationale for doing so was that, while we will still need disciplinary science, to fully understand and provide advice on '*the major challenges facing human society in the 21st century*', we also need to take '*more holistic approaches that will integrate knowledge from individual disciplines*' [72]. The new journal '*deals with the state of the planet and its expected evolution. It publishes papers that emphasize the Earth as an interactive system under the influence of the human enterprise. It provides science-based knowledge on risks and opportunities related to environmental changes*' [72]. The paradigm has changed.

I have no doubt that Humboldt would have been cheered by these developments, although I know that they remain anathema to some of my more traditionally minded (and commonly older) geological contemporaries – often the same ones who resist the notion that so-called 'small' changes in $CO_2$ can affect our climate. It is inevitable that with change comes resistance, which often turns out to represent simply an inability to keep up. Many of the changes have been so recent, and the literature is advancing so quickly, that only the younger generation of geologists trained in Earth Sciences and Earth System Science can be expected to be fully aware of the dramatic progress that has been made and continues to be made. Even then, awareness may be only partial, not least because many of the advances are in sub-fields like palynology, geochemistry, and biogeochemistry, which demand specialized knowledge. The general public is likely to be even less aware of these various new developments and of their significance, not least because many new advances remain for the time being hidden away behind a pay-wall in top-of-the line scientific journals readily available only to academics (a good reason to write a book about them).

One new concept has been much more in the public eye than many of the specialized elements we address in this book. It is the proposal by the English scientist and inventor James Ephraim Lovelock (1919–) that the living and non-living parts of the Earth's surface form a complex interacting system that can be thought of as a single organism, and that the biosphere has a regulatory effect on the Earth's environment that acts to sustain life. First putting forward this proposal in a scientific paper in 1969 [73], he later named it after *Gaia*, the Greek goddess or personification of the Earth [74]. In 1974, Lovelock fleshed out the concept with American microbiologist Lynn Margulis (1938–2011) [75]. In her 1999 book *The Symbiotic Planet* Margulis defined Gaia as '*the series of interacting ecosystems that compose a single huge ecosystem at the Earth's surface*', and '*an emergent property of interaction among organisms*' [76]. Lovelock popularized the concept in his 1979 book *Gaia: A New Look at Life on Earth* [77]. Harking back to both Lovelock and Vernadsky, Euan Nisbett of Royal Holloway College wrote: '*The planet shapes life, but life also shapes the planet. The maintenance of surface temperature is managed by the air: hence as life controls the composition of the air and the atmospheric greenhouse, then life sets the surface temperature*' [78]. If Lovelock and Nisbett are right, then $CO_2$ plays an important role in keeping our climate within the natural envelopes described in this book.

The Gaia concept has been explored through a series of conferences, starting in 1985. Whilst the concept of studying the Earth as a living whole has taken some hits, Lovelock's genius has been recognized by numerous awards, not least Fellowship of the Royal Society in 1974, and the Geological Society's Wollaston Medal in 2006. His PhD student Andy Watson, also elected a Fellow of the Royal Society, and Watson's PhD student Tim Lenton, expanded on Lovelock's ideas, summarizing the present state of play in their 2011 book *Revolutions that Made the Earth* [79]. I once met Lovelock in passing – he was on the committee that interviewed me for the post of director of the Institute of Oceanographic Sciences Deacon Laboratory in 1987. I must have pressed the right buttons on the day.

Does life shape the Earth then, or is it shaped by plate tectonic processes, including volcanic activity or mountain building, with its attendant weathering? Mike Leeder (1947–) of the University of East Anglia, and the Geological Society's Lyell medallist for 1992, was in no doubt [80]: '*Tectonics, climate and sea level are the dominant controls on the nature and distribution of sedimentary environments ... The processes of global tectonics that cause widespread mountain belt and continental plateau uplift have produced numerous "severe" events during Earth history: these often random workings-out of the plate tectonic cycle define the state of "Cybertectonic Earth" (from cyber, after the Greek kubernan: to steer or govern). This has worked within a usually zonally arranged series of climate belts and within the framework provided by biological evolution. At certain times* *in the geological past, a combination of factors arose as a response to continental uplift and have acted to cause certain Earth surface conditions and variables (notably mean global surface temperature, atmospheric pCO$_2$, pO$_2$) to have varied by very large amounts (x2 – x10) compared with the present value. These large fluctuations, such as those responsible for the Neoproterozoic, Late Palaeozoic and Late Tertiary glaciations, lead one to doubt the reality of homeostatic control of surface conditions as proposed in Lovelock's Gaia hypothesis. The Gaia kubernētēs (steersman) had a weak hand on the helm, frequently unable to prevent vast areas of the globe experiencing rapid fluctuations in environmental conditions and inimical conditions to life for very long periods during Neoproterozoic and Phanerozoic times. At the same time, biogenic and abiogenic processes have proved capable of returning Earth to states of mean stability, although biogeochemical cycling models seem alarmingly ad hoc and largely untestable as scientific hypotheses in any true geological sense*' [80].

Leeder argued that Lovelock's concept of the Earth as a self-regulating entity was unsatisfactory because it '*operated without reference to, and entirely independent of, the activities of plate tectonics*' [80]. Instead, Leeder's 'cybertectonic Earth' and Lovelock's biogeochemical 'Gaia' had to work together. '*It can be argued*' Leeder suggested '*that this combination of tectonics and biogeochemistry is the great fulfilment of the Huttonian philosophical scheme*', Hutton having first introduced the idea of a mobile Earth that was a '*superorganism whose proper study is physiology*' [80]. '*Modern sedimentary (and other) geologists*' Leeder said '*can thrive only if they study, or are at least aware of, not only the rocks and sediment under the surface, but also the host of related disciplines that deal with the material on and above Earth*' [80]. Humboldt would have liked that. And Lyell would have seen in it an endorsement for his call for comprehensive palaeoenvironmental analysis.

Leeder's 2007 analysis was very slightly off the mark in that he relied on Dave Rowley's 2002 suggestion that rates of plate construction had not changed for the past 180 Ma [81]. But I have produced other lines of evidence (variations through time in sea level, CO$_2$, sulphur, and calcite-versus aragonite deposition reflecting changes in ocean chemistry), which show that rates of spreading in the Cretaceous were faster than in the Cenozoic. Müller was right and Rowley wrong (Chapter 5 and above). Hence, Leeder's dismissal of the Vail curve of changes in sea level through time [9], which was linked to changes in the rates of sea floor spreading, now seems suspect. Nevertheless, Leeder's notion of a cybertectonic Earth has distinct attractions for understanding past climate change, especially when linked to biogeochemical cycles – provided we accept in addition the occasional catastrophe imposed by asteroids of the kind that ended the Cretaceous and wiped out the dinosaurs.

## 10.6 Meteorite Impacts

You only have to look at the Moon through binoculars to see that its surface is pitted with craters from giant impacts with passing asteroids. And many of us will have seen the meteor showers that grace our skies from time to time. Most visitors to Arizona, like me, will have peered into 'Meteor Crater' a giant hole in the ground some 1200 m across and 170 m deep, 69 km east of Flagstaff. Many readers will recall seeing the 1998 sci-fi disaster movie *Armageddon*, in which NASA sent Bruce Willis into space to deflect a giant asteroid from its path towards Earth. And let's not forget Ted Nield's book *Incoming* [82]. So, we should not be startled by the notion that asteroids have hit the Earth a few times in its history. Fortunately, the frequency of impact has declined very substantially with time.

Numerous massive impacts would have affected Earth's climate, at least for short periods. Georges Cuvier would have doubtless latched on to asteroid impacts as one of the missing engines for the catastrophes that he thought punctuated geological time. But they might have posed a conundrum for Charles Lyell and his doctrine of gradualism or uniformitarianism. His ignorance of their existence explains the fact that they do not disturb the pages of his *Principles*. Indeed, collisions of asteroids with the Earth do not even feature in Arthur Holmes's 1965 magnum opus *Principles of Physical Geology*, more than a century after Lyell's *Principles* were published.

We should not be too surprised at this oversight, because it was not until 1963 that Eugene (Gene) Merle Shoemaker (1928–1997) (Box 10.4) proved conclusively that Meteor Crater was indeed an impact crater [83]. The crater was initially named Canyon Diablo Crater and thought to be the result of a volcanic steam explosion. Daniel Barringer (1860–1929) correctly identified it as a meteorite impact structure in 1903, and it was renamed 'Barringer Crater' in his honour, though its more common name remains Meteor Crater.

Knowing what shock features to look for, geologists began searching methodically for meteor craters, discovering more than 50 by 1970. Support for their identification as meteorite strikes – where the term 'meteorite' covers everything from comets to asteroids – came from the Apollo Moon landings, starting in 1969. Because the lack of erosion on the moon allows craters to last indefinitely, it was possible to identify the rate of cratering, which likely applied to the Earth, too [84]. Fewer craters are visible on Earth. Plate tectonic processes, weathering, and burial by

## Box 10.4 Eugene Shoemaker

Whilst studying Meteor Crater for a PhD at Princeton in 1960, Shoemaker found coesite and stishovite in the ground there – rare varieties of silica formed when quartz has been severely shocked. Such 'shocked quartz' is now recognized as one of the metamorphic products of impact events. Shoemaker went on to work for the US Geological Survey, where he pioneered the field of astrogeology, founding the Survey's Astrogeology Research Program in 1961. Shoemaker was in an ideal position to advise NASA about its Lunar Ranger missions to the Moon, and at one point trained as an astronaut, although he was eventually disqualified on medical grounds. Arriving at Caltech in 1969, he began a systematic search for asteroids, discovering the Apollo Asteroids. While there, he proposed that asteroid strikes on Earth had likely been 'common' on the geological time-scale, and would have caused sudden geological changes. Previously, impact craters were thought to be volcanic in origin – even on the Moon. In 1993 Shoemaker co-discovered a comet, named 'Shoemaker–Levy 9', which provided scientists with a first opportunity to observe a cometary impact on a planet, when it slammed into Jupiter in 1994, leaving a massive scar. This helped to emphasize what extraterrestrial objects might be able to do if they hit the Earth. Shoemaker was awarded the Barringer Medal in1984 and the US National Medal of Science in 1992. In 1999 some of his ashes were taken to the moon and buried there by the Lunar Prospector space probe.

sediments have obliterated their traces. Those most easily found tend to be young, like Meteor Crater, which is 50 Ka old. Buried ones can only be identified by geophysical survey. Large circular structures tend to be a giveaway.

One buried crater, the 180 km-wide 'Chicxulub Crater' lies beneath the northern edge of Mexico's Yucatan Peninsula and its adjacent continental shelf and slope [85]. One of the largest impact structures on Earth, it represents a collision with a bolide (the Greek for 'missile') at least 11 km in diameter – about the size of Manhattan – which took place at the Cretaceous–Tertiary (or K–T) boundary 65 Ma ago [86]. It was discovered in 1990.

The astonishing notion that there had been a major meteorite impact at this boundary stemmed from research that Walter Alvarez (1940–) was doing on magnetic reversals in deep-sea limestones in Italy. There, he found a widespread clay layer right at the K–T boundary. Knowing that this was when the dinosaurs went extinct, he wondered what the clay layer meant, and discussed the matter with his father, Luis Walter Alvarez (1911–1988). Luis was a Nobel Prize-winning physicist from the University of California at Berkeley, who worked on the Manhattan Project during World War II, and later used a bubble chamber to discover new fundamental particles. He persuaded colleagues from the Lawrence Berkeley Laboratory to use neutron activation to analyse the clay layer. They made one of science's greatest discoveries when in 1980 they found that the clay contained abundant iridium, a chemical element common in meteorites but not on Earth [87]. Later the clay was found to contain soot, glassy spherules, shocked quartz, microscopic diamonds and other materials that formed under high temperature and pressure [88]. The researchers deduced that a meteorite impact had brought the Cretaceous to a close. The crater was only found 10 years later. The immense cloud of dust and gas from the impact would have blocked sunlight, inhibited photosynthesis, and cooled the atmosphere for a decade. It apparently affected the climate sufficiently to cause a major extinction event that wiped out the dinosaurs and other creatures, including the ammonites.

This discovery upset those of a Lyellian bent [89]. Palaeontologists were especially unhappy at the intrusion of geochemists into their cosy uniformitarian world. Lyell had noted the gap in continuity of fossils across the K–T boundary, but assumed it represented just one of those annoying gaps in the geological record. Darwin, too, marvelled at the sudden disappearance of the ammonites then, but agreed with Lyell's interpretation. Sadly both Lyell and Darwin were wrong. The paradigm had shifted. Catastrophes did happen.

A review of the evidence in 2010 showed that the global ejecta layer and the extinction event coincided. Moreover the ecological patterns in the fossil record agreed with modelled environmental perturbations (darkness and cooling). Reviewers concluded that the impact triggered the mass extinction [90]. Later, in 2013, Paul Renne from the Berkeley Geochronology Centre presented argon isotopic data that established that the impact and the extinction coincided to within 32 Ka – a very small 'error window' in geological terms [91]. Renne and colleagues suggested that the subsequent perturbation of atmospheric carbon at the boundary probably lasted less than 5 Ka, but that recovery of the major ocean basins took much longer. The impact most likely triggered a shift in the state of ecosystems that were already under stress [91]. The meteorite impact vaporized carbonate sediments at the impact site, leading to a rise in atmospheric $CO_2$ that led to warming once the impact dust had fallen from the atmosphere. Oxygen isotopes in phosphatic remains (fish bones) show a decrease indicative of a warming of about 5 °C in a geological section spanning c. 100 Ka [92].

Not everyone agreed with the Alvarez's claim, not least because the massive eruption of flood basalts in India's Deccan Traps occurred at about the same time and might well have had a similar catastrophic effect [93]. But it is not easy to see how the long eruptive period of the Deccan Traps – extending over about 1 Ma and spanning the K–T boundary – fits with the tightly constrained evidence for a very short extinction event. Even so, the Trap eruptions, which probably produced 10 million km$^3$ of lava at rates of up to or even more than 1 million km$^3$/year [93], may have affected the global ecosystem enough for it to have succumbed more easily to the effects of the impact (more on that later). The Deccan eruption was a response to plate tectonic processes. It immediately preceded and was possibly related to the opening of the Arabian Sea.

Were there multiple impacts at the K–T boundary – an asteroid shower, or impacts from bits of a fragmented asteroid? Two impact craters of the right age have been identified – the 24 km diameter Boltysh Crater in the Ukraine, and the 20 km diameter Silverpit Crater in the North Sea [86]. Others as yet unidentified may be hidden beneath the sediments of the deep ocean floor.

The Alvarezes were not the first to suggest that major bolide impacts might have caused biological extinctions, either. That honour goes to Digby Johns McLaren (1919–2004) (Box 10.5).

Studying the Devonian around the world, McClaren saw that the Late Devonian Frasnian–Famennian boundary (374.5 Ma ago) was knife-edge sharp, synchronous globally, and accompanied by extinction of 50% of the biomass. In his presidential address to the Palaeontological Society of America in 1969 [94], he argued that the only explanation for such a thing was the impact of a giant meteorite. In the

---

**Box 10.5 Digby McLaren**

Digby McClaren was born in Carrickfergus in Northern Ireland, brought up in Yorkshire in England and spent his working life with the Geological Survey of Canada (GSC), specializing in studies of the Devonian. By 1973, he was Director-General of the GSC, a post he held until 1980. McLaren was honoured in many ways for his scientific contributions, not least by being made a fellow of the Royal Society of Canada in 1968, and becoming its president from 1987 to 1990. Amongst his several honours he was made a fellow of London's Royal Society in 1979, and the Geological Society of London awarded him its Coke Medal in1985 and made him an honorary fellow in 1989. The Digby McLaren Medal of the International Commission on Stratigraphy is named after him.

---

words of his biographer '*the members of the Society were left in a state of shock; the general consensus was that he must have lost his marbles*' [95].

McLaren's revolutionary mechanism for explaining mass extinctions was partly inspired by the research of Robert Dietz, of the Navy Electronics Laboratory (whom we met in Chapter 5), which showed in 1964 that the giant circular Sudbury structure in Ontario was an astrobleme (an impact structure) [96]. McClaren had made one of those giant intellectual leaps, and the reward came when an iridium anomaly was discovered at the Frasnian–Famennian boundary in Australia's Canning Basin [97]. In his address as the retiring president of the Geological Society of America, in October 1982, he reviewed progress in the search for bolides at boundaries, concluding that it was highly probable that a large body impact had caused the extinctions of the Late Devonian and end Cretaceous, possible that such a mechanism accounted for the extinctions in the Late Ordovician and Late Triassic, and somewhat likely that such a mechanism accounted for the Late Permian extinction [98]. Could it be that meteorite impacts might be quite common? Drawing on evidence from Gene Shoemaker, McLaren said '*several 1-km-wide objects might be expected to arrive every million years, whereas larger objects of about 10 km in diameter should arrive at an interval of between 60 and 100 m.y., or even every 50 m.y. …. [and] there is the possibility of the relatively rare arrival of a body as much as 20 km in diameter*' [98]. The main effect of such an impact would be a massive ejection of dust into the stratosphere, which would remain in place for months (if not years), reflecting sunlight and seriously cooling the planet. Given that 72% of the surface of the planet is covered by the ocean it seemed likely that 70% of the meteorite strikes would also be in oceanic areas and thus be difficult to detect.

The publications of Shoemaker, Dietz, and McClaren may well have influenced the great Harold Urey, who suggested in 1973 that '*it does seem possible and even probable that a comet in collision with the Earth destroyed the dinosaurs and initiated the Tertiary division of geologic time*' [99].

Back in 1998, however, Tony Hallam reminded us that really good evidence for major meteorite impacts, in the form of iridium layers of global extent and shocked quartz coincident with extinction, is only available for the K–T boundary event [100]. Glassy spherules or tektites formed by meteorite strikes have been found in the geological record, but not in association with other major extinctions. That lack of a direct association between impacts on the one hand and extinctions on the other has commonly led geologists to look to other causes of extinctions, notably massive volcanic eruptions like those of large igneous provinces.

Before we leave the topic of extraterrestrial impacts, we should note in passing the notion that the Sun's orbit around the galactic centre every 250–300 Ma might also affect Earth's climate from time to time. One of those giving some early thought to that prospect was Herbert Friedman of the US National Research Council, in 1985 [101]. Friedman noted that the Sun undulates above and below the galactic plane with a period of 27 Ma, providing the possibility of collision with dense clouds of intergalactic dust in the disk of the Milky Way. As it circles the galactic nucleus, the Sun also drifts slowly through the dust clouds in the galaxy's spiral arms. Collision of our solar system with a dust cloud would slow the solar wind, and collision of dust with the Sun's surface might raise its temperature very slightly. Both effects have the potential to change Earth's climate a little on a regular basis. Friedman recognized that '*it might seem far-fetched .... to dwell .... on [such] exotic suggestions*' [101]. James Pollack, of NASA's Ames Research Centre, agreed, stating his opinion that interstellar dust clouds were far too thin to affect Earth's climate [102]. Besides, present data do not show 300 Ma cycles or 150 Ma cycles back beyond around 650 Ma ago [18]. Furthermore, the occurrence of the short Ordovician–Silurian glaciation 440 Ma ago is an anomaly in the 300 Ma cycle mode, and is not easily explainable by the extraterrestrial argument.

Henrik Svensmark of the National Space Institute of the Technical University of Denmark proposed a slightly different 'galactic' hypothesis relating life and climate on Earth to the incidence of galactic cosmic rays arriving from supernovae. His basic argument is, first, that galactic cosmic rays impacting the outer atmosphere can lead to the formation of low clouds whose high surface albedo contributes to cooling; second, that the solar system has experienced many large short-term increases in the flux of such rays from nearby supernovae; and, third, that these increases may reflect the passage of the Sun through the spiral arms of the galaxy [103]. He claimed to have found a link between his calculated rates of supernova production and Earth's climate and marine biodiversity over the past 510 Ma. He explained an inverse correspondence between his supernova rates and our planetary $CO_2$ levels by calling upon enhanced bio-productivity in oceans that were better fertilized under cold conditions [103]. Svensmark calculated that there were substantial peaks in supernova production over the past 15 Ma, and 160–110 Ma ago, 230–210 Ma ago, 310–250 Ma ago, 400–360 Ma ago, 450–415 Ma ago, and 510–490 Ma ago, with the largest occurring 310–250 Ma ago. Smaller episodes of supernova production lay between these major peaks. Comparing his supernova production rates to changes in sea level over the past 50 Ma, suggested to Svensmark that there was a crude

link between those rates and changes of sea level of 25 m or more. We are not dealing here with precise correlation. As Svensmark points out: '*The aim here is not to try to achieve a perfect covariance by further statistical iterations, but to illustrate that, in the absence of any other explanation for them, the fast sea level falls are just what are to be expected as signals of closer-than-usual SN detonations*' [103]. Nevertheless, geologists will be well aware that there are also close links between changes in sea level and plate tectonic processes including basin subsidence. The link between Svensmark's supernova production rates and changes between warm and cool climate through time over the past 510 Ma are similarly crude, as are the links to palaeotemperature as represented by $\partial^{18}O$ values through time for the past 200 Ma [103].

Svensmark assessed the relation between cosmic ray flux and biodiversity by assuming that the evolution of marine biodiversity is a function of the combination of the supernova production rate and sea level. However, that approach ignores, first, the well recognized close association between sea level and numbers of marine species, and second, the well recognized biological reality of marked extinction events, for which there is an adequate geological explanation (e.g. end Permian) [104]. Svensmark argued that a major fall in supernova production rates at end Permian time (c. 250 Ma ago) should have led to major and sudden warming as the solar system '*left the very active Norma spiral arm*' of the galaxy [103]. Equally, we would expect an abrupt effect on climate from the Siberian Trap volcanism. The issue becomes one of rates. How rapid would the transition be as the solar system exited a galactic arm, compared with the suddenness of a major igneous event on Earth? Svensmark's time bins last 8 Ma, so his end Permian galactic transition could have lasted up to that length of time, while in contrast the Siberian Trap eruptions were massive and short-lived. Svensmark claims to find an association between his supernova production rates and palaeoproductivity as represented by $\partial^{13}C$ values (both of which, according to him, should increase in cold periods); in addition, he thought that $CO_2$ should be drawn down from the atmosphere by that high productivity (without considering the roles of volcanism as a $CO_2$ source or weathering as a $CO_2$ sink). Svensmark showed $CO_2$ as decreasing from 450 Ma ago to around 280 Ma ago, in agreement with Berner (Figure 9.7) and Royer (Figure 9.13). But we know that this pattern can be readily attributed to planetary processes, notably the gradual rise in land plants, which absorb $CO_2$, rather than extraterrestrial control. Similarly, much of the rise in $^{13}C$ in marine carbonates in the late Carboniferous, which Svensmark saw as representing the influence of cosmic rays cooling the planet, most likely representing the tying up in coal deposits of the $^{12}C$ favoured by plants. In summary, it appears that Svensmark has worked hard to make the case for galactic cosmic rays as an important control on Earth's climate and life, while ignoring the more plausible geological explanations based on Earth's surface processes for those same variations.

As we can see from the analyses of Earth history by the likes of Fischer, Worsley, Frakes, Vaughan, Hallam, Leeder, Lenton, Watson, and Müller, we can find most of the answers we need by examining the behaviour of the Earth itself.

## 10.7 Massive Volcanic Eruptions and Biological Extinctions

By the mid-1980s Jack Sepkoski (1948–1999), a palaeontologist from the University of Chicago, had identified a number of biological extinctions through time over the past 250 Ma, and he and a colleague, David Raup, speculated that they were part of a cycle recurring with an interval of 26 Ma [105]. One of these (the K–T boundary) appeared related to a meteorite impact, as spelled out above, and they speculated that all or some of the others might be, too. I note in passing that David Raup subsequently considered

that the cycle he had identified might be a sort of statistical fluke [89]. Nevertheless, it does replicate what Müller and others have found recently in the global records of $CO_2$ and sea-floor spreading [44].

Could these extinctions be related instead to volcanic eruptions? Vincent Courtillot, from the Institut de Physique du Globe, in Paris, thought so [106]. It was he who drew attention to the possible linkage between the K–T boundary and the eruption of the flood basalts of the Deccan Traps in India [93]. In 1994 he reported a link between nine of Sepkoski's 10 extinction events of the past 300 Ma and some of the 12 known examples of large continental flood basalt provinces from that period. By 1994, the term 'Large Igneous Province' (or LIP) was widely applied to the areas of massive eruptions that produced flood basalts both on the continents and in the oceans (e.g. the submarine Ontong–Java Plateau, east of New Guinea and north of the Solomon Islands). Courtillot concluded that flood basalt events provide the most likely explanation for extinctions, and that the Deccan Traps eruptions had set in motion the extinction event that culminated with the meteorite impact at the K–T boundary [106]. He particularly favoured the eruption of the flood basalts of the Siberian Traps to explain the massive extinction at the Permo–Triassic (P–T) boundary. Tony

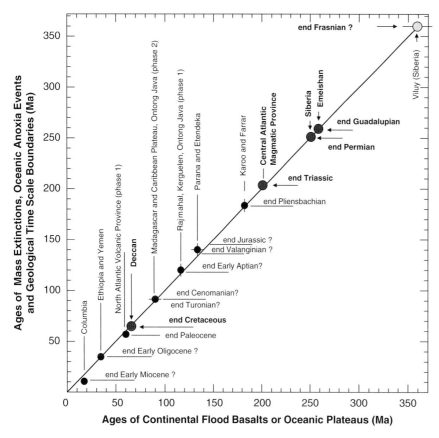

**Figure 10.7** Links between large igneous provinces and biological extinctions. *Source:* From figure 1 in Ref. [107].

Hallam agreed that the end Permian extinction correlated with eruption of the Siberian Traps [100].

As the timings of LIPs became better established, Courtillot found yet more evidence for a linkage between 16 massive igneous events and biological extinctions, naming four such events in particular as having an important causal connection – the Deccan Traps (K–T boundary), the Siberian Traps (P–T boundary), the Central Atlantic Magmatic Province (CAMP) (Triassic–Jurassic [T–J] boundary), and the Ethiopian and Yemen Traps c. 30 Ma ago (Figure 10.7) [107]. By 2007, he had refined this picture, noting that the four most recent large mass extinctions of the Phanerozoic associated with major flood basalt eruptions were the late Permian (Guadalupian) event at 258 Ma ago, the end Permian event at 250 Ma ago, the end Triassic event at 200 Ma ago, and the end Cretaceous event at 65 Ma ago [108]. Given the subsequent slight adjustments to the geological time scale, in 2018 Benton listed the volcanically related major extinctions of the Permian to early Mesozoic as the Capitanian (260 Ma ago), associated with the Emeishan Traps in China; the Permian–Triassic (252 Ma ago) and the Smithian–Spathian (249 Ma ago), associated with the Siberian Traps; the Carnian (232 Ma ago), associated with the Wrangellia basalts of western North America; the end of the Triassic (201 Ma ago), associated with the CAMP; and the early Toarcian (183 Ma ago), associated with the Karoo–Ferrar basalts of South Africa and Antarctica [109]. This was a time of globally warm temperatures, with no icecaps, and the single supercontinent, Pangaea.

Recent acquisition of a high resolution magnetostratigraphy, biostratigraphy, and carbon isotope stratigraphy for the Toarcian Stage in the Llanbedr Borehole of Cardigan Bay, Wales, demonstrates that there is a precise correlation between the Toarcian Oceanic Anoxic Event 183 Ma ago and the basalt lava sequence of the Karoo–Ferrar Large Igneous Province, which links the Toarcian climatic and environmental perturbation directly to this episode of major volcanic activity [110].

Just how big were these massive volcanic eruptions? At their maximum extent, the flood basalts of India's Deccan Traps probably covered some 1.5 million km$^2$ – about the size of modern India. These large volumes of magma typically erupt through long fissures over geologically short periods in places well away from the boundaries of the major tectonic plates. They are not like volcanoes, which represent point sources of eruptive material. They are attributed to the activity of 'mantle plumes': vertical plumes of hot lava arising from places deep in the Earth's mantle and enabling the eruption of large volumes of material over a relatively small area (Figure 5.10). The plumes are thought to arise from periodic instabilities in

the thermal boundary layer just above the Earth's core–mantle boundary, especially around the edges of the 'large low shear-wave velocity provinces' (LLSVPs) labelled *Jason* and *Tuzo* (Figure 5.10 and Chapter 5), for example as described by Torsvik and Cocks [69]. Such eruptions would have ejected volcanic ash and sulphurous aerosols high into the air, cooling the climate for decades or even centuries. Not surprisingly, these places have come to be termed 'hot-spots'.

How exactly did these eruptions occur? Gerta Keller of Princeton and colleagues estimated in 2012 that the volcanic flows of the Deccan Traps took place in a number of pulses, each of which could have been as large as 10 000 km$^3$ and erupted within 100 years [111]. In comparison, one of the largest historical eruptions of a single basaltic volcano, Laki, in Iceland, in 1783, produced a mere 15 km$^3$ of lava in a single year. Just one Deccan flow would have represented 667 Lakis! We have no human memory of such enormous events as plateau basalt eruptions. Our experience is of single volcanoes, like Laki, that may eject up to, say, 12 km$^3$ of lava, are active in a major way for less than a year, and cause the climate to cool by up to 0.5 °C for a period of no more than a year or so. LIPs are not on our radar.

The amount of carbon and sulphur dioxides emitted from one Deccan pulse is likely to have been as large as the amounts emitted by the asteroid impact in Yucatan [111]. The sulphur dioxide (SO$_2$) would have risen to the stratosphere, combining with water to form droplets of sulphuric acid that reflected solar energy. Cooling from any one eruptive phase would have been short-lived, because sulphuric acid droplets get rained out of the upper atmosphere within a couple of years following major eruptions. In contrast, the CO$_2$ would have stayed in the atmosphere for several thousands of years, contributing to long-term warming. It would also have tended to acidify the oceans, extending the range of extinction from terrestrial to oceanic.

Keller and colleagues argue that '*none of the "big five" mass extinctions was brought about by a single simultaneous event causing sudden environmental collapse. All are characterized by prolonged periods of high stress before and after mass extinctions, and three (end-Permian, end-Devonian, end-Ordovician) show multiple extinction phases, sometimes separated by hundreds of thousands of years*' [111]. Careful examination of the Cretaceous boundary extinction suggested to them that the simple impact-kill scenario was inadequate, not least because many species '*groups died out gradually or decreased in diversity and abundance well before the boundary, including dinosaurs*' [111]. She and her colleagues reiterated these conclusions following a multidisciplinary international conference in 2013 on 'Volcanism, Impacts and Mass Extinctions: Causes and Effects' [112]. More recently, however, using Bayesian statistics to model the evolutionary dynamics of speciation and extinction of Mesozoic dinosaurs, Sakamoto and colleagues found abundant evidence for a significant decline in dinosaur speciation over a period of at least 40 Ma throughout the Late Cretaceous, which would have made dinosaurs '*vulnerable to extinction and unable to respond quickly to and recover from the final catastrophic event*' [113]. They attributed the decline to a combination of factors including the break-up of Pangaea (limiting free movement), intense prolonged volcanism, fluctuations in sea levels and interactions with other animal groups, especially mammals, that were rapidly expanding and flourishing [113].

Tony Hallam pointed to the close association between extinctions and other possible major causes of environmental change, notably major marine regressions caused by falling sea level [100]. More important than any other factor in causing extinctions, in his view, were major transgressions caused by rising sea level associated with climatic warming and the associated spread of anoxic bottom waters onto shallow continental shelves. Intensive research cast serious doubt on the possibility that meteorite impacts had anything to do with the claim of a 26 Ma periodicity in extinction through time. Gerta Keller and colleagues agreed with Hallam that as well as volcanism and bolide impacts '*sea level and climate changes (warming and cooling), ocean acidification, ocean anoxia, and atmospheric changes have to be considered in any extinction scenario to understand the causes and consequences of mass extinctions*' [112].

Was a link between LIPs and biodiversity justified? In 2010, David Kidder and Tom Worsley drew attention to a dozen or more examples of a tripartite link between LIPs, biological extinctions, and geologically brief (<1 Ma) periods of exceptional warmth (their hothouse intervals) (Figure 10.3) [26, 27]. In 2013, statistical analyses by Rampino and Prokoph showed that both fossil diversity and the ages of LIPs show cycles of around 62 and 140 Ma, with an additional weaker 30–35 Ma cycle over the past 135 Ma [29]. Those new data suggested a link to Fischer and Arthur's 32 Ma cycles in Earth's history, but recent analyses by Müller and Dutkiewicz suggest that there is indeed a plate tectonic underpinning for a 26 Ma long cycle [31].

Although biological diversity seems to have been least at times of massive flood basalts, in 2010, Peter Schulte of Germany's Universität Erlangen-Nürnberg, and colleagues argued that, whilst the evidence for an association between major igneous provinces and biological extinctions was reasonably sound, not all flood basalts were associated with extinctions or decreases in diversity. This led them to present the case that it was unlikely that volcanism associated with the Deccan Traps somehow destabilized the biosphere making it more likely to collapse with the meteorite impact at the K–T boundary [114]. Their interpretation

meshes well with the observations of Sakamoto on the gradual decline of the dinosaurs over at least 40 Ma prior to the K–T boundary [113].

Let us now examine the evidence for two major extinctions known to be associated with massive eruptions of plateau basalts: the end Permian extinction at the boundary with the Triassic some 252 Ma ago, and the end Triassic extinction at the boundary with the Jurassic c. 201 Ma ago.

The extinction at the Permian–Triassic (P–T) boundary some 252 Ma ago has recently been the subject of extensive research carried out through the International Geological Correlation Programme (IGCP) of the International Union of Geological Sciences (IUGS) and UNESCO, now renamed the International Geoscience Programme (same acronym – IGCP). This extinction has been comprehensively reviewed in recent years for example by Chen et al. (in 2014) [115], Cui and Kump (in 2015) [116], Komar and Zeebe (in 2016) [117], Stordal et al. (in 2017) [118], Baresel et al. (in 2017) [119], and Song et al. (in 2018) [120].

The extinction coincides with the emplacement of the Siberian Traps, a vast area of plateau basalts in Siberia's Tunguska Basin. The Traps covered an area of some two to five million $km^2$ at the surface, but represent an enormous volume, because much of the associated lava solidified as 50–250 m thick basalt 'sills' that embedded themselves horizontally between layers of sediment deep within the basin and extended laterally for several hundred kilometers [116]. Vertical sheets ('dykes') or tubes of lava, now solidified, connected the sills with their deep magma source, with each other, and with the surface (Figure 10.8) [118]. The vertical pipes, or tubes, were about 11 km across [116]. Volcanism was likely to have been episodic.

Gases emitted with the lavas included mostly those formed by contact metamorphism – the heating of the basin's sedimentary rocks by contact with the sills, rather than by degassing of the lava [116, 118]. The sills intruded into and baked a mixture of evaporites and organic rich country rocks, releasing gases such as $CO_2$, methane ($CH_4$), chloromethane ($CH_3Cl$), hydrogen sulphide ($H_2S$) and sulphur dioxide ($SO_2$), some being trapped within the sediments, some seeping to the surface, and some being blasted into the atmosphere through explosion pipes [116, 118, 121–123]. Moreover, the Trap magmas contained anomalously high amounts of sulphur, chlorine, and fluorine. Ejection of large loads of such chemicals into the atmosphere may have contributed to serious deterioration in the global environment at the end of the Permian, contributing to extinction [123]. Stordal and colleagues estimated the atmospheric composition and temperature effects of these various emissions, using for the first time estimates of the volumes of gases emitted by the metamosphism of the country rocks. Like the eruptions, the emission of gases

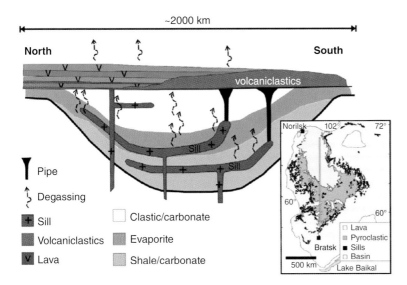

**Figure 10.8** Schematic north–south cross section along line in inset map across the Siberian Traps showing the two dominant sources of carbon gas: (i) mantle carbon from basalts, and (ii) sedimentary carbon from metamorphic contact aureoles around intrusive sills. The sedimentary carbon was partly degassed from pipe structures rooted in contact aureoles. *Source:* From figure 1 in Ref. [118].

was likely to have been episodic, leading to episodes of acid rain (caused by the emission of $SO_2$), and of destruction of the ozone layer (caused by the emission of chlorine compounds). Focusing on individual large-scale volcanic events (e.g. the formation of a single event over a period of a century) Stordal and colleagues calculated that each event could have led to global temperature increase of 7 °C, which would have had severe effects on precipitation and the capacity for adaptation of many biological species [118].

Considering the entire trap volume, we can calculate that the total emission of $CO_2$ would likely have been around >100 000 $GtCO_2$ [122], which would have led to long term warming, while the emission of $SO_2$ would have led to decadal scale cooling from a veil of sulphuric acid in the stratosphere, which would rain out fairly rapidly [116]. At the same time the addition of sulphuric acid droplets and $CO_2$ to the ocean would have caused temporary acidification [119]. Marine carbonate deposits show a negative excursion of 4‰ in $\partial^{13}C$, indicating the massive release of $^{12}C$-rich carbon, presumably from the metamorphism of the organically enriched sediments of the Tunguska Basin [116].

According to Cui and Kump, prior to the event, data and models indicate that Pangaea experienced a warm climate characterized by increasing aridity, with strong seasonal contrasts between very hot summers (>40–50 °C) and cold winters (<−15 °C) plus wet monsoonal circulation along the Tethyan coast in the east contrasting with a very dry interior [116]. Tropical sea surface temperatures were between 22 and 25 °C and atmospheric $pCO_2$ may have reached 500–4000 ppm, whilst during the event those temperatures rose to 30–35 °C and $CO_2$ to perhaps 8000 ppm [116]. That led to a global rise in temperature of 5–9 °C

according to $\partial^{18}O$ data, suggesting a high climate sensitivity [116]. However, that warming was interrupted by a brief cool event. The intense cooling induced by the emission of $SO_2$ is recognized by an abrupt decline in sea surface temperature at the boundary [119]. The cooling probably led to the formation of ice and accounted for a brief global regression, which may have lasted for some 30 Ka [115] or slightly longer [119]. It was associated with a sixfold decrease in the rate of supply of sediment from the coast to deep waters, suggestive of the development of a brief cool dry climate [119].

The ensuing rise in $CO_2$ and temperature would have temporarily increased the rate of chemical weathering on land, supplying abundant phosphate to the ocean to induce eutrophication (nutrient enrichment leading to excessive growth of marine algae, decomposition, and the spread of oxygen poor waters) [116]. However, modelling by Komar and Zeebe suggests that the amount of carbon emitted by the Traps was not itself sufficient to cause all of the observed temperature rise [117]. In addition, the death of land plants would have decreased the supply of organic matter to the ocean from land, and the death of marine organisms would have lessened primary productivity, both factors working together to shut down the marine biological pump that would otherwise have transferred $CO_2$ from the ocean surface into the deep ocean as plankton died and their remains sank to the seabed [117]. Shutting down the biological pump meant that much more of the emitted $CO_2$ stayed in the air to cause more extreme warming [117].

What effect did the end Permian event have on life? In 2011, David Retallack of the University of Oregon and colleagues noticed that the Late Permian mass extinction was followed by an unusually prolonged recovery through the Early Triassic. Citing new records from Australia's Sydney Basin, they found five successive spikes of unusually high atmospheric $CO_2$, estimated from stomatal indices in fossil leaves, along with signs of deep chemical weathering. These 'greenhouse crises' coincided with unusually wet and warm climates at a palaeolatitude of 61°S. Between these crises were long periods of cool dry climate with low $CO_2$. These patterns, they felt, '*may account for the persistence of low diversity and small size in Early Triassic plants and animals*' [121]. What might have caused these periodic events? They thought it might be '*Extraordinary atmospheric injections of isotopically light carbon, perhaps from thermal metamorphism of coal by feeder dikes to Siberian Trap lavas*' [121]. It may be significant that Lenton et al. found much the same spikes in carbon burial and $\partial^{13}C$ during the transition from the Ordovician glaciation into the Silurian [124]. Conceivably these Ordovician–Silurian and Permian–Triassic oscillations both represent the instability of a climate system in adjustment between different stable states.

The end Permian extinction was Earth's most catastrophic loss of biodiversity, when some 57% of all biological families, 83% of all genera, and 87% of marine genera (including 96% of marine species) became extinct (the latter probably because of heating, ocean acidification and anoxia reducing the habitable depth zone) [125, 126]. Insect life declined dramatically too, as did large land animals (amphibians, reptiles, and proto-mammals), though land plants suffered probably no more than a 50% loss in species diversity. Most plants and animals suffer major physiological damage at sustained temperatures of 35–40 °C. Extreme temperatures and drought killed animals and insects, whilst acid rain killed off forests and damaged soils, leading to a 10 Ma-long 'coal gap' because there were no trees [109]. Stable gymnosperm-dominated Permian floras were replaced by lycopods and ferns in the Early Triassic [109]. So much biodiversity was lost that the recovery of land-dwelling life took significantly longer than after any other extinction event, possibly 10–30 million years [125]. Temperatures on land likely rose to around 40 °C [126]. Understanding the extinction has led to Benton's standard model for mass extinctions induced by plateau basalts associated with LIPs, and involving interactions between multiple stressors (Figure 10.9) [109]. As an example of the combined effect of different stressors, think about the effect of restriction of the habitable zone in the ocean by a rise in anoxia from below and lethal warming from above, comprising a 'double whammy' [109].

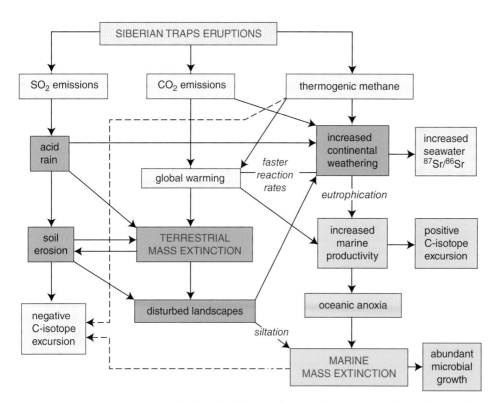

**Figure 10.9** Generic model for volcanically induced extinctions, based on the likely environmental consequences of the Siberian Traps eruptions, showing the flows of consequences of global warming and acid rain. Causal links are indicated by solid arrows, and possible second-order controls on the negative carbonate C-isotope excursion are indicated by dashed lines. *Source:* From figure 1 in Ref. [109].

The extinction killed >90% of marine species in less than 100 000 years and was followed by a slow recovery through the Triassic that has been documented in detail by Song and others [120]. They show that marine ecosystems dominated by non-motile animals (such as brachiopods, bryozoans, hydrozoans, corals, echinoderms, foraminifera, radiolaria, and sponges) before the extinction shifted to ones dominated by free-swimming 'nektonic' groups including pelagic predators such as cephalopods, reptiles and bony fishes. In the Triassic ocean '*animals at higher trophic levels recovered faster than those at lower levels, … [and] top down rebuilding of the marine ecosystem was still underway in the latest Triassic ~50 Ma after the extinction*' [120]. The diversity of the non-motile animals fell from more than 500 genera in the Late Permian to fewer than 100 in the Early Triassic. The faunal shift changed the structure of the marine ecosystem to one dominated by nekton (free-swimmers), which climbed from 14% of genera in the latest Permian to around 67% 2 Ma years later. The proportion of non-motile animals then recovered gradually, reaching 70% by the Rhaetian (latest Triassic). To some extent this pattern may reflect the free-moving ability of the nekton (free-swimmers) as opposed to the benthos (bottom-dwellers) [120]. Song and colleagues noted that the recovery of the structure of the marine ecosystem referred to above, which took some 50 Ma, was much slower than the rate of recovery of taxonomic diversity, which recovered within about 5 Ma [120].

The end Permian extinction brought about the end of the Palaeozoic (old life times), and the entry of dominant new life forms in the Mesozoic (middle life times). The Cenozoic – (new life times) – followed the mass extinction at the K–T boundary.

The study by Song et al. '*reaffirms the importance of protecting global ecosystem diversity because, once it is destroyed, restoration requires dozens of million years, much longer than human history*' [120].

Could a bolide impact have caused the extinction or, perhaps, triggered the formation of the Siberian Traps? Searches for impact craters dated to the end Permian have found one in central Brazil. Eric Tohver of the University of Western Australia suggested that, although the 40 km diameter crater is quite small, the bolide impact may have disturbed underlying sediments rich in organic matter and methane across a radius of 700–3000 km. That seismic disruption could have supplied vast amounts of methane to the atmosphere in a brief period [127]. Assuming that some 1600 Gt of methane were released (equivalent to 135 000 Gt of $CO_2$), that would be more than enough to explain massive warming leading to an extinction.

There is also some evidence for another impact crater of approximately the right age off northwest Australia [128].

We certainly do need something large to explain an extinction that wiped out 90% of terrestrial life and 70% of marine life on Earth. But is it realistic to call on bolide impacts? Investigating a claim for a massive bolide impact at the Permian–Triassic boundary, Christian Koeberl of the University of Vienna, and colleagues, concluded in 2002 that none of the evidence provided '*even a vague suggestion of an impact event at the P-T boundary*' [129]. The Alvarez's, too, found no evidence for an iridium spike like that of the K–T boundary in clays at the end of the Permian.

The major extinction at the end of the Triassic is also thought to have resulted from a mass outpouring of lava – the flood-basalts and associated sills and dykes of the Central Atlantic Magmatic Province, covering area about the size of Australia, which developed before Pangaea split apart and is now found in separate pieces in Venezuela and Amazonia (Brazil), the southeastern United States, Europe, and Northwest Africa. In March 2013, Terrence Blackburn of MIT and Washington's Carnegie Institute used new uranium-lead ages from zircon crystals taken from ancient lavas in North America and Morocco to demonstrate that the start of the volcanism coincided with the extinction at the end of the Triassic c. 202 Ma ago [130]. Over 1 million $km^3$ of lava poured out within less than 30 Ka, enough to smother an area the same size as the lower 48 states of the USA to a depth of more than 100 m, changing the global climate and the global ecosystem and paving the way for the dinosaurs to dominate our planet [130]. Smaller episodes of volcanism followed about 60 Ka, 270 Ka and 620 Ka after that first phase.

In 2018, Heimdal and others confirmed that the main phase of the eruption did indeed take place at the boundary, proving the synchroneity of the main intrusive phase and the end Triassic mass extinction [131]. They found that subsurface lava intrusions had formed sills like those found beneath the Siberian Traps, and comprised up to 20% of the stratal layers in the sedimentary basins of the Amazonian area. The sills intruded into extensive deposits of carbonate and evaporite, along with organic-rich shales and petroleum reservoirs, much as was found in Siberia. Heimdal concluded that thermal metamorphism of the country rocks by the hot intrusions would have generated some 80 000 Gt $CO_2$, along with substantial amounts of $SO_2$ and chlorine compounds from the evaporites [131], These gases probably entered the atmosphere through explosion pipes, as in the case of the Siberian Traps, but the extensive jungle in the Amazonian region precluded their detection.

Examining the densities of stomata in fossilized leaves from a range of species on either side of the Triassic–Jurassic boundary, Jenny McElwain from the University of Sheffield found lower densities above the boundary than below it [132]. She and her team deduced that the $CO_2$ concentration in the air was about 600 ppm before the boundary and between 2100 and 2400 ppm after it – enough to cause a rise in mean global temperature of 4 °C, which likely interfered with the efficiency of large leaves to photosynthesize, leading to the extinction of 95% of land plants. The probable origin for the rise in $CO_2$ was extensive volcanism as Pangaea began to break-up.

Bas van der Schootbrugge of Goethe University found that terrestrial vegetation in Germany and Sweden had also been significantly affected by volcanic activity in the central Atlantic province at the end of the Triassic [133]. A fern-dominated association typical of disturbed ecosystems replaced Gymnosperm forests, and the associated sediments contained little charcoal but abundant polycyclic hydrocarbons, suggesting incomplete combustion of organic matter by flood basalts. This severe and abrupt shift in vegetation is unlikely to have been triggered by an increase in greenhouse gases alone. It probably resulted also from the emission of pollutants like $SO_2$ along with toxic aromatic hydrocarbons [133]. In the marine realm, Song et al. point out that benthic marine organisms maintained their growing diversity until the end of the Triassic, which means that the extinction is best viewed as an abrupt termination of a ~50 Ma trend [120].

It therefore seems most likely that the end Triassic extinction followed much the same pattern as the end Permian extinction, with initial ejection of $SO_2$ and dust leading to temporary cooling due to shading from the Sun, $SO_2$ producing acid rain, chlorine compounds damaging the ozone layer and allowing the penetration of harmful ultra-violet radiation, and $CO_2$ causing significant warming along with ocean acidification, with the effects of the warming emerging after the decay of the short term cooling caused by dust and sulphate aerosols. The mounting evidence from studies such as these is leading towards a '*more unifying view of causes and effects of large igneous provinces*' [119].

The research examined here confirms that there are longlasting cycles in our planet's climate. They seem to relate mainly to processes taking place deep within the Earth. There is a growing body of evidence that mantle plumes generate hot spots (like Iceland, for example), flood basalts in LIPs (like the Deccan Traps), and regional uplift and rifting. Rampino and Prokoph go so far as to suggest that '*plumes may act as a pacemaker for changes in sea level, climate and biodiversity*' and that we are looking at the exciting '*possibility of a unification of geologic processes, related in part to changes in the deep mantle*' [29]. Tectonic–volcanic cycles related to plumes and plate processes drive the slow carbon cycle, controlling the supply of $CO_2$ from volcanic vents and its eventual removal from the atmosphere through the weathering of newly up-thrust mountains. A fast carbon cycle uses biogeochemical processes to maintain climates suitable for life. Periodicity in the record

suggests a variation on Lyell's uniformitarianism that would accommodate Cuvier's catastrophism. Steady and periodic processes were interrupted by Cuverian catastrophes in the form of massive meteorite impacts like that which ended the reign of the dinosaurs at the end of the Cretaceous, as well as by occasional extended eruptions of flood basalts. There is no convincing evidence for asteroid impacts causing other major biological extinctions within the past 540 Ma.

The operations of plate tectonics and plumes evidently cause changes in the abundance of $CO_2$, which goes on to play a primary role as a climate regulator for the planet [134]. Tectonics, volcanism, weathering, $CO_2$, biology, and climate are inextricably linked. Their interactions kept Earth's climate cycling between fairly narrow limits of both temperature and $CO_2$ through Phanerozoic time. Levels of $CO_2$ were higher for a given temperature early in the period, when the Sun was fainter, and lower later in the period, when the Sun's output was stronger. For the most part, the climate was warm, resting in a greenhouse state, with occasional falls to glacial conditions and occasional rises to hothouse conditions. The limits of the natural envelope of Earth's climate system for the bulk of Phanerozoic time ranged from 180 ppm $CO_2$ and an average global temperature of around 11 °C in peak glacial conditions at the low end, to somewhere between 4500 and 8500 ppm $CO_2$ and between 30 and 32 °C in peak hothouse conditions at the high end.

## 10.8 An Outrageous Hypothesis: Snowball Earth

One of the most surprising features of Earth's climate is a suite of glaciations between 750 and 540 Ma ago in the late Proterozoic (see Frontispiece for Geological Time Scale). They differ dramatically from subsequent glaciations in seeming to have covered the entire Earth, converting it in effect into a giant snowball. Reviewing the evidence, science writer Gabrielle Walker tells us that with temperatures averaging −40 °C the ocean surface was frozen, and with no exposed ocean there was no evaporation and therefore no clouds [135]. Microbes survived in refugia. Snowball conditions ended with global warming triggered by the slow build-up of $CO_2$ in the atmosphere, from volcanic activity, and when the ice melted it did so with a bang, lurching to +40 °C. This was Earth's most extreme shock – the worst catastrophe in our planet's history, lasting hundreds of thousands of centuries [135].

When it first came out, the Snowball Earth concept was regarded as outrageous. But, as a concept it shares the merit of Wegener's Continental Drift theory (Chapter 5), by explaining several disparate facts and puzzles under one umbrella. And as the Wegener example showed, it may be unwise to dismiss extraordinary ideas out of hand. Snowball Earth was difficult to accept because it described a planet different in almost every characteristic from the one we see today, and seemed to violate Lyell's principle of 'uniformitarianism'. But uniformitarianism was dislodged from its perch long ago, not least when Walter and Luis Alvarez proposed that the Cretaceous was brought to an end by a giant asteroid strike. Time does not flow smoothly. Nothing had prepared the world of geological intuition and training for something as outrageous as Snowball Earth. But there are times when you just have to think outside the box. Conventional wisdom can be wrong. And Snowball Earth was well outside the box of convention, with '*oceans that freeze over completely, even in the tropics and at the equator. An ice age that lasted for millions of years. A planet that then plunged from the coldest temperatures it had ever experienced into an immense hothouse within just a few centuries. Carbon dioxide levels hundreds of times higher than have ever been seen in the geological record. Rock weathering rates like nothing on Earth today*' [135]. No wonder the hypothesis attracted vehement criticism.

The saga began with the observations of Brian Harland, leader of Cambridge University's Cambridge Arctic Shelf Programme (CASP), which formed in 1975 as a successor to the Cambridge Spitzbergen Expeditions, which Harland began in 1948. Harland first drew attention to the global distribution of late Proterozoic glacial deposits in 1964, postulating that they extended into the tropics [136]. He considered that there would have been open water in tropical regions [137], which would make for what Dan Schrag and Paul Hoffman called a 'Slushball Earth' rather than a 'Snowball Earth' [138]; they coined the term to refer to a climate model solution by Hyde and others [139] that was close to Harland's conception. The survival of assemblages of marine microbiota does suggest the existence of open water over ice-free continental shelves, at least seasonally [140].

This book is not the place for a comprehensive review of the Snowball Earth hypothesis, which would require a whole volume to itself. Instead I will rely on selected research papers including reviews like that of Fairchild and Kennedy [141] and Hoffman et al. [142], which provide satisfying syntheses whilst recognizing the need for further research. Earth System Science approaches, considering interactions and feedbacks between multiple variables, have helped to evaluate the various conjectures relating to the Snowball Earth glaciations by making testable predictions.

As Hambrey and Harland pointed out in 1981 [143], the Neoproterozoic glacial units in any one region typically comprise one, two, or three discrete glacial formations, grouped within three major glacial periods: Sturtian (717–659 Ma ago), with its type locality in the Sturt Gorge near Adelaide in South Australia; Marinoan (645–635 Ma ago), with its type location in Marino Rocks near Adelaide; and the short-lived Gaskiers (c. 580 Ma ago), with its type location in Newfoundland, Canada [140–142]. The first two, but not the latter are claimed to be Snowball events [140].

The glaciations took place on the supercontinent of Rodinia, which, as mentioned in Chapter 5, was assembled about 1100 Ma ago and broke up about 700 Ma ago during or just before the Sturtian glaciation (Figure 10.10) [69, 142, 144]. Rodinia was centred on the Equator, with no land at either pole, and its dispersal was accompanied by massive volcanism in LIPs [69], one of which (the Franklin

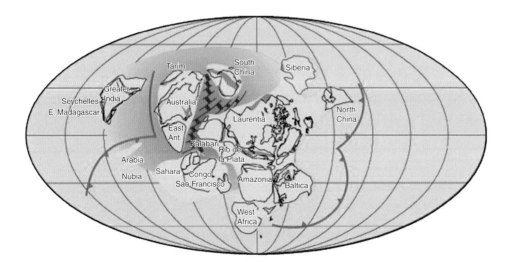

**Figure 10.10**  Cartoon showing the breakup of Rodinia c. 750 Ma ago. Green = active margin with subduction; brown shading = area of mantle superplume beneath Rodinia; purple = spreading ridge with fracture zones; orange = continental rift. The mantle superplume led to widespread continental rifting between c. 825 and 740 Ma. The first major break-up event occurred along the western margin of Laurentia (as shown). *Source:* From figure 9g in Ref. [144].

LIP in Arctic Canada) covered part of Rodinia just before the Sturtian glaciation [142]. Rapid weathering of its plateau basalts likely contributed to reducing $CO_2$ in the atmosphere. The climate would have been cooler anyway, because the Sun was 6–7% dimmer at the time. Moreover, in the absence of vascular land plants the '*continental surface albedos were presumably higher than present day albedos*' [142], which would have contributed to cooling, especially at night (as in modern deserts). As Rodinia broke up, the continental interiors would have gained more moisture, increasing the efficiency of chemical weathering and hence contributing to the lowering of $CO_2$ [142].

Having major glaciations centred on an equatorial continent '*contrasts starkly with the Phanerozoic norm of glacial eras starting on a polar continent*' [141]. The glaciations are thought to have originated due to the growth of sea-ice in the polar regions, with its spread towards the tropics encouraged by cool conditions driven by low atmospheric $CO_2$, low solar luminosity (7% less than today), and the high albedo of the growing ice surface [142]. Freezing seawater added to the base of the sea ice would have thickened it from below, creating 'sea glaciers' up to c. 750 m thick [142], hence similar in thickness to some of modern Antarctica's surrounding ice shelves.

A carbonate sediment cap typically overlies each glacial deposit. The presence of the carbonates, which have no analogue in the younger glacial sequences of the Phanerozoic, is indicative of a major perturbation in the global carbon cycle [141, 143]. Whereas most of the carbonates of the Phanerozoic originated through the accumulation of skeletal remains, such skeletal forming organisms had not yet developed in Proterozoic times, so the cap carbonates are all precipitates suggesting high levels of saturation of seawater in carbonate ions and a tendency to form in warm shallow waters where conditions encouraged seawater to evaporate [141]. Many of the cap carbonates are similar in character to deposits that formed, or are forming today, in the cold conditions of the Antarctic Dry Valleys [141]. They are commonly about 1–5 m-thick dolomite deposits, separated from the underlying glacial strata by a sharp contact indicating an abrupt cessation of glacial conditions. The sharp contact suggests that the carbonates, deposited in shallow water, represent a marine transgression during a post-glacial rise in sea level as the former ice sheets melted. The carbonates may have accumulated over periods as long as about 1 Ma [141].

Joe Kirschvink was the first to suggest, in 1992, that the low latitude of Rodinia most likely facilitated the development of its ice cover at a time of low atmospheric $CO_2$, when Earth's albedo increased under extremely cold conditions, and that the key to escape from those conditions was a steady rise in volcanically sourced $CO_2$, which provided

enough warmth to eventually overcome the cooling effects of the albedo [145]. The process is now thought to work as follows [142]: ice starts to form under cold conditions with low $CO_2$; once ice has covered the Earth, volcanic activity causes $CO_2$ to increase in the atmosphere over time to very high levels, warming the atmosphere; ice melts very quickly once the atmosphere has become hot; chemical weathering of the land surface (e.g. by acid rain under hot conditions) facilitates slow recovery to the initial conditions, and the process can restart [141, 142, 146]. This oscillating process is known as hysteresis [142], a topic explored in Chapter 6 (see Figure 6.19). As Tim Lenton points out: '*The resulting oscillation would be typical of the behaviour of a system in which a fast positive feedback – in this case the ice-albedo feedback – interacts with a slow negative feedback process, in this case the silicate weathering feedback*' [146].

As Hoffman explained [142], the ice edge line changed with latitude as a function of changing solar or $CO_2$ radiative forcing. In response to lower forcing, the ice edge migrated equator-wards. This reduced sinks for carbon, allowing normal volcanic out-gassing to drive atmospheric $CO_2$ higher over millions to tens of millions of years until it reached a deglaciation threshold, whereupon the tropical ocean began to open. Decreasing ice-albedo feedback then drove the ice edge rapidly poleward (in ~2 Ka). High $CO_2$ combined with low surface albedo created a torrid greenhouse climate. The resulting intense silicate weathering and carbon burial then lowered atmospheric $CO_2$ (in $10^7$ years) to the threshold for the reestablishment of a polar ice cap. The hysteresis loop model predicts that Snowball glaciations were long-lived, began synchronously at low latitudes, and ended synchronously at all latitudes under extreme $CO_2$ radiative forcing. The ocean is expected to have experienced severe acidification and de-acidification in response to the $CO_2$ hysteresis.

Kirschvink also pointed to the association of the glaciations first with oxygen depletion in the adjacent ocean, due to the lack of stirring by the action of winds, and, second, with banded iron formations, last seen much earlier in the Precambrian prior to the so-called great oxidation event around two billion years ago [145]. The association of the glacial events with positive carbon isotope signals indicating an enrichment of sediments in $^{12}C$ is taken as evidence for the enhanced burial of organic carbon under anoxic conditions [140]. The combination of banded iron formations (indicating oxygenation) and the deposition and trapping of marine organic matter (suggesting reducing conditions) might be explained by the development of well oxygenated deep 'glacial' bottom waters below a well developed but shallower oxygen minimum zone [140], like the one we find in the modern eastern Pacific Ocean or Arabian Sea (although there we do not have enough dissolved iron to create modern banded ironstones). However, Hoffman and colleagues suggest that iron formations may have developed where upwelling water rich in dissolved iron met oxygenated surface water immediately beneath the ice glacier. That would explain why the iron formations are found near ice-grounding lines in semi-restricted basins along continental edges [142]. In support of that concept, Fairchild and colleagues note that the banded iron formations are both thin and geographically restricted, quite unlike the very considerable masses of banded ironstones typical of the older Archaean periods of the Precambrian [141].

The sea ice may have been thinnest in low latitude embayments, providing sufficient light penetration for sub-ice photosynthesis to continue, supporting both cyanobacteria (unicellular prokaryotes – lacking a nucleus) and primitive multi-celled organisms (eukaryotes – organisms with a nucleus contained within a cell surrounded by a membrane) such as red algae [142]. Fresh surface meltwater could have penetrated sea glaciers and coastal sea ice through 'moulins' (vertical channels carved by relatively warm surface meltwater) [142], much as happens today on the Greenland Ice Sheet. Bacteria and eukaryotes could also have survived in meltwater ponds atop the ice in tropical regions where the ice thinned through sublimation where evaporation exceeded precipitation [142, 147]. Was the ocean completely ice-covered? Corsetti and colleagues thought not. Because they found that the biota from pre-glacial, syn-glacial, and post-glacial units were more or less identical, they deduced that a completely ice-covered ocean was unlikely. If they were correct, then we did not need to call on refugia to explain the survival of the biota through glacial times [148].

Hoffman tested Kirschvink's model by collecting sedimentological, stratigraphic, and geochemical evidence from key locations, notably Namibia [149, 150]. He argued that carbon isotopes showed that the glaciations were periods of low organic productivity, the ocean being isolated from the atmosphere by ice. Subsequently, the novel work of work of Bao and others demonstrated that both the cap carbonates and glaciers were deposited under atmospheres rich in $CO_2$ [151, 152]. Modelling shows that glaciation was initiated at low $CO_2$ times. $CO_2$ then rose due to volcanic emissions and its normal reduction by dissolution in the ocean was limited by the ice cover [153]. $CO_2$ concentrations at glacial terminations are thought to have been substantially higher than today, reaching 10 000–100 000 ppm [151, 152] '*giving rise to torrid greenhouse aftermaths*' [142]. Ice cover of the land surface was not ubiquitous; away from the ice the evidence favours a cold, hyper-arid, tropical continent, though one where spring snowmelt encouraged the formation of lakes [153], hence the analogy with the

dry valleys of Antarctica. It is significant that using his novel $^{17}O$ technique, Bao could also confirm that there were very high $CO_2$ levels in the early Cambrian [151].

The origin of both the cap carbonates and the $\partial^{13}C$ signal are still matters of speculation representing '*a fascinating intellectual battleground*' [140]. However, given the presence of carbonate sediments with Mg/Ca ratios >2, the seas of the Neoproterozoic would have given rise to aragonite as the primary carbonate mineral precipitate [154]. Shields and Veizer noted in 2002 that whilst the prolonged and anomalously high $\partial^{13}C$ signal of these times could represent anomalously high rates of burial of organic matter, this was difficult to prove [155]. If rates of accumulation and storage of organic matter had increased, as the carbon isotope data suggested, then atmospheric oxygen levels would likely have risen as a consequence [155]. Producing one of the latest compilations of $\partial^{13}C$ from the Neoproterozoic, Saltzman, and Thomas concluded in 2012 that there was no doubt that this period was a time of very large climatic swings, which are represented by the carbon isotope oscillations [156]. These large negative isotopic excursions are radically different from the positive isotopic excursions characterizing the Hirnantian (Ordovician) glaciations that we explored in Chapter 9 [156]. Difficulties in interpreting the $\partial^{13}C$ record reflect the fact that we know very little about the carbon cycle in times prior to the evolution of hard shelled organisms in the Cambrian, and, furthermore, '*many features of the present carbon cycle are linked to the present state of biotic evolution*' [156]. As a result, '*Considerably more research is necessary ... before we can confidently evaluate whether the Neoproterozoic carbon isotope record can indeed be interpreted as reflecting the functioning of an unfamiliar carbon cycle or reflects diagenetic processes*' [156].

The message from studies by Lovelock, Margulis, and Nisbett is that life does play a vital role in regulating the climate. Lenton and Watson seem to agree. In 2004, they reminded us that '*biological colonization of the land surface began in the Neoproterozoic*' around 1000 Ma ago [157]. They hypothesized '*that this colonization involved selective weathering of P [phosphorus] from rocks, as well as an amplification of overall weathering rates*' [157]. The '*increase in the weathering flux of P to the ocean would have caused a rise in atmospheric $O_2$ in the Neoproterozoic. This in turn may have provided a necessary condition for the evolution of animals with hard skeletons seen in the [subsequent] "Cambrian explosion." Increased weathering of silicate rocks [by advancing life on land] would also have caused a decline in atmospheric $CO_2$, which could have been a causal factor in the Neoproterozoic glaciations*' [157]. They pointed out that the evolution of large metazoan life (eukaryotes) depended on sufficient oxygen being available, and the appearance of the soft-bodied Ediacaran animals c. 570 Ma ago suggested that oxygen was already at more than 0.1–0.3 times present atmospheric levels then, while yet more would have been required for formation of the skeletons of the Cambrian period that followed. Evidence for Proterozoic life is abundant, for example in the 1170–950 Ma old Paranoá Group and the 950–700 Ma old São Francisco Supergroup in Brazil, which contain abundant microfossils (mostly cyanobacteria and acritarchs – organic-walled microfossils of uncertain origin) preserved within carbonaceous (i.e. organic-rich) black chert (the silica probably from marine organic remains) in flat-laminated and high-relief stromatolites and representing a wide spectrum of environments from nearshore lagoonal to supratidal, subtidal and offshore marine settings [158]. Evidence from geochemistry indicates, further, that prior to about 800 Ma ago, phosphorus supply was limited. It then underwent a fundamental shift, signalled by an increase in the burial of phosphorus in shallow marine environments [159]. An increase in its supply suggests that phosphorus was being supplied from the land to the coastal ocean by chemical weathering on land, perhaps stimulated by the movement of life from the ocean onto the land, but perhaps driven by weathering).

Turning to the carbon isotopes, Lenton and Watson noted that the $\partial^{13}C$ signal was strongly positive during nonglacial periods before and after glaciations (Figure 10.11). They interpreted this signal as indicating a high rate of organic carbon burial, consistent with a drawdown in atmospheric $CO_2$ as a result of increased chemical weathering on land, which also led to a supply of nutrients to the ocean to support an increase in productivity there. Strontium isotopes also suggested an increase in weathering, with $^{87}Sr/^{86}Sr$ ratios increasing through the Neoproterozoic (Figure 10.11). Although there was no direct evidence of colonization of the land surface by photosynthesizing organisms '*there are possible microfossils and indications from carbon isotopes of highly productive ... microbial communities on land after 1200 Ma [ago], and protein sequence analysis suggests that green algae and the major lineages of fungi had diverged by ~1000 Ma and this allows for the possibility that lichens (from fungal symbiosis with either cyanobacteria or green algae) could have evolved thereafte*r' [157]. Selective weathering of phosphorus by lichens would have increased the biological availability of this element and its supply to the ocean, stimulating productivity there and the preservation of organic remains under anoxic conditions in marine sediments. The resulting decline in $CO_2$ and temperature would have suppressed chemical weathering. We can see the analogy in the Palaeozoic, when '*the evolution of vascular plants is thought to have enhanced silicate weathering and organic carbon burial, leading to lower $CO_2$ and higher $O_2$ concentrations, peaking in the Permo-Carboniferous*' [157]. Land plants were thought to have originated about 420 Ma ago (the age

of their earliest fossils), but recent data suggests they originated 100 Ma earlier in earliest Cambrian times. Those ancestors – embryophytes – likely had rudimentary pores and roots, so could process soil and carbon dioxide and influence Earth's biogeochemistry [160]; they probably had less advanced precursors originating in the Neoproterozoic.

In 2011, Lenton and Watson elaborated on this hypothesis in their book *Revolutions that Made the Earth* [79]. Basically, they argued that over time plants transformed the global environment, creating their own moist environment through evapotranspiration, and creating their own soils. The first land plants likely comprised lichens, symbionts in which fungi use organic acids to dissolve rock silicates and extract essential nutrients like iron and phosphorus for supply to algae (plants), which take in $CO_2$ to produce sugars via photosynthesis for fungi to use. Lichens can also trap dust, beginning the formation of soils. Fungi are thought to have evolved around 900 Ma ago, and the colonization of the land surface may have begun around 800 Ma ago, explaining the rise in sedimentary phosphate and clay minerals (both products of chemical weathering) from then onwards [79]. Microbial mats may have started the colonization of the land, along the coasts, around 850 Ma ago. These authors speculate that the cycle of glaciations may have continued through into the Ordovician, explaining the origin of the glaciation there [79] (see Chapter 9).

In 2014, Lenton produced a further update of what we may call the 'Lenton and Watson' hypothesis, explained in Figure 10.11. His team argued that '*The traditional view is that a rise in atmospheric oxygen concentrations led to the oxygenation of the ocean, thus triggering the evolution of animals. We argue instead that the evolution of increasingly complex eukaryotes, including the first animals, could have oxygenated the ocean without requiring an increase in atmospheric oxygen*' [161]. They imagine large eukaryotes sinking rapidly through the water column, transferring their decomposition and consumption of oxygen to the seabed, which would have allowed overlying waters to accumulate more oxygen, eventually leading to the oxygenation of parts of the deep ocean after the Gaskiers glaciation (Figure 10.11b). And '*deep convection of oxygenated high latitude surface waters could ... have ventilated parts of the Proterozoic deep ocea*n' [161]. This notion of the oxygenation of deep water conflicts with the suggestions of Fairchild [141] and of Hoffman [142], and deserves further research.

That there were large non-shelled organisms in the Neoproterozoic we know from the fauna from the Ediacarian period, which extends from the end of the Cryogenian Period (635 Ma ago) to the start of the Cambrian Period at 541 Ma ago, and which included the enigmatic *Dicksonia*. The Cryogenian Period includes the two major glaciations – Sturtian and Marinoan, and extends from 720 to

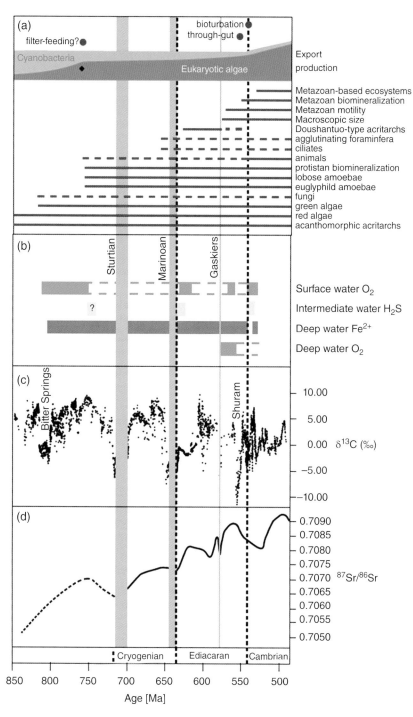

**Figure 10.11** Timeline of biological and environmental change spanning the Cryogenian, Ediacaran and Cambrian periods. (a) Biological evidence: red dots = key animal traits; blue diamond = first substantial occurrence of eukaryotic biomarkers; red lines = first appearance of key eukaryotic phenomena, traits, and clades (dashed lines indicate more debatable evidence). (b) Summary of ocean redox conditions at different depths (question mark indicates possible intervals of intermediate-water euxinia). (c) Secular variation in the carbon isotope composition of marine carbonates. (d) Strontium isotope composition of seawater. *Source:* From figure 1 in Ref. [157].

635 Ma ago. An even older creature than *Dicksonia* – a sort of comb jelly – was recently found in China and dates back to 600 Ma ago.

One wonders if the oscillating $\partial^{13}C$ record of the Neoproterozoic evident from Figure 10.11c could be interpreted simply as a series of pulses of chemical weathering on land leading, first, to the supply of nutrients stimulating the production of $^{12}C$-rich oceanic organic matter in the ocean, resulting in the $\partial^{13}C$ signal in sediments becoming progressively more negative (downward) in Figure 10.11c). That process drew down atmospheric $CO_2$ leading to cooling that in some instances was sufficient to cause glaciation. During the glaciations, the gradual accumulation of volcanically generated $CO_2$ eventually warmed the environment and provided progressively more $^{13}C$ for inclusion in sediments, making the $\partial^{13}C$ signal more positive (upward) (Figure 10.11c), during a period when life slowly reclaimed the land surface. Eventually an equilibrium state was reached at a carbon isotopic composition of 5–10‰ $\partial^{13}C$. Chemical weathering, enhanced by warm and probably more humid conditions (due to evaporation of water vapour from the warmed ocean surface), then grew and the cycle repeated itself. Figure 10.11c shows two events that fit with this pattern, but which were not extreme enough to produce glaciations, namely the Bitter Springs event about 810 Ma ago, and the Shuram event at about 555 Ma ago. A remarkably similar negative $\partial^{13}C$ excursion occurred in the Early Cambrian at about 528 Ma ago, also not associated with a major glaciation [156]. Nevertheless, the Early Cambrian was also a period of extremely high $CO_2$ [151], and the $\partial^{13}C$ excursion and high $CO_2$ at the time may have the same explanation as those of the Neoproterozoic. Given the atmospheric $CO_2$ data, it seems more likely that the $\partial^{13}C$ excursions with which they are associated have a biogenic rather than a diagenetic explanation.

Rodinia's equatorial position would have attracted moisture via the Hadley circulation, and that moisture would have penetrated inland especially when the supercontinent began to break up into a number of fragments. Here we can see the potential for an increased weathering flux driven by warmth and humidity, as well as increased weatherability in the sense of Kump and Arthur (i.e. a large area of freshly exposed silicate rock) (see Chapter 9). It has long been known that there is a profound gap in the stratigraphic record ('The Great Unconformity'), below the base of the Cambrian. Keller and colleagues showed in 2019 that it is associated with large global excursions in oxygen and hafnium isotopes that suggest a massive Late Neoproterozoic erosion event or events [162]. These events coincide quite closely with the Neoproterozoic glaciations, which begs the question – did the glaciations erode the rocks on land, or was the erosion driven by the tropical weathering and weatherability of the fragmenting Rodinian supercontinent. If the latter, then the largely physico-chemical erosion *prior* to glaciation could have supplied sufficient phosphorus to the ocean to stimulate the growth of marine plants, leading to an accumulation of $^{13}C$-poor organic-rich sediment whose accumulation would have pulled down atmospheric $CO_2$, leading to glaciation. A series of such erosional events could have created the series of $\partial^{13}C$ anomalies of Figure 10.11 (Bitter Springs, Sturtian, Marinoan, Gaskiers, and Shuram), without the need to postulate that life on land had played any role (a reasonable assumption bearing in mind the origin of most land plants in the mid Cambrian (see Chapter 9)). The Great Unconformity could be the result of cumulative erosional events like those described here, enhanced by the effect of yet more erosion during the actual glaciations.

We also need to consider the possibility that the $\partial^{13}C$ cycles of Figure 10.11 represent at least in part cycles in sea-floor spreading of the kind identified by Müller, with a period of 26 Ma, since Mg/Ca ratios indicate similar oscillations in the Proterozoic to those identified in the Phanerozoic [31, 44]. They could represent the control of ocean chemistry by plate tectonic processes, perhaps, as Lenton and Watson suggest, modified by life emerging from the ocean onto the land surface before the generation of recognizable land plants.

In addition, we need to consider alternatives for the supply of phosphorus. Gernon and colleagues argued in 2016 that it came from plate tectonic processes involved in the break-up of Rodinia, during which time '*the oceans gained massive amounts of alkalinity, culminating in the deposition of massive cap carbonates on deglaciation*' [163]. Some of these changes may have come about from changes in terrestrial runoff, but others may have resulted from submarine volcanism creating large volumes of glass fragments (hyaloclastics) that were readily hydrated to form palagonite on shallow spreading ridges. Dissolution or alteration of palagonite under marine conditions would have led, first, to supersaturation of the ocean in calcium and magnesium ions, which would help to explain the volume of cap carbonates formed, and, second, to an abundant supply of phosphorus to fuel marine productivity, which would account for the increasing signal of $^{12}C$ enrichment of sediments through the burial of marine organic remains as glaciations approached [163]. This would also help to explain the high concentrations of silica ($SiO_2$) in the thin, banded ironstone formations. There is abundant evidence for elements of Rodinia's break-up occurring before the Sturtian, Marinoan, and Gaskiers glaciations. Subaerial weathering alone is unable to account for the volume of the cap carbonates. The marine weathering, on the other hand, would have continued even under a glaciated ocean, building up calcium and magnesium concentrations to the required level to form the cap carbonates on deglaciation [163]. Long term outgassing of volcanogenic $CO_2$ would explain the rise in $\partial^{13}C$ towards the cap carbonates. Gersson's model '*provides a mechanism for producing high dissolved phosphorus levels that, in the absence of biological removal, could persist over ~10-100 Myr … over repeated cycles*' [163].This mechanism may have driven the increase in primary productivity that, through subsequent decomposition of dead organic remains, led to a large rise in oxygen levels in the wake of Snowball Earth events, in due course oxidizing the Ediacarian ocean sufficiently to facilitate the emergence of multicellular organisms [163] like those of the Burgess Shale.

To draw this section to a close there is one final reminder that life itself may play a significant role in shaping the Earth System. Georg Feulner and colleagues observed is in 2015 that prior to the Neoproterozoic glaciations of the Cryogenian, Earth experienced one billion years of climatic stability – a time known as the 'boring billion'. This suggested to them that something other than just the drawdown of $CO_2$ must have triggered snowball Earth, and they thought that the answer may lie within the ocean [164]. They realized that the gradual evolution of rise of the eukaryotic algae had caused these creatures to become ecologically prominent by 800–750 Ma ago, just prior to the Sturtian glaciation. Now, eukaryotic algae produce a particular chemical (dimethylsulfoniopropionate or DMSP for short), which, on cell death, is converted to an organic aerosol – dimethyl sulfide (or DMS) – which is readily oxidized to produce sulphur compounds that can serve as nuclei for cloud condensation. Feulner thought it likely that when eukaryotes such as the planktonic haptophyte algae rose to dominance they would have contributed progressively larger amounts of cloud condensation nuclei, stimulating the production of low-level clouds over the ocean that would have cooled the planet. The production of sulphur aerosols like DMS also leads to the formation of acid rain, which, falling on land, would have accentuated chemical weathering (e.g. of easily weatherable plateau basalts), hence contributing to the drawdown of atmospheric $CO_2$. Using a climate model they showed that with a modern cloud cover an atmospheric $CO_2$ abundance of less than 110 ppm was sufficient to freeze areas of the open ocean. Their analysis suggested that '*Earth would have been exceedingly unlikely to enter a snowball regime for Neoproterozoic boundary conditions before DMSP-producing eukaryotic algae rose to global ecological significance*' [164]. The magnitude of the rise in diversity and abundance of eukaryotic algae observed in the microfossil record, they thought, made a rise in cloud condensation nuclei in the Cryogenian plausible, leading to a lowering of temperature of up to 10 °C, provided that atmospheric $CO_2$ levels were lower than 110 ppm. This 'cloud cooling' concept does not mitigate the need for $CO_2$ to be drawn down (e.g. by weathering or the accumulation of organic remains on the deep ocean floor) to trigger the onset of a Snowball Earth event, but it does reduce the amount of $CO_2$ drawdown required. We will see other examples of the role of life in shaping the Earth System later in this book.

It's now time to explore further how numerical general circulation models may aid our understanding of past climate change, and especially the role of $CO_2$ in the Cenozoic. With that improved understanding in mind, it is time, too, to examine some case histories of the role of $CO_2$ in our climate system. We will do so in Chapter 11.

# References

**1** Umbgrove, J.H.F. (1947). *The Pulse of the Earth*, 358 pp. The Hague: Martinus Nijhoff.

**2** Fischer, A.G. and Arthur, M.A. (1977). Secular variations in the pelagic realm. In: *Deep Water Carbonate Environments* (ed. H.E. Cook and P. Enos). *Society of Economic Palaeontologists and Mineralogists Special Publication* 25: 19–50.

**3** Beall, A.O., and Fischer, A.G. (1969) Sedimentology, *Initial Reports Deep Sea Drilling Project* **1**, Washington D.C., 521–593.

**4** Fischer, A.G. (1981). Climatic oscillations in the biosphere. In: *Biotic Crises in Ecological and Evolutionary Time* (ed. M.H. Nitecki), 103–131. New York: Academic Press.

**5** Fischer, A.G. (1984). The two Phanerozoic cycles. In: *Catastrophes and Earth History: The New Uniformitarianism* (eds. W.A. Berggren and J.A. Van Couvering), 129–150. Princeton University Press.

**6** Ager, D.V. (1993). *The Nature of the Stratigraphical Record*, 3e, 151 pp. Chichester: Wiley.

**7** Holland, H.D. (1978). *The Chemistry of the Atmosphere*, 351 pp. New York: Wiley.

**8** Worsley, T.R., Moody, J.B., and Nance, R.D. (1985). Proterozoic to recent tectonic tuning of biogeochemical cycles. In: *The Carbon Cycle and Atmospheric $CO_2$: Natural Variations Archean to Present*, Geophysical Monograph **32** (eds. E.T. Sundquist and W.S. Broecker), 561–572. Washington D.C.: American Geophysical Union.

**9** Vail, P.R., Mitchum, R.W., and Thompson, S. (1977). Seismic stratigraphy and global changes of sea level 4. Global cycles of relative changes of sea level. *American Association of Petroleum Geologists Memoir* 26: 82–97.

**10** Wilson, J.T. (1966). Did the Atlantic close and then re-open? *Nature* 211: 676–681.

**11** Wilson, J.T. (1968). Static or mobile Earth: the current scientific revolution. *Proceedings of the American Philosophical Society* 112: 309–320.

**12** Haq, B.U. (1984). Paleoceanography: a synoptic overview of 200 million years of Earth history. In: *Marine Geology* and Oceanography of Arabian Sea and Coastal Pakistan* (eds. B.U. Haq and J.D. Milliman), 201–232. London: Van Nostrand Reinhold.

**13** Sheridan, R.E. (1986). Pulsation tectonics as the control of North Atlantic palaeoceanography. In *North Atlantic Palaeoceanography* (ed. C.P. Summerhayes and N.J. Shackleton). *Geological Society Special Publicaiton* 21: 255–275.

**14** Sheridan, R.E. (1987). Pulsation tectonics as the control of long-term stratigraphic cycles. *Paleoceanography* 2: 97–118,

**15** Frakes, L.A., Francis, J.E., and Syktus, J.I. (1992). *Climate Modes of the Phanerozoic – The History of the Earth's Climate over the Past 600 Million Years*, 274 pp. Cambridge University Press.

**16** Budyko, M.I. and Ronov, A.B. (1979). Chemical evolution of the atmosphere in the Phanerozoic. *Geochemistry International* 16 (3): 1–9; translated from *Geokhimiya* **5**, 1979, 643-653.

**17** Vaughan, A.P.M. (2007). Climate and geology – A Phanerozoic perspective. In: *Deep-Time Perspectives on Climate Change: Marrying the Signal from Computer Models and Biological Proxies*, Micropalaeontological Society Special Publicaiton (eds. M. Williams, A.M. Haywood, F.J. Gregory and D.N. Schmidt), 5–59. London: The Geological Society.

**18** Royer, D.L., Berner, R.A., Montañez, I.P. et al. (2004). $CO_2$ as a primary driver of Phanerozoic climate. *GSA Today* 14 (3): 4–10. https://doi.org/10.1130/1052-5173(2004)014<4:CAAPDO>2.0.CO;2.

**19** Royer, D.L. (2006). $CO_2$ – forced climate thresholds during the Phanerozoic. *Geochimica Cosmochimica Acta* 70: 5665–5675.

**20** Montañez, I.P. and Poulsen, C.J. (2013). The late Paleozoic ice age: an evolving paradigm. *Annual Reviews of Earth and Planetary Science* 41: 629–656.

**21** Goddéris, Y., Donnadieu, Y., Carretier, S. et al. (2017). Onset and ending of the late Palaeozoic ice age triggered by tectonically paced rock weathering. *Nature Geoscience* 10: 382–386.

**22** Montañez, I.P. (2016). A late Paleozoic climate window of opportunity. *Proceedings of the National Academy of Sciences of the United States of America* 113 (9): 2334–2336.

**23** Goddéris, Y., Donnadieu, Y., Le Hir, G. et al. (2014). The role of palaeogeography in the Phanerozoic history of atmospheric $CO_2$ and climate. *Earth-Science Reviews* 128: 122–138.

**24** Rugenstein, J.K.C., Ibarra, D.E., and von Blanckenburg, F. (2019). Neogene cooling driven by land surface reactivity rather than increased weathering fluxes. *Nature* 571: 99–102.

**25** Korte, C., Hesselbo, S.P., Ullmann, C.V. et al. (2015). Jurassic climate mode governed by ocean gateway. *Nature Communications* 6, Article number: 10015. https://doi.org/10.1038/ncomms10015.

**26** Kidder, D.L. and Worsley, T.R. (2010). Phanerozoic Large Igneous Provinces (LIPs), HEATT (Haline Euxinic Acidic Thermal Transgression) episodes, and mass extinctions. *Palaeogeography, Palaeoclimatology, Palaeoecology* 295: 162–191. https://doi.org/10.1016/j.palaeo.2010.05.036.

**27** Kidder, D.L. and Worsley, T.R. (2012). A human-induced hothouse climate? *Geology Today* 22 (2): 4–11.

**28** Hay, W.W. (2013). *Experimenting on a Small Planet – A Scholarly Entertainment*, 983 pp. New York: Springer (also an e-book).

**29** Rampino, M.R. and Prokoph, A. (2013). Are mantle plumes periodic? *EOS* 94 (12): 113–114.

**30** Rampino, M.R. (2015). Disc dark matter in the Galaxy and potential cycles of extraterrestrial impacts, mass extinctions and geological events. *Monthly Notices of the Royal Astronomical Society* 448: 1816–1820.

**31** Müller, R.D. and Dutkiewicz, A. (2018). Oceanic crustal carbon cycle drives 26-million-year atmospheric carbon dioxide periodicities. *Scientific Advances* 4: eaaq0500, 7pp.

**32** Arthur, M.A. (1982) The carbon cycle—controls on atmospheric CO2 and climate in the geologic past, pp 55-67, in *Climate In Earth History: Studies in Geophysics* (ed W.H. Berger and J.C. Crowell), Report of Panel on Pre-Pleistocene Climates, for the Geophysics Study Committee of the Geophysics Research Board of the Commission on Physical Sciences, Mathematics, and Applications of the US National Research Council, based on a meeting in Toronto in 1980, National Academy Press, Washington D.C., ISBN: 0-309-10784-9, 212pp.

**33** Walker, J.C.G. (1986). Global geochemical cycles of carbon, sulfur and oxygen. *Marine Geology* 70: 159–174.

**34** Müller, R.D., Sdrolias, M., Gaina, C. et al. (2008). Long-term sea-level fluctuations driven by ocean basin dynamics. *Science* 319: 1357–1362.

**35** Sandberg, P.A. (1983). An oscillating trend in Phanerozoic nonskeletal carbonate mineralogy. *Nature* 305: 19–22.

**36** Stanley, S.M. and Hardie, L.A. (1998). Secular oscillations in the carbonate mineralogy of reef-building and sediment-producing organisms driven by tectonically forced shifts in seawater chemistry. *Palaeogeography, Palaeoclimatology, Palaeoecology* 144: 3–19.

**37** Hardie, L.A. (1996). Secular variation in seawater chemistry: an explanation for the coupled secular variation in the mineralogies of marine limestones and potash evaporites over the past 600 m.y. *Geology* 24: 279–283.

**38** Holland, H.D. (2003). The geological history of seawater. In: *The Oceans and Marine Geochemistry*, Treatise on

Geochemistry (ed. H.D. Holland and K.K. Turekian), vol. 6 (ed. H. Elderfield), 583–625. Oxford: Elsevier-Pergamon.

39 Haq, B.U., Hardenbol, J., and Vail, P.R. (1987). Chronology of fluctuating sea levels since the Triassic. *Science* 235: 1156–1167.

40 Hallam, A. (1984). Pre-Quaternary sea level changes. *Annual Reviews of Earth and Planetary Science* 12: 205–243.

41 Elderfield, H. (2010). Seawater chemistry and climate. *Science* 327: 1092–1093.

42 Zalasiewicz, J. and Williams, M. (2012). *The Goldilocks Planet – The Four Billion Year Story of Earth's Climate*, 303 pp. Oxford University Press.

43 Coggon, R.M., Teagle, D.A.H., Smith-Duque, C.E. et al. (2010). Reconstructing past seawater Mg/Ca and Sr/Ca from mid-ocean ridge flank calcium carbonate veins. *Science* 327: 1114–1117.

44 Müller, R.D., Dutkiewicz, A., Seton, M., and Gaina, C. (2013). Seawater chemistry driven by supercontinent assembly, breakup, and dispersal. *Geology* 41 (8): 907–910.

45 Rowley, D.B. (2008). Extrapolating oceanic age distributions: lessons from the Pacific region. *Journal of Geology* 116: 587–598.

46 Müller, R.D., Dutkiewicz, A., Seton, M., and Gaina, C. (2014). Seawater chemistry driven by supercontinent assembly, breakup and dispersal. *Geology Forum* https://doi.org/10.1130/G35636Y.1.

47 Pälike, H., Lyle, M.W., Nishi, H. et al. (2012). A Cenozoic record of the equatorial Pacific carbonate compensation depth. *Nature* 488: 609–615.

48 Schlanger, S.O. and Jenkyns, H.C. (1976). Cretaceous oceanic anoxic events: causes and consequences. *Geologie en Mijnbouw* 55: 179–184.

49 Arthur, M.A., Schlanger, S.O., and Jenkyns, H.C. (1987). The Cenomanian-Turonian oceanic anoxic event, II, palaeoceanographic controls on organic-matter production and preservation. In: *Marine Petroleum Source Rocks* (ed. J. Brooks and A.J. Fleet). *Geological Society of London Special Publicaiton* 26: 401–420.

50 Jenkyns, H.C. (2010). Geochemistry of oceanic anoxic events. *Geochemistry Geophysics Geosystems* 11: Q03004. https://doi.org/10.1029/2009GC002788.

51 Trabucho-Alexandre, J., Hay, W.W., and De Boer, P.L. (2012). Phanerozoic environments of black shale deposition and the Wilson cycle. *Solid Earth* 3: 29–42.

52 Summerhayes, C.P. (1987). Organic-rich Cretaceous sediments from the North Atlantic. In: *Marine Petroleum Source Rocks* (ed. J. Brooks and A.J. Fleet). *Geological Society of London Special Publication* 26: 301–316.

53 Brass, G.W., Southam, J.R., and Peterson, W.H. (1982). Warm saline bottom waters in the ancient ocean. *Nature* 296: 620–623.

54 Friedrich, O., Erbacher, J., Moriya, K. et al. (2008). Warm saline intermediate waters in Cretaceous tropical Atlantic Ocean. *Nature Geoscience* 1: 453–457.

55 Topper, R.P.M., Trabucho-Alexandre, J., Tuenter, E., and Meijer, P.T. (2011). A regional ocean circulation model for the mid-Cretaceous North Atlantic basin: implications for black shale formation. *Climate of the Past* 7: 277–297.

56 Summerhayes, C.P., and Masran, T.C. (1984) Organic facies of Cretaceous sediments from Deep Sea Drilling Project sites 535 and 540, eastern Gulf of Mexico, *Initial Reports DSDP* 77, Washington D.C., 451–457

57 Trabucho Alexandre, J., Van Gilst, R.I., Rodriguez-Lopez, J.P., and De Boer, P.L. (2011). The sedimentary expression of oceanic anoxic event 1b in the North Atlantic. *Sedimentology* 58: 1217–1246.

58 Stow, D.A.V., and Dean, W.E. (1984) Middle Cretaceous black shales at site 530 in the southeastern Angola Basin, *Initial Reports DSDP* 75, Washington D.C., 809–817.

59 Arthur, M.A., Dean, W.E., and Schlanger, S.O. (1985). Variations in the global carbon cycle during the Cretaceous related to climate, volcanism, and changes in atmospheric $CO_2$. In: *The Carbon Cycle and Atmospheric CO2: Natural Variations Archean to Present*, Geophysical Monograph 32 (eds. E.T. Sundquist and W.S. Broecker), 504–529. Washington D.C.: American Geophysical Union.

60 Summerhayes, C.P., and Masran, T.S. (1983) Organic facies of Cretaceous and Jurassic sediments from Deep Sea Drilling Project site 534 in the Blake-Bahama Basin, western North Atlantic, *Initial Reports DSDP* 76, Washington D.C., 469–480.

61 Bralower, T.J. and Thierstein, H.R. (1984). Low-productivity and slow deep-water circulation in mid-Cretaceous oceans. *Geology* 12: 614–618.

62 Keller, C.E., Hochuli, P.A., Weissert, H. et al. (2011). Volcanically induced climate warming and floral change preceded the onset of OAE1 (arly Cretaceous). *Palaeogeography, Palaeoclimatology, Palaeoecology* 305: 43–49.

63 Pogge von Strandmann, P.A.E., Jenkyns, H.C., and Woodfine, R.G. (2013). Lithium isotope evidence for enhanced weathering during Oceanic Anoxic Event 2. *Nature Geoscience* 6 https://doi.org/10.1038/NGEO1875.

64 Montañez, I.P., Norris, R.D., Algeo, T. et al. (2011). *Understanding Earth's Deep Past: Lessons for Our Climate Future*, 194 pp. Washington, DC: National Academies Press.

65 Foster, G.L. and Rohling, E.J. (2013). Relationship between sea level and climate forcing by $CO_2$ on geological timescales. *Proceedings of the National Academy of Sciences of the United States of America* 110 (4): 1209–1214.

66 Van de Wal, R.S.W., De Boer, B., Lourens, L.J. et al. (2011). Reconstruction of a continuous high-resolution

$CO_2$ record over the past 20 million years. *Climate of the Past* 7: 1459–1469.

67 Parkinson, D.N. and Summerhayes, C.P. (1985). Synchronous global sequence boundaries. *American Association of Petroleum Geologists Bulletin* 69 (5): 685–687.

68 Summerhayes, C.P. (1986). Sealevel curves based on seismic stratigraphy: their chronostratigraphic significance. *Palaeogeography, Palaeoclimatology, Palaeoecology* 57: 27–42.

69 Torsvik, T.H. and Cocks, L.R.M. (2017). *Earth History and Palaeogeography*, 317 pp. Cambridge University Press.

70 Steffen, W., Sanderson, A., Tyson, P.D. et al. (2004). *Global Change and the Earth System: A Planet Under Pressure*, Global Change – The IGBP Series, 336 pp. Berlin: Springer.

71 http://en.wikipedia.org/wiki/Earth_system_science (December 2013).

72 Brasseur, G.P. and Van der Pluijm, B. (2013). Earth's future: navigating the science of the Anthropocene. *Earth's Future* 1: 1–2. https://doi.org/10.1002/2013 EF000221.

73 Lovelock, J.E. and Giffin, C.E. (1969). Planetary atmospheres: compositional and other changes associated with the presence of life. *Advances in the Astronautical Sciences* 25: 179–193.

74 Lovelock, J.E. (1972). Gaia as seen through the atmosphere. *Atmospheric Environment* 6 (8): 579–580.

75 Lovelock, J.E. and Margulis, L. (1974). Homeostatic tendencies of the Earth's atmosphere. *Origins of Life* 5 (1–2): 93–103.

76 Margulis, L. (1999). *The Symbiotic Planet: A New Look at Evolution*. London: Phoenix.

77 Lovelock, J. (1979). *Gaia: A New Look at Life on Earth*. Oxford: Oxford University Press.

78 Nisbett, E.G. and Fowler, C.M.R. (2005). The early history of life. In: *Biogeochemistry*, Treatise on Geochemistry (eds H.D. Holland and K.K. Turekian), vol. 8 (ed. W.H. Schlesinger), 1–39. Oxford: Elsevier-Pergamon.

79 Lenton, T.M. and Watson, A. (2011). *Revolutions That Made the Earth*, 423 pp. Oxford University Press.

80 Leeder, M. (2007). Cybertectonic Earth and Gaia's weak hand: sedimentary geology, sedimentary cycling and the Earth system. *Journal of the Geological Society of London* 164: 277–296.

81 Rowley, D.B. (2002). Rate of plate creation and destruction: 180 Ma to present. *Bulletin of the Geological Society of America* 114 (8): 927–933.

82 Nield, T. (2011). *Incoming! or Why We Should Stop Worrying and Learn to Love the Meteorite*. London: Granta Books.

83 http://en.wikipedia.org/wiki/Eugene_Merle_Shoemaker (December 2013).

84 http://en.wikipedia.org/wiki/Impact_crater (December 2013).

85 http://en.wikipedia.org/wiki/Chicxulub_Crater (December 2013).

86 http://en.wikipedia.org/wiki/Cretaceous–Paleogene_boundary (December 2013).

87 Alvarez, L.W., Alvarez, W., Asaro, F., and Michel, H.V. (1980). Extraterrestrial cause for the Cretaceous-Tertiary extinction: experiment and theory. *Science* 208 (4448): 1095–1108.

88 http://en.wikipedia.org/wiki/Luis_Walter_Alvarez (December 2013).

89 Kolbert, E. (2014). *The Sixth Extinction*, 319 pp. London: Bloomsbury Press.

90 Schulte, P., Alegret, L., Arenillas, I. et al. (2010). The Chicxulub asteroid impact and mass extinction at the Cretaceous-Paleogene boundary. *Science* 327 (5970): 1214–1218.

91 Renne, P.R., Deino, A.L., Hilgen, F.J. et al. (2013). Time scales of critical events around the Cretaceous-Paleogene boundary. *Science* 339 (6120): 684–687.

92 MacLeod, K.G., Quinton, P.C., Sepúlveda, J., and Negra, M.H. (2018). Postimpact earliest Paleogene warming shown by fish debris oxygen isotopes (El Kef, Tunisia). *Science* 360 (6391): 1467–1469.

93 Courtillot, V., Besse, J., Vandamme, D. et al. (1986). Deccan flood basalts at the Cretaceous-Tertiary boundary. *Earth and Planetary Science Letters* 80: 361–174.

94 McLaren, D.J. (1970). Time, life, and boundaries. *Journal of Paleontolology. (Presidential Address)* 44 (5): 801–813.

95 Hattersley-Smith, G. (2007). Digby Johns McLaren, 1 December 1919–8 December 2004: elected FRS 1979. *Biographical Memoirs Fellows of the Royal Society* 53 https://doi.org/10.1098/rsbm.2007.0007.

96 Dietz, R.S. (1962). Sudbury structure as an astrobleme. *Journal of Geology* 72: 412–434.

97 McLaren, D.J. and Goodfellow, W.D. (1990). Geological and biological consequences of giant impacts. *Annual Reviews Earth Planetary Science* 18: 123–171.

98 McLaren, D.J. (1983). Bolides and biostratigraphy. *Bulletin of the Geological Society of America* 94 (3): 313–324.

99 Urey, H.C. (1973). Cometary collisions and geological periods. *Nature* 242: 32–33.

100 Hallam, A. (1998) Mass extinctions in Phanerozoic time, *Special Publication* **140**, Geological Society of London, 259–274.

101 Friedman, H. (1985) The science of global change – an overview, in *Global Change* (eds T.F. Malone and J.G. Roederer), Proc. ICSU Sympos. Ottowa, September 25, 1984, ICSU Press and Cambridge Univ. Press, 20–52.

102 Pollack, J. (1982) Solar, astronomical, and atmospheric effects on climate, pp 68-76, in *Climate In Earth History: Studies in Geophysics* (ed W.H. Berger and J.C. Crowell), Report of Panel on Pre-Pleistocene Climates, for the Geophysics Study Committee of the Geophysics Research Board of the Commission on Physical Sciences, Mathematics, and Applications of the US National Research Council, based on a meeting in Toronto in 1980, National Academy Press, Washington D.C., ISBN: 0-309-10784-9, 212 pp.

103 Svensmark, H. (2012). Evidence of nearby supernovae affecting life on Earth. *Monthly Notices of the Royal Astronomical Society* 423: 1234–1253.

104 Roberts, G.G. and Mannion, P.D. (2019). Timing and periodicity of Phanerozoic marine biodiversity and environmental change. *Scientific Reports* 9: 6116.

105 Raup, D. and Sepkoski, J.J. (1984). Periodicity of extinctions in the geologic past. *Proceedings of the National Academy of Sciences of the United States of America* 81 (3): 801–805.

106 Courtillot, V. (1994). Mass extinctions in the last 300 million years: one impact and seven flood basalts? *Israel Journal of Earth Science* 43: 255–266.

107 Courtillot, V.E. and Renne, P.R. (2003). On the ages of flood basalt events. *Comptes Rendus Geoscience* 335: 113–140.

108 Courtillot, V.E. and Olsen, P. (2007). Mantle plumes link magnetic superchrons to Phanerozoic mass depletion events. *Earth and Planetary Science Letters* 260: 495–504.

109 Benton, M.J. (2018). Hyperthermal-driven mass extinctions: killing models during the Permian–Triassic mass extinction. *Philosophical Transactions. Series A, Mathematical, Physical, and Engineering Sciences* 376: 20170076. doi:https://doi.org/10.1098/rsta.2017.0076.

110 Xu, W., MacNiocaill, C., Ruhl, M. et al. (2018). Magnetostratigraphy of the Toarcian Stage (Lower Jurassic) of the Llanbedr (MochrasFarm) Borehole, Wales: basis for a global standard and implications for volcanic forcing of palaeoenvironmental change. *Journal of the Geological Society of London* 175: 594–604.

111 Keller, G., Armstrong, H., Garper, D. et al. (2012). Volcanism, impacts and mass extinctions. *Geoscientist* 22 (10): 10–15.

112 Keller, G., Kerr, A.C., and MacLeod, N. (2013). Exploring the causes of mass extinction events. *EOS* 94 (22): 200.

113 Sakamoto, M., Benton, M.J., and Venditti, C. (2016). Dinosaurs in decline tens of millions of years before their final extinction. *Proceedings of the National Academy of Sciences of the United States of America* 113: 5036–5040.

114 Schulte, P. et al. (2010). Cretaceous extinctions: multiple causes; response to letters by Archibald et al., Keller et al., and Courtillot and Fluteau. *Science* 328, Letters: 975–976.

115 Chen, Z.-Q., Algoe, T.J., and Bottjer, D.K. (2014). Global review of the Permian-Triassic mass extinctio and subsequent recovery: part 1. *Earth Science Reviews* 137: 1–5.

116 Cui, Y. and Kump, L.R. (2015). Global warming and the end Permian extinction event: proxy and modeling perspectives. *Earth Science Reviews* 149: 5–22.

117 Komar, N. and Zeebe, R.E. (2016). Calcium and calcium isotope changes during carbon cycle perturbations at the end-Permian. *Paleoceanography* 31: 115–130. https://doi.org/10.1002/2015PA002834.

118 Stordal, F., Svensen, H.H., Aarnes, I., and Roscher, M. (2017). Global temperature response to century-scale degassing from the Siberian Traps Large igneous province. *Palaeogeography, Palaeoclimatology, Palaeoecology* 471: 96–107.

119 Baresel, B., Bucher, H., Bagherpour, B. et al. (2017). Timing of global regression and microbial bloom linked with the Permian-Triassic boundary mass extinction: implications for driving mechanisms. *Nature Scientific Reports* 7: 43630. https://doi.org/10.1038/srep43630.

120 Song, H., Wignall, P.B., and Dunhill, A. (2018). Decoupled taxonomic and ecological recoveries from the Permo–Triassic extinction. *Scientific Advances* 4: eaat5091.

121 Retallack, G.J., Sheldon, N.D., Carr, P.F. et al. (2011). Multiple early Triassic greenhouse crises impeded recovery from late Permian mass extinction. *Palaeogeography, Palaeoclimatology, Palaeoecology* 308: 233–251.

122 Svensen, H., Planke, S., Polozov, A.G. et al. (2008). Siberian gas venting and the end-Permian environmental crisis. *Earth and Planetary Science Letters* 277 (3–4) https://doi.org/10.1016/j.epsl.2008.11.015.

123 Black, B.A., Elkins-Tanton, L.T., Rowe, C., and Peate, I.U. (2012). Magnitude and consequences of volatile release from the Siberian Traps. *Earth and Planetary Science Letters* 317-318: 363–373.

124 Lenton, T.M., Dahl, T.W., Daines, S.J. et al. (2016). Earliest land plants created modern levels of atmospheric oxygen. *Proceedings of the National Academy of Sciences of the United States of America* 113 (35): 9704–9709.

125 Benton, M.J. (2005). *When Life Nearly Died: The Greatest Mass Extinction of All Time*. London: Thames & Hudson. ISBN 978-0-500-28573-2.

126 Sun, Y., Joachimski, M.M., Wignall, P.B. et al. (2012). Lethally hot temperatures during the early Triassic Greenhouse. *Science* 338 (6105): 366–370.

127 Tohver, E., Cawood, P.A., Riccomini, C. et al. (2013). Shaking a methane fizz: seismicity from the Araguianha

impact event and the Permian-Triassic global carbon isotope record. *Palaeogeography, Palaeoclimatology, Palaeoecology* 387: 66–75.

**128** Becker, L., Poreda, R.J., Hunt, A.G. et al. (2001). Impact event at the Permian-Triassic boundary: evidence from extraterrestrial noble gases in fullerines. *Science* 291: 1530–1533.

**129** Koeberl, C., Gilmour, I., Reimold, W.U. et al. (2002). End-Permian catastrophe by bolide impact: evidence of a gigantic release of sulfur from the mantle: comment and reply. *Geology* 30: 855–856.

**130** Blackburn, T.J., Olsen, P.E., Bowring, S.A. et al. (2013). Zircon U-Pb geochronology links the end-Triassic extinction with the Central Atlantic Magmatic Province. *Science* 340 (6135): 941–945.

**131** Heimdal, T.K., Svensen, H.H., Ramezani, J. et al. (2018). Large-scale sill emplacement in Brazil as a trigger for the end-Triassic crisis. *Nature Scientific Reports* 8 (141) https://doi.org/10.1038/s41598-017-18629-8.

**132** McElwain, J.C., Beerling, D.J., and Woodward, F.I. (1999). Fossil plants and global warming at the Triassic–Jurassic boundary. *Science* 285: 1386–1389.

**133** Van Der Schootbrugge, B., Quan, T.M., Lindström, S. et al. (2009). Floral changes across the Triassic/Jurassic boundary linked to flood basalt volcanism. *Nature Geoscience* 2: 589–594.

**134** Walker, J.C.G., Hays, P.B., and Kasting, J.F. (1981). A negative feedback mechanism for the long term stabilization of earth's surface temperature. *Journal of Geophysical Research* 86: 9976–9982.

**135** Walker, G. (2003). *Snowball Earth: The Story of the Great Global Catastrophe That Spawned Life as We Know It*, 269 pp. London: Bloomsbury.

**136** Harland, W.B. (1964). Critical evidence for a great infra-Cambrian glaciation. *Geologische Rundschau* 54: 45–61.

**137** Harland, W.B. and Rudwick, M.J.S. (1964). The great infra-Cambrian ice age. *Scientific American* 211 (2): 28–36.

**138** Schrag, D.P. and Hoffman, P. (2001). Life, geology and snowball Earth. *Nature* 409: 306.

**139** Hyde, W.T., Crowley, T.J., Baum, S.K., and Peltier, W.R. (2000). Neoproterozoic 'snowball Earth' simulations with a coupled climate/ice-sheet model. *Nature* 405: 425–429.

**140** Spence, G.H., Le Heron, D.P., and Fairchild, I.J. (2016). Sedimentological perspectives on climatic, atmospheric and environmental change in the Neoproterozoic Era. *Sedimentology* 63: 253–306.

**141** Fairchild, I.J. and Kennedy, M.J. (2007). Neoproterozoic glaciation in the Earth system. *Journal of the Geological Society* 164: 895–921. https://doi.org/10.1144/0016-76492006-191.

**142** Hoffman, P., Abbot, D.S., Ashkenazy, A. et al. (2017). Snowball Earth climate dynamics and Cryogenian geology-geobiology. *Scientific Advances* 3: e1600983, 43 pp.

**143** Hambrey, M.J. and Harland, W.B. (eds.) (1981). *Earth's Pre-Pleistocene Glacial Record*. Cambridge: Cambridge University Press.

**144** Li, Z.X. et al. (2008). Assembly, configuration, and break-up history of Rodinia: a synthesis. *Precambrian Research* 160 (1–2): 179–121.

**145** Kirschvink, J.L. (1992). Late Proterozoic low-latitude global glaciation: The Snowball Earth. In: *The Proterozoic Biosphere: A Multidisciplinary Study* (eds. J.W. Schopf, C. Klein and D. Des Maris), 51–52. Cambridge University Press.

**146** Lenton, T.M. (2016). *Earth System Science: A Very Short Introduction*, 153 pp. Oxford University Press.

**147** Abbot, D.S., Voigt, A., Li, D. et al. (2013). Robust elements of Snowball Earth atmospheric circulation and oases for life. *Journal of Geophysical Research: Atmospheres* 118: 1–11.

**148** Corsetti, F.A., Awramik, S.M., and Pierce, D. (2003). A complex microbiota from snowball Earth times: Microfossils from the Neoproterozoic Kingston Peak Formation, Death Valley, USA. *Proceedings of the National Academy of Sciences of the United States of America* 100 (8): 4399–4404.

**149** Hoffman, P., Kauffman, W., and Halverson, G.P. (1998). Comings and goings of global glaciations on a Neoproterozoic tropical platform in Namibia. *GSA Today* 8 (5): 1–9.

**150** Hoffman, P.F. and Schrag, D.P. (2002). The snowball Earth hypothesis: testing the limits of global change. *Terra Nova* 14: 129–155.

**151** Bao, H., Lyons, J.R., and Zhou, C. (2008). Triple oxygen isotope evidence for elevated $CO_2$ levels after a Neoproterozoic glaciation. *Nature* 453: 504–506.

**152** Bao, H., Fairchild, I.J., Wynn, P.M., and Spötl, C. (2009). Stretching the envelope of past surface environments: Neoproterozoic glacial lakes from Svalbard. *Science* 323: 119–122.

**153** Fairchild, I.J., Fleming, E.J., Bao, H. et al. (2016). Continental carbonate facies of a Neoproterozoic panglaciation, north-east Svalbard. *Sedimentology* 63: 443–497.

**154** Hardie, L.A. (2003). Secular variations in Precambrian seawater chemistry and the timing of Precambrian aragonite seas and calcite seas. *Geology* 31 (9): 785–788.

**155** Shields, G. and Veizer, J. (2002). Precambrian marine carbonate carbonate isotope database: version 1.1. *Geochemistry, Geophysics, Geosystems* 3 (6) https://doi.org/10.1029/2001GC000266.

**156** Saltzman, M.R. and Thomas, E. (2012). Carbon isotope stratigraphy. In: *The Geologic Time Scale 2012* (ed. F. Gradstein). Elsevier https://doi.org/10.1016/B978-0-444-59425-9.00011-1.

**157** Lenton, T. and Watson, A. (2004). Biotic enhancement of weathering, atmospheric oxygen and carbon dioxide in the Neoproterozoic. *Geophysical Research Letters* 31: L05202. https://doi.org/10.1029/2003GL018802.

**158** Fairchild, T.R., Schopf, J.W., Shen-Miller, J. et al. (1996). Recent discoveries of Proterozoic microfossils in South-Central Brazil. *Precambrian Research* 80 (1–2): 125–152.

**159** Reinhard, C.T., Planavsky, N.J., Benjamin, C. et al. (2017). Evolution of the global phosphorus cycle. *Nature* 541: 386–389.

**160** Morris, J.L., Puttick, M.N., Clark, J.W. et al. (2018). The timescale of early land plant evolution. *Proceedings of the National Academy of Sciences of the United States of America* 115 (10): E2274–E2283.

**161** Lenton, T.M., Boyle, R.A., Poulton, S.W. et al. (2014). Co-evolution of eukaryotes and ocean oxygenation in the Neoproterozoic era. *Nature Geoscience* 7 (4): 257–265. ISSN 1752-0894.

**162** Keller, C.B., Husson, J.M., Mitchell, R.N. et al. (2019). Neoproterozoic glacial origin of the Great Unconformity. *Proceedings of the National Academy of Sciences of the United States of America* 116 (4): 1136–1145.

**163** Gernon, T.M., Hincks, T.K., Tyrrell, T. et al. (2016). Snowball Earth ocean chemistry driven by extensive ridge volcanism during Rodinia breakup. *Nature Geoscience* 9: 242–248.

**164** Feulner, G., Hallmann, C., and Kienert, H. (2015). Snowball cooling after algal rise. *Nature Geoscience* 8: 659–662.

# 11

# Numerical Climate Models and Case Histories

**LEARNING OUTCOMES**

- Understand and be able to explain what we can learn from numerical models and geological data about Cretaceous climate and its influence by $CO_2$.
- Explain Climate Sensitivity and Earth System Sensitivity and the difference between them, bearing feedbacks in mind.
- Ignoring hyperthermal events, be able to explain the various arguments for the control of $CO_2$ and climate from end Cretaceous to end Eocene time by: plate tectonics, weathering, ocean circulation, the changing nature of marine plankton, and changes to the biological pump.
- Understand and be able to explain the origin of the Antarctic ice sheet 34 Ma ago.
- Understand and be able to explain the nature and origin of Eocene hyperthermal events, including the Palaeocene-Eocene Thermal Maximum (PETM).
- Understand and be able to explain the variability of the Miocene climate in relation to ocean gateways, orbital insolation and fluctuating $CO_2$.
- Understand and be able to explain the origin and effect of the Mid Pliocene climatic optimum.

## 11.1 CO$_2$ and General Circulation Models

Having established that fluctuations in $CO_2$ are directly related to changing temperature through the operation of the slow carbon cycle, it is time to examine the effect of integrating $CO_2$ into a General Circulation Model (GMC) of the climate system, of the kind introduced in Chapter 6.

Realizing that $CO_2$ must have played an important role in the climate story, Eric Barron and Warren Washington, whom we met in Chapter 6, modified their GCM to consider the effect of increased levels of $CO_2$ in the air. At the 1984 Tarpon Springs meeting, they reported that the addition of four times the then present day value (in 1984) of $CO_2$ increased the globally averaged surface temperature of the Cretaceous by an additional 3.6 °C [1]. Added to the warming of 4.8 °C produced by geographical change, that gave a total global average temperature rise of 8.4 °C, which was close to that estimated from $\partial^{18}O$ data. While this seemed convincing, the tropical temperatures in the model output were a bit too high. Barron thought that might be because their GCM was primarily an atmospheric model. Lacking a dynamic ocean component, it missed the role of the oceans in poleward heat transport. This defect would take a while to correct. Meanwhile, working with Steve

Schneider, Barron used his GCM for the Cretaceous to show that it was unlikely that winter temperatures rose above freezing in mid-continent at high latitudes [2], confirming what was known from geological data and fossils.

By 1995, advances in computing power enabled Barron to use a new GCM, GENESIS, which incorporated a dynamic ocean component as well as a dynamic atmosphere, to assess the effect of ocean heat transport [3]. He ran the model with different combinations of ocean heat transport and $CO_2$ to see what best matched the observed distribution of Cretaceous temperatures. He also reviewed the latest data on Cretaceous temperatures to satisfy himself that the polar regions were free of ice on land, even if they were seasonally below freezing. The model's results confirmed that: '*Higher carbon dioxide is required to achieve sufficient polar warmth, and greater oceanic heat flux is required to prevent the tropical oceans from overheating under conditions of higher carbon dioxide*' [3]. The model that best matched observations contained four times the current value of $CO_2$ (340 ppm) (Figure 11.1). In effect, we have come full circle, back to the picture painted in 1899 by T.C. Chamberlin.

Was all well? Barron knew that past ocean temperatures estimated from $\partial^{18}O$ data might be inaccurate because the $\partial^{18}O$ ratio of seawater in an ice-free ocean

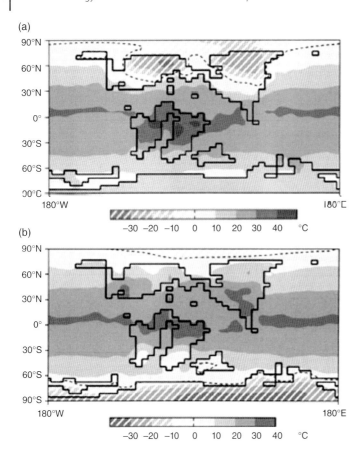

**Figure 11.1** Mid Cretaceous modelled temperatures for $4 \times CO_2$: (a) December, January, and February and (b) June, July, and August. Warming is evident everywhere, especially at high latitudes. DJF Arctic warming locally exceeds 40 °C. High latitude continental interiors remain below freezing in winter. Tropical ocean surface temperatures increase by 1.5–2.0 °C. Tropical continental regions experience greater warming. The polar warming approaches palaeoclimate observations, but Arctic temperatures remain too cold. Mid-latitude temperatures are close to realistic. Tropical temperatures are close to exceeding the range of observed values. *Source:* From figure 6 in Ref. [3].

might vary with salinity, and the $\partial^{18}O$ measurements were made on a wide range of organisms living at different depths and thus representing different water temperatures in the top 100 m of the water column. To adjust for those possibilities, his team compared the outputs of GENESIS with palaeotemperatures adjusted for the effects of salinity and water depth. This improved the match between the data and the model outputs [4]. Temperatures predicted by the model agreed well with isotopic temperatures for low latitudes, and with the low part of the range of isotopic temperatures from high latitudes. One problem remained: the modelled temperatures from high latitudes were lower than the extremes of the range of temperatures determined from $\partial^{18}O$ data at those latitudes. Whilst the '*results suggest that the mid-Cretaceous climate may have been as warm or warmer than that simulated in a mid-Cretaceous $4 \times CO_2$ experiment*', Barron recommended that additional $\partial^{18}O$ data from high latitudes be collected to resolve the discrepancy [4]. This is an example of how GCM outputs can identify places where new data should be collected.

There was an additional problem. Extensive continental interiors, especially in northeast Asia (Siberia) were

modelled as below freezing despite plant and other evidence to the contrary [5]. Bill Hay and colleagues thought about this for quite some time before realizing in 2018 that the $\partial^{18}O$ signal from marine plankton could represent either ice on land, *or* an increase in the volume of groundwater reservoirs on land. The problem of a warm Arctic with abundant flora could not be explained by freezing in winter (even by simulating increases in $CO_2$), but it could be explained if the topography was low and there were extensive wetlands (rather like Finland or northern Canada today). By assuming that $CO_2$ reached 8 times present values and that 50–75% of the surface provided water vapour as a supplementary greenhouse gas on the continent, they found that the regional temperature gradients produced conditions compatible with fossil and sedimentological evidence. The interiors were not excessively cold [5].

Could different proxy data tell us if the $4 \times$ modern $CO_2$ level was a reasonable estimate for the Cretaceous climate? Bob Spicer and Judy Parrish argued that high $CO_2$ might have helped to maintain the polar warmth necessary to sustain vegetation in the Arctic during the Late Cretaceous and Palaeocene [6]. They thought that it was '*unlikely ...*

*that levels of 4 × present $CO_2$ were exceeded because above this level many plants with conventional $C_3$ photosynthetic pathways close their stomata and productivity is thereby reduced'* [6].

Others followed Barron's lead. For instance, in 1992, geologists from the Chevron oil company applied a GCM to the Late Jurassic, to see what difference might arise between an atmosphere containing the pre-industrial level of 280 ppm $CO_2$, and one containing 1120 ppm, a level suggested for the Jurassic by Berner's models and by compilations of $CO_2$-proxy data [7, 8]. Their GCM comprised a dynamic atmosphere coupled to a static ocean, so missed the effects of ocean heat transport. The climate simulation with 1120 ppm best matched the distribution of palaeoclimate data. Coals occurred where rainfall was high, evaporites where rainfall was low. Warming was greatest over the high latitude oceans and least over the tropics. Sea ice was restricted to offshore seas at high latitudes. Strong summer monsoon rains developed over South East Asia. The Trade Winds brought heavy winter rainfall to eastern Gondwana and heavy summer rains to the tropical margins of Tethys.

A comparable approach was taken, also in 1992, in an investigation of the climate of the Kimmeridgian (155–150 Ma), an oil rich part the Late Jurassic period, by modeller Paul Valdes, whom we met in Chapter 6, and geologist Bruce Sellwood (1946–2007), professor of applied sedimentology at Reading University [9]. Their GCM coupled the atmosphere to a static ocean with temperatures ranging from 27 °C at the equator to 0 °C at the poles. Like Chevron, they varied the $CO_2$ concentration in the air – in their case from 350 to 1080 ppm – to see what difference this made. The most accurate reflection of the palaeoclimate data came from the simulation with high $CO_2$. The average global temperature was 20 °C. Their results replicated those of Chevron, but they also found the southwestern United States to be arid, southern Europe to be seasonally arid, and winter temperatures over Siberia and southeastern Gondwana to have been below 0 °C, in agreement with the palaeoclimate data.

Models go on developing, and data go on accumulating. Fifteen years later, Sellwood and Valdes applied an updated GCM to palaeoreconstructions for the Late Triassic, Late Jurassic, and Early and Late Cretaceous, validating the models with palaeoclimate data [10]. *'Compared to the present',* they observed *'the Mesozoic Earth was an alien world with a greenhouse climate … dense forests grew close to both poles but experienced months-long daylight in warm summers, and months-long darkness in cold, sometimes snowy, winters. Ocean depths were warm (8°C or more to the ocean floor) and reefs, with corals, grew*

*10° of latitude further north and south than at present time. The whole Earth was warmer than now by several degrees centigrade, generating high atmospheric humidity and a greatly enhanced hydrological cycle'* [10]. Much of the rainfall was focused over the oceans, leaving major expanses of desert on the continents. Polar ice sheets were absent, but local mountain glaciers could not be ruled out on Antarctica in the Cretaceous. There may have been sea ice in the Arctic, especially in the Cretaceous. During the Triassic and Jurassic, atmospheric $CO_2$ was at least four times present values. Some discrepancies between data and models suggested problems with either the models or the proxy data; for instance, model simulations of continental interiors were too cold by 15 °C or more, and real ocean temperatures could have been warmer. Otherwise the model performed well.

At this point we might recall Hasegawa's finding, mentioned in Chapter 6, that during the Cretaceous the northern margin of the Hadley Cell (Figure 6.17) advanced poleward along with global warming during the Early and Late Cretaceous to latitudes between about 32° and 40°N, but that during the exceptionally hot 'super-greenhouse' of the Middle Cretaceous the northern edge of the Hadley Cell shrank equatorward to latitudes of between 22° and 30°N [11], contrary to the expectation of Bill Hay [12]. Hasegawa and his team thought that these results *'suggest the existence of a threshold in atmospheric $CO_2$ level and/or global temperature, beyond which the Hadley circulation shrinks drastically'* [11]. Levels of $CO_2$ may have reached about 1500 ppm during the extremely warm Middle Cretaceous, compared with 500–1000 ppm during the Late Cretaceous.

Hasegawa supported his interpretation of events by citing recent observational and modelling studies suggesting *'that present-day Hadley circulation is expanding poleward in response to the increasing atmospheric $CO_2$ level and consequent global warming'* [11]. He also noted that *'Such a relationship between the width of the Hadley circulation and global temperature and/or atmospheric $CO_2$ levels has also been reported from paleoclimate records of glacial–interglacial transitions'* [11]. These changes in width of the Hadley circulation are related to changes in the Equator-to-pole temperature gradient, with a wider Hadley Cell during interglacial than glacial periods. Deep-ocean circulation was sluggish in the Mid Cretaceous super-greenhouse period, and more vigorous before and after it, indicating an atmosphere–ocean link. The reason may have been enhanced vertical mixing in the high latitude ocean due to enhanced storminess in the Mid Cretaceous, suppressing the formation of deep water there [11], along lines also suggested by Bill Hay [12]. The causal links are not yet clear.

## 11.2  Climate Sensitivity

Models calculate global temperatures by using as boundary conditions the desired $CO_2$ level (say 2 × present) and something called the 'radiative forcing value' calculated from experimental work on how $CO_2$ absorbs and re-radiates infrared radiation. For example, the radiative forcing value was $1.46\,W\,m^{-2}$ for the increase from 1750 to 1998, and would be $3.7\,W\,m^{-2}$ for a doubling of $CO_2$ [13, 14]. The rise in temperature for a given rise in $CO_2$ is known as the 'climate sensitivity'. For a comprehensive view on what climate sensitivity was in the geological past, we turn to the PALEOSENS Project, comprising a large group of climatologists and palaeoclimatologists who reported in 2012 [15]. Over the past 65 Ma the climate sensitivity calculated from past temperatures and $CO_2$ levels implied a warming of 2.2–4.8 °C for a doubling of atmospheric $CO_2$ [15]. That agrees reasonably well with calculations for the modern climate by the Intergovernmental Panel on Climate Change (IPCC), of an equilibrium climate sensitivity of 1.5–4.5 °C [16].

Calculations of climate sensitivity usually take into account only relatively fast changes in climate, including factors like snow melt, the exchange of $CO_2$ between the air and the ocean, and the behaviour of clouds and water vapour. But, from the geological perspective, we also must consider factors that change slowly, like ice sheet decay and the behaviour of the longer-term carbon cycle, including the rates of dissolution of deep-sea carbonates. This 'Earth System Sensitivity' will differ from the climate sensitivity. It should govern our estimates of long-term changes in sea level and of the amounts of $CO_2$ in the air that would be required to stabilize climate change. Studies of the Last Glacial Maximum suggest the Earth System Sensitivity ranged from 1.4 °C to more than 6.5 °C [15, 17]. We'll explore that more later on.

Jim Hansen of NASA, whom we came across in Chapter 8, argued that the global temperature change of 5 °C between the peak of the last Ice Age and the Holocene implied an equilibrium climate sensitivity of ¾ °C for each Watt of forcing [18]. As a doubling of $CO_2$ amounts to a forcing of about $4\,W\,m^{-2}$, the equilibrium sensitivity of ¾ $W\,m^{-2}$ implies that temperature would rise by 3 °C for such a doubling. Hansen pointed out that if the Earth were a black body without climate feedbacks, the equilibrium response to a forcing of $4\,W\,m^{-2}$ would be a rise in temperature of 1.2 °C. That value is more than doubled by feedbacks from the increase in water vapour caused by warming, from the decrease in albedo caused by the loss of snowfields and sea ice, and from other lesser effects. Hansen agreed that whilst the sensitivity of climate to fast feedbacks is about 3 °C for a doubling of $CO_2$, this would be amplified by the effect of slow feedbacks, including

changes in ice sheet size and the long duration of $CO_2$ in the atmosphere [19].

This topic continues to be one of active research. For instance, one 2012 study of recent global warming suggested that the climate sensitivity to a doubling of $CO_2$ should be much closer to 2 °C [20]. Another, by Dana Royer, Mark Pagani and David Beerling, used Cretaceous and Eocene samples to suggest that climate sensitivity then was higher than the 3 °C suggested by several numerical climate models, and may have exceeded 6 °C in glacial times [21]. Royer and his colleagues considered that '*Climate models probably do not capture the full suite of positive climate feedbacks that amplify global temperatures during some globally warm periods, as well as other characteristic features of warm climates such as low meridional temperature gradients*' [21]; these models erred on the low side as a result. This may reflect '*the long-standing failure of GCMs in underestimating high-latitude temperatures and overestimating latitudinal temperature gradients*' [21], possibly by missing biological feedbacks. Such feedbacks would include '*emissions of reactive biogenic gases from terrestrial ecosystems …. [with] substantial impacts on atmospheric chemistry … in particular …. the tropospheric concentrations of GHGs [greenhouse gases] such as ozone, methane and nitrous oxide … Climate feedbacks of these trace GHGs are missing from most pre-Pleistocene climate modeling investigations*' [21].

As we saw in Chapter 8, studies of climate sensitivity began long ago. Arrhenius began this work in the 1890s. Manabe and colleagues continued it in the 1960s. And Jim Hansen published his estimate of it in 1981. Whilst there has been much discussion since then about this seemingly elusive value, in 2018, Cox and colleagues used the variability of temperature about the long-term historical warming trend to calculate that the equilibrium climate sensitivity was indeed 2.8 °C [22], just as Hansen had calculated back in 1981. However, as Proistosescu and Huybers showed in 2017, the problem with estimating climate sensitivity is that the historical estimates of equilibrium climate sensitivity are derived assuming a linear radiative response to warming and ignore '*the radiative response to warming from an evolving contribution of interannual to centennial modes of radiative response … which display stronger amplifying feedbacks and ultimately contribute 28 to 68% of equilibrium warming, yet they comprise only 1 to 7% of current warming*' [23]. Taking palaeoclimate data into account (i.e. centennial modes), rather than just the instantaneous climate sensitivity reliant on modern climate data, yielded an equilbrium climate sensitivity of 2–4.5 °C in agreement with model-derived estimates [23].

In 2019, Farnsworth and colleagues used an ensemble of climate model simulations covering the period 150–35 Ma ago to show that climate sensitivity to $CO_2$ doubling ranged from ~3.5–5.5 °C, with the variation reflecting the additional influence on forcing, first, of changes in continental position and ocean area; second, of increasing solar luminosity through time; and, third, of non-linear feedbacks from changing water vapour and clouds [24]. '*Changes in palaeogeography [especially the area of ocean, which affects Earth's albedo] have a substantial effect on regional climate that must be considered when interpreting long proxy records*' [24]. Climate sensitivity at present seems low because of the low $CO_2$ baseline, the smaller area of ocean compared with that of Cretaceous time, and the presence of ice, which offsets the temperature increase that would otherwise be expected from increasing solar luminosity [24]. These findings are consistent with others that found '*higher climate sensitivity (~4°C per $CO_2$ doubling) in the relatively warm early Eocene climatic optimum than in the relatively cool late Eocene (~3°C per $CO_2$ doubling)*' [24]. Temperatures rose from the earliest to the latest Cretaceous as continental flooding increased then fell substantially through the Palaeocene as sea level fell (see sea level curve in Figure 5.13); climate sensitivity followed suit [24].

Even Farnsworth's estimate may be on the low side. For example, for the most recent period of global warming, the Middle Pliocene, Isabel Montañez and Richard Norris point to evidence suggesting that climate sensitivity may have been as high as 7 °C to 9.6 °C $\pm$ 1.4 °C per doubling of $CO_2$. Why so high? '*Long-term feedbacks operating at accelerated timescales (decadal to centennial) promoted by global warming can substantially magnify an initial temperature increase*' [25]. Their estimate of this Earth System Sensitivity is in broad agreement with other geologically based estimates, which differ from estimates of 'instantaneous climate sensitivity' based on considering just the present relationship between warming and $CO_2$.

These two scientists pointed out in 2011 that because current GCMs cannot reproduce the exceptional warmth of high latitudes during all past warm periods without invoking unreasonably high levels of $CO_2$, they must be unable '*to fully capture the processes and feedbacks governing heat transport and retention or the processes that might generate heat in the polar regions under elevated atmospheric greenhouse gases*' [25]. They are close, but not close enough. There is more work to do.

One of the gases missing from models is the dimethylsulfide produced by phytoplankton. This creates nuclei around which water vapour condenses, encouraging the production of low-level clouds that reflect solar energy. Did heat stress from the warm Cretaceous ocean reduce the ability of phytoplankton to produce this gas, thus reducing cloud cover and warming the atmosphere – especially at the poles? Our understanding of these 'known unknowns' is poor, reducing confidence in climate sensitivity values derived from GCMs applied to palaeoclimate problems [21]. We need more robust GCMs to address this issue. Other trace greenhouse gases may have been equally important in the past, including methane ($CH_4$), nitrous oxide ($N_2O$), and ozone ($O_3$) [26]. It is difficult to apply this realization to the geological past, because we lack suitable geochemical or biological proxies for such gases back beyond the age range of ice cores.

Recently, David Beerling (1965–) (Box 11.1) of the University of Sheffield, and colleagues, used an Earth Systems model to evaluate how terrestrial, climatic and atmospheric processes might have interacted in the past to

---

**Box 11.1 David Beerling**

Studying at the University of Wales' College of Cardiff, David Beerling obtained his BSc and PhD degrees in botany. His current fields of interest include plant evolution, geobiology and global change. He is the director of the Leverhulme Centre for Climate Change Mitigation and Sorby Professor of Natural Sciences in the Department of Animal and Plant Sciences at the University of Sheffield, UK, where his group integrates evidence from fossils, experiments, and theoretical models to investigate fundamental questions concerning the conquest of the land by plants and the role of terrestrial ecosystems in shaping Earth's global ecology, climate and atmospheric composition. Their discoveries are improving our understanding of current global climate change issues. Beerling and his team discovered evidence for a substantial increase in the atmospheric $CO_2$ concentration and 'super-greenhouse' conditions across the Triassic–Jurassic boundary, based on analyses of fossil stomata and leaf morphology from Greenland, which linked a catastrophic extinction event with the break-up of Pangaea [28], and he extended that research through collaboration with Bob Berner (Box 9.3). Beerling is a leader in experimental palaeobiology, through which he helped to determine that the establishment of rootless early land plants in soils was promoted by their partnership with soil fungi, and that the development of this association helped to influence the climate of the Palaeozoic. His 2007 book *The Emerald Planet: How Plants Changed Earth History* [29] won international plaudits and formed the basis for a three-part BBC 2 television series *How to Grow a Planet*. He was awarded a Philip Leverhulme Prize in 2001, a Royal Society Wolfson Fellowship in 2014, and elected as a fellow of the Royal Society in 2014. His latest book is *Making Eden* [30].

regulate atmospheric chemistry [27]. Their model includes a climate model coupled to a tropospheric chemistry-transport model and to a model simulating the biogeography of terrestrial vegetation. They applied it to determine the likely concentration of greenhouse gases in the air and their influence on the planetary energy budget in the Early Eocene (55 Ma ago) and the Late Cretaceous (90 Ma ago). The $CO_2$ content of the air was fixed at 4× present levels, with the addition of a model with 2× present levels for the Early Eocene. Pre-industrial levels of the trace gases were derived from studies of ice cores. Methane concentrations calculated for these two periods were high (2580–3614 ppb – parts per billion), double the current concentration of around 1800 ppb. That reflects the warm moist climates of those times and the increased abundance of the wetlands from which much methane comes. Nitrous oxides would also have been more abundant at these times due to the greater extent and decomposition of plant remains. Surface ozone concentrations were calculated to be 60–70% higher than today. These increases provided additional radiative forcing at times of rapid warming and rising $CO_2$. Compared with the global warming produced by $CO_2$ alone, the extra forcing produced **additional warming** of 1.4 °C and 2.7 °C for the 2× and 4× $CO_2$ Eocene simulations, and 2.2 °C for the Cretaceous [27].

How close to reality were the results? The model reproduced the mean annual land temperatures suggested by leaf fossils better than it did mean ocean temperatures. This might reflect a cool bias in proxy leaf data, or a warm bias in proxy ocean data. Alternatively, the model may be missing some amplifying feedback process. It failed to reproduce accurately the Equator-to-pole temperature gradient in the ocean, and hence ocean heat transport, perhaps partly because of the relatively poor resolution of the ocean component. Nevertheless, this exercise was an interesting test of the way in which the different components of the Earth System can be represented by mathematical calculations in such a way as to simulate the complex operation of the climate system [27]. Models stimulate ideas.

Having established that modern GCMs can help us to understand how fluctuations in $CO_2$ and related gases may have affected past climates, we now turn to look at the details of the Cenozoic, because its climate history is the most relevant for understanding the period in which we now live. Readers keen to explore past climate modelling further may find helpful the 2007 review by Williams and others [31].

## 11.3 $CO_2$ and Climate in the Early Cenozoic

By 2008, having absorbed all of the new information about the role of $CO_2$ in the climate system and the effects of mountain building on climate, Dennis Kent of Rutgers University and his colleague Giovanni Muttoni of the University of Milan felt able to explain both the increase of $CO_2$ and temperature from the Late Cretaceous to the Middle Eocene climatic optimum at around 50 Ma ago, and the subsequent decline of both $CO_2$ and temperature towards the present day [32]. As a starting assumption, they accepted Dave Rowley's view that the rate of production of new seafloor (hence $CO_2$) was constant over the past 180 Ma [33], which we now know is probably wrong (see Chapters 5 and 10). Were there other ways of changing $CO_2$ emissions through time, they wondered? Focusing on Tethys, the ocean between India and Asia, they assumed that shallow carbonate reefs festooned the Indian and Asian margins, and that pelagic carbonate oozes mantled the deep-sea floor between them. As India drifted north, the floor of Tethys and its carbonate load were subducted beneath Asia, where they melted and gave off volcanic gases rich in $CO_2$. Eventually, the carbonate reefs were subducted too, with similar results. This output of $CO_2$ was augmented by $CO_2$ and other gases erupted from India's Deccan Traps near the Cretaceous boundary 65 Ma ago. India collided with Asia about 50 Ma ago, switching off this 'subduction factory'. By then the Deccan Traps had drifted into the tropical rain belt, where chemical weathering of the basalts soaked up $CO_2$ from the atmosphere. By the Oligocene, some 30 Ma ago, the Himalaya started to rise as the collision of India with Tibet continued. Mountain building in a humid environment led to chemical weathering of silicate rocks in the mountains, soaking up more $CO_2$.

Kent and Muttoni's concept provides a convincing mechanism for the increase in $CO_2$ from the end of the Cretaceous to a peak some 50 Ma ago, and its subsequent decline, through increased chemical weathering. This picture is consistent with the change in the depth of the CCD with time [34]. The changing $CO_2$ explains the associated rise and decline in global temperature, for which there is no other plausible explanation. This interpretation is consistent with that of Froehlich and Misra, who argued that much of the rise in $CO_2$ in the Paleogene (the early Cenozoic) was due to the scarcity of newly uplifted, fresh, weatherable silicate rocks in a hothouse climate, when the land surface was largely a peneplain (a low relief plain formed by protracted erosion at a time of tectonic stability), providing relatively little rock exposure to encourage the kind of weathering that would extract large volumes of $CO_2$ from the air [35]. That was followed by the fall in $CO_2$ in the Neogene (the later Cenozoic) due to increased mountain building (e.g. as the Himalayan mountains rose) and its associated physical and chemical weathering from the beginning of the Oligocene onwards [35]. In effect the Early Cenozoic peneplain led to a temporary 'runaway' increase in the $CO_2$ concentration of the ocean–atmosphere system, which accentuated warming in the Eocene. Geochemical evidence supporting their argument comes from the absence in the Early to Middle Eocene of increases in the ratios of the strontium isotopes ($^{87}Sr/^{86}Sr$), which started to rise strongly from 40 Ma ago, and the lithium isotopes ($\partial^7Li$), which started to rise strongly from 50 Ma ago as mountain uplift began [35].

In 2013, Kent and Muttoni extended their conceptual model of $CO_2$ and climate back by another 30 Ma into the Early Cretaceous, 120 Ma ago [36] Once again they accepted Rowley's (probably false) assumption that sea floor spreading rates had been constant over the past 120 Ma [37]. They also assumed that rates of emission of $CO_2$ from spreading centres, mantle plumes and arc volcanoes were like those of the present day – about 0.26 Gt $CO_2$/year (see Chapter 8). Reconstructing past continental positions, they documented the times and places of the collisions of Arabia with Iran, India with Tibet, and New Guinea with Indonesia, and of the southeastward migration of fragments of Asia to form South East Asia. This enabled them to see where these components had been in relation to the equatorial humid zone, through time. They assumed that subduction and volcanic activity associated with India's northward march began 120 Ma ago, supplying $CO_2$ to the atmosphere from that time on. The rate of emission of $CO_2$ would have depended on the amount of carbonate-rich sediment available. As pelagic carbonates are most abundant in the tropics, the rate of $CO_2$ emission would have increased as plate motion brought carbonate sediment north from tropical regions of Tethys into the Asian subduction zone. They calculated that the subducted load of carbonate would have risen from 60 Megatonnes (Mt)/year at 120 Ma ago to 220 Mt/year at 80 Ma ago, and dropped to virtually zero at collision time (50 Ma ago). For the collision of Arabia with Iran they calculated that the subducted load of carbonate would have varied between 23 Mt/year at 70 Ma ago and 35 Mt/year at 20 Ma ago [36].

Assuming a relatively low rate of recycling of $CO_2$ from carbonate sediments into $CO_2$ emissions, Kent and Muttoni thought that their subduction mechanism would have made a modest contribution of $CO_2$ to the warm atmospheres of Cretaceous and Eocene times, contrary to the conclusion of their 2008 paper. Might emissions from Large Igneous Provinces have helped to increase global $CO_2$ levels? They thought not, since both the geological evidence and numerical models suggested that excess $CO_2$ supplied from such relatively short-term sources would be adsorbed within one million years by negative feedback processes [36]. If $CO_2$ emissions from seafloor spreading were more or less constant through time, and if the subduction factory added only relatively small

amounts of $CO_2$, might variations in carbon sinks play the key role on controlling the concentration of atmospheric $CO_2$? Investigating that question, Kent and Muttoni found that the weathering of continental basalts, which today make up less than 5% of land area, could account for between one third and one half of the consumption of $CO_2$ by the weathering of silicate minerals during the period in question (Figure 11.2) [36]. Knowing that chemical weathering is strongest under warm and humid conditions typical of the tropical regions between 5°S and 5°N, they plotted the area of continental crust passing through that region between 120 Ma and the present, finding that it averaged about 12 million km$^2$. Within that area, the most weatherable silicate rocks – the basalts – were particularly abundant between 50 and 40 Ma ago, with a subsequent rise from 30 Ma ago to the present. They calculated that the rate of consumption of $CO_2$ in the tropics began rising 65 Ma ago to a peak 45 Ma ago, dropped back to former levels about 35 Ma ago, then rose rather steadily to the present (Figure 11.3). Significant consumption of $CO_2$ took place in South East Asia and New Guinea as well as over India's Deccan Trap basalts and the Ethiopian Trap basalts. Silicate weathering under relatively tropical conditions in the rising Himalaya, as well as in the 4000 m high mountains of Borneo and the nearly 5000 m high mountains of New Guinea might also have contributed to further $CO_2$ drawdown, they suggested [36].

What is to stop Kent and Muttoni's proposed tropical chemical weathering from having a runaway effect on atmospheric $CO_2$? Their common-sense answer was that under tropical conditions chemical weathering leads to the development of thick soils deficient in cations, which are likely to retard further weathering of the underlying bedrock. Nevertheless, continued or periodic uplift, like that which has characterized South East Asia and New Guinea, would cause protective soil coverings to be shed from time to time, allowing $CO_2$ to continue to be taken up. Rains from prolonged La Niña conditions would further exacerbate chemical weathering in those areas. Continued weathering of basaltic terrains in equatorial regions might well draw down sufficient atmospheric $CO_2$ over time to lead to glacial conditions [36].

As I pointed out in Chapters 5 and 10, geochemical and sea level evidence strongly suggests that Rowley was wrong [33, 37], and sea floor spreading rates did decline, as Müller suggested [38], from 80 Ma ago to 50 Ma ago, after which they changed little. So $CO_2$ supply would have been high but declining over that 30 Ma period, providing an additional source of $CO_2$ through the Late Cretaceous and Palaeocene to be added to Kent and Muttoni's calculations [32, 36].

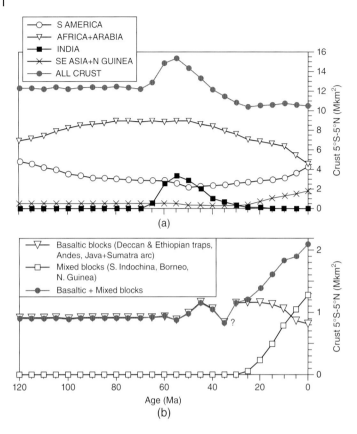

(a)

(b)

**Figure 11.2** Weathering of continental basalts through time: (a) Estimates of land area within equatorial humid belt (5°S–5°N) as a function of time since 120 Ma ago. (b) Estimates of most weatherable land areas of volcanic-arc provinces (Java, Sumatra, Andes), large basaltic provinces (Deccan Traps, Ethiopian Traps), and mixed igneous-metamorphic provinces (South Indochina, Borneo, New Guinea) in the equatorial humid belt (5°S–5°N) as a function of time from 120 Ma ago to present. *Source:* From figure 6 in Ref. [36].

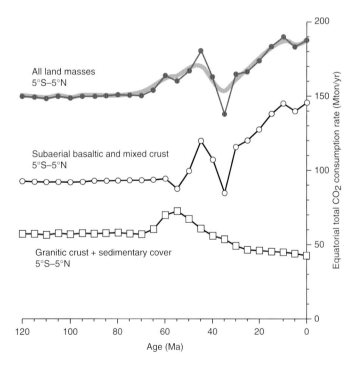

**Figure 11.3** Extraction of $CO_2$ through time by chemical weathering. Total $CO_2$ consumption rates from silicate weathering since 120 Ma ago of land areas in equatorial humid belt (5°S–5°N) obtained by multiplying a nominal $CO_2$ consumption rate of 100 t $CO_2$/year/km$^2$ for basaltic provinces and 50 t $CO_2$/year/km$^2$ for mixed basaltic-metamorphic provinces and 5 t $CO_2$/year/km$^2$ for remaining continental land areas to the corresponding cumulative distribution curves in Figure 11.2. These total consumption rates should be divided by 2 for net $CO_2$ consumption because half of the $CO_2$ consumed by silicate weathering is returned to the atmosphere–ocean during carbonate precipitation. *Source:* From figure 7 in Ref. [36].

Should we not be seeing significant emissions of $CO_2$ today, driven by the subduction of carbonate-rich sediments? No, because – according to geochemist John Edmond – plate motions today give rise to relatively little subduction of carbonate-rich oceanic sediments, most of which are found in the Atlantic where subduction is uncommon [39]. It is partly this lack of carbonate subduction that has driven $CO_2$ levels slowly down through the Cenozoic to as low as 170–180 ppm during Pleistocene glacial periods. Any future closure of the Atlantic, with accompanying subduction of its rich load of carbonate sediments, would push atmospheric $CO_2$ levels back up.

Might there be other reasons for the decline in $CO_2$ from the Mid Cretaceous to the Oligocene? Bill Hay (who we met in Box 6.9) provides two plausible explanations [12]. Firstly, oceanic productivity fell at the Cretaceous–Tertiary boundary, when most calcareous plankton were wiped out by the asteroid impact. Whilst that lack of photosynthesizing marine plankton allowed the concentration of $CO_2$ in the air to increase in the Palaeocene, by the Eocene, '*evolution had done its work and there were new species of calcareous plankton back at work, making deep-sea carbonate ooze and lowering atmospheric $CO_2$ again*' [12]. Secondly, the warm seas of the Cretaceous and early Cenozoic were dominated by haline circulation, with warm salty waters from tropical margins descending to abyssal depths, as in today's Mediterranean. But global cooling in the Late Eocene changed ocean circulation to thermo-haline, with the deep ocean becoming filled with cold bottom water sinking at the poles. Cold polar water dissolves much more $CO_2$ than warm water, so that change accentuated the decline in the concentration of $CO_2$ in the air by drawing it down into the deep ocean [12].

Another angle comes from Eleanor John of the University of Cardiff, and colleagues, who reminded us in 2013 that the warmth of Eocene surface waters would have affected the operation of the biological pump and the cycling of carbon and related nutrients [40]. Using $\partial^{13}C$ values in planktonic foraminifera from Tanzania and Mexico, John and colleagues found that the vertical carbon isotope profile of the Eocene tropical ocean was steeper and larger than it is today. This suggests that organic matter was being decomposed and recycled then at much shallower depths than today. The warm water raised metabolic rates, sped the recycling of organic matter, and reduced the rate of burial of organic matter, thus minimizing the loss of $CO_2$ to the deep ocean and keeping Eocene atmospheric $CO_2$ levels high. This is an important conclusion, given that some 90% of Phanerozoic climate was in the warm greenhouse state. Others had also concluded that oceanic metabolism played a key role in controlling $CO_2$ levels in Phanerozoic climates [41, 42].

It seems highly likely that the changes in atmospheric $CO_2$ with time through the early to mid Cenozoic may reflect changes in $CO_2$ supply and demand induced by plate tectonics, as suggested by Kent and Muttoni, modified by: first, a decline in seafloor spreading (and $CO_2$ emissions) between 80 and 50 Ma ago, as suggested by Müller; second, changes in the population of oceanic plankton, as suggested by Hay; third, the changing metabolism of marine plankton in moving from warm to cool conditions, as implied by John and others; and fourth, the drawing down of $CO_2$ from the atmosphere as deep water circulation cooled, as suggested by Hay.

In 2015, Inglis and colleagues used the $TEX_{86}$ proxy to reconstruct sea surface temperature for the Eocene. This proxy is '*based on the distribution of marine isoprenoidal glycerol dialkyl glycerol tetraether lipids*' extracted from the remains of marine plankton [43]. Using that technique, Inglis found that towards the late Eocene there was significant cooling (c. 6 °C) at high latitudes (>55°), and in the tropics (c. 2.5 °C). Given that their modelling suggested that less than 10% of the cooling could have been caused by the reorganization of continental fragments and ocean gateways, they concluded that declining atmospheric $CO_2$ had to be the most likely driver of the cooling towards the Oligocene icehouse [43].

In 2016, Anagnostou and colleagues used the boron isotope ($\partial^{11}B$) composition of planktonic foraminifera to show that during the Early Eocene Climate Optimum 53–51 Ma ago, when mean annual surface air temperatures were over 10 °C warmer than in pre-industrial times, $CO_2$ concentrations reached c. 1400 ppm before declining by c. 50% by the end of Eocene time (Figure 11.4). This was sufficient to drive the cooling to the Oligocene boundary, and suggested a climate sensitivity of c. 2.1–4.6 °C for a doubling of $CO_2$ [44].

Despite the warm conditions of the Eocene, Sean Gulick and colleagues noted that '*marine geological and geophysical data from the continental shelf seaward of the Aurora subglacial basin [show] that marine-terminating glaciers existed at the Sabrina Coast by the early to middle Eocene epoch, [which] … implies the existence of substantial ice volume in the Aurora subglacial basin before continental-scale ice sheets were established about 34 million years ago*' [45]. The basin lies in Antarctica's Wilkes Land, south of Australia, between 105 and 130°E (Figure 7.5). These results show that we can expect a dynamic response from the East Antarctic ice sheet under conditions warmer than today's.

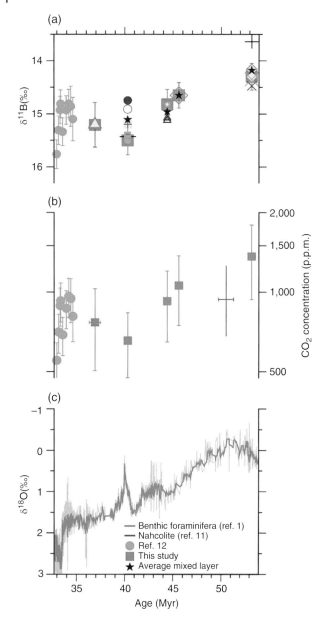

**Figure 11.4** New atmospheric $CO_2$ reconstructions from shallow planktonic foraminiferal $\partial^{11}B$: (a) $\partial^{11}B$ of shallower foraminifera; (b) Atmospheric $CO_2$ calculated from $\partial^{11}B$; (c) Benthic foraminiferal $\partial^{18}O$ representing bottom water temperature. Error bars are shown in (a) and (b). *Source:* From figure 3 in Ref. [44].

## 11.4 The First Great Ice Sheet

As we saw in Chapter 7, Jim Kennett attributed the development of the first major ice sheet on Antarctica at the Eocene–Oligocene boundary 34 Ma ago to the breaking away of Australia and South America from Antarctica leading to the formation of the Antarctic Circumpolar Current, which isolated the continent

from the warm air and warm water of the tropics. This is the 'isolation hypothesis'.

By 1994, Jim Zachos (Box 11.2), and his colleagues had used oxygen isotopes to show a warming of the Southern Ocean into the Early Eocene, followed by cooling into the Early Oligocene [46].

Knowing that Mike Arthur and his colleagues had used carbon isotopes in organic matter from phytoplankton in 1991 to suggest that $CO_2$ had declined from six to three times present values between the Middle Eocene and Early Oligocene [48], Zachos speculated that the cooling at that time was probably related to the fall in $CO_2$ [46]. But there was a problem: the Early Eocene tropical temperatures appeared to be like those of today, which would seem unlikely in a high $CO_2$ world. To explain this discrepancy Zachos hypothesized that some mitigating process must have operated to take heat away from the tropics at that time, citing increased poleward transport of heat by air and ocean currents. Zachos's suggestion is consistent with the results of the modelling of Eocene climate in 2000 by Karen Bice and colleagues [49].

By 2001, as we saw in Figure 7.7, Zachos and his team had established a global pattern for bottom water temperatures [50], confirming a rise from about 8 °C at 70 Ma ago to 12 °C at about 50 Ma ago, then a decline back to about 8 °C by 42 Ma ago, and to about 5 °C near the end of the Eocene, followed by a steep drop to about 0.5 °C at the Eocene–Oligocene boundary at 34 Ma ago. $CO_2$ showed a similar pattern, rising from a low of about 400 ppm near 70 Ma ago to a peak at about 1100 ppm between 45 and 50 Ma ago, before falling to near 400 ppm by 28 Ma ago, close to the

**Box 11.2   James C. Zachos**

Following his PhD in geological oceanography from the University of Rhode Island in 1988, Jim Zachos eventually joined the University of California Santa Cruz, where he is currently a professor and chair of the Department of Earth and Planetary Sciences. He specializes in studies of the biological, chemical, and climatic evolution of Late Cretaceous and Cenozoic oceans, using isotopic analyses to reconstruct past changes in ocean temperature and circulation, continental ice-volume, productivity, and carbon cycling [47]. In 2004, he served as the co-chief scientist of the Ocean Drilling Program's Leg 208 expedition to the south Atlantic, in 2016 he was awarded the Milutin Milankovic Medal by the European Geosciences Union, and in 2017 he became a member of the US National Academy of Sciences.

Eocene–Oligocene boundary [51]. The relationship was close enough to suggest a causal connection.

Growing understanding of the role of $CO_2$ in the climate system eventually led to a reappraisal of the Antarctic 'isolation hypothesis' in 2003, by Rob DeConto of the geosciences department of the University of Massachusetts at Amherst, and Dave Pollard from the Earth and Environmental Systems Institute of Penn State University [52]. DeConto and Pollard simulated '*the glacial inception and early growth of the East Antarctic Ice Sheet using a general circulation model with coupled components for atmosphere, ocean, ice sheet and sediment, and which incorporates palaeogeography, greenhouse gas, changing orbital parameters, and varying ocean heat transport*' [52]. In their model, '*declining Cenozoic $CO_2$ first leads to the*

*formation of small, highly dynamic ice caps on high Antarctic plateaux. ... [Then] cooling due to declining $pCO_2$ would have gradually lowered annual snowline elevations until they intersected extensive regions of high Antarctic topography. Once some threshold was reached, feedbacks related to snow/ice-albedo and ice-sheet height/mass-balance could have initiated rapid ice-sheet growth during orbital periods favourable for the accumulation of glacial ice, ... [with the ice caps] eventually coalescing into a continental-scale East Antarctic Ice Sheet*' [52]. That threshold is what we might now see as a 'tipping point' – one that pushed the world into a new climate state (Figure 11.5) [53]. The Eocene–Oligocene boundary is marked by an oxygen isotope ($\partial^{18}O$) event referred to as Oi-1 [52] (see caption to Figure 11.5).

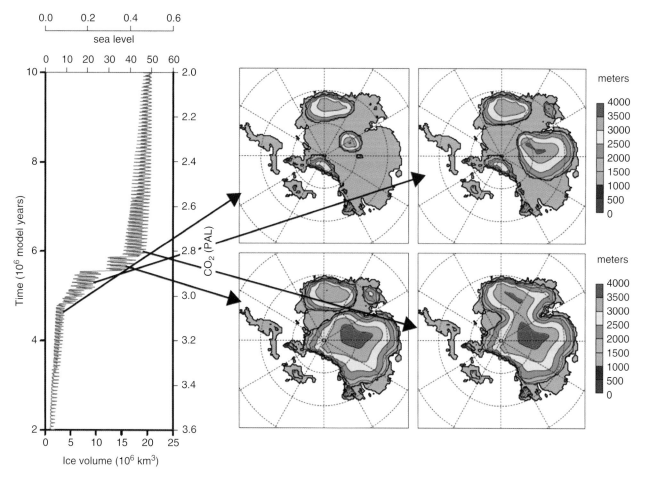

**Figure 11.5** Ice volume (left) and corresponding ice-sheet geometries (right) simulated by a coupled GCM-ice sheet model in response to a slow decline in atmospheric $CO_2$ and idealized orbital cyclicity across the E/O (Eocene–Oligocene) boundary. The sudden, two-step jump in ice volume (left panel) corresponds to the Oi-1 event. The left panel shows simulated ice volume (red line), extrapolated to an equivalent change in sea level and the mean isotopic composition of the ocean (top). Arbitrary model years (left axis) and corresponding, prescribed atmospheric $CO_2$ (right axis) are also labelled. $CO_2$ is shown as the multiplicative of pre-industrial (280 ppmv) levels. Ice-sheet geometries (right panels) show ice-sheet thickness in metres. Black arrows correlate the simulated geometric evolution of the ice sheet through the Oi-1 event. PAL = preindustrial atmospheric level at 280 ppm. Diagram modified from Ref. [52]. *Source:* From figure 8.14 in Ref. [53].

According to the DeConto and Pollard simulation, the transition from small ice caps on the Mühlig–Hofmann Mountains and Gamburtsev Mountains to a full ice sheet covering East Antarctica would have taken around one million years and resulted in a fall in global sea level of 40–50 m. They also simulated the effect of an increase in the strength of the Antarctic Circumpolar Current, but found it unlikely to have caused as rapid a change as that resulting from the decline in atmospheric $CO_2$, concluding '*the opening of Drake Passage can only be a potential trigger for glacial inception when atmospheric $CO_2$ is within a relatively narrow range .... reinforcing the importance of pCO$_2$ as a fundamental boundary condition for Cenozoic climate change*' [52]. Using a more advanced model in 2005, they found that an increase of more than eight times the pre-industrial atmospheric level of $CO_2$, equivalent to a local temperature increase of 15–20 °C, would be needed to make the East Antarctic Ice Sheet retreat significantly, something that is not likely to happen within the next 100 years [54].

The greenhouse to icehouse transition from a relatively deglaciated state to one characterized by a small Antarctic ice sheet is thought to have occurred within about 200–300 Ka. It seems to have occurred in a series of steps (marked in red on Figure 11.5), each with an increase of ice volume and a lowering of sea level [52].

Milankovitch variation may also have played its part in developing the ice sheet over time, as suggested by Michel Crucifix, with periods of high eccentricity accentuating the effect of precession on insolation and causing the ice sheet to contract during the Early Oligocene, and periods of low eccentricity allowing for phases of expansion, changes that also affected sea level and are reflected in $\partial^{18}O$ values in the shells of marine organisms [55].

Independent support for the ideas of DeConto and Pollard came from two different sources. First, when $CO_2$ is abundant in the atmosphere it is also abundant in the ocean, which thus becomes slightly more acidic. As a result, the carbonate compensation depth, or CCD, rises, as we saw in Chapter 9 (e.g. see Figure 9.3). Helen Coxall of the Southampton (now National) Oceanography Centre pointed out, with colleagues, that the CCD deepened by 1 km in the latest Eocene, indicating a fall in atmospheric $CO_2$ [56]. This deepening was more rapid than previously thought, taking place in two 40 Ka long jumps, each synchronous with the stepwise onset of the growth of the Antarctic ice-sheet. Coxall thought that the fall in $CO_2$ had led to cooling that preconditioned Antarctica for glaciation, which was then initiated in response to a period of cool summers favoured by declining orbital insolation; the changes in $\partial^{18}O$ composition across the Eocene/Oligocene boundary were too large to be explained by the growth of the ice-sheet alone. And she knew that recent research by

Mark Pagani and others, based on $\partial^{13}C$ analyses of alkenones from deep marine sediments, suggested that the decline in $CO_2$ across the Eocene–Oligocene boundary had been larger than former estimates [57]. Later research by Pagani et al. showed that $CO_2$ fell from 1000 ppm to around 600 ppm between 40 and 33 Ma ago [58]. Furthermore, Lear had documented a fall in bottom water temperature starting in the Late Eocene, consistent with that fall in $CO_2$ [59]. Carbon and oxygen isotopes both show a stepwise fall across the transition: first came Eocene–Oligocene Transition 1 (EOT-1) at 34–33.9 Ma ago; second came an event labelled Oi-1 at 33.7 Ma ago [60]. What role did atmospheric $CO_2$ play in this transition? In 2016, Armstrong McKay and colleagues demonstrated that '*that the rapidity of the two steps in the benthic $\partial^{13}C$ paleorecord can best be attributed to the net sequestration of ~1000 [Gt] of organic carbon during the EOT through processes such as permafrost and peatland expansion. Antarctic permafrost soil carbon must have either been sequestered by advancing ice sheets or eroded and oxidized prior to the EOT*' [61]. McKay's model tended to agree with Pagani's conclusions [58].

We could argue that whilst $CO_2$ clearly provides an important control on temperature through time, the development of glaciation also relies on the combination of lowered $CO_2$ – hence cooling – with movements of the continents that place landmasses at the pole at the right time – echoing Lyell. But, as we saw in Chapter 7, land at the poles is not enough by itself, there having been no extensive glaciation when Antarctica lay at the South Pole between 92 and 34 Ma ago. Even so, as we learned in Chapter 7, there probably were ice caps and small glaciers on East Antarctica, at least during the Eocene, providing nuclei for later ice sheets.

Did an ice sheet form rapidly on Antarctica as Pollard and DeConto suggest [52, 54]? Reviewing the results of shallow offshore drilling by the SHALDRIL project east of the Antarctic Peninsula, John Anderson and Julia Wellner suggested in 2011 that the onset of glacial conditions at the Eocene–Oligocene boundary was not as rapid as had been implied [62]. Anderson considered that '*This observation is more consistent with the Kennett hypothesis, which pinned glaciation on the opening of the Drake Passage .... it is not one single thing, like carbon dioxide, that is driving ice sheet evolution*' [63]. Had he gone out on a shaky limb? Observations on glacial onset based on SHALDRIL are compromised by the short fragmentary nature of the cores: ~10 m from 38 to 36 Ma ago, and 10 m from 32 to 24 Ma ago.

Support for the rapidity of the climate change across the Eocene–Oligocene boundary comes from Ocean Drilling Program (ODP) Leg 199 in the tropical Pacific [56]. There, as we saw earlier, Helen Coxall showed that the CCD deepened abruptly across the Eocene–Oligocene boundary (see also Figure 9.3), consistent with rapid $CO_2$ fall and global cooling.

A team led by Miriam Katz of Rensselaer Polytechnic in New York later confirmed that picture [64]. Integrating $\partial^{18}O$ and Mg/Ca values from benthic foraminifera with stratigraphic data from continental margins, they constructed a record of temperature, ice volume and sea level changes to show that the transition across the Eocene–Oligocene boundary happened in three steps (as Coxall had indicated) with the influence of ice volume continually increasing (Figure 11.6). The initial Antarctic ice sheet by the time of the Oi-1 event (Figure 11.6) was 25% larger than at present, while sea level fell by 67 m. Katz's team attributed the shift from greenhouse to icehouse climate as a response to a combination of factors including the opening of seaways around the continent, the global fall in $CO_2$, and long cool summers driven by declining orbital insolation. Large fluctuations in ice volume continued through the Oligocene [64]. At the same time as the land ice formed, or even before as temperatures began to fall, it seems likely that sea ice would have started to grow around Antarctica. That in turn would have created cold bottom water, which originates when the formation of sea ice excretes salt, making cold surface waters dense enough to sink and take their signal down into deep water. The sinking of Antarctic Bottom Water around the continent would have increased as sea ice began to form due to cooling fostered by both the decline in $CO_2$ and the development of the Antarctic Circumpolar Current.

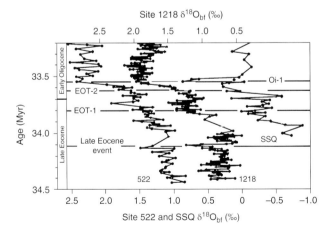

**Figure 11.6** Comparisons of $\partial^{18}O$ in benthic foraminifera from (at left) ODP Site 522 (S Atlantic), (in centre) ODP Site 1218 (Equatorial Pacific) and (at right) Saint Stevens Quarry, Alabama. Note definition of a series of $\partial^{18}O$ events: Eocene–Oligocene Transitions 1 and 2 (EOT-1 and EOT-2) that comprise the Eocene–Oligocene climate transition, which culminated in the Oi-1 event referenced in the caption to Figure 11.5. *Source:* From figure 2 in Ref. [64].

In 2009, Paul Pearson of Cardiff University and colleagues provided the matching $CO_2$ pattern, which showed a slight dip across the boundary between 34 Ma and 33.5 Ma, associated with the fall in $\partial^{18}O$ indicative of a growth in ice volume [65]. The dip in $CO_2$ was followed by a brief recovery in $CO_2$ between 33.5 and 33.2 Ma, after which $CO_2$ continued to fall. Evidently, once the large ice sheet had formed it was resistant to climate forcing by the brief rise in $CO_2$, though some reduction in ice volume was likely from the associated rise in temperature [65]. The later more detailed profile of $CO_2$ across the boundary, from Pagani et al. in 2011, illustrates more clearly the magnitude of that change [58].

As Michel Crucifix pointed out, there was also an important biotic change across the Eocene–Oligocene boundary, where there was a palaeontological extinction event known as the 'Grande Coupure' [55]. He reminds us that '*studies of species abundance and taxa have revealed that the biological activity in the southern ocean increased remarkably as early as 39.1Ma [ago], that is 5 Ma before [the] Oi-1 [event]*' [55]. He speculated that this increase may have been triggered by changes in ocean circulation resulting from the slow opening of the Drake Passage. The increase in production would have caused more $CO_2$ to be buried in sediments, thus drawing $CO_2$ down from the atmosphere and preconditioning Antarctica for glaciation [55]. Although this scenario is admittedly hypothetical, it illustrates Humboldt's point that 'everything is connected'.

Comparison of the global $\partial^{18}O$ curve representing bottom water temperatures and ice volume (Figure 7.7) [50] and the $CO_2$ curve [51] for the whole Cenozoic (Figure 9.16) suggests that the link between temperature, ice volume and $CO_2$ across the Eocene–Oligocene boundary is reasonably robust (cf. Figure 7.10). However, the continued fall in $CO_2$ (Figure 9.16), takes place as the ice sheet *decreases* towards the Middle Miocene and as bottom water temperatures slow their rate of fall (Figure 7.10) [66]. We deal with these post-Eocene–Oligocene changes in later sections of this chapter.

Cramer's temperature reconstruction (Figure 7.10) indicates '*differences between deep ocean cooling and continental ice growth in the Late Cenozoic: cooling occurred gradually in the Middle-Late Eocene and Late Miocene-Pliocene while ice growth occurred rapidly in the earliest Oligocene, Middle Miocene, and Plio-Pleistocene. These differences are consistent with climate models that imply that temperatures, set by the $CO_2$ "steady state", should change only gradually on timescales >2 Myr, but growth of continental ice sheets may be rapid in response to climate thresholds due to feedbacks that are not yet fully understood*' [66]. This tells us that although $CO_2$ and plate tectonics (opening or closing ocean gateways) set the broad climate scene behind the global cooling from the Mid Cretaceous to the present, from the mid-Cenozoic onwards climate changes at finer scales were then imposed by the vagaries of ice growth and decay. Indeed, once an ice sheet forms it may become self-stabilizing not least because as it grows upwards its surface cools by around 10 °C /km. But it can also grow and decline around its edges by advancing and retreating across the continental shelf [67].

Whilst initial ice growth was connected to falling temperatures driven by falling $CO_2$, ice growth and decay at the >2 Ma scale is only crudely connected to the fall in $CO_2$ from the Eocene–Oligocene boundary onwards, presumably because the temperatures also respond to changing insolation [66]. In addition, we have to bear in mind the changing nature of sources of moisture and the changing extent of sea ice, which regulated the ocean to atmosphere exchange of heat and $CO_2$, as Australia moved away from Antarctica from the basal Oligocene into the Middle Miocene.

Katz and Coxall both agree that the effects of orbital variations were superimposed on the effects on climate of plate tectonics and $CO_2$ during the Cenozoic. In the late 1990s, studies of the variability of $\partial^{18}O$ in deep-sea cores showed that between 34 and 15 Ma ago orbital cycles caused fluctuations in global temperatures and the volumes of ice at high latitudes. Off Cape Roberts in the western Ross Sea, in 2001, Tim Naish and others found numerous cyclic variations linking the extent of the East Antarctic Ice Sheet to orbital cycles during the Oligocene–Miocene transition (24.4–23.7 Ma ago) [68]. By 2007, that picture was extended to fluctuations between 34 Ma and 17 Ma ago, with the 40 Ka obliquity cycle and 100 Ka eccentricity cycle being particularly prominent [69].

Heiko Pälike and his colleagues from the UK's National Oceanography Centre tell us that orbital cycles formed '*the heartbeat of the Oligocene climate system*' [70]. Their research was based on a 13 Ma long record of variations in $\partial^{13}C$, representing the carbon cycle, and $\partial^{18}O$ representing ocean temperature and ice volume, from a deep-sea drill core from the equatorial Pacific. Results confirmed that the climate system responds to intricate orbital variations, which induce a fundamental relationship between solar forcing, the carbon cycle, and glacial events. Periodically recurring glacial and carbon cycle events showed that the heartbeat comprised cycles of orbital eccentricity that were 405, 127, and 96 Ka long, and a 1.2 Ma cycle in obliquity. The orbitally modulated variations in the carbon cycle induced changes in deep ocean acidity manifest as changes in the amount of $CaCO_3$ in bottom sediments. Pushing their analogy to the human body further, Pälike concluded '*Earth seems to "breathe" on time scales ranging from the annual to the orbital. We hypothesize that in all cases these cycles are driven by the expansion and contraction of biosphere productivity in response to changes in solar insolation*' [70].

Investigating the onset of glaciation at the Eocene–Oligocene (E–O) boundary, Pälike's team found from numerical models that '*by imposing a gradual decrease in atmospheric $CO_2$ levels some time before the E-O transition ….we obtain a rapid onset of glaciation at the time that it is observed in our records. We find that onset of glaciation is independent of the exact timing of $CO_2$ reduction and is triggered by astronomical forcing as soon as atmospheric $CO_2$ levels are close to a threshold value. Our model results therefore confirm the view [of DeConto and Pollard] that a decrease in atmospheric $CO_2$ is a possible mechanism to explain the record across the E-O transition*' [70].

By 2016, a team including DeConto and Pollard had concluded from the latest drilling in the western Ross Sea that an Antarctic ice sheet formed as $CO_2$ fell below c. 750 ppm [71]. The initial ice sheet was small, and at $CO_2$ levels of ≥600 ppm was highly responsive to local forcing by orbital insolation as shown by glacial–interglacial cycles dating to 34–31 Ma ago. A more stable ice sheet reaching the coastline formed at c. 32.8 Ma ago when $CO_2$ levels fell below c. 600 ppm [71]. The authors concluded '*that the partial pressure of atmospheric $CO_2$ was the primary influence on the overall climate state and variability of [ice sheet] volume, including its sensitivity to orbital forcing, which implies a close linkage between carbon cycle dynamics and [ice sheet] evolution on both long-period and short-period orbital time scales. Amplification of the long-period eccentricity component … tracks the establishment of low-latitude $\partial^{13}C$ variability with a 405-kyr periodicity*' [71]. The coherence and phasing between glacial cycles and marine $\partial^{13}C$ records shows that carbon cycle feedbacks contributed to $CO_2$ changes and amplified short- and long-period eccentricity-paced glacial–interglacial cycles in the Early Oligocene, just as in the Northern Hemisphere during the Pleistocene. The data suggest that the ice sheet should display threshold-like behaviour in response to long-term trends in atmospheric $CO_2$ levels; its tendency to melt '*increases dramatically between 600–750ppm*' [71].

The formation of the ice sheet undoubtedly influenced local circulation, facilitating the formation of Antarctic Bottom Water and Antarctic Intermediate Water, which expanded through the world ocean to further influence circulation and climate as we saw in Chapter 7. Ice too had its role to play.

## 11.5 Hyperthermal Events

Hyperthermal events are particularly helpful for our understanding of modern climate change, because they represent natural examples of abrupt climate change [72]. In that respect, Eocene times are especially fascinating, because the warming phase of the Early Eocene Climatic Optimum (EECO), which peaked at about 50 Ma ago and represented the warmest period of the Cenozoic, was punctuated by six prominent but short-lived warming events linked to the release of $^{13}$C-depleted $CO_2$ and known as the Early Eocene hyperthermals (Figure 11.7) [73]. Generally lasting about 40 Ka, they involved rapid redistributions of carbon between Earth's surface reservoirs. From the almost linear relationship between $\partial^{13}$C and $\partial^{18}$O during these events, Lauretano and colleagues deduced that there must have been an intimate connection between changing oceanic and global temperature and $^{12}$C-rich marine organic

carbon [73]. Carbon was supplied to the air as $CO_2$ by periodic rapid release of dissolved organic carbon, due to oxidation of deep waters, then rapidly reabsorbed back by the ocean [74]. These brief warm events were also marked by ocean acidification and the dissolution of deep sea carbonates indicative of a supply of excess $CO_2$ to the ocean. They are also identified in sediments on land, for example in the Bighorn Basin of Wyoming. One of these hyperthermal events, Eocene Thermal Maximum 2 (ETM2; Figure 11.7), can be seen even in the Arctic, where sea surface temperatures rose by 3–5 °C, accompanied by eutrophicaton of the surface waters, which led to oxygen poor

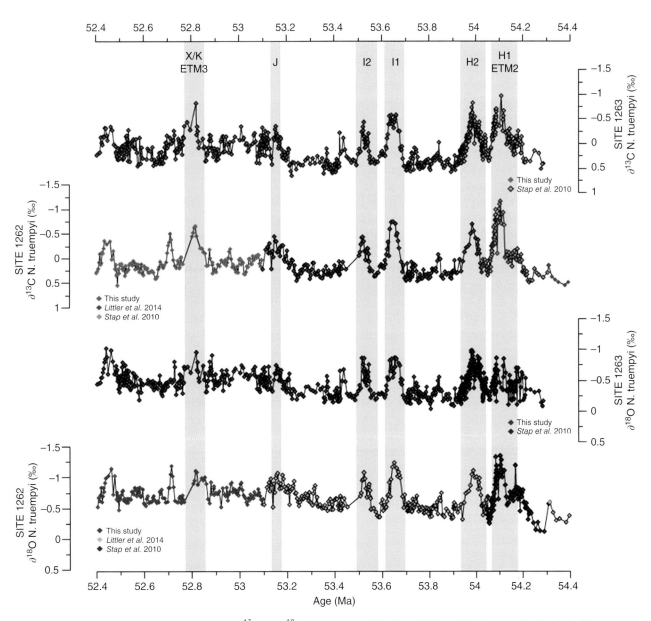

**Figure 11.7** Benthic foraminiferal N. truempyi $\partial^{13}$C and $\partial^{18}$O records from ODP Sites 1263 and 1262 on the flanks of the Walvis Ridge in the South Atlantic, plotted versus Age (Ma). Highlighted intervals represent the position of the Early Eocene hyperthermal events. *Source:* From figure 4 in Ref. [73].

conditions in the photic zone [75]. Temperatures during the coldest months remained above 8 °C in the Arctic at the time, possibly due to cloudy conditions reducing winter cooling. De Conto and Pollard suggest that the $CO_2$ came from the decomposition of soil carbon in polar permafrost [76], but most authors, like Lauretano [73] favour a marine source for the carbon.

Wavelet analyses of the $\partial^{13}C$ and $\partial^{18}O$ records show that spectral power is concentrated at the 405 and 100 Ka long orbital cycles of eccentricity, with maxima in $^{12}C$ in phase with maxima (representing warming) in the eccentricity cycles [73]. Did the hyperthermals die out as the peak of Eocene warming was approached at 50 Ma ago? It appears not: '*a series of [smaller] negative stable isotope excursions continued throughout the EECO*' attesting to the continuance of episodes of carbon release [73]. Kirtland Turner at al [77] suggest that the orbital forcing mechanism operating in the Early Eocene was not exceptional, but similar to that operating in the Oligocene [70] and Miocene [68, 69] when cyclic variations in the carbon cycle were clearly paced by forcing, especially in the eccentricity bands. Thus the six events can be seen as the more extreme variants of a large number of events with a spacing of about 100 Ka, which were most probably paced by the cycles in the eccentricity of the Earth's orbit [74]. During these transient events the deep ocean warmed by 2–4 °C, and the deposition of sedimentary carbonate was reduced by an increase in deep ocean acidity. Sexton suggested that the warmth of the Eocene ocean led to depletion of the deep ocean in oxygen, which led to a build-up of dissolved organic carbon from the decay of sinking organic matter [74]. Periodic cooling at high latitudes would have supplied the deep ocean with more dissolved oxygen, resulting in the oxidation of the carbon in the abyssal ocean reservoir and its release to the atmosphere to cause warming. These oceanic processes would have been rapid, given the relatively short overturning time of the ocean (300–1000 years) in relation to the long cycles in eccentricity (100 Ka) and tilt (40 Ka). These events were associated with abrupt and extreme climate change, an accelerated hydrological cycle, and ocean acidification. The rates of negative feedback processes like chemical weathering were too slow to restore the global carbon cycle to a steady state in a short space of time [25].

Richard Zeebe and colleagues observed in 2017 that the sedimentary cycles of the Palaeocene and Eocene are expressions of changes in global climate and carbon cycling paced by astronomical forcing [78]. As is well known, maxima in eccentricity lead to large amplitudes in the precession signal, likely corresponding to more intense wet and dry seasons, whereas eccentricity minima result in small precession amplitudes, likely corresponding to more equable climates, which, Zeebe suggests, favour enhanced long-term burial of organic carbon. This in turn leads to the isotopic signature of atmospheric and oceanic carbon being richer in $\partial^{13}C$ during eccentricity minima [78]. The Late Palaeocene and Eocene orbital signals are superimposed on multi-million year warming and declining $\partial^{13}C$ trends. Zeebe concluded that – assuming a baseline $CO_2$ of 1000 ppm – variations associated with the 400 Ka eccentricity cycle could be on the order of ±200 ppm $CO_2$, which would have substantially amplified the orbital forcing [78]. Such a variation is twice as much as experienced during the glacial to interglacial fluctuations of the last 400 Ka, suggesting that '*the carbon cycle of a warm world is more sensitive to astronomical influence than is that of a cooler one*' [79].

Pollen records from Eocene Lake Messel, in Germany, demonstrate, further, that the climate then also experienced variation in the sub-Milankovitch range indicative of solar variability, and natural fluctuations at the millennial scale typical of what would be expected from oscillations like ENSO (the El Niño–Southern Oscillation) [80]. Variations in pollen indicate the prevalence of short-term cyclic changes in the composition of vegetation that were driven by changes in precipitation and lake level. Clearly the drivers of climate variability typical of the Quaternary are of long-standing geologically.

Among the early Cenozoic hyperthermals was one at the Palaeocene–Eocene boundary that was much larger and longer than the rest – the Palaeocene–Eocene Thermal Maximum (or PETM). The data from the PETM carry an important message for us. While the event is not an exact analogue of what is happening to Earth's climate today, because the rate of growth in $CO_2$ then was slower than it is now, and because the continental configuration was slightly different, we can use it, like the Pliocene, as a **case history** to suggest what might happen to our climate if we keep emitting greenhouse gases at an increasing rate.

## 11.6 Case History – The Palaeocene–Eocene Boundary

Convincing evidence of the role of $CO_2$ in raising temperature comes from the end of the Palaeocene, about 56 Ma ago, when there was a sudden short-lived warming event in which temperatures rose by about 5–6 °C globally and by as much as 8 °C at the poles. This event, the PETM, was superimposed upon the underlying warming of the Early Cenozoic (Figure 7.7). It caused one of the largest extinctions of deep-sea benthic organisms in the last 90 Ma, acidified the surface ocean as well as the deep ocean and stands out as a time of significant changes in the terrestrial biota, including the first appearance of the ancestors of modern hoofed mammals and major changes in the taxonomic composition of floras in the Bighorn Basin of Wyoming, with rapid reorganization of plant communities in response to climatic and environmental change [81]. For example, tropical forest plant communities in Colombia and Venezuela survived the heightened temperatures, increasing rapidly in diversity, with a set of new taxa, mostly angiosperms, added to the existing low diversity stock of Palaeocene flora [82].

Discovered as recently as 1991 by Kennett and Stott [83], the PETM has been analysed in detail by Jim Zachos and colleagues [50], who interpreted their $\partial^{13}C$ data to suggest that it was accompanied by a major release of more than 2000 Gigatons (Gt) of $^{12}C$-rich carbon as $CO_2$ into the ocean

and air (Figure 11.8) [84]. Its signal appears everywhere [85, 86], including the Arctic, where ocean temperatures near the North Pole rose from 18 °C to >23 °C, at the same time as nearby lands warmed from 17 to 25 °C [87]. Appy Sluijs, from Utrecht University, and colleagues, found evidence for a rise in temperature of 7 °C as far south as the Tasman Sea at a palaeolatitude of 65°S [88]. Rises of this magnitude seem typical of continental margin rather than open ocean settings, where the rise was less. The Tasman site is far enough south to suggest that the warming there may represent polar amplification [88]. Maximum temperatures there ranged between 29 and 34 °C, indicating a low Equator-to-pole thermal gradient. Field evidence in the form of transgressive sedimentary deposits showed that the warming led to a rise in sea-level, with thermal expansion of the ocean contributing around 5 m, and the melting of all or part of the relatively small Antarctic mountain ice caps that may have existed at that time contributing some

**Figure 11.8** The Palaeocene–Eocene Thermal Maximum (PETM) as recorded in benthic foraminiferal isotopic records. A rapid decrease in carbon isotope ratios (top panel) indicates a large increase in atmospheric greenhouse gases ($CO_2$ and $CH_4$) coincident with 5 °C global warming (centre panel). Much of the added $CO_2$ would have been absorbed by the ocean, thereby lowering seawater pH and causing widespread dissolution of seafloor carbonates (lower panel), manifest as a transient reduction in the carbonate ($CaCO_3$) content of sediments. The ocean's carbonate saturation horizon representing the CCD rapidly shoaled more than 2 km, then gradually recovered as buffering processes slowly restored the chemical balance of the ocean. *Source:* From figure 6.2 in Ref. [84].

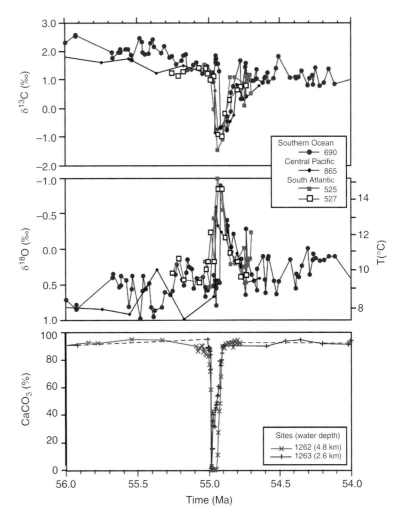

additional amount, likely less than 10 m [89]. The existence of similar though smaller such events in the Early Eocene suggests that the PETM was not a unique event and that its initiation may be linked to orbital eccentricity [81]. Subsequent considerations of all the available data suggest that the total amount of carbon emitted and sedimented during the entire duration of the PETM (170 Ka) was about 1000 Gt [72].

How fast was the carbon emitted? The distribution of fossil remains across the Palaeocene–Eocene boundary suggests that the onset of the event took place in a geological instant – less than 500 years [90]. Ying Cui of Penn State University, and colleagues, found that the peak rate of addition of carbon to the Earth system during the PETM in Svalbard was probably in the range 0.3–1.7 GtC/year [91], which is much less than the current rate of carbon emissions of c. 10 GtC/year [25, 92]. One recent high- resolution carbon isotope record from the Bighorn Basin in Wyoming suggested that rates of carbon release during the PETM approached those associated with modern carbon emissions, making the PETM a likely analogue for modern conditions [93]. However, recognizing that geological methods make it difficult to determine the age of release of carbon precisely, in 2018 Turner combined data and modelling studies to conclude that the carbon emissions occurred over just a few thousand years at a rate about 10 times slower than present rates [92]. That agrees well with the 2016 finding of Richard Zeebe and colleagues, who calculated that the initial carbon release during the onset of the PETM occurred over at least 4000 years, which constrains the maximum release rate to less than 1.1 GtC/year [94]. That was significantly less than the modern rate of emission of 10 GtC/year (in 2014). Given that the carbon release rate at the PETM was the highest of the past 66 Ma, these authors deduced that we are now in a 'no analogue' state that '*represents a fundamental challenge in constraining future climate projections*' [94].

Where did the $CO_2$ come from? Gerald Dickens of Rice University suggested that a massive disturbance of the seabed led to the rapid release of methane ($CH_4$), quickly oxidized to $CO_2$ [95]. But evidence is emerging that the warming associated with the $\partial^{13}C$ excursion may have begun a short while before the excursion itself took place, and may have triggered the event [96]. Good evidence for such a precursor recently came from studies of the PETM in the Bighorn Basin of Wyoming, where Ross Secord of the University of Nebraska, and colleagues, found continental warming of about 5 °C prior to the event [97]. Given that the precursor warming is unrelated to changes in marine carbon ($\partial^{13}C$), their results suggest that there were at least two sources of warming, the earlier of which was

unlikely to be marine methane. One possible source of the atmospheric carbon was mantle-derived $CO_2$ released during volcanic activity within the North Atlantic Igneous Province (a Large Igneous Province associated with massive extrusions of basalt) [96, 97] However, mantle-derived $CO_2$ has far too high a $^{13}C$ composition to provide a realistic source. The size of the shift in the $\partial^{13}C$ ratio measured in the marine carbonates would have required the transfer into the atmosphere of an amount of terrestrial carbon equivalent to most of today's terrestrial biosphere including soils [98]. Rather than a volcanic source, then, it seems more likely that we are looking at the release of a massive amount of sea-floor methane, which is significantly enriched in $^{12}C$. As Dickens suggested [95], it may have come from the ice-methane hydrates known as clathrates, which grow in the sediments of the continental slope. These hydrates form where molecules of water under high pressure in the cold pores of the sediment form cages of ice around molecules of methane gas released into the sediment by the decomposition of organic matter. One possibility is that warming destabilized these clathrates, leading to a burst of methane that rapidly converted to $CO_2$ in the atmosphere [96].

Not everyone agreed with Dickens. John Higgins and Daniel Schrag, of Harvard, argued that factors such as the magnitude of the warming, the rise in tropical sea surface temperature, the abrupt warming of the deep sea, the ocean acidification and the extent of benthic extinctions at the time meant that the size of the addition of carbon was larger than could be accounted for by the methane hypothesis [99]. They concluded that oxidation of 5000 Gigatons of organic carbon was required. Logical sources included contact metamorphism associated with intrusions of a large igneous province into organic-rich sediments, or a peat-land fire of global dimensions, or desiccation of a large epicontinental sea. Nevertheless, in 2018 Turner pointed out that constraints are still needed on the duration of the onset, the mass of carbon released, and the total duration of the release, before we can calculate the isotopic composition of the initial carbon input and thus detect its most probable source [92].

In his 2010 book *Challenged by Carbon*, Bryan Lovell, from the Earth Sciences Department of Cambridge University, suggested possible cause for the timing of the PETM [100]. At the Palaeocene–Eocene boundary, the hot plume of magma beneath the volcanic island of Iceland produced a lateral pulse of magmatic fluid that pushed eastward beneath Scotland, lifting the terrain. This igneous-tectonic process is likely to have destabilized the adjacent continental slopes of the northeast Atlantic, making the soft sediments slump and releasing methane gas from frozen methane hydrates, which rapidly oxidized to $CO_2$. $CO_2$ levels in the atmosphere were already abundant, but the additional $CO_2$ warmed it even more. Eventually this lateral pulse of magmatic activity subsided, ceasing the destabilization of the margin.

Lovell's imaginative concept draws on research by his Cambridge colleagues, who used three-dimensional seismic reflection data to build a picture of a buried Late Palaeocene landscape beneath the seabed north of Scotland [101]. This landscape, they contended, was uplifted by between 600 and 1200 m due to the passage of the subterranean pulse of magma from Iceland, before being reburied after about 1 Ma. The uplift took place at the time of the PETM, which is why Lovell connected it to the possible release of methane from gas hydrates. The igneous activity to which he refers is likely to have been associated with the million years or so of massive flood basalt volcanism in East Greenland that created the North Atlantic Igneous Province, began just before and ended just after the PETM, and marked a key stage in the separation of Greenland from western Europe [102]. Magma interacting with sedimentary basins full of Greenland's carbon-rich sedimentary rocks could have provided an excess of $CO_2$ at this time [103]. This linkage may perhaps explain Secord's observation that a $CO_2$-induced warming event, perhaps related to Atlantic volcanic activity, immediately preceded the main PETM event [97]. It has also recently been proposed that both the PETM and the Eocene hyperthermal events can be explained by the warming driven by orbital forcing that triggered decomposition of organic carbon in polar permafrost [76].

As we know from Chapter 9, increasing the amount of $CO_2$ in the ocean makes it slightly more acid, dissolving deep-sea carbonates and raising the CCD to shallower depths. Not surprisingly, then, Zachos found that calcium carbonate disappeared from the deep sea during the warming event, which lasted 170 Ka [104]. Whereas the CCD occurs in the Atlantic today at water depths around 5 km (see Chapter 9), during the PETM it shoaled to 2.5 km (e.g. see the disappearance of $CaCO_3$ from the deep-sea floor in the bottom panel of Figure 11.8). It took at least 100 Ka for the CCD to fall back to its original level (Figure 11.8), consistent with the residence time the average $CO_2$ molecule spends in the air [104]. A good summary of Zachos's findings appears in the October 2011 issue of National Geographic [105].

A side effect of making the ocean more acid and raising the CCD was to kill off the benthic foraminiferal species of the deep sea floor, as well as causing most shallow water coral reefs to vanish. Coral reefs did not become widespread again until the Middle Eocene about 49 Ma ago [25]. For more information on ocean acidification see Box 9.1

and the growing literature on the geological record of ocean acidification [106], and on its cause of crises in ancient coral reefs [107]. The surface ocean became acidified by 0.3 pH units globally [108], in addition to the acidification of the deep ocean that raised the CCD. Warming and acidification of surface waters changed the biogeography and population of marine plankton [83, 109]. For instance, in 2017, Friehling and colleagues used a variety of palaeothermometers to show that sea surface temperatures along the Nigerian coast rose to >36 °C during the PETM, associated with a massive drop in dinoflagellate abundance and diversity, indicating '*that the base of tropical food webs is vulnerable to rapid warming*' [110]. Frieling and others concluded in 2018 that there must have been a '*general absence of eukaryotic surface-dwelling microplankton during peak PETM warmth in the eastern equatorial Atlantic, most likely caused by heat stress*' [111].

Do we know everything we need to about the PETM event? Not yet. Richard Zeebe of the University of Hawaii, along with Jim Zachos and Gerald Dickens, used the record of dissolution of deep-sea carbonates through the PETM to suggest that the release of carbon must have been pulsed rather than instantaneous [112]. Atmospheric $CO_2$ would have increased to 1700 ppm from a base level of 1000 ppm. Given the IPCC's accepted range of climate sensitivity as being a rise of 1.5–4.5 °C for a doubling of $CO_2$ [16], this 700 ppm increase should have caused at most 3.5 °C of warming – less than observed. Zeebe and his team concluded that: '*our results imply a fundamental gap in our understanding of the amplitude of global warming associated with large and abrupt climate perturbations. This gap needs to be filled to confidently predict future climate change*' [112]. This begs the question – did they give enough consideration to slow feedback elements of the climate system, such as changes in terrestrial ecosystems? Furthermore, subsequent analysis suggests a total emission of 10 000 Gt, which may well have taken $CO_2$ to higher levels.

By 2019, we had learned a great deal more about the PETM from a study of ODP site 959 in the Gulf of Guinea and other sites, led by Joost Frieling of the Laboratory of Palaeobotany and Palynology of Utrecht University. They confirmed that about 2 °C of global warming preceded the onset of the PETM by some 2000 years (Figure 11.9) [113]. During the PETM the surface waters in the Gulf of Guinea experienced a further warming of 1.9 °C, while at other sites an initial pre-PETM warming of 2 °C was followed by a further warming of 3 °C [113]. Frieling's team found that the PETM was associated at this and other sites by a significant increase in barium (Ba). They interpreted this chemical signal to indicate the release of carbon from the destabilization of frozen methane hydrates on continental margins, which in turn allowed the massive release of

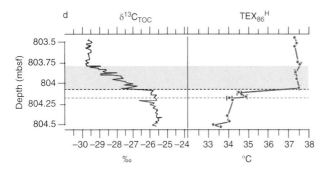

**Figure 11.9** PETM at Ocean Drilling Program Site 959, Equatorial Atlantic, showing the onset of warming with, at left, stable carbon isotope ratios of total organic carbon ($\delta^{13}C_{TOC}$), and, at right, sea surface temperatures derived from analyses of the TetraEtherindeX (TEX), labelled $TEX_{86}^{H}$. The pink dashed line shows the onset of the pre-PETM warming. The grey shaded band represents a turbiditic organic-lean layer at the base of the PETM (which is marked by the dashed black line). *Source:* From figure 2(d) in Ref. [113].

dissolved Ba from the pore waters previously trapped in sediments below the frozen methane hydrate zone [113]. Frieling attributed the initial 2 °C warming to a rise in $CO_2$ most likely from volcanism associated with the North Atlantic Igneous Province. He and colleagues had previously suggested that hydrothermal vents in the Norwegian Sea might have been a source for the bulk of the $CO_2$ in the 50 Ka *after* the initiation of the PETM, but could not have provided the trigger [114].

A subsequent 2019 study, by Shelby Lyons from Penn. State University, provided further information [115]. She and her team realized that $CO_2$ inputs continued long after the initial rapid onset, creating the main body of the Carbon Isotope Event. Their careful analysis showed that much of this extra $^{12}C$-rich carbon had been reworked from coastal sediments, creating an order of magnitude increase in the delivery of fossil carbon to the oceans that began 10–20 Ka after the onset of the event. The oxidation of this remobilised carbon released between $10^2$ and $10^4$ Gigatonnes of carbon as $CO_2$ during the body of the PETM, which sustained the elevated atmospheric $CO_2$ levels through the carbon isotope excursion, thus contributing to the delayed recovery of the climate system for many thousands of years. Lyons attributed the erosion of the sedimentary carbon to warming-induced enhancement of the hydrological cycle [115]. But one additional factor is likely to have been the associated rise in sea level of 12–15 m.

Kiehl and colleagues reminded us that warming led to the evaporation of significant amounts of water vapour, which dramatically influenced the hydrological cycle. Dry regions such as the interior of the American continent became drier, wet regions like the East Asian region got wetter, and the seasonal climate and hydrological cycle

became more extreme [116]. For example, rainfall increased dramatically in certain regions, such as northern Spain, where '*extreme floods and channel mobility quickly denuded surrounding soil-mantled landscapes, plausibly enhanced by regional vegetation decline, and exported enormous quantities of terrigenous material towards the ocean*' [117]. Evidence from drilling on the Lomonosov Ridge in the Arctic Ocean shows that laminations developed in the sediments during the PETM, which suggests that the water column became stratified, leading to the bottom waters becoming anoxic, which suggests in turn that there was a substantial increase in the run-off of freshwater and associated soil carbon from land [118]. Gabriel Bowen and David Beerling agreed in 2004 that the release of carbon would have had distinct regional effects [119]. Additional warmth, along with increased rainfall and enhanced $CO_2$, would have encouraged vigorous plant growth and an increase in the turnover of soil organic matter, doubling the rate of fast carbon cycling. Higher rainfall and temperature would have expanded wetlands, leading to an increase in atmospheric methane – another greenhouse gas. And, as Lyons has now indicated, erosion of sedimentary carbon played a key role in maintaining the event for long after its inception [115]. There is an implication here for what may happen with today's rise in $CO_2$ [115].

Why did it take around 100 Ka for conditions to return to normal? Isn't the residence time of $CO_2$ in the ocean short? That is a popular misconception, as we saw in Chapter 8. As David Archer of the University of Chicago pointed out, '*The carbon cycle of the biosphere will take a long time to completely neutralize and sequester anthropogenic $CO_2$*' [120]. He calculated '*that 17–33% of the fossil fuel carbon will still reside in the atmosphere 1 kyr from now, decreasing to 10–15% at 10 kyr, and 7% at 100 kyr. The mean lifetime of fossil fuel $CO_2$ is about 30–35 kyr*' [120]. Nevertheless, Lyons' analysis shows that persistence of the carbon isotopic anomaly can be attributed to the large addition of $CO_2$ from oxidized eroded carbon well after the initiation of the event [115].

The only way to take the $CO_2$ out of the ocean naturally is by the gradual accumulation in sediment of the organic carbon and calcium carbonate remains of marine organisms and the remains of terrestrial vegetable matter brought down to the coast by rivers, which is extremely slow. In effect, then, the PETM gives us a realistic illustration of what happens naturally when a pulse of carbon is added to the atmosphere. If the PETM is anything to go by, we should expect our current even more rapid additions to lead to significant warming, a slightly more acidic ocean, destruction of the carbonate-based organisms of deep water, a profound decrease in coral reefs, a substantial rise in sea level, and a long recovery time.

Foster explains the recovery from hyperthermal events (also called hypsithermals) such as the PETM as follows: first, the carbon added during a hyperthermal dissolves in the ocean, lowering the pH, which is partly neutralized via reaction with deep ocean carbonate sediments; second, warmer temperatures and higher rainfall enhance the rate of silicate weathering, which slowly draws down atmospheric $CO_2$; third, a sedimentary sink for $CO_2$ develops, as warm water stores less oxygen, and enhanced run-off supplies more nutrients to the ocean, a combination that enhances the burial of organic matter; by trapping marine organic matter rich in $^{12}C$, these processes reverse the negative $\partial^{13}C$ excursion [72]. Foster points out that much the same happened, although in more extreme form, during the Oceanic Anoxic Events of the Cretaceous. How effective is silicate weathering as a negative feedback part of this equation? Kump points out that while it would have been extremely limited in the warming event at the end of the Permian, because of the vast extent of the arid interior of the Pangaean supercontinent, that restriction did not apply to the early Cenozoic, when the continents were all separated [121].

The PETM event is not the only one of its kind. Anthony Cohen of the Open University in the UK, and colleagues, found another one in the Early Jurassic Toarcian period (178–183 Ma) [122]. Like the PETM, it was associated with a major $\partial^{13}C$ excursion, significant biotic extinctions, severe global warming, and an enhanced hydrological cycle. It was also accompanied by widespread seawater anoxia, forming a prominent Oceanic Anoxic Event. In Britain, the rich deposit of organic matter formed by this event crops out in the cliffs of Whitby, Yorkshire, as the Jet Rock – a black layer of fossil wood from the monkey-puzzle tree. It was once popular as jewellery. In mourning for Prince Albert, Queen Victoria declared that only jet jewellery could be worn at court for a year [123]. Like the PETM, the Toarcian event is most readily be explained '*by the abrupt, large-scale dissociation of methane hydrate that followed a period of more gradual environmental change linked to the emplacement of a large igneous province*' [122].

Could the PETM be related, like the end-Cretaceous event, to a bolide impact? Recent work shows that there was a small bolide impact at the Palaeocene–Eocene boundary, manifest as a 12.7 km wide crater at the Marquez Dome of central Texas. However, the crater is a tenth the size of the one associated with the impact that ended the Cretaceous, suggesting that the impact was not large enough to explain the extensive nature of the PETM [124]. Schaller and Fung argue that the impact may have induced an acceleration of the volcanism around Greenland that is thought to be the trigger for the PETM [124]. However, Zeebe and others present several convincing arguments against the bolide impact hypothesis [125]. More likely, given the situation of the PETM as one of a number of similar, albeit shorter hypsithermals, is that it is an extreme example of one of these natural events, in which (as suggested by Lyons) [115] there was tendency for runaway conditions to develop that accessed excess carbon from coastal regions during a rise in sea level, which contributed to this event being at least four times as long as its Palaeocene and Eocene 'contemporaries'.

Can we model the PETM? The simple answer is not yet. As Paul Valdes explains, the background climate state of the period was characterized by an extremely flat temperature gradient between the Equator and the poles [126]. Climate models have so far been unable to simulate the intense warmth of the polar regions of the time. According to Valdes '*Not being able to start from a realistic global temperature distribution for the Late Palaeocene makes it unrealistic to simulate further abrupt warming associated with the Palaeocene–Eocene Thermal Maximum*' [126]. More worryingly, '*similarly flat latitudinal temperature gradients are a common feature of extreme warm climates of the past, suggesting that IPCC-type, complex climate models may not be well suited to simulating climate dynamics during the past, extremely warm periods*' [126].

## 11.7   Case History – The Mid Miocene Climatic Optimum

What were the respective roles of plate tectonics, orbital changes, and $CO_2$ in modifying the climate of the later Cenozoic? Plate tectonics must have played some role, as it did in the Early Cenozoic [36]. For example, Australia and New Guinea began colliding with the island arcs of South East Asia, constricting the link between the Indian and Pacific Oceans. Numerical modelling suggests that this constriction would have affected the location of deep-water formation, oceanic heat transport and sea surface temperatures at both high and low latitudes. Those changes would likely also have affected the marine carbon cycle and atmospheric $CO_2$ [127]. That collision may help to explain the origin of the Oligocene–Miocene boundary around 23 Ma ago. The northward movement of Australia also widened the Southern Ocean between Tasmania and Antarctica at roughly the same time as the Drake Passage opened between Tierra del Fuego and the Antarctic Peninsula. That widening strengthened the Antarctic Circumpolar Current, further isolating Antarctica thermally and helping to keep it cool, which may be why Southern Hemisphere ice sheets came to be more or less permanent features of the climate system from 34 Ma onwards.

What about the role of $CO_2$, which declined quite rapidly from levels near 750 ppm at the Eocene–Oligocene boundary to about 400 ppm by the end of the Oligocene (Figure 9.16) [51, 57, 128]? Cramer and colleagues tell us that Oligocene bottom water temperatures stayed more or less flat at about 6 °C, while there was a persistent but dynamic Antarctic ice sheet (see Figure 7.10) [66]. Bottom water temperatures increased slightly to about 7 °C, 26 Ma ago, and stayed warm across the Oligocene-Miocene boundary until about 16 Ma ago (Figure 7.10) as the ice sheet decreased in volume towards the Middle Miocene. The existence of some high $CO_2$ values at about 26 Ma ago (Figure 9.16) [51] suggests that the warming then may have had something to do with increasing $CO_2$. But $CO_2$ then fell to lowish values across the Oligocene-Miocene boundary [51], while bottom temperatures stayed on the warm side (Figure 7.10) [66]. High-resolution $\partial^{18}O$ data at the Oligocene-Miocene boundary around 23 Ma ago revealed a 2 Ma long cold event coinciding with a fall in sea level, interpreted as a period of growth in the Antarctic ice sheet [127]. The boundary event correlated with minima in the amplitudes of both the low frequency (400 Ka) eccentricity cycle and the high frequency (41 Ka) obliquity cycle, which sustained unusually cold summers. Whilst this orbital 'event' was transient, it may have been enough to expand the Antarctic ice sheet to dimensions like those of the Last Glacial Maximum for some 200 Ka [129]. Orbital variations are readily identifiable for the Oligocene [70] and the Miocene [68].

During the Middle Miocene there was a warm event lasting from 18 to 14 Ma ago, when bottom water temperatures reached 7 °C (Figure 7.10) [66] This so-called 'Mid Miocene Climatic Optimum' was associated with high values of $CO_2$. Mark Pagani's original alkenone data from 2005 had suggested that $CO_2$ levels were close to 200–300 ppm during the Miocene [57]. By 2009, Aradhna Tripati from Cambridge University had found that the Mid Miocene warm period was characterized by $CO_2$ values ranging between 350 and 450 ppm. Tripati and her team used a new technique – '*boron/calcium ratios in foraminifera*' to show that '*during the Middle Miocene, when temperatures were ~3° to 6°C warmer and sea level was 25 to 40 meters higher than at present, pCO₂ appears to have been similar to modern levels*' [130]. In the later Miocene and Pliocene '*Decreases in pCO₂ were apparently synchronous with major episodes of glacial expansion ....[~14 to 10 million years ago] and ... [~3.3 to 2.4 million years ago]*' [130]. The more recent data from Zhang (Figure 9.15) [131], and from Beerling and Royer (Figure 9.16) [51] showed that Mid Miocene $CO_2$ levels were in fact closer to 400–600 ppm.

There is widespread evidence for the Mid Miocene warming event on land between about 18 and 14 Ma ago, as we saw in Chapters 6 and 7. Antarctica warmed significantly, with temperatures in the Ross Sea region averaging 10 °C in January, a reduction in sea ice, and a proliferation of woody plants on land [132]. Mg/Ca ratios from planktonic and benthic foraminifera show that warm waters prevailed around Antarctica [133]. ANDRILL data suggest that the East Antarctic Ice Sheet may have retreated into the Transantarctic Mountains at this time. Drill cores confirm that the ice sheet fluctuated dynamically during this period [127]. The warming coincided with the eruption of the Columbia River Plateau basalts – a Large Igneous Province that was at its most vigorous between 17 and 14 Ma ago. These eruptions may have contributed to the elevated $CO_2$ of the time [134].

Puzzling over apparent inconsistencies between quite sizeable changes in the size of the Miocene ice sheet and relatively small changes in $CO_2$ (between values close to pre-industrial levels of c. 280 ppm, or less, to maxima of c. 500 ppm), Edward Gasson of the Climate Research Center of the University of Massachussetts at Amherst and his team used an advanced numerical model coupling climate and ice sheet behaviour to assess the evolution of the climate system, the ice sheet and sea level during the Early to Middle Miocene [135]. Drilling in the Ross Sea, they observed, provides evidence for four advances and retreats of the ice sheet in this period, associated with changes in benthic $\partial^{18}O$ and sea level. For their colder interval simulations they used an atmospheric $CO_2$ of 280 ppm and an astronomical configuration favourable for Antarctic glaciation (low obliquity, high eccentricity, perihelion during boreal summer). For their warmer interval simulations they used an atmospheric $CO_2$ of 500 ppm and an astronomical configuration favourable for Antarctic deglaciation (high obliquity, high eccentricity, perihelion during austral summer) and 2 °C of ocean warming. For their high $CO_2$ simulations they used the warmer climate simulations but with atmospheric $CO_2$ raised to 840 ppm. And for each climate simulation they used two submarine topographies, one with deeper water than the other. '*Bedrock topography is ... important in determining the magnitude of East Antarctic ice sheet retreat for the high $CO_2$ experiments*' [135]. As a result, '*One scenario [has] large-scale retreat into East Antarctic subglacial basins; the second scenario has limited retreat of the East Antarctic ice sheet, with variability largely a result of the expansion of a terrestrial West Antarctic ice sheet*' [135]. Although both scenarios satisfy constraints on ice volume, only the shallow topography satisfied observations from drilling in the Ross Sea, suggesting that much of the change in ice volume came from fluctuations in the volume of ice on East rather than West Antarctica [135]. Both the 280 ppm and 500 ppm condition satisfy the observational data, but the 840 ppm model does not. However even at that level of $CO_2$ the Antarctic ice sheet survived, though much reduced '*supporting the idea that there was a core of stable ice on Antarctica throughout the warm intervals of the early to mid-Miocene*' [135].

The climate cooled sharply starting about 14 Ma ago, with bottom waters beginning a long temperature decline (Figure 7.10) [66, 128]. This pronounced cooling led to strengthening of the Antarctic Circumpolar Current and enlargement of the Antarctic ice sheet [130]. The cooling went along with a slow decline of $CO_2$ levels to between 200 and 350 ppm between 12 and 5 Ma ago (Figure 9.16). In contrast, mid-latitude Pacific Ocean surface waters stayed warm [136]. Sea level also rose at the time [130]. Stable isotope measurements from the western equatorial Pacific show that over this period the thermocline gradually became shallower, which eventually led to the stronger coupling between $CO_2$, sea surface temperature and climate typical of the Pliocene and Pleistocene [136]. Between 8 and 2.5 Ma ago, bottom water temperatures fell from around 6 °C to less than 2 °C [66]. $CO_2$ levels fell in parallel [137].

The post-Mid Miocene cooling starting abruptly about 13.8 Ma ago saw the establishment of the stable, modern Antarctic ice sheet, and was a critical step in moving towards the Pleistocene and modern era with its bipolar ice sheets [138]. This cooling step and its accompanying Antarctic ice growth was accompanied by a substantial positive $\partial^{13}C$ excursion caused probably by the burial of $^{12}C$-rich organic matter through enhanced upwelling and associated productivity driven by an increase in wind strength; as a result, $pCO_2$ fell by 20–80 ppm from a starting level near 330 ppm [138]. These results highlight '*the enigma of a warmer Miocene world with, at times, less ice than today but only slightly higher levels of atmospheric $pCO_2$*' [138]. '*Further global temperature records are required to solve the Miocene paleoclimate enigma … [which may have arisen because] climate sensitivity to changing $pCO_2$ is greater than previously thought*' [138]. In Antarctica itself, tundra conditions had become established, for example in the McMurdo Dry Valleys, during the climatic optimum, and disappeared with a cooling of at least 8 °C between 14.07 and 13.85 Ma ago [139]. Lakes were well developed, with plants and insects around their margins. The tundra conditions are represented, for example, by freeze-dried mosses than can, like museum specimens, be rehydrated [139].

Despite the cooling, there was still only one ice sheet – on Antarctica – and the Late Miocene climate was in many regions warmer and/or wetter than today, with warm temperate forests in the circum-Arctic region, and grasslands in many of what are now desert regions (Arabia, Sahara) [140]. However climate influences were changing, for example with the uplift of the Himalaya, the Tibetan Plateau, the Andes, the Rockies, the East African Plateaus,

and the Alps. Also, unlike today, the Indonesian seaway and the Central American seaway were open. Common sense and climate models agree that the atmospheric and oceanic circulation would have been quite different in the Late Miocene from what it is now. Using a fully coupled atmosphere–ocean–vegetation GCM model with two $CO_2$ levels (280 and 400 ppm), Bradshaw and colleagues found that during the Late Miocene cool, dry air spread across the North Atlantic into Eurasia, with regional warming in areas of low topography, and reductions in monsoonal precipitation [140]. They concluded that '*late Miocene atmospheric $CO_2$ concentrations were [likely] generally nearer the high end of the range of reconstructions, and that palaeogeography play[ed] a much more localised role in determining late Miocene temperatures*' [140].

This is a useful point at which to remind ourselves that '*a monsoon is a regional weather regime characterised by a seasonal reversal of wind direction, generally caused by the seasonal cycle of the Sun back and through the equator*' [55]. Monsoons are strongest when summer insolation is highest (e.g. at perihelion in June in the Northern Hemisphere). They will also strengthen when obliquity is high, because that strengthens the meridional temperature gradient between low and high latitudes, especially in Africa because of the large land area, but less so in South Asia [55]. Astronomical forcing also influences ice sheet growth and decay, the ocean's alkalinity, sea surface temperature, ocean circulation, continental temperature and vegetation, the supply of nutrients to the ocean and marine productivity [55].

Monsoonal circulation is responsible for the deposition of organic rich layers named sapropels in the deeps of the Eastern Mediterranean. They accumulated there over the past 13.5 Ma after the closure of the eastern Tethys, which formerly extended into the Mediterranean. The sapropels lasted between 3 and 5 Ka and recurred at intervals of about 21 Ka, reflecting control by precession. High summer precession brings rains to East Africa, swelling the outflow of the Nile, which forms a freshwater lid on the eastern Mediterranean, causing stagnation of the bottom waters, which, between times, are oxygenated by salty water sinking from the surface. Sapropels are associated with eccentricity maxima, which enhance the precession signal [55].

Returning to the Miocene, whilst it still experienced strong connections between $CO_2$ and temperature, as in the early Cenozoic, for two main reasons these links did not explain all the changes found in palaeoclimate records. First, plate tectonics caused gateways to open or close, changing ocean circulation and thus the distribution of heat around the globe. Second, as Foster and Rohling pointed out [141], the relationship between $CO_2$ and temperature is disrupted at times when there are large ice sheets, which alter the thermal state of the Earth System, as we saw in Chapter 10.

This analysis of the relation of $CO_2$ to the picture of global change in deep ocean bottom water derives from stacking together a number of oxygen isotopic and Mg/Ca records from deep sea cores [50, 66]. But it does not tell us what was going on region-by-region at the surface. Sam Savin, whom we met in Chapter 7, found in 1990 that most of the cooling was confined to high latitudes and deep bottom waters. In contrast, the equatorial region warmed, and mid-latitude waters showed no change [142]. This constitutes something of a paradox. Savin found no evidence for global refrigeration at the time of the onset of new Antarctic ice growth 14 Ma ago. Yet we know from the $CO_2$ data of Pagani, Zhang, and Beerling and Royer that this ice growth took place with only a small change in $CO_2$. What, then, made the ice grow?

Much of Savin's picture of the cooling of high southern latitudes with time may have to do with the stabilizing effect on global climate of the existence of an Antarctic ice sheet. Undoubtedly the ice sheet helped to keep the surrounding air and water cool. But as the Southern Ocean widened the Antarctic Circumpolar Current migrated north, expanding the area of cold water south of the Polar Front. The circulation of the Southern Ocean of the time was probably much like it is today, with seasonal sea ice and seasonal production of Antarctic Bottom Water around the Antarctic coast; northward movement of cold surface water driven by the westerly wind; replacement of surface water by the upwelling of Circum-Polar Deep Water; and the sinking of cool Intermediate Water at the Polar Front. These processes are balanced, today by the sinking of cold North Atlantic Deep Water in the Nordic Seas. But in the Oligocene and Miocene, there was no major north polar ice sheet, and in any case the Fram Strait between Svalbard and Greenland was closed until 12 Ma ago, all of which suggests that the water from the north balancing the outward flow of cold Antarctic Bottom Water and Intermediate Water from the south would have been warmer than it is today. The ocean and atmosphere processes associated with Antarctica and its growing Southern Ocean would have kept southern high latitudes cool and stable, but growing colder as the Southern Ocean widened. Changes from time to time might be expected as a result of changes in orbital forcing, but global temperature and $CO_2$ may not have changed much because of the absence of Northern Hemisphere ice sheets and the corresponding weakness of the supply of North Atlantic Deep Water.

The key change driven by the growing Southern Ocean would have been, as Savin saw, gradual intensification of the Equator-to-pole thermal gradient, leading to a more vigorous thermohaline circulation, stronger winds and enhanced upwelling – hence cooling – on continental margins [142], including around Antarctica. But he missed another factor – a change in the source of deep water, as we shall see.

As in all of the developments we have examined, ideas have changed with time. Let's look at the Late Miocene. At the Tarpon Springs meeting in 1984, palaeontologist Edith Vincent and geochemist Wolfgang Berger, of Scripps, pointed out that this cooling step occurred within a substantial 'excursion' of carbon isotopes towards heavier values (higher $\partial^{13}C$) signifying the excess trapping of $^{12}C$-rich organic matter in sediments [143]. The timing, they thought, coincided with the widespread deposition of diatomaceous deposits on the Pacific margin, especially the shales rich in diatomaceous silica and organic matter, along with phosphorite, of California's Monterey Formation. That association led them to propose that global cooling had increased the equator-to-pole thermal gradient, which strengthened coastal winds that in turn enhanced upwelling, productivity, the development of the oxygen minimum zone, and the sedimentation and entrapment of organic matter, so trapping $^{12}C$-rich material. The massive production, deposition, and entrapment of organic carbon in sediments combined to pull $CO_2$ out of the atmosphere, they thought, which in turn contributed to a further lowering of global temperature through the greenhouse mechanism. This cycle would have been broken when the nutrients in the upwelling waters, especially phosphorus, became exhausted, causing production to decline [143]. However, we now know that the positive $\partial^{13}C$ excursion ended around 16–15 Ma ago, whilst the peak accumulation of organic matter in the Monterey Formation was focussed later, between 14.3 and 10.6 Ma ago [144].

By 2009, with the benefit of access to a global set of deep ocean drill cores, Vincent and Berger's 'Monterey Hypothesis' was found wanting by a team led by Liselotte Diester-Haas of the University of Saarland, in Saarbrücken, Germany [145]. Diester-Haas used multiple proxies of climate change to analyse the sediment records of Middle Miocene age in deep-sea drill cores from the Atlantic, which crossed the prominent positive Mid Miocene excursion in $\partial^{13}C$ between 17.5 and 13.5 Ma ago, at the heart of the Mid Miocene Climatic Optimum (MCO). Her team made two significant discoveries. First, marine productivity in the Atlantic was not related to the $\partial^{13}C$ excursion, so the excursion could not represent a global marine productivity event of the kind envisaged by Vincent and Berger. Second, the $\partial^{13}C$ values of bulk organic matter in deep marine sediments derived from ocean surface plankton paralleled those in benthic foraminifera that drew their carbon from dissolved organic carbon in bottom waters. The absence of a surface to seabed gradient in carbon isotopes made it highly unlikely that some hypothetical marine productivity event had led to a significant change in the $CO_2$ content of the air during this period. How, then, did the ocean come to be enriched in $^{13}C$ at this time? The answer, thought Diester-Haas, must lie on land. The $^{13}C$ excursion in the ocean coincided with the Mid Miocene Climate Optimum, when warm conditions prevailed and $^{12}C$-rich vegetation spread to high latitudes. Evergreen forests extended north to 45°N in North America and to 52°N in Europe, leading to major deposits of brown coal, or lignite, worldwide. $CO_2$ was relatively abundant at the time, possibly, as mentioned above, due to the eruption of the Columbia River Plateau Basalts [134]. The trapping of $^{12}C$ in terrestrial organic matter enriched the ocean and its sediments in $^{13}C$ [145].

Global cooling and growth of the Antarctic ice sheet at 14 Ma ago followed the Mid Miocene Climatic Optimum (MCO). Grasslands replaced the forests, and the deposition of brown coal ceased [146]. The Monterey Hypothesis failed to explain the time lag between the onset of the $\partial^{13}C$ excursion at 17.5 Ma ago, and the global cooling, which followed at around 14 Ma ago. Not only that, but the hypothesis also failed to explain the significant changes in atmospheric $CO_2$ [145, 146].

Like Savin, Diester-Haass called on a major reorganization of ocean circulation to explain the Late Miocene cooling [145]. We can imagine such a reorganization resulting from the continued reorganization of landmasses and ocean gateways as the continents continued to move. One of the key changes has been suggested by differences in the $\partial^{13}C$ ratios of benthic foraminifera from different ocean basins. These indicate that in the Early Miocene the Southern Ocean received warm saline deep water from a tropical source, probably in the Indonesian region, and that the flow of this water ceased in the Mid Miocene, perhaps due to the collision of New Guinea with the Indonesian islands, thus reducing meridional heat transport and possibly triggering expansion of the Antarctic ice sheet [137].

We still do need an explanation for the enrichment of the Monterey Formation in organic remains. Even though it is rich in organic matter, the rate of accumulation of organic matter, it now turns out, was rather low [144]. The modern environment encourages the preservation of organic matter on the California margin for three reasons: first, the eastern Pacific is old water poor in oxygen, so that there is a well-developed oxygen minimum zone; second, run-off of silt and sand is low because the interior is arid, so organic matter is not diluted; and third, the coastal

offshore basins are enclosed, with restricted circulation leading to anoxic conditions. Much the same conditions may have applied in the Late Miocene. However, whilst those conditions were also necessary in the Late Miocene to account for the creation of the Monterey Formation, there must have been a global reorganization of subsurface waters at the time to account for the fact that multiple phosphorite deposits dating to around 11 Ma ago formed not only in the California Borderland, but also on the shallow continental margins off Morocco, off Cape Town, and on the Chatham Rise off the South Island of New Zealand, for example [147], for which we need a global explanation.

Astronomical forcing also played an important role in the dynamics of the Miocene ice sheet and Miocene climate, as we see from the work of Ann Holbourn of Kiel's Christian-Albrechts University and colleagues [148]. Figure 11.10(b) shows the profile of Miocene warming based on the $\partial^{18}O$ signal. Warming was associated with a raised CCD indicating an increase in atmospheric $CO_2$ (Figure 9.3). That stimulated accumulation of $^{12}C$-rich organic matter on land [145] driving up the $\partial^{13}C$ signal in the ocean (Figure 11.10a). The MCO onset event (wide orange strip at 16.9–16.7 Ma in Figure 11.10) was associated with a very long period of high eccentricity (Figure 11.10(c)). The sharp changes in $\partial^{18}O$ and $\partial^{13}C$ at the onset of the MCO are '*contemporaneous with a massive increase in carbonate dissolution, demonstrat[ing] that abrupt warming was coupled to an intense perturbation of the carbon cycle*' [148].

The recovery in $\partial^{13}C$ at c.16.7 Ma, c.250 Ka after the beginning of the MCO (Figure 11.10), '*marks the onset of the first carbon isotope maximum within the long-lasting "Monterey Excursion", ... [and] the results lend support to the notion that atmospheric $pCO_2$ variations drove profound changes in the global carbon reservoir through the MCO, implying a delicate balance between changing $CO_2$ fluxes, rates of silicate weathering, and global carbon sequestration*' [148]. Temperatures during the MCO were probably 3–8 °C higher than pre-industrial levels, $CO_2$ values reached 500 ppm, and the Antarctic ice sheet was much reduced. Holbourn thought that the MCO was kicked off by the volcanic eruptions and $CO_2$ emissions associated with the emplacement of the Columbia River Basalts (a Large Igneous Province event) [148]. Warming then affected the hydrological cycle, impacting rates of silicate weathering on land and the marine nutrient and carbon budgets, prolonging the MCO. A brief, well-defined peak warming event occurred at 15.6 Ma ago (Figure 11.10), seemingly paced by orbital change (possibly by precession) [148]. Equally, I might add that there was another slightly less

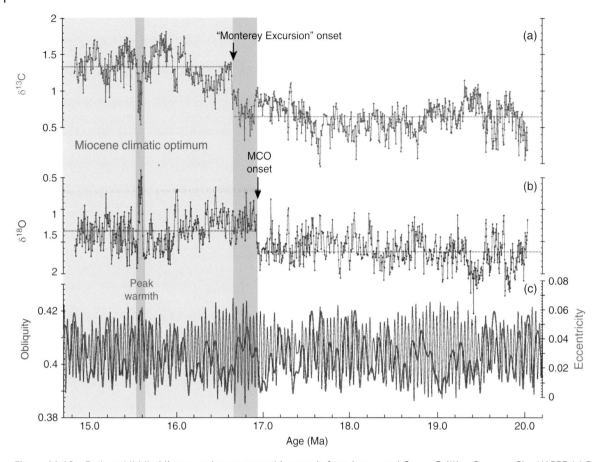

**Figure 11.10** Early to Middle Miocene paleoceanographic records from Integrated Ocean Drilling Program Site U1337. (a) Benthic foraminiferal $\partial^{13}C$ (in ‰ versus Vienna Peedee belemnite, VPDB). (b) Benthic foraminiferal $\partial^{18}O$ (in ‰ versus VPDB). (c) Eccentricity and obliquity from Laskar et al. (2004). Dashed lines in (a) and (b) indicate mean $\partial^{18}O$ and $\partial^{13}C$ values. Darker orange shading marks peak warming episodes. MCO = Miocene Climatic Optimum. *Source:* From figure 1 in Ref. [148].

extreme event at about 15.95 Ma, also tied to an eccentricity peak (Figure 11.10).

Levy and others analysing the behaviour of the Mid Miocene Antarctic ice sheet as represented in the ANDRILL cores from the Ross Sea agreed in 2016 that episodes of ice sheet advance recognized in the sedimentary record coincided with minima in eccentricity and $CO_2$ and were correlated with falls in sea level (of 30–60 m) and increased production of cold bottom water [149]. $CO_2$ levels ranged between 280 and 500 ppm. During warm intervals, the sediments were rich in palynomorphs indicative of pollen-bearing plants on land, and Levy's team found indications that surface water in the Ross Sea was some 6–10 °C warmer than it is today and that the Ross Ice Shelf was absent. He speculated that '*whereas orbital variations were the primary driver of glacial cycles, atmospheric $CO_2$ variations modulated the extent of ice sheet advance and retreat*' [149]. Their studies '*suggest that polar climate and the [Antarctic Ice Sheet] were highly sensitive to relatively small changes in atmospheric $CO_2$ during the early to mid-Miocene*' [149]. During periods of

peak warmth, summer land temperatures were at least 10 °C, tundra vegetation characterized the coast, and the ice sheet retreated inland. They inferred that '*sustained levels of atmospheric $CO_2$ >400 ppm may represent a stability threshold for marine-based portions of the ... ice sheets*' [149].

In 2013 Holbourn presented data from the late part of the MCO (16.4–8.4 Ma ago), showing that this event ended abruptly at about 13.8 Ma ago with a sharp increase in $\partial^{18}O$ and $\partial^{13}C$, indicative of cooling and a fall in atmospheric $CO_2$ (Figure 11.11) [150]. The figure illustrates the modulation of the eccentricity and obliquity cycles ascertained by Laskar [151] and mentioned in Chapter 3. The $\partial^{13}C$ signal closely follows the 400 Ka eccentricity cycle, whilst the $\partial^{18}O$ signal closely follows the 100 Ka eccentricity cycle and 41 ka obliquity cycle. The obliquity cycle has its most prominent effect from ~14.6 to 14.1 Ma and from ~9.8 to 9.2 Ma, when high-amplitude variability in obliquity coincides with 100 Ka variability in eccentricity (a configuration occurring only every 2.4 Ma). The $\partial^{18}O$ signal also experiences a series of downward steps (arrows in Figure 11.11)

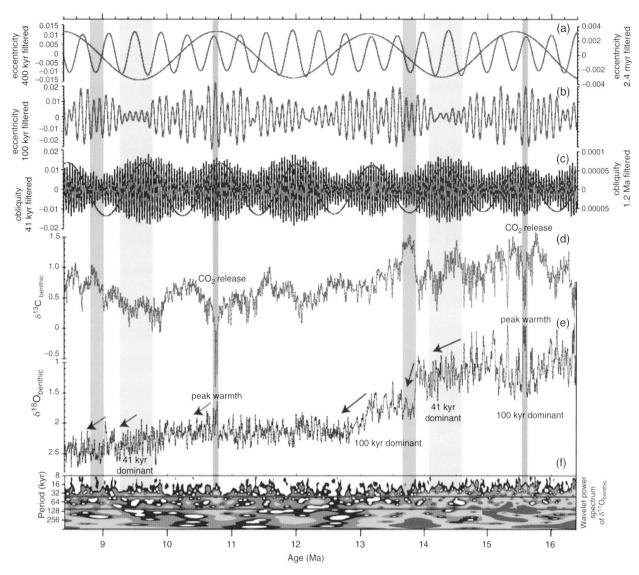

**Figure 11.11** Comparison of benthic $\partial^{18}O$ and $\partial^{13}C$ data in ODP Site 1146 with orbital variability: (a) 2.4 Ma eccentricity and 400 Ka eccentricity; (b) 100 Ka eccentricity; (c) 1.2 Ma obliquity and 41 Ka obliquity; (d) Benthic $\partial^{13}C$; (e) Benthic $\partial^{18}O$; (f) Wavelet power spectrum of benthic $\partial^{18}O$. Orange bands mark transient warmings and negative $\partial^{13}C$ excursions (increases in $CO_2$) at eccentricity maxima (100 and 400 Ka). Light blue shading highlights periods of low (100 Ka) eccentricity variability and high (41 Ka) obliquity variability, whilst $\partial^{18}O$ exhibits mainly 41 Ka variability. Darker blue bands mark cooling episodes concomitant with nodes in obliquity. Arrows indicate cooling trends. *Source:* From figure 7 in Ref. [150].

attributed to progressive deep-water cooling and/or glaciation episodes after the MCO [150]. The post-MCO cooling was interrupted by a brief warming event at ~10.8–10.7 Ma during a peak insolation event forced by maxima in 100 and 400 Ka eccentricity (Figure 11.11), similar to the event at 15.6 Ma that is also seen in Figure 11.10 and is associated with a large decline in $\partial^{13}C$. Bearing in mind that '*studies of Pliocene-Pleistocene climate variations have shown ... that changes in the ocean/atmosphere $CO_2$ exchange, latitudinal temperature gradient, meridional overturning circulation, and ice sheet hysteresis also exert major control on climate development through intricate internal feedback processes*'

Holbourn and colleagues agreed that the response of the ocean-atmosphere system to orbital forcing was non-linear: '*other triggering factors [than simply changes in orbital properties, or temperature] were also required to force the climate across critical thresholds*' [150].

Holbourn's $CO_2$ records across the MCO (Figures 11.10 and 11.11) are based on the notion that the $\partial^{13}C$ signal in deep ocean sediments represents atmospheric $CO_2$. Independent evidence for the actual atmospheric concentration of $CO_2$ at the time comes from Tripati's use of boron/calcium ratios [130] and from a new boron isotope method applied in 2014 by Greenop and colleagues to an

ODP core from the Indian Ocean [152]. Greenop's study showed pronounced variability of $CO_2$ between 300 and 500 ppm on a roughly 100 Ka timescale during the MCO between 17 and 15.5 Ma ago. Noticing that rather larger concentrations of $CO_2$ seemed necessary to account for changes in ocean temperature than was the case for the Pleistocene glaciations, Greenop and colleagues deduced that this was because during the Miocene there was no Northern Hemisphere ice sheet, which reduced the size of the reservoir of ice that would be sensitive to subtle variations in atmospheric $CO_2$ [152]. They considered that the intervals of extreme warmth in the MCO were brought about by a combination of high tilt and high $CO_2$ in warm summers generated by high obliquity.

In 2014, Holbourn was a member of a team led by Karlos Kochann that analysed a group of deep-ocean cores through the MCO in the eastern equatorial Pacific Ocean. Their sediment records depicted prominent cyclicity with periods of 100 and 400 Ka in both $\partial^{13}C$ and $CaCO_3$ records, '*pointing to a tight coupling between the marine carbon cycle and climate variations*' [153]. They also found that '*the lysocline behaved in highly dynamic manner throughout the MCO, with >75% carbonate loss occurring at paleodepths ranging from ~3.4 to ~4 km … Carbonate dissolution maxima coincide with warm phases ($\partial^{18}O$ minima) and $\partial^{13}C$ decreases*' [153]. This showed that climate–carbon cycle feedbacks differed from the pattern for late Pleistocene glacial–interglacial change, where dissolution maxima correspond to $\partial^{13}C$ maxima and $\partial^{18}O$ minima. Instead, the carbonate dissolution cycles of the MCO were more similar to the patterns from Palaeogene hyperthermals (see Section 11.4), which are also characterized by enhanced $CaCO_3$ dissolution coupled with low $\partial^{13}C$ and $\partial^{18}O$ in phase with eccentricity maxima. This may suggest a stronger link to temperature and humidity driven silicate weathering cycles on land than to changes in ice volume, and to '*radically different climate-carbon cycle feedbacks during greenhouse climate modes*' [153]. During Palaeogene hyperthermals it is thought that temperature increase forces the release into the atmosphere of large amounts of isotopically light $CO_2$ from high latitude soils at times when polar ice is reduced or disappears, thus contributing to elevated $CO_2$, which is transferred to the oceans, making deep water more acid and raising the lysocline and CCD [153]. Greenop's boron isotope-based $CO_2$ record from the MCO [152] supports the notion of $CO_2$-forcing of high amplitude carbonate dissolution and climate cycles, recognizing that the ultimate driver is orbital eccentricity [153]. This feedback process between terrestrial and marine counterparts of the carbon cycle differs radically from the late Pleistocene glacial–interglacial pattern in which ice provides a key feedback medium [153].

The marked decrease in the variability of $\partial^{18}O$ amplitude after ~12.9 Ma (Figure 11.11) suggests that a substantially drier, colder, and more stable climate became established over Antarctica, probably connected to a change from a warm- to a cold-based ice sheet. Decreased sensitivity to eccentricity forcing and damped $\partial^{18}O$ variability suggest that moisture transport became, as it is today, the dominant control on Antarctic ice sheet growth as Antarctica became increasingly glaciated and with a more permanent ice cover [150]. The waxing and waning of the climate, representing glacial expansion and the cooling of deep water, was associated with a sea level rise of 20–30 m.

Comparing the $\partial^{13}C$ signals across the inception of the MCO (Figure 11.10) with those across Oceanic Anoxic Event-1a (OAE-1a), Holbourn saw that both experienced an initial fall in $\partial^{13}C$ followed by a sharp recovery. The difference between the two events was the much larger $\partial^{13}C$ change in OAE-1a, and its association with very widespread black shales indicating the sequestration of much larger amounts of organic matter in sediments than during the MCO [148].

What has not previously been considered in thinking about the age of this global event is the effect of the opening of Fram Strait around 12 Ma ago (Chapter 5). The opening of the Strait would have strengthened the supply of cold North Atlantic Deep Water, moving Atlantic circulation from its Oligocene–Miocene state towards its modern state. This would have been accompanied, due to the increasing size of the Antarctic ice sheet, by an increased northward flow of cold Antarctic Bottom Water, Antarctic Intermediate Water (with a core at about 800–1000 m), and Subantarctic Mode Water (with a core at 200–300 m). These last two water masses originating in the Southern Ocean take nutrients throughout the world's oceans. Upwelling currents tap into them to bring nutrients to the surface to provide up to 75% of global productivity between 30°S and 30°N [147]. Development of Fram Strait is likely to have intensified inter-basin oceanic gradients between the Atlantic and Pacific Oceans after ~12 or ~13 Ma ago, probably as a result of North Atlantic Deep Water formation in response to the opening of the Strait.

Another possibility, also based on the control of ocean chemistry by the opening or closing of ocean gateways, and suggested by Ian Dalziel, was the full development of a deep water Antarctic Circumpolar Current consequent upon the opening of appropriate gaps in the Neogene sill bounding the northern side of the Scotia Sea [154].

To understand the effect of these vast changes in vertical circulation we must recall that at present the limited supply of cold oxygenated deep water from the north in both the Indian and North Pacific Oceans leads their deep waters to become old, oxygen poor and nutrient rich. The

Mid Miocene of the North Pacific would have been similar, which would help to explain its unusually deep oxygen minimum zone during the time of formation of the Monterey Formation [155]. Much the same would have been true of the Late Miocene North Atlantic before the Fram Strait opened. Opening of the Strait would have helped in due course (possibly by c. 10.8–11 Ma ago) to stir up the nutrient-rich deep water of the world ocean, making more nutrients available globally to upwelling systems and hence stimulating productivity leading to the formation of widespread phosphorites. Eventually the supply of subsurface nutrients, especially phosphorus, would have declined, ending the global formation of continental margin phosphorites.

The cooling of the Late Miocene increased the Equator-to-pole thermal gradient and strengthened coastal winds, which in turn increased coastal upwelling, the supply of nutrients, productivity, depletion of oxygen in the oxygen minimum zone, and the thickening of that zone. These interconnected processes help to explain the extension of the oxygen minimum zone into deeper water on the continental margin west of California during deposition of the Monterey Formation [155]. Vertical upward displacement of nutrients by strengthening deep waters, combined with strengthened upwelling, is likely to have affected the global ocean as well as the seas off California.

The newly available data suggest that Savin was wrong to imply that the global climate did not cool in the Late Miocene. Indeed, as we saw in Chapters 6 and 7 there is ample evidence on land for a global cooling at that time. However, the tropics may not have experienced significant cooling.

Most of the increase in ice volume in the Late Miocene occurred on East Antarctica [137]. Whilst seismic studies also pointed to an expansion of ice caps on West Antarctica, several lines of evidence showed that there was no major ice sheet on West Antarctic or on the Antarctic Peninsula from the Mid Miocene to the early part of the Late Miocene. A large and thick West Antarctic Ice Sheet advanced up to the edge of the continental shelf during the latest Miocene and Early Pliocene, and the Antarctic Peninsula Ice Sheet expanded at the same time. While those changes were slowly taking place, the ice sheets were expanding and contracting more rapidly in response to periodic orbital changes in insolation [137].

An intriguing resolution of the Mid Miocene climate paradox came in 2014 from new research by the Alfred Wegener Institute for Polar and Marine Research (AWI), in Bremerhaven. Gregor Knorr and Gerrit Lohmann used numerical model simulations of the Miocene climate to find out why, when the Antarctic ice sheet grew to its present size around 14 Ma ago, some parts of the world, including parts of the Southern Ocean, became warmer [156].

The expansion of the ice sheet increased the elevation of Antarctica. The resulting lapse rate (a fall of 6–10 °C/1000 m) cooled the surface by up to 22 °C in the high central part of the ice sheet. Katabatic winds spilling down to the coast from the cold, high interior favoured the development and northward export of sea ice, cooling the surface waters in the Ross Sea region. In contrast the coastal wind patterns for the Weddell Sea region were primarily onshore, leading to the import of warmer waters from the north, accompanied by the retreat of the sea ice. These changes in the wind pattern served to increase the rate of deep-water formation in the Weddell Sea and to decrease it in the Ross Sea region. Superimposed on these regional changes, which included local warming of the Southern Ocean, was a fall in temperature associated with the continued decline in $CO_2$. High latitude air cooled by around 6 °C in both hemispheres as sea ice, and thus albedo (reflection of solar energy), increased. The decline in $CO_2$ and associated fall in global temperature had less effect on the wind field of the Southern Ocean than did the growth of the ice sheet, which, in effect, caused pronounced regional change. Hence the Mid Miocene $\partial^{18}O$ shift reflected the growth of the ice sheet, a regional event that encouraged regional warming of the Southern Ocean in the Atlantic sector during an overall global cooling trend in sea surface temperature and bottom water temperature driven by the global decline in $CO_2$.

As Holbourn and her team advised, we also need to bear in mind the role of the size and thickness of the ice sheet as setting thresholds that must be reached so as to sustain ice growth. For instance, '*during the Miocene climatic optimum, when the latitudinal temperature gradient was relatively low and polar temperatures elevated, small ice sheets remained highly susceptible to summer radiative forcing. In contrast, prolonged periods of low summer insolation would inhibit melting and favor ice growth ... as Antarctica became thermally more insulated following major ice sheet expansion and concomitant intensification of the Antarctic Circumpolar Current, moisture transfer from lower latitudes became reduced and further ice accumulation decreased. The marked decrease in $\partial^{18}O$ amplitude variability after ~12.9 Ma in Site 1146 [Figure 11.11] suggests that a substantially drier, colder, and more stable climate became established over the Antarctic continent that likely reflected a long-term change from warm- to cold-based ice grounding after that time. Decreased sensitivity to eccentricity forcing and dampened $\partial^{18}O$ variability suggest that moisture transport became, as it is today ... the dominant control on Antarctic ice sheet growth*' [148].

Evidently, understanding climate change in the Miocene requires careful attention to the respective roles of: plate tectonics and its effects on ocean circulation; the changing volume and elevation of Antarctic ice and its effects on the wind field and supply of moisture; and the overall decline in $CO_2$. Changing geographic gateways forced changes on global ocean and atmospheric circulation. The climate was generally stable, probably due to the influence of stable oceanographic processes driven by the Antarctic thermostat, which kept southern high latitude temperatures cool, along the lines suggested by Ralph Keeling and Martin Visbeck [157]. The thermostat works by strengthening the flow of Antarctic Bottom Water into the ocean interior and cooling the deep ocean when other sources of deep water are too warm, and by turning off or reducing the supply of Antarctic Bottom Water when deep bottom waters are too cold, thus allowing the deep ocean to warm again through the input of warm deep water from elsewhere, eventually achieving a steady state. In due course, plate tectonic processes switched off the supply of warm tropically sourced deep water, allowing cold deep waters to predominate and making oceanic and atmospheric circulation more vigorous. That led temperatures at high latitudes to cool enough to enable a bigger ice sheet to grow. The widening Southern Ocean played a key role in these developments. With the opening of Fram Strait between Svalbard and Greenland from 12 Ma ago onwards, the ocean became preconditioned for the development of ice in the Northern Hemisphere.

We will now examine a case history for the Pliocene, to see what it can tell us about the kinds of changes that increases in $CO_2$ may lead to in the future.

## 11.8 Case History – The Pliocene

As we saw in Chapter 7, the cooling trend of the late Cenozoic was interrupted during the Pliocene (5.33–2.58 Ma ago) by conditions warm enough to melt Antarctica's Ross Ice Shelf and deposit diatomaceous ooze there. In their 2001 compilation of $\partial^{18}O$ data for the past 65 Ma, Zachos and his colleagues reported that: '*The Early Pliocene is marked by a subtle warming trend until 3.2 Ma*' (Figure 7.7) [50]. We can also see this slight warming in the $\partial^{18}O$ data in Figure 9.3, while Tripati highlights the warmth between 3.6 and 3.4 Ma ago [130]. The available geological and geochemical data were comprehensively reviewed in 2011 by a team led by Ulrich Salzmann of Northumbria University in Newcastle-upon-Tyne, who found that the global mean surface temperatures of the Pliocene were 2–3 °C warmer than they are today, and accentuated at high latitudes, so reducing the Equator-to-pole thermal gradient [158]. '*The Pliocene world ... was not only warmer but on most continents also wetter than today, leading to an expansion of tropical savannas and forests in Australia and Africa at the expense of deserts*' [158], they concluded.

Arctic temperatures were estimated from Pliocene tree rings in fossil wood from Ellesmere Island in Arctic Canada [159]. For each tree ring, oxygen isotopes provided mean annual temperatures, carbon isotopes provided growing season temperatures, and hydrogen isotopes provided relative humidity. Growing season temperatures averaged 15.8 ± 5 °C, varying by up to 4 °C from year to year. They were about 12 °C warmer than today's summer temperatures. Mean annual temperatures averaged −1.4 ± 4 °C, varying by up to 2 °C from year to year. They were around 18 °C warmer than today's average. These 'fossil temperatures' are typical of boreal forests growing today at 15–20° of latitude *south* of Ellesmere Island. There would have been almost none of the tundra that today marks the Canadian and Siberian Arctic; '*instead the polar sun rose across [a] ... well-nigh endless green Pliocene forest*' [123].

Coring in Lake El'gygytgyn in northeast Arctic Russia recently showed that Pliocene summer temperatures there were about 8 °C warmer 3.6–3.4 Ma ago than today, at a time when atmospheric $pCO_2$ was about 400 ppm [160]. Arctic summers stayed warm until 2.2 Ma ago, following the onset of Northern Hemisphere glaciation 2.6 Ma ago, confirming that Arctic cooling was not sufficient to support large ice sheets until the Early Pleistocene [160]. Evidently, mid Pliocene temperatures were significantly warmer at high latitudes and in the Arctic than globally – a sign of polar amplification of the global warming signal of the times [161].

In contrast, a permanent ice sheet covered the Antarctic Peninsula at this time [162]. Even so, the fossil seashells of Cockburn Island in the northwestern Weddell Sea east of the Antarctic Peninsula tell us that '*during warm intervals of the Pliocene there was little or no seasonal ice in the northern part of the Weddell Sea*', where the large pectinid scallop *Chlamys* flourished [123]. Close relatives of this species now live further north, on the coast of Patagonia. An increase in biological productivity suggests that there was a reduction in sea ice in the Southern Ocean during the Pliocene, and silicoflagellate fossil assemblages suggest that at times the Pliocene Southern Ocean was about 5 °C warmer than it is today [137].

Conditions began to cool in the late Pliocene as the Antarctic ice sheets expanded towards their present state between about 3–2.5 Ma ago. This is when the West Antarctic ice streams advanced into the Ross Sea. Ice shelves began to develop, along with offshore floating extensions of glacier termini, replacing relatively warm and often land-based ice margins [163]. Even so, the picture was not simple. Ice streams along the Peninsula repeatedly advanced to the shelf break throughout the late Pliocene. '*Enhanced progradation of the continental shelf and slope started all around Antarctica at ca. 3 Ma*' [137], which is when the Antarctic ice sheets became primarily of the cold-mode type, more or less at the same time as the build-up of major ice sheets in the Northern Hemisphere.

Given the warmth of the mid Pliocene event, we should not be surprised at the evidence for sea level having been much higher than it is today. Ken Miller of Rutgers University, whom we met in Chapter 5, examined a global spread of data with his team to suggest that peak sea level between 2.7 and 3.2 Ma ago was 22 m ± 10 m higher that it is today [164]. It may even have reached as high as 36 m above today's level [25], although a recent (2019) report shows that 3.26–3.0 Ma ago, when temperatures were 2–3 °C warmer than pre-industrial, it averaged 16.2 m above today's level (range 5.6–19.2 m), and 4.4–4.0 Ma ago – in the Pliocene Climatic Optimum – when temperatures were 4 °C warmer than pre-industrial, it averaged 23.5 m (range 9–26.7 m) above today's level [165]. These large rises help to explain why the Pliocene coastal cliff line in the eastern United States sits tens of kilometres inland as the Orangeburg Scarp between Florida and North Carolina. Much the same was seen in New Zealand, where Naish and Wilson recorded sea-level oscillations of 10–30 m paced by obliquity [166].

As there was no large Northern Hemisphere ice sheet in the Pliocene, Miller's estimate of a peak Pliocene sea level of +22 m above today's level requires the melting of the West Antarctic Ice Sheet (5 m of sea level) and a significant part of the East Antarctic Ice Sheet (15+ m of sea level). Pliocene palaeoclimate data show that warm and cool periods alternated with one another, and that while sea level was higher than at present in the warm periods, it was close to present sea level in the cool periods. The ~20 m range varied with orbitally induced ice volume changes that took place mainly in Antarctica [164, 167].

Maureen Raymo of Boston University was concerned that these estimates of sea level rise failed to take into account the kinds of isostatic adjustments made as ice was removed from landmasses, so she and her team used a numerical model to investigate such effects [168]. Their results implied that the actual eustatic change in the mid Pliocene warm period may have been significantly less than measurements of fossil sea level features implied. They suggested that '*much needed constraints on Pliocene [sea-level], and therefore ice volume, can only be achieved with a large global matrix of palaeoshoreline data, which does not yet exist, evaluated within the context of model predictions … Indeed, results described here can be used to target diagnostic regions where evidence for palaeoshorelines should be sought*' [168].

Miller agreed that glacial isostatic adjustment (GIA – land rising when ice is removed) complicates estimates of past sea level at each site, which is partly why he cited a potential error range of ±10 m [164]. But GIA should not affect the average sea level calculated for multiple global sites.

Eelco Rohling from the University of Southampton estimated with colleagues that an equilibrium concentration of 387 ppm $CO_2$ in the atmosphere (the level in 2009) would have led to a sea level of 25 ± 5 m above present during interglacials of the Pliocene warm period [169], which is well within the range of Miller's estimates. It should be noted that the global average hides the range of Pliocene sea levels measured at different coastal sites, which is affected first by GIA (the vertical 'rebound' effect), and second by the lateral redistribution of material in the subcrust. For instance, when ice loads an area, pushing down the crust (e.g. over Hudsons Bay), material in the hot, viscous asthenosphere below the crust moves laterally, causing a compensatory bulge up to a few thousand kilometres away (e.g. in the Caribbean). When ice loading goes, that effect does too [168, 170].

Sea level rise implies ice sheet collapse. What direct evidence is there for that on or around Antarctica? The evidence comes from drilling through the 80 m thick McMurdo Ice Shelf and into the underlying sea floor, as described in Chapter 7 [171]. It confirmed that in the mid Pliocene, when global temperatures were 2–3 °C warmer than today and atmospheric $CO_2$ was around 400 ppm, there was no Ross Ice Shelf and most likely no West Antarctic Ice Sheet. Dave Pollard and Rob Deconto simulated the ice sheet cycles recorded in the drill core, and estimated a rise of up to 7 m in sea level, most of it from the West Antarctic Ice Sheet [172]. Their subsequent modelling of the mid Pliocene retreats of the Antarctic ice sheets indicates an even greater loss of ice, with up to about 20 m of sea level rise coming from additional melting of the East Antarctic Ice Sheet during the warmest intervals of the Pliocene [173]. Further drilling, by the International Ocean Drilling Program (IPOD) some 300 km off the coast of Adélie Land in 2009, provides evidence for partial melting of the massive East Antarctic ice sheet during the Pliocene. Carys Cook of Imperial College London, and colleagues, used ratios of neodymium to strontium isotopes to show that rocks from the Wilkes Sub-Basin, a large, low-lying part of East Antarctica (Figure 7.5) had been eroded in the mid Pliocene in the absence of ice [174, 175]. Offshore drilling off Wilkes Land in 2011 provided further evidence for ice retreat, showing that sea surface temperatures rose periodically above 3 °C during the early Pliocene climatic optimum between 5.2 and 3.3 Ma ago in response to eccentricity cycles [176]. Figure 11.12 shows what Antarctica may have looked like in warm Pliocene times [177].

My take on the complementary data emerging from the studies of Miller, Rohling, De Conto, Pollard, and Cook is

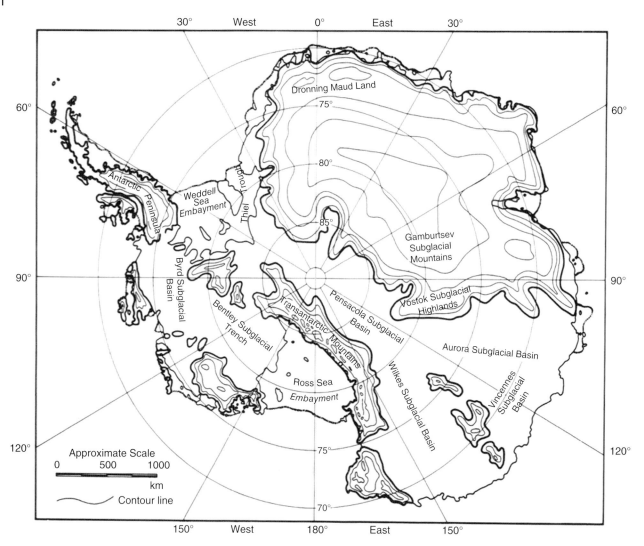

**Figure 11.12** Limited distribution of glacial ice during peak Pliocene warm times. *Source:* From Denton et al. in Tingey 1991 (Ref. [177]).

that we are converging on the idea that during Pliocene interglacial warm periods average global sea level may well have reached on the order of 20 m above today's level for periods of, say, 1–2000 years during each 40 Ka cycle. Raymo may be right to assume that sea levels varied widely from place to place, due to GIAs [168], but it is difficult to argue with the observed convergence. The data imply that neither the West nor the East Antarctic Ice Sheets may be stable under warm conditions like those of today.

Knowing what we now do about the relation of carbon dioxide to temperature, it is hardly surprising that a 2011 compilation of proxy data for $CO_2$ by David Beerling and Dana Royer showed that the Pliocene was characterized by values reaching almost 450 ppm [51]. And Salzmann's team noted that *'Elevated atmospheric $CO_2$-concentrations, ranging from 330–425 ppmv during warm interglacials have been quoted as one of the main reasons for higher global temperatures during the Pliocene'* [158], which nicely connects

the raised $CO_2$ values to Miller's evidence for raised sea-levels during Pliocene interglacials. Earlier, Tripati had recorded values reaching almost 325 ppm from a much more limited data set [130]. Marcus Badger, from the University of Bristol, and colleagues, confirmed in 2013, that Pliocene $CO_2$ values also followed cycles of about 41 Ka, indicative of orbital control [178]. These $CO_2$ variations were small, about $\pm 40$ ppm, suggesting that the climate was quite stable, but with fluctuations in ice volume, temperature and $CO_2$ driven by or linked in some tight way to orbital changes. Fluctuations of $CO_2$ in association with orbitally driven climate change are expected, because warm waters release $CO_2$, whilst cold waters absorb it.

The Pliocene warm period has engaged the attention of the climate modelling community because, as Alan Haywood of the British Antarctic Survey and Mark Williams of the University of Leicester reminded us in 2005, *'The mid-Pliocene is the last time in geological history when our planet's climate was significantly warmer, for a prolonged period, than it is today*' [179]. Not only that, but as Salzmann's team pointed out, the Pliocene could almost form an analogue for today's climate, given that the land geography, ocean bathymetry and marine and terrestrial biology were all quite similar to today's [158]. According to Zalasiewicz and Williams it '*looks more and more like a potential late twenty-first century climate scenario on Earth*' [123], much as Mikhail Budyko had warned. As a result, a concerted effort is being made both to assemble Pliocene temperature data and to model the Pliocene climate. Even so, the Pliocene cannot be a direct analogue for our present and future climate because, first, the rate of rise of temperature today is faster than during the Pliocene, and, second, our trajectory of $CO_2$ increase will take it far above anything measured in Pliocene samples.

Amongst the leading figures in modelling the Pliocene climate is palaeontologist Harry Dowsett of the US Geological Survey (USGS) in Reston, Virginia, who is addressing the question of Pliocene warmth through the USGS Pliocene Research Interpretation and Synoptic Mapping (PRISM) project, which started in the early 1990s [180]. He and his international team fill the PRISM database with data for the warm Pliocene world of 3 Ma ago. They used their data to map Pliocene sea surface temperatures globally, and found a good fit between the data and the output of an Atmosphere-Ocean GMC [180]. The model outputs confirmed that equatorial temperatures were higher than today; polar temperatures were much higher than today; and polar ice sheets were smaller, especially in Greenland. The agreement between the 'real' data estimated from proxies for temperature and the 'simulated' data in outputs from the model showed that the palaeoclimate modellers are getting better at their job. There were some discrepancies, for example in the North Atlantic and Arctic, where the use of modern bathymetry for the Pliocene might not have been appropriate. To test that possibility, the model was re-run with slight deepening of the Greenland–Scotland Ridge, a barrier to the northward movement of warm water from the Atlantic to the Arctic [181]. The modelled Pliocene ocean currents increased the poleward heat transport, so increasing Arctic sea surface temperatures. This provides a possible mechanism for warming the Arctic and for explaining the discrepancy between 'real' and 'simulated' data.

The subject is advancing apace. Back in 2007, Dowsett's PRISM database indicated significant Pliocene warming in the polar regions but not in the tropics [182]. Applying those data in a numerical climate model, Alan Haywood and his colleagues suggested that the warming of the polar regions must reflect an increase in the flow of currents taking heat from the Equator to the poles, and not an increase

in $CO_2$ in the atmosphere, which would have warmed the tropics as well as the poles [183].

Was Dowsett's tropical data accurate? Kira Trillium Lawrence of Lafayette College in the USA, and colleagues, used alkenones from marine phytoplankton to show that tropical temperatures in the mid Pliocene had in fact warmed considerably relative to modern conditions [184]. This finding confirmed that there had been global warming then, most likely related to the increase in $CO_2$. Lawrence and her team found that the pronounced cooling between the mid Pliocene and the Pleistocene preceded the onset of significant Northern Hemisphere glaciation. The onset of the glaciation seems to have been the culmination of more gradual changes in the Earth's climate, possibly related to changes taking place in the Southern Hemisphere. The global cooling accentuated the regional cooling in areas of coastal upwelling, like the Benguela Current off Namibia, where the sea surface temperature dropped 10 °C over this period. In the eastern equatorial Pacific, surface temperatures were as warm as they are today (around 28 °C) in the western Pacific warm pool off New Guinea, suggesting that El Niño-like conditions dominated Pacific equatorial circulation. We will consider the implications of expanded El Niño conditions in a moment.

These data-to-model comparisons for past time slices enable modellers to test the GCMs used to predict how the climate may change in the future, by adding certain likely forcings such as $CO_2$ to the atmosphere. Carrying out these tests on the Pliocene is especially important because its temperatures and $CO_2$ levels are so similar to those forecast for the future Earth. Geological data, then, have become a critical part of the effort to improve the numerical models used to suggest what range of future climate conditions we may face with continued greenhouse gas emissions.

Recent studies of Pliocene temperatures in relation to the content of $CO_2$ in the atmosphere have raised questions about the climate sensitivity – the response of mean global temperature to a doubling of $CO_2$. As we saw earlier in this chapter, the sensitivity proposed by the IPCC was a 1.5–4.5 °C warming for a doubling of $CO_2$ [16]. But as geochemist Mark Pagani and colleagues pointed out in *Nature* in 2010 '*this value incorporates only relatively rapid feedbacks such as changes in atmospheric water vapour concentrations, and the distributions of sea ice, clouds and aerosols ... [it excludes] the effects of long-term feedbacks such as changes in continental ice-sheet extent, terrestrial ecosystems and the production of greenhouse gases other than $CO_2$*' [185]. Examining the data from the Pliocene, they found that '*only a relatively small rise in atmospheric $CO_2$ levels was associated with substantial global warming*' [185]. In other words, the climate system appears to have been more sensitive than the IPCC supposed to an increase in atmospheric $CO_2$.

Pagani and his team concluded that the climate sensitivity of Earth's ice-ocean–atmosphere system was '*significantly higher over the past five million years than estimated from fast feedbacks alone*' [185]. Their concern was echoed by a team led by Daniel Lunt, of the University of Bristol, who used a coupled Atmosphere–Ocean GMC to simulate the climate of the mid Pliocene warm period and compared the outputs with proxy records of mid Pliocene sea surface temperature [186]. They estimated '*that the response of the Earth system to elevated atmospheric carbon dioxide concentrations is 30–50% greater than the response based on those fast-adjusting components of the climate system that are used traditionally to estimate climate sensitivity*' [186]. They concluded, '*targets for the long-term stabilization of atmospheric greenhouse-gas concentrations aimed at preventing a dangerous human interference with the climate system should take into account this higher sensitivity of the Earth system*' [186].

Warming is not the only factor of interest. More evaporation in a warmer world will both change the hydrological cycle, and provide a positive feedback to warming, as water vapour is a greenhouse gas. Not surprisingly, then, climate models of global warming predict an intensified hydrological cycle and, on a global scale, enhanced precipitation [25]. Increased atmospheric water vapour will enhanced the transfer of latent heat from the tropics to the poles, thus sustaining polar warmth, melting polar ice (especially sea ice) decreasing albedo and reinforcing greenhouse conditions [25].

Both global warming and the hydrological cycle will have been affected by the presence or absence of El Niño-like conditions in the Pacific. During the Pliocene the Pacific Ocean was characterized by a mean state resembling El-Niño-like conditions, which contributed to overall warming. Palaeoclimate proxies like Mg/Ca ratios and the alkenone $U^{K}_{37}$ index tell us that there was only a very slight or negligible temperature gradient between the eastern and western equatorial Pacific from 5 to 2 Ma ago, with surface waters averaging about 28 °C, after which the western Pacific became warmer (averaging 30 °C) and the eastern Pacific became cooler (averaging 23 °C) [25]. The emission of heat during El Niño events raises global temperatures above the mean temperature today, and is likely to have done so in the past. El Niño events bring rains to the western coasts of North, Central and South America and aridity to Australia, southern Africa and northeastern Brazil. Abundant warm surface waters also tend to keep $CO_2$ in the air rather than dissolved in the ocean.

Weak Trade Winds are a key feature of El Niño events, so, as pointed out by Ana Ravelo of the University of California, Santa Cruz, the low (1.5 °C) temperature difference from east to west across the Pacific, compared with

the 5 °C difference there today, implies that the Trade Winds were weaker in the mid Pliocene and therefore that upwelling may have been suppressed along the Peruvian and Californian margins [187]. The palaeoclimate records are not of high enough resolution to distinguish individual El Niño events. All they can tell us is that the mean state of the Pacific at this time resembled warm El Niño conditions. Ravelo and colleagues wrote: '*The oceanic processes that cause change in the long-term mean surface temperature pattern through the Pliocene are thought to be different from the rapid air-sea processes that generate interannual variability and El Niño events in today's climate*' [187]. Whatever the conditions were that kept the Pliocene warm, they ended by 3 Ma ago and were replaced by conditions like those of the present by 2 Ma ago.

Alexey Fedorov, from Yale, and colleagues thought that expanded hurricane activity might have played a role in sustaining the 'permanent' El Niño state typical of the warm Pliocene, which was about 2–3 °C warmer than today [188]. Fedorov and his team pointed out that tropical cyclones (hurricanes) increase vertical mixing by deepening the ocean mixed layer to depths of 120–200 m in their wake, in effect pumping heat from the surface down into the interior. Applying a GCM with the boundary conditions for the Pliocene, they found that tropical cyclones would have been much more common and longer lasting globally then than they are today. They felt confident that their model was correct, since with modern boundary conditions it correctly simulated the present distribution of tropical cyclones. They concluded that the expanded warm pool enhanced hurricane activity in the subtropical Pacific, which led to stronger vertical mixing, which warmed deep parcels of water moving east in the equatorial undercurrent, which then warmed the eastern tropical Pacific as well as deepening the tropical thermocline. This positive feedback sustained permanent El Niño conditions and strong hurricane activity across the equatorial Pacific [188]. The implication of Fedorov's findings is that a warmer ocean will generate more tropical cyclones, increase the frequency of El Niño events and warm the ocean, much as happened in the Pliocene tropics and subtropics [189]. And warm oceans emit $CO_2$.

The transition from warm Pliocene to cool Ice Age conditions began at least 4 Ma ago, with marked cooling from 3.5 Ma ago [187]. But the timing of the establishment of Ice Age conditions was different in high compared with low latitudes. Mean ice volume has been greater than today for the past 2.5 Ma, but coastal upwelling systems have been cooler than today for only the past 1.6 Ma, implying that the cause of cooling at low latitude must be somewhat independent of the cause of changes in the size of ice sheets at high latitudes [187]. Ravelo explained the difference by calling on changes in the conditions of subsurface water. For example, where the thermocline is deep today, as in the eastern equatorial Indian Ocean, upwelled water is warm, whilst where the thermocline is shallow, as off Namibia or California, upwelled water is cold. Similarly, today the thermocline in the eastern equatorial Pacific may be either deep, during warm El Niño events, or shallow, during cold La Niña events. Changes in the mean state of the Pacific in the Pliocene could thus reflect a shift from El Niño-dominated (warm) to La Niña-dominated (cold) conditions. Thus the cooling of upwelling regions with time during the Pliocene could represent shoaling of the thermocline rather than an increase in wind strength; indeed, the palaeoenvironmental evidence favours that explanation [187]. Possibly the same applied to the Late Miocene.

Pliocene warming is a front line research topic of some urgency, and new results emerge every year. In 2012, Dan Lunt and his colleagues refined their picture [190]. Evidence had accumulated for large fluctuations in ice cover on Greenland and West Antarctica, which were likely free of ice during the warmest periods, as were parts of East Antarctic over the Aurora and Wilkes sub-glacial basins. In addition the Arctic Ocean may have been seasonally free of sea ice. Melting ice and snow would have exposed land and ocean; thus less solar energy would have been reflected, heating the Earth. Melting sea ice would have exposed more ocean as a source of $CO_2$ for the atmosphere. Rising sea levels flooded the continental margins, providing an even greater oceanic surface area for exchange of $CO_2$ with the atmosphere. These changes help to explain why atmospheric concentrations of $CO_2$ rose possibly up to 450 ppm. The warming ocean would have encouraged evaporation, increasing water vapour – another greenhouse gas – in the air. At this time East African highlands were higher by about 500 m, whilst the mountains along the western side of the Americas were lower; the effect was to decrease the amount of cool land. Lowering the mountains also changed the location of the Northern Hemisphere's jet stream.

Is there a smoking gun – some single change that accounts for the Pliocene warming? Lunt and his colleagues calculated that 48% of the global average Pliocene temperature rise of 3.3 °C was due to $CO_2$, 21% was caused by the changes in mountain height, 10% was from the loss of sea and land ice lowering the albedo, and 21% was due to vegetation changes – increased 'greening' of the Arctic (meaning an increase in Arctic vegetation as snow decreased) further lowering the albedo there [190]. Warming was highest in the Polar Regions – the phenomenon known as polar amplification – due to the loss of ice, the gain of vegetation and the consequent decrease in albedo. The combined increases in $CO_2$ and water vapour

contributed 61% of the total surface temperature change. Partitioning the responsibility for aspects of the warming still does not provide us with its cause, but we do need to explain the rise in $CO_2$.

Evidently, the Pliocene warming was not like that at the PETM. We do not need to call on methane clathrates or volcanic activity to explain it. Perhaps the answer ultimately comes down to changes in global tectonics. The gradual emergence of the isthmus of Panama closed the Central American Seaway, severely limiting the equatorial exchange of seawater between the Atlantic and the Pacific Oceans at around 4.7 Ma ago, and cutting it off completely by around 2.5 Ma ago [191]. Restricting the east to west movement of equatorial surface waters across the Atlantic forced them to move north and south, taking more heat towards the poles. This gradual process may have primed the Earth system to warm. An increase in the amount of solar radiation due to subtle changes in orbital eccentricity and axial tilt may have further primed the pump. Gradual warming would have stimulated a rise in $CO_2$ and water vapour that fed back into further temperature rises [190].

Michael Sarnthein from the University of Kiel, favoured the plate tectonic solution [192]. He and his colleagues noted that the severe deterioration of climate towards the close of the Pliocene occurred in three steps between 3.2 and 2.7 Ma (Figure 11.13). Data from deep ocean drill cores suggested clear linkages between the onset of Northern Hemisphere glaciation and three steps in the final closure of the Central American Seaway, which they deduced from rising salinity differences between the Caribbean Sea and the East Pacific. They concluded that each closing event strengthened the poleward transport of salt and heat in the Atlantic, which warmed the temperate regions. At the same time there was an increase in the northward transport of moisture in the air, which increased precipitation and run-off in northern Eurasia. That lowered salinity in the Arctic, which increased sea ice and so increased albedo, cooling the Arctic. The combination of more salty water in the Norwegian Greenland Sea and Arctic cooling enhanced the sinking of North Atlantic Deep Water and thus strengthened the Atlantic Meridional Overturning Circulation, which in turn drew more warm salty water north via the Gulf Stream to sustain warming around the North Atlantic.

Their new evidence also showed that closing the Central American Seaway increased sea level in the North Pacific, which doubled the through-flow of cool water from the Bering Strait through the Arctic to the East Greenland Current (EGC), ultimately cooling the Labrador Sea by 6 °C. Once the EGC had been established, it '*led to robust thermal isolation of East Greenland from the poleward heat transport further east through the North Atlantic and*

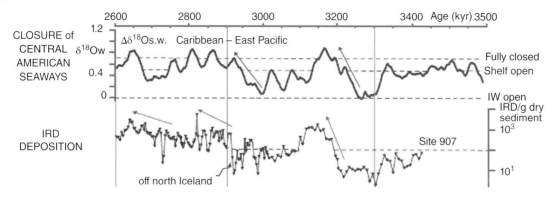

**Figure 11.13** Closing the Central American Seaway linked to episodes of ice rafting in the North Atlantic. The closing of the seaway is estimated from the difference in salinity between Caribbean ODP site 999 and East Pacific ODP site 1241, deduced from variations in $\partial^{18}O$ and Mg/Ca-based sea surface temperatures in *Globigerinoides sacculifer*, a species growing at 50–100 m water depth. A $\partial^{18}O$ gradient of 0.6–0.7 equates to a sea surface salinity gradient of 1.2–1.8 salinity units, which corresponds to full closure of the seaway. Salinity increases in the Caribbean and decreases in the East Pacific with closure. IRD = ice rafted debris abundance at Site 907 north of Iceland. IW = Intermediate Water connection. Arrows indicate closures centred on (i) 3.2 Ma; (ii) 2.9 Ma, and (iii) 2.65 Ma ago, after which connection was essentially closed. Sea Surface Temperatures in the Irminger Current and North Atlantic Current increased with closures, by 2–3 °C, while deep-water temperatures decreased by 1.5–2.0 °C. Warming mainly affected temperature at temperate sites (west of Ireland) and subpolar sites (south of Iceland). At same time flow increased from the Pacific to the Arctic and thermally isolated Greenland, leading to the onset of major Northern Hemisphere glaciation near 2.8 Ma ago. *Source:* From figure 14 in Ref. [192].

*Norwegian Currents. Since this time, the East Greenland Current] formed a barrier important to promote the growth of continental ice*' [192]. As a result, whilst most of the Northern Hemisphere warmed after the closure of the Seaway, Greenland experienced cooling and accelerated snowfall. Formation of the Greenland Ice Sheet enhanced the polar high-pressure cell, which accelerated the northerly winds along its eastern margin, further strengthening the current. Once the ice sheet was established, about 3.18–3.12 Ma ago, and especially after 2.9 Ma ago, it formed a nucleus for the development of Quaternary glaciations in the Northern Hemisphere [192]. The increased run-off to the Arctic and the resulting increase in sea ice further contributed to the eventual development of Northern Hemisphere ice sheets from about 2.8 Ma ago. While Sarnthein may be right about intensification of glaciation on Greenland, actual development of glacial conditions there may be as old as 18 Ma (Chapter 7).

Sarnthein's argument does not require a significant rise in $CO_2$ to explain the warming of the mid Pliocene. Instead, it regards the warming of the North Atlantic region as sufficient to influence the global average temperature. The warming caused $CO_2$ to be released from surface waters, providing positive feedback and further warming. In due course, the increased heat transport to the north was overcompensated for by the flow of low salinity cold water through the Arctic from the North Pacific, inducing dramatic cooling and freshening of the EGC and the onset of major Northern Hemisphere glaciation [192].

So, Sarnthein's 2009 model explains both the warming of the Pliocene and the ensuing cooling that led to the initiation

of the Pleistocene Ice Age, as spelled out subsequently in his comprehensive 2013 review of these matters [193]. Cronin agreed that closing the isthmus of Panama intensified the global thermohaline circulation system and strengthened production of North Atlantic Deep Water, pre-conditioning the high latitudes of the North Atlantic for a later build-up of large ice sheets [194]. Closure induced greater transport of moisture from the equator to the pole in the Atlantic, thus enhancing the run-off of rivers into the Arctic and thereby encouraging the formation of sea-ice. The Trade Winds intensified, strengthening upwelling along the African margins and thus enhancing productivity and further draw-down of $CO_2$, driving further cooling. Nonetheless, as Cronin reminded us, we still have no specific causal connection between the closure of the isthmus and the development of the Northern Hemisphere ice sheets. Central America was not the only seaway to close. We also have the closing of the connection between the Mediterranean and the Atlantic, when Spain met Africa around 5–6 Ma ago and shut off supplies of warm water from the Med.

Back in 2007, Ana Ravelo and her colleagues had disagreed that closure of the Central American Seaway triggered the development of the Ice Age 2.6 Ma ago [187]. But she had accepted that the closure did occur just before a steep decline in deep ocean temperature, which suggested a connection between tectonics and climate. There was no significant dispute about the timing of glaciation. Shackleton was aware back in the mid-1980s that North Atlantic bottom water like that of today first formed at around 3.5 Ma ago, and that major ice-rafting developed over the Rockall Plateau west of the UK around 2.5 Ma ago [195, 196]. And he and Kennett had already deduced from Southern Ocean studies in 1975 that the onset of the Northern Hemisphere glaciation must have taken place 2.6 Ma ago [197].

Montañez and colleagues thought that it was the decline in $CO_2$ from around 450 ppm to the 200 or so ppm typical of the Pleistocene glacials that '*most probably accounted for the initiation and growth of Northern Hemisphere ice sheets at around 3 Ma*' [25]. Whilst that agrees with the analysis of Barry Saltzman, as we shall see in Chapter 13, it begs the question: what was driving the $CO_2$, if not ocean temperatures? Ravelo and colleagues argued that because mid Pliocene $CO_2$ levels were only slightly higher than they are today, whilst temperatures were 2–3 °C warmer, global cooling from the Pliocene into the Ice Age can only be explained by increasing albedo from the formation of polar ice and/or decreases in water vapour [187]. The change from 'permanent' El Niño to more modern conditions might explain about 1 °C of cooling, whilst the shallowing of the thermocline and cooling of the sea surface temperatures in upwelling currents along continental margins might explain another 0.6 °C of cooling, leaving unidentified processes or factors to explain the remaining ~2 °C of cooling [187]. Ravelo and her team considered as one possibility the closing of the Central American Seaway. While it seemed unlikely to them that the resulting northward transport of heat would have contributed to generating an Ice Age, they thought it possible that the closure would have stimulated shallowing of the thermocline, making upwelling currents much colder, decreasing water vapour and so cooling the climate. Demise of 'permanent' El Niño conditions in the Pacific would have cooled North America, providing conditions under which ice sheets could grow there. This in turn would have increased albedo, enhanced the Equator-to-pole thermal gradient, strengthened winds, and so further enhanced upwelling and thence further cooling in a positive feedback loop. Increased circulation and ocean stratification would have contributed to more $CO_2$ being sequestered in the ocean, adding to cooling.

Matteo Willeit of the Potsdam Institute for Climate Impact Research and colleagues agreed that declining $CO_2$ helped to trigger the formation of the Northern Hemisphere ice sheets. They calculated that between 3.2–2.4 Ma ago $CO_2$ declined from 375 to 425 ppm to 275–300 ppm [198]. This long term decline was accompanied by a relatively abrupt intensification of the Northern Hemisphere glaciation around 2.7 Ma ago, as a result of the '*threshold behaviour of the ice sheets response to gradual $CO_2$ decrease and orbital forcing*' [198]. Before 2.7 Ma ago, Greenland was ice-free during summer insolation maxima and only partly covered by ice during periods of minimal summer insolation [198]. Willeit found that $CO_2$ also varied coherently with the 41 Ka obliquity cycle.

Further evidence that plate tectonics, rather than $CO_2$, may have played a role in initiating this final cooling comes from Cyrus Karas of the Leibnitz Institute of Marine Sciences. Karas and his colleagues suggested in 2009 that the continued northward motion of New Guinea restricted flow through the Indonesian 'gateway' between the Pacific and Indian Oceans. That cut off the supply of warm water from the central and South Pacific, replacing it with cool water from the North Pacific between 4 and 3 Ma ago [199]. Karas et al. used '$\partial^{18}O$ *and Mg/Ca ratios of planktonic foraminifera to reconstruct the thermal structure of the eastern tropical Indian Ocean*' to show that '*subsurface waters freshened and cooled by about 4°C between 3.5 and 2.95 Myr ago*' [199]. That would have caused the thermocline in the Indian Ocean to shallow and cool, eventually cooling the Benguela Current on the warm water route from the Indian Ocean to the North Atlantic, and possibly also leading to the development of the eastern equatorial Pacific cold tongue [199]. These changes would have stimulated or acted as further feedbacks to global cooling by reducing poleward heat transport and so pre-conditioning the growth of ice sheets [200]. Considering all the evidence, Michel Crucifix concluded that '*the transition into the glacial–interglacial regime is best seen as a progressive phenomenon which emerged from complex dynamics involving low and high latitudes*' [55].

These arguments beg the question, what made $CO_2$ decline? In 2016, David Lang from the National Oceanography Centre, Southampton, and colleagues showed that the Northern Hemisphere was preconditioned for significant continental glaciation between 3.3 and 2.6 Ma ago by cooling of the Arctic and enhanced storage of $CO_2$ in abyssal waters [201]. These things came about as a consequence of the long-term expansion of the Antarctic Ice Sheet and coastal sea ice, the development of stratification of the ocean at high latitudes in both hemispheres from 2.7 Ma ago, and increased transport of heat from the Northern Hemisphere to the deep Pacific Ocean during warm stages. Once those preconditions were met, incursions of southern component waters into the deep North Atlantic during cold stages helped to amplify glacial cycles through increasing sequestration of $CO_2$ in the deep Atlantic, leading to tighter coupling of inter-hemispheric climate on orbital time scales from 2.72 Ma onwards [201].

On a final note, regarding Hutton's dictum on the use of the past to suggest what might happen in the future, sea level expert Ken Miller and his team were quick to point out that when our future climate reaches Pliocene $CO_2$ levels (we are almost there), and the climate stays warm long enough for equilibrium conditions to be reached, we could in the long term be facing a sea level rise of a similar magnitude. Even if global sea level only rose by just 12 m to the low end of Miller's estimated rise, it would still be extremely uncomfortable for coastal communities. The latest data from Grant and colleagues in 2019 confirms that sea level varied by $13 \pm 5$ m over glacial–interglacial cycles during the mid to late Pliocene (3.3–2.5 Ma ago) in phase with orbital precession [202].

While these Pliocene changes are broadly consistent with computer climate model outputs for that period [203], they are much larger than we have experienced since 1900 despite the similarity in $CO_2$ levels. This is because the high concentrations of $CO_2$ in the Pliocene were maintained for millennia, allowing atmosphere, ice sheets and oceans to come to equilibrium. Our present climate is not in equilibrium with the current rapid rise in $CO_2$. Hence, assuming that $CO_2$ levels like those of today persist, it may still take a long time – perhaps between 500 and 2500 years – for the equilibrium level of sea level to be reached [141].

Looking at all the evidence displayed above it seems to me that in the Pliocene, as in the Miocene, there is ample room for slowly changing plate tectonic processes to have set the scene for climate change. I find the arguments convincing for the closure of the Panama seaway having first warmed the polar regions by the north–south deflection of formerly east–west tropical currents. The resulting warming of the global water surface would have decreased the ocean's ability to take up $CO_2$, leaving more of it in the atmosphere, reinforcing warming via positive feedback. Eventually, the change in sea level of the North Pacific would have cooled the EGC, isolating Greenland. Whilst warm water was pushing north in the North Atlantic Current, it was being countered by cold water moving south in the EGC. Eventually, with the cooling of the Labrador Sea, a source of cool deep water, conditions would have approached those that we have today, with the sinking of North Atlantic Deep Water fed from the Labrador and Norwegian–Greenland Seas, further cooling the north and leading to the development of the other Northern Hemisphere ice sheets. The process would have been accentuated by the cooling in the south driven by the shrinkage of the Indonesian through-flow from the Pacific into the Indian Ocean.

Clearly, there is ample evidence for the primacy of plate tectonic process in being the ultimate drivers of climate change through modulations in the supply of $CO_2$ via volcanoes, through the uplift of mountains stimulating increases in physical and chemical weathering, and through the opening or closing of ocean gateways. Rises in $CO_2$ may cause temperature change, as in the Eocene, or be caused by temperature change, as in the Pliocene. In both cases the $CO_2$ and temperature tend to move in lock step through positive feedback. $CO_2$ may be either chicken or egg, depending on the plate tectonic circumstances. That does not mean that the Pliocene is useless as a case history in comparison with today's climate change. Indeed, the Pliocene provides a concrete example of what happens when $CO_2$ rises and reinforces an ongoing rise in temperature that was triggered by some other mechanism (like the closing of the Panama seaway).

Given this background, we now have an opportunity to reflect on the prevalence of ocean acidification (Box 9.1) in past times. There is no simple way to measure past ocean pH, but we can infer past episodes of ocean acidification from new geochemical tools, such as (i) the boron isotopic composition ($\partial^{11}B$) of marine carbonates, which represents pH; (ii) the trace element (such as B, U, and Zn)-to-calcium ratio of foraminiferal shells, which represents the ambient concentration of carbonate ions ($CO_3^{2-}$), a key element in ocean acidification (see Figure 9.2); and (iii) the stable carbon isotope ($\partial^{13}C$) composition of organic molecules (alkenones), which represents the amount of aqueous $CO_2$ in the surface ocean [106]. As we saw in Chapter 9, ocean acidification can also be inferred from changes in the CCD, but care is needed because ultimately the accumulation of sediment rich in $CaCO_3$ may represent the degree of saturation of the ocean with respect to $CaCO_3$, rather than the pH per se. Honisch's team found that the covariation of ocean acidification, warming and oxygen depletion precludes attributing extinctions to a single cause. But she also discovered that the extent of ocean acidification is rate-dependent, which is not readily apparent from Figure 9.2. Slower rates of $CO_2$ supply may lead to a smaller degree of saturation of seawater in $CaCO_3$, to the extent that pH and carbonate saturation may be almost decoupled for very slowly increasing atmospheric $CO_2$. Conversely, during rapid changes in atmospheric $CO_2$, both surface ocean pH and carbonate saturation decline in tandem, with the rate of decrease in saturation being an order of magnitude larger for a rapid $CO_2$ increase than a slow one over 100 Ka. In effect the capacity of the ocean's buffering system keeps pH from becoming more acidic when the rate of $CO_2$ supply is low, but not when the rate of supply is rapid. Hence, we can expect ocean acidification events (where rapid changes in $CO_2$ rapidly lower pH) at time intervals of c. 10 Ka or less, as at the present [106]. Investigating such past events, Honisch found that ocean acidification was most like the present (state P-0) at the PETM and the Triassic–Jurassic boundary; was partly similar to the present at the last deglaciation (state P-1); was unlike the present at the end of the Cretaceous (state P-3); and was at some intermediate state (P-2) at the Permian–Triassic boundary and during the oceanic anoxic events in the Toarcian (Jurassic), and Cretaceous (OAE-1a and OAE-2). She found that pH levels in Jurassic and Cretaceous time may have hovered around 7.6, and probably also declined 0.25–0.45 units during the the PETM, when the CCD rose by c. 2 km to less than 1.5 km as against its present depth of close to 4 km [106]. It appears, counterintuitively, that the $CaCO_3$ saturation state of the ocean may be regulated primarily by the extent of weathering on long time scales, not by the immediate partial pressure of $CO_2$ in the atmosphere. Even so, during periods of long slow change in $CO_2$, there may have been pulses lasting 20 Ka or less during which the rate of supply was faster, leading to acidification events. '*Because of the decoupling between pH and saturation on long time scales ... extended intervals of elevated [atmospheric] $CO_2$*

*such as the middle Miocene, Oligocene, and Cretaceous can be firmly ruled out as future-relevant analogs [i.e. rapid acidification events like the present]*' [106]. For the early Jurassic and beyond we are handicapped by the absence of deep-sea records, most Jurassic and older deep ocean crust having been subducted.

In this chapter we have examined a few aspects of climate change in broad-brush terms, and the history of the Cenozoic in some detail, along with three case histories, from the PETM, the Mid Miocene and the Pliocene, to illustrate the role of $CO_2$ in changing the climate. Now it is time to turn our attention to the Pleistocene Ice Age and its aftermath, the post-glacial Holocene and the Anthropocene, the period in which we now live.

## References

**1** Barron, E.J. and Washington, W.M. (1985). Warm Cretaceous climates: high atmospheric $CO_2$ as a plausible mechanism. In: *Carbon Cycle and Atmospheric $CO_2$: Natural Variations Archean to Present, Geophysical Monograph Series* **32** (eds. E.T. Sundquist and W.S. Broecker), 546–553. Washington, D.C.: American Geophysical Union.

**2** Schneider, S.H., Thompson, S.L., and Barron, E.J. (1985). Mid-Cretaceous continental surface temperatures: are high $CO_2$ concentrations needed to simulate above-freezing winter conditions. In: *Carbon Cycle and Atmospheric $CO_2$: Natural Variations Archean to Present, Geophysical Monograph Series* **32** (eds. E.T. Sundquist and W.S. Broecker), 554–559. Washington, D.C.: American Geophysical Union.

**3** Barron, E.J., Fawcett, P.J., Peterson, W.H. et al. (1995). A "simulation" of mid-Cretaceous climate. *Paleoceanography* 10 (2): 953–962.

**4** Poulson, C.J., Barron, E.J., Peterson, W.H., and Wilson, P.A. (1999). A reinterpretation of mid-Cretaceous shallow marine temperatures through model-data comparison. *Paleocanography* 14 (6): 679–697.

**5** Hay, W.W., DeConto, R.M., and De Boer, P.L. (2018). Possible solutions to several enigmas of Cretaceous climate. *International Journal of Earth Sciences* https://doi.org/10.1007/s00531-018-1670-2.

**6** Spicer, R.A. and Parrish, J.T. (1990). Late Cretaceous-Early Tertiary palaeoclimates of northern high latitudes: a quantitative view. *Journal of the Geological Society of London* 147: 329–341.

**7** Moore, G.T., Hayashida, D.N., Ross, C.A., and Jacobson, S.R. (1992). Paleoclimate of the Kimmeridgian/Tithonian (Late Jurassic) world: I. results using a general circulation model. *Palaeogeography Palaeoclimatology Palaeoecology* 93: 113–150.

**8** Moore, G.T., Sloan, L.C., Hayashida, D.N., and Umrigar, N.P. (1992). Paleoclimate of the Kimmeridgian/Tithonian (Late Jurassic) world: II. Sensitivity tests comparing three different paleotopographic settings. *Palaeogeography Palaeoclimatology Palaeoecology* 95: 229–252.

**9** Valdes, P.J. and Sellwood, B.W. (1992). A palaeoclimate model for the Kimmeridgian. *Palaeogeography Palaeoclimatology Palaeoecology* 95: 47–72.

**10** Sellwood, B.W. and Valdes, P.J. (2007). Mesozoic climates. In: *Deep-Time Perspectives on Climate Change: Marrying the Signal from Computer Models and Biological Proxies*, Micropalaeontological *Society. Specoa; Publocatopm* (eds. M. Williams, A.M. Haywood, F.J. Gregory and D.N. Schmidt), 201–224. London: The Geological Society.

**11** Hasegawa, H., Tada, R., Jiang, X. et al. (2012). Drastic shrinking of the Hadley circulation during the mid-Cretaceous Supergreenhouse. *Climate of the Past* 8: 1323–1337.

**12** Hay, W.W. (2013). *Experimenting on a Small Planet: A Scholarly Entertainment*, 983 pp. New York: Springer.

**13** Ramaswamy, V., Boucher, O., Haigh, J., et al. (2001) Radiative forcing of climate change, *IPCC Third Assessment Report*, Chapter 6, 349–416.

**14** Myhre, G., Highwood, E.J., Shine, K.P., and Stordal, F. (1998). New estimates of radiative forcing due to well mixed greenhouse gases. *Geophysical Research Letters* 25 (14): 2715–2718.

**15** PALEOSENS Project Members (2012). Making sense of palaeoclimate sensitivity. *Nature* 491: 683–691.

**16** IPCC (2013) *Climate Change 2013: The Physical Science Basis, Contribution of Working Group I to the Fifth Assessment Report of the Intergovernmental Panel on Climate Change* [Stocker, T.F., Qin, D., Plattner, G.-K., Tignor, M., Allen, S.K., Boschung, J., Nauels, A., Xia, Y.,

Bex, V., and Midgely, P.M., (eds.)]. Cambridge University Press, Cambridge, United Kingdom and New York, NY, USA, 1535pp.

**17** Schmittner, A., Urban, N.M., Shakun, J.D. et al. (2011). Climate sensitivity estimated from temperature reconstructions of the Last Glacial Maximum. *Science* 334: 1385–1388.

**18** Hansen, J.E. and Sato, M. (2012). Paleoclimate implications for human-made climate change. In: *Climate Change: Inferences from Paleoclimate and Regional Aspects* (eds. A. Berger, F. Mesinger and D. Šijački), 21–48. Springer https://doi.org/10.1007/978-3-7091-0973-1_2.

**19** Hansen, J., Sato, M., Russell, G., and Kharecha, P. (2013). Climate sensitivity, sea level and atmospheric carbon dioxide, In: *Warm Climates of the Past – A Lesson for the Future* (ed. D.J. Lunt, H. Elderfield, R. Pancost and A. Ridgwell). *Philosophical Transactions of the Royal Society A: Mathematical, Physical and Engineering Sciences* 371: 2001, 20120294.

**20** Ring, M.R., Lindner, D., Cross, E.F., and Schlesinger, M.E. (2012). Causes of the global warming observed since the 19th century. *Atmospheric and Climate Sciences* 2: 401–415.

**21** Royer, D.L., Pagani, M., and Beerling, D.J. (2012). Geobiological constraints on Earth System Sensitivity to $CO_2$ during the Cretaceous and Cenozoic. *Geobiology* 10 (4) https://doi.org/10.1111/j.1472-4669.2012.00320.x.

**22** Cox, P.M., Huntingford, C., and Williamson, M.S. (2018). Emergent constraint on equilibrium climate sensitivity from global temperature variability. *Nature* 553: 319–322.

**23** Proistosescu, C. and Huybers, P.J. (2017). Slow climate mode reconciles historical and model-based estimates of climate sensitivity. *Science Advances* 3: e1602821.

**24** Farnsworth, A., Lunt, D., O'Brien, C.L. et al. (2019). Climate sensitivity on geological timescales controlled by non-linear feedbacks and ocean circulation. *Geophysical Research Letters* https://doi.org/10.1029/2019GL083574.

**25** Montañez, I.P., Norris, R.D., Algeo, T. et al. (2011). *Understanding Earth's Deep Past: Lessons for Our Climate Future*, 194 pp. Washington, D.C.: National Academies. Press.

**26** Ramanathan, V. (1998). Trace-gas greenhouse effect and global warming: underlying principles and outstanding issues: Volvo Environmental Prize Lecture 1997. *Ambio* 27 (3): 187–197.

**27** Beerling, D.J., Fox, A., Stevenson, D.S., and Valdes, P.J. (2011). Enhanced chemistry-climate feedbacks in past greenhouse worlds. *Proceedings of the National Academy of Sciences of the United States of America* 108 (24): 9770–9775.

**28** McElwain, J.C., Beerling, D.J., and Woodward, F.I. (1999). Fossil plants and global warming at the Triassic-Jurassic

boundary. *Science* 285 (5432): 1386–1390. https://doi.org/10.1126/science.285.5432.1386.

**29** Beerling, D. (2007). *The Emerald Planet: How Plants Changed Earth History*. Oxford University Press.

**30** Beerling, D. (2019). *Making Eden*. Oxford University Press.

**31** Williams, M., Haywood, A.M., Gregory, F.J., and Schmidt, D.N. (eds.) (2007). *Deep-Time Perspectives on Climate Change: Marrying the Signal from Computer Models and Biological Proxies*, 589 pp, Micropal. Soc. Spec. Publ. London: The Geological Society.

**32** Kent, D.V. and Muttoni, G. (2008). Equatorial convergence of India and Early Cenozoic climate trends. *Proceedings of the National Academy of Sciences* 105 (42): 16065–16070.

**33** Rowley, D.B. (2002). Rate of plate creation and destruction: 180 Ma to present. *Geological Society of America Bulletin* 114: 927–933.

**34** Pälike, H. et al. (2012). A Cenozoic record of the equatorial Pacific carbonate compensation depth. *Nature* 488: 609–615.

**35** Froehlich, P. and Misra, S. (2014). Was the Late Paleocene-Early Eocene hot because Earth was flat? *Oceanography* 27 (1): 36–49.

**36** Kent, D.V, and Muttoni, G. (2013). Modulation of Late Cretaceous and Cenozoic climate by variable drawdown of atmospheric $pCO_2$ from weathering of basaltic provinces on continents drifting through the equatorial humid belt. *Climate of the Past* 9: 525–546.

**37** Rowley, D.B. (2008). Extrapolating oceanic age distributions: lessons from the Pacific region. *Journal of Geology* 116: 587–598.

**38** Müller, R.D., Sdrolias, M., Gaina, C. et al. (2008). Long-term sea-level fluctuations driven by ocean basin dynamics. *Science* 319: 1357–1362.

**39** Edmond, J.M. and Huh, Y. (2003). Non-steady state carbonate recycling and implications for the evolution of atmospheric $pCO_2$. *Earth and Planetary Science Letters* 216: 125–139.

**40** John, E.H., Pearson, P.N., Coxall, H.K. et al. (2013). Warm ocean processes and carbon cycling in the Eocene, In: *Warm Climates of the Past – A Lesson for the Future* (ed. D.J. Lunt, H. Elderfield, R. Pancost and A. Ridgwell). *Philosophical Transactions of the Royal Society A: Mathematical, Physical and Engineering Sciences* 371 (2001): 20130099.

**41** Olivarez, L.A. and Lyle, M.W. (2006). Missing organic carbon in Eocene marine sediments: is metabolism the biological feedback that maintains end-member climates? *Paleoceanography* 21: PA2007. https://doi.org/10.1029/2005PA001230.

**42** Stanley, S.M. (2010). Relation of Phanerozoic stable isotope excursions to climate, bacterial metabolism, and major extinctions. *Proceedings of the National Academy of Sciences* 107: 19185–19189.

43  Inglis, G.N., Farnsworth, A., Lunt, D. et al. (2015). Descent toward the Icehouse: Eocene sea surface cooling inferred from GDGT distributions. *Paleoceanography* 30: 1000–1020. https://doi.org/10.1002/2014PA002723.

44  Anagnostou, E., John, E.H., and Edgar, K.M. (2016). Changing atmospheric $CO_2$ concentration was the primary driver of early Cenozoic climate. *Nature* 533 (7603): 380–384.

45  Gulick, S.P.S., Shevenell, A.E., Montelli, A. et al. (eds.) (2017). Initiation and long-term instability of the East Antarctic Ice Sheet. *Nature* 552: 225–229.

46  Zachos, J.C., Stott, L.D., and Lohmann, K.C. (1994). Evolution of Early Cenozoic marine temperatures. *Paleoceanography* 9 (2): 353–387.

47  eps.ucsc.edu/faculty/Profiles/singleton.php?&singleton=true&cruz_id=jzachos (December 2013).

48  Arthur, M.A., Hinga, K.R., Pilson, M.E.Q. et al. (1991). Estimates of $pCO_2$ for the last 120 Ma based on the $\partial^{13}C$ of marine phytoplanktonic organic matter. *Eos* 72 (17), Spring Mtg. Suppl., p166.

49  Bice, K.L., Scotese, C.R., Seidov, D., and Barron, E.J. (2000). Quantifying the role of geographic change in Cenozoic ocean heat transport using uncoupled atmosphere and ocean models. *Palaeogeography, Palaeoclimatology, Palaeoecology* 161: 295–310.

50  Zachos, J., Pagani, M., Sloan, L. et al. (2001). Trends, rhythms, and aberrations in global climate 65 Ma to present. *Science* 292: 686–693.

51  Beerling, D.J. and Royer, D.L. (2011). Convergent Cenozoic $CO_2$ history. *Nature Geoscience* 4: 418–420.

52  DeConto, R.M. and Pollard, D. (2003). Rapid Cenozoic glaciation of Antarctica induced by declining atmospheric $CO_2$. *Nature* 421: 245–249.

53  Francis, J.E., Marenssi, S., Levy, R. et al. (2009). From Greenhouse to Icehouse – The Eocene/Oligocene in Antarctica. In: *Antarctic Climate Evolution*, Developments in Earth and Environmental Sciences **8** (eds. F. Florindo and M. Siegert), 309–368. The Netherlands: Elsevier. ISBN: 978-0-444-52847-6.

54  Pollard, D. and DeConto, R.M. (2005). Hysteresis in Cenozoic Antarctic ice-sheet variations. *Global and Planetary Change* 45: 9–21.

55  Crucifix, M. (2019). Pleistocene Glaciations. In: *Climate Changes in the Holocene, Impacts and Human Adaptation* (ed. E. Chiotis), 77–106. CRC Press publishers, Taylor and Francis. ISBN: 97808153938.

56  Coxall, H., Wilson, P.A., Palike, H. et al. (2005). Rapid stepwise onset of Antarctic glaciation and deeper calcite compensation in the Pacific Ocean. *Nature* 433: 53–57.

57  Pagani, M., Zachos, J.C., Freeman, K.H. et al. (2005). Marked decline in atmospheric carbon dioxide concentrations during the Paleogene. *Science* 309: 600–603.

58  Pagani, M. et al. (2011). The role of carbon dioxide during the onset of Antarctic glaciation. *Science* 334: 1261–1264.

59  Lear, C.H., Elderfield, H., and Wilson, P.A. (2000). Cenozoic deep-sea temperatures and global ice volumes from Mg/Ca in benthic foraminiferal calcite. *Science* 287: 269–272.

60  Coxall, H.K. and Wilson, P.A. (2011). Early Oligocene glaciation and productivity in the eastern equatorial Pacific: Insights into global carbon cycling. *Paleoceanography* 26: PA2221. https://doi.org/10.1029/2010PA002021.

61  Armstrong McKay, D.I., Tyrrell, T., and Wilson, P.A. (2016). Global carbon cycle perturbation across the Eocene-Oligocene climate transition. *Paleoceanography* 31: 311–329. https://doi.org/10.1002/2015PA002818.

62  Anderson, J.B. and Wellner, J.S. (2011). *Tectonic, Climate, and Cryospheric Evolution of the Antarctic Peninsula*, Special Publications **63**, 218 pp. American. Geophysical Union.

63  EOS (2013). Tectonic, Climate, and Cryospheric Evolution of the Antarctic Peninsula. *Eos* 94 (23): 210.

64  Katz, M.E., Miller, K.G., Wright, J.D. et al. (2008). Stepwise transition from the Eocene greenhouse to the Oligocene icehouse. *Nature Geoscience* 1: 329–334.

65  Pearson, P.N., Foster, G.L., and Wade, B.S. (2009). Atmospheric carbon dioxide through the Eocene-Oligocene climate transition. *Nature* 461: 1110–1114.

66  Cramer, B.S., Miller, K.G., Barrett, P.J., and Wright, J.D. (2011). Late Cretaceous–Neogene trends in deep ocean temperature and continental ice volume: reconciling records of benthic foraminiferal geochemistry ($\partial^{18}O$ and Mg/Ca) with sea level history. *Journal of Geophysical Research* 116: C12023, 23 pp.

67  Mawbey, E.M. and Lear, C.H. (2013). Carbon cycle feedbacks during the Oligocene–Miocene transient glaciation. *Geology* https://doi.org/10.1130/G34422.1.

68  Naish, T.R., Woolfe, K.J., Barrett, P.J. et al. (2001). Orbitally induced oscillations in the East Antarctic ice sheet at the Oligocene/Miocene boundary. *Nature* 423: 719–723.

69  Barrett, P.J. (2007). Cenozoic climate and sea level history from glacimarine strata off the Victoria Land Coast, Cape Roberts Project, Antarctica, In: *Glacial Processes and Products* (ed. Hambrey M.J., Christoffersen P., Glasser N.F., and Hubbart B.). *International Association of Sedimentologists. Special Publication* 39: 259–287.

70  Pälike, H., Norris, R.D., Herrie, J.O. et al. (2006). The heartbeat of the Oligocene climate system. *Science* 314: 1894–1898.

**71** Galeotti, S., DeConto, R., Naish, T. et al. (2016). Antarctic ice sheet variability across the Eocene-Oligocene boundary climate transition. *Science* https://doi.org/10.1126/science.aab0669.

**72** Foster, G.L., Hull, P., Lunt, D.J., and Zachos, J.C. (2018). Placing our current 'hyperthermal' in the context of rapid climate change in our geological past. *Philosophical Transactions of the Royal Society A: Mathematical, Physical and Engineering Sciences* 376: 20170086.

**73** Lauretano, V., Littler, K., Polling, M. et al. (2015). Frequency, magnitude and character of hyperthermal events at the onset of the Early Eocene Climatic Optimum. *Climate of the Past* 11: 1313–1324.

**74** Sexton, P.F., Norris, R.D., Wilson, P.A. et al. (2011). Eocene global warming events driven by ventilation of oceanic dissolved organic carbon. *Nature* 471: 349–352.

**75** Sluijs, A., Schuten, S., Donders, T.H. et al. (2009). Warm and wet conditions in the Arctic region during Eocene Thermal Maximum 2. *Nature Geoscience* 2: 777–780.

**76** DeConto, R.M., Galeotti, S., and Pagani, M. (2012). Past extreme warming events linked to massive carbon release from thawing permafrost. *Nature* 484: 87–91.

**77** Kirtland Turner, S., Sexton, P.F., Charles, C.D., and Norris, R.D. (2014). Persistence of carbon release events through the peak of early Eocene global warmth. *Nature Geoscience* 12: 1–17. https://doi.org/10.1038/ngeo2240.

**78** Zeebe, R.E., Westerhold, T., Littler, K., and Zachos, J.C. (2017). Orbital forcing of the Paleocene and Eocene carbon cycle. *Paleoceanography* 32 https://doi.org/10.1002/2016PA003054.

**79** Meyers, S. (2017). Cracking the palaeoclimate code. *Nature* 546: 219–220.

**80** Lenz, O.K., Wilde, V., and Riegel, W. (2017). ENSO- and solar-driven sub-Milankovitch cyclicity in the Palaeogene greenhouse world; world; high-resolution pollen records from Eocene Lake Messel, Germany. *Journal of the Geological Society of London* 174 (1): 110–128. https://doi.org/10.1144/jgs2016-046.

**81** Sluijs, A., Bowen, G.J., Brinkhuis, H. et al. (2007). The Palaeocene–Eocene Thermal Maximum super greenhouse: biotic and geochemical signatures, age models and mechanisms of global change. In: *Deep-Time Perspectives on Climate Change: Marrying the Signal from Computer Models and Biological Proxies*, The Micropalaeontological Society, Special Publications (eds. M. Williams, A.M. Haywood, F.J. Gregory and D.N. Schmidt), 323–349. London: The Geological Society.

**82** Jaramillo, C., Ochoa, D., Contreras, L. et al. (2010). Effects of rapid global warming at the Paleocene-Eocene boundary on neotropical vegetation. *Science* 330 (6006): 957–961.

**83** Kennett, J.P. and Stott, L.D. (1991). Abrupt deep-sea warming, palaeoceanographic changes and benthic extinctions at the end of the Palaeocene. *Nature* 353: 319–322.

**84** Jansen, E., Overpeck, J., et al. (2007) *Paleoclimate*. Chapter 6 in the 4th Assessment Report of the IPCC, Figure 6.2. Published with permission from the IPCC Secretariat.

**85** Dypvik, H., Riber, L., Burca, F. et al. (2011). The Paleocene–Eocene thermal maximum (PETM) in Svalbard — clay mineral and geochemical signals. *Palaeogeography Palaeoclimatology Palaeoecology* 302: 156–169.

**86** Handley, L., Crouch, E.M., and Pancost, R.D. (2011). A New Zealand record of sea level rise and environmental change during the Paleocene–Eocene Thermal Maximum. *Palaeogeography Palaeoclimatology Palaeoecology* 305: 185–200.

**87** Miller, G.H., Brigham-Grette, J., Alley, R.B. et al. (2010). Temperature and precipitation history of the Arctic. *Quaternary Science Reviews* 29: 1679–1715.

**88** Sluijs, A., Bijl, P.K., Schouten, S. et al. (2011). Southern ocean warming, sea level and hydrological change during the Paleocene-Eocene thermal maximum. *Climate of the Past* 7: 47–61.

**89** Sluijs, A., Brinkhuis, H., Crouch, E.M. et al. (2008). Eustatic variations during the Paleocene–Eocene greenhouse world. *Paleoceanography* 23 https://doi.org/10.1029/2008PA001615.

**90** Zachos, J.C., Bohaty, S.M., John, C.M. et al. (2010). The Palaeocene–Eocene carbon isotope excursion: constraints from individual planktonic foraminifer record. *Philosophical Transactions of the Royal Society A: Mathematical, Physical and Engineering Sciences* 365: 1829–1842.

**91** Cui, Y., Kump, L.R., Ridgwell, A.J. et al. (2011). Slow release of fossil carbon during the Palaeocene–Eocene Thermal Maximum. *Nature Geoscience* 4 https://doi.org/10.1038/NGEO1179.

**92** Turner, S.K. (2018). Constraints on the onset duration of the Paleocene–Eocene Thermal Maximum. *Philosophical Transactions of the Royal Society A: Mathematical, Physical and Engineering Sciences* 376: 20170082. https://doi.org/10.1098/rsta.2017.0082.

**93** Bowen, G.J., Maibauer, B.J., Kraus, M.J. et al. (2014). Two massive, rapid releases of carbon during the onset of the Palaeocene–Eocene thermal maximum. *Nature Geoscience* https://doi.org/10.1038/NGEO2316.

**94** Zeebe, R.E., Ridgwell, A., and Zachos, J.C. (2016). Anthropogenic carbon release rate unprecedented during the past 66 million years. *Nature Geoscience* 9: 325–329.

**95** Dickens, G.R. (2011). Down the rabbit hole: toward appropriate discussion of methane release from gas hydrate systems during the Paleocene-Eocene thermal maximum and other past hyperthermal events. *Climate of the Past* 7: 831–846.

96 Sluijs, A., Brinkhuis, H., Schouten, S. et al. (2007). Environmental precursors to rapid light carbon injection at the Palaeocene/Eocene Boundary. *Nature* 450: 1218–1221.

97 Secord, R., Gingerich, P.D., Lohmann, K.C., and McLeod, K.G. (2010). Continental warming preceding the Palaeocene–Eocene Thermal Maximum. *Nature* 467: 955–958.

98 Sundquist, E.T. and Visser, K. (2005). The geologic history of the carbon cycle. In: *Biogeochemistry*, *Treatise on Geochemistry* (eds H.D. Holland and K.K.), vol. 8 (ed. W.H. Schlesinger), 425–513. Turekian: Elsevier-Pergamon. Oxford.

99 Higgins, J.A. and Schrag, D.P. (2006). Beyond methane: towards a theory for the Paleocene-Eocene Thermal Maximum. *Earth and Planetary Science Letters* 245: 523–537.

100 Lovell, B. (2010). *Challenged by Carbon*. Cambridge University Press.

101 Hartley, R.A., Roberts, G.G., White, N., and Richardson, C. (2011). Transient convective uplift of an ancient buried landscape. *Nature Geoscience* 4: 563–565.

102 Storey, M., Duncan, R.A., and Swisher, C.C. (2007). Paleocene-Eocene Thermal Maximum and the opening of the Northeast Atlantic. *Science* 316: 587–589.

103 Svensen, H., Planke, S., Melthe-Sørenssen, A. et al. (2004). Release of methane from a volcanic basin as a mechanism for initial Eocene global warming. *Nature* 429: 542–545.

104 Zachos, J.C., Dickens, G.R., and Zeebe, R.E. (2008). An Early Cenozoic perspective on greenhouse warming and carbon-cycle dynamics. *Nature* 451: 279–283.

105 Kunzig, R. (2011). Hothouse Earth: world without ice. *National Geographic* 220: 90.

106 Honisch, B., Ridgwell, A., Schmidt, D.N. et al. (2012). The geological record of ocean acidification. *Science* 335: 1058–1063.

107 Kiessling, W. and Simpson, C. (2011). On the potential for ocean acidification to be a general cause of ancient reef crises. *Global Change Biology* 17: 56–67. https://doi.org/10.1111/j.1365-2486.2010.02204.x.

108 Babila, T.L., Penman, D.E., Hönisch, B. et al. (2018). Capturing the global signature of surface ocean acidification during the Palaeocene–Eocene Thermal Maximum. *Philosophical transactions. Series A, Mathematical, physical, and engineering sciences* https://doi.org/10.1098/rsta.2017.0072.

109 Kelly, D.C., Bralower, T.J., and Zachos, J.C. (1998). Evolutionary consequences on the latest Paleocene thermal maximum for tropical planktonic foraminifera.

*Palaeogeography Palaeoclimatology Palaeoecology* 141: 139–161. https://doi.org/10.1016/S0031-0182 (98)00017-0.

110 Frieling, J., Gebhardt, H., and Huber, M. (2017). Extreme warmth and heat-stressed plankton in the tropics during the Paleocene-Eocene Thermal Maximum. *Science Advances* 3: e1600891.

111 Frieling, J., Reichart, G.-J., Middelburg, J.J. et al. (2018). Tropical Atlantic climate and ecosystem regime shifts during the Paleocene–Eocene Thermal Maximum. *Climate of the Past* 14: 39–55.

112 Zeebe, R., Zachos, J.C., and Dickens, G.R. (2009). Carbon dioxide forcing alone insufficient to explain Palaeocene–Eocene Thermal Maximum warming. *Nature Geoscience* https://doi.org/10.1038/NGEO578.

113 Frieling, J., Peterse, F., Lunt, D.J. et al. (2019). Widespread warming before and elevated barium burial during the Paleocene-Eocene Thermal Maximum: Evidence for methane hydrate release? *Paleoceanography and Paleoclimatology* 34 https://doi.org/10.1029/2018PA003425.

114 Frieling, J., Svensen, H.H., Planke, S. et al. (2016). Thermogenic methane release as a cause for the longduration of the PETM. *Proceedings of the National Academy of Sciences* 113 (43): 12059–12064.

115 Lyons, S.L., Baczynski, A.A., and Babila, T.L. (2019). Palaeocene–Eocene Thermal Maximum prolonged by fossil carbon oxidation. *Nature Geoscience* 12: 54–60.

116 Kiehl, J.T., Shields, C.A., Synder, M.A. et al. (2018). Greenhouse- and orbital forced climate extremes during the early Eocene. *Philosophical Transactions of the Royal Society A: Mathematical, Physical and Engineering Sciences* 376: 20170085. https://doi.org/10.1098/rsta.2017.0085.

117 Chen, C., Guerit, L., Foreman, B.Z. et al. (2018). Estimating regional flood discharge during Palaeocene–Eocene global warming. *Nature research, Scientific Reports* 8: 13391.

118 Sluijs, A., Schouten, S., Pagani, M. et al. (2006). Subtropical Arctic Ocean temperatures during the Palaeocene/Eocene Thermal Maximum. *Nature* 441: 610–613.

119 Bowen, G.J., Beerling, D.J., Koch, P.L. et al. (2004). A humic climate state during the Paleocene/Eocene Thermal Maximum. *Nature* 432: 495–499.

120 Archer, D. (2005). Fate of fossil fuel $CO_2$ in geologic time. *Journal of Geophysical Research* 110: C09S05. https://doi.org/10.1029/2004JC002625.

121 Kump, L.R. (2018). Prolonged Late Permian–Early Triassic hyperthermal: failure of climate regulation? *Philosophical transactions. Series A, Mathematical, physical, and engineering sciences* 376: 20170078. https://doi.org/10.1098/rsta.2017.0078.

**122** Cohen, A.S., Coe, A.L., and Kemp, D.B. (2007). The Late Palaeocene–Early Eocene and Toarcian (Early Jurassic) carbon isotope excursions: a comparison of their time scales, associated environmental changes, causes and consequences. *Journal of the Geological Society of London* 164 (6): 1092–1108.

**123** Zalasiewicz, J. and Williams, M. (2012). *The Goldilocks Planet – The Four Billion Year Story of Earth's Climate*, 303 pp. Oxford University Press.

**124** Schaller, M.F. and Kung, M.K. (2018). The extraterrestrial impact evidence at the Palaeocene–Eocene boundary and sequence of environmental change on the continental shelf. *Philosophical Transactions of the Royal Society A: Mathematical, Physical and Engineering Sciences* 376: 20170081. https://doi.org/10.1098/rsta.2017.0081.

**125** Zeebe, R.E., Dickens, G.R., Ridgwell, A. et al. (2014). Onset of carbon isotope excursion at the Paleocene-Eocene thermal maximum took millennia, not 13 years. *Proceedings of the National Academy of Sciences* 111 (12): E1062–E1063. https://doi.org/10.1073/pnas.1321177111.

**126** Valdes, P. (2011). Built for stability. *Nature Geoscience* 4: 414–416.

**127** Wilson, G.S., Pekar, S.F., Naish, T.R. et al. (2009). The Oligocene–Miocene boundary – Antarctic climate response to orbital forcing. In: *Antarctic Climate Evolution, Developments in Earth and Environmental Sciences* **8** (eds. F. Florindo and M. Siegert), 369–400. Amsterdam: Elsevier.

**128** Pagani, M., Arthur, M.A., and Freeman, K.H. (1999). Miocene evolution of atmospheric carbon dioxide. *Paleoceanography* 14: 273–292.

**129** Sluijs, A., Bowen, G.J., Brinkhuis, H. et al. (2007). The Palaeocene–Eocene Thermal Maximum super greenhouse: biotic and geochemical signatures, age models and mechanisms of global change. In: *Deep-Time Perspectives on Climate Change: Marrying the Signal from Computer Models and Biological Proxies, Micropal. Soc. Spec. Publ* (eds. M. Williams, A.M. Haywood, F.J. Gregory and D.N. Schmidt), 23–349. London: The Geological Society.

**130** Tripati, A.K., Roberts, C.D., and Eagle, R.A. (2009). Coupling of $CO_2$ and ice sheet stability over major climate transitions of the last 20 million years. *Science* 326: 1394–1397.

**131** Zhang, Y.G., Pagani, M., Lu, A. et al. (2013). A 40-million-year history of atmospheric $CO_2$, in *Warm Climates of the Past – A Lesson for the Future* (ed. D.J. Lunt, H. Elderfield, R. Pancost and A. Ridgwell). *Philosophical Transactions of the Royal Society A: Mathematical, Physical and Engineering Sciences* 371 (2001): 20130096, 20 pp.

**132** Warny, S., Askin, R.A., Hannah, M.J. et al. (2009). Palynomorphs from a sediment core reveal a sudden remarkably warm Antarctica during the Middle Miocene. *Geology* 37 (10): 865–960.

**133** Shevenell, A.E. and Bohaty, S.M. (2012). Southern exposure: new paleoclimate insights from Southern Ocean and Antarctic margin sediments. *Oceanography* 25 (3): 106–117.

**134** Foster, G.L., Lear, C.H., and Rae, J.W.B. (2012). The evolution of $pCO_2$, ice volume and climate during the Middle Miocene. *Earth and Planetary Science Letters* 341-344: 243–254.

**135** Gasson, E., DeConto, R.M., Pollard, D., and Levy, R.H. (2016). Dynamic Antarctic ice sheet during the early to mid-Miocene. *Proceedings of the National Academy of Sciences* 113 (13): 3459–3464.

**136** LaRiviere, J.P., Ravelo, A.C., Vrimmins, A. et al. (2012). Late Miocene decoupling of oceanic warmth and atmospheric carbon dioxide forcing. *Nature* 486: 97–100.

**137** Haywood, A.M., Smellie, J.L., Ashworth, A.C. et al. (2009). Middle Miocene to Pliocene history of Antarctica and the Southern Ocean, in *Antarctic Climate Change* (ed. F. Florindo and M. Siegert). *Developments in Earth and Environmental Sciences* 8: 401–463.

**138** Badger, M.P.S., Lear, C.H., Pancost, R.D. et al. (2013). $CO_2$ drawdown following the middle Miocene expansion of the Antarctic Ice Sheet. *Paleoceanography* 28: 42–53.

**139** Lewis, A.R., Marchant, D.R., Ashworth, A.C. et al. (2008). Mid-Miocene cooling and the extinction of tundra in continental Antarctica. *Proceedings of the National Academy of Sciences* 105 (31): 10676–10680.

**140** Bradshaw, C.D., Lunt, D.J., Flecker, R. et al. (2012). The relative roles of $CO_2$ and palaeogeography in determining late Miocene climate: results from a terrestrial model-data comparison. *Climate of the Past* 8: 1257–1128.

**141** Foster, G.L. and Rohling, E.J. (2013). Relationship between sea level and climate forcing by $CO_2$ on geological timescales. *Proceedings of the National Academy of Sciences of the United States of America* 110 (4): 1209–1214.

**142** Savin, S.M. and Woodruff, F. (1990). Isotopic evidence for temperature and productivity in the Tethyan Ocean. In: *Phosphate Deposits of the World, Vol.3, Neogene to Modern Phosphorites, Int. Geol. Congr. Project* **156** (eds. W.C. Burnett and S.R. Riggs), 241–259. Phosphorites: Cambridge University Press.

**143** Vincent, E. and Berger, W.H. (1985). Carbon dioxide and polar cooling in the Miocene: The Monterey Hypothesis. In: *The Carbon Cycle and Atmospheric $CO_2$: Natural*

*Variations Archean to Present, Geophys. Monogr.* **32** (eds. E.T. Sundquist and W.S. Broecker), 455–468. Washington D.C.: American Geophysical Union.

144 Föllmi, K.B., de Kaenel, E., Stille, P. et al. (2005). Phosphogenesis and organic-carbon preservation in the Miocene Monterey Formation at Naples Beach, California – The Monterey hypothesis revisited. *GSA Bulletin* 117 (5-6): 589–619.

145 Diester-Haass, L., Billups, K., Gröcke, D.R. et al. (2009). Mid-Miocene paleoproductivity in the Atlantic Ocean and implications for the global carbon cycle. *Paleoceanography* 24 (1): PA1209, 19 pp.

146 Bertler, N.A.N. and Barrett, P.J. (2010). Vanishing polar ice sheets. In: *Changing Climates, Earth Systems and Society, International Year of Planet Earth* (ed. J. Dodson), 49–83. New York: Springer.

147 Summerhayes, C.P. (2016). Upwelling. In: *Encyclopedia of Marine Geosciences* (eds. J. Harff, M. Meschede, S. Petersen and J. Thiede), 900–912. Springer.

148 Holbourn, A., Kuhnt, W., Kochhann, K.G.D. et al. (2015). Global perturbation of the carbon cycle at the onset of the Miocene Climatic Optimum. *Geology* 43 (2): 123–126.

149 Levy, R., Harwood, D., Florindo, F. et al. (2016). Antarctic ice sheet sensitivity to atmospheric $CO_2$ variations in the early to mid-Miocene. *Proceedings of the National Academy of Sciences* 113 (13): 3453–3458.

150 Holbourn, A.E., Kuhnt, W., Clemens, S.C. et al. (2013). Middle to late Miocene stepwise climate cooling: evidence from a high resolution deep-water isotope curve spanning 8 million years. *Paleoceanography* 28: 688–699.

151 Laskar, J., Robutel, P., Joutel, F. et al. (2004). A long-term numerical solution for the insolation quantities of the Earth. *Astronomy & Astrophysics* 428: 261–285.

152 Greenop, R., Foster, G.L., Wilson, P.A., and Lear, C.H. (2014). Middle Miocene climate instability associated with high-amplitude CO2 variability. *Paleoceanography* 29: 845–853. https://doi.org/10.1002/2014PA002653.

153 Kochhann, K.G.D., Holbourn, A., Kuhnt, W. et al. (2016). Eccentricity pacing of eastern equatorial Pacific carbonate dissolution cycles during the Miocene Climatic Optimum. *Paleoceanography* 31: 1176–1192. https://doi.org/10.1002/2016PA002988.

154 Dalziel, I.W.D., Lawver, L.A., Pearce, J.A. et al. (2013). A potential barrier to deep Antarctic circumpolar flow until the late Miocene. *Geology* 41 (9): 947–950.

155 Summerhayes, C.P. (1981). Oceanographic controls on organic matter in the Miocene Monterey Formation, offshore California, in *The Monterey Formation and Related Siliceous Rocks of California* (ed. R.E. Garrison, and R.G. Douglas). *Society of Economic Petrologists and Paleontologists, Pacific Section* 15: 213–219.

156 Knorr, G. and Lohmann, G. (2014). Climate warming during Antarctic ice sheet expansion at the Middle Miocene transition. *Nature Geoscience* 7: 376–381.

157 Keeling, R.F. and Visbeck, M. (2011). On the linkage between Antarctic surface water stratification and global deep-water temperature. *Journal of Climate* 24: 3545–3557.

158 Salzmann, U., Williams, M., Haywood, A.M. et al. (2011). Climate and environment of a Pliocene warm world. *Palaeogeography Palaeoclimatology Palaeoecology* 309: 1–8.

159 Csank, A.Z., Patterson, W.P., Eglington, B.M. et al. (2011). Climate variability in the Early Pliocene Arctic: annually resolved evidence from stable isotope values of sub-fossil wood, Ellesmere Island, Canada. *Palaeogeography Palaeoclimatology Palaeoecology* 308: 339–349.

160 Brigham-Grette, J., Melles, M. Minyuk, P., et al. (2013) Pliocene warmth, polar amplification, and stepped Pleistocene cooling recorded in NE Arctic Russia, *Science, Science Express Index online*, May 9 2013, DOI: 10.1126/science.1233137.

161 Dowsett, H.J., Robinson, M.M., Haywood, A.M. et al. (2012). Assessing confidence in Pliocene sea surface temperatures to evaluate predictive models. *Nature Climate Change* 2: 365–371. https://doi.org/10.1038/nclimate1455.

162 Salzmann, U., Riding, J.B., Nelson, A.E., and Smellie, J.L. (2011). How likely was a green Antarctic Peninsula during warm Pliocene interglacials? A critical reassessment based on new palynofloras from James Ross Island. *Palaeogeography Palaeoclimatology Palaeoecology* 309 (2011): 73–82.

163 Naish, T., Carter, L., Wolff, E. et al. (2009). Late Pliocene-Pleistocene Antarctic climate variability at orbital and suborbital scale: ice sheet, ocean and atmospheric interactions. In: *Antarctic Climate Change, Developments in Earth and Environmental Sciences* **8** (eds. F. Florindo and M. Siegert), 465–529. Amsterdam: Elsevier.

164 Miller, K.G., Wright, J.D., Browning, J.V. et al. (2008). High tide of the warm Pliocene: implications of global sea level for Antarctic deglaciation. *Geology* 40 (5): 407–410.

165 Dumitru, O.A., Austermann, J., Polyak, V.J. et al. (2019). Constraints on global mean sea level during Pliocene warmth. *Nature* https://doi.org/10.1038/s41586-019-1543-2.

166 Naish, T.R. and Wilson, G.S. (2009). Constraints on the amplitude of Mid-Pliocene (3.6–2.4 Ma) eustatic sea-level fluctuations from the New Zealand shallow-marine sediment record. *Philosophical Transactions of*

*the Royal Society A: Mathematical, Physical and Engineering Sciences* 367: 169–187.

**167** Dwyer, G.S. and Chandler, M.A. (2009). Mid-Pliocene sea level and continental ice volume based on coupled benthic Mg/Ca palaeotemperatures and oxygen isotopes. *Philosophical Transactions of the Royal Society A: Mathematical, Physical and Engineering Sciences* 367: 157–168.

**168** Raymo, M.E., Mitrovica, J.X., O'Leary, M.J. et al. (2011). Departures from eustacy in Pliocene sea-level records. *Nature Geoscience* 4: 328–332.

**169** Rohling, E.J., Grant, K., Bolshaw, M. et al. (2009). Antarctic temperature and global sea level closely coupled over the past five glacial cycles. *Nature Geoscience* 2: 500–504.

**170** Stocchi, P., Escutia, C., Houben, A.J.P. et al. (2013). Relative sea-level rise around East Antarctica during Oligocene glaciation. *Nature Geoscience* 6: 380–384.

**171** Naish, T.R. et al. (2009). Obliquity-paced Pliocene West Antarctic Ice Sheet oscillations. *Nature* 458: 322–329.

**172** Pollard, D. and DeConto, R.M. (2009). Modelling West Antarctic Ice Sheet growth and collapse through the past five million years. *Nature* 458: 329–333.

**173** DeConto, R.M., and Pollard, D. (2013) Pliocene retreat of Greenland and Antarctic ice sheet margins, *Abstract, AGU Fall Meeting*, San Francisco 9–13 December 2013.

**174** Cook, C.P., Van de Flierdt, T., Williams, T. et al. (2013). Dynamic behaviour of the East Antarctic Ice Sheet during Pliocene warmth. *Nature Geoscience* 6: 765–759.

**175** Gramling, C. (2013) East Antarctica's ice sheet not as stable as thought, *Science NOW*, 21 July.

**176** Hansen, M.A., Passchier, S., Khim, B.-K. et al. (2015). Threshold behavior of a marine-based sector of the East Antarctic Ice Sheet in response to early Pliocene ocean warming. *Paleoceanography* 30: 789–801. https://doi.org/10.1002/2014PA002704.

**177** Denton, G.H., Prentice, M.L., and Burckle, L.H. (1991). Cainozoic history of the Antarctic ice sheet. In: *Geology of Antarctica* (ed. R.J. Tingey), 365–433. Oxford University Press.

**178** Badger, M.P.S., Schmidt, D.N., Mackensen, A., and Dancost, R.D. (2013). High-resolution alkenone palaeobarometry indicates relatively stable $pCO_2$ during the Pliocene (3.3–2.8 Ma), In: *Warm Climates of the Past – A Lesson for the Future* (ed. D.J. Lunt, H. Elderfield, R. Pancost and A. Ridgwell). *Philosophical Transactions of the Royal Society A: Mathematical, Physical and Engineering Sciences* 371 (2001): 20130094, 16 pp.

**179** Haywood, A. and Williams, M. (2005). The climate of the future: clues from three million years ago. *Geology Today* 21 (4): 138–143.

**180** Dowsett, H.J., Haywood, A.M., Valdes, P.J. et al. (2011). Sea surface temperatures of the mid-Piacenzian warm period: a comparison of PRISM3 and HadCM3. *Palaeogeography Palaeoclimatology Palaeoecology* 309 (2011): 83–91.

**181** Robinson, M.M., Valdes, P.J., Haywood, A.M. et al. (2011). Bathymetric controls on Pliocene North Atlantic and Arctic sea surface temperature and deepwater production. *Palaeogeography Palaeoclimatology Palaeoecology* 309: 92–97.

**182** Dowsett, H.J. (2007). The PRISM palaeoclimate reconstruction and Pliocene sea-surface temperature. In: *Deep-Time Perspectives on Climate Change: Marrying the Signal from Computer Models and Biological Proxies, Micropal. Soc. Spec. Publ* (eds. M. Williams, A.M. Haywood, F.J. Gregory and D.N. Schmidt), 459–480. London: The Geological Society.

**183** Haywood, A.M., Valdes, P.J., Hill, D.J., and Williams, M. (2007). The Mid-Pliocene warm period: a test-bed for integrating data and models. In: *Deep-Time Perspectives on Climate Change: Marrying the Signal from Computer Models and Biological Proxies, Micropal. Soc. Spec. Publ* (eds. M. Williams, A.M. Haywood, F.J. Gregory and D.N. Schmidt), 43–457. London: The Geological Society.

**184** Lawrence, K.T., Herbert, T.D., Dekens, P.S., and Ravelo, A.C. (2007). The application of the alkenone organic proxy to the study of Plio-Pleistocene climate. In: *Deep-Time Perspectives on Climate Change: Marrying the Signal from Computer Models and Biological Proxies, Micropal. Soc. Spec. Publ* (eds. M. Williams, A.M. Haywood, F.J. Gregory and D.N. Schmidt), 539–562. London: The Geological Society.

**185** Pagani, M., Liu, Z., LaRiviere, J., and Ravelo, A.C. (2010). High Earth-system climate sensitivity determined from Pliocene carbon dioxide concentrations. *Nature Geoscience* 3: 27–30.

**186** Lunt, D.J., Haywood, A.M., Schmidt, G.A. et al. (2010). Earth system sensitivity inferred from Pliocene modelling and data. *Nature Geoscience* 3: 60–64.

**187** Ravelo, A.C., Billups, K., Dekens, P.S. et al. (2007). Onto the ice ages: proxy evidence for the onset of Northern Hemisphere glaciation. In: *Deep-Time Perspectives on Climate Change: Marrying the Signal from Computer Models and Biological Proxies, Micropal. Soc. Spec. Publ* (eds. M. Williams, A.M. Haywood, F.J. Gregory and D.N. Schmidt), 563–573. London: The Geological Society.

**188** Fedorov, A.V., Brierley, C.M., and Emanuel, K. (2010). Tropical cyclones and permanent El Niño in the Early Pliocene Epoch. *Nature* 463: 1066–1070.

**189** Sriver, R.L. (2010). Tropical cyclones in the mix. *Nature* 463: 1032–1033.

**190** Lunt, D.J., Haywood, A.M., Schmidt, G.A. et al. (2012). On the causes of mid-Pliocene warmth and polar amplification. *Earth and Planetary Science Letters* 321-322: 128–138.

**191** Schmidt, D.N. (2007). The closure history of the Central American Seaway: evidence from isotopes and fossils to models and molecules. In: *Deep-Time Perspectives on Climate Change: Marrying the Signal from Computer Models and Biological Proxies*, *Micropal. Soc. Spec. Publ* (eds. M. Williams, A.M. Haywood, F.J. Gregory and D.N. Schmidt), 427–442. London: The Geological Society.

**192** Sarnthein, M., Bartoli, G., Prange, M. et al. (2009). Mid-Pliocene shifts in ocean overturning circulation and the onset of Quaternary-style climates. *Climate of the Past* 5: 269–283.

**193** Sarnthein, M. (2013). Transition from Late Neogene to Early Quaternary Environments. In: *The Encyclopedia of Quaternary Science*, vol. 2 (ed. S.A. Elias), 151–166. Amsterdam: Elsevier.

**194** Cronin, T.M. (2010). *Paleoclimates: Understanding Climate Change Past and Present*, 441 pp. New York: Columbia University Press.

**195** Shackleton, N.J., and Hall, M.A. (1984) Oxygen and carbon isotope stratigraphy of Deep Sea Drilling Project hole 552A: Plio-Pleistocene glacial history, in *Init. Repts. DSDP* **81**, US Govt. Print. Off. Washington D.C., 599–609.

**196** Shackleton, N.J., Backman, J., Zimmerman, H. et al. (1984). Oxygen isotope calibration of the onset of ice-rafting and history of glaciation in the North Atlantic region. *Nature* 307: 620–623.

**197** Shackleton, N.J., and Kennett, J.P. (1975) Late Cenozoic oxygen and carbon isotopic changes at DSDP Site 284: implications for glacial history of the Northern Hemisphere and Antarctica, *Init. Rep. DSDP* **24**, US Govt. Print. Off. Washington D.C., 801–807.

**198** Willeit, M., Ganopolski, A., Calov, R. et al. (2016). The role of $CO_2$ decline for the onset of Northern Hemisphere glaciation. *Quaternary Science Reviews* 119: 22–34.

**199** Karas, C., Nurnberg, D., Gupta, A. et al. (2009). Mid-Pliocene climate change amplified by a switch in Indonesian subsurface throughflow. *Nature Geoscience* 2: 434–438.

**200** Cane, M.A. and Molnar, P. (2001). Closing of the Indonesian Seaway as a Precursor to East African Aridification around 3–4 million years ago. *Nature* 411 (6834): 157–162.

**201** Lang, D.C., Bailey, I., and Wilson, P.A. (2016). Incursions of southern-sourced water into the deep North Atlantic during late Pliocene glacial intensification. *Nature Geoscience* 9: 375–379.

**202** Grant, G.R., Naish, T.R., Dunbar, G.B. et al. (2019). The amplitude and origin of sea-level variability during the Pliocene epoch. *Nature* https://doi.org/10.1038/s41586-019-1619-z.

**203** Haywood, A.M., Hill, D.J., Dolan, A.M. et al. (2013). Large-scale features of Pliocene climate: results from Pliocene model Intercomparison project. *Climate of the Past* 9: 191–209.

# 12

# Solving the Ice Age Mystery – The Deep Ocean Solution

---

**LEARNING OUTCOMES**

- Understand and be able to explain how the precession, eccentricity, and obliquity signal changed over the past 12 Ka, what is likely to happen to them over the next 100 Ka, and what that means for our future climate.
- Understand and be able to explain the origin of the global signal of climate change in deep ocean sediments, and what oxygen isotopes tell us about the relative contributions of water temperature and ice volume.
- Be able to explain the significance of the 1976 paper in Science by Hays, Imbrie, and Shackleton.
- Understand and be able to describe the Mid Pleistocene Transition, and to explain its possible origins.
- Understand and be able to describe the Meridional Overturning Circulation (MOC), its significance for climate, for instance in connecting the two polar regions, and the effects of its variation through time.
- Be able to explain the role of airborne dust in causing climatic change between glacial and interglacial time.
- Understand and be able to explain the possible origins of the millennial variability of the last glacial period.
- Understand and be able to explain the differences and similarities between the interglacial climates of Marine Isotope Stages (MIS) 1, 5, 11, and 19.
- Understand and be able to explain how and why global sea level varied in relation to climate over the past 130 Ka.
- Understand and be able to explain the existence of 'natural climatic envelopes'.

---

## 12.1 Astronomical Drivers

By the time we get to the Pleistocene, the Earth has cooled to its lowest point since the Carboniferous glaciation 300 Ma ago. During the Pleistocene much of the variation in the Earth's climate is due not to the plate tectonic processes that were important in older times, but to celestial mechanics. This led to cycles of 100 Ka in the eccentricity of the Earth's orbit, of 41 Ka in the Earth's axial tilt (or 'obliquity'), which dominates radiation at high latitudes, and of 23 and 19 Ka in the precession of the equinoxes, which dominates radiation at low latitudes. And as we saw in Chapter 6, Milutin Milankovitch set out the basis for our understanding of this process between 1920 and 1941 [1], expanding on the work of James Croll that we examined in Chapter 3.

In 1945, Frederick Zeuner (1905–1963), of London's Institute of Archaeology, tested Milankovitch's theory by examining how it applied to what was known of the Pleistocene period on land [2]. Finding a close match between the sequence of Ice Age strata and the variations in insolation, he concluded '*no objection can be raised against the astronomical theory of the glacial and interglacial*

*phases of the Pleistocene*' [2]. Amongst those intrigued by what controlled the changes between glacial and interglacial periods was Richard Foster Flint (1901–1976) of Yale University, who along with other honours would be awarded the Prestwich Medal by the Geological Society of London in 1972 for his contributions to understanding the Ice Age. Flint was one of the most influential figures in Quaternary science in the twentieth century, much admired for his seminal 1957 text *Glacial and Pleistocene Geology* [3]. But his book concluded '*the geometric scheme of distribution of insolation heating must be considered inadequate in itself to explain the Pleistocene climatic changes*' [3].

He was wrong. Things have changed since then. As Mike Walker of the University of Wales, and John Lowe of Royal Holloway College London pointed out in 2007, Flint's approach was rooted in glacial geology. Since his day, those investigating Quaternary science have moved '*away from Flint's somewhat narrow glacial-geological paradigm towards the multi- and inter-disciplinary approach to the study of recent Earth history that is practiced today*' [3]. With this new approach we can now analyse the '*rich and often readily accessible Quaternary record .... at a level of*

*Palaeoclimatology: From Snowball Earth to the Anthropocene*, First Edition. Colin P. Summerhayes.
© 2020 John Wiley & Sons Ltd. Published 2020 by John Wiley & Sons Ltd.

**Figure 12.1** André Berger.

---

**Box 12.1  André Berger**

Climatologist André Berger has a master's degree in meteorology from MIT (1971) and a PhD from the Catholic University of Louvain, Belgium (1973). He is renowned for contributing to the renaissance of Milankovitch's theory of climate change, for making major contributions to simulating future climate change and for working on the first Earth model of intermediate complexity. He was professor of meteorology and climatology at Louvain, and then director of the Institute of Astronomy and Geophysics Georges Lemaître from 1978 to 2001, where he now has Emeritus status. He has served as president or chairman of several national and international scientific organizations or committees and was honorary president of the European Geosciences Union. He was on the steering committee for the International Geosphere–Biosphere Programme (IGBP), and initiated the Palaeoclimate Intercomparison Modelling Project (PMIP). He has received many honours for his discoveries, including the Milutin Milankovitch Medal of the European Geophysical Society, and in 1996 was made a knight of the realm by King Albert II.

---

*detail not normally possible for older geological periods*' [3]. Milankovitch theory has come to stay [4].

Milankovitch lacked computers. André Léon Georges Chevalier Berger (1942–) (Figure 12.1, Box 12.1) used them to refine Milankovitch's calculations [5–8]. Figure 12.2 provides an introduction to Berger's findings, as explained below.

Astronomers are confident in the robustness of their calculations of the variation of the Earth's orbit not only for millions of years back through time (with the best results for the past 20 Ma), but also for millions of years into the future [9], which enables the calculation of insolation not only for the past, but also for future times. Croll used the astronomical information available in his day to show how eccentricity was expected to change over the next one million years (Figure 3.2), and Berger used modern astronomical information to show how all three Milankovitch cycles are expected to change over the next 100 Ka and what that will do to insolation in summer at 65°N (Figure 12.2). Here we have the basis for coarse-grained climate prediction.

Examining Figure 12.2 we can see that at present eccentricity is small and getting smaller with time. We know that eccentricity modulates precession, which has also been declining in intensity with time; both are now close to what they were 350–400 Ka ago. In contrast, the obliquity signal remains quite strong. While precession increased slightly over the past 10 Ka, eccentricity and obliquity both decreased. The net result was a fall in Northern Hemisphere insolation, which is predicted to remain low for the next 40 Ka before a further fall between 45 and 55 Ka (Figure 12.2). At times precession and obliquity may coincide, whilst at other times they may be asynchronous (Figure 12.2).

Berger demonstrates that obliquity, which determines seasonality, is synchronous in both hemispheres and strongest at high latitudes [10]. Consequently, we would expect insolation at high latitudes in both hemispheres to follow broadly similar patterns. Indeed, we find that the poles are connected not only by the synchroneity of the annual obliquity signal, but also by global changes in sea level, by the global ocean thermohaline conveyor system (which is driven by the sinking of cold dense surface water in both polar regions – more on that later), by atmospheric $CO_2$, which is well-mixed globally and is largely driven by warming and cooling of the polar oceans (cold water absorbs $CO_2$, warm water emits it), and by atmospheric water vapour (which increases with oceanic warming and decreases with oceanic cooling). Hence we should expect that deep ocean sediments, whether or not they are close to the poles, should contain a signal of global climate change. Novel studies of thousands of deep ocean sediments collected by piston corers since the 1940s and from hundreds of deep-ocean drilling sites since 1968 show that this expectation is justified, as do very long ice cores from both poles, which have been largely collected in the 2000s. Our understanding of Earth's Ice Age climate is thus relatively new.

Studies of these various cores has enormously expanded the science of the Quaternary, which comprises the Pleistocene, starting at 2.6 Ma ago, and the Holocene – the last 11 700 years [3]. To help move the field forward, scientists formed the International Quaternary Union (INQUA) in 1928 [3]. Several scientific journals emerged to meet

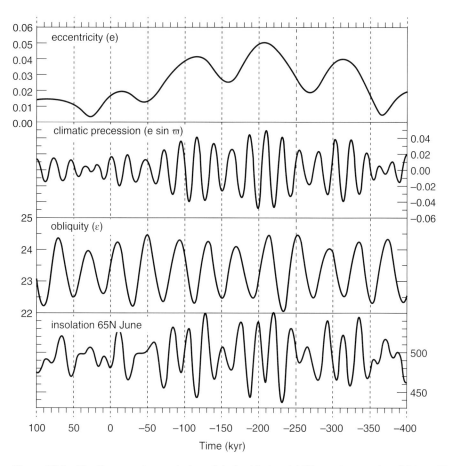

**Figure 12.2**   The Berger astronomical model of orbital variability past, present and future. These curves appear in numerous formats in several publications by André Berger and Marie-France Loutre, and are reproduced here with the permission of André Berger. Obliquity is expressed in degrees of tilt of the Earth's axis. Insolation at the summer solstice at 65°N is expressed in W/m$^2$.

their needs – *Quaternary Research* (1971), *Boreas* (1972), *Quaternary Science Reviews* (1982), *Journal of Quaternary Science* (1985), *Quaternary International* (1989) and *The Holocene* (1991). Researchers also make good use of journals like *Paleoceanography* (renamed *Paleoceanography and Paleoclimatology* in 2018) and *Palaeogeography, Palaeoecology, and Palaeoclimatology*, as well as the four volume *Encyclopedia of Quaternary Science* (2006) [11].

## 12.2   An Ice Age Climate Signal Emerges from the Deep Ocean

As we saw in Chapter 7, the first to exploit the new technology of piston coring to examine the history of climate recorded in deep-sea sediments was Gustaf Arrhenius. In 1952 he attributed alternations between carbonate-rich and carbonate-poor sediments in east Pacific cores to changes in the 'aggressiveness' of polar bottom waters. During the Ice Age, he thought, large volumes of bottom water were derived from the Polar Regions. Being cold,

they carried large amounts of dissolved $CO_2$ enabling them to dissolve, or to prevent the deposition of, deep-sea carbonates. Bottom waters of intervening warm periods carried less dissolved $CO_2$, so were less 'aggressive'. At the time, Arrhenius, like Flint, dismissed Milankovitch's ideas [12], although he later adopted them.

At about the same time, in the early 1950s, Lamont began its routine collection of long piston cores from the world's oceans. David Ericson, who was in charge of the new Lamont core store, found that although the distribution of the planktonic foraminiferan *Globorotalia menardii* indicated warm conditions, it was also influenced by ocean currents. Another species, *Globigerina pachyderma*, was a cold-water indicator, as were *Globigerina inflata* and *Globigerina bulloides*. Changes from warm to cold were also indicated by changes in the coiling direction of *Globorotalia truncatulinoides*. From the distribution of these species down core, Ericson built a Quaternary stratigraphy incorporating the Holocene, the last glaciation and the previous interglacial [13].

Ericson's and Arrhenius's interests were shared by Cesar Emiliani, whom we met in Chapters 6 and 7 (see Box 7.2). In 1955, Emiliani analysed the oxygen isotopes in planktonic foraminifera collected from eight piston cores from the Swedish Deep-Sea Expedition and four from the Lamont core store. Finding fluctuations in the $\partial^{18}$O ratio with time, he interpreted them as representing the variations in climate between glacial and interglacial periods [14]. By analysing the same species of surface dwelling foraminiferan, he eliminated the effect of metabolic differences between different species. The resulting variations were due to differences in either the temperature or the isotopic composition of seawater, the latter representing the amount of water tied up as ice on land. Emiliani '*guessed that 60% of the signal was due to the temperature effect, 40% to the ice effect*' [15]. Following Zeuner's reasoning [2], Emiliani thought that the variations in $\partial^{18}$O with time represented changing insolation [14]. He invented the nomenclature that is still in use today, in which each warm or cold period is identified as a marine isotope stage (MIS). MISs with even numbers are cold stages, those with odd numbers are warm stages. Some can be subdivided into sub-stages (e.g. 5a, 5b, 5c).

In March 1961, Emiliani teamed up with Flint to link the Pleistocene record in continental and deep-sea sediments [16]. They considered that MIS 1 was the Holocene, MIS 2 was the Main Würm glacial stage on land, MIS 3 was the Early to Main Würm glacial interval on land, MIS 4 was the Early Würm on land; and MIS 5 was the last interglacial. Thinking about the possible drivers for the Ice Age, they favoured a model of glaciation based on Milankovitch's concept that ice accumulated during cool summers driven by insolation in the Northern Hemisphere. Evidently, Flint had changed his mind about Milankovitch since 1957. They discounted Plass's idea that the ice ages were in some way controlled by atmospheric $CO_2$, because they thought, wrongly, that Revelle and Suess [17] had concluded, based on studies of $^{14}$C, that atmospheric $CO_2$ would be rapidly taken up by the ocean. It would be a while before the role of $CO_2$ in the Ice Age climate would be fully understood.

Before Emiliani's isotopic analyses of deep-sea cores it was thought that there were only four major glacial periods during the Ice Age [2]. Emiliani found more than twice that! While that shocked classical Quaternary geologists, physicist Nick Shackleton (Box 7.3) agreed with Emiliani [18]. The fly in the ointment was the inadequacy of methods for accurately dating Emiliani's cores. Emiliani guessed that his main $\partial^{18}$O cycles were 41 Ka long, but they were later found to last about 100 Ka.

Subtle differences in the way that Ericson and Emiliani interpreted their data meant that they disagreed, for a while, about the precise sequence of cold and warm events. When they did agree, we had two independent means of mapping changes between glacial and interglacial periods

**Figure 12.3** John Imbrie.

in deep sea cores – one palaeontological, from microscopic fossils of marine plankton (microfossils), and the other geochemical, from oxygen isotopes in those same fossils.

By the mid-1960s, two new figures occupied centre stage: Nick Shackleton (Box 7.3), and the micropalaeontologist and stratigrapher John Imbrie (1925–2016), of Brown University (Figure 12.3) (Box 12.2). Our understanding of Ice Age climate owes a great deal to these two men.

Using his refined mass-spectrometric technique (see Chapter 7), Shackleton showed by 1967 that a distinctive $\partial^{18}$O pattern is evident in the benthic foraminifera that live on the seabed, which are bathed in cold oxygen-rich bottom water sinking from the surface in the polar regions [21]. Most of this water is colder than 4 °C, and variations in the $\partial^{18}$O ratio in these organisms tell us more about changing polar surface water temperature hence changing ice volume. Shackleton found that planktonic foraminifera growing in surface waters displayed a similar signal. He calculated that about 66% of the $\partial^{18}$O shift between glacial and interglacial periods was due to changes in ice volume, not to the influence of temperature, which was the opposite of what Emiliani had concluded. This revelation caused a paradigm shift in our understanding of $\partial^{18}$O ratios in the service of palaeothermometry. Urey had assumed that the oxygen isotopic composition of seawater would be invariant [22]. Clearly it was not. Today, the $\partial^{18}$O ratios of benthic foraminifera are taken as representing the temperatures of polar surface waters.

By 1969, Imbrie realized that because the total assemblage of planktonic foraminifera in surface waters should reflect the environment in which they lived, and multivariate statistical analyses could quantify that relationship,

Imbrie obtained a BA from Princeton in 1948 after serving with the US 10th Mountain Division in Italy during the Second World War. Having obtained a PhD from Yale in 1951, he taught at Columbia University until 1967. Joining Brown University, he held the Henry L. Doherty Chair in Oceanography, and later held Emeritus status there. He pioneered the use of computers to demonstrate the relation of assemblages of plankton to the temperature of surface waters, thus providing palaeoceanography with one of its key tools. His book *Ice Ages: Solving the Mystery* [19], written with his daughter Katherine, won the 1976 Phi Beta Kappa Prize. Imbrie was co-author with Hays and Shackleton of the 1976 paper in *Science* that linked Milankovitch variations to the sediment record [20]. He was elected to the US National Academy of Sciences in 1978, and received the Maurice Ewing Medal of the American Geophysical Union in 1986, the Twenhofel Medal of the Society of Sedimentary Geology and the Lyell Medal of the Geological Society of London in 1991 and the Vetlesen Prize in 1996.

statistical analyses of faunal assemblages down core could be used to ascertain past climate change [23]. Reanalysing the Caribbean cores analysed by Ericson and Emiliani, he and Nilva Kipp showed that the temperatures of surface waters of glacial periods there fell by just 2 °C, not 6 °C. Because the fluctuations in their data agreed with those determined by Emiliani from isotopes, it was clear that *G. menardii*, which Ericson used to identify warm periods, fluctuated in ways unrelated to temperature, explaining the discrepancy between Ericson's and Emiliani's data [24, 25]. That was a breakthrough.

By the late 1960s, as the Deep Sea Drilling Project got underway, analyses of $\partial^{18}O$ changes down piston cores from different parts of the ocean showed that sediments could be routinely subdivided into Emiliani's MISs representing glacial and interglacial periods. These stages coincided with intervals defined by Imbrie's assemblages of microfossils and could be correlated from one core to another over vast oceanic distances, suggesting planetary control.

Was that planetary control the same as Milankovitch's astronomically controlled insolation? To find out, geologists needed closely spaced dates down core. In the late 1960s-early 1970s microfossils could tell us about environmental change from cold to warm and back, but not about the ages of cold or warm stages. As we saw in Chapter 6, radiocarbon dating was useful, but only in sediments less

than 50 Ka old. Layers of volcanic ash older than 100 Ka could be dated by the potassium-argon (K–Ar) method. Changes in the Earth's magnetic field could be used to date specific sedimentary horizons, but only in sediments older than 780 Ka ago. Together, these independent techniques provided a crude means of dating sediment layers in cores. Over time, new techniques would become available. A major breakthrough in radiocarbon ($^{14}C$) dating came about in 1977, when a new technique – accelerator mass spectrometry (AMS) – enabled us to count $^{14}C$ atoms, as opposed to measuring $^{14}C$ decay. AMS can date samples as small as a single pin-head sized microfossil. The technique is fast and cheap, but still limited to sediments less than about 50 Ka old.

Determining the ages of the glacial and interglacial sedimentary stages identified by $\partial^{18}O$ and microfossil analyses became a major objective of the international Climate Long Range Investigation, Mapping and Prediction (CLIMAP) project. Founded by Imbrie, Shackleton and others, CLIMAP began in spring 1971 as part of the International Decade of Ocean Exploration [26]. The project aimed to establish average boundary conditions for the Last Glacial Maximum (LGM) at 18 Ka ago. Those conditions included the geography of the continents, the albedo of land and ice surfaces, the extent and elevation of permanent ice, and the sea surface temperature. Modellers would use those conditions in atmospheric General Circulation Models (GCMs) to map the climate of the LGM. CLIMAP would then test model outputs against palaeoclimate data. The first simulation was for August 18 Ka ago [26]. Sea surface temperature values were derived from $\partial^{18}O$ data and from Imbrie's statistical analyses of planktonic faunal assemblages. The extent of sea ice in the polar regions was estimated from the presence or absence of diatomaceous sediments, absence indicating the presence of sea ice.

The CLIMAP data showed that extensive cooling at the poles and an expanded area of land and sea ice steepened the thermal gradient between the Equator and the poles, thus strengthening the winds. In the Southern Ocean, the Antarctic Polar Front moved north, along with Antarctic sea ice. The Subtropical Front moved far enough north to limit the passage of warm Indian Ocean water around South Africa into the South Atlantic. This cooled the South Atlantic, and created a closed anticlockwise gyre in the Indian Ocean. Upwelling increased where it is found today, along continental margins and along the Equator, as the winds that drove it increased in strength. On land, grasslands, steppes, deserts, and ice spread at the expense of forests, increasing the Earth's albedo. Not only was there more ice, there was correspondingly less ocean area as sea levels fell. That fact, and the cooler temperatures, meant that there was less evaporation, so the climate was drier, which, together with the stronger winds, meant there was more dust in the air.

Palaeoclimatologists noticed that the wiggles in the $\partial^{18}O$ curves down sediment cores seemed to mimic the wiggles in the patterns of Earth's insolation through time. If the match was real, then it offered an opportunity to date the age of the sediments from the pattern of wiggles in the oxygen isotope data. Starting with just a few radiometric dates as tie points, the CLIMAP scientists assumed that the rates of sedimentation in different MISs were constant downcore. That enabled them to estimate the age of each wiggle on the curve of variation in $\partial^{18}O$ back through time. Applying spectral analysis to the $\partial^{18}O$ curve dated in this way, James Hays (1938–), of Lamont, was able to demonstrate in a landmark paper published in 1976 with Imbrie and Shackleton [20], that in cores where sedimentation was undisturbed, the variations in $\partial^{18}O$ over the past 450 Ka accurately mimicked the orbital signals calculated by André Berger (Figure 12.2), with spectral climate peaks at intervals of close to 100, 41, 23, and 19 Ka. This was particularly fascinating in that Berger had calculated from orbital data that the precession signal should have two components, one at about 23 Ka and one at about 19 Ka [27]. '*This split is playing a fundamental role in the explanation of the 100 000-year cycle found in geological record, as this period is often assumed to originate from a nonlinear response of the climate system to these two precessional periods*' Berger observed [28]. Not only that, but Hays' core data confirmed that the climate changes in the Northern Hemisphere were essentially synchronous with those observed in the Southern Hemisphere, as would be expected from the synchronicity of the obliquity signal at both poles [29]. The correlation between the astronomical variables and $\partial^{18}O$ told Hays and his team that '*... changes in the earth's orbital geometry are the fundamental cause of the succession of Quaternary Ice Ages*' [20]. The three Milankovitch mechanisms – eccentricity, tilt, and precession – worked in unison as the 'pacemaker of the ice ages'. In 2010, Hays was awarded the Milankovitch Medal of the European Geosciences Union for his many contributions to palaeoclimatic understanding.

The fact that the same pattern of wiggles occurred everywhere meant that even cores without precise radiometric dates could be dated by reference to a standard $\partial^{18}O$ curve derived by merging data from several well-dated cores. Wiggle matching offered an incredible opportunity to date intervals of time as small as about 1000 years long. Patterns of tree rings back through time offer much the same possibility, provided they come from much the same area and so experienced more or less the same climate changes through time. While this is very helpful for the past 11 Ka, tree rings do not provide us with a lengthy and globally distributed database of the kind provided by deep-sea cores. Wiggle matching of oxygen isotope curves from core to core does contain the assumption that the sediment section has not been disturbed by either erosion or the lateral introduction of material by turbidity currents, but these possibilities are identified or eliminated by comparing individual cores with the global standard. Indeed, wiggle matching can identify how much of the sedimentary record has been removed by local erosion or added, for example by turbidity currents'! Nowadays, wiggle matching has enabled palaeoclimatologists to push the $\partial^{18}O$-calibrated time scale back into the Oligocene, about 30 Ma ago [15]. An amazing advance!

This incredible breakthrough refined the resolution of the geological time scale beyond anything previously imaginable, except where annual layering was preserved in tree rings, corals, stalagmites, or lake sediments, most of which did not allow dating back beyond about 2000 years ago. Palaeoclimatologists who started their careers in the 1980s and later take these advances for granted, but they were astonishing developments to the geologists of my age group. As Mike Leeder pointed out in 2011, this new understanding was '*arguably as big an earth sciences discovery as that of plate tectonics*' [30]. Mike Walker and John Lowe agreed, citing the 1976 Hays paper as '*perhaps the most important Quaternary paper of the past 50 years*' [3]. It definitively overturned the long-held perception that there had been four major Quaternary glacial periods [31], by showing there had been many more cycles of climate, ice volume and sea level in the Late Cenozoic, and that these cycles formed in response to variations in Earth's orbital parameters. As Nick McCave and Harry Elderfield point out in Shackleton's obituary '*This clear recognition of orbital control is also now revolutionizing the whole of stratigraphy (the study of geological strata) because it provides in principle a means of correlating beds at separated parts of the Earth to a precision of 20000 years at a time of hundreds of millions of years ago, and of determining precise "orbitally tuned" age-calibrated stratigraphies back to about 250 Ma ago*' [32].

That was not all. Given that a large fraction of the climate was governed by celestial mechanics, and that Berger's data projected Earth's orbital properties and insolation far into the future (Figure 12.2), Hays, Imbrie, and Shackleton deduced that '*the long term trend over the next 20,000 years is towards extensive Northern Hemisphere glaciation and a cooler climate*' [20]. That's not exactly the picture we have today, but it's close. Berger's data shows that because the Earth's orbit is at present close to circular, the present warm Holocene interglacial should last some 30–50 Ka (Figure 12.2), of which we have already experienced 10 Ka [33, 34]. So, we should have 20 Ka or so more of relatively warm climate before the next glacial period. Imbrie's 1979 classic *Ice Ages: Solving the Mystery* ably summarizes these various dramatic developments [19], as, more recently, does Berger's 2012 review [28].

In the 1980s, the CLIMAP project was succeeded by the SPECMAP (Spectral Mapping) project, designed by Imbrie, Shackleton and others to produce continuous time series of Ice Age climate change from deep-sea sediments, and to facilitate studying their spectral properties. A key SPECMAP achievement was publication of a time-scale for the last 780 Ka, based on a $\partial^{18}O$ reference curve compiled by stacking together planktonic foraminiferal $\partial^{18}O$ records from five low- and middle-latitude sites. Stacking avoids local 'noise' interfering with the underlying signals [35]. Known as the SPECMAP stack, this curve, which was tuned (phase-locked) to the oscillations of precession and obliquity, provided a continuous geological time scale for the Late Pleistocene, the divisions of which were accurate to within ±5000 years, an astoundingly high resolution for

geological records. The SPECMAP $\partial^{18}O$ stack was improved and extended over time [36, 37]. In 2005, Lorraine Lisiecki, then at Brown University, and Maureen Raymo, then at Boston University, replaced it with a stack made from combinations of benthic foraminiferal data (Figure 12.4) [38, 39], which show less variability than planktonic data. Over 100 MISs have been identified, going back some 6.6 Ma. The stack provided a 'type section' against which new core measurements could be compared.

The LGM, or MIS 2, extended from 30 to 15 Ka ago. On average, compared with today, it was originally thought to be 5–6 °C colder globally. It was much colder in the Arctic, with temperatures in central Greenland depressed by as much as 20 °C [40]. At that time, much Arctic land lay beneath continental ice sheets, and the Arctic Ocean was mantled by continuous sea ice and entrapped icebergs. The lack of northward transport of warm, salty water during the winters made them exceptionally cold. Polar desert replaced tundra. Ice volume peaked about 21 Ka ago, after

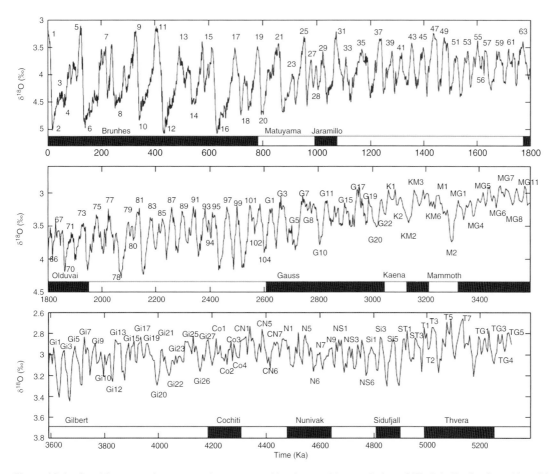

**Figure 12.4** Benthic oxygen isotope stack, constructed by the graphic correlation of 57 globally distributed benthic $\partial^{18}O$ records covering 5.3 Ma. Note that the scale of the vertical axis changes from panel to panel. From this stack, a number of new marine isotope stages were identified in the early Pliocene. Marine isotope stages are identified by MIS numbers back to 2.6 Ma; before that, the lettering refers to the name of the magnetic chron in which the isotope peaks appear (e.g. Si = Sidufjall, Co = Cochiti etc.). *Source:* From figure 4 in Ref. [38].

which rising insolation caused ice sheets and glaciers to melt back, with most coastlines becoming ice-free before 13 Ka ago [40]. We will discuss deglaciation later.

Although the variation in orbital and axial properties was smooth (Figure 12.2), the climate signal in ice cores was saw-toothed (Figure 12.4), something pointed out by Broecker and Van Donk back in 1970 [41]. The saw-tooth shape represents the slow growth of ice followed by rapid deglaciation, calling for strong positive feedback mechanisms to accelerate melting. We will examine that process more closely in Chapter 13.

CLIMAP's maps of sea surface temperature for the winter and summer of the LGM (21.5–18 Ka ago) were later updated by the Glacial Atlantic Mapping Project (GLAMAP) [42], which reconstructed the glacial Atlantic Ocean from 275 deep-sea sediment cores [43]. During the northern winter, sea ice extended south as far as about 50°N, close to the latitude of Cork, on the south coast of Ireland. During the northern summer, warm surface waters moved northwards into the Norwegian–Greenland Sea much as they do today, but between ice sheets on both sides, meeting sea ice at about latitude 70°N. As today, there was a return flow from the Arctic down the east side of Greenland. A proto-Gulf Stream swept eastward from Labrador to Lisbon where the sea surface temperature averaged about 16 °C.

Parallel with GLAMAP, a comparable project began to examine Environmental Processes of the last Ice Age: Land, Oceans, and Glaciers (EPILOG). It focused on the 21 Ka interval, this being the age of minimal summer insolation at 65°N, and the time when ice sheets reached their maximal volume, as represented by sea level fall [44]. GLAMAP and EPILOG were followed by the Multiproxy Approach for the Reconstruction of the Glacial Ocean project (MARGO) [45]. These various studies and others concluded that the last deglaciation began between 22 and 18 Ka ago. Two schools of thought emerged, one suggesting that deglaciation began in the Southern Hemisphere, with surface and deep-ocean warming followed by tropical sea surface temperatures and by atmospheric $CO_2$, while the other suggested that it began in the Northern Hemisphere where summer insolation at high northern latitudes was the trigger for ice sheet decay and sea-level rise [46]. We explore these issues in Chapter 13.

The MARGO Project team constructed maps of sea surface temperature to provide constraints on ocean cooling at the LGM [47]. Like CLIMAP, they found that the strongest mean annual cooling (−10 °C) occurred in the mid-latitude North Atlantic, extending into the western Mediterranean, but unlike CLIMAP they found that this cooling was most pronounced in the east. Indeed, most ocean basins had cooler eastern than western sides. This eastern cooling was probably due to increased coastal upwelling of cold water forced by stronger coastal winds. For reasons not then fully understood (in 2009), existing GCMs for the LGM failed to replicate this eastern cooling. In contrast with CLIMAP, the MARGO team found that conditions were ice-free in the summer in the Nordic seas, and that the tropics were on average 1.7 °C cooler than CLIMAP had thought. In the Southern Ocean the Polar Front shifted north from near 60°S to 45°S, associated with a cooling of 2 °C down to 6 °C in the austral winter.

One of the key features evident from the stack of $\partial^{18}O$ data (Figure 12.4), noted in 1976 by Shackleton and Opdyke [48], is that climate variability has grown with time [22]. In the earlier part of the Pleistocene, the signal comprised relatively small glacial to interglacial changes in which signals of precession (22 Ka cycles) and obliquity (41 Ka cycles) predominated. The signals became much larger at about 900 Ka, after which a signal with a spacing of 100 Ka intervals predominated. The 100 Ka signal itself became larger with time especially from around 430 Ka (MIS-11) onwards. The change at about 900 Ka ago formed the Mid Pleistocene Transition, or MPT. What did it represent?

Harry Elderfield and colleagues used Mg/Ca ratios to establish what part of the $\partial^{18}O$ signal at the Transition was due to ice volume rather than water temperature [49]. Changes in ice volume from glacial to interglacial were much smaller before the transition than since, presumably because the older ice sheets were smaller in area and/or in thickness. The transition was a sudden jump, not the result of a long-term trend towards increased ice volume and colder temperatures. Elderfield's team concluded that it represented '*an abrupt reorganization of the climate system*' [49]. The trigger seemed to be a brief period of anomalously low summer insolation in the Southern Hemisphere during the warm MIS-23 (see Figure 12.4). Low insolation cooled the climate, suppressing the melting of ice that had formed previously in cold MIS-24, allowing unusually extended ice growth in the following cold MIS-22, at 900 Ka ago, to yield a very large ice sheet associated with a lowering of sea level of about 120 m [49].

Investigating the behaviours of ice volume and temperature, Elderfield's team confirmed that ice volume followed a saw-toothed pattern, growing steadily from low amounts during interglacials to high amounts during glacials, then suddenly retreating. In contrast, bottom water temperatures followed a square wave pattern, falling to a certain level as ice volume grew, then staying more or less constant before rising again as ice volume decreased. The temperatures of bottom-waters during glacial periods remained constant at −1.5 to −2.0 °C, because once the temperature of surface waters in the source region fell to about the freezing point of salt water (−2.0 °C), it would fall no further. Bottom water temperatures warmed to 3 °C during interglacial periods.

Prior to the Transition, sea level fell to 70 m below present levels. The drop by a further 50 m at the Transition exposed continental shelf sediments to erosion, transferring marine organic matter rich in $^{12}$C to the deep sea and lowering the $\partial^{13}$C ratio of bottom water and benthic organisms. Elderfield's team calculated that about half of the fall in $\partial^{13}$C at 900 Ka ago was due to this change in carbon reservoirs, with the other half coming from a reduction in the influence of North Atlantic Deep Water (NADW) [49].

What caused the Transition? Michel Crucifix reminded us that the Transition involved mainly a marked cooling of glacial periods, whilst interglacials kept broadly the same temperatures through the past 2 Ma, and contained much the same amounts of $CO_2$ as each other [29]. Conceivably the transition was triggered by changes in some internal feedbacks. For example, as the volume of ice increased across the Transition, the supply of aeolian dust, represented by the rates of accumulation of sedimentary iron and terrestrial leaf waxes, doubled in the Southern Ocean [50]. The increase in the dust supply tells us that the surrounding lands dried out as the globe cooled. The increase in iron helped to cool the globe further, via positive feedback, because an increase in iron as a key nutrient stimulates productivity, drawing $CO_2$ from the atmosphere, as we see in more detail below [50]. This interpretation is supported by an increase in the sedimentation of opal, representing diatom productivity, at the same time. The rise in productivity drove a 30 ppm reduction in atmospheric $CO_2$ across the Transition. By driving a descent into deep, cold, glacial periods, the insolation/dust/$CO_2$ feedback may have initiated the strong 100 Ka periodicity that characterized subsequent climate change. On the global scale, an increase in the supply of dust also coincided with the start of the major Northern Hemisphere glaciation at about 2.6 Ma ago, which drew down $CO_2$ in much the same way.

Alternatively, for instance, the mechanical erosion of soils and sediments (known as the 'regolith') beneath the Laurentide Ice Sheet eventually exposed hard crystalline basement on which ice would have tended to stick rather than flow, hence encouraging the growth of thicker ice sheets, which, because of their height and coldness, may have resisted decay due to insolation on the 40 Ka time scale of the Early Pleistocene [51]. Willeit and colleagues agreed that the removal of the regolith was in part responsible for the cooling of the Quaternary, but they also showed that a long-term decrease in $CO_2$ was necessary for the initiation of the Northern Hemisphere glaciation and the increase in the amplitude of glacial–interglacial variations. The timing of the Mid Pleistocene Transition depended on *both* the decline in $CO_2$ and the removal of the regolith [52].

So, what drove the reduction in $CO_2$? The latest explanation of the Mid Pleistocene Transition comes from Farmer and colleagues in 2019. There was no shift in orbital characteristics at this time, but trace elements (Ba/Ca and Cd/Ca ratios) and Neodymium isotopes in planktonic foraminiferal skeletons show that carbonate ion saturation decreased and phosphate concentration increased significantly across the transition. These changes indicate that the carbon storage in the deep Atlantic increased by ~50Gt then, in response to a profound change in the Atlantic Meridional Overturning Circulation (MOC) that involved a 20% reduction in the supply of NADW to the abyssal South Atlantic. The weakening of the overturning circulation and the increase in carbon storage in the deep ocean drew down atmospheric $CO_2$, which cooled the atmosphere sufficiently to expand terrestrial ice volume [53]. The mechanism is thought to have operated along the following lines: first, by about 1 Ma ago, the continued cooling associated with the expansion of the East Antarctic ice sheet would have expanded the area of sea ice in both hemispheres; second, the expansion of the area of sea ice in the Southern Ocean reduced the contact of surface water with the atmosphere, leading to less loss of $CO_2$; third, the extraction of salt from the expanding sea ice increased the formation of Antarctic Bottom Water (AABW); fourth, expansion of sea ice in the north reduced the supply of NADW, which in turn reduced the supply of nutrients and $CO_2$ to the surface in the Southern Ocean. As a result, the storage of nutrients and $CO_2$ increased in the deep ocean [53]. Farmer's conclusions match those of Lear and colleagues in 2016, who found that '*the respired carbon content of glacial Atlantic deep water increased across the MPT. Increased dominance of corrosive bottom waters during glacial intervals would have raised mean ocean alkalinity and lowered atmospheric pCO₂*' [54]. Lear thought as did Farmer, that '*increased sea ice cover or ocean stratification during glacial times may have reduced $CO_2$ outgassing in the Southern Ocean, providing an additional mechanism for reducing glacial atmospheric pCO₂*' [54].

The notion of using a single stack of $\partial^{18}$O values to represent Earth's recent glacial history would have seemed odd to Joseph Adhémar and James Croll, whom we met in Chapter 3, because they thought that cooling related to precession would alternate between the two hemispheres. But, as we now know, glaciers, and ice sheets in Patagonia and Antarctica advance and retreat at more or less the same times as those in the Northern Hemisphere. Why? The answer lies in those feedbacks that Croll first introduced us to. Antarctica is an ice-covered continent surrounded by ocean – there is nowhere for its land ice to expand into. When Antarctic ice is at a maximum, global ice can only increase by growing on the Northern Hemisphere continents. That growth lowers sea level, exposing Antarctica's continental shelf and so providing space for yet more ice growth. Sea level links ice growth on Antarctica to that on the northern continents. Thus as mentioned earlier, glaciation tends to become more or less synchronous in both hemispheres, linked by changes in global elements such as atmospheric $CO_2$, deep-ocean circulation, and sea-level.

The latest view of the temperature of the LGM is that it was probably $4° \pm 0.8°C$ cooler globally than the modern pre-industrial climate [55], and $9°C$ colder in Antarctica

The growing literature on orbital variations and their record in cores from ice, sediments, corals, and stalactites fuelled intensive discussion about the precise mechanisms underlying the climate changes of the Ice Age, which we review in Chapter 13.

## 12.3 Flip-Flops in the Conveyor

Palaeoceanographic studies radically changed our understanding of the variability of Ice Age climate and the role of the ocean in climate change. One key result was the realization that the circulation of the ocean had different stable states for glacials and for interglacials. During interglacials like the Holocene, in which we live now, ocean circulation was much as it is today (Figure 12.5) [56].

Wally Broecker originally referred to the buoyancy-driven parts of this process (Figure 12.5) as the Thermohaline Conveyor Belt (from 'thermo', meaning heat and 'haline', meaning salt), which moves heat and salt around the globe [57]. The modern term for this is the MOC. However, as pointed out by Marshall and Speer in 2012 [56], Wally's concept neglected the wind driven contribution to the Meridional Overturning Circulation (MOC) that affects surface water in the Southern Ocean, as shown in Figure 12.6.

Oceanographers now see the process operating as follows. Warm salty surface water is drawn towards the Arctic through the northern branch of the Gulf Stream, losing heat to the atmosphere en route. This heat warms northwest Europe. By the time the salty water reaches the Norwegian–Greenland Sea, it has cooled to the point of becoming dense enough to sink and form NADW, which moves south towards Antarctica and fills the mid-water depths of the Atlantic, Indian, and Pacific Oceans. The strong westerly winds blowing around Antarctica towards the east force these northern-sourced deep waters to the surface through the process of upwelling. The newly upwelled surface waters then return to the North Atlantic

to close the cycle through two pathways. First, under the influence of Antarctica's coastal easterly winds some of them move south onto the Antarctic continental shelf, where the excretion of salt from sea ice forming at the ocean's surface makes that water dense enough to sink to the deep ocean floor as deep, cold AABW, which moves back to the north through the Atlantic, Indian, and Pacific Oceans. Because cold water dissolves larger amounts of oxygen from the atmosphere than does warm water, these deep waters are rich in oxygen and aerate the bottom of the world's oceans. Second, much of the rest of the deep water eventually mixes up to the surface in the Pacific and Indian Oceans, where it becomes entrained in the major surface currents that move west from the North Pacific through the Indonesian archipelago, across the Indian Ocean, down the East African coast in the Agulhas Current, and across the South Atlantic to the Equator in the Benguela Current, gaining salt and heat along the way. This water ends up feeding in to the southern end of the Gulf Stream to repeat the cycle. There is a third pathway not shown in Figure 12.5. Northward moving surface water in the Southern Ocean eventually sinks at the Polar Front near 60°S to form Antarctic Intermediate Water that circulates through the world ocean at depths of 600–1000 m, while Subantarctic Mode Water sinks somewhat further north and circulates through much of the ocean at depths of 200–300 m.

The upwelling process around Antarctica brings to the surface deep water enriched in nutrients and $CO_2$ from the decomposition of sinking organic remains brought there by the biological pump. These nutrients are pumped through the ocean in Intermediate and Mode Water and are brought to the surface by wind-driven upwelling along the Equator and along primarily the western margins of the continents, to feed the world's major oceanic fisheries.

The processes described in Figures 12.5 and 12.6 are key elements of the climate variability of the past 2.6 Ma, and much discussed in this and subsequent chapters, so it is advisable to become familiar with them. During glacial times the upper cell involving the production of deep water (green north to south arrows in both figures) is switched off. For more details on the Atlantic MOC see references by Marshall [56, 58].

The MOC is geologically young. It did not exist before the opening of the Drake Passage. Robbie Toggweiler from Princeton and H. Bjornsson from Iceland used experiments with an ocean model in 2001 to show that, prior to the opening of the passage, ocean temperature should have been symmetric about the Equator, with meridional overturning being driven by deep water formation at the poles in both hemispheres [59]. With the passage open, the overturning took the form of an interhemispheric conveyor with deep-water formation primarily in the Northern

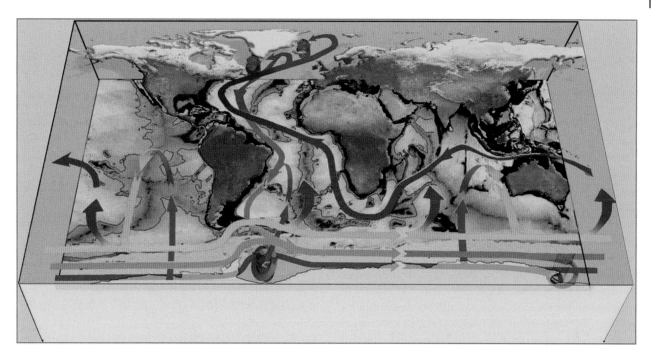

**Figure 12.5** Meridional overturning circulation, showing the directions and depths of cold, salty, oxygen-rich deep currents, warm surface currents, and vertical connections from deep to shallow and vice versa. Warm surface water moves from the Pacific through the Indian Ocean (red arrow) becoming saltier with time (purple to blue in North Atlantic). It becomes cold enough to sink to form North Atlantic Deep Water, which returns south at depth (blue becoming green). This water joins the Circumpolar Deep Water around Antarctica (green). Where it reaches the surface close to the continent it absorbs salt excreted by sea ice, and become dense enough to sink (green circular arrows (Weddell Sea and Ross Sea) to form Antarctic Deep Water (blue). This circulates around the continent and to the north, where it mixes upwards (blue to green to yellow arrows) and reconnects with the Circumpolar Deep Water. These are all density-driven processes. The strong westerly winds around the continent drive the upwelling of Circumpolar Deep Water, which reaches the surface and is driven north (yellow to red arrows). *Source:* From box 1 in Ref. [56].

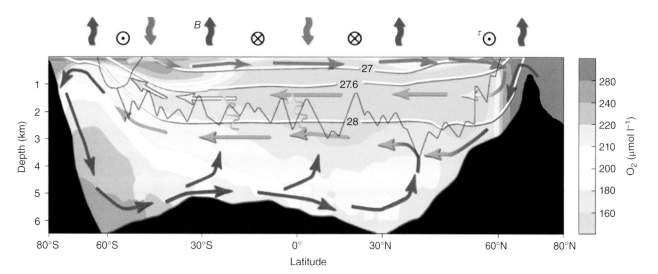

**Figure 12.6** Vertical profile of the Atlantic Meridional Overturning Circulation (AMOC) through the Atlantic (based on the plan view in Figure 12.5). Colours represent the zonally averaged oxygen distribution (yellows = low values, hence older water; purples = high values, hence recently ventilated water). Directions of currents based on density are shown by coloured arrows: red = less dense mode and thermocline waters; yellow = upper deep waters; green = lower deep waters including North Atlantic Deep Water (NADW); blue = bottom waters. Low-latitude, wind-driven shallow cells are not indicated. Density surfaces are in white, with the 27.6 kg m$^{-3}$ surface (thin black line) as the rough divide between a deeper south-moving cell and an upper (surface) return cell. The jagged thin black line is roughly the depth of the Mid-Atlantic Ridge and the Scotia Ridge (just downstream of Drake Passage) in the Southern Ocean, which can act as barriers to lateral movement at depth. Wiggly red and blue arrows show where surface waters are being made less dense (blue = rising subsurface water), or more dense (red = sinking surface water). Circles with dots = axis of westerly winds; those with x = easterly. Vertical squiggly green arrows = mixing associated with topography. *Source:* From figure 1 in Ref. [56].

Hemisphere. The conveyor made temperatures rise in the Northern Hemisphere and fall in the Southern Hemisphere, as the ocean transported heat north across the Equator especially in the Atlantic. The high salt content of the warm surface water allowed northern waters to become dense when cooled, thus driving the return flow at depth as NADW. While salinity differences are obviously important in driving the conveyor, Toggweiler's model showed that the conveyor could not be entirely driven by buoyancy. The westerly winds funnelled through Drake Passage do more than 'set the stage' for the work of the buoyancy forces in the North Atlantic; they are an indispensable part of the conveyor circulation, because they drive the upwelling of deep water around Antarctica to bring NADW to the surface as Circumpolar Deep Water [59].

In glacial times, the power of the overturning circulation was much reduced, as we can see from the geochemistry of microfossils from cores collected at different water depths. In 1982, Ed Boyle from MIT and Lloyd Keigwin from Woods Hole Oceanographic Institution (WHOI) found that the shells of benthic foraminifera contained variations down-core in the ratios of cadmium (Cd) to calcium (Ca) [60]. The ocean distribution of Cd follows that of nutrients, so the Cd/Ca ratio in these shells records variations in the nutrient content of bottom waters. NADW turned out to be relatively poor in Cd compared with deep water of southern origin. The Cd/Ca evidence told Boyle and Keigwin that the intensity of the northern source relative to the southern one diminished by a factor of two during severe glaciations. $\partial^{13}C$ values also document the distribution of nutrients, being low in old waters containing abundant dissolved organic carbon rich in $^{12}C$. Both Cd/Ca and $\partial^{13}C$ distributions suggest a strong stratification in the North Atlantic at the LGM, with a low nutrient, high $\partial^{13}C$ water mass (Glacial North Atlantic Intermediate Water) occupying depths down to around 2000 m, above a high nutrient, low $\partial^{13}C$ water mass of southern origin [61]. Laurent Labeyrie from the Centre National de la Recherche Scientifique (CNRS) laboratory in Gif-sur-Yvette, southwest of Paris, and a winner of the European Geophysical Union's Hans Oeschger Medal in 2005, confirmed that the overturning circulation of the glacial Atlantic was shallower and weaker than today's [62]. Production of NADW was much reduced during the LGM.

With the development of vast ice sheets on western Europe and North America, the sea's surface froze during the winter as far south at times as 40°N – the latitude of Boston in the west and Lisbon in the east. With the freezing over of the Norwegian–Greenland Sea there was no longer a significant source for NADW [63, 64]. Boyle realized that this icing up prevented deep-water formation [65]. The Cd/Ca signal showed strong periodicity at 41 ka, synchronous

with changes in the tilt of the Earth's axis, confirming that northern hemisphere ice cover (at least in the Norwegian–Greenland Sea) was controlled by insolation at high latitudes. Increasing tilt produced higher summer insolation and less ice cover [65].

The icing up of the North Atlantic north of 40°N at the LGM switched off the branch of the Gulf Stream that extended into the Nordic Seas, and deflected the Gulf Stream east between New York and Lisbon. Sea ice also extended 10° closer to the equator in the Southern Hemisphere, putting a lid on Southern Ocean processes. With the growth of ice on land, sea level fell 120–130 m. The growth of land ice and sea ice increased Earth's albedo significantly, helping to cool the planet.

## 12.4  Ice Age CO$_2$ Signal Hidden on Deep Sea Floor

Could carbon isotopes from marine sediments tell us about the abundance of $CO_2$ in the Ice Age atmosphere? Wally Broecker suggested that the atmospheric $CO_2$ signal could be represented by the difference in $\partial^{13}C$ ratios between surface planktonic foraminifera and bottom dwelling benthic foraminifera, an idea followed up by Nick Shackleton and colleagues in 1983 [66]. The variations they detected in $CO_2$ by this means matched those found in ice cores by Oeschger, as mentioned in Chapter 9. Clearly, $CO_2$ rose and fell with rises and falls in temperature during the Ice Age. How reliable was the association between estimated $CO_2$ and the $\partial^{18}O$ values used to estimate temperature? In 1985, working with Nick Pisias from Oregon State University to analyse the $\partial^{13}C$ and $\partial^{18}O$ profiles through the past 340 Ka in a core from 3091 m depth in the Pacific, Shackleton found that $CO_2$ closely followed temperature [67]. It slightly lagged solar insolation in June at 65°N, led the response in ice volume (documented by variations in $\partial^{18}O$) by about 2500 years, and was closely linked to variations in axial tilt, which dominate the insolation signal at middle to high latitudes. They concluded that the $CO_2$ signal was forced by high latitude insolation through '*a mechanism at present not fully understood*' [67] – probably the effect of insolation on ocean circulation. As changes in $CO_2$ led changes in ice volume in the North Atlantic, the $CO_2$ must have contributed to the forcing of changes in ice volume there. Although it was a forcing factor there, as we shall see later it was not a primary forcing factor in the Southern Ocean.

These conclusions were consistent with the proposal by Wally Broecker and Tsung-Hung Peng in their 1982 classic *Tracers in the Sea* that, since the ocean contains about 60 times more carbon than the atmosphere, the glacial-to-interglacial change in atmospheric $CO_2$ content must have been driven by changes in ocean chemistry [68]. Broecker seems to have been the first to suggest that a glacial increase in the strength of the biological pump drove down $CO_2$ levels. His initial ideas about the biological pump revolutionized the field of chemical oceanography [69].

Broecker was also part of a team that in 2016 considered how the chemistry of the deep ocean might have changed during the last glaciation, and what that meant regarding atmospheric $CO_2$, which declined markedly about 70 Ka ago as the Earth approached the LGM. Evidence for increased storage of carbon in the deep ocean had been elusive until then. The team used the Ba/Ca ratios of Atlantic benthic foraminifera to reconstruct a sharp decline in the carbonate ion concentration of the deep Atlantic between 80 and 45 Ka ago. '*This drop implies that the deep Atlantic carbon inventory increased by at least 50 Gt around the same time as the amount of atmospheric carbon dropped by about 60 Gt*' [70]. The team inferred that this carbon sequestration coincided with a weakening of the Atlantic MOC, due to weakening of the supply of NADW and a corresponding expansion of AABW throughout the deep Atlantic (expressed as upward vertical migration of the blue and green arrows in Figure 12.6), which must have '*played an important role in reductions of atmospheric $CO_2$ concentrations during the last glaciation, by increasing the carbon storage in the deep Atlantic*' [70].

Broecker's various ideas got John Martin (1935–1993) thinking. Director of the Moss Landing Marine Laboratories in California, and crippled by polio when he was 19, Martin set himself the task of figuring out the role of phytoplankton – the grass of the sea – in the global climate system. Phytoplankton use $CO_2$ for photosynthesis. When their remains sink to the seafloor and decompose, the $CO_2$ is returned to deep ocean waters or trapped in sediment so can no longer contribute to warming the planet. In order to determine how much plankton sank to the seafloor in a given time, Martin organized the Vertical Transport and Exchange of Oceanic Particulate Program (VERTEX) in 1981, placing sediment traps across the North Pacific to sample the flux and composition of settling particulates. Amongst other things, he discovered that the parts of the ocean that are high in nutrients but low in chlorophyll were depleted in iron (Fe) [71]. Joseph Hart, an English scientist, had speculated in the 1930s that this might be the case, but was unable to prove it. Martin proposed that iron (Fe) was a limiting nutrient, and that production of phytoplankton could be limited by its supply in airborne dust [72].

During the LGM, the supply of dust (hence Fe) was 50 times greater than today, enhancing productivity enough to draw $CO_2$ out of the air. Lack of Fe-rich dust during interglacials slowed productivity, leaving $CO_2$ in the air.

Martin suggested testing his hypothesis through Fe-enrichment experiments. Several such experiments were carried out, the first in 1993, although none at the scale of an entire ocean. The early experiments found that the excess organic matter created by the addition of Fe was recycled in the water column; it did not settle to the seabed as Martin had imagined [73]. That changed in 2012, when Victor Smetacek of Germany's Alfred Wegener Institute for Polar and Marine Research and colleagues performed a five-week long Fe-fertilization experiment in the Antarctic Circumpolar Current, and discovered that at least half of the diatom bloom caused by the fertilization sank far below 1000 m, with much being likely to have reached the deep sea floor [74]. This confirmed the view that '*iron-fertilized diatom blooms may sequester carbon for timescales of centuries in ocean bottom water and for longer in the sediment*' [74]. Here was support for the geo-engineering notion that iron fertilization of the ocean could transport $CO_2$ out of the surface waters and into the deeps, thus drawing down atmospheric $CO_2$.

The abundance of $CO_2$ in the air between glacial and interglacial times was also governed by the presence or absence of sea ice [75]. Growing sea-ice placed a lid on polar surface waters, preventing them from absorbing $CO_2$ from the air. As a result, when sea-ice expanded $CO_2$ would gradually accumulate in the air, causing it to warm. Melting of sea ice exposed the cold dark surface of the ocean, enabling it to absorb $CO_2$ from the air and contribute to eventual cooling.

## 12.5 A Surprise Millennial Signal Emerges

In the 1970s marine geologists dredged large angular boulders of continental rocks like granite from the Mid-Atlantic Ridge in the North Atlantic, and some people unwisely took this to mean that the ridge was made of continental rock [76]. By 1977, Bill Ruddiman, of Lamont Geological Observatory, had mapped wide swathes of ice-rafted glacial debris over much of the North Atlantic sea floor north of a line connecting Boston with Lisbon, and it was realized that the angular boulders were also ice-rafted [77–79]. More detailed studies of piston cores from the North Atlantic by German geologist Hartmut Heinrich (1952–) of the Deutsches Hydrographisches Institut in Hamburg, found ice-rafted debris (IRD) concentrated in six layers deposited half a precession unit (11 Ka) apart [80]. Heinrich's discovery, in 1988, aroused intense interest, and an international team formed under the leadership of Gerard Clark Bond (1940–2005), of Lamont, to investigate these sediments, which Bond's group named 'Heinrich Layers' [81].

Heinrich Layers formed between 10 and 60 Ka ago, apparently in response to surging within the Canadian ice sheet, which led to it breaking up into myriads of icebergs – an armada carrying rocks across the ocean. Charles Lyell would have been pleased, since this was the mechanism he had proposed back in the 1830s to explain the widespread distribution of glacial debris across western Europe. However, as we saw in Chapter 2, it is not reasonable to call on drifting icebergs to cover Europe – there we must call on a grounded ice sheet to explain the origin of the erratic boulders and associated boulder clay. Along with the melting icebergs came vast volumes of cold, fresh water, cooling the North Atlantic. Each event lasted around 750 years, and began suddenly – within about a decade. These outbreaks reached the European margin off Portugal, as I discovered in 1995, when I went to sea with Nick Shackleton on the maiden voyage of the new French research vessel *Marion Dufresne,* with Yves Lancelot as Chief Scientist. My goal was to collect samples for a project on 'Northeast Atlantic Palaeoceanography and Climate Change' that I had cooked up with geochemist John Thomson, from my institute at Wormley in Surrey. We cored the Portuguese continental margin en route from the Azores to Marseilles. I can heartily recommend French marine research expeditions to those with a taste for cordon-bleu cuisine and fine wines, a fitting reward for weeks of being away from home in very basic surroundings on a frequently storm-tossed sea.

Much to our delight, Thomson and I found in the *Marion Dufresne's* 40 m long giant piston cores certain layers rich in magnesium derived from dolomite rock deposited as IRD by the armadas of melting icebergs from Canada [82, 83]. Later I spent a short sabbatical at Lancelot's laboratory at Aix-en-Provence, working up some of the data from our long cores. I remember his kindness in loaning me a car to help me get around during my month there. It was also a pleasure to meet other prominent French palaeoclimatologists during my stay, amongst them Edith Vincent and Edouard Bard.

The periodic irruption of ice rafting in Heinrich events showed that the climate of the Ice Age was variable at the millennial scale, as well as at Milankovitch's orbital frequencies. Bond was keen to find out more about these millennial-scale processes, and in 1995, with Rusty Lotti, carried out close-spaced analyses down two cores collected west of Ireland and spanning the past 9–38 Ka [84]. Between each of the Heinrich Layers they found yet more layers of IRD, but of lesser magnitude. They deduced that iceberg calving had recurred at intervals of 2000–3000 years. Examining the rock grains in these layers showed that whilst the carbonate-rich ones came from Canada and were concentrated in Heinrich Layers, red

(haematite)-stained rock grains from multiple sources were also common, along with grains of volcanic glass that came from Iceland or Jan Mayen Island, in the other millennial scale layers.

By 1997, yet more detailed studies by Bond and his team suggested that red (haematite)-stained rock grains in these layers came from East Greenland or Svalbard [85]. The ice-rafting took place on average every $1536 \pm 563$ years. This cyclicity persisted into the Holocene, where the frequency was $1374 \pm 502$ years, statistically indistinguishable from the cyclicity of the glacial period [86]. In due course Bond was able to demonstrate that the same cycles appeared in Greenland ice cores [86], and might be related to solar cycles, something we discuss in Chapter 15. Averaging the Holocene and glacial signals gave a frequency of $1470 \pm 532$ years, which, Bond thought, '*reflects the presence of a pervasive, at least quasi-periodic, climate cycle occurring independently of the glacial–interglacial climate state*' [85]. Furthermore '*ocean circulation ... [is implicated] as a major factor in forcing the climate signal and in amplifying it during the last glaciation ... with a single mechanism – an oscillating ocean surface circulation, we can explain at once the synchronous ocean surface coolings, changes in IRD [ice-rafted debris] and foraminiferal concentrations, and changes in petrologic tracers*' [85]. Laurent Labeyrie's French team distilled this jargon-rich waffle down to conclude that a 'climate oscillator' caused the millennial oscillations [87].

In 2019, Zielhofer and colleagues showed that Bond's cycles can also be seen for example at Lake Sidi Ali in Morocco's Middle Atlas Mountain region inland from the westernmost Mediterranean (Figure 12.7) [88]. Cooling induced by the fall in summer insolation was associated with wet winters over the past 5000 years, whereas higher summer insolation in previous years was associated with dry winters (compare Figures 12.7d and f). Note the shift in Bond Cycles from higher to lower amplitude at the same time (c. 5000 years BP) (Figure 12.7). The Sahara dried out at about the same time, probably owing to a southward shift in the Inter-Tropical Convergence Zone (ITCZ).

Back in 1999, Bond's team confirmed the persistence of the 1470-year cycle back to 80 Ka ago [89]. Why weren't each of these ice-rafting events associated with massive iceberg outbreaks of the kind forming Heinrich Events? Bond thought that after a massive purge of ice during a Heinrich Event it took a few thousand years for the ice at source to grow back and reach the unstable conditions needed for another massive discharge [89]. Noting that Heinrich Events got closer together between MIS 4 and the LGM, he thought that this might reflect deterioration of global climate after the last interglacial, with the recovery time for the ice in Hudson Strait taking progressively less

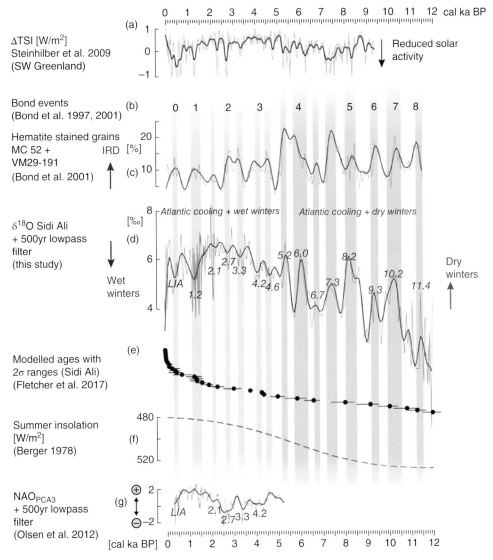

**Figure 12.7** Holocene North Atlantic ice-rafted debris record versus Western Mediterranean (Sidi Ali) winter rain record. (a) Total solar irradiance; (b) numbered Holocene Bond events 0–8; (c) Bond's Ice-rafted debris (IRD) record based on haematite-stained grains of cores from the subpolar North Atlantic – the black line shows results of a low-pass filter (500 years) removing centennial variability; (d) Improved Sidi Ali $\delta^{18}O$ record from closely related species *Fabaeformiscandona* sp. and *Candona* sp. – the grey line represents the original data, and the black line shows results of a low-pass filter (500 years); blue/red numbers and pale blue/orange bars indicate North Atlantic cooling events and wet/dry winters in the Western Mediterranean; (e) Modelled ages with 2 standard deviation ranges; (f) Summer insolation (65°N, June – note reversed axis); (g) Palaeo–North Atlantic Oscillation (NAO) record with a 500 year low-pass filter. See original reference for data sources. *Source:* From figure 3 in Ref. [88].

time as the climate cooled. Heinrich Events were not strictly periodic, and the oscillations were internal to the system.

Bond's team found signs of oceanic cooling leading up to Heinrich Events [89], including a southward extension of polar surface waters and disruption of the North Atlantic Current, the branch of the Gulf Stream that transports warm waters north. Mark Maslin, of the Environment Change Research Centre at University College London, suggested that some of Bond's 1500 year cooling events may have produced ice surges from Iceland and East

Greenland that were large enough to raise sea level to the extent that it undercut and destabilized the edges of the Laurentide ice sheet, so precipitating a full-scale Heinrich Event [90].

This rather begs the question – what caused the surges? In 2017, Jeremy Bassis proposed that it was the same mechanism that destabilizes Antarctic ice shelves today, namely subsurface ocean warming associated with variations in the overturning circulation, which increases underwater melting along the calving face of ice shelves, triggering rapid retreat of the ice margin and increased discharge of icebergs [91].

On millennial timescales, isostatic adjustment related to loss of ice mass causes the underlying bed to lift, isolating the subsurface grounding line from subsurface warming and allowing the ice sheet to re-advance until it is poised for another Heinrich Event [91]. We discuss this mechanism in more detail later.

Following up Bond's work, Bill Curry and his team from the WHOI, on Cape Cod, used $\partial^{18}O$, $\partial^{13}C$, and Cd/Ca ratios from benthic foraminifera from a core near Iceland to show that these cooling events occurred at more or less the same time as decreases in the production of NADW and cooling of surface waters in the western equatorial Atlantic [92]. After each Heinrich Event, warm, saline surface water re-entered the area, and thermohaline circulation resumed [92]. Labeyrie and his French team agreed that millennial ice rafting events were associated with cooling of the surface waters and southward extension of cold and low salinity Arctic waters, but found that the widespread low-salinity melt water accompanying these events delayed resumption of the production of deep water by several hundred years by forming a lid on the ocean [87].

Analysing one of the cores collected on the Iberia margin by the 1995 *Marion DuFresne* cruise, Shackleton discovered that the $\partial^{18}O$ oxygen isotopic signal from the *planktonic* foraminifera from between 12 and 50 Ka ago faithfully followed that from the Greenland GRIP ice core, which told his team that as the north polar region cooled, so did sea surface temperatures in the northeastern Atlantic off Portugal [93]. Much to his surprise, however, the $\partial^{18}O$ oxygen isotope signal in the *benthic* foraminifera followed that from Antarctica's Vostok ice core, where temperatures warmed as those in Greenland cooled. Evidently northern and southern temperature signals were out of phase, with the benthic signals representing the influx of AABW and, in turn, fluctuations in Antarctic ice volume [93]. Subsequently, by 2007, Belen Martrat, working with Shackleton on a suite of the *Marian DuFresne* cores from off Portugal, discovered that the millennial-scale variations in sea surface temperature typical of the last glacial period [93] recurred in each of the four main glacial periods back to 420 Ka ago [94]. They also confirmed the persistence of the out-of-phase link between water from northern and southern sources, cold episodes in the north occurring when the predominant source for deep water became southern (i.e. Antarctic) rather than northern. They found that '*Decreases in the $C_{26}OH$ ratio over the deglaciations trace the early arrival of NADW [North Atlantic Deep Water] from northern latitudes. During the warmest and largely ice-free portions of the interglacials, the benthic $\partial^{13}C$ record exhibits high values for nutrient-depleted waters arriving from northern locations. Maintenance of low $C_{26}OH$ ratios and high benthic $\partial^{13}C$ indicates conditions of high deep-water ventilation by NADW ...*

*[periods of] well-defined increases in the $C_{26}OH$ ratio, recording lower oxygenation and a slowdown in the intensity of bottom currents ... ended abruptly in harsh drops in [sea surface temperature] preceded by steep decreases in both $C_{26}OH$ and benthic $\partial^{13}C$ ratios, indicating a reinvigoration of the deep currents ... caused by the inflow of southern-sourced deep waters (AABW)[Antarctic Bottom Water] ... After the swift cold spell within a few centuries, a sharp movement back to warm SST [sea surface temperature] was marked by sudden decreases in deep water ventilation (increase in $C_{26}OH$ ratio) and a restraint in AABW flow (increase in benthic $\partial^{13}C$)*' [94]. Evidently, '*cold episodes after relatively warm and largely ice-free periods occurred when the dominant element of deep ocean circulation changed from North Atlantic Deep Water (NADW) to Antarctic Bottom Water (AABW). These changes occurred a few centuries before the subsequent generation of icebergs*' [94].

In a key divergence from the findings of Bond and others [84–86], Martrat and his team observed that '*The mean ~1500-year time spacing of the events of the last glacial is not observed in previous climate cycles. This is a further indication that a time scale of this nature is not a universal characteristic of the climate system*' [94]. They also observed that '*Whatever mechanisms were responsible, the combination of triggers, amplifiers, and sources of persistence during each climate cycle was never identical, and the abundance, length, and magnitude of the events varied accordingly. For example, the last ice age was characterized by abundant multicentennial shifts, but cold spells were equally or even more severe in the three previous climate cycles*' [94]. Indeed, they went on, '*The finding that SST [sea surface temperature] shifts become more abundant as climate cycles approach the present calls for a pressing need to understand the mechanisms of rapid climate change*' [94].

In 2014, a team led by Stephen Obrochta of the University of Tokyo expanded on Martrat's 2007 findings about the spacing of millennial events in prior glacial periods, by analysing deep-sea drilling sites from the central North Atlantic close to 50°N [95]. Obrochta's team found that millennial variability in the last glacial period (MIS 2–4) (30–65 Ma ago) was similar to that in MIS 8, but different from that in MIS 6 (135–185 Ma ago), where there was much less millennial variability [95]. They attributed that lack of variability to the location of the drill site at a place that was heavily influenced during MIS 6 by more or less continual sea ice, hence by a more or less continuous input of glacial meltwater and IRD. The drill site was a full 9° north of Martrat's sites off Lisbon, which lay well south of the normally expected sea ice cover. Modernizing the age scale for the drill core, Obrochta also found that the ice-rafting record did not follow Bond's proposed 1500 year cycle, but rather comprised distinct 1000 and 2000 year components, and, furthermore, that '*No systematic relationship exists between [ice-rafted grain] peaks and Greenland climate*' [95]. Indeed, it appears from recent work that '*little evidence exists for actual 1500-yr intervals of climate variability in either Greenland records* [96] *or the [ice-rafted grain] record from well-dated intervals of Site 609*' [95]. We will explore this millennial scale variability more in Chapter 13.

Changes at the millennial scale, like Heinrich Events, Bond Cycles and the Younger Dryas cold event at the end of the last glaciation (the latter event discussed in section 12.7 and chapter 13), are visible in the reconstructions of past sea surface temperatures made for the Mediterranean and North Atlantic from alkenone palaeothermometry [97]. The Canary Current transports these millennial signals south into the tropics along the coast of northwest Africa, where, as a result, the coldest time at the sea surface in the past 80 Ka (−12 °C) was not at the LGM at around 20 Ka ago, but during Heinrich Event 2, just before the LGM, and Heinrich Event-1 just after the LGM [98]. Surface currents also transported these signals, including all those seen in Greenland ice cores for the past 50 Ka, into the Mediterranean and affected the climate there too [99]. As off northwest Africa, the temperatures of these events were colder in the Mediterranean than were those of the LGM. As seen in Figure 12.7, these signals are also found in Atlantic Mediterranean lakes [88].

Although the armadas of icebergs in the Heinrich Events of the northern North Atlantic floated along in cold surface waters (as one might expect from the abundance of melting ice), careful recent analyses show that all six Heinrich Events of the last 70 Ka are associated with a rapid warming of surface waters by 2–4 °C that took place over just a few thousand years further south in the mid-latitude North Atlantic, in response to a northward expansion of the subtropical gyre [100]. At the time, then, we had '*an antiphased (seesaw) pattern in [sea surface temperatures] ... between the midlatitude (warm) and northern North Atlantic (cold)*' [100]. At the same time, Greenland experienced its longest and coldest stadials at the end of so-called Bond cycles [100].

Heinrich Events represent a large injection of freshwater into the ocean from an ice store on land. Not surprisingly, this gave rise to a sea level rise at far field locations of up to 15 m [101]. This would have affected North Atlantic hydrography, lessening the production of NADW, reducing northward heat flow and resulting in Antarctic warming [95]. Much the same pattern is suggested for MIS-8 as for MISs 2–4 [95].

Some of the millennial changes were quite fast, and, as mentioned above, coincided with large rapid climate changes recorded in ice cores from Greenland. The discovery of these large and sudden changes in piston cores and ice cores was a revelation, proving that the glacial period was far from being as stable as was once supposed. It now appeared

that slow steady changes in the climate of that time led eventually to 'tipping points' at which the climate changed to a different state, before eventually tipping back to its previous one [90]. Some of the most pronounced changes, especially around the northern North Atlantic, occurred within a few decades or even just a few years. These sudden step-like transitions would have had significant effects on human life at the time, making it prudent for us to reflect on what caused them in the past and what might do so in the future as global warming continues. Was this 'flickering' between one state and another typical just of glacial periods, or might it occur also in interglacials like the one we are now living through? We revisit that question in Chapters 13 and 14, when we explore these rapid changes in some detail, including their relationship to solar variability.

Before concluding this section we should note the conclusion of Julian Dowdeswell, of Scott Polar Research Institute, concerning the behaviour of ice sheet margins. The response of an ice sheet to a climatic event or a rise in sea level is not necessarily uniform from one ice margin to another [102]. Random processes could have led to surges in Northern Hemisphere ice sheets during the last glacial period, which means that even if some external forcing agent were involved there may be a random component to major Heinrich Events or their lesser ice-rafting equivalents. Readers wanting more detail on the competing models to explain periodic surging by ice sheets may find it useful to consult Cronin's *Paleoclimates* [46].

Regarding Labeyrie's mention of centennial variability being controlled by a climate oscillator [87], Michel Crucifix explains, '*the oscillator is a dynamical system that has a globally attracting limit cycle. In more simple terms it oscillates even in the absence of an external drive*' [103]. The system jumps back and forth between two stable states (end members) that limit the cycle. The two stable states provide the limits to the cycle, which operates rather like the hysteresis diagram of Figure 6.19. For instance, insolation (Figure 12.2) follows a limit cycle, being bounded by well-defined extremes (stable end members) that, in turn, keeps temperature and ice volume, represented by $\partial^{18}O$, within well-defined limits (Figure 12.4) along with other properties dictated ultimately by insolation, such as $\partial^{13}C$. As Crucifix explains [103], Heinrich events can also be seen as an example of the operation of a limit cycle. We will explore oscillators more in Chapter 13.

## 12.6 Ice Age Productivity

Prior to my *Marion DuFresne* cruise I had been trying to test the hypothesis that the increase in the steepness of the thermal gradient between the Equator and the poles during glacial times increased the strength of the Trade Winds and so enhanced upwelling on continental margins. I did so in 1995, using a piston core that I had collected back in 1973 from the continental slope off Namibia when I was Chief Scientist on one leg of *RV Chain* cruise 115 between Dakar and Cape Town [104]. I wanted to know if the upwelling associated with the Benguela Current changed with time, and if so how and when. The answer required assembling a multidisciplinary team.

To assess the temperature history I needed alkenone data. Fortunately, I knew Geoff Eglinton well, having first met him in the late 1970s, when we were both members of the Organic Geochemistry Panel advising the Deep Sea Drilling Project. Geoff agreed to provide the alkenone data we needed to determine sea surface temperatures over the past 70 Ka [104]. To our surprise, we found that sea surface temperatures were coldest and most productive during MIS-3 (60–24 Ka ago), a warm interstadial during the last glacial period. We deduced that the alongshore Trade Winds had been strongest during stage 3, thus driving more upwelling of cold, nutrient-rich, and highly productive water. In the globally colder isotopic stages above and below (MIS-2 and MIS-4), Benguela waters were slightly warmer and slightly less productive. Whilst this could indicate reduced wind strength, the evidence suggested that wind directions may have changed, there being more winds rich in desert dust blowing directly offshore, and fewer of the alongshore Trades that drove upwelling currents. In contrast, today's sea surface along that margin is very much warmer and less productive than it was in glacial times, though upwelling still prevails there and surface waters are still highly productive – this is one of the world's great fishing grounds. Our organic carbon signal fluctuated through time on a cycle of about 22 Ka, evidently driven by variations in the precession of the Earth's orbit.

Several other researchers were extracting climate signals from piston cores from close by in the southeastern Atlantic at the same time, and we pooled our resources to show that Heinrich Events, when icebergs were most abundant in the North Atlantic, were represented by warming oceanographic signals in the South Atlantic [105]. We explore the reasons for that unexpected hemispheric climatic connection in Chapter 13.

Looking at the Benguela Current system in rather more detail, a group of German researchers used the alkenone method to show that the warming characteristic of the LGM began before it, most probably in response to a change in the winds that allowed subtropical surface waters to move south down the coast from Angola [106]. That coincided with conditions less favourable for upwelling, which helps to explain the decrease in organic carbon accumulation we found on the continental slope off Walvis Bay.

Timothy Herbert of Brown University found much the same thing off southern California – just a slight cooling at the LGM close to the coast. In both of these environments the cores from the open ocean further offshore contained temperature profiles typical of those seen in the global SPECMAP stack, with the coldest sea surface temperatures at the time of maximum ice volume – the LGM [97]. Along the Benguela and California coast, then, upwelling was stronger and more productive than during the Holocene during glacials, including at the LGM, but was not as strong or productive at that maximum as it was in the interstadial period (MIS 3).

Was our finding that upwelling had decreased during peak glacial times (MIS 2) typical, I wondered? Yes. As Sigman and Haug explained, the coastal upwelling zones off California and Mexico in the north and off Peru in the south were less productive during the recent glacial period [69]. Sigman and Haug attribute this to the effect of continental cooling (and a large North American ice sheet in the case of the California Current) on the winds that currently drive coastal upwelling. Upwelling associated with monsoonal circulation in the Somali Current of the western Indian Ocean also decreased, because the cooling of the Tibetan Plateau weakened the southwest monsoonal winds. In contrast, upwelling was strengthened in the equatorial Indian Ocean where the northeast monsoonal winds remained strong; the same applied in the South China Sea. It was also increased in the eastern equatorial Pacific [107], and in the equatorial Atlantic [108].

Yet another geochemical technique helped to ascertain the history of productivity in the Southern Ocean during the Ice Age. Because the element thorium (Th) is rapidly adsorbed from the ocean onto sinking particles, and there is little lateral transport of dissolved Th from its site of production to its site of deposition, an isotope of thorium ($^{230}$Th) can be used as a proxy for the vertical downward flux of sediment. Use of this technique helped to determine the vertical fluxes of opal, barium, organic carbon and other proxies for palaeoproductivity in the Southern Ocean [109]. Compared with the Holocene, productivity was lower south of the Polar Front and higher north of the Polar Front during the LGM in the Atlantic and Indian Ocean sectors [110, 111]. The main planktonic organisms in these cold waters are siliceous diatoms. Diatom production shifted north as temperatures cooled. Whilst this applied in the Atlantic and Indian Ocean sectors, it did not apply in the Pacific sector, where productivity was lower in the LGM than in the Holocene.

The northward shift in productivity reflects northward migration of the oceanic fronts and their accompanying sea ice during glacial times. The absence of high productivity in the Pacific sector was probably due to its excessive distance from the westerly sources of dust that transported Fe to fertilize the ocean. The lower productivity of Antarctic waters during the recent glacial period was most likely due to decreasing supply of nutrient-rich deep water to the surface, resulting from the diminished supply of NADW, which would have driven a relative fall in atmospheric $CO_2$. In addition the prevailing westerly winds shifted northwards as the Hadley Cell shrank when glaciation increased, reducing upwelling in the Antarctic coastal sector. Besides that, more extensive cover of sea ice in the Southern Ocean limited the exposure of the ocean to the air, contributing further to a fall in atmospheric $CO_2$ [110–112]. At the LGM, sea ice was double its present extent both in winter and in summer [112]. Sigman and Haug suggest that salinity stratification associated with sea-ice was a major limiting factor on $CO_2$ exchange with the air during glacial times [69]. Whether there was a net change in total productivity of the Southern Ocean from the LGM to the Holocene remains a topic for debate [86]. It seems more likely that marine productivity stayed the same but underwent a lateral shift from south to north in the LGM. Regardless of what happened in the polar regions, studies of $^{230}$Th in the equatorial Pacific show little change in productivity from glacial to interglacial [109].

## 12.7 Observations on Deglaciation and Past Interglacials

The last deglaciation was the most massive change in Earth's climate in the past 25 Ka. The Northern Hemisphere ice sheets began to melt back around 21 Ka ago, as insolation and $CO_2$ began to rise. Rising seas contributed to the rapid decay of those ice sheets, encouraging an increase in the rate of flow of ice streams draining the interior, thus thinning the ice sheets and facilitating their collapse [113]. Increased melting formed large melt-water lakes on the southern fringes of the ice sheets, especially in North America where Lake Agassiz covered an area about the size of the Black Sea. Its remnants today form Lakes Winnipeg and Manitoba. The sudden drainage of the lake put a freshwater cap on the North Atlantic, shutting down the northern arm of the MOC. This cap was probably responsible for the Younger Dryas cold period or stadial interrupting the deglaciation between 12 800 and 11 500 years ago, which caused temperatures to drop 5 °C in the United Kingdom, for instance (more on that in Chapter 13). A further drainage from the lake gave rise to a brief cooling 8200 years ago, which we examine in Chapter 14. Other melt-water pulses occurred at around 14 200 and 11 000 years ago [114]. In many respects, the cold Younger Dryas period represented a temporary return to the glacial circulation pattern of reduced NADW. Surprising though it may seem, it was primarily a Northern Hemisphere phenomenon, although the alkenone data show sea surface temperatures fell by some 12 °C off western North America, and cooling of the same age in the South China Sea, the Indian Ocean and the South Atlantic [97]. The puzzle of how Lake Agassiz drained into the ocean was eventually solved. From gravels and a regional erosion plain in northern Canada, Julian Murton of the University of Sussex, and colleagues, showed in 2010 that it discharged into the Arctic Ocean along the path of the Mackenzie River [115].

Temperatures derived from new alkenone data can also be used to cheque the temperatures obtained for the LGM by CLIMAP researchers. The alkenone data show that the surface ocean was cooler then than CLIMAP researchers thought, but that the tropics cooled much less than the high latitudes, perhaps by only 1 °C [97].

Are past interglacials analogues for the Holocene – the interglacial we are now living in? The simple answer is: no. Interglacials are not all alike. Back in 2003, it seemed to André Berger that the modulating effect of the roughly 400 Ka cycle of eccentricity meant that the interglacial most similar to our own was that from roughly 400 Ka ago, during MIS 11 [116] (Figure 12.4). Analyses of the $\partial^{18}O$ ratios in samples of the right-coiling planktonic foraminifera *Neogloboquadrina pachyderma* from stage 11 in deep-sea drill cores in the northeast Atlantic show that sea surface temperatures varied by less than $\pm 1$ °C from the long term mean for at least 30 Ka during that stage [117]. The near-circular orbit of the Earth at the time prevented the 20 Ka precession signal from having much effect within this isotope stage. In effect, the Milankovitch cycle 'missed a beat', thus prolonging the interglacial to close on 50 Ka. MIS-11 was in effect about two precession cycles long, instead of one.

As André Droxler of Rice University in Houston, and his colleagues, pointed out in 2003, MIS 11 and the present interglacial are similar because their orbital variables are almost identical: '*both interglacials correspond to times when the eccentricity of the Earth orbit was at its minimum, so that the amplitude of the precessional cycle was damped*' [118] (see Figure 12.2). The strongest and longest Pleistocene interglacial, stage 11 (Figure 12.4) had prolonged intense warmth, sea level stands possibly up to 13–20 m above present levels [119], and significant poleward penetration of warm waters. It lasted twice as long as subsequent interglacial stages. The Holocene is likely to be just as long, as we shall see in later Chapters. In Chapter 13 we will look at possible explanations for these patterns.

The warming in stage 11 was important for the establishment of coral reefs. Wolf Berger from Scripps and Gerold

Wefer from the University of Bremen used deep-sea drill core data to show that the western Pacific warm pool of surface water expanded dramatically some 400 Ka ago, helping to explain the growth of Australia's Great Barrier Reef [120]. At that time, shallow carbonate platforms grew to the point where they clogged the flow of surface water through the Indonesian islands between the Pacific and Indian Oceans. The warming also triggered the establishment of other barrier reefs like that off Belize [118]. These reefs grew when the large rise in sea level at the end of the previous glacial maximum flooded fluvial plains extensively, preventing the former supply of river-borne silt and sand from reaching offshore reef sites. It is something of a paradox, then, that Wolfgang Kiessling and colleagues discovered in 2012 that during the last interglacial (125 Ka ago) warming was strong enough to decimate corals in the equatorial zone, and displace species away from the equator [121].

'*Will such warm conditions be replicated as the Holocene continues*'? asked Droxler and his colleagues [118]. Yes, they concluded, '*... we can expect another ~20,000 years of interglacial conditions, independent of any anthropogenic forcing*' [118]. Is MIS 11 an exact analogue for the Holocene? Not according to David Hodell, then of the University of Florida at Gainsville, and colleagues, who found maxima in the $\partial^{13}C$ of the planktonic foraminifera *G. bulloides*, and in fragmented foraminiferal remains in stage 11 sediments from a deep-sea drill core from the Cape Basin off South Africa [122]. The same indicators in sediments from the last 100 Ka and the Holocene have much lower values, and the carbonate compensation depth was 600 m shallower in stage 11 than is it now. These patterns suggest a lowering in the concentration of carbonate ions in the ocean in MIS 11 times, possibly related to the massive building of barrier reefs in shallow waters then. Although atmospheric and oceanic feedbacks are not operating in exactly the same way today as they were then, a comparison of sea surface temperatures and of $\partial^{18}O$ ratios from benthic foraminifera in the southeast Atlantic shows that the 11.7 Ka of the Holocene are comparable to the first 12 Ka of MIS-1 [123].

What about the last interglacial, the Eemian, during MIS 5, which began at around 135 Ka ago and lasted until around 110 Ka ago (Figure 12.4)? Alkenone and Mg/Ca data from marine sediments suggest that it was warmer than the late Holocene by up to 3 °C – consistent with stage 5 experiencing significantly higher orbital insolation [97], as seen in Figure 12.2. These new data improve on the CLIMAP data, which suggested that there was little difference between stage 5 and today. It now seems likely that stage 5 was at least as warm as the early Holocene climatic optimum, because insolation was much higher then than it is today (Figure 12.2). We can extract more evidence of Eemian climate change from pollen and lake records in central Europe,

loess sediments from central China, and marine sediment cores from the eastern subtropical Atlantic. These data provide evidence for a single sudden cool event in the middle Eemian at about 120–122 Ka ago, showing that short, sharp cold periods can occur in interglacials.

Compared with today, global ice volumes were smaller and solar radiation was 13% stronger over the Arctic in summer during the Eemian interglacial. According to Gifford Miller and his team, sea ice and permafrost were vastly reduced, boreal forest expanded to the Arctic shore, and most Northern Hemisphere glaciers melted [40]. Summer temperature anomalies over Arctic lands were 4–5 °C above present values, especially in the Atlantic sector. Northern Canada and parts of Greenland were 5 °C warmer than today in summer, but Alaska and Siberia were only about 2 °C warmer. Interpreting marine data is complicated by the stratification of the Arctic Ocean, which commonly has a cool relatively fresh cap ($<-1$ °C) overlying warmer subsurface waters ($>1$ °C).

In 2011, André Berger and his colleague Qui Zhen Yin revisited the question of which interglacial was most similar to the Holocene, using a model, of intermediate complexity (LOVECLIM) and a 'factor separation' technique to evaluate the individual contributions of insolation at 65°N on the one hand and of greenhouse gases on the other hand [124]. They found that the insolation pattern of MIS-1 6–12 Ka ago ($CO_{2eq}$ at 264 ppm and insolation at 529 W/m$^2$) was more similar to MIS-19 780 Ka ago ($CO_{2eq}$ at 265 ppm and insolation at 533 W/m$^2$) than to MIS-11 405 Ka ago ($CO_{2eq}$ at 286 ppm and insolation at 522 W/m$^2$) (see Figure 2.4 for timing of MISs). '*Both MIS-1 and MIS-19 [were] underinsulated over the whole globe during boreal summer with a deep minimum over the low latitudes around the summer solstice, and [were] overinsulated during boreal winter with a maximum over the South Pole (a signature of a much lower eccentricity and a slightly larger obliquity as compared to the reference)*' [124]. In contrast, the obliquity of MIS-11 was smaller, and it showed a '*much cooler Earth in boreal summer*', making MIS-19 the better analogue for MIS-1 [124]. Consistent with observations the model confirmed that MIS-9, MIS-5 and MIS-11 were the three warmest interglacials, and MIS-17 and MIS 13 were the two coolest ones (much as suggested by the $\partial^{18}O$ data of Figure 12.4). The deep-sea records confirm that MIS-7 was the coldest interglacial of the past 430 Ka (Figure 12.4), although not quite as cold as MIS 13 and 17; the cooling effect of MIS-7's insolation and the weakness of its greenhouse warming contributed to this picture [124].

Comparing the relative contributions of greenhouse gases and insolation to the variation in annual mean temperature between the nine interglacials back to MIS-19, Yin and Berger found that greenhouse gases explained most of the variance globally, whilst insolation played a more important role in the northern high latitudes and greenhouse gases played a more important role in the southern high latitudes [124] (where the vast Southern Ocean is a major source and sink for $CO_2$).

Yin and Berger also analysed the distribution of tree cover in relation to insolation and $CO_2$, finding a correlation with $CO_2$, but a high positive correlation between tree cover, precession, and precipitation, reflecting the increase in monsoonal strength at times of high precession. In some interglacials the tree fraction increased over North Africa, but in others it decreased, with '*a systematic increase of the desert in North Africa peaking at MIS-11*' [124]. The next most desert-like Sahara occurred in MIS-1 (the Holocene). Obliquity has a greater influence than precession on precipitation and tree cover in the high latitudes [124].

Sea ice cover in interglacials in the Northern Hemisphere is correlated linearly with local summer temperature, which is negatively correlated with obliquity and eccentricity and mostly reflects insolation rather than the effect of greenhouse gases [124]. In the Southern Hemisphere the variation in winter is mostly explained by greenhouse gas concentration because insolation is low in the southern winter, whilst the summer sea ice cover is controlled by both $CO_2$ and insolation. '*The difference in the importance of [greenhouse gases] and insolation between the Arctic and Southern Ocean sea ice area is mainly due to their different geographical configuration, which leads to the climate responding more to local and seasonal forcing (insolation) over [the] Arctic but more to global and annual forcing [greenhouse gases] over the Southern Ocean*' [124].

In summary '*the insolation-induced global annual mean temperature is significantly correlated with obliquity … as a direct consequence of the influence of obliquity on both daily and annual insolations in the high latitudes of both hemispheres and of the strong climate feedbacks there. This correlation is much larger in the southern high latitudes than in the northern ones where the importance of precession increases in relationship with the vegetation feedbacks [which are] very well expressed in the tree fraction*' [124].

A final point to bear in mind is that interglacials as warm as the present one occurred for only about 10% of the time in the late Quaternary [125]. The climate of the past 800 Ka was predominantly cold (as implied by the $\partial^{18}O$ data in Figure 12.4). We live in geologically unusual times.

## 12.8  Sea Level

Rising sea level is one of the most highly visible results of a warming world, driven by the melting of ice on land and the expansion of seawater as it warms. These changes are termed 'eustatic' [125]. Although changes in sea level provide us with yet another proxy for past climate change – especially for ice volume – the relation between climate and sea level is not simple, as we saw in Chapter 11. As R. Lawrence Edwards of the University of Minnesota and his colleagues remind us [126], sea level can also change as the result of tectonic uplift or sinking of the Earth's surface, or isostatic changes through which land sinks beneath ice sheets but rises around their periphery to form a 'fore-bulge' – a process that reverses when the ice sheets melt. Only eustatic change is truly global. Tectonic and isostatic adjustments cause local or regional changes that complicate the extraction of a global sea level signal, as we saw in Chapter 10. Lyell knew all about tectonic effects, having observed that the so-called Temple at Serapis, in Italy, had first been partly drowned and then uplifted. These competing signals must be unravelled to separate the local from the global signal, as we will see in some detail in Chapter 14.

There is also the question of the rates of isostatic adjustment. For example, the Scandinavian ice sheet had melted away by about 6000 years ago, but Scandinavia is still slowly rising. So too is Scotland, which lost its ice long ago. In contrast, southern England, the southern edge of the Baltic, and the west coasts of Germany and the Netherlands, which were on the fore-bulge around the European ice sheets, are still slowly subsiding. Similarly, the parts of the northernmost USA and Canada that lay beneath the Laurentide ice sheet are now rising, whilst the southern states of the USA, which formed the fore-bulge area outside that ice sheet, are slowly sinking. Within the area of the former Laurentide ice sheet, its core region – Hudson's Bay – is slowly rising, although much of it is still depressed below sea level.

Past sea levels can be determined directly by using the $^{14}C$ or other radiometric techniques to date carbonates like reefs or other features that formed at or very close to sea level [126]. These techniques include U/Th dating, which involves calculating ages from radioactive decay relationships between $^{238}U$, $^{234}U$, and $^{230}Th$ isotopes; this is also known as $^{230}Th$ dating. A further check on accuracy can be obtained from U/Pa dating, in which ages are calculated from the relationships between Uranium-235 ($^{235}U$) and its daughter isotope, Protoactinium-231 ($^{231}Pa$). U/Th and U/Pa dating extend the range of $^{14}C$ dating (maximum 50 Ka) to 250 Ka ($^{230}Pa$) and 600 Ka ($^{230}Th$). These techniques, like AMS $^{14}C$ dating, came into their own after the mid-1980s, with the development of mass spectrometric measurements that reduced sample size, and increased the speed and precision of analyses. Even so, despite the accuracy of the dates, all estimates of past sea level come with some uncertainty.

In this section, we focus on how high sea level may have been during past warm interglacials. The data available to Edwards in 2003 suggested that sea levels were up to 20 m above today's level in MIS-11 (400 Ka ago); up to 29 m above in MIS-9 (330 Ka ago); up to 9 m above in MIS-7 (240 Ka ago); and around 5 m ± 3 m in MIS-5, the last interglacial (100 Ka ago) [126] (for timings see Figure 12.4). The rise in sea level from a low point of about 130 m below present during the LGM is known in some detail, thanks to comparable data from the New Guinea's Huon Peninsula, Tahiti, South East Asia's Sunda Shelf, and northwest Australia's Bonaparte Gulf. These are far-field sites remote from polar ice sheets, so isostatic adjustments are unimportant, and the data reflect a true global signal.

These estimates were refined for the last interglacial (MIS-5) by a team led by Robert Kopp, of Princeton, who in 2009 compiled a large number of indicators of local sea level change and applied a statistical approach for estimating global sea level [127]. They found a 95% probability that global sea level peaked at least 6.6 m higher than today, and a 67% probability that it exceeded 8 m, but only a 33% likelihood that it exceeded 9.4 m. Rates of sea level rise could have varied between about 56–92 cm per century. For comparison the present rate of sea level rise is currently around 33 cm per century. The last interglacial was only slightly warmer than the present – by about 2 °C. Achieving a sea level rise in excess of 6.6 m higher than present '*is likely to have required major melting of both the Greenland and West Antarctic ice sheets*' they concluded [127].

In 2016, Michael O'Leary and colleagues found that during the last interglacial mean sea level remained relatively stable at about 3–4 m above present sea level between 127 and 119 Ka ago, then rose rather abruptly to about 9 m above present at the end of the interglacial about 118 Ka ago [128]. O' Leary inferred that a critical climate threshold was crossed during the last years of the interglacial, causing '*the catastrophic collapse of polar ice sheets and substantial sea level rise*' [128]. We need to factor possible changes like this into our view of what is possible in a warming world.

Eelco Rohling from the University of Southampton used data from the Red Sea to suggest, like Kopp, that during the last interglacial, sea level reached a mean position of +6 m, with individual short-term peak positions up to about +9 m compared with today's level. The rates of rise of sea level, according to Rohling and colleagues, were about 1.6 m per century, which '*would correspond to disappearance of an ice sheet the size of Greenland in roughly four centuries*' [129]. This rate of rise occurred when global mean temperature was 2 °C higher than today. Kurt Lambeck (1941–) (Box 12.3) of the Australian National University evaluated sea level for the last interglacial by using tectonically stable

---

**Box 12.3   Kurt Lambeck**

Kurt Lambeck, professor of geophysics at the Australian National University in Canberra, was born in Utrecht in the Netherlands. From 2006 to 2010 he was President of the Australian Academy of Science. He has been honoured with several awards, among them fellowship in the French and US Academies of Science and London's Royal Society, the international Balzan Price (2012), and the Wollaston Medal of the Geological Society of London (2013).

---

sites in the 'far-field', estimating that it was 5.5–9 m above today's level [130], consistent with the findings of both Kopp and Rohling.

Maureen Raymo of Boston University, and one of Kopp's team, Jerry Mitrovica from Harvard, looked into the contentious suggestion that Pleistocene shoreline features on the tectonically stable islands of Bermuda and the Bahamas were more than 20 m higher than today in MIS-11, some 400 Ka ago [131]. They found both sites to be located on the outer edge of the peripheral bulge of the Laurentide ice sheet. To account for post-glacial crustal subsidence at these sites, the elevations of these shoreline features were adjusted by about 10 m, which reduced global eustatic sea level rise to ~6–13 m above today's level in the second half of MIS-11. The rise was caused by prolonged warmth leading to collapse of both the Greenland and West Antarctic Ice Sheets. Given that the likely maximum rises in sea level for the melting of the Greenland and West Antarctic Ice Sheets are 7 and 5 m respectively, the estimated rise of 6–13 m suggests that changes in the volume of the East Antarctic Ice Sheet must have been minor.

Roland Gehrels, of the University of Plymouth, drew attention to other flaws in the analysis of sea-level change, focusing on the rise of sea level since the LGM [132]. Fairbanks' classic paper on sea level rise, published in 1989, and based largely on data from Barbados, suggested that sea level was 120 m below present at the LGM [133]. Gehrels cited three possible sources of error in Fairbanks' data. First, Barbados lies in an active tectonic setting on the edge of the Caribbean Plate. Even slight tectonic changes could have affected the absolute amount of sea-level change registered on the island. Second, Barbados lay on the trailing edge of the glacial fore-bulge pushed up around the margins of the Laurentian ice sheet, and the collapse of that feature would have created further vertical change. Third, Fairbanks' curve included data from other islands (Martinique, Bahamas, Puerto Rico, and St Croix), '*thereby introducing errors resulting from differential isostatic movements and regional sea-level variations*' [132].

Focusing on far-field sites, Yokoyama, working with Lambeck, calculated in 2000 that global sea level was as low as 130–135 m below present levels at the LGM [134]. Claire Waelbroeck and colleagues provided much the same picture, with a maximal lowering to –135 m at the LGM [135]. They also calculated that sea level fell to about −125 m in MIS 6 (140 Ka ago) and MIS 10 (345 Ka ago), and to about −110 m during MIS 8 (250 Ka ago). Their values differ significantly from those derived by Nick Shackleton [136], and provide good reasons for discounting Shackleton's data. Lambeck recently estimated sea level lowering at the LGM as −134 m [137], explaining that this was a measure of grounded ice volume, including ice grounded on continental shelves. Along far-field continental margins the LGM sea levels would generally be less than this due to isostatic/gravitational effects, while in mid-oceans they would exceed this [138].

When it was published in 1989, the Barbados sea-level curve [133] gained a great deal of attention because it showed evidence for episodes of very rapid sea level rise, most notably the event known as melt-water pulse 1a, dated to about 14 Ka ago, when sea levels rose by 15–25 m at rates of over 40 mm/year [132]. The jury is still out with respect to the source of the melt-water pulse, which could have originated in surges of the Laurentian or the Antarctic ice sheet, or from the discharge of large glacial lakes.

Clearly, sea level has changed through time in response to the waxing and waning of ice sheets, and measurements of past sea level can be used as a proxy for ice volume change. In order to refine these calculations further, there is much still to learn about regional variations in sea level, which depend on local tectonics and on glacial isostatic adjustments of the Earth's surface to the addition or removal of large masses of ice. Knowing that '*regional sea-level changes resulting from polar ice melt can depart by up to 30% from the global mean*', Gehrels concluded that '*regional sea-level variability precludes the use of the term "eustasy" in the traditional sense (i.e. global average sea-level change). The recognition that "eustasy" is only a concept should ... lead to [improved] regional sea-level predictions*' [132].

As we saw in Chapter 11, one way to avoid problems created by using past shorelines to establish past sea levels is to use the $\partial^{18}O$ composition of seawater, which is related to both ocean temperature and ice volume, both of which are, in effect, global. Subtracting the temperature signal enables us to determine ice volume and hence sea level [139, 140].

Using stable oxygen isotope analyses of planktonic foraminifera and bulk sediments from the Red Sea, Eelco Rohling and his team developed a relative sea level record for the past 520 Ka [119]. It shows a striking similarity to the record of Antarctic temperature, a relationship that remains the same regardless of whether the climate system is shifting towards glaciation or deglaciation, and which does not drift as one moves back through time. As this is a robust relationship within the climate system, it could be applied in estimating the effects of future climate change (see Chapter 16).

Jacqueline Austermann of Harvard, and colleagues, agreed with the revisions to the Fairbanks model of sea-level change [141]. Their model confirmed that at the LGM sea level should have been lowered to about −130 m, not −120 m as Fairbanks thought. That left a significant volume of ice in the Northern Hemisphere unaccounted for; it appeared from sea level data that more ice must have melted than had been available in the ice sheets of Laurentia (North America) and Fennoscandia. A joint team from Germany's Alfred Wegener Institute and the Korean Polar Research Institute discovered in 2013 that furrows that arise when large ice sheets become grounded on the seabed are widespread on the seabed off the coast of northeast Siberia. The team estimated that the furrows represent the former existence of an Arctic ice sheet that covered an area at least as large as Scandinavia and was up to 1200 m thick. This previously missing ice may well explain the accounting discrepancy [142].

The study of sea level is worth an entire book. Readers wishing to probe further might like to start with a 2010 compilation of global data entitled *Understanding Sea-Level Rise and Variability* [143].

## 12.9 Natural Climatic Envelopes

Returning to the astronomical calculations, it is now abundantly clear that regular changes in the Earth's orbit and axial tilt cause the amount of insolation we receive to vary within narrow limits, a discovery as influential in its own way as plate tectonic theory. The limits define a 'natural envelope' in which the maxima and minima are seldom if ever exceeded. These limits in turn apply to global temperature, which varied over the narrow global range of 4–5 °C between glacial and interglacial times. Back in 1982, Wolf Berger realized that this was '*a striking phenomenon, important especially for the survival of higher organisms*' [144]. That Ice Age natural envelope was superimposed on a background climate whose extremes varied within another natural envelope in which, as we saw at the end of Chapter 9, the variation in $CO_2$ was driven by plate tectonic processes including the emission of $CO_2$ from volcanoes, and its extraction by weathering, especially in mountainous areas, and by sedimentation in growing ocean basins. The natural envelope of $CO_2$ from 1000 to 200 ppm

(Chapter 9) was only occasionally exceeded when unusual hothouse conditions prevailed, as we saw in Chapter 10.

To summarize, our view of Pleistocene climate changed dramatically from the mid-1960s onwards when piston corers and deep ocean drilling enabled us to study for the first time the climate history recorded over the 66% of the Earth's surface covered by water depths of more than 200 m. Application of novel palaeontological and geochemical techniques showed that Earth had experienced many more substantial variations in climate than was apparent from studies of glaciation on land, where the advances of later glaciers and ice sheets removed the records of earlier ones. The realization that changes in orbital insolation were intimately linked to changes in temperature and ice volume enabled palaeoclimatologists to tune their signals of climate change to orbital changes, thus deriving a novel method for dating core horizons to an unheard of accuracy of ±2000 years, over periods of more than one million years. Furthermore, as clockwork variations in orbital insolation could be projected into the future, it became possible to estimate the extent, duration, and timing of the next glaciation.

The growing global array of deep ocean cores enabled comparisons of glacial and interglacial conditions caused by changes in the extent of sea ice and in the ocean's Meridional Overturning Circulation. These comparisons confirmed the validity of orbital cycles, and highlighted the saw-toothed pattern of actual climate change, reflecting the slow build-up of ice sheets and their rapid eventual demise, a pattern suggesting that once warming caused melting to reach some critical rate, land ice reservoirs collapsed to produce a glacial termination [144]. Carbon isotopes could be used to estimate the amount of $CO_2$ in the air, showing that $CO_2$ varied with temperature, presumably because carbon reservoirs slowly built up in peat beds, rain forest debris, fine-grained organic-rich sediments, and deep ocean waters as ice accumulated, before decaying rapidly and releasing $CO_2$ as ice melted and the climate warmed [144].

$CO_2$ provided one positive feedback, affecting temperature. Sea ice provided another, firstly through its effect on albedo, and secondly through governing the exchange of $CO_2$ between ocean and atmosphere [144]. Dust provided a third, increasing fertilization of the ocean with iron in glacial times, thus enhancing $CO_2$ draw-down; its absence in interglacials had the opposite effect. Water vapour provided a fourth, following $CO_2$, influencing temperature, and governing change in the water cycle. Sea level provided a fifth, increasing, or decreasing the area of ocean available for exchanging $CO_2$ and water vapour between ocean and atmosphere. Sea levels in past interglacials may have been as high as 9 m above today's level.

A millennial level of natural variability became apparent from concentrations of IRD. Large glacial outbreaks formed Heinrich Events; small ones formed the roughly 1500-year Bond Cycles. They corresponded with cold periods in the North Atlantic and warm periods in the South Atlantic. The millennial fluctuations seemed to reflect periods of instability, or flickering, as the climate changed gradually from one stable state (interglacial) to another (glacial). They were present in all glacial periods dating back to 400 Ka ago, but not with the same roughly 1500-year periodicity that had seemed to characterize the latest glacial period.

During the latest deglacial transition, outbreaks from massive glacial lakes flooded the northern oceans from time to time, causing for example the northern freeze known as the Younger Dryas, the cooling centred on 8200 years ago, and possibly also the brief periods of exceptionally rapid rise in sea level during the latest deglaciation.

In the next chapter we will look at the exciting discoveries that the drilling of ice cores was revealing, and compare those with the amazing discoveries that were emerging from studies of the ocean floor. The history of fossil $CO_2$ in ice cores enables us to further explore the relationship between $CO_2$ and the curious 100 Ka climate cycle. We will also examine possible mechanisms for glacial–interglacial climate change.

# References

**1** Milankovitch, M. (1941). Kanon der erdbestrahlung and seine andwendung auf das eiszeitproblem, *Special Publication Vol.* **133**, Royal Serbian Academy Belgrade, 633 pp; English translation published by Israel Program for Scientific Translations, U.S. Dept of Commerce, 1969.

**2** Zeuner, F.E. (1945). *The Pleistocene Period – Its Climate, Chronology and Faunal Successions*, 447 pp. London: Hutchinson Scientific and Technical (2nd Ed, 1959).

**3** Walker, M. and Lowe, J. (2007). Quaternary science 2007: a 50-year retrospective. *Journal of the Geological Society of London* 164: 1073–1092.

**4** Weedon, G. (2003). *Time-Series Analysis and Cyclostratigraphy*, 259 pp (on line). Cambridge University Press.

**5** Berger, A. (1976). Long-term variations of daily and monthly insolation during the Last Ice Age. *Eos* 57 (4): 254.

6 Berger, A. (1976). Obliquity and general precession for the last 5 000 000 years. *Astronomy and Astrophysics* 51: 127–135.

7 Berger, A. (1977). Support for the astronomical theory of climatic change. *Nature* 268: 44–45.

8 Berger, A. (1978). Long-term variations of caloric insolation resulting from the Earth's orbital elements. *Quaternary Research* 9: 139–167.

9 Laskar, J., Robutel, P., Joutel, F. et al. (2011). A long-term numerical solution for the insolationquantities of the Earth. *Astronomy and Astrophysics* 428: 261–285.

10 Berger, A., Loutre, M.-F., and Yin, Q. (2010). Total irradiation during any time interval of the year using elliptic integrals. *Quaternary Science Reviews* 29: 1968–1982.

11 Elias, S. (2006). *Encyclopedia of Quaternary Science*. Amsterdam: Elsevier.

12 Arrhenius, G. (1952). Sediment Cores from the East Pacific, *Reports of the Swedish Deep Sea Expedition* 5, 227 pp.

13 Ericson, D.B. (1953). Sediments of the Atlantic Ocean, *Technical Report on Submarine Geology* 1, Lamont Geological Observatory, Palisades, New York, Columbia University, 34 pp.

14 Emiliani, C. (1955). Pleistocene temperatures. *Journal of Geology* 63: 538–578.

15 Hay, W.W. (2013). *Experimenting on a Small Planet: A Scholarly Entertainment*, 983 pp. New York: Springer.

16 Emiliani, C. and Flint, R.F. (1963). The Pleistocene record. In: *The Sea: Ideas and Observations on Progress in the Study of the Seas, Vol. 3, the Earth Beneath the Sea – History* (ed. M.N. Hill), 888–927. London: Wiley.

17 Revelle, R. and Suess, H.E. (1957). Carbon dioxide exchange between atmosphere and ocean and the question of an increase of atmospheric $CO_2$ during the past decades. *Tellus* 9 (1): 18–27.

18 Shackleton, N.J. and Turner, C. (1967). Correlation between marine and terrestrial Pleistocene successions. *Nature* 216: 1079–1082.

19 Imbrie, J. and Imbrie, K. (1979). *Ice Aged: Solving the Mystery*. Berkeley Heights, N.J: Enslow.

20 Hays, J.D., Imbrie, J., and Shackleton, N.J. (1976). Variations in the earth's orbit: pacemaker of the ice ages. *Science* 194 (4270): 1121–1132.

21 Shackleton, N.J. (1967). Oxygen isotope analyses and Pleistocene temperatures re-assessed. *Nature* 215: 15–17.

22 Lea, D.W. (2003). Elemental and isotopic proxies of past ocean temperatures. In: *The Oceans and Marine Geochemistry, Treatise on Geochemistry* (eds H.D. Holland and K.K. Turekian), vol. 6 (ed. H. Elderfield), 365–390. Oxford: Elsevier-Pergamon.

23 Imbrie, J. and Kipp, N.G. (1969). Quantitative interpretation of late Pleistocene climate based on planktonic foraminiferal assemblages in Atlantic cores (abstract), *Geol. Soc. Amer. Mtg. Program for 1969, pt.* **7**, 113.

24 Imbrie, J. and Kipp, N.G. (1971). A new micropaleontological method for quantitative paleoclimatology: application to a late Pleistocene Caribbean core. In: *Cenozoic Glacial Ages* (ed. K.K. Turekian), 71–181. New Haven: Yale University Press.

25 Imbrie, J., Van Donk, J., and Kipp, N.G. (1973). Paleoclimatic investigation of a Caribbean core: comparison of isotopic and faunal methods. *Quaternary Research* 3: 10–38.

26 CLIMAP Project Members (1976). The surface of the ice age Earth. *Science* 191 (4232): 1131–1137. [Members included A. McIntyre, T.C. Moore, B. Andersen, W. Balsam, A. Bé, C.Brunner, J. Cooley, T. Crowley, G. Denton, J. Gardner, K. Geitzenauer, J.D. Hays, W. Hutson, J. Imbrie, G. Irving, T. Kellog, J. Kennett, N. Kipp, G. Kukla, H. Kukla, J. Lozano, B. Luz, S. Mangion, R.K. Mathews, P. Mayewski, B. Molfino, D. Ninkovich, N. Opdyke, W. Prell, J.Robertson, W.F. Ruddiman, H. Sachs, T. Saito, N. Shackleton, H. Thierstein, and P. Thompson].

27 Berger, A. (1978). Long-term variations of daily insolation and Quaternary climatic changes. *Journal of Atmospheric Science* 35 (12): 2362–2367.

28 Berger, A. (2012). A brief history of the astronomical theories of paleoclimates. In: *Climate Change, Inferences from Paleoclimate and Regional Aspects* (eds. A. Berger, D. Mesinger and D. Sijacki), 107–129. Springer.

29 Crucifix, M. (2019). Pleistocene glaciations. In: *Climate Change in the Holocene, Impacts and Human Adaptation* (ed. E. Chiotis). CRC Publishers, Taylor and Francis https://doi.org/10.1201/9781351260244.

30 Leeder, M. (2011). *Sedimentology and Sedimentary Basins – from Turbulence to Tectonics*, 2e, 768 pp. Chichester: Wiley Blackwell.

31 Flint, R.F. (1971). *Glacial and Quaternary Geology*. New York: Wiley.

32 McCave, I.N. and Elderfield, H. (2011). Sir Nicholas John Shackleton. *Biographical Memoirs of Fellows of the Royal Society* 57: 435–462. https://doi.org/10.2307/41412890.

33 Berger A. and Loutre M.F. (1994). Astronomical forcing through geological time, Spec. Publs. **19**, Int. Assn. Sedimentol., 15–24

34 Berger, A. and Loutre, M.F. (2002). An exceptionally long interglacial ahead? *Science* 297: 1287–1288.

35 Imbrie, J., Hays, J.D., Martinson, D.G. et al. (1984). The orbital theory of Pleistocene climate: support from a revised chronology of the marine $\partial^{18}O$ record. In: *Milankovitch and Climate*, Part I (eds. A.L. Berger, J. Imbrie, J. Hays, et al.), 269–305. Dordrecht: D. Reidel.

36 Imbrie, J., Boyle, E.A., Clemens, S.C. et al. (1992). On the structure and origin of major glaciation cycles. I.

Linear responses to Milankovitch forcing. *Paleoceanography* 7: 701–738.

37 Imbrie, J., Berger, A., Boyle, E.A. et al. (1993). On the structure and origin of major glaciation cycles. II. The 100,000-year cycle. *Paleoceanography* 8: 699–735.

38 Lisiecki, L.E. and Raymo, M.E. (2005). A Pliocene-Pleistocene stack of 57 globally distributed benthic $\delta^{18}O$ records. *Paleoceanography* 20 https://doi.org/10.1029/2004PA001071, 17 pp.

39 http://lorraine-lisiecki.com/stack.html (December 2013).

40 Miller, G.H., Brigham-Grette, J., Alley, R.B. et al. (2010). Temperature and precipitation history of the Arctic. *Quaternary Science Reviews* 29: 1679–1715.

41 Broecker, W.S. and Van Donk, J. (1970). Insolation changes, ice volumes, and the O-18 record in deep-sea cores. *Reviews of Geophysics and Space Physics* 8: 169–197.

42 Pflaumann, U., Sarnthein, M., Chapman, M. et al. (2003). Glacial North Atlantic: sea-surface conditions reconstructed by GLAMAP 2000. *Paleoceanography* 18: 1065.

43 Sarnthein, M., Gersonde, R., Niebler, S. et al. (2003). Overview of glacial Atlantic Ocean mapping (GLAMAP 2000). *Paleoceanography* 18: 1030.

44 Mix, A.C., Bard, E., and Schneider, R. (2001). Environmental processes of the last ice age: land, oceans, and glaciers (EPILOG). *Quaternary Science Reviews* 20: 627–657.

45 Kucera, M., Rosell-Melé, A., Schneider, R. et al. (2005). Multiproxy approach for the reconstruction of the glacial ocean surface (MARGO). *Quaternary Science Reviews* 24: 813–819.

46 Cronin, T.M. (2010). *Paleoclimates: Understanding Climate Change Past and Present*, 441 pp. New York: Columbia University Press.

47 MARGO Project Members (2009). Constraints on the magnitude and patterns of ocean cooling at the Last Glacial Maximum. *Nature Geoscience* 2: 127–132.

48 Shackleton, N.J. and Opdyke, N. (1976). Oxygen isotope and paleomagnetic stratigraphy of Pacific core V28-239 late Pliocene to latest Pleistocene, in Investigation of Late Quaternary Paleoceanography and Paleoclimatology (ed. R. M. Cline and J. D. Hays). *Geological Society of America Memoirs* 145: 449–464.

49 Elderfield, H., Ferretti, P., Greaves, M. et al. (2012). Evolution of ocean temperature and ice volume through the mid-Pleistocene climate transition. *Science* 337: 704–709.

50 Martinez-Garcia, A., Rosell-Melé, A., Jaccard, S.L. et al. (2011). Southern Ocean dust-climate coupling over the past four million years. *Nature* 476: 312–315.

51 Clark, P.U., Archer, D., Pollard, D. et al. (2006). The middle Pleistocene transition: characteristics.

Mechanisms, and implications for long-term changes in atmospheric pCO$_2$. *Quaternary Science Reviews* 25: 3150–3184.

52 Willeit, M., Ganopolski, A., Calov, R., and Brovkin, V. (2019). Mid-Pleistocene transition in glacial cycles explained by declining CO$_2$ and regolith removal. *Science Advances* 5: eaav7337.

53 Farmer, J.R., Hönisch, B., Haynes, L.L. et al. (2019). Deep Atlantic Ocean carbon storage and the rise of 100,000-year glacial cycles. *Nature Geoscience* 12: 355–360.

54 Lear, C.H., Billups, K., Rickaby, R.E.M. et al. (2016). Breathing more deeply: deep ocean carbon storage during the mid-Pleistocene climate transition. *Geology* 44 (12): 1035–1038.

55 Annan, J.D. and Hargreaves, J.C. (2013). A new global reconstruction of temperature changes at the Last Glacial Maximum. *Climate of the Past* 9: 367–376.

56 Marshall, J. and Speer, K. (2012). Closure of the meridional overturning circulation through Southern Ocean upwelling. *Nature Geoscience* 5: 171–180.

57 Broecker, W.S. (1987). The great ocean conveyor. *Natural History Magazine* 97: 74–82.

58 Buckley, M.W. and Marshall, J. (2016). Observations, inferences, and mechanisms of Atlantic meridional overturning circulation variability: a review. *Reviews of Geophysics* 54 https://doi.org/10.1002/2015RG000493.

59 Bjornsson, H. and Toggweiler, J.R. (2001). The climatic influence of Drake Passage. In: *The Oceans and Rapid Climate Change: Past, Present and Future, Geophysical Monograph* **126** (eds. D. Seidov, B.J. Haupt and M. Maslin), 243–259. Washington, DC: American Geophysical Union.

60 Boyle, E.A. and Keigwin, L.D. (1982). Deep circulation of the North Atlantic over the last 200,000 years: geochemical evidence. *Science* 218 (4574): 784–787.

61 Lynch-Stieglitz, J. (2003). Tracers of past ocean circulation. In: *The Oceans and Marine Geochemistry, Treatise on Geochemistry* (eds H.D. Holland and K.K. Turekian), vol. 6 (ed. H. Elderfield), 433–451. Oxford: Elsevier-Pergamon.

62 Labeyrie, L.D. (1992). Changes in the vertical structure of the North Atlantic Ocean between glacial and modern times. *Quaternary Science Reviews* 11: 401–413.

63 McIntyre, A., Kipp, N.G., Bé, A.W.H. et al. (1976). Glacial North Atlantic 18,000 years ago; a CLIMAP reconstruction. *Geological Society of America Memoirs* 145: 43–76.

64 CLIMAP (1981). Seasonal reconstructions of the Earth's surface at the last glacial maximum, *Map Series, Technical Report* **MC-36**, Geological Society of America, Boulder, Colorado.

65 Boyle, E.A. (1984). Cadmium in benthic foraminifera and abyssal hydrography: evidence for a 41 kyr obliquity

cycle. *Geophsical Monograph*. **29**, AGU Washington D.C., 360–368.

66 Shackleton, N.J., Hall, M.A., Line, J., and Cang, S. (1983). Carbon isotope data in core V19-30 confirm reduced carbon dioxide concentration of the ice age atmosphere. *Nature* 306: 319–322.

67 Shackleton, N.J. and Pisias, N.G. (1985). Atmospheric carbon dioxide, orbital forcing, and climate. In: *The Carbon Cycle and Atmospheric CO$_2$: Natural Variations Archean to Present, Geophysical Monograph* **32** (eds. E.T. Sundquist and W.S. Broecker), 303–317. Washington, DC: American Geophysical Union.

68 Broecker, W.S. and Peng, T.H. (1982). *Tracers in the Sea*, 690 pp. LDGO, Columbia University.

69 Sigman, D.M. and Haug, G.H. (2003). Biological pump in the past. In: *The Oceans and Marine Geochemistry, Treatise on Geochemistry* (eds H.D. Holland and K.K. Turekian), vol. 6 (ed. H. Elderfield), 491–528. Oxford: Elsevier-Pergamon.

70 Yu, J., Menviel, L., Jin, Z.D. et al. (2016). Sequestration of carbon in the deep Atlantic during the last glaciation. *Nature Geoscience* 9: 319–324.

71 Martin, J.H. and Fitzwater, S.E. (1988). Iron deficiency limits phytoplankton growth in the North-East Pacific Subarctic. *Nature* 331 (6154): 341.

72 Martin, J.J. (1990). Glacial-interglacial CO$_2$ change: the iron hypothesis. *Paleoceanography* 5 (1): 1–13.

73 De La Rocha, C.L. (2003). The biological pump. In: *The Oceans and Marine Geochemistry, Treatise on Geochemistry* (eds H.D. Holland and K.K. Turekian), vol. 6 (ed. H. Elderfield), 83–111. Oxford: Elsevier-Pergamon.

74 Smetacek, V., Klaas, C., Strass, V.H. et al. (2012). Deep carbon export from a Southern Ocean iron-fertilized diatom bloom. *Nature* 487: 313–319.

75 Frakes, L.A., Francis, J.E., and Syktus, J.I. (1992). *Climate Modes of the Phanerozoic – The History of the Earth's Climate Over the Past 600 Million Years*, 274 pp. Cambridge University Press.

76 Meyerhoff, A.A. and Meyerhoff, H.A. (1972). The new global tectonics: major inconsistencies. *Bulletin of the American Association of Petroleum Geologists* 56 (2): 269–336.

77 Ruddiman, W.F. (1977). Late Quaternary deposition of ice-rafted sand in the subpolar North Atlantic (Lat. 40° to 65°N). *Bulletin Geological Society of America* 88: 1813–1827.

78 Ruddiman, W.F. and McIntyre, A. (1977). Late Quaternary surface ocean kinematics and climate change in the high-latitude North Atlantic. *Journal of Geophysical Research* 82 (27): 3877–3887.

79 Ruddiman, W.F. (1977). North Atlantic ice rafting: a major change at 75,000 yr B.P. *Science* 196: 1208–1211.

80 Heinrich, H. (1988). Origin and consequences of cyclic ice rafting in the Northeast Atlantic Ocean during the past 130,000 years. *Quaternary Research* 29: 142–152.

81 Bond, G.C., Heinrich, H., Broecker, W.S. et al. (1992). Evidence for massive discharges of icebergs into the North Atlantic Ocean during the last glacial period. *Nature* 360: 245–249.

82 Thomson, J., Nixon, S., Summerhayes, C.P. et al. (1999). Implications for sediment changes on the Iberian margin over the last two glacial/interglacial transitions from ($^{230}$Th-excess) systematics, Earth and Plan. *Science Letters* 165: 255–270.

83 Thomson, J., Nixon, S., Summerhayes, C.P. et al. (2000). Enhanced productivity on the Iberian margin during glacial/interglacial transitions revealed by barium and diatoms. *Journal of the Geological Society of London* 157: 667–677.

84 Bond, G.C. and Lotti, R. (1995). Iceberg discharges into the North Atlantic on millennial time scales during the last glaciation. *Science* 267: 1005–1010.

85 Bond, G., Showers, W., Cheseby, M. et al. (1997). A pervasive millennial-scale cycle in North Atlantic Holocene and glacial climates. *Science* 278: 1257–1266.

86 Bond, G.C., Kromer, B., Beer, J. et al. (2001). Persistent solar influence on North Atlandtic climate during the Holocene. *Science* 294: 2130–2136.

87 Labeyrie, L., Leclaire, H., Waelbroeck, C. et al. (1999). Temporal variability of the surface and deep waters of the north west Atlantic Ocean at orbital and millennial scales. In: *Mechanisms of Global Climate Change at Millennial Time Scales, Geophysical Monograph* **112** (eds. P.U. Clark, R.S. Webb and L.D. Keigwin), 77–98. Washington, DC: American Geophysical Union.

88 Zeilhofer, C., Köhler, A., Mishke, S. et al. (2019). Western Mediterranean hydro-climatic consequences of Holocene ice-rafted debris (Bond) events. *Climate of the Past* 15: 463–475.

89 Bond, G.C., Showers, W., Elliot, M. et al. (1999). The North Atlantic's 1-2kyr climate rhythm: relation to Heinrich Events, Dansgaard/Oeschger Cyclcs and the Little Ice Age. In: *Mechanisms of Global Climate Change at Millennial Time Scales, Geophysical Monograph* **112** (eds. P.U. Clark, R.S. Webb and L.D. Keigwin), 35–58. Washington, DC: American Geophysical Union.

90 Maslin, M., Seidov, D., and Lowe, J. (2001). Synthesis of the nature and causes of rapid climate transitions during the Quaternary. In: *The Oceans and Rapid Climate Change: Past, Present and Future, Geophysical Monograph* **126** (eds. D. Seidov, B.J. Haupt and M. Maslin), 9–52. Washington, DC: American Geophysical Union.

91 Bassis, J.N., Petersen, S.V., and Cathles, L.M. (2017). Heinrich events triggered by ocean forcing and modulated by isostatic adjustment. *Nature* 542: 332–334.

**92** Curry, W.B., Marchitto, T.M., McManus, J.F. et al. (1999). Millennial-scale changes in ventilation of the thermocline, intermediate, and deep waters of the glacial North Atlantic. In: *Mechanisms of Global Climate Change at Millennial Time Scales*, *Geophysical Monograph* **112** (eds. P.U. Clark, R.S. Webb and L.D. Keigwin), 59–76. Washington, DC: American Geophysical Union.

**93** Shackleton, N.J., Hall, M.A., and Vincent, E. (2000). Phase relationships between millennial-scale events 64,000–24,000 years ago. *Paleoceanography* 15: 565–569.

**94** Martrat, B., Grimalt, J.O., Shackleton, N.J. et al. (2007). Four climate cycles of recurring deep and surface water destabilizations on the Iberian margin. *Science* 317: 502–507.

**95** Obrochta, S.P., Crowley, T.J., Channell, J.E.T. et al. (2014). Climate variability and ice-sheet dynamics during the last three glaciations. *Earth and Planetary Science Letters* 406: 198–212.

**96** Ditlevsen, P.D., Andersen, K.K., and Svensson, A. (2007). The DO-climate events are probably noise induced: statistical investigation of the claimed 1470 years cycle. *Climate of the Past* 3 (1): 129–134. http://dx.doi.org/10.5194/cp-3-129-2007.

**97** Herbert, T.D. (2003). Alkenone paleotemperature determinations. In: *The Oceans and Marine Geochemistry*, *Treatise on Geochemistry* (eds H.D. Holland and K.K. Turekian), vol. 6 (ed. H. Elderfield), 391–432. Oxford: Elsevier-Pergamon.

**98** Zhao, M., Beveridge, N.A.S., Shackleton, N.J., and Sarnthein, M. (1995). Molecular stratigraphy of cores off Northwest Africa: sea surface temperature history over the last 80ka. *Paleoceanography* 10: 661–675.

**99** Cacho, I., Grimalt, J.O., and Canals, M. (2002). Response of the western Mediterranean Sea to rapid climatic variability during the last 50,000 years: a molecular biomarker approach. *Journal of Marine Systems* 33–34: 253–272.

**100** Naafs, B.D.A., Hefter, J., Grutzner, J., and Stein, R. (2013). Warming of surface waters in the mid-latitude North Atlantic during Heinrich events. *Paleoceanography* 28: 153–163.

**101** Yokoyama, Y., Esat, T.M., and Lambeck, K. (2001). Coupled climate and sea-level changes deduced from Huon Peninsula coral terraces of the last ice age. *Earth and Planetary Science Letters* 193 (3–4): 579–587. http://dx.doi.org/10.1016/S0012-821.

**102** Dowdeswell, J.A., Elverhoi, A., Andrews, J.T., and Hebbeln, D. (1999). Asynchronous deposition of ice-rafted layers in the Nordic seas and North Atlantic Ocean. *Nature* 400: 348–351.

**103** Crucifix, M. (2012). Oscillators and relaxation phenomena in Pleistocene climate theory. *Philosophical Transactions. Series A, Mathematical, Physical, and Engineering Sciences* 370: 1140–1165.

**104** Summerhayes, C.P., Kroon, D., Rosell-Mele, A. et al. (1995). Variability in the Benguela Current upwelling system over the past 70,000 years. *Progress in Oceanography* 35: 207–251.

**105** Little, M.G., Schneider, R.R., Kroon, D. et al. (1997). Trade wind forcing of upwelling, seasonality, and Heinrich events as a response to sub-Milankovitch climate variability. *Paleoceanography* 12 (4): 568–576.

**106** Kirst, G., Schneider, R.R., Muller, P.J. et al. (1999). Late Quaternary temperature variability in the Benguela current system derived from alkenones. *Quaternary Research* 52: 92–103.

**107** Pedersen, T.F. (1983). Increased productivity in the eastern equatorial Pacific during the Last Glacial Maximum (19 000–14 000 yr B.P.). *Geology* 11: 16–19.

**108** Lyle, M. (1988). Climatically forced organic carbon burial in equatorial Atlantic and Pacific Oceans. *Nature* 335: 529–532.

**109** Anderson, R.F. (2003). Chemical tracers of particle transport. In: *The Oceans and Marine Geochemistry*, *Treatise on Geochemistry* (eds H.D. Holland and K.K. Turekian), vol. 6 (ed. H. Elderfield), 247–291. Oxford: Elsevier-Pergamon.

**110** Crosta, X., Pichon, J.J., and Burckle, L.H. (1998). Application of modern analogue technique to marine antarctic diatoms: reconstruction of maximum sea-ice extent at the Last Glacial Maximum. *Paleoceanography* 13 (3): 284–297.

**111** Crosta, X., Sturm, A., Armand, L., and Pichon, J.J. (2004). Late Qaternary Sea ice history in the Indian sector of the Southern Ocean as recorded by diatom assemblages. *Marine Micropaleontology* 50: 209–223.

**112** Gersonde, R., Crosta, X., Abelmann, A., and Armand, L. (2005). Sea-surface temperature and sea ice distribution of the Southern Ocean at the EPILOG Last Glacial Maximum – a Circum-Antarctic view based on siliceous microfossil records. *Quaternary Science Reviews* 24: 869–896.

**113** Denton, G.H. and Hughes, T.J. (1986). *The Last Great Ice Sheets*, 484 pp. New York: Wiley.

**114** Blanchon, P. (2011). Meltwater pulses. In: *Encyclopedia of Modern Coral Reefs* (ed. D. Hopley), 683–690. New York: Springer.

**115** Murton, J.B., Bateman, M.D., Dallimore, S.R. et al. (2010). Identification of Younger Dryas outburst flood path from Lake Agassiz to the Arctic Ocean. *Nature* 464: 740–743.

**116** Berger, A. (2009). Astronomical theory of climate change. In: *Encyclopedia of Paleoclimatology and Ancient Environments* (ed. V. Gornitz), 51–57. Dordecht: Springer.

117 McManus, J., Oppo, D., Cullen, J., and Healey, S. (2003). Marine isotope stage 11 (MIS 11): analog for Holocene and future climate? In: *Earth's Climate and Orbital Eccentricity: The Marine Isotope Stage 11 Question*, *Geophysical Monograph* **137** (eds. A. Droxler, R.Z. Poore and L.H. Burckle), 69–85. Washington, DC: American Geophysical Union.

118 Droxler, A., Alley, R.B., Howard, W.R. et al. (2003). Introduction: unique and exceptionally long interglacial marine isotope stage 11: window into earth warm future climate. In: *Earth's Climate and Orbital Eccentricity: The Marine Isotope Stage 11 Question*, *Geophysical Monograph* **137** (eds. A. Droxler, R.Z. Poore and L.H. Burckle), 1–14. Washington, DC: American Geophysical Union.

119 Rohling, E.J., Grant, K., Bolshaw, M. et al. (2009). Antarctic temperature and global sea level closely coupled over the past five glacial cycles. *Nature Geoscience* 2: 500–504.

120 Berger, W.H. and Wefer, G. (2003). On the dynamics of the ice ages: stage 11 paradox, mid-Brunhes climate shift, and 100-ky cycle. In: *Earth's Climate and Orbital Eccentricity: The Marine Isotope Stage 11 Question*, *Geophysical Monograph* **137** (eds. A. Droxler, R.Z. Poore and L.H. Burckle), 41–59. Washington, DC: American Geophysical Union.

121 Kiessling, W., Simpson, C., Beck, B. et al. (2012). Equatorial decline of reef corals during the last Pleistocene interglacial. *PNAS* 109 (52): 21378–21383.

122 Hodell, D.A., Kanfoush, S.L., Venz, K.A. et al. (2003). The mid-Brunhes transition in ODP sites 1089 and 1090 (subantarctic South Atlantic). In: *Earth's Climate and Orbital Eccentricity: The Marine Isotope Stage 11 Question*, *Geophysical Monograph* **137** (eds. A. Droxler, R.Z. Poore and L.H. Burckle), 113–129. Washington, DC: American Geophysical Union.

123 Dickson, A.J., Beer, C.J., Dempsey, C. et al. (2009). Oceanic forcing of the Marine Isotope Stage 11 interglacial. *Nature Geoscience* 2: 428–433.

124 Yin, Q.Z. and Berger, A. (2011). Individual contribution of insolation and CO$_2$ to the interglacial climates of the past 800,000 years. *Climate Dynamics* 38 (3): 7009–7724.

125 Montañez, I.P., Norris, R.D., Algeo, T. et al. (2011). *Understanding Earth's Deep Past: Lessons for Our Climate Future*, 194. Washington, DC: National Academies Press.

126 Edwards, R.L., Cutler, K.B., Cheng, H., and Gallup, C.D. (2003). Geochemical evidence for Quaternary Sea-level changes. In: *The Oceans and Marine Geochemistry*, *Treatise on Geochemistry* (eds H.D. Holland and K.K. Turekian), vol. 6 (ed. H. Elderfield), 343–364. Oxford: Elsevier-Pergamon.

127 Kopp, R.E., Simons, F.J., Mitrovica, J.X. et al. (2009). Probabilistic assessment of sea level during the last interglacial stage. *Nature* 462: 863–868.

128 O'Leary, M.J., Hearty, P.J., Thompson, W.G. et al. (2016). Ice sheet collapse following a prolonged period of stable sea level during the last interglacial. *Nature Geoscience* 6: 796–800.

129 Rohling, E.J., Grant, K., Hemleben, C. et al. (2008). High rates of sea-level rise during the last interglacial period. *Nature Geoscience* 1: 38–42.

130 Dutton, A. and Lambeck, K. (2012). Ice volume and sea level during the last interglacial. *Science* 337 (6091): 216–219.

131 Raymo, M.E. and Mitrovica, J.X. (2012). Collapse of polar ice sheets during the stage 11 interglacial. *Nature* 483: 453–456.

132 Gehrels, R. (2010). Sea-level changes since the Last Glacial Maximum: an appraisal of the IPCC Fourth Assessment Report. *Journal of Quaternary Science* 25 (1): 26–38.

133 Fairbanks, R.G. (1989). A 17,000-year glacio-eustatic sea level record: influence of glacial melting rates on the Younger Dryas event and deep-ocean circulation. *Nature* 342: 637–642.

134 Yokoyama, Y., Lambeck, K., de Deckker, P. et al. (2000). Timing of the Last Glacial Maximum from observed sea-level minima. *Nature* 406: 713–716.

135 Waelbroeck, C., Labeyrie, L., Michel, E. et al. (2002). Sea level and deep water temperature changes derived from benthic foraminifera isotopic records. *Quaternary Science Reviews* 21: 295–305.

136 Shackleton, N.J. (2000). The 100,000-year ice-age cycle identified and found to lag temperature, carbon dioxide, and orbital eccentricity. *Science* 289: 1897–1902.

137 Lambeck, K., Yokoyama, Y., and Purcell, A. (2002). Into and out of the Last glacial Maximum: sea level change during oxygen isotope stages 3-2. *Quaternary Science Reviews* 21: 343–360.

138 Personal communication (e-mail via James Scourse, 30 September 2013).

139 Chappell, J. and Shackleton, N.J. (1986). Oxygen isotopes and sea level. *Nature* 324: 137–140.

140 Elderfield, H. and Ganssen, G. (2000). Past temperature and $\partial^{18}O$ of surface ocean waters inferred from foraminiferal Mg/Ca ratios. *Nature* 405: 442–445.

141 Austermann, J., Mitrovica, J., Ltychev, K., and Milne, G.A. (2013). Barbados-based estimate of ice volume at Last Glacial Maximum affected by subducted plate. *Nature Geoscience* 6: 553–557.

142 Niessen, F., Hong, J.K., Hegewald, A., et al. (2013). Repeated Pleistocene glaciation of the east Siberian continental margin, *Nature Geoscience* October 2013, DOI: 10.1038/NGEO1904.

**143** Church, J.A., Woodworth, P.L., Aarup, T., and Wilson, W.S. (2010). *Understanding Sea-Level Rise and Variability*, 428 pp. Chichester and Oxford: Wiley Blackwell.

**144** Berger, W.H. 1982. Climate Steps in Ocean History – Lessons from the Pleistocene, pp 43–54 in *Climate In Earth History: Studies in Geophysics* (ed W.H. Berger and J.C. Crowell), Report of Panel on Pre-Pleistocene Climates, for the Geophysics Study Committee of the Geophysics Research Board of the Commission on Physical Sciences, Mathematics, and Applications of the US National Research Council, based on a meeting in Toronto in 1980, National Academy Press, Washington D.C., ISBN: 0-309-10784-9, 212 pp (free PDF from: http://www.nap.edu/catalog/11798.html).

# 13

# Solving the Ice Age Mystery – The Ice Core Tale

---

**LEARNING OUTCOMES**

- Know the basic details of the nature and distribution of the major ice sheets in Pleistocene times, including the arguments for one in northeast Siberia.
- Know the pattern of temperature change in Antarctica over the past 400 Ka and its relation to sea level, and be able to explain what the temperatures and sea levels of the past four large interglacials tell us about future climate change.
- Understand and be able to explain the bipolar seesaw, including the origins of Greenland's Dansgaard–Oeschger Events and their warm Antarctic counterparts.
- Understand and be able to describe the relationship between $CO_2$ and temperature in Antarctic ice cores, and the origins of the changing relationship between the two between periods of increasing and decreasing insolation.
- Describe the natural envelopes for temperature and $CO_2$ for the past 400 Ka and explain the feedbacks controlling the boundaries of these envelopes.
- Describe and explain the possible origin of the Younger Dryas in relation to the Antarctic Cold Reversal (ACR).
- Understand and be able to explain (in terms of emergent properties) the significance of Saltzman's unification of the two major theories of ice ages (the $CO_2$ theory and the Milankovitch theory) with a third theory based on the role of internal instability.
- Understand and be able to explain the ways in which north and south polar regions are more or less directly connected.

---

## 13.1 The Great Ice Sheets

This book is not the place for a detailed review of the Northern Hemisphere ice sheets, but it is useful background to know where they were. Here we focus on the Arctic ice sheet, because the Antarctic ice sheet was described in Chapter 7. The vast Laurentide ice sheet covered all of Canada and a bordering strip of the USA (Figure 13.1) [1]. Large lateral moraines, including those forming Long Island and Cape Cod, mark its extreme southern edge. Ice covered about half of Alaska, all of the Canadian islands, and Greenland. Aside from the ice sheet in Greenland, a tiny remnant of the Laurentide Ice Sheet remains in the Barnes Ice Cap on Baffin Island. The Laurentide ice sheet was much like Antarctica is today, covering a similar area (almost 13 million km²) to about the same maximum thickness (about 4 km). Its height and surface albedo kept the region cold, and it formed a significant barrier to atmospheric circulation. A massive and equally thick Fennoscandian ice sheet of about 8.5 million km² covered Scotland, Ireland, Scandinavia and most of the Barents Sea between Sweden and the islands of Svalbard, Franz Josef Land and Nova Zemlya, extending south to a line running from just north of London in the west to just south of Berlin and Moscow in the east (Figure 13.1).

The ice sheets of both poles extended to the edges of the continental shelf when sea level was low. Floating ice shelves probably bordered the Arctic ice sheets, as they do around Antarctica today. During the winters at least, the Arctic ice sheets would have been flanked by sea ice on the one hand and tundra on the other hand. The weight of the great Arctic ice sheets depressed the underlying crustal rocks by as much as 700–800 m [2]. Temperatures close to the edges of the ice sheets were probably about 10 °C lower than today on average, and 15–20 °C lower in winter. Because cold air holds little moisture, conditions became very dry over the ice sheets in both polar regions and may have been about 50% drier than usual in the regions near the ice sheets, resulting in a significant increase in the atmospheric transport of dust, which is recorded in ice cores from Greenland and Antarctica.

*Palaeoclimatology: From Snowball Earth to the Anthropocene*, First Edition. Colin P. Summerhayes.
© 2020 John Wiley & Sons Ltd. Published 2020 by John Wiley & Sons Ltd.

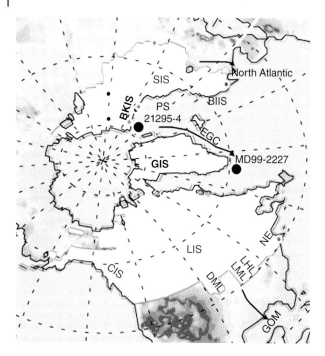

**Figure 13.1** The large ice sheets of the Northern Hemisphere at the Last Glacial Maximum. Cordilleran (CIS), Laurentide (LIS), British Isles (BIIS), Scandinavian (SIS) and Barents–Kara (BKIS) ice sheets labelled. Also Lake Michigan (LML), Lake Huron (LHL), and Des Moines (DML) ice lobes, and the New England margin (NE). Small dots = dated points. Large dots = specific ocean cores. Black arrows = directions taken by meltwater flow from BKIS via EGC (East Greenland Current) and from LIS to Gulf of Mexico (GOM). *Source:* From figure 1 in Ref. [1].

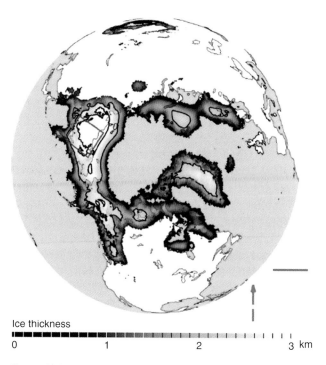

Ice thickness

0 1 2 3 km

**Figure 13.2** Possible Laurentide-Eurasian ice sheet configuration incorporating eastern Siberia, based on numerical modelling experiments. Colours represent ice thickness. *Source:* From figure 2 in Ref. [3].

Figure 13.1 excludes the recently identified northeast Siberian ice sheet (see Chapter 12), which is depicted in Figure 13.2 [3]. As Zhang et al. pointed out in 2018, '*The debate surrounding the existence and history of this enigmatic NE Siberian ice sheet highlights fundamental gaps in our current understanding of the mechanisms of glacial climate evolution*' [3]. To address that issue, Zhang and colleagues '*combine[d] climate and ice sheet simulations to demonstrate how [the dynamics of] ice-vegetation-atmosphere-ocean [interactions] can lead to two ice sheet configurations: the well-known Laurentide-Eurasian configuration with large ice sheets over North America and NW Eurasia, and a circum-Arctic configuration with large ice sheets over NE Siberia and the Canadian Rockies*' [3]. The climate system can swing between the two different patterns depicted in Figures 13.1 and 13.2. Further research is needed to ascertain why NE Siberia was glaciated in some glacial periods and not in others; the answers may lie in feedbacks between ice, vegetation, atmosphere and ocean '*triggered by different combinations [of] ... orbital parameters and greenhouse gas levels*' [3].

The Eurasian ice sheet (Figure 13.3) was the third largest ice mass during the Last Glacial Maximum and previous glacial maxima [4]. It extended laterally over about 4500 km

and was responsible for about 20 m of sea level change. During the deglaciation following the Last Glacial Maximum a '*post-glacial hydrological network ... form[ed] the "Fleuve-Manche" mega-catchment which had an area of 2.5 x 10^6 km^2 and drained the present-day Vistula, Elbe, Rhine and Thames rivers through the Seine Estuary*' [4], which formed the proto English Channel (Figure 13.4). '*During the Bølling/Allerød [warm] oscillation [14.9–12.9 Ka ago], two major proglacial lakes formed in the Baltic and White seas, buffering meltwater pulses from eastern Fennoscandia [and persisting] through to the Younger Dryas [cold stadial 12.9–11.7 Ka ago]. ... [As the ice melted] these massive proglacial freshwater lakes [then] flooded into the North Atlantic Ocean*'. The Baltic ice lake discharged westwards through the North Sea, and the White Sea ice lake discharged northwards through the Barents Sea (Figure 13.3) [3]. The catastrophic release of large volumes of water from these two lakes would have disturbed North Atlantic circulation starting around 11.7 Ka ago [5]. During the Younger Dryas cold stadial 12.9–11.7 Ka ago, remnant ice across Svalbard, Franz Josef Land, Novaya Zemlya, Fennoscandia and Scotland experienced a short-lived re-advance. As warming continued, most ice disappeared by 8.7 Ka ago, although the isostatic recovery of the land continues today [4]. And there still are glacial remnants for example on Norway (which boasts the large Jostedal Glacier – the largest in western Europe – and almost

**Figure 13.3** Major drainage routes (blue arrows) of the Eurasian ice sheet complex at the Last Glacial Maximum, showing in brown the locations of major trough mouth fans (TMF): PB = Porcupine Bank; BDF = Barra and Donegal Fans; RB = Rosemary Bank; NSF = North Sea Fan; Bj = Bjørnøyrenna Fan. Purple lines = extent of ice during the Younger Dryas. Key islands around the Barents Sea include: Svalbard at top left (with purple line showing Younger Dryas ice extent; Franz Josef Land – the group of small islands east of Svalbard; and Novaya Zemlya, the long thin island on the eastern edge of the Barents Sea shelf next to the Kara Sea (which was not ice covered at the Last Glacial Maximum).
*Source:* From figure 1 in Ref. [4].

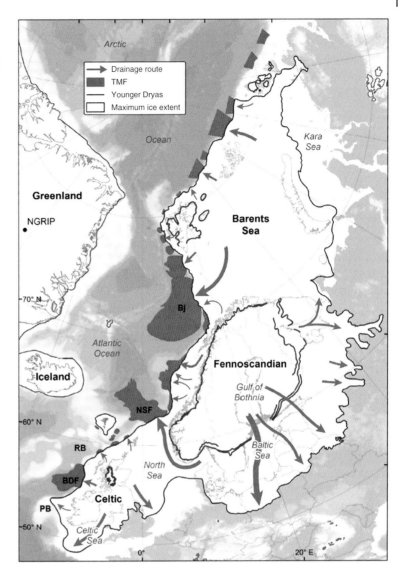

30 smaller ones) and on Svalbard (which boasts eight small ice caps and a large number of glaciers).

## 13.2  The Greenland Story

Palaeoclimatologists were keen to drill through the Greenland and Antarctic ice sheets, which – like ocean or lake sediments – were expected to operate like gigantic tape recorders, with successive snowfalls capturing details of climate change through time. Confirmation that the palaeoceanographers were right in their identifications of Ice Age climate changes began to emerge in the very early 1980s from studies of the oxygen isotopes down ice cores from Greenland. At the 4th Maurice Ewing Symposium in Palisades, New York, in 1982, the Danish and Swiss teams working on those cores presented their findings, led by Willi Dansgaard (1922–2011) (Figure 13.5). Dansgaard was

professor of geophysics at the University of Copenhagen, a member of the Academies of Science of Denmark, Sweden, and Iceland, and soon to be a winner of the 1995 Swedish Crafoord Prize.

Initially, Dansgaard worked on the 1390 m long Camp Century core, drilled in northeast Greenland in 1966 by the US Army Cold Regions Research and Engineering Laboratory. This was the first drilling through an ice sheet. He then led the drilling party for the DYE-3 ice core, drilled 1400 km away in South Greenland in the 1970s. The DYE-3[1] core was a product of the Greenland Ice Sheet Project (GISP) run by the Danes, Americans, and Swiss

---

[1] The DYE sites were part of the USA's network of Distant Early Warning radar stations, the call sign of which was DYE. DYE station 3 (DYE-3) was located on the Greenland ice sheet at an altitude of 8700 ft at 65°11′N 43°50′W.

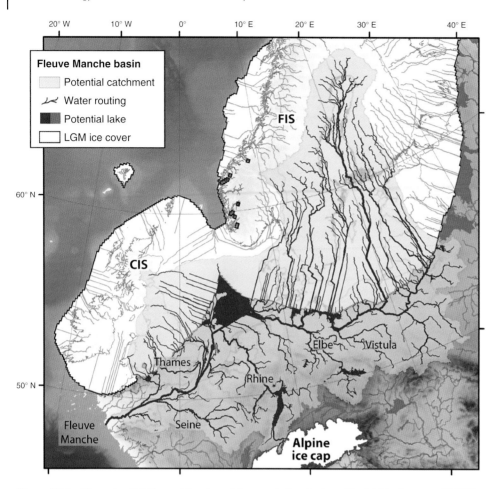

**Figure 13.4** The potential Fleuve Manche catchment during the Last Glacial Maximum at 22.73 Ka ago, when the Celtic (CIS) and Fennoscandian (FIS) ice sheets were coalesced and around their peak extents. Over half of the catchment basin (53%) was ice-covered, draining meltwater from the Eurasian Ice Sheet Complex (EISC) and Alpine ice cap. *Source:* From figure 13 in Ref. [4].

**Figure 13.5** Willi Dansgaard.

from 1971 to 1981. In 1979 the drill reached bedrock at 2038 m and retrieved ice possibly as old as the last interglacial, around 130 Ka ago [6].

Much to their amazement, Dansgaard and the Swiss ice scientist Hans Oeschger (see Chapter 9) found in their Greenland ice cores evidence for large, abrupt variations in $\partial^{18}O$ ratios within the last glaciation. They interpreted these changes as caused by fast changes in the latitude of the polar front. Abrupt rises in temperatures of up to 6 °C took place in as little as 2.5 years, remained stable for 500–1000 years, then plunged [7]. The fluctuations seemed to follow a cycle of about 2550 years, now named 'Dansgaard–Oeschger Cycles', and there was some slight indication of a cycle with a frequency near the 1470 years cycle that Bond had identified in North Atlantic sediment cores, as mentioned in Chapter 12 [8, 9]. Dansgaard and Oeschger thought that these cycles followed some feedback mechanism within the climate system, rather than some as yet unknown solar cycle [6]. The cycles became colder with each cool event, the coldest being associated with ice-rafting in a Heinrich Event, after which conditions warmed and the process repeated itself. Oeschger and Dansgaard also showed that the $\partial^{18}O$ temperature record from the last 60 Ka of the DYE-3 core paralleled the $\partial^{18}O$ record from Lake Gerzensee in Switzerland [10]. This convinced them that advances and retreats of cold Atlantic surface water were influencing the climate of the region.

GISP was followed by GRIP, the international Greenland Ice Core Project, which drilled a 3038 m core to basement at Summit in central Greenland between 1989 and 1992. As with the GISP core, the ice layers at the bottom of the hole were contorted due to squeezing by the weight of the ice sheet. To avoid distortions created by ice flow they drilled another hole, this time in North Greenland at a site where the basal topography was flat, making distortions less likely. The North GRIP (NGRIP) team obtained a core 3085 m long between 1999 and 2003. At its base were 5000 undisturbed years of the last interglacial – the Eemian. Dansgaard–Oeschger Events were well represented in the

GRIP Summit core [11], reinforcing the conclusion from ocean core studies that the Ice Age climate oscillated between two stable states. The temperature increases during these events ranged between 5 and 15 °C in and around Greenland [12].

By 1993, it was obvious to Gerard Bond and his team that millennial variations in ice-rafted debris from North Atlantic cores were synchronous with Dansgaard–Oeschger events in Greenland ice, and that this relationship held up over the past 90 Ka [13] (Figure 13.6) [14]. Making progressively more detailed studies in 1997 and 1999 they found that the millennial cycles were bundled into cooling cycles lasting around 10–15 Ka, with asymmetric saw-tooth shapes due to fast warm beginnings and slow cooling endings [8, 9]. The most massive discharges of icebergs into the North Atlantic (Heinrich Events) happened at the end of each such bundle, followed by warm Dansgaard–Oeschger events (Figure 13.6). As we saw in Chapter 12, it seemed likely that a climate oscillator within the ocean–atmosphere system periodically injected cold

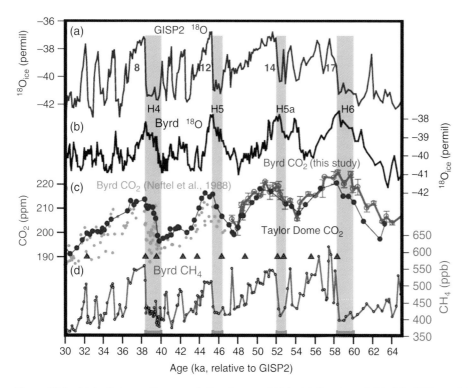

**Figure 13.6** Links Between Dansgaard–Oeschger and Heinrich Events and $CO_2$ (65–30 Ka ago). (a) $\delta^{18}O_{ice}$ from GISP-2 as a proxy for surface temperature. Bold numbers = Dansgaard–Oeschger warm events. (b) $\delta^{18}O_{ice}$ from Byrd station (Byrd ice core), and Antarctic warm events. (c) Atmospheric $CO_2$. Small solid circles back to 47 Ka are existing data from Byrd station; open circles prior to 47 Ka are new data from Byrd; large solid dots are data from Taylor Dome; triangles are control points for the synchronization of gas ages between Byrd and Taylor Dome records. (d) $CH_4$ from Byrd ice core. Ages are adjusted to synchronize abrupt increases of $CH_4$ and Greenland temperature. Vertical blue shaded bars represent cold Heinrich events. Note that atmospheric $CO_2$ rose several thousand years before abrupt warming in Greenland associated with Dansgaard–Oeschger events 8, 12, 14, 17, the four large warm events that follow Heinrich events. The $CO_2$ rise predates Heinrich events associated with these D–O events, and terminated at the onset of Greenland warming for each of these events. Atmospheric $CO_2$ is strongly correlated with the Antarctic isotopic temperature proxy, with an average time lag of 720 ± 370 year (mean ± 1σ) during the time interval studied. *Source:* From figure 1 in Ref. [14].

fresh water to the Nordic Seas [8]. Those injections weakened or shut down the thermohaline circulation bringing warm surface waters from the south. Iceberg outbreaks occurred once a threshold was crossed [9].

The longest and warmest Dansgaard–Oeschger events were associated with elevated amounts of methane (Figure 13.6), because warmth encouraged the formation of tropical wetlands from which $CH_4$ was emitted. Initially it appeared from Greenland ice cores that these events were also associated with high levels of $CO_2$, but those are now known to be artefacts caused by acids deposited from volcanic eruptions interacting with carbonate-rich dust blown onto the ice from Canada [15]. Such reactions do not affect Antarctic ice, which contains less dust and virtually no carbonate.

Enter well-known palaeoclimatologist Richard Alley (1957–) of Pennsylvania State University (Box 13.1).

Summarizing the results of an AGU Chapman Conference on 'Mechanisms of Millennial-Scale Global Climate Change' that took place in Snowbird, Utah, in June 1998, Alley reminded us that although the Dansgaard–Oeschger oscillation was primarily an oceanic process centred on the North Atlantic, winds transported this climate signal rapidly to surrounding areas, including the monsoonal areas of Africa and Asia. The signal is also identified in cores from the Cariaco Basin off Venezuela, the Arabian Sea, and the Santa Barbara Basin off California [17]. Although the oscillation seemed to be periodic, there was much variability in the spacing of its signal. Variability of the kind represented by Dansgaard–Oeschger events has subsequently been detected in older glacial periods (MISs 6 and 8) (see Chapter 12), suggesting that the climate of glacial periods is predisposed to unstable behaviour [18, 19]: more on that later.

Understanding Greenland's history is important because its ice sheet contains the equivalent of 7.4 m of sea level rise.

We now know from analyses of isotopes at the surface of buried bedrock near Greenland's summit that Greenland lost 90% of its ice sheet for extended periods during the Pleistocene [20]. All that it was left with was a small remnant ice sheet and some glacial ice in the east Greenland highlands. Such conditions likely applied for periods of 8–10 Ka during the large interglacial periods of the past 400 Ka [20].

## 13.3 Antarctic Ice

Another set of stunning discoveries emerged from drilling into the Antarctic ice sheet. The first results were not inspiring – a 2164 m long core drilled to bedrock by the Americans in Marie Byrd Land in 1968 (the Byrd ice core - see data in Figure 13.6) reached ice almost 90 Ka old. Major success came from a 3623 m long core from the USSR's Vostok Station[2], 3490 m up on the Polar Plateau (Figure 7.5), where the Russians began drilling in the 1970s; cores from depths greater than 1000 m were not drilled until 1984, and the oldest ice was cored in 1996. The coring reached ice 420 Ka old and revealed four glacial cycles. Many of the geochemical analyses of this core were carried out by French scientists led by Claude Lorius (1932–) (Figure 13.7) (Box 13.2) at the Laboratoire de Glaciologie et de Géophysique de

**Figure 13.7** Claude Lorius.

---

[2]Vostok is Russian for east, and the station was named for the Vostok spacecraft that flew Yuri Gagarin into space in April 1961 as the world's first spaceman.

---

**Box 13.1   Richard Alley**

Plaeoclimatologist Richard Alley is a member of the US National Academy of Sciences (elected 2008), a foreign fellow of the Royal Society (2014), and a recipient of the Louis Agassiz Medal of the European Geosciences Union (2005), of the Seligman Crystal of the International Glaciological Society (2005), and of the Geological Society of London's Wollaston Medal (2017). Alley is well known for his climate change books: *The Two-Mile Time Machine: Ice Cores, Abrupt Climate Change, and Our Future* (2000) [16], and *Earth: The Operator's Manual* (2011) [5], which was turned into a TV documentary series.

Lorius was one of three young Frenchmen who were the first scientists to winter over on the Polar Plateau during the International Geophysical Year of 1957–1958. He was one of the first to demonstrate a relationship between hydrogen isotopes and temperature in Antarctic ice, thus discovering a new palaeothermometer, $\partial D$ or $\partial^2 D$ (the ratio in parts per thousand, or ‰, between hydrogen and its isotope deuterium) [21]. For his work on ice cores, Lorius has been showered with medals and prizes, including the Blue Planet Prize in 2008. He and Jouzel were jointly awarded France's highest scientific award, the Gold Medal of the Centre National de la Recherche Scientifique (CNRS) in 2002. Amongst other awards France named him a Commandeur of the Légion d'Honneur in 2009, and he is a member of the French Academy of Sciences. Along with Dansgaard and Oeschger, he received the Tyler Prize for Environmental Achievement in 1996.

l'Environnement in St Martin d'Hères, near Grenoble, and by Jean Jouzel (1947–) of the Laboratoire de Géochimie Isotopique of the Laboratoire d'Océanographie Dynamique et de Climatologie (LODYC) at Gif-sur-Yvette, near Paris.

Landmark papers announced the French and Russian results. In August 1985, Lorius and others presented the Vostok temperature record for the past 150 Ka, deduced from analyses of $\partial^{18}O$ of the earliest (hence shallowest and youngest) cores retrieved [22]. This was the first ice core record extending through both the last glacial and the last interglacial, something not achieved in Greenland. In October 1987, Jouzel and others presented the temperature record for the past 160 Ka, deduced from analyses of hydrogen isotopes [23]. The isotopic data showed that the Last Glacial Maximum in Antarctica was about 9 °C colder than the early Holocene, whilst the last interglacial around 140 Ka ago was around 2–3 °C warmer. The last glacial period contained several warm and cool intervals, repeating on a cycle of 40 Ka, typical of the behaviour of Earth's axial tilt. In due course, Petit's work showed that the core provided a record of four major interglacials at 125, 240, 320, and 420 Ka ago (Figure 13.8) [24, 25]

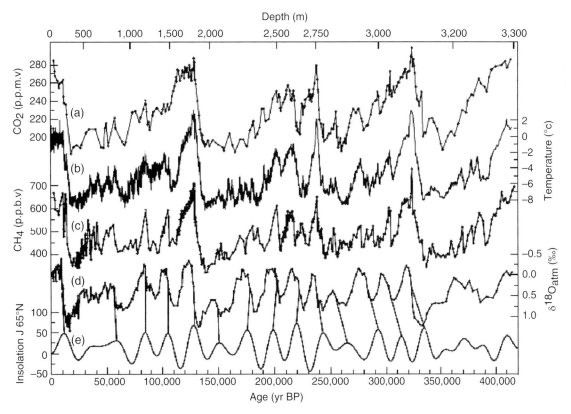

**Figure 13.8** Vostok: the climate of the past 400 000 Years: (a) $CO_2$; (b) isotopic-based temperature of the atmosphere; (c) $CH_4$; (d) $\partial^{18}O_{atm}$; (e) mid-June insolation at 65 °N (in W/m²). *Source:* From figure 3 in Ref. [24].

Vostok's astounding success was followed by even greater penetration back through time obtained by the European Project for Ice Coring in Antarctica (EPICA). EPICA drilled two holes on opposite sides of the Polar Plateau. One with a low rate of accumulation was drilled at the Franco–Italian Concordia base on Dome C, 560 km from Vostok and at a slightly lower altitude of 3233 m (Figure 7.5). It reached a depth of 3270 m in December 2004, and recovered ice as old as 800 Ka, containing eight glacial cycles [26] (Figure 13.9) [27]. The other, with a faster rate of accumulation, was drilled at Germany's Kohnen summer station in Dronning Maud Land, at a height of 2892 m. It reached a depth of 2774 m in January 2006, and ice at a depth of 2416 m was 150 Ka old [28]. This is the EPICA Dronning Maud Land (EDML) core. The Japanese also drilled a long core, on Dome F, the site of Japan's Dome Fuji (DF) station and one of the highest points of Dronning Maud Land at 3810 m (Figure 7.5). Drilling began in August 1995, achieved a depth of 3035 m in the southern winter of 2006/7, and reached back 720 Ka. The Japanese results confirmed findings made in the Vostok and EPICA cores [29].

I visited the Kohnen station on the Polar Plateau in late November 2004, as a guest of the Alfred Wegener Institute for Marine and Polar Research (AWI). At 2892 m above sea level, the station lies almost on the Greenwich Meridian at 75°S, and a mere 1035 miles from the South Pole. We landed in a twin-engined Dornier and wandered around outside the orange-coloured laboratories and workshops made from shipping containers and placed on stilts to prevent the snow piling up against them (Figure 13.10). Nearby was the drilling pit, but the drilling season would not start for another week, so the place was deserted and the drill rig was dismantled. Due to the altitude the temperature was a cold −22°C, which seemed more like −30°C due to the slight wind chill from a light 5 knot breeze. Around us the vast, silent, uninterrupted white expanse of the Polar Plateau swept away in all directions (Figure 13.10). What a lonely spot! Seen close up, the surface was roughened by 30 cm high ridges of hard ice – 'sastrugi' – carved by the wind (Figure 13.10). They made the surface look like an old-fashioned washboard. Taking off to fly to our base, the German Neumayer station down on the coast, was a nightmare. The plane had so much fuel aboard, and the air was so thin at that altitude, that it took the pilot 13 attempts at full throttle, plus taxiing back to the start, all the while rattling and banging over the sastrugi, before we finally lifted off. What it must have been like for Scott and his men towing heavily loaded sledges over such terrain, I cannot imagine. But I can empathize with his remark on reaching the pole in January 1912: '*Great God, this is an awful place*'. At least I had the benefit of a rapid escape route, and a radio.

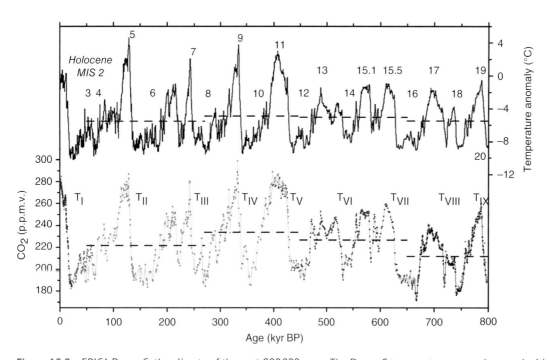

**Figure 13.9** EPICA Dome C: the climate of the past 800 000 years. The Dome C temperature anomaly record with respect to the mean temperature of the last millennium (based on deuterium isotope data). Data for $CO_2$ are from Taylor Dome (25–75 Ka ago), Vostok (25–425 Ka ago), and Dome C. Horizontal lines are the mean values of temperature and $CO_2$ for the time periods 799–650, 650–450, 450–270 and 270–50 Ka ago. $T_{1,2,3}$ etc. = glacial terminations; MIS numbers are Marine Isotope Stages. *Source:* From figure 2 in Ref. [27].

**Figure 13.10**   Twin engine Dornier with skis sits on the vast empty plain of the Polar Plateau next to Kohnen Station, on crusty ice carved into sastrugi. Inset – author well wrapped up at Kohnen Station in early southern summer (27 November 2004) a week before the ice drilling season started.

The long cores from Vostok, Fuji and Dome C confirmed the picture from deep-ocean drilling. There were five major interglacial periods spaced around 100 Ka apart over the past 400 Ka [30]. We are living in the last of them, named the Holocene Epoch. Each of the four prior interglacials was thought to be about 2–3 °C warmer than the present, and sea level was higher than it is today by between 4 and 9 m [31], as mentioned in Chapter 12. A re-examination of the last interglacial, in 2017, showed that its mean global sea surface temperature temperatures were 5.5 ± 0.3 °C warmer than the mean from 1870 to 1889, and indistinguishable from the 1995–2014 mean [32].

In the 1990s, the mean global difference between the Ice Age and the present was thought to be about 5 °C. Recent research shows that it was closer to 4 °C [33, 34], possibly even 3.5 °C [35]. In Antarctica the difference was thought to be 9 °C below present, based on interpretation of isotopic values, until Kurt Cuffey of the University of California,

Berkeley, and colleagues used borehole temperatures from the West Antarctic Ice Sheet (WAIS) Divide core in 2016 to show that the Last Glacial Maximum was actually significantly cooler at 11.3 ± 1.8 °C below present, and that the net warming from the Last Glacial Maximum to the middle Holocene thermal maximum was possibly as large as 13.7 °C [34]. Because '*cited estimates for deglacial warming at locations in East Antarctic are smaller, from 7°C to 9.3°C ... it is possible that there exists a real regional difference [between East Antarctica and] West Antarctica, where the climate is more strongly influenced by the proximal ocean*' [34]. Nevertheless those values derive from isotopic data, which may be inaccurate.

Antarctic warming during the deglaciation was mostly completed by 15 Ka ago, several millennia earlier than in the Northern Hemisphere [34]. It coincided with glacial recession in mountain ranges in Patagonia and New Zealand. Cuffey and colleagues thought that the pattern of deglaciation reflected changes in insolation, $CO_2$, and ocean heat transport through the Atlantic Meridional Overturning Circulation (AMOC; Figures 12.5 and 12.6). Greenland did not reach interglacial temperatures for another four millennia, at least in part because Atlantic heat transport was reduced between 19 and 14.7 Ka ago, which cooled Greenland and warmed Antarctica [34]. Decreasing ice sheet extent (hence reduced albedo) and increasing greenhouse gas concentrations ($CO_2$ and $CH_4$) account for about 90% of the direct global-averaged climate forcing, with respective magnitudes of ~2.9 W m$^{-2}$ and 2.2 W m$^{-2}$ [34]. Cuffey argued that the contribution of greenhouse gases to Antarctic warming from 18.5 to 15 Ka ago was ~65% [34], roughly agreeing with French estimates [24, 25].

The French geochemists found that there was more airborne dust and sea salt during cold periods. Nearby dry areas either expanded at the time or were subject to stronger winds that took more of their dust offshore, while Southern Ocean winds strengthened, whipping up waves and enhancing the supply of sea salt to the air. Analyses of sulphate showed no long-term link between volcanism and climate in Antarctica [36, 37].

By 2009, Tim Naish of Victoria University of Wellington had accumulated a great deal of information from ice cores, continental margin drill cores, and deep ocean drilling cores from the Southern Ocean, about Ice Age climate change in and around Antarctica [38]. He and his team found an interlude of anomalous continental-scale warmth dated to about 1 Ma ago and correlating with MIS 31 (see Figure 12.4). It was represented by foraminiferal ooze and coccolith-bearing assemblages in cores from the Weddell Sea and Prydz Bay, along with bioclastic limestone in the Ross Sea. Sea surface temperatures locally warmed by 4–6 °C then, apparently due to an extended interval of unusually warm or long summers forced by orbital variations. Following it, as seen in marine sediment cores (Chapter 12), and ice cores, was the Mid Pleistocene Transition (MPT) between 900 and 700 Ka ago, when orbital cycles changed in frequency from an older set dominated by 41 Ka cycles to a younger set dominated by roughly 100 Ka cycles (see Figures 12.4 and 13.9). After the MPT, evidence for substantial discharge of melt-water disappeared, suggesting that the ice sheet had become more dry-based: a sign of cooling. Naish's team described the transition as '*one of the most poorly understood events in palaeoclimatology*' [38]. While that was true back in 2009, by 2012 Elderfield had come up with an explanation for it [39], as we saw in Chapter 12.

Naish's team found that during glacial periods, Antarctic sea ice advanced further north, sea surface temperatures fell by about 6 °C, oceanic frontal zones migrated north by 5–10° of latitude, and zonal winds intensified, strengthening

the Antarctic Circumpolar Current (ACC) [38]. These processes intensified between 3000 and 7000 years before maximum ice volume was reached and the ice sheet covered the continental shelf. Expansion and contraction of the ice sheet across the full width of the continental shelf contributed 15–20 m to the global changes in sea level of the past 700 Ka. During the Last Glacial Maximum, ice built up around the continental margins by up to several hundred metres due to increased snowfall, but diminished in the interior by 100–150 m due to lower rates of snowfall there. The dust blown into the region by high winds during glacials supplied nutrients (especially iron, Fe) to the Southern Ocean, increasing its productivity and helping to draw down the content of $CO_2$ in the atmosphere. Equally, the cooling of the surface waters meant that more $CO_2$ could dissolve in them.

## 13.4 Seesaws

The fine-scale signals of climate change within the last glaciation that Dansgaard and Oeschger found in Greenland ice cores are also identified in Antarctic ice cores, where they are much less abrupt, as Ahn and Brook showed in 2007 (Figure 13.6) [14, 38]. To compare these signals, the cores from the two poles had to be matched in time. What clever means could be used to do that? The answer – methane! Methane gas is rapidly mixed throughout the atmosphere, so a prominent methane signal will be trapped in ice at both poles at the same time. Superimposing the patterns of wiggles in the methane signal through time in the cores from both hemispheres enabled the relationships in time between the temperature signals from the two regions to be fixed [40–42]. Slow warming in Antarctica seemed to precede each sudden warming in Greenland (Figure 13.6). Greenland warming rose rapidly as Antarctic warming reached a peak and began to cool. Greenland then cooled rapidly as the Antarctic slowly started to warm again. What was going on? The answer was what Wally Broecker and Thomas Stocker called the 'bipolar seesaw' [43, 44]. Later, we will look in detail at the relative timing of these Northern and Southern Hemisphere events.

This out-of-phase relationship between the climates of the two hemispheres is one of the most unexpected finds in the study of millennial-scale climate change [45], and one of the most exciting additions to the Thermohaline Conveyor Belt (or Meridional Overturning Circulation) paradigm [46]. Initial research suggested that the bipolar seesaw works as follows. Warming in the north leads to iceberg outbreaks in the North Atlantic. As the bergs melt they cool the surface ocean and flood it with a freshwater lid. That stops or reduces the sinking of formerly dense salty surface water to produce North Atlantic Deep Water (NADW), a process that in effect pulled warm salty Gulf Stream to the north to cool and sink to form NADW (see Figure 12.6). Instead, because heat was no longer being pulled north it accumulated in the South Atlantic and spread through the Southern Ocean. Meanwhile, deep, northward-moving cold Antarctic Bottom Water (AABW) (see Figure 12.6) displaced slightly warmer south-moving NADW as the main source of cross-equatorial flow at depth (this does not necessarily imply an increase in AABW production – merely a change in the balance between AABW and NADW at depth). The hypothesis then posits that once the northern iceberg armada shrinks, the freshwater lid disappears, and warm salty surface waters return north enabling NADW production to resume, which pulls yet more warm salty Gulf Stream water north, enabling the cycle to repeat itself. Meanwhile, in the south the warming of the Southern Ocean eventually melts sea ice and ice shelves, making the Antarctic ice sheet surge, which cools and freshens surface waters, slowing AABW production. The net result is a continual off-and-on reversal of cross-equatorial flow – an 'internal oscillator'. No external forcing is required.

Wally Broecker thought that the seesaw was all down to salt, which changes the density of ocean water [47]. The NADW exports salt from the North Atlantic, via the sinking of water with a relatively high salt content, made denser by cooling in the Norwegian–Greenland and Labrador Seas. When little or no NADW is produced, salt transported north by the Gulf Stream builds up in the northernmost Atlantic to the point where eventually the NADW has to switch back on to remove it. In due course, this export lowers the salt content of the North Atlantic again to the point where the NADW once more switches off, because surface waters are no longer dense enough to sink. Hence, deep-water processes are at the root of the bipolar seesaw. Other amplifiers also come into play. As the North Atlantic cools, the sea ice front moves south, increasing the Earth's albedo and thus causing further cooling. Expansion of the areal cover of sea ice limits the formation of NADW. Cooling steepens the thermal gradient between the Equator and the Pole, strengthening winds and lofting dust and sea salt into the air to reflect solar energy and further amplify cooling. In due course, salt build-up turns NADW production back on.

Dan Seidov and his team from Penn State used an advanced ocean model to explore the chicken-and-egg implications of these findings. They concluded that while the NADW is the main driver of the Meridional Overturning Circulation (Figures 12.5 and 12.6), in some circumstances the Southern Ocean could overpower that Northern Hemisphere driver and become a major player in long-term

climate change [48]. They agreed that deep ocean circulation drove the seesaw.

Mark Maslin and colleagues explained the seesaw in terms of 'heat piracy' [45]. During warm periods the North Atlantic 'steals' heat from the Southern Hemisphere, sustaining a strong Gulf Stream and preventing ice build-up in the north. This is northern heat piracy. During a Heinrich Event, when melt water covers the North Atlantic, heat is transported south across the Equator, warming the southern oceans. The South Atlantic 'steals' heat from the north. This is southern heat piracy. Under 'normal' glacial conditions the production of AABW and NADW are in balance and there is no heat piracy.

The abrupt nature of each warming in Greenland suggests that the climate there moved from cool to warm within a decade. This implies that the arrival of heat from the south pushed local temperature over a threshold, a 'tipping point', beyond which it rose suddenly. The incoming heat maintained the new warm stable state around Greenland for a time, during which the south cooled as it lost heat to the north. The cool waters from the south eventually arrived in the north and tipped the northern system back below the threshold, returning it to its original cold state. The threshold for stopping NADW formation is much higher than the threshold for restarting it, because of a phenomenon termed 'hysteresis', a Greek word meaning 'a coming short' or 'deficiency' (e.g. see Figure 6.19). We use it in palaeoclimate studies to indicate where an effect is lagging behind its cause. A larger response than would be expected from gradual linear forcing is referred to as an overshoot [45]. Overshoots are difficult to reverse even when the forcing subsequently declines below the original threshold. Hysteresis typically gives rise to loop-like behaviour of the kind described in the discussion on oscillators in Chapter 12. For instance, a gradually increasing freshwater lid eventually gets to the point where North Atlantic overturning represented by the production of NADW abruptly stops: a high threshold value, in this case abundant freshwater at the surface, has been passed – we have an overshoot. Before NADW production can restart, freshwater has to decline to a far smaller area than the one at which it stopped that production [49].

Two researchers from Germany's Potsdam Institute for Climate Impact Research, Andrey Ganopolski and Stefan Rahmstorf, used a coupled ocean–atmosphere climate model to see how stable the Meridional Overturning Circulation was, and to explore possible mechanisms for rapid change [50]. Whilst for the modern climate they found two stable modes for Atlantic circulation – the 'warm' and the 'off' mode, they found only one – 'cold' – for glacial times. The 'cold' mode has deep-water formation south of Iceland and relatively shallow overturning, with weak outflow to the South Atlantic. The 'warm' mode, with deep-water formation north of Iceland, has deeper overturning and strong outflow to the South Atlantic. It is unstable under glacial conditions, where it can occur for short periods when the flux of fresh water to the North Atlantic is low. Temporary transitions from the 'warm' to the 'cold' mode and back explain the observed Dansgaard–Oeschger cycles. The model outputs confirmed that changes in freshwater supply in the North Atlantic could drive warming in the south. Given the non-linear threshold response of the North Atlantic circulation to forcing, a weak climate cycle can trigger large-amplitude episodic warm events. Rapid transitions can occur because the sites for deep-water formation between the 'cold' mode (south of Iceland) and the 'warm' mode (north of Iceland) are geographically far apart.

As mentioned in Chapter 12, I was part of a group that identified the bipolar seesaw effect in sediment cores from the South Atlantic back in 1997. We found that cold Heinrich Events in the North Atlantic were represented as warm events in the South Atlantic [51]. This is consistent with what one would expect if the thermohaline circulation collapsed at those times [49], as explained above.

Is Bond's internal oscillator, which appears to be an element of Broecker's bipolar seesaw, truly periodic, or does it just seem that way? Richard Alley was one of those to puzzle over that question. In 1999, he and some others thought that the extent to which Bond's cycle was truly regular or merely seemed so remained to be seen [17]. Some thought that it was a free oscillation possibly related to the El Niño-Southern Oscillation characteristic of the Pacific Ocean [52]. The length of the cycle suggested a major role for the ocean, rather than control by changes in the ice or the atmosphere [17]. By 2001, Alley and his team had an answer [53]. Analysing temperature records from Greenland ice cores they found peaks in periodicity at 1500, 3000, and 4500 years, diminishing in amplitude from 1500 to 4500 years. This pattern is typical of 'stochastic resonance', meaning that it is largely random or non-deterministic. Financial markets use stochastic models to represent the seemingly random behaviour of the stock market. Bond's sediment cores showed the same pattern. Alley's team suggested that weak periodicity combined with statistical noise in an internal oceanic oscillator made the climate switch from one mode to another with an **apparent** periodicity of about 1500 years, driven by random fluctuations in freshwater supply. We will revisit this question again, both below and in Chapters 14 and 15. But for the moment it is worth recalling that Bond's supposed 1470 year cycle is roughly equivalent to the overturning age of the ocean. Vikings sailed their long ships on the c. 1000 year old

fossil North Atlantic surface water now welling up to the surface in the North Pacific.

In 2011, Stephen Barker of Cardiff University set out with colleagues to test the notion that Dansgaard–Oeschger fluctuations in climate persisted back through time beyond the limit of the Greenland ice cores, and that the events in Greenland and Antarctica were connected through the AMOC, the part of the Thermohaline Conveyor that links the two poles (Figure 12.5) [54]. Using the bipolar seesaw model, they constructed an 800 Ka long synthetic record of climate variability. It agreed with what was known of climate change in Greenland for the past 100 Ka and with what was known of climate change from a well-dated Chinese cave speleothem record covering the past 400 Ka, providing confidence that this same climatic behaviour probably extended back through time for the full 800 Ka record seen in Antarctic ice cores. Mark Siddall of the University of Bristol carried out a similar exercise [55], probing the Dome C record to see if older intervals than the last glacial period contained similar millennial-scale events. Over the past 500 Ka, Siddall and colleagues found clusters of millennial events recurring with a period of about 21 Ka, like that of orbital precession, especially during periods of intermediate ice volume when sea level was lowered by 40–80 m. Millennial variability was absent during periods of high ice volume, probably because the southward extension of large ice sheets and widening of the belt of sea ice during glacial maxima affect the zonality of atmospheric circulation, impeding the development of millennial variability. Intermediate amounts of land ice would not have had the same effect, allowing changes in precession to force changes in sea ice through zonal changes in atmospheric circulation and surface air temperature. The lack of a consistent correlation between millennial variability and ice-rafted debris of the kind one might expect from Bond's theory suggested to Siddall that a different cause must be sought for Bond cycles other than iceberg release into the North Atlantic. As discussed at length in Chapter 12, marine sediment cores have now been found that show that there was indeed millennial variation in all four of the recent major mid-glacial periods.

Recently, Tzedkis and colleagues found that millennial/centennial variability was also common in the last interglacial [56]. Although the amplitude of the variability was substantially less than it was in periods of intermediate ice volume, it was somewhat higher than in the Holocene. Episodic incursions of cold surface water down the Portuguese margin, probably reflecting reorganization of ocean surface circulation at times of minor disruptions to the AMOC, correlated with periods of aridity identified from pollen data from southern Europe. The cold water incursions may represent periods of warming that led to the melting of Greenland ice, which put a cold freshwater lid on the northeastern Atlantic [56].

A high-resolution analysis of variability down the Dome Fuji and Dome C ice cores by Kawamura and colleagues recently confirmed that millennial Antarctic warming events were typical of intermediate glacial stages [57], much as found by Barker [54] and Siddall [55]. These same fluctuations are also identified in all mid-glacial intervals sampled in sediment cores, as noted in Chapter 12.

To investigate possible causes for the Antarctic warming events and their relation to Greenland, Kawamura used an Atmosphere–Ocean General Circulation Model (GCM). Part of his investigation involved an experiment in which the North Atlantic was 'hosed' with fresh water, the idea being that this would put a lid on the area where NADW formed, thus causing the AMOC (Figure 12.6) to cease transporting heat there, with knock-on effects further south. The hosing was applied for 500 years, then turned off. As hosing began, sea ice formed rapidly in the northern North Atlantic where NADW would normally form. That led to cooling in Greenland, and the slowing of northward transport by the AMOC. At the same time a slow warming began into an Antarctic Isotope Maximum at a rate of ~1 °C/500 year. Clearly the rapid changes in the north were connected to slow changes in the south, suggesting that Northern Hemisphere change may drive Southern Hemisphere change. When the hosing was turned off, northward transport slowly began in the North Atlantic, and sea ice slowly began to diminish. 200 years after hosing ceased there was a marked fall in sea ice and a sudden rise in northward transport in the AMOC and in Greenland temperatures. Meanwhile, the rise in Antarctic warming flattened at around 400 model years, showed a diffuse peak at around 550 model years, then a very slight decline before cooling more rapidly by about 850 model years as warming in Greenland and AMOC transport reached their peak highs and sea ice reached its minimum [57]. What we learn from this experiment is that the periodic addition of freshwater to the cold surface water of the North Atlantic under mid-glacial climatic conditions maintains extensive sea ice cover in the region of formation of deep water, leading to weakening of the AMOC, consistent with the ice-core data, and the weakening of the AMOC gradually warms the southern South Atlantic and the Southern Ocean [57]. However, very different rates are involved: the shut down of the AMOC in the north is rapid, the warming in the south is slow. Kawamura's results confirm that variations in ice rafting in the North Atlantic were much smaller during interglacials than in glacials (see Chapter 12). And his modelling experiments also showed that applying the same amount of fresh water during interglacial simulations as during mid-glacial simulations was insufficient to form extensive sea ice in the North Atlantic, so had a much lesser impact on the AMOC [57], consistent with observations from sediment cores [56].

Figure 13.6 shows that the millennial warmings during Antarctic Isotope Maxima were associated with increase of about 20 ppm in $CO_2$ (probably emitted from the warming Southern Ocean) and about 150 ppb in methane (probably emitted from expanding tropical wetlands as equatorial precipitation increased during stronger monsoons). These variations in $CO_2$ were built in to Kawamura's model [57].

Kawamura concluded from his experiments that '*the model results are consistent with paleoclimatic data and theory, which propose that the climate change signal from the northern North Atlantic rapidly propagates through the ocean to midlatitudes of the South Atlantic*' [57]. Furthermore '*results are consistent with the hypotheses that the climatic oscillation may have been caused by weak forcing [so-called stochastic resonance] and/or by the coupling of northern ice sheets and AMOC*' [57]. In addition, he noted that '*we emphasize the importance of greenhouse gas forcing as a determining factor in the stability of AMOC and global climate*' [57]. However, his results still leave open the question of what causes the shut down of the AMOC in the north. They do not conclusively prove that Northern Hemisphere change leads Southern Hemisphere change. Nevertheless, his '*results suggest that, if a large freshwater flux from the Greenland ice sheet occurs in the future, AMOC weakening could be amplified*' [57], which seems wholly reasonable

As ever, science marches on. Examining the Dansgaard–Oeschger events in MISs 3 (at 28–60 Ka ago) and 5 (at 73.5–123 Ka ago) in Greenland, and their counterparts in Antarctica, Emilie Capron of the Institut Pierre-Simon Laplace of the Laboratoire des Sciences du Climat et de l'Environnement at Gif-sur-Yvette, in France, found in 2010 that these sudden warm events were preceded by a short warm period lasting less than 100 years, and were followed by a brief and low amplitude 'rebound' event [58]. Moreover, the duration of individual Dansgaard–Oeschger events varied in relation to sea level: the longer ones were associated with the greatest sea level rise, hence the largest amount of ice melt. Her team's exceptionally high-resolution studies confirm that more research is needed to quantify the influence of insolation, ice-sheet volume, sea ice and the hydrological cycle on variability at the sub-millennial scale – the precursor and rebound events [58]. While Dansgaard–Oeschger events and their Antarctic counterparts are evidently linked through changes in the AMOC, the link is not as simple as was supposed at first. To obtain a better understanding of the response of Antarctica to sub-millennial scale variability (precursor and rebound events) we need to quantify the influence of insolation, ice-sheet volume, sea ice and the hydrological cycle on that variability [58].

In 2018, Christo Buizert of Oregon State University and colleagues capitalized on the existence of a new high-resolution ice core from West Antarctica to carry out one of the latest analyses of Antarctica's climate history in relation to Dansgaard–Oeschger (DO) events [59]. They drew on data from two Greenland summit ice cores (GISP2 and GRIP), and five Antarctic ice cores: European Programme for Ice Coring in Antarctica (EPICA) Dome C (EDC); EDML; Dome Fuji (DF); and Talos Dome (TALDICE) (TAL), all from East Antarctica; and the West Antarctic Ice Sheet Divide core (WDC) (Figure 13.11). The cores from the two poles were synchronized using their methane signals, and dated with reference to volcanic ash deposits of known age.

Summarizing what was then known about Dangaard–Oeschger [DO] Events and the oceanic seesaw, Buizert and his colleague Andreas Schmittner of Oregon State University explained in 2015 that '*Not all DO cycles are created equal, as their duration, recurrence time and magnitude varies through time*' [60], as we can see from Figure 13.11. DO events seem to be grouped into the so-called Bond cycles about 7 Ka long, separated by Heinrich Events, with successive DO events within each Bond cycle

decreasing in duration and amplitude. The variability of DO events is greatest at times of intermediate ice volume (like MIS-3, between c. 50 and 25 Ka ago (Figure 13.11), and they are absent when ice volume is low – during interglacials [60]. Plotting the duration of Greenland DO interstadials (warm periods) against Antarctic temperatures for the same time period (represented by $\partial^{18}O$), Buizert and Schmittner found that DO warm interstadials were long when Antarctic temperatures were high, and short when Antarctic temperatures were cold [60] (Figure 13.12a). Later, Ed Brook and Christo Buizert also found that the warmer the Antarctic temperature, the higher the $CO_2$ concentration in ice core bubbles (Figure 13.12b) [35, 60].

Buizert and Schmittner proposed that '*the timing characteristics of the DO cycles are controlled by [Southern Hemisphere] climate through the influence of [Southern Ocean] processes on the strength and stability of the [Atlantic Meridional Overturning Circulation] AMOC. During periods of [Southern Hemisphere] (high-latitude) warmth, the vigorous interstadial AMOC mode is more stable. This stability allows the interstadial mode to persist [for] longer periods of time, which is reflected by long interstadial durations. With cooling [Southern Hemisphere] high-latitude temperatures, the interstadial AMOC mode becomes increasingly unstable, resulting in shorter interstadial durations*' [60].

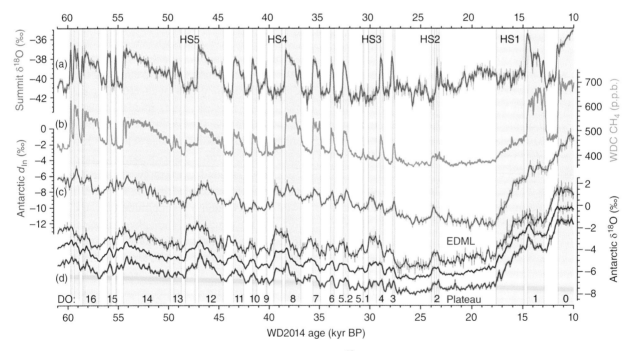

**Figure 13.11** Records of abrupt glacial climate variability. (a) Ice-core $\partial^{18}O$ (average of GISP2 and GRIP) at the Greenland summit. (b) WDC methane in ppb. (c) Antarctic five-core (WDC, EDC, EDML, TAL and DF) average $d_{ln}$ anomaly. (d) Antarctic $\partial^{18}O$ anomaly at EDML (blue), Antarctic Plateau average of DF and EDC (red) and five-core average (black). All records are synchronized to WDC2014 chronology; Antarctic data are shown as anomalies relative to the present, with the EMDL and Antarctic Plateau lines offset by +1.4 and −1.3‰, respectively, for clarity. DO interstadial periods are marked in grey and numbered, and Heinrich stadials (HS1 – HS5) are marked in blue. Isotope ratios are on the VSMOW (Vienna Standard Mean Ocean Water) scale. *Source:* From figure 1 in Ref. [59].

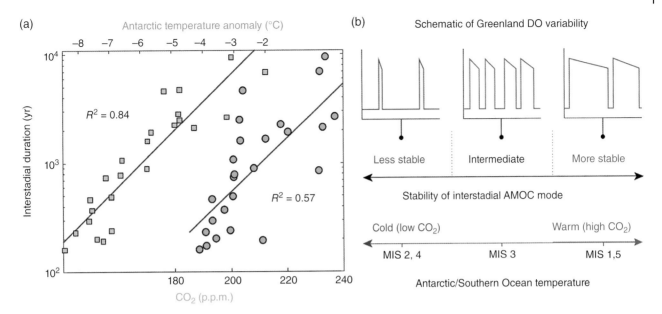

**Figure 13.12** Dependence of millennial-scale variability on the background climate state. (a) Dansgaard–Oeschger interstadial duration (logarithmic scale) from Greenland ice core records plotted against Antarctic temperature and atmospheric $CO_2$, with the coefficient of determination $R^2$ listed for each case. (b) Schematic of Greenland Dansgaard–Oeschger variability during various states of the background climate. At left, during cold (low-$CO_2$) periods (e.g. Marine Isotope Stages [MIS] 2 and 4), Dansgaard–Oeschger interstadials are infrequent and short, suggesting that the interstadial AMOC mode is unstable. In the middle, during intermediate climates (e.g. MIS 3), Dansgaard–Oeschger interstadials are frequent and of medium duration, resulting in high event frequency. At right, during warm (high $CO_2$) periods (e.g. MIS 5), Dansgaard–Oeschger interstadials are long, resulting in a lower event frequency, suggesting that the interstadial AMOC mode is very stable. *Source:* From figure 3 in Ref. [35], modified from figures 2 and 4 in primary Ref. [60].

Conversely, when the Southern Hemisphere climate is coldest (e.g. MIS-2) the AMOC mode is too unstable to persist, leading DO events to be more or less absent (Figure 13.12). However, while the Southern Hemisphere high latitude climate sets the background condition of AMOC stability under which the DO-cycle operates, this does not mean that the Southern Hemisphere high latitudes control the timing of any individual event. The pattern of recurrence of DO events suggests a deterministic control, '*contrary to the suggestion by Ditlevsen* [61] *that DO recurrence times are purely stochastic*' [60]. Buizert and Schmittner also found '*no evidence that DO recurrence times cluster around multiples of 1470 years*' [60]. Indeed the 1470 year cycle is not found in the most current Greenland ice core chronology [60, 62]. Buizert and Schmittner concluded that the climate system contains a 'Goldilocks Zone' of intermediate stability of the interstadial AMOC mode that allows a rapid sequence of DO events to develop (Figure 13.12b). Modelling confirmed that when the Southern Hemisphere high latitude was warm, the AMOC was more stable, less likely to collapse, and thus persisted for long periods [60]. The seesaw seems to work as follows: the first DO of a 7 Ka Bond cycle is preceded by a Heinrich Event in which iceberg-delivered freshwater suppresses the formation of NADW; this reduces northward heat transport via the AMOC, allowing heat to build up in the Southern Ocean; the warmed Southern Ocean leads to a more stable AMOC, which transfers heat to the Northern Hemisphere creating a long interstadial DO event; the gradual cooling of the Southern Hemisphere through each Bond cycle leads to progressive shortening and cooling of DO interstadials, and increasing instability of the AMOC, eventually leading to the collapse of Northern Hemisphere ice margins and the creation of another Heinrich Event, precipitating the next phase of the cycle [60]. The Bond cycle could be viewed as a damped oscillator in which the heat accumulated in the Southern Hemisphere during Heinrich stadials is reduced through a series of DO oscillations of diminishing amplitude and duration [60]. The gradual nature of Antarctic climate change probably represents buffering of the seesaw changes by a large heat reservoir – probably the Southern Ocean and ocean interior [59, 60].

That still begs the question – what caused iceberg outbreaks? It seems likely that Heinrich Events occur when warm water penetrates beneath an ice shelf, causing it to collapse (Figure 13.13) [63]; a process described by Shepherd and colleagues [64]. Much the same process is likely around the edges of Northern Hemisphere ice sheets. Sea ice then retreats rapidly, leading to abrupt warming at

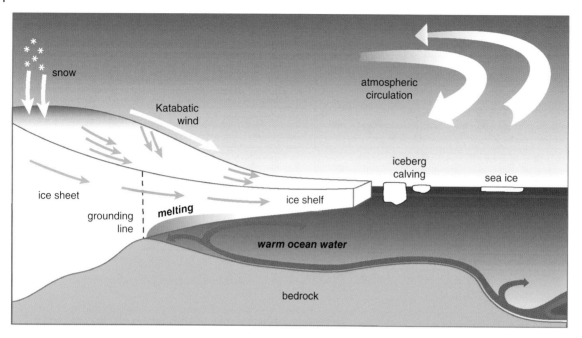

**Figure 13.13** A schematic cross section through Antarctica and the adjacent Southern Ocean showing the displacement and thinning of inland ice towards the coast leading to the formation of ice shelves, which are then eroded from beneath by warm ocean water welling up from below, which weakens their buttressing effect allowing more ice to move from land to ocean. Surface waters, which freeze to form sea ice in winter, would typically reach temperatures of −1.8 °C, while bottom waters would be at 1 °C or slightly more. *Source:* From Ref. [63].

the onset of a D–O cycle. Gradual cooling through the ensuing warm interstadial period then encourages the re-growth of the ice shelf and its associated sea ice, driving the climate back into cold stadial conditions [65]. The cycle then repeats.

Tas Van Ommen pointed out that '*this picture leaves questions of cause and effect unresolved ... Potential drivers of AMOC variations include changes in freshwater balance, sea ice or ice shelves, any of which can initially affect deep-water formation in the north or the south. Buizert and colleagues suggest a straightforward interpretation of the north–south lag: the Northern Hemisphere AMOC shift triggers changes in the Southern Hemisphere, rather than the other way around. But, as others have noted and the authors recognize, the concept of a "trigger" may be poorly framed in a system of closely coupled oscillations – the system may be preconditioned for a state change by remote, as yet unidentified factors*' [66]. It's the old story of the chicken and the egg – which came first?

To answer that question, we may have some help from an investigation of a sediment core from the western side of the Reykjanes Ridge, just south of Iceland, by Rasmussen and colleagues in 2016. They found that Intermediate Water down to a depth of 1730 m warmed steadily through all stadials and Heinrich Events between 65 and 25 Ka ago [67]. Within the Nordic seas, Rasmussen noted that '*During stadials, large numbers of melting icebergs and sea ice*

*created a stratified water column composed of a relatively thin layer of cold, low saline surface water overlying a denser intermediate water mass, which was gradually warming. Similar conditions probably existed in the [Ice-Rafted Debris] belt during Heinrich events. In the Nordic seas, the abrupt warming has been attributed to a rapid surfacing of the warm intermediate water, which broke the stratification and restored convection*' [67]. Their ridge core showed that gradual warming began with the abrupt rise in ice-rafting, which showed that the increased melting of icebergs was caused by the warming. Warming led to a higher input of meltwater and stratified the upper ocean, preventing convection. The patterns of warming reflect what is known of the temperature of the Gulf Stream system, which in turn represents what was happening in the source region in the South Atlantic. The warming seen in the cores was coincident with warming seen in Antarctic ice cores, and occurred 211 years after the start of the interstadial (the same delay seen by Buizert – see earlier). Evidently, '*the D–O warmings in the open North Atlantic were gradual and in phase with the gradual warmings in the Antarctic ice cores and in the southern and central Atlantic ... [and] were out of phase with the abrupt warmings in the Greenland ice cores, in the Nordic seas, and areas in the North Atlantic [that were] strongly affected by meltwater during stadials. This implies that the hinge line between areas showing gradual warming and areas showing abrupt warming was displaced far to the north close to the Greenland-Scotland Ridge'* [67]. Thus, the main cause for D–O events seems to be the 'turn-on' and 'turn-off' of a slow-down in convection in the Nordic Seas, with the South Atlantic remaining passive. Rasmussen concluded that rather than the Atlantic being the arena for a 'seesaw' mechanism, we should see it as the arena for a 'push-and-pull' mechanism [67]. In the 'pull' phase, during warm interstadial periods, more and more warm salty water is 'pulled' north through the Atlantic to become dense and sink as it cools in the cold Nordic seas. In the 'push' phase, during cold stadial periods, when Nordic seas ice-up and convection stops, warm water from the South Atlantic pushes northward via the Gulf Stream **beneath** the sea ice and gradually warms the Nordic seas from below until their ice melts and the 'pull' phase re-starts. The 'pull' is driven by the outflow of the deep cold, dense convective water from the Nordic Seas over the Iceland–Scotland Ridge, which currently entrains 75% of the inflow to the Nordic Seas from the south. This outflow creates a sea-level gradient (barotropic pressure gradient) across the ridge, which accounts for the 'pull' [67].

Recent work on stable oxygen isotopes from marine diatoms from the northeastern North Pacific shows that the region experienced abrupt intrusions of relatively fresh surface water coincidental with some of the Atlantic Heinrich Events and probably derived from meltwater originating from the Cordilleran Ice Sheet [68]. Maier and colleagues argued that northwardly directed warm waters caused the melting of parts of the ice sheet, adding freshwater to the North Pacific at the same time as some of the North Atlantic Heinrich Events [68]. Evidently the process in the Pacific was far less extreme than in the Atlantic.

Clearly the ocean provides a major route for connecting the poles. What about the atmosphere? In 2018, having noticed that there was a spatial pattern to the Antarctic response to DO warming or cooling, Buizert and colleagues used this to explore the relationship between ice core signals and changes in wind patterns [59]. Using principal component analysis they found that two modes of variability explain 96% of the variance in the Antarctic records. First of these, explaining 83% of the variance, was the classic bipolar thermal signal with a 200 year lag behind Greenland warming typical of the ocean response. The second main component, explaining 13% of the variance was quite different, being more or less synchronous with Northern Hemisphere change. This second mode is likely to represent an atmospheric teleconnection [59]. To better understand the operation of the atmospheric mode, Buizert used the logarithmic description of the deuterium excess $d_{\ln}$ (Figure 13.11), which preserves information on moisture, hence the incidence of moist winds coming off the sea. This signal lags behind Northern Hemisphere climate

by just 8–9 years for DO warming and cooling events, presumably in response to an equatorward shift to the north by the Southern Hemisphere westerlies in response to Northern Hemisphere warming, and a poleward shift to the south in response to Northern Hemisphere cooling [59]. The $d_{\ln}$ response is largest for the interior plateau sites, presumably because they lie farthest from the coast. Fluctuations in time in the $d_{\ln}$ response reflect the expansions and contractions in the Southern Annular Mode of atmospheric circulation around the continent. The West Antarctic core has the weakest response to shifts in the Southern Annular Mode, both in these palaeo-data and today. In contrast, maritime cooling that is probably forced by the extensive sea ice cover of the adjacent Weddell Sea influences the EDML core. Evidently, then, there is good evidence for Antarctic climate responding to both oceanic and atmospheric forcing on different timescales, and for displaying regional variability depending on geographic constraints such as proximity to the ocean and proximity to large areas of sea ice [59]. The continent does not operate as one simple system: different parts respond to the same forcings in different ways, and different forcings are present to varying extents in different parts of the continent.

Confirming the impact and timing of atmospheric processes on Antarctica, Bradley Markle and colleagues *'use[d] a high-resolution deuterium-excess record from West Antarctica to show that the latitude of the mean moisture source for Antarctic precipitation changed in phase with abrupt shifts in Northern Hemisphere climate, and significantly before Antarctic temperature change. This provides direct evidence that Southern Hemisphere mid-latitude storm tracks shifted within decades of abrupt changes in the North Atlantic, in parallel with meridional [atmospheric] migrations of the intertropical convergence zone'* [69].

Trying to enhance understanding of the operation of the bipolar seesaw, in 2018 Joel Pedro and colleagues put forward the idea that it was in fact part of a broader coupling between the two hemispheres and involving all the oceans, the poles and the atmosphere [70]. The preceding discussions in this section had already begun to make this apparent. Pedro and colleagues drew four main conclusions: first, changes in Atlantic heat transport invoked by the thermal seesaw are partially compensated by opposing changes in heat transport by the global atmosphere and Pacific Ocean. This compensation is an integral part of inter-hemispheric coupling, influencing the global pattern of climate anomalies; second, while this coupling involves the presence of a heat reservoir, its location is the global interior ocean north of the ACC, not – as commonly assumed – the Southern Ocean south of the ACC; third, the process driving Antarctic warming during Antarctic Isotopic Maxima (AIM) events is an increase in poleward atmospheric heat and moisture transport following sea ice retreat and surface warming over the Southern Ocean; and fourth, this Antarctic sea ice retreat is driven by eddies transporting heat south across the ACC, and amplified by feedback from the changing albedo as the area of sea ice decreases. The lag of Antarctic warming after the collapse of the AMOC reflects the time required for heat to accumulate in the upper 1500 m of the ocean interior north of the ACC before it can be mixed southwards across this dynamic barrier by eddies [70].

It is counterintuitive that when the AMOC collapses due to freshwater input and cooling in the northern North Atlantic and Arctic, the heat content of the global ocean increases. In the North Atlantic that increase takes place sub-surface beneath the shallow layer of surface cooling. The increase is attributed to expansion of sea ice in the north, reducing heat loss to the atmosphere, and to reduction in the production of cold NADW. The resulting warm anomaly is advected south in the Western Boundary Undercurrent (blue arrow pointing south off the east coast of North America in Figure 12.5). At the same time heat increases in the South Atlantic as the thermocline deepens by 50–100 m and there is reduced northward advection of heat in the uppermost limb of the AMOC (Figure 12.6) [70]. The heat anomaly in the South Atlantic is propagated east in the subsurface by the ACC (Figure 12.5), and after 300 years close to two thirds of the total increase in global heat content is found in the subsurface of the combined Indian and Pacific Basins; in contrast, the Southern Ocean south of the ACC is the region of the global ocean with the slowest rate of change of heat content [70]. When the AMOC resumes, northern sea ice melts away, convection resumes, and more heat is transferred from the subsurface ocean to the northern polar atmosphere; as a result the ocean interior cools. There is, again, very little change south of the ACC. Resumption of the AMOC is associated with a decrease in global ocean heat content.

'*These results challenge the common assumption that the Southern Ocean [south of the ACC] is a major heat reservoir during AMOC variations*' [70]. On the contrary, it is the area with the smallest change in heat content. '*Hence the "seesawing" of heat … is between the global ocean and the combined surface ocean and atmosphere at northern high latitudes*' [70]. The atmosphere is involved, too, in that following the collapse of the AMOC, '*the Atlantic ITCZ migrates south, which affects the wind field over the entire South Atlantic gyre. [The changes in the wind] also favour thermocline deepening and surface warming*' [70]. In addition, the Southern Hemisphere westerlies strengthen and shift southward, creating a positive anomaly in the Southern Annular Mode (the wind pattern around Antarctica), which increases wind stress by 10% south of

the ACC, and increases the flow of the ACC through Drake Passage by 10%. The southward eddy driven heat flux across the ACC increases by 20%, bringing warmer water to the sea ice zone and melting the sea ice. The warmer air over the ocean then transports heat and moisture south to warm Antarctica. Together these effects drive a warming of 0.1–0.2 °C/century in Antarctica. The same mechanisms operate in reverse to cool the region, giving us Antarctic warming at the onset of Greenland stadials (cool periods) and cooling at the onset of Greenland interstadials (warm periods) [70]. The southward shift of the inter-tropical convergence zone (ITCZ) weakens the southeasterly Trade Winds and the wind stress over the Benguela Current off Namibia, thus weakening upwelling there. Much the same is happening with modern-day warming [70].

A key aspect of these new messages is that once an oscillator gets going (the bipolar seesaw) it is self-sustaining to the point where it is not possible to determine which came first – the egg or the chicken.

The existence of Dansgaard–Oeschger and Heinrich Events tells us that slow, gradual changes in climate forcing can lead to abrupt climate shifts. Now, there's a good reason to study the processes and interactions within our climate system!

## 13.5 $CO_2$ in the Ice Age Atmosphere

So much for temperature; now what about $CO_2$ in the Ice Age air? As we saw in Chapter 9, early work by Oeschger and Dansgaard on $CO_2$ in ice cores showed that the air carried 180–200 ppm $CO_2$ at the Last Glacial Maximum, and 260–300 ppm in the pre-industrial era [10, 71]. By 1984, these two scientists showed that there was a good correlation between the $CO_2$ in bubbles of fossil air and the $\partial^{18}O$ composition of the ice, an indicator of temperature. As warming increased, so did atmospheric $CO_2$. They concluded that the rise in $CO_2$ at the end of the last glaciation contributed to the rise in temperature [10, 71].

With their results from Vostok, the French reached much the same conclusion by 1987. Jean-Marc Barnola of the Laboratoire de Glaciologie et Géophysique de l'Environnement, at St. Martin d'Heres showed with colleagues that $CO_2$ shifted from 190–280 ppm as conditions warmed at the ends of glacial periods, confirming a strong link between the climate cycle and the carbon cycle [72]. Cristophe Genthon of the Laboratoire de Géochimie Isotopique within the Laboratoire d'Océanographie Dynamique et Climatologie at Gif sur Yvette concluded that $CO_2$ was the dominant factor mediating between the two hemispheres with their different insolation forcing [73]. If Genthon and colleagues were right, then the 100 Ka glacial–interglacial cycle might originate from variations in $CO_2$ rather than from processes associated with the growth and decay of ice sheets suggested by Imbrie's SPECMAP project (indeed, Genthon's suggestion would cause Imbrie to modify the SPECMAP concept).

Claude Lorius summarized the Vostok data at a symposium in Vancouver in August 1987 [36, 37]. $CO_2$ in Greenland fluctuated between 190 and 200 ppm in cold periods and 260–280 ppm in warm periods. The results '*provide the first direct evidence of a close association between atmospheric $CO_2$ and climatic (temperature) changes in a glacial-interglacial time scale*' said Lorius [36, 37]. He went on to say that '*a simple linear multivariate analysis suggests that $CO_2$ changes may have accounted for more than 50% of the Vostok temperature variability, the remaining part being associated with orbital forcing ... The close correlation between $CO_2$ and temperature records and their spectral characteristics supports the idea that climatic changes could be triggered by an insolation input, with the relatively weak orbital forcing strongly amplified by possibly orbitally induced $CO_2$ changes*' [36, 37]. So, fluctuations in ice volume might not, after all, reflect just orbital forcing – even though that was the primary driver of change. $CO_2$ acting as a greenhouse gas must play some role. That is not surprising, because as the ocean is warmed by rising insolation it will emit $CO_2$, which will then feed back onto temperature. However, Petit found in 1999 that there seemed to be a lag between temperature and $CO_2$ peaks, with '*The $CO_2$ decrease [lagging] the temperature decrease by several kyr*' [24]. Nevertheless, he qualified that observation by noting that the mean resolution of the $CO_2$ profile was about 1500 years, increasing to 6000 years in fractured zones and in the bottom part of the record [24]. He went on to report that although recent work on glacial terminations had suggested that $CO_2$ concentration lagged warming by just $600 \pm 400$ years, '*considering the large gas-age/ice-age uncertainty (1,000 years, or even more if we consider the accumulation-rate uncertainty), we feel that it is premature to infer the sign of the phase relationship between $CO_2$ and temperature at the start of terminations*' [24]. Note, first, that this caveat regarding **gas-age/ice-age uncertainty** appears to be little known or commonly ignored, and, second, that the gas-age is based on a model whose assumptions (two decades later) appear be suspect, as we shall see. Evidently Petit saw his data (Figure 13.8) as a preliminary result requiring more work. Nevertheless, he and his team concluded from their data that '*our results support the idea that greenhouse gases have contributed significantly to the glacial–interglacial change. This correlation, together with the uniquely elevated concentrations of these gases today, is of relevance with respect to the continuing debate on the future of Earth's climate*' [24].

Aside from the close similarity between the trends of $CO_2$, $CH_4$ and local temperature calculated from hydrogen isotopes ($\partial^2 H$) in the Vostok core (Figure 13.8), a conspicuous feature of these trends was their 'saw-toothed' pattern, like that seen in deep ocean sediment cores (Figure 12.4). Since insolation, the main heat source, has a sinusoidal character, the saw-toothed pattern had to reflect slow ice growth and rapid decay [15]. By 2008 a composite analysis of $CO_2$ for the past 800 Ka was available (Figure 13.9) [74], as was data for methane [75]. These confirmed the stability of the saw-toothed $CO_2$, and temperature patterns back through time. That begged the question – which parts of the saw-tooth represented mostly the insolation signal, and which the ice signal?

Both at Vostok and Dome C, maximum values of $CO_2$ and $CH_4$ reached 280–300 ppm and 650–700 ppb, respectively, in interglacials, and fell to 180 ppm and 320–350 ppb in glacials. These limits described what I call the 'natural envelope' for these gases (Figures 13.8 and 13.9). That natural envelope coincides in turn with the natural envelope for Ice Age temperatures (Figures 12.4 and 13.6) (see also Chapter 12), providing compelling evidence that the carbon cycle was intimately connected to the climate system throughout the 2.6 Ma of the Pleistocene period, most probably through connecting feedback processes. Paul Falkowski of Rutgers University and his co-workers agreed, noting that '*The remarkable consistency of the upper and lower limits of the glacial–interglacial atmospheric $CO_2$ concentrations, and the apparent fine control over periods of many thousands of years around those limits, suggest strong feedbacks that constrain the sink strengths in both the oceans and terrestrial ecosystem*' [76]. How close is the relationship between $CO_2$ and temperature with time? Not surprisingly, the parallel oscillations between the two (Figure 13.9) [27], are reflected in statistics showing a moderately good correlation (correlation coefficient ~0.7) [24, 25, 74].

Is the $CO_2$ signal reliable back through time down core? Yes. First, over the past 800 Ka the values of $CO_2$ in the Dome C ice core have remained within the 'natural envelope' of 180–280 ppm as shown by the University of Bern's Dieter Lüthi and colleagues (Figure 13.9) [27]. A careful check on those data in 2015, by Bernhard Bereiter from the same university, and his team, found that an analytical artefact had very slightly affected $CO_2$ results from the deepest 200 m of the EPICA Dome C core, but that his had not significantly affected the overall pattern shown in Figure 13.9 [77]. Second, the consistency of the correlation between $CO_2$ and temperature, regardless of age down core, rules out the prospect of gas migration after compaction of the firn [74, 78]. Third, this pattern is consistent from one long core site to another (Vostok, Kohnen, Fuji, and Dome C). Fourth, measurements of $CO_2$ covering the past 1000 years in Antarctic ice cores from Law Dome, Siple Dome, South Pole, and Kohnen station, show the same $CO_2$ patterns despite being very different in location, altitude, and rate of accumulation. That rules out anything that might affect the concentration of $CO_2$ after compaction of the firn has stopped gas migration [78]. For example, Law Dome is a coastal site with a mean annual temperature of $-19\,°C$, a relatively high concentration of some impurities, and a high rate of accumulation, while Dome C is a high-altitude inland site with a mean annual temperature of $-44.6\,°C$, low accumulation and low concentrations of impurities. All of these sites show more or less no change from a peak of 280 ppm until the late 1700s or early 1800s, after which their $CO_2$ abundances rise exponentially to merge imperceptibly with those from Keeling's air measurements begun in the late 1950s (Figure 9.11).

It is of particular interest for studies of future climate change, as well as for understanding past climate change, that the modest 0.7 correlation coefficient between $CO_2$ and temperature shows that this is not a simple one-to-one relationship. That piqued the interest of Barry Saltzman (1931–2001) of Yale (Box 13.3).

Working with Mikhail Verbitsky in 1994, Saltzman had the innovative idea of plotting $CO_2$ data from Jouzel's Vostok ice core against ice volume data from Imbrie's

SPECMAP $\partial^{18}O$ data set, and against sea surface temperature data from a North Atlantic core [84, 85]. Plotting the data in 2000 year time steps, they expected to find that $CO_2$ decreased as temperatures cooled and ice volume grew, while $CO_2$ increased as temperatures warmed and ice volume shrank. While that was broadly true, the data deviated from the expected pattern. Starting from a fully glacial cold mode, $CO_2$ increased while the temperature stayed constant, then the temperature jumped while $CO_2$ stayed constant until a fully warm interglacial mode was reached. $CO_2$ then decreased while temperature stayed constant, then temperature decreased while $CO_2$ stayed constant until the cool mode was reached again. The smoothed pattern is illustrated in Figure 13.14. This pattern forms a 'hysteresis loop', something described in Chapter 6 (Figure 6.19). As both temperature and ice volume lagged behind $CO_2$, Saltzman assumed that the change in $CO_2$ caused the changes in the other two properties. He thought he had discovered an oscillation within the climate system between two stable states (glacial and interglacial) terminated by internal feedbacks, and that $CO_2$ was the key forcing agent in this system. His novel findings are summarized in his 2002 book *Dynamical Paleoclimatology* [86]. As we shall see, his analysis may be distorted by the assumption that Petit was right [24, 25], and that there was a significant lag between rising temperature and rising $CO_2$. We explore his ideas in some depth later.

Stimulated by Saltzman's research, in 1999 Hubertus Fischer, at Scripps, carried out a more detailed study of the relationship between $CO_2$ and temperature in ice cores [87]. In the Vostok and Taylor Dome cores, Fischer and

---

**Box 13.3  Barry Saltzman**

Saltzman's work in 1962 on chaos theory led to the development of the 'Saltzman-Lorenz attractor', but in 1980 he turned his attention to climate change and spent two decades at Yale developing models and theories of how ice sheets, winds, ocean currents, $CO_2$ concentration and other factors work together to make the climate oscillate in a 100 Ka cycle [79, 80]. Perhaps not surprisingly, given his earlier involvement in the development of chaos theory, with its emphasis on attractors, he became widely known for developing a numerical approach to modelling past climate change. Amongst other things, Saltzman's models predicted the possibility of 1–2 Ka millennial oscillations of the climate system many years before widespread interest in this topic developed in the palaeoclimate community [79]. He thought that a long-term tectonically forced decrease in atmospheric $CO_2$ could lead to a bifurcation of the system from steady state to a near 100 Ka auto-oscillation, a subject he explored with Kirk Allen Maasch [81–83]. For his research on how the climate system works, the American Meteorological Society awarded Saltzman its highest honour, the Carl Gustaf Rossby Research Medal, in 1998.

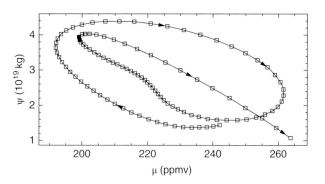

**Figure 13.14** Smoothed trajectory of the interrelation between ice mass (in kg×$10^{19}$) against $CO_2$ (in ppmv) in 2000 year time steps, demonstrating the near 100 Ka-period oscillations, beginning 218 Ka ago at 226 ppmv $CO_2$ and $1.5 \times 10^{19}$ kg of ice volume, and ending at time 0 (the present) (at lower right) with 264 ppmv $CO_2$ and $1.1 \times 10^{19}$ kg. The main inflexion points lie (i) in the outer ring close to 193 ppmv $CO_2$ and $4 \times 10^{19}$ kg, at about 170 Ka ago; (ii) in the outer ring close to 262 ppmv $CO_2$ and $2.4 \times 10^{19}$ kg, at about 130 Ka ago; and (iii) in the inner ring close to 200 ppmv $CO_2$ and $4 \times 10^{19}$ kg, at about 35 Ka ago. *Source:* From figure 7 in Ref. [84].

colleagues found that whilst $CO_2$ and temperature rose more or less together at glacial terminations, the temperature dropped more sharply than the $CO_2$ as interglacials ended. This is also evident from the work of Petit on Vostok (Figure 13.8), and from the work of Lüthi on Dome C (Figure 13.9). And the same is true for changes in temperature *within* glacials, as we can see from comparing the $\partial^{18}O$ signal, representing temperature, with the $CO_2$ signal, in Figure 13.6. As Fischer put it: '*high carbon dioxide concentrations can be sustained for thousands of years during glaciations*'[87]. He recommended that '*the carbon cycle–climate relation should be separated into (at least) a deglaciation and a glaciation mode*' [87]. In the deglaciation mode, when orbital insolation increased and led to warming, '*a net transfer of carbon from the ocean to the atmosphere*' directly connected temperature and $CO_2$ as warming exuded $CO_2$ from the ocean [87]. But when insolation decreased and led to cooling, several processes maintained abundant $CO_2$ in the air until temperatures reached their lowest point. First, the reduced area of the ocean as sea level fell on cooling meant that there was less area of ocean to absorb $CO_2$, even though more $CO_2$ would be expected to dissolve in the cooler surface water. Second, the growth of sea ice in cold periods further limited the uptake of $CO_2$ by surface waters. Third, a decline in terrestrial vegetation in glacial times (as tundra expanded) kept more $CO_2$ in the air, as did decomposing marine organic matter when falling sea level exposed organic remains on continental shelves. Eventually those influences waned, and once temperatures were at their lowest, land sources of $CO_2$ ceased supplying $CO_2$ to the air, and the cooler ocean away from sea ice areas took up more $CO_2$, reducing the amount in the air. Conditions moved towards equilibrium.

Back in 1961, Plass had predicted a further disconnect between temperature and $CO_2$ that should be evident during deglaciation in the Arctic. Whilst increasing insolation warmed mid-latitudes and tropics, releasing $CO_2$ from a warming ocean, in the Arctic the increased insolation energy went into melting ice sheets, keeping the surroundings cold despite the increase of $CO_2$ in the globally circulating air [88].

These various observations confirm that the relationship between temperature and $CO_2$ during the Ice Age is not simple. In effect, the climate sensitivity differs between glaciation and deglaciation modes. Paul Falkowski and his team reached much the same conclusions [76]. They called for further investigation, because understanding the carbon cycle is essential for deciding whether or not we can '*distinguish between anthropogenic perturbations and natural variability in biogeochemical cycles and climate*', and, further, to determine '*the sensitivity of Earth's climate to changes in atmospheric $CO_2$*' [76].

This raised a question: was there evidence that $CO_2$ and temperature changed in lock step as the world warmed? Or were there leads and lags in the system? If you have ever walked on fresh snow, you will recall the crunching sound your footsteps make. That's the sound of the snow being compressed and causing pockets of trapped air to collapse. Without human intervention, the deposition of more and more snow eventually compresses earlier layers into dense snow, or 'firn'. The air between the snow particles in the firn can still mix with air at the surface down to depths of 50 or 100 m, at which point the firn is converted to glacial ice and the air is locked in as bubbles. Comparing the properties of the air bubbles with those of the surrounding ice requires an adjustment to the time scale for the air bubbles that takes into account the younger age of the trapped air than of the surrounding ice. From such calculations it was thought that the rise in $CO_2$ lagged behind the rise in temperature by 600–1900 years [74]. This estimated lag was a topic of controversy, firstly because its calculation was based on assumptions that might or might not be accurate about the 'lock-in depth' at which air was finally trapped, and secondly because basic physics showed that temperature and $CO_2$ should rise together in a warming ocean. As Ed Brook said, we have to know – '*Does $CO_2$ drive climate cycles or is it a feedback in the system that contributes to warming?*' [89]. If the model of firn compaction used to determine the difference between the gas age and the age of its surrounding ice was wrong, the lag could be more, less, or even non-existent.

The controversy was resolved in 2013 with the latest exciting development in studies of ice core $CO_2$ by an international team led by Frédéric Parrenin of the Laboratoire de Glaciologie et Géophysique de l'Environnement, in Grenoble [90]. Parrenin and his team used nitrogen isotopes to establish the difference between the gas age and the age of the surrounding ice in several Antarctic ice cores. The ratio of $^{15}N/^{14}N$ (expressed as $\partial^{15}N$) in the air bubbles is enriched in firn due to gravitational settling, and depends on the thickness of the firn [90]. Given this ratio, the offset in depth between the gas and the ice of the same age can be determined, along with the amount of time represented. Using this new technique, Parrenin's team found that $CO_2$ concentrations and Antarctic temperatures were tightly coupled throughout the last deglaciation, within a quoted uncertainty of less than $\pm 200$ years [89, 90]. The correlation between $CO_2$ and Antarctic temperature using this new gas-age chronology was very high, at 0.993, suggesting that the rise in $CO_2$ did contribute to much of the rise in temperature in Antarctica during the last deglaciation, even at its onset, more, or, less as Lorius had initially suggested. Parrenin's data agreed closely with those of Joel Pedro and colleagues [91], even though both laboratories had used largely independent methods and data. This suggested that there was a very tight coupling between regional Antarctic temperatures and $CO_2$ during deglaciation [89].

The new data were confirmed by further analyses by Beeman and others from Parrenin's laboratory in 2019, which incorporated data from the latest long Antarctic ice core from the West Antarctic Divide (Figure 13.15) [92]. Beeman and colleagues detected four major common break points during the last glacial termination in both the temperature and $CO_2$ time series. The phasing between $CO_2$ and Antarctic climate was small but variable, ranging from a centennial-scale $CO_2$ lead, to synchrony, to a small centennial-scale lead by Antarctic climate. Synchrony between $CO_2$ and temperature was within the 95% uncertainty range for the four major changes except the end of glacial termination 1 (T1) at 11.5 Ka ago when Antarctic temperature ($\partial^{18}O$) appeared to have led $CO_2$ by 532 years (range 337–629 years) [92], much as noted

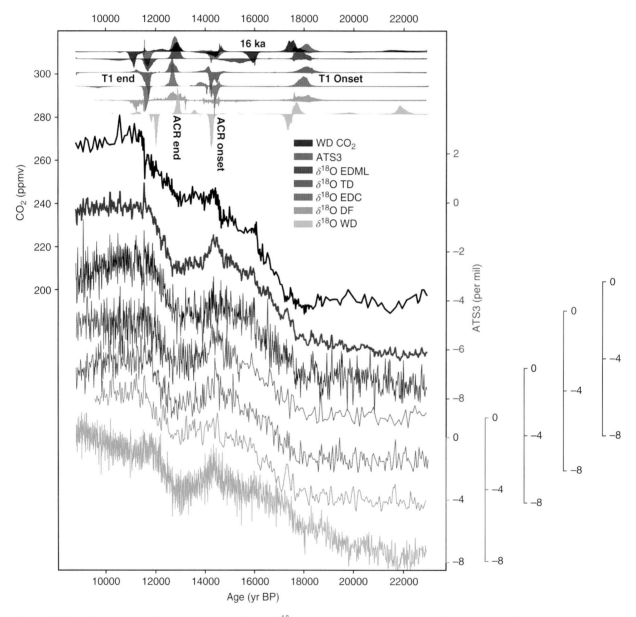

**Figure 13.15** Atmospheric $CO_2$ (black) and individual $\partial^{18}O$ records placed on a common timescale, with the normalized histograms of probable change points (eight points) for each ice core used in the ATS3 temperature stack; the locations of the drill sites are shown in the top right corner. Histograms are plotted downward-oriented when the rate of change decreases and upward-oriented when it increases (same colours, y-axis not shown). Probabilities are normalized so that the integrated probability for a given histogram sums to 1. In four distinct time intervals, both series show concurrent probable change points. *Source:* From figure 5 in Ref. [92].

by Parrenin [90]. However, if the true $CO_2$ change point occurred slightly closer to 11.5 Ka ago, the phasing would be reduced to 174 years (range 65–280 years) and thus lie within the 95% uncertainty range. At the T1 onset the temperature may have led $CO_2$ by 570 years (range 127–751 years) [92], whereas Parrenin had found no significant age difference [90]. Even there, given the increased uncertainty of the age determination prior to 16 000 years ago, Beeman and colleagues could not exclude the possibility of synchrony at the 95% confidence level, which indeed appears to be the case for the record from DF [92].The phasing at the onset and closure of the Antarctic Cold Reversal (ACR) was in essence synchronous (within ±50–60 years) as found earlier by Parrenin [90].

We can see these various interactions clearly in Figure 13.16, published in 2010 by Barbara Stenni, then at the University of Trieste in Italy, which explores the relationships between Greenland and the TALDICE ice core from the Talus Dome in the Ross Sea region [93]. Talus Dome data was also used in compiling Figure 13.6, and in both figures methane was used to synchronize the ice core data. Figure 13.16 confirms that '*warming during the transition from glacial to interglacial conditions was markedly different between the hemispheres, a pattern attributed to the thermal bipolar see-saw*' [93]. It also confirms the close relationship between temperature change in Antarctica and the $CO_2$ record, although these $CO_2$ data predate the revision to the dating of fossil $CO_2$ in Antarctic ice cores by Parrenin [90] and by Beeman [92] and represented in Figure 13.15. Figure 13.16 confirms the synchronous relationship between warm Antarctic Isotope Maxima (AIM 0, 1, and 2) and cold Dansgaard–Oeschger Events (DO 0, 1, and 2), and between subsequent Antarctic cooling and Greenland warming, one example of which is the synchronous relationship between '*the cooling of the Antarctic Cold Reversal ... [and] the Bølling–Allerød warming in the northern hemisphere 14,700 years ago*' [93]. The onset of the ACR coincides with '*the exceptionally large Melt Water Pulse 1a (MWP1a) ... at around 14.6 [Ka ago, leading to] a global sea-level rise between 10 and 20 m in a few centuries*' [93] (Figure 13.16). Part of that sea level rise most likely came from the melting of land ice in Antarctica. As Stenni noted '*A massive freshwater release in the high-latitude Southern Ocean is expected to shut down convection, reduce southward heat transport and increase sea ice and associated albedo feedbacks, inducing high-latitude Southern Ocean cooling*' [93]. The Bølling–Allerød warming is seen as due to a recovery of the AMOC (and production of NADW) that brought heat to the north following cold Heinrich Event 1

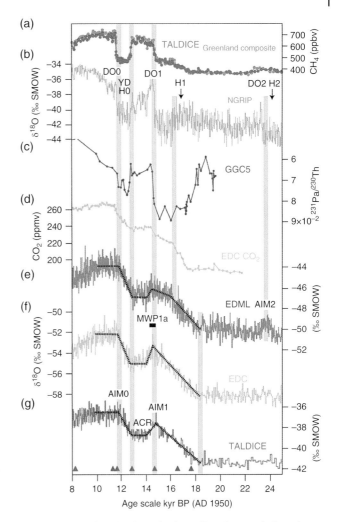

**Figure 13.16** Compilation of palaeoclimatic records from ice and marine cores to depict the bipolar sequence of events during the last glacial termination. (a) $CH_4$ compiled from Greenland ice cores and TALDICE; (b) $\partial^{18}O$ from Greenland's NGRIP ice core; (c) $^{231}Pa/^{230}Th$ from marine core GGC5 from Bermuda Rise in the deep western subtropical Atlantic, taken as a proxy for the Atlantic Meridional Overturning Circulation (AMOC) strength; (d) $CO_2$ from the EPICA Dome C ice core; (e) $\partial^{18}O$ from EDML (EPICA Dronning Maul Land) ice core; (f) $\partial^{18}O$ from EPICA Dome C; (g) $\partial^{18}O$ from TALDICE. The dotted lines are the ramps obtained with the Rampfit and Breakfit software. The Younger Dryas (YD) and Dansgaard–Oeschger (DO) interstadials are indicated. The black horizontal bar corresponds to Meltwater Pulse (MWP) 1a. Arrows represent the ages of Heinrich layers 1 and 2 (the younger Dryas is regarded as Heinrich Layer 0, or H0). Triangles indicate synchronization $CH_4$ tie points. Grey vertical bars correspond to the Antarctic Isotope Maximum (AIM) 2 event; the start of the deglaciation (18.2 ± 0.7 Ka ago); the slowing of the warming at EDML (16.0 ± 0.2 Ka ago), the AIM-1 event (14.7 ± 0.3 Ka ago), the end of the Antarctic Cold Reversal (ACR) (12.7 ± 0.3 Ka ago) and the AIM-0 (11.9 ± 0.3 Ka ago) event as inferred from $\partial^{18}O$ ice-core records. *Source:* From figure 2 in Ref. [93].

(H1), and a corresponding decrease in the production of AABW caused by the influx of fresh melt water to the Southern Ocean, as expected from the bipolar seesaw [93].

Stenni and colleagues also found distinct differences in temperature history between different Antarctic regions *'pointing to differences in the climate evolution of the Indo-Pacific and Atlantic sectors of Antarctica'* [93]. For instance, in the Atlantic sector, they found *'that the rate of warming slowed between 16,000 and 14,500 years ago [see dashed line in Figure 13.16e], parallel with the deceleration of the rise in atmospheric carbon dioxide concentrations [Figure 13.16d] and with a slight cooling over Greenland [Figure 13.16b]'* [93]. This slowing and deceleration coincided with weakening of the strength of the AMOC (Figure 13.16c). This suggests that the bipolar seesaw is expressed differently in the two regions, probably because of differences in basin scale circulation [93].

The changes across the ACR and at the end of the termination (T1) corresponded with changes in methane ($CH_4$) (Figure 13.16) probably driven by changes in the tropics. Similarly, the onset of warming at the beginning of the termination (T1) appeared to be more or less synchronous with a gradual rise in $CH_4$. There were also times when $CO_2$ and temperature were decoupled, e.g. at 16 and 14Ka ago. Evidently, the mechanisms of coupling between temperature and $CO_2$ are complex, perhaps (i) due to modulation by external forcing or background conditions that impact heat transfer and oceanic circulation (hence $CO_2$ release), (ii) internal feedbacks that change the response times of the two series, or (iii) mechanisms that provoke similar response in both temperature and $CO_2$ but with different lags [92]. Beeman and his team noted that averaging cores from across West and East Antarctica in this study may obscure or be affected by regional differences, for instance the effect of ice sheet elevation on temperature; we may note for instance that East Antarctica's DF core is far away from the WAIS Divide core, at a much higher altitude (3700 m as opposed to 1797 m), and exposed to a different wind system (the WAIS region receives warm moist air funnelled down the Antarctic Peninsula towards the Ross Sea as part of the Amundsen Sea Low Pressure Cell) [92]. The collection of further long cores from West Antarctic would help to address this issue, as would the acquisition of high-resolution $CO_2$ records through the beginning of the Holocene (12.5–11.5 Ka) [92].

These various authors theorized that insolation warmed the ocean, which simultaneously released $CO_2$ to the air, providing an immediate positive feedback that enhanced further warming. The same warming released water vapour, another greenhouse gas, which acted in concert with the $CO_2$. For the most part there was no significant delay between warming and $CO_2$ emission in the Antarctic.

The link between $CO_2$ and temperature at glacial terminations is not a coincidence, it is causal and tied to insolation. Physics rules! We have moved far beyond the era of inadequate age models of ice core gases that led to an impression that $CO_2$ lagged temperature by thousands of years [24]. However, we have to remember that different controls operated during periods of glacial growth, when insolation drove falls in temperature that were not followed immediately by $CO_2$, as explained earlier.

Independent support for these conclusions came from Joel Pedro of the Antarctic Climate and Ecosystems Cooperative Research Centre at the University of Tasmania, Hobart, and colleagues (Figure 13.17) [91]. Using multiple Antarctic ice cores with both high and low rates of accumulation to refine estimates of the lock-in depth of air bubbles in the firn, Pedro and colleagues found in 2012 that during the last deglaciation *'the increase in $CO_2$ likely lagged the increase in regional Antarctic temperature by less than 400 yr and that even a short lead of $CO_2$ over temperature cannot be excluded. This result ... implies a faster coupling between temperature and $CO_2$ than previous estimates, which had permitted up to millennial-scale lags'* [91]. Pedro agreed with Parrenin: *'Mounting evidence attributes a large component of the deglacial $CO_2$ increase to release of old $CO_2$ from the deep Southern Ocean through changes in its biogeochemistry and physical circulation'* [91].

Yet another international team, led by Jeremy Shakun of Harvard, agreed in 2012 that: *'the covariation of carbon dioxide ($CO_2$) concentration and temperature in Antarctic ice-core records suggests a close link between $CO_2$ and climate during the Pleistocene ice ages'* [33]. Their analysis, which was not as refined as Parrenin's, suggested that *'The role and relative importance of $CO_2$ in producing these climate changes remains unclear ... in part because the ice-core deuterium record [used as a proxy for temperature] reflects local rather than global temperature'* [33]. Using a record of *global* temperature constructed from 80 proxy records, they showed *'that temperature is correlated with and generally lags $CO_2$ [globally] during the last ... deglaciation'* [33]. This is *'consistent with $CO_2$ acting as a primary driver of global warming, although its continuing increase is presumably a feedback from changes in other aspects of the climate system'* [33].

Clearly, we need to differentiate local from global signals, much as Al Fischer recommended (Chapter 10). Examining Shakun's results, Pedro explained that $CO_2$ led the Northern Hemisphere temperature reconstruction by $720 \pm 330$ years, led the global temperature reconstruction by $460 \pm 340$ years, and lagged behind the Southern Hemisphere temperature reconstruction by $620 \pm 660$ years [91]. Given the uncertainties in dating, they thought that the $620 \pm 660$ year lag reported by Shakun's team for the

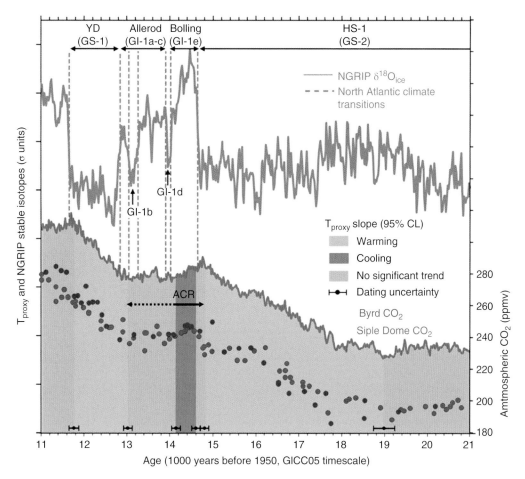

**Figure 13.17** Synchronous proxy temperature ($T_{proxy}$) and atmospheric $CO_2$ signals in the last deglaciation at Byrd and Siple Coring Sites, displaying the bipolar seesaw. Significant warming and cooling trends in $T_{proxy}$ are represented by shaded vertical bands. Climate in the North Atlantic region is represented by the NorthGRIP ice core $\partial^{18}O$ record at top. Changes in the slope of Antarctic $T_{proxy}$ are synchronous with climate transitions in the North Atlantic (vertical dashed lines), within relative dating uncertainties (horizontal error bars). The deglacial increase in $CO_2$ occurs in two steps, starting at 19 and 13 ka and corresponding to significant warming trends in $T_{proxy}$. A pause in the $CO_2$ rise is aligned with a break in the Antarctic warming trend during the Antarctic Cold Reversal (ACR). Within the core of the Antarctic Cold Reversal, significant cooling in $T_{proxy}$ (dark shaded band) coincides with an apparent decrease in $CO_2$. Fast-acting inter-hemispheric coupling mechanisms linking Antarctica, Greenland and the Southern Ocean are required to satisfy these timing constraints. *Source:* From figure 3 in Ref. [91].

Southern Hemisphere was '*not inconsistent*' with their own conclusion that the lag in the Antarctic was less than 400 years and that $CO_2$ might even lead temperature change [91]. That being so, Shakun's result for the Southern Hemisphere is not inconsistent with Parrenin's finding in 2013 [90] and Beeman's finding in 2019 [92] that during the last deglaciation $CO_2$ and temperature changed largely synchronously in the Southern Hemisphere. That begs the question, *what delayed warming in the north?*

Examined in detail, the parallel curves of $CO_2$ and temperature in Antarctic ice cores through the last deglaciation display two warming trends interrupted by the cooling of the ACR between 14.7–13.1 Ka ago (Figures 13.16 and 13.17) [91]. Pedro's team concluded that '*Evidence from Southern Ocean marine sediment cores directly links each of*

*these two warming steps with release of $CO_2$ accumulated in the deep Southern Ocean during the last glacial period .... [including] pulses in upwelling (represented by opal fluxes) ... and coincident with negative excursions in atmospheric $\partial^{13}C$ ... while cores ... identify a source of old ($^{14}C$-depleted) carbon-rich water that dissipated over ... corresponding intervals*' [91]. Pedro proposed '*that the increases in [wind-driven] upwelling were responsible for the simultaneous delivery of both sequestered heat and $CO_2$ to the atmosphere around Antarctica*' [91].

Upwelling also enhanced the upper ocean's concentration of nutrients, since these had been sequestered in deep-ocean bottom waters, along with $CO_2$, by the biological pump during glacial periods. As these nutrient-rich waters welled up to the surface around Antarctica in response to increasing wind strength and a decrease in sea ice cover, they were then subducted north of the Polar Front by the sinking of surface water to form Intermediate Water and Subantarctic Mode Water. These two water masses circulate through the world ocean and provide the feedstock for continental margin upwelling. Studies of sediment cores from the eastern equatorial Pacific upwelling system show that this nutrient supply, originating from southern sources, increased away from glacial maxima and peaked at glacial terminations [94], as did the associated $CO_2$. Equally important was the intensity of the Trade Wind system governing equatorial upwelling, which increased in response to the retreat of the Laurentide Ice Sheet [94].

Can we be certain where the $CO_2$ came from during deglaciations? It was popular to assume that as the $CO_2$ was very depleted in $^{14}C$, it must have been out of touch with the atmosphere for long enough for its atmosphere-derived $^{14}C$ to decay, hence the logical source was upwelling in the Southern Ocean [95]. But, careful work by Hain and colleagues in 2014 found that the relationship between $CO_2$ and $^{14}C$ in ice cores suggested a more complex solution. They recalled that the $^{14}C$ content of the atmosphere declined as the Earth's magnetic field increased over the past 40 Ka (more on that process in Chapter 15). Furthermore the repeated stalling and starting of the production of NADW (e.g. in the Younger Dryas event) would have led to periods of less and more $^{14}C$ extracted with $CO_2$ from the atmosphere [95].

Figure 13.6 shows how $CO_2$ behaves over small time steps, like individual Dansgaard–Oeschger events during the bipolar seesaw. In 2000, analysing an ice core from Antarctica's Taylor Dome, Andreas Indermühle and colleagues from the University of Bern's Climate and Environmental Physics group showed that $CO_2$ rose by about 20 ppm as the environment warmed and fell by the same amount as it cooled, during climate changes between 60 and 20 Ka ago [96], more or less as shown in Figure 13.6. In 2007, comparing the Greenland's GISP2 core with Antarctica's Byrd core, Jinho Ahn and Ed Brook from Oregon State University confirmed that $CO_2$ rose along with temperature in the Antarctic predecessor to a Greenland Dansgaard–Oeschger event (Figure 13.6) [14]. As temperature and $CO_2$ reached their peak in the south and started to decline, a Dansgaard–Oeschger event suddenly began in the north. The decline in $CO_2$ was less rapid than the decline in temperature in the south, much as found by Saltzman and by Fischer (see earlier). Ahn and Brook argued that the close relationship between $CO_2$ and Antarctic temperatures at both the coarse scale of orbital change (Figure 13.8) and at the fine scale of Dansgaard–Oeschger change (Figure 13.6), and the fact that the pattern is saw-toothed at both scales, reinforces the argument

that atmospheric $CO_2$ must be governed by physical processes that control surface air and ocean temperatures in the Southern Ocean. These processes likely included an increase in warming (driven by some forcing factor – explored below), which caused a decline in sea ice cover, allowing winds to interact with the ocean surface, causing upwelling, thus overturning southern deep water, decreasing the stratification of the Southern Ocean, and bringing $CO_2$ to the surface at the same time as ocean warming.

Comparing the climate records from Greenland with those of Antarctica (Figure 13.16 and 13.17), Pedro's team found that '*there is little or no time lag separating the major millennial and sub millennial climate transitions in Greenland … from the onsets and ends of the warming and cooling trends in Antarctica*' [91], suggesting that there must be rapid coupling between Antarctic temperatures, Greenland temperatures and atmospheric $CO_2$. They concluded that '*The ice core observations point to a tightly coupled system operating with little or no time delay between the onsets/terminations of North Atlantic climate stages and near simultaneous trend changes in both Antarctic temperature and atmospheric $CO_2$ …. [which] lends support to the current concept of an atmospheric teleconnection between the northern and southern high latitudes which forces wind-driven $CO_2$ release from the Southern Ocean [via increased upwelling]*' [91]. To put it another way, $CO_2$ and temperature vary in the same way at the glacial–interglacial scale and at the millennial scale; they rise together first in the south.

While these studies showed that the distribution of $CO_2$, like temperature, responded to the bipolar seesaw, they did not identify what warmed the Southern Ocean to set the seesaw off. As we saw earlier, initial arguments suggested the warming of the Southern Ocean was influenced by fluctuations in the production of NADW. To test that possibility, Andreas Schmittner of Oregon State University and Eric Galbraith from Princeton applied a coupled ocean–atmosphere-sea ice-biosphere model, forcing it by varying freshwater inputs to the North Atlantic [97]. Much as expected, adding freshwater stopped the circulation, reduced the northward transport of heat, cooled the North Atlantic and warmed the Southern Ocean, as subsequently shown by Kawamura [57]. It also reduced the transport of salt from the North Atlantic to the Southern Ocean via NADW, reducing the stratification of the Southern Ocean and allowing more Circumpolar Deep Water to reach the surface to release $CO_2$, thus increasing atmospheric $CO_2$ concentrations in concert with Southern Ocean (and Antarctic) temperatures. These changes overrode the ability of the biological pump to store $CO_2$ in deep water.

Did the Southern Ocean contain enough $CO_2$ during the last glaciation to cause the rapid rises in atmospheric $CO_2$ seen during the subsequent deglaciation? Yes. Luke Skinner of the Godwin Laboratory of Cambridge University showed in 2010 that the deep waters of the Southern Ocean during the Last Glacial Maximum were depleted in $^{14}C$, which increased as $CO_2$ was degassed from the ocean during the deglaciation, and diluted the $CO_2$ containing excess $^{14}C$ that had accumulated in the air during the glacial period [98]. Skinner and colleagues data confirmed that the Atlantic sector of the Southern Ocean contained a poorly ventilated deep carbon pool during the Last Glacial Maximum. That reservoir sequestered $CO_2$, reducing its concentration in the air. The increase in $CO_2$ and decrease in $^{14}C$ in the air at deglaciation went along with a decrease in sea ice estimated from the increased supply of sea salt to coastal Antarctic sites. Shrinkage of sea ice from its maximal extent in the Last Glacial Maximum allowed westerly winds to act directly on the ocean, causing Circum-Polar Deep Water to well up from below and provide old $CO_2$ to the air [99]. Models and data agree with that interpretation. Whilst Schmittner and Galbraith's model seems to confirm that this process was driven by what happened in the north (the freshwater lid model) [97], we must not neglect the possibility the ultimate trigger for the seesaw lay in the south. More on that later.

In 2019, MacGilchrist and colleagues proposed an exciting modification to the conventional framework for the subpolar Southern Ocean carbon cycle [100]. They reminded us that the understanding of this cycle has traditionally been considered in terms of the two-dimensional latitude-depth circulation framework represented in Figure 12.6, in which the most conspicuous feature is vertical overturning, which brings up carbon from depth. Some of this carbon may be transferred to the atmosphere, and some may be transferred back into deep water through the formation of AABW. However another key feature of the Meridional Overturning Circulation is the horizontal transport of ocean properties from west to east by the ACC, as shown in Figure 12.5. Furthermore, within the Southern Ocean the circulation south of the ACC is characterized by large horizontal gyres shown in Figure 13.18 [101], but not visible at the global scale of Figure 12.5. These include the Weddell Gyre between Africa and Antarctica, and the Ross Gyre between New Zealand and Antarctica [101], early versions of which – at a time just prior to formation of the ACC – are depicted in Figure 7.9 to show that these are long-standing features of Southern Ocean circulation. Calculating the amount of transport of carbon both vertically and horizontally for the Weddell Gyre, MacGilchrist found that it was overwhelmingly associated with horizontal transport in Circumpolar Deep Water [100]. Much of

this material is not old carbon welled up from depth, but new carbon formed by local open-ocean productivity within the ACC system at the northern boundary of the gyre. Much the same processes operate within the Ross Gyre (Figure 13.18) [100]. Hence their deduction '*the present-day rate of carbon uptake in the subpolar Southern Ocean is set by the [modern] open-ocean biological pump and the horizontal gyre circulation*' [100]. Until now these '*dynamics ... have been broadly absent from historical perspectives on the carbon cycle in the subpolar Southern Ocean ... Changes in a proxy signal are commonly interpreted to reflect changes in the overturning circulation ... [but] our results indicate that processes far removed from the overturning circulation critically affect the biogeochemical signature of deep waters, fundamentally undermining [traditional] interpretation[s]*' [100]. For more information on the modern circulation of the Southern Ocean and its relation to climate see Rintoul [101] and Meredith and Brandon [102].

Another fascinating new insight about carbon storage, which helps us to understand the tight linkage between temperature and $CO_2$ in ice cores, also emerged in 2019, from Khatiwala and colleagues [103]. It has been commonly thought that the decrease in atmospheric $CO_2$ in glacial periods was due to the increased efficiency of the biological pump, which takes planktonic remains into the deep ocean, combined with increased productivity driven by an increase in the supply of dust containing iron, and the formation of a sea ice lid preventing the outgassing of old $CO_2$. Khatiwala argued that this understanding was inadequate, being based on a method of estimating respired $CO_2$ that uses 'apparent oxygen utilization', which overestimates the inventory of respired $CO_2$ in the modern ocean by as much as 50% [103]. Khatiwala pointed out that at high latitudes poleward-moving surface waters experience heat loss and carbon gain from the atmosphere, but because the air-sea exchange of $CO_2$ is slow (and further hindered by the existence of sea ice), the carbon gain by the ocean is incomplete, so that polar waters sinking to the interior are undersaturated – depleted in carbon relative to what they would be if they had been in equilibrium with the atmosphere. This disequilibrium reduces carbon storage by the ocean. Biology, in contrast, causes carbon to increase in the subsurface. When upwelling and mixing bring old biogenic $CO_2$ to the surface, slow gas exchange and a sea ice lid once again inhibit exchange, causing oversaturation of carbon in the subsiding surface water. In this case the disequilibrium with the atmosphere enhances carbon storage in the ocean. These disequilibria are likely to have been exacerbated during glacial conditions. Khatiwala's findings have important implications. First, even large changes in ocean circulation or sea ice seem to have substantially smaller effects than previously thought on changes in atmospheric

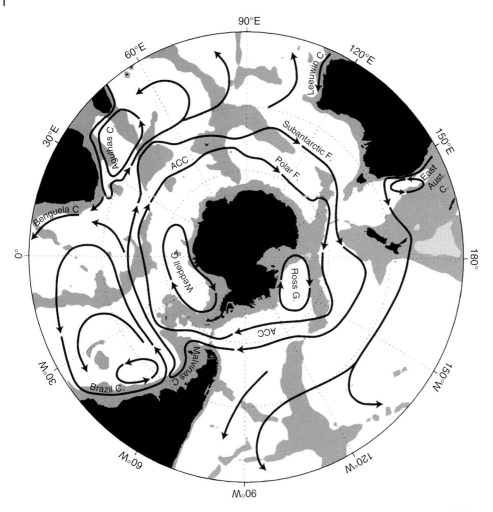

**Figure 13.18** Schematic map of major currents in the Southern Hemisphere oceans south of 20°S, showing depths shallower than 3500 m as shaded. The two major cores of the ACC are the Subantarctic Front and the Polar Front. F = Front; C = Current; G = gyre. *Source:* From figure 4.6.1 in Ref. [101].

$CO_2$ abundance between warm and cold periods. Second, changes in biological production and respiration, especially in response to dust (Fe) flux, have a larger overall effect than previously thought. Finally, cooling (the effect of temperature on $CO_2$ solubility) causes the largest change in atmospheric $CO_2$ between present and glacial conditions [103]. Khatiwala concluded that '*Spatial variations in temperature increase [the carbon disequilibrium] by reducing undersaturation, and thus amplify the impact of ... cooler temperatures, which accounts for about half of the total glacial-interglacial $CO_2$ change. This may explain the tight coupling of $CO_2$ with Antarctic temperatures observed in ice cores*' [103]. This 2019 finding should influence future climate modelling.

Examining the record of temperature and $CO_2$ over the past 800 000 years, Yin and Berger realized in 2010 that during the last 1 Ma, the amplitude of glacial–interglacial climate cycles had increased substantially after the

Mid-Brunhes Event about 430 Ka ago' [104] (see Figure 13.9). They determined that the later interglacials were warmer primarily because of increased global mean temperatures during Northern Hemisphere winters, caused by increased insolation during this season, relative to the interglacials that preceded the Mid-Brunhes Event, in conjunction with increased atmospheric greenhouse-gas concentrations. By 2011, they were able to determine the respective roles of $CO_2$ and insolation in determining the temperatures of each interglacial, finding that $CO_2$ tended to play a more important role than insolation in the Southern Hemisphere, while insolation tended to play a more important role in the Northern Hemisphere [105]. During boreal winter (austral summer) about 60% of the warming was due to greenhouse gases and 30% to insolation [104]. As Pedro observed, $CO_2$ tended to change with temperature in the south, so provided an important feedback there, but preceded temperature change in the north, where insolation played a larger role in melting northern hemisphere ice than did $CO_2$ [91]. This inter-hemispheric difference may reflect the vast size of the Southern Ocean (a source of $CO_2$ and water vapour in the south) compared with the extremely limited ocean area in the north polar region, much of it covered with sea ice and hence having high albedo.

## 13.6  The Ultimate Climate Flicker – The Younger Dryas Event

As we saw in Chapter 12, one of the best-known features of the warming from the last glacial maximum to the Holocene was its interruption, in the Northern Hemisphere, by the 1300 year long cold period known as the Younger Dryas (Figures 13.16 and 13.17), named after a white Arctic flower *Dryas octopetala,* which spread southwards at the time. This cold period, before temperatures plunged almost back to the levels of the Last Glacial Maximum, was preceded by an initial warm period – the Bølling–Allerød interstadial – and followed by rapid warming to the start of the Holocene. Mapping these changes in detail has only been possible through the advent of ice-cores, although marine sediment cores, despite their lower resolution in time, have also been useful.

One of the biggest surprises to emerge from a comparison of the Greenland and Antarctic ice cores was that the warm Bølling/Allerød interstadial and cold Younger Dryas stadial of Greenland ice cores were missing from Antarctica (Figures 13.16 and 13.17) [106]. Even so, both of these events are visible in Antarctic ice cores, for example at Dome C, where they are represented not by temperature but by methane ($CH_4$) (Figures 13.8c and 13.16a). Jeff Severinghaus, of Scripps, and colleagues, showed that the $CH_4$ began to rise within 0–30 years of the beginning of the temperature rise at the end of the Younger Dryas event [107]. Plants decaying under low oxygen (anaerobic) conditions emit methane, mostly from wetlands. These were thought to lie mostly in the Northern Hemisphere, where there is more land, although some scientists suggested that the rapidity of the $CH_4$ rise reflected sudden release from large volumes of methane stored in marine and continental sediments [15]. Although the Arctic was commonly cited as a possible source, Bill Ruddiman and Maureen Raymo showed from the gradients of $CH_4$ between the hemispheres that most of the methane came from wetlands in the northern tropics, controlled by the monsoons and their response to changes in insolation at 30°N [108]. Regardless of its origin, once emitted $CH_4$ spread rapidly everywhere.

In Antarctica, a slight cooling, the ACR, coincided with Greenland's warm Bølling/Allerød interstadial, and was followed by warming throughout the period represented by Greenland's cold Younger Dryas (Figures 13.16 and 13.17). In addition, the rise in temperature in Antarctica, from 18 Ka ago, coincided with flat or declining temperature in Greenland (Figures 13.16 and 13.17). As we saw earlier, these opposing trends mimic the bipolar seesaw, which likely reflected changes within the AMOC.

How might we account for the cooling in the Younger Dryas? As we saw above and in Chapter 12, the prevailing theory, proposed by Wally Broecker and colleagues back in 1989, was that the Younger Dryas came about because of a sudden shut-down of the global ocean conveyor, when melting ice in North America allowed an enormous glacial lake, Lake Agassiz, to drain to the North Atlantic via the St Lawrence seaway, providing a temporary freshwater lid that stopped the formation of deep water in the Norwegian-Greenland Sea [109]. While Lev Tarasov and Richard Peltier of the University of Toronto agreed with that notion in principle, they showed in 2005 that the discharge most likely ran down the Mackenzie River into the Arctic and then into the Atlantic [110]. Peltier is the holder of several awards for his contributions to climate science, including the Milutin Milankovitch Medal of the European Geosciences Union, in 2008, and was elected a Fellow of the Royal Society of Canada in 1986.

This is not the place to go into detail about the history of Lake Agassiz, but, as we saw in Chapter 12, outwash gravels have been found that confirm Tarasov and Peltier's interpretation of the route. It now seems likely that there were also many smaller discharges from the lake between 17 and 8 Ka ago, several of which can be correlated with $\partial^{18}O$ events in Greenland ice cores [19]. The most recent such event occurred 8.2 Ka ago (see Chapter 14). I discount the recently proposed notion that the cooling of the Younger Dryas was the result of a major bolide impact around 12.9 Ka ago [111], since the evidence for it is controversial.

Analyses of the GRIP and GISP-2 ice core records shows that the warm Bølling–Allerød interstadial arose remarkably quickly at 14.7 Ka ago, and the transition from the cold of the Younger Dryas to the warmth of the Holocene was also extremely quick with about half the warming concentrated into about 15 years [45] (cf. Figures 13.16 and 13.17). As Stenni and colleagues point out '*the end of the YD in the North shows the same characteristics of a typical abrupt DO [Dansgaard–Oeschger] warming event (DO-0), associated with the early Antarctic Holocene optimum [Antarctic Isotope Maximum-0, or AIM-0] in a thermal bipolar see-saw pattern. The ACR-AIM-0 transition starts synchronously at 12.7 ± 12.9 [Ka ago] in [the] TALDICE-EDC-EDML [ice cores], synchronously with the precisely dated glacier retreat in New Zealand, demonstrating a coherent picture for the end of deglaciation between mid- and high latitudes of the Southern Hemisphere*' [93].

It was once thought that the Younger Dryas was present in sedimentary sequences in the Southern Hemisphere [19]. But in 2007, the Waiho Loop, a large moraine associated with New Zealand's Franz Josef Glacier, was re-dated using beryllium ($^{10}$Be) and chlorine ($^{36}$Cl) isotopes produced in the upper atmosphere, and shown to be older than the Younger Dryas period [112]. Further examination of glacial moraines in New Zealand's Southern Alps in 2010 showed that the glaciers there advanced during the ACR. They were older than the Younger Dryas [113]. Michael Kaplan of Lamont and colleagues used maps of landforms, high-precision $^{10}$Be dating of the period of exposure of boulder surfaces, and reconstructions of former snowlines and ice extents, to show that New Zealand glaciers melted back during the Younger Dryas [114]. Glacier resurgence in New Zealand peaked 13 Ka ago, coincident with the ACR, probably in response to northward migration of the southern Subtropical Front bringing cold Southern Ocean water close to New Zealand [113]. Like others, Kaplan argued that sea ice forming in the Northern Hemisphere in the Younger Dryas curtailed Atlantic meridional overturning, leading to stronger winds over the Southern Ocean that increased upwelling, releasing $CO_2$ from the Southern Ocean that warmed the Southern Hemisphere whilst the Northern Hemisphere cooled [114]. Other signs of the disconnect between the Younger Dryas and the Southern Hemisphere include warming of sea surface temperatures in the Tasman Sea right through the Younger Dryas into the Holocene [112], and the advance of glaciers in Patagonia, after the Younger Dryas [115].

To conclude this section, two observations spring to mind. First, as Cronin noted, Broecker's 1989 paper on the Younger Dryas '*generated a worldwide search for evidence for the Younger Dryas and other abrupt climate reversals. More generally, it shifted emphasis from research on orbital-scale climate dynamics to suborbital timescales, especially the abrupt onset and termination of millennial events, which remain directly relevant today as a reminder about the vulnerability of the climate system to abrupt changes*' [19]. Second, as Eric Sundquist reminded us, the melting of the continental ice sheets of the Northern Hemisphere took place well after the rise of $CO_2$ and associated warming had begun in the south. Thus the role of those ice sheets as amplifiers of cooling via the albedo effect must have been less important than the role of $CO_2$ as an amplifier of global warming at the time [15]. The presence of the melting ice in the Northern Hemisphere soaked up heat, keeping the atmosphere cool there despite the global rise in atmospheric $CO_2$ and the associated warming in the Southern Hemisphere. We explore these issues in the next section.

## 13.7 Problems in the Milankovitch Garden

Palaeoclimate indicators in ice cores replicate the cyclicity seen in marine sediment cores and expected from the orbital periodicities calculated by Milankovitch: 100 Ka cycles of eccentricity; 41 Ka cycles of axial tilt; and 23 Ka and 19 Ka cycles in precession (see Chapters 3 and 12). While many saw this as indicating direct control of Earth's climate by orbital forcing, which dictates the amount of insolation received from the Sun at the top of the atmosphere, the change in eccentricity changes insolation by less than 0.2%, which is insufficient to explain the magnitude of the 100 Ka cycles seen in the proxy climate data. Many, like Frakes, wondered how could this tiny change lead to major cyclic glaciations [116]?

The puzzle is even more enigmatic than that simple question might imply. André Berger, noted that the most important theoretical period of orbital eccentricity, 400 Ka, was weak before 1 Ma ago, but strengthened towards the present, whilst the strength of the theoretically observed 100 Ka signal decreased after 900 Ka ago. Paradoxically, the weakening of this 100 Ka period in the astronomical calculations coincided with the strengthening of that same period in the paleoclimate records. '*This implies that the 100 ka period found in paleoclimatic records cannot, by any means, be considered to be linearly related to the eccentricity*' said Berger [117]. Feedbacks must be at work to give us the strong roughly 100 Ka signal that we see in the record, something Croll had also noticed.

This paradox applies also to the long interglacial of MIS 11, about 400 Ka ago. Simple 100 Ka orbital forcing cannot explain this long-lived event, as pointed out in 2003 by Wolf Berger from Scripps, and Gerold Wefer from the University of Bremen, amongst others [118]. Its existence suggests that the climate system must be responding in addition to internal fluctuations, not just orbital change. Along with Marie-France Loutre, André Berger argued from an analysis of orbital parameters in 2003 that the low amplitude of insolation change during both MIS 11 and the Holocene made the climate more sensitive to changes in the concentration of $CO_2$ in the atmosphere at those times than at times of higher amplitude of the insolation signal [119]. Considering that possibility for the record of MIS-11 in the Vostok ice core, Dominic Reynaud and colleagues interpreted the results of a climate model in 2003 to suggest that insolation alone could not have been responsible for the duration of this stage. Atmospheric $CO_2$ must have made an important contribution to the sustained warming [120]. In 2012, an Indian research team led by Das Sharma agreed [121]. Using advanced statistical techniques, they found that atmospheric $CO_2$ was the driving signal for change in MIS-11, whilst all the other climate proxies (like sea surface temperature and the carbon isotopic composition of organic carbon) were responses.

There is another complication. As Thomas Cronin reminded us [19], the interglacial peaks of the late Pleistocene do not repeat with a strict 100 Ka beat – they can be anywhere from 82 to 123 Ka apart. Some palaeoclimatologists have argued that forcing by precession influenced climate more than eccentricity did, and that the 100 Ka peaks occurred at times of low precession, recurring every fourth or fifth precession cycle [122]. Others suggested that variations in obliquity drove the 100 Ka cycles [123, 124], making the 100 Ka cycle a non-linear response to the 41 Ka obliquity cycle, a suggestion that remains controversial according to André Berger [125]. As Brook and Buizert pointed out in 2018, accurate dating of speleothems by the U/Th method '*suggests that glacial terminations are actually spaced by four or five precession cycles*' which must thus be the predominant drivers of glacial terminations [35]. Tzedakis and colleagues considered that the main driver for climate change was the 41 Ka obliquity signal (which had driven the glacial–inerglacial change of the Early Pleistocene), but that several obliquity signals had been skipped during the last 1 Ma leading to a deglaciation signal approaching 100 Ka. Evidently the deglaciation threshold had risen (for reasons not yet clear), giving rise to longer glacial periods and the accumulation of larger ice sheets [126].

We can get a sense of that rise in threshold from Figure 13.19. Peter Huybers observed in 2007 that the character of the $\partial^{18}O$ record in sediment cores changed regularly and progressively following an increasing trend throughout the Ice Age, the signals being smaller 2 Ma ago and growing steadily with time (Figure 13.19a). Figure 13.19 shows that Pleistocene glacial variability is better described by a trend, or progression, than by any single transition. Huybers deduced that the cyclicity at each point in the record must have been derived from the same underlying mechanism, and therefore that eccentricity did not pace the glacial cycles [127]. The origin of the gradual increase with time was unclear, perhaps reflecting gradual cooling through slow progressive loss of $CO_2$ (much as suggested by Saltzman [86]) ultimately reflecting control by plate tectonic processes. As yet we do not have an ice core long enough to check that possibility.

Huybers also discovered that a remarkable 33 out of 35 deglaciation features occurred when axial tilt (obliquity) was anomalously large [127]. During the early Pleistocene, deglaciations occurred at 40 Ka intervals with nearly every obliquity cycle, whilst in the late Pleistocene deglaciations skipped one or two obliquity beats, providing interglacials with a spacing of 80–120 Ka and averaging about 100 Ka.

Aside from the thorny 100 Ka issue, we were faced with another of those puzzling chicken-and-egg problems. Milankovitch realized that in order for us to compare insolation with climate we needed to know how much insolation reached particular latitudes, especially polar latitudes, since temperatures there have the greatest effect on the preservation or melting of snow, and hence the presence or absence of ice sheets. His calculations and later improvements upon them by André Berger and Marie-France Loutre [128] suggested that if insolation were high in the Northern Hemisphere it would generally be low in the Southern Hemisphere. As Peter Huybers of Harvard explained, the climates of the two hemispheres should be disconnected: '*One implication of orbital geometry is that at the time when precession aligns Earth's closest approach to the Sun (perihelion) with Northern Hemisphere summer, Earth is furthest away from the Sun (at aphelion) during the Southern Hemisphere summer*' [129]. Obviously, this difference between the hemispheres has a greater effect when the orbit is elliptical than when it is circular. Does this matter? Paradoxically, despite the implication that the climates of the two hemispheres should not follow each other exactly, Southern Hemisphere climate proxies do tend to follow closely the changing intensity of Northern Hemisphere insolation, as pointed out by Huybers and Denton [129] (Figure 13.20). This reflects the fact that when insolation is high in the Northern Hemisphere summer (21 June), it is also high (though only about one third of the amplitude) in the Southern Hemisphere spring (10 September), but the southern summer lasts longer that the northern. So the south still does get a fair amount of heat even in the Northern Hemisphere winter. Equally, we now know that – unlike precession – obliquity is synchronous in the two hemispheres [130]. The observation that precession was opposed in each hemisphere led to the notion that northern insolation controlled southern climate, probably through the Thermohaline Conveyor – something we have seen arguments for already in this chapter. However, we have to remember that precession is relatively weak at high latitudes. We will examine the issue of synchroneity in more detail later on.

As we saw earlier, a third key question concerned the role of $CO_2$ during the Ice Age. Did it lead or lag temperature? Did it drive climate change, causing temperatures to rise as we saw it doing in past ages, or did it just act as a feedback to climate change, enhancing the effect of an already rising temperature? Perhaps insolation caused temperature and $CO_2$ to rise simultaneously globally, by causing a warming ocean to carry less $CO_2$. Teasing out the answer was complicated by the fact that $CO_2$ seemed to change with temperature change in the Antarctic but ahead of temperature change in the Arctic as discussed

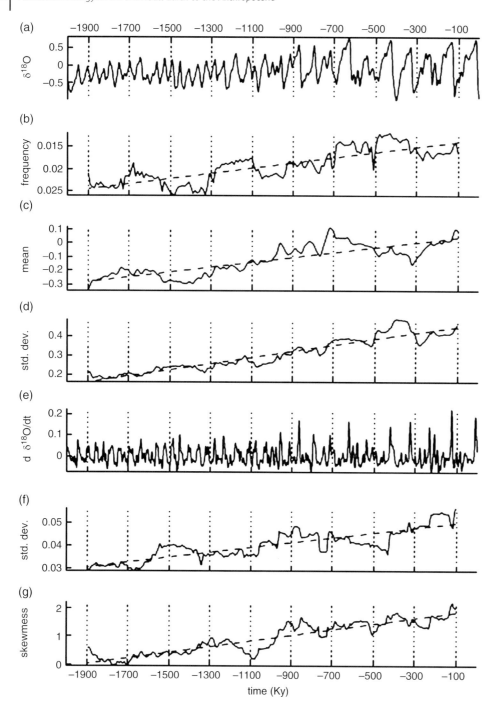

**Figure 13.19** The progression of Pleistocene glacial variability: (a) the average $\partial^{18}O$ record oriented so that up corresponds to more ice volume, (b) the first moment of the power spectrum (i.e. the weighted average of the frequency), (c) the mean value and (d) the standard deviation of the $\partial^{18}O$ record, (e) the time derivative of the $\partial^{18}O$ record in %/Ka, and the associated (f) standard deviation and (g) skewness. Skewness of the rate of change in $\partial^{18}O$ indicates the asymmetry between rates of glaciation and deglaciation. The dashed line in each panel indicates the least-squares best fit to each trend. *Source:* From figure 8 in Ref. [127].

above in relation to the findings of Parrenin [90], Beeman [92] and Pedro [91].

Then there was the question of the role of ice. Were changes in the great ice sheets solely driven by changing temperature (under the influenced of insolation), or did

their own internal dynamics play a key role in driving change?

Back in 2002, Barry Saltzman was not surprised that nobody yet had a definitive answer to the problem of how the Pleistocene Ice Age worked. He reminded us that

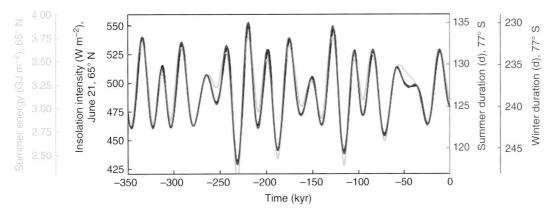

**Figure 13.20** Different measures of insolation covary with one another. Examples are the summer energy at 65 °N using a threshold of 400 Wm² (yellow), the diurnal average insolation intensity at the top of the atmosphere for the summer solstice at 65 °N (black), the duration of the summer at 77 °S (red, measured as the number of days whose diurnal average insolation exceeds 250 Wm²) and the number of winter days (hidden beneath the summer duration curve, measured as days below 250 Wm²; note the reversed y axis). *Source:* From figure 1 in Ref. [129].

*'Earth's climate system is delicately poised near the freezing point of water, allowing relatively slow but major fluctuations in the proportion of surficial ice to liquid water that involve only small perturbations to the global energy cycle. The low levels of energy flux involved, occurring in a complex, heterogeneous, nonlinear, nonequilibrium system, pose a class of problems that is as difficult and important as those more commonly treated in modern physics'* [86]. Putting it another way, he went on: *'The paleoclimatological (e.g., ice-age) problem .... is, in short, a problem in ultraslow, complex evolution in which the rates of change, and the fluxes of mass, momentum, and energy that accompany and drive them, are too small to be calculable directly or in some cases even measurable, though we are sure they occurred and are still occurring'* [86]. A thorny problem indeed.

In 2016, Andrey Ganopolski and colleagues noted that whilst the rapid growth of Northern Hemisphere ice sheets was commonly attributed to reduced summer insolation in boreal settings, such insolation was already near its minimum yet there were no signs of a new ice age [131]. To explain this paradox, they proposed that the anticipated *'glacial inception was narrowly missed before the beginning of the Industrial Revolution ... due to the combined effect of relatively high late-Holocene $CO_2$ concentrations and the low orbital eccentricity of the Earth'* [131]. By relatively high $CO_2$ concentrations they meant the pre-industrial level of 280 ppm; models using 240 ppm did lead to a glaciation. Their analysis suggested that even in the absence of anthropogenic emissions of $CO_2$ the current interglacial would probably last for another 50 000 years, whilst *'moderate anthropogenic cumulative $CO_2$ emissions of 1,000 to 1,500 gigatonnes of carbon will postpone the next glacial inception by at least 100,000 year'* [131].

Despite appearances to the contrary, then, for some considerable time everything has not been completely rosy in the Milankovitch garden, and much time and effort has been spent by many palaeoclimatologists on attempts to resolve these complex issues [132]. In this book we are only going to look at a few of these attempts to give you some idea of where current thinking is headed. For more comprehensive reviews readers might like to consult additional palaeoclimate textbooks by Cronin [19], Ruddiman [133], or Hay [134] or other texts in the reading list in Appendix 1.

## 13.8 The Mechanics of Change

To address these various questions we'll start our tour back in the CLIMAP era, in the mid-1970s. Inevitably, the rapid accumulation of new data from seabed cores spawned efforts to see if the new generation of atmospheric GCMs could be used to simulate the climate of the Ice Age [99]. In 1976, W. Lawrence Gates (1928–), then at Oregon State University, used the CLIMAP data set along with an atmospheric GCM to simulate the July climate in the Last Glacial Maximum, at 18 Ka ago [135]. Compared with the present, he found the Last Glacial Maximum was cooler and drier, especially over the Northern Hemisphere, due to enhanced anticyclonic circulation over the major ice sheets, and reduced summer monsoonal circulation. The mid-latitude westerlies were stronger. There was not much change in global cloudiness or relative humidity, but summer precipitation was 20% below that for today's July. Manabe and Hahn carried out a similar exercise in 1977. Comparing modern and Ice Age conditions they found that the tropics were much drier in the Ice Age. This was because the continents cooled more than the oceans, reducing the flow of air between them. Increased continental albedo weakened the Asian monsoon in their Ice Age simulation [136].

At that time, we still lacked data on atmospheric $CO_2$. The arrival of $CO_2$ data at the end of the 1970s changed the picture. Jim Hansen and his space physicists from NASA were the first to apply climate models to assess the effects of changing $CO_2$ on the climate of the Last Glacial Maximum as mapped by the CLIMAP team. In a paper for the 1982 Maurice Ewing Symposium in New York, Hansen explained, '*records of past climate provide a valuable means to test our understanding of climate feedback mechanisms, even in the absence of a complete understanding of what caused the climate change*' [137]. His team simulated the climate of the Last Glacial Maximum at 18 Ka ago by incorporating climate boundary conditions, including land ice, sea ice, and sea surface temperature, from the CLIMAP project. They assessed the effects of different feedbacks on the planetary radiation balance, testing the sensitivity of the model to forcing by different parameters. Global surface temperature was 3.6 °C cooler than today, and coolest in the polar regions. A fall in $CO_2$ to 200 ppm at 18 Ka ago (from an assumed interglacial level of 300 ppm), caused about 0.6 °C of the estimated global cooling, the rest coming from reductions in water vapour and clouds, a growth in land and sea ice (hence albedo), and a decrease in vegetation cover (also affecting albedo) [137].

Hansen's results suggested an overall climate sensitivity ranging from 2.5–5 °C, and averaging about 4 °C, for a doubling of atmospheric $CO_2$ (implicit was the accompanying rise in water vapour and fall in albedo). In recent years, with more data available, that estimate dropped to about 3 °C, as we saw in Chapter 11. Given the slow nature of ocean circulation, Hansen calculated that '*the response time of surface temperature to a change of climate forcing is of the order of 100 years*' [137]. Thus any change would be around for a long time beyond the initial forcing. Regardless of the initial response time, the recovery from a major warming event is thought to be ~100 000 years, as we saw for the PETM case history in Chapter 11.

One of Saltzman's early numerical modelling exercises on this topic, in 1984, using a dynamical model, not a GCM, showed that the 100 Ka cycles could be explained by dynamical variations in land ice, marine ice, and the mean climate (represented by ocean temperature), and feedbacks between them, paced by orbital variations in insolation [138]. In a nutshell, orbital cycles could modulate and pace a self-oscillating climate system. Way back in 1980, Saltzman was one of the early proponents of the notion that $CO_2$ played a key role in such a self-oscillating system [139].

At the Tarpon Springs meeting in 1984, two ocean modellers from Princeton, Robbie Toggweiler and Jorge Louis Sarmiento, tested the idea that the proposed glacial to interglacial changes in $CO_2$ were related to changes in the nutrient content of high latitude surface waters [140]. During glacial periods, when the Thermohaline Conveyor was switched off or reduced, the deep ocean accumulated nutrients and dissolved $CO_2$ through the action of the solubility pump (cold water holds more $CO_2$) and the biological pump (sinking organic matter dissolves in the deep ocean, enriching deep waters in $CO_2$), (see Chapter 9). These deep-ocean waters were unable to reach the atmosphere because the surface of the Southern Ocean was capped by a stable thermocline topped with sea ice. Much as we saw from later studies, Toggweiler and Sarmiento's model showed that in the deglaciation, the supply of NADW would be enhanced, the Southern Ocean thermocline – which had inhibited deep convection – would be disrupted, sea ice would melt, and $CO_2$ would be released from the deep ocean. They concluded that '*climate forcing of atmospheric $CO_2$ probably played a pivotal role in amplifying orbital forcing which paced the climate change*' [140], thus providing independent support to the Shackleton and Pisias 1985 model of glacial to interglacial climate change [141], and to Saltzman's concept. These ideas have a long pedigree.

By 1987, Syukuro Manabe, from the Geophysical Fluid Dynamics Laboratory at Princeton was ready to use a GCM combining the atmosphere with a mixed layer ocean to examine the effects of expanded continental ice, reduced atmospheric $CO_2$, and changes in land albedo on the climate of the Last Glacial Maximum [142]. Working with A. J. Broccoli, he found that expanded continental ice and reduced $CO_2$ had a substantial impact on global mean temperature. Increasing albedo from growing Northern Hemisphere ice sheets cooled that region. The Antarctic ice sheet hardly affected the picture because it already occupied most of the available land space, so could not grow much. Reduced $CO_2$ cooled both hemispheres. The presence of the northern ice sheets caused substantial change to atmospheric circulation in winter, amplifying the westerlies near the ice sheets. The sea surface temperature in the model reduced in much the same way as shown in the CLIMAP data, with greater cooling in the Northern than in the Southern Hemisphere. Half of the global 1.9 °C cooling in the sea surface temperatures of the Last Glacial Maximum was attributable to the $CO_2$ effect, which was larger in the Southern Hemisphere. Reduction in sea surface temperature was greatest near the margin of the sea ice. It did not decrease further poleward, because temperatures there were already close to freezing. Air temperatures were especially cold over both land and sea ice, and there was a larger area of cold air in the north than in the south, reflecting the larger area of Northern Hemisphere land and ice. Cooling due to a decline in $CO_2$ was less than 1 °C in the tropics, and much greater at high latitudes, mainly during the cold season. Broccoli and Manabe's study '*supports the hypothesis that glacial–interglacial variations in $CO_2$ concentration may provide a linkage between the two hemispheres*' [142]. Their model suggested that the presence of large ice sheets alone was insufficient to explain the glacial climate of the Southern Hemisphere, where the cooling was most likely driven by the fall in atmospheric $CO_2$.

By 1989, Wally Broecker from Lamont and George Denton, a glaciologist from the University of Maine at Orono, had entered the ring [143]. In addition to the puzzle that forcing by precession and obliquity seemed to produce bigger signals than eccentricity, they wanted to know '*What might explain the asymmetric shape of the 100 ky cycle, with its rapid start and slow decay?*' [143]. Long intervals of gradual increase in $\partial^{18}O$ in marine cores (signifying cooling) ended abruptly with a rapid decrease in $\partial^{18}O$ (signifying warming) at what they called 'glacial terminations' as the world changed from a glacial to an interglacial climate. Explaining how those rapid terminations were produced at 100 Ka intervals posed a difficult challenge.

Broecker and Denton thought that the answers to their questions lay in the ocean, and that glacial to interglacial transitions involved major reorganizations of the ocean–atmosphere system. Basically, the climate oscillated between two stable states of operation, which caused changes in the greenhouse gas content of the atmosphere and the albedo of the planet's surface. Only in this way could they account for the rapidity of the glacial terminations, the apparent synchroneity between the two hemispheres and the large variations in air temperatures and dust concentrations. They thought that the connection between insolation and climate was driven by orbital change leading to changing freshwater input, and the impact of freshwater on ocean salinity, hence density and ocean transport [143]. Their arguments for the role of the Thermohaline Conveyor (Meridional Overturning Circulation) have been rehearsed in detail earlier in this chapter, having been adopted by later researchers.

Their research suggested that no *one element changed enough by itself* to cause a glacial termination: changes in insolation were too small, as were decreases in albedo caused by reducing the area of sea ice, the 80 ppm increase in atmospheric $CO_2$ emitted from areas of the ocean formerly covered by sea ice, and the decrease in dust caused by weakening westerly winds. Whether there were more or fewer clouds may have been a factor, but not a major one. All these things together provided positive feedback, making glacial to interglacial change much faster and more

extensive than it would otherwise have been, so accounting, in their minds, for the asymmetry of the 100 Ka peaks in the record. They saw no basis for rejecting Saltzman's idea that the changes in mode from glacial to interglacial were part of a self-sustaining internal oscillation that operated even in the absence of orbital changes, but which, in the presence of orbital changes (especially at low temperatures) were paced by those changes.

Over the next decade, palaeoclimatologist Maureen Raymo, of Lamont, noticed that the excess ice typical of 100 Ka climate cycles tended to accumulate when July insolation at 65°N was unusually low for more than a full precessional cycle (21 Ka), and that once established it did not last beyond the next precessional maximum in summer insolation: '*Thus, the timing of the growth and decay of large 100-kyr ice sheets, as depicted in the deep sea $\partial^{18}O$ record, is strongly (and semipredictably) influenced by eccentricity through its modulation of the orbital precession component of Northern Hemisphere summer insolation*' [144]. Referring to studies of ice sheet decay by Oerlemans [2] and Pollard [145], she deduced that the rapid collapse of the ice sheet at the precessional high point might be triggered as some threshold was passed, such as a critical level of isostatic adjustment of the bedrock beneath the growing ice [122]. Nevertheless, she also thought that '*the global carbon cycle is likely to be a critical component of the mechanism(s) controlling the 100 kyr cycles*' [122]. We explore her argument for isostatic adjustment later.

While the behaviour of the global Thermohaline Conveyor got a lot of attention for providing a means of connecting the two hemispheres via the ocean, by 1999 Mark Cane and Amy Clement at Lamont drew attention '*to a part of the system that is known from the modern climate record to be capable of organising global scale climate events: the tropics*' [146]. Behind that goal was the growing realization that El Niño events were driven by an internal climate oscillator and not by external (solar) forcing. During an El Niño event the Pacific Ocean warms. It then cools during the following La Niña event, in a pattern that repeats on a 2–7 year cycle and has a global reach [147], as we saw in Chapter 11. El Niños warm the globe by spreading warm water across the surface of the Pacific, whilst La Niñas cool the globe by making those same waters cold enough to absorb heat. A key feature in this system is the oscillation of the 'warm pool' in the western Pacific, near New Guinea. The strong easterly Trade Winds of a La Niña pile warm water up in the west to feed the warm pool; the weak Trades of an El Niño allow the warm pool to collapse, and its warm surface water sloshes back to the east.

Cane and Clement wanted to know if the tropical Pacific climate could change *on its own,* with no influence from higher latitudes [146]. Their model showed that non-linear interactions in the tropical Pacific could generate variations in sea surface temperature on both orbital and millennial time scales through essentially the same physics as involved in El Niño-La Niña events. They suggested that '*global scale millennial and glacial cycles may be initiated from the tropical Pacific*' [52]. Low orbital insolation cooled the tropics by between 2 and 5 °C, creating a cooler more La Niña-like system in the tropical Pacific. The changes in sea surface temperature altered the character and location of atmospheric convection, which altered global climate through long-range connections (teleconnections). Through the natural oscillations in the system, the cold (La Niña-like) phase would tend to cool the Earth further by increasing low cloud cover, so increasing planetary albedo, and by reducing atmospheric water vapour – a greenhouse gas. This would increase glaciation over North America. The warm (El Niño-like) phase would do the opposite. As non-linear processes drove this millennial variability, peaks would tend to cluster in broad time bands rather than at particular time intervals [146].

They argued that there had been too much emphasis on fluctuations in NADW as a primary control of glacial to interglacial climate change [52]. Centres of tropical convection tend to lie over the warmest water. Moving the locus of that water alters the convection, which changes the impacts on distant locations. This works for the fast 2–7 year scale El Niño-La Niña changes. There was no intrinsic reason why it should not work on longer time scales, with similar global effects, and be accentuated by global changes triggered by orbital variations. Given the non-linear nature of the system, they argued that, whilst their model runs showed peaks near 1500 years, like those found by Bond, such peaks might well be artefacts of the short record. The longer the record in both models and palaeoclimate data, the more one expects a broad band of peaks originating from non-linear processes. They went on to point out that '*in common with other non-linear systems .... the tropical Pacific ocean-atmosphere may exhibit regime-like behaviours which persist far longer than any obvious intrinsic physical timescale*' [52]. It may vary on millennial timescales independent of influences from elsewhere.

The Pacific climate does, in fact, vary on longer timescales than the short wavelength El Niño–La Niña cycle. That cycle is superimposed on a longer wavelength oscillation, the Pacific Decadal Oscillation. In the positive phase of the Pacific Decadal Oscillation, the equatorial region is warm (like an El Niño) and the Gulf of Alaska is cool, whilst in the negative phase the equatorial region is cool (like a La Niña) and the Gulf of Alaska is warm. Each of these phases can last from 10 to 25 years [148]. Cane and Clement were suggesting that oscillations like these, if

extended enough in time, could have a substantial and possibly lasting effect on the global climate system.

Haug and colleagues agreed that the Pacific had been overlooked in considering what drove glaciations, especially its role in providing a moisture source for the North American ice sheet. Examining alkenone unsaturation ratios and diatom oxygen isotope ratios from a sediment core in the western subarctic Pacific Ocean, they found that 2.7 Ma ago late summer sea surface temperatures there rose in response to an increase in stratification, whilst winter sea surface temperatures cooled, winter floating ice became more abundant and global climate descended into glacial conditions. They suggested that the '*summer warming extended into the autumn, providing water vapour to northern North America, where it precipitated and accumulated as snow, and thus allowed the initiation of Northern Hemisphere glaciation*' [149]. Clearly there is much to be said for giving serious consideration to the role of the North Pacific in feeding the Laurentide Ice Sheet, and the separate role of the North Atlantic in feeding moisture to the great European ice sheets.

Imbrie, Ruddiman and Shackleton, three of the CLIMAP scientists, also had views on the mechanisms of climate change. Initially, in devising the SPECMAP model of 1984, Imbrie and his co-workers followed Milankovitch in accepting that summer insolation in the Northern Hemisphere forced the growth and decay of ice sheets directly. This was consistent with the original CLIMAP view published by Hays, Imbrie and Shackleton in 1976 [150], which we examined in Chapter 12. Their conclusions were published before we knew about the history of $CO_2$ in ice cores. Recognizing what ice cores were telling us, Imbrie changed his mind in the final SPECMAP models of 1992–1993. He proposed that Northern Hemisphere summer insolation triggered a train of climatic responses that were transmitted via deep water to the Southern Hemisphere, where they caused changes in $CO_2$ and other feedbacks that then affected Northern Hemisphere ice sheets. Direct forcing of the Northern Hemisphere ice sheets had changed to indirect.

Shackleton supported that interpretation. In 2000, he used the $\partial^{18}O$ record from Vostok to separate the ice volume component from the ocean temperature component of the $\partial^{18}O$ signal of marine sediments. He showed that atmospheric $CO_2$, Vostok air temperature and deep-water temperature were all in phase with orbital eccentricity, whilst ice volume lagged these three variables. '*The coherences and phases in the 100-ky band strongly suggest that atmospheric $CO_2$ has a direct and immediate control on deep water temperature (presumably with high latitude air temperature as an intermediary)*' he wrote [151]. '*Hence, the 100,000-year [ice] cycle does not arise from ice sheet* *dynamics; instead, it is probably the response of the global carbon cycle that generates the eccentricity signal [in ice volume] by causing changes in atmospheric carbon dioxide concentrations'* [151]. Ice volume then responds to these changes. Finally, *'The effect of orbital eccentricity probably enters the paleoclimatic record through an influence on the concentration of atmospheric $CO_2$'* [151]. This supported Saltzman's model.

Bill Ruddiman thought this was unnecessarily complicated [152]. He felt that although the 100 Ka cycle of eccentricity was too small to achieve much change by itself, it grew in importance by accentuating the patterns of insolation caused by variations in precession and obliquity (tilt). It accentuated the summer (mid-July) insolation that forced Northern Hemisphere ice sheets at the 41 Ka obliquity period, helping to change ice volume, sea surface temperature, dust supply, and the production of NADW, which produced a strong positive $CO_2$ feedback that further amplified ice-volume changes. It also accentuated insolation at the roughly 21 Ka precession period, focused in the tropics, which influenced wetlands through monsoonal changes that drove fast feedbacks in methane. Clearly, Ruddiman thought that $CO_2$ was not a primary driver of ice sheet change in the Northern Hemisphere. Instead he saw $CO_2$ as providing positive feedback to accentuate the climate signal. These effects were stronger at the Mid Pleistocene Transition where *'gradual global cooling allowed ice sheets to survive during weak precession insolation maxima and grow large enough during 41,000-year ice-volume maxima to generate strong positive $CO_2$ feedback'* [152].

Ruddiman highlighted the importance of combining Milankovitch's three orbital signals to generate the 100 Ka cyclicity: *'each broad eccentricity maximum at the 100,000-year cycle spans 2 or 3 individual insolation maxima at the precessional cycle. During the last several deglaciations, the climate system response has latched onto one or other of these precession maxima in creating the observed termination. As a result, all terminations occur at or very near even multiples of 4 or 5 precession cycles (90 ,000-115,000 years). The specific precession maximum chosen by the climate system depends on ... close alignment with a nearby obliquity maximum. Modulation of precession by the longer-term 400,000-year eccentricity cycle also plays a role: it makes all precession maxima in a particular 100,000-year cycle either weaker or stronger, thereby affecting which peak is chosen for the termination'* [152]. As a result, some interglacials started almost in phase with eccentricity (e.g. Termination I, at 20 Ka ago), whilst others might lead eccentricity by 17 000 years or more (e.g. Termination II, at 135 Ka ago).

There is no doubt that the ocean did play a key role in storing and releasing $CO_2$ during the Ice Age. Thanks to the 2007 discovery that the ratio of boron to calcium (B/Ca) in benthic foraminifera is directly related to the concentration of carbonate ions ($CO_3^{2-}$) in bottom waters [153], we can now use that ratio to show changes in the concentration of carbonate ions ($CO_3^{2-}$) in oceanic deep waters over the past 25 Ka [154]. The data show that the biological pump caused large amounts of $CO_2$ to be stored in the deep glacial ocean, thus helping to reduce the levels of $CO_2$ in the air during the Last Glacial Maximum.

The decrease in $CO^2$ in the atmosphere (by about 60 Gt) during the last glacial period was matched by an increase in the deep Atlantic carbon inventory (by about 50 Gt), which coincided with the shoaling of the Atlantic Meridional Overturning Circulation (AMOC), but led to a decline in deep Atlantic $CO_3^{2-}$ [154]. This decline is thought to have resulted from an increase in dissolved inorganic carbon (DIC) [154]. Under 'normal circumstances' *'$CO_2$ sequestration in the deep ocean across MIS 5a-MIS 4 would inevitably raise deep-water acidity, lower seawater [$CO_3^{2-}$], and consequently intensify deep-sea $CaCO_3$ dissolution'* [154]. But, while there are indications that the CCD shoaled somewhat in the glacial Atlantic, it did not in the Pacific or Indian Oceans. Evidently, the increased dissolution of carbon in the glacial ocean was accommodated by the increased size of the Atlantic's cold deep water pool, caused by the reduction in the supply of North Atlantic Deep Water, rather than by any massive change in the CCD [154].

On deglaciation, firstly $CO_2$ was released back into the air from the rising deep water in the Southern Ocean as sea ice declined, exposing the ocean to the atmosphere, and secondly the declining supply of iron-rich dust reduced ocean productivity, thus lessening the extraction of $CO_2$ from the air.

Was it reasonable to consider the ocean as the main source of change in $CO_2$? In 1996, Guy Munhoven and Louise François of the University of Liege in Belgium suggested that not enough attention had been paid to the possibility that variations in rock weathering might have altered $CO_2$ levels between glacial and interglacial periods [155]. They used the ratio of germanium to silicon (Ge/Si) in marine sediments as an indicator of the consumption of $CO_2$ by the weathering of silicate rocks, noting that the Ge/Si ratio was lowest where the rate of weathering was highest, at the Last Glacial Maximum. This is because increased cycles of freezing and thawing enhance mechanical weathering, so increasing the surface area of exposed rock that can be subjected to chemical weathering, and thus supplying more dissolved silicon to the ocean. This kind of weathering is most efficient in mountains, and might have been enough to significantly lower atmospheric $CO_2$. They calculated that $CO_2$ consumption by rock weathering in glacials could have reduced atmospheric $CO_2$ by 50–60 ppm Previously, rock weathering was dismissed as too slow to influence glacial–interglacial variations in $CO_2$, but the existence of the Ge/Si signal in marine sediments suggests that this perception was misguided.

Criticizing palaeoclimatologists for focusing on a search for changes in *external* forcing to explain palaeoclimatic variability, Barry Saltzman thought that much of it could have resulted from internal instability within a system forced steadily by the thermal gradient between equator and pole [86]. As a scientist steeped in chaos theory, he was puzzled by the fact that although the character of orbital forcing was unchanged over the full Pleistocene period and before, there were signs of significant instability, including the rather sudden onset of the Pleistocene glacial epoch at about 2.5 Ma, the Mid Pleistocene Transition at about 900 Ka, and the dominance of 100 Ka cyclicity in the past 400 Ka. He concluded that: '*the main variations of planetary ice mass do not represent a linear response to the known orbitally induced radiative forcing, having a temporal spectrum that is much different than that of the forcing ... Although it is possible that the orbital external forcings may be a necessary condition for the observed ice variation ... they cannot be a sufficient condition*' [86]. This greatly interested him, because other modellers assumed that external forcing controlled the 100 Ka oscillations, with the implication that if those forcings were removed then the 100 Ka oscillations would vanish. Those other models, he emphasized, could not account for the transitions at 2.5 Ma or 900 Ka. He declared that '*If .... the tectonic decrease in $CO_2$ over the Late Cenozoic provided the threshold state for the initiation of the major ice build ups, it would then deserve recognition as the "cause" of the ice epoch and its oscillations*' [86]. These were profound and counterintuitive insights.

Saltzman thought that Ice Age variability resulted from a mix of both external forced and internal free effects, and that the broad spectral peak centred on 100 Ka resulted from an internally driven fluctuation caused by instability, rather than the response to an externally driven forcing like that at the 41 Ka period. He interpreted Ice Age changes with time as showing that there was strong free variability within the climate system, part of which might be described as 'climatic turbulence' stimulated when the tectonically forced value of $CO_2$ achieved a critically low range of values. Recognizing that the system was complicated and the solution to the problem would not be easy, the intrepid Saltzman set out '*to develop [a] quantitative theory of climate in which all relative forcings, feedbacks, and competitive physical factors are taken into account simultaneously. That is, we consider the explanations of variations in*

*the climate system as a problem in mathematical physics, in which the basic conservation laws for mass, momentum, and energy are expressed in symbolic forms so that the power of mathematical deductive logic can be used to extract quantitative relationships .... It is the purpose of theory*' he continued '*to provide a predictive connection between the known external forcing ... and the observed internal behaviour*' [86].

His dynamical system model of the Northern Hemisphere assumed that atmospheric $CO_2$ linearly decreased from 350 ppm at 5 Ma ago to 250 ppm in preindustrial times in response to tectonic forcing, and that the only other forcing came from orbitally induced variations in summer insolation at high northern latitudes. '*In essence*' he explained, his model unified '*the two major theories of the ice ages: the $CO_2$ theory (in which longwave radiation is altered by the greenhouse effect) and the Milankovitch theory (in which the distribution of shortwave radiation is altered by Earth-orbital changes), supplemented by a new third major theory resting on the possible role of internal instability*' [86]. His numerical model showed that the global ice mass over the past 5 Ma responded with a significant jump at 2.5 Ma, coincident with the $\partial^{18}O$ record from ocean drilling cores. Hence, '*the imposition of the slow tectonic forcing of $CO_2$ transforms what would otherwise be the chaotic, intermittent ... [distribution of ice volume through time] into an organized sequence of clearly defined regimes separated by well-developed transitions*' [86]. One regime started 900 Ka ago with emergence of 100 Ka variability.

Saltzman's model suggests that the increase of Pleistocene ice volume at 2.5 Ma is an emergent property of the systematic decline in $CO_2$ with time. He thought that the 100 Ka fluctuations were associated with and probably driven by internally generated $CO_2$ fluctuations, whilst the 20–40 Ka fluctuations were driven by externally imposed orbital variations and the associated instability of the ice sheet, leading to calving [81]. Calving instabilities contributed to the rapid deglaciation of the Laurentide and Fennoscandian ice sheets by causing an earlier collapse of ice mass than predicted by the insolation and $CO_2$ variations alone. His model's predictions have been validated by testing against the data from Vostok and Dome C. André Berger, too, use a numerical model to demonstrate that given a linear decline in $CO_2$, a shift from glacial cycles dominated by periodicity of 41 Ka to cycles dominated by periodicity of 100 Ka took place about 1 Ma ago at the Mid Pleistocene Transition (MPT) [156].

Elderfield was not convinced that they were right. Having found that the Mid Pleistocene Transition was associated with a particular pattern of insolation, he concluded that '*Data of $CO_2$ .... are as yet too sparse to determine the respective roles of temperature and the carbon system*' in causing the Mid Pleistocene Transition [39]. Bärbel

Hönisch of Lamont, and her colleagues, agreed [157]. Using boron isotopes in planktonic foraminiferal shells as a proxy for the partial pressure of $CO_2$ in ocean surface waters from the past 2.1 Ma, and comparing glacial and interglacial values before and after the MPT, they found that whilst $CO_2$ was slightly higher in the *glacial* periods from before the MPT, it was *not in the interglacial* periods. While this confirmed a close linkage between atmospheric $CO_2$ concentration and global climate, it militated against long-term drawdown of atmospheric $CO_2$ as being the ultimate cause of the MPT.

Huybers, too, investigated the Mid Pleistocene Transition [127, 129]. The pacing of deglaciations by obliquity throughout the Pleistocene, mentioned earlier, plus the regular and progressive increase in the properties of glacial cyclicity, which followed a trend from smaller to larger signals over the past 2 Ma (Figure 13.19), suggested that there was no sudden onset of 100 Ka cyclicity. Hence the MPT must be an artefact. In particular it was becoming apparent that whilst eccentricity and precession signals showed no significant trend with time over the past 1 Ma, the amplitude of the obliquity signal steadily increased. It was the obliquity signal that drove the periodicity of the Ice Age climate signal prior to the MPT.

Arguments can easily develop, for instance about Earth System Sensitivity (ESS). In 2016, Carolyn Snyder noted that over the past 2 Ma, global temperature gradually cooled until roughly 1.2 Ma ago, then stalled, somewhere near the Mid Pleistocene Transition. That change in trend preceded the increase in the maximum size of the ice sheets around 0.9 Ma ago, and the change from 40 to 100 Ka cyclicity with its attendant increase in the amplitude of the glacial–interglacial signal [158]. Comparing her new temperature reconstruction with the calculated radiative forcing from greenhouse gases, Snyder estimated that the ESS over millennial timescales to a doubling of $CO_2$ was 9 °C (ranging from 7 to 13 °C). She then calculated that this meant that '*stabilization at today's greenhouse gas levels may already commit Earth to an eventual total warming of 5 °C (range 3 to 7 °C) over the next few millennia as ice sheets, vegetation and atmospheric dust continue to respond to global warming*' [158]. Her estimate is close to the maximum documented in Chapter 11. However, in 2017, Gavin Schmidt and others concluded that her methodology did not reliably estimate the ESS, and that the amount of committed warming had to be substantially less than she had calculated [159]. The chief problem with Snyder's analysis, they thought, was that it '*was based on the assumption that greenhouse gases were solely responsible for long-term global-mean glacial–interglacial temperature changes*' [159]. Although greenhouse gases do have a large role, '*quantifying that role is difficult because of simultaneous changes in*

*many factors that also influence the energy balance of Earth (such as the extent of the ice sheets, snow cover, vegetation, dust load and cloud cover*' [159]. Moreover, even the role of orbital forcing (the ultimate trigger) is '*enhanced by fast and slow feedbacks that involve the ice albedo, clouds, the carbon cycle, vegetation, and so on*' making attribution difficult [159]. The correlation between $CO_2$ and temperature '*conflates the sensitivity of the climate to $CO_2$ and the response of the carbon cycle to variations in temperature and ice sheet extent*' [159] (and in sea ice, and sea level and land vegetation cover for example). Finally, Schmidt observed that a warming of 3–7 °C seemed implausible, because there is a limit on the rate at which the areal coverage of current ice sheets can change [159]. Snyder responded that her calculation of ESS did summarize the behaviour of Earth's climate system including ice sheets, vegetation and dust as internal feedbacks; in effect it '*summarize[d] the past aggregate, correlational relationships among those feedbacks*' [160]. In that sense, she noted, her calculation of ESS was in line with previous calculations for the Pleistocene and Cenozoic. In her mind the debate was not about calculating past ESS, but about applying the results of that calculation to estimating future climate change based on current levels of $CO_2$ [160]; '*states warmer than the present day may have different ESS*' [160]. One take home message here is that calculations based on what happened during the Pleistocene may not apply (or not apply well) to what our climate may do in the future.

Despite Saltzman's valiant attempts to find out how the Ice Ages worked, there was still much to consider. For instance, his model did not examine the role of sea ice. Yet the changing distribution of sea ice evidently played a key role in governing the exchange of $CO_2$ between air and ocean, and in the switching on and off of the supply of NADW [161], as we saw earlier. These arguments usually apply to the interactions of sea ice and the production of NADW. But Ralph Keeling from Scripps and Britton Stephens from the University of Colorado argued that the ocean Thermohaline Conveyor could also be destabilized by the influence of Antarctic sea ice [162]. They supposed '*that changes in the freshwater budget of high southern latitudes may provide the link between Antarctic warming and sudden Greenland climate changes associated with long-lived D/O events … [and that] the duration of the short-lived interstadial events would be linked to the timescale for NADW to propagate from the North Atlantic to high southern latitudes*' [162].

Trond Dokken of the Bjerknes Centre for Climate Research, Bergen, Norway agreed that northern sea ice played an important role in the development of Dansgaard–Oeschger events [163]. New sediment core data showed Dokken and colleagues that warm subsurface Atlantic water had flowed in to the Nordic seas beneath the sea ice and its associated freshwater cap. Eventually that warm water destabilized the cool surface system and its associated sea ice, venting heat to the air and warming the region abruptly by as much as 10 °C. This warmth then gradually melted the Fennoscandian ice sheet, recreating the freshwater cap on the Nordic seas, allowing sea ice to reform. Dokken's hypothesis avoids ad hoc proposals for the periodic arrival of a freshwater cap.

Confirmation that variations in atmospheric $CO_2$ were most likely related to variations in the ocean came from an Earth system model of intermediate complexity (CLIMBER-2) run by Brovkin and others in 2007 [164]. This model showed that the change from interglacial to glacial reduced Atlantic thermohaline circulation, caused the thermocline to shallow, allowed southern deep waters to penetrate further north and drew down atmospheric $CO_2$ by 43 ppm. Upwelling and dust fertilized the Southern Ocean in the Atlantic and Indian Ocean sector north of the Polar Front, drawing down a further 37 ppm of $CO_2$. There was an accompanying decrease in the cooled terrestrial biosphere, thus diminishing photosynthesis, which increased $CO_2$ in the atmosphere by 15 ppm, and an increase in ocean salinity resulting from the conversion of water to ice, which led to a further rise of 12 ppm A decrease in deposition of $CaCO_3$ in shallow water following the fall of sea level drew down atmospheric $CO_2$ by another 12 ppm These various mechanisms explained 65 ppm (more than two thirds) of the fall in $CO_2$ during glacial times, suggesting that the model captured reasonably well the effects of reorganization of biogeochemistry in the Atlantic Ocean. The rest of the fall might be explained by less well-known processes including changes in terrestrial weathering, and iron-fertilization of the sub-Antarctic Pacific Ocean.

Despite the intriguing outputs of Saltzman's model, Michael Crucifix pointed to its '*failure to reproduce the steadily increasing trend in CO₂ concentration during marine isotope stage (MIS) 11 …. [suggesting that] some stabilizing mechanisms may have been ignored in this model*' [165]. He reminded us that the correct length for MIS-11 (two precession cycles) had been predicted by a model devised in 2001 by Didier Paillard of the French Laboratoire des Sciences du Climat et l'Environnement [166]. Paillard's model featured three possible climate regimes (glacial, mild glacial, interglacial) to which the climate system was successively attracted, depending on insolation and ice volume. For his pioneering ideas on the response of Quaternary climate system dynamics and the carbon cycle to Milankovitch forcing, Paillard was awarded the Milutin Milankovitch Medal of the European Geosciences Union in 2013.

Reviewing progress up to the mid-2000s in the development of our understanding of ice age variability, I am reminded of Crowley and North's summary of the position back in 1991: '*A diversity of ideas exists as to the origin of the ice-age CO₂ fluctuations …. Working on the carbon cycle on this time-scale is like trying to piece together a giant puzzle for which some of the pieces are missing and some of the rules not thoroughly understood. But the progress that has been made … is impressive, and we are optimistic that a revisitation of this subject in a few years will indicate considerable advances over what has been presented*' [99]. In this book we are examining those advances.

Part of the problem, as Michael Crucifix explained in 2009, was that '*There is presently no comprehensive model … capable of representing the interactions between the slow components of the climate system satisfactorily enough to predict the evolution of ice volume and greenhouse gas concentrations over several glacial–interglacial cycles*' [165].

Significant further progress has been made since the mid-2000s, as we see further on. This stems in particular from results emerging from the completed analyses of the long Antarctic ice cores collected from Dome F in 2007, Dome C in 2004, Dronning Maud Land in 2007, and the West Antarctic Divide in 2011. These cores have not been available for long, so it is not surprising that our understanding of the operation of the Ice Age climate system has evolved rapidly quite recently. Ice cores from Greenland were available for longer, but did not go much past the last interglacial and were unreliable for $CO_2$.

In 2010, Lorraine Lisiecki of the University of California at Santa Barbara carried out a statistical analysis of the links between eccentricity and the 100 Ka glacial cycle [167]. She showed that the 100 Ka cycles are indeed paced by eccentricity, but that, paradoxically, strong eccentricity is associated with weak power in the 100 Ka glacial cycle. She argued that strong forcing by precession disrupted the internal climate feedbacks that drive the 100 Ka glacial cycle. Her findings '*support the hypothesis that internally driven climate feedbacks are the source of the 100,000-year climate variations*' [167]. These internal feedbacks must be phase-locked to eccentricity, and vary slowly over long periods, as do the carbon cycle and the ice sheets. Glacial terminations are driven by precession and obliquity after large ice sheets develop, and are paced by eccentricity (which affects the amplitude of precession). Lisiecki agreed that the 100 Ka cycle was likely to originate from processes associated with the carbon cycle, as Saltzman had suggested.

What might make the carbon cycle vary in that way? In 2008, Robbie Toggweiler of NOAA's Geophysical Research Laboratory in Princeton, thought that the answer might lie in the Southern Ocean [168]. There the effect of salinity on the overturning and mixing of the Southern Ocean grows as polar waters approach the freezing point, with overturning being particularly strong at temperatures between 1 and 3 °C, and weak at temperatures below 0 °C, when sea ice tends to form. Given that constraint, he saw that as $CO_2$ and temperatures fell during the late Cenozoic, temperatures would eventually reach the critical point where polar waters became prone to overturning and mixing. Where those temperatures were relatively warm (1–3 °C), strong overturning and mixing would supply deep-water $CO_2$ to the air; where they cooled to less than 0 °C, reduction of overturning and mixing would encourage $CO_2$ to dissolve in the ocean and be carried to the depths. In warm times, when $CO_2$ was released to the air, the deficit of carbonate ($CO_3^{2-}$) ions in the deep ocean would switch to an excess, enhancing the burial of $CaCO_3$. This would release more $CO_2$, leading to more warming and more overturning, which would lead to more $CO_2$ release and more $CaCO_3$ burial and so on, converting a relatively minor overturning fluctuation into a major transition. Over a period of 50 Ka, the excess of $CO_3^{2-}$ ions in the deep ocean would be erased by $CaCO_3$ deposition, lowering the partial pressure of $CO_2$ in the ocean and making more $CO_2$ from the air dissolve in the ocean to begin the cycle again. In Toggweiler's model, the booms and busts for atmospheric $CO_2$ took 50 Ka each, together making up a whole 100 Ka cycle. In the 'on' state, when the $CO_2$ was high, the Southern Ocean and Antarctica were warm. This process could have triggered the global seesaw. Toggweiler concluded that '*most of the 100,000-year temporal variability in the ocean is a greenhouse response to $CO_2$ cycles from the south, as suggested by Shackleton (2000) ... taken together the Northern and Southern Hemispheres would seem to have dominant influences and dominant periods of variability that are basically independent: precession and tilt make the ice sheets grow and shrink in the north; the internal mechanism warms and cools the south. The greenhouse effect from the internal mechanism in the south transmits some of the 100,000-year southern variability to the northern ocean and the northern ice sheets*' [168].

In 2018, Chandranath Basak and colleagues asked how old $CO_2$ stored in deep bottom waters could escape to the atmosphere during the deglaciation. They used neodymium isotopes extracted from deep-sea cores to show that during the Last Glacial Maximum Ross Sea Bottom Water (which then extended northwards to the region of the surface Subantarctic Front) was sharply isolated from the overlying Circumpolar Deep Water. The ocean was strongly stratified, trapping old carbon in deep water. Early in the deglaciation phase the warming of the Southern Hemisphere warmed the Southern Ocean, encouraging sea ice retreat and the southward shifting of the westerlies.

This led to deeper mixing between surface and deep water, as well as the wind-driven upwelling of deep water, processes that broke down the former stratification and allowed carbon-rich deep water to reach the surface. This led to the release of *old* dissolved $CO_2$ to the atmosphere, which, in turn, enhanced deglacial warming. The warming process and the southerly migration of the westerly wind belt also encouraged the southward penetration of NADW into the circum-polar region as the Atlantic Meridional Overturning Circulation strengthened [169].

Basak's findings mesh well with those of Rae and others, who used boron isotope data from deep-sea corals near Antarctica to show that over the past 40 Ka deep-ocean pH was lower (i.e. more acidic), representing more $CO_2$ storage, during colder periods when the ocean was more stratified, and higher (i.e. less acidic), representing loss of $CO_2$ from the ocean to the atmosphere, during warmer periods when the ocean was less stratified (and wind-driven upwelling was stronger) [170]. Rae reminded us that Antarctica's deep-ocean bottom waters were formed by the excretion of salt during the formation of sea ice, which made cold surface water dense enough to sink. This meant that when conditions were favourable for wind-driven upwelling, the upwelling process brought salt as well as $CO_2$ into the upper ocean. This vertical transfer of salt, and its subsequent transfer through surface currents to the North Atlantic, may have helped to re-initiate the formation of NADW [170]. Resumption of the formation of NADW, pulling Gulf Stream water northwards, may have been a key to warming the Northern Hemisphere. Part of the control of changes of climate with time could thus be flipping between these modes of stratification and connectivity in the Southern Ocean, especially during the midglacial conditions that gave rise to Dansgaard–Oeschger events [170].

How independent were the Northern and Southern Hemispheres? In 2007, Kenji Kawamura of Japan's Tohoku University decided with colleagues to use the ratio of oxygen to nitrogen molecules in fossil air from the DF and Vostok ice cores as a proxy for local summer insolation (stronger insolation diminishes the $O_2$ concentration), allowing them to examine the phase relationships between climate records from the ice cores and changes in insolation [171]. They found that southern summer insolation was out of phase with Antarctic climate change, and interpreted that to mean that Antarctic climate change on orbital timescales must be paced by northern summer insolation and its effects on northern ice sheets and the northern oceans. They concluded that '*Northern Hemisphere summer insolation triggered the last four deglaciations*' [171]. Like Saltzman and Fischer, they thought that Antarctic cooling into past glacial periods began earlier by several millennia than the corresponding $CO_2$ falls. That led them to suggest '*that post-interglacial cooling began in the Northern Hemisphere with ice area growth and was transferred to Antarctica quickly through modulation of poleward heat transport and methane concentration decrease – before the reduced $CO_2$ forcing, or the sea level drop caused by northern ice volume growth, became significant*' [171]. By their reckoning, then, $CO_2$ was an amplifier of orbital input, not a primary driver of change. However, as we learned earlier, at least one of their assumptions, that $CO_2$ preceded warming, is now known to be wrong, as shown by Parrenin [90], Beeman [92] and Pedro [91] (Figures 13.15–13.17).

Were they right to focus on southern summer insolation? Examining the month-by-month changes in Southern Hemisphere insolation, Peter Huybers, together with George Denton of the University of Maine, found in contrast that some aspects of that insolation in fact did co-vary with the pattern of insolation in the Northern Hemisphere, contrary to expectation [129, 172] (Figure 13.20). As usual, the devil is in the detail. In principle, when summer insolation was highest at 65°N due to Earth being at perihelion (closest to the Sun), summer insolation at 77°S should have been weak due to Earth being at aphelion (furthest from the Sun). But Huybers and Denton realized from Kepler's second law that although summers have less intense insolation when Earth is at aphelion, they are *longer* and the associated winters will be shorter than the average. Calculating for the Southern Hemisphere the number of summer days experiencing daily average insolation of more than $250\,\mathrm{W\,m}^{-2}$ and the number of winter days experiencing daily average insolation of less than $250\,\mathrm{W\,m}^{-2}$, they found that Northern Hemisphere insolation covaried positively with the duration of southern summers and negatively with the duration of southern winters over the past 350 Ka (Figure 13.22) [129, 172]. Spring insolation intensity at high southern latitudes also varied closely with the duration of the southern summer (Figure 13.20) [129, 172]. Hence, orbitally driven changes in the south could coincide with different orbitally driven changes in the north, both causing warming at essentially the same time Evidently, when trying to assess the relationship between insolation and climate in different hemispheres, the insolation of all seasons has to be considered, along with their duration. Milankovitch knew that, but dismissed Antarctica as too cold for changes in southern insolation to influence its volume of ice [129]. He was wrong. By focusing on annual insolation, Kawamura's team had missed the fact that the duration of southern summer, and the pattern of spring insolation in the south, might explain what they saw [171]. Their focus was too narrow, as shown by Huybers [129, 172] (Figure 13.20).

Huybers and Denton speculated '*that the increasing summer and decreasing winter durations caused by the alignment of aphelion with southern summer solstice coordinates the effects of summer radiation balance, winter sea ice and atmospheric $CO_2$ so as to increase Antarctic temperature. Variations in sea ice and $CO_2$ may also explain why climate variations similar to those in Antarctica are observed in mid-latitude southern marine and continental environments*' [129]. This freed southern data from northern forcing at precession and obliquity timescales. Northern climate would respond to summer insolation intensity à la Milankovitch, whilst southern climate would respond to the duration of summer and winter seasons, perhaps also reflecting the contrasting distributions of land and sea and ice sheets in the two hemispheres [129, 172]. The net result was that the two hemispheres operated in sync (Figure 13.20). And as mentioned in Chapter 12, the two hemispheres were also linked by changes in sea level, which would influence the ocean-atmospheric exchange of $CO_2$ through changes in ocean area.

One key difference between the two hemispheres, apart from the larger amount of land in the north, was the fact that the Antarctic ice sheet covered an entire continent. Aside from the narrow continental shelf, there was no room for expansion on land in glacials. As a result the area and volume of its ice was relatively stable compared with the Northern Hemisphere ice sheets, which waxed and waned considerably with changing insolation, thus affording significant opportunities for changes in both elevation and albedo to affect northern temperatures. Independent hemispheric responses to insolation at orbital timescales would imply that there was no need to invoke causality in explaining lead–lag relationships between the hemispheres. Nevertheless, Huybers and Denton went on to say that '*An Antarctic response to local changes in insolation is consistent with hypotheses calling on terminations to be triggered by changes in southern insolation … If a long summer and a short winter lead to a decrease in production and extent of Antarctic sea ice, they may also increase the outgassing of $CO_2$ from the Southern Ocean by decreasing near-surface stratification … Once the northern ice sheets are sufficiently large to become unstable, the combination of a long southern summer and an intense northern summer may be the one-two punch that leads to the collapse of northern ice sheets*' [129].

Nevertheless, Huybers agreed that the beat of Northern Hemisphere insolation could influence Antarctic climate through the transfer of heat across the equator either in the atmosphere or in the ocean [172]. Regardless of that possibility, there was no doubt that $CO_2$ amplified temperatures measured in Antarctic ice cores by about half, was sourced mainly from the Southern Ocean, and was a good candidate for orchestrating global climate change, since it was well mixed through the atmosphere. Deciding between the various options would require more research on precisely how the annual climate signal became fixed in Antarctic ice [172]. Some confusion is inevitable when attempting to relate global climate signals to either Northern or Summer Hemisphere forcing, because '*many aspects of the insolation forcing have essentially identical variability*' [172] (e.g. see Figure 13.20).

By 2009, Eric Wolff, then at the British Antarctic Survey, agreed that the Southern Hemisphere was in the driving seat for glacial terminations. Wolff and colleagues noticed that '*the initial stages of glacial terminations are indistinguishable from the warming stage of events in Antarctica known as Antarctic Isotopic Maxima, which occur frequently during glacial periods*' [173] (e.g. see Figure 13.6). Those warmings, which are associated with increasing $CO_2$, are directly associated with Dansgaard–Oeschger events in the north (Figure 13.6). The Antarctic warmings begin to reverse with the onset of the warm Dansgaard–Oeschger events (Figure 13.6). Wolff and his team argued that glacial terminations were in effect an extreme variety of this relationship, in which there was no reversal of the Antarctic warming. As that warming continued, a full deglaciation became inevitable. In these findings there was both an implication and a question. The implication was that if the millennial AIM were identical with Antarctic warmings at the start of deglaciations, then the timings of the deglaciations were probably not orbitally controlled. The question was – why did some Antarctic warming events not reverse, instead leading to terminations?

Wolff noticed that before each termination, the global climate reached a cold extreme, when ice sheets and sea ice had their largest extent (e.g. see Figures 13.8, 13.9). This excessive cold may have disabled the system from producing a Dansgaard–Oeschger event following an Antarctic warming. Eventually a Dansgaard–Oeschger event did occur, but too late to prevent the Antarctic warming from continuing virtually unchecked. In effect '*terminations are caused by southern warming that runs away because the north cannot produce a DO event*' [173]. The seesaw was temporarily switched off for long enough for Antarctic warming to swamp the system.

In an independent review in 2010, Daniel Sigman and Gerald Haug confirmed the importance of the Southern Ocean in controlling Ice Age $CO_2$ [174]. Knowing that the modern Southern Ocean releases into the air old $CO_2$ from upwelling deep water, they surmised that this 'leak' was suppressed during glacial periods, thus increasing the storage of $CO_2$ within the deep ocean. This made the deep ocean more acid, causing deep-ocean carbonates to dissolve, which in turn made the global ocean more alkaline, increasing the solubility of $CO_2$ in seawater and so driving

more uptake of $CO_2$, hence driving further cooling through positive feedback, much as Toggweiler had suggested. The ocean must drive these changes because it is by far the largest reservoir of $CO_2$ on the planet. Sea ice played a key role by limiting the 'leak' of $CO_2$ to the atmosphere in cold periods. A decrease in the flow of NADW served to reduce oxygenation of carbon-rich southern sourced deep water. Thus both northerly *and* southerly processes exacerbated the build-up of $CO_2$ at abyssal depths. Surface water productivity also played a role, especially in the sub-Antarctic, where high productivity and efficient grazing of phytoplankton by zooplankton led to massive increases in siliceous ooze composed mostly of diatoms in sediments from glacial periods. The biological pump was efficiently transferring abundant $CO_2$ directly to the deep sea. The excessive production may reflect an increase in iron fertilization stemming from an enhanced influx of wind-blown dust in glacial times. What we end up with is robust coupling of atmospheric $CO_2$ to climate cycles driven largely by changes in the ocean [174].

In 2013, Feng He of the University of Wisconsin-Madison set out to test the Huybers and Denton model of independent hemispheric response to forcing. Feng He and colleagues used a coupled atmosphere–ocean GCM to identify the impacts of forcing on air temperature from changes in orbits, $CO_2$, ice sheets and the AMOC connecting the two hemispheres via the ocean (Figure 12.6) [175]. They interpreted their results to suggest that rising insolation in spring and summer in the Northern Hemisphere initiated the last deglaciation and controlled the timing and the magnitude of the evolution of surface temperature in the Southern Hemisphere, which would appear to support the Milankovitch model. They concluded that the orbitally induced retreat of the Northern Hemisphere ice sheets stimulated changes in the AMOC that prompted deglacial warming in the Southern Hemisphere and its subsequent lead over Northern Hemisphere temperature. $CO_2$ rising with the Southern Ocean warming provided a critical feedback encouraging global warming and deglaciation. Other researchers disagreed, as we see below.

For example, Kawamura showed that the increase in the AMOC in the North Atlantic occurred more or less at the same time as the decline began in Antarctic warming, and well before the warming in Greenland, which – since the AMOC directly connects both northern and southern hemispheres might tend to suggest that the drive for change came from the south [171].

In 2012, Joel Pedro and his team suggested two possible mechanisms for the last deglaciation, one involving fast connection through the atmosphere, the other a slower connection through the ocean [91]. Their atmospheric model starts with an orbitally induced increase in northern summer insolation initiating local warming and retreat of the Northern Hemisphere ice-sheets. Melting supplies freshwater to the North Atlantic, weakening the AMOC and leading to warming of North Atlantic subsurface waters. Those warm waters destabilize ice shelves, driving further ice retreat and releasing more freshwater. The surface cooling stimulated by freshwater makes sea ice expand, cooling the air and pushing the ITCZ south, which displaces southward and strengthens the Southern Hemisphere's mid-latitude westerly winds. These winds generated more upwelling, which draws up warm deep waters to release heat and old $CO_2$, further warming the air. In this atmospheric scenario, North Atlantic cooling and the release of $CO_2$ from the high latitude Southern Ocean are almost simultaneous.

In contrast, Pedro's ocean pathway invokes the bipolar seesaw along the lines suggested by Broecker [43], as discussed earlier. Weakening the AMOC reduces northward heat transport, allowing heat to accumulate in the south.

Whatever the solution turns out to be in terms of atmospheric versus oceanic connections, Pedro noted that '*the ice core observations point to a tightly-coupled system operating with little or no time delay between the onset/terminations of North Atlantic climate stages and a near-simultaneous trend change in both Antarctic temperature and atmospheric $CO_2$*' [91].

Thomas Crowley agreed that the ocean route played a significant role, introducing a novel additional idea. Switching off the thermohaline conveyor by reducing the formation of NADW meant that heat entering the South Atlantic from the Indian Ocean around South Africa would not be able to escape to the north via the Gulf Stream, as it did in warm times. Instead it would turn south along the South American coast in the Brazil Current, thus pumping heat into the Southern Ocean [176].

Jeremy Shakun and his team found in 2012 that their observed temperature variations in Antarctic ice cores closely matched variations in the strength of the AMOC, interpreting this to '*suggest that ocean circulation changes driven primarily by freshwater flux, rather than by direct forcing from greenhouse gases or orbits, are plausible causes of the hemispheric differences in temperature change seen in the proxy records*' [33]. The fact that their Southern Hemisphere temperature stack led Northern Hemisphere temperatures during the deglaciation supported their '*inference that AMOC-driven internal heat redistributions explain the Antarctic temperature lead and global temperature lag relative to $CO_2$*' [33]. What then triggered deglacial warming? They found that '*substantial temperature change at all latitudes … as well as a net global warming of about 0.3°C … precedes the initial increase in $CO_2$ concentration at 17.5 kyr ago, suggesting that $CO_2$ did not initiate deglacial warming*' [33]. They went on to '*suggest that these

spatiotemporal patterns of temperature change are consistent with warming at northern mid to high latitudes, leading to a reduction in the AMOC at ~19 kyr ago, being the trigger for the global deglacial warming that followed*' [33]. The trigger may have been '*rising boreal insolation driving northern warming*' [33]. Then the Northern Hemisphere ice sheets retreated, and the resulting influx of freshwater reduced the AMOC and thus warmed the Southern Hemisphere through the bipolar seesaw, leading to the release of old $CO_2$. The observed pattern '*is difficult to reconcile with hypotheses invoking a southern high latitude trigger for deglaciation*' [33]. Even so, they concluded '*$CO_2$ [was] a key mechanism of global warming during the last deglaciation*' [33]. But, every hypothesis has to stand up to scrutiny based on new data, and Parrenin's results [90, 92] (e.g. Figure 13.15) now show that the core of Shakun's model – the assumption that $CO_2$ lagged the Antarctic rise in temperature – is suspect for most of the deglacial period.

The correspondence between bipolar seesaw oscillations and changes in atmospheric $CO_2$ during glacial terminations back through time suggested to Stephen Barker and his team in 2011 that the bipolar seesaw played a key role in the mechanism of deglaciation through positive feedbacks associated with increasing $CO_2$ [54]. '*With the supercritical size of continental ice sheets as a possible precondition, and in combination with the right insolation forcing and ice albedo feedbacks, the $CO_2$ rise associated with an oscillation of the bipolar seesaw could provide the necessary additional forcing to promote deglaciation*' they felt [54]. Thus, the mechanism of glacial termination might result from the timely and necessary interaction between millennial variations (Dansgaard–Oeschger events) and orbital time scale variations, much as Wolff had suggested.

Frédéric Parrenin and his team disagree with Shakun and other Meridional Overturning enthusiasts. The tight correlation they discovered between $CO_2$ and temperature in Antarctica (Figure 13.15) meant that '*invoking changes in the strength of the Antarctic meridional overturning circulation is no longer required to explain the [previously apparent] lead of [Antarctic temperature] over [atmospheric] $CO_2$*' [90, 92]. Parrenin's team suggested that '*given the importance of the Southern Ocean in carbon cycle processes … one should not exclude the possibility that [atmospheric] $CO_2$ and [Antarctic temperature] are interconnected through another common mechanism such as a relationship between sea ice cover and ocean stratification*' [90]. This latest finding lends support to the Huybers and Denton model in which southern spring insolation and the length of southern summers controls climate change in the south at the same time that summer insolation at 65°N controls what happens in the north (see Figure 13.20) [129, 172]. Toggweiler and Wolff would no doubt agree.

A recent twist in the complex tale of the evolution of Ice Age climate theory came in 2013 from a team led by Ayako Abe-Ouchi of the University of Tokyo [177], and including Maureen Raymo, whom we met earlier arguing for a role for glacial rebound. They used numerical physical models in combination with a GCM to assess the relative importance of internal mechanisms that might drive the 100 Ka glacial cycles. These included delayed bedrock rebound (i.e. glacial isostatic adjustment), the calving of icebergs from ice sheet margins, variations in $CO_2$, and feedback from aeolian dust and the oceans. The advantage of their method was that '*The ice-sheet model with the climate parameterization ... can represent fast feedbacks, such as water vapour, cloud and sea-ice feedbacks ... [as well as] slow feedbacks, such as albedo/temperature/ice-sheet and lapse rate/temperature/ice sheet feedbacks*' [177]. Lapse rate was included because in moist air it is about 6.5 °C per 1000 m, making ice sheets cooler as they grow upwards and warmer as they shrink downwards. Their rising and shrinking may be functions not only of precipitation, but also of isostatic adjustment, when the ice mass depresses the underlying crust. Abe-Ouchi's approach thus took into consideration variables that had been missing in previous theoretical analyses, including extent of sea ice, processes governing growth and decay of ice sheets, and isostatic adjustments to ice sheets as the increase or decrease mass of ice either depressed or raised the underlying land.

The team calculated ice-sheet variation for the past 400 Ka forced by insolation and atmospheric $CO_2$ content, validated the results using proxy palaeoclimate data and conducted sensitivity experiments to investigate the possible mechanisms controlling the 100 Ka glacial cycles. Their model took into account the fact that as ice sheets thicken they depress the land beneath them, which may lower their tops into areas of warmer air, increasing the amount of melt and the area exposed to melting. Ice sheets also flow, which may widen their area and lower their height, thus exposing them to melting. These various processes become effective late in the glacial cycle, because it takes a long time for ice sheets to grow to the point where they will both lose height and spread. By incorporating these and other feedbacks, the model '*realistically simulates the sawtooth characteristics of glacial cycles, the timing of the terminations and the amplitude of the Northern Hemisphere ice-volume variations ... as well as their geographical patterns at the Last Glacial Maximum and the subsequent deglaciation*' [177]. Amongst other things, the team found that '*The ~100-kyr periodicity, the sawtooth pattern and the timing of the terminations are reproduced ... [in sensitivity experiments] with constant $CO_2$ levels ... [and that] the crucial mechanism for the ~100-kyr cycles is the delayed glacial isostatic rebound, which keeps the ice elevation low, and, therefore, the ice ablation high, when the ice retreats*' [177].

Abe-Ouchi's team found that the relationship between ice volume and temperature for the North American (Laurentide) ice sheet followed a hysteresis loop, with ice sheet volume first declining gradually as temperature anomalies increased from −5 to 0 °C, then declining rapidly as temperatures increased to +2 °C. Re-growth happened rapidly only after temperature anomalies fell to zero and declined below it. When the ice sheet was large enough in extent, its rapid disintegration was triggered by just a modest increase in insolation, enhanced by low elevation due to the delayed isostatic response, calving into pro-glacial lakes, increasing $CO_2$ concentrations (amplifying warming), dust feedback, and basal sliding as water made its way to the base of the ice sheet. Thus '*insolation and internal feedbacks between the climate, the ice sheets and the lithosphere–asthenosphere system explain the 100,000-year periodicity ... The larger the ice sheet grows and extends towards lower latitudes, the smaller is the insolation required to make the mass balance negative. Therefore, once a large ice sheet is established, a moderate increase in insolation is sufficient to trigger a negative mass balance, leading to an almost complete retreat of the ice sheet within several thousand years. This fast retreat is governed mainly by rapid ablation due to the lowered surface elevation resulting from delayed isostatic rebound, which is the lithosphere–asthenosphere response. Carbon dioxide is involved, but is not determinative, in the evolution of the 100,000-year glacial cycles*' [177]. The observation about $CO_2$ is fair enough in that, as we now know, it started increasing long before the warming and collapse of the Northern Hemisphere ice sheets. The possibility emerges, then, that $CO_2$ output is dominantly a Southern Hemisphere phenomenon and has its greatest effect on climate change there, whilst the events in the north are much more closely controlled by the dynamics of ice sheet response to northern (boreal) insolation forcing. I am reminded of Plass's observation from the 1950s that Arctic temperatures would remain low while ice sheets melted, even though $CO_2$ was rising, simply because melting ice absorbs huge amounts of energy. Hence we should not expect a direct link between Northern Hemisphere ice and $CO_2$, given that the rise in $CO_2$ is driven from the Southern Ocean.

The European ice sheet behaved differently from the North American one because it was thinner, less extensive, and located in a warmer climate [178]. It also followed a hysteresis loop, with ice sheet volume declining rapidly between −2 °C and 0 °C compared with modern conditions, and re-growing rapidly only after temperature anomalies fell to −1 °C and declined further. It responded mainly to insolation following the obliquity cycle of

~40 kyr. The difference from its North American equivalent was caused in part because summers in Europe are warmer than in North America. Under these conditions the European ice sheet could not sustain large ice volumes for long. Obliquity was much less important for the North American ice sheet, where eccentricity modulated the amplitude of the precession signal by causing critical changes in summer insolation to create the 100 Ka cycle. Abe-Ouchi's team reached the remarkable conclusion that '*the 100-kyr glacial cycle exists only because of the unique geographic and climatological setting of the North American ice sheet with respect to received insolation*' [177].

Others were working along similar lines. In 2011, Andrey Ganopolski and Reinhard Calov of the Potsdam Institute for Climate Impact Research in Germany used an Earth System Model of Intermediate Complexity to show that 100 Ka variations in ice volume and in the timing of the terminations of glacial periods could be simulated as non-linear responses of the climate-cryosphere system to orbital forcing, provided that $CO_2$ levels were below those typical of interglacial periods [179]. Like Abe-Ouchi, they attributed the existence of long glacial cycles mainly to the behaviour of the Laurentide ice sheet of North America. Although this behaviour could be simulated without any variation in $CO_2$, they found that variations in $CO_2$ with time amplified the 100 Ka cycles. They also found that the development of the 100 Ka pattern depended on former ice sheets having cleared northern North America of sediments that might have enhanced sliding at the base of the ice sheet. It was important in the model for the ice sheet to be sitting on rock. Their model showed that ice formed when summer insolation fell below a certain threshold allowing ice to remain all year. Rapid terminations of the ice sheets were strongly related to their coverage by dust, which reduced albedo, enhanced melting (ablation) and thus amplified the response of the ice sheet to rising insolation.

The solid Earth also seems to have been important in another way, through $CO_2$ provided by volcanic eruptions. Comparing volcanic activity against the deuterium ($\partial^2 D$) proxy for temperature in the EPICA Dome C ice core, a team including Jean Jouzel from Gif-sur-Yvette found no clear evidence in 2004 for a close relationship between climate change and volcanism over the past 45 Ka [180]. But, a later more detailed examination by Peter Huybers and Charlie Langmuir from Harvard, in 2009, showed that subaerial volcanism increased globally by two to six times above background levels during the last deglaciation 12–7 Ka ago [181]. That rise was consistent with an increase of 40 ppm in atmospheric $CO_2$ during the second half of the last deglaciation. Huybers and Langmuir suggested that the glacial isostatic adjustments associated with shrinkage of ice caps decompressed the mantle in deglaciating regions.

That increased the number and intensity of volcanic eruptions, which raised atmospheric $CO_2$, which warmed the atmosphere and caused more deglaciation, and so on. They concluded that '*Such a positive feedback may contribute to the rapid passage from glacial to interglacial periods*' [181]. Their hypothesis provides us with a further modification to the commonly accepted notion that glacial to interglacial variations in $CO_2$ are primarily attributable to oceanic processes. Along similar lines, Maya Tolstoy observed in 2015 that when the pull of other planets made Earth's orbit most eccentric, Earth's shape would have been sufficiently distorted to enhance submarine volcanism and the exhalation of $CO_2$ especially at mid-ocean ridge crests. That may have helped to enhance glacial–interglacial cycles and may help to explain the dominance of the 100 Ka cycle [182]. Confirmation of an association between 100 Ka climate cyclicity and large volcanic eruptions came subsequently from the distribution of tephra (volcanic ash) at an ocean drilling site on the Izu-Bonin Arc, a prominent ridge extending south from Japan into the northwest Pacific [183]. '*A spectral analysis of the dataset yielded a statistically significant spectral peak at the ~100 kyr period, which dominates the global climate cycles since the Middle Pleistocene … volcanism peaks after the glacial maximum and <13 ± 2 kyr before the $\delta^{18}O$ minimum at the glacial/interglacial transition … [and] the correlation is especially good for the last 0.7 Myr*' [183]. The correlation was weaker for the period 0.7–1.1 Ma, during the Middle Pleistocene Transition (MPT), mainly because the 100 Ka periodicity in the $\delta^{18}O$ record from benthic foraminifera diminished, whilst the tephra record maintained its strong Ka beat [183].

The story is almost complete. The latest data emerging from West Antarctica provides us with a yet more refined view of climate change. In 2011 the members of the WAIS Divide team drilled the 3405 m long WAIS Divide Ice Core (WDC) in central West Antarctica at an altitude of 1766 m, recovering ice as old as 68 Ka [184]. Initial data appeared in 2013. The key thing to bear in mind is that the long East Antarctic cores discussed so far (from Vostok, Fuji, Dome C and Dronning Maud Land) came from sites high up on the Polar Plateau that are isolated from the influence of changes of circulation and sea ice within the Southern Ocean, whereas the West Antarctic Divide site is lower, hence warmer, and the recipient of moist maritime air. It thus preserves a clearer record of changes in ocean circulation and sea ice. Warming began at the WDC site 20 Ka ago, at least 2 Ka before significant warming in East Antarctica [184]. At the same time, sea-salt sodium (Na) increased, suggesting a decline in the cover of sea ice. This agrees with data from a marine core from the southwest Atlantic showing that sea ice began to retreat shortly before 22 Ka ago [185]. The warming began *before* the decrease in the AMOC that had been called upon to explain Southern Ocean warming. Its likely cause was local orbital forcing, with annual insolation at 65°S increasing 1% between 22 and 18 Ka ago. Furthermore '*The increase in integrated summer insolation, where summer is defined as days with insolation above a threshold of 275W/m², is greater than the total annual increase ... Thus, the increase comes in summer, when it is most likely to be absorbed by low-albedo open water. The summer duration also begins increasing at 23 kyr ago*' [184]. Longer summers and shorter winters will likely have melted sea ice and warmed the ocean. Decreasing sea ice would have decreased albedo, causing further warming.

Can we tie the Northern and Southern Hemisphere pictures together now? The WAIS team summarized the picture as follows [184]: '*While the abrupt onset of East Antarctic warming, increasing CO₂ and decreasing AMOC 18 kyr ago has supported the view that deglaciation in the Southern Hemisphere is primarily a response to changes in the Northern Hemisphere ... the evidence of warming in West Antarctica and corresponding evidence for sea-ice decline in the [southeast] Atlantic show that climate changes were ongoing in the Southern Ocean before 18 kyr ago, supporting an important role for local orbital forcing. Warming in the high latitudes of both hemispheres before 18 kyr ago implies little change in the interhemispheric temperature gradient that largely determines the position of the intertropical con-*

*vergence zone and the position and intensity of the mid-latitude westerlies*' [184]. They proposed that '*when Northern Hemisphere cooling occurred ~18 kyr ago, coupled with an already-warming Southern Hemisphere, the intertropical convergence zone and mid-latitude westerlies shifted southwards in response. The increased wind stress in the Southern Ocean drove upwelling, venting of CO₂ from the deep ocean, and warming in both West Antarctica and East Antarctica*' [184]. This new WAIS core thus confirmed an active role for the Southern Hemisphere in initiating global deglaciation, whilst also confirming the influence of Northern Hemisphere processes on the Southern Hemisphere. Everything is connected, even more so when we recall Trond Dokken's observation that warm water from the south penetrates beneath sea ice covered Nordic seas and eventually destabilizes them (suggesting a southern driver for northern change) [163].

By 2015, the WAIS Divide Project Members had taken advantage of the high resolution of the WAIS Divide core and of Greenland's NGRIP core to examine in fine detail the relationship between Antarctic warm events (AIMs) and Greenland's Dansgaard–Oeschger events [186]. Their results dealing with abrupt climate variability, summarized by Ed Brook and Christo Buizert in 2018 [35], appear in Figure 13.21. Analysing the relationships between Greenland and Antarctica, the WAIS Divide team showed

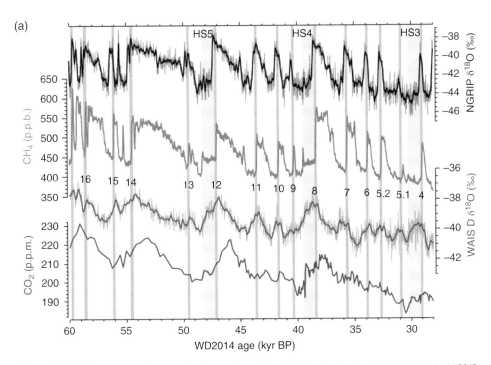

**Figure 13.21** From top to bottom, the traces show Greenland water isotope ratios from the NGRIP core, atmospheric CH₄ from the Antarctic WAIS Divide ice core, Antarctic ice core water isotope ratios from the WAIS Divide core (WDC) and atmospheric CO₂ from a multi-core Antarctic compilation. Water isotope ratios are measured relative to Vienna Standard Mean Ocean Water (VSMOW). Blue bars show the approximate timing of Heinrich Stadials 3–5. Numbers in black indicate AIM events. *Source:* From figure 2 in Ref. [35].

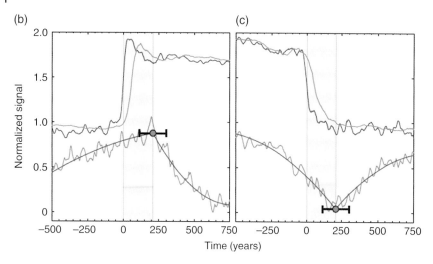

**Figure 13.22** Interpolar phasing of abrupt climate change. Using the data from Figure 13.9, the stacked NGRIP isotope (blue) and WDC methane records (green) are aligned with the stacked WDC isotope data (orange) at the abrupt Northern Hemisphere transitions (yellow vertical lines), and averaged to obtain the shared climatic signal for warming events (b) and cooling events (c). The Antarctic response of the bipolar seesaw is delayed by around two centuries behind their abrupt Northern Hemisphere equivalent events. *Source:* from figure 2 in Ref. [186].

that '*on average, abrupt Greenland warming leads the corresponding Antarctic cooling onset by 218±92 years ... for Dansgaard – Oeschger events, including the Bølling event; Greenland cooling leads the corresponding onset of Antarctic warming by 208±96 years*' [186] (Figure 13.22). The data suggest that abrupt phases of the Dansgaard–Oeschger events are initiated in the Northern Hemisphere, presumably in the North Atlantic. Hence, these '*results demonstrate a north-to-south directionality of the abrupt climatic signal, which is propagated to the Southern Hemisphere high latitudes by oceanic rather than atmospheric processes. The similar interpolar phasing of warming and cooling transitions suggests that the transfer time of the climatic signal is independent of the AMOC background state ... [the] findings confirm a central role for ocean circulation in the bipolar seesaw and provide clear criteria for assessing hypotheses and model simulations of Dansgaard–Oeschger dynamics*' [186].

And yet – Hain and colleagues argue that the continuation of wind-driven upwelling in the Southern Ocean at times when the production of NADW stalled would have acted to deepen the global thermocline, which would have encouraged the re-initiation of the production of that deep water. They base their argument on the concentration of $^{14}C$ in $CO_2$, because re-initiation of the production of NADW mixes new $^{14}C$ into the deep ocean. The most abrupt $^{14}C$ declines precede the timings of NADW onset by a few centuries, most probably because Southern Ocean upwelling was producing buoyant Intermediate and Mode Water that thickened the upper cell of the Atlantic Meridional Circulation when NADW production stopped (e.g. in the Younger Dryas), hence causing the thermocline

to deepen (Figure 12.6). The thermocline would become shallower when the production of dense NADW resumed. Hain argues that such thickening of the thermocline may be fundamental to the physical mechanism of the bipolar seesaw. The production of more buoyant water in the south would create a sea level gradient leading down to the north, exerting pressure on the North Atlantic to resume NADW production so as to rectify the physical imbalance [95]. The roughly 500 year change in $^{14}C$ suggests the time scale for this Southern Ocean influence [95].

So, by the beginning of the twenty-first century, much as hoped by Crowley and North [99], palaeoclimatologists had brought us close to a definitive explanation of the variability of the Pleistocene ice age. Access to very recently obtained long cores from the Antarctic ice sheet were critical to this growth in our understanding, and much of the data emerging from these cores has yet to make it into the literature available to the wider public. These data have been obtained at a considerable cost, much like those obtained from deep sea drilling. Without the results from drill cores through the ice sheets and the ocean floor our understanding of the Earth's climate evolution would be dim indeed.

Looking at these various developments, we can now see that $CO_2$ and the circulation of the Southern Ocean play key roles within the glacial–interglacial climate system. Rising spring insolation and the long duration of the summer in the Southern Hemisphere caused the Southern Ocean to lose its sea ice, to warm and to simultaneously release $CO_2$, which further accentuated the warming. The ocean began to carry less $CO_2$ as it warmed and as the rate of production of deep water decreased. Atmospheric $CO_2$ mixed rapidly to the Northern Hemisphere, whose rise in temperature lagged that in the south because of the thermal inertia of the Northern Hemisphere ice sheets. The AMOC connected the two hemispheres, feeding 'old' $CO_2$ south from depth to emerge at the surface of the Southern Ocean (Figures 12.5 and 12.6). The strengthening of the AMOC, previously called upon to explain Southern Ocean warming, began *after* warming began in the south, indicating the importance of local southern factors. Tropical warming and associated humidity increased wetlands, which provided methane to further enhance warming. Global warming enhanced oceanic evaporation, providing water vapour to stimulate further warming. Rising and warming seas provide a larger surface area from which to supply increasing volumes of both water vapour and $CO_2$. Volcanic eruptions associated with glacial isostatic adjustment as ice melted in the north, and with Earth's changing eccentricity, provided yet more $CO_2$ during the deglaciation. The *three* increasing greenhouse gases in the atmosphere ($CO_2$, $H_2O$, $CH_4$) added to warming, which was further accentuated by rising insolation in the Northern Hemisphere and by the eventual disappearance of Northern Hemisphere ice, which reduced albedo, and in due course we arrived at a warm interglacial, aided by the northward penetration of warm subsurface water from the south that eventually destabilized the frozen ocean surface. The disappearance of Northern Hemisphere ice sheets was largely controlled by the response of ice to local insolation, combined with the decay of the growing ice sheet as it sank into warmer air and spread widely across its surroundings. Continued production of buoyant surface and intermediate water by Southern Ocean upwelling when the production of NADW was switched off may have thickened the thermocline enough to re-initiate NADW production.

In contrast, during cold conditions, $CO_2$ was drawn down from the atmosphere into the ocean by dissolving in cold surface waters and was taken into deep water by enhanced deep-water production. Intensified chemical and physical weathering in mountainous areas further drew down atmospheric $CO_2$. A low equilibrium level of atmospheric $CO_2$ was reached at glacial maxima, when sea level was at its lowest, reducing ocean areas, and sea ice was most abundant, further reducing the area of ocean exposed to the atmosphere. At that time expanded ice sheets reduced the area covered by vegetation, a potential source of $CO_2$ and $CH_4$. The area of tropical wetlands, a major source of $CH_4$, was also reduced. Nevertheless, exchange of $CO_2$ between ocean and atmosphere continued, especially where strengthening winds intensified upwelling along the Equator, beneath the westerly winds of the Southern Ocean, and along continental margins,

bringing *old* $CO_2$ to the surface. Falling seas and cold conditions reduced the potential for the production of water vapour. These various factors combined to cool the planet. Eventually insolation rose, and the warming cycle began over again, leading from glaciation to deglaciation.

It would appear that there is no simple answer to the chicken and egg question of what initiated flip-flops in the overturning circulation. There is evidence for Dansgaard–Oeschger events preceding their Antarctic counterparts by some 200 years, but other evidence has emerged suggesting that the events in the north were triggered by a thickening of the ocean's thermocline driven from the south, and that northern sea ice was destabilized by northward moving warm southern waters. Once an oscillation starts it becomes impossible to trace its origin.

We have learned one important lesson in addition. Temperature per se is not absolutely dependent on $CO_2$. It is also governed by insolation, by albedo (e.g. the extent of ice sheets and sea ice), by the local melting of land ice, and by the movement of warm currents. In contrast, whilst $CO_2$ is dependent on factors like the warmth of seawater (the solution pump), it is also affected by productivity (the biological pump), the distribution of carbonate (the CCD), the presence or absence of sea ice, the area of the ocean (which changes with sea level), and the amount and type of vegetation on land. Under *normal* interglacial conditions $CO_2$ reaches an equilibrium level (about 280 ppm), driven by the various feedbacks within the climate system, before declining insolation begins to drive the system back towards cooler conditions. That decline decreases the conditions that lead to high concentrations of atmospheric $CO_2$, which in turn aids cooling. The rate of decline in temperature in encroaching glaciation exceeds the rate of decline in $CO_2$, being driven more by falling insolation and increasing albedo due to the growing extent of land and sea ice, the last of which prevents $CO_2$ decline in polar regions. Eventually $CO_2$ reaches a new equilibrium level (about 180 ppm), driven by the feedbacks within the climate system (for example the dissolution of $CO_2$ in increasingly cold surface water). These various processes account for the fluctuation of $CO_2$ during the Ice Age within a *natural envelope* of 180–280 ppm

Changes in $CO_2$ tend to follow AIM events closely (Figure 13.21), as well as during deglaciation, where temperature and $CO_2$ are essentially synchronous (Figures 13.15 and 13.23) [35]. The high-resolution $CO_2$ data from the West Antarctic Divide core show that rapid $CO_2$ changes are superimposed on the slower millennial change in $CO_2$ during the deglaciation (Figure 13.23). Ones of particular interest, of about 10 ppm, occur in association with Northern Hemisphere warming at the start of the Bølling–Allerod interstadial, and at the end of the Younger Dryas

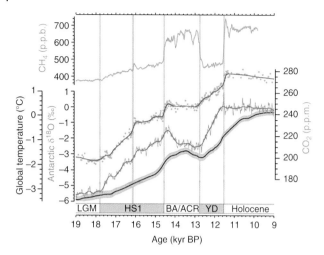

**Figure 13.23** shows atmospheric $CO_2$ change from the WAIS Divide (Antarctica) ice core for the period 19–9 Ka ago (red), global temperature reconstruction (black, with grey shading indicating uncertainty), the east Antarctic oxygen isotope stack (the water $^{18}O/^{16}O$ isotope ratio anomaly relative to the present) (blue), and the atmospheric $CH_4$ record from the WAIS Divide ice core (green). Vertical yellow bars show the timing of major inflection points in the $CO_2$ record. LGM = Last Glacial Maximum; HS1 = Heinrich Stadial 1 (Northern Hemisphere); BA = Bølling–Allerød warming in the Northern Hemisphere; ACR = Antarctic Cold Reversal; and YD = Younger Dryas (Northern Hemisphere). *Source:* From box 2 in Ref. [35].

stadial, which are also associated with rises in methane ($CH_4$) (Figure 13.23). These shifts appear to be related to warming of the sea surface in the north [35]. A third increased rate of $CO_2$ rise, also associated with a small rise in $CH_4$, is associated with Heinrich stadial 1, a Northern Hemisphere cold event. This shift may be related to cooling-induced southward shifts in the ITCZ and its effect on tropical hydrology [35].

Neither insolation nor $CO_2$ are the sole keys to understanding the 100 Ka cycles. Both hemispheres are connected through the ocean and the atmosphere, which means that each can influence the other, for example through the AMOC. Aspects of the insolation characteristics of each hemisphere mean that both can warm or cool at more or less the same time, despite their separate drivers. With the latest data in hand from Frédéric Parrenin [90, 92] and Joel Pedro [91] (Figures 13.15 and 13.17), $CO_2$ can no longer be seen as merely a follower of temperature during the Ice Age. At least around Antarctica, the two vary in lock-step during deglaciations and during the Antarctic equivalents of Dansgaard–Oeschger events (Figures 13.15–13.17). In a complex climate system characterized by both internal oscillations and external forcings it remains difficult to see what the trigger may be for both millennial variations and the growth and termination of ice sheets (the chicken and

egg problem) [66]. It seems likely that the dynamics of the great Laurentide ice sheet may play a key role. When it gets too big, it spreads out, loses height, and becomes more susceptible to rapid removal by small increases in northern insolation combined with warmth brought north by the ocean, independently of global $CO_2$ levels, which play a much larger role in controlling climate change in the Southern Hemisphere. Nevertheless it is equally obvious that the AMOC and the bipolar seesaw have '*important role[s] in the machinery of glacial cycles, and should be considered in answering questions that are commonly placed in the "orbital" realm, such as … [what causes] the 100 kyr cycle and interhemisperic symmetry at the obliquity and precession timescales?*' [35].

We have also learned that isostatic adjustments during deglaciation in the north stimulate volcanic activity that provides new $CO_2$ unrelated to plate tectonic processes, and that volcanic activity on the same 100 Ka time scale may be triggered by the Earth's response to changing eccentricity. We also now know from Ge/Si ratios that chemical weathering can proceed fast enough to draw down $CO_2$ in glacial times. Hence $CO_2$ can have both a primary role (via volcanic activity and weathering) and a secondary role (via emerging from the ocean as it warms, then enhancing warming through positive feedback), with water vapour and methane adding to the feedback. We understand the complexities much better than we used to.

$CO_2$ and temperature follow the same saw-toothed pattern at orbital scales (Figures 13.8 and 13.9) and at Dansgaard–Oeschger scales (Figure 13.6). $CO_2$ is not operating independently of temperature. Both are responding at the same time to some common external forcing – which drives the upwelling that releases old $CO_2$ from the Southern Ocean. Contrary to initial ideas, that forcing kicks off the process in the south, possibly during long southern summers. What would account for the same patterns at orbital and millennial scales? We can expect there to be harmonics in the orbital signal that would lead to millennial scale variation between the main orbital periods. Alternatively, once the oscillations are set up, as Broecker initially suggested, they may simply persist with no external driver. As Brooks and Buizert point out, there clearly is a link between orbital (especially precessional) and millennial signals [35]. $CO_2$ and temperature operate together at both time scales.

The detailed examination carried out in previous chapters and here of the relationship between $CO_2$ and temperature in past geological times and during the Pleistocene Ice Age confirms, first, that there are times when the emission of $CO_2$ in response to volcanic activity driven by plate tectonics or by plume eruptions causes changes in temperature, and, second, that there are times when changes in temperature (e.g. driven by orbital change) cause changes in $CO_2$, as we have seen during the Ice Age. In both cases temperature and $CO_2$ act as feedbacks to enhance each other, within a natural envelope representing the operation of a natural thermostat.

Having established with a reasonable degree of confidence how the Ice Age climate system works, we turn now to a detailed examination of climate change in the latest interglacial period, the Holocene, starting 11 700 years ago, in Chapter 14.

## References

**1** Carlson, W.E. and Winsor, K. (2012). Northern Hemisphere ice-sheet responses to past climate warming. *Nature Geoscience* 5 (9): 607–613.

**2** Oerlemans, J. (1982). Glacial cycles and ice-sheet modelling. *Climate Change* 4: 353–374.

**3** Zhang, Z., Yan, Q., Farmer, E.J. et al. (2018). Instability of Northeast Siberian ice sheet during glacials. *Climate of the Past Discussions* https://doi.org/10.5194/cp-2018-79.

**4** Patton, H., Hubbard, A., Andreassen, K. et al. (2017). Deglaciation of the Eurasian ice sheet complex. *Quaternary Science Reviews* 169: 148–172.

**5** Alley, R. (2011). *Earth: The Operator's Manual*. New York and London: W.W. Norton & Co.

**6** Dansgaard, W., Johnsen, S.J., Clausen H.B., et al. (1984) North Atlantic climatic oscillations revealed by deep Greenland ice cores, in *Climate Processes and Climate Sensitivity* (eds J.E. Hansen and T. Takahashi), *Geophysical Monograph* **29**, American Geophysical Union, Washington D.C., 288–298.

**7** Steffensen, J.P., Andersen, K.K., Bigler, M. et al. (2008). High-resolution Greenland ice core data show abrupt climate change happens in few years. *Science* 321: 680–684.

**8** Bond, G., Showers, W., Cheseby, M. et al. (1997). A pervasive millennial-scale cycle in North Atlantic Holocene and glacial climates. *Science* 278: 1257–1266.

**9** Bond, G.C., Showers, W., Elliot, M. et al. (1999). The North Atlantic's 1-2kyr climate rhythm: relation to Heinrich Events, Dansgaard/Oeschger cycles and the Little Ice Age. In: *Mechanisms of Global Climate Change at Millennial Time Scales*, *Geophysical Monograph* **112** (eds. P.U. Clark, R.S. Webb and L.D. Keigwin), 35–58. Washington, D.C.: American Geophysical Union.

**10** Oeschger, H., Beer, J., Siegenthaler, U. et al. (1984). Late glacial climate history from ice cores. In: *Climate Processes and Climate Sensitivity*, *Geophysical Monograph* **29** (eds. J.E. Hansen and T. Takahashi), 299–306. Washington D.C.: American Geophysical Union.

**11** Dansgaard, W., Johnsen, S.J., Clausen, H.B. et al. (1993). Evidence for general instability of past climate from a 250 kyr ice core record. *Nature* 364: 218–220.

**12** Jouzel, J. (2006). Water stable isotopes: atmospheric composition and applications in polar ice core studies.
In: *The Atmosphere, Treatise on Geochemistry* (eds H.D. Holland and K.K. Turekian), vol. 4 (ed. R.F. Keeling), 213–243. Oxford: Elsevier-Pergamon.

**13** Bond, G.C., Broecker, W.S., Johnsen, S. et al. (1993). Correlations between climate records from North Atlantic sediments and Greenland ice. *Nature* 365: 143–147.

**14** Ahn, J. and Brook, E.J. (2007). Atmospheric $CO_2$ and climate from 65 to 30 ka B.P. *Geophysical Research Letters* 34: L10703, GL029551.

**15** Sundquist, E.T. and Visser, K. (2005). The geologic history of the carbon cycle. In: *Biogeochemistry, Treatise on Geochemistry* (eds. H.D. Holland and K.K. Turekian), vol. 8 (ed. W.H. Schlesinger), 425–513. Oxford: Elsevier-Pergamon.

**16** Alley, R. (2000). *The Two-Mile Time Machine: Ice Cores, Abrupt Climate Change, and Our Future*. Princeton University Press.

**17** Alley, R.B., Clark, P.U., Keigwin, L.D., and Webb, R.S. (1999). Making sense of millennial-scale climate change. In: *Mechanisms of Global Climate Change at Millennial Time Scales*, *Geophysical Monograph* **112** (eds. P.U. Clark, R.S. Webb and L.D. Keigwin), 385–394. Washington, D.C.: American Geophysical Union.

**18** Siddall, M., Stocker, T.F., Blunier, T. et al. (2007). Marine Isotope Stage (MIS) 8 millennial variability stratigraphically identical to MIS 3. *Paleoceanography* 22: PA1208.

**19** Cronin, T.M. (2010). *Paleoclimates – Understanding Climate Change Past and Present*, 441 pp. Columbia University Press.

**20** Schaefer, J.M., Finkel, R.C., Balco, G., and Alley, R.B. (2017). Greenland was nearly ice-free for extended periods during the Pleistocene. *Nature* 540: 252–255.

**21** Lorius, C., and Merlivat, L. (1977) Distribution of mean surface stable isotope values in East Antarctica; observed change with depth in a coastal area, in *Isotopes and Impurities in Snow and Ice*. Proc. Grenoble Sympos. 1975, *IAHS* **118**, 125-137.

**22** Lorius, C., Jouzel, J., Ritz, C. et al. (1985). A 150,000-year climatic record from Antarctic ice. *Nature* 316: 591–596.

**23** Jouzel, J., Lorius, C., Petit, J.R. et al. (1987). Vostok ice core: a continuous isotope temperature record over the last climatic cycle (160,000 years). *Nature* 329: 403–408.

24 Petit, J.R., Jouzel, J., Raynaud, D. et al. (1999). Climate and atmospheric history of the past 420,000 years from the Vostok ice core, Antarctica. *Nature* 399: 429–436.

25 Petit, J.R., Basile, I., Leruyuet, A. et al. (1997). Four climate cycles in Vostok ice core. *Nature* 387: 359–360.

26 EPICA community members (56 authors) (2004). Eight glacial cycles from an Antarctic ice core. *Nature* 429: 623–628.

27 Lüthi, D., Le Floch, M., Bereiter, B. et al. (2008). High resolution carbon dioxide concentration record 650 000–800 000 years before present. *Nature* 453. 379–382.

28 Oerter, H., Drücker, C., Kipfstuhl, S., and Wilhelms, F. (2008). Kohnen Station – the drilling camp for the EPICA deep ice core in Dronning Maud Land. *Polarforschung* 78 (1–2): 1–23.

29 Watanabe, O., Kamiyama, K., Motoyama, H. et al. (1999). The paleoclimate record in the ice core at Dome Fuji Station, East Antarctica. *Annals of Glaciology* 29 (1): 176–178.

30 Jouzel, J., Masson-Delmotte, V., Cattani, O. et al. (2007). Orbital and millennial Antarctic climate variability over the past 800 000 years. *Science* 317 (5839): 793–796.

31 Kopp, R.E., Simons, F.J., Mitrovica, J.X. et al. (2009). Probabilistic assessment of sea level during the last interglacial stage. *Nature* 462: 863–868.

32 Hoffman, J.S., Clark, P.U., Parnell, A.C., and He, F. (2017). Regional and global sea-surface temperatures during the last interglaciation. *Science* 355: 276–279.

33 Shakun, J.D., Clark, P.U., He, F. et al. (2012). Global warming preceded by increasing carbon dioxide concentrations during the last deglaciation. *Nature* 484: 49–54. https://doi.org/10.1038/nature10915.

34 Cuffey, K.M., Clow, G.D., Steig, E.J. et al. (2016). Deglacial temperature history of West Antarctica. *Proceedings of the National Academy of Sciences* 113 (50): 14249–14254.

35 Brook, E. and Buizert, C. (2018). Antarctic and global climate history viewed from ice cores. *Nature* 558: 200–208.

36 Lorius, C., Barnola, J.-M., Legrand, M. et al. (1989). Long-term climatic and environmental records from Antarctic ice. In: *Understanding Climate Change, Geophysical Monograph* **52** (eds. A. Berger, R.E. Dickinson and J.W. Kidson), 11–16. Washington DC: American Geophysical Union, and *IUGG* 7.

37 Lorius, C., Barkov, N.I., Jouzel, J. et al. (1988). Antarctic ice core: $CO_2$ and climatic change over the last climatic cycle. *Eos* 69 (26): 681, 683-684.

38 Naish, T., Carter, L., Wolff, E. et al. (2009). Late Pliocene-Pleistocene Antarctic climate variability at orbital and suborbital scale: ice sheet, ocean and atmospheric interactions. In: *Antarctic Climate Change, Developments in Earth and Environmental Sciences* **8** (eds. F. Florindo and M. Siegert), 465–529. Amsterdam: Elsevier.

39 Elderfield, H., Perretti, P., Greaves, M. et al. (2012). Evolution of ocean temperature and ice volume through the mid-Pleistocene climate transition. *Science* 337: 704–709.

40 Blunier, T., Chappellaz, J., Schwander, J. et al. (1998). Asynchrony of Antarctic and Greenland climate change during the last glacial period. *Nature* 394: 739–743.

41 EPICA community members (86 authors) (2006). One-to-one coupling of glacial climate variability in Greenland and Antarctica. *Nature* 444: 195–1998.

42 Blunier, T., Spahni, R., Barnola, J.-M. et al. (2007). Synchronization of ice core records via atmospheric gases. *Climate of the Past* 3: 325–330.

43 Broecker, W.S. (1998). Paleocean circulation during the last deglaciation: a bipolar seesaw? *Paleoceanography* 13: 119–121.

44 Stocker, T.F. (1998). The seesaw effect. *Science* 282: 61–62.

45 Maslin, M., Seidov, D., and Lowe, J. (2001). Synthesis of the nature and causes of rapid climate transitions during the Quaternary. In: *The Oceans and Rapid Climate Change: Past, Present and Future, Geophysical Monograph* **126** (eds. D. Seidov, B.J. Haupt and M. Maslin), 9–52. Washington D.C.: American Geophysical Union.

46 Barron, E.J. and Seidov, D. (2001). Ocean currents of change: introduction. In: *The Oceans and Rapid Climate Change: Past, Present and Future, Geophysical Monograph* **126** (eds. D. Seidov, B.J. Haupt and M. Maslin), 1–5. Washington D.C.: American Geophysical Union.

47 Broecker, W.S. (2001). The big climate amplifier: ocean circulation-sea ice-storminess-dustiness-albedo. In: *The Oceans and Rapid Climate Change: Past, Present and Future, Geophysical Monograph* **126** (eds. D. Seidov, B.J. Haupt and M. Maslin), 53–56. Washington D.C.: American Geophysical Union.

48 Seidov, D., Haupt, B.J., Barron, E.J., and Maslin, M. (2001). Ocean bi-polar seesaw and climate: southern versus northern meltwater impacts. In: *The Oceans and Rapid Climate Change: Past, Present and Future, Geophysical Monograph* **126** (eds. D. Seidov, B.J. Haupt and M. Maslin), 147–167. Washington D.C.: American Geophysical Union.

49 Stocker, T.F., Knutti, R., and Plattner, G.-K. (2001). The future of the thermohaline circulation – a perspective. In: *The Oceans and Rapid Climate Change: Past, Present and Future, Geophysical Monograph* **126** (eds. D. Seidov, B.J. Haupt and M. Maslin), 277–293. Washington D.C.: American Geophysical Union.

50 Ganopolski, A. and Rahmstorf, S. (2001). Stability and variability of the thermohaline circulation of the past and future: a study with a coupled model of intermediate

complexity. In: *The Oceans and Rapid Climate Change: Past, Present and Future*, *Geophysical Monograph* **126** (eds. D. Seidov, B.J. Haupt and M. Maslin), 261–275. Washington D.C.: American Geophysical Union.

51 Little, M.G., Schneider, R.R., Kroon, D. et al. (1997). Trade wind forcing of upwelling, seasonality, and Heinrich events as a response to sub-Milankovitch climate variability. *Paleoceanography* 12 (4): 568–576.

52 Cane, M. and Clement, A.C. (1999). A role for the tropical Pacific coupled ocean-atmosphere system on Milankovitch and millennial timescales: Part II: global impacts. In: *Mechanisms of Global Climate Change at Millennial Time Scales*, *Geophysical Monograph* **112** (eds. P.U. Clark, R.S. Webb and L.D. Keigwin), 373–383. Washington D.C.: American Geophysical Union.

53 Alley, R.B., Anandakrishnan, S., Jung, P., and Clough, A. (2001). Stochastic resonance in the North Atlantic: further insights. In: *The Oceans and Rapid Climate Change: Past, Present and Future*, *Geophysical Monograph* **126** (eds. D. Seidov, B.J. Haupt and M. Maslin), 57–68. Washington D.C.: American Geophysical Union.

54 Barker, S., Knorr, G., Edwards, R.L. et al. (2011). 800 000 years of abrupt climate variability. *Science* 334: 347–351.

55 Siddall, M., Rohling, E.J., Blunier, T., and Spahni, R. (2010). Patterns of millennial variability over the last 500 ka. *Climate of the Past* 6: 295–303.

56 Tzedakis, P.C., Drysdale, R.N., Margari, V. et al. (2018). Enhanced climate instability in the North Atlantic and southern Europe during the last interglacial. *Nature Communications* 9: 4235. https://doi.org/10.1038/s41467-018-06683.

57 Kawamura, K., Abe-Ouchi, A., Motoyama, H. et al. (2017). State dependence of climatic instability over the past 720,000 years from Antarctic ice cores and climate modeling. *Science Advances* 3 (2): e1600446. https://doi.org/10.1126/sciadv.1600446.

58 Capron, E., Landais, A., Chappellaz, J. et al. (2010). Millennial and sub-millennial scale climatic variations recorded in polar ice cores over the last glacial period. *Climate of the Past* 6: 345–365.

59 Buizert, C., Sigl, M., Severi, M. et al. (2018). Abrupt ice-age shifts in southern westerly winds and Antarctic climate forced from the north. *Nature* 563: 681–685.

60 Buizert, C. and Schmittnet, A. (2015). Southern Ocean control of glacial AMOC stability and Dansgaard–Oeschger interstadial duration. *Paleoceanography* 30: 1595–1612. https://doi.org/10.1002/2015PA002795.

61 Ditlevsen, P.D., Kristensen, M.S., and Andersen, K.K. (2005). The recurrence time of Dansgaard–Oeschger events and limits on the possible periodic component. *Journal of Climate* 18 (14): 2594–2603. https://doi.org/10.1175/jcli3437.1.

62 Ditlevsen, P.D., Andersen, K.K., and Svensson, A. (2007). The DO-climate events are probably noise induced: statistical investigation of the claimed 1470 years cycle. *Climate of the Past* 3 (1): 129–134.

63 National Academies of Sciences, Engineering, and Medicine (NASEM) (2015). *A Strategic Vision for NSF Investment in Antarctic and Southern Ocean Research*. Washington, DC: The National Academies Press.

64 Shepherd, A., Fricker, H.A., and Farrell, S.L. (2018). Trends and Connections Across the Antarctic Cryosphere. *Nature* 558: 223–232.

65 Petersen, S.V., Schrag, D.P., and Clark, P.U. (2013). A new mechanism for Dansgaard–Oeschger cycles. *Paleoceanography* 28: 24–30. https://doi.org/10.1029/2012PA002364.

66 Van Ommen, T. (2015). Northern push for the bipolar see-saw. *Nature* 520: 630–631.

67 Rasmussen, T.L., Thomsen, E., and Moros, M. (2016). North Atlantic warming during Dansgaard–Oeschger events synchronous with Antarctic warming and out-of-phase with Greenland climate. *Scientific Reports* 6: 20535. https://doi.org/10.1038/srep20535.

68 Maier, E., Zhang, X., Abelmann, A. et al. (2018). North Pacific freshwater events linked to changes in glacial ocean circulation. *Nature* 559: 241–245.

69 Markle, B.R., Steig, E.J., and Buizert, C. (2016). Global atmospheric teleconnections during Dansgaard–Oeschger events. *Nature Geoscience* 10: 36–40.

70 Pedro, J.B., Jochum, M., Buizert, C. et al. (2018). Beyond the bipolar seesaw: toward a process understanding of interhemispheric coupling. *Quaternary Science Reviews* 192: 27–46.

71 Oeschger, H., Stauffer, B., Finkel, R., and Langway, C.C. (1985). Variations of the $CO_2$ concentration of occluded air and of anions and dust in polar ice cores. In: *The Carbon Cycle and Atmospheric CO2: Natural Variations Archean to Present*, *Geophysical Monograph* **32** (eds. E.T. Sundquist and W.S. Broecker), 132–142. Washington D.C.: American Geophysical Union.

72 Barnola, J.M., Raynaud, D., Korotkevitch, Y.S., and Lorius, C. (1987). Vostok ice core: a 160,000 year record of atmospheric $CO_2$. *Nature* 329: 408–414.

73 Genthon, C., Barnola, J.M., Raynaud, D. et al. (1987). Vostok ice core: the climate response to $CO_2$ and orbital forcing changes over the last climate cycle (160,000 years). *Nature* 329: 414–418.

74 Siegenthaler, U., Stocker, T.F., Monnin, E. et al. (2005). Stable carbon cycle-climate relationship during the late Pleistocene. *Science* 310: 1313–1317.

75 Loulergue, L., Schilt, A., Spahni, R. et al. (2008). Orbital and millennial-scale features of atmospheric $CH_4$ over the past 800,000 years. *Nature* 453: 383–386.

**76** Falkowski, P., Scholes, R.J., Boyle, E. et al. (2000). The global carbon cycle: a test of our knowledge of Earth as a system. *Science* 290: 291–296.

**77** Bereiter, B., Eggleston, S., Schmitt, J. et al. (2015). Revision of the EPICA Dome C $CO_2$ record from 800 to 600 kyr before present. *Geophysical Research Letters* 42: 542–549.

**78** Wolff, E.W. (2011). Greenhouse gases in the Earth System: a palaeoclimate perspective. *Philosophical transactions. Series A, Mathematical, physical, and engineering sciences* 369: 2133–2147.

**79** Yale (2001). Professor Barry Saltzman, a pioneer in the study of the atmosphere and climate, dies. *Yale Bulletin and Calendar* 29 (18).

**80** Maasch, K.A., Oglesby, R.J., and Fournier, A. (2005). Barry Saltzman and the theory of climate. *Journal of Climate* 18: 2142–2150.

**81** Maasch, K.A. and Saltzman, B. (1990). A low-order dynamical model of global climatic variability over the full Pleistocene. *Journal of Geophysical Research* 95: 1955–1963.

**82** Saltzman, B. and Maasch, K. (1990). A first-order global model of late Cenozoic climate change. *Transactions of the Royal Society of Edinburgh* 81: 315–325.

**83** Saltzman, B. and Maasch, K. (1991). A first-order global model of late Cenozoic climate change ii: a simplification of $CO_2$ dynamics. *Climate Dynamics* 5: 201–210.

**84** Saltzman, B. and Verbitsky, M. (1994a). Late Pleistocene climate trajectory in the phase space of global ice, ocean state, and $CO_2$: observations and theory. *Paleoceanography* 9: 767–779.

**85** Saltzman, B. and Verbitsky, M. (1994b). $CO_2$ and the glacial cycles. *Nature* 367: 418.

**86** Saltzman, B. (2002). *Dynamical Paleoclimatology: Generalized Theory of Global Climate Change*, 354 pp. London: Academic Press.

**87** Fischer, H., Wahlen, M., Smith, J. et al. (1999). Ice core records of atmospheric $CO_2$ around the last three glacial terminations. *Science* 283: 1712–1714. https://doi.org/10.1126/science.283.5408.1712.

**88** Plass, G.N. (1961). The influence of infrared absorptive molecules on the climate. *New York Academy of Sciences* 95: 61–71.

**89** Brook, E.J. (2013). Leads and lags at the end of the last ice age. *Science* 339: 1042–1043.

**90** Parrenin, F., Masson-Delmotte, V., Köhler, P. et al. (2013). Synchronous change of atmospheric $CO_2$ and Antarctic temperature during the last deglacial warming. *Science* 339: 1060–1063.

**91** Pedro, J.B., Rasmussen, S.O., and Van Ommen, T.D. (2012). Tightened constraints on the time-lag between Antarctic temperature and $CO_2$ during the last deglaciation. *Climate of the Past* 8: 1213–1221.

**92** Beeman, J.C., Gest, L., Parrenin, F. et al. (2019). Antarctic temperature and $CO_2$: near-synchrony yet variable phasing during the last deglaciation. *Climate of the Past* 15: 913–926.

**93** Stenni, B., Buiron, D., Frezzotti, M. et al. (2010). Expression of the bipolar see-saw in Antarctic climate records during the last deglaciation. *Nature Geoscience* 4: 46–49.

**94** Jakob, K.A., Wilson, P.A., Bahr, A. et al. (2016). Plio-Pleistocene glacial-interglacial productivity changes in the eastern equatorial Pacific upwelling system. *Paleoceanography* 31: 453–470. https://doi.org/10.1002/2015PA002899.

**95** Hain, M.P., Sigman, D.M., and Haug, G.H. (2014). Distinct roles of the Southern Ocean and North Atlantic in the deglacial atmospheric radiocarbon decline. *Earth and Planetary Science Letters* 394: 198–208.

**96** Indermühle, A., Monnin, E., Stauffer, B. et al. (2000). Atmospheric $CO_2$ concentration from 60 to 20 kyr BP from the Taylor Dome ice core, Antarctica. *Geophysical Research Letters* 27 (5): 735–738.

**97** Schmittner, A. and Galbraith, E.D. (2008). Glacial greenhouse-gas fluctuations controlled by ocean circulation changes. *Nature* 456: 373–376.

**98** Skinner, L.C., Fallon, S., Waelbroeck, C. et al. (2010). Ventilation of the deep Southern Ocean and deglacial $CO_2$ rise. *Science* 328: 1147–1151.

**99** Crowley, T.J. and North, G.R. (2011). *Paleoclimatology, Oxford Monographs on Geology and Geophysics* **18**, 339 pp. Oxford University Press.

**100** MacGilchrist, G.A., Naveira-Garabato, A.C., Brown, P.J. et al. (2019). Reframing the carbon cycle of the subpolar Southern Ocean. *Science Advances* 5: eaav6410.

**101** Rintoul, S.R., Hughes, C.W., and Olbers, D. (2001). The Antarctic circumpolar current system. In: *Ocean Circulation and Climate; Observing and Modelling the Global Ocean, International Geophysics Series* **77** (eds. G. Siedler, J. Church and J. Gould), 271–302. Academic Press.

**102** Meredith, M.P. and Brandon, M.A. (2017). Oceanography and sea ice in the Southern Ocean. In: *Sea Ice*, 3e (ed. D.N. Thomas), 216–238. Chichester: Wiley.

**103** Khatiwala, S., Schmittner, A., and Muglia, J. (2019). Air-sea disequilibrium enhances ocean carbon storage during glacial period. *Science Advances* 5, eaaw4981 12.

**104** Yin, Q.Z. and Berger, A. (2010). Insolation and $CO_2$ contribution to the interglacial climate before and after the mid-Brunhes event. *Nature Geoscience* 3: 243–246.

**105** Yin, Q.Z. and Berger, A. (2011). Individual contribution of insolation and $CO_2$ to the interglacial climates of the past 800 000 years. *Climate Dynamics* 38 (3): 7009–7724.

**106** Monnin, E., Indermuhle, A., Dallenbach, A. et al. (2001). Atmospheric $CO_2$ concentrations over the last glacial termination. *Science* 291: 112–114.

**107** Severinghaus, J.P., Sowers, T., Brook, E. et al. (1998). Timing of abrupt climate change at the end of the Younger Dryas interval from thermally fractionated gases in polar ice. *Nature* 391: 141–146.

**108** Ruddiman, W.F. and Raymo, M.E. (2003). A methane-based time scale for Vostok ice. *Quaternary Science Reviews* 22 (2–4): 141–155.

**109** Broecker, W.S., Kennett, J.P., Flower, B.P. et al. (1989). Routing of meltwater from the Laurentide ice sheet during the Younger Dryas cold episode. *Nature* 341: 318–321.

**110** Tarasov, L. and Peltier, W.R. (2005). Arctic freshwater forcing of the Younger Dryas cold reversal. *Nature* 435: 662–665.

**111** Firestone, R.B., West, A., Kennett, J.P. et al. (2007). Evidence for extraterrestrial impact 12,900 years ago that contributed to the megafaunal extinctions and Younger Dryas cooling. *Proceedings of the National Academy of Sciences of the United States of America* 104: 16016–16021.

**112** Barrows, T.T., Lehman, S.J., Fifield, L.K., and De Deckker, P. (2007). Absence of cooling in New Zealand and the adjacent ocean during the Younger Dryas Chronozone. *Science* 318: 86–89.

**113** Putnam, A.E., Denton, G.H., Schaefer, J.M. et al. (2010). Glacier advance in southern middle-latitudes during the Antarctic Cold Reversal. *Nature Geoscience* 3: 700–704.

**114** Kaplan, M.R., Schaefer, J.M., Denton, G.H. et al. (2010). Glacier retreat in New Zealand during the Younger Dryas Stadial. *Nature* 467: 194–197.

**115** Ackert, R.P., Becker, R.A., Singer, B.S. et al. (2008). Patagonian glacier response during the late Glacial-Holocene transition. *Science* 321: 392–395.

**116** Frakes, L.A., Francis, J.E., and Syktus, J.I. (1992). *Climate Modes of the Phanerozoic*, 274 pp. Cambridge University Press.

**117** Berger, A. (2009). Astronomical theory of climate change. In: *Encyclopedia of Paleoclimatology and Ancient Environments* (ed. V. Gornitz), 51–57. Dordrecht: Springer.

**118** Berger, W.H. and Wefer, G. (2003). On the dynamics of the ice ages: stage 11 paradox, mid-Brunhes climate shift, and 100-ky cycle. In: *Earth's Climate and Orbital Eccentricity: The Marine Isotope Stage 11 Question*, *Geophysical Monograph* **137** (eds. A. Droxler, R.Z. Poore and L.H. Burckle), 41–59. Washington, D.C.: American Geophysical Union.

**119** Berger, A. and Loutre, M.-F. (2003). Climate 400 000 years ago, a key to the future? In: *Earth's Climate and Orbital Eccentricity: The Marine Isotope Stage 11 Question*, *Geophysical Monograph* **137** (eds. A. Droxler, R.Z. Poore and L.H. Burckle), 17–26. Washington, DC: American Geophysical Union.

**120** Raynaud, D., Loutre, M.F., Ritz, C. et al. (2003). Marine isotope stage (MIS) 11 in the Vostok ice core: $CO_2$ forcing and stability of East Antarctica. In: *Earth's Climate and Orbital Eccentricity: The Marine Isotope Stage 11 Question*, *Geophysical Monograph* **137** (eds. A. Droxler, R.Z. Poore and L.H. Burckle), 27–40. Washington, D.C.: American Geophysical Union.

**121** Das Sharma, S., Ramesh, D.S., Bapanayya, C., and Raiju, P.A. (2012). Sea surface temperatures in cooler climate stages bear more similarity with atmospheric $CO_2$ forcing. *Journal of Geophysical Research – Atmospheres* 117 (D13): D13110. https://doi.org/10.1029/2012JD017725.

**122** Ridgwell, A.J., Watson, A.J., and Raymo, M.E. (1999). Is the spectral signature of the 100 kyr glacial cycle consistent with a Milankovitch origin? *Paleoceanography* 14 (4): 437–440.

**123** Masson-Delmotte, V., Kageyama, M., Baconnot, P. et al. (2006). Past and future polar amplification of climate change: climate model intercomparisons and ice-core constraints. *Climate Dynamics* 27: 437–440.

**124** Huybers, P. and Wunsch, C. (2005). Obliquity pacing of the late Pleistocene glacial terminations. *Nature* 434: 491–494.

**125** Berger, A. (2012). A brief history of the astronomical theories of paleoclimates. In: *Climate Change, Inferences from Paleoclimate and Regional Aspects* (eds. A. Berger, D. Mesinger and D. Sijacki), 107–129. Springer.

**126** Tzedakis, P.C., Crucifix, M., Mitsui, T., and Wolff, E.W. (2017). A simple rule to determine which insolation cycles lead to interglacials. *Nature* 542: 427–432.

**127** Huybers, H. (2007). Glacial variability over the last two million years: an extended depth-derived age model, continuous obliquity pacing, and the Pleistocene progression. *Quaternary Science Reviews* 26: 37–55.

**128** Berger, A. and Loutre, M.-F. (1991). Insolation values for the climate of the last 10 million years. *Quaternary Science Reviews* 10: 297–317.

**129** Huybers, P. and Denton, G. (2008). Antarctic temperature at orbital timescales controlled by local summer duration. *Nature Geoscience* 1: 787–792.

**130** Crucifix, M. (2019). Pleistocene glaciations. Chapter 3. In: *Climate Change in the Holocene, Impacts and Human Adaptation* (ed. E. Chiotis), 77–106. CRC Publishers, Taylor and Francis https://doi.org/10.1201/9781351 260244.

**131** Ganopolski, A., Winkelmann, R., and Schellnhuber, H.J. (2016). Critical insolation–$CO_2$ relation for diagnosing past and future glacial inception. *Nature* 529 (7585): 200–203. https://doi.org/10.1038/nature18452.

**132** Elkibbi, M. and Rial, J.A. (2001). An outsider's review of the astronomical theory of the climate: is the eccentricity-driven insolation the main driver of the ice ages? *Earth-Science Reviews* 56: 161–177.

**133** Ruddiman, W.F. (2013). *Earth's Climate, Past and Future*, 445 pp, 3e. New York: W.H. Freeman and Co.

**134** Hay, W.W. (2013). *Experimenting on a Small Planet – A Scholarly Entertainment*, 983 pp. New York: Springer.

**135** Gates, W.L. (1976). The numerical simulation of ice-age climate with a global general circulation model. *Journal of the Atmospheric Sciences* 33: 1844–1873.

**136** Manabe, S. and Hahn, D. (1977). Simulation of the tropical climate of an ice age. *Journal of Geophysical Research* 82 (C27): 3889–3911.

**137** Hansen, J.E., Lacis, A., Rind, D. et al. (1984). Climate sensitivity: analysis of feedback mechanisms. In: *Climate Processes and Climate Sensitivity*, *Geophysical Monograph* **29** (eds. J.E. Hansen and T. Takahashi), 130–163. Washington D.C.: American Geophysical Union.

**138** Saltzman, B., Hansen, A.G., and Maasch, K.A. (1984). The late Quaternary glaciations as the response of a three-component feedback system to earth orbital forcing. *Journal of the Atmospheric Sciences* 41: 3380–3389.

**139** Saltzman, B. and Moritz, R. (1980). A time-dependent climatic feedback system involving sea-ice extent, ocean temperature, and $CO_2$. *Tellus* 32: 93–118.

**140** Toggweiler, J.R., and Sarmiento, J.L. (1985) Glacial to interglacial changes in atmospheric carbon dioxide: the critical role of ocean surface water in high latitudes, in *The Carbon Cycle and Atmospheric $CO_2$: Natural Variations Archaean to Present* (eds E.T. Sundquist and W.S. Broecker), *Geophys. Monog.* 32, Proc. Chapman Conf., Tarpon Springs, Florida, January 9–13, 1984,163–184.

**141** Shackleton, N.J. and Pisias, N.G. (1985). Atmospheric carbon dioxide, orbital forcing, and climate. In: *The Carbon Cycle and Atmospheric $CO_2$: Natural Variations Archean to Present*, *Geophysical Monograph* **32** (eds. E.T. Sundquist and W.S. Broecker), 303–317. Washington D.C.: American Geophysical Union.

**142** Broccoli, A.J. and Manabe, S. (1987). The influence of continental ice, atmospheric $CO_2$, and land albedo on the climate of the Last Glacial Maximum. *Climate Dynamics* 1: 87–99.

**143** Broecker, W.S. and Denton, G.H. (1989). The role of ocean-atmosphere reorganization in glacial cycles. *Geochimica et Cosmochimica Acta* 53: 2465–2501.

**144** Raymo, M.E. (1997). The timing of major climate terminations. *Paleoceanography* 12 (4): 577–585.

**145** Pollard, D. (1982). A simple ice sheet model yields realistic 100 kyr glacial cycles. *Nature* 296: 334–338.

**146** Clement, A.C. and Cane, M. (1999). A role for the tropical Pacific coupled ocean-atmosphere system on Milankovitch and millennial timescales: Part I: a modelling study of tropical Pacific variability. In: *Mechanisms of Global Climate Change at Millennial Time Scales*, *Geophysical Monograph* **112** (eds. P.U. Clark, R.S. Webb and L.D. Keigwin), 363–371. American Geophysical Union.

**147** Kenyon, J. and Hegerl, G.C. (2010). Influence of modes of climate variability on global precipitation extremes. *Journal of Climate* 23: 6248–6262.

**148** Chavez, F.P., Ryan, J., Lluch-Cota, S.E., and Ñiquen, M. (2003). From anchovies to sardines and back: multidecadal change in the Pacific Ocean. *Science* 299: 217–221.

**149** Haug, G.H., Ganopolski, A., Sigman, D.M. et al. (2005). North Pacific seasonality and the glaciation of North America 2.7 million years ago. *Nature* 433: 821–825.

**150** Hays, J.D., Imbrie, J., and Shackleton, N.J. (1976). Variations in the Earth's orbit: pacemaker of the ice ages. *Science* 194 (4270): 1121–1132.

**151** Shackleton, N.J. (2000). The 100,000-year ice-age cycle identified and found to lag temperature, carbon dioxide, and orbital eccentricity. *Science* 289: 1897–1902.

**152** Ruddiman, W.F. (2003). Orbital insolation, ice volume, and greenhouse gases. *Quaternary Science Reviews* 22: 1597–1629.

**153** Yu, J.M. and Elderfield, H. (2007). Benthic foraminiferal B/Ca ratios reflect deep water carbonate saturation state. *Earth and Planetary Science Letters* 258: 73–86.

**154** Yu, J.M., Menviel, L., Jin, Z.D. et al. (2016). Sequestration of carbon in the deep Atlantic during the last glaciation. *Nature Geoscience* 9: 319–323.

**155** Munhoven, G. and François, L.M. (1996). Glacial and interglacial variability of atmospheric $CO_2$ due to changing continental silicate rock weathering: a model study. *Journal of Geophysical Research* 101 (D16): 21423–21437.

**156** Berger, A., Li, X.S., and Loutre, M.F. (1999). Modelling Northern Hemisphere ice volume over the last 3 Ma. *Quaternary Science Reviews* 18: 1–11.

**157** Hönisch, B., Hemming, N.G., Archer, D. et al. (2009). Atmospheric carbon dioxide concentration across the mid-Pleistocene transition. *Science* 324: 1551–1554.

**158** Snyder, C.W. (2016). Evolution of global temperature over the past two million years. *Nature* 538 (7624): 226–228. https://doi.org/10.1038/nature19798.

**159** Schmidt, G.A., Severinghaus, J., Abe-Ouchi, A. et al. (2017). Overestimate of committed warming. *Nature* 547: E16–E17.

**160** Snyder, C. (2017). Snyder replies. *Nature* 547: E17–W18.

**161** Sigman, D.M. and Haug, G.H. (2003). Biological pump in the past. In: *The Oceans and Marine Geochemistry, Treatise on Geochemistry* (ed. H.D. Holland and K.K. Turekian), vol. 6 (ed. H. Elderfield), 491–528. Oxford: Elsevier-Pergamon.

**162** Keeling, R. and Stephens, B.B. (2001). Antarctic Sea ice and the control of Pleistocene climate instability. *Paleoceanography* 16: 112–131, and *Paleoceanog.* 16 (3): 330–334.

**163** Dokken, T., Nisancioglu, K.H., Li, C. et al. (2013). Dansgaard–Oeschger cycles: interactions between ocean and sea ice intrinsic to the Nordic Seas. *Paleoceanography* 28 (3): 491–502.

**164** Brovkin, V., Ganopolski, A., Archer, D., and Rahmstorf, S. (2007). Lowering of glacial atmospheric $CO_2$ in response to changes in oceanic circulation and marine biogeochemistry. *Paleoceanography* 22: PΛ4202. https://doi.org/10.1029/2006PA001380.

**165** Crucifix, M. (2009). Modeling the climate of the Holocene. In: *Natural Climate Variability and Global Warming: A Holocene Perspective* (eds. R.W. Battarbee and H.A. Binney), 98–122. Chichester and Oxford: Wiley Blackwell.

**166** Paillard, D. (2001). Glacial cycles: towards a new paradigm. *Reviews of Geophysics* 39: 325–346.

**167** Lisiecki, L.E. (2010). Links between eccentricity forcing and the 100 000-year glacial cycle. *Nature Geoscience* 4: 349–352.

**168** Toggweiler, J.R. (2008). Origin of the 100,000-year timescale in Antarctic temperatures and atmospheric $CO_2$. *Paleoceanography* 23: PA2211. https://doi.org/10.1029/2006PA001405.

**169** Chandranath Basak, C., Henning Fröllje, H., Lamy, F. et al. (2018). Breakup of last glacial deep stratification in the South Pacific. *Science* 359: 900–904.

**170** Rae, J.W.B., Burke, A., Robinson, L.F. et al. (2018). CO2 storage and release in the deep Southern Ocean on millennial to centennial timescales. *Nature* 562: 569–573.

**171** Kawamura, K., Parrenin, F., Lisiecki, L. et al. (2007). Northern Hemisphere forcing of climatic cycles in Antarctica over the past 360 000 years. *Nature* 448: 972–977.

**172** Huybers, P. (2009). Antarctica's orbital beat. *Science* 325: 1085–1086.

**173** Wolff, E.W., Fischer, H., and Röthlisberger, R. (2009). Glacial terminations as southern warmings without northern control. *Nature Geoscience* 2: 206–209.

**174** Sigman, D.M., Hain, M.P., and Haug, G.H. (2010). The polar ocean and glacial cycles in atmospheric $CO_2$ concentration. *Nature* 466: 47–55.

**175** He, F., Shakun, J.D., Clark, P.U. et al. (2013). Northern Hemisphere forcing of Southern Hemisphere climate during the last deglaciation. *Nature* 494: 81–85.

**176** Crowley, T.J. (2011). Proximal trigger for late glacial Antarctic circulation and $CO_2$ changes. *PAGES News* 19 (2): 70–71.

**177** Abe-Ouchi, A., Saito, F., Kawamura, K. et al. (2013). Insolation-driven 100 000-year glacial cycles and hysteresis of ice-sheet volume. *Nature* 500: 190–194.

**178** Marshall, S.J. (2013). Solution proposed for ice-age mystery. *Nature* 500: 159–160.

**179** Ganopolski, A. and Calov, R. (2011). The role of orbital forcing, carbon dioxide and regolith in 100 kyr glacial cycles. *Climate of the Past* 7: 1415–1425.

**180** Castellano, E., Becagli, S., Jouzel, J. et al. (2004). Volcanic eruption frequency in the last 45 ky as recorded in EPICA-Dome-C ice core (East Antarctica) and its relationship to climate changes. *Global and Planetary Change* 42: 195–205.

**181** Huybers, P. and Langmuir, C. (2009). Feedback between deglaciation, volcanism and atmospheric $CO_2$. *Earth and Planetary Science Letters* 286: 479–491.

**182** Tolstoy, M. (2015). Mid-ocean ridge eruptions as a climate valve. *Geophysical Research Letters* 42 (5): 1346–1351.

**183** Schindlbeck, J.C., Jegen, M., Freundt, A. et al. (2018). 100- kyr cyclicity in volcanic ash emplacement: evidence from a 1.1 Myr tephra record from the NW Pacific. *Scientific Reports* 8: 4440.

**184** WAIS Divide Project Members (2015). Precise interpolar phasing of abrupt climate change during the last ice age. *Nature* 520: 661–665.

**185** Collins, L.G., Pike, J., Allen, C.S., and Hodgson, D.A. (2012). High-resolution reconstruction of Southwest Atlantic Sea-ice and its role in the carbon cycle during marine isotope stages 3 and 2. *Paleoceanography* 27: PA3217.

**186** WAIS Divide Project Members (42 authors) (2013). Onset of deglacial warming in West Antarctica driven by local orbital forcing. *Nature* 500: 440–446.

# 14

# The Holocene Interglacial

## 14.1 Holocene Climate Change

For most of the past 12 000 years we have been living in the Holocene, which has taken us up to the Anthropocene of very recent time (see Chapter 15). If we want to get some idea of how our climate may evolve into the future, we need to see how it developed over the 11 700 years since the end of the Younger Dryas cold event. A recent review by John Birks, of the Bjerknes Centre for Climate Research of the University of Bergen, explored the early stirrings of inquisitiveness about Holocene climate change [1]. Natural historians noted back in the late eighteenth century *'the impressive occurrence of large fossil trunks and stumps (megafossils) of pine trees buried in peat bogs in northwest Europe'* [1]. For example, a Mr. H. Maxwell observed in 1815 that *'one of the greatest enigmas of natural science is presented in the remains of pine forest buried under a dismal treeless expanse on the Moor of Rannoch, and on the Highland hills up to and beyond 2000 feet altitude'* [1]. Similar changes in Denmark were attributed in the late nineteenth century to changes in moisture and to cooling in the post-glacial period. Norwegian botanist Axel Blytt (1843–1898) interpreted tree layers in peat bogs and changes from dark, humified peat to pale, fresh peat as evidence for alternations between dry and wet periods. Swedish botanist

Rutger Sernander (1866–1944) added ideas about summer temperature changes to propose the famous four Blytt–Sernander periods of post glacial time: Boreal (warm, dry) from 10 000 to 8000 years ago; Atlantic (warmest, wet) from 8000 to 5000 years ago; Sub-Boreal (warm, dry) from 5000 to 2500 years ago; and Sub-Atlantic (cool, wet) from 2500 years ago to the present. These were preceded by a Pre-Boreal period (cool, sub-Arctic) prior to 10 000 years ago. This Eurocentric scheme *'became the dominant paradigm for Holocene climate history'* during the early part of the twentieth century [1].

Birks also reminded us that another late nineteenth century scientist, Swedish botanist Gunnar Andersson (1865–1928), whilst searching peat bogs, had discovered plant fossils including hazel nuts, well north of their current range, suggesting that the climate had once been warmer than today. This led Andersson to present the idea of a gradually rising temperature curve reaching a long early to mid Holocene period[1] of temperature warmer than today

[1] In 2019, the International Commission on Stratigraphy formally agreed that the Holocene should be subdivided into three units: Greenlandian (11 700–8200 years before present, or BP), Northgrippian (8200–4200 years BP), and Meghalayan (4200 years BP to present). For convenience here I refer to these broadly as early, middle (or mid) and late.

*Palaeoclimatology: From Snowball Earth to the Anthropocene*, First Edition. Colin P. Summerhayes.
© 2020 John Wiley & Sons Ltd. Published 2020 by John Wiley & Sons Ltd.

(a thermal maximum), followed by a subsequent decline. The notion of a Holocene thermal maximum seems to have arisen independently in the minds of other natural historians, including the Scottish palaeontologist Thomas F. Jamieson (1829–1913), based on his study of the molluscan fauna of mid Holocene estuarine clays in Scotland [1].

Early in the twentieth century, a Swedish geologist, Lennart Von Post (1884–1951) proposed that '*pollen analysis ... [be used] as a technique for relative dating and for reconstructing past vegetation and past climate*' [1]. In contrast to megafossils and macrofossils, it could provide a continuous record of changing vegetation and climate. Von Post integrated the Blytt and Salander phases of climate with Andersson's notion of gradual climate change and the findings of his own pollen analyses. This led to pollen analysis being used as a key means of determining Holocene geological time and climate change prior to about 1960. Quoting Ed Deevey, Birks told us that '*Von Post's simple idea that a series of changes in pollen proportions in accumulating peat was a four-dimensional look at vegetation, must rank with the double-helix as one of the most productive suggestions of modern times*' [1]. So it should be no surprise that for his contributions Von Post was awarded the Vega Medal of the Swedish Anthropological and Geographical Society in 1944.

Holocene climate change was the stuff of particular fascination for H.H. Lamb, whom we met in Chapter 8. In his 1966 book *The Changing Climate* (written in 1964) Lamb produced a chronicle of climate change for the Holocene [2]. The first version of this chronicle, dated 1959, began with the disappearance of the last major ice sheet from Scandinavia between 10 and 9 Ka ago, followed by warming to a post-glacial 'climatic optimum' of 6–4 Ka ago, when he thought that world temperatures were 2–3 °C warmer than now. The warming roughly coincided with a 'sub-pluvial' period in North Africa between 7.0 and 4.4 Ka ago, when there were settlements in the Sahara. As well as archaeological evidence for the presence of humans there, the development of major river systems draining from the interior down to the Atlantic coast of the Sahara, now almost completely dry, attests to a formerly much more humid climate. These rivers include the Oued Souss (reaching the coast at about 30° 30′N) and the Oued Dra (reaching the coast at about 28° 30′N), which are largely dry, and the Seguia del Hamra (reaching the coast at 27° 30′N), which is almost always dry [3]. The 'climatic optimum' was well known to students of the Holocene, and was recognized in deep-sea cores by Emiliani in 1955, for instance [4].

Lamb found that subsequent cooling and increased rainfall in Europe were followed by a drier and warmer climate during the Roman Era (250 BCE – 400 CE), and by a second, Medieval, climatic optimum between 400 and 1200 CE, with a peak around 800–1000 CE. The climate then cooled into a period that Lamb referred to as the Little Ice Age (a term first introduced by Matthes in 1939, and used to describe an '*epoch of renewed but moderate glaciation which followed the warmest part of the Holocene*' [5, 6]). The cold winters of this period, which Lamb dated to between 1550 and 1850 CE, were immortalized by Pieter Brueghel the Elder in his February 1565 painting *Hunters in the Snow*. We now know that following the Holocene Climatic Optimum, glaciers began to readvance. George Denton of the University of Maine referred to this new period of glacial advance (giving rise to the Little Ice Age of Matthes) as the 'Neoglaciation' [7, 8]. It incorporates the narrower range of Lamb's Little Ice Age. We will explore these changes of the past 2000 years in some detail in Chapter 15.

André Berger's astronomical calculations show that insolation is dominated by obliquity at high latitudes, although precession does play an important subordinate role there especially when eccentricity is high (Figure 12.2). It is also clear from Figure 12.2 that during the Holocene insolation peaked around 9–10 Ka ago. Yin and Berger confirmed that annual irradiance was basically the same then at high latitudes in both hemispheres, increasing to a maximum between 11.7 and 7.5 Ka ago with a peak around 9 Ka ago [9]. For the Northern Hemisphere, this peak 9 Ka ago was a product of high obliquity and of the Sun being at perihelion in Northern Hemisphere summer in response to precession (see Figures 12.2 and 14.1) [10]. In contrast, Figure 14.1 shows that the Southern Hemisphere summer experienced less insolation [10]. Although that might suggest that the Southern Hemisphere should have been cool, Figure 13.22 explains that it was in fact warm because the southern summer (being at aphelion), was several days longer than the northern hemisphere one [11, 12]. These basic astronomical data help to explain what caused the mid Holocene climatic optimum. The optimum was somewhat earlier in the south than in the north because the melting of the great Northern Hemisphere ice sheets at the beginning of the Holocene used up available solar energy, keeping the region cool until the ice sheets had gone.

As we know from Chapter 6, astronomical calculations of the influence of the other planets on the Earth's orbit and axial tilt enable us to determine insolation well out into the future (Figure 12.2). Berger's and Loutre's calculations [13] show that the Earth should have cooled from 10 000 years ago to the present, and will remain cool for the next 5000 years (Figure 12.2). Given their data, we can confidently predict that the present low level of summer

Insolation anomaly (W/m²)

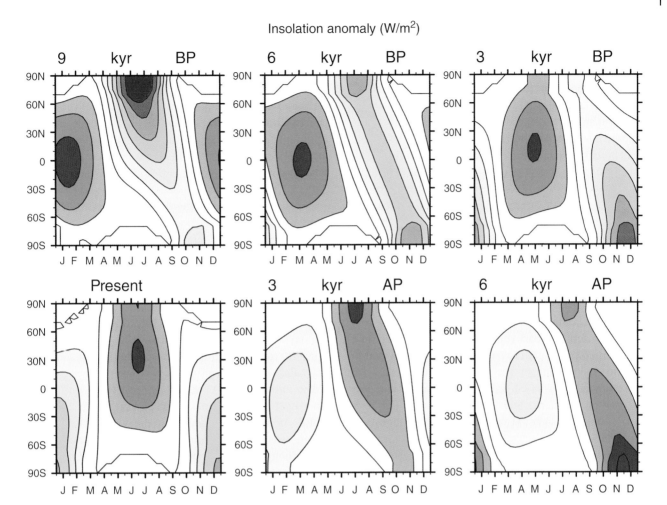

**Figure 14.1** Distribution of shortwave radiation (insolation) received from the Sun at the top of the atmosphere between 9 Ka ago (Before Present, or BP) and 6 Ka into the future (After Present, or AP). A mean distribution of insolation assuming no eccentricity and a mean obliquity of 23° 20′ was subtracted from the annual insolation in order to highlight the effects of changes in precession and obliquity. Precession redistributes heat across the seasons (positive anomalies around July 9 Ka ago in the north and around January at present in the south). The decrease in obliquity during the Holocene reduces summer insolation in both hemispheres from 9 Ka ago onwards. *Source:* From figure 4.4 in Ref. [10].

insolation at 65°N, which has persisted for the past 2000 years (Figure 12.2), will remain more or less unchanged for at least another 5000 years. And as we know from the work of Huybers [11, 12] (see Figure 13.20) the patterns of warming and cooling in both hemispheres are intimately related. As we learned in Chapter 13, the two hemispheres are also connected through global changes in sea level, which are triggered by changes in ice volume driven by changes in insolation, and through the Atlantic Meridional Overturning Circulation (AMOC), which takes warm salty water north at the surface and returns cold salty

water south at depth as North Atlantic Deep Water (Figures 12.5 and 12.6).

Figure 14.1 shows how insolation changed at both poles and in the tropics as Earth moved from the Holocene thermal maximum 9 Ka ago to the present [10]. Precession caused a gradual redistribution of insolation away from the Northern Hemisphere as perihelion in summer moved towards the Southern Hemisphere. The few thousand years after present show just slight changes in insolation (consistent with Figure 12.2), with progressively decreased insolation (cooling) in the south as precession moves Earth back towards aphelion (cold) in the southern summer (Figure 14.1).

Lamb's work was succeeded by the CLIMAP project to map the climate of the Last Glacial Maximum (see Chapter 12), which in turn stimulated formation of a successor project, COHMAP, to look at the climate of the Holocene. Starting out as 'Climates of the Holocene – Mapping Based on Pollen Data', COHMAP quickly became global both in its scope and in the climate proxies it considered. Renamed the 'Cooperative Holocene Mapping Project', it ran from 1977 to 1995 and was a turning point for the palaeoclimate community studying the Holocene. It tied models and proxy data of the Holocene together for the first time, and stimulated international collaboration in Holocene research [1]. Key organizers were John Kutzbach, whom we met in Chapter 6, and his colleagues Tom Webb, and Herb Wright. They aimed to use the Community Climate Model, an atmospheric General Circulation Model of the US National Centre for Atmospheric Research (NCAR), to simulate past climates in 3000 year long time slices centred on 18, 15, 12, 9, 6, 3 Ka ago and 0 Ka, and to compare the outputs of their model simulations with palaeoclimate maps based on data from pollen, lake levels, pack-rat middens, and marine plankton [14, 15]. These simulations ignored events of short duration, like Bond cycles.

Berger and Loutre's solar radiation parameters were input to the COHMAP model as an external forcing variable. Other boundary conditions within the model provided internal climate forcing from changes in aerosol loading (dust), ice volume, sea surface temperature (SST) and atmospheric $CO_2$. Those conditions were not allowed to change between 18 and 15 Ka ago. After 15 Ka ago, dust was decreased to around present levels by about 12 Ka ago, and ice volume to around present levels by 9 Ka ago (there was still some ice on North America and Scandinavia at that time). SST increased to present levels and $CO_2$ to preindustrial levels between 15 and 9 Ka ago, with $CO_2$ increasing again in modern times [14].

COHMAP's results, which started appearing in 1988, were stunning [14, 15]. The combination of climate data and models enabled palaeoclimatologists to understand for the first time what caused global climate changes to occur at different times between the last glacial and the present, and how these changes were distributed across different regions. Previously, the role of the Earth's orbital variations had been discounted, the full impact of changes in ice sheets had not been understood, and climate models had not been well enough developed to test the influence of these and other factors.

Changes in the precession and tilt cycles made seasonality increase in the Northern Hemisphere and decrease in the Southern Hemisphere between the 15 and 9 Ka ago time slices. By 9 Ka ago, the Northern Hemisphere received 8% more solar radiation in July and 8% less in January compared with today. After 9 Ka ago, these extremes diminished towards modern values. These changes increased the thermal contrast between land and ocean, so causing strong monsoons during the Holocene thermal optimum between 12 and 6 Ka ago in the northern tropics and subtropics, particularly in Africa and Asia, and raising lake levels in regions that are arid today, such as the Sahara. Crocodiles and hippos expanded into the Saharan region, and people settled on the lake shores there. At the same time, summers became warm and dry in the northern interiors.

After 6 Ka ago, as summer insolation continued to decrease (Figure 12.2, 14.1), temperatures fell over the land, monsoon rains weakened, deserts expanded and the present climate regime developed. In the southern tropics, the changed insolation (more in July, less in January) had the opposite effect, with less seasonality and less intense rains in tropical South America, southern Africa and Australia.

The initial COHMAP model suggested that during the glacial period (the 18 and 15 Ka ago time slices), the large North American ice sheet split the westerly winter jet stream over North America into a northerly branch running along the Arctic shore then down the Labrador Sea between Canada and Greenland, and a southerly branch running over the American southwest and joining the northerly branch over the mid Atlantic, before heading for Europe. A later, more advanced version of the model, along with a mixed layer ocean model and an ice sheet reconstruction that gave a lower height to the Laurentide ice sheet, found that the ice sheet did not split the jet stream at 18 Ka ago [16]. Apart from that, the results seemed robust, with good coherence between the outputs from the earlier and later versions of the model and between both models and the palaeoclimate data.

The COHMAP data and models showed that there is no simple description of Holocene climate. Orbital change forced climate changes that varied across the globe not only in magnitude but also spatially and in time. Patterns of variation in both precipitation and seasonality changed along with temperature. Lamb's European-based climate reconstruction was not globally representative.

As a result of the increase in insolation, which peaked around 10 Ka ago (Figure 12.2), and the subsequent absorption of heat to melt the Northern Hemisphere ice sheets, temperatures remained cool in Europe until about 9 Ka ago, when the combination of moderately high insolation and much reduced ice sheets created the thermal Holocene Climatic Optimum between 9 and 5 Ka ago; it is also known as the 'Holocene Thermal Maximum' and the 'hypsithermal', and was delayed in relation to the insolation maximum. In the Arctic, at 70°N, June insolation 11 Ka ago was about 45 W/m$^2$ larger than today, and by 4 Ka ago it had dropped to about 15 W/m$^2$ larger than today [17]. During the optimum, temperatures were up to 4°C above later Holocene levels there, but while northwest Europe warmed, southern Europe cooled and there was little or no change in the tropics. This so-called 'event' was not globally uniform in either magnitude or timing.

Overall, the decrease in insolation from 10 Ka to the present cooled the Northern Hemisphere significantly. For example, SSTs off Cap Blanc, Mauretania, declined by between 4 and 6°C during that period. There the input of dust rose abruptly at around 5.5 Ka ago when the African Humid Period came to an abrupt end [18, 19]. This is an example of a 'tipping point': an abrupt response emerging from a gradual change. During the African Humid Period around 6 Ka ago a wide belt of enlarged lakes had occupied the tropics between 32°N and 18°S, from which we can infer that the Intertropical Convergence Zone (ITCZ) then lay much further north than it does now, greatly enhancing the monsoonal transport of moisture into the tropical continents at that time [20]. As the climate cooled, the ITCZ moved south and the lakes disappeared.

As more data poured in from tree rings, corals, ice cores, stalactites, and sediment cores from lakes and oceans, the Holocene climate record was further refined. In 2000, Richard Alley used ice isotopic ratios to calculate a profile of late glacial and Holocene temperature from the Greenland's GISP2 ice core (his figure 1), which showed an irregular pattern of warming and cooling about an *increasing warming trend* towards the present from the end of the Younger Dryas period [21]. More recently, Sigfus Johnsen of the University of Copenhagen's Niels Bohr Institute and his team reassessed Greenland's Holocene climatic history using ice core data from six sites (Camp Century, Dye-3, GRIP, GISP2, Renland, and NorthGrip) [22]. His team found that oxygen isotopes in the GRIP and NorthGRIP ice cores fluctuated by about 1‰ about trend lines that showed a *very slight decline* over the past 10 Ka, following a strong increase in trend indicative of the rapid warming of the earliest Holocene at the end of the Younger Dryas cold interval (Figure 14.2). The isotopic data also clearly showed up a major cold event centred on about 8.2 Ka ago (Figure 14.2). Subtle differences between the GRIP and NorthGRIP records reflects the fact that NorthGRIP lies at a slightly lower altitude than GRIP, and some 4° of latitude further north (at 75°N), so would be subject to a different climate.

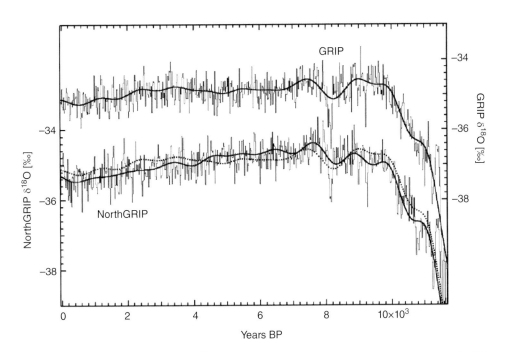

**Figure 14.2** Twenty-year resolution $\partial^{18}O$ profiles from the GRIP and NorthGRIP ice cores for the Holocene. The solid curves represent a 1 Ka low-pass filtered version of the data sets to show the long-term trends. The dotted curve is the GRIP 1-Ka low-pass data shown together with the NorthGRIP data to highlight the slight difference between them. The long-term trends are similar for both cores except for the interval from 8.5 to 4.2 Ka, the climatic optimum, where the NorthGRIP data are 0.6 more positive than in the GRIP profile. *Source:* From figure 7 in Ref. [22].

Although it is likely that the rapid ups and downs in the isotopic record in Figure 14.2 represent alternation between warm and cool periods throughout the Holocene from 10 Ka ago to the present, it would be misleading to suppose that the isotopic data could be translated simply as a temperature record. An indication of the likely actual temperatures at the surface through time comes from the application, to the isotopic record, of the borehole temperatures, which preserve surface temperatures as snow continues to accumulate [22]. These data suggest that Greenland's climate warmed to about 7.5 Ka ago then remained static until a cooling trend (Denton's 'neoglacial') set in about 4 Ka ago [22].

However, by 2009 that interpretation had been called into question by Bo Vinther of the University of Copenhagen's Niels Bohr Institute [23]. Examining ice cores from Greenland and Arctic Canada, Vinther's team saw that changes in altitude and past thinning of the ice sheet, both caused by warming, had shaped the $\partial^{18}O$ record that earlier researchers had used for their temperature reconstructions. '*Contrary to the earlier interpretation of $\partial^{18}O$ evidence from ice cores*' Vinther said, '*our new temperature history reveals a pronounced Holocene climatic optimum in Greenland coinciding with maximum thinning near the GIS [Greenland Ice Sheet] margins*' [23]. Their new record of Greenland's temperature history (Figure 14.3) shows a steep warming of around 6 °C between 12 and 10 Ka ago, followed by a slight rise by a further 0.5 °C to a peak around 8 Ka ago (with a sharp, brief, cooling event [recognized globally] at around 8.2 Ka ago), then a steady fall of around 2.5 °C to between 1600 and 1850 CE (the Little Ice Age), followed by a rise into the modern era. Evidently, the broad pattern of rise and fall in temperature for Greenland generally followed the pattern of Northern Hemisphere summer insolation.

Incidentally, Vinther's data show that the Greenland ice sheet responds much more rapidly to warming than had formerly been supposed, making it '*entirely possible that a future temperature increase of a few degrees Celsius in Greenland will result in GIS [Greenland Ice Sheet] mass loss and contribution to sea level change larger than previously projected*' [23].

The fall in Greenland temperature over the past 4 Ka was further confirmed by a team led by Takuro Kobashi of the National Institute of Polar Research in Tokyo, and including Bo Vinther and Sigfus Johnsen of the Niels Bohr Institute in Copenhagen. Analysing the isotopic composition of nitrogen and argon in air trapped in the GISP2 ice core from near the summit of the ice sheet they found a long term 1.5 °C cooling over that period, leading to a current average of −29.9 °C for 2001–2010 [24]. Superimposed

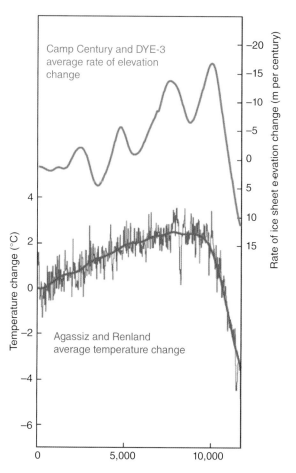

**Figure 14.3** Greenland temperatures adjusted for changing elevation. Temperature change in Greenland derived from Agassiz and Renland $\partial^{18}O$ records, and average rate of elevation change of ice sheets at the DYE-3 and Camp Century drill sites. Elevation changes distort the temperature records for these drill sites. *Source:* From figure 2c in Ref. [23].

on the cooling trend were brief centennial events reaching as much as 2 °C above or below the trend. One of the largest was a 2-century-long warming event centred on CE 700, and separated by a similarly long cooling event from the warmth of the Medieval Warm Period of CE 1050–1200 [24] (for more on the climate of the past 2000 years see Chapter 15).

The existence of an early to mid Holocene thermal maximum is well documented in proxy records of SST from the high latitude North Atlantic and the Nordic Seas, where polar amplification is clear [25]. The SST maximum there was forced by the summer insolation maximum, and is not simply due to advection of SST anomalies from further south. Superimposed on the SST trends is variability at the century or millennial scale, the amplitude of which increased after the mid Holocene. There is very little persistence in the main frequencies of variability through the Holocene. The increased climatic variability affected primarily wintertime conditions, such as sea ice, and was consistent with the onset of neoglacial conditions in Europe. According to Eystein Jansen and his team, '*Increased sea ice cover following the reduced summer insolation may have put in place amplification mechanisms leading to stronger ocean temperature variability .... In the absence of a clear attribution of this variability to external forcings (e.g. solar, volcanic) ... it appears most likely that the century to millennial scale variability is primarily caused by the long time-scale internal dynamics of the climate system*' [25]. We will look at this variability in more detail later on.

The Arctic tree line also changed, with boreal forests advancing to near the Arctic coast between 9 and 7 Ka ago, during the Holocene climatic optimum, when temperatures there were 2.5–7 °C warmer than today, before retreating to its present position between 4 and 3 Ka ago [26]. The retreat, leaving large numbers of ancient tree stumps on the tundra of northern Siberia, coincided with declining summer insolation, cooling Arctic Ocean water and neoglaciation. The establishment of forests contributed to Holocene warming in northern Eurasia by lowering high latitude albedo and influencing the position of the Arctic Front [26]. The southward progression of the tundra would have raised the albedo, providing positive feedback to enhance cooling.

Polar amplification comes about for a variety of reasons, not least the reduction in sea ice and land ice and snow cover, which reduces albedo and allows the exposed ocean to absorb heat, which is then given up to the atmosphere, which causes more sea ice to be lost and so on, through positive feedback. Gifford Miller of the Institute of Arctic and Alpine Research of the University of Colorado, Boulder, explored this phenomenon in some detail in the Arctic in 2010 [27]. He and his team found that Arctic temperature change exceeded the Northern Hemisphere average by a factor of 3 to 4. For example, warming compared to today during the Holocene Thermal Maximum was $+1.7 \pm 0.8$ °C (Arctic), $+0.5 \pm 0.3$ °C (Northern Hemisphere) and $0 \pm 0.5$ °C (global). During the last interglacial it was $+5 \pm 1$ °C (Arctic) and $+1 \pm 1$ °C (global and Northern Hemisphere). And during the mid Pliocene it was $+12 \pm 3$ °C (Arctic) and $+4 \pm 2$ °C (global and Northern Hemisphere) [27].

By 2017, the Arctic's Holocene temperature history had been re-evaluated in some detail by Benoit Lecavalier and colleagues using an ice core from the Agassiz ice cap on Ellesmere Island. Their thermal history reconstruction found that the Holocene thermal maximum had been earlier and 4–5 °C warmer than previous reconstructions

suggested (with reference to the preindustrial value in 1750), and that temperatures regularly exceeded contemporary values for a period of ~3000 years then [28]. Their results showed that '*air temperatures in this region are now at their warmest in the past 6,800–7,800 [years], and that the recent rate of temperature change is unprecedented over the entire Holocene*' [28]. The early Holocene warming caused northwest Greenland to lose a 1 km thickness from its ice sheet [28].

A comprehensive recent survey of Holocene climate variability, based on some 50 globally distributed palaeoclimate records from a wide range of environments extending from Greenland to Antarctica, was carried out in 2004 by a team led by Paul Mayewski of the Climate Change Institute of the University of Maine at Orono [29]. In recognition of his extensive research on climate change in Greenland and Antarctica, Mayewski was awarded the Seligman Crystal of the International Glaciological Society in 2009, and the science medal of the Scientific Committee on Antarctic Research in 2006, and has an Antarctic peak named after him.

Mayewski's team found clear signs of cooling in line with the decline in summer insolation in the Northern Hemisphere from about 7400 years ago to the present [29]. Norwegian glaciers advanced, the Swedish tree line moved southwards, and temperatures (from $\partial^{18}O$) fell in Soreq Cave, Israel. Africa's Lake Victoria and Ethiopia's Lake Ahhe began shrinking, and the Trade Winds began to weaken over Venezuela's Cariaco Basin. Peru's Huascaran ice cap and Antarctica's Taylor Dome began cooling, as did SSTs in Namibia's Benguela Current.

In 2006, Stephan Lorenz, then at the Max-Planck-Institut für Meteorologie, Hamburg, used alkenone data and a coupled ocean–atmosphere circulation model forced by orbital changes, to examine global patterns of SST for the past 7000 years [30]. As in COHMAP, the patterns proved to be heterogeneous. Whilst the higher latitudes cooled over time, the tropics warmed slightly. In the North Atlantic region many aspects of the regional climate are dictated by the behaviour of the North Atlantic Oscillation (NAO), measured from the difference in the air pressure between the high latitude Iceland low-pressure centre and the low latitude Azores high-pressure centre. When the pressure difference is large, the NAO is positive, westerly winds are strong, Europe has cool summers and mild, wet winters and the Mediterranean area is dry and cool. This pressure difference has decreased towards the present, making the NAO more negative, which weakened the westerlies, making European winters colder and dryer, and the Mediterranean warmer and wetter. This trend is consistent with the observation that the Little Ice Age late in the Holocene in Europe was characterized by very cold winters (more on this in Chapter 15). The trend in the NAO operated *against* the trend in orbital insolation, which *increased* in winter in the Northern Hemisphere – a good example of how local effects can override Milankovitch variations. Overall, summer insolation in the Northern Hemisphere over the past 7000 years decreased by more than 30 W/m² at middle and northern latitudes, whilst winter insolation increased by about 25 W/m² at low latitudes (Figure 14.1). Lorenz's study noted that '*Northern Hemisphere summer cooling during the Holocene [meaning the last 7,000 years] is of the same order of magnitude as the warming trend over the last 100 years*' [30] (my emphasis).

Several more extensive studies of Holocene climate change were later carried out by Heinz Wanner (1945–), the former Director of the Oeschger Centre for Climate Change Research at the University of Bern in Switzerland, which was named after the ice core expert Hans Oeschger. Among other things, in 2011 Wanner and his colleagues used high resolution records of Holocene climate change to produce a Holocene Climate Atlas containing 100 anomaly maps representing 100 year averages of climatic conditions for the last 10 Ka [31]. In recognition of his contributions to climate change research, he was awarded the Vautrin Lud Prize – regarded unofficially as the Nobel Prize in Geography – in 2006.

In 2008, Wanner and his colleagues published a comprehensive review of mid to late Holocene climate change spanning the last 6000 years [32]. They chose this period because (as Wanner later showed in 2011) '*the boundary conditions of the climate system did not change dramatically*' during it [32]. The large continental Northern Hemisphere ice sheets had melted, so there were no large outflows of fresh water from melting ice sheets, nor any major changes in sea level. Plus there were abundant, detailed regional palaeoclimatic proxy records. The team mapped the distribution of temperature and precipitation through time, supplementing the data with results from GCMs and Earth System Models of Intermediate Complexity (EMICs) fed with data on the agents forcing climate change: orbital variations, solar variations, large volcanic eruptions, and changes in land cover and greenhouse gases. Their goal was '*to establish a comprehensive explanatory framework for climate changes from the Mid-Holocene (MH) to pre-industrial time*' [32]. One of their sources of information was the international Palaeoclimate Modelling Intercomparison Project (PMIP), which started in the early 1990s in an effort to improve palaeoclimate models. Another source was the IGBP's Palaeovegetation Mapping Project (BIOME 6000), which provides a global dataset derived from pollen and plant fossils for use as a

benchmark against which to test the outputs from palaeoclimate models.

In 2013, Wanner was part of another team that produced a detailed analysis of the global climate of the past 2000 years [33] as a contribution to the IGBP's Past Global Changes (PAGES) programme, a successor to COHMAP. One of the PAGES subgroups, the '2k network', aims to produce a global array of regional climate reconstructions for the past 2000 years. It coordinates with the NOAA World Data Center for Paleoclimatology to maintain a benchmark database of proxy climate records for that period [33]. By 2013, the PAGES 2k dataset included 511 time series of tree rings, pollen, corals, lake and marine sediments, glacier ice, speleothems, and historical documents recording changes in processes sensitive to variations in temperature. Resolution is annual, enabling the team to examine multi-decadal variability by focusing on 30 year mean temperatures. The PAGES 2k team showed that the long-term continental cooling driven by the fall in insolation continued right through the past 2000 years, during which '*all regions experienced a long-term cooling trend followed by recent warming during the 20th century*' [33].

Wanner's various studies, made at much higher resolution than the COHMAP studies, confirmed that decreasing insolation in the Northern Hemisphere summer led not only to Northern Hemisphere cooling, but also to a southward shift of the summer position of the ITCZ and a weakening of the Northern Hemisphere summer monsoon systems in Africa and Asia, associated with increasing dryness and desertification [32, 34]. The southward shift in the ITCZ and the weakening of the monsoons came about as a result of the interplay between decreasing insolation in the Northern Hemisphere summer and increasing insolation in the Southern Hemisphere summer (see Figure 14.1). Insolation in the Northern Hemisphere summer declined by more than insolation in the Southern Hemisphere summer increased, so the cooling associated with it had a wider effect on tropical systems than did the warming associated with Southern Hemisphere insolation.

Wanner's analyses of changes in insolation over the past 6 Ka help to explain not only changes in climate with time, but also variations between regions, as demonstrated in Figure 14.4 [32, 35], which expands on the insolation data shown in Figure 14.1 From these data it is clear that the continual annual decrease in the Northern Hemisphere insolation at high latitudes (Figure 14.4d) was mostly due to the change in June (northern summer) (Figure 14.4a). Much the same pattern was apparent in the annual insolation data for the high latitudes of the Southern Hemisphere (Figure 14.4d), largely reflecting the Southern Hemisphere change in December (southern summer) (Figure 14.4b). These data suggest that there should have been cooling

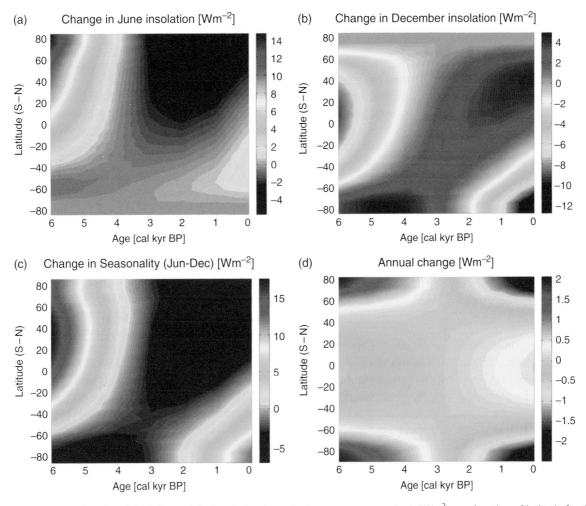

**Figure 14.4** Calculated deviations of the insolation from the long-term mean values (W/m$^2$) as a function of latitude for the past 6000 years: (a) June (boreal summer); (b) December (austral summer); (c) seasonality (difference between June and December); (d) annual mean. *Source:* From figure 6 in Ref. [35].

over the past 3 Ka at *both* poles (Figure 14.4d), in the appropriate seasons (Figure 14.4c). The tropics, meanwhile, show signs of slightly increased insolation (implying slight warming), over most of the past 4 Ka (Figure 14.4d), following the significant decline in Northern Hemisphere summer insolation (Figure 14.4a) that led to the demise of the African Pluvial Period.

The cooling of the Northern Hemisphere summer with time increased the activity of the El Niño–La Niña system in the Pacific up to around 1300 CE, since when that activity has fluctuated significantly. The cooling also led to development of an increasingly negative atmospheric NAO between 6 and 2 Ka ago, followed by a weak reversal [32]. As mentioned earlier, the negative phase of the NAO is associated with colder winters over Europe, and a warmer drier climate over the Mediterranean. SSTs declined in the North Atlantic and the Norwegian Sea, along with

southward retreat of the Arctic tree line implying declining summer temperatures [32]. Glaciers advanced across the Arctic, in the Alps and in the Western Cordillera of North America, while decreasing in the Western Cordillera of South America. In the Southern Hemisphere, lake levels were low from 6 to 4.5 Ka ago, following which they increased towards the present, the opposite of the trend in the Northern Hemisphere subtropics. This seems to reflect an intensification of northward migrating westerlies, consistent with an increase in upwelling along the Pacific coast of South America [32].

The spectacular decrease of vegetation in the Sahara between 6 and 4 Ka ago was '*related to a positive atmosphere-vegetation feedback, triggered by comparatively slow changes in orbital forcing*' [32]. As Wanner explained, '*Due to a decrease in the intensity of the African monsoon, related to the decrease in summer insolation, precipitation decreases in the Sahara during the Holocene. This induces a decrease in the vegetation cover, and thus an additional cooling [caused by an increase in albedo, or reflectivity] and reduction of precipitation that amplifies the initial decrease in vegetation cover. The amplification is particularly strong when a threshold is crossed, leading to a rapid desertification and ... fast changes'* [32]. This provides a good example of a gradual change in insolation causing knock-on effects in the climate system that – together with the insolation – caused a tipping point to be crossed that led to a rather abrupt shift in climate.

In marine records, the coherent long term cooling averaging about 1 °C over the past 9–10 Ka was not confined to the North Atlantic Ocean [36]. It also extended to the Mediterranean [37–39], which is consistent with glacial re-advance in Iceland, with cooling in Greenland ice cores, and with pollen data in Europe and North America that indicate southward migration of cool spruce forest [40]. Tim Herbert pointed out [40] that this appeared to be a regional rather than global pattern, because alkenone data from the Indian Ocean, South China Sea and western tropical Atlantic showed very slight warming from the early Holocene to the present [41], while data from the western margin of North America showed no trend at all during last 9 Ka, apart from minor millennial oscillations of about 1 °C. The observed tropical warming might be expected from the slight rise in insolation seen in the tropics in Figure 14.4d.

I noticed one important caveat to the report of the PAGES 2k programme: whilst Antarctica showed the same cooling as the other continents, it did not – in the data available to the PAGES group – show the recent warming of the twentieth century [33]. That perception is clearly wrong, as subsequent data show, because some parts of Antarctica, especially the Antarctic Peninsula, but also to a lesser extent West Antarctica, do show this recent warming [42, 43]. It is evident in an ice core from the ice cap of James Ross Island (JRI) in the western Weddell Sea, for example (Figure 14.5) [44, 45], and at an Ocean Drilling Programme site (1089) in the Palmer Deep just south of Anvers Island [46, 47]. We will revisit these data in Chapter 15.

Studying Antarctic Peninsula climate changes in relation to insolation and to other regional temperature profiles, and with particular reference to ODP site 1098 in the Palmer Deep on the continental shelf west of the Peninsula near 65°S, Amelia Shevenell of University College London and her team found that '*surface ocean temperatures at the continental margin of the western Antarctic Peninsula cooled by 3-4°C over the past 12,000 years, tracking the Holocene decline of local (65°S) spring insolation'* [46]. This

pattern of insolation is related to the duration of the southern summer (much as seen in Figure 13.20), and would account for the decline in temperature at JRI (Figure 14.5). Much the same decline in temperature is seen in the Ross Sea and the adjacent southwest and southeast sectors of the Pacific Ocean [46]. As we saw in Chapter 13, there are ample reasons for invoking local orbital control on Antarctic climate rather than forcing by the orbital patterns of the Northern Hemisphere, where summer insolation at 65°N followed that same decrease [48]. Shevenell drew upon the known teleconnections in the modern climate system between the tropical Pacific and the Antarctic coast to suggest that these connections strengthened during the Holocene in association with changes in the position of the main westerly wind belt [46]. The Southern Ocean westerlies play a key role in warming and cooling locally, as well as in regulating the emission of $CO_2$ from the Southern Ocean, or its subduction in Antarctica Bottom Water and Intermediate Water. Shevenell's ODP core showed a significant warm event extending from about 2000–500 years ago, which may be related to the local changes in insolation observed by Wanner (Figure 14.4b and d).

Based on the declining insolation in the north, and its effects, Wanner subdivided the Holocene into three periods [34]. First was an early deglaciation phase, between 11.7 and 7 Ka ago, characterized by high summer insolation in the Northern Hemisphere, a cool temperate climate near the melting ice sheets in North America and Eurasia, and strong monsoonal activity in Africa and Asia. Second came the Holocene Thermal Optimum, between 7 and 4.2 Ka ago, with high summer temperatures in mid- and high-latitude parts of the Northern Hemisphere, and active but weakening monsoonal systems at low latitudes. Third, there was a neoglacial period, with falling summer temperatures in the Northern Hemisphere, terminating with the sharp rise in global temperature as we entered the modern era. These divisions map onto the three new stratigraphic divisions of the Holocene promulgated by the International Union of Geological Sciences in 2018: Greenlandian (base at 11.7 Ka ago); Northgrippian (base at 8.3 Ka ago); and Meghalayan (base at 4.2 Ka ago) [49].

In 2013, Shaun Marcott from Oregon State University (OSU) and colleagues presented their new reconstruction of global temperature change for the Holocene [36]. Most of their data came from marine sediment cores, but they tell basically the same story as Wanner and his colleagues. Combining their data into a single global temperature stack, Marcott's team found a rise in temperature of around 0.6 °C between 11.3 and 10 Ka ago, with Early

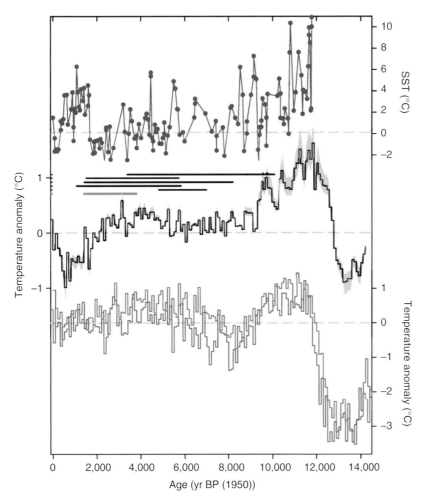

**Figure 14.5**   Holocene temperature history of the Antarctic Peninsula. Top panel: sea surface temperature (SST) reconstruction offshore from the western Antarctic Peninsula. Middle Panel: James Ross Island ice-core temperature reconstruction relative to the 1961–1990 mean, in 100 year averages, with grey shading indicating the standard error of the calibration. Horizontal bars show intervals when open water was present in the adjacent Prince Gustav Channel. Lower Panel: temperature reconstructions from the Dome C (uppermost) and Dronning Maud Land (lowermost) ice cores from East Antarctica. *Source:* From figure 3 in Ref. [44].

Holocene warmth remaining more or less stable between 10 and 7.0 Ka ago. This stable period was followed by 0.7 °C of cooling, largely biased by a 2 °C cooling of the North Atlantic, and culminating in the coldest temperatures of the Holocene in the Little Ice Age about 200 years ago. After that, there was a sudden rise of 1 °C to the warm temperatures of the modern era. The global stack was most similar to the data from the Northern Hemisphere between 30°N and 90°N, suggesting a primary control by Northern Hemisphere summer insolation. The tropics (30°N–30°S) warmed by 0.3 °C from 11.3 to 5 Ka ago, then cooled by a similar amount to around 250 years ago, before warming sharply to modern values. Southern Hemisphere temperatures varied more. The narrowly defined mid Holocene Thermal Optimum between 7 Ka ago and 4 Ka ago in the Northern

Hemisphere, identified by Lamb and Wanner, was not readily evident in Marcott's global data.

Criticism of Marcott's temperature profile stemmed from the fact that as the bulk of the Holocene data represented averages for time intervals of 200 years, the same should have applied to the modern era, whereas Marcott used instead the full instrumental temperature record. Setting that aside, his findings from 11.7 Ka ago to the beginning of the last 200 years are consistent with multiple other sources of proxy data from the Holocene.

In 2012, using a global atmosphere-ocean-vegetation model to analyse the variations in the timing and magnitude of the Holocene thermal maximum and their dependence on various forcings over the past 9 Ka, Renssen and colleagues found that the warmest conditions occurred at high latitudes in both hemispheres, reaching 5 °C above

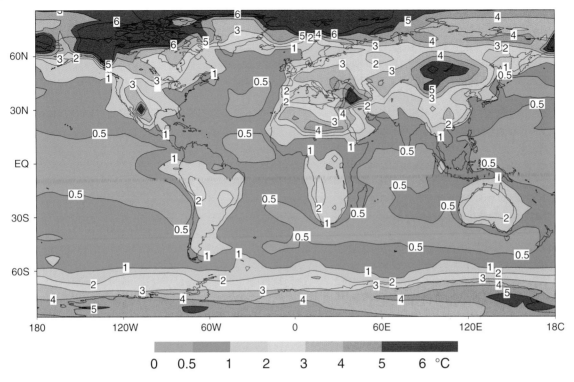

**Figure 14.6** Simulated maximum positive surface temperature anomaly in the Holocene as calculated by the OGMELTICE model, relative to the pre-industrial mean, based on monthly mean results. *Source:* From Fig. 3(a) in Ref. [48].

the pre-industrial level, while the smallest temperature signal (<0.5 °C) occurred over the tropical oceans (Figure 14.6) [48]. The major forcing factor was the orbital signal, which was largest at high latitudes, further enhanced by polar amplification. The timing of the thermal maximum was earliest (before 8 Ka BP) in regions not affected by the remnant Laurentian Ice Sheet. Close to the ice sheet the signal was delayed by 2–3 Ka. These timings were consistent with the evidence from proxies.

Marcott's data show that the Northern Hemisphere signal of climate change tends to dominate the Holocene global signal, as it does today [36]. In part the predominance of that signal is due to the greater area of land in the Northern Hemisphere. Land tends to heat up and cool down much faster than water, and water dominates the Southern Hemisphere. The global cooling had an impact on the North Atlantic Current, a northern branch of the Gulf Stream that takes heat to western Europe. Less heat was transported northwards in the neoglacial, helping Europe to cool [36].

Marcott's graphs were also questioned in 2014 by an international team led by Zhengyu Liu of Peking University, who argued that the cooling trend was opposite to the expected and simulated global warming trend due to retreating ice sheets and rising atmospheric $CO_2$,

creating what his team referred to as 'The Holocene temperature conundrum' [50]. One of their criticisms of Marcott's data, much of which came from planktonic remains from marine sediment cores, was that these remains might be biased towards summer temperatures. They also raised the possibility that there might be a similar seasonal bias in the terrestrial pollen and plant records used by other investigators. They considered the effects of orbital change to be insignificant compared to the effects of greenhouse gases and changes in ice coverage. However, they did note that reduced summer insolation had weakened the summer monsoon, which could have caused regional drying and an increase in the supply of dust that operated as a solar reflector that could cause a cooling trend, which might help to explain the discrepancy between models that simulated Holocene warming, and observations that suggested cooling [50]. Were the models or the observations biased?

In 2017, Jonathan Baker and colleagues agreed that climate models showed continual global and Eurasian climate warming, in contrast to what observations suggested. Baker tested this paradox by extracting temperature records for winter climate from $\partial^{18}O$ in stalagmites in a cave at 54.2°N in the Urals, which documented warming throughout the Holocene in agreement with climate models [51].

They attributed this pattern to: first, retreat of the Northern Hemisphere ice sheets (to 7 Ka ago); second to rising atmospheric greenhouse gas concentrations (since 8 Ka ago); and, third, to rising winter insolation over the same period. They concluded that winter warming dominated Holocene temperature evolution in the continental interior of Eurasia, away from coastal locations [51]. In summary, then, one might make the argument that – during the Holocene – Earth's orbital cycles caused cooling in the Northern Hemisphere summer (as observed by Marcott) and warming during the winter (as observed by Baker). However, careful examination of the orbital data (Figure 14.1) shows that if insolation was the main driver of temperature in the Northern Hemisphere, the high latitudes (>60°N) should have been dominated by warmth peaking in the summer and declining towards the present; summers should have been cool and autumns slightly warmer by 3 Ka ago, and the cooling should have grown slightly to the present whilst winters would have been very slightly warmer. Overall, then, the high latitude picture based on orbital data (Figure 14.1) suggests a climate of cooling summers and more or less static to slightly warmer winters. Baker argued that Eurasian warming is also apparent from pollen-based annual temperatures from northern Europe (but these are from summer-influenced vegetation), shrub and pine pollen abundance in the Caspian Sea (also a summer vegetation index), ice wedge $\partial^{18}O$ in the Siberian Arctic and borehole temperature from the southern Urals. Is the $\partial^{18}O$ data trustworthy? Baker himself points out that the $\partial^{18}O$ reflects moisture sources for the westerly winds that drive local climate, and agrees that his case rests on the notion that '*the proxy archives capture winter climate variability*' [51]. They may indeed capture climate variability, but since the trend opposes what is predicted from orbital insolation, does it reflect temperature or some other climatic property? Assuming that the $\partial^{18}O$ represents temperature, he suggests that forcings other than orbital insolation must explain the (apparent) wintertime warming. These might include the slightly rising abundance of $CO_2$ since 8 Ka ago, and slightly rising $CH_4$ since 3 Ka ago, discussed later. Prior to that, the warming signal was paced by the decline in the northern Hemisphere ice sheets, until about 7.5 Ka ago. The decline of the ice sheets enabled the jet stream to migrate north, which in turn allowed warm air from the south to enter the central Eurasian region.

These new data require us to recognize the importance of regional change driven by geography. Clearly, central Eurasia, far from the ocean, behaved differently from the Atlantic Ocean. Extrapolating from the Atlantic to central Eurasia seems unwise, as is the assumption that all climate change will reflect orbital change. Statements about modern temperature rise in mid continental regions must take this background into account. Nevertheless, it is equally clear from non-vegetational data that cooling after the Holocene thermal maximum [48] pushed the ITCZ south (drying out the Sahara) and that Denton's neoglacial led to advances in both glaciers and sea ice during the past 3000 years (see below), observations that call into question the conclusions of both Liu [50] and Baker [51].

Evidence that cooling predominated in the late Holocene comes from both polar regions. In the Arctic, Gifford Miller's team found that sea ice first decreased during the early Holocene, as insolation increased, then increased in the late Holocene, as insolation declined [17], a pattern confirmed by studies of sediment cores [52]. Likewise, the tree line expanded northwards to as much as 200 km beyond its current position, before beginning to retreat southward starting 3–4 Ka ago, leaving tundra in its wake [26]. Permafrost followed much the same pattern, melting south of the Arctic Circle, then refreezing after 3 Ka ago. Summer temperature anomalies along the northern margins of Eurasia in the thermal optimum ranged from 1 to 3 °C above today's, and SSTs were up to 4–5 °C warmer than today's. As elsewhere, cooling began between 6 and 3 Ka ago. Most Arctic mountain glaciers and ice caps expanded during the 'neoglaciation' of the late Holocene, as did the Greenland ice sheet. This cooling trend culminated in most places in the mid-nineteenth century [17, 53, 54]. Much the same broad-brush picture of Arctic sea ice was true of the Antarctic (Figure 14.7). There, Xavier Crosta found evidence for minimal distribution of sea ice as Holocene temperatures rose in the summer months between 9 and 4 Ka ago, followed by a substantial expansion as temperatures fell over the next 3000 years [55], driven by the decline in insolation (Figure 14.4d). These patterns are consistent with Renssen's model (Figure 14.6) [48].

We also now have new information about the behaviour of the West Antarctic Ice Sheet. While previous reconstructions assumed that it retreated progressively throughout the Holocene, Kingslake, and colleagues showed in 2018 that over this period its grounding line (the point at which the ice sheet is no longer in contact with the substrate and becomes a floating ice shelf) retreated several hundred kilometres inland to a minimum position between 10.2–9.7 Ka ago, during the thermal maximum of the Southern Hemisphere, then re-advanced to its present position [56]. The re-advance is attributed not solely to temperature change but also to isostatic rebound in which the loss of the weight of ice caused the underlying seabed to rise, which in turn caused the grounding line to advance seawards once more. However, the present grounding line is still 1000 km or so inland of where it was at the last Glacial Maximum [56].

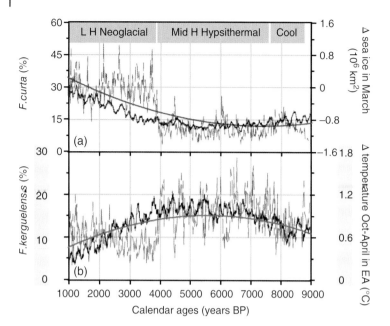

**Figure 14.7** Sea ice area around Antarctica. Relative abundances of (a) the *Fragilariopsis curta* group (blue dashed line) and (b) the *Fragilariopsis kerguelensis* group (red dashed line) versus calendar ages in core MD03-2601 close to the coast of Adélie Land, Antarctica (see Figure 7.5). *F. curta* is a sea-ice diatom, whose abundance represents denser and longer-lasting coverage of the ocean by sea ice. *F. kerguelensis*, in contrast, is most abundant along the Polar Front where there is less sea ice and temperatures are warmer. Polynomial regressions indicate the first-order evolution for the *F. curta* group (a, solid smooth line) and the *F. kerguelensis* group (b, solid smooth line), compared to simulated time series of the March sea ice area (a, black zig-zag line) and the October–April temperature for East Antarctica over the last 9000 years (b, black zig-zag line), plotted as deviation from the preindustrial mean. The data show warming from a cool early Holocene into a warm MH hypsithermal with low sea ice, then into a cool late Holocene neoglacial with abundant sea ice. *Source:* From figure 4 in Ref. [55].

The retreat of ice shelves is due largely to melting from beneath (Figure 13.13), which in turn pulls ice from the interior to the coast by removing the buttressing effect of the ice shelf on the movement of ice streams from the interior. Hillenbrand and colleagues use sedimentary evidence to show that the upwelling of warm circum-polar deep water onto the continental shelf of the Amundsen Sea *'forced deglaciation of this sector from at least 10,400 years ago until 7,500 years ago [the Holocene thermal optimum in the Southern Hemisphere] – when an ice-shelf collapse may have caused rapid ice-sheet thinning further upstream'* [57].

The work of Mayweski [29], Wanner [34], and Marcott [36] confirms that the continued existence of ice sheets in the Northern Hemisphere until around 7 Ka ago prevented temperatures from rising far despite high insolation between 10 and 7 Ka ago. The disappearance of the great ice sheets, combined with moderately high insolation, created the Holocene Thermal Optimum there between 7.0 and 4.2 Ka. Then neoglacial conditions set in, with glacier advances across the Northern Hemisphere and the re-formation of ice shelves along the northern coast of Ellesmere Island in northern Canada [58]. The most recent advance of glaciers, in the Little Ice Age, was the most extensive in all areas, making it the coldest episode of the Holocene

[58]. The cooling of the North Atlantic after the thermal optimum may represent weakening of the Atlantic Meridional Overturning Circulation, associated with the increase in Northern Hemisphere winter insolation [59]. Volcanic activity continued at variable levels throughout the Holocene, with no significant trend, making forcing from this source an unlikely driver of the overall cooling in the climate system [29, 34, 36].

At first glance, recently published pollen data seem to provide a slightly different picture of Northern Hemisphere Holocene climate from that described above and represented by the global stack of marine data. In 2018, Marsicek and colleagues showed that pollen data largely from middle latitudes (30–60°N) in Europe and North America indicated sharp warming from 11 to 7.5 Ka ago, then a relaxation of warming to a thermal plateau 6–2.5 Ka ago, then a significant cooling from 2 Ka ago to about 1900, followed by a steep warming [60]. This pattern is more or less consistent with what we see in the patterns of annual insolation for those latitudes (Figure 14.1): summer temperatures (June, July, August) would have risen from the end of the Younger Dryas to substantial warmth by 9 Ka ago, following which there would have been somewhat less warming by 6 Ka ago, cooler conditions by 3 Ka ago, then relatively cool conditions by the present. However, the mid-latitude pollen patterns on the continents reflect to some degree the fact that ice persisted at the northern edges of these regions until around 7 Ka ago, enhancing the development of cool conditions even though the rest of the globe was warming. As Marsicek explained – '*temperature depression caused by remnant ice sheets extended across North America and Europe*' [60]. Part of the explanation for subtle differences between marine records and continental pollen records lies in the response of the latter not only to insolation but also to the length of the growing season, which offset the decline in maximum warmth [60]. Both the marine and continental records exhibit centennial and millennial variations, which are in phase with one another, including the prominent cooling event centred on about 8.2 Ka ago and representing the effects of a glacial lake outflow; the temperature increase at about 5.5 Ka ago – which correlates with the start of Saharan aridity; and the Little Ice Age.

The pattern of climate change leading into the Late Holocene neoglacial and Little Ice Age reflects the pattern of insolation calculated by André Berger and Marie-France Loutre [13, 61] (as seen for example in Figures 12.2 and 13.20). Insolation peaked in the Northern Hemisphere 11 Ka ago, then declined to the present, with a substantial flattening of the rate of decline over the past 1000 years or so. Insolation is calculated to stay just as low for at least another 1000 years [13, 61] (Figure 12.2). Taken together, the astronomical and the palaeoclimate data suggest that the world should have continued to cool, so it is more than a little surprising that: '*in the brief interval of less than two centuries, the Northern Hemisphere (at least) has experienced the warmest and the coldest extremes of the late Holocene*' [58]. Recent warming bucks the trend imposed by orbital forcing, which should have kept our climate in a cool neoglacial state. We will explore this issue further in Chapter 15.

## 14.2 The Role of Greenhouse Gases – Carbon Dioxide and Methane

Did changes in $CO_2$ and $CH_4$ have a significant role to play in driving the climate patterns of the Holocene? Reference to Figure 14.8 will help us to answer that question. Eric Sundquist of the US Geological Survey at Woods Hole reminded us in 2005 that measurements of fossil atmospheric $CO_2$ from the Taylor Dome ice core, from near McMurdo Sound in the Ross Sea, showed that $CO_2$ declined from around 270 ppm at 10.5 Ka ago to 260 ppm at 8 Ka ago, and then increased to values near 285 ppm by CE 1000 [62]. Much the same picture emerged in 1999 from ice cores at

Law Dome, in the Australian sector of Antarctica, and at the Russian Vostok station on the Polar Plateau (Figure 14.8a and b) [63], and Figure 14.9a) [64]. Andreas Indermühle from the University of Bern, in Switzerland, and colleagues, explained the initial decrease in $CO_2$ in Antarctica as representing the expansion of vegetation into formerly glaciated areas in the Northern Hemisphere (hence absorption of $CO_2$ by plants), which was followed by a subsequent increase in $CO_2$ caused by gradual release of carbon in response to global cooling and drying, the cooling being a response to the orbitally controlled decline in insolation [65]. Confirming the association between early Holocene $CO_2$ decline and expanding Arctic vegetation, Glen Macdonald of the University of California at Los Angeles, and colleagues, found in 2006 that circum-Arctic peatlands began to develop rapidly as ice and snow began to disappear 16.6 Ka ago and expanded between 12 and 8 Ka ago when insolation was at its highest, drawing down $CO_2$ during the early Holocene [66].

Methane in ice cores from Taylor Dome and from Dome C changed over the same period, but in a different way, as shown by Jérôme Chappelaz of the Laboratoire de Glaciologie et Géophysique de l'Environnement of the University of Grenoble, and colleagues, in 1997. $CH_4$ declined from values near 700 ppb at 10.5 Ka ago to values between 550 and 600 ppb at 5 Ka ago, before increasing to values near 700 ppb by CE 1000 [67] (see Figure 14.8d). Many scientists attributed the late Holocene increase in $CH_4$ values to an expansion of boreal wetlands. Not surprisingly, as most wetlands are found in the Northern Hemisphere, the Holocene data showed a gradient in $CH_4$ from north to south, with between 30 and 50 ppb more in air from the north. The addition of recent anthropogenic sources of $CH_4$ in the north increased this gradient by a factor of three [62]. The $CH_4$ data showed a sudden short-term drop at c. 8 Ka ago (Figure 14.8d), which corresponded to the widespread but short-lived cooling event caused by a major flood of freshwater into the North Atlantic from a melting event in the Laurentide ice sheet.

The rise in atmospheric $CO_2$ values from around 260 ppm at about 8 Ka ago to 285 ppm 200–400 years ago (Figure 14.8a and b, and 14.9a) might be expected to have led to a small warming. Instead, cooling continued (Figure 14.9e), showing that the forcing by declining Northern Hemisphere insolation had a much greater effect on climate than did forcing by the 25 ppm addition of $CO_2$. Much the same argument applies also to the rise in methane ($CH_4$), which increased by about 160 ppb towards the pre-industrial era [29] – mostly over the past 3 Ka, despite the continued cooling [32] (Figure 14.9e). It is worth recalling at this point the divergence of temperature and $CO_2$ typical of the re-entry to glacial conditions from the peaks of past interglacials

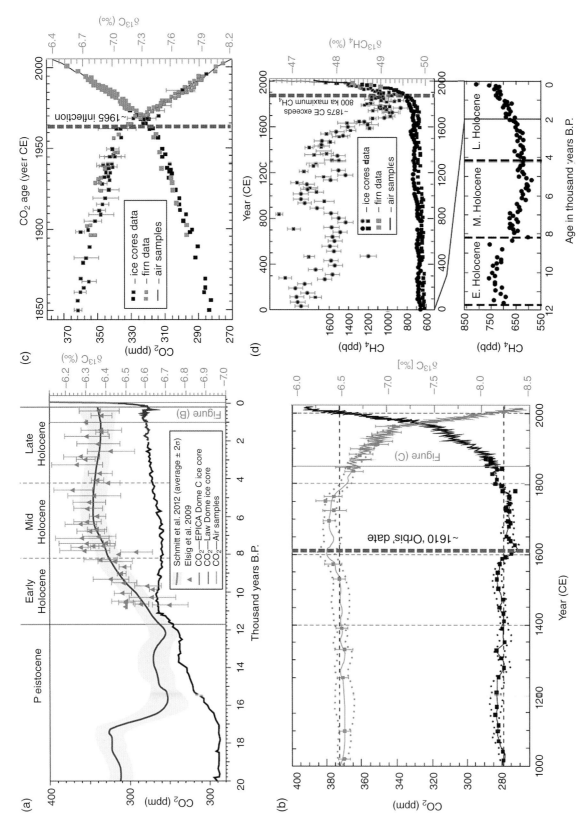

**Figure 14.8** Perturbations of the carbon cycle evidenced by glaciochemical $CO_2$ and $CH_4$ concentrations and carbon isotopic ratios. (a) Atmospheric $CO_2$ from the Antarctic Law Dome and EPICA Dome C ice cores and $\partial^{13}C$ from atmospheric $CO_2$. (b) $CO_2$ concentration and $\partial^{13}C$ from the Law Dome ice cores for the time interval indicated by the green rectangle in (a), showing a 10 ppm dip in $CO_2$ recognized as the Orbis event (which we will discuss in Chapter 15). (c) $CO_2$ concentration and $\partial^{13}C$ from atmospheric $CO_2$ from the Law Dome ice core, firn data, and air samples for the time period indicated by the green rectangle in (b), showing inflexions at ~1965 CE. (d) Antarctic (squares) and Greenland (circles) ice core and firn records for $CH_4$ concentration and $\partial^{13}C$ for atmospheric $CH_4$ for the past two millennia (top) with Greenland ice core $CH_4$ data for the Holocene (bottom). *Source:* From figure 5 Ref. [63], which contains the references to data sources.

**Figure 14.9** Holocene records of fossil-bound $\partial^{15}N$ and biogenic opal flux from the Southern Ocean, compared with atmospheric $CO_2$ and with climate- and circulation-related records from the Northern Hemisphere. (a) Atmospheric $CO_2$ concentration composite. (b) $\partial^{15}N$ records from the Antarctic and Polar Frontal Zones of the Southern Ocean. Diatom-bound $\partial^{15}N$ records are shown in green, from the Pacific Antarctic (circles), the Indian Antarctic (triangles) and the Indian Polar Frontal Zone (diamonds). The compilation of coral-bound $\partial^{15}N$ data (Drake Passage) is shown in blue. (c) Fossil-bound $\partial^{15}N$ records from the Subantarctic Zone of the Southern Ocean. Red, foraminifera-bound $\partial^{15}N$ record from ODP Site 1090 in the Atlantic; dark blue, coral-bound $\partial^{15}N$ compilation from the Drake Passage; light blue, coral-bound $\partial^{15}N$ compilation from south of Tasmania. (d) Thorium-normalized opal flux records from two cores just inside the Polar Front (circles and triangles), and one core at the Subantarctic Front (diamonds). (e) Temperature anomaly from 30 to 90° N relative to the 1961–1990 instrumental mean. (f) Mean grain size of the sortable silt fraction, a proxy for near-bottom flow speed, from the Gardar Drift at 2620 m water depth in the North Atlantic (orange) and mean relative grain size from a stacked record from 13 sediment cores located south of Iceland (grey), indicative of the relative strength of the Iceland–Scotland overflow and thus North Atlantic Deep Water formation. Data in (a) and (e) are shown for the period before 1900. The right-hand axis in (c) (SAZ $\partial^{15}N$ coral-bound) has been scaled by two-thirds for plotting purposes. *Source:* From figure 2 in Ref. [64], which contains references to data sources.

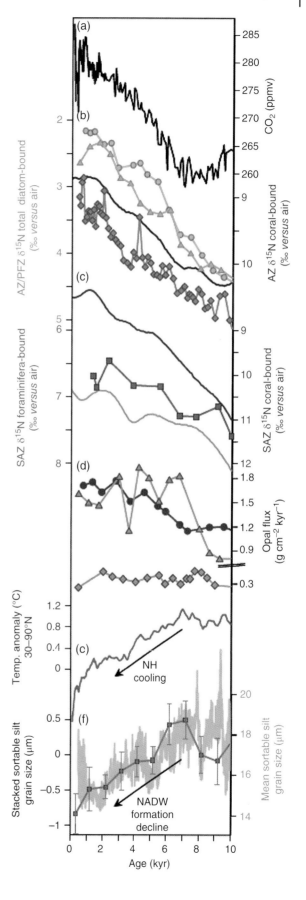

that we saw in Chapter 13, with temperature, driven primarily by insolation, falling faster than $CO_2$ [68, 69]. As in those cases, the lack of correlation between $CO_2$ or $CH_4$ and temperature between the early and late Holocene is likely to have much to do not only with the decrease in insolation per se, which drives temperature, but also with its effect on albedo, through the growth in sea ice, land ice, and snow cover into the neoglacial. This temporary lack of correlation changed as we moved into the modern era, when massive increases in the two gases caused a sharp increase in temperatures unprecedented in the previous millennium [32, 33, 36, 58], as we see in more detail in Chapter 15.

This seems like a good point to review the ideas of Bill Ruddiman concerning the effect humans may have had on climate since the beginning of the Holocene [70]. Ruddiman argued that the fall in insolation during the Holocene should have led to the formation of new ice sheets by now [71]. He suggested that '*$CO_2$ and $CH_4$ concentrations should [also] have fallen steadily from 11,000 years ago until now*' [71]. Indeed they did start to fall from the beginning of the Holocene, but then '*$CO_2$ and $CH_4$ began anomalous increases at 8000 and 5000 years ago, respectively*' [71] (e.g. see Figures 14.8a and d). He thought that these rises were most probably due to human activities, and, further, that those increases had prevented the occurrence of a new glaciation. He attributed the slow, small rise of $CO_2$ over the past 8 Ka or so to forest clearance and the development of agriculture [71, 72].

Birks reminded us that there is some independent support for that idea, in that the distribution of charcoal, an index of the existence of fires, parallels the increase in $CO_2$ concentrations, suggesting that biomass burning could possibly have been a cause for the rise in $CO_2$ [1]. Nevertheless, as spelled out below, others suggests that the rise most probably had several causes, including changes in calcite compensation in the ocean, changes in SST, and the post-glacial build-up of coral reefs, and was not caused by major changes in the storage of carbon on land.

Investigating the rise in atmospheric $CO_2$ over the past 8 Ka, Wally Broecker and colleagues had found in 1999 that it was reflected in the declining carbonate ion concentration of the deep ocean (Figure 14.10) [73]. Broecker explained the pattern of $CO_2$ as follows (Figure 14.11): an early Holocene increase in terrestrial biomass, caused atmospheric $CO_2$ to fall, which increased the carbonate ion concentration in the ocean, so deepening the carbonate compensation depth (CCD) and causing more accumulation of carbonate sediment on the deep-sea floor, which in turn increased the carbonate ion concentration and led to the ocean emitting $CO_2$.

If Broecker's scenario were correct, then we would expect no early to late change in the $\partial^{13}C$ of atmospheric

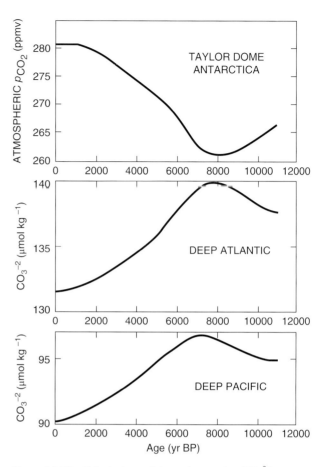

**Figure 14.10** Calculations of the carbonate ion ($CO_3^{2-}$) concentration of the deep ocean based on the $CO_2$ data of Indermühle from the Taylor Dome ice core [65]. *Source:* From figure 10 in Ref. [73].

$CO_2$ [73], and that is indeed what we find (see Figure 14.8a). In 2006, Broecker and Thomas Stocker found that the latest measurements of $\partial^{13}C$ in $CO_2$ from Antarctic ice suggested that the main changes in atmospheric $CO_2$ over this period were the result of changes in the world's oceans [74]. They reasoned that as the amount of $CO_2$ in the air had increased, then so too must the amount of $CO_2$ in the ocean. In other words much more $CO_2$ had been emitted than could be accounted for by deforestation alone [74]. Instead they proposed, much as Broecker had done in 1999 [73] (Figure 14.10), that the $CO_2$ rise was triggered by the ocean's response to the extraction of $CO_2$ from the ocean to create the early Holocene increase in forest cover (Figure 14.11). Removing that $CO_2$ from the ocean lowered early Holocene $CO_2$ in the air and increased the carbonate ion concentration of ocean water, hence deepening the CCD and causing calcium carbonate to accumulate. That drew down the carbonate ion concentration, making the CCD rise and increasing the $CO_2$ content of ocean water, which fed back to increasing the $CO_2$ in the air. According

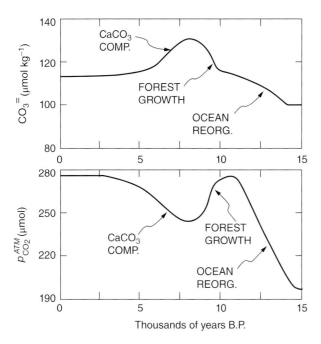

**Figure 14.11** Broecker's scenario to explain the Holocene $CO_2$ record from Antarctica's Byrd ice core, showing that a 0.35% rise in the $\partial^{13}C$ of oceanic $CO_2$ would lower the $CO_2$ content of the atmosphere by c. 25 ppm. $CaCO_3$ compensation would then lower the carbonate ion concentration of the deep sea, thus raising the $CO_2$ content of the atmosphere. *Source:* From figure 12 in Ref. [73].

to Broecker, then, the $CO_2$ change through the Holocene is a story of natural ocean chemistry responding to the impacts of declining insolation, not to anthropogenic deforestation.

Support for Broecker's position came from the discovery in 2007, mentioned in Chapter 13, that the ratio of boron to calcium (B/Ca) in species of benthic foraminifera is directly related to the concentration of carbonate ions ($CO_3^{2-}$) in bottom water, a finding that has greatly improved our understanding of the behaviour of the carbonate system in the ocean [75, 76]. Jimin Yu and colleagues used the B/Ca ratios in deep-sea cores to show that the concentration of carbonate ions ($CO_3^{2-}$) in the deep waters of the Pacific and the Indian Oceans declined during the Holocene [76], as Broecker had suggested (Figure 14.10). They thought it likely that the build-up of coral reefs during the Holocene caused the whole ocean concentration of $CO_3^{2-}$ to decline. That decline reduced the ocean's alkalinity, causing the solubility of $CO_2$ in the ocean to decline, so contributing to the 20 ppm rise in atmospheric $CO_2$ over the past 8 Ka. Much the same conclusion was reached in 2012 by numerical modelling [77]. Similarly, Ciais concluded in 2013 that the oceans may have contributed most or all of this growth in $CO_2$ [78], as they did in the preceding glacial-to-interglacial transition [79]. Along the same lines, Studer and colleagues noted in 2018 that changes in the nitrogen isotope

composition of organisms from the Southern Ocean were consistent with a weakening of the biological pump that stores $CO_2$ in deep water, thus possibly accounting for much of the Holocene rise in atmospheric $CO_2$ [64] (see Figure 14.9).

Ruddiman was not deterred. First, he revised the start date for the $CO_2$ rise from 8 Ka ago to 7 Ka ago [80]. Second, he published a comparison of the behaviour of $CO_2$ in the Holocene interglacial with the *averaged* behaviour of $CO_2$ in previous interglacials (red symbols in Figure 14.12b), which made the Holocene appear unique. However, close inspection of his data show that $CO_2$ increased throughout one interglacial stage (MIS-15), and underwent temporary increases after an initial decline in others (MISs −7, −9, and −11), which would tend to suggest that the Holocene

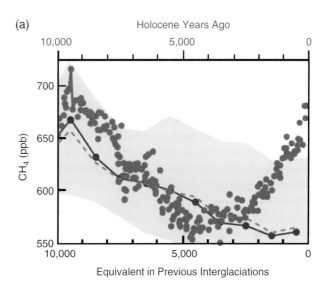

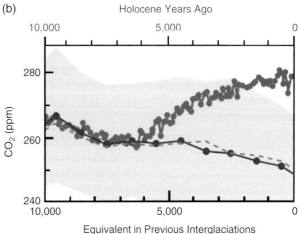

**Figure 14.12** Comparison of Holocene trends (red) to stacked averages for previous interglaciations (blue). The light blue shading shows one standard deviation. (a) = $CH_4$; (b) = $CO_2$; from Dome C (as in Figure 14.8). Stage 15 is excluded in the average shown by the solid blue line and included in the trend shown by the dashed line. *Source:* From figure 3 in Ref. [80].

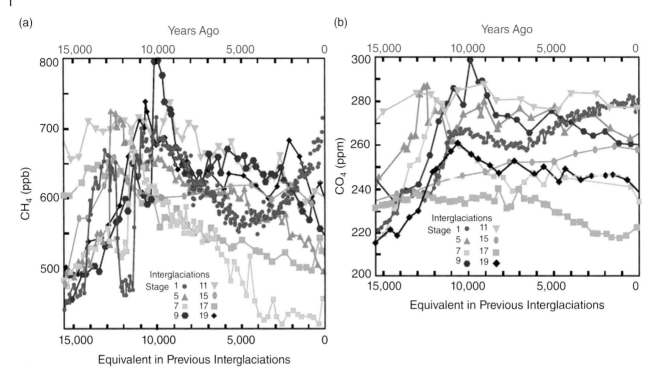

**Figure 14.13** Holocene climate-related indices (in red) compared to previous interglaciations. (a) = $CH_4$ signals; (b) = $CO_2$ signals; from the Dome C ice core in Antarctica. *Source:* From figure 2 in Ref. [80].

pattern of $CO_2$ change (red symbols in Figure 14.13b) was not entirely unique.

Bearing in mind Broecker's arguments, Ruddiman agreed that 17 of the 20 ppm rise in $CO_2$ through the Holocene originated from the oceans, but argued that this was the result of feedbacks triggered by initial carbon emissions from changes in land use (deforestation) [80]. In contrast, Zalasiewicz and colleagues argued that the human population at the start of and during most of the Holocene was simply far too small to have generated the amount of deforestation required to create the $CO_2$ anomaly from 8000 years ago as required by Ruddiman's hypothesis [81]. In support of that argument, Zalasiewicz showed that population growth through the Holocene was extremely low and that as a result so was the growth in the greenhouse gases $CO_2$ and $CH_4$ (Figure 14.14) [81]. Comparing the $CO_2$ profile from Figure 14.9a with the rate of population change illustrated in Figure 14.14 suggests that there was no significant connection between population and $CO_2$ change from around 7000 years ago to about 1800. In any case, the spread of agriculture and associated deforestation was not a synchronous global phenomenon. It began slowly some 10 000 years ago in the Middle East, never occurred in Australia, and did not occur in a large way in the Americas until long after the migrations of early peoples there. Nevertheless, the ArchaeoGLOBE Project team recently came up with a revised assessment of historical land use whose patterns lend some support to Ruddiman's arguments. Extensive agriculture and pastoralism began to grow around 7 Ka ago, becoming widespread by 2 Ka ago; intensive agriculture began to grow around 6 Ka ago, becoming widespread around 1 Ka ago; and foraging remained steady from 10 to 6 Ka ago then declined to very low levels by about 200 years ago [82]. The declines in the rates of growth in agriculture and pastoralism from about 3 Ka ago [82] seem to match the decrease in the rate of the $CO_2$ rise from 3.5 Ka ago onwards (Figure 14.9a). The question remains – to what extent were these agricultural and pastoral changes connected to the $CO_2$ record? Equally one has to question the robustness of the ArchaeoGLOBE data, which its own members agreed have some distinct regional deficiencies [82].

What else might $CO_2$ be connected to, if not population? Broecker and Stocker suggested it was a gradual change in ocean processes [74]. In that context we see that global sea level, which had been rising fast while the last Northern Hemisphere ice sheets melted away, slowed between 8.2 and 6.7 Ka ago along with the final phase of North American deglaciation [83, 84] (Figure 14.15), which is when the rate of $CO_2$ fall flattened (Figure 14.9a). There followed a further progressive decrease in the rate of sea level rise, of which c. 3 m occurred between 6.7 and 4.2 Ka ago, with a further rise of ≤1 m up to the onset of the recent rise about 100–150 years ago (Figure 14.15). Evidently,

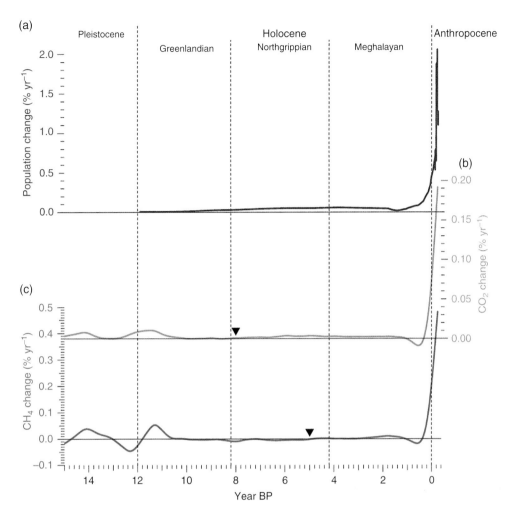

**Figure 14.14** Average values of relative change to (a) global human population, (b) atmospheric $CO_2$ concentration, and (c) $CH_4$ concentration over the last 20 000 years. Age is in thousands of years before 1950. *Source:* From figure 1 in Ref. [81], which contains references to the data sources.

ocean area must have been nearly constant from 2.5 Ka ago until that time [83]. When sea level was rising rapidly, the area of the surface ocean was increasing rapidly, representing a growing sink for atmospheric $CO_2$. But, once sea level had reached its equilibrium position at around 7 Ka ago there was very little further change in the area of the ocean, which therefore no longer formed a growing sink for $CO_2$. At the same time we would expect commensurate changes in surface and deep ocean circulation, much as suggested by Broecker and Stocker [74]. In conclusion it seems highly likely that the $CO_2$ profile for the Holocene reflects the interplay between the atmosphere, ice, ocean, and marine and land plants, rather than any externally imposed human impacts.

Reviewing the various arguments for and against Ruddiman's hypothesis, Michel Crucifix concluded in 2009 that '*the early anthropogenic theory implies – if it is correct – that there was a bifurcation point during the past 6000*

*years during which the climate system hesitated before opting for a glacial inception or staying interglacial. The anthropogenic perturbation gave it the necessary kick to opt for a long interglacial*' [10]. The key word there is *if*. However, the oceanic evidence cited above makes Ruddiman's $CO_2$ argument seem increasingly untenable.

Before examining Ruddiman's theory further, we should recall Berger's calculations suggesting that, given what we know about future insolation (Figure 12.2), the Holocene interglacial may prove to last as long as 30 Ka or more, like Marine Isotope Stage 11, which occurred some 400 Ka ago [85, 86]. By using an Earth Model of Intermediate Complexity, Michel Crucifix was able to show that given Berger's and Loutre's projected data on insolation there should be no glaciation for the next 50 Ka even if $CO_2$ were to fall to mid-glacial levels of 240 ppm (Figure 14.16) [10]. The model suggests that glaciation could have occurred if $CO_2$ had fallen to 210 ppm as early as 7 Ka ago (Figure 14.16).

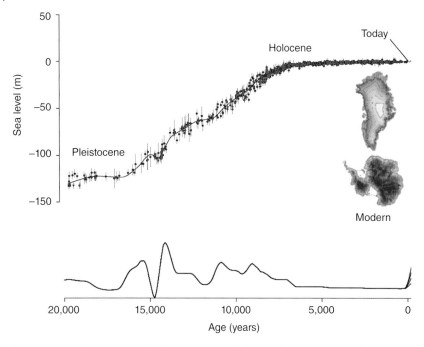

**Figure 14.15** Long-term global mean sea-level change for the past 20 000 years (black line) based on palaeo sea level records (black dots with depth uncertainties shown by blue vertical lines). *Source:* From figure 2a in Ref. [84].

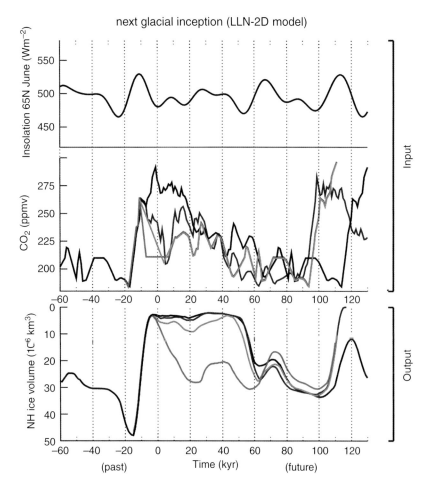

**Figure 14.16** Top panel = Northern Hemisphere insolation; middle panel = input of four prescribed $CO_2$ profiles modelled on Vostok data for the long interglacial of MIS-11; lower panel = output prediction of the next glacial inception derived from an atmosphere-ocean-ice sheet climate model using the four prescribed $CO_2$ scenarios (red, green, blue, black lines). Glaciation would only occur early (red line in lower panel) if $CO_2$ fell to 210 ppm 7 Ka ago (red line in middle panel). *Source:* From figure 4.5 in Ref. [10].

Clearly it did not fall that far (Figure 14.9a). Hence Ruddiman's argument that the small rise in $CO_2$ in the late Holocene has helped to stave off the arrival of the next glaciation seems implausible.

Noting that most interglacials have lasted no more than about 10 Ka, Crucifix answers the question '*why [are we] not currently in the process of glacial inception?*' [87] by pointing out that the Holocene differs from previous interglacials in having exceptionally low eccentricity (Figure 12.2), which means that it has not experienced the sharp decline in insolation that explains, for example, the penultimate glacial inception 120 Ka ago. '*Consequently the effect of climatic precession is damped and the movement towards the glacial inception is much less rapid*' [87]. In effect this makes the Holocene a borderline case. Whilst the insolation circumstances of the Holocene are like those of MIS-19, the latter experienced $CO_2$ values below 250 ppm when glaciation began. The bounce-back of $CO_2$ in the mid Holocene took values well above that 210 ppm threshold [87]. As to what the climate will look like in the not too distant future based just on analyses of changing patterns of insolation, we can see from Figure 14.1 that the climate 3 and 6 Ka from now will be somewhat less extreme than it has been in the late Holocene, but with a slightly cooler Northern Hemisphere polar summer 3 Ka from now, and a cooler Southern Hemisphere polar summer 6 Ka from now. These patterns reflect the continuing low values for eccentricity and precession [10].

In 2009, a team led by Joachim Elsig from the University of Bern, and including Thomas Stocker, used carbon isotopes to help to establish the most likely origin for the Holocene $CO_2$ pattern. They concluded that '*the evolution of atmospheric $CO_2$ over the past 11 kyr is dominated by an early Holocene increase in land biosphere [especially the take up of $CO_2$ by growing peatlands] and changes in the marine-carbonate cycle*' [88]. The changes in the marine carbonate cycle included, on the one hand dissolution of deep ocean carbonates in response to the take up of $CO_2$ by land plants in the early Holocene, and on the other hand the development of coral reefs in the late Holocene, which changed the carbonate ion balance in seawater and released $CO_2$ [88].

Turning to the methane part of his argument, Ruddiman attributed the small rise in $CH_4$ that began about 5 Ka ago (Figures 14.8d, 14.12a and 14.13a) to the increasing culture of rice in Asian wetlands [71, 72, 80]. A numerical simulation of the climate system by Joy Singarayer, Paul Valdes and colleagues in 2011 found that this rise in methane was most likely due to a natural expansion in wetlands [89]. That, in turn, was related to changes in orbital precession and its control over insolation, especially at low latitudes. As Eric Wolff explained: '*Insolation reaches its maximum during the part of the precession cycle when the elliptical orbit of the Earth takes the planet closest to the Sun during northern summer. The result is a stronger monsoon in Asia and other regions, with more summer precipitation, and consequently greater wetland areas and methane production by soil-dwelling microorganisms*' [90]. The methane increase of the past 5 Ka departed from that pattern, increasing when northern summer insolation was on the wane. The reason seemed to be that during the late Holocene there was an increase in wetland sources in South America that outweighed the expected decreases in Eurasia and East Asia. The South American source was reacting to Southern Hemisphere insolation with a different phase from that in the Northern Hemisphere [90]. Hence there was no need to call upon human intervention as Ruddiman had.

Nevertheless, other recent data do provide Ruddiman with some support. In 2012, Celia Sapart from Utrecht University, and colleagues, reviewed variations in the abundance of atmospheric methane and its carbon isotopic composition for the past 2000 years [91]. Their new high-resolution data came from the recent North Greenland Eemian Ice Drilling programme (NEEM) and EUROCORE ice cores from the Summit site in central Greenland. These data confirmed that methane increased by about 70 ppb from 100 BCE–1800 CE, accompanied by fluctuations in its $\partial^{13}C$ composition (Figure 14.8d, upper panel). There was a good correlation between methane emissions and anthropogenic land use, implying a strong anthropogenic control on those emissions over the past 2000 years. After 1800 CE, $CH_4$ rose abruptly, accompanied by a rapid rise in its $\partial^{13}C$ composition, consistent with increased emissions from fossil fuel burning associated with the onset of industrialization. Three centennial-scale falls in $\partial^{13}C$ concentration between 100 BCE and 1600 CE (Figure 14.8d, upper panel) were attributed to periods of biomass burning for clearing or maintaining land for crops. Clearly, human activities must have played an important role in governing the long-term increase in $CH_4$ over this period [91]. However, at this time the population was very much larger than it was when the shift in $CO_2$ began 7 Ka ago (Figure 14.12b).

Do the arguments of Sapart apply all the way back to 5 Ka ago when the slow rise in the $CH_4$ signal began (Figure 14.8d lower panel and Figure 14.12a), as Ruddiman would have it? The difficulty in resolving that issue lies in the fact that there were comparable (albeit temporary) reversals in the methane signal during MISs −9 and −19 (Figure 14.13a). Furthermore, the rise in $CO_2$ slowed right about the time that the rise of $CH_4$ began (compare Figures 14.12a and b), whereas one might expect them to move in parallel if growth in agriculture is the source. The analysts of the Vostok core, led by Jean Jouzel, pointed out in 2013 that in Marine Isotope Stage 11, the 420 Ka interglacial sampled in the ice cores from Vostok and EPICA Dome C, '*the highest levels [of methane] were observed at the very beginning of the interglacial period, then, rather similarly for the Holocene, decreased for 5,000 years, and then began to increase. This increase could obviously not be attributed to human activity ....and it weakened Ruddiman's argument, which was also not obviously supported by the records of concentrations of carbon dioxide recorded at Vostok and at Dome C throughout that interglacial period*' [92]. Jouzel regarded this as crucially important information, because the long interglacial of stage 11 was thought to be an excellent analogue for the Holocene.

Much has been made in the popular press of the possibility that continued warming may melt the Arctic permafrost, releasing large amounts of methane trapped within or beneath it. Eric Wolff pointed out [90] that the modelling by Singarayer and others [89] showed that, during the last interglacial, atmospheric methane decreased slightly at a time when temperatures globally and in the Arctic were a few degrees warmer than at present. This suggests that a large influx of additional emissions from methane hydrates currently trapped in permafrost may be unlikely with continued warming in the immediate future [90]. The existence of a natural envelope for methane, like that for $CO_2$, over the past 400 Ka of the Vostok core (Figure 13.8) suggests that continued warming to the levels typical of past interglacials will not release large extra amounts of $CH_4$.

Before we leave this topic we should return to the subject of volcanoes and their emission of $CO_2$. As we saw in Chapter 13, volcanism increased two to six times above background levels during the last deglaciation between 12 and 7 Ka ago, probably due to decompression of the Earth's mantle in deglaciating regions [93]. The increase in $CO_2$ from these increased volcanic emissions provided a positive feedback, helping to warm the deglaciating world, and may have accounted for the addition of 40 ppm $CO_2$ to the atmosphere. Huybers and Langmuir pointed out that '*Conversely, waning volcanic activity during the Holocene would contribute to cooling and reglaciation, thus tending to suppress [further] volcanic activity and promote the onset of an ice age*' [93]. However, there is no evidence to suggest that volcanic activity contributed to the cooling trend of the late Holocene [32].

Having a strong interest in predicting what the effect on our climate and sea level might be from further emissions of $CO_2$, Jim Hansen, of NASA, was keen to see how much warmer recent interglacial periods were than the Holocene [94]. He found that the temperatures of the major interglacials of the past 450 Ka tended to be close to or above those of the Holocene, whilst the smaller interglacials of the preceding 400 Ka tended to be cooler than the Holocene. The large warm interglacials, he thought, had '*moved into a regime in which there was less summer ice around the*

*Antarctic and Greenland land masses, there was summer melting on the lowest elevation of the ice sheets, and there was summer melting on the ice shelves, which thus largely disappeared. In this regime, we expect warming on the top of the ice sheet to be more than twice global mean warming*' [94]. Why? Because melting of the sea ice and ice shelves created large areas of warm open ocean that reduced albedo and thus affected temperature year-round, something unlikely to have happened in the smaller interglacials prior to 450 Ka ago.

Hansen went on to suggest that the relative stability of Holocene climate came about because orbital controls kept global temperature just below the level required to melt enough ice to decrease albedo in the same way. His calculations suggested that during the peak warm interglacials at 420 and 120 Ka ago, the global climate was only about 1 °C warmer than the Holocene climatic optimum, thus about 2 °C warmer than the late (pre-industrial) Holocene [94]. For that reason, Hansen and his colleague Sato considered that, with modern temperature rise having returned Earth's mean temperature close to the Holocene maximum, the planet is poised to experience the strong amplifying polar feedbacks that likely led to the warmest interglacials of the past 450 Ka. The continued decline in extent of September sea ice in the Arctic in the modern era is one sign that Earth's climate is already on that path. Melting of all or parts of the Greenland and West Antarctic Ice Sheets would be expected to follow, albeit much more slowly.

Undoubtedly there have been human influences on climate through time in the Holocene, not least through fire, forest clearance, agriculture, wiping out large mammals, and changes to the hydrological cycle through the damming of rivers. Even so, the vast majority of the Holocene changes we examine in this chapter seem to have come about naturally, rather than being affected by human activities. In Chapter 15 we will explore in more detail processes happening in the last 2000 years of the Holocene when the human population approached and then exceeded 1 billion people, eventually generating enough waste to affect the ice, ocean, and atmosphere.

Sceptics of the influence of $CO_2$ on climate point to the fact that the decline in temperature over the latter part of the Holocene accompanied a 25 ppm rise in $CO_2$, taking this as evidence against $CO_2$ as a driver of modern climate change. Looking at just temperature and $CO_2$ is not the most appropriate way to check the influence of $CO_2$ on climate, however, since global temperature is driven by other factors too, such as change in albedo, for instance, which may be caused both by changes in vegetation and by changes in ice cover, along with changes in insolation, and the balance between all of them. To get the right answer, all forcings have to be considered simultaneously. Climate change is not a single-issue game featuring just two players: $CO_2$ and temperature.

## 14.3 Climate Variability

As we saw in Chapter 12, Gerard Bond discovered that the deposition of ice-rafted debris during the last glacial period seemed to follow a climate cycle of roughly 1500 years that continued into the Holocene (Figure 12.7) [95]. Although Bond's cycles in ice-rafted debris were very much more prominent during the last glacial period dating back to 80 Ka ago than they were during the Holocene, the amount of variability in any one grain type (Icelandic glass or red-stained grains) was about the same in both periods; only the absolute abundance decreased [95]. That suggested that the cycles were independent of orbital forcing and carried on regardless of glacial state. Some periodic or quasi-periodic forcing agent must cause ice streams to grow or shrink, so increasing or decreasing the rates of discharge of icebergs and thus freshening surface waters in the North Atlantic. During the glacial periods of the Ice Age some threshold (or tipping point) was eventually passed, beyond which there was a major iceberg discharge and cooling event, from which the system eventually recovered, passing back across the threshold as the system warmed up. Such tipping points were not exceeded during the Holocene interglacial, because with the melting of the major ice sheets on North America and Scandinavia there was no longer enough mass of cold ice to respond in such a way in the Northern Hemisphere. Consequently, the cycles in the Holocene had very much lower amplitude than those in the last glacial period.

Looked at more closely, the length of Bond cycles varied from a low of $1328 \pm 539$ years between 31 and 43 Ka ago, and a high of $1795 \pm 425$ years between 64 and 79 Ka ago, with the Holocene cycle lasting $1374 \pm 502$ years [96]. Although there is a difference of 400 years between the extremes of these averages, the standard deviations about the means all overlap to some degree making it difficult to differentiate between them statistically. Spectral analysis of the record of red (haematite) stained grains revealed a slightly different picture, with peaks at intervals of around 1.8 and 4.7 Ka [95, 96]. Ignoring the apparent cycle of 4.7 Ka, Bond felt confident that his team had identified a natural millennial-scale cycle of 1–2 Ka, and he thought that the Little Ice Age, which we'll come to later, was the most recent cold phase of that cycle [96]. However, we do need to bear in mind that as reported in Chapters 12 and 13 subsequent analyses of millennial and centennial variability in prior interglacials have not confirmed the persistence of a 1500 year cycle – the cycle exists but its length is variable.

Bond was not alone in his thinking. In 1999, Giancarlo Bianchi and Nick McCave, of Cambridge University, confirmed that sediments from a Holocene core taken south of Iceland also followed a quasi-periodic 1500 year cycle [97]. Peter De Menocal found much the same signal in marine sediments off west Africa [18]. The supposed 1500 year climate signal was also identified in 2011 in Holocene cave formations known as speleothems in Israel [98]. Speleothems comprise precipitates of calcium carbonate, and most people know them as stalactites (growing down from the cave ceiling, like icicles), and stalagmites (growing up from the cave floor). Their internal layering is a reflection of the history of local rainfall, and the individual layers may be analysed for their $\partial^{18}O$ characteristics, which provide climate signals through time.

In 2012, a team led by Philippe Sorrel found similar variability in their examination of Holocene records of high energy estuarine and coastal sediments from the south coast of the English Channel [99]. '*High storm activity occurred periodically with a frequency of about 1,500 years, closely related to cold and windy periods diagnosed earlier*', Sorrel's team concluded [99]. These various oscillations, linked to Bond cycles, appeared within a range of different spectral signatures ranging from a 2500 year cycle to a 1000 year cycle. There was no consistent correlation between spectral maxima in records of storminess and solar irradiation, making solar activity seem an unlikely cause of millennial-scale variability. Rather, to Sorrel, the storminess reflected a natural periodic cooling of the North Atlantic.

Also in 2012, Dennis Darby of Old Dominion University, Virginia, found Bond's 1500 year cycle in an 8000 year record of ice-rafted iron-rich grains in sediments from the Arctic [100]. These grains were rafted to the Alaskan coast from Russia's Kara Sea during strongly positive phases of the Arctic Oscillation (AO). Darby and colleagues used the sediment record to document an 8000 year history of the AO. Recognizing that there was no 1500 year solar cycle (see Chapter 15 for a detailed discussion on solar cycles), they attributed the forcing to internal variability within the climate system, or to an indirect response to solar forcing at low latitudes [100]. Low latitude palaeoclimate records do show significant linear solar forcing, suggesting that the El Niño–Southern Oscillation system in the Pacific acts as a mediator of the solar influence on the climate system's low latitude heat engine, rather than on the high latitude one [101]. That may explain why Bond's ice-rafting events correlate well with SSTs from the low latitude Atlantic. Solar heating of the tropics would create a stronger equator-to-pole thermal gradient, strengthening winds and storms, and in due course leading to southwardly directed outbreaks of ice rafting [100]. The link would be indirect. As yet the precise mechanism for propagating low-latitude forcing through the climate system to high latitudes is unknown [101].

Other researchers also realized that superimposed on the significant changes of temperature that were driven by orbital changes in insolation through the Holocene, there were slight variations on the millennial scale of the kind – although not the magnitude – identified by Dansgaard and Oeschger in ice records and by Bond in marine sediment records from the last glacial period. For instance, Vinther's data showed that temperatures on the Greenland Ice Sheet varied on the millennial time scale by up to 1 °C both above and below the background Holocene temperature trend [23]. Amongst these variations was a warm period centred near 1000 CE that might represent Lamb's Mediaeval Warm Epoch, followed by a subsequent cold period more or less coinciding with Lamb's Little Ice Age. Alley, too, identified in Greenland ice a warm period centred on about CE 1030, followed by a cold one reaching maximum lows between around 1650 and 1850 CE [21].

Working with Mayewski and others, Alley also identified a significant short-lived cooling event at 8.2 Ka ago [102], most likely caused by a North American glacial lake draining into the adjacent North Atlantic, as mentioned earlier. Vinther calculated the amount of cooling in that event to be around 2 °C [23]. Careful scrutiny of palaeoclimate records from the early Holocene by Eelco Rohling and Heiko Pälike of the UK's National Oceanography Centre in Southampton showed that the 8.2 Ka event was the peak of a cooling event that started at 8.6 Ka and lasted 400–600 years [103]. Rohling and Pälike concluded that whilst the peak event may well have been caused by outflow from glacial Lake Ojibway-Agassiz, its regional and even global extent was difficult to estimate because of its occurrence within a longish cool period of the Holocene. Nevertheless there was good evidence to suggest that the freshwater lid formed over the North Atlantic by discharge from the lake did cause the Meridional Overturning Circulation to slow at about 8.2 Ka ago, with global effects [104].

Mayewski's team's analysis of a global data set provided an opportunity to obtain a global picture of millennial events in the Holocene [29]. They identified six of what they called 'rapid climate change' events at 9–8 Ka ago, 6–5 Ka ago, 4.2–3.8 Ka ago, 3.5–2.5 Ka ago, 1.2–1.0 Ka ago and 600–150 years ago. Polar cooling, tropical aridity, and major changes in atmospheric circulation characterized most of these events, whilst the 6–5 Ka ago event marked the end of the Holocene humid period in tropical Africa, beginning a trend towards tropical aridity. Several of these events coincided with major disruptions of civilization, suggesting the impact of climatic variability on past civilizations.

Like Alley and Vinther, Mayewski documented the brief cooling event at 8.2 Ka, which lay within his 9–8 Ka ago

period [29]. At that time there were still remnants of the ice sheets on North America and Scandinavia. Conditions were cool over much of the Northern Hemisphere. There were major episodes of ice-rafting, stronger winds over the North Atlantic and Siberia, outbreaks of polar air over the Aegean Sea, and glacier advances in Scandinavia and North America. In contrast, glaciers retreated in the Alps, as the air became drier there. In the tropics the monsoon weakened and there were widespread droughts. In the Southern Hemisphere, SSTs warmed around southern Africa, and grounded ice retreated in the Ross Sea.

Most of the other global events reviewed by Mayewski shared the twinning of cool poles with dry topics that also characterized cool periods during the Ice Age. Evidence for the events at 4.2–3.8 and 1.2–1.0 Ka ago (the latter equivalent to 800–1000 CE) appeared in fewer of the records, but their widespread distribution and synchrony suggested the operation of global connections. At these latter times winds weakened over the North Atlantic and Siberia, and temperatures fell in North America and Eurasia [9].

The most recent of these events covered the end of the Holocene from around 600 years ago (or 1400 CE) to the beginning of the modern era [29]. It differed from previous events in being characterized by cool poles and wet tropics. There were rather fewer records to draw upon than one might expect for this relatively recent period, leading Mayewski to complain: '*Unfortunately, determining the nature and duration of later stages of this interval is difficult because high-resolution records for this time are relatively scarce and because several records are missing recent sections as an artifact of sampling. Moreover, interpretation is complicated by potential anthropogenic influences*' [29]. As a consequence his team investigated this event only from 600 to 150 years ago (1400 to 1850 CE). In the Northern Hemisphere, this cool event had the fastest and strongest onset of any of the Holocene events, with glaciers advancing and westerlies strengthening. Whilst Venezuela, Haiti and Florida became more arid, tropical Africa became more humid, and monsoon rains increased in India. Whilst parts of the Antarctic Peninsula warmed, East Antarctic cooled. Glaciers advanced in New Zealand, rainfall increased in Chile, and southern Africa became cool and dry. This event corresponds to the Little Ice Age, of which more in Chapter 15.

Heinz Wanner and his group also identified six 'cold relapses' in their collection of global samples from the Holocene, with peaks at 8.2, 6.3, 4.7, 2.7, 1.55 Ka and 550 years ago [34]. Each of these peaks lay amid a period spanning about 500 years. One might expect these cool peaks to correlate reasonably well with the cool events identified by Mayewski and his team, and with the cold peaks of Bond Cycles. Table 14.1 compares the ranges of Wanner's and Bond's cold peaks (both measured from

**Table 14.1** Comparison of the ranges of Wanner's and Bond's cold peaks [34], with Mayewski's cold events [29].

| Bond cycles (Ka) | Mayewski events (Ka) | Wanner peaks (Ka) | Wanner glacial advances (Ka) |
|---|---|---|---|
| 6: 9.6–9.2 | | | |
| 5a: 8.6–8.25 | 9–8 | 8.55–8.0 | 8.6–8.1 |
| 5b: 7.8–7.2 | | | 7.8–7.5 |
| 4: 5.95–5.1 | 6–5 | 6.45–5.9 | |
| 3: 4.6–3.8 | 4.2–3.8 | 4.8–4.6 | 4.5–4.0 |
| 2: 3.4–2.65 | 3.5–2.5 | 3.35–2.45 | 3.6–3.1 |
| 1: 1.6–1.0 | 1.2–1.0 | 1.8–1.5 | 1.8–1.0 |
| 0: 700–100 years | 600–150 years | 850–150 years | 600–150 years |

Wanner's Figure 3 [34]) and Mayewski's cold events [29], in Ka before present.

The comparison looks close for all three for Bond cycles 5a, 5b, 2, and 0. It is less good for Bond cycle 4, where Wanner's cool period does not closely match the other two. Moreover, Mayewski's 6–5 Ka event is not strictly speaking a cold event, rather the end of the Saharan pluvial period. The comparison is also less good for Bond cycle 3, which does not fit Wanner's cool period, although it does match Wanner's glacial advance. Bond cycle 1 fits with Wanner's glacial advance, but only overlaps slightly with Mayewski's and Wanner's cool periods, which themselves fail to overlap. Neither Wanner nor Mayewski identified Bond cycle 6 as a cool period.

Discrepancies like these may arise because of inadequacies in the chronology of the core samples used. Can we rely on Mayewski's and Wanner's data? They all come from published sources. One of these was subsequently found to be suspect – Alley's 2000 AD Greenland data profile [21], which Vinther later showed to be misleading as a guide to Greenland temperature [23]. Nevertheless, the discrepancies are likely to be real and to reflect the geographical and temporal heterogeneity of the climate signal globally. For instance Mayewski observed advances in glaciers over North America and Scandinavia during his 9–8 Ka ago event, at the same time that glaciers were retreating in the Alps, and each of Wanner's events showed a high degree of spatial and temporal variability. Selecting the boundaries of a global cool period is likely to be a more subjective process than identifying a Bond cycle or a glacial advance. Discrepancies between the North American and European climate are not unexpected, because the North American ice sheet took much longer to melt away than did the Scandinavian and British ice sheets. Clearly, some climate signals tend to be heterogeneous rather than homogeneous.

What about the rapidity of change? Contrary to Mayewski's conclusions, with regard to the past 6 Ka [29], Wanner's PAGES team did not '*find any time period for which a rapid or dramatic climate transition appears even in a majority of … time series*', although rapid shifts were recognized at certain times and in particular regions [33]. However, Wanner did agree that there had been a rapid and short-term climate change event at 8.2 Ka [34].

What are we to make of the various warm and cool periods that are superimposed on the gradual cooling driven by the orbital decline in insolation throughout the Holocene? If we are to extrapolate into the future what we know of the past, we need to be sure the cycles are real and to understand what caused them. MIT oceanographer Carl Wunsch (1941–) called Bond's 1500 year cycle into question on the grounds that it may be a simple alias of inadequately sampled seasonal cycles [105]. When Wunsch removed this signal from ice core and marine core data, climate variability appeared as a continuous process, suggesting that finding a narrowly defined 1500 year cyclicity in the data set may not represent actual millennial events. His analysis supported the idea that Heinrich Events and Dansgaard–Oeschger Events may be quasi-periodic and driven by many possible influences.

Richard Alley disagreed. Since the periodicity was evident in a wide variety of analyses, regardless of sampling interval and other details of the analyses, Alley thought it could not be an alias of any shorter periodicity, such as the annual cycle, as Wunsch had suggested [106]. Wanner and his PAGES 2k colleagues, on the other hand, were inclined to agree with Wunsch, noting '*There is thus scant evidence for consistent periodicities and it seems likely that much of the higher frequency variability observed is due to internal variability or complex feedback processes that would not be expected to show strict spectral coherence*' [33]. Wanner's view of Bond Cycles was that '*the origin of these cycles remains unknown*' [32], and he went on to question the relevance of Bond Cycles for the Southern Hemisphere.

Bianchi and McCave agreed with Bond that the 1500 year cycles probably represented some manifestation of the internal circulation of the ocean of unknown cause, but they did not rule out the possibility of some modulation of climate behaviour by Earth's orbital properties [97]. Bond and his team dismissed that possibility. Millennial-scale climate change could arise from harmonics and combinations of the three main orbital periodicities, but cycles originating in those ways were mostly longer than the observed 1500 year cycle [95].

How might the internal circulation of the ocean have changed? The most likely culprit was the AMOC, which transports warm salty water to high latitudes, where they cool, sink, and return southwards at depth (Figures 12.5

and 12.6). David Thornally, Harry Elderfield and Nick McCave investigated this possibility by using Mg/Ca and $\partial^{18}O$ ratios measured in foraminifera from a sediment core taken in 1938 m of water close to Iceland [107]. The temperatures of near surface waters oscillated between about 10–11 °C over the past 10 Ka, whilst their salinity slowly increased. Subsurface waters from below the thermocline showed much greater variability in both salinity and temperature, depending on whether this water was drawn from the cold, fresh subpolar gyre or the warm, saline subtropical gyre. From 12 to 8.4 Ka ago the North Atlantic was well stratified, with fresh surface water – probably from melting ice – overlying warm, saline tropical gyre water. Then there was a switch to well-mixed waters, followed by an oscillation between stratified and well-mixed waters roughly every 1500 years, attributable to changes in ocean dynamics, much as suggested by Debret and colleagues in 2007 [108]. This appeared to be very much a North Atlantic phenomenon. A strong link is likely between the behaviour of the oceanic AMOC and the atmospheric NAO, which is in turn linked to the behaviour of the atmospheric AO. And as we saw earlier, the 1500 year cycle also shows up in the Arctic where ice rafting occurred during positive phases of the AO [100]. These linkages help to explain why Bond cycles tend to be focussed around the North Atlantic rather than elsewhere.

Greenhouse gases were an equally unlikely cause of millennial change. Neither $CO_2$ nor $CH_4$ showed sufficient variability at the millennial scale during the Holocene to have caused the observed millennial cooling events, nor did volcanic eruptions [29, 34].

Can we blame multi-centennial internal oscillations in the ocean for these millennial events? There is ample evidence for the occurrence of natural oscillations on the decadal scale, for instance in the shape of the Pacific Decadal Oscillation (PDO), or the Atlantic Multi-Decadal Oscillation (AMO) [32]. When the PDO is positive, we get warm conditions in the central tropical Pacific and up the American west coast to Alaska, along with cold conditions in the northwest Pacific. The opposite occurs during the negative phase. Analyses of tree rings data from the region show that the PDO signal is recognizable at least back to 1470 CE, but is not a persistently dominant feature. The AMO is recognized from SST patterns in the North Atlantic, which warm by around 0.2 °C in the positive phase, and decline by the same amount in the negative phase, at intervals of around 20 years – much like the timescale of the PDO. Part of the global warming in the 1930s is likely to have been caused by the positive mode of the AMO, as is the American dustbowl of the same era [10]. These quasi-periodic oscillations supply much of the background high frequency *noise* in the climate spectrum. Similar variability, on time scales of between 35 and 120 years, has been detected in the AMOC. Shorter-term variations attributable to El Niño are superimposed on these larger scale cycles.

Long-period oscillations may arise within the ocean system through non-linear processes affecting either *advection* – the transport of heat and salt, or *convection* – the vigorous vertical mixing of water that occurs when denser water lies atop lighter water [10]. Much of what we know about advection started with Henry (Hank) Stommel (1920–1992), who was appointed to the US National Academy of Science in 1962, and awarded the US National Medal of Science in 1989. His 1961 box model of the thermohaline circulation of the oceans showed that it could feature a sharp decrease in advective transport that we might think of as a shutdown of the Thermohaline Conveyor [109]. The role of oceanic convection is to restore gravitational stability. As Crucifix explained [10], it is a self-maintained process – the exchange of heat between the ocean and the atmosphere can make surface waters denser, promoting further convection. Deep convection occurs today in the Norwegian–Greenland Sea, the Labrador Sea, and the Weddell and Ross Seas. Modelling by Stommel [109] and others shows that convective instability may induce repeated stops and starts of convection, which may explain the abrupt warming and cooling observed by Bond in the Norwegian–Greenland Sea during the Holocene. During a convective shutdown, sea ice advances southwards, maintaining the shutdown and pushing the system into a cold state. Such a state may become persistent in the presence of external forcing, for example in the form of a decrease in solar radiation, or cooling caused by a volcanic eruption. The probability of such events occurring would tend to increase through the Holocene with the long-term cooling brought about by the persistent decrease in orbital insolation during summer in the Northern Hemisphere. Convective instability and the inception of a temporary cold state may also have prolonged the cooling 8.2 Ka ago caused by the discharge of freshwater from Lake Agassiz, which would have taken place in a few years at most, whilst the 8.2 Ka event lasted at least 100 years [10]. While more or less complete shutdowns of the conveyor did occur in glacial times, they do not seem to have taken place during the Holocene, unless perhaps temporarily during the 8.2 Ka ago cool event. According to Crucifix '*A refined version of Stommel's model incorporating realistic propagation times suggests that advective processes may also cause sustained oscillations … associated with the propagation of temperature and salinity anomalies through the conveyor belt, and their periods range between two and four millennia*' [10] (Figure 14.17). Such advective oscillations may explain Bond's millennial cycles [10]. Once again we find ourselves looking at interactions and feedbacks within the climate system.

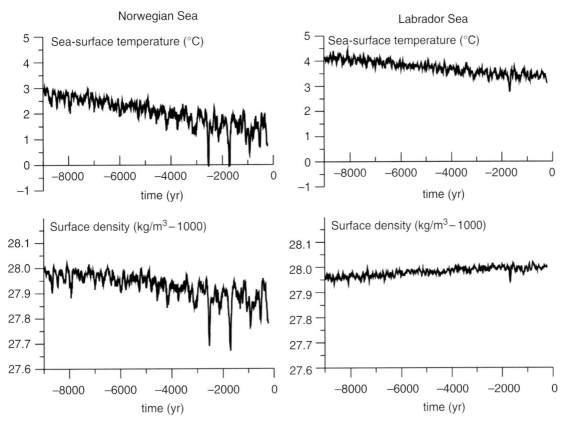

**Figure 14.17** Sea surface temperature and surface water density in the Norwegian Sea predicted by an atmosphere–ocean–vegetation climate model of intermediate complexity and taking into account orbital forcing whilst leaving other boundary conditions (including land and sea ice) as at today. Convective shutdowns occur with increasing frequency with increased cooling, notably at around 2400 years and 1800 years ago. They were triggered by stochastic advances in sea ice maintained by cyclonic anomalies in atmospheric circulation that were caused by the cooling. *Source:* From figure 4.6 in Ref. [10].

Pulling these various bits of information together, the basic theme of Holocene climate is one of initially high insolation leading to the demise of the Northern Hemisphere ice sheets, apart from Greenland, followed by an early to mid Holocene climatic optimum in the Northern Hemisphere as the influence of ice diminished, followed by declining insolation leading to the development of a neoglacial period characterized by glacial advances, culminating in the Little Ice Age. Most of these changes were either typical of, or most strongly developed in, the Northern Hemisphere. In contrast, the tropics tended to warm slightly. The rise in insolation during Termination 1, and the accompanying early Holocene thermal maximum, also made the Antarctic ice sheet shrink away from its maximum extent at the edge of the continental shelf to approximately its present location. The present continental shelf is gouged by deep channels representing the locations of the ice streams that drained the formerly more extensive ice sheet and deposited their sedimentary loads in giant submarine fans on the continental slope.

Superimposed on this fairly large overall cooling (Figure 14.9e) were small climate variations on the millennial or centennial scale [19]. Some distinct oscillations or climate steps appear to be widespread, for example across the North Atlantic region, notably at 8.2, 5.5–5.3, and 2.5 Ka ago, where they punctuated the Holocene decline in temperature. In that region, the coldest points of each of these millennial steps in SST were 2–4 °C colder than the warmest part [19]. The event at 8.2 Ka ago was the most sudden and striking of the Holocene, bringing cool dry conditions to the North Atlantic region. It is picked up in ice cores from as far away as Greenland and Antarctica. That particular event represents a massive flood of glacial lake water into the Arctic and North Atlantic. There was a broader and less intense shift to colder and drier conditions at 5.5 Ka ago, and a similar shift at 4 Ka ago that coincided with the collapse of a number of early civilizations [19]. While it is possible that solar forcing played a role in causing these cycles, it may have merely exacerbated the effects of oscillations caused by natural variability in oceanic advection and convection, as suggested by Michel Crucifix

[10]. That being said, it is more likely that the effects of advection and convection will be focused in specific regions, like the North Atlantic and Arctic provinces, rather than globally.

In 2017, Pepijn Bakker and colleagues puzzled over the fact that based on proxy-based indicators of climate change, global climate models systematically underestimated Holocene climatic variability on centennial to multi-millennial timescales. One of the several explanations for this discrepancy could be that '*global climate models do not consider [potentially important] slow climate feedbacks related to the carbon cycle or interactions between ice sheets and climate*' [110]. The authors showed that fluctuations in the discharge of ice from Antarctica, caused by quite small changes in subsurface ocean temperature (cf. Figure 13.13), could amplify multi-centennial climate variability both regionally and globally, and may have driven some Holocene climate fluctuations on centennial to multi-millennial timescales [110]. For much the same reasons, the inadequacy of the treatment of ice sheet–climate interactions in the models for the Holocene may explain current disparities between observed Southern Ocean temperature and sea level and their simulated values [110].

Decadal change is also apparent throughout much of the Holocene. Mads Knudsen of Aarhus University in Denmark and colleagues observed in 2011 that '*distinct, ~55- to 70-year oscillations characterized the North Atlantic ocean–atmosphere variability over the past 8,000 years*' [111]. These appear to be tied to '*a quasi-persistent ~55- to 70-year [Atlantic Multidecadal Oscillation] (AMO), linked to internal ocean-atmosphere variability*', with no direct link to solar variability [111]. We will explore in Chapter 15 the significance of the AMO for more recent climate change.

Crucifix summed up the end Holocene situation as follows: '*whatever caused the mid-Holocene $CO_2$ trend, there is no doubt about the fact that we are now en route for an exceptionally long interglacial. $CO_2$ concentration is now above 400 ppm and is quite likely to double at least its value since the pre-industrial period. It will take many tens of thousands of years before the anthropogenic shock is absorbed by the climate system, first via alkalinity changes (at the expense of corals and other calcigeniferous organisms), and then by the slow dynamics of silicate weathering (Archer and Brovkin 2008 [112]). We are likely to miss orbital windows for the next glacial inception during 100 ka at least, and perhaps even more*' [87].

However, before concluding we must turn to Chapter 15 to see what role the Sun may have played in driving millennial or centennial climate change, and to explore in some detail climate change over the last 2000 years and beyond.

# References

**1** Birks, H.J.B. (2009). Holocene climate research – progress, paradigms, and problems. In: *Natural Climate Variability and Global Warming: A Holocene Perspective* (eds. R.W. Battarbee and H.A. Binney), 7–57. Chichester and Oxford: Wiley Blackwell.

**2** Lamb, H.H. (1966). *The Changing Climate*, 236 pp. London: Methuen.

**3** Summerhayes, C.P., Milliman, J.D., Briggs, S.R. et al. (1976). Northwest African shelf sediments: influence of climate and sedimentary processes. *Journal of Geology* 84: 277–300.

**4** Emiliani, C. (1955). Pleistocene temperatures. *Journal of Geology* 63 (6): 538–578.

**5** Matthes, F.E. (1939). Report of committee on glaciation, April 1939. *Transactions of the American Geophysical Union* 20: 518–523.

**6** Crowley, T.J. and North, G.R. (2011). *Paleoclimatology, Oxford Monographs on Geology and Geophysics* **18**, 339 pp. Oxford University Press.

**7** Porter, S.C. and Denton, G.H. (1967). Chronology of neoglaciation in the North American Cordillera. *American Journal of Science* 265: 177–210.

**8** Denton, G.H. and Karlen, W. (1973). Holocene climatic variations – their pattern and possible causes. *Quaternary Research* 3: 155–205.

**9** Yin, Q. and Berger, A. (2015). Interglacial analogies of the Holocene and its natural near future. *Quaternary Science Reviews* 120: 28–46.

**10** Crucifix, M. (2009). Modeling the climate of the Holocene. In: *Natural Climate Variability and Global Warming: A Holocene Perspective* (eds. R.W. Battarbee and H.A. Binney), 98–122. Chichester and Oxford: Wiley Blackwell.

**11** Huybers, P. and Denton, G. (2008). Antarctic temperature at orbital timescales controlled by local summer duration. *Nature Geoscience* 1: 787–792.

**12** Huybers, P. (2009). Antarctica's orbital beat. *Science* 325: 1085–1086.

**13** Loutre, M.-F. and Berger, A. (2000). Future climate changes: are we entering an exceptionally long interglacial? *Climatic Change* 46: 61–90.

**14** COHMAP Members (33 authors) (1988). Climatic changes of the last 18,000 years: observations and model simulations. *Science* 241: 1043–1052.

**15** Wright, H.E., Kutzbach, J.E., Webb, T. et al. (1993). *Global Climates Since the Last Glacial Maximum*, 569 pp. Minneapolis: University Minnesota Press.

**16** Kutzbach, J.E., Gallimore, R., Harrison, S.P. et al. (1998). Climate and biome simulations for the past 21,000 years. *Quaternary Science Reviews* 17: 473–506.

**17** Miller, G.H., Geiersdottir, A., Zhong, Y. et al. (2012). Abrupt onset of the Little Ice Age triggered by volcanism and sustained by sea-ice/ocean feedbacks. *Geophysical Research Letters* https://doi.org/10.1029/2011GL050168.

**18** De Menocal, P., Ortiz, J., Guilderson, T., and Sarnthein, M. (2000). Coherent high- and low- latitude climate variability during the Holocene warm period. *Science* 288: 2198–2202.

**19** Maslin, M., Seidov, D., and Lowe, J. (2001). Synthesis of the nature and causes of rapid climate transitions during the Quaternary. In: *The Oceans and Rapid Climate Change: Past, Present and Future, Geophys. Monogr.* **126** (eds. D. Seidov, B.J. Haupt and M. Maslin), 9–52. Wshington D.C.: American Geophysical Union.

**20** Frakes, L.A., Francis, J.E., and Syktus, J.I. (1992). *Climate Modes of the Phanerozoic – The History of the Earth's Climate Over the Past 600 Million Years*, 274 pp. Cambridge University Press.

**21** Alley, R.B. (2000). The Younger Dryas cold interval as viewed from Central Greenland. *Quaternary Science Reviews* 19: 213–226.

**22** Johnsen, S.J., Dahl-Jensen, D., Gundestrup, N. et al. (2001). Oxygen isotope and palaeotemperature records from six Greenland ice-core stations: Camp Century, Dye-3, GRIP, GISP2, Renland and NorthGRIP. *Journal of Quaternary Science* 16: 299–307.

**23** Vinther, B.M., Buchardt, S.L., Clausen, H.B. et al. (2009). Holocene thinning of the Greenland ice sheet. *Nature* 461: 385–388.

**24** Kobashi, T., Kawamura, K., Severinghaus, J.P. et al. (2011). High variability of Greenland surface temperature over the past 4000 years estimated from trapped air in an icc core. *Geophysical Research Letters* 38 https://doi.org/10.1029/2011GL049444.

**25** Jansen, E., Andersson, C., Moros, M. et al. (2009). The early to mid-Holocene thermal optimum in the North Atlantic. In: *Natural Climate Variability and Global Warming: A Holocene Perspective* (eds. R.W. Battarbee and H.A. Binney), 123–137. Chichester and Oxford: Wiley Blackwell.

**26** MacDonald, G.M., Velichko, A.A., Kremenetski, C.V. et al. (2000). Holocene Treeline history and climate change across Northern Eurasia. *Quaternary Research* 53: 302–311.

**27** Miller, G.H., Alley, R.B., Brigham-Grette, J. et al. (2010). Arctic amplification: can the past constrain the future?

**28** Lecavaliera, B.S., Fisher, D.A., and Milne, G.A. (2017). High Arctic Holocene temperature record from the Agassiz ice cap and Greenland ice sheet evolution. *Proceedings of the National Academy of Sciences of the United States of America* 114 (23): 5952–5957.

**29** Mayewski, P.A., Rohling, E., Stager, J.C. et al. (2004). Holocene climatic variability. *Quaternary Research* 62: 243–255.

**30** Lorenz, S.J., Kim, J.-H., Rimbu, N. et al. (2006). Orbitally driven insolation forcing on Holocene climate trends: evidence from alkenone data and climate modeling. *Paleoceanography* 21 https://doi.org/10.1029/2005PA001152.

**31** HOCLAT A Web-based Holocene Climate Atlas. https://www.oeschger.unibe.ch/research/projects_and_databases/web_based_holocene_climate_atlas_hoclat/index_eng.html (December 2013)

**32** Wanner, H., Beer, J., Bütikofer, J. et al. (2008). Mid-to late Holocene climate change; an overview. *Quaternary Science Reviews* 27: 1791–1828.

**33** PAGES 2k Consortium (78 authors) (2013). Continental-scale temperature variability during the last two millennia. *Nature Geoscience* 6: 339–346.

**34** Wanner, H., Solomina, O., Grosjean, M. et al. (2011). Structure and origin of Holocene cold events. *Quaternary Science Reviews* 30: 3109–3123.

**35** Beer, J. and Wanner, H. (2012). Corrigendum to "Mid- to late Holocene climate change - an overview" [Quaternary Sci. Rev. 27 (2008) 1791-1828]. *Quaternary Science Reviews* 51: 93–94.

**36** Marcott, S.A., Shakun, J.D., Clark, P.U., and Mix, A.C. (2013). A reconstruction of regional and global temperature for the past 11,300 years. *Science* 339: 1198–1201.

**37** Zhao, M., Beveridge, N.A.S., Shackleton, N.J., and Sarnthein, M. (1995). Molecular stratigraphy of cores off Northwest Africa: sea surface temperature history over the last 80 ka. *Paleoceanography* 10: 661–675.

**38** Calvo, E., Grimalt, J., and Jansen, E. (2002). High resolution $U^k_{37}$ sea surface temperature reconstruction in the Norwegian Sea during the Holocene. *Quaternary Science Reviews* 21: 1385–1394.

**39** Marchal, O., Cacho, I., Stocker, T.F. et al. (2000). Apparent long-term cooling of the sea surface in the Northeast Atlantic and Mediterranean during the Holocene. *Quaternary Science Reviews* 21: 455–483.

**40** Herbert, T.D. (2003). Alkenone paleotemperature determinations. In: *The Oceans and Marine Geochemistry, Treatise on Geochemistry* (eds H.D. Holland and K.K. Turekian), vol. 6 (ed. H. Elderfield), 391–432. Oxford: Elsevier-Pergamon.

*Quaternary Science Reviews*: 1–12. https://doi.org/10.1016/j.quascirev.2010.02.008.

**41** Bard, E., Rostek, F., and Sonzogni, C. (1997). Interhemispheric synchrony of the last deglaciation inferred from alkenone paleothermometry. *Nature* 385: 707–710.

**42** Turner, J., Bindschadler, R.A., Convey, P. et al. (2009). *Antarctic Climate Change and the Environment*, Scientific Committee on Antarctic Research, Cambridge, ISBN 978-0-948277-22-1, 526 pp; https://www.scar.org/library/scar-publications/occasional-publications/3508-antarctic-climate-change-and-the-environment-1

**43** Stenni, B., Curran, M.A.J., Abram, N.J. et al. (2017). Antarctic climate variability on regional and continental scales over the last 2000 years. *Climate of the Past* 13: 1609–1634.

**44** Mulvaney, R., Abram, N.J., Hindmarsh, R.C.A. et al. (2012). Recent Antarctic Peninsula warming relative to Holocene climate and ice-shelf history. *Nature* 489: 141–145.

**45** Abrams, N.J., Mulvaney, R., Wolff, E.J. et al. (2013). Acceleration of snow melt in an Antarctic Peninsula ice core during the twentieth century. *Nature Geoscience* 6: 404–411.

**46** Shevenell, A.E., Ingalls, A.E., Domack, E.W., and Kelly, C. (2011). Holocene Southern Ocean surface temperature variability west of the Antarctic Peninsula. *Nature* 470: 250–254.

**47** Etourneau, J., Collins, L.G., Willmott, V. et al. (2013). Holocene climate variations in the western Antarctic Peninsula: evidence for sea ice extent predominantly controlled by changes in insolation and ENSO variability. *Climate of the Past* 9: 1431–1446.

**48** Renssen, H., Goosse, H., Fichelet, T. et al. (2005). Holocene climate evolution in the high-latitude Southern Hemisphere simulated by a coupled atmosphere-sea ice-ocean-vegetation model. *The Holocene* 15 (7): 951–984.

**49** IUGS International Union of Geological Sciences (2018). http://www.stratigraphy.org/ICSchart/ChronostratChart2018-08.jpg.

**50** Liu, Z., Zhu, J., Rosenthal, Y. et al. (2014). The Holocene temperature conundrum. *Proceedings of the National Academy of Sciences of the United States of America* 11 (2014): E3501–E3505.

**51** Baker, J.L., Lachniet, M.S., Chervyatsova, O. et al. (2017). Holocene warming in western continental Eurasia driven by glacial retreat and greenhouse forcing. *Nature Geoscience* 10: 430. https://doi.org/10.1038/NGEO2953.

**52** Stein, R., Fahl, K., Schade, I. et al. (2017). Holocene variability in sea ice cover, primary production, and Pacific-Water inflow and climate change in the Chukchi and East Siberian Seas (Arctic Ocean). *Journal of Quaternary Science* 32 (3): 362–379.

**53** Alley, R.B., Andrews, J.T., Brigham-Grette, J. et al. (2013). History of the Greenland Ice Sheet: paleoclimatic insights. *Quaternary Science Reviews* 29: 1728–1756.

**54** Öberg, L. and Kullman, L. (2011). Recent glacier recession – a new source of postglacial treeline and climate history in the Swedish Scandes. *Landscape Online* 26: 1–38.

**55** Crosta, X., Debret, M., Denis, D. et al. (2007). Holocene long- and short-term climate changes off Adelie Land, East Antarctica. *Geochemistry, Geophysics, Geosystems* 8 (11): 15. https://doi.org/10.1029/2007GC001718.

**56** Kingslake, J., Scherer, R.P., Albrecht, T. et al. (2018). Extensive retreat and re-advance of the West Antarctic Ice Sheet during the Holocene. *Nature* 558 (7710): 430–434.

**57** Hillenbrand, C.-D., Smith, J.A., Hodell, D.A. et al. (2017). West Antarctic Ice Sheet retreat driven by Holocene warm water incursions. *Nature* 547: 43–48.

**58** Bradley, R.S. (2009). Holocene perspectives on future climate change. In: *Natural Climate Variability and Global Warming: A Holocene Perspective* (eds. R.W. Battarbee and H.A. Binney), 254–268. Chichester and Oxford: Wiley Blackwell.

**59** Hoogakker, B.A.A., Chapman, M.R., McCave, I.N. et al. (2011). Dynamics of North Atlantic deep water masses during the Holocene. *Paleoceanography* 26: PA4214.

**60** Marsicek, J., Shuman, B.N., Bartlein, P.J. et al. (2018). Reconciling divergent trends and millennial variations in Holocene temperatures. *Nature* 554: 92–96.

**61** Berger, A. and Loutre, M.-F. (2002). An exceptionally long interglacial ahead. *Science* 297: 1287–1288.

**62** Sundquist, E.T. and Visser, K. (2005). The geologic history of the carbon cycle. In: *Biogeochemistry, Treatise on Geochemistry* (eds H.D. Holland and K.K. Turekian), vol. 8 (ed. W.H. Schlesinger), 425–513. Oxford: Elsevier-Pergamon.

**63** Waters, C.N., Zalasiewicz, J., Summerhayes, C. et al. (2016). The Anthropocene is functionally and stratigraphically distinct from the Holocene. *Science* 351 (6269): 10. https://doi.org/10.1126/science.aad2622.

**64** Studer, A.S., Sigman, D.M., Martínez-García, A. et al. (2018). Increased nutrient supply to the Southern Ocean during the Holocene and its implications for the pre-industrial atmospheric $CO_2$ rise. *Nature Geoscience* https://doi.org/10.1038/s41561-018-0191-8.

**65** Indermühle, A., Stocker, T.F., Joos, F. et al. (1999). Holocene carbon-cycle dynamics based on $CO_2$ trapped in ice at Taylor Dome, Antarctica. *Nature* 398: 121–126.

**66** MacDonald, G.M., Beilman, D.W., Kremenetski, K.V. et al. (2006). Rapid early development of circumarctic peatlands and atmospheric $CH_4$ and $CO_2$ variations. *Science* 314: 285–288.

**67** Chappelaz, J., Blunier, T., Kints, S. et al. (1997). Changes in the atmospheric CH$_4$ gradient between Greenland and Antarctica during the Holocene. *Journal of Geophysical Research – Atmospheres* 102: 15987–15997.

**68** Falkowski, P., Scholes, R.J., Boyle, E. et al. (2000). The global carbon cycle: a test of our knowledge of Earth as a system. *Science* 290: 291–296.

**69** Fischer, H., Wahlen, M., Smith, J. et al. (1999). Ice core records of atmospheric CO$_2$ around the last three glacial terminations. *Science* 283: 1712–1714. https://doi.org/10.1126/science.283.5408.1712.

**70** Ruddiman, W.F. and Thomson, J.S. (2001). The case for human causes of increased atmospheric CH$_4$. *Quaternary Science Reviews* 20: 1769–1777.

**71** Ruddiman, W.F. (2003). Orbital insolation, ice volume, and greenhouse gases. *Quaternary Science Reviews* 22: 1597–1629.

**72** Ruddiman, W.F. (2005). *Plows, Plagues and Petroleum: How Humans Took Control of Climate*, 202 pp. Princeton University Press.

**73** Broecker, W.S., Clark, E., McCorkle, D.C. et al. (1999). Evidence for a reduction in the carbonate ion content of the deep sea during the course of the Holocene. *Paleoceanography* 14 (6): 744–752.

**74** Broecker, W.S. and Stocker, T.F. (2006). The Holocene CO2 rise: anthropogenic or natural? *Eos* 87 (3): 27.

**75** Yu, J.M. and Elderfield, H. (2007). Benthic foraminiferal B/Ca ratios reflect deep water carbonate saturation state. *Earth and Planetary Science Letters* 258: 73–86.

**76** Yu, J.M., Anderson, R.F., and Rohling, E.J. (2013). Deep ocean carbonate chemistry and glacial-interglacial atmospheric CO$_2$. *Oceanography* 27 (1): 16–25.

**77** Menviel, L. and Joos, F. (2012). Toward explaining the Holocene carbon dioxide and carbon isotope records: results from transient ocean carbon cycle-climate simulations. *Paleoceanography* 27 https://doi.org/10.1029/2011pa002224.

**78** Ciais, P., Sabine, C., Bala, G. et al. (2013). Carbon and other biogeochemical cycles. In: *Climate Change 2013: The Physical Science Basis, Contribution of Working Group I to the Fifth Assessment Report of the Intergovernmental Panel on Climate Change* (eds. T.F. Stocker, D. Qin, G.-K. Plattner, et al.), 465–570. Cambridge and New York: Cambridge University Press.

**79** Skinner, L.C., Fallon, S., Waelbroeck, C. et al. (2010). Ventilation of the deep Southern Ocean and deglacial CO$_2$ rise. *Science* 328: 1147–1151.

**80** Ruddiman, W.F., Fuller, D.Q., Kutzbach, J.E. et al. (2016). Late Holocene climate: Natural or anthropogenic? *Reviews of Geophysics* 54 https://doi.org/10.1002/2015RG000503.

**81** Zalasiewicz, J., Waters, C., Head, M.J., Poirier, C., Summerhayes, C.P. et al. (2019). A formal Anthropocene is compatible with but distinct from its diachronous anthropogenic counterparts: a response to W.F. Ruddiman's "three-flaws in defining a formal Anthropocene". Progress in Physical Geography. PPG-18–130.

**82** Stephens, L., Fuller, D., Boivin, N. et al. (2019). Archaeological assessment reveals Earth's early transformation through land use. *Science* 365 (6456): 897–902. https://doi.org/10.1126/science.aax1192.

**83** Lambeck, K., Rouby, H., Purcell, A. et al. (2014). Sea level and global ice volumes from the Last Glacial Maximum to the Holocene. *Proceedings of the National Academy of Sciences of the United States of America* 111: 15296–15303.

**84** Clark, P.U., Shakun, J.D., Marcott, S.A. et al. (2016). Consequences of twenty-first-century policy for multi-millennial climate and sea-level change. *Nature Climate Change* 6: 360–369.

**85** Berger, A. (2009). Astronomical theory of climate change. In: *Encyclopedia of Paleoclimatology and Ancient Environments* (ed. V. Gornitz), 51–57. Dordecht: Springer.

**86** Berger, A. and Loutre, M.-F. (2003). Climate 400,000 years ago, a key to the future? In: *Earth's Climate and Orbital Eccentricity: The Marine Isotope Stage 11 Question, Geophys. Monogr.* **137** (eds. A. Droxler, R.Z. Poore and L.H. Burckle), 17–26. Washington D.C.: American Geophysical Union.

**87** Crucifix, M. (2019). Pleistocene glaciations. In: *Climate Change in the Holocene, Impacts and Human Adaptation* (ed. E. Chiotis). CRC Publishers, Taylor and Francis https://doi.org/10.1201/9781351260244.

**88** Elsig, J., Schmitt, J., Leuenberger, D. et al. (2009). Stable isotope constraints on Holocene carbon cycle changes from an Antarctic ice core. *Nature* 461: 507–510.

**89** Singarayer, J.S., Valdes, P.J., Friedlingstein, P. et al. (2011). Late Holocene methane caused by orbitally controlled increase in tropical sources. *Nature* 470: 82–82.

**90** Wolff, E.W. (2010). Methane and monsoons. *Nature* 470: 49–50.

**91** Sapart, C.J., Monteil, G., Prokopiou, M. et al. (2012). Natural and anthropogenic variations in methane sources during the past two millennia. *Nature* 490: 85–88.

**92** Jouzel, J., Lorius, C., and Raynaud, D. (2013). *The White Planet: The Evolution and Future of our Frozen World*, 306 pp. Princeton University Press.

**93** Huybers, P. and Langmuir, C. (2009). Feedback between deglaciation, volcanism and atmospheric CO$_2$. *Earth and Planetary Science Letters* 286: 479–491.

**94** Hansen, J.E. and Sato, M. (2012). Paleoclimate implications for human-made climate change. In: *Climate Change: Inferences from Paleoclimate and Regional Aspects* (eds. A. Berger, F. Mesinger and D. Šijački), 21–48. Springer https://doi.org/10.1007/978-3-7091-0973-1_2.

95 Bond, G., Showers, W., Cheseby, M. et al. (1997). A pervasive millennial-scale cycle in North Atlantic Holocene and glacial climates. *Science* 278: 1257–1266.

96 Bond, G.C., Showers, W., Elliot, M. et al. (1999). The North Atlantic's 1–2 kyr climate rhythm: relation to Heinrich events, Dansgaard/Oeschger cycles and the Little Ice Age. In: *Mechanisms of Global Climate Change at Millennial Time Scales, Geophys. Monogr.* **112** (eds. P.U. Clark, R.S. Webb and L.D. Keigwin), 35–58. Washington, D.C.: American Geophysical Union.

97 Bianchi, C. and McCave, I.N. (1999). Holocene periodicity in North Atlantic climate and deep-ocean flow south of Iceland. *Nature* 397: 515–517.

98 Bar-Matthews, M. and Ayalon, A. (2011). Mid-Holocene climate variations revealed by high-resolution speleothem records from Soreq Cave, Israel and their correlation with cultural changes. *The Holocene* 21 (1): 163–171.

99 Sorrel, P., Debret, M., Billeaud, I. et al. (2012). Persistent non-solar forcing of Holocene storm dynamics in coastal sedimentary archives. *Nature Geoscience* 5: 892–896.

100 Darby, D.A., Ortiz, J.D., Grosch, C.E., and Lund, S.P. (2012). 1,500-year cycle in the Arctic oscillation identified in Holocene Arctic sea-ice drift. *Nature Geoscience* 5: 897–900.

101 Marchitto, T.M., Muschelere, R., Ortiz, J.D. et al. (2010). Dynamical response of the tropical Pacific Ocean to solar forcing during the early Holocene. *Science* 330: 1378–1381.

102 Alley, R.B., Mayewski, P.A., Sowers, T. et al. (1997). Holocene climatic instability: a prominent widespread event 8200 years ago. *Geology* 25: 483–486.

103 Rohling, E.J. and Pälike, H. (2005). Centennial-scale climate cooling with a sudden cold event around 8,200 years ago. *Nature* 434: 975–979.

104 Cronin, T.M. (2010). *Paleoclimates – Understanding Climate Change Past and Present*, 441 pp. Columbia University Press.

105 Wunsch, C. (2000). On sharp spectral lines in the climate record and the millennial peak. *Paleoceanography* 15: 417–424.

106 Alley, R.B., Anandakrishnan, S., Jung, P., and Clough, A. (2001). Stochastic resonance in the North Atlantic: further insights. In: *The Oceans and Rapid Climate Change: Past, Present and Future, Geophys. Monogr.* **126** (eds. D. Seidov, B.J. Haupt and M. Maslin), 57–68. Washinton DC: American Geophysical Union.

107 Thornally, D.J.R., Elderfield, H., and McCave, I.N. (2009). Holocene oscillations in temperature and salinity of the surface subpolar North Atlantic. *Nature* 457: 711–714.

108 Debret, M., Bout-Roumazeilles, V., Grousset, F. et al. (2007). The origin of the 1500-year climate cycles in Holocene North Atlantic records. *Climate of the Past* 3: 569–575.

109 Stommel, H. (1961). Thermohaline convection with two stable regimes of flow. *Tellus* 13: 224–230.

110 Bakker, P., Clark, P.U., Golledge, N.R. et al. (2017). Centennial-scale Holocene climate variations amplified by Antarctic Ice Sheet discharge. *Nature* 54: 72–76.

111 Knudsen, M.F., Seidenkrantz, M.-S., Jacobsen, B.H., and Kuijpers, A. (2011). Tracking the Atlantic Multidecadal Oscillation through the last 8,000 years. *Nature Communications* 2 (1): 178. https://doi.org/10.1038/ncomms1186.

112 Archer, D. and Brovkin, V. (2008). The millennial atmospheric lifetime of anthropogenic CO2. *Climatic Change* 90 (3): 283–297.

# 15

# The Late Holocene and the Anthropocene

---

**LEARNING OUTCOMES**

- Understand and be able to explain the relationship between global temperatures, Greenland temperatures, aerosols, greenhouse gases, and the Atlantic Multidecadal Oscillation (AMO) for the past 100 years.
- Understand and be able to describe the climatic variations of the Medieval Warm Period and Little Ice Age (LIA) in relation to forcing by orbital change, volcanic change, and solar change.
- Be able to compare and contrast Lamb's view of climate change over the past 1000 years with that of modern palaeoclimatologists (e.g. Mann, Ljungqvist, Wanner, Vinther, Moberg, Mayewski, and the PAGES 2kyr group).
- Understand and be able to describe and explain climate change in the Arctic (including Greenland) and in the Antarctic over the past 2000 years, and why these two regions differ.
- Understand and be able to explain the evidence for solar variability over the past 1000 years, naming the main solar cycles and their effects.
- Know and be able to explain how these solar cycles relate to cooling, precipitation history, ice rafting, and the Atlantic Meridional Overturning Circulation (AMOC).
- Understand and be able to explain what solar cycles suggest about climate change over the next 150 years.
- Understand and be able to provide examples of volcanic activity affecting temperature, tree growth and glacier growth over the past 1000 years.
- Understand and be able to explain the evidence for the rise in sea level from about 1850 onwards in relation to the patterns of ice loss or gain from Greenland, Antarctica, and regions of mountain glaciers.
- Understand and be able to explain the origins of concern about the possible future rate of ice loss from East and West Antarctica.
- Understand and be able to explain the reasons for the 'pause' in temperature rise between 2000 and 2013, and what this means regarding the relationship between $CO_2$ and temperature.
- Explain what $CO_2$ and climate-related changes we can expect to see in the Anthropocene.

---

## 15.1 The Medieval Warm Period and the Little Ice Age

The best known, or perhaps the most widely discussed, climatic events during the late Holocene are the Medieval Warm Period and the LIA. We can get a modern perspective on these events from Takuro Kobashi from the National Institute of Polar Research in Tokyo (Figure 15.1). Kobashi found that the average modern snow temperature at the GISP2 site near the summit of the Greenland ice sheet, was −29.9 °C (for the period 2001–2010), which was not significantly different statistically from, though slightly above, the −30.7 °C ± 1.0 °C average for the past 4 Ka [1]. Nevertheless, the modern temperature was part of a trend

of rising temperature following the nadir of the LIA at about −33.5 °C close to 1750 CE. Over the past 2 Ka the warmest temperatures, rising some 2 °C above the average, occurred around 750 CE. They were followed by a brief but sharp cooling to about −33 °C, after which the temperatures rose to a broad period of warmth between 900 and 1250 CE – the Medieval Warm Period – when temperatures reached close to those of 1950 for periods of about 20 years at about 950 CE and 1140 CE. Temperatures were generally higher 4–3 Ka ago, consistent with the decline in insolation since then (as noted in Chapter 14). Following the LIA, there was a slow recovery to about −31.5 °C by 1920, then a sharp rise to present temperatures in a brief warming event between 1930 and 1950. Temperatures then cooled back to

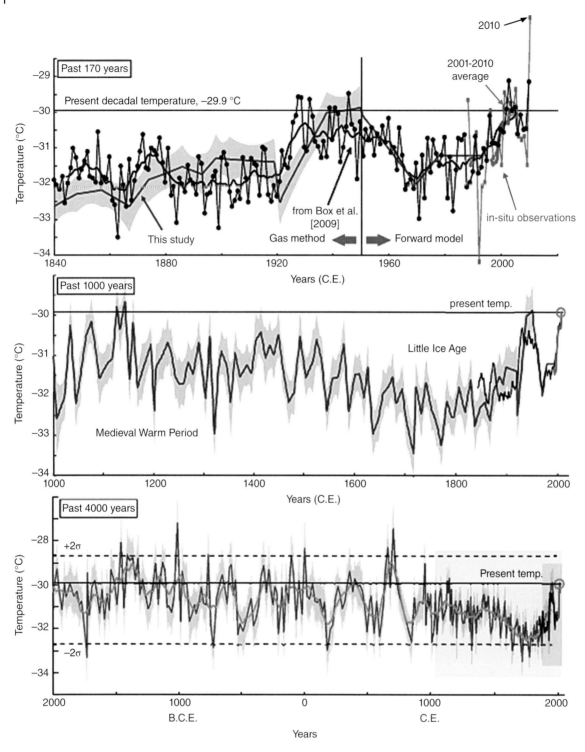

**Figure 15.1** Top: Air temperature over the past 170 years (1840–2010), and (middle and lower panels) reconstructed Greenland snow surface temperatures for the past 4000 years. The thick blue line and blue band represents the reconstructed Greenland temperature and one standard deviation error, respectively. Thick and thin black lines are the inversion-adjusted reconstructed 'Summit' annual air temperatures and 10-year moving average temperatures, respectively. Thin and thick red lines are the inversion adjusted annual and ten year moving average air temperature records, respectively. Middle: Past 1000 years of Greenland temperature. Thick blue line and band are the same as above. Black and red lines are the 'Summit' and air temperature decadal average temperatures, as above. Bottom: Past 4000 years of Greenland temperature. Thick blue line and band are the same as above. Thick green line represents 100 year moving averages. Black and red lines are the 'Summit' and air temperature decadal average temperature, respectively. Blue and pink rectangles represent the periods of 1000–2010 CE (middle) and 1840–2010 CE (top), respectively. Present temperature is calculated from the inversion adjusted air temperature decadal average (2001–2010) as – 29.9 °C (top). Present temperature and ± 2 standard deviations are illustrated by lines in the plots. Green circles are the current decadal average temperature, as above (–29.9 °C, for 2001–2010). *Source:* From figure 1 in Ref. [1].

−32°C by 1970 before rising to present values by about 2005. These local Greenland patterns followed the global one. The ice core record correlated well with observations of air temperature during the instrumental period (post-1840), so can be taken as representative of actual air temperatures during the pre-instrumental measurement period [1].

Corroboration of the medieval warming pattern comes from records of sea surface temperature (SST) from the southeast Greenland continental shelf, where the warmest surface water conditions occurred between 1000 and 1200 CE [2]. This was followed by a cool phase lasting to 1890 CE, associated with the LIA and with more extensive sea ice. Much colder temperatures were apparent between 500 BCE and 600 CE [2], though that is not apparent from the air temperatures recorded in Greenland ice (Figure 15.1).

In the modern era, since 1870, these local Greenland patterns follow the SST fluctuations of the AMO, as we can see by comparing Figure 15.1 with Figure 15.2. As Mads Knudsen pointed out, the AMO index began increasing from about 1920, and was in a positive phase from 1925 to 1965, coinciding with warming in North Atlantic SSTs (Figure 15.2) [3]. This was also a period of anomalous warming globally (Figure 15.2a), and in Greenland (Figure 15.1). Both the global and the Greenland patterns may well have been influenced by this natural oscillation in the climate system, as well as by the Pacific Decadal Oscillation (PDO), which also moved into its positive (warm) phase between 1920 and 1945 [4], likely providing yet a further push to the global rise in temperature. These natural oscillations are important features of the climate system, and it is intriguing to see that the oscillations of the two oceans are roughly (though not completely) in phase. Knudsen noted that the AMO began moving into a second positive phase in about 1990, and thought that this '*may have accentuated global warming in this [latest] period*' [3].

**Figure 15.2** Global and North Atlantic climate trends over the past ~150 years: (a) global annual mean sea-surface temperature (SST) anomalies for the period 1870–2008; (b) annual mean North Atlantic SST anomalies for the period 1870–2008; (c) the Atlantic Multidecadal Oscillation (AMO) index for the period 1870–2008. Five year running means are shown by heavy black lines with fill in all panels. *Source:* From figure 1 in Ref. [3].

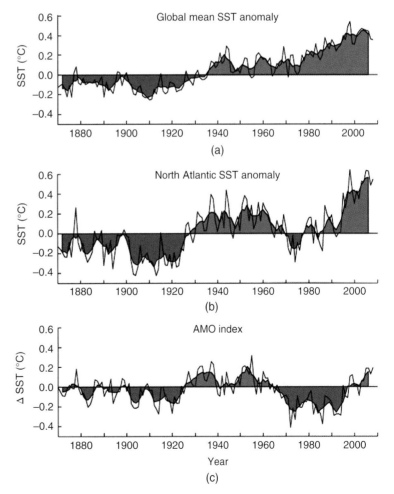

In contrast, the PDO began moving into its second positive phase in about 1970, and was back in its negative (cold) phase by 1998 [4], a state that it maintained until about 2013, coincident with the pause in global warming.

Although these fluctuations may well be natural, or have a natural component, Booth and colleagues remind us that we have also to bear in mind research that suggests important roles for aerosols reflecting either industrial activity or volcanic emissions. In 2012, Booth used '*a state-of-the-art Earth system climate model to show that aerosol emissions and periods of volcanic activity explain 76 per cent of the simulated multidecadal variance in detrended 1860–2005 North Atlantic sea surface temperatures*' [5]. His results show that aerosols were least abundant in the years 1880–1900, 1930–1965, and 1985–2000. This pattern mimics the warming seen in Greenland (Figure 15.1) and in the North Atlantic (Figure 15.2), which suggests the possibility that aerosols play an important role in governing North Atlantic temperatures [5].

As mentioned in Chapter 14, these events fascinated Hubert Lamb [6]. In 1959 he pointed to a climatic optimum in medieval times with a peak between 800 and 1000 CE, followed by cooling into what he called the Little Ice Age between 1550 and 1850 CE [6]. By 1963 he was referring to the climatic optimum as '*The early medieval warm epoch*' [7, 8], which was later more commonly referred to as the Medieval Warm Period, the Medieval Climatic Optimum, or the Medieval Climate Anomaly (MCA) of 1000–1200 CE. In 1964, he presented a comprehensive view of temperature change through time to the Southampton meeting of the British Association for the Advancement of Science (Figure 15.3) [9]. Entitled *Temperatures (°C) prevailing in central England, 50 year averages*, his illustration included graphs for average annual, summer and winter temperatures since CE 900 in central England. He grouped the data into 50 year averages to eliminate high frequency noise caused by decadal variability. The graph of annual averages shows a smooth rise from 900 CE to a broad peak between 1140 and 1270 CE representing his *Early Middle Ages Warm Epoch*. The following LIA was represented by low values in ~1470 and between 1570 and 1670 CE, slight rises in ~1520 and between 1720 and 1870, ending with a rise to 1950. Much the same figure appeared in 1988 in Lamb's book *Climate, History and the Modern World* [10].

Lamb's medieval data were gleaned mostly from records of pollen types, with their assumed relationship to past temperatures, rather than from sources that modern palaeoclimatologists would regard as reliable – like the oxygen or hydrogen isotope data or alkenone data or Mg/Ca ratios that we have discussed in previous chapters. Not only were his original data based on data representing central England, but also much of his subsequent data was from western Europe. Although in his later work he was attempting to build a global picture, his focus tended to be Eurocentric, not least because that's where he got most of his data. Much of Lamb's original work was done in the late 1950s and 1960s. Aside from having few climate proxies to work with, Lamb had many fewer records than are now available, besides which they suffered from having poorer resolution in time (modern geochronology offers far higher resolution records).

During Lamb's medieval peak, the Vikings colonized the southern coast of Greenland and reached the shores of North America, and wine was produced in England as far north as York (almost 54°N). Lamb thought that summer temperatures in England were about 1–2 °C higher than when he was writing about climate in the late 1950s to early 1960s. Cooling then set in, although Lamb's research suggested that the decline from the 1300s onwards was not uniform and there was a partial recovery from it in the period 1440–1500 CE. A warming at that time seems slightly odd in retrospect, because it coincides with the Spörer sunspot minimum (1450–1550 CE). We will look at the effects of sunspots in more detail later.

Lamb concluded that – compared with 1960 values – temperatures were raised 2–3 °C in the warm epochs of the Holocene and lowered 1–2 °C in the LIA, which he thought ranged from 1430 to 1850 AD, and was coldest from 1550 to 1700 CE [9]. He associated the cooling with weakening of both atmospheric circulation and incident solar radiation, and thought that the latter may have been caused by a veil of dust from volcanic activity. We have to remember that Lamb's temperature data were mainly from central England, where the mean monthly data series began in 1659, and the mean daily data series in 1772. Prior to those dates he had to use proxies – mainly estimates based on pollen, and mostly from England [9], which makes the accuracy of his pre-1659 estimates questionable.

During the LIA the Thames froze over four times in the 1500s, eight times in the 1600s, and six times in the 1700s, encouraging people to hold *frost fairs* on the ice [6]. The average temperature for January in central England in the 1780s was 2.5 °C lower than it was in the 1920s and 1930s, though summers in southern England were not much different in the 1780s from what they were in the 1950s (sunspots were at about the same level in those two times, as we see later). The Gulf Stream was further south in 1780–1820 than it is now, and ice was more extensive in northern seas. Nevertheless, there were warm periods within the LIA in the 1630s, 1730s, 1770s, and 1840s. Superimposed on these broad patterns were large oscillations lasting on the order of 20–60 years affecting ice extent and winter temperatures especially in Europe, where, for instance, there were cold winters with much ice in the 1880s and early 1890s. I have

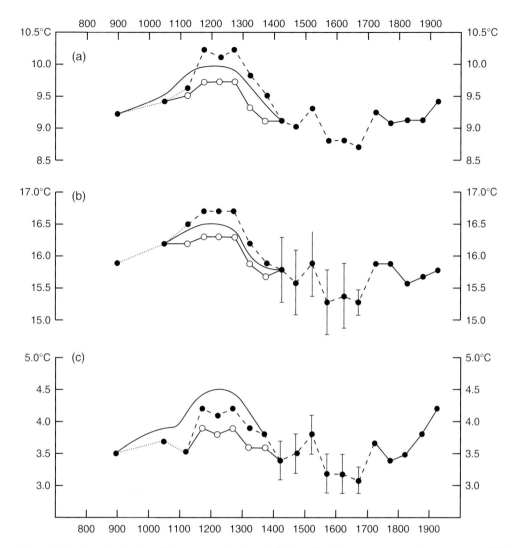

**Figure 15.3** Lamb's Graphs of Temperature in Central England in 50 year averages of temperatures (°C): (a) annual; (b) summer (July and August); (c) winter (December, January, and February). The ranges indicated by vertical bars are three times the standard error of the estimates. Heavy solid line (from 1680 to 1950) = observed values. Line connecting open circles = unadjusted values based on meteorological evidence. Dashed line = preferred values including temperatures adjusted to fit botanical indications. Fine dotted line (pre-1150 CE) = connects points corresponding to 100–200 year means indicated by sparse data. Thin solid line (pre 1400) = Lamb's preferred option. *Source:* From figure 10, in Ref. [9].

seen photographs of the Thames frozen over in London in February 1896, but there were hardly any cold winters in 1896–1897 and 1939–1940.

Lamb thought that global temperatures fell 1–2 °C in the LIA. His observation that the LIA was primarily a wintertime phenomenon has been confirmed by more modern data [11]. Eduardo Moreno-Chamarro of the Max Planck Institute for Meteorology and colleagues used climate reconstructions to show in 2017 that cooling was strongly amplified during LIA winters due to the development of persistent atmospheric blocking conditions [12]. These exceptional wintertime conditions arose from the expansion of sea ice and subsequent reduced loss of heat to the atmosphere in the Nordic and Barents Seas, which was driven by a multicentennial reduction in the northward transport of oceanic heat by the subpolar gyre, and which led in turn to persistent southward advection of cold air from the Barents Sea and western Russia towards western Europe [12]. These anomalous conditions were largely decoupled from European atmospheric variability in the summer, which closely tracks variations of the mean temperature of the Northern Hemisphere. So, while direct radiative forcing drives variability in European summer, the extreme variation in winter (which is much larger than that of the Northern Hemisphere mean) has a specific regional source. It is not in itself a direct reflection of solar forcing, nor does it represent weakened northward heat transport by the AMOC or a persistent North Atlantic Oscillation (NAO) in the atmosphere [12]. Moreno concluded that the subpolar gyre itself is sustained by its own dynamics through the redistribution of ocean salinity and heat. Freshening of the ocean surface during winter weakened the gyre and led to the observed regional effects [12].

Lamb's graph of annual average temperatures for the past 1000 years for central England (Figure 15.3) became iconic following its reproduction more or less unchanged in Chapter 7 of the First Assessment Report of the Intergovernmental Panel on Climate Change (IPCC) in 1990, with the misleading title *Schematic diagrams of global temperature variations since … the last thousand years* (Figure 15.4) [13]. Although the caption cited no specific source for the figure, the report refers to a 1988 paper by Lamb that is the probable source [14]. Not surprisingly it is remarkably similar to the figure in his 1988 book [10] and his 1964 Southampton paper [9]. Given that caption, readers could be forgiven for taking it as a representation of global temperature, rather than what it was in fact – the temperature of central England.

How widespread was Lamb's 'Early Middle Ages Warm Epoch' (now the Medieval Warm Period) and what was its magnitude? This is a question that has come to bedevil discussions about global warming. Many of those who think the medieval warm period was warm everywhere seem to have forgotten, if they ever knew, that Lamb saw it as asymmetric over the Northern Hemisphere [10]. While medieval warming appeared to have been widespread across Europe, North America and the European Arctic, he found that it was cold in China and Japan, where there was warming from 650 to 850 CE when it was cold in Europe [10]. Even within the Atlantic sector there were variations from one part to another, with the warm phase passing its peak in Greenland, Iceland, and eastern Europe in the twelfth century, but continuing in western Europe to reach a maximum about 1300 CE [10]. Clearly, medieval warming could

not be considered global even based just on Northern Hemisphere data. That does not mean it was not triggered by a global event. It means that processes internal to the Earth's climate system interfere with such signals, so they don't have a uniform global impact, as we show later.

By 1995, Lamb's view that the medieval warm epoch might not be global had gained some traction amongst the climate cognoscenti. By then, a great many proxy data for temperature over the past 1000 years had been obtained from isotopic analyses and assessments of changes in tree rings [15]. In Chapter 3 of the IPCC's Second Assessment Report, published in 1996, lead author Neville Nicholls from the Australian Bureau of Meteorology Research Centre, and colleagues, concluded that the evidence for a Medieval Warm Period between the ninth and fourteenth centuries was geographically limited and equivocal [15]. Evidence for it was clear for parts of Europe. Elsewhere there was no such evidence, or warmer conditions prevailed, but at different times. More and better-calibrated proxy records were needed to obtain a clearer picture. Bearing that in mind, Nicholls concluded that '*it is not possible to say whether, on a hemispheric scale, temperatures declined from the 11-12th to the 16-17th century. Nor, therefore, is it possible to conclude that global temperatures in the Medieval Warm Period were comparable to the warm decades of the late 20th century*' [15]. The emphasis in the last sentence should lie on the word 'global', as it seemed that European temperatures in the Medieval Warm Period may have been slightly warmer than they were in the cool 1960s when Lamb first produced his graphs.

With regard to the LIA, Nicholls concluded that '*the climate of the last few centuries was more spatially and temporally complex than this simple concept implies …. It was a period of both warm and cold climatic anomalies that varied in importance geographically*'. Nevertheless, '*despite the spatial and temporal complexity, it does appear that much of the world was cooler in the few centuries prior to the present century*' [15]. Lamb, too, documented the climate of the LIA, finding that temperatures in China closely paralleled those in Europe, and most of North America was also cold, a pattern unlike that for the period around 1000 CE [7]. He found almost no cooling in the tropics, and rather mild conditions in Antarctica, but his analysis suffered from being data limited.

Nicholls and his IPCC team from 1996 drew heavily on three papers published between 1992 and 1995 by climatologist Raymond S. Bradley (1948–) of the University of Massachusetts at Amherst, and Lamb's colleague Phil Jones from the East Anglia's Climate Research Unit (CRU) [16–18]. Jones was listed as a key contributor to Nicholls' IPCC report, and Bradley also contributed to it [15]. In due course, Bradley was to rise to the position of Research

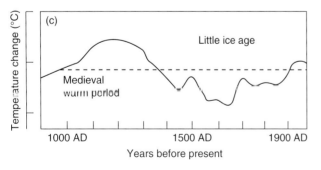

**Figure 15.4** The figure the IPCC got wrong – Lamb's graph of 'English' temperature taken as global. Captioned 'Schematic diagrams of global temperature variations since the last thousand years'. The dashed line nominally represents conditions near the beginning of the twentieth century'. *Source:* From figure 7.1c, in Ref. [13].

Director of the University of Massachusetts Amherst's Climate System Research Centre, and Jones to the position of Director of the CRU. Jones would later star in the 'Climategate' saga of the winter of 2009–2010.

By the early 2000s, the LIA was seen as divisible into two main parts: an early LIA phase, or Medieval Cold Period, between about 1250–1550 CE, and the main LIA lasting from 1550 to 1850, with peak cooling between 1750 and 1850 [19, 20] (see Figure 15.1 middle panel). Crowley and North observed that the main LIA '*consisted of two main cold stages of about a century's length ...in the seventeenth and nineteenth centuries ... [and that] the coldest decades occurred in the mid-late 1600s, the early 1800s, and the late 1800s*' [21] (see Figure 15.1 middle panel). The start of the early LIA phase coincided with pronounced glacial advance in the European Alps, an expansion of sea ice around Iceland, and a return to pluvial conditions following medieval drought in tropical Africa [19]. Crowley and North pointed out that '*The Little Ice Age is considered to end around 1890, although different authors might choose slightly different dates*' [21]. For instance, in eastern Africa it ended with an abrupt switch to wet conditions in the early 1800s [19], while Lamb clearly fancied 1850 [6]. Later on we will assess evidence for the end to the LIA.

Lamb associated the LIA with a weakening of both atmospheric circulation and incident radiation, possibly affected by volcanic dust [6]. Gerard Bond disagreed because his team had identified the LIA as one of his 1500 year ice-rafting cycles [22]. Linking the LIA to the cold phase of Bond's persistent climate cycle ruled out volcanism as a cause, since volcanism would not be expected to vary in a continuous and regular manner. Wanner agreed with Bond, because the distribution of large climate-changing eruptions over the past 6 Ka was '*highly inhomogeneous*' [23]. Nevertheless, Wanner reminded us '*more large tropical volcanic eruptions [of the kind likely to cool the climate for a year or two] have occurred during certain intervals of the last millennium, i.e. between AD 1200 and 1350 or around AD 1700 and 1800, than at other times during the Holocene. These maxima of volcanic activity happen to coincide with both low orbitally induced insolation in the [Northern Hemisphere] and an unusual concentration of solar activity minima. Therefore, it seems plausible that the cold intervals of the past millennium, including the Little Ice Age, might be attributed to a combination of orbital, volcanic and solar forcing*' [23]. We will consider volcanic forcing later.

To be more explicit, Wanner attributed the cooling of the LIA to three factors: the continued decrease in insolation attributable to orbital forcing; occasional sunspot minima representing declines in solar activity and which we now know were driven by the 208 year Suess–De Vries cycle

(of which more later); plus the output of some large tropical volcanoes [23]. For the past 2000 year period, the PAGES team had identified five short periods of cooling attributable to either volcanic activity or its combination with solar activity at: 1251–1310 CE (initial volcanic activity followed by the solar low of the Wolf Minimum); 1431–1520 CE (initial volcanic activity within the solar low of the Spörer Minimum); 1581–1610 CE (volcanic activity during a sunspot high); 1641–1700 CE (combined volcanic activity within the Maunder Minimum in sunspots); and 1791–1820 CE (volcanic activity followed by and overlapping with the Dalton Minimum in sunspots) [24] (compare those dates with the temperature profile of Figure 15.1 middle panel).

The PAGES team agreed that declining insolation was an important forcing factor. The cooling of the continents over the past 2 Ka was '*consistent with the cooling of global sea surface temperatures from year 1 to 1800 CE [Common Era, or AD] exhibited in the PAGES Ocean2k synthesis*' [24]. The continents cooled at about 0.2 °C/1000 years over that period. In mid- to high latitudes of the Northern Hemisphere this was due to a decrease in orbitally driven local summer insolation (Figure 14.1), whilst in the Southern Hemisphere it was a delayed response to the decrease in spring insolation modulated by the Southern Ocean's thermal inertia.

What temperature changes could be discerned in the Medieval Warm Period and the LIA? As we saw earlier, Lamb concluded in the mid-1960s that – compared with then present values – temperatures were raised 2–3 °C in the warm epochs of the Holocene and lowered 1–2 °C in the LIA, which was a time of glacial advance in virtually all the world's mountain regions, and of advancing sea ice around Iceland [9]. Were his temperature suggestions reasonable? No. His database at the time was skewed heavily towards the UK in particular and western Europe in general. By the year 2000, Bard thought that the world had cooled by 0.5–1.0 °C during the LIA, and that the Medieval Warm Period could have been as warm as the mid-twentieth century [25]. While that appeared true for the Northern Hemisphere, would it hold globally? We shall see.

Jan Esper of the Swiss Federal Research Institute, and colleagues, was keen to test the idea that tree rings could provide an accurate picture of climate change over the past 1000 years. Analysing tree rings from high elevation and middle to high latitude sites in the Northern Hemisphere, he showed in 2002 that temperatures were above average in the Medieval Warm Period (900–1300 CE), although there were local differences in the timing of peak warmth then, indicating a high degree of spatial variability in the Medieval Warm Period signal [26]. Low temperatures characterized the LIA between 1200 and 1850 CE, followed by a warming trend. Esper wondered how reliable tree ring data were, and in a later paper showed that differences of up to 0.5 °C could be derived by different research teams from much the same tree ring data, due to differences in the methods used to calculate temperature. Great care was required in interpreting results derived by different teams. Ideally all researchers should follow the same methodology [27]. Caveat emptor.

By 2005, Anders Moberg from the Department of Meteorology of Stockholm University, and colleagues, found that moderately high temperatures like those observed in the twentieth century before 1990 occurred in the Northern Hemisphere around 1000 to 1100 CE in the Medieval Warm Period, and minimum temperatures that were about 0.7 °C below the average of 1961–1990 occurred around 1600 CE during the LIA [28]. Jasper Kirby, from CERN in Switzerland, confirmed that temperatures in the Medieval Warm Period in Europe were about 1.7 °C warmer than at their minimum in the LIA [29]. The evidence emerging from modern palaeoclimate investigations did not support Lamb's contention that the temperatures of the Medieval Warm Period in the Northern Hemisphere had been 1–2 °C higher than in the mid-twentieth century [9].

More recently, in 2012 a team from Stockholm University led by Fredrik Ljungqvist used a large set of temperature-sensitive proxies along with historical records to assess the temperature history of the Northern Hemisphere back to about 800 AD (Figure 15.5) [30]. They found widespread positive temperature anomalies from the ninth to eleventh centuries comparable to the twentieth century mean, and widespread negative anomalies from the sixteenth to eighteenth centuries. Most of their data were from Europe, Greenland, North America, and China. More data were needed from interior Asia, North Africa, and the Middle East. Temperatures varied regionally: '*almost all of North America, western Europe and much of central and eastern Asia warmed from the 17th to the 18th century but not Greenland, eastern Europe and northwestern Asia. Notable cooling occurred from the 18th to 19th century in northern Europe and much of Asia except in the south to southwest. This cooling caused the 19th century to be the coldest over much of northwestern Eurasia*' [30]. Once again, the focus of these several studies was the Northern Hemisphere.

These recent studies replicated work done by Raymond Bradley's group at Amherst. Bradley worked closely with Michael Mann, who joined Bradley's lab as a post-doc in 1996 to work on taking '*a new statistical approach to reconstructing global patterns of annual temperature … based on the calibration of multiproxy data networks by the dominant patterns of temperature variability in the instrumental record*' [31], as a contribution to the 'Analysis of Rapid and Recent Climate Change' project sponsored by the National

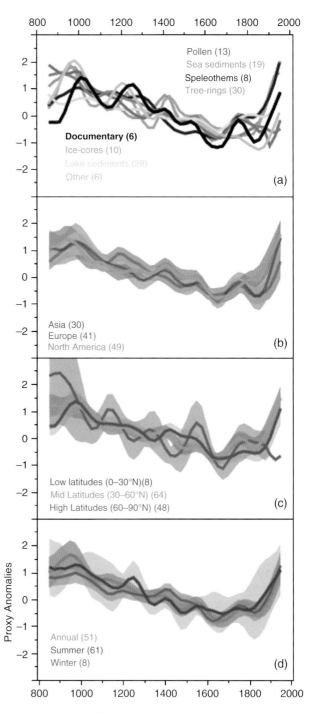

**Figure 15.5** Mean time-series of centennial proxy anomalies separated by: (a) data type, (b) continents, (c) latitude, (d) seasonality of signal. The curves in (c–d) show the mean confidence intervals (±2σ). The numbers in parentheses indicate the number of proxies in each category. *Source:* From figure 4, in Ref. [30].

Science Foundation (NSF) and the National Oceanic and Atmospheric Administration (NOAA). The thinking behind this study went as follows: '*If a faithful empirical description of climate variability could be obtained for the past several centuries, a more confident estimation could be made of the role of different external forcings [such as volcanic activity, solar irradiance and greenhouse gases] and internal sources of variability on past and recent climate*' [31]. A worthy goal.

Mann's initial results, published in 1998 with Bradley and with Malcolm Hughes of the Laboratory of Tree Ring Research of the University of Arizona, reconstructed surface temperature patterns for the Northern Hemisphere over the past 600 years [31]. They resembled the patterns published by Bradley and Jones in the early 1990s [16–18], showing '*pronounced cold periods during the mid-seventeenth and nineteenth centuries, and somewhat warmer intervals during the mid-sixteenth and late eighteenth centuries, with almost all years before the twentieth century well below the twentieth-century climatological mean*' [31]. Furthermore, '*the years 1990, 1995 and 1997 .... each show anomalies that are greater than any other year back to 1400*' [31]. Solar forcing proved to be especially important in and around the Maunder Minimum, and in association with the warming through the nineteenth century. Forcing by greenhouse gases worked jointly with solar forcing for most of the past 200 years, while both were increasing, then greenhouse gas forcing increased further to become more significant as solar forcing levelled off after the mid-twentieth century. Explosive volcanism had only a spasmodic influence on climate, for example with the widespread cooling in 1816 caused by the eruption of Tambora. The authors hoped '*It may soon be possible to faithfully reconstruct mean global temperatures back over the entire millennium, resolving, for example the enigmatic medieval period*' [31].

Enigmatic? Yes, because they were referring to a 1994 study by Malcolm Hughes, of the University of Arizona, and Henry Diaz, of NOAA, which found that whilst some areas of the globe were warmer in summer during the medieval era than they are today (Scandinavia, China, the Sierra Nevada, the Rockies, and Tasmania), others experienced similar temperatures at later times (southeast USA, the Mediterranean and parts of South America) [32]. '*Taken together*', they concluded, '*the available evidence does not support a global Medieval Warm Period, although more support for such a phenomenon could be drawn*' perhaps from high elevation records [32]. This was not a new problem; it had been identified by Lamb in 1988 [10], investigated in the early 1990s by Bradley [16–18], and highlighted in the 1995 IPCC report by Nicholls [15]. Mann was the new kid on the block, within a team that had been carrying out this kind of research for several years.

By 1999 the Mann, Bradley, and Hughes had extended their temperature reconstruction for the Northern Hemisphere back a full 1000 years [33], producing a graph that became known as the 'Hockey Stick'. It showed minor variability about a trend of steadily declining temperature from 1000 CE through to about 1900 (the handle of the stick), after which there was a sharp upward kick (the blade) to the high temperature of 1998. Reappearing as Figure 2.20 in Chapter 2 of the IPCC's third assessment report in 2001 [34], this graph became just as iconic as Lamb's diagram in the IPCC report of 1990 (Figure 15.3). But in fact it differs very little from similar analyses carried out subsequently by others (e.g. Figure 15.5).

Had Mann demolished the Medieval Warm Period? No. On the contrary, he said '*Our reconstruction thus supports the notion of relatively warm hemispheric conditions earlier in the millennium*' [33]. But in comparison with Lamb's graph (Figures 15.3 and 15.4), his results reduced the amplitude of that medieval warming for the Northern Hemisphere. This reversal of received wisdom by a young upstart created controversy in the minds of those who failed to appreciate that Lamb's graph was not global, even though Mann and his team correctly noted that – with respect to the Medieval Warm Period – Lamb had recognized that it contained both cooler and warmer periods [33]. We should expect a difference between the Lamb and Mann curves because Mann's reconstruction was for the entire Northern Hemisphere, whereas Lamb's was primarily for the UK, which he thought representative of western Europe. Like others referred to earlier, Mann saw his Northern Hemisphere temperature curve as a response to declining orbitally driven insolation, pointing out that '*the recent warming is especially striking if viewed as defying a long-term cooling trend associated with astronomical forcing*' [33].

Critiques of Mann's 1999 paper led the US National Research Council (NRC), the working arm of the US National Academies, to appoint a committee chaired by Gerald North of Texas A&M University to review the topic of *Surface Temperature Reconstructions for the Last 2,000 Years* [35]. As one test of Mann's hockey stick, presented in updated form in 2003 by Mann and Phil Jones from East Anglia's CRU [36], North's report in 2006 [35] provided a comparison with two other independent climate reconstructions back to 900 AD, one from Anders Moberg of Stockholm University [28, 37] and one from Jan Esper of the Swiss Federal Institute of Technology [26]. Despite their independent construction, all showed much the same thing, with the medieval warming of the Northern Hemisphere just about reaching temperatures typical of the 1950s, although the Mann and Jones curve peaked around 0.2 °C lower in the medieval period than did the other two. By 2007, Eystein Jansen of the University of Bergen in Norway, and Jonathan Overpeck of the University of Arizona, had expanded this comparison [38]. They used 12 independent reconstructions, mostly back to 800 CE to show that peak temperatures for the medieval period lay 0.1°–0.2 °C below those typical of 1950. North's report concluded, like Nicholls in 1996 [15], that relatively warm conditions were centred near 1000 CE in the Medieval Warm Period, and relatively cold conditions were centred around 1700 in the LIA, especially in the Northern Hemisphere. Evidence for a LIA between 1500 and 1850 was widespread, but evidence for a Medieval Warm Period was limited. The timing and duration of warm periods varied from region to region, and the magnitude and geographic extent of the warmth were uncertain [35]. Jansen and Overpeck reached much the same conclusion in 2007 [38]. And the 2013 PAGES report agreed with them [24], as did Ljungqvist's study in 2012 (Figure 15.5) [30]. Given the parallels available at the time, North's panel agreed with Mann that the late twentieth century warmth of the Northern Hemisphere was unprecedented for at least the last 1000 years [35].

One of the criticisms associated with Mann's analyses, which were based on tree rings, concerned the fact that some tree ring data suggested that temperatures had fallen in recent decades. Rosanne D'Arrigo, of Lamont, and colleagues, explained that this anomalous pattern, known as the 'divergence problem', is attributable to several causes including '*temperature-induced drought stress, nonlinear thresholds or time-dependent responses to recent warming, delayed snowmelt and related changes in seasonality, and differential growth/climate relationships inferred for maximum, minimum and mean temperatures*' [39]. The key is the changing balance between precipitation and temperature, both of which affect ring patterns. At the International Polar Year Conference in Montreal in April 2012, Trevor Porter reported on his observations of 'divergence' in tree rings from White Spruce trees in the Yukon Territory [40]. In one relatively small area of boreal forest, the trees of about half the sites responded negatively to summer temperature increase in the period 1930–2007, while those from the other half responded positively. Yet both shared a common response prior to 1930. A wider geographic examination confirmed that most White Spruce sites reacted positively to temperature increases over the period between 1300 CE and 1930. Replying to a question that I posed as the moderator of his oral session, Porter explained that the leading hypothesis for the decline seen in some of his studied trees was localized drought stress, and that the region had become significantly drier in recent decades, with precipitation down to 200–400 mm/year. However, not all tree sites were equally affected. Both positive and negative responses were found even within a few kilometres of one another. The reason may lie in the soil moisture levels. Local changes in permafrost, with areas of melting sited close to areas of permanently frozen ground depending on the slope of the ground in relation to the Sun, may account

for extreme local variation in tree response, but factors such as the density of stands of trees and the amounts of organic material supplying nutrients to the surface layer may also play a part. Much the same result came from a recent study of boreal trees by Reich and colleagues in 2018, which confirmed that climate warming influences photosynthesis via thermal effects and by altering soil moisture [41]. They found that '*the effects of climate warming flip from positive to negative as southern boreal forests transition from rainy to modestly dry periods during the growing season ... photosynthesis decreased during dry spells, and did so more sharply in warmed plants than in plants at ambient temperatures*' [41]. Clearly, great care needs to be taken, especially in boreal areas, in selecting the trees that most appropriately reflect the actual temperature conditions through time, rather than those responding to water stress.

In a 2008 review of global climate change in the mid to late Holocene, Wanner confirmed that the time of the medieval warm peak was not simultaneous around the globe [23]. Amongst other things, this convinced him that external forcing had not caused the transition between the Medieval Warm Period and LIA. In contrast, the glacial maxima at the time of the LIA were reasonably synchronous around the world [23]. By 2013, as a member of the PAGES team, he had firmed up his opinions [24]. The PAGES team's 2013 compilation of 511 data sets for the past 2000 years proved that '*at multi-decadal to centennial scales temperature variability shows distinctly different regional patterns, with more similarity within each hemisphere than between them. There were no globally synchronous multi-decadal warm or cold intervals that define a worldwide Medieval Warm Period or Little Ice Age, but all reconstructions show generally cold conditions between 1580 and 1880* CE *[Common Era, or* AD*], punctuated in some regions by warm decades during the 18th century. The transition to these colder conditions occurred earlier in the Arctic, Europe and Asia than in North America or the Southern Hemisphere regions*' [24].

In response to the recommendations of North's NRC report, in 2008 Mann and colleagues reconstructed surface temperature at both hemispheric and global scales for much of the last 2000 years, using an expanded set of proxy data [42]. They concluded that '*the hemispheric-scale warmth of the past decade for the NH [Northern Hemisphere] is likely anomalous in the context of not just the past 1,000 years ... but longer... [and] appears to hold for at least the past 1,300 years*'. Moreover, '*this conclusion can be extended back to at least the past 1,700 years if tree ring data are used*' [42]. The picture was '*less definitive for the SH [Southern Hemisphere] and globe, which we attribute to large uncertainties arising from the sparser available proxy data in the*

*SH*' [42]. Much the same picture has emerged from the studies of Moberg and PAGES, as mentioned earlier.

In 2009, Mann and his team extended their analysis. Using a network of 1000 climate proxy records from various sites around the world they concluded '*The Medieval period is found to display warmth that matches or exceeds that of the past decade in some regions, but which falls well below recent levels globally ... [and] The coldest temperatures of the Little Ice Age are observed .... over the extratropical Northern Hemisphere continents*' [43]. Wanner and his PAGES team reached much the same conclusions [23, 24]. How could this variation be explained?

Mann noticed that La Niña-like conditions, making the tropics cool, tended to predominate during the medieval period in the tropical Pacific, most noticeably between 950 and 1100 CE. At the same time SSTs warmed significantly in the North Atlantic and the North Pacific. This pattern is typical of the negative phase of the Pacific Decadal Oscillation. Mann thought that the effects of orbital, solar, volcanic, and greenhouse forcings were being modulated regionally through oceanic responses [43]. The medieval period showed enhanced warmth over the interior of North America and the Eurasian Arctic, and cooling over central Eurasia '*suggestive of the positive phase of the [atmospheric] North Atlantic Oscillation (NAO) and the closely related Arctic Oscillation (AO) sea-level pressure (SLP) pattern*' [43]. It would seem that medieval warmth was associated with the positive phase of the NAO, which was in turn associated with relatively strong solar forcing. The positive phase of the NAO enhanced the westerlies, creating cool summers, and mild winters with frequent rain in northwest Europe, snow in Scandinavia, and dry conditions in the Mediterranean. In contrast, the LIA was associated with a predominantly negative phase of the NAO, driven by weak solar forcing. That suppressed the strength of the westerlies, bringing hot dry summers and very cold winters to northwest Europe and rain to the Mediterranean. This helps to explain why the LIA was mainly a winter phenomenon [11], as Lamb had noted.

Rather like Mann, both the PAGES group [24] and Marcott [44] found 'global' medieval temperatures around 1000 CE to be around 0.1 °C lower than the 1961–1991 reference period. Those recent studies vindicate Mann's earlier conclusion that the Medieval Warm Period, though real, was not a major global climatic phenomenon. Thus in terms of absolute 'global' warming, the Medieval Warm Period is no more than a small upward blip on the orbitally driven downward curve of Holocene cooling. That picture emerges time and time again, for example as shown in Figures 15.5 and 15.6 [45]. Moreover, a 2019 study by Neukom and others confirms that there is no compelling evidence for a globally coherent Medieval Warm Period or LIA [46].

In 2016, Luterbacher and colleagues carried out a detailed examination of European summer temperatures for the past 2000 years [47]. They found that '*the mean 20th century European summer temperature was not significantly different from some earlier centuries, including the 1st, 2nd, 8th and 10th centuries CE. The 1st century [and] 10th century may even have been slightly warmer than the [mean] 20th century, but the difference is not statistically significant. Comparing each 50 yr period with the 1951–2000 period reveals a similar pattern. Recent summers, however, have been unusually warm in the context of the last two millennia and there are no 30 yr periods in either reconstruction that exceed the mean average European summer temperature of the last 3 decades (1986–2015CE)*' [47]. In the period since 1850 the coolest summers were centred on 1910, following which temperatures warmed to a peak between 1940 and 1950, then cooled before rising again from about 1975, much as seen in Figures 8.13 and 15.1. During the 2000 year period there was substantial variability on both decadal and multi-decadal timescales.

We turn now to some more detailed regional analyses of climate change. For example, studies of the Great Barrier Reef and coral islands in the tropical southwest Pacific led Erica Hendry of the Australian National University,

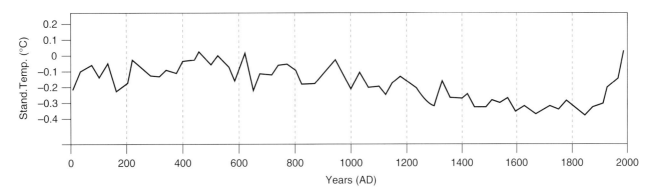

**Figure 15.6** Global temperature for the last 2000 years, represented by 30 year means. The zero on the temperature scale represents the average from 1950 to 2000, which is also the last data point. *Source:* From figure 5, in Ref. [45].

Canberra, to conclude that during the LIA local surface waters were relatively warm and saline, and that '*Cooling and abrupt freshening of the tropical southwestern Pacific coincided with the weakening of atmospheric circulation at the end of the Little Ice Age, when glaciers worldwide began to retreat*' [48]. She suggested that the Equator-to-pole temperature gradient steepened during the LIA, strengthening the easterly Trade Winds, which pushed warm surface waters west towards the reefs. These warm waters would also have been more subject to evaporation, possibly providing a source for the moisture that helped polar glaciers to grow between 1600 and 1860. On the other side of the Pacific, analysis of a coastal sediment core from San Francisco Bay showed Mary McGann of the US Geological Survey that the Medieval Warm Period was warm and dry on the California coast, whilst the LIA was humid with an influx of freshwater [49]. A 500 year record of alkenone palaeotemperatures ($U^{k'}_{37}$) from deep-sea cores along the Californian margin showed no dip in SST during the LIA, probably because almost the entire core was of that age [50]. In contrast, the tops of the cores carried evidence of recent warming.

Turning to the Arctic, in 2001, researchers used $\partial^{18}O$ analyses on fossil shells from lake sediments in the Arctic's Alaska Range to extract a record of growing-season temperatures for the past 2000 years. They found three periods of comparable warmth: 0–300 CE (Roman Warm Period), 850–1200 CE (Medieval Warm Period), and post 1800 [51]. The LIA cooling peaked in 1700 CE with a climate 1.7 °C colder than in the year 2000. Another prominent cooling event centred on c. 600 CE corresponded to the European Dark Ages. Both cold periods were times of significant glacial advance in Alaska, and were wetter than the warmer periods. Later work by Michael Loso of Alaska Pacific University in Anchorage, on a varved glacial lake from southern Alaska, showed that temperatures over the past 1500 years varied by about 1.1 °C with the maximum in the late twentieth century [52]. The Medieval Warm Period there was cooler than the climate of recent decades.

In 2016, Ulf Büntgen and colleagues confirmed that the Dark Ages period from 536 to 650 AD was particularly cold, quite possibly triggered or exacerbated by a series of large volcanic eruptions in 536, 540, and 547 CE [53]. They called this period the Late Antique LIA, found it to span most of the Northern Hemisphere, and related it to: the establishment of the Justinian plague, the transformation of the eastern Roman Empire, migrations out of the Asian steppe and Arabian Peninsula, the spread of Slavic-speaking peoples, and political upheavals in China [53].

A team led by Kristin Werner of the Leibnitz Institute for Marine Science used climate proxies from a marine sediment core from the Svalbard continental margin on the east side of Fram Strait to show, in 2011, that warm water inflow from the Atlantic led to seasonally ice-free conditions during the Medieval Warm Period at about 80°N [54]. Summer SSTs reached 4.4 °C between 650 and 1400 AD. The site was at or close to the sea ice margin in the LIA after 1400 CE, and was covered by extensive sea ice and icebergs after 1730. The coldest part of the LIA on Svalbard occurred between 1760 and 1900. Warm Atlantic water inflow increased after about 1860 CE. Trond Dokken of the Bjerknes Centre for Climate Research, Bergen, Norway, and colleagues, noted in 2013 that warm subsurface Atlantic water periodically flows into Nordic seas beneath the sea ice, destabilizing it and rapidly warming the region, leading to formation of a freshwater cap associated initially with ice rafting. They suggested that the interplay between warm inflow and local sea ice played an important role in the development of Dansgaard–Oeschger events, as we saw in Chapter 13 [55]. It may have helped control Bond Cycles, and may be operating now, promoting development of an ice-free Arctic [55]. These various data imply that the AMOC (Figures 12.5 and 12.6) was strong during the Medieval Warm Period, taking warm water north, and weak during the LIA, when a freshwater lid and sea ice covered the Norwegian–Greenland Sea, which is where the cooling and sinking of dense salty water from the south in the absence of sea ice would normally pull more such warm water north, so strengthening the AMOC.

On the west side of Fram Strait, analyses of winter season $\partial^{18}O$ ratios and borehole temperatures in Greenland ice cores suggested to Bo Vinther and colleagues, in 2010 that '*temperatures during the warmest intervals of the Medieval Warm Period were as warm as or slightly warmer than present day Greenland temperatures*' [56]. Takuro Kobashi recognized that his team's Greenland temperature reconstruction (Figure 15.1) [1] differed from the reconstruction of Arctic summer air temperature over the past 2 Ka, which showed a long cooling trend ending with pronounced warming (like that of Figure 8.13). They attributed this to three factors: first, the Greenland ice core record provides a mean annual palaeotemperature rather than a summer temperature, whereas Arctic temperatures may represent summer maxima; second, Greenland's temperature may be affected by cloud cover and wind speed; and, third, Greenland is not representative of the whole Arctic, at least in part because Greenland is a large block of ice reaching high altitudes (3694 m) so has its own microclimate [1].

Reviewing Arctic temperature records, Miller and his team found that the most consistent records of medieval warming came from the North Atlantic sector, but the evidence for medieval warmth elsewhere in the Arctic was less clear. This regional variability meant that '*the

*Arctic as a whole was not anomalously warm throughout Medieval time*' [57].

In 2006, Paul Mayewski and his colleague, Kirk Allen Maasch, used the abundances of the ions of sodium ($Na^+$) and calcium ($Ca^{2+}$) in ice cores to map polar climate change over the past 2000 years [58]. Stronger winds blow more dust and sea salt into the interiors of ice sheets, so changes in those components seen in a network of cores over an ice sheet provide information about changes in wind speed and direction, hence changes in atmospheric circulation, for comparison with changes in temperature deduced from $\partial^{18}O$ or $\partial^2 D$ measurements. The warm temperatures typical elsewhere of the Medieval Warm Period (800–1400 CE) were associated with lower wind strengths representing weaker westerlies, whilst the cool temperatures typical of the LIA (1401 to 1850 CE) were associated with higher wind strengths typical of stronger westerlies (Figure 15.7).

Comparing the behaviour of the two polar regions, Mayewski and Maasch found that the changes in the winds associated with the warming of the Medieval Warm Period and with the cooling associated with the LIA seemed to begin 300–400 years earlier in the south than in the north, and that the strong winds were much stronger in the north [58]. Furthermore, whilst the peak of the LIA appeared to coincide with the Maunder Sunspot Minimum in solar activity in the north, it did not in the south. For those

reasons they advised against applying this terminology to the Antarctic. In contrast with that lag in time between Northern and Southern Hemisphere events, they observed that present day warming was unusual in occurring simultaneously in both hemispheres – there was no lag. Their observation that the cooling between the Medieval Warm Period and the LIA began earlier in the south than in the north begs the question – is this another example of the climate seesaw, with warming starting first in the Southern Ocean and being transmitted to the north through the AMOC? The time-scale is about right.

In 2009, I worked with Mayewski and others on a review of the state of the Antarctic and Southern Ocean ecosystem [59]. Amongst other things we found that the climate of West Antarctica, represented by an ice core from Siple Dome in the Ross Sea area (location in Figure 7.5), began to warm around 6 Ka ago, in parallel with the local rise in insolation at 60°S and the gentle global rise in $CO_2$. That warming steepened after around 1750 CE, following the curve of rising $CO_2$. Over the past 1200 years the concentrations of calcium and sea salt increased at Siple Dome, telling us that the westerly winds were getting stronger there, bringing in more dust from surrounding continents and warm air from the open ocean. In contrast, in East Antarctica – represented by an ice core from Law Dome – we saw a cooling trend continuing from around

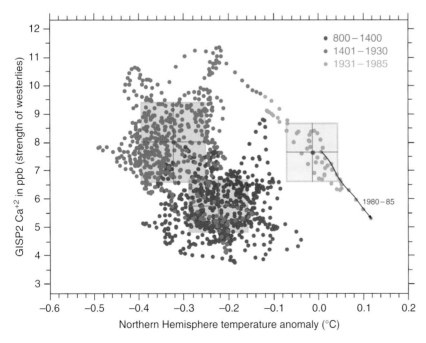

**Figure 15.7** Phase diagram for Northern Hemisphere temperature versus an ice core proxy for Northern Hemisphere westerlies (calcium ions in the GISP ice core). Cluster over box at lower right centre = 800–1400 CE; cluster over box at upper left centre = 1401–1930 CE; cluster over box at far right and along the line labelled 1931–1985 = 1931–1985 CE. Shaded boxes represent the mean ±1σ for each period. Black line = 1980–1985, the last five years of the data record. Clearly climatic conditions since 1930 diverge from those of both the Medieval Warm Period and Little ice Age. *Source:* From figure 4, in Ref. [58].

1300 CE to the present. The opposing climatic trends in East and West Antarctica tell us about the development of the wind system around the continent. In a perfect world the westerly winds would run in a circle around the South Pole. But the Antarctic part of the world is far from perfect. The biggest ice cube in the world, the enormous upstanding mass of East Antarctica, which reaches a height of 4092 m, is offset to the east of the pole (Figure 7.5), distorting the wind pattern and allowing the formation of a giant eddy, the Amundsen Sea low pressure centre, over the much lower topography of West Antarctica in the lee of East Antarctica's giant ice cube [60]. This 'Amundsen Sea Low' pulls warm air from the north down the Antarctic Peninsula, and carries warm, salt-laden marine air over West Antarctica's Siple Dome. Intensification of the Amundsen Sea Low with time accounts for the difference in climate regimes between East and West Antarctica. Antarctic studies highlight the difference in climate between the high-standing block of East Antarctica (averaging 3000 m high) and the much lower block of West Antarctica (averaging 1800 m high) [59, 60]. Warm wet air brought in by northerly winds sucked in by the Amundsen

Sea Low makes precipitation abundant in West Antarctica, while precipitation is as low as that in a desert over the high and extremely cold interior plateau of East Antarctica. Antarctica as a whole is also colder than the Arctic for a number of other reasons, spelled out by Ramanathan [61] (see Chapter 8).

In 2002, Joseph Souney and colleagues (including Mayewski) published an oxygen isotope profile from Law Dome for the past 700 years (back to 1300 CE), which while it did show a decline of about 1 °C between 1300 and 1800 CE, then showed a rise of about 1.5 °C to 1900, followed by a fall of about 1 °C to c. 1930, then a rise of c. 0.5 °C to about 1980 [62]. This pattern replicates the one found by Dorthe Dahl-Jensen and colleagues in 1999 [63], and looks typical of the pattern seen in East Antarctic cores by Stenni [64] (Figure 15.8).

In 2013, Mayewski and Maasch and colleagues examined West Antarctica's sensitivity to natural and human-forced climate change over the Holocene [65]. Comparing ice cores from Siple Dome in West Antarctic and Taylor Dome in East Antarctica (on the western edge of the Ross Sea – see Figure 7.5), they suggested that global warming over the

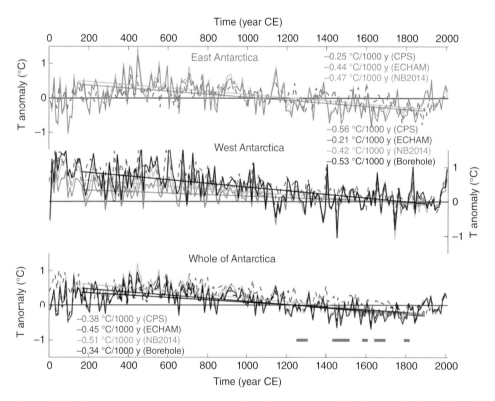

**Figure 15.8** Composite temperature reconstructions (T anomalies in °C referenced to the 1900–1990 CE period) for East, West and the whole of Antarctica using 10 year averages and different temperature scaling approaches represented by dotted coloured lines; grey lines; coloured lines; and black lines, with linear trends calculated between 165 and 1900 CE. The grey horizontal segments correspond to volcanic–solar downturn intervals as defined in PAGES 2k Consortium (2013) [24] and corresponding to the following periods: 1251–1310, 1431–1520, 1581–1610, 1641–1700, and 1791–1820 CE. *Source:* From figure 8 in Ref. [64].

past 150 years strengthened the Southern Ocean westerlies and made them migrate towards Antarctica. This change in zonal wind flow warmed West Antarctic and the Antarctic Peninsula, while cooling East Antarctica, much as described by SCAR scientists in their reports on Antarctic Climate Change and the Environment in 2009 [59, 60]. Nevertheless, things do change, and whilst the northern part of the Antarctic Peninsula had warmed by 0.5 °C annually and 1 °C in winter following the 1950s, this trend reversed in the late 1990s [66]. Temperatures decreased in response to an influx of cold, east-to-southeasterly winds, caused by more cyclonic conditions in the northern Weddell Sea, which were forced by a strengthening mid-latitude jet. These circulation changes brought more sea ice towards the east coast of the Peninsula (on the western edge of the Weddell Sea – see Figure 7.5), amplifying the cooling effects. These local *'temperature changes ... are not primarily associated with the drivers of global temperature change but, rather, reflect the extreme natural internal variability of the regional atmospheric circulation'* [66]. The increase in sea ice prevented cruise ships, like the one I was lecturing on, from accessing the east side of the Peninsula via Antarctic Sound, although we had transited the Sound to visit Paulet Island in the Weddell Sea in December 2010. By December 2012 the eastern end of the Sound was blocked by sea ice, preventing access to the island, and in 2015 the build-up of sea ice prevented me from accessing Port Lockroy, on the western side of the Peninsula[1].

There is no clear sign at either James Ross Island (JRI) in the western Weddell Sea, just east of the Antarctic Peninsula, or in the Palmer Deep on the continental shelf south of Anvers Island in the Bellingshausen Sea, just west of the Peninsula, of the LIA or the Medieval Warm Period (Figure 15.9) (see Figure 7.5 for broad geographic outline).

**Figure 15.9** Temperature history of Antarctic Peninsula over 2000 Years. Lower panel = the James Ross Island (JRI) ice core temperature reconstruction with 100 year averaging (heavy line) and 10 year averages (grey ranges) relative to the 1961–1990 mean (dashed line). Warming by 1.56 °C over the past 100 years (red line) is highly unusual in the context of natural variability. Middle panel = sea surface temperature (SST) record from Ocean Drilling Program site 1098 in Bellingshausen Sea west of the Antarctic Peninsula. Upper panel = reconstructed Northern Hemisphere temperature anomaly relative to 1961–1990 mean, with envelope showing 95% confidence interval. *Source:* From figure 4, in Ref. [67].

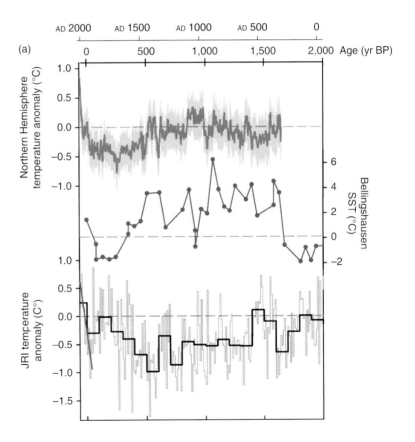

[1]According to a WMO committee of experts, the highest temperature for the "Antarctic continent" defined as the main continental landmass and adjoining islands was the temperature extreme of 17.5 °C (63.5 °F) recorded on 24 March 2015 at the Argentine Research Base Esperanza located near the northern tip of the Antarctic Peninsula, and reported on 1 March 2017 in EOS: https://eos.org/features/evaluating-highest-temperature-extremes-in-the-antarctic. According to the Guardian newspaper (17 Feb 2020), a new record of 18.3 °C was reached at the same base in February 2020; it awaits verification by the WMO committee.

Indeed, between 1671 and 1777 at the heart of the LIA in the Northern Hemisphere, temperatures on JRI rose as rapidly as 1.5 °C/century [67]. '*The [recent] warming by 1.56 °C over the past 100 yr (redline in [Figure 15.9]) is highly unusual in the context of natural variability*' Mulvaney noted [67].

Jennifer Pike of Cardiff University, and colleagues, concluded from their study of Palmer Deep, which is one of the deep channels cut by ice streams across the continental shelf west of the Antarctic Peninsula, that warming there from about 2200 years ago was driven by peak summer insolation at 60°S and an increase in the intensity of La Niña events, which drove an increasing discharge of glacial ice from the land [68]. What was not clear was the cause of the increase in La Niña events in the equatorial Pacific, although as we shall see later, such increases were consistent with the development of the Medieval Warm Period and may have had a solar link. The warming seen on JRI east of the Peninsula (Figure 15.9) [67], and the warming evident in the Palmer Deep west of the Peninsula [68], was typical of the past 2 Ka, despite an increase in the extent of sea ice. That expansion in sea ice had been going on throughout the Holocene as temperatures cooled around Antarctica in response to the decline in insolation around Antarctica during the Holocene [69]. It was something of a paradox, then, as Johan Etourneau noted in 2013, that during the past 2 Ka, when temperatures around the Peninsula had started to warm, the sea ice had continued to expand [69]. Climate models suggested an answer. They showed that '*orbital forcing induces a greater seasonality in the position of the Westerlies over the course of the Holocene ... [with] equatorward displacement of the southern westerlies wind belt and surface temperature cooling during the spring season ... [leading to] longer spring presence of sea ice ... [followed by] southward migration of the Westerlies and greatest surface temperature warming during summer and autumn*' [69]. These seasonal contrasts between a cold winter and a warm summer may have been enhanced towards the late Holocene by the combination of decreasing insolation with an increasing frequency of El Niño and La Niña events in the equatorial Pacific. Those events are part of the so called Southern Oscillation, in which tropical signals are transported south by Rossby Waves in the atmosphere, eventually impacting West Antarctica and the Antarctic Peninsula [69].

The difference between East and West Antarctica is obvious from a comparison of the JRI data (Figure 15.9) and data from the past 1000 years from ice cores in Dronning Maud Land in East Antarctica (Figure 15.10) (Figure 7.5 shows the location of Dronning Maud Land). As pointed out by Wolfgang Graf and his team in 2002, the Dronning

Maud Land data show a decline in temperature (represented by $\partial^{18}O$) from the Medieval Warm Period (1025–1175) towards a period of lower temperatures between 1275 and 1875, punctuated by warm intervals, like the one at 1575 [70]. Even though that period of generally lower temperature broadly covered the period of the LIA, they considered that '*the Little Ice Age cold period ... is not present in Dronning Maud Land ... [instead] this period 1650–1850 is characterized [there] by strong fluctuations of $^{18}O$ contents around average values*' [70] (Figure 15.10). Comparing their Dronning Maud Land data with data from the same period in the Weddell Sea region they found '*that $^{18}O$ records can in some periods be positively correlated and in others negatively correlated, indicating a complex climatic history in time and space*' [70].

Antarctic ice core science has moved on considerably since Graf's 2002 paper, and we can now make comprehensive comparisons between different Antarctic regions for the past 2000 years. Barbara Stenni and colleagues used several methods to demonstrate these regional differences in 2017 [64]. Figure 15.8 shows that a cooling trend was common to all reconstruction regions regardless of the method used: the period between 1200 and 1900 CE formed the coldest interval; whilst the warmest interval lay between 300 and 1000 CE. Although West Antarctica (with an average altitude of 1800 m) was clearly warmer than East Antarctica (with an average altitude of 3000 m), the cooling trend between 1200 and 1900 CE was stronger in West Antarctica. There is some evidence in these composite data for the warming trend of the industrial period (Figure 15.8), which according to regional details shown by

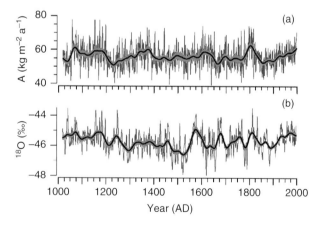

**Figure 15.10** Composite 1000 year records of (a) accumulation rates and (b) $^{18}O$ contents from Dronning Maud Land in East Antarctica. Each record consists of three individual records before 1800 and of 16 records thereafter. Also shown are the smoothed records (by use of a 60 year Gaussian low-pass filter) and the 95% confidence intervals of these records. *Source:* From figure 4 in Ref. [70].

Stenni et al. (their Figure 4) [64], seems to have begun in about 1925. Within West Antarctica the warming of the past 100 years appeared strongest on the Antarctic Peninsula, which sticks up well to the north of the rest of Antarctica, most of which lies south of 70°S (see Figure 7.5), and where '*the last 100-year trend [is] unusual in the context of natural century-scale trends [in Antarctica] over the last 2000 years*' [64].

The West Antarctic Ice Sheet (WAIS) and Dronning Maud Land also showed signs of warming over the last 100 years, but not as strongly as on the Peninsula [64]. And Elizabeth Thomas of the British Antarctic Survey also showed, in 2017, that '*in the Antarctic Peninsula region the late 20th century snowfall increase is [also] unusual in the context of the past 300 years*' [71]. The Stenni and Thomas data sets [64, 71] suggest that the Peninsula is responding to global climate trends that have not yet affected the bulk of the continent except locally and weakly. Furthermore, the lack of any distinctive Medieval Warm Period or LIA in the bulk data (Figure 15.8) tends to confirm that whilst Antarctic temperature is responding to declining insolation, it is less susceptible to other external forcing (e.g. solar or volcanic), something we examine further later on. In its 2013 report on climate change for the past 2 Ka, the PAGES team averaged data for each continent. By so doing, they missed the differentiation between East and West Antarctica noted by Mayewski [65] and later reported by Stenni [64].

Further evidence about Antarctic climate came in 2018 from the new ice core on Roosevelt Island at the mouth of the Ross Sea. The evidence agrees with that from a core from the nearby Siple Dome. Bertler and colleagues showed that for most of the past 2.7 Ka the eastern Ross Sea (adjacent to Marie Byrd Land and including the Roosevelt Island and Siple Dome cores – for general location see Figure 7.5) was warming, with increased snow accumulation and probably a decrease in nearby sea ice, whilst West Antarctica cooled and the western Ross Sea (adjacent to the Transantarctic Mountains and including the Taylor Dome and Talos Dome [TALDICE] cores) showed no significant temperature trend.[2] Bertler referred to this pattern as the Ross Sea Dipole [72]. The eastern Ross Sea warming continued through the LIA, when West Antarctica and the western Ross Sea experienced colder than average temperatures. The Roosevelt Island and Siple Dome cores show no warming or cooling associated with the Medieval Warm

Period and LIA [72]. From the seventeenth century onwards, this pattern changed, with all three regions warming. Bertler observed that the accumulation of snow at Roosevelt Island correlated negatively with sea ice off the coast. More sea ice means less moist air, hence less snow. Thus, accumulation at the core site reflects the presence or absence of a coastal polynya; when it is present, sea ice is exported and moist marine air brings snow to the site. Increasing sea ice correlates with cooler surface air temperature, which in turn correlates with a positive phase of the Southern Annular Mode (or SAM) (the main ring-like mode of atmospheric circulation around Antarctica, otherwise known as the Antarctic Oscillation). During its positive phase, the westerly wind belt contracts towards the continent. This demonstrates that climate change in Antarctica must also be considered from the regional perspective.

Studying the regional complexity of the climate signal in the Antarctic region, Julie Jones and colleagues confirmed in 2016 that it is difficult to separate the effects of global linear trends in climate from the large inter-annual to decadal variability typical of the region: '*Most observed trends, however, are not unusual when compared with Antarctic palaeoclimate records of the past two centuries*' [73]. The one parameter that does have a global signal is the leading mode of atmospheric circulation variability of the Southern Hemisphere's high latitudes – the SAM – which is a measure of the mid-to-high-latitude atmospheric pressure gradient and reflects the strength and position of the circumpolar westerly winds. Global warming has strengthened the westerlies and increased the SAM index, pushing it into a positive mode. General Circulation Models that include greenhouse gas forcing are not compatible with the other observed trends, suggesting either '*that natural variability overwhelms the forced response in the observations, … [or that] the models may not fully represent this natural variability or may overestimate the magnitude of the forced response*' [73]. Most climate records '*support the conclusion that most of the recent changes for any single variable largely result from natural variability and are not unprecedented over the past two centuries*' [73]. Nevertheless, Jones's team concluded that '*the accelerating melting and calving of Antarctic ice shelves could have a pronounced influence on the recent and future evolution of the high-latitude Southern Ocean*'. For most climate indicators, '*the sparsity of observations and proxy data is a clear limitation, especially in the*

---

[2]For readers who find the use of eastern and western confusing in the Antarctic context, it may help to refer to Figure 7.5. East and West Antarctica are so named because they lie mostly either east or west of the Greenwich meridian. But moving along a circle of latitude, e.g. across the Ross Sea towards Marie Byrd Land and away from the Transantarctic Mountains, you are moving towards

the east, hence from the Western Ross Sea towards the Eastern Ross Sea. Equally confusingly, the winds that blow the Antarctic Circumpolar Current to the east are labelled westerlies, because they blow from the west, whilst the coastal winds that blow to the west are labelled easterlies because they come from the east!

*ocean, and ... averaging climate properties over the entire Antarctic or Southern Ocean potentially aliases the regional differences*' [73]. The different oceanic sectors (Atlantic, Pacific, Indian) experience different regional forcing and behave in different ways from one another.

Contradicting both Turner [66] and Jones [73], a more detailed investigation by Neil Swart in 2018 showed that '*the observed changes [in Southern Ocean climate and oceanography] are inconsistent with [either] the internal variability or the response to natural forcing alone. Rather, the observed changes are primarily attributable to human-induced greenhouse gas increases, with a secondary role for stratospheric ozone depletion*' [74]. Using observational data and climate models they showed that greenhouse gases played the dominant role in governing both temperature and salinity since 1950, consistent with the understanding that increasing greenhouse gases are the principal driver of climate warming and of recent anomalous heat uptake by the ocean. The Ozone Hole, they found, was responsible for the cooling observed north of 40°S and for warming to the south. Swart's team '*expect[s] to see continued warming and freshening of the Southern Ocean over the coming decades, despite the mitigating effects of ozone recovery*' [74]. One of the side effects of the warming south of 60°S is freshening and warming of Antarctic Bottom Water, the rates of which increased between 2007 and 2016 compared with 1994 to 2007 [75].

The evidently complex global distribution of climatic patterns during both the Medieval Warm Period and LIA underscores the perceptive observation of Damon and Peristykh in 2004 that because '*climate is complex involving the immense thermohaline currents of the oceans, global circulation of air masses, cyclical phenomena like the El Niño-La Niña, quasi-decadal circulation of the oceans, aerosols injected into the stratosphere by volcanoes, etc ... – even during a climatic event like the Little Ice Age only regionally sensitive areas will be observably cooler*' [76] (my bold highlight). Jonathan Cowie also reminded us that we would be unwise to expect a uniformly global effect, given that the Northern Hemisphere is predominantly land while the Southern Hemisphere is dominantly ocean, and both react differently to incoming solar energy [77]. More recent data, from Neukom in 2019 suggested that whilst the modern warming is effectively synchronous in both hemispheres, the Medieval Warm Period, the Roman Warm Period, the Dark Ages Cold Period and the LIA were primarily evident in particular regions rather than being globally synchronous [46]. Indeed, '*the coldest epoch of the last millennium – the putative Little Ice Age – is most likely to have experienced the coldest temperatures during the fifteenth century in the central and eastern Pacific Ocean, during the seventeenth century in northwestern Europe and*

*southeastern North America, and during the mid-nineteenth century over most of the remaining regions*' [46]. Neukom observed that '*This lack of spatiotemporal coherence [of all events but that of modern times] indicates that preindustrial forcing was not sufficient to produce globally synchronous extreme temperatures at multidecadal and centennial time-scales. By contrast, we find that the warmest period of the past two millennia occurred during the twentieth century for more than 98 per cent of the globe. This provides strong evidence that anthropogenic global warming is not only unparalleled in terms of absolute temperatures, but also unprecedented in spatial consistency within the context of the past 2,000 years*' [46], in agreement with what we learned earlier from Wanner, Mayewski and others.

Bard and his colleagues agreed in 2000 that climate models would not necessarily show the Medieval Warm Period and LIA as global, because advective processes that moved heat around in the atmosphere meant that not every location would experience cooling at exactly the same time [25]. For the same reason we should expect to see regional variability in the signal of the LIA. For instance, there is some evidence that during the LIA the Gulf Stream's normal flow was reduced by 10%. That would have reduced the flow of heat to northwest Europe, and may explain why the LIA was so pronounced there [78]. Michael Schlesinger and Natalia Andronova of the University of Illinois at Urbana calculated that the average global temperature was around 0.34 °C lower in the Maunder Sunspot Minimum of the 1600s than immediately before it, resulting in a shift towards the negative phase of the atmospheric Arctic Oscillation and NAO that led to higher pressure over the Arctic and lower pressure over the mid-latitude North Atlantic, weakening the polar vortex and reducing the transport of warm air from the oceans to the continents, thus cooling Europe and eastern North America by around 1–2°C [79]. As mentioned earlier, during the negative phase of the NAO, Europe experiences hot dry summers and cold dry winters, whilst the Mediterranean region is wet. This helps to explain why the LIA seemed mostly like a winter phenomenon [11]. The AMOC would have weakened at the same time, transporting less heat north.

It is not entirely surprising that the atmospheric $CO_2$ levels during the LIA were very slightly depressed (Figure 14.8b), but the question is – why? This depression has been attributed to increased storage of carbon in plants on land, although it is not clear whether it represented either the response of the terrestrial biosphere to cold temperatures, or vegetation re-growing on abandoned land [80]. Carbonyl sulphide can be used to represent primary productivity, and in 2016 Rubino and colleagues used it to show that such productivity declined during the LIA, most likely due to temperature change, not the re-growth of vegetation. The changing $CO_2$ flux was most probably driven by the high-latitude (especially Arctic) Northern Hemisphere response to temperature [80]. The unique fall in $CO_2$ in about 1590 has been attributed to reforestation of the Americas following the 90% decline in the indigenous population caused by disease brought by Europeans [81].

These various analyses help to confirm that whilst Lamb's suggested temperatures for medieval warming and the LIA may seem plausible for western Europe, they are not plausible for the globe as a whole, nor are his maxima (from proxies) for the Medieval Warm Period likely to be representative. The evidence seems to suggest that solar variability may have played a role in forcing climate change during the late Holocene, and we will explore that possibility next.

## 15.2 Solar Activity and Cosmic Rays

The evidence suggests that an overall fall in orbital insolation (Figures 12.2 and 14.1) cooled the world from the mid Holocene Climatic Optimum into the neoglacial climate of the late Holocene (e.g. Figures 14.1, 15.5, 15.6) [82], despite evidence for local winter warming, and that superimposed on that cooling trend were brief warming or cooling events (like the Medieval Warm Period or the LIA) that appeared unrelated to the broad, slow changes in orbital insolation. Did such variations represent fluctuations in the activity of the Sun itself? Could the same centennial or millennial forcing be responsible for the warming of the modern era? To address these questions we must first examine variations in the output of the Sun over the past 2000 years.

The Sun's face suffers from spots. They come and go in a roughly 11 year cycle, the Schwabe Cycle, named after German astronomer Heinrich Schwabe (1789–1875), in which insolation varies by a tiny 0.1%, or 0.25 Watts/m$^2$ (W/m$^2$) [83]. When sunspots are abundant our climate is very slightly warmer; when they are fewer it is very slightly cooler. Back in 1801, the astronomer William Herschel noticed that the price of wheat rose when the number of sunspots fell [84]. The French historian Emmanuel Le Roy Ladurie (1929–) found a similar association between sunspots and the French wine harvest [85]. He also found that the dates of French grape harvests between 1370 and 1879 were inversely proportional to April–September temperatures in Paris [86–88]. The LIA cooling was not manifest in his Parisian summer records, confirming that in Europe, at least, the LIA was most effective in winter [11].

The sunspot cycle varies in both length and amplitude. Its length ranges from 9.7 to 11.8 years. The longer cycles are cooler, whilst shorter ones are warmer [89]. The amplitude of the cycle is modulated by three other cycles. The Gleissberg Cycle, named after German astronomer Wolfgang Gleissberg (1903–1986), lasts 88 years and ranges from 70 to 100 years. The Suess–De Vries Cycle, named after Hans Suess and Dutch physicist Hessel De Vries (1916–1959), lasts 208 years and ranges from 170 to 260 years. The Hallstatt Cycle, named after a cool and wet period in Europe when glaciers advanced substantially, lasts 2300 years. One of the first to appreciate the modulation of the sunspot cycle was American astronomer John (Jack) Eddy (1931–2009), who pointed out that '*the long-term envelope of sunspot activity carries the indelible signature of slow changes in solar radiation which surely affect our climate*' [90]. In 1976, he drew attention to the period of zero sunspots between 1645 and 1715 known as the Maunder Minimum, after English astronomer Edward Walter Maunder (1851–1928) [91]. In 1987, the US National Academy of Sciences awarded Eddy the Arctowski Medal for his work on long solar cycles.

Observations of the sunspot cycle through telescopes began in 1601. Back beyond that we rely on proxy measurements of solar activity in the form of the radioactive nuclides of carbon ($^{14}$C), and beryllium ($^{10}$Be), with half-lives of 5730 years and 1.39 Ma, respectively. Bombardment by cosmic rays creates these radioisotopes in the upper atmosphere. They are least abundant when there are many sunspots, the Sun's output is high, and the solar wind deflects galactic cosmic rays away from our outer atmosphere (Figure 15.11) [92]. The climate is correspondingly warm. Cosmic rays are also deflected when the Earth's internal magnetic field is strong (compare A and B in Figure 15.12) [92]. For example, to determine the effects of solar variability on $^{14}$C Steinhilber had first to extract from the $^{14}$C data in Figure 15.11b the effect of changes in the Earth's geomagnetic dipole field strength. Subtracting the effect of the magnetic field on the abundance of $^{14}$C gave the production rate of $^{14}$C (Figure 15.11c), which is a reflection of solar activity. The normalized records of $^{14}$C and $^{10}$Be (Figure 15.11d) represent solar output [92]. These signals are trapped in Earth materials, like tree rings ($^{14}$C) and ice cores ($^{10}$Be in ice, and $^{14}$C in bubbles of fossil air). Single living trees can provide annual rings extending back 1000 or more years. Ancient trees preserved in sediments and overlapping in age can provide a continuous annual record dating back several thousand years. A treasure trove of these trees in the flood deposits of rivers like the Rhine and the Danube gets us back to 11 919 years ago [93], and shows that $^{14}$C declined steadily through the Holocene until about 1500 years ago [92] (see Figure 15.11b).

$^{14}$C production would have been higher ~40 Ka ago, when Earth's geomagnetic field almost disappeared for a few thousand years during the Laschamp geomagnetic event [94]. Because this field deflects cosmic rays, its growing

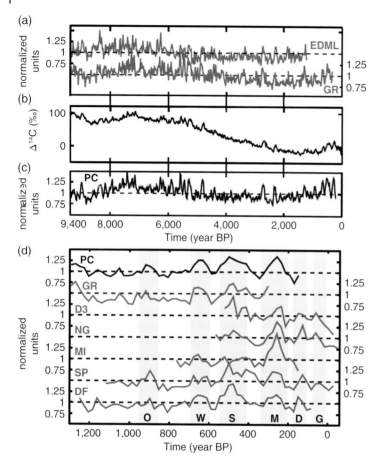

**Figure 15.11** ${}^{10}$Be concentration records are green (Greenland) and red (Antarctica) and ${}^{14}$C is black. Time is given as year before present (BP where present refers to 1950 CE). All records are mean normalized (divided by the mean) 22 year averages. (a) New ${}^{10}$Be record from the Antarctic EDML ice core (red) and of the existing ${}^{10}$Be record from the GRIP (GR) ice core (green); (b) $\Delta^{14}$C, the deviation of the atmospheric ${}^{14}$C/${}^{12}$C ratio from a standard value, measured in tree rings; (c) ${}^{14}$C production rate p${}^{14}$C (PC), calculated with a box-diffusion carbon cycle model from $\Delta^{14}$C over the last 9400 years (takes out the effect of the increasing magnetic field); (d) ${}^{14}$C production rate p${}^{14}$C (PC) and ${}^{10}$Be concentrations for ice cores (GR = GRIP; D3 = Dye-3; NG = NorthGRIP; MI = Milcent; SP = South Pole; DF = Dome Fuji) over the last 1200 years. Grand solar minima (cool periods) are marked by yellow bands: O = Oort; W = Wolf; S = Spörer; M = Maunder; D = Dalton; G = Gleissberg. *Source:* From figure 2 in Ref. [92].

strength led to the production of progressively less ${}^{14}$C in the upper atmosphere over the last 40 Ka (e.g. see Figure 15.11b). When we examine the reported magnetic field strength (Figure 15.12a), that rise in field strength is evident from 6 Ka ago to a broad peak centred about 2 Ka ago, followed by a decline. As explained by Korte and Muscheler in 2012 (their figure 3) [95], there is uncertainty about the field strength from 9.5–6 Ka ago, shown by the grey shading in Figure 15.12a. This implies that the underlying background shape of the cosmic intensity graph in Figure 15.12b may be uncertain as a guide to Earth's temperature, which is likely anyway as it runs directly counter to what we know of Holocene temperature and its relation to orbital insolation. In any case, as Korte and Muscheler put it, '*geomagnetic field models have to be improved further before they can provide a truly robust means to eliminate the influence of geomagnetic variability in cosmogenic radionuclide production studies*' [95]. What we can be sure of are the centennial scale fluctuations on that cosmic intensity curve, which reflect changes in solar intensity derived from the ${}^{14}$C

and ${}^{10}$Be data. These data, then, are proxies for solar cycles (Figures 15.12b–d) [92, 96, 97], bearing in mind the following caveats.

Excursions in ${}^{14}$C are not a precise guide to solar change, because the distribution of carbon is influenced by exchange with organisms and the ocean, which damps the solar signal [93, 94]. The ${}^{14}$C data extend up to about 1955, but after that are distorted by the effects of atomic weapons tests [98]. In addition, as pointed out in 2009 by Quan Hua of the Australian Nuclear Science and Technology Organization, '*the large decrease in atmospheric ${}^{14}$C after c.AD 1900 is mainly due to the continuous release of ${}^{14}$C-free CO$_2$ to the atmosphere as a consequence of the combustion of fossil fuels since AD 1850, a process known as the Suess effect … The combustion of fossil fuels in the [prior] period c.AD 1850–1900 was too small to cause an obvious decrease in atmospheric ${}^{14}$C*' [98]. So it is wise to ignore ${}^{14}$C data after 1900.

How reliable is ${}^{10}$Be, then, as measure of solar activity? Joel Pedro observed in 2012 that ${}^{10}$Be is not influenced by

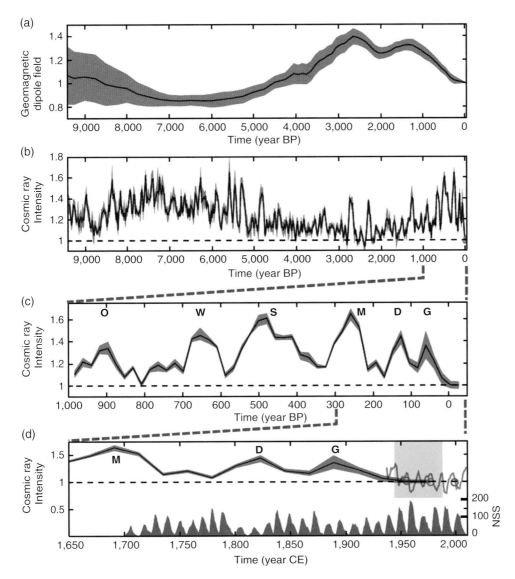

**Figure 15.12** (a) Geomagnetic dipole field strength relative to today; (b) cosmic radiation based on the first principal component of several radionuclide records, at 22 year averages, over the last 8000 years. Time is given as years BP. Grey band represents standard deviation of the individual radionuclide records. Black dashed line represents average cosmic ray intensity for 1944–1988 CE; (c) same as (b), for the past millennium. Grand solar minima are: O = Oort; W = Wolf; S = Spörer; M = Maunder; D = Dalton; G = Gleissberg. (d) Same as (c) for the past 350 years. Time is given as year CE. Red circles and green curve are 22 year averages and yearly averages of cosmic ray intensity. At the bottom in red are the annual sunspot numbers. *Source:* From figure 3 in Ref. [92].

the carbon cycle, but '*interpretation of the $^{10}$Be record is hindered by limited understanding of the physical processes governing its atmospheric transport and deposition to the ice sheets*' [99]. The $^{10}$Be signals tend to record patterns of atmospheric circulation, leading Pedro and colleagues to advise caution in the use of $^{10}$Be from *single* sites in solar forcing reconstructions [99]. Also, as Friedhelm Steinhilber of the Swiss Federal Institute of Aquatic Science and Technology reminded us, the $^{10}$Be signal in ice cores is affected by '*the variation of the snow*

*accumulation rate*' [92]. Although Steinhilber assumed that the rates of snow accumulation in Greenland and Antarctica '*have been rather stable during the Holocene so that $^{10}$Be concentrations and fluxes yield similar results*' [92], this is an assumption and ignores Pedro's observations. Evidently, then, although similar processes in the atmosphere produce both radionuclides, their subsequent geochemical behaviour differs. Hence it is no simple matter to use $^{14}$C and $^{10}$Be from ice cores or tree rings to reconstruct the past intensity of solar activity.

Nevertheless, it is clear from Bond's data (Figure 12.7) that there is a link during the Holocene between fluctuations in $^{14}C$ and $^{10}Be$ (representing the abundance of cosmic rays and hence the strength of the solar wind and/or the strength of the Earth's shielding magnetic field, or both) and the climate as reflected by ice rafting: decreased solar activity leads to cooling as well as to more cosmic rays, hence to more of those isotopes and to more ice rafting (compare Figures 12.7, 15.11, and 15.12) [92].

There is solid backing for Steinhilber's conclusions. Studies of radionuclides in the atmosphere in relation to climate date back at least to 1980, when Minze Stuiver of the University of Washington discovered that the $^{14}C$ in the fossil $CO_2$ from ice cores increased during the Maunder Sunspot Minimum, confirming that solar output was weak then [96, 97]. The solar connection was confirmed when Dansgaard and Oeschger found in 1984 that $^{10}Be$ was also abundant during that event [100, 101]. Stuiver found similar increases in $^{14}C$ at the times of the Wolf Sunspot Minimum (1280 to 1345 CE) and the Spörer Sunspot Minimum (1420 to 1540 CE). Analyses of $^{14}C$ and $^{10}Be$ by Manfred Schüssler and Dieter Schmitt of the Max-Planck Institute for Solar System Research at Katlenburg-Lindau, in 2004, found maximum values coinciding with the Oort Sunspot Minimum (1040–1080 CE), and the Dalton Sunspot Minimum (1790–1820 CE) [102]. Another sunspot minimum, the Gleissberg Minimum covered the period 1890–1910 [92]. Low $^{14}C$ and $^{10}Be$ values suggesting high sunspot activity occurred in medieval times (1100–1250 CE) and the modern era (1960–1990) [96, 97], all of these earlier data confirming the picture shown in Figures 15.11 and 15.12.

By 1998, Stuiver and his colleagues had extended the variation of $^{14}C$ and the solar signal back to 24 Ka ago [103]. Two years later, Eduard Bard, of the Université d'Aix-Marseille, and colleagues, confirmed Stuiver's findings for solar activity for the past 1200 years, finding radionuclide highs associated with the Maunder, Dalton, and Gleissberg sunspot minima [25]. The radionuclide high associated with the Maunder Minimum lay within a rather long period of low solar irradiance between 1450 and 1750 CE. In between the solar minima were solar maxima (Figures 15.11 and 15.12) representing warm periods, including one with values slightly higher than today centred on 1200 CE (between the Wolf and Spörer minima). Bard concluded that the radionuclide data supported the idea that variations in solar output contributed to the Medieval Warm Period and the LIA [25]. In 2007, Raimund Muscheler of Lund University in Sweden refined Bard's findings, identifying solar maxima tied to warming at 1100–1200 CE (the Medieval Warm Period), at 1750–1800,

with a peak at 1790, and at 1960 CE with somewhat lesser maxima at 1370, 1550–1630, and 1850–1870 CE [104]. Intervening solar minima were associated with cool periods typical of the LIA.

The centennial peaks and troughs of Steinhilber's graphs of cosmic intensity (Figure 15.12b–d) can be taken as the inverse of solar intensity, and thus the inverse of Total Solar Irradiance (TSI), which increases with increasing numbers of sunspots up to a sunspot number of about 150 [105]. These peaks and troughs fluctuate within the limits of what I call the 'natural envelope' of solar variation through recent time (Figures 15.11c and d) [92]. They suggest that the Roman Warm Period lasted with some fluctuation from c. 250 BCE to c. 600 CE and was followed by an intense cold period representing the European Dark Ages (c. 600–800 CE), then the warmth of the Medieval Warm Period (c. 850–1250 CE), followed by the cold of the LIA (1250–1750 CE), with the cooling continuing with lesser intensity through the Dalton Minimum (1820 CE) and the Gleissberg Minimum (1900 CE). If we can take the baseline cosmic intensity (dashed horizontal line in Figure 15.12b) as representative of the baseline global temperature for the last 2000 years, this would imply that the Roman Warm Period was slightly warmer than the Medieval Warm Period, whilst the Dark Ages were not as cold as the LIA, both of which assumptions seem to reflect what is known of European temperatures. Steinhilber's data from 1950 onwards suggest more or less stable temperatures (green line in Figure 15.12d), and that recent TSI (1944–1988 CE) was not far different from what it was in the Medieval Warm Period. Given that the cosmic intensity of the Medieval Warm Period was not much different from that of some of the warm periods within the LIA (Figure 15.12c), one might wonder why it was warmer. Possibly the answer lies in its duration, since it lasted for some 200 years. But, recalling Neukom's analysis, this may turn out to be a regional effect [46].

What do sunspot data themselves tell us about late Holocene climate change? Previous analyses of sunspot activity suggested in 1998 that sunspot numbers had been somewhat higher during the twentieth century than in the nineteenth [106]. But a careful reappraisal in 2014 of the way in which sunspot numbers had been recorded over time showed Frédéric Clette of the Royal Belgian Observatory and colleagues that the apparent broad peak of sunspot activity that had been thought to exist in the middle and later parts of the twentieth century [106] was instead an artefact of a formerly inadequate way of recording sunspots [107]. The revised sunspot data showed Clette that between 1749 and the present there was hardly any difference between the peaks of solar activity in 1780–1790,

1840–1870, and 1980–1990 (Figure 15.13). This suggests that the warming characteristic of the late twentieth century (e.g. 1980 and 1990) was not driven by rising solar activity, contrary to the suggestion by Hoyt and Schatten in 1993 [108], which we now know to have been based on inadequate sunspot data for the period prior to about 1900 [107]. Figure 15.14 shows some of the latest corrections to sunspot numbers [109], which are consistent with the reconstruction in Figure 15.13 [107].

Can we discern discrete sunspot cycles in the proxy data? Yes. Analysing the spectrum of $^{14}$C and $^{10}$Be data, Stuiver identified the 2300 year long Hallstatt solar cycle in 1998 [103]. By 2004, Raimund Muscheler and colleagues showed that the 208 year Suess–De Vries Cycle, which had persisted for at least the past 50 Ka, caused the Maunder, Spörer, Wolf and Oort Solar Minima [94]. At the same time, Paul Damon and Alexie Peristykh of the University of Arizona identified not only the 208 year Suess–De Vries

Cycle and the 88 year Gleissberg Cycle, but also their respective 'overtones' at 104 and 44 years in Greenland ice cores [76]. Overtones recur at half the interval of major peaks. They also found 'combination tones', where signals of different wavelength interfere with one another to accentuate or diminish the underlying signal. Modification of the 88 year Gleissberg cycle by the 208 year Suess cycle produced combination tones at 152 years and 61.8 years. *The latter is most probably the origin of the apparent 60 year cycle observed weakly in much modern climate data.* Damon and Peristyk also found the 2300 year Hallstatt Cycle in the ice core data. Heinz Wanner and his team carried out similar spectral analyses in 2008, finding peaks in solar variability at 208, 150, 104, and 88 years [23]. Steinhilber's team also confirmed the existence of the Suess–De Vries Cycle, which they found to vary with a period of about 2200 years (the Hallstatt cycle) [92]. The highest amplitudes of the Suess–De Vries Cycle occurred

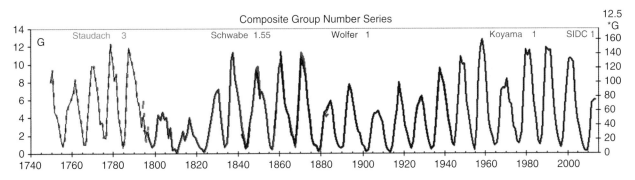

**Figure 15.13** A composite record of the yearly mean number of sunspot groups, back to 1749. The right-hand scale shows a quantity 12.5 times the average group count, as an 'equivalent' GN. *Source:* From figure 28 in Ref. [107].

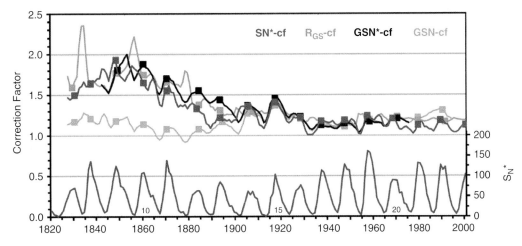

**Figure 15.14** Three year running averages for 1826–2000 of correction-factor (CF) time series for four sunspot number time series. Years of maxima are indicated by square symbols and cycle numbers are given at the bottom of the figure, which shows the latest sunspot numbers. *Source:* From figure 5 in Ref. [109].

when the Hallstatt cycle was at a minimum – at 8200, 5500, 2500, and 500 years ago – and the lowest occurred during grand solar minima (like the Maunder Minimum) [92]. Steinhilber also identified the Eddy Cycle (~1,000 years) and an unnamed cycle (~350 years), amongst other, less significant cycles [92].

How did these solar variations relate to climate records? In 1993, Michel Magny, of the Laboratoire de Chrono-Environnement in Besançon, found that lake levels in Europe were high, indicating cool and wet conditions, when $^{14}$C increased by 5% (=low solar activity), and that glaciers advanced and the tree-line descended when $^{14}$C increased above 10%, (= even lower solar activity) [110]. Cooling and high lake levels occurred during the Maunder and Spörer sunspot minima. There was an earlier 'little ice age' of high lake levels with high $^{14}$C at around 750 BCE. This was the cool and wet 'Hallstattzeit', or Hubert Lamb's 'early iron-age cold epoch'. The cooling of the LIA with its $^{14}$C peaks at 1500 and 1700 CE, and the cooling of the Hallstattzeit with its $^{14}$C peak at 750 BCE, are about 2300 years apart (see Figure 15.12b) – the period of the Hallstatt solar cycle.

Between 1999 and 2007, Magny and his colleagues used Stuiver's $^{14}$C data to extend their analysis of the relationship between $^{14}$C peaks and mid-European lake levels back to the start of the Holocene (Figure 15.15) [111–114]. High lake levels [113, 114] indicative of cooling and high precipitation corresponded with peaks in $^{14}$C and $^{10}$Be indicating low solar output. The largest positive $^{14}$C excursions and

lake high stands of the past 3000 years were those at 750 BCE (the Hallstattzeit) and 1500 CE (LIA). The larger of the lake high stands corresponded to peaks in Bond cycles of ice-rafting (Figure 15.15), hence to significant regional cooling as well as low solar activity.

Complementing Magny's work, Dirk Verschuren of Ghent University and Dan Charman, then at the University of Plymouth, found evidence in 2009 for a decrease in monsoonal activity associated with these cool events, suggesting that the Intertropical Convergence Zone (ITCZ) extended less far north in cool periods (solar minima) than in warm ones [19]. That agreed with the results of Steinhilber's examination of data from the Dongge Cave in southern China, where low $\partial^{18}$O (indicative of warming) corresponded to strong Asian monsoons and high solar irradiance [92].

Luterbacher and colleagues thought that their data showed a link between European summer temperatures and solar activity (Figure 15.16) [47]. But their 2016 estimates of solar activity were based on the work of Schmidt and others in 2011 [115], which predated the revision of sunspot data by Clette in 2014 [107] and Cliver [109], which would have shown much less of a rise in the modelled temperatures shown in green and purple in Figure 15.16. The observation-based data (blue and red lines) show that the temperature rise in 1950 matched what was seen in Greenland (Figure 15.1) and fits with the AMO (Figure 15.2). The subsequent rise from 2000 onwards (not shown in Figure 15.16) would significantly exceed the

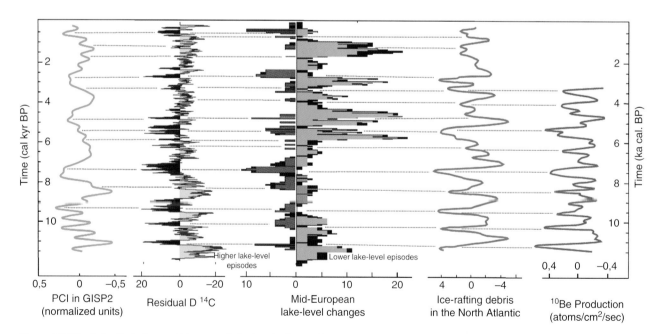

**Figure 15.15** Solar Variation and European Lake Levels. Comparison between the Polar Circulation Index (PCI) at GISP2, the atmospheric residual $^{14}$C variations, the Greenland $^{10}$Be record, the mid-European phases of higher lake-level, and the ice-rafting debris (IRD) events in the North Atlantic Ocean. *Source:* From figure 6, in Ref. [113].

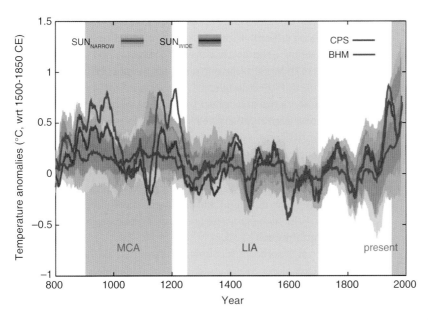

**Figure 15.16** Simulated and reconstructed European *summer* land temperature anomalies (with respect to 1500–1850 CE) for the last 1200 years, smoothed with a 31 years moving average filter. Reconstructed temperatures are shown in blue (BHM = Bayesian hierarchical modeling) and red (CPS = composite - plus-scaling) over the spread of model runs. Simulations are distinguished by taking solar forcing into account: forcing was estimated as either stronger (purple) or weaker (green), with darker shading accounting for 50% and lighter accounting for 80% of the spread). Orange shading indicates the extent of the Medieval Climate Anomaly (MCA) and the present warming, whilst blue shading outlines the extent of the Little Ice Age (LIA). *Source:* From figure 3 in Ref. [47].

levels of warming of the Medieval period. As an aside, it is not clear why Luterbacher chose to show the LIA as ending in 1700, when most observers have it ending around 1850, and the red and blue observational data show a substantial rise from about 1900. Nevertheless Luterbacher concluded that '*changes in external forcing have had a pronounced influence on past European summer temperature variations ... [with] summer temperatures during the [period] 1986–2015 [being] anomalously high*' [47], presumably owing to increasing emission of greenhouse gases especially since 1950.[3]

Analysing the distribution of stomata in fossil pine needles collected from a lake 1311 m up on the slopes of Mount Rainier (Washington State), Lenny Kouwenberg and a team from Utrecht University found in 2005 that $CO_2$ varied in parallel with temperature over the past 1200 years (Figure 15.17) [116]. There was slightly more $CO_2$ in warm periods than cold periods. The centennial scale trends reached 300 ppm in the Medieval Warm Period (about 970 CE) and almost 320 ppm during the warm period between the Wolf and Spörer Sunspot Minima (Figure 15.17). That does not mean that $CO_2$ drove the warming, but it does suggest that warming by solar activity enhanced the emission of $CO_2$ from the ocean, which then provided a positive feedback to the ongoing warming,

much as we saw occurring during the last glacial period (e.g. see Figure 13.6).

Comparing Figures 15.17, 14.8b, and 9.11, we see that the stomatal data from Washington State are more variable than the Law Dome data from East Antarctica (location in Figure 7.5) for the past 1000 years, which averaged 280 ppm (range 275–285 ppm). This is at least in part because the Northern Hemisphere $CO_2$ patterns are much more seasonally variable that those in Antarctica (see Figure 8.7), and the maximal $CO_2$ values of the Northern Hemisphere tend to be higher than those of South Pole by about 8 ppm This pattern helps to explain some of the differences between Figures 15.12, 14.8b, and 9.11. However, more recent analyses, of the WAIS ice core by Ahn and colleagues in 2012 [117] show that $CO_2$ values there were slightly higher than those at East Antarctica's Law Dome, averaging about 285 ppm for the period 1100–1600 CE (Figure 15.18). A broad $CO_2$ peak in West Antarctica between 1250 and 1050 CE coincides with the warming of the Indo-Pacific warm pool (Figure 15.18), suggesting the continuation back through time of the close link between Pacific equatorial temperatures and those of West Antarctica that is known to apply also today. Note that whilst the West Antarctic $CO_2$ data fall to a broad low during the LIA they do not show the brief sharp fall in $CO_2$ dated to about 1620 seen at Law

---

[3]More than 50% of all fossil fuel ever burned has been burned since1950.

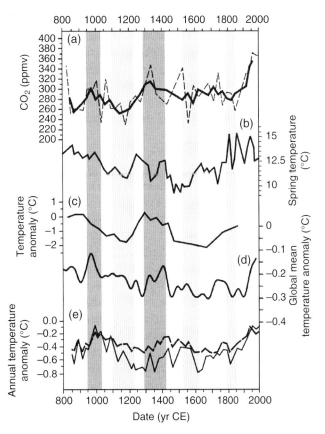

**Figure 15.17** Comparison between $CO_2$ and Northern Hemisphere climate records. (a) $CO_2$ concentration from stomatal counts on *Tsuga heterophylla* needles from Jay Bath lake. Thin dashed line connects means of 3–5 needles per sample depth; thick line = three-point moving average to emphasize centennial scale trends. Dark grey bands = periods of high $CO_2$ concentration; light grey bands = periods of low $CO_2$. (b) Sea-surface temperatures of Chesapeake Bay. (c) Sea-surface temperature cold season anomalies offshore West Africa. (d) Global mean temperatures (45 years running average) from multiproxy records based on 1961–1990 reference period. (e) Summer temperature anomalies in two sets of tree ring records from the Northern Hemisphere. *Source:* From figure 4 in Ref. [116], which contains the sources of the data.

Dome (Figure 15.18). The greater variability of the Northern Hemisphere temperature record compared to that of the Pacific warm pool is evident in the data (Figure 15.18). This Northern Hemisphere thermal variability may help to explain the variability of the Northern Hemisphere stomatal record of $CO_2$ (Figure 15.17). The relative smoothness of the Antarctic $CO_2$ record (e.g. Figure 9.11) must owe much to the fact that Antarctic $CO_2$ values are naturally far less variable than Northern Hemisphere ones (Figure 8.7), but the possibility also exists that because ice accumulates very slowly, the $CO_2$ signal in ice cores may be somewhat smeared. Investigating that possibility, Ahn and colleagues found that smearing likely had its greatest effect in subduing divergences, like the negative Law Dome peak at about

1620, changing the value by as much as c. 7 ppm [117]. It was difficult to see evidence of smearing of the signal before or after that event (between 1500 and 1750 CE) [117]. We also have to consider the possibility that in part the difference between Antarctic and Northern Hemisphere $CO_2$ values may reflect the seasonal growth of the sampled Northern Hemisphere leaves early in the year, when $CO_2$ is at its most abundant, whilst Antarctic values represent annual averages. Nevertheless, it is difficult to compare stomatal reconstructions like that of Figure 15.17 with Antarctic ones (Figure 15.18) because of the high variability (30 ppm range) and the high uncertainty (represented by the high standard deviation of the data) of the stomatal information [118].

It is somewhat counterintuitive that as the Norwegian–Greenland Sea warmed in recent decades, the northern North Atlantic off the southern end of Greenland cooled. Investigating the past 800 years of Greenland's temperature history, Kobashi and his colleagues may have found an explanation. In 2013, they noticed that with more solar activity, the local surface waters would have warmed, making them less dense, so less likely to sink, hence reducing the overturning circulation and pulling less Gulf Stream water northwards, paradoxically cooling the northern North Atlantic (which is what we see in the ocean today) and southernmost Greenland [119]. That could explain why Greenland showed less pronounced warming than did the Northern Hemisphere as a whole in the latter half of the twentieth century [119]. Kobashi also observed, that during periods of lower solar activity, the AMOC strengthened, warming the North Atlantic and Greenland, because cooling the ocean off Greenland made salty northward moving Gulf Stream Water dense enough to sink. In effect Kobashi found a slight negative correlation with solar activity. Bond, in contrast, found a strong positive correlation with solar activity, with lower activity corresponding to stronger ice rafting (Figure 12.7), much like Magny's correlation of lower solar activity with heavier rainfall (Figure 15.15). In evaluating these apparently paradoxical findings we have to remember that the cold East Greenland Current, which moves south along the east coast, separates Greenland from the bulk of the Norwegian–Greenland Sea. East of the current, warm water from the Gulf Stream moves up into the Arctic through the Norwegian–Greenland Sea, warming the west coasts of Britain and Norway as it does so. Many of Bond's samples come from areas to the east of the East Greenland Current. Here, then, we have regional grounds for differentiating between the climate regimes of Greenland and the Arctic, much as noted by Moffa-Sanchez in 2019 [120].

Could quasi-regular variations in the Sun's output explain Bond's cycles? In 1997, Bond and his team thought

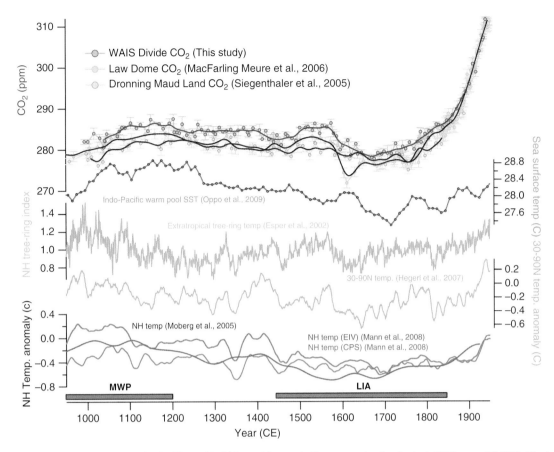

**Figure 15.18** (top) Atmospheric $CO_2$ and (middle and bottom) climate proxies for the last 1000 years. WAIS Divide data (red) are compared with those from the Law Dome (blue) and EPICA Dronning Maud Land (black). To facilitate comparison of temperature and $CO_2$ records, a Gaussian filter ($1\sigma$ = 10 years) was applied to interpolated annual data sets (darker lines). *Source:* From figure 6 in Ref. [117].

not – because there was no known solar cycle with a 1500 year period. But, working with specialists in solar signals – Raimund Muscheler and Jürg Beer – Bond found by 2001 that the ratio of red-coated (iron-stained) rock grains to total ice-rafted grains in the Holocene correlated with fluctuations in $^{14}$C and $^{10}$Be, which were abundant when the Sun cooled [121], as shown in Figures 12.7 and 15.15. Red ice-rafted grains had a more northeasterly source, suggesting that cooling forced ice southwest. The correlation coefficients were not high, at 0.44 for $^{14}$C and 0.56 for $^{10}$Be. While the linkage convinced Bond that small cyclic temperature changes, with a period of around 1500 years, might indeed be paced by changes in solar luminosity in the Holocene, he recognized that the large errors in the ages estimated from his marine sediment cores made the correlations with solar output fuzzy. Nevertheless, he considered that the signal was likely to be global, because there was a clear link between strong solar minima, cooler ice-bearing waters in the North Atlantic, and reduced monsoonal activity and rainfall in the tropics. Evidently, what he called '*the enigmatic, at best quasi-periodic, "1500-year" cycle*' must be '*a pervasive feature of the climate*

*system .... linked to variations in solar irradiance*' [121]. Bond and his team knew that General Circulation Models implied that '*at times of reduced solar irradiance, the downward-propagating effects triggered by changes in stratospheric ozone lead to cooling of the high northern latitude atmosphere, a slight southward shift of the northern subtropical jet, and a decrease in the Northern Hadley circulation*', which would increase North Atlantic drift ice, cool the ocean surface and the atmosphere above Greenland, and reduce precipitation at low latitudes [121]. This cooling and freshening of the northern seas would have temporarily reduced the production of North Atlantic Deep Water (hence pulling less warm water north from the Gulf Stream). In conclusion, they said, '*If forcing of North Atlantic drift ice and surface hydrography is fundamentally linked to the Sun and begins in the stratosphere, then atmospheric dynamics and their link to the ocean's circulation are much more important for interpreting centennial and millennial time sales of climate variability than has been assumed*' [121]. Equally, that influence offers the prospect of direct solar signals being distorted by oceanic and atmospheric circulation patterns.

In 2004, Damon and Peristykh set out to test Bond's theory of a more or less direct solar link to climate. Among a number of small peaks of millennial duration in the $^{14}$C record in Holocene tree rings they found one at 1433 years – about the same interval as the 1470 year peak in Bond Cycles [76]. But this peak was '*too insignificant compared to the noise level to be counted as real*' [76]. Their analysis of the $^{14}$C spectrum identified peaks spaced roughly 800 years apart over the past 9000 years. Half of these were larger than the others and spaced around 1600 years apart. These peaks were emergent properties of the system, created by a variety of combination tones and overtones rather than by primary solar forcing signals. Nevertheless, they do suggest that emergent millennial variation in the centennial solar cycle at approximately 1600 year intervals might possibly account for the ice rafting events, even if that millennial variation did not reflect specific independent solar signals with a regular beat.

Holger Braun of the Heidelberg Academy of Sciences and colleagues found in 2005, that the 88 year Gleissberg and 208 year Suess–De Vries solar cycles could have led to periodic inputs of freshwater to the North Atlantic in such a way that '*superposition of the two .... cycles, together with strongly nonlinear dynamics and the long characteristic timescale of the thermohaline circulation*' could have combined to provide the robust 1470 year climate cycle underpinning Dansgaard–Oeschger events during the last glaciation [122]. Thus, despite the absence of a major 1470 year solar cycle, solar forcing could have triggered those events through non-linear connections within the climate system. Although that was the case for the last glaciation, they could not get it to apply to the Holocene, possibly because of the large differences in the thermohaline circulation between glacial and Holocene times.

A subsequent more sophisticated statistical analysis of Bond's Holocene data by a French group in 2007 suggested that his millennial cycles were – as Bond had first thought – most probably caused by internal oceanic oscillations [123]. They did find evidence in Bond's data for solar cycles, but with periodicities of 1000 (the Eddy Cycle) and 2500 years (the Hallstatt Cycle), not 1500 years [123].

Nevertheless, in 2014, analysing $\partial^{18}$O and Mg/Ca ratios from a deep-water sediment core off Iceland, Paola Moffa Sanchez of Cardiff University and colleagues did find a correlation between solar output, climate, and variations in ocean circulation for the period from 818 to 1780 CE [124]. They inferred '*that the hydrographic changes probably reflect variability in the strength of the subpolar gyre associated with changes in atmospheric circulation. Specifically ... low solar irradiance promotes the development of frequent and persistent atmospheric blocking events, in which a quasi-stationary high-pressure system in the eastern North Atlantic*

*modifies the flow of the westerly winds ... this process could have contributed to the consecutive cold winters documented in Europe during the Little Ice Age*' [124]. The blocking events are mid-latitude weather systems in which a quasi-stationary high-pressure system over the North Atlantic modifies the flow of westerly winds by blocking or diverting their path. Blocking events derive from meanders in the jet stream, mostly in winter in association with a negative phase of the NAO. They keep western Europe cold by blocking the northeastward transport of warm maritime air from the southwest and replacing it with cold southerly transport of air from Scandinavia or Russia. This pattern caused the cold European winters of 1963, 2009, 2010, and 2013. Lamb had already called on blocking to explain the severe winters of the Maunder Minimum in the LIA [125]. Similar cold air outbreaks from the Arctic induced by atmospheric dynamics, not by climate change, help to explain cold winter episodes in the United States and Canada.

In 2005, Kirk Maasch and colleagues comparing their climate data with $^{14}$C and $^{10}$Be records confirmed that the major cooling event of the past 2000 years occurred during the period of low solar activity that embraces the LIA [126]. Much like Magny, they concluded that there was '*a first-order relationship between a variable Sun and climate. The relationship is seen on a global scale*' [126].

Given these relationships, it is not entirely clear why the cool periods identified by Wanner and Mayewski and examined in Chapter 14 [127, 128], do not correlate more closely with the $^{14}$C, $^{10}$Be and lake high-stand data and Bond cycles. Three of Mayewski's long cool periods (600–150 years, 2.5–3.5 Ka, and 5–6 Ka ago) correlate with major positive $^{14}$C excursions, but one (8–9 Ka ago) corresponds to a major depletion in $^{14}$C [128]. All four of his events seem to have been selected based on the abundance of potassium (K$^+$) ions in the GISP-2 ice core. The event at 8–9 Ka ago occurred when volcanic activity was abundant, which may explain its high K$^+$ signal.

Considering these various analyses, it seems highly likely that most of the short cool periods of the Holocene represent periods of low amplitude in century-scale solar cycles. These cool periods occurred irregularly within a cycle having a broad envelope of 1–2 Ka, much as Bond had suggested in moving away from a precisely defined 1500 year signal [22, 121]. They led to high European lake levels, as well as North Atlantic ice rafting and associated freshwater outbreaks during Bond Cycles, which slowed the Thermohaline Conveyor system temporarily. That in turn would have transmitted Northern Hemisphere cooling signals slowly to the Southern Ocean via the bipolar seesaw, thereby making Southern Hemisphere responses to the same forcing events out of phase with their Northern

Hemisphere counterparts, and thus making more difficult the identification of originally global signals of solar output. Wanner reminded us that given the different hemispheric responses to solar changes, we should not expect uniform global warming or cooling – there is and ought to be significant regional heterogeneity [23]. The incidence of large volcanic eruptions complicates matters by contributing additional cooling at times, as we shall see later.

Should we be confident of the results emerging from examinations of the increasing flood of proxy solar data? By the late 1980s, Lamb knew that variations in solar activity might have controlled the climate of the past 2000 years, but he was not convinced that it had, because the signal of climate change was too heterogeneous [10]. In his day, not only were the signals of climate change geographically heterogeneous, but also the timings of so-called climatic events did not coincide precisely with the extent of known solar events. Even as recently as 2012, Gerald North of Texas A&M University, and colleagues, agreed that current palaeoclimatic evidence for solar forcing was limited – it was difficult to find clear indications of the impact of solar variations on climate [129]. At that time they felt that most records did not provide the necessary resolution or signal strength to detect a weak solar signal even if it was present. Other forcing factors like volcanic eruptions or oceanic oscillations might be operating on similar time scales, thus complicating signal detection at certain frequencies. For instance, more than 90% of the variance in temperatures in the Central England Temperature record since about 1650 CE can be accounted for by volcanic activity or by fluctuations in the AMO. That begs the question – to what extent were those oscillations linked to solar activity? For the Pacific, North agreed that there might be a link between peak sunspot years and cool La Niña events. He and his team called for more interdisciplinary research and better linkages between data and models as the basis for recognizing solar signals in the palaeoclimate record [129]. Looking at the distribution of key radionuclides ($^{14}$C and $^{10}$Be), Cronin observed, in 2010, that *'geomagnetic reconstructions vary greatly depending on the region and source of measurements, that equally complex obstacles surround reconstruction of long-term modulation of solar energy flux by solar wind and other processes, and that these factors preclude a simple quantification of millennial and short-term irradiance changes from nuclear records ... [thus] a great deal of caution is needed when interpreting paleoclimate records before A.D. 1600 because reconstructed radionuclide records are not direct measures of solar activity'* [130]. Korte and Muscheler expressed much the same caveat in 2012 [95]. Presumably, Neukom would agree [46].

Despite those conclusions, it seems to me that we now have, in Steinhilber's 2012 data (Figure 15.12) [92], and

Clette's review of sunspot data since 1749 (Figure 15.13) [107], an up to date view of solar variability with time, along with stronger evidence in Moffa-Sanchez's 2014 report [124], for a clear relationship between local climate and solar forcing, in the North Atlantic, as well as good evidence in Magny's data (Figure 15.15) for a direct link between solar activity and European lake levels. If the association between declining solar activity and cooling proves to be robust, then the recent decline in solar forcing since 1990 (Figure 15.13) suggests a return to more severe European winters – unless that is countered by anthropogenic global warming.

Horst-Joachim Lüdecke of the University of Applied Sciences in Saarbrücken, Germany, and Carl-Otto Weiss of the Physikalisch-Technische Bundesanstalt in Braunschweig, Germany, provide a somewhat different perspective. Carrying out a spectral analysis of a global temperature curve that they constructed for the past 2000 years, they found evidence for probable solar cycles with periods of c. 1000 years (Eddy Cycle), 460 years, and 190 years (the latter being, they thought, the Suess–De Vries Cycle) [131]. Their temperature maxima represented the Roman, Medieval and modern warm periods, which have been widely noted by others. They suggested that much of the rise in temperature since 1900 must be solar and driven by the Suess–De Vries Cycle. Attributing at least some of the rise in temperature since 1900 to solar effects accords with what we see from the sunspot numbers, which rise from a low peak in 1910 to a high peak in 1960 (Figure 15.13). However, sunspot peaks then flattened and have been in decline since 1990 (Figure 15.13), whilst temperatures have gone on rising. Hence there is a current disconnect between solar activity and global average temperature, which Lüdecke and Weiss either failed to note or failed to comment upon. In passing, I note that the Antarctic data these authors used to derive their global temperature record did not include the rise in temperature for the late twentieth century seen in West Antarctica (Figure 15.8). Nevertheless, projecting the sine waves of their three main solar cycles into the future suggested to the two scientists that solar outputs should decline from now to 2050 CE, rise slightly to 2130 CE, then decline again to 2200 CE, with the implication that global temperatures should follow that same path [131]. Some Russian scientists offer a similar scenario [132].

In 2013 Steinhilber and Beer provided an alternative view, based on spectral information obtained from a 9.4 Ka long reconstruction of Holocene solar activity, and assuming that the same periodicities found in the past will also recur in the future (meaning the next 500 years). They noted that whilst their methods applied to the past record seemed *'to predict the trend in solar activity reasonably well'* they were less successful at predicting the amplitude of those solar signals [133]. For the future, their methods predicted (for 2100 CE) a solar minimum like that of the Dalton Minimum of the early 1800s or the Gleissberg Minimum of the end of the nineteenth century (roughly half the amplitude of the Maunder Minimum – see Figure 15.12). This minimum would last around 50–100 years, after which solar activity would slowly increase with some oscillations, reaching something like the sunspot level of about the 1920s by 2500 CE. They pointed out that future cosmic ray intensity would depend not only on changing solar activity but also on the nature of the Earth's magnetic field, whose future activity is uncertain [133].

There is a discrepancy between the Lüdecke and Weiss solar activity forecast (decline to 2050 then rise to 2130) and the Steinhilber and Beer forecast (decline to 2100 then a slight rise). In any case there is no evidence in these predictions for a new Maunder Minimum, hence no suggestion that temperatures for the predicted period would replicate those of the depth of the LIA. Lüdecke and colleagues commented, further, that the association of the twentieth century rise in temperature with solar activity '*does not rule out a warming by anthropogenic influences such as an increase of atmospheric CO₂*' [134].

This seems an appropriate point to comment on the theory proposed by Henrik Svensmark (whom we met in Chapter 10) that a weakening of the solar wind and/or the Earth's magnetic field leads to an increase in cosmic rays that create ions in the upper atmosphere, which causes the growth of cloud condensation nuclei, leading to the production of clouds that serve to cool the climate [135–137]. If this hypothesis is correct, then temperatures should have cooled substantially when the Earth's magnetic field was at its weakest 40 Ka ago during the Laschamp geomagnetic reversal. But, examination of the $\partial^{18}O$ proxy temperature for Greenland's GRIP ice core, collected in 2004, shows no evidence that Greenland cooled along with that reversal [138]. Instead, the GRIP data show that the cold of 40 Ka ago was more or less identical with that for the preceding and following decades [138]. Furthermore, Petit's and Luthi's data from Antarctic ice cores show that there was a warm peak 40 Ka ago (Figures 13.8 and 13.9). An evaluation of the effect of cosmic rays on climate, by Mike Lockwood (1954–) of the Meteorology Department of the UK's University of Reading concluded in 2012 that galactic cosmic rays provide an '*increasingly inadequate explanation of observations*' [139]. A review of solar influences on climate by Gray and others in 2010 agreed that current data do not support the postulated link between cloud cover and cosmic rays [140]. And a further critique of Svensmark's hypothesis offered by Bomin Sun and Raymond Bradley in 2004, concluded '*that there is a lack of evidence to support the GCR (galactic cosmic ray)-cloud hypothesis*' [141].

From the geological perspective it should also be pointed out that there seems to be no relationship between climate and the 183 reversals of the Earth's magnetic field that characterize the past 83 million years, contrary to what Svensmark's hypothesis might suggest. Nevertheless, a recent Japanese study by Ueno and colleagues did find a possible link between a cooling event in China and the Matuyama–Brunhes magnetic reversal 783–776 Ka ago [142]. Careful examination of their published data shows that there was a brief small amplitude cooling event at that time, which they interpreted as due to relaxation of the Earth's magnetic field resulting in an increase in galactic cosmic rays that amplified low cloud cover. However, this event, which would have occurred during the warm interglacial event labelled MIS-19, is not discernible in the global $\partial^{18}O$ record (Figure 12.4), so cosmic rays must have had a trivial effect compared to that of orbital precession and obliquity. Clearly, if galactic cosmic rays do affect the climate they do so in ways that are largely obscured by the effects of orbital and axial variability.

## 15.3 Volcanoes and Climate

Can the relative effects of volcanic and solar activity be disentangled? Andrew Schurer and colleagues from the University of Edinburgh attempted to do so for the past millennium across the Northern Hemisphere. They concluded that '*Although solar forcing may be relatively unimportant for large-scale climate change, it could still play a significant role in regional and seasonal variability, owing to its influence on climate dynamics, an influence that is strongly diminished when averaging annually and over the whole Northern Hemisphere*' [143]. Instead they thought that the changes of the past millennium were influenced more by volcanic eruptions and changes in greenhouse gas concentrations than by changes in solar output. When I examined their data, I found 12 major volcanic eruptions within the solar minima between 800 and 2000 CE, and 15 within the intervening solar maxima. I therefore remain unconvinced that volcanic activity was a significant control of centennial change through the last millennium, though I would expect it to have had effects at the decadal level or less. Recognizing that volcanic activity generally leads to cooling lasting no more than a few months to two years, it seems far more likely to me that Schurer's team underestimated the solar effect and overestimated the volcanic effect.

Undoubtedly, eruptive events did influence the climate of the past 1000 years. For instance, Rosanne D'Arrigo of Lamont, and colleagues, found some connections between low-latitude volcanic eruptions and tropical climate over the past 400 years [144]. Amongst the eruptions causing temporary global cooling were those of Tambora in 1815 and Krakatoa in 1883. The stratospheric veil of dust and sulphuric acid droplets caused by these eruptions scattered incoming solar radiation and led to fine red sunsets, which may have influenced the paintings of J.W.M. Turner following the eruption of Tambora, and of Edvard Munsch after the eruption of Krakatoa. Tambora exploded during the Dalton Sunspot Minimum, and temporarily contributed to its cooling effects. That combination led to 1816 being labelled the 'year without a summer', when cooling and excessive rains made the harvests fail in western Europe. Similar effects may have resulted from the eruptions in 1902 and 1911/12 that coincided with the Gleissberg Sunspot Minimum. Even so, my calculations suggest that it is unwise to assume that this association weakens the argument for the importance of solar activity. After all, large eruptions are short-lived, whilst solar minima are not, as we saw earlier.

Until 2013, the source of the volcanic activity of the period 1251–1310, identified from ash in polar ice, was a mystery. That year, we learned that in 1257 CE there was a colossal eruption of the Salamas volcano on Lombok Island in Indonesia. It was among the largest eruptions of the Holocene, with an ash column that most likely reached an altitude of 43 km, depositing 40 km³ of volcanic ash [145]. That eruption may help to explain the rapidity with which the LIA set in. A team headed by Gifford Miller of the University of Boulder, Colorado found in 2012 that '*precisely dated records of ice-cap growth from Arctic Canada and Iceland [show] that Little Ice Age summer cold and ice growth began abruptly between 1275 and 1300 CE, followed by a substantial intensification [in] 1430–1455 CE. Intervals of sudden ice growth coincide with two of the most volcanically perturbed half centuries of the past millennium*' [57]. Sea ice cover also expanded north of Iceland at these times. '*The persistence of cold summers is best explained by consequent sea-ice/ocean feedbacks during a hemispheric summer insolation minimum – large changes in solar irradiance are not required*' [57]. Miller viewed the underlying (and continuing) low orbital insolation of the late Holocene (Figure 12.2) as the main contributor to the coolness of the LIA. In his view, volcanism and sunspot minima exacerbated what would have been a cold period in any case. Once the sea ice formed, and after the volcanic activity ceased, low insolation and sunspot minima helped to maintain a large area of sea ice that suppressed regional summer temperatures for centuries: '*an explanation of the Little Ice Age does not require a solar trigger*' [57], he said. Nor does it require a volcanic one, although the recovery of the climate system after a very large eruption may take a decade due to the persistence of local cold anomalies in the

ocean. Orbital insolation is quite enough by itself to make cold summers persist, and, clearly, solar minima aid in that process even if they are not the triggers for it [57].

Mayewski and Maasch observed that the wind system intensified abruptly both in Greenland and at Siple Dome around 1400 CE, close to the beginning of the LIA. While this could suggest the global imprint of a solar signal, there was no abrupt solar event at the time [58]. Instead, this intensification could reflect a global response to the atmospheric signal of the major volcanic episodes identified by Miller and his team in the Arctic as taking place between 1430 and 1455 [57].

By 2015, Sigl and colleagues were using new measurements of aerosol loading from polar ice cores to suggest that '*large eruptions in the tropics and high latitudes were primary drivers of interannual-to-decadal temperature variability in the Northern Hemisphere*' over the past 2500 years [146]. Overall, they found that '*cooling was proportional to the magnitude of volcanic forcing and persisted for up to ten years after some of the largest eruptive episodes*' [146] (Figure 15.19). Volcanic activity led to reduced Northern Hemisphere tree growth, and temporary thermal responses averaged a cooling of $-0.6\pm0.2\,°C$ there, following the largest tropical eruptions, and $-0.4\pm0.4\,°C$ following the largest Northern Hemisphere eruptions, with the strongest response (exceeding Tambora in 1815) reaching $-1.1\pm0.6\,°C$. Sigl and colleagues dated the volcanic episode identified by Miller as taking place most likely in 1458 (Figure 15.19). Examining Sigl's record of temperature for the past 2500 years confirms that Northern Hemisphere temperatures were warmer in the Roman than the Medieval Period and that recent warming was warmer than that of the Medieval period by the late twentieth century (Figure 15.19). Temperatures in the Dark Ages were cool, but not as cold as those of the LIA. The overall variability of the temperature profile is most likely solar, as described in earlier sections, superimposed on the overall cooling driven by Earth's orbital and axial insolation, and with volcanic activity punctuating the overall temperature profile with discrete, relatively short-lived cooling events in which the major cooling happened immediately after the eruption, then tailed off rapidly. Nevertheless, fossil wood found buried at the front of the Piancabella glacier proved to have an age of 1040–1280 CE indicating that the treeline in the Medieval Warm Period was 200 m higher than in the mid-twentieth century [147], suggesting at least a local link to solar variability.

Sigl and colleagues also examined in some detail the relationships between glacier length, temperature, solar activity and volcanic activity in Alpine ice [148]. They found a strong association since 1800 between periods of glacial advance, cool summer temperature, and periods of

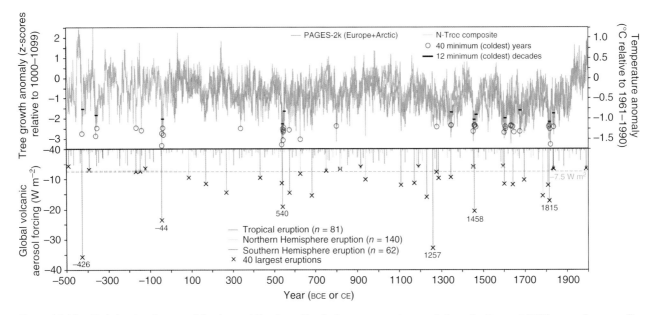

**Figure 15.19** Global volcanic aerosol forcing and Northern Hemisphere temperature variations for the past 2500 years: (top panel) 2500 year record of tree growth anomalies relative to 1000–1099 CE (N-Tree)(green), and reconstructed summer temperature anomalies for Europe and the Arctic (orange), indicating (o) the 40 coldest single years and the 12 coldest decades; (lower panel) Reconstructed global volcanic aerosol forcing from bipolar sulphate records from tropical (bipolar), Northern Hemisphere, and Southern Hemisphere eruptions, indicating (x) the 40 largest volcanic signals and the ages for events with atmospheric sulphate loading exceeding that of Tambora's (1815). *Source:* From figure 3 in Ref. [146].

strong volcanic activity in 1807–1820 (includes Tambora in 1815), 1831–1854, 1883–1893 (includes Krakatoa in 1883), 1913–1924 and 1967–1982 (Figure 15.20). Between those periods the glaciers were in retreat. Extending their analysis back through time they found significant glacial advance associated with periods of strong volcanic activity in 1585–1600, and at 1640, following which there was periodic glacial retreat (Figure 15.20). Although Sigl did not find an obvious connection between glacier length and the solar fluctuations of the Spörer, Maunder and Dalton Sunspot Minima, comparison of Figures 15.12d and 15.20 does suggest that glacier growth did begin within the Dalton Minimum before the large eruptions of the early 1800s, and that there was significant re-growth of the *Mer de Glace* during the Maunder Minimum.

This 'modern' history of glacial change in the Alps is reminiscent of the Holocene changes there. Thomas Stocker of Switzerland's University of Bern, whom we met earlier discussing the bipolar seesaw, produced with colleagues a listing of multi-century glacier fluctuations in the Swiss Alps [149], which proved to be remarkably similar to glacier fluctuations in Scandinavia [150] (Table 15.1). Each recession was documented from $^{14}$C dating of the remains of trees that grew at higher elevation between glacial episodes. The existence of these remains shows that during these recessions the glaciers were smaller than the 1985 reference set, and the treeline was higher [149]. Joerin argued in 2006 that

the pattern of glacial retreat changed during the Holocene, with glaciers at their smallest extent early in the Holocene and up to 7 Ka ago, experiencing 500 year long recessions interrupted by short (200 year) advances, and much larger glaciers and more frequent advances after 3.3 Ka ago peaking in the LIA. He noted that '*The trend [towards increased glacier extent] is in line with a continuous decrease of summer insolation during the Holocene ... [and] consistent with a long-term reduction of sea surface temperature in the North Atlantic*' [149]. The trend is also consistent with other palaeoclimate data sets discussed in Chapter 14.

Although we might infer from Sigl's analyses that glacial ice is more susceptible to the influence of volcanic aerosols on temperature than is tree growth, which is a more seasonal phenomenon and more responsive to solar influence, the data from Joerin [149] indicates overall susceptibility to forcing by orbital change, while the data from Scapozza [147], suggests susceptibility to solar forcing, as does the link between the glacial advance of the *Mer de Glace* during the Maunder Minimum, mentioned earlier.

## 15.4 Sea Level

Svetlana Jevrejeva of the Proudman Marine Laboratory in Liverpool showed in 2008 and 2014 that once tectonic changes and glacial isostatic adjustments have been taken

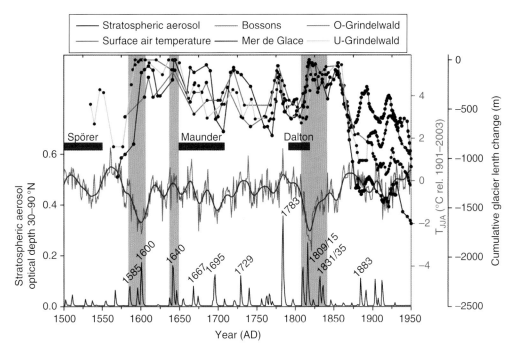

**Figure 15.20** Cumulative glacier length changes for the four glaciers Bossons, Mer de Glace, Oberer (O-) Grindelwald, and Unterer (U-) Grindelwald, with black dots marking years with observations, tree-ring reconstructed Alpine summer (JJA) temperatures, minima in solar activity, and volcanic aerosol forcing, from ᴄᴇ 1500 to 1950. Grey shading marks time periods with increased volcanic aerosol forcing. *Source:* From figure 9 in Ref. [148].

**Table 15.1** Glacier *recessions* during the Holocene. Times are BP (before present), where ¹⁴C dates are measured from 0 at 1950. The results from Joerin in 2006 [149] improve on the chronology of Hormes in 2001 [150].

| Scandinavian [150] | Alpine [150] | Alpine [149] |
| --- | --- | --- |
| 10–8.5 Ka | same | 9.85–9.6 Ka |
| | | 9.3–8.65 Ka |
| | | 8.55–8.05 Ka |
| 7.9–7.5 Ka | no record | 7.77–7.55 Ka |
| 7.2–7.65 Ka | same | 7.45–6.55 Ka |
| 6.1–5.9 Ka | same | 6.15–5.95 Ka |
| 5.8–5.5 Ka | same | 5.7–5.5 Ka |
| 5.2–5.05 Ka | same | 5.2–4.4 Ka |
| 4.9–4.5 Ka | same | overlap with above |
| 4.2–3.4 Ka | same | 4.3–3.4 Ka |
| no record | 3.64–3.36 Ka | overlap with above |
| 3.2–3.05 Ka | no record | no record |
| 2.8–2.7 Ka | same | 2.8–2.7 Ka |
| 2.05–1.9 Ka | same | 2.15–1.85 Ka |
| 1.6–1.2 Ka | same | 1.4–1.2 Ka |
| 1.–0.7 Ka | no record | no record |

into account, sea levels have been rising throughout the past 200 years along with the rise in temperature typical of the modern era [151, 152]. Later, in 2017, Dangendorf and colleagues reassessed sea level rise to produce an improved reconstruction of sea level (Figure 15.21) [153]. Their sea level data seem to show the influence of the AMO, with a rise towards 1940 followed by a fall to 1960, then a further rise, more or less in line with global temperature.

A decline in sea level from 1800 to 1850 in Jevrejeva's data [152] reflects the increase in glacier length over this period found by Sigl (Figure 15.20), the effect of which on sea level was calculated by Leclercq and colleagues (Figure 15.22) [154]. In addition to these long records, we now have available a precisely calibrated continuous record of global sea level change between 1992 and the present, based on tide gauge and satellite data [155]. Average global sea levels are currently higher than at any point within the past c. 115 Ka, since the termination of the last interglacial [156].

Sea level rose from about 1850 onwards in response to glacier retreat (Figure 15.20), ice melt (increase in water mass), and thermal expansion (increase in water volume), all of which respond to global warming. As John Church of the Centre for Australian Weather and Climate Research, pointed out in 2010, a 1000 m column of seawater expands by about 1 or 2 cm for every 0.1 °C of warming [157].

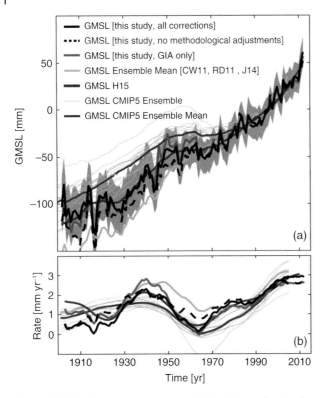

**Figure 15.21** Time series and rates of Global Mean Sea Level (GSML) change during the period 1902–2012: (a) Revised GMSL reconstruction in comparison with previous estimates (CW11; RD11; J14; H15 – see original text for references) and with modeling attempts based on historical CMIP5 models (see original text for reference). The solid black line is the preferred reconstruction, with gray shading marking its 1σ errors. The dotted black line represents a GMSL reconstruction with all corrections, but without methodological adjustments such as area weighting and the use of a common mean. (b) The corresponding rates calculated with a singular spectrum analysis using an embedding dimension of 15 years; trends have been removed to shows the natural oscillations within the climate system. *Source:* From figure 6 in Ref. [153].

According to Konrad Steffen of the University of Colorado at Boulder, and colleagues, thermal expansion of ocean water accounted for about 0.4 mm/year of sea level rise for the past four or five decades, rising to about 1.5 mm/year during the 2000s, whilst melting glaciers and ice caps contributed a rise of 0.3–0.45 mm/year over the past 50 years, rising to 0.8 mm/year during the 2000s [158]. While that left an unexplained contribution of about 1 mm/year that most probably came from the large ice sheets, John Church explained that this apparent 'enigma' in resolving the sea-level budget (the relationship between actual measurement and likely contribution from different sources), has now been largely resolved '*by combining... revised estimates for upper ocean thermal expansion and glacier and ice-cap contributions with reasonable but more poorly known*

estimates of contribution from deep-ocean thermal expansion and the Greenland and Antarctic ice sheet contributions' [159]. For his work on sea level, the Intergovernmental Oceanographic Commission awarded Church the 2006 Roger Revelle Medal.

Salt marshes are useful for estimating past sea levels. John Church and his team noted in 2008 that '*changes in the local sea level estimated from sediment cores collected in salt marshes reveal an increase in the rate of sea-level rise in the western and eastern Atlantic Ocean during the 19th and early 20th centuries .... consistent with the few long tide-gauge records from Europe and North America*' [160]. In 2010, Roland Gehrels of the University of Plymouth, and President of the INQUA Commission on Coastal and Marine Processes for 2011–2015, recalled that '*The AR4 [the IPCC's 4th Assessment Report] states with "high confidence" that the onset of modern rates of sea-level rise occurred between 1850 and 1950*' [161]. He went on to observe that '*The timing of the change in trend has been more precisely determined in proxy records from salt marshes in eastern North America ... and New Zealand ... and occurred between 1880 and 1920 .... This acceleration is also visible in the long tide-gauge record of Brest ... In Iceland, evidence from salt marsh sediments shows that sea level started rising rapidly between 1800 and 1840, possibly as a result of regional ocean warming, and continued to the present, without showing an inflexion at the beginning of the 20th century ... An acceleration of sea-level rise starting in the early 1800s has been proposed on the basis of the oldest sea-level measurements in the world at Amsterdam ... The first sea-level acceleration within the 20th century occurred in the 1930s ... and is demonstrated by model-dependent global reconstructions of tide-gauge data*' [161].

Church observed that '*sea-level reconstructions based on salt-marsh sediments from eastern North America and Europe show that centennial variability during the late Holocene was probably of the order of 0.2 m or less ... [in agreement with] micro-atoll studies from Australia [which] also limit rates of Holocene sea-level change to 0.1–0.2 m per century*' [159]. Kurt Lambeck of the Australian National University (Box 12.3) concurred, noting that '*The absence of pre 20th century sea level fluctuations in proxy records from North America suggests that the 20th-century sea-level acceleration was unprecedented in (at least) the past millennium*' [162]. Seeing that there was some evidence from the Baltic for small changes of less than 1 m during the late Holocene, he went on to observe that '*coral evidence from the Great Barrier Reef, ... or from the Abrolhos Islands of Western Australia .... does not require oscillations during the Late Holocene and suggests that what is seen in the Baltic is a regional rather than a global signal ... Baltic levels are*

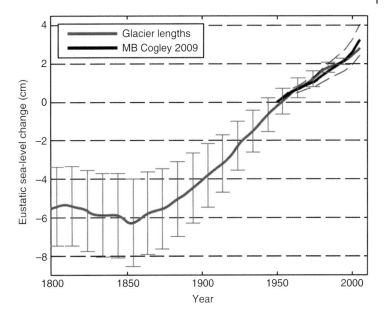

**Figure 15.22** Reconstructed contribution of glaciers and icecaps to sea-level rise. The cumulative mass balance record is given in black; the reconstructed sea-level change from the glacier length records is shown in red. The range of the upper and lower estimates is indicated with dashed black lines for the mass balance record and with red uncertainty bars for the reconstruction from glacier lengths. *Source:* From figure 6 in Ref. [154].

strongly correlated with meteorological conditions ... and it is possible that what is recorded... is atmospherically forced' [162]. With reference to the Baltic we should recall that this is an area where the land is rising due to glacial isostatic adjustment following the removal of the Scandinavian ice sheet. In the western Pacific, sea levels rise with La Niña events, which push warm surface waters to the west, and decline with El Niño events, when weak Trade Winds allow the warm water from the west to flood back towards the east. The same happens within the fluctuations associated with the PDO. These natural changes help to obscure a global sea level signal in that region.

Phil Woodworth of the Proudman Oceanographic Laboratory in Liverpool, and winner of the 2010 Vening Meinesz Medal of the European Geosciences Union, explained with colleagues in 2011 that records of sea-level change from salt-marshes overlap with tide gauge records but have lower resolution, preventing examination of inter-annual variability [163]. They concluded that '*The most we can say with regard to findings from this relatively new field is that salt marsh ... and tide gauge ... data appear to reflect similar features of acceleration between the 19th and 20th centuries, with important details to be resolved by further research*' [163].

Andrew Kemp of the University of Pennsylvania, and colleagues, used data from salt-marsh sediments on the Atlantic coast of North Carolina, and similar environments, corrected for glacial isostatic adjustment, to show that '*North Carolina sea level was stable from at least BC 100 until AD 950. Sea-level rose at a rate of 0.6 mm/y from about* AD *950 to 1400 as a consequence of Medieval warmth ... followed*

by a further period of stable, or slightly falling, sea-level that persisted until the late 19th century... due to cooler temperatures associated with the Little Ice Age. A second increase in the rate of sea-level rise occurred between around AD 1880-1920; in North Carolina the mean rate was 2.1 mm/y, in response to 20th century warming. This historical rate of rise was greater than any other persistent century-scale trend during the past 2100 y' [164].

Using a comparable model, Aslak Grinsted from the Arctic Centre of Finland's University of Lapland, and colleagues, calculated in 2010 that sea levels were some 10 cm below today's average level (for 1980–1999) at about 600 CE, reached 12–20 cm above present average levels in the Medieval Warm Period, and fell to about 19–26 cm below present levels in the LIA [165]. The levels estimated for the 200 year long Medieval Warm Period reflected the long time it takes to melt substantial amounts of land ice, and imply that it takes 100 years or more for sea level to reach an equilibrium position with respect to temperature rise. Grinsted's team used two independent estimates of global temperature as inputs to their model, and found that the 2005 palaeotemperature estimates of Moberg [28] gave a more realistic result than did the 2004 estimates of Jones and Mann [166].

Much the same results emerged from another modelling experiment, carried out by Jonathan Gregory from the Meteorology Department of the University of Reading, with colleagues from the Hadley Centre for Climate Prediction and Research of the UK's Meteorological Office [167]. They found that sea level fell during the sixteenth and seventeenth centuries due to ocean cooling associated with the LIA. With increased warming at the end of the Maunder Sunspot Minimum, and fewer volcanoes, the model showed sea level rising in the eighteenth century, before falling in the early nineteenth century with cooling associated with the Dalton (sunspot) Minimum (1790–1830), and the gigantic Tambora volcanic eruption of 1815–1816 (see Figure 15.20). Sea level rose back to former levels after about 1820, as sunspots increased following the Dalton Minimum, and volcanic activity decreased. Gregory concluded '*that sea level did not go outside natural variability [shown in Figure 14.15] until the twentieth century*' [167].

Nevertheless we do have to be cautious about model outputs. Gehrels reminded us not to interpret them too literally, not only for past times, but also for the present. They cannot yet replicate accurately either observed trends or the magnitude of inter-annual to decadal variability of modern times [161]. More research is needed to reconcile observations with models; their outputs are plausible within statistical limits, and they can act as guides to our understanding.

P.W. Leclercq and colleagues took a different approach to assessing sea level change through time, by calculating the contribution of melting glacier and ice caps to sea level rise for the period 1800–2005 (Figure 15.22) [154]. Global glacier length grew slightly between 1800 and 1850, leading to a small decrease in sea level, but then shrank more or less linearly to the present, which helps to account for the rise in sea level from 1850 onwards. According to Leclercq, 35–50% of the observed rise in sea level since 1800 was due to the melting of glaciers and ice caps [154]. In a companion paper, Leclercq and colleagues used changing glacier length to estimate the change in global temperature over the same period, finding that temperature estimates from glacial length followed global temperature profiles reasonably well [168]. They concluded that '*the rise of the global temperature and the temperatures of both hemispheres started between 1830 and 1850 and continued uninterrupted into the 20th century. The high global averaged temperatures of the period between 1980 and 2000 are unprecedented in at least the last 400 years. In addition, the rate of temperature increase over the period 1980–2000 is the highest of the period 1600–2000*' [168]. The most rapid glacial retreat in the Alps occurred between about 1860 and 1875, by which time the glaciers had lost 80% of their length (Figure 15.20) [148]. Volcanic eruptions between 1883 and 1920, probably combined with the effect of the Gleissberg Sunspot Minimum of 1890–1920, interrupted that trend, which then continued towards the present (Figure 15.20).

Given greenhouse gas theory, one would certainly expect to see not only an increase in temperature since 1900, but also an accompanying acceleration of the rate of sea level rise. According to Steffen the rate of rise averaged 1.7 mm/year from 1870 to 2001, with an increase in the rate of rise over this period from 1 mm/year prior to the 1930s to about 2.5 mm/year in the late 1950s [158]. Large volcanic eruptions between 1963 and 1991 caused temporary cooling that lowered the rate of rise briefly to less than 2 mm/year. Against this backdrop, the rate of sea level rise since the mid-1980s is unusual. Church noted that data from satellite altimeters '*indicate that global average sea level has been rising at 3.1 ± 0.4 mm [per year] ... [which] is faster (by almost a factor of two) than the average rate of rise during the 20th century, which, in turn, was an order of magnitude larger than the rate of rise over the two millennia prior to the 18th century*' [160]. Jevrejeva showed that '*Despite strong low frequency variability (larger than 60 years) the rate of sea level rise is increasing with time*' [152]. Current rates of rise since 1993 from tide gauges agreed with satellite data, at c. 3.2 mm/year, compared with c. 1.9 mm/year for earlier in the twentieth century [152]. Roland Gehrels confirmed that '*Rapid rates of sea-level rise during the 20th century, as recorded in many places around the global, represent a significant departure from late Holocene trends of sea-level change*' [161]. And Kurt Lambeck agreed, as we saw above. Nevertheless, we must recall that these cited rises are global averages, and that regional pictures will differ, especially around coasts like Canada, Scotland, and the Baltic, which are rising due to glacial isostatic adjustment following the removal of their glacial ice load. Even the global signal shows variations in rate of rise (Figure 15.21b) [153], with a more rapid rise between 1920 and 1940, a flattening when global temperatures were cool in the 1950s–1960s, then a faster rise in the 1990s–2000s at about the same rate as in the 1920s–1940s, when temperatures were also rising quite fast (e.g. see Figure 15.1). The two periods of moderate acceleration in sea level rise, one between about 1920 and 1945, and another post 1980, both with similar rates approaching 3 mm/year, may well relate to the influence on sea level as well as on global temperature of the warming associated with the AMO (Figure 15.2) and the PDO. These natural variations introduce small amounts of structure into the overall curve of rising sea level. It is not surprising that temperature and sea level are connected. Church and colleagues explore these changes in some detail [156]. Regardless of rates of change the overall trend is upwards (Figure 5.21).

These various lines of evidence show that after the Last Glacial Maximum sea level rose quite rapidly until about 7 Ka ago, by which time the major Northern Hemisphere ice sheets had mostly melted (Figure 14.15). The rise then slowed down and was extremely slow over the past 7 Ka [160]. Comparison of recent sea level with that of 2 Ka ago, deduced from the positions of Roman fish tanks at the coast, suggests that there was little change over that period until the start of the nineteenth century. This picture was influenced by Fairbanks's development of a global sea level curve based on data from Barbados [169], as discussed in Chapter 12. Even so, careful examination of the rationale behind the construction of that curve led Roland Gehrels to conclude '*The Holocene part of the Barbados curve [produced by Fairbanks 1989] is … without much merit*' [161]. A key part of the story, as we saw in Chapter 12, is the change to the Earth's crust when the ice load is removed. Land that was formerly depressed by ice sheets begins to rise, whilst land that was forced to bulge up around the ice-laden crust begins to sink. This siphons water from the tropics towards the poles to fill the space left by the former peripheral bulge. Similarly, the rise in sea level after the Last Glacial Maximum drowned the continental shelves, which sank under the extra weight of water. Water was siphoned from the open ocean into the space created by this sinking. That siphoning lowered sea level in tropical and mid-latitudes, explaining why Holocene shorelines there are commonly found several metres above present sea level. Fairbanks' Caribbean studies are suspect because the islands (like Barbados) lay on the peripheral bulge around the Laurentide Ice Sheet.

In response to these various effects, models suggest that Holocene sea level should have fallen by 0.28–0.36 mm/year along with the fall in insolation and accompanying global cooling. It did not do so. There is some evidence that the Greenland Ice Sheet and small Northern Hemisphere glaciers were growing in the late Holocene (the neoglacial), which implies – if sea level were more or less constant – that ice must have been melting in Antarctica, for which there is some evidence, for example from the Ross Sea [161]. Getting precise data on late Holocene sea level change, says Gehrels, '*requires an interdisciplinary approach by glaciologists, GIA [glacial isostatic adjustment] modelers, sea-level scientists and space geodesists*' [161]. Care is also needed in measuring recent regional changes in sea level where coasts are subsiding, as they are in many estuarine and deltaic environments due to the weight of sediments brought down by rivers like the Rhine. Even so, the vast majority of scientists making sea level measurements agree that sea level began to rise in the late 1800s and that the rate has increased since then, being especially high in 1930–1950 and 1990–present [153].

In 2009, Eelco Rohling and colleagues used their analysis of the relationship between Antarctic temperature and sea level, and the relationship between $CO_2$ and temperature in Antarctic ice cores [170], to conclude that a high $CO_2$ value like that of today's world could lead over a few hundred years to an equilibrium rise in temperature of up to 16 °C, and an equilibrium rise in sea level of around $25 \pm 5$ m, like that observed for the Pliocene [171] (see Chapter 11). In contrast, climate models suggested that the relationship between $CO_2$ and temperature may diverge from the 'natural and linear pattern' that they observed for the Pliocene, and that a modern $CO_2$ value would equate to an Antarctic temperature of 5 °C, suggesting a sea-level rise of up to only 5 m [171]. The discrepancy between end points of 5 m or 25 m, may arise either because the natural linear relationship may remain so with increasing $CO_2$, or because the high-side model projections for the relationship between temperature and $CO_2$ do not use a high enough climate sensitivity (i.e. the rise in temperature for a doubling of $CO_2$ may be much higher than 3 °C, as some have suggested). In any case the upper limit of the projections (25 m) is based not on models but on actual observations (see Chapter 11). What we need to know is how long such a change might take.

Rohling later extended his study of the relationship between $CO_2$ and sea level back through time for 40 Ma, as we saw in Chapter 10 [172]. Given modern atmospheric concentrations of $CO_2$, and knowing the long-term equilibrium relationship between $CO_2$ and sea level, he estimated that without any further rise in $CO_2$ the long term equilibrium rise in sea level compared with the present would reach between +9 and +31 m, averaging +24 m, over some 5–25 centuries. Rises of $CO_2$ above the modern level will cause larger long-term change.

Stefan Rahmstorf of the Potsdam Institute for Climate Impact Research of Germany's Potsdam University near Berlin, and Martin Vermeer of Helsinki University of Technology also had ideas about future sea level change. In 2007, Rahmstorf found a linear relationship between global temperature and the rate of rise of sea level rise, which suggested that sea level could rise by up to about 1.4 m by 2100, given greenhouse gas emissions following 'business as usual' [173]. By 2009, Vermeer and Rahmstorf had refined that calculation to suggest that the rise might be up to 1.9 m [174]. Kemp and others used these semi-empirical relationships to convert temperatures estimated from proxies by Michael Mann and colleagues in 2008 [42] into past sea levels for comparison with the observations from North Carolina [164]. They found that the estimated sea levels diverged from those measured in salt marsh sediments prior to 1000 CE, possibly because global temperatures estimated from proxies for the period from 500 to 1100 CE were too high by about 0.2 °C.

The estimates of Rahmstorf [173] and his colleague Vermeer [174] are more than double the maximal sea level rise of about 80 cm suggested for 2100 by the IPCC in its fourth and fifth Assessment Reports. In contrast, Jim Hansen of NASA considered that Rahmstorf's various projections failed to consider non-linearities in the climate system, and that the ongoing rise in temperature could destabilize ice sheets and lead to a cumulative rise in sea level of 5 m by 2100 [175]. Tad Pfeffer of the Institute of Arctic and Alpine Research of the University of Colorado at Boulder, and colleagues, disagreed, suggesting that such a rapid rate of rise was physically untenable given what was known about the behaviour of ice sheets. Considering all the available data they opted instead for a maximum rise of 2 m, and a likely rise of up to 0.8 m by 2100 [176].

Hansen thought that Pfeffer and his team were too conservative in their estimation of the amount of forcing likely from growing emissions of greenhouse gases, which by mid-century was expected to lead to extensive summer melting in a long melt season in Greenland, and to the loss of the ice shelves that prevented inland ice streams from discharging land ice rapidly at the coast of Antarctica [177]. Ice loss from the Pine Island Glacier (really a 50 km wide ice stream as wide as the English Channel between Folkestone and Boulogne) and its sister glaciers draining West Antarctica is significant and increasing. It is attributed not to surface warming but to the penetration onto the continental shelf beneath the glacier of Circumpolar Deep Water with a temperature of about 1 °C, which carries sufficient warmth to thin the glacier from beneath, making it discharge faster [177] (Figure 13.13). The ice streams represented by the glaciers draining from West Antarctica into the Amundsen Sea Embayment are all moving faster, posing the risk of eventual collapse of the WAIS and a possible sea level rise near 3 m [59, 60]. Evidence from sediment cores shows that incursions of warm circumpolar deep water onto the continental shelf of the Amundsen Sea caused retreat of the ice shelf to begin in about 1940, after a previous bout of retreat during the Holocene thermal maximum (Chapter 14), was followed by glacial advance [178].

In 2019, Eric Rignot of the Jet Propulsion Laboratory of the California Institute of Technology and colleagues showed that the rate of loss of land ice from Antarctica increased from 40 ± 9 Gt/year in 1979–1990 to 252 ± 26 Gt/year in 2009–2017 [179]. Most came from the coast of the Bellingshausen Sea (for location see Figure 7.5) through melting from beneath the ice shelves there (Figure 13.13). The Bellingshausen Sea lies adjacent to Marie Byrd Land, beneath whose ice are some active volcanoes [180, 181]. However, these are randomly distributed whilst the ice melt is clearly related to the ice shelves along the coast and concentrated in the ice streams that drain directly to them (Pine Island, Thwaites, etc). The loss is rapid, and '*In the Amundsen and Bellingshausen regions, some ice shelves have lost up to 18% of their thickness in less than two decades*' [182].

In addition, recent analyses suggest that the ice of the Wilkes Subglacial Basin beneath Wilkes Land in East Antarctica (see Figure 7.5) is also at risk of collapse, providing the prospect of an additional rise of sea level of 3–4 m [183]. In 2019, Rignot showed that the Wilkes Basin lost 51 ± 13 Gt/year [179] The risk will increase markedly when erosion by warm seawater beneath the seaward edge of the ice shelf there removes a volume of ice equivalent to a sea level rise of 8 cm. That will be enough to melt the ice plug that currently stops incursion of warm water into the Wilkes Basin. There are signs of past active erosion of continental bedrock from within the now ice-covered Wilkes Basin, indicating previous retreat of the ice sheet there by several hundred kilometres inland. That melting is likely to have added between 3 and 10 m to global sea level [184]. New evidence from sedimentological and geochemical records in offshore deep ocean drill cores shows that the ice margin thinned or retreated along the edge of the Wilkes Basin during warm interglacials of the Late Pleistocene (especially Marine Isotope Stages 5, 9, and 11), when Antarctic temperatures were 2 °C warmer than pre-industrial temperatures over periods of 2500 years or more [185]. This evidence shows that East Antarctic melting contributed to the sea level rises at those times, and indicates that even modest warming does affect the East Antarctic ice sheet [185].

The Totten Glacier, behind the Totten Ice Shelf, drains East Antarctica's Aurora Subglacial Basin with a volume equivalent to at least 3.5 m of global sea level rise [186]. Geological analyses of the interior behind the Totten Ice Shelf shows that it too has been subject to repeated erosion in the Sabrina Subglacial Basin, caused by advance and retreat of the Totten Glacier [187]. And it is clear that warm water is already being funnelled beneath the Totten Ice Shelf (cf. Figure 13.13) through a newly discovered deep channel [188], which will destabilize its buttressing effect. This deep underwater trough extends all the way from the edge of the Totten Ice Shelf to the grounding line 125 km inland, and as deep as 2.7 km below sea level. According to Jane Qiu, very recent research by ice-probing radar from aircraft shows that beneath the Totten Glacier catchment basin lies a 1100-km-long canyon – the longest in the world, and almost as deep as the Grand Canyon – which connects with the coast. Moreover, some 21% of the Totten glacier catchment is more than 1 km below sea level, and could easily flood under the right conditions. '*This deeply contoured landscape could allow warming waters from offshore to quickly reach and erode the ice*' [189].

Clearly, then, the great East Antarctic Ice Sheet is capable of a dynamic response to fairly small amounts of warming. High-resolution records of ice-rafting in the adjacent Southern Ocean show that small changes in subsurface ocean temperature during the Holocene increased discharge of icebergs around Antarctica, which amplified multi-centennial variability in the climate system both regionally and globally [190], providing further evidence for the dynamic nature of the Antarctic ice sheet. Ice shelves are already being progressively lost down the Antarctic Peninsula [191], with a large piece of the Larsen C ice shelf disappearing in 2017.

Nevertheless, Jeremy Shakun and colleagues found in 2018 that ice had been present in the East Antarctic sector of the Ross Sea for at least 8 Ma, despite the warming of the Pliocene and the large warm interglacials of the past 400 Ka. They deduced that the land-based sectors of the ice sheet that drain into the Ross Sea had been stable for this entire period, and that any melting of the ice sheet must have been focused on marine ice margins [192]. Those would most likely include the retreats in the Aurora Subglacial Basin mentioned above.

Down the western side of the Antarctic Peninsula, hundreds of glaciers between 63 and 70°S have been steadily shrinking in recent decades. This is due primarily to the warming of mid-depth water that wells up onto the continental shelf beneath glacier tongues [193]. In the south, glaciers that terminate in this warmer water have undergone substantial retreat, whilst in the north, where the glaciers terminate in cooler water, there has been less retreat. Mid-ocean warming since the 1990s in the south coincides with accelerated glacier retreat [193].

Greenland also continued to melt through the summer of 2016, with only one year (2012) in 37 years of satellite observations of the Greenland ice sheet (1979–2016) showing earlier onset of the annual spring melt [194]. Greenland loses ice not only by sublimation and surface melting, but also by the melting back of glaciers that terminate at the coast [195]. Based on LANDSAT and ASTER satellite data, in 2016 coastal terminating glaciers lost 60.6 km$^2$, the largest loss since 2012; 22 glaciers retreated, losing 100.8 km$^2$, and 11 advanced, gaining 40.9 km$^2$ [194]. Regional effects are important. GRACE data show that 70% of the ice loss in Greenland for the period 2003–2013 (280 ± 58 Gt/year) came from the southeast (40%) and northwest (30%), driven by ice dynamics; elsewhere losses were driven more by changes in surface mass balance (sublimation and melting) [196]. Khan and colleagues considered in 2016 that estimates of ice loss from Greenland (based on the GRACE satellite data of Velicogna [196]) may underestimate actual losses (hence contribution to sea level rise) by failing to take into account the isostatic rise of land due to ice loss [197]. Adjusting for those factors, they concluded that Greenland had provided 4.6 m of sea level rise since the Last Glacial Maximum, or 44% more than previously estimated. Greenland has been losing ice at an increasing rate

as the Arctic has warmed, losing since 1988 about the same amount as Antarctica [198].

The current global rise in sea level is attributed to about 33% each from thermal expansion of the ocean, the melting of mountain glaciers, and the melting of ice caps and ice sheets on land. A group led by Alex Gardner of the University of Alberta showed in 2011 that the ice caps of the Canadian Archipelago on Baffin Island and Ellesmere Island had lost 61 ± 7 Gt/year in 2004–2006 and 2007–2009, in response to summer warming [199]. This was the single largest contributor to sea level rise (0.17 ± 0.02 mm/year) outside Greenland and Antarctica.

Returning to Robert Kopp of Princeton, whom we met in Chapter 12, we are reminded that '*Incorporating a large database of palaeoclimatic constraints … highlights the vulnerability of ice sheets to even relatively low levels of sustained global warming*' [200]. Kopp's team found that with just 2–3 °C of warming above present levels in the last interglacial, sea levels peaked at least 6.6 m higher than today. Analysing the change in sea level over the past 2.5 Ka, Kopp and colleagues found in 2016 that in relation to its level in 1950, sea level was at its lowest (about 15 cm below) during the LIA, at its highest (about 5 cm below) in the Medieval Warm Period, and low again (about 12.5 cm below) around 200 BCE [201]. Although it is challenging to disentangle natural and anthropogenic factors in late-nineteenth and early-twentieth century sea level rise, because the LIA came to its end as coal consumption rose in the nineteenth century, based on his study of the past 2500 years of sea level change Kopp estimated that, without warming, twentieth century global-mean sea level would have been limited to −0.4 to +0.8 mm/year – leaving between 40 and 130% of the observed rise attributable to the effects of twentieth-century global warming [201]. The global-mean sea level signal of warming emerged at the 95% probability level by 1970. This means that the ~3 mm/year of global-mean sea level rise since the early 1990s has brought the world outside the realm of late Holocene experience.

In 2010, an IPCC workshop on sea level change included the observation that '*ice sheets are capable of highly nonlinear dynamical behaviour that could contribute significantly to short-term sea level rise (to 2100), and may also produce a long-term commitment (e.g. centuries-long) to substantial (many metres) of sea level*' [202]. John Church agreed, noting in 2011 that whilst '*major deficiencies in our understanding remain … perhaps the major challenge is the response of ice sheets, particularly those parts grounded below sea-level*' [203]. We do know that rates of loss of ice from Greenland and Antarctica are increasing with time [179, 198]. But more research on the range of rates of ice sheet decay under different thermal regimes is needed before we can answer the question 'how much and how fast?' with confidence. Meanwhile, engineers responsible for the Thames Barrier, which protects London from storm surges, plan to redesign it to withstand a further rise of 2 m.

## 15.5 The End of the Little Ice Age

We can use Steinhilber's cosmic intensity (Figure 15.12) [92] to define the most likely boundaries of solar-driven warm and cool periods. Accepting that the Medieval Warm Period lay between the Oort and Wolf sunspot minima when cosmic intensity was <1.25, then it lasted roughly 200 years from 1070 to 1270 CE. Using that same cut-off we can define the LIA as lasting from 1270 to 1920 CE, and comprising the Wolf, Spörer, Maunder, Dalton and Gleissberg sunspot minima plus their intervening warm periods, of which there were four (at 1350–1390, 1560–1640, 1730–1790, and 1840–1870), when cosmic intensity was <1.25 and the climate was likely to have been as warm as it was in the Medieval Warm Period. Whilst many scientists (e.g. Folland) [13] regard 1850 as a rough guide to the end of the LIA (end of the Dalton sunspot minimum, with its nadir at about 1820), we might reasonably extend that to about 1920 (end of the Gleissberg sunspot minimum), following which there was a continued slight rise in sunspot activity to the peak solar maximum of the modern era between 1960 and 1990 (Figure 15.14) [204]. The key question is – has the LIA come to a natural end, or have we simply been living in a post Gleissberg sunspot maximum, which was merely another of those temporary warm intervals of the LIA and is now in decline? The latter possibility seems to be the view of both Steinhilber [133] and Lüdecke [134] and their colleagues, with their predictions of a lessening of solar output between now and 2050–2100 (broadly estimated).

Many studies confirm that modern temperatures are the highest they have been since about 1900, and that sunspots rose to a recent maximum in the period between 1960 and 1990. But how reliable is the link? A 2007 review by Jasper Kirby, from the European Centre for Nuclear Research (CERN) in Switzerland, confirmed that sunspots as represented by $^{14}$C and $^{10}$Be were about as abundant in the Medieval Warm Period as they were in the late twentieth century [28], much as found by Steinhilber and his team in 2012 [92]. Nevertheless, the devil as usual is in the detail, and the pattern of temperature change for the late twentieth century does not match the sunspot record in the same way that it did in earlier times. For instance, whilst global temperature fell from 1880 to a low in 1910, during the Gleissberg solar minimum, then rose to a peak in 1940

following the solar data [205], the solar data and temperature data then diverged (Figure 15.23) [206]. In effect, temperature fell when it should have been rising according to solar activity (1950 and 1960), and rose when it should have been falling (1990 through to the present) [206, 207]. Meanwhile $CO_2$ continued to rise exponentially – something that began slowly in the late 1700s with the increased burning of coal at the start of the Industrial Revolution (Figure 9.11). It is these convergences ($CO_2$ and temperature) and divergences (temperature and solar data) that lead almost all scientists working with climate data to suggest that we are now living in an era when $CO_2$ is playing a more important role than solar activity in driving global temperature.

It seems most likely that the shape of the global temperature curve since 1900 (Figure 8.13) reflects a combination of the solar activity signal, plus irregular volcanic activity, plus the temporary effects of short-term (two to six year) internal oscillations like El Niño-La Niña events, plus longer term (10–30 year) fluctuations in the NAO and the PDO [208], superimposed on a trend driven by greenhouse gas forcing that increased with time [205]. As we saw in Chapter 8, aerosols from human activities also affected global temperature, reflecting solar energy and keeping temperatures low, especially in the late 1950s and 60s, as demonstrated by Budyko [209], and by Smith [210]. The increase in reflective aerosols was driven by the massive rise in industrial output that accompanied and followed the World War II, and was most effective before the introduction of increasingly comprehensive clean air acts by several industrial nations between 1950 and 1970 [205].

Results emerging from global numerical modelling reinforce the conclusion that the divergence of empirically observed temperature and solar signals after 1940 is most likely due to anthropogenic activities. In 2009, Caspar Amman of the National Center for Climate Research in Boulder, Colorado, (NCAR), used a coupled ocean–atmosphere Global Climate Model (GCM) to see if solar irradiance patterns determined from $^{10}$Be, along with data on past explosive volcanic emissions (represented by volcanic aerosols in ice cores), might explain the variations seen in climate data from the Northern Hemisphere since 850 CE [211]. Amman and colleagues concluded that the climate of the Northern Hemisphere for the past 1150 years was directly attributable to small fluctuations in solar output and explosive volcanic output up until about 1870 [211]. It goes without saying that those changes were themselves superimposed on the low global temperature background of the neoglacial, driven by declining insolation. Those same climatic conditions should have continued with little change – a slight warming of perhaps 0.2 °C to the 1940s

**Figure 15.23** The divergence of temperature and solar data. The graph shows annual averages of solar activity indices and global surface temperature since the year 1950. $S(t)$ is total solar irradiance measured by satellites (only since 1978). Climax CRF is the cosmic ray flux measured at Climax in Colorado. The aa index is a geomagnetic index prepared by the International Service of Geomagnetic Indices. Tglobe UEA is the global surface temperature anomalies from HadCRUT3 with its total uncertainty at the 95% level. NRF stands for the net radiative forcing of the Sun (i.e. $S(t)$ divided by 4 and multiplied by 0.7). Only the Tglobe curve has an upward trend (0.11 °C per decade, $r = 0.87$, since 1950). All other curves oscillate about a trend that is basically flat or, in the case of $S(t)$ slightly declining. Large dips of the Tglobe curve occurred just after major volcanic eruptions (e.g. 1963, 1982, and 1991). *Source:* From figure 3, in Ref. [206].

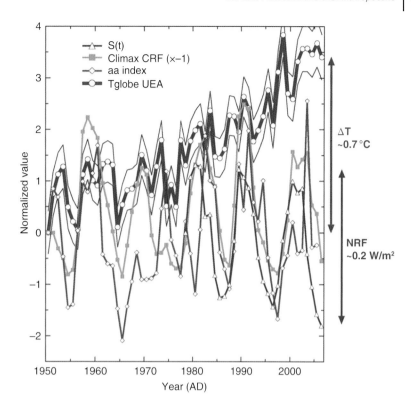

and 1950s, driven by solar factors, followed by a decrease of about 0.1 °C to the present (meaning 2007) as solar energy declined, interrupted by brief episodes of cooling due to small increases in explosive volcanism from eruptions like that of Mount Pinatubo in 1991, which cooled the world by 0.2 °C in 1992. Superimposed on this natural background (driven by orbital and axial insolation, solar variability and volcanic activity), was a progressive warming from about 1900 driven by increases in human-added greenhouse gases. Amman deduced that '*the 20th century warming is not a reflection of a rebound from the Little Ice Age cool period, but it is largely caused by anthropogenic forcing*' [211]. This anthropogenic increase is what has taken late twentieth century and early twenty-first century temperatures above the levels of the Medieval Warm Period.

Hugues Goosse and colleagues carried out a similar modelling exercise, in 2009 [212]. Considering the effects of changes in land use, solar irradiance, aerosols, volcanoes and greenhouse gases, their model confirmed that solar forcing explained the high temperatures of the 1100s, 1200s, and early 1900s, and the low temperatures of the 1500s and late 1700s to early 1800s. Large volcanic forcing explained short but particularly cold periods in 1258, 1452, 1600, 1641, and 1815–1816. Land use changes were associated with a long-term cooling trend, whilst cooling associated with increasing (industrial) aerosols was confined to

the late twentieth century. Greenhouse gas forcing induced increasing warming during the past 150 years. Responses to forcing were faster and larger in the Northern Hemisphere, and lower and slower in the Southern Hemisphere because of its large area of ocean. The high heat capacity of the Southern Ocean damped the signals of global warming or cooling. In addition, upwelling brought to the surface of the Southern Ocean old cold ocean water that had acquired its characteristics decades to centuries earlier, which resulted in a more complex interaction between forcing and response than we see in the north [212]. This helps to explain Neukom's observations about the regional nature of climate change [46].

Goosse and his team also used model simulations to investigate the response of particular parts of the climate system to external forcing. Their results '*indicate a counterintuitive tendency towards El Niño (warm eastern and central tropical Pacific) conditions in response to negative radiative forcing (past explosive tropical volcanic eruptions or decreases in solar irradiance), and a tendency for La Niña-like conditions [cool tropical Pacific] in response to positive radiative forcing (i.e., increases in solar irradiance). This prediction matches available evidence from tropical Pacific coral records*' [212]. This counterintuitive finding explains '*why the tropical Pacific appears to have been in a cold La Niña-like state during the so-called "Medieval Warm*

*Period" and a warm El Niño-like state during the "Little Ice Age"'* [212].

In a 2010 review of solar influences on climate, Lesley Gray from Reading University's Meteorology Department supported Amman's conclusions [140]. Gray and colleagues found that although the surface response of greenhouse gases and solar forcing might be similar, the greenhouse gas response in the stratosphere is the opposite (cooling) of what would be expected from solar forcing (warming). Model simulations of all known forcings showed that whilst much of the global warming in the first half of the twentieth century was natural and much was due to solar activity (a slow rise in sunspot activity), this did not apply in the second half of the twentieth century or the beginning of the twenty-first century, when solar forcing played at most a weak role in current temperature trends [140], as also shown by Edouard Bard [206] (Figure 15.23). A panel of the US National Academy of Sciences in 2012 agreed that changes in TSI were unrelated to the increase in global temperature over the past 50 years [129].

Evidence compiled from the Arctic by Gifford Miller and his team showed in 2012 that warming began in the late nineteenth century in most palaeoclimate records [57]. During the last millennium, orbital forcing caused midsummer insolation to fall about $1 \text{W/m}^2$ at 75°N and about $2 \text{W/m}^2$ at 90°N. Because orbital insolation was weakening '*additional forcing was needed in the 20th century to give the same summertime temperatures as achieved in the Medieval Warm Period*' [57]. That 'additional forcing' must have been large, since in 2012 Miller was seeing vegetation emerging from beneath ice caps on Baffin Island that expanded 1600 years ago, lasted through the Medieval Warm Period, and began melting back early in the twentieth century [57]. Similarly, the percentage of summer melting of the Agassiz Ice Cap in the Canadian High Arctic decreased along with the orbital/axial insolation trend through the Holocene, but increased significantly during the twentieth century to the point where current rates of melting are greater than at any time in the past 1700 years. Tree ring data show that Arctic summer temperatures now are the most favourable for tree growth within the past 4 Ka. Across the Arctic, lakes were dominated by a first-order cooling trend for most of the past 2 Ka, because of declining orbital/axial insolation, but this trend reversed in the twentieth century *despite* continued reduction of summer insolation. The warmest 50 year interval in the 2 Ka composite record occurred between 1959 and 2000 CE [57]. Miller concluded, '*The strong warming trend of the past century across the Arctic, and of the past 50 years in particular, stands in stark contrast to the*

*first-order Holocene cooling trend, and is very likely a result of increased greenhouse gases that are a direct consequence of anthropogenic activities*' [213]. Given a consistent polar amplification of global temperatures in the Arctic by a factor of 3–4 in recent times (see Chapter 14), Miller suggested that '*Arctic warming will continue to greatly exceed the global average over the coming century, with concomitant reductions in terrestrial ice masses and, consequently, an increasing rate of sea level rise*' [213].

As I pointed out elsewhere [214], the '*last remnants of the Laurentide Ice Sheet include the Barnes and Penny Ice Caps on Baffin Island. Both are in retreat, and it is estimated that the Barnes Ice Cap will have melted away over the next 300 years [according to Gilbert* [215]]. *Only in the three warmest interglaciations of the past 2.6 Ma has the Barnes Ice Cap been as small as it is presently. After 2000 years of little change in its dimensions, recent observations show that the ice cap is now losing mass at all elevations, despite the continued decrease in summer insolation, and this mass loss is most likely in response to the exceptionally warm 21st century Arctic climate*' [214].

In 2019 Miller's group updated their research on Baffin Island, finding evidence that '*pre-Holocene radiocarbon dates on plants collected at the margins of 30 ice caps in Arctic Canada suggest those locations were covered by ice for >40 Ka, but are now ice free*' [216]. They also found that it was unlikely that any of these sites had been ice-free during the Holocene thermal optimum, leading them to '*suggest that summer warmth of the past century exceeds now any century in 115,000 years*' [216].

A prior review of Arctic temperature data in 2003, by Igor Polyakov and colleagues, suggested that the rise in Arctic temperatures from 1900 to 1940 had been followed after a temperature dip by a second rise to similar levels since about 1970 [217]. At the time (2003), they interpreted those changes to be responses to a natural cycle unrelated to greenhouse gas emissions. The idea that Arctic climate change is subject to a natural 60 year cycle has been popular with some Russian authors [218]. In contrast, American investigators Kevin Wood of the University of Washington and James Overland of NOAA, writing in 2010, attributed the early twentieth century warming to increasing southerly winds bringing warm air north from the Atlantic in the form of '*a random climate excursion imposed on top of the steadily rising global mean temperature*' [219]. Those winds were associated with the arrival of anomalous high pressure over Europe and a deepening of the Iceland low, implying strengthening of the NAO. The following year, Overland and Wood, with Muyin Wang, pointed out that the pattern of sea ice during the early twentieth century warm event was quite different from that of the late twentieth century warming [220]. Surface air temperatures during the earlier event were warmest in the Atlantic sector of the Arctic. The distribution of sea ice in August 1938 reflected that pattern, disappearing all along the European and Russian coasts but remaining along the coast of Canada, the USA and Greenland, where the Northwest Passage remained closed [220]. The modern distribution of sea ice is quite different. Sea ice is being lost from all Arctic margins except the northern edges of Greenland and the Canadian Arctic islands, leaving the Northwest Passage open. These differences confirm that different forcing factors were responsible for the two warming periods. Polyakov's most recent contribution, in 2012, showed remarkable warming of the Arctic from 1990 to 2008, associated with a dramatic loss of sea ice since 1979, which he attributed to polar amplification of warming [221]. Acosta Navarro and colleagues have attributed this amplification to the reduction in European sulphate aerosols since 1980 in response to air quality regulations, which may have contributed an unexpected additional 0.5 °C to regional temperatures [222]. Turning back for a moment to the supposed 60 year cycle, this could be the product of a combination tone caused by modification of the 88 year Gleissberg cycle by the 208 year Suess cycle, as mentioned earlier [76].

Examining data on the flow of sea ice through Fram Strait since c. 1580 CE, Knud Lassen and Peter Thejll of the Danish Meteorological Institute found in 2005 that it varied in sync with approximately 80 year fluctuations in solar intensity, which suggests a connection to the 88 year Gleissberg solar cycle [223].

The variation in Arctic sea ice through time provides a useful backdrop to these investigations. In 2008, Christophe Kinnard of the University of Ottawa published time series of maximum and minimum Arctic sea-ice extent from 1870 to 2003 [224]. As seen in Figure 15.24 [225, 226], the area of winter ice was essentially constant from 1870 until 1950, after which it steadily declined, and the area of summer ice fell slightly from 1900 to 1950, then began a steep decline. The slightly low summer sea ice area between 1920 and 1945 corresponds to the prior period of Arctic warming described by Wood and Overland [219, 220], during which a different climatic forcing than that which operated during the late twentieth century was in effect. That early twentieth century warm event is also seen in Greenland (Figure 15.1). We should bear in mind that that some 30–40% of the decline in sea ice area was likely due to atmospheric effects rather than global warming [227, 228]. At the same time as the area of sea ice cover diminished, the area of multi-year ice (>1 year old) shrank to 22%

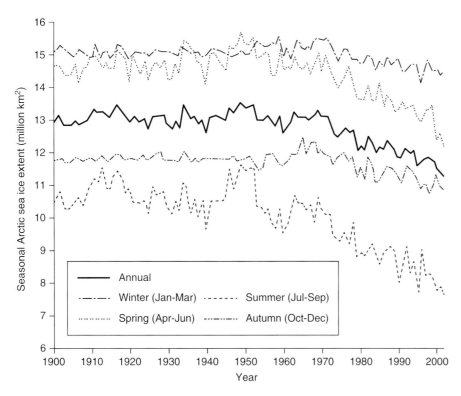

**Figure 15.24** Arctic sea ice area from 1900 onwards. *Source:* Extended from figure 16.14 in [225], From figure 6.1 in [226].

of the total compared with 45% in 1985 [194], with the thickest ice piled against the north coast of Greenland and the Canadian islands. The rest is mostly 1 m thick and easily broken up by winds and currents. Although much Arctic sea-ice loss can be related to atmospheric forcing [227, 228], ice reductions are also caused by the shoaling of warm intermediate-depth Atlantic Water, which has made the eastern Eurasian Basin of the Arctic Ocean similar to the western one [229]. '*The associated enhanced release of oceanic heat has reduced winter sea-ice formation at a rate now comparable to losses from atmospheric thermodynamic forcing ... This encroaching "atlantification" of the Eurasian Basin represents an essential step toward a new Arctic climate state, with a substantially greater role for Atlantic inflows*' [229].

In 2017, Gagné and colleagues found a subtle difference from the Arctic sea anomaly pattern shown in Figure 15.24. Updated observational datasets without climatological infilling show that there was a slight (5%) increase in sea ice concentration in the Eastern Arctic between 1950 and 1975. This turns out to correlate with a global decline in temperature, which was probably caused by increasing anthropogenic aerosols and natural forcing [230] (e.g. the negative, cold, phase of the PDO and/or the Atlantic Meridional Oscillation – see Figure 15.2).

In 2011, Kinnard and colleagues used a range of climate proxies to extend the study of Arctic sea ice back to 1450 years before present [231]. Remarkably, they found that '*both the duration and magnitude of the current decline in sea ice seem to be unprecedented for the past 1,450 years*' [231]. Not only has sea ice shrunk in area, it has also thinned, making the volume of sea ice decline radically in recent years. The area covered by thin 1 year ice expanded greatly at the expense of older and thicker ice. Because the 1 year ice tends to melt each season, this has meant a great expansion in the area covered by seasonal ice. The extremely low sea ice extent observed since the mid-1990s is well below the range of natural variability inferred by the team's reconstructions. Nevertheless, the extent of sea ice was also low before 1200 CE, in association with the warm conditions of the Medieval Warm Period, although not as low as it is today. As far as they could tell, the driver for declining sea ice seemed to be enhanced advection of warm Atlantic water to the Arctic. The warming in medieval times and at the end of the twentieth century was related to an increase in the positive index of the NAO, which pushed the westerly winds north between Scotland and Iceland, funnelling warm air and ocean water towards the Arctic. There was also a period of warming and reduced sea ice in the middle of the LIA about 1470–1520, likely driven by the same mechanism. Given the present state of knowledge, Kinnard and his team thought that anthropogenic 'greenhouse gas'

warming must explain the record warming and sea ice loss of recent decades, since those earlier solar drivers of change were lacking [231]. The trend towards an increasing area of seasonal (as opposed to permanent) sea ice shows no sign of diminishing, though the area fluctuates from year to year (e.g. the September 2019 area was only slightly less than that in 2012, which was an all-time low).

A 2013 study by Martin Tingley and Peter Huybers of Harvard confirmed that the magnitude and frequency of recent warm extreme temperatures of high northern latitudes were the highest for 600 years and that the high latitude summers of 2005, 2007, 2010, and 2011 were warmer than those of all prior years back to 1400 CE [232]. These extremes greatly exceeded those expected about a stationary climate, but were consistent with variability about an increasing mean temperature trend.

A synthesis of proxy temperature records for the past 2000 years from north of latitude 60°N, by Darrell Kauffman, from Northern Arizona University, and colleagues, concluded in 2009 that, excluding short-term, low-amplitude events like the Medieval Warm Period and the LIA, the long-term trend comprised gradual cooling down to the end of the nineteenth century [233]. That cooling then reversed, with four of the five warmest decades of the 2000 year long reconstruction occurring between 1950 and 2000 [233]. Twentieth century warming took temperatures above anything that their proxy data set revealed in the previous 19 centuries. This is consistent with the fact that over the past 60 years the Arctic has warmed by more than 2 °C – more than double the global average warming for the same period – due to polar amplification driven by positive feedback from melting sea ice [234].

Anne Bjune of the Bjerknes Centre for Climate Research at the University of Bergen observed a similar pattern [235]. From studies of pollen from lakes in Fennoscandia and on the Kola Peninsula, she and her colleagues found in 2009 that the mean July temperatures were about 0.2 °C above present between 0 and 1100 CE and fell to about 0.2 °C below present in the LIA. Abrupt warming occurred at about 1900 CE, and the twentieth century was the warmest century since 1100 CE. They were unable to detect a Medieval Warm Period.

Bo Vinther and colleagues reported from an analysis of winter season $\partial^{18}$O and borehole temperatures in three Greenland Ice Cores that '*the warming that commenced in the early 20th century has brought Greenland temperatures to a level matching the warmest periods of the Medieval Warm Period some 900–1300 years ago. After a cold spell in the 1970s and 1980s Greenland temperatures have increased rapidly and present day temperatures have just about reached the same level as during the warm period in 20th century. This result implies that further warming of present day Greenland climate will result in temperature conditions that are warmer than anything seen in the past 1400 years'* [56] (e.g. Figure 15.1). As we saw earlier, the surrounding Arctic has warmed more than Greenland. Nevertheless, a surface temperature reconstruction from the recently drilled NEEM (North Greenland Eemian Ice Drilling) ice core shows warming by $2.7 \pm 0.33\,°C$ in the period 1982–2011 compared with the long-term average for 1900–1970 ($-28.55 \pm 0.29\,°C$) [236]. The warming is due to increased downward long-wave heat flux, which is underestimated by 17% in atmospheric re-analyses. Greenland experienced a sharp rise in surface air temperature starting in 1993, with 2001–2010 being the warmest decade since the onset of meteorological measurements, surpassing the generally warm 1920s–1930s by $0.2\,°C$ [237]. 2010 was exceptionally warm, particularly in West Greenland, and was associated with a record melt over the Greenland Ice Sheet. This was associated with a very negative NAO in 2010 and 2011. In the past 20 years, SSTs off southwest Greenland have warmed by c. $0.5\,°C$ in winter and c. $1\,°C$ in summer [237]. These changes are likely to result from development of the positive phase of the AMOC since the mid 1990s (Figure 15.2). Current coastal surface air temperatures are comparable to the mean surface air temperatures of the mid Holocene 4000–6000 years ago, following which temperatures declined at about $0.4\,°C/1000$ years [237, 238].

Incidentally, we owe much of our knowledge of Greenland to the pioneering efforts of two prominent female scientists, Dorthe Dahl-Jensen (1958–) (Figure 15.25) (Box 15.1) from Denmark, and Valérie Masson-Delmotte (1971–) (Figure 15.26) (Box 15.2) from France.

| Box 15.1    Dorthe Dahl-Jensen |
|---|
| Dorthe Dahl-Jensen (1958–) is professor of ice physics at the Niels Bohr Institute of the University of Copenhagen. She heads the university's Centre for Ice and Climate, where researchers focus on ice core data to understand the climate of the past, present, and future. She led the North Greenland Eemian Ice Drilling project (NEEM) – the Danish International Polar Year's ice core drilling program on the Greenland ice sheet. A member of the Royal Danish Academy of Sciences and Letters, she has been awarded the European Union's Descartes Prize (2008), the Swedish Vega medal (2008), and the Louis Agassiz Medal of the European Geosciences Union (2014), amongst others, for outstanding scientific contributions in polar glaciology and climate studies centred on Greenland ice cores. |

**Figure 15.26**    Valérie Masson-Delmotte.

In Fram Strait, between Svalbard and Greenland at about 80°N, summer SSTs increased significantly after about 1800 CE, reaching a maximum of $6\,°C$ in recent years, which was warmer than conditions in the Medieval Warm Period [54]. Even so, surface water stayed cool, and sea ice and ice-rafted debris continued to be abundant well into the twentieth century, probably because of an increase in glacial melt-water accompanied by icebergs. Evidence from planktonic foraminifera suggests the persistence of a thick, cold and fresh surface water layer overlying warm, saline Atlantic Water that penetrates the Arctic in the subsurface, much as suggested by Dokken and colleagues [55].

Exciting new finds confirm the dramatic warming of the Arctic. Miller's 2013 examination of rooted tundra plants

**Figure 15.25**    Dorthe Dahl-Jensen (http://www.thesolutionsjournal.com/author/dorthe-dahl-jensen).

exposed by receding cold-based ice caps in eastern Canada shows that 5000 years of regional summertime cooling has been reversed, taking the average summer temperatures of the past 100 years to levels higher than in any summer period for more than 44 Ka [239]. That includes the peak warmth of the early Holocene when Arctic summer insolation was 9% above modern levels. Summers cooled by some 2.7 °C over the past 5000 years, until the reversal of the modern era. Miller's findings show that in the Arctic '*anthropogenic emissions of greenhouse gases have now resulted in unprecedented recent summer warmth that is well outside the range of that attributable to natural climate variability*' [239].

Independent confirmation of the unprecedented nature of modern Arctic warming comes from a study in northern Scandinavia by Jan Esper and colleagues (Figure 15.27) [240]. Like others, Esper observed that the climatic trend for the past 2000 years was one of cooling driven by the Holocene decline in insolation. His study of maximum latewood density in trees indicated a decline of ~0.31 °C/1000 years from 138 BCE to 1900 CE, a signal that is missing from published tree ring proxy records of climate change. Peak warmth occurred in Roman and Medieval times, alternating with severe cold conditions in the fourth and fourteenth centuries. A decline in orbital

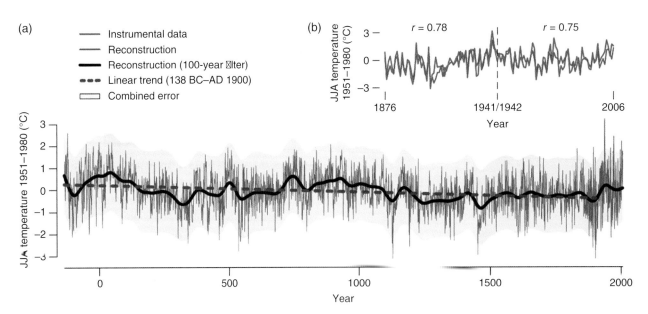

**Figure 15.27** Northern Hemisphere temperature reconstruction based on maximum density (MXD) tree ring data from northern Scandinavia. Summer (June, July, August) (JJA) temperature reconstruction and fit with regional instrumental data. (a), Reconstruction extending back to 138 BCE highlighting cool and warm periods on decadal to centennial scales (black curve, 100 year spline filter) amid variation between extremely cool and warm summers (jagged dark data), and a long-term cooling trend of –0.31 °C/1000 years from 138 BCE to 1900 CE (dashed line), within an overall uncertainty area (grey) representing two standard deviations (2σ). (b), MXD trend (darker line) compared with measured JJA temperatures (light line) from 1876 to 2006. Correlation between MXD and instrumental data is 0.77. Lighter jagged values in (a) post-1876 indicate measured air temperatures from (b). *Source:* From figure 2, in Ref. [240].

summer insolation by ~6 W/m² in 2000 years made the Roman period (21–50 CE) 1.05 °C warmer than the 1951–1980 mean, and ~2 °C warmer than the coldest period (CE 1451–1480), which was −1.19 °C colder than the mean, and ~0.5 °C warmer than the maximum twentieth century warmth of 1921–1950. Temperature rose sharply into the modern era from 1900, cutting across the orbitally driven decline [240].

Similar evidence is emerging from other regions. For instance, Lonnie Thompson, of the Byrd Polar Research Centre of Ohio State University, and his team, found that glaciers at Quelccaya, 5670 m up in the Peruvian Andes, declined rapidly in ice cover between 1983 and 2003 [241], just like those of other low latitude ice caps and glaciers (Figure 15.28) [242]. These retreats reverse advances that began at the end of the Holocene climatic optimum, which ended some 5 Ka ago. This is an abrupt climatic transformation. As Michael Zemp of the University of Zurich points out, these data are in marked contrast '*to statements repeatedly made in the grey literature claiming that (1) glacier retreat or mass loss could not be substantively evidenced globally … or that (2) glaciers are globally*

*not retreating but advancing*' [242]. Using the World Glacier Monitoring Service's comprehensive database, and adding their own satellite observations, Zemp and his team reported in 2019 that the global mass loss of mountain glacier ice has increased significantly in the last 30 years and is contributing almost 1 mm of sea level rise per year [243]. '*The present glacier mass loss [between 1961 and 2016] is equivalent to the sea level contribution of the Greenland Ice Sheet, clearly exceeds the loss from the Antarctic Ice Sheet, and accounts for 25–30 per cent of the total observed sea level rise*' [243]. Those conclusions are supported by satellite studies of Himalayan glaciers showing that they have lost as much as a quarter of their mass in the last 40 years and that the rate of loss doubled after the years 2000 at the same time as the average temperature there increased by 1 °C [244].

In contrast with the Holocene neoglacial, when warming in the southern Hemisphere made glaciers there retreat whilst cooling in the north caused them to advance, glaciers in both hemispheres retreated synchronously post 1850–1900 with the warming that began then [242, 243, 245–248](Figure 15.28).

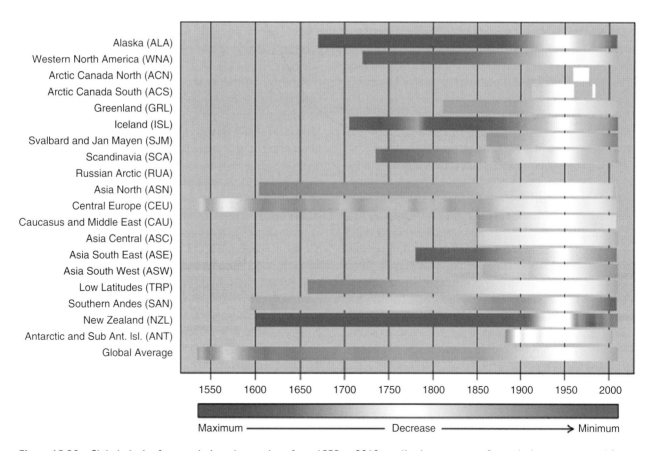

**Figure 15.28** Global glacier front variation observations from 1535 to 2010: qualitative summary of cumulative mean annual front variations. The colours range from dark blue for maximum extents (+2.5 km) to dark red for minimum extents (−1.6 km) relative to the extent in 1950 as a common reference (i.e. 0 km in white). *Source:* From figure 6a in Ref. [242].

Nevertheless, regional conditions do mean that in some places glaciers are locally advancing, for example in New Zealand, where Andrew Mackintosh and colleagues found 58 glaciers advanced in the Southern Alps between 1983 and 2008 [249]. These advances were caused not by increased precipitation but by colder temperatures associated with anomalous southerly winds reflecting changes in the structure of extratropical atmospheric circulation over the South Pacific. This may be related to the southward migration of the Polar Front by about 60 km (1° of latitude) between 1992 and 2007 discovered by Sokolov and Rintoul and attributed to global warming [250].

Paul Spence and colleagues observed that the strengthening and southward shifting of the Southern Hemisphere's westerly winds since the 1950s could produce an intense warming of subsurface coastal waters that exceeded 2 °C at 200–700 m depth [251]. Their model helped to explain how the movement of the winds diminished the down-welling that kept surface waters cold on the continental shelf of the Antarctic Peninsula, enabling the boundary between cold surface water and warm deeper layers to move up enough to flood the shelf with warm water that could have attacked the under sides of floating ice shelves and moved their grounding lines landwards (Figure 13.13).

As shown in Figure 15.20, Sigl and colleagues found that 80% of the decline in glacier length in the Alps had occurred between 1800 and the mid-1870s. That was before black carbon – representative of industrial pollution – began to build up on the surface of the ice [148]. Hence the decline in glacial length was a response to some other forcing – either decline in volcanic aerosols or increase in solar activity or increase in greenhouse gases.

Mayewski and Maasch found that the climate conditions for the modern era (1931–1980) in the Northern Hemisphere differ from those of both the Medieval Warm Period and the LIA, having variable wind strengths associated with significantly higher temperatures (Figure 15.7) [58]. Much the same is true for the Southern Hemisphere, except that there today's winds tend to be stronger than in the Medieval or LIA climate eras, whilst the temperatures are not much different from those times [58]. Evidently the modern climate in both polar regions differs from what it was in the past, but the climates in the two regions also differ from one another, with stronger winds and less warming in the south. To a polar expert this is no surprise. After all, the Arctic has an ocean surrounded by land, whilst the Antarctic has a continent surrounded by ocean. The wall of winds around Antarctica keeps warm air out; there is no comparable wall around the Arctic.

In Europe, studies of tree rings, published in 2011 and 2013, confirm that the recent warming at the end of the twentieth century is unprecedented in eastern Europe over

the past 1000 years and in central Europe over the past 2500 years [252, 253].

Ljungqvist's team from the University of Stockholm found that the warmth of the Northern Hemisphere in the ninth to eleventh centuries was comparable to that of the twentieth century mean (Figure 15.5) [30]. But the rate of warming from the nineteenth to the twentieth century was by far the fastest between any two centuries in the past 1200 years, and was '*unprecedented in the context of the last 1200 yr*' [30]. They also noted that analyses of instrumental data [254] showed that the last decade of the twentieth century was much warmer than the twentieth century mean nearly everywhere over Northern Hemisphere land areas, '*thus providing evidence that the long-term, large-scale, NH [Northern Hemisphere] warming that began in the 17th century and accelerated in the 20th century has continued unabated*' [30].

Turning to the Antarctic, we see that sea ice remained more or less stable, with a slight tendency for a year-on-year increase from 1978, when satellite measurements of sea ice began, until 2016, when the sea ice coverage reduced suddenly back to what it had been in 1978. It has stayed low since then. The precise cause is not known, but Turner, and Walsh both consider the earlier growth to have probably resulted from a strengthening of the westerly winds forced by the growth of the ozone hole, rather than from global warming [227, 255]. The stability of Antarctic sea ice, as compared with that in the Arctic, is explained by such factors as: (i) melting of coastal ice shelves freshening the sea surface, stratifying it and making it more likely to freeze [256]; (ii) the increase in reflective sea ice area keeping surface air and water cool close to the Antarctic coast [256]; and (iii) strengthening offshore winds blowing surface water (and sea ice) further to the north, increasing the sea ice area [257] (that helps to explain why Antarctic sea ice shows different patterns depending on the region, growing in most sectors but decreasing in the Bellingshausen and Amundsen Seas); (iv) linkage of Antarctic climate to cooling in the equatorial Pacific, attributed to the negative phase of the PDO [258]; and (v) pronounced upwelling of cool deep water driven by the strong westerly winds keeps surface waters cool, a process lacking in the Arctic. Wind-driven northward transport of freshwater in the form of melting sea ice helps to determine the mean salinity distribution and density structure of the Southern Ocean, which has critical consequences for the global climate by affecting the exchange of heat, carbon and nutrients between the deep ocean, surface waters and the atmosphere [259].

In Antarctica there is growing evidence that inter-annual variability is primarily governed by the Antarctic Oscillation and the (tropical) inter-decadal Pacific Oscillation modes of atmospheric variability, with some connection also to the decadal behaviour of the Indian Ocean Dipole [260].

Looking at the global picture, the PAGES group analysing the climate of the past 2000 years reported in 2013 that '*Recent warming reversed the long term cooling [of the Holocene]; during the last 30-year period (1971–2000 CE), the area-weighted average reconstructed temperature was likely higher than anytime in nearly 1400 years*' [24]. Evidently, they concluded, '*The global warming that has occurred since the end of the 19th century reversed a persistent long-term cooling trend*' [24].

Like the PAGES team, Sean Marcott and colleagues found that global temperatures for 2000–2009 were warmer than 72% of the stacked Holocene record, but had not yet exceeded the highs of the early Holocene between 10 and 5 Ka ago. In contrast, the decadal mean global temperature for 1900–1909 (the Gleissberg Minimum) was cooler than 95% of the Holocene temperature distribution. '*Global temperature, therefore, has risen from near the coldest to the warmest levels of the Holocene within the past century, reversing the cooling trend that began ~5000 yr B.P.*' [44].

Unfortunately, our instrumental records cover only about the past 150 years, so they are too short to assess anthropogenic climate change comprehensively [261]. As Abram and colleagues pointed out, greater clarity emerges from the data when results from coastal oceanic upwelling regions are separated from non-upwelling regions: enhanced oceanic upwelling plausibly explains recent cooling trends at some upwelling sites, consistent with the idea that global warming could, in some locations, strengthen the surface winds that drive upwelling and the supply of cold water to the surface along certain continental margins (Peru/Chile; California; Namibia; Morocco) and in the Antarctic Circumpolar Current system. In the latter case, the strengthening of westerly winds over the Southern Ocean caused both enhanced upwelling and the northward advection of any surface warming signal, and so delayed development of sustained industrial era warming over Antarctica [261]. Similarly, sustained ocean warming in the North Atlantic is delayed or characterized by cooling due to the slowdown of the AMOC. The finding that warming began in Antarctica during the mid-nineteenth century when forcing by rising greenhouse gases was small '*suggests that Earth's surface temperature may respond to even small increases in greenhouse gas forcing more rapidly than previously thought*' [261]. And as we saw earlier, significant Antarctic warming since 1900 CE is apparent on the Antarctic Peninsula, the WAIS and the Dronning Maud Land coast [64]. However, Abram's conclusions about the influence of greenhouse gas emissions in the nineteenth century have been called into question, with Lewis suggesting that such influence was unlikely prior to the 1870s [262].

Armour and colleagues agreed with Abram that delayed warming south of the Antarctic Circumpolar Current, in contrast with substantial warming along its northern flank, can be attributed to the Southern Ocean's meridional overturning circulation in which wind-driven upwelling of cold deep-ocean water south of the current drives that cool water north and damps warming around the continent [263]. As those cool waters are moved north, they warm and are then subducted along the northern flank of the current to form Antarctic Intermediate Water and Subantarctic Mode Water, in a process that stores their heat in the subsurface.

The present climate is also an anomaly as far as $CO_2$ is concerned. Hans Oeschger's team in Switzerland showed that the pre-industrial values of about 280 ppm in fossil air from shallow ice cores began to rise exponentially following the mid-1700s, driven by the industrial revolution (Chapter 9). Those rising values mesh with the values measured in background air, providing a seamless transition between modern air and fossil air (Figure 9.12) [264, 265]. There is no analogue deeper in the Antarctic ice core record for the changes in $CO_2$ seen since the nineteenth century. The fastest rate of increase in $CO_2$ seen during the last glacial termination was 20 ppm/1000 years; $CO_2$ rose by the same amount in the 11 years before 2010, or 100 times as fast [265]. Changes like that led Nobel chemist Paul Crutzen to describe as the 'Anthropocene' the period of time since the late eighteenth century when humans began to impact the Earth's climate [266]. This anomalous rise in $CO_2$ is closely connected to the rise in temperature since 1900 (Figure 15.29).

According to Richard Zeebe, the present emission rate of $CO_2$ is unprecedented for the late Cenozoic, and puts us in a 'no-analog' state that '*represents a fundamental challenge in constraining future climate projections*' [270]. $CO_2$ is now at a level of 411 ppm, which is 131 ppm above the maximum of past interglacials, an increase of 47%. The 1.2 °C average temperature anomaly for 2016 (relative to 1860–1900) was 0.4 °C higher than the average global temperature estimated for the Holocene thermal optimum of 9000–6000 years ago [44]. Earth's climate is now close to the warmest temperatures of the last interglacial, which were above those typical of the mid twentieth century by 0.5° ± 0.3 °C globally [271]. The recent temperature highs of 2014 through 2016 had a negligible likelihood of occurrence in the absence of anthropogenic global warming [272] and 2019 was only fractionally cooler than 2016.

In 2016, Adolf Stips and colleagues used a novel approach based on information flow to explore whether the link between temperature and $CO_2$ in the modern era was due to causation or correlation. Their research suggested that greenhouse gases were indeed the main drivers of recent

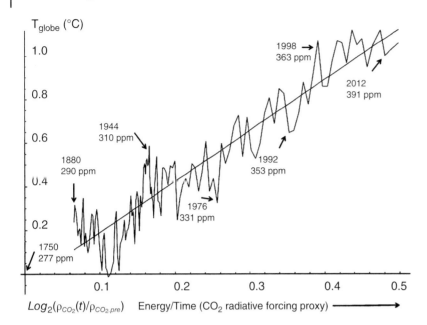

**Figure 15.29** Global temperature (from NASA, for 1880–2013) plotted as a logarithmic function of $CO_2$ representing radiative forcing. Fluctuations represent natural variability. The slope of 2.33 °C/$CO_2$ doubling is the 'effective climate sensitivity' (based on figure 1 in Lovejoy (2015) [267], which is based on Lovejoy (2014a) [268]). The post-war cooling was likely due not only to a negative PDO but also to the accumulation of industrial aerosols prior to the development of Clean Air Acts; the 'pause' from 1998 to 2012 is thought to be the result of a further negative phase of the PDO (e.g. Lovejoy 2014b [269]). Warming has risen since this graph ended in 2013, to progressively higher values between 2014 and 2019.

warming [273]. There was evidence for minor forcing from aerosols, but insignificant forcing from natural variability (e.g. PDO or AMO), solar activity or volcanoes; although volcanic activity can be important in the short term it has no long-lasting effect over the past 150 years. Significant links indicative of a causal relation between $CO_2$ and temperature emerge from the data beginning in about 1960 [273]. Extending their investigation back through the 800 000 years of the EPICA Dome C ice core, Stips found evidence for temperature causing changes in $CO_2$, but not for the reanalysis of the EPICA $CO_2$ data made by Parrenin in 2013. In effect this discrepancy reinforces Parrenin's discovery that the dating of $CO_2$ by earlier methods was inadequate (see Chapter 13). However, Stips' expedition into the past was spoilt because his team failed to consider the roles of orbital forcing and the enhancement of its effects by fast and slow feedbacks that involve the ice albedo, clouds, the carbon cycle, vegetation, and so on, sometimes resulting in hysteresis behaviour.

Methane too has seen a significant (162%) rise, from 700 ppb in past interglacials [274] to an annual average of about 1900 ppb in recent years [275, 276] (Figures 13.8 and 14.8d). Methane emission was at a plateau from about 2000 to 2007, but since then has increased at rates of up to c. 12.5 ppb/year [276]. The current changes are beginning to exceed the largest events in the past millennium [276]. Along with the rise in $CH_4$ abundance since 2007, the $\partial^{13}C$ of its carbon became significantly more negative (Figure 14.8d), suggesting that much of the increase in the supply of methane is biogenic, '*led from emissions in the tropics and Southern Hemisphere, where the isotopically depleted biogenic sources are primarily microbial emissions from wetlands and ruminants*' [276]. Fossil fuels have declined as a proportion of the total methane budget. The increase coincides with the unusual meteorological conditions over tropical wetlands, which have experience increased warming and more intense wet seasons in recent years. Nisbet and colleagues wonder if this is a trend driven by climbing climate change, or merely the result of a decadal weather oscillation: '*if methane growth continues, and is indeed driven by biogenic emissions, the present increase is already becoming exceptional, beyond the largest events in the last millennium*' [276].

According to Raimund Muscheler and his team the most recent minimum of the 208 year Suess De Vries sunspot cycle was combined with a minimum in the Gleissberg sunspot cycle at about 1900 [94]. That implies that the corresponding maximum in the Suess–De Vries Cycle should have occurred between 1985 and 2030, more or less as suggested by Lüdecke [131, 134], which may account for the peak in sunspot activity at about 1980–1990 and its subsequent decline [207]. Assuming that Muscheler and his team are right, then the next cold minimum in the Suess–De Vries and Gleissberg cycles should occur between about

2070 and 2120 (my calculation), more or less as suggested by Steinhilber and Beer [133]. Similarly, assuming that the 2300 year Hallstatt solar cycle is real, and that it was manifest as a solar minimum and cold period in 750–500 BCE, then the minimum should have recurred about 1500 CE in the LIA, which may explain the rather low temperatures of that period, and should recur again at about 3800–4000 CE. If Bond was right and there is a 1500 year cycle of which the last major cooling peak was ~500 years ago during the Maunder Minimum, then the next similar cooling peak should occur about 1000 years from now, with a warm peak in between at about 2350 CE. Time will tell. In any case the data of Berger's Belgian team remind us that that these solar (and volcanic) variables are superimposed on a virtually flat neoglacial pattern of orbital/axial insolation (Figure 12.2), which will continue for at least the next 1000 years and which provides the underpinning or infrastructure of the LIA and the present climate (Figure 12.2). Unless the Sun increases its output in some extraordinary manner, unlike its Holocene behaviour, the only thing that will keep us as warm as we are now (or warmer) is the increased concentration of greenhouse gases in the atmosphere [277].

To conclude this section we examine the flattening of the warming signal between 2000 and 2013, which has attracted much attention, although it was followed by a sharp rise to a new peak in 2016, driven by a large El Niño event (Figure 8.13). As we have seen in this book, one should not expect there to be one-to-one relationship between $CO_2$ and temperature all the time. $CO_2$ is not the only thing to affect temperature and vice versa. Global temperature can also be affected, for instance by internal oceanic oscillations like El Niño–La Niña, and the PDO. As mentioned in Chapter 13, during the positive phase of the PDO the equatorial region is warm (as it is in an El Niño) and the Gulf of Alaska is cool, whilst in the negative phase the equatorial region is cool (as it is in a La Niña) and the Gulf of Alaska is warm. Each of these PDO phases can last from 10 to 25 years. During a La Niña, and during the negative phase of the PDO, cold water spreads out over the central and eastern Pacific Ocean. This 'new' large area of cold water cools the air, bringing down the global average temperature [208, 278–280]. Did the negative phase of the PDO between 2000 and 2013 help to stabilize global temperature by preventing a rise that would otherwise have occurred? Oceanographers like Yan and others writing in 2016 consider that the cooling of the equatorial Pacific in one of the long phases of the PDO is at least partly responsible for the global warming slowdown of 2000–2013 [281]. According to Knudsen and colleagues we also have to consider the global impact of the AMO during which the North Atlantic warms by about 0.2 °C in the positive (strong) phase, and

cools by the same amount in the negative (weak) phase at intervals of about 50–70 years (Figure 15.2) [3]. The positive phase of the AMO may have contributed to the American 'dust bowl' conditions of the 1930s [3]. It is not to be confused with the NAO, which is mainly an atmospheric phenomenon driven by changes in air pressure between the Iceland low and the Azores high pressure cells. These and other quasi-periodic internal oscillations like the PDO and the El Niño–Southern Oscillation (ENSO) supply much of the background high frequency noise in the climate spectrum (Figure 8.13). If these interpretations are correct, then reversion to a long period of the positive phase of the PDO should lead to an increase in the global warming signal. We should note in passing that even when the global climate warming signal was flat, the Arctic continued to warm, not least because the continued supply of heat melted its sea ice, so reducing the Arctic's albedo.

The apparent stillstand in temperature began in 2000, when the PDO went negative [208]. It did not begin in 1998, when a massive El Niño event temporarily increased global warming. That was followed by an equally large La Niña and associated cooling. With the continued rise in $CO_2$ beyond the 400 ppm level first reached in May 2014 [282], the stillstand has effectively ended, with the past five years (2015 through 2019) being the warmest since records began. This makes the recent 'pause' just as temporary as the 'hiatus' or flattening in temperature rise was between 1950 and 1970, when the PDO was last in its negative (i.e. cool) phase [208].

An additional perspective on these observations of change in the surface ocean comes from an analysis of the role of deep ocean circulation. A comprehensive analysis of ocean warming showed in 2014 that at the same time as the 'hiatus' started, more heat began moving into the deeps of the Atlantic and Southern Oceans. According to Chen and Tung, cooling associated with this deep heat sequestration usually lasts 20–35 years, after which surface warming should accelerate [283]. In 2018, they expanded on that interpretation [284]. They started by recalling that during the last glaciation the AMOC weakened when excess freshwater was supplied to the North Atlantic, resulting in cooling. However, the recent weakening of the AMOC (1975–1998) led to rapid surface warming, not cooling (Figure 15.30). At that time, the AMOC's 'normal' role of transporting surface heat north (Figures 12.5 and 12.6), changed to storing heat in the deep North Atlantic. Between '*the mid 1990s to the early 2000s, the AMOC stored about half of excess heat globally, contributing to the global-warming slowdown*' [284]. Since 2004, the AMOC and oceanic heat uptake have weakened. These variations in regional behaviour are best explained by multi-decadal variability rather than any anthropogenically forced trend (Figure 15.30) [284].

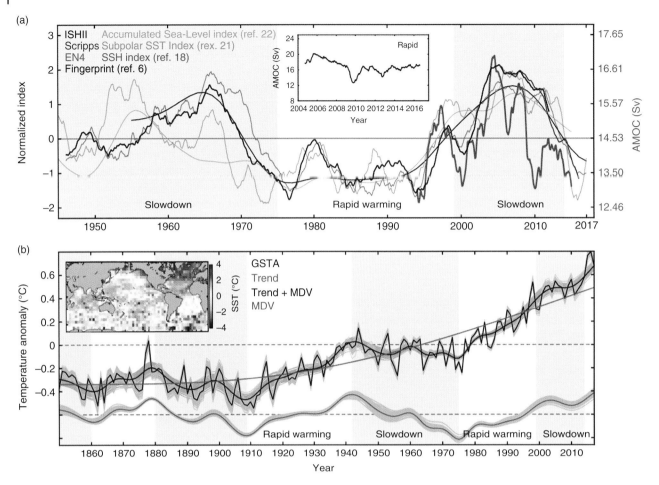

**Figure 15.30** Variations in Atlantic Meridional Overturning Circulation (AMOC) and the Global mean Surface Temperature Anomaly (GSTA): (a) Mid and subpolar latitude AMOC strength at 41°N (red, two year running mean, Sverdrup scale shown on the right, where 1 Sv = 1 000 000 m³/s), based on the ISHII (dark blue) [285] and Scripps (purple) datasets, with a two year running mean. The green curve is the subpolar salinity. The AMOC fingerprint (dark blue) and the accumulated sea-level index (turquoise) were smoothed with 10 year and 7 year low-pass filters, respectively. The subpolar gyre sea surface temperature (SST) index in orange is also a two year running mean. The inset shows the AMOC at 26°N; (b) Shows the GSTA from the HadCRUT4.6 data set (black), the nonlinear secular trend (close to the 100 year linear trend) (brown) and variation about the trend for timescales longer than decadal (multidecadal variability [MDV], in red). The inset shows the SST spatial pattern associated with MDV, obtained by regressing SST onto its time series. The blue curve is the smoothed version of GSTA obtained as the sum of the secular trend and MDV. The faint lines around the solid lines show the range of uncertainty of the data. *Source:* From figure 3 in Ref. [284].

By 2017 the AMOC decline was ending, probably leading to an AMOC minimum that may last about 20 years based on past experience. Chen and Tung suggest that this new minimum will reduce the uptake of heat by the ocean and lead to a period of rapid global surface warming [284]. A comparison of Figures 15.1, 15.2, and 15.30 should help in understanding climate change in the modern era.

Independent verification of Chen and Tung's conclusions comes from a 2019 study of ocean heat content from 1871 to the present by Laure Zanna and colleagues. Zanna pointed out that '*Most of the excess energy stored in the climate system due to anthropogenic greenhouse gas emissions has been taken up [cumulatively] by the oceans, leading to thermal expansion and sea-level rise. The oceans thus have an important role in*

*the Earth's energy imbalance*' [286]. Zanna showed that surface ocean heat content increased in two phases, with a first increase from 1910 to 1960, then a flattening before a further rise from 1975 to 2017. The rises and flattenings in ocean heat content for the Atlantic correspond to the pattern of the AMOC shown in Figure 15.30a. Deep water (>2000 m) shows much the same pattern of heat storage but with a delay of some 20 years [286]. It is significant that the rise in ocean heat content was the same in 1990–2015 as it was in 1921–1946 (the early period of global warming – see Figure 15.30b). This pattern mimics the patterns of sea level rise and global temperature rise for the modern era (from 1900 onwards) (e.g. see Figure 8.13). Clearly, monitoring ocean heat content is a key to predicting global and regional change in climate and sea level [286]. As mentioned earlier, these changes (e.g. Figure 15.30) also mimic the distribution of aerosols. As Booth pointed out (see earlier), aerosols were most abundant in the periods 1890–1920 and 1955–1975 when Atlantic and Greenland temperatures were lowest [5].

A careful study of temperature trends in the deep ocean, by Gebbie and Huybers, in 2019, showed that whilst those at all depths in the Atlantic reflected modern warming, those in the deep Pacific showed that it was still adjusting to the cooling imposed by the LIA, a finding corroborated by comparing modern physical oceanographic measurements with those from the Challenger Expedition of 1872–76 [287]. The discrepancy between the two oceans occurs because the waters of the Atlantic are derived from the main sinking in the meridional overturning circulation, which takes place from both ends of the Atlantic, whilst the deep waters of the Pacific are far from these sources (at the end of the line, as it were) and so are oldest, which in this case also means coldest. The subsurface of the Pacific Ocean is thus in disequilibrium. Gebbie and Huybers argue that the global ocean stored more heat during the Medieval Warm Period than today, not because surface temperature was greater, but because the deep ocean had more time to adjust to surface anomalies [287] (the MWP lasted for some 200 years).

The changes in the AMOC observed by Chen and Tung (Figure 15.30) are also related to changes in the NAO (a measure of the atmospheric pressure difference between Iceland and the Azores), both being positive (or negative) at the same time [288]. If Chen and Tung are right about the AMOC leading to a period of warming over the next two decades, then northward transport of heat through the Atlantic, stimulated by the NAO, will likely lead to a substantial further decline in the area of Arctic sea ice, augmenting the decline due to anthropogenic warming [288]. However, since 1990 the NAO has weakened slightly, making such a prediction somewhat uncertain.

The weakening of the AMOC has weakened the Gulf Stream (or its ability to carry heat northwards at the surface). But this behaviour is anomalous as far as the other major western boundary currents are concerned. Yang and colleagues demonstrated in 2016 that all of these currents apart from the Gulf Stream are intensifying and shifting towards the poles in response to an intensification and poleward shift of near-surface ocean winds driven by global warming [289]. This includes the Kuroshio (off Japan), the Brazil Current, the East Australian Current and the Agulhas Current (off South Africa), all of which carry heat polewards. The northward shifts in the western boundary currents are consistent with the poleward shift of the Hadley Cell, the expansion of the tropical belt, and the poleward shift of the subtropical dry zones [289]. It is

counterintuitive that despite the weakening of the Gulf Stream, Chen and Tung find evidence to suggest a weak AMOC will lead to greater warming in the northern North Atlantic (Figure 15.30) over the next two decades where we have recently seen cooling [284].

Further research has also shown that that the 'apparent' recent slow-down in global warming was not as slow as had been thought [290]. It becomes less extreme when Arctic data are incorporated into the analysis of global trends, since the global stillstand in temperature is not seen there.

It has also been suggested that the flattening of the curve of temperature rise between 2000 and 2013 may reflect a slight increase in volcanic aerosols emitted into the upper atmosphere [291], for example from the eruption of Soufrière Hills in Montserrat and Tavurvur in Papua New Guinea, both in 2006 [292]. There were also growing emissions of aerosols from the rapid increase in coal burning to support industrial expansion in India and China [293]. China's coal consumption more than doubled from 2002 to 2007, accounting for 77% of the rise in global coal use, and increasing global sulphur emissions by 26% [293]. Volcanoes and coal-burning power stations emit sulphur dioxide. Combined with water it forms sulphuric acid, whose droplets in the stratosphere reflect solar radiation. These increases in aerosols may help to prevent global temperature from rising as fast as it might do otherwise.

Finally, we should note the suggestion from André Berger that we have neglected to consider the fact that the rapid melting of Arctic sea ice since 1978 will have absorbed a great deal of heat that would otherwise have gone into warming the atmosphere [294]. This would also help to explain the so-called 'pause' in warming. It seems reasonable to suppose that the 'pause' represents the combined effects of surface ocean cooling, deep water storage of heat, loss of heat due to the melting of Arctic sea ice, and aerosol loading from Chinese and Indian coal burning plus the occasional volcanic eruption. These observations help to underline the fact that we should not expect to see a simplistic one-to-one relationship between $CO_2$ and temperature.

In summary, we are now in the 'modern warm period'. The IPCC's 5th assessment, in 2013, concluded that the warming from 1900 to about 1945 (Figure 8.13) was attributable in part to an increase in solar energy caused by development of the latest peak in the Suess–De Vries solar cycle (Figure 15.14), and in part, amongst other things, to a gradually growing contribution from climbing emissions of $CO_2$ (Figure 9.12), balanced against cooling effects from growing aerosols [295]. A disconnect between solar output and global warming began in about 1950. Solar energy increased towards the double peak of the solar cycle between 1960 and 1990 [139, 204], (Figure 15.13) whilst global temperatures flattened (Figure 8.13), possibly due to cooling induced by a growing load of industrial aerosols between 1945 and 1970 [209, 210]. Thereafter, temperatures rose whilst solar energy remained constant (Figure 15.23) [206]. Temperatures continued to rise after the solar peak in 1990, when solar output began its present decline (Figures 15.13 and 15.14) [204].

Greenhouse gases started to increase exponentially from about 1776, when the industrial revolution took off with the patenting of James Watts' steam engine (Figure 14.8b). The rise in greenhouse gases, due especially to the burning of coal, began to contribute significantly to the rise in temperature from about 1900 onwards (overlapping with the slight rise in solar energy), increasing eventually to the point where they became the main drivers of warming after 1950, accentuated by water vapour evaporated from the warming ocean [295]. However, some of that warming from 1900 to 1945 may also have been caused by the natural move towards the positive phases of both the AMO and the Pacidic Decadal Oscillation (e.g. Figure 15.2). The global temperatures of the mid-twentieth century may not have been much different from those of the mainly European Medieval Warm Period. But present temperatures are significantly higher, especially in the Arctic, where the shrinkage of sea ice makes the ocean and lower atmosphere warm much faster than elsewhere. These warming trends cut across both the long-term trend of cooling driven by declining insolation, and the recent decline in sunspot activity representative of solar output.

In a nutshell, then, orbital insolation (Figure 12.2) tells us we should still be in the neoglacial LIA of the Holocene and experiencing periodic minor warming and cooling due mainly to predictable fluctuations in solar energy modified by occasional volcanic outpourings. Solar data (Figure 15.13) tell us that our temperatures from the late twentieth and early twenty-first centuries should be no greater than they were during the LIA at around 1780 or 1860. Given that context, the actual warming since 1950 is unprecedented in the past 2000 years, as is the shrinkage of glaciers and the rise in sea level (which are proxies for temperature). On the basis of the information reviewed above we could assume, based just on solar data, that the LIA came to its end following the Gleissberg Sunspot Minimum (i.e. around 1920). However, it would also appear that the LIA is the manifestation of the current low point in orbital/axial insolation, and that the sunspot minima following the Medieval Warm Period merely accentuated that low point. Both Steinhilber [133] and Ludecke [131, 134] agree that we will soon face a further sunspot minimum in the 2050–2100 era, which may descend to something like the depths of the Dalton or Gleissberg solar minima, which belonged within the late phase of the LIA. Hence it could be argued

that we are merely experiencing another of those brief warm interludes within the LIA identified by the pattern of cosmic intensity in Figure 15.12. If the present solar maximum lasts from 1920 (end of Gleissberg Minimum) to 2050 or 2100 it will be either 130 or 180 years long, which is somewhat shorter than the usual 208 year long Suess–De Vries cycle, but not much different from the warm period between the Spörer and Maunder Minima (Figure 15.12). Accepting that, we have to explain why our current temperature is about 1 °C above Late Holocene temperatures, and in fact at or close to that of the Holocene thermal optimum. Nobody has yet provided a more plausible explanation that that this rise is due to our emissions of greenhouse gases [295].

## 15.6   The Anthropocene

The idea that we are no longer living in the Holocene came from Nobel atmospheric chemist Paul Crutzen, in 2000. His proposal that the Anthropocene should be declared as a new geological time unit led to the formation of the Anthropocene Working Group as a body of the Subcommission on Quaternary Stratigraphy, which reports to the main body responsible for naming geological units, the International Commission on Stratigraphy. Professor Jan Zalasiewicz of the University of Leicester in the UK leads the working group. Under his leadership, the group's members have recently comprehensively reviewed the case for the Anthropocene as a geological time unit [296]. Initially, Crutzen suggested that the Anthropocene might have begun with the Industrial Revolution. But careful evaluation of a wide range of indicators suggest that the major break with the Holocene can be dated to about 1950, when post-war industrial activity ramped up globally contributing to major changes in the character, composition, and behaviour of the atmosphere, the ocean, the cryosphere and the land surface [296]. The change from the Holocene to the Anthropocene can be seen in a wide range of geological materials, making it feasible to identify a base for the new unit. Research is ongoing to identify that base in a wide variety of environments, prior to the case being made formally to the International Commission. Included in the case as support for the transition are the rise in global temperature by 1.2 °C since 1900, the cooling of the stratosphere, the warming of the ocean to progressively greater depths, the rise in sea level, the shrinkage of Arctic sea ice, the loss of substantial masses of land ice (hence reducing Earth's albedo) from Greenland, Antarctica and regions of mountain glaciers, the increase in ocean acidity, and the rise in greenhouse gases, notably $CO_2$, $CH_4$ and $N_2O$. In keeping with these developments we see plants insects and animals moving poleward of their customary ranges, especially in the Northern Hemisphere, along with decreasing natural diversity, increasing volumes of domesticated animals and widespread intercontinental distribution of species [297, 298]. These changes are all regarded as the products of increasing human impact on the planet from a global population that reached 2.5 billion in c. 1950, and 7.5 billion by c. 2015, and is projected to reach 9–10 billion by c. 2050. Humans have become equal to a geological force, radically changing the planet's ocean, atmosphere, ice, land surface, and biota.

These radical changes have made the Anthropocene the target for integrated research of holistic Humboldtian style through Earth System Science, a topic introduced in Chapter 8, and which reflects the increasing need to understand the workings of planet Earth by climbing out of our discipline silos and taking a systems view of life, an approach championed by the International Geosphere Biosphere Program (IGBP) [299], and by such visionaries as Fritjof Capra [300]. For more on Earth System Science, see Lenton (2016) [301]. And for more on the Anthropocene see Ellis (2018) [302] as well as Waters (2016) [303].

Stepping back to take a long view, Hubertus Fischer and colleagues pointed out in 2018 that the several intervals of the past 3.5 Ma when climate conditions were warmer than the pre-industrial Holocene provide us with insights into potential future climate impacts and ecosystem feedbacks over the centennial to millennial scales that are NOT covered by the current generation of climate models. Investigating that history tells him that there is a '*low risk of runaway greenhouse feedbacks for global warming of no more than 2°C [although] substantial regional impacts can occur [such as] a shift in climate zones and the spatial distribution of land and ocean ecosystems [along with] substantial reductions of the Greenland and Antarctic ice sheets, with sea-level increases of at least several metres on millennial timescales*' [304]. Comparing palaeo observations with results from climate models suggests that '*due to the lack of certain feedback processes [such as interactive ice sheets, cloud processes and biogeochemical feedbacks] … model-based climate projections may underestimate long-term warming in response to future radiative forcing by as much as a factor of two, and thus may also underestimate centennial to millennial scale sea-level rise*' [304]. We have been warned.

As Ed Brook and Christo Buizert noted in 2018, '*Perhaps the most fundamental message from ice cores is … The strong correlation of variations in [greenhouse] gases with climate proxies over the last 800 kyr verifies the importance of the*

*greenhouse effect in global climate. [Furthermore] the temperature records …confirm the theory of polar amplification [with its implication for substantial loss of ice with quite small increases in 'global' warmth]* [305].

The main drivers for climate change in the Anthropocine will be, first, continued anthropogenic greenhouse gas emissions [306–309]. The upward trajectory of change will be modified, second, by variations in solar output and occasional volcanic activity, which will create short-term slight warming and cooling above and below the underlying upward trend of warming. On the longer timescales of orbital change, as we saw earlier, Ganopolski calculated that greenhouse gas emissions are now abundant enough to keep the Earth from descending into another glaciation for 50 Ka from now, and that a moderate further addition of $CO_2$ is likely to keep the Earth from moving into another glacial period for 100 Ka from now [309].

In 2018, considering the possible climatic trajectories for the future, Steffen and colleagues illustrated possible scenarios in which Earth's climate could reach progressively the temperatures of the Eemian interglacial, the mid Pliocene, the mid Miocene, and, eventually the mid Eocene [308]. Burke and colleagues assessed those possibilities in relation to the emission scenarios of the IPCC, concluding that under the maximum emission scenario (RCP8.5)[4] Earth's climate could resemble that of the mid Pliocene by 2030 and the mid Eocene by 2050; under a more moderate emission scenario (RCP4.5) climate would stabilize at Pliocene-like levels by 2040 [310]. Climates like those of the Pliocene and Eocene would emerge first in mid-continental regions, then expand outwards. These studies suggest that we are moving quite rapidly towards climate states well beyond those typical of the Holocene, the period in which our current agricultural civilization arose [308, 309].

We have turned up the thermostat, which will bring many unintended consequences. For example, it is not certain what continued warming will do to plant productivity. Up until now, we have unintentionally provided plants with increasing concentrations of plant food – $CO_2$, which has already led to some greening of the planet. This greening represents an increase in historical global primary productivity of $31 \pm 5\%$ over the twentieth century [311]. In 2016 Zhu and colleagues used satellite data to show that there had been persistent and widespread greening (involving an increase in the length of the growing season), over 25–50% of the global vegetated area, with less than 4% of that area showing browning [312]. Fertilization by increasing $CO_2$ explained 70% of this greening, while the rest was

[4]RCP = representative concentration pathway, which is a greenhouse gas concentration (not emissions) trajectory

labelled after its expected forcing value in 2100, where (say) RCP8.5 = 8.5W/m$^2$.

explained by nitrogen deposition (9%), climate change (e.g. change in warming and/or precipitation) (8%) and changing land cover (including intensity of land management, fertilization, irrigation, forestry and grazing) (4%) [312]. The $CO_2$ effect was most prominent in the tropics, while the global warming effect was most prominent at high latitudes; changes in land use had their greatest effect in southeast China and the eastern United States [312]. The increased uptake of $CO_2$ by plants caused a recent pause in the rate of growth of $CO_2$ in the atmosphere [313].

Experimenting with the response of plants to rising $CO_2$, Lombardozzi and colleagues found in 2018 that such increases tended to make leaves thicker, ultimately making photosynthesis less effective, so reducing a plant's ability to act as a carbon sink [314]. When plants take in $CO_2$, the carbon is used to create triose phosphate, which is later converted to sugar. But under high conditions of $CO_2$ plants hit a limit in the production of triose phosphate, so stop absorbing increasing amounts of $CO_2$. Hitting this limit will speed $CO_2$ rise. Whether further increases in $CO_2$ will increase plant growth substantially thus remains to be seen, since plant health depends not only on $CO_2$ and plant chemistry, but also on the supply of nutrients (which can become exhausted if not renewed) and on water.

We now know that the emission of $CO_2$ during Palaeocene–Eocene Thermal Maximum was considerably prolonged by the erosion of carbon from coastal sediments, which was oxidized and converted to atmospheric $CO_2$ [315]. As explained in Chapter 11, Lyons suggested that the erosion was the result of enhanced rainfall induced by evaporation of water vapour in a warming world [315], but I also suggested in Chapter 11 that it may in addition reflect the erosional effects of the rise in sea level of 10–15 m that accompanied the warming at the time. Such erosional sources of carbon have not yet been factored in to the Global Climate Models of what effect increasing greenhouse gas concentrations may have on our planet in future.

We should also recall that Kiessling pointed out in 2012 that the demise of equatorial coral reefs and their displacement poleward during the last interglacial suggests that continued warming will affect these delicate environments [316]. That message was recently reinforced by William Leggat of the University of Newcastle, Australia, and colleagues, who pointed out in 2019 that '*severe-heatwave-induced mortality events should be considered as a distinct biological phenomenon from bleaching events on coral reefs … heatwave conditions result in an immediate heat-induced mortality of the coral colony, rapid coral skeletal dissolution, and the loss of the three dimensional reef structure*' [317]. Leggat thought that such heatwave mortality events associated with rapid reef decay, would likely become more frequent as the intensity of marine heatwaves increases [317].

The land ice on Greenland and Antarctica will continue to shrink, as will the associated sea ice in the Arctic and Southern Oceans, and glacial ice in mountain areas, the latter creating problems for local water supplies. Rob De Conto and Dave Pollard calculated in 2016 that '*Antarctica has the potential to contribute more than a metre of sea-level rise by 2100 and more than 15 metres by 2500, if emissions continue unabated*' [318]. The future behaviour of the ice sheet will represent a combination of the effects of both long-term anthropogenic warming and multi-centennial natural variability [190]. Antarctica may lose one third of its sea ice by 2100 [319]. And there is the prospect of an ice-free Arctic Ocean by around 2050 [320]. From time to time these trends will be temporarily interrupted by the effects of natural oscillations within the climate system, for instance like that which led to an abrupt decline in ice loss from Greenland starting in mid-2013, driven by a positive trend in the NAO [321]. That situation did not last, and '*The next time [the North Atlantic Oscillation] turns strongly negative, [surface mass balance] will trend strongly negative over west and especially southwest Greenland … [and] we infer that within two decades this part of the [Greenland Ice Sheet] will become a major contributor to sea level rise*' [321]. Indeed, as Mads Knudsen observed, '*A return from a warm to a cold [Atlantic Meridional Oscillation] AMO phase could temporarily mask the effects of anthropogenic global warming, and thus lead to possible underestimation of future warming if the variability of the AMO is not taken into account*' [3].

Nicholas Golledge and colleagues reminded us in 2019 that '*simulations of the Greenland and Antarctic ice sheets constrained by satellite-based measurements of recent changes in ice mass [suggest] that increasing meltwater from Greenland will lead to substantial slowing of the Atlantic overturning circulation, and that meltwater from Antarctica will trap warm water below the sea surface, creating a positive feedback that increases Antarctic ice loss [by melting ice shelves from beneath]*' [322]. These processes may well increase sea level by a further quarter of a metre by 2100 [322]. And the rise will not stop there.

Already we can see climate change around Antarctica changing the area of sea ice there. In 2016, following a slow but very slight increase in sea ice area since satellite measurements of it began in 1978, the sea ice area suddenly declined by 28%. Loss was greatest in the Weddell and Ross Seas and continued [323]. According to the US National Snow and Ice Data Centre the area was just as low in the southern summer of 2018 as it was in 2017. Will this state continue?

Finally, it is important to recognize that the magnitude of melting of the Greenland ice sheet is '*exceptional over at least the last 350 years*' [324]. Luke Trusel and colleagues demonstrated in 2018 that '*the initiation of increases in [Greenland Ice Sheet] melting closely follow the onset of industrial-era Arctic warming in the mid-1800s, but that the magnitude of [that] melting has only recently emerged beyond the range of natural variability*' [324]. Because surface melting responds in a non-linear way to increasing summer air temperature, continued atmospheric warming will lead to rapid increases in melting, run-off and sea-level rise [324].

Will we see a sudden massive injection of methane from the decomposition of organic matter stored in melting Arctic permafrost? Wolff argues that we may not, because we did not see anything like that in the warmer major interglacials of the past 400 Ka [265]. In 2015, Schuur and colleagues found evidence suggesting that methane emissions from that source were and would likely continue to be 'gradual and prolonged in a warming climate' [325]. Earlier I pointed out that methane, one of the strongest greenhouse gases, is only emitted when decomposition takes place under reducing conditions. Decomposition under oxidizing conditions produces $CO_2$, which has a weaker impact. We do not yet know what the ratio will be in future between reducing and oxidizing conditions in the permafrost melt zone.

Will future changes be as rapid as, say, Dansgaard–Oeschger Events (10 °C rise in 5–50 years)? That seems unlikely given that much of the millennial variability of that kind occurred under different (i.e. glacial) conditions including lower sea level, larger ice sheets and lower $CO_2$ levels [305]. Are the '$CH_4$ time bomb' scenarios of a runaway positive carbon feedback for the Arctic likely? That, too, seems unlikely given that '*the ice core records contain no indications of very large bursts of CH4 during [warm] interglacial periods*' [305]. Did the WAIS collapse during the last interglacial when conditions were slightly warmer than today (as it did during the warm Pliocene)? We need longer cores in West Antarctica to answer that question [305]

Nevertheless, climate expert Veerabhadran Ramanathan and colleagues tell us that '*global warming will happen faster than we think*' [326]. They cite three reasons: first, emissions are rising, which could mean temperature rising faster than the 0.2 °C/decade of recent years; second, air pollution (i.e. reflective aerosols) is declining faster than anticipated; and, third, natural climate cycles are changing to favour warming (e.g. AMOC weakening and PDO becoming more positive), which could warm Earth to 2 °C by 2045 [326].

In recent years the melting of Arctic ice has freshened the surface of the North Atlantic, leading to a slowing of the AMOC that has cooled the North Atlantic off the southern tip of Greenland. Paradoxically, a study of the behaviour of the AMOC during the intermediate stages of the last glaciation (see Chapter 13) showed that with natural global warming this circulation becomes quite stable and leads to long periods of warm climate in the Northern Hemisphere [327]. Buizert and Schmittner deduced that '*The future evolution of the AMOC [Atlantic Meridional Overturning Circulation] may thus depend on the competition between a short-term transient weakening driven by North Atlantic surface warming/freshening, and a long-term increase in ... strength driven by circulation changes in the [warming] Southern Ocean*' [327]. Chen and Tung showed that a weakened AMOC is likely to lead to enhanced warming over the next two decades [284]. However, Liu and colleagues pointed out in 2017 that '*the AMOC is in an unstable regime susceptible [to] large changes in response to perturbations*' [328]. Applying a climate model, they showed that an abrupt doubling of $CO_2$ from its level in 1990 (with associated warming) could cause the AMOC to collapse within 300 years, which would cool the North Atlantic and surrounding areas and increase Arctic sea ice [328]. Their research indicates the need for improvements in reducing the biases in models.

This is not the place for a lengthy analysis of future trends, a topic I have addressed elsewhere [308, 329]. But it should be obvious that the pace of warming will vary depending on local circumstances. Just to cite one little-appreciated example, '*Conifers are not moving north across Siberia as rapidly as one might expect from global warming. It appears that the growth of shallow-rooted larch trees on top of soil hardened by permafrost preserves ice below ground from melting, making it impossible for deeper rooted conifers to move north. This kind of "vegetation-climate lag" was once thought to last no more than a few centuries, but now appears to be persistent for millennia* [330]. *Given continued or persistent warming on the multi-millennial scale, we can expect a slow northward migration of larch and then conifer and spruce into what is now Arctic tundra, darkening the surface, which will have a further positive feedback effect on warming*' [329].

In summary, this chapter confirmed that the climate continued to cool to neoglacial conditions over the past 2000 years in response to the decrease in summer orbital and axial insolation (Figure 12.2). Superimposed on that cooling trend were several short-term warming and cooling events lasting 50–200 years and attributable to variations in solar output, modified by shorter term coolings lasting up to a decade or so and caused by large volcanic eruptions. This pattern was modified by natural internal variations (oscillations) within the ocean–atmosphere system – like the PDO, the NAO and the ENSO. As the trend in orbital insolation is expected to continue more or less flat for at least another 1000 years, Earth's natural climate should therefore remain in a cool state rather like that of the LIA, with occasional warmings and coolings. If nations choose to persist with business as usual, causing greenhouse gas emissions to rise significantly, then the trajectory of future change is likely to take us past the warmth of the four large interglacials of the past 400 Ka, past the warmth of Pliocene times, when sea level was at least 10 m higher than today, to the even warmer conditions of mid Miocene times or, at worst, the heat typical of the mid-Eocene 50 Ma ago, when there were no ice sheets [308]. At each of these times the sea level was significantly higher than it is now.

## References

**1** Kobashi, T., Kawamura, K., Severinghaus, J.P. et al. (2011). High variability of Greenland surface temperature over the past 4000 years estimated from trapped air in an ice core. *Geophysical Research Letters* 38: L21501. https://doi.org/10.1029/2011GL049444.

**2** Miettinen, A., Divine, D.V., Husum, K. et al. (2015). Exceptional ocean surface conditions on the SE Greenland shelf during the Medieval Climate Anomaly. *Paleoceanography* 30: 1657–1674. https://doi.org/10.1002/2015PA002849.

**3** Knudsen1, M.F., Seidenkrantz, M.-S., Jacobsen, B.H., and Kuijpers, A. (2011). Tracking the Atlantic Multidecadal Oscillation through the last 8,000 years. *Nature Communications* 2 (1): 178. https://doi.org/10.1038/ncomms1186.

**4** Chavez, F.P., Ryan, J., Lluch-Cota, S.E., and Ñiquen, M. (2003). From anchovies to sardines and back: multidecadal change in the Pacific Ocean. *Science* 299: 217–221.

**5** Booth, B.B.B., Dunstone, N.J., Halloran, P.R. et al. (2012). Aerosols implicated as a prime driver of twentieth-century North Atlantic climate variability. *Nature* 484: 228–232.

**6** Lamb, H.H. (1959). Our changing climate, past and present. *Weather* 14: 299–318. Reprinted in: Lamb, H.H. (ed) 1966 The Changing Climate, 236 pp, Methuen, London.

**7** Lamb, H.H. (1964). The role of the atmosphere and oceans in relation to climate changes and the growth of ice sheets on land. In: *1966, The Changing Climate* (ed. H.H. Lamb), 140–156. London: Methuen.

**8** Lamb, H.H. (1965). The Early Medieval Warm Epoch and its sequel. *Palaeogeography, Palaeoclimatology, Palaeoecology* 1: 13–37.

**9** Lamb, H.H. (1964). Britain's climate in the past, Unpublished lecture to Section X of the British Association for the Advancement of Science, Southampton; reprinted in. In: *The Changing Climate – Selected Papers* (ed. H.H. Lamb), 170–195. London: Methuen.

**10** Lamb, H.H. (1988). *Climate, History and the Modern World*, 433 pp. London: Routledge, (second edition 1995).

**11** Pfister, C. (1985), *CLIMHIST: A Weather Data Bank for Central Europe*, 125–186. Bern: Meteotest.

**12** Moreno-Chamarro, E., Zanchettin, D., Lohmann, K. et al. (2017). Winter amplification of the European Little Ice Age cooling by the subpolar gyre. *Scientific Reports* 7: 9981. https://doi.org/10.1038/s41598-017-07969-0.

**13** Folland, C.K., Karl, T.R., Vinnikov, K.Y. et al. (1990). Observed climate variations and change. In: *Report of Working Group 1, Climate Change*, the IPCC First Assessment Report (eds. J.T. Houghton, G.J. Jenkins and J.J. Ephraums), 195–238. Cambridge University Press.

**14** Lamb, H.H. (1988). Climate and life during the Middle Ages, studied especially in the mountains of Europe, in section 4 of. In: *Weather, Climate and Human Affairs*, 40–74. London: Routledge.

**15** Nicholls, N., Gruza, G.V., Jouzel, J. et al. (1996). Observed climate variability and change. In: *Report of Working Group 1*, IPCC Second Assessment Report (eds. J.T. Houghton, L.G. Meira-Filho, B.A. Callander, et al.), 133–192. Cambridge University Press.

**16** Bradley, R.S. and Jones, P.D. (1993). Little Ice Age summer temperature variations; their nature and relevance to recent global warming trends. *The Holocene* 3: 367–376.

**17** Bradley, R.S. and Jones, P.D. (1995). Recent developments in studies of climate since AD 1500. In: *Climate Since AD 1500*, 2e (eds. R.S. Bradley and P.D. Jones), 666–679. London: Routledge.

**18** Jones, P.D. and Bradley, R.S. (1992). Climatic variations over the last 500 years. In: *Climate Since AD 1500* (eds. R.S. Bradley and P.D. Jones), 649–665. London: Routledge.

**19** Verschuren, D. and Charman, D.J. (2009). Latitudinal linkages in late Holocene moisture-balance variation. In: *Natural Climate Variability and Global Warming: A Holocene Perspective* (eds. R.W. Battarbee and H.A. Binney), 189–231. Chichester and Oxford: Wiley Blackwell.

**20** De Menocal, P., Ortiz, J., Guilderson, T., and Sarnthein, M. (2000). Coherent high- and low- latitude climate variability during the Holocene warm period. *Science* 288: 2198–2202.

21 Crowley, T.J. and North, G.R. (2011). *Paleoclimatology, Oxford Monographs on Geology and Geophysics*, 339 pp, vol. 18. Oxford University Press.

22 Bond, G.C., Showers, W., Elliot, M. et al. (1999). The North Atlantic's 1-2kyr climate rhythm: relation to Heinrich events, Dansgaard/Oeschger cycles and the Little Ice Age. In: *Mechanisms of Global Climate Change at Millennial Time Scales, Geophys. Monogr.* **112** (eds. P.U. Clark, R.S. Webb and L.D. Keigwin), 35–58. American Geophysical Union.

23 Wanner, H., Beer, J., Bütikofer, J. et al. (2008). Mid-to late Holocene climate change; an overview. *Quaternary Science Reviews* 27: 1791–1828.

24 PAGES 2k Consortium (2013). Continental-scale temperature variability during the last two millennia. *Nature Geoscience* 6: 339–346.

25 Bard, E., Raisbeck, G., Yiou, F., and Jouzel, J. (2000). Solar irradiance during the last 1200 years based on cosmogenic nuclides. *Tellus* 52B: 985–992.

26 Esper, J., Cook, E.R., and Schweingruber, F.H. (2002). Low frequency signals in long tree-ring chronologies for reconstructing past temperature variability. *Science* 295: 2250–2252.

27 Esper, J., Frank, D.C., Wilson, R.S., and Briffa, K.R. (2005). Effect of scaling and regression on reconstructed temperature amplitude for the past millennium. *Geophysical Research Letters* 32: L07711.

28 Moberg, A., Sonechkin, D.M., Holmgren, K. et al. (2005). Highly variable Northern Hemisphere temperatures reconstructed from low- and high-resolution proxy data. *Nature* 433: 613–618.

29 Kirby, J. (2007). Cosmic rays and climate. *Surveys in Geophysics* 28: 333–375. https://doi.org/10.1007/s10712-008-9030-6.

30 Ljungqvist, F.C., Krusic, P.J., Brattström, G., and Sundqvist, H.S. (2012). Northern Hemisphere temperature patterns in the last 12 centuries. *Climate of the Past* 8: 227–249.

31 Mann, M.E., Bradley, R.S., and Hughes, M.K. (1998). Global-scale temperature patterns and climate forcing over the past six centuries. *Nature* 392: 779–787.

32 Hughes, M.K. and Diaz, H.F. (1994). Was there a 'Medieval Warm Period', and if so, where and when? *Climatic Change* 26: 109–142.

33 Mann, M.E., Bradley, R.S., and Hughes, M.K. (1999). Northern Hemisphere temperatures during the past millennium: inferences, uncertainties, and limitations. *Geophysical Research Letters* 26 (6): 759–762.

34 Folland, C.K., Karl, T.R., Christy, J.R. et al. (2001). Observed climate variability and change. In: *Climate Change 2001, the Scientific Basis*, Contribution of Working Group 1 to the IPCC Third Assessment Report (eds. J.T. Houghton, Y. Ding, D.J. Griggs, et al.), 101–181. Cambridge University Press.

35 North, G.R., Biondi, F., Bloomfield, P. et al. (2006). *Surface Temperature Reconstructions For The Last 2,000 Years*, 145 pp, NRC Report. Washington D.C.: National Academies Press http://www.nap.edu/catalog/11676.html.

36 Mann, M.E. and Jones, P.D. (2003). 2,000 year hemispheric multi-proxy temperature reconstructions, *IGBP PAGES/World Data Center for Paleoclimatology Data Contribution Series* **2003-051**, NOAA/NGDC Paleoclimatology Program, Boulder, CO.

37 Moberg, A., Sonechkin, D.M., Holmgren, K. et al. (2005). 2,000-year Northern Hemisphere temperature reconstruction, *IGBP PAGES/World Data Center for Paleoclimatology Data. Contribution Series* **2005-019**, NOAA/NGDC Paleoclimatology Program, Boulder, CO.

38 Jansen, E., Overpeck, J., Briffa, K.R. et al. (2007). Palaeoclimate. In: *Climate Change 2007, the Physical Science Base*, Contribution of Working Group I to the IPCC Fourth Assessment Report (eds. S. Solomon, D. Qin, M. Manning, et al.), 433–497.

39 D'Arrigo, R.S., Wilson, R., Liepert, B., and Cherubini, P. (2008). On the 'divergence problem' in northern forests: a review of the tree-ring evidence and possible causes. *Global and Planetary Change* 60: 289–305.

40 Porter, T.J. and Pisaric, M.F.J. (2012). Abstract, 3rd International Polar Year Conference (From Knowledge to Action), Montreal, Canada, April 22-27, 2012 (unpublished).

41 Reich, P.B., Sendall, K.M., Stefanski, A. et al. (2018). Effects of climate warming on photosynthesis in boreal tree species depend on soil moisture. *Nature* 562 (7726): 263–267.

42 Mann, M.E., Zhang, Z., Hughes, M.K. et al. (2008). Proxy-based reconstructions of hemispheric and global temperature variations over the past two millennia. *Proceedings of the National Academy Sciences of the United States of America* 105 (36): 13252–13257.

43 Mann, M.E., Zhang, Z., Rutherford, S. et al. (2009). Global signatures and dynamical origins of the Little Ice Age and Medieval Climate Anomaly. *Science* 326: 1256–1260.

44 Marcott, S.A., Shakun, J.D., Clark, P.U., and Mix, A.C. (2013). A reconstruction of regional and global temperature for the past 11,300 years. *Science* 339: 1198–1201.

45 Wanner, H., Mercolli, L., Grosjean, M., and Ritzl, S.P. (2014). Holocene climate variability and change – a data based review. *Journal of the Geological Society* 172 (2): 254–263.

46 Neukom, R., Steiger, N., Gómez-Navarro, J.J. et al. (2019). No evidence for globally coherent warm and cold periods over the preindustrial Common Era. *Nature* 571: 550–554.

47 Luterbacher, J., Werner, J.P., Smerdon, J.E. et al. (2016). European summer temperatures since Roman times.

*Environmental Research Letters* https://doi.org/10.1088/1748-9326/11/2/024001.

**48** Hendry, E.J., Gagan, M.K., Alibert, C.A. et al. (2002). Abrupt decrease in tropical Pacific sea surface salinity at end of Little Ice Age. *Science* 295: 1511–1514.

**49** McGann, M. (2008). High resolution foraminiferal, isotopic and trace element records from Holocene estuarine deposits of San Francisco Bay, California. *Journal of Coastal Research* 24 (5): 1092–1109.

**50** Zhao, M., Eglinton, G., Read, G.G., and Schimmelmann, A. (2000). An Alkenone (U$^{k'}_{37}$) quasi Annual sea surface temperature record (AD1440 to 1940) using varved sediments from the Santa Barbara Basin. *Organic Geochemistry* 31: 903–917.

**51** Hu, F.S., Ito, E., Brown, T.A. et al. (2001). Pronounced climatic variations in Alaska during the last two millennia. *Proceedings of the National Academy Sciences of the United States of America* 98 (19): 10552–10556.

**52** Loso, M.G. (2009). Summer temperatures during the Medieval Warm Period and Little Ice Age inferred from varved proglacial lake sediments in southern Alaska. *Journal of Paleolimnology* 41: 117–128.

**53** Büntgen, U., Myglan, V.S., Ljungqvist, F.C. et al. (2016). Cooling and societal change during the Late Antique Little Ice Age from 536 to around 660 AD. *Nature Geoscience* 9: 231–235.

**54** Werner, K., Spielhagen, R.F., Bauch, D. et al. (2011). Atlantic water advection to the eastern Fram Strait – multi-proxy evidence for Holocene variability. *Palaeogeography, Palaeoclimatology, Palaeoecology* 308: 264–276.

**55** Dokken, T., Nisancioglu, K.H., Li, C. et al. (2013). Dansgaard-Oeschger cycles: interactions between ocean and sea ice intrinsic to the Nordic Seas. *Paleoceanography* 28 (3): 491–502.

**56** Vinther, B.M., Jones, P.D., Briffa, K.R. et al. (2010). Climatic signals in multiple highly resolved stable isotope records from Greenland. *Quaternary Science Reviews* 29: 522–538.

**57** Miller, G.H., Geiersdottir, A., Zhong, Y. et al. (2012). Abrupt onset of the Little Ice Age triggered by volcanism and sustained by sea-ice/ocean feedbacks. *Geophysical Research Letters* https://doi.org/10.1029/2011GL050168.

**58** Mayewski, P.A. and Maasch, K.A. (2006). Recent warming inconsistent with natural association between temperature and atmospheric circulation over the last 2000 years. *Climate of the Past Discussions* 2: 1–29.

**59** Mayewski, P.A., Meredith, M.P., Summerhayes, C.P. et al. (2009). State of the Antarctic and Southern Ocean climate system. *Reviews of Geophysics* 47: RG1003. https://doi.org/10.1029/2007RG000231. 38 pp.

**60** Turner, J., Bindschadler, R.A., Convey, P. et al. (2009). *Antarctic Climate Change and the Environment*, 526 pp.

Cambridge: Scientific Committee on Antarctic Research, ISBN 978–0–948277-22-1: http://www.scar.org/publications/occasionals/acce.html.

**61** Ramanathan, V., Lian, M.S., and Cess, R.D. (1979). Increased atmospheric $CO_2$: zonal and seasonal estimates of the effect on the radiation energy balance and surface temperature. *Journal of Geophysical Research* 84 (C8): 4949–4958.

**62** Souney, J.M., Mayewski, P.A., Goodwin, I.D. et al. (2002). A 700-year record of atmospheric circulation developed froom the Law Dome ice core, East Antarctica. *Journal of Geophysical Research* 107 (D22): 4608. https://doi.org/10.1029/2002JD002104.

**63** Dahl-Jansen, D., Morgan, V.I., and Elcheikh, A. (1999). Monte Carlo inverse modeling of the Law Dome (Antarctica) temperature profile. *Annals of Glaciology* 29: 145–150.

**64** Stenni, B., Curran, M.A.J., Abram, N.J. et al. (2017). Antarctic climate variability on regional and continental scales over the last 2000 years. *Climate of the Past* 13: 1609–1634.

**65** Mayewski, P.A., Maasch, K.A., Dixon, D. et al. (2013). West Antarctica's sensitivity to natural and human forced climate change over the Holocene. *Journal of Quaternary Science* 18 (1): 40–48.

**66** Turner, J., Lu, H., White, I. et al. (2016). Absence of 21st century warming on Antarctic Peninsula consistent with natural variability. *Nature* 535: 411–415.

**67** Mulvaney, R., Abram, N.J., Hindmarsh, R.C.A. et al. (2012). Recent Antarctic Peninsula warming relative to Holocene climate and ice-shelf history. *Nature* 489: 141–145.

**68** Pike, J., Swann, G.E.A., Leng, M.J., and Snelling, A.M. (2013). Glacial eischarge along the West Antarctic Peninsula during the Holocene. *Nature Geoscience* 6: 199–202.

**69** Etourneau, J., Collins, L.G., Willmott, V. et al. (2013). Holocene climate variations in the western Antarctic Peninsula: evidence for sea ice extent predominantly controlled by changes in insolation and ENSO variability. *Climate of the Past* 9: 1431–1446.

**70** Graf, W., Oerter, H., Reinwarth, O. et al. (2002). Stable-isotope records from Dronning Maud Land, Antarctica. *Annals of Glaciology* 35: 195–201.

**71** Thomas, E.R., van Wessem, J.M., Roberts, J. et al. (2017). Regional Antarctic snow accumulation over the past 1000 years. *Climate of the Past* 13: 1491–1513.

**72** Bertler, N.A.N., Conway, H., Dahl-Jensen, D. et al. (2018). The Ross Sea Dipole – temperature, snow accumulation and sea ice variability in the Ross Sea region, Antarctica, over the past 2,700 Years. *Climate of the Past* 14: 193–214.

**73** Jones, J.M., Gille, S.T., Goosse, H. et al. (2016). Assessing recent trends in high-latitude Southern Hemisphere surface climate. *Nature Climate Change* 6: 917–926.

74 Swart, N.C., Gille, S.T., Fyfe, J.C., and Gillett, N.P. (2018). Recent Southern Ocean warming and freshening driven by greenhouse gas emissions and ozone depletion. *Nature Geoscience* 11 (11) https://doi.org/10.1038/s41561-018-0226-1.

75 Menezes, V.V., Macdonald,1, A.M., and Schatzman, C. (2017). Accelerated freshening of Antarctic Bottom Water over the last decade in the Southern Indian Ocean. *Science Advances* 3: e1601426.

76 Damon, P. and Peristykh, A.N. (2004). Solar and climatic implications of the centennial and millennial periodicities in atmospheric Δ14C variations. In: *Solar Variability and Its Effects on Climate, Geophysi. Monogr.* **114** (eds. J.M. Pap and P. Fox), 237–249. American Geophysical Union.

77 Cowie, J. (2013). *Climate Change: Biological and Human Aspects*, 558 pp, 2e. Cambridge University Press.

78 Lund, D.C., Lynch-Stieglitz, J., and Curry, W.B. (2006). Gulf Stream density structure and transport during the past millennium. *Nature* 444: 601–604.

79 Schlesinger, M.E. and Andronova, N.G. (2004). Has the sun changed climate? modeling the effect of solar variability on climate. In: *Solar Variability and its Effects on Climate, Geophysical Monograph* **114** (eds. J.M. Pap and P. Fox), 261–282. American Geophysical Union.

80 Rubino, M., Etheridge, D.M., Trudinger, C.M. et al. (2016). Low atmospheric $CO_2$ levels during the Little Ice Age due to cooling-induced terrestrial uptake. *Nature Geoscience* 9 (9): 691–694.

81 Koch, A., Brierley, C., Maslin, M.M., and Lewis, S.L. (2019). Earth system impacts of the European arrival and Great Dying in the Americas after 1492. *Quaternary Science Reviews* 207: 13–36.

82 Berger, A. and Loutre, M.-F. (2002). An exceptionally long interglacial ahead. *Science* 297: 1287–1288.

83 Beer, J. and Van Geel, B. (2009). Holocene climate change and the evidence for solar and other forcings. In: *Natural Climate Variability and Global Warming: A Holocene Perspective* (eds. R.W. Battarbee and H.A. Binney), 138–162. Chichester and Oxford: Wiley Blackwell.

84 Herschel, W. (1801). Observations tending to investigate the nature of the sun in order to find the causes or symptoms of its variable emission of light and heat. *Philosophical Transactions Royal Society of London* 91: 265–318.

85 Le Roy Ladurie, E. (1988). *Times of Feast, Times of Famine: a History of Climate Since the Year 1000*, 438 pp. Farrar Straus and Giroux.

86 Le Roy Ladurie, E. and Baulant, M. (1980). Grape Harvests from the fifteenth through the nineteenth centuries. *Journal of Interdisciplinary History* 10: 839–849.

87 Chuine, I., Yiou, P., Viovy, N. et al. (2004). Historical phenology: grape ripening as a past climate indicator. *Nature* 432 (7015): 289–290.

88 Le Roy Ladurie, E. (2005). *Comptes Rendus Biologies* 328 (3): 213–222.

89 Friis-Christensen, F. and Lassen, L. (1991). Length of the solar cycle: an indicator of solar activity closely associated with climate. *Science* 254: 698–700.

90 Eddy, J.A. (1977). Climate and the changing sun. *Climatic Change* 1: 173–190.

91 Eddy, J.A. (1976). The Maunder minimum. *Science* 192 (4245): 1189–1202.

92 Steinhilber, F., Abreu, J.A., Beer, J. et al. (2012). 9,400 years of cosmic radiation and.

93 Broecker, W. (2006). Radiocarbon. In: *The Atmosphere, Treatise on Geochemistry* (eds H.D. Holland and K.K. Turekian), vol. 4 (ed. R.F. Keeling), 245–260. Oxford: Elsevier-Pergamon.

94 Muscheler, R., Beer, J., and Kubik, P.W. (2004). Long-term solar variability and climate change based on radionuclide data from ice cores. In: *Solar Variability and Its Effects on Climate, Geophysical Monograph* **114** (eds. J.M. Pap and P. Fox), 221–235. American Geophysical Union.

95 Korte, M. and Muscheler, R. (2012). Centennial to millennial geomagnetic field variations. *Journal of Space Weather and Space Climate* 2: A08. https://doi.org/10.1051/swsc/2012006.

96 Stuiver, M. and Quay, P.D. (1980). Changes in atmospheric carbon-14 attributed to a variable sun. *Science* 207: 11–19.

97 Stuiver, M. and Quay, P.D. (1981). Atmospheric $^{14}$C changes resulting from fossil fuel $CO_2$ release and cosmic ray flux variability. *Earth and Planetary Science Letters* 53: 349–362.

98 Hua, Q. (2009). Radiocarbon: a chronological tool for the recent past. *Quaternary Geochronology* 4: 378–390.

99 Pedro, J.B., McConnell, J.R., and Von Ommen, T.D. (2012). Antarctica and Greenland during the neutron monitor era. *Earth and Planetary Science Letters* 355-356: 174–186.

100 Dansgaard, W., Johnsen, S.J., Clausen, H.B. et al. (1984). North Atlantic climatic oscillations revealed by deep Greenland ice cores. In: *Climate Processes and Climate Sensitivity, Geophysical Monograph* **29** (eds. J.E. Hansen and T. Takahashi), 288–298. Washington D.C.: American Geophysical Union.

101 Oeschger, H., Beer, J., Siegenthaler, U. et al. (1984). Late glacial climate history from ice cores. In: *Climate Processes and Climate Sensitivity, Geophysical Monograph* **29** (eds. J.E. Hansen and T. Takahashi), 299–306. Washington D.C.: American Geophysical Union.

102 Schüssler, M. and Schmitt, D. (2004). Theoretical models of solar magnetic variability. In: *Solar Variability and its Effects on Climate, Geophysical Monograph* **114** (eds. J.M. Pap and P. Fox), 33–49. American Geophysical Union.

103 Stuiver, M., Reimer, P.J., Bard, E. et al. (1998). Intcal 98 radiocarbon age calibration, 24000–0 cal BP. *Radiocarbon* 40 (3): 1041–1083.

104 Muscheler, R., Snowball, I., Jos, F. et al. (2007). Reply to the comment by Bard et al. on "Solar activity during the last 1000 yr inferred from radionuclide records". *Quaternary Science Reviews* 26: 2301–2308.

105 Hempelmann, A. and Weber, W. (2012). Correlation between the sunspot number, the total solarirradiance, and the terrestrial insolation. *Solar Physics* 277: 417–430.

106 Hoyt, D.V. and Schatten, K.H. (1998). Group sunspot numbers: a new solar activity reconstruction. *Solar Physics* 179: 189–219. (Reprinted with figures in Solar Phys. 181, 491, 1998).

107 Clette, F., Svalgaard, L., Vaquero, J.M., and Cliver, E.W. (2014). Revisiting the Sunspot number. *Space Science Review* https://doi.org/10.1007/s11214-014-0074-2.

108 Hoyt, D.V. and Schatten, K.H. (1993). A discussion of plausible solar irradiance variations, 1700-1992. *Journal of Geophysical Research* 98 (A11): 18895–18906.

109 Cliver, E.W. (2017). Sunspot number recalibration: the ~1840–1920 anomaly in the observer normalization factors of the group sunspot number. *Journal of Space Weather and Space Climate* 7 (A12) https://doi.org/10.1051/swsc/2017010.

110 Magny, M. (1993). Solar influences on Holocene climatic changes illustrated by correlations between past lake level fluctuations and the atmospheric $^{14}$C record. *Quaternary Research* 40: 1–9.

111 Magny, M. (1999). Lake-level fluctuations in the Jura and French Subalpine Ranges associated with ice-rafting events in the North Atlantic and variations in the polar atmospheric circulation. *Quaternaire* 10: 61–64.

112 Magny, M. (2004). Holocene climatic variability as reflected by mid-European lake-level fluctuations, and its probable impact on prehistoric human settlements. *Quaternary International* 113: 65–79.

113 Magny, M. (2007). West-Central Europe In: *Lake Level Studies, Encyclopedia of Quaternary Science* (ed. S. Elias), 1389–1399. Elsevier.

114 Magny, M., De Beaulieu, J.-L., Drescher-Schneider, R. et al. (2007). Holocene climate changes in the central Mediterranean as recorded by lake-level fluctuations at Lake Accesa (Tuscany, Italy). *Quaternary Science Reviews* 26 (13–14): 1736–1758.

115 Schmidt, G.A., Jungclaus, J.H., Amman, C.M. et al. (2011). Climate forcing reconstructions for use in PMIP simulations of the last millennium (v.1.0). *Geoscientific Model Development* 4: 33–45.

116 Kouwenberg, L., Wagner, R., Kurschner, W., and Visscher, H. (2005). Atmospheric CO2 fluctuations during the last millennium reconstructed by stomatal frequency analysis of *Tsuga heterophylla* needles. *Geology* 33 (1): 33–36.

117 Ahn, J., Brook, E.J., Mitchell, L. et al. (2012). Atmospheric $CO_2$ over the last 1000 years: a high-resolution record from the West Antarctic Ice Sheet (WAIS) divide ice core. *Global Biogeochemical Cycles* 26 https://doi.org/10.1029/2011GB004247.

118 Van Hoof, T.B., Wagner-Cremer, F., Kürschner, W.M., and Visscher, H. (2008). A role for atmospheric $CO_2$ in preindustrial climate forcing. *Proceedings of the National Academy of Sciences of the United States of America* 105: 15815–15818.

119 Kobashi, T., Shindell, D.T., Kodera, K. et al. (2013). On the origin of multidecadal to centennial Greenland temperature anomalies over the past 800 yr. *Climate of the Past* 9: 583–596.

120 Moffa-Sánchez, P., Moreno-Chamarro, E., Reynolds, D.J. et al. (2019). Variability in the northern North Atlantic and Arctic Oceans across the last two millennia: a review. *Paleoceanography and Paleoclimatology* 34. https://doi.org doi: 10.1029/2018PA003508.

121 Bond, G.C., Kromer, B., Beer, J. et al. (2001). Persistent solar influence on North Atlantic climate during the Holocene. *Science* 294: 2130–2152.

122 Braun, H., Christl, M., Rahmstorf, S. et al. (2005). Possible solar origin of the 1,470-year glacial climate cycle demonstrated in a coupled model. *Nature* 438: 208–211.

123 Debret, M., Bout-Roumazeilles, V., Grousset, F. et al. (2007). The origin of the 1500-year climate cycles in Holocene North Atlantic records. *Climate of the Past* 3: 569–575.

124 Moffa-Sanchez, P., Born, A., Hall, I.R. et al. (2014). Solar forcing of North Atlantic surface temperature and salinity over the past millennium. *Nature Geoscience* 7: 275–278.

125 Lamb, H. (1979). Climatic variation and changes in the wind and ocean circulation: The Little Ice Age in the Northeast Atlantic. *Quaternary Research* 11 (1).

126 Maasch, K.A., Mayewski, P.A., Rohling, E.J. et al. (2005). A 2000-year context for modern climate change. *Geografiska Annaler* 87A: 7–15.

127 Wanner, H., Solomina, O., Grosjean, M. et al. (2011). Structure and origin of Holocene cold events. *Quaternary Science Reviews* 30: 3109–3123.

128 Mayewski, P.A., Rohling, E., Stager, J.C. et al. (2004). Holocene climatic variability. *Quaternary Research* 62: 243–255.

**129** North, G.R., Baker, D.N., Bradley, R.S. et al. (2012). *The Effects of Solar Variability on Earth's Climate*, 58 pp, National Research Council Report. Washington D.C.: National Academies Press.

**130** Cronin, T.M. (2010). *Paleoclimates – Understanding Climate Change Past and Present*. Columbia University Press.

**131** Lüdecke, H.-J. and Weiss, C.-O. (2017). Harmonic analysis of worldwide temperature proxies for 2000 years. *The Open Atmospheric Science Journal* 11: 44–53.

**132** Babich, V.V., Darin, A.V., Kalugin, I.A., and Snolyaninova, L.G. (2016). Climate prediction for the extratropical northern hemisphere for the next 500 years based on periodic natural processes. *Russian Meteorology and Hydrology* 41 (9): 593–600. http://dx.doi.org/10.3103/S1068373916090016.

**133** Steinhilber, F. and Beer, J. (2013). Prediction of solar activity for the next 500 years. *Journal of Geophysical Research: Space Physics* 118: 1861–1867. http://dx.doi.org/10.1002/jgra.50210.

**134** Lüdecke, H.-J., Hempelmann, A., and Weiss, C.-O. (2013). Multi-periodic climate dynamics: spectral analysis of long-term instrumental and proxy temperature records. *Climate of the Past* 9: 447–452.

**135** Svensmark, H. and Friis-Christensen, E. (1997). Variation of cosmic ray flux and global cloud coverage – a missing link in solar-climate relationships. *Journal of Atmospheric and Terrestrial Physics* 59 (11): 1225–1232.

**136** Svensmark, H. (2015). Cosmic rays, clouds and climate. *Europhysics News* 46 (2): 26–29. https://doi.org/10.1051/epn/2015204.

**137** Svensmark, H., Enghoff, M.B., Shaviv, N.J., and Svensmark, J. (2017). Increased ionization supports growth of aerosols into cloud condensation nuclei. *Nature Communications* 8 https://doi.org/10.1038/s41467-017-02082-2.

**138** North Greenland Ice Core Project Members (2004). High-resolution record of Northern Hemisphere climate extending into the last interglacial period. *Nature* 431: 147–151.

**139** Lockwood, M. (2012). Solar influence on global and regional climates. *Surveys in Geophysics* 33 (3–4): 505–534.

**140** Gray, L.J., Beer, J., Geller, M. et al. (2010). Solar influences on climate. *Reviews of Geophysics* 48: RG4001.

**141** Sun, B. and Bradley, R.S. (2004). Reply to comment by N. D. Marsh and H. Svensmark on "Solar influences on cosmic rays and cloud formation: a reassessment". *Journal of Geophysical Research: Atmospheres* 109 (D14) https://doi.org/10.1029/2003JD004479.

**142** Ueno, Y., Hyodo, M., Yang, T., and Katoh, S. (2019). Intensified East Asian winter monsoon during the last geomagnetic reversal transition. *Scientific Reports* https://doi.org/10.1038/s41598-019-45466-8.

**143** Schurer, A.P., Tett, S.F.B., and Hegerle, G.C. (2014). Small influence of solar variability on climate over the past millennium. *Nature Geoscience* 7: 104–108. https://doi.org/10.1038/NGEO2040.

**144** D'Arrigo, R., Wilson, R., and Tudhope, A. (2008). The impact of volcanic forcing on tropicaltemperatures during the past four centuries. *Nature Geoscience* 2: 51–56.

**145** Lavigne, F., Degeal, J.-P., Komorowski, J.-C. et al. (2013). Source of the great A.D. 1257 mystery eruption unveiled, Samalas volcano, Rinjani Volcanic Complex, Indonesia. *Proceedings of the National Academy of Sciences of the United States of America*, Published online before print September 30, 2013, doi:https://doi.org/10.1073/pnas.1307520110.

**146** Sigl, M., Winstrup, M., McConnell, J.R. et al. (2015). Timing and climate forcing of volcaniceruptions for the past 2,500 years. *Nature* 523: 543–549.

**147** Scapozza, C., Lambiel, C., Reynard, E. et al. (2010). Radiocarbon Dating of fossil wood remains buried by the Piancabella Rock Glacier, Blenio Valley (Ticino, Southern Swiss Alps): implications for Rock Glacier, treeline and climate history. *Permafrost and Periglacial Processes* 21 (1): 90–96.

**148** Sigl, M., Abram, N.J., Gabrieli, J. et al. (2018). 19th century glacier retreat in the Alps preceded the emergence of industrial black carbon deposition on high-alpine glaciers. *The Cryosphere* 12: 3311–3331.

**149** Joerin, U.E., Stocker, T.F., and Schlüchter, C. (2006). Multicentury glacier fluctuations in the Swiss Alps during the Holocene. *The Holocene* 16 (5): 697–704.

**150** Hormes, A., Müller, B.U., and Schlüchter, C. (2001). The Alps with little ice: evidence for eight Holocene phases of reduced glacier extent in the Central Swiss Alps. *The Holocene* 11 (3): 255–265.

**151** Jevrejeva, S., Moore, J.C., Grinsted, A., and Woodworth, P.L. (2008). Recent global sea level acceleration started over 200 years ago. *Geophysical Research Letters* 35: L08715. https://doi.org/10.1029/2008GL033611.

**152** Jevrejeva, S., Moor, J.C., Grinsted, A. et al. (2014). Trends and acceleration in global and regional sea levels since 1807. *Global and Planetary Change* 113: 11–22.

**153** Dangendorf, S., Marcos, M., Woppelmann, G. et al. (2017). Reassessment of 20th century global mean sea level rise. *PNAS* 114 (23): 5946–5951.

**154** Leclercq, P.W., Oerlemans, J., and Cogley, J.G. (2011). Estimating the glacier contribution to sea-level rise for the period 1800–2005. *Surveys in Geophysics* 32: 519–535.

155 Pugh, D. and Woodworth, P. (2014). *Sea-Level Science: Understanding Tides, Surges, Tsunamis and Mean Sea-Level Changes*, 395 pp. Cambridge: Cambridge University Press.

156 CHU Church, J.A., Clark, P.U., Cazenave, A. et al. (2013). Sea level change. In: *Climate Change 2013: The Physical Science Basis, Contribution of Working Group I to the Fifth Assessment Report of the Intergovernmental Panel on Climate Change* (eds. T.F. Stocker, D. Qin, G.-K. Plattner, et al.), 1137–1216. Cambridge and New York: Cambridge University Press.

157 Church, J.A., Roemmich, D., Domingues, C.M. et al. (2010). Ocean temperature and salinity contributions to global and regional sea-level change. In: *Understanding Sea-Level Rise and Variability* (eds. J.A. Church, P.L. Woodworth, T. Aarup and W.S. Wilson), 143–176. Chichester and Oxford: Wiley Blackwell.

158 Steffen, K., Thomas, R.H., Rignot, E. et al. (2010). Cryospheric Contributions to Sea-Level Rise and Variability. In: *Understanding Sea-Level Rise and Variability* (eds. J.A. Church, P.L. Woodworth, T. Aarup and W.S. Wilson), 177–225. Chichester and Oxford: Wiley Blackwell.

159 Church, J.A., Aarup, T., Woodworth, P.L. et al. (2010). Sea-level rise and variability: synthesis and outlook for the future. In: *Understanding Sea-Level Rise and Variability* (eds. J.A. Church, P.L. Woodworth, T. Aarup and W.S. Wilson), 402–419. Chichester and Oxford: Wiley Blackwell.

160 Church, J.A., White, N.J., Aarup, T. et al. (2008). Understanding global sea levels: past, present and future. *Sustainability Science* 3 (1) https://doi.org/10.1007/s11625-008-0042-4.

161 Gehrels, R. (2010). Sea-level changes since the Last Glacial Maximum: an appraisal of the IPCC Fourth Assessment Report. *Journal of Quaternary Science* 25 (1): 26–38.

162 Lambeck, K., Woodroffe, C.D., Antonioli, F. et al. (2010). Paleoenvironmental records, geophysical modeling, and reconstruction of sea-level; trends and variability on centennial and longer timescales. In: *Understanding Sea-Level Rise and Variability* (eds. J.A. Church, P.L. Woodworth, T. Aarup and W.S. Wilson), 61–121. Chichester and Oxford: Wiley Blackwell.

163 Woodworth, P.L., Gehrels, W.R., and Nerem, R.S. (2011). Nineteenth and twentieth century changes in sea level. *Oceanography* 24 (2): 80–93.

164 Kemp, A.C., Horton, B.J., Donnelly, J.P. et al. (2011). Climate-related sea-level variations over the past two millennia. *Proceedings of the National Academy of the United States of America* 108 (27): 11017–11022.

165 Grinsted, A., Moore, J.C., and Jevrejeva, S. (2010). Reconstructing sea level from paleo and projected temperatures 200 to 2100 AD. *Climate Dynamics* 34 (4): 461–472. https://doi.org/10.1007/s00382-008-0507-2.

166 Jones, P.D. and Mann, M.E. (2004). Climate over past millennia. *Reviews of Geophysics* 42: RG2002. https://doi.org/10.1029/2003RG000143.

167 Gregory, J.M., Lowe, J.A., and Tett, S.F.B. (2006). Global-mean sea level changes over the last half-millennium. *Journal of Climate* 19: 4576–4591.

168 Leclercq, P.W. and Oerlemans, J. (2012). Global and hemispheric temperature reconstruction from glacier length fluctuations. *Climate Dynamics* 38: 1065–1079.

169 Fairbanks, R.G. (1989). A 17,000-year glacio-eustatic sea level record: influence of glacial melting rates on the Younger Dryas event and deep-ocean circulation. *Nature* 342: 637–642.

170 Rohling, E.J., Grant, K., Bolshaw, M. et al. (2009). Antarctic temperature and global sea level closely coupled over the past five glacial cycles. *Nature Geoscience* 2: 500–503.

171 Rohling, E.J., Grant, K., Hemleben, C. et al. (2008). High rates of sea-level rise during the last interglacial period. *Nature Geoscience* 1: 38–42.

172 Foster, G.L. and Rohling, E.J. (2013). Relationship between sea level and climate forcing by $CO_2$ on geological timescales. *Proceedings of the National Academy Sciences of the United States of America* 110 (4): 1209–1214.

173 Rahmstorf, S. (2007). A semi-empirical approach to projecting future sea-level rise. *Science* 315: 368–370.

174 Vermeer, M. and Rahmstorf, S. (2009). Global Sea level linked to global temperature. *Proceedings of the National Academy of Sciencesof the United States of America* 106: 21527–21532.

175 Hansen, J.E. and Sato, M. (2012). Paleoclimate implications for human-made climate change. In: *Climate Change: Inferences from Paleoclimate and Regional Aspects* (eds. A. Berger, F. Mesinger and D. Šijački), 21–48. Dordrecht: Springer https://doi.org/10.1007/978-3-7091-0973-1_2.

176 Pfeffer, W.T., Harper, J.T., and O'Neel, S. (2008). Kinematic constraints on glacier contributions to 21st century sea level rise. *Science* 321: 1340–1343.

177 Jacobs, S.S., Jenkins, A., Giulivi, C., and Dutrieux, P. (2011). Stronger ocean circulation and increased melting under Pine Island Glacier ice shelf. *Nature Geoscience* https://doi.org/10.1038/ngeo1188.

178 Hillenbrand, C.-D., Smith, J.A., Hodell, D.A. et al. (2017). West Antarctic Ice Sheet retreat driven by Holocene warm water incursions. *Nature* 547: 43–48.

179 Rignot, E., Mouginot, J., Scheuchl, B. et al. (2019). Four decades of Antarctic Ice Sheet mass balance from 1979–2017. *Proceedings of the National Academy of*

*Sciences of the United States of America* https://doi.org/10.1073/pnas.1812883116.

**180** Fisher, A.T., Mankoff, K.D., Tulaczyk, S.M. et al. (2015). High geothermal heat flux measured below the West Antarctic Ice Sheet. *Science Advances* 1 (6): e1500093.

**181** Schroeder, D.M., Blankenship, D.D., Young, D.A., and Quartini, E. (2014). Evidence for elevated and spatially variable geothermal flux beneath the West Antarctic Ice Sheet. *Proceedings of the National Academy of Sciences of the United States of America* 111 (25): 9070–9072.

**182** Paolo, F.S., Fricker, H.A., and Padman, L. (2015). Volume loss from Antarctic ice shelves is accelerating. *Science* 348 (6232): 327–331.

**183** Mengel, M. and Levermann, A. (2014). Ice plug prevents irreversible discharge from East Antarctica. *Nature Climate Change* online 4 May 2014. doi:https://doi.org/10.1038/nclimate2226.

**184** Cook, A.J., van der Flierdt, T., Williams, T. et al. (2013). Dynamic behaviour of the East Antarctic ice sheet during Pliocene warmth. *Nature Geoscience* 6: 765–769.

**185** Wilson, D.J., Bertram, R.A., Needham, E.F. et al. (2018). Ice loss from the East Antarctic Ice Sheet during late Pleistocene interglacials. *Nature* 561: 383–386.

**186** Greenbaum, J.S., Blankenship, D.D., Young, D.A. et al. (2015). Ocean access to a cavity beneath Totten Glacier in East Antarctica. *Nature Geoscience* 8: 294–298.

**187** Aitken, A.R.A., Roberts, J.L., van Ommen, T.D. et al. (2016). Repeated large-scale retreat and advance of Totten Glacier indicated by inland bed erosion. *Nature* 533: 385–389.

**188** Rintoul, S.R., Silvano, A., Pena-Molino, B. et al. (2016). Ocean heat drives rapid basal melt of the Totten Ice Shelf. *Science Advances* 2: e1601610.

**189** Qiu, J. (2017). The threat beneath Antarctica. *Nature* 544: 152–154.

**190** Bakker, P., Clark, P., Golledge, N.R. et al. (2017). Centennial-scale Holocene climate variations amplified by Antarctic Ice Sheet discharge. *Nature* 541: 72–76.

**191** Cook, A.J. and Vaughan, D.G. (2010). Overview of areal changes of the ice shelves on the Antarctic Peninsula over the past 50 years. *The Cryosphere* 4 (10): 77–98.

**192** Shakun, J.D., Corbett, L.B., Bierman, P.R. et al. (2018). Minimal East Antarctic Ice Sheet retreat onto land during the past eight million years. *Nature* 558: 284–287.

**193** Cook, A.J., Holland, P.R., Meredith, M.P. et al. (2016). Ocean forcing of glacier retreat in the western Antarctic Peninsula. *Science* 353 (6296): 283–286.

**194** Richter-Menge, J., Overland, J. E., and Mathis, J. T. (eds.) (2016). *Arctic Report Card 2016*. http://arctic.noaa.gov/Report-Card.

**195** Hanna, E., Navarro, F.J., Pattyn, F. et al. (2013). Ice sheet mass balance and climate change. *Nature* 498: 51–59.

**196** Velicogna, I., Sutterley, T.S., and van den Broeke, M.R. (2014). Regional acceleration in ice mass loss from Greenland and Antarctica using GRACE time-variable gravity data. *Journal of Geophysical Research Space Physics* 41: 8130–8137.

**197** Khan, S.A., Sasgen, I., Bevis, M. et al. (2016). Geodetic measurements reveal similarities between post–Last Glacial Maximum and present-day mass loss from the Greenland ice sheet. *Science Advances* 2: e1600931.

**198** Mouginot, J., Rignot, E., Bjork, A.A. et al. (2019). Forty-six years of Greenland Ice Sheet mass balance from 1972 to 2018. *Proceedings of the National Academy of Sciences of the United States of America* 116 (19): 9239–9244.

**199** Gardner, A.S., Moholdt, G., Wouters, B. et al. (2011). Sharply increased mass loss from glaciers and ice caps in the Canadian Arctic Archipelago. *Nature* 473: 357–360.

**200** Kopp, R.E., Simons, F.J., Mitrovica, J.X. et al. (2009). Probabilistic assessment of sea level during the last interglacial stage. *Nature* 462: 863–868.

**201** Kopp, R.E., Kemp, A.C., Bittermann, K. et al. (2016). Temperature-driven global sea-level variability in the Common Era. *Proceedings of the National Academy of Sciences of the United States of America* 113 (11): E1434–E1441.

**202** Stocker, T., Dahe, Q., Plattner, G.-K. et al. (eds) (2010). *Workshop on Sea Level Rise and Ice Sheet Instability*. Report of IPCC Workshop, Kuala Lumpur, Malaysia, 21–24 June 2010, IPCC, WG1 Technical Support Unit, University of Bern, 227 pp.

**203** Church, J.A., Gregory, J.M., White, N.J. et al. (2011). Understanding and projecting sea level change. *Oceanography* 24 (2): 130–143. https://doi.org/10.5670/oceanog.2011.33.

**204** Lockwood, M. (2010). Solar change and climate: an update in the light of the current exceptional solar minimum. *Proceeding of the Royal Society A Mathematical, Physical and Engineering Sciences* 466: 303–329.

**205** Hansen, J., Sato, M., and Ruedy, R. (2013). Global temperature update through 2012. http://www.nasa.gov/pdf/719139main_2012_GISTEMP_summary.pdf.

**206** Bard, E. and Delaygue, G. (2008). Comment on "Are there connections between the Earth's magnetic field and climate?" by V. Courtillot, Y. Gallet, J.-L. Le Mouël, F. Fluteau, A. Genevey EPSL 253, 328, 2007. *Earth and Planetary Science Letters* 265: 302–307.

**207** Lockwood, M. and Fröhlich, C. (2007). Recent oppositely directed trends in solar climate forcings and the global mean surface air temperature. *Proceedings of the Royal Society A* https://doi.org/10.1098/rspa.2007.1880.

**208** Trenberth, K.E. and Fasullo, J.T. (2013). An apparent hiatus in global warming. *Earth's Future* https://doi.org/10.1002/2013EF000165, 14 pp.

**209** Budyko, M.I. (1977). On present-day climatic changes. *Tellus* 29: 193–204.

**210** Smith, S.K., Van Aardenne, J., Klimont, Z. et al. (2011). Anthropogenic sulfur dioxide emissions:1850-2005. *Atmospheric Chemistry and Physics* 11: 1101–1116.

**211** Amman, C.M., Joos, F., Schimel, D.S. et al. (2007). Solar influence on climate during the past millennium: results from transient simulations with the NCAR climate system model. *Proceedings of the National Academy Sciences of the United States of America* 104 (10): 3713–3718.

**212** Goosse, H., Mann, M.E., and Renssen, H. (2009). Climate of the past millennium: combining proxy data and model simulations. In: *Natural Climate Variability and Global Warming: A Holocene Perspective* (eds. R.W. Battarbee and H.A. Binney), 163–188. Chichester and Oxford: Wiley Blackwell.

**213** Miller, G.H., Alley, R.B., Brigham-Grette, J. et al. (2010). Arctic amplification: can the past constrain the future? *Quaternary Science Reviews*: 1–12. https://doi.org/10.1016/j.quascirev.2010.02.008.

**214** Summerhayes, C.P. (2019). Ice. In: *The Anthropocene as a geological time unit* (eds. J. Zalasiewicz, C. Waters, M. Williams and C.P. Summerhayes), 218–233. Cambridge: Cambridge University Press.

**215** Gilbert, A., Flowers, G.E., Miller, G.H. et al. (2017). Projected demise of Barnes Ice Cap: evidence of an unusually warm 21st century Arctic. *Geophysical Research Letters* 44 (6): 2810–2816.

**216** Pendleton, S.L., Miller, G.H., Lifton, N. et al. (2019). Rapidly receding Arctic Canada glaciers revealing landscapes continuously ice-covered for more than 40,000 years. *Nature Communications* 10: 445. http://doi.org/10.1038/s41467-019-08307-w.

**217** Polyakov, I.V., Bekryaev, R.V., Alekseev, G.V. et al. (2003). Variability and trends of air temperature and pressure in the maritime Arctic, 1875–2000. *American Meteorological Society* 16: 2067–2077.

**218** Frolov, I.E., Gudkovich, Z.M., Karklin, V.P. et al. (2009). *Climate Change In Eurasian Arctic Seas: Centennial Ice Cover Observations*, 160 pp. Praxis, Chichester: Springer-Praxis Books in Geophysical Sciences.

**219** Wood, K.R. and Overland, J.E. (2010). Early 20th century Arctic warming in retrospect. *International Journal of Climatology* 30: 1269–1279.

**220** Overland, J.E., Wood, K.R., and Wang, M. (2011). Warm Arctic - cold continents: climate impacts of the newly open Arctic Sea. *Polar Research* 30: 15787. https://doi.org/10.3402/polar.v30i0.

**221** Polyakov, I.V., Walsh, J.E., and Kwok, R. (2012). Recent changes of Arctic multiyear sea ice coverage and the likely causes. *American Meteorological Society*. In Box Insights and Innovations 93: 145–151.

**222** Acosta Navarro, J.C., Varma, V., Riipinen, I. et al. (2016). Amplification of Arctic warming by past air pollution reductions in Europe. *Nature Geoscience* 9: 277–281.

**223** Lassen, K. and Thejll, P. (2005). *Multi-decadal variation of the East Greenland Sea-Ice Extent: AD 1500–2000*, 13 pp, Scientific Report 05–02. Danish Meteorological Institute.

**224** Kinnard, C., Zdanowicz, C.M., Koerner, R.M., and Fisher, D.A. (2008). A changing Arctic seasonal ice zone: observations from 1870–2003 and possible oceanographic consequences. *Geophysical Research Letters* 35: L02507. https://doi.org/10.1029/2007GL032507.

**225** ACIA (2005). *Arctic Climate Impact Assessment*, 1042 pp. Cambridge: Cambridge University Press.

**226** Wadhams, P. (2016). *A Farewell to Ice: A Report from the Arctic*, 240 pp. Penguin Random House, UK: Allen Lane.

**227** Walsh, J.E. (2009). A comparison of Arctic and Antarctic climate change, present and future. *Antarctic Science* 21 (3): 179–188.

**228** Ding, Q., Schweiger, A., L'Heureux, M. et al. (2017). Influence of high-latitude atmospheric circulation changes on summertime Arctic sea ice. *Nature Climate Change* 7: 289–295.

**229** Polyakov, I.V., Pnyushkov, A.V., Alkire, M.B. et al. (2017). Greater role for Atlantic inflows on sea-ice loss in the Eurasian Basin of the Arctic Ocean. *Science* 356: 285–291.

**230** Gagné, M.E., Fyfe, J.C., Gillett, N.P. et al. (2017). Aerosol-driven increase in Arctic sea ice over the middle of the 20th century. *Geophysical Research Letters* https://doi.org/10.1002/2016GL071941.

**231** Kinnard, C., Zdanowicz, C.M., Fisher, D.A. et al. (2011). Reconstructed changes in Arctic sea ice over the past 1,450 years. *Nature* 479: 509–512.

**232** Tingley, M.P. and Huybers, P. (2013). Recent temperature extremes at high northern latitudes unprecedented in the past 600 years. *Nature* 496: 201–205.

**233** Kauffman, D., Schneider, D.P., McKay, N.P. et al. (2009). Recent warming reverses long-term Arctic cooling. *Science* 325: 1236–1239.

**234** Walsh, J.E. (2013). Melting ice – what is happening to Arctic Sea ice, and what does it mean for us? *Oceanography* 26 (2): 171–181.

**235** Bjune, A.E., Seppä, H., and Birks, H.J.B. (2009). Quantitative summer-temperature reconstructions for the last 2000 years based on pollen-stratigraphical data from northern Fennoscandia. *Journal of Paleolimnology* 41: 43–56.

236 Orsi, A.J., Kawamura, K., Masson-Delmotte, V. et al. (2017). The recent warming trend in North Greenland. *Geophysical Research Letters* https://doi.org/10.1002/2016GL072212.

237 Masson-Delmotte, V., Swingedouw, D., Landais, A. et al. (2012). Greenland climate change: from the past to the future. *Wiley Interdisciplinary Reviews. Climate Change* https://doi.org/10.1002/wcc.186.

238 Vinther, B.M., Buchardt, S.L., Clausen, H.B. et al. (2009). Holocene thinning of the Greenland ice sheet. *Nature* 461: 385–388.

239 Miller, G.H., Lehman, S.J., Refsnider, K.A. et al. (2013). Unprecedented recent summer warmth in Arctic Canada. *Geophysical Research Letters* 40: 5745–5751.

240 Esper, J., Frank, D.C., Timonen, M. et al. (2012). Orbital forcing of tree-ring data. *Nature Climate Change* 2: 862–866.

241 Thompson, L.G., Mosley-Thompson, E., Brecher, H. et al. (2006). Abrupt tropical climate change: past and present. *Proceedings of the National Academy Sciences of the United States of America* 103: 10536–10543.

242 Zemp, M., Frey, H., Gärtner-Roer, I. et al. (2015). Historically unprecedented global glacier decline in the early 21st century. *Journal of Glaciology* 61 (228): 745–762. https://doi.org/10.3189/2015JoG15J017.

243 Zemp, M., Huss, M., Thibert, E. et al. (2019). Global glacier mass changesand their contributions to sealevel rise from 1961 to 2016. *Nature* 568: 382–386.

244 Maurer, J.M., Schaefer, J.M., Rupper, S., and Corley, A. (2019). Acceleration of ice loss across the Himalayas over the last 40 years. *Science Advances* 5 (6) https://doi.org/10.1126/sciadv.aav7266.

245 Masiokas, M.H., Rivera, A., Espizua, L.E. et al. (2009). Glacier fluctuations in extratropical South America during the past 1000 years. *Palaeogeography, Palaeoclimatology, Palaeoecology* 281: 242–268.

246 Glasser, N.F., Harrison, S., Jansson, K.N. et al. (2011). Global sea-level contribution from the Patagonian Icefields since the Little Ice Age maximum. *Nature Geoscience* 4.

247 Mernild, S.H., Lipscomb, W.H., Bahr, D.B. et al. (2013). Global glacier retreat: a revised assessment of committed mass losses and sampling uncertainties. *The Cryosphere Discussions* 7: 1987–2005.

248 Kaser, G., Cogley, J.G., Dyurgerov, M.B. et al. (2006). Mass balance of glaciers and ice caps: consensus estimates for 1961–2004. *Geophysical Research Letters* 33: L19501. https://doi.org/10.1029/2006GL027511.

249 Mackintosh, A.N., Anderson, B.M., Lorrey, A.M. et al. (2017). Regional cooling caused recent New Zealand glacier advances in a period of global warming. *Nature Communications* 8: 14202. https://doi.org/10.1038/ncomms14202.

250 Sokolov, S. and Rintoul, S.R. (2009). Circumpolar structure and distribution of the Antarctic Circumpolar Current fronts: 2. Variability and relationship to sea surface height. *Journal of Geophysical Research* 114: C11019. https://doi.org/10.1029/2008JC005248.

251 Spence, P., Griffies, S.M., England, M.H. et al. (2014). Rapid subsurface warming and circulation changes of Antarctic coastal waters by poleward shifting winds. *Geophysical Research Letters* 41: 4601–4610. https://doi.org/10.1002/2014GL060613.

252 Büntgen, U., Tegel, W., Nicolussi, K. et al. (2011). 2500 years of European climate variability and human susceptibility. *Science* 331: 578–582.

253 Büntgen, U., Kyncl, T., Ginzler, C. et al. (2013). Filling the eastern European gap in millennium-long temperature reconstructions. *Proceedings of the National Academy of Science of the United States of America* 110 (5): 1773–1778.

254 Brohan, P., Kennedy, J.J., Harris, I. et al. (2006). Uncertainty estimates in regional and global observed temperature changes: a new dataset from 1850. *Journal of Geophysical Research* 111 (D12): 106. https://doi.org/10.1029/2005jd006548.

255 Turner, J., Comiso, J.C., Marshall, G.J. et al. (2009). Non-annular atmospheric circulation change induced by stratospheric ozone depletion and its role in the recent increase of Antarctic sea ice extent. *Geophysical Research Letters* 36: L08502. https://doi.org/10.1029/GL037524.

256 Bintanja, R., van Oldenborgh, G.J., Drijfhout, S.S. et al. (2013). Important role for ocean warming and increased ice-shelf melt in Antarctic sea-ice expansion. *Nature Geoscience* 6: 376–379.

257 Holland, P.R. and Kwok, R. (2012). Wind-driven trends in Antarctic sea-ice drift. *Nature Geoscience* 5: 872–875.

258 Meehl, G.A., Arblaster, J.M., Bitz, C.M. et al. (2016). Antarctic sea-ice expansion between 2000 and 2014 driven by tropical Pacific decadal climate variability. *Nature Geoscience* 9: 590–595.

259 Haumann, F.A., Gruber, N., Münnich, M. et al. (2016). Sea-ice transport driving Southern Ocean salinity and its recent trends. *Nature* 537: 89–92. https://doi.org/10.1038/nature19101.

260 Ekaykin, A.E., Vladimirova, D.O., Lipenkov, V.Y., and Masson-Delmotte, V. (2016). Climatic variability in Princess Elizabeth Land (East Antarctica) over the last 350 years. *Climate of the Past Discussions* https://doi.org/10.5194/cp-2016-76.

**261** Abram, N.J., McGregor, H.V., Tierney, J.E. et al. (2016). Early onset of industrial-era warming across the oceans and continents. *Nature* 536: 411–418.

**262** Lewis, N. (2016). Was early onset industrial-era warming anthropogenic, as Abram et al. claim? Climate Audit, August 31st [http://www.climateaudit.info/tipjar.html]

**263** Armour, K.C., Marshall, J., Scott, J.R. et al. (2016). Southern Ocean warming delayed by circumpolar upwelling and equatorward transport. *Nature Geoscience* 9: 549–554.

**264** Friedli, II., Lotscher, H., Oeschger, H. et al. (1986). Ice core record of the $^{13}C/^{12}C$ ratio of atmospheric $CO_2$ in the past two centuries. *Nature* 324: 237–241.

**265** Wolff, E.W. (2011). Greenhouse gases in the Earth system: a palaeoclimate perspective. *Philosophical Transactions of the Royal Society A Mathematical, Physical and Engineering Sciences* 369: 2133–2147.

**266** Crutzen, P.J. and Stoermer, E.F. (2000). The "Anthropocene". *International Geosphere-Biosphere Newsletter* 41: 17–18.

**267** Lovejoy, S. (2015). Climate Closure. *Eos* 96 https://doi.org/10.1029/2015EO037499.

**268** Lovejoy, S. (2014a). Scaling fluctuation analysis and statistical hypothesis testing of anthropogenic warming. *Climate Dynamics* 42: 2339–2351. https://doi.org/10.1007/s00382-014-2128-2.

**269** Lovejoy, S. (2014b). Return periods of global climate fluctuations and the pause. *Geophysical Research Letters* 41: 4704–4710. https://doi.org/10.1002/2014GL060478.

**270** Zeebe, R.E., Ridgwell, A., and Zachos, J.C. (2016). Anthropogenic carbon release rate unprecedented during the past 66 million years. *Nature Geoscience* 9: 325–329.

**271** Hoffman, J., Clark, P.U., Parnell, A.C., and He, F. (2017). Regional and global sea-surface temperatures during the last interglaciation. *Science* 335 (6322): 276–279.

**272** Mann, M.E., Miller, S.K., Rahmstorf, S. et al. (2017). Record temperature streak bears anthropogenic fingerprint. *Geophysical Research Letters* 44 https://doi.org/10.1002/2017GL074056.

**273** Stips, A., Macias, D., Coughlan, C. et al. (2016). On the causal structure between $CO_2$ and global temperature. *Nature Scientific Reports* 6: 21691. https://doi.org/10.1038/srep21691.

**274** Loulergue, L., Schilt, A., Spahni, R. et al. (2008). Orbital and millennial-scale features of atmospheric $CH_4$ over the past 800,000 years. *Nature* 453: 383–386. (15 May 2008) | doi:https://doi.org/10.1038/nature06950.

**275** Blasing, T.J. (2016). Recent greenhouse gas concentrations. *Carbon Dioxide Information Analysis Center* https://doi.org/10.3334/CDIAC/atg.032, http://cdiac.ornl.gov/pns/current_ghg.html.

**276** Nisbet, E.G., Dlugokencky, E.J., Manning, M.R. et al. (2016). Rising atmospheric methane: 2007–2014 growth and isotopic shift. *Global Biogeochemical Cycles* 30: 1356–1370. https://doi.org/10.1002/2016GB005406.

**277** Summerhayes, C.P. (2017). Comment on "The Medieval Quiet Period" – implications arising from models of solar irradiance. *The Holocene* 27 (2): 315–316. https://doi.org/10.1177/0959683616658532.

**278** Met Office (2013). *The Recent Pause in Global Warming (2): What Are the Potential Causes?*, 22 pp. Exeter: Met Office.

**279** Kosaka, Y. and Xie, S.-P. (2013). Recent global-warming hiatus tied to equatorial Pacific surface cooling. *Nature* 501: 403–407.

**280** England, M.H., McGregor, S., Spence, P. et al. (2014). Recent intensification of wind-driven circulation in the Pacific and the ongoing warming hiatus. *Nature Climate Change* 4 (3): 222–227.

**281** Yan, X.-H., Boyer, T., Trenberth, K. et al. (2016). The global warming hiatus: slowdown or redistribution? *Earth's Future* 4 https://doi.org/10.1002/2016EF000417.

**282** Schmidt, G.A., Shindell, D.T., and Tsigaridis, K. (2014). Reconciling warming trends. *Nature Geoscience* 7: 158–160.

**283** Chen, X. and Tung, K.-K. (2014). Varying planetary heat sink led to global-warming slowdown and acceleration. *Science* 345: 897–903.

**284** Chen, X. and Tung, K.-K. (2018). Global surface warming enhanced by weak Atlantic overturning circulation. *Nature* 559: 387–391.

**285** Ishii, M. and Kimoto, M. (2009). Reevaluation of historical ocean heat content variations with time-varying XBT and MBT depth bias corrections. *Journal of Oceanography* 65 (3): 287–299.

**286** Zannaa, L., Khatiwalab, S., Gregoryc, J.M. et al. (2019). Global reconstruction of historical ocean heat storage and transport. *Proceedings of the National Academy of Sciences of the United States of America* https://doi.org/10.1073/pnas.1808838115.

**287** Gebbie, G. and Huybers, P. (2019). The Little Ice Age and 20th-century deep Pacific cooling. *Science* 363: 70–74.

**288** Delworth, T.L., Zeng, F., Vecchi, G.A. et al. (2016). The North Atlantic Oscillation as a driver of rapid climate change in the Northern Hemisphere. *Nature Geoscience* 9: 509–912.

**289** Yang, H., Lohmann, G., Wei, W. et al. (2016). Intensification and poleward shift of subtropical Western boundary currents in a warming climate. *Journal of Geophysical Research, Oceans* 121: 4928–4945. https://doi.org/10.1002/2015JC011513.

**290** Cowtan, K. and Way, R. (2013). Coverage bias in the HadCRUT4 temperature series and its impact on recent temperature trends. *Quarterly Journal of the Royal Meteorological Society Part B* 140 (683): 1935–1944.

291 Santer, B.D., Bonfils, C., Painter, J.F. et al. (2014). Volcanic contribution to decadal changes in tropospheric temperature. *Nature Geoscience* 7: 185–189.

292 Solomon, S., Daniel, J.S., Neely, R.R. et al. (2011). The persistently variable "background" stratospheric aerosol layer and global climate change. *Science* 333: 866–870.

293 Kaufmann, R.K., Kauppi, H., Mann, M.L., and Stock, J.H. (2011). Reconciling anthropogenic climate change with observed temperature 1998-2008. *Proceedings of the National Academy of Sciences of the United States of America* 108 (29): 11790–11793. https://doi.org/10.1073/pnas.1102467108.

294 Berger, A., Yin, Q., Nifenecker, H., and Poitou, J. (2017). Slowdown of global surface air temperature increase and acceleration of ice melting. *Earth's Future* 5: 811–822.

295 Myhre, G., Shindell, D., Bréon, F.-M. et al. (2013). anthropogenic and natural radiative forcing. In: *Climate Change 2013: The Physical Science Basis, Contribution of Working Group I to the Fifth Assessment Report of the Intergovernmental Panel on Climate Change* (eds. T.F. Stocker, D. Qin, G.K. Plattner, et al.), 1137–1216. New York: Cambridge University Press.

296 Zalasiewicz, J., Waters, C., Williams, M., and Summerhayes, C.P. (2019). *The Anthropocene as a Geological Time Unit – A Guide to the Scientific Evidence and Current Debate*, 361 pp. Cambridge: Cambridge University Press.

297 Juniper, T. (2013). *What Has Nature Ever Done for Us?* London: Profile Books.

298 Beerling, D. *Making Eden*, 257 pp. Oxford University Press.

299 Steffen, W., Sanderson, A., Tyson, P.D. et al. (2004). *Global Change and the Earth System: A Planet Under Pressure*, 336 pp, Global Change – the IGBP Series. Berlin: Springer.

300 Capra, F. and Luisi, P.L. (2014). *The Systems View of Life – A Unifyng Vision*, 498 pp. Cambridge: Cambridge University Press.

301 Lenton, T. (2016). *Earth System Science – A Very Short Introduction*, 153 pp. Oxford: Oxfrd University Press.

302 Ellis, E.C. (2018). *Anthropocene – A Very Short Introduction*. Oxford: Oxford University Press, 183 pp.

303 Waters, C.N., Zalasiewicz, J., Summerhayes, C. et al. (2016). The Anthropocene is functionally and stratigraphically distinct from the Holocene. *Science* 351 (6269): 10. https://doi.org/10.1126/science.aad2622.

304 Fischer, H., Meissner, K.J., Mix, A.C. et al. (2018). Palaeoclimate constraints on the impact of 2°C anthropogenic warming and beyond. *Nature Geoscience* 11: 474–485.

305 Brook, E. and Buizert, C. (2018). Antarctic and global climate history viewed from ice cores. *Nature* 558: 200–208.

306 Clark, P.U. et al. (2016). Consequences of twenty-first-century policy for multi-millennial climate and sea-level change. *Nature Climate Change* https://doi.org/10.1038/nclimate2923.

307 Hansen, J., Sato, M., Hearty, P. et al. (2016). Ice melt, sea level rise and superstorms: evidence from paleoclimate data, climate modeling, and modern observations that 2_C global warming could be dangerous. *Atmospheric Chemistry and Physics* 16: 3761–3812.

308 Steffen, W., Rockström, J., Richardson, K. et al. (2018). Trajectories of the Earth System in the Anthropocene. *Proceedings of the National Academy of Sciences of the United States of America* 115 (33): 8252–8259.

309 Ganopolski, A., Winkelmann, R., and Schellnhuber, H.J. (2016). Critical insolation–$CO_2$ relation for diagnosing past and future glacial inception. *Nature* 529: 200–203. https://doi.org/10.1038/nature16494.

310 Burke, K.D., Williams, J.W., Chandler, M.A. et al. (2018). Pliocene and Eocene provide best analogs for near-future climates. *Proceedings of the National Academy of Sciences of the United States of America* 115 (52): 13288–13293.

311 Campbell, J.E., Berry, J.A., Seibt, U. et al. (2017). Large historical growth in global terrestrial gross primary production. *Nature* 544: 84–87.

312 Zhu, Z., Piao, S., Myneni, R.B. et al. (2016). Greening of the Earth and its drivers. *Nature Climate Change* 6 (8) https://doi.org/10.1038/NCLIMATE3004.

313 Keenan, T.F., Prentice, I.C., Canadell, J.G. et al. (2016). Recent pause in the growth rate of atmospheric $CO_2$ due to enhanced terrestrial carbon uptake. *Nature Communications* 7: 13428. https://doi.org/10.1038/ncomms13428.

314 Lombardozzi, D.L., Smith, N.G., Cheng, S.J. et al. (2018). Triose phosphate limitation in photosyhnthesis models reduces leaf photosynthesis and global terrestrial carbon storage. *Environmental Research Letters* 13: 074025. https://doi.org/10.1088/1748-9326/aacf68.

315 Lyons, S.L., Baczynski, A.A., and Babila, T.L. (2019). Palaeocene–Eocene thermal maximum prolonged by fossil carbon oxidation. *Nature Geoscience* 12: 54–60.

316 Kiessling, W., Simpson, C., Beck, B. et al. (2012). Equatorial decline of reef corals during the last Pleistocene interglacial. *Proceedings of the National Academy of Sciences of the United States of America* 109 (52): 21378–21383.

317 Leggat, W.P., Camp, E.F., Suggett, D.J. et al. (2019). Rapid coral decay is associated with marine heatwave mortality events on reef. *Current Biology* 29: 1–8.

**318** DeConto, R.M. and Pollard, D. (2016). Contribution of Antarctica to past and future sea-level rise. *Nature* 531: 591–597.

**319** Bracegirdle, T.J., Connolley, W.M., and Turner, J. (2008). Antarctic climate change over the twenty first century. *Journal of Geophysical Research* 113: D03103. https://doi.org/10.1029/2007JD008933.

**320** Swart, N. (2017). Natural causes of Arctic sea-ice loss. *Nature Climate Change* 7: 239–241.

**321** Bevis, M., Harig, C., Khan, S.A. et al. (2019). Accelerating changes in ice mass within Greenland, and the ice sheet's sensitivity to atmospheric forcing. *Proceedings of the National Academy of Sciences of the United States of America* https://doi.org/10.1073/pnas.1806562116.

**322** Golledge, N.R., Keller, E.D., Gomez, N. et al. (2019). Global environmental consequences of twenty-first-century ice-sheet melt. *Nature* 566: 65–72.

**323** Turner, J., Phillips, T., Marshall, G.J. et al. (2017). Unprecedented springtime retreat of Antarctic Sea ice in 2016. *Geophysical Research Letters* 44 https://doi.org/10.1002/2017GL073656.

**324** Trusel, L.D., Das, S.B., and Osman, M.B. (2018). Nonlinear rise in Greenland runoff in response to post-industrial Arctic warming. *Nature* 564 (7734) https://doi.org/10.1038/s41586-018-0752-4.

**325** Schuur, E.A.G., McGuire, A.D., Schädel, C. et al. (2015). Climate change and the permafrost carbon feedback. *Nature* 520: 171–179.

**326** Xu, Y., Ramanathan, V., and Victor, D.G. (2018). Global warming will happen faster than we think. *Nature* 564: 30–31.

**327** Buizert, C. and Schmittnet, A. (2015). Southern Ocean control of glacial AMOC stability and Dansgaard-Oeschger interstadial duration. *Paleoceanography* 30: 1595–1612. https://doi.org/10.1002/2015PA002795.

**328** Liu, W., Xie, S.-P., Liu, Z., and Zhu, J. (2017). Overlooked possibility of a collapsed Atlantic Meridional Overturning Circulation in warming climate. *Science Advances* 3: e1601666.

**329** Summerhayes, C.P. (2019). Climate. In: *The Anthropocene as a Geological Time Unit* (eds. J. Zalasiewicz, C. Waters, M. Williams and C.P. Summerhayes), 200–218. Cambridge: Cambridge University Press.

**330** Herzschuh, U., Birks, J.B., Laepple, T. et al. (2016). Glacial legacies on interglacial vegetation at the Pliocene-Pleistocene transition in NE Asia. *Nature Communications* https://doi.org/10.1038/ncomms11967.

# 16

# Putting It All Together

## 16.1 A Fast Evolving Subject

Following the explorers of Earth's climate as they progressively uncovered more and more of its complex workings has been a fascinating journey. Starting in the late eighteenth century this field of enquiry expanded slowly at first, and then with gathering pace as scientists began to find ways of measuring greenhouse gases not only in our present atmosphere, but also in fossil air from ice cores dating back 800 000 years. By various clever means they also found out how to estimate the likely $CO_2$ content of past atmospheres for millions of years back beyond the reach of ice cores.

Establishing this atmospheric history and its relation to our climate demanded that we develop new ways of sampling the Earth, including the record concealed in sediments beneath the 72% of the planet covered by ocean floor, and in the ice of polar regions. We benefited from the invention of new seabed sampling technologies in the form of cores and drills, of ice drilling technology, and of new analytical techniques in the laboratory. Ratios of one chemical element or isotope to another added to the tools we use as proxies for past climate conditions. Fossil remains representing different forms of climate-sensitive life were our initial tools, along with climate-sensitive sediments. Our fossil armoury expanded through the discovery of mathematical transfer functions that relate the assemblage of marine plankton to the temperature of the water mass from which it came. Numerical models proved useful in stimulating our imaginations, in linking widely separated data points, and in enabling us to test ideas about past climates. Thanks to advances in dating Earth materials we can now analyse climate change at ever-smaller intervals of time.

In this unfolding of the flower of climate change, whole new subdivisions of science arose – palaeoclimatology, palaeoceanography, biogeochemistry, and Earth System Science – along with new journals to house the research papers emerging from the growing communities of scientists in these new fields. We now know much more than we did even a decade ago about what makes our climate variable, and why.

Not only the scope, but also the rate of discovery continues to increase. Palaeoclimate studies began to rise exponentially in the 1940s. In this book I refer to no more than 10 research papers or books per decade from 1750 to 1940. The numbers rose fast to 383 for the decade 2000–2009 and to 679 for the decade 2010–2019 (by end September 2019). These are just a small subset of the fast growing publications on past climates, but the pattern is representative.

Much of the rise in output reflects the growing application in World War II and after of echo-sounding to map the deep ocean floor; the invention of the piston corer in 1947; the development of palaeomagnetic analyses and magnetic mapping of the deep ocean floor in the early 1950s; the expansion of interest in the workings of the planet from the International Geophysical Year of 1957–1958 onwards, including the opening up of Antarctica to research; the evolution of plate tectonic theory in the early 1960s; the advent of deep ocean drilling in 1968; and widespread drilling into ice sheets and glaciers from about 1980 on; the measurement of various of Earth's properties from satellites; and the application of computing to numerical analysis and modelling.

As the opportunities to study newly sampled materials grew, so did the numbers of young scientists keen to make names for themselves by working at this intellectual frontier. Yet more people became involved as the number of analytical techniques required to unravel the complex strands of climate change grew. Before 1970 each palaeoclimate paper to which I refer averaged 1.2 authors. In the 1970s it doubled. By the beginning of the twenty-first century it had doubled again. I calculate that there were at least 30 times more scientists publishing on past climate change in the early 2000s than in the 1950s. This accelerating rate reflects not only the inherent intellectual

*Palaeoclimatology: From Snowball Earth to the Anthropocene*, First Edition. Colin P. Summerhayes.
© 2020 John Wiley & Sons Ltd. Published 2020 by John Wiley & Sons Ltd.

fascination of the topic, but also the growing need to understand just how variable Earth's climate is, as background for making predictions about its future behaviour. Research on past climate change now occupies a few thousand scientists worldwide.

The pattern of the science has changed too. Nowadays, through Earth System Science, we take a much more integrative, holistic approach to understanding climate change – following Humboldt's example [1] (see Chapter 2). The study of past climates brings together astronomers calculating Earth's orbit and axial tilt, astrophysicists studying the behaviour of the Sun, palaeobotanists, and micropalaeontologists studying how organisms and their ecology change with climate and time, geochemists with their battery of elements and isotopes, geophysicists determining where land and ocean were in relation to one another through time, oceanographers, meteorologists, and plain old geologists with their intimate knowledge of Earth's surface processes – volcanic activity, weathering, sedimentation, and the endless recycling implicit in plate tectonics, mountain building and continental drift. It is not enough to consider just one or two variables in isolation, like the relation between temperature and $CO_2$. Many factors can affect this interaction. All have to be considered as we try to work out how Spaceship Earth operates. Otherwise one day we may find it uninhabitable.

The scientists whose work I reviewed in this book did not take statements about climate change on trust. They produced, examined and interpreted the evidence from the geological record or from fields of study pertinent to interpreting that record. Many of their findings are so recent that they have not yet made it into standard undergraduate textbooks or teaching courses. They tend to be available only from scientific journals that are not readily accessible to the general public because they are hidden behind a 'pay-wall' as pointed out in Chapter 1. Do we know everything we need to know? No. As Wally Broecker reminds us, science is '*a constant struggle to understand more fully and more accurately how the world really works*' [2]. There is always more to do. But we do know much more now than we did in any previous decade about how our climate system works.

## 16.2 Natural Envelopes of Climate Change – Earth's Thermostat

What has our journey uncovered? We learned that for the past 800 Ma Earth's climate oscillated within certain limits forming a natural envelope. Within that envelope Earth experienced alternating warm and cold periods driven by internal or external forces. Some of those forces are extraterrestrial. Ultimately the output from our star, the Sun, drives our planetary climate. And like all main sequence stars of its type, the Sun's output is increasing slowly with time, by about 7% over the past 800 Ma covered in this book. Internal changes within the Sun cause its output to fluctuate cyclically by small amounts on decadal, centennial, and millennial scales. The amount of that solar energy the Earth receives also varies, due to subtle but regular changes in the orbit of the Earth around the Sun, and the tilt of the Earth's axis, operating on time-scales of tens-to-hundreds-of-thousands-of-years. We can detect these solar and orbital changes by various means because their climate signals are trapped in sediments, tree rings, corals, stalagmites, and ice.

Our climate is a by-product of those extraterrestrial effects mixed with internal effects, which include the balance, on a time scale of millions of years, between the supply of $CO_2$ provided by volcanic activity stimulated by plate tectonic processes (a source), and the extraction of $CO_2$ by weathering and its sequestration in sedimentary carbonates and organic matter (a sink). On much shorter time scales our climate is modulated by internal oscillations involving transfers of energy by wind and ocean currents, which mostly cause regional changes at the multi-annual to multi-decadal scale (for example the Pacific Decadal Oscillation and the Atlantic Multidecadal Oscillation), but also work at the scale of glacial to interglacial change (for instance through the Meridional Overturning Circulation of the global ocean). The ocean accentuates the effects of these various changes in two ways. It transfers heat and salt slowly around the globe, and, with its vast store of dissolved $CO_2$, it plays an important feedback role in releasing $CO_2$ as its waters well up to the surface; warm water releases $CO_2$ to the air, and cold water extracts it from the air.

These natural operations are interrupted from time to time by 'catastrophic' events, including short-term major volcanic eruptions like that of mount Pinatubo in 1991, which cooled the climate slightly for between one and two years, and occasional massive outpourings of flood basalts over periods of a few millions of years, which cause mass extinctions of life at intervals of several tens of millions of years. Much rarer are large asteroid impacts like that which killed off the dinosaurs at the end of Cretaceous time. Catastrophic events tend to be both irregular in time and unpredictable, unlike the present mass extinction documented by Elizabeth Kolbert in her 2014 book *The Sixth Extinction – An Unnatural History* [3].

Plate tectonics play an important role. The changing positions of the continents through time modulate the record of change within the natural envelope, most notably by moving land into more tropical or more polar positions, by breaking apart land masses to expose new coasts to moisture, and by changing the positions of the seaways that provide routes for ocean currents and ultimate sources of moisture.

Sea level rose and fell as mid-ocean ridges grew and declined at the sites of sea-floor spreading. Growing mid-ocean ridges pumped out $CO_2$, as did volcanoes in subduction zones around plate edges. Plate tectonic forces also governed the formation of mountains and plateaus whose rise accentuated the physical and chemical weathering that extracted $CO_2$ from the air. The rises and falls in atmospheric $CO_2$ were manifest through ocean chemistry, because a rise in atmospheric $CO_2$ increases the concentration of $CO_2$ in the ocean. That makes the oceans more acidic, causes deep-ocean carbonate sediments to dissolve and hence raises the level of the carbonate compensation depth or CCD.

Just as climate affects life, so too life affects climate, given that plants 'eat' $CO_2$. The rapid growth and evolution of land plants in Devonian time lowered atmospheric $CO_2$ enough to lead to the Carboniferous glaciation, aided by the trapping of plant remains in swamps, which eventually turned to coal – a carbon sink. Just as plants affect $CO_2$, so $CO_2$ can affect plants, and the gradual decrease in $CO_2$ with time over the past 450 Ma eventually led to the evolution and expansion of the so-called C4 plants, many of them grasses, which extract $CO_2$ more efficiently from the air than do the so-called C3 plants [4].

Keeping Earth's climate within certain limits reflects the operation of a natural thermostat. Consider chemical weathering. As temperatures rise, chemical reactions proceed faster. Faster chemical weathering draws down more $CO_2$, a primary warming agent. The climate cools, which slows the rate of chemical weathering. Chemical weathering also progresses faster in regions where physical weathering breaks rocks apart. Once mountains are worn down to form peneplains, few minerals are exposed and chemical weathering slows. Something similar takes place in the tropics where intense rain exacerbates chemical weathering, forming thick deposits of minerals like kaolinite (the basis for 'china clay'), and clayey deposits like laterite. These form a weathered cover slowing further weathering of the underlying bedrock. Chemical weathering is also more widespread at times when there is more exposed and weatherable rock. Fresh basalt is highly weatherable, and when widespread (as in the case of plateau basalts) may help to draw down levels of $CO_2$.

In his Gaia theory, Lovelock cited an experiment named 'Daisy World' to show how plants may unconsciously adapt to keep the climate suitable for life. In earlier times, when the Sun's output was low, there would be more black daisies, which absorbed heat. As the Sun's output grew, the black daisies would be replaced gradually and unconsciously by white daisies that reflected heat. The influence of the plants kept the climate in a range suitable for daisies, via thermostatic control.

The changes from glacial to interglacial within the recent Ice Age provide further examples of thermostatic control via orbital variation in insolation. These fluctuations ensured that temperature, sea level and $CO_2$ fluctuated within rather narrow limits ($CO_2$ between 180 and 280 ppm for instance).

It has become increasingly clear since early in the nineteenth century that because natural gases like $CO_2$, $H_2O$-vapour, $CH_4$, and $N_2O$ both absorb and re-emit infrared radiation, the planetary atmosphere containing them benefits from a natural greenhouse effect that keeps the Earth some 32 °C warmer than it would be otherwise, and hence suitable for life. Adding more of any one of these gases to the atmosphere will enhance that effect. $CO_2$ may be added by out-gassing from volcanoes and metamorphism, by the decomposition of organic matter, by the rising to the surface of deep ocean water rich in $CO_2$, and by the warming of the ocean – as warm water holds less $CO_2$ than cold water. $CO_2$ may be extracted from the atmosphere by the chemical weathering of silicate and carbonate minerals in rocks, by photosynthesis, by dissolving in cold surface water, by the sinking of that water into the deep ocean, by the marine biological pump (in which sinking dead and decomposing plankton release into deep water the $CO_2$ absorbed near the ocean's surface), and by eventual incorporation into carbonate sediment and sedimentary organic matter. The effects of $CO_2$ in the air are exacerbated by water vapour, which increases by evaporation as temperature rises, providing a positive feedback, and which decreases as temperature falls. Warmth reduces the proportion of $CO_2$ the ocean can hold and increases the proportion of water vapour the air can carry. Disconnects between temperature and $CO_2$ can arise, for example when volcanoes make $CO_2$ abundant in the absence of mountains that might otherwise be subject to chemical weathering, or when sea ice prevents exchange of $CO_2$ between the ocean and atmosphere, or where the melting of large volumes of land ice soaks up heat from the atmosphere and keeps the local environment cold regardless of $CO_2$ in the air. Fluctuations in the intensity of the greenhouse effect with time result naturally from the changing balance between $CO_2$ sources and sinks.

We must also bear in mind that the main atmospheric gases, oxygen and nitrogen, have no effect on infrared radiation. Although water vapour averages just 0.4% of the atmosphere and $CO_2$ is 0.04%, these two gases comprise virtually 90% and 10% respectively of the part of the atmosphere that affects infrared radiation. Hence the effect of today's roughly 410 ppm $CO_2$ is thus vastly out of proportion to that small amount, once oxygen and nitrogen have been discounted.

It should be self-evident from the preceding chapters that it is eminently possible to understand not only how the climate has fluctuated with time, but also what forced it to do so in a particular way at any particular time. Our knowledge of those fluctuations and their causes is weaker the further we go back through time, as the geological record becomes progressively less complete, especially as much of the ocean floor older than the Cretaceous has been subducted into the Earth's mantle.

Many attempts have been made using infrared radiative transfer theory to calculate climate sensitivity – the amount by which the temperature will rise if atmospheric $CO_2$ doubles. From the geological perspective it seems abundantly clear that the instantaneous climate sensitivity (i.e. what happens next week if we double $CO_2$ now) is an inadequate guide, because the ocean and the ice are out of equilibrium with the atmosphere. Ice absorbs masses of energy in order to melt, which it does very slowly. The ocean absorbs heat at its surface then transfers it to the deeps, also very slowly. If we take these very slow feedbacks into consideration, then it is clear that a small change in $CO_2$ may in fact have an unexpectedly large effect on temperature (unexpected, that is, if one has forgotten the importance of these slow natural feedback processes). If we stopped pumping $CO_2$ into the atmosphere now, these slow feedback processes would continue on the 100+ year timescale. Ice would continue melting and sea level would continue rising, until equilibrium with the atmosphere was reached. The thermostat operates at geological pace.

The acid test of a plausible theory is its power of prediction. Many scientists, starting with Arrhenius in the 1890s, forecast that the continued burning of fossil fuels would raise temperature. Forecasts made in the late 1970s and 1980s about what temperatures we might expect at the end of the twentieth century turned out to be remarkably prescient. Forecasts seemed to some people to go off track with the 'pause' in warming between 2000 and 2013, but with the benefit of hindsight that turns out to be because not enough knowledge of how the natural system works was built in to the numerical forecasting models. One has to take into account in addition natural oscillations (e.g. Pacific Decadal Oscillation, Atlantic Multidecadal Oscillation), which are themselves difficult to forecast precisely, along with the fact that the summer melting of 4 million km$^2$ of Arctic sea ice absorbs enormous amounts of heat, and that much additional heat has also sunk into the deep ocean, penetrating first to moderate depths and now reaching well below 2000 m. Incorporating these new understandings of the operation of the natural world into numerical climate models will improve their performance.

There has been some reluctance to incorporate solar change into numerical climate models, given the small changes in incoming radiation with which solar change is associated. However, given the geological evidence for solar-terrestrial linkages, we may find that climate scientists have neglected solar terrestrial processes in the atmosphere that may enhance those signals at the Earth's surface. Time will tell.

## 16.3 Evolving Knowledge

As previous chapters show, many theories came and went to explain the workings of the climate system through the ages. The science progressed. Agassiz's ice sheet theory displaced Lyell's iceberg theory. Yet we now know that armadas of icebergs did play a key role in millennial change, especially in the North Atlantic during the Ice Age. The savants of the eighteenth century and the geologists of the early nineteenth century knew that the Earth's climate had cooled through what they called Tertiary time and we now call the Cenozoic Era – the past 65 Ma. Charles Lyell thought that the cooling might have come about by the continents moving into polar regions. But that would not have been enough. Antarctica reached the South Pole 90 Ma ago, but remained free of an ice sheet for a further 56 Ma!

Wegener gave us moving continents, but verification had to await the development of palaeomagnetism in the early 1950s and plate tectonics in the late 1960s. Mapping continental positions helped to put palaeoclimatology on a firmer footing, but did not explain why the Cretaceous was hot and the late Cenozoic was cold. In the late 1800s Arrhenius offered an elegant explanation for the warming and cooling of the Ice Age, based on Tyndall's discoveries about the absorptive properties of $CO_2$. Calling on the findings of the then recent Challenger Expedition of 1872–1876, Chamberlin added a new dimension to Arrhenius's theory by observing that the amount of $CO_2$ in the atmosphere would fluctuate depending on the temperature of the sea's surface, since cold water held much more dissolved gas than warm water. Arrhenius had considered only the air. Chamberlin extended his analysis to cover much of geological time. We had a theory: adding or subtracting $CO_2$ could change the climate. But did it do so in fact? In the absence of fossil $CO_2$, speculation ruled.

The necessary tests depended on advances in spectroscopy that gave us the complete spectra of the main greenhouse gases – $CO_2$ and water vapour – and on knowing how $CO_2$ was distributed in the ocean and exchanged with the atmosphere. Some of what we needed to know had appeared by the mid-1950s, enabling Plass to resurrect Chamberlin's theory. But in the absence of hard data on past levels of $CO_2$, geologists took a conservative approach to the causes of past climate change.

Things began to change at the end of the 1970s. Soviet scientists Alexander Ronov in geology and Mikhail Budyko in climatology showed that periods of warm climate occurred when volcanic activity was common and $CO_2$ was likely to have been abundant in the air. Hans Oeschger's team found $CO_2$ in bubbles of fossil air trapped in ice cores, and Minze Stuiver used carbon isotopes from tree rings to estimate likely past levels of atmospheric $CO_2$ for the Holocene.

By the early 1980s, geochemists like Bob Garrels, James Walker, and Bob Berner took advantage of the increasing power of computers to model the behaviour of the carbon cycle. Refined over the years, carbon cycle models confirm the interdependence of $CO_2$ and temperature back through time. These various developments led Al Fischer to identify two great tectono-climatic cycles with a periodicity of about 300 Ma, driven by convection in the Earth's mantle, which led to cyclic changes in the abundance of $CO_2$ in the air between a 'greenhouse state' when $CO_2$ was abundant (Cretaceous and early Cenozoic), and an 'icehouse state' when $CO_2$ was depleted (late Carboniferous and late Cenozoic). The great cooling of the Cenozoic took place when volcanism declined, sea level fell, chemical weathering increased along with mountain uplift, and $CO_2$ declined as a result. Fischer's tectono-climatic cycle concept gave the Arrhenius–Chamberlin–Plass model a meaningful geological context and a significant degree of respectability. Other geologists rapidly followed Fischer's lead, amongst them Thomas Worsley, who observed that the correlation in time between prolonged episodes of mountain building and icehouse climate '*is prima facie evidence for orogenically driven $CO_2$ drawdown and carbon burial*' [5]. The development of a unifying world-view of the evolution of our climate system was beginning to materialize.

By the late-1980s we knew that the numbers of pores (stomata) on leaves declined as the percentage of $CO_2$ increased in the atmosphere. David Beerling and Dana Royer [6] and others confirmed that there is a rather good correlation between $CO_2$ and temperature over the past 450 Ma since land plants evolved. $CO_2$ was abundant in the warm early Palaeozoic, declined to near present-day levels in the late Carboniferous glaciation, rose again during the Mesozoic, then declined to the levels typical of the Pleistocene Ice Age [7]. Independent confirmation came from the carbon isotopes and alkenones in the remains of plankton. Like temperature, $CO_2$ has fluctuated within a fairly narrow natural envelope [7].

Careful analysis suggests that the slow rise in both $CO_2$ and temperature from the end of the Cretaceous into the Middle Eocene 50 Ma ago and their subsequent slow fall was a response to plate tectonics, especially the northward migration of India and its collision with Asia 50 Ma ago. Subduction of carbonate sediment at the Asian margin provided $CO_2$ via volcanoes to heat the planet, a process that ceased with the collision and the associated rise and chemical weathering of the Himalaya. The chemical weathering of recently erupted plateau basalts around the Indian Ocean also played a part in reducing $CO_2$ in the air and hence cooling the late Cenozoic.

By 34 Ma ago, the decline of $CO_2$ in the atmosphere had cooled the planet to the extent that an ice sheet could form on Antarctica, a process encouraged by the physical isolation of Antarctica behind the Antarctic Circumpolar Current, which developed as Australia and South America broke their links with the southern continent. The growing ice sheet enhanced the Earth's albedo, cooling the climate further. Positive feedback also played a role. As declining $CO_2$ caused temperatures to fall, the Southern Ocean cooled to the point where it could take up more $CO_2$ from the air, and less water vapour evaporated from the ocean surface. The cooling of the Southern Ocean made surface waters dense enough to sink, taking newly dissolved $CO_2$ with them into the ocean depths, drawing down yet more $CO_2$ in a 'slow runaway' process leading in due course to the latest Ice Age.

A convincing link between $CO_2$ and temperature comes from the Palaeocene-Eocene boundary 56 Ma ago, when a major injection of carbon into the air made temperatures rise suddenly by 4–6 °C. The amount of $CO_2$ emitted acidified the ocean, dissolved deep ocean carbonates and killed off organisms on the deep sea floor. The warming made sea level rise by at least 10 m, and the Earth System took 100 000 years to recover. This was a time of massive eruption of flood basalts in the northeast Atlantic volcanic province. A magmatic event on that scale, especially the intrusion of basalt sills into carbon-rich rocks, could have supplied much of the $CO_2$. However, earthquakes associated with eruptions could also have destabilized the continental slope, releasing $CH_4$ previously trapped as methane hydrates. Methane is a powerful greenhouse gas in its own right, but is rapidly oxidized to $CO_2$ in the atmosphere. The precise trigger for this event is still unclear. It seems to be one of a series of short-lived Eocene hyperthermal events that may have been initiated by orbital changes of the kind that later drove the warm interglacial periods of the Ice Age.

Warming was also associated with abundant $CO_2$ in the mid Pliocene, when $CO_2$ rose to about 450 ppm, temperatures rose globally by 2–3 °C (and by up to 18 °C in the Arctic), the Southern Ocean was about 5 °C warmer than it is today, Antarctica's huge Ross Ice Shelf disappeared, and sea level may have risen by 10–20 m. Like the shorter-lived event 56 Ma ago, this may be a rough analogue for our future climate. As there was limited land ice in the Northern Hemisphere, much of West Antarctica and parts of East Antarctica must have melted away. The warming may have been triggered by plate tectonic processes, when the gradual closure of the central American seaway between 4.7 and 2.5 Ma ago forced to the north warm equatorial waters that formerly flowed from the Atlantic to the

Pacific. An expansion of warm surface water globally in this way would have put more $CO_2$ and water vapour into the air to warm the planet. As in the case of Antarctica, this is another of the outcomes of opening and closing ocean gateways and redirecting ocean currents, and in so doing modifying the effects of the primary driver for our climate, the Sun. The climate, tectonics and the carbon cycle are intimately linked. As Mike Leeder pointed out, we are dealing with a *'Cybertectonic Earth'*, and the *'combination of tectonics and biogeochemistry is the great fulfilment of the Huttonian philosophical scheme'* [8].

Our knowledge of the way in which orbital and axial changes might affect the Ice Age climate grew through the 1800s, but was difficult to apply in the field because it was impossible then to subdivide Pleistocene geological sequences on land with sufficiently high resolution. Early in the twentieth century Milankovitch improved our understanding of what the climate record *should* look like based on orbital theory, but we had to wait until Carbon-14 dating arrived in the 1950s to date the Pleistocene geological record to test his ideas. Applying André Berger's refinement of Milankovitch's calculations in the 1970s, Hays, Imbrie and Shackleton discovered that deep-ocean sediments faithfully record the orbital beat [9]. Ice grew when insolation was weak, and melted when insolation was strong. We could map the extraterrestrial signal through time, a really exciting breakthrough.

Human civilization developed in the Holocene, one of the short warm interludes of a 2.6 Ma-long Ice Age that was predominately cold – some 3–4 °C colder than the mid-twentieth century average. Following the Last Glacial Maximum 20 Ka ago, rising insolation warmed the planet, melting the great Northern Hemisphere ice sheets. Their continued melting, while insolation was high at the start of the Holocene, absorbed solar energy, preventing the Northern Hemisphere from warming to the extent suggested by the amount of solar energy received. As a result, the warming signal did not emerge there until 6000–7000 years ago, when the ice sheets melted away. By then the insolation was already in decline. After a relatively short mid Holocene warm period temperatures fell towards the neoglacial conditions that culminated in the cold temperatures of the Little Ice Age (LIA). Celestial mechanics tells us that low insolation and cool conditions should continue for at least another 1000 years.

With insolation providing the primary driving force for glacial to interglacial cycles, and with a low overall level of planetary $CO_2$, the role of $CO_2$ in the Ice Age climate was secondary. Its role only came to the fore when Oeschger's team found $CO_2$ in bubbles of fossil air in ice cores in 1978. A decade later French scientists found that $CO_2$ and temperature fluctuated more or less in concert down the Vostok

ice core, a pattern extended to 800 Ka in the Dome C ice core in 2006. Carbon isotopes enabled estimates of atmospheric $CO_2$ to be made from marine sediment cores, which showed the same pattern. Palaeoclimate scientists agreed that insolation at high latitudes warmed the ocean, which then emitted $CO_2$, further enhancing temperature. The warming enhanced evaporation, supplying water vapour, causing further warming, and so on. $CO_2$ had to be a secondary forcing factor for ice volume, following the primary drivers – orbital and axial change. As far as cooling was concerned, Wally Broecker thought that falling insolation cooled the polar regions, increased the strength of winds and so stimulated the biological pump that drew $CO_2$ into the ocean interior. John Martin took that notion one step further, proposing in 1990 that the stronger winds of glacial periods transported iron-rich dust that enhanced oceanic productivity and so drew $CO_2$ out of the atmosphere, accelerating cooling.

The evidence tells us that just as changes in the amount of $CO_2$ affect temperature by volcanic sources of $CO_2$ being out of phase with weathering sinks for $CO_2$, so do changes in temperature affect $CO_2$, as in the Ice Age when orbital and axial change drove temperature changes, which were reinforced by feedback from rising or falling $CO_2$. In that respect temperature and $CO_2$ are almost inseparable non-identical twins. Nevertheless, the correlation between $CO_2$ and temperatures is not one-to-one. Whilst $CO_2$ and temperature rise more or less together as ice sheets decay into interglacials, they diverge as temperatures fall more rapidly than atmospheric $CO_2$ when orbital forcing declines and ice sheets begin to grow. There is an obvious explanation. As insolation decreases and temperatures fall, sea ice grows over the polar oceans, increasing the Earth's albedo and exacerbating cooling. Even though a colder ocean absorbs $CO_2$ from the atmosphere, the fall in sea level and the growth in sea ice reduce the area of sea available for that task. Meanwhile, on land, terrestrial vegetation declines as snow and ice grow, releasing $CO_2$ to the air through decomposition. So, whilst temperature can keep up with declining insolation, $CO_2$ cannot. $CO_2$ is part of hysteresis loop in which there is a dynamic lag between input and output. During an Ice Age, it moves with temperature on warming, but lags temperature on cooling.

As $CO_2$ emerges from the ocean when it warms, it was difficult to understand why ice cores showed an apparent time-lag of about 1000 years between warming and the rise in $CO_2$. That mystery was solved in 2013, when Frédéric Parrenin used new isotopic techniques to establish that in Antarctica the abundance of $CO_2$ in the atmosphere increased at the same time as the temperature during the last deglaciation. The major source for this $CO_2$ was the Southern Ocean. Northern seas did not warm at the same time, because the energy supplied by rising insolation there was spent mostly on melting the Northern Hemisphere ice sheets, a process that kept the northern regions cold despite the rise in global $CO_2$ exhaled from the Southern Ocean. This is yet another example of the disconnect that may occur at times for obvious physical reasons between temperature and $CO_2$.

It had long seemed odd that the ice sheets of the Northern and Southern Hemispheres waxed and waned more or less in concert despite opposing patterns of average insolation. However, it is not the average insolation that matters. The intensity of northern summer insolation correlates not only with the length of the summer period in the south, but also with the intensity of spring insolation in the south. We now realize that we do not need to call on forcing by Northern Hemisphere insolation to explain the link in climate response between the two hemispheres.

The last glacial period was punctuated by small periodic warmings and coolings of millennial duration that were gentler in the south and larger and more abrupt in the north. The Antarctic warm events seemed to precede the Greenland ones, reversing as the northern events (known as Dansgaard–Oeschger Events) started. Eric Wolff argued that terminations of the main glacial periods were like Antarctic warm events in which the following warm event in Greenland failed to develop. That allowed the Antarctic warming to continue, to the point that deglaciation began in the north to match that taking place in the south. The north was preconditioned not to develop a Dansgaard–Oeschger warming event, by a combination of low insolation and maximal volume and extent of land ice in the Northern Hemisphere. If Wolff is right, then glacial terminations are periods of runaway southern warming, and the Southern Hemisphere is in the driving seat for glacial terminations. However, rather than seeking either a northern or a southern trigger for the bipolar seesaw pattern, it would seem wiser to recognize that we are seeing a natural oscillation in which neither pole is in the driving seat.

The Northern and Southern Hemispheres are connected through both the atmosphere and the ocean. The rapid circulation of the air means that both polar regions get the same signals of change in $CO_2$ or $CH_4$ at almost the same time. The slow circulation of the ocean connects the poles via the Atlantic Meridional Overturning Circulation (AMOC), the Atlantic branch of the global ocean's Thermohaline Conveyor that transports heat and salt around the world. Rising summer insolation in the north eventually triggers the melting of the Laurentide ice sheet, which at its maximum extent becomes lower due to depression of the Earth's crust caused by isostatic adjustment, and wider, due to ice flow, making it more susceptible to orbitally induced warming. The Scandinavian ice sheet

melts faster because it is smaller and thinner. The fast retreat of both ice sheets covers the northern North Atlantic with cold fresh water on which sea ice forms. That inhibits the re-start of the Atlantic Meridional Oceanic Circulation. Temperatures stay low, despite the rising insolation and the rising heat and $CO_2$ supplied from the south, because most of the available solar energy that is not reflected goes to melt northern land ice. Where sea ice is at its maximum extent it reaches a line from Brest to Newfoundland, causing winter North Atlantic region temperatures to fall as low as those in central Siberia. Increasing insolation and the import of heat from the south eventually melts northern sea ice, allowing the AMOC to re-start. That transports $CO_2$-rich deep water to the south where the old $CO_2$ emerges through the upwelling of Circumpolar Deep Water, to be exchanged with the atmosphere and further reinforce global warming. Under this scenario, the AMOC helps to maintain glacial (AMOC-off or reduced) states versus interglacial (AMOC-on) states, but it was not in the driving seat for deglaciation. The seesaw between the two poles at the millennial scale formed the Dansgaard–Oeschger events and their Antarctic counterparts during periods of intermediate glaciation.

In summary, insolation is the primary driver for glacial to interglacial climate change, where $CO_2$ plays a critical role as an amplifier of its effect, being emitted from the Southern Ocean directly as temperatures increase. $CO_2$ makes temperatures increase beyond what insolation could achieve, thereby increasing the supply of water vapour. The process stops when insolation declines. Methane plays an ancillary role, being given off by wetlands as monsoonal activity grows in interglacial times. $CO_2$, in this context, is no longer the handmaiden of temperature, following where temperature leads as was once thought. It is an equal partner.

Given that insolation oscillates between well-defined limits, and that the climate system marches to the insolation beat during the Ice Age, temperature too oscillates between well-defined limits. So too do the greenhouse gases like $CO_2$ and methane. The natural envelope of $CO_2$ during the Ice Age was 280–180 ppm The rise of $CO_2$ to 416 ppm in May 2019 is so far outside this natural envelope that it should give us pause for thought. However, it is still within the natural envelope of the past 450 Ma (now thought to be 2000–200 ppm [7]). Whilst Ruddiman suggested that the slight rise in $CO_2$ that began about 8 Ka ago might have been caused by early human land clearances, the jury is still out on that question. In any case it was not enough of a rise to significantly offset the cooling driven by the decline in insolation that led to the neoglacial conditions of the late Holocene. Would the declining insolation have led to a glaciation in the absence of that $CO_2$ rise, as

Ruddiman implies? André Berger's calculations of declining insolation suggest not.

Changes in sea level associated with melting ice reinforced warming as sea level rose, and cooling as sea level fell. Rising sea level increases the area of ocean that can exchange $CO_2$ with the atmosphere, whilst falling sea level decreases it. And a wider ocean area provides a larger surface for the evaporation of water vapour.

Volcanic eruptions played a minor role in the Ice Age climate story. They increased as the land rose when ice was removed, and decreased as ice was added. Similarly, chemical weathering on land played a minor part in the Ice Age, with its influence being more regional than global. Cool periods in western Europe in the Holocene were associated with increased rainfall, which implies more chemical weathering then. Whilst weathering may have played a role in $CO_2$ drawdown within the oscillations of the Ice Age, its contribution would have been dwarfed by ocean–atmosphere interactions, because of the large size of the ocean $CO_2$ reservoir.

Superimposed on the multi-thousand year variations in insolation caused by regular changes in the Earth's orbit and axis are much shorter and smaller centennial to millennial variations in solar output. Orbital change induced 3–4 °C of temperature change between glacial and interglacial times, while changes in solar output induced changes of about 0.5–1.0 °C, mostly at the regional scale. A key modifier of the 11-year sunspot cycle is the ~208-year Suess–De Vries Cycle, whose variation drove the development of the grand solar minima of the past 2000 years. It is likely that most of the centennial cool periods of the Holocene represent weak solar output leading to more rainfall and higher lake levels in western Europe, decreased monsoonal activity in the tropics, and increased drift ice in the north Atlantic. Drift ice cycles between 1000 and 2000 years long probably arose from combinations of different solar cycles, mediated by patterns of ocean circulation.

A solar maximum peaking at about 1100 CE lay behind the development of the Medieval Warm Period (1000–1200 CE), which was followed by the LIA. The LIA was not permanently cold. It experienced warm periods nearly as warm as the Medieval Warm Period in the 1630s, 1730s, 1770s, and 1840s, associated with solar maxima. The coldest periods of the LIA occurred during grand solar minima, and it was coldest in winter. Its summers were quite warm. The intensity of the LIA may owe its origin to the 2300-year Hallstatt solar cycle.

The effects of solar variability on the North Atlantic are stronger than they are elsewhere because of interactions between the warm Atlantic and the icy Arctic. As a result the signal of the Medieval Warm Period is stronger around the North Atlantic than elsewhere, and it is difficult to identify either the Medieval Warm Period or the LIA in and around the Southern Ocean. The oceanographic signals of those events in the north are transmitted south over periods of some hundreds of years through the ocean's subsurface via the AMOC. That makes it difficult to find ubiquitous and distinct global records of any short-term event having a small solar signal, like the Medieval Warm Period, which lasted no more than about 200 years. One Holocene signal that is completely global is the short, sharp cooling at 8.2 Ka ago, which was most probably caused by the draining of a vast Arctic glacial lake. This large signal rapidly reached the Southern Ocean via the air.

Roughly the same intensity of solar activity characterized the Medieval Warm Period and the mid-twentieth century. Whilst both global temperature and sunspot activity rose slightly from 1900 to about 1940, the two then diverged. Sunspot activity rose to a maximum in 1960–1990 then fell to intermediate levels, while temperatures fell to a low in 1950 and stayed low until about 1965–1970 before rising to the high levels typical of the 2000s. The two signals became disconnected in about 1940. Soviet data suggest that the rise in aerosols from increasing industrial activity broke the connection.

$CO_2$ is now rising by about 20 ppm/decade. As far as we can tell from ice core studies, the last time it rose fast was over a period of 1000 years during the last deglaciation. It is now rising 100 times as fast as it did then. The dramatic rise in $CO_2$ from the 1800s onwards made it an increasingly important agent of climate change in the late twentieth century at a time when solar output flattened. The lack of corresponding rise in temperature between 1950 and 1970 most likely reflects the growing abundance of dirty air filled with reflective aerosols as post-war industrial output grew. Regulation later cut the output of aerosols: for instance the UK's Clean Air Act of 1956 was promulgated in response to the London smog of 1952 (more on that below), and was updated in 1968 and 1993. The US Clean Air Act was promulgated in 1970 and amended in 1977 and 1990. Other countries too have enacted Clean Air Acts, for example New Zealand in 1972. Collectively, by the late 1960s to early 1970s these Acts significantly cut the global aerosol load that had cooled the atmosphere and hid the warming due to rising $CO_2$. The removal of shading aerosols explains part of the global rise in temperature since about 1970. An increase in aerosols from the growing burning of coal in the industrial revolutions of China and India may have helped to hide the warming that should have continued during the early part of the present century due to growing outputs of greenhouse gases.

Nevertheless, we must remember that radiative forcing is a function of all human emissions, which sum to 'equivalent $CO_2$' or $CO_2$e, which in 2016 was 527 ppm, close to double the maximum $CO_2$ of 280 ppm in ice cores.

The large additional climate forcing provided by increasing emissions and $H_2O$-vapour recently shrank the Baffin Island ice cap, exposing vegetation that last saw the light of day before the Medieval Warm Period. Other data on plant roots now being uncovered by melting ice in the Canadian Arctic suggest that conditions are now the warmest for the Holocene. Climate signals from across the Arctic confirm that conditions are now the warmest for the past 2000 years. The one exception is Greenland, probably because its climate is partly insulated, from the warming of the Norwegian–Greenland Sea, by the cold East Greenland Current.

Consistent with the current warming, satellite data show that both Greenland and Antarctica are losing land ice at increasing rates.

Further confirmation that the Arctic is changing radically comes also from its sea ice. The duration and magnitude of its decline is unprecedented since records began ~1500 years ago. In contrast, although sea ice around Antarctica increased very slightly since satellite measurements began (in 1978) it recently fell to levels below those of 1978. The extent of sea ice there is a function of the southward migration of the Polar Front due to global warming plus the increased intensity of surface winds caused by the development of the ozone hole [10]. In addition, as Jinlun Zhang of the Polar Science Centre of the Applied Physics Laboratory at the University of Washington pointed out in 2007, warming and freshening the Southern Ocean's surface increases its stratification, which prevents warm deep ocean water from rising to melt sea ice at the surface, thus helping to increase the potential of the surface to freeze [11]. Furthermore, the melting of ice shelves from beneath by warm water welling up onto the continental shelf supplies more cold freshwater near the coast, and freshwater freezes more readily than salt water. In addition, the stronger coastal winds push floes together in areas where winds converge, causing sea ice to thicken in the Weddell, Bellingshausen, Amundsen, and Ross Seas, thus increasing sea ice volume [12]. But the picture is not uniform. Changing wind patterns have made sea ice grow in the Indian Ocean sector and shrink in the Pacific Ocean sector.

Similar signs that the warming at the end of the twentieth century exceeds that of the past 2000 years come from studies of tree rings in Europe, from global collections of marine sediment cores, from a variety of other Earth materials collected globally, and from the shrinkage of the Quelccaya ice cap in the Peruvian Andes, which is now reversing changes that began some 5000 years ago.

These changes also appear in numerical palaeoclimate models fed with data on insolation, solar output, volcanic activity and greenhouse gases. Their outputs show that natural factors including the decline in Holocene insolation, plus the cooling effects of occasional large volcanic eruptions like that of Mt Pinatubo in 1991, plus the ups and downs of the solar cycle should have cooled the climate slightly since 1900. Instead it has warmed. The only plausible explanation is the addition of greenhouse gases, particularly $CO_2$ and methane from human activities, plus water vapour evaporated from the warming ocean, plus the effect of decreasing albedo attributable to the loss of Arctic sea ice and the greening of the Arctic. We do not need the models to tell us that. They merely confirm what is obvious from the palaeoclimate and modern data.

## 16.4   Where Is Climate Headed?

My analysis of the data presented in this book is that over the long term solar output will continue gradually rising as it has done for the past four billion years, and $CO_2$ will continue falling slightly in such a way that the climate will remain within limits suitable for life, much as described by Foster and colleagues [7]. Over the hundreds of millions of years timescale, plate tectonic processes are likely to regroup continental fragments into some new configuration like a Rodinia or a Pangaea. Already the conjunction of Australia and New Guinea with southeast Asia is underway, as is the conjunction of Europe and Africa. Whether or not the Atlantic may close again, as its predecessors have done in the past remains to be seen. If it does, then subduction of the carbonate sediments of the deep Atlantic will provide abundant $CO_2$, leading to a temporary rise in $CO_2$ that over time will be balanced by chemical weathering. Over the hundreds of thousands of years timescale, orbital and axial change is likely to continue much as before, as is solar change on timescales of millennia (e.g. the 2300 year Hallstatt cycle), centuries (e.g. the Suess–DeVries cycle), and decades (e.g. the Gleissberg Cycle), along with their combination tones and half tones. Plant life will play its own role in evolving to meet the changing levels of $CO_2$ and temperature.

On the timescale of human interest, orbital and axial properties will tend to keep the climate close to the neoglacial state for some 5000 years to come. It is the neoglacial state that brought us the LIA, possibly exacerbated by the manifestation of the 2300-year Hallstatt solar cycle. We could consider, then, that we should still be in the LIA as a continuing manifestation of the Holocene neoglacial.

The LIA itself was not uniform. It was punctuated by periods of solar warmth the latest of which peaked in 1780. While some scientists say that the LIA ended in 1850, the solar intensity was no different from that of 1780 in 1860, and that in turn was no different from what it was in 1980–1990. That observation confirms, to my mind, that if we had not added greenhouse gases to the atmosphere we might still be in the LIA.

What might we expect if the present decline in solar activity leads to another grand solar minimum? The published projections discussed in Chapter 15 suggest that we might get such an event somewhere in the 2050–2100 time frame. If we assume that the Suess–DeVries cycle reached its minimum somewhere around 1890, and a maximum around 1980–1990, then its next minimum might be expected around 2100. If that coincides with a Gleissberg solar minimum the climate might as cool (in the absence of $CO_2$) as it was in 1900. However, based on observations by Ljungqvist (Figure 15.4), we might expect the temperature of the Northern Hemisphere during such a minimum to fall by up to ~0.7 °C at most, and global temperature to fall much less, perhaps by 0.2–0.3 °C. By 2100, according to the IPCC, our emissions are likely to have taken the global temperature to between 1 and 4 °C above today's level [13]. Subtracting 0.3 °C caused by a possible grand solar minimum would reduce that rise to between 0.7 and 3.7 °C, values that are still substantial. Hence the arrival of such a solar minimum would not stop global warming – it would merely make conditions slightly less warm than they would otherwise be. Gerald Meehl of the US National Centre for Atmospheric Research, and colleagues, reached much the same conclusion in May 2013 [14]. Knowing that solar variability fluctuates within a narrow natural envelope suggests that it will not push us into icy conditions.

Beside the effects of variations in solar output, we would expect the future trajectory of global warming to be modified slightly upward or downward by internal oscillations like that of the El-Niño (warm)–La Niña (cool) couplet, the Pacific Decadal Oscillation, and the Atlantic Multidecadal Oscillation (positive phases = warm; negative phases = cool), as well as by cooling attributable to the occasional volcano capable of putting large volumes of reflective dust and sulphuric acid into the stratosphere. In addition, we are likely to see some slight cooling from any increase in the burning of coal and the associated output of aerosols caused by continued industrial expansion and transportation in the developing world, including India and China as they continue to industrialize.

Global heating will continue to slowly melt ice in the Arctic and to warm the ocean, with surface heat being taken slowly into the deep ocean. These processes will contribute to rising sea level. Measures of changing heat storage in the ocean, melting ice, snow and permafrost, and rising sea level provide independent measures of global warming **and should take precedence over just the average global temperature signal in our analysis of how the climate is changing**. Other independent measures of global warming include the extent to which animals, insects and plants are increasing their range polewards.

By how much might sea level rise as the ocean warms and land ice melts? In September 2019, the IPCC concluded that the rate of rise of sea level had increased to 3.6 mm/year between 2006 and 2015, a rate unprecedented over the last century, mostly due to the increasing loss of ice from Greenland and Antarctica [13]. By 2100, continued warming of the ocean and increased melting of land ice is projected to increase global sea level by 0.43 m (range 0.29–0.59 m) under the IPCC's low emissions scenario and by 0.84 m (range 0.61–1.10 m) under the high emissions scenario, with further increases beyond 2100 due especially to loss of ice from Antarctica. This means that local sea level rise events that historically occurred once per century may recur at least annually by 2100. The increasing frequency of high water levels can have severe impacts at the local level especially for low-lying areas and islands [13]. We have to bear in mind the disequilibrium in the climate system, which means that ice melt and sea level rise will lag a long way behind temperature rise. Looking at the geological evidence we can see that with the rise in temperature of 2–3 °C above today's levels in recent interglacial periods, sea level was between 4 and 9 m higher than it is today. And when warming persisted for much longer than the length of an interglacial period, sea levels for comparable rises in temperatures rose by up to ~15 m during the Palaeocene-Eocene Thermal Maximum, and possibly by as much as 22 m in the mid-Pliocene warm period. With the current rise in $CO_2$ to levels above 400 ppm, and of $CO_2e$ above 500 ppm, it is not stretching things much to suggest we could see a long-term rise in sea level of >9 m above the present in the 200–500 year timeframe if temperatures rise by 2 °C or more above 1900 levels [15]. Sea level rises of the order of 9 m would require the melting of significant volumes of land ice. But already there are signs that both the West Antarctic Ice Sheet in the vicinity of the Amundsen Sea, and the East Antarctic Ice Sheet in the Wilkes Basin are at risk of substantial decline [16, 17].

As a side effect, the rise in $CO_2$ is also gradually acidifying the ocean, which will lead to the saturation of the surface ocean in $CaCO_3$ falling below the level needed to sustain the building of aragonite skeletons by marine plankton. This will affect the Southern Ocean first, to the detriment of the organisms like pteropods (sea butterflies) at the base of the food chain [18]. The deleterious effects of ocean acidification are evident from the Palaeocene–Eocene Thermal Maximum 56 Ma ago, an event from which it took the world 100 000 years or more to recover.

The prospect of rapid climate change is of significant concern to scientists and policy makers alike. In 2013, James C. White of the University of Colorado at Boulder led a team of experts to evaluate the need to understand and monitor abrupt climate change and its impacts, on behalf of the US National Academy of Science [19]. And in 2011, on behalf of the Geological Society of London, I attended two meetings of experts convened by the UK government's Chief Scientific Advisor, Sir John Beddington, to consider the current evidence for and views on potential thresholds or 'tipping points' in the climate system. Both the US and UK groups agreed with the general consensus in the wider climate science community that there are likely to be tipping points in the climate system beyond which the rate of change may accelerate, but that there is a lot of uncertainty about what they are, when they might occur, and whether or not they might be preceded by warning signals.

Is the past climate record any guide to us as to what these tipping points may be? Possibly not, because the world now is not the same as it was when rapid changes happened in the past. For example, the rapid changes typical of Dansgaard–Oeschger Events took place during the intermediate stages of glaciations, when there were major ice sheets on North America and Scandinavia. Will further warming switch off the Thermohaline Conveyor, which brings warm air to western Europe via the Gulf Stream and its northern branches? According to Richard Alley [20], most numerical models suggest that this circulation may slow but not stop. Some models suggest that the melting of the Greenland ice could put a freshwater lid on the northern North Atlantic that could make the conveyor stop. Alley suggests that: '*even though a large, rapid, high-impact event seems unlikely based on most of the literature, the nonzero possibility and the potentially large impacts motivate further research*' [20].

One of the worries about continued global warming is its effect on possible supplies of methane ($CH_4$), which is a more potent greenhouse gas than $CO_2$. Methane is abundant on the deep sea floor, where the gas is trapped in cages of ice known as clathrates (see Chapter 10). While these clathrates are currently stable, their stability depends on the relation between their pressure and their temperature. As the deep ocean warms, the stability of the methane clathrate field may be compromised [21]. Melting clathrates would destabilize the sediments in which they sit, causing submarine slides and slumps that would release large quantities of methane. The volumes of gas trapped in these sediments are enormous, amounting to several trillion tonnes. However, the likelihood of marine clathrates melting over the next couple of hundred years is remote [21].

More likely is the emission of methane from the melting of Arctic permafrost, which is thought to contain about 1400 GtC [22]. Arctic expeditions on land and at sea find methane bubbling up from melting permafrost [22]. When we examine ice cores we find that methane values ranged from 350 to 400 ppb during glacials to 600–800 ppb during interglacials [23]. These limits define methane's natural envelope for the Ice Age. Even though past interglacials were 2–3 °C warmer than the present, their methane values did not rise above 800 ppb. We could infer that a comparable warming in the near future might not be accompanied by significantly increased emissions of methane [23]. Nevertheless, present $CH_4$ values (~1850 ppb) are already double the highest levels of past interglacials. Continued emissions of $CH_4$ from our activities, added to a likely increase in natural emissions from melting permafrost, must be of concern. The potential size of this concern is underlined by the fact that about 25% of the land in the northern hemisphere is underlain by permafrost of a few metres to more than 1 km in thickness. In some places in Siberia, natural gas is trapped beneath the permafrost, and might be released if the permafrost melts. Nevertheless, in 2019 the IPCC found '*low agreement whether northern permafrost regions are currently releasing additional net methane and $CO_2$ due to thaw*' [13].

Will the ice sheets on Greenland and West Antarctic melt rapidly? Parts of the West Antarctic Ice Sheet are now in a state of irreversible decline. There is also a risk that if sea level rises a few centimetres more we may lose ice from the Wilkes Basin of East Antarctica. We know that the Ross Ice Shelf melted away during the warm Pliocene. Numerical models suggest that it may also have done so during warm interglacials, along with much of West Antarctica [24]. But the timescale for such melting is likely to be a few hundred years, not decades. Even so, there is no reason to be complacent, since the melting of all Greenland's ice would raise sea level by about 7 m, whilst the melting of West Antarctica could add a further 5–6 m. At most we appear to be looking at a rise of 1–2 m by 2100.

Taking into consideration what we know of past climate, my view is that we are not in for rapid change in the next few decades. We face slow incrementally accelerating change, which to most people will seem imperceptible – at least for a while. If nothing is done, we will be subject to what I call a 'creeping catastrophe'. By that I mean that the effect by 2100 would be the equivalent of a catastrophe if it had happened all at once. Global warming is not like a tsunami.

Will we get over global warming quickly if we stop emissions now, or soon? No. There is a popular misconception that our additions of $CO_2$ to the air fall out within just a few years. In fact, David Archer of the University of Chicago calculates that '*The carbon cycle of the biosphere will take a long time to completely neutralize and sequester anthropogenic $CO_2$*' [25]. If we stopped putting $CO_2$ into the atmosphere tomorrow: '*17–33% of the fossil fuel carbon will still reside in the atmosphere 1 kyr from now, decreasing to 10–15% at 10 kyr, and 7% at 100 kyr. The mean lifetime of fossil fuel $CO_2$ is about 30–35 kyr*' [25]. This is comparable to '*the 10 kyr lifetime of nuclear waste … [which] seems quite relevant to public perception of nuclear energy decisions today. A better approximation of the lifetime of fossil fuel $CO_2$ for public discussion might be "300 years, plus 25% that lasts forever"*' [25]. Along with Andrey Ganopolski of the Potsdam Institute for Climate Impact Research, Archer took that analysis one step further. Estimating that 25% of present anthropogenic emissions will remain in the atmosphere for thousands of years, and about 7% will remain beyond 100 Ka, they concluded that it may take 500 Ka before the $CO_2$ falls low enough for the next glacial period to start [26].

We are now in the Anthropocene, a period during which human activity has become prevalent enough for it to operate like a geological force on the planetary environment [27]. We are following a trajectory of change out of the Holocene, in which the trapping of heat has now raised global temperature by 1–1.2 °C above what it was in 1900, to at least the level of the Holocene climatic optimum. Where will that trajectory take us next in relation to the temperatures of 1900: to 2–3 °C above, as typical of recent large interglacials or of the Pliocene; to 3–6 °C above, as typical of the Mid Miocene, or to 10 °C above, as typical of the Mid Eocene? [28].

## 16.5 Some Final Remarks

As my colleagues and I pointed out in writing the 2013 addendum to the statement on climate change of the Geological Society of London: '*These various geologically based considerations lend strength to the argument that continued emissions of $CO_2$ will drive further rises in both temperature and sea level. Given that the Earth system takes a long time to reach equilibrium in the face of change, the present changes are likely to continue long beyond 2100*' [29]. Furthermore, to end with the concluding phrase of the original GSL statement '*In the light of the evidence presented here it is reasonable to conclude that emitting further large amounts of $CO_2$ into the atmosphere over time is likely to be unwise, uncomfortable though that fact may be*' [30].

That statement is not alarmist. It is a simple statement of fact. Nevertheless, there is a significant implication in our continued and expanding use of fossil fuels and in the resulting expansion in emissions of $CO_2$ and their accumulation in the air and in the ocean. This book is not the place for a comprehensive review of potential impacts, nor of the means of adapting to or mitigating them. Nevertheless, some comments seem in order based on what the past tells us. If we continue to emit $CO_2$ in growing amounts the world will warm over the next 50–100 years, even if some of that warming is offset by a decrease in solar output that will, by its nature, be temporary. Even if such a temporary cooling effect should arise, it will not stop $CO_2$ from further increasing ocean acidification, with deleterious effects on the base of the food chain.

Continued warming will melt more glaciers, ice caps, and parts of the big ice sheets, raising sea level further. Low lying coastal communities will be affected. Many coastal cities may find themselves having to seek advice from the Dutch to stop the sea invading public places. Governments are already considering what coastal engineers must do to protect coasts from ongoing sea level rise in vulnerable areas. They can't all be protected, and some coastal land will be lost. In the developing world where coastal engineering is unaffordable, migration seems inevitable.

The effects of warming on land areas will be diverse. Cooling brings rain to middle latitudes. Warming does the opposite. Areas that are now dry may get dryer, whilst wet areas may get wetter. The reverse tends to be true in monsoonal areas. That geologically based conclusion is echoed by the reports of the IPCC and other organizations studying potential impacts. The drying of already dry areas like Australia, southern Africa, the south-western USA and the Mediterranean has begun. It will have deleterious effects on water supplies and agriculture, leading to migration on a scale that the military sees as a potential threat [31]. Warming also has an upside. It will push north the potential for wheat growing in Canada and Siberia. More $CO_2$ in the air will promote faster growth for some plants, but the accompanying warming may prove deleterious once specific thresholds are passed. How plants respond depends not just on $CO_2$, but also on heat and water.

My conclusions are based on the work of a great many palaeoclimatologists from a wide range of countries, as should be evident from the multitude of references to the scientific literature in each chapter. Their work tells us that the argument by a few scientists that the present warming is a natural response of the climate system to the end of the LIA is wrong. Amongst those sceptics of the human influence on global warming was Australian geologist Dr Robert Carter (1942–2016), who argued that we should accept what he called his '*null hypothesis*' of climate change, which is that '*global climate changes are presumed to be natural unless and until specific evidence is forthcoming for human causation*' [32]. However, Carter later acknowledged that '*The possibility of human-caused global warming nonetheless remains, because carbon dioxide is indubitably a greenhouse gas. The major unknown is the actual value of climate sensitivity … [although in his view] climate sensitivity is significantly less than argued by the IPCC'* [33]. In effect that was an admission that human causation had taken effect, albeit to what he considered an unknown and possibly trivial extent.

We now know, from the palaeotemperature profiles emerging from the outstanding and comprehensive studies of the likes of Vinther, Wanner, Mayewski, Moberg, Esper, Gifford Miller, Marcott and others, that it is highly improbable that the rapid rise in temperature since 1900 to the levels seen during the last decade of the twentieth century and those of the beginning of the twenty-first century is the result of a supposed natural 'recovery' to supposed 'normal values' from the cold depths of the LIA. The rise has been greater and faster than any seen during the 11 700 years of the Holocene, apart from the 'return to normal' after the short-lived 8.2 Ka cold event, and it is happening despite a lessening of the Sun's output since 1990. What should be considered 'normal' in the present case, as pointed out by Esper and many others, are the underlying cool temperatures of the past 2000 years of the Holocene neoglacial. Those temperatures are the core that runs through the various data within a range (the 'natural envelope') extending from the peak of medieval warming to the cold of the Maunder Minimum. That cold core temperature is what I would call the 'norm', and it is colder than what we are experiencing now. We have moved away from that cold average, and out of the influence of the natural envelope of orbital insolation for the late Holocene, out of the natural envelope of solar output for the past 2000 years, and out of the natural envelope of variability in atmospheric $CO_2$ for the past 800 Ka into a new domain, driven by our own emissions. **We do now have what Carter was seeking – specific evidence that [a good deal of] the present global change is unnatural**.

If I had been writing a book just about the climate of the Holocene, I could have gone into much greater detail about how unstable the hydrological cycle was over the past 11 700 years, and how that impacted upon early civilizations. Drought may well have been a contributory factor, for example, in the collapse of the Mayan civilization in central America. That level of detail was not my aim. It is well covered elsewhere [34]. Here it is sufficient for me to remind you of Magny's research that linked excessive rainfall in western Europe to lows in solar activity that caused widespread cooling. Long periods of floods or droughts will have been inimical to the development of populations anywhere on the globe. They have happened in the past and will undoubtedly happen again, and they will be just as disruptive or even more so, given that they are likely to affect much larger populations than those of the Maya. Many of those affected will not be city dwellers, but small-scale agriculturalists. There will be little or no opportunity for them to move to greener pastures. As Gwynne Dyer pointed out – conflicts will become more likely [31]. That is the published view of the US military, for example [35].

Before the Pleistocene Ice Age, there were larger fluctuations in $CO_2$ than we have experienced in the past 2.6 Ma. These too were constrained within a natural envelope since the rise of the land plants some 450 Ma ago. During most of that time levels of $CO_2$ in the air did not rise much above about 2000 ppm [7]. At the low end, given the continuity of photosynthesizing plants, they most likely did not fall below about 180 ppm, the lowest values of the Ice Age. This broad natural envelope reflects the interaction between changing rates of supply of $CO_2$, from volcanism and related processes in the Earth's interior, and its rates of extraction by chemical weathering, photosynthesis, and the sedimentation of marine carbonates. Lovelock's notion that Gaian processes maintain Earth's temperature at a suitable level (within a natural envelope) most likely applies also to the abundances of the atmospheric gases, though in his case he argued that life was providing the main control, whereas here we may be dealing more with a largely inanimate geochemical thermostat [8].

Even though amounts of $CO_2$ may have been higher in past times, rates of change were generally slower than over the past 2.6 Ma. For example, $CO_2$ rose by 600 ppm from 65 to 50 Ma ago in the Early Cenozoic [6], at a rate of 40 ppm/Ma, then fell by the same amount at the same rate between 50 and 35 Ma ago. Over the same period average global temperatures rose by about 10 °C, then fell by the same amount, whilst bottom water temperatures rose by between 2 °C [37] and 4 °C [37], then cooled by between 4 °C [36] and 7 °C [36]. These small, slow changes are typical of Earth in its equilibrium state with no ice sheets, although the somewhat larger cooling of the late Eocene bottom water represents a period when glacial ice was beginning to form on Antarctica. The doubling of $CO_2$ from 65 to 50 Ma ago caused a temperature rise of 8 °C, indicating a high equilibrium climate sensitivity. Much the same sensitivity applied from 50 to 35 Ma ago. We have come across similar estimates of climate sensitivity elsewhere in this book (and see the penultimate paragraph of this chapter).

Since we were not around to observe and measure what was going on, we have to use the principles of historical geology (as invented by Lyell) to use our knowledge of processes now at work to infer processes that we could not see from results that have been preserved, just as Darwin did in studying evolution from the fossil record. In taking that step we must be willing to '*recognize how much work the most insignificant processes can accomplish with enough time*' [38]. Given enough time, tiny changes in $CO_2$ can cause significant changes in climate. Changes in $CO_2$ can cause large changes in the short term that become smeared out as slow ocean circulation and weathering take effect and restore the planetary equilibrium. Historical reasoning is demanded. We use it to infer the history of stars, the history of evolution, the history of our planet, and, now, the history of our climate. That history is part of our probing of deep time. It provides us with a wider appreciation of possibilities for the future. It enriches our world-view.

This is a story about 'emergence' – the emergence of the recognition that greenhouse gases play one of the key roles in controlling our climate. There are parallels with the story of the emergence from the late 1960s on, with the advent of powerful microscopic and genetic technologies, of the notion that all plants and animals are symbiotic assemblages of organisms (think of the mitochondria in animal cells or the plastoids of plant cells, which were once free living bacteria) [39]. Emergence of the greenhouse gas theory of climate control challenges geological orthodoxy by contradicting the dogma that the effect of greenhouse gases is insignificant or that human emissions can have no effect on the climate. The emergence of the infrared radiative transfer theory of climate change and the serial endosymbiosis theory of life both provide radical revisions of scientific thought in the era since 1960. The data concerning them are freely available from the life work of scholars worldwide. Margulis, the inventor of the endosymbiosis theory of life is also, with Lovelock, one of the developers of the Gaia theory, which argues that just as our bodies maintain a relatively stable internal temperature despite changing external conditions, so the Earth System keeps the planet's temperature stable (within certain natural limits), by modifying external conditions through negative feedback – e.g. lowering $CO_2$ and hence temperature to counter the rise in the Sun's output through time. As Margulis puts it '*Gaia is an emergent property of interaction among organisms, the ... planet ... and an energy source, the Sun ... It is a convenient name for an Earthwide phenomenon: the regulation of temperature, acidity/alkalinity, and gas composition ... [by] a series of interacting ecosystems that compose a single huge ecosystem at the Earth's surface*' [39]. Margulis regarded the entire planetary surface as 'metastable' because it is far from chemical equilibrium – otherwise we would not have reactive gases like methane in the atmosphere. Nevertheless, Leeder is right to point out that Gaia theory fails to incorporate the non-physical facts of plate tectonics, which is why he calls for a merging of plate tectonic and Gaian theory to understand the operations of the Earth's surface, including its ocean and atmosphere [8]. To fully comprehend how the world works now '*requires geologists, geochemists, atmospheric chemists, and even*

*meteorologists to understand science outside of their own fields ... [including] biology, especially microbiology ... [actions] that people in related fields are loath to take*' [39]. I concur.

The concept of natural envelopes for climate that I promote here fits well with the notions of Hutton and Lyell that Earth cycles on within certain well-defined limits. But there is also plenty of room for Cuverian catastrophes, notably in the form of the lengthy and massive eruptions of flood basalts in Large Igneous Provinces from time to time, and the occasional asteroid impact. Whilst these pushed Earth's climate temporarily out of its natural climate envelope and caused mass extinctions of life, things 'rapidly' (in geological terms) returned to 'normal' afterwards.

## 16.6   What Can Be Done?

Will we be able to do anything about the changes we are now making to the climate system? We did successfully take a stand against the emission of ozone-destroying substances, but they were mainly produced by a very few chemical companies in a very few countries, and it turned out after much discussion that the companies had alternatives on the production line [40]. Problem solved. An easy win.

Global warming is not like that. We all contribute to it, by heating or cooling our homes, driving cars, travelling on planes and trains, and consuming the products of manufacture and agriculture from a globalized industry. The energy consumed comes not from a few chemical companies in a few countries, but from globally distributed gas, oil, and coal companies located in many countries and managed either privately or nationally. Those activities are supported by a massive infrastructure stretching from coal mines and oil and gas wells through pipelines, tankers, and bulk carriers to power stations and petrol station forecourts. This complexity and ubiquity is what makes the solution so much more difficult than closing the ozone hole.

The general tendency of governments faced with environmental problems is to wait until a catastrophe happens, at which point people clamour for change and governments finally act. Take the River Thames for example. For centuries the river was used as London's sewer. This became more and more of a problem as the city's population grew, until, in the words of Prime Minister Benjamin Disraeli (1804–1881), the river had become '*A Stygian pool reeking with ineffable and unbearable horror*' [41]. The year 1858 was labelled the '*Great Stink*'. Members of Parliament contemplated moving the House of Commons upstream to Hampton Court, and London's law courts prepared to move out to Oxford. Something had to be done. Parliament already had a solution to hand, having set up a Metropolitan Board of Works in 1855 to consider how to clean up the river. Civil engineer Joseph Bazalgette (1819–1891) devised a scheme to take sewage through tunnels to two main out-falls well downstream. His network of 1100 miles of local sewers and 165 miles of main sewers was described by the Observer newspaper of the time as '*the most extensive and wonderful work of modern times*' [41]. It solved the problem, and is still in use today. Peter Ackroyd describes the enormous task of disposing of London's human waste as '*the city's secret industry*' [41]. That industry is now at work everywhere in all cities, ensuring public health. We do not think twice about it. Out of sight, out of mind! But it took a virtual catastrophe to start the ball rolling. And no doubt there was a large initial investment.

The story of smog is much the same. The burning of coal in London polluted the city with smoke, and was a cause of complaint to Elizabeth I, amongst others. In 1661, diarist and writer John Evelyn (1620–1706) wrote a treatise on the problem, entitled '*Fumifigium*', or '*The Inconvenience of the Air and the Smoak of London*', lamenting the condition of the city covered by '*a Hellish and dismal Cloud of SEA-COAL*' [41]. Peter Ackroyd reminds us that: '*Victorian fog is the world's most famous meteorological phenomenon. It was everywhere, in Gothic drama and in private correspondence, in scientific correspondence and in "Bleak House" (1852-53) ... Gas lights were turned on throughout the day in order to afford some interior light .... The street lamps seemed points of flame in the swirling miasma*' [41]. The worst of the fogs were the London '*smogs*' of the 1950s, which I remember well. You could hardly see your hand in front of your face at mid-day on occasion. Thousands died of respiratory problems. Although public disquiet led to the UK passing a Clean Air Act in 1956, a severe smog in 1962 '*killed sixty people in three days; there was "nil visibility" on the roads, shipping "at a standstill", trains cancelled*' [41]. Something had to be done, and was, by means of a more extensive Clean Air Act in 1968. Further improvements followed. The costs were accepted. Smog was stopped.

These simple examples from one city in one country represent many similar public health initiatives around the world. They illustrate the truth of the words of Jorgen Randers in his global forecast for the next 40 years to 2052 [42]: '*Experience shows that it is hard for democratic, free-market economies to make proactive decisions to increase voluntary investments before they are unavoidable. It is much simpler [to do so] after crisis has struck and there is an externally imposed threat of destroyed infrastructure and livelihoods ... Solutions will come on line much later than optimal – at least in those parts of the world where the majority favors the market. Collective solutions will not be used until it is overabundantly clear that private solutions (based on individual initiatives in an unrestrained market) will not suffice*' [42]. Perceptive words.

We are now living through the global equivalent of London's '*Great Stink*' of 1858 or the vile smogs that preceded the Clean Air Acts of the 1950s and 1960s. The difference is that the scale is now global not local, and that the carbon dioxide that we are adding to the atmosphere is colourless and odourless. Not being able to see it or smell it, it's hard to believe it's really there or that it can possibly be in any way a danger to us. $CO_2$ is an integral component of the natural environment, so can it be classified as a pollutant? Yes, in the same way that many toxic chemicals, like arsenic, or mercury, are naturally present in the environment in small amounts, but become pollutants when we dump concentrations of them into small areas. In the case of $CO_2$ that area is the whole atmosphere and the upper layers of the ocean. I use the word pollutant here in the sense of it being a substance that has undesirable side effects when added to the environment. Note the use of the word 'added'. It is only our 'additions' of $CO_2$, not the natural background of $CO_2$, that pollute. And even here the description is moot, because plants are using the added $CO_2$ to 'green the planet'. What we tend to forget is that the atmosphere is very thin – most of its air occurs within about 10 km of Earth's surface, a distance an athlete can run in just over 25 minutes; in relative terms it is equivalent to the veneer on a piece of furniture.

Whilst a target of no more than 450 ppm of combined greenhouse gas emissions in the atmosphere has been deemed likely to hold global warming to an average of 2 °C above preindustrial levels, countering it could require an 80% or more cut in $CO_2$ emissions by 2050. And in December 2015 the UN's IPCC recommended a target of 1.5 °C. As Randers pointed out [42], given the $CO_2$ levels we have today (meaning 2012), it seems highly unlikely that we shall meet these targets. Indeed, the voluntary pledges made at a New York Heads of State meeting in September 2019 (and especially those pledges from the industrialized nations) were still above the levels required to meet those targets. $CO_2$ goes on climbing. 450 ppm $CO_2$ is just around the corner. Nevertheless, improvements are being made. Energy efficiency is likely to increase by a further 30% by 2050, but by then there will be 2.5 billion more people on the planet wanting to use progressively more energy [42]. Randers forecasts that energy use will grow by 50% in that time period. If there were no change in energy sources, that would imply an increase in $CO_2$ emissions by 50%. But, Randers expects a levelling off in the use of oil, which should peak by 2025 and then decline. He expects the use of coal and gas to peak by 2040 and then decline [42]. We shall see. The Stockholm Resilience Centre seems to think we can meet these targets by thinking and acting smarter [43].

Energy demand will increasingly be met by renewable energy sources, because they are technically feasible and their costs continue to fall, making them increasingly competitive. As an example of the rapid rate of falling costs, Chris Goodall cites solar panels, whose costs declined from about 100 $ per Watt in 1975 to about 35 cents by 2016, and could be down to 10 cents per Watt by about 2025 [44]. In due course, solar panels may give way to solar roof tiles, which could be fitted to every house [45]. If Randers is right, the world's consumption of fossil fuels could be in steep decline by 2052 [42]. Others, like Jeremy Leggett, broadly agree [45]. In the short-term Randers sees a major swing to gas, which has already begun in the USA and western Europe. Replacing coal with natural gas in power stations will reduce the amount of $CO_2$ emitted by two-thirds, and is a big step towards a low-carbon future.

Whilst this is a good thing in the short-term, it postpones the inevitable shift away from burning fossil fuels to eventual use of direct or indirect solar power. The road to moving there directly will remain blocked for some time by a combination of costs, the inadequacies of storage (we need new battery technology), vested interests, the emphasis on short-term profits, and the weakness of governments in the face of powerful lobbies [42, 45]. Long-term thinking is rare in our predominantly market-driven modern economies. Long-term action is even rarer. But, by mid-century the problems caused by global warming will loom large, renewable energy technologies will be much cheaper, and carbon capture and storage may have begun, which may lead governments to take more serious action than now seems possible.

As US economist William Nordhaus of Yale University points out, the economic case for action seems beyond dispute. Nordhaus accepts that we do not know everything about global warming with 100% certainty, but good scientists are never 100% sure about any empirical phenomenon. *'The advice of climate science contrarians is to ignore the dangers in [what he calls] the Climate Casino. To heed that advice is a perilous gamble ... Those who burn fossil fuels'* he reminds us *'are enjoying an economic subsidy – in effect, they are grazing on the global commons and not paying for what they eat. Raising the carbon price would correct for the implicit subsidy on the use of carbon fuels'* [46]. Companies who think otherwise, he opines *'are really looking out for their profits and not for the public welfare'* [46]. Much of the confusion about global warming in the public's mind, he reminds us, comes from disinformation supplied by lobbyists keen to 'sow doubt', a tactic well-documented by Naomi Oreskes and Erik Conway [47], and used in the past by those keen to prevent the public from appreciating the strong statistical link between cigarette smoking and cancer. Indeed, the battle against global warming has many parallels with the battle against smoking – both require people to fight against their freedoms. And yet those same people will obey traffic lights in the interests of the greater good (and their own personal safety).

We also face the problem that these days the variety of economics taught globally is Neoclassical Economics, in which the environment is considered to be an externality – something that has no value. If the environment (air, water, ocean) has no value, then you can dump what you like in it, at zero cost. This kind of economics underpins the 'Tragedy of the Commons': nobody owns the atmosphere or the ocean, so we use them as dumping grounds. The polluter does not pay. The public bears the clean-up cost. Happily, there are alternative economic approaches. The students of Manchester University recently published *The Econocracy: The Perils of Leaving Economics to the Experts* (meaning the neoclassical economists) [48]. Kate Raworth takes a similar approach in her book *Doughnut Economics* [49], as does Clair Brown in *Buddhist Economics* [50]. We should pay attention.

I hope that *Palaeoclimatology* contributes constructively to the discussion of these major issues by bringing to the fore what we have learned from this 200-year old science about the history of climate change. It has a much to tell us about what our future may look like as we continue to spew $CO_2$ into the air. One can always argue with particular pieces of evidence and with the outputs of computer models of the climate system, but as my friend Bryan Lovell is fond of pointing out – *'you can't argue with a rock'*. As long as he is prepared for ice to be included in the definition of rock, I wholeheartedly agree. Rocks carry an integral climate message from the past. They have a stirring tale to tell that should make us pause in our onward rush. They happen, not by accident, to carry the same message that we hear by other routes and based on other sciences, from the IPCC. Action is urgently needed. Policies are being developed to address the matter, though we are in danger of doing too little too late.

Aside from examining the geological evidence to see if current theories of climate change are making testable predictions, we also have to challenge the self-styled global warming sceptics to present coherent and comprehensive hypotheses of their own, the predictions from which can be tested against the observable record. For example, if anyone disagrees with the conclusion that a rise in $CO_2$ caused the rise in temperature at the Palaeocene–Eocene boundary, or that a fall in $CO_2$ led to the cooling of the Cenozoic, they must come up with plausible and testable alternatives. As Nate Silver reminds us, we cannot just *'rummage through fact and theory alike for argumentative and ideological convenience*'; rather we must '*weigh the strength of the new evidence against the overall strength of the theory*' [51]. Cherry-picking the data to support a particular argument is not on. And belief is out of the question.

Where are we now? As Gavin Foster and colleagues observed in 2018 '*Given that the rate of carbon addition during our 'anthropogenic hyperthermal' eclipses that of the PETM, at the very least we are likely looking at a potential future with a more severe impact of life on Earth than any climate change event in the last 56 Myr. Exactly how severe, however, remains perhaps one of the most pressing of the "unknown unknowns"*' [52].

Science is an iterative process – there is always some progress at the leading edge that improves the picture. For our purposes I highlight the very latest (2019) study of climate sensitivity, by Jiang Zhu of the University of Michigan at Ann Arbor, and colleagues. Zhu, like others before, knew that climate models of past times commonly run cold – that is they seem unable to replicate the warmth of periods like the mid Eocene, the Mid Miocene and the Mid Pliocene. Zhu's new simulations of Eocene times were able to replicate the warmth of the Mid Eocene 50 Ma ago and that of the Palaeocene–Eocene Thermal Maximum, by using a model in which there was strong short-wave feedback from clouds. They summarized their results dryly as follows: '*simulations exhibit increasing equilibrium climate sensitivity with warming and suggest an Eocene sensitivity of more than 6.6°C, much greater than the present-day value (4.2°C)*' [53]. In essence, without going into the detail, their results show that warming affects the small-scale physical processes within clouds, which act to reduce both cloud cover and the opacity of clouds, thus increasing the penetration of short-wave solar radiation to the Earth's surface and hence enhancing warming. Assuming that they are correct, this means that current warming will beget future warming by weakening the protective capacity of clouds. A warming world will warm faster because climate sensitivity will increase with warming [53]. If indeed the equilibrium sensitivity is as large as they suggest (and as others have pointed out elsewhere in this book, based on consideration of the palaeoclimate record), then we can expect higher temperatures, more melting ice, and higher sea levels than the IPCC suggests for 2100 and periods beyond; the amounts of increase are yet to be calculated.

Let me conclude with two observations from biologist Lynn Margulis: '*we cannot put an end to nature; we can only pose a threat to ourselves*' … [and] '*runaway populations … always collapse*' [39]. There is no Planet B.

# References

**1** Jackson, S.T. (2019). Humboldt for the Anthropocene. Perspectives essay for insights. *Science* 365 (I645): 1074–1075.

**2** Broecker, W. (2010). *The Great Ocean Conveyor: Discovering the Trigger for Abrupt Climate Change*, 154 pp. Princeton: Princeton Univeristy Press.

**3** Kolbert, E. (2014). *The Sixth Extinction – An Unnatural History*, 319 pp. London: Bloomsbury.

**4** Beerling, D. (2007). *The Emerald Planet: How Plants Changed Earth's History*, 288 pp. Oxford University Press.

**5** Kidder, D.L. and Worsley, T.R. A human-induced hothouse climate. *GSA Today* 22 (2): 4  11.

**6** Beerling, D.J. and Royer, D.L. (2011). Convergent Cenozoic $CO_2$ history. *Nature Geoscience* 4: 418–420.

**7** Foster, G.L., Royer, D.L., and Lunt, F.J. (2017). Future climate forcing potentially without precedent in the last 420 million years. *Nature Communications* 8 (14845) https://doi.org/10.1038/ncomms14845.

**8** Leeder, M. (2007). Cybertectonic Earth and Gaia's weak hand: sedimentary geology, sedimentary cycling and the Earth system. *Journal of the Geological Society of London* 164: 277–296.

**9** Hays, J.D., Imbrie, J., and Shackleton, N.J. (1976). Variations in the earth's orbit: pacemaker of the ice ages. *Science* 194 (4270): 1121–1132.

**10** Maksym, T., Stammerjohn, S.E., Ackley, S., and Massom, R. (2012). Antarctic sea ice – a polar opposite. *Oceanography* 25 (3): 140–151.

**11** Zhang, J. (2007). Increasing Antarctic sea ice under warming atmospheric and oceanic conditions. *Journal of Climate* 20: 2525–2529.

**12** Zhang, J. (2014). Modeling the impact of wind intensification on Antarctic sea ice volume. *Journal of Climate* 27: 202–214. https://doi.org/10.1175/JCLI-D-12-00139.1.

**13** IPCC (2019). Summary for policymakers. In: *IPCC Special Report on the Ocean and Cryosphere in a Changing Climate* (eds. H.-O. Pörtner, D.C. Roberts, V. Masson-Delmotte, et al.). Intergovernmental Panel on Climate Change. In press.

**14** Meehl, G.A., Arblaster, J.M., and March, D.R. (2013). Could a future "Grand Solar Minimum" like the Maunder Minimum stop global warming? *Geophysical Research Letters* 9: 1789–1793.

**15** Foster, G.L. and Rohling, E.J. (2013). Relationship between sea level and climate forcing by $CO_2$ on geological timescales. *Proceedings of the National Academy of Sciences of the United States of America* 110 (4): 1209–1214.

16 Pritchard, H.D., Ligtenberg, S.R.M., Fricker, H.A. et al. (2012). Antarctic ice-sheet loss driven by basal melting of ice shelves. *Nature* 484: 502–505.

17 Mengel, M. and Levermann, A. (2014). Ice plug prevents irreversible discharge from East Antarctica, *Nature Climate Change online* 4: May 2014. doi:10.1038/nclimate2226.

18 Orr, J.C., Fabry, V.J., Aumont, O. et al. (2005). Anthropogenic ocean acidification over the twenty-first century and its impact on calcifying organisms. *Nature* 437: 681–686. https://doi.org/10.1038/nature04095.

19 White, J.C., Alley, R.B., Archer, D.E. et al. (2013). *Abrupt Impacts of Climate Change – Anticipating Surprises*, 188 pp. Washington D.C.: National Academies Press.

20 Alley, R.B. (2007). Wally was right: predictive ability of the North Atlantic "Conveyor Belt" hypothesis for abrupt climate change. *Annual Reviews of Earth and Planetary Science* 35: 241–272.

21 Hay, W.W. (2013). *Experimenting on a Small Planet: A Scholarly Entertainment*, 983 pp. New York: Springer.

22 Shakhova, N., Semiletov, I., Salyuk, A., Kosmach, D. (2008). Anomalies of methane in the atmosphere over the East Siberian Shelf: is there any sign of methane leakage from shallow shelf hydrates? *Geophysical Research Abstracts* 10, EGU General Assembly 2008-A-01526.

23 Wolff, E.W. (2011). Greenhouse gases in the Earth system: a palaeoclimate perspective. *Philosophical Transactions. Series A, Mathematical, Physical, and Engineering Sciences* 369: 2133–2147.

24 Pollard, D. and De Conto, R.M. (2009). Modelling West Antarctic ice sheet growth and collapse through the past five million years. *Nature* 458: 329–333.

25 Archer, D. (2005). Fate of fossil fuel $CO_2$ in geologic time. *Journal of Geophysical Research* 110: C09S05. https://doi.org/10.1029/2004JC002625.

26 Archer, D. and Ganopolski, A. (2005). A movable trigger: fossil fuel $CO_2$ and the onset of the next glaciation. *Geochemistry Geophysics Geosystems* 6: Q05003. https://doi.org/10.1029/2004GC000891, 7 pp.

27 Waters, C.N., Zalasiewicz, J., Summerhayes, C. et al. (2016). The Anthropocene is functionally and stratigraphically distinct from the Holocene. *Science* 351 (6269) https://doi.org/10.1126/science.aad2622, 10 pp.

28 Steffen, W., Rockström, J., Richardson, K. et al. (2018). Trajectories of the Earth system in the Anthropocene. *Proceedings of the National Academy of Sciences of the United States of America* 115 (33): 8252–8259.

29 Geological Society of London (2013). *Addendum to Geological Society Statement on Climate Change*, http://www.geolsoc.org.uk/climatechange.

30 Geological Society of London (2010). *Geological Society Statement on Climate Change*, https://www.geolsoc.org.uk/climaterecord

31 Dwyer, G. (2008). *Climate Wars: The Fight for Survival as the World Overheats*, 267 pp. London: Random House.

32 Carter, R.M. (2010). *Climate: The Counter Consensus*, 315 pp. Stacey International.

33 Carter, R.M., and Courtillot, V. (2013). Letter to Dr B Peiser of the Global Warming Policy Foundation, dated 14/02/2013 and published on the science page of the GWPF web site. https://www.thegwpf.org/geological-perspective-global-warming

34 Diamond, J. (2005). *Collapse: How Societies Choose to Fail or Survive*, 575 pp. Allen Lane: Penguin Books, London.

35 U.S. Department of Defense (2015). *Report to Congressional Inquiry on National Security Implications of Climate-Related Risks and a Changing Climate.* U S Department of Defense RefID: 8-6475571, 14pp.

36 Cramer, B.S., Miller, K.G., Barrett, P.J., and Wright, J.D. (2011). Late Cretaceous–Neogene trends in deep ocean temperature and continental ice volume: reconciling records of benthic foraminiferal geochemistry ($\partial^{18}O$ and Mg/Ca) with sea level history. *Journal of Geophysical Research* 116: C12023, 23 pp.

37 Zachos, J., Pagani, M., Sloan, L. et al. (2001). Trends, rhythms and aberrations in global climate 65 Ma to present. *Science* 292: 686–693.

38 Gould, S.J. (1984). Worm for a Century and all seasons. Chapter 10 in *Hen's Teeth and Horse's Toes. – Further Reflections on Natural History*. Pelican Books, 413pp

39 Margulis, L. (1998). *Symbiotic Planet – A New Look at Evolution*, 146 pp. Basic Books.

40 Oppenheimer, M., Oreskes, N., Jamieson, D. et al. (2019). *Discerning Experts – The Practices of Scientific Assessment for Environmental Policy*, 281 pp. Chicago University Press.

41 Ackroyd, P. (2001). *London – The Biography*, 822 pp. Random House, London: Vintage.

42 Randers, J. (2012). *2052: A Report to the Club of Rome Commemorating the 40th Anniversary of "The Limits to Growth"*, 392 pp. Vermont: Chelsea Green Publishing.

43 Randers, J., Rockström, J., Stoknes, P.E. et al. (2018). *Transformation Is Feasible: How to Achieve the Sustainable Development Goals within Planetary Boundaries*, 59 pp. Stockholm Resilience Centre.

44 Goodall, C. (2016). *The Switch*, 274 pp. Profile Books.

45 Leggett, J. (2014). *The Energy of Nations: Risk Blindness and the Road to Renaissance*, 252 pp. Routledge, London: Earthscan.

46 Nordhaus, W. (2013). *The Climate Casino: Risk, Uncertainty, and Economics for a Warming World*, 378 pp. Yale University Press.

**47** Oreskes, N. and Conway, E.M. (2010). *Merchants of Doubt*, 355 pp. London: Bloomsbury Press.

**48** Earle, J., Moran, C., and Ward-Perkins, Z. (2017). *The Econocracy – The Perils of Leaving Economics to the Experts*, 212 pp. Manchester University Press.

**49** Raworth, K. (2017). *Doughnut Economics*, 372 pp. Penguin, Random House.

**50** Brown, C. *Buddhist Economics*, 203 pp. London: Bloomsbury Press.

**51** Silver, N. (2013). *The Signal and the Noise: The Art and Science of Prediction*, 534 pp. London: Penguin Books.

**52** Foster, G.L., Hull, P., Lunt, D.J., and Zachos, J.C. (2018). Placing our current 'hyperthermal' in the context of rapid climate change in our geological past. *Philosophical Transactions of the Royal Society. Series A, Mathematical, Physical, and Engineering Sciences* 376 20170086.

**53** Zhu, J., Poulsen, C.J., and Tierney, J.E. (2019). Simulation of Eocene extreme warmth and high climate sensitivity through cloud feedbacks. *Science Advances* 5: eaax1874.

# Appendix 1

# Further Reading

Alley, R.B. (2000). *The Two-Mile Time Machine: Ice Cores, Abrupt Climate Change, and Our Future*, 229 pp. Princeton: Princeton University Press.

Alley, R.B. (2011). *Earth: The Operators' Manual*, 479 pp. New York: Norton.

Alverson, K., Bradley, R.S., and Pedersen, T. (2003). *Paleoclimate, Global Change and the Future*, 220 pp. London: Springer.

Archer, D. (2007). *Global Warming: Understanding the Forecast*, 194 pp. Oxford: Blackwell.

Archer, D. and Pierrehumbert, R. (2011). *The Warming Papers: The Scientific Foundation for the Climate Change Forecast*, 419 pp. Chichester: Wiley Blackwell.

Battarbee, R.W. and Binney, H.A. (eds.) (2008). *Natural Climate Variability and Global Warming: A Holocene Perspective*, 276 pp. Chichester: Wiley Blackwell.

Beering, D. (2019). *Making Eden*, 257 pp. Oxford University Press.

Beerling, D. (2007). *The Emerald Planet*, 288 pp. Oxford University Press.

Bender, M.L. (2013). *Paleoclimate*, Princeton Primers in Climate, 306 pp. Princeton University Press.

Berner, R.A. (2004). *The Phanerozoic Carbon Cycle: $CO_2$ and $O_2$*, 144 pp. Oxford University Press.

Broecker, W. (2010). *The Great Ocean Conveyor: Discovering the Trigger for Abrupt Climate Change*, 154 pp. Princeton: Princeton University Press.

Capra, F. and Luisi, P.L. (2014). *The Systems View of Life: A Unifying Vision*, 498 pp. Cambridge University Press.

Cowie, J. (2013). *Climate Change: Biological and Human Aspects*, 558 pp, 2e. Cambridge: Cambridge University Press.

Cronin, T.M. (2010). *Paleoclimates*, 441 pp. New York: Columbia University Press.

Euzen, A., Gaill, F., Lacroix, D., and Cury, P. (2017). *The Ocean Revealed*, 321 pp. Paris: CNRS Éditions. ISBN: 978-2-271-11907-0.

Hay, W.W. (2013). *Experimenting on a Small Planet: A Scholarly Entertainment*, 983 pp. London: Springer.

Jouzel, J., Lorius, C., and Raynaud, D. (2013). *The White Planet: The Evolution and Future of our Frozen World*, 306 pp. Princeton: Princeton University Press.

Köppen, W. and Wegener, A. (1924) Die Klimate der geologischen Vorzeit, Borntraeger Publishing House, Berlin, 255 pp. Reprinted in 2015 by the same publishing house, as well a translation into English by Bernard Oelkers (Bremen).

Kump, L.R., Kasting, J.F., and Crane, R.G. (2013). *The Earth System*, 462 pp, 3e. Pearson, Ltd.

Langmuir, C.H. and Broecker, W. (2012). *How to Build a Habitable Planet*, 718 pp. Princeton University Press.

Lenton, T. (2016). *Earth System Science: A Very Short Introduction*, 153 pp. Oxford University Press.

Lenton, T. and Watson, A. (2011). *Revolutions that Made the Earth*, 423 pp. Oxford University Press.

Lewis, S.L. and Maslin, M.A. (2018). *The Human Planet: How We Created the Anthropocene*, 465 pp. London: Pelican Books, Penguin.

Lovell, B. (2010). *Challenged by Carbon: The Oil Industry and Climate Change*, 212 pp. Cambridge University Press.

Macdougall, D. (2006). *Frozen Earth: The Once and Future Story of Ice Ages*, 256 pp. Berkeley: California University Press.

Mann, M.E. (2012). *The Hockey Stick and the Climate Wars: Dispatches from the Front Lines*, 395 pp. New York: Columbia University Press.

Masson-Delmotte, V. (2011). *Climat: Le Vrai et le Faux*, 103 pp. Paris: Éditions Le Pommier.

Mayewski, P.A. and White, F. (2002). *The Ice Chronicles: The Quest to Understand Global Climate Change*, 233 pp. Hanover: University Press of New England.

North, G.R., Biondi, F., Bloomfield, P. et al. (2006). *Surface Temperature Reconstructions for the Last 2000 Years*, 145 pp. Washington, D.C.: National Academies Press.

Oreskes, N. and Conway, E.M. (2010). *Merchants of Doubt*, 355 pp. London: Bloomsbury Press.

Pollack, H. (2009). *A World Without Ice*, 290 pp. New York: Avery-Penguin.

*Palaeoclimatology: From Snowball Earth to the Anthropocene*, First Edition. Colin P. Summerhayes.
© 2020 John Wiley & Sons Ltd. Published 2020 by John Wiley & Sons Ltd.

Richardson, K., Steffen, W., and Liverman, D. (2013). *Climate Change: Global Risks, Challenges and Decisions*, 501 pp. Cambridge University Press.

Ruddiman, W.F. (2005). *Plows, Plagues and Petroleum*, 202 pp. Princeton: Princeton University Press.

Ruddiman, W.F. (2014). *Earth's Climate: Past and Future*, 445 pp, 3e. New York: W.H. Freeman.

Steffen, W., Sanderson, A., Tyson, P.D. et al. (2004). *Global Change and the Earth System: A Planet Under Pressure*, 336 pp. Springer, Berlin: Global Change – the IGBP Series.

Turner, J. and Marshall, G.J. (2011). *Climate Change in the Polar Regions*, 434 pp. Cambridge University Press.

White, S., Pfister, C., and Mauelshagen, F. (eds.) (2018). *The Palgrave Handbook of Climate History*, 656 pp. London: Palgrave/Macmillan.

Zalasiewicz, J. and Williams, M. (2012). *The Goldilocks Planet: The Four Billion Year Story of Earth's Climate*, 303 pp. Oxford: Oxford University Press.

Zalasiewicz, J., Waters, C.N., Williams, M., and Summerhayes, C.P. (2019). *The Anthropocene as a Geological Time Unit*, 361pp. Cambridge University Press.

# Appendix 2

## List of Figure Sources and Attributions

**Figure 2.1** Portrait by François-Hubert Drouais, Musée Buffon, Montbard, 1753.Wikimedia – in the public domain.

**Figure 2.2** Portrait by Sir Henry Raeburn in the Scottish National Portrait Gallery. Wikimedia – in the public domain.

**Figure 2.3** Portrait by François-Andre Vincent. In 'The Gallery of Portraits: with Memoirs' vol. 2, C. Knight, London. Wikimedia – in the public domain.

**Figure 2.4** Portrait by Friedrich Georg Weitsch. In McCrory, D. (2010) 'Nature's Interpreter – The Life and Times of Alexander von Humboldt'. Lutterworth Press, Cambridge. Wikimedia – in the public domain.

**Figure 2.5** Extract from Humboldt's Tableau Physique in 'Geographie der Pflanzen in den Tropen-Landern' (1805). Wikimedia – in the public domain.

**Figure 2.6** Extract from the Frontispiece to Bonney, T.G. (1895) 'Charles Lyell and Modern Geology'. Macmillan, New York. In compliance with the conditions specified by Project Gutenberg (http://www.gutenberg.net).

**Figure 2.7** From Figures 14 (A) and 15 (B) in Lyell, C. (1875) 'Principles of Geology'. 12th Ed., v.1, John Murray, London; (also reprinted as Figure 1 in Summerhayes, C.P. (1990) Palaeoclimates. J. Geol. Soc. London 147, 315–320). In the public domain.

**Figure 2.8** Portrait engraved by Samuel Cousins, 1833, after the 1832 portrait by Thomas Phillips. Published with permission from the Oxford University Museum of Natural History.

**Figure 2.9** Darwin at 40; lithograph by T. H. Maguire 1849. Creative Commons Atribution only license CC by 4.0; http://creativecommons.org/licenses/by/4.0/.

**Figure 2.10 (a) and (b)** Photographed by the author.

**Figure 2.11** Published with permission from the Archives of the Museum of Comparative Zoology, Ernst Mayr Library, Harvard University.

**Figure 2.12** Image from the Geological Society Library, reference GSL/POR/50/21-1, with the permission of the Geological Society of London.

**Figure 2.13** From the frontispiece to Newbigin, M.I. & Flett, J.S. (1917) 'James Geikie, The Man and the Geologist'. Oliver and Boyd, Edinburgh. In compliance with the conditions specified by Project Gutenberg (http://www.gutenberg.net).

**Figure 2.14** From Figure 7 from Clements, D. (2012) The Geology of London. Geologists Association Guide no 68. With permission from the Geologists' Association.

**Figure 3.1** Extract from a photograph in Irons, J.C. (1896) 'Autobiographical Sketch of James Croll, with Memoirs of his Life and Work'. Edward Stanford, London. Out of copyright.

**Figure 3.2** From Croll, J. (1875) 'Climate and Time in their Geological Relations: a Theory of Secular Changes of the Earth's Climate'. Appleton and Co., New York; following p. 312. Out of copyright.

**Figure 4.1** Engraving by Ambroise Tardieu. Wikimedia – in the public domain.

**Figure 4.2** Photo by Daniel*D, File: Chamoix 2007 100 0022.jpg. Permission under the terms of the GNU Free Documentation License, Version 1.2 (Free Software Foundation).

**Figure 4.3** From Figure 6, in Bard, E., (2004) 'Greenhouse effect and ice ages: historical perspective'. Comptes Rendus Geoscicnce **336**. Reproduced with permission of Elsevier.

**Figure 4.4** Licensed under the Creative Commons Attribution-Share Alike 2.0 Generic license. Out of copyright

**Figure 4.5** Photograph from web site of Tyndall's old school, Leighlinbridge National School, County Carlow, Eire, courtesy of headmaster Mr John Threadgold. Also available at www.bing.com/. Out of copyright.

**Figure 4.6** From Figure 2 in Tyndall, J. (1864) 'The Absorption and Radiation of Heat by Gaseous and Liquid Matter'. London, Edinburgh and Dublin Philosophical Magazine and Journal of Science vol. XXVIII, 81–106, (reproduced as Figure 14 in Part V of Tyndall, J. (1872) Contributions to Molecular Physics in the Domain of Radiant Heat – a Series of Memoirs Published in the 'Philosophical Transactions' and 'Philosophical Magazine', with Additions. Longmans, Green and Co., London). Out of copyright.

**Figure 4.7** Extract from a copy by Tägtström of an original oil painting by R.Berg. Reproduced with permission from the Centre for History of Science of the Royal Swedish Academy of Science.

**Figure 4.8** Reproduced with permission from the Archives and Records Management Office, University of Wisconsin - Madison.

**Figure 4.9** Reproduced with permission from the Geological Survey of Austria, Vienna.

**Figure 5.1** Reproduced with permission from the Alfred Wegener Institute for Marine and Polar Research.

**Figure 5.2** From Figure 23 in Wegener, A. (1920) 'Die Entstehung der Kontinente und Ozeane'. 2nd ed., Die Wissenschaft, Band 66. Vieweg und Sohn, Braunschweig. Out of copyright.

**Figure 5.3** From Figure 4 in Köppen, W. and Wegener, A. (1924) 'Die Klimate der Geologischen Vorzeit'. Borntraeger, Berlin. Reproduced with permission from E. Schweizerbart'sche Verlagsbuchhandlung OHG.

**Figure 5.4** Reproduced with permission from the School of Geosciences, University of Edinburgh.

**Figure 5.5** From Figure 262 in Holmes, A. (1944) 'Principles of Physical Geology'. Thomas Nelson & Sons, London.

**Figure 5.6** Reproduced with permission from the Department of Geosciences, Princeton University.

**Figure 5.7** Reproduced with permission from the University of Toronto Mississauga archives.

**Figure 5.8** Courtesy of Geological Society of London at www.geolsoc.org.uk/Plate-Tectonics/Chap2-What-is-a-Plate. Open source Wikipedia.

**Figure 5.9** From Figure 2 in Meschede, M. (2016) 'Plate Tectonics'. In Harff, J., Meschede, M., Petersen, S., and Thiede, J., (eds.) Encyclopedia of Marine Geosciences. Springer, Dordrecht, 676–680. With permission from Springer.

**Figure 5.10** From Figure 1 in Meschede, M. (2016) 'Plate Tectonics'. In Harff, J., Meschede, M., Petersen, S., and Thiede, J., (eds.) Encyclopedia of Marine Geosciences. Springer, Dordrecht, 676–680. With permission from Springer.

**Figure 5.11** Reproduced with permission from the United States Geological Survey. Public domain.

**Figure 5.12** From Maps 31 and 16 in Smith, D.G., Smith, A.G., and Funnell, B.M. (1994) 'Atlas of Mesozoic and Cenozoic Coastlines'. Reproduced with permission from Cambridge University Press via PLS Clear.

**Figure 5.13** Prepared by Robert A Rohde and made available through Global Warming Art at http://commons.wikimedia.org/wiki/File: Phanerozoic_Sea_Level.png. Licensed under the Creative Commons Attribution-GNU Free documentation License 1.2.

**Figure 6.1** From Figure 8, in Köppen, W., and Wegener, A. (1924) 'Die Klimate der Geologischen Vorzeit'. Borntraeger, Berlin. Reproduced with permission from E. Schweizerbart'sche Verlagsbuchhandlung OHG.

**Figure 6.2** From Figure 4 in Lewis, J.M. (1996) Winds over the world sea: Maury and Köppen. Bulletin of the American Meteorological Society **77** (5), 935–952. Also at www.bing.com; assume out of copyright.

**Figure 6.3** From Köppen, W. (1918) 'Klassifikation der Klimate nach Temperatur, Niederschlag und Jahreslauf'. Petermann's Geographische Mitteilungen 64(9.12), 243–248. Supplied with permission by Walter Obermiller of Schweizerbart.

**Figure 6.4** From Figure 7 in Köppen, W. (1931) 'Grundriss der Klimakunde'. Walter de Gruyter and Co., Berlin and Leipzig. Out of copyright.

**Figure 6.5** From Figure 1 in Köppen, W. and Wegener, A. (1924) 'Die Klimate der geologischen Vorzeit'. Borntraeger, Berlin, 1–255. Reproduced with permission from E. Schweizerbart'sche Verlagsbuchhandlung OHG.

**Figure 6.6** From Figure 2 in Summerhayes, C.P. (1990) Palaeoclimates. J. Geol. Soc. London, **147**, 315–320, based on Figure 1 in Scotese, C.R., and Summerhayes, C.P. (1986) Computer model of paleoclimate predicts coastal upwelling in the Mesozoic and Cenozoic. Geobyte **1** (3), 28–44 and 94. Reproduced with the permission of the Geological Society of London.

**Figure 6.7** From the portrait by Paja Jovanovic, 1943, reproduced with the permission of John Imbrie.

**Figure 6.8** Based on Figure 48 in Milankovitch, M. (1941) 'Kanon der erdbestrahlung und seine anwendung auf das eiszeitenproblem, Special Publications **132**, Section of Mathematics and Natural Sciences **33**. Belgrade: Königliche Serbische Akademie, reproduced as Figure 3 in Grubic, A. (2006) The astronomic theory of climatic changes of Milutin Milankovitch. Episodes **29** (3), 197–203. Out of copyright.

**Figure 6.9** Provided by Prof. Saswati Bandyopadhyay, Geological Studies Unit, Indian Statistical Institute, Kolkata

**Figure 6.10** From Figure 1a in Robinson, P.L. (1973) Palaeoclimatology and continental drift. In Implications of Continental Drift to the Earth Sciences, Vol. 1 (eds. D.H. Tarling and S.K. Runcorn). Academic Press, London, pp 451–476. Reproduced with permission from Elsevier.

**Figure 6.11** Photograph provided by Judith Parrish.

**Figure. 6.12** From Figure 6 in Parrish, J.T., Bradshaw, M.T., Brakel, A.T., et al. (1996) 'Palaeoclimatology of Australia during the Pangean interval'. Palaeoclimates, **1**, 241–281 (Adapted from Parrish, J.T., and Curtis, R.L. (1982) 'Atmospheric circulation, upwelling and organic-rich rocks in the Mesozoic and Cenozoic'. Palaeogeography, Palaeoclimatology and Palaeoecology **40**, 31–66). Published with the permission of Taylor and Francis.

**Figure 6.13** From Figure 3, in Parrish, J.T., and Curtis, R.L. (1982) 'Atmospheric circulation, upwelling and organic-rich rocks in the Mesozoic and Cenozoic'. Palaeogeography, Palaeoclimatology and Palaeoecology **40**, 31–66). Published with the permission of Elsevier.

**Figure 6.14** Photograph supplied by Dame Jane Francis.

**Figure 6.15** From Figure 5A in Francis, J.E. and Poole, I. (2002) 'Cretaceous and early Tertiary climates of Antarctica: evidence from fossil wood'. Palaeogeography, Palaeoclimatology, Palaeoecology 182, 47–64. Published with the permission of Palaeogeography, Palaeoclimatology, Palaeoecology.

**Figure 6.16** Reproduced with the permission of the Ocean Drilling Program Office of Texas A & M University.

**Figure 6.17** From Figure 2.46a in Steffen, W., Sanderson, A., Tyson, P.D., et al. (2004) 'Global Change and the Earth System: A Planet Under Pressure'. Global Change – the IGBP Series. Springer, Berlin, 336pp. Published with permission of Springer.

**Figure 6.18** From Figure 4.1 in Crucifix, M. (2009) 'Modeling the Climate of the Holocene, in Natural Climate Variability and Global Warming: A Holocene Perspective'. Battarbee, R.W. and Binney, H.A. (eds.), Wiley-Blackwell, Chichester and Oxford, 98–122. Published with permission from WILEY.

**Figure 6.19** From Figure 1 in Crucifix, M., (2016) 'Tipping Ice Ages'. PAGES Magazine 24(1) 6–7. With permission from the PAGES Office.

**Figure 6.20** From Figure 6.20 in Steffen, W., Sanderson, A., Tyson, P.D., et al. (2004) 'Global Change and the Earth System: A Planet Under Pressure'. Global Change – the IGBP Series. Springer, Berlin, 336pp. With permission from Springer.

**Figure 7.1** Photograph supplied by Jim Kennett.

**Figure 7.2** Photograph by Robert Ginsburg. Reproduced with permission from the Archives of the Rosenstiel School of Marine and Atmospheric Science, University of Miami.

**Figure 7.3** From McCave, I.N., and Elderfield, H. (2011) 'Sir Nicholas, John Shackleton. 23 June 1937- 24 January 2006'. Biographical Memoirs of Fellows of the Royal Society. Pp 1–28. doi:10.1098/rsbm.2011.0005. Reproduced with the permission of the Godwin Laboratory, Cambridge University.

**Figure 7.4** With the permission of the Godwin Laboratory, University of Cambridge.

**Figure 7.5** Reproduced with the permission of Tony Phillips, British Antarctic Survey.

**Figure 7.6** Photograph supplied by Peter Barrett

**Figure 7.7** From Figure 8.12 in Francis, J.E., Marenssi, S., Levy, R., et al (2009) 'From Greenhouse to Icehouse – the Eocene/Oligocene in Antarctica'. In: Antarctic Climate Evolution (eds. F Florindo and M. Siegert). Developments in Earth and Environmental Sciences **8**. Elsevier, Amsterdam, pp 309–368. Published with the permission of Elsevier.

**Figure 7.8** Author photograph.

**Figure 7.9** From Figure 8.11 in Francis, J.E., Marenssi, S., Levy, R., et al. (2009) 'From Greenhouse to Icehouse – the Eocene/Oligocene in Antarctica, in Antarctic Climate Evolution', (eds F. Florindo and M. Siegert), Developments in Earth and Environmental Sciences 8, Elsevier, Amsterdam. 309–368. Published with the permission of Elsevier.

**Figure 7.10** From Figure 9, in Cramer, B.S., Miller, K.G., Barrett, P.J., and Wright, J.D. (2011) 'Late Cretaceous-Neogene trends in deep ocean temperature and continental ice volume: reconciling records of benthic foraminiferal geochemistry ($\partial^{18}O$ and Mg/Ca) with sea level history'. Journal of Geophysical Research **116**, C12023. Published with the permission of Wiley.

**Figure 7.11** Photograph supplied by Jörn Thiede.

**Figure 8.1** From Schmidt, G. (2010) 'Taking the measure of the Greenhouse Effect'. Science Briefs. Goddard Institute for Space Studies, NASA (https://www.giss.nasa.gov/research/briefs/schmidt_05). NASA public domain.

**Figure 8.2** From http://ozonedepletiontheory.info/what-is-radiation.html. Copyright Peter I. Ward. Permission supplied October 2019.

**Figure 8.3** Source unknown.

**Figure 8.4** Published with the permission of Special Collections and Archives, University of California, San Diego.

**Figure 8.5** Published with the permission of Special Collections and Archives, University of California, San Diego.

**Figure 8.6** Published with the permission of Special Collections and Archives, University of California, San Diego.

**Figure 8.7** Published with the permission of Scripps Institution of Oceanography.

**Figure 8.8** Published with the permission of Scripps Institution of Oceanography.

**Figure 8.9** Published with the permission of Scripps Institution of Oceanography.

Nature. This is an open access article distributed under the terms of the Creative Commons CC BY license.

**Figure 9.24** From Figure 2.22 in Torsvik, T.H., Cocks, L.R.M., (2017) 'Earth History and Palaeogeography', Cambridge Uni. Press. 317pp. With permission of Cambridge University Press and the authors.

**Figure 9.25** From Figure 24.5 in Vandenbroucke, T.R.A., Armstrong, H.A., Williams, M., et al. (2013) Chapter 24 'Late Ordovician zooplankton maps and the climate of the Early Palaeozoic Icehouse'. Geological Society of London, Memoirs 38, 399–405. With the permission of the Geological Society of London.

**Figure 10.1** Photograph supplied by Jim Kennett.

**Figure 10.2** From Figure 1 in Vaughan, A.P.M. (2007) 'Climate and geology – a Phanerozoic perspective'. In: Deep-Time Perspectives on Climate change: Marrying the Signal from Computer Models and Biological Proxies (eds. M. Williams, A.M. Haywood, F.J. Gregory, and D.N. Schmidt). Micropalaeontological Society Special Publication. The Geological Society of London, pp. 5–59. Reproduced with the permission of the Geological Society of London.

**Figure 10.3** From Figure 3 in Kidder, D.L., and Worsley, T.R. (2012) 'A human-induced hothouse climate'? Geology Today 22 (2), 4–11. With permission from WILEY.

**Figure 10.4** From Figure 3 in Müller, R.D., Dutkiewicz, A., Seton, M., and C. Gaina, C, (2013) 'Seawater chemistry driven by supercontinent assembly, breakup, and dispersal'. Geology, 41(8) 907–910. With the permission of the Geological Society of America.

**Figure 10.5** From Figure 3 in Foster, G.L., and Rohling, E.J. (2013) Relationship between sea level and climate forcing by $CO_2$ on geological timescales, Proceedings of the National Academy of Sciences. 110 (4). With the permission of the National Academy of Sciences.

**Figure 10.6** From Figure 1 in Van de Wal, R.S.W., De Boer, B., Lourens, L.J., et al (2011) 'Reconstruction of a continuous high-resolution $CO_2$ record over the past 20 million years'. Climate of the Past **7**, 1459–1469. Published with permission of Climate of the Past under the Creative Commons Attribution 3.0 License.

**Figure 10.7** From Figure 1 in Courtillot, V.E. and Renne, P.R. (2003) 'On the ages of flood basalt events'. C. R. Geoscience 335, 113–140. Published with permission of Elsevier France.

**Figure 10.8** From Figure 1 in Stordal, F., Svensen, H.H., Aarnes, I. and Roscher, M. (2017) Global temperature response to century-scale degassing from the Siberian Traps Large igneous province. Palaeogeography, Palaeoclimatology, Palaeoecology 471, 96–107.With permission from Elsevier.

**Figure 10.9** From Figure 1 in Benton M.J. (2018) 'Hyperthermal-driven mass extinctions: killing models during the Permian–Triassic mass extinction'. Philosophical Transactions of the Royal Society A 376: 20170076. With permission from the Royal Society under the terms of the Creative Commons Attribution License http://creativecommons.org/licenses/by/4.0/.

**Figure 10.10** From Figure 9(g) in Li et al. (2008) 'Assembly, configuration, and break-up history of Rodinia: A synthesis'. Precambrian Research 160, Issues 1–2, 179–21. With permission from Elsevier.

**Figure 10.11** From Figure 1 in Lenton, T.M, Boyle, R.A., Poulton, S.W., et al. (2014) 'Co-evolution of eukaryotes and ocean oxygenation in the Neoproterozoic era'. Nature Geoscience 7 (4), 257 – 265. With permission from Springer Nature.

**Figure 11.1** From Figure 6 in Barron, E.J., Fawcett, P.J., Peterson, W.H., et al (1995) 'A 'simulation' of mid-Cretaceous climate'. Paleoceanography **10** (2) 953–962. Published with permission of AGU/Wiley.

**Figure 11.2** From Figure 6 in Kent, D.V., and Muttoni, G., (2013) 'Modulation of late Cretaceous and Cenozoic climate by variable drawdown of atmospheric $pCO_2$ from weathering of basalt provinces on continents drifting through the equatorial humid belt'. Climate of the Past **9**, 525–546. Published with the permission of Climate of the Past under the Creative Commons Attribution 3.0 License.

**Figure 11.3** From Figure 7 in Kent, D.V., and Muttoni, G., (2013) 'Modulation of late Cretaceous and Cenozoic climate by variable drawdown of atmospheric $pCO_2$ from weathering of basalt provinces on continents drifting through the equatorial humid belt'. Climate of the Past **9**, 525–546. Published with the permission of Climate of the Past under the Creative Commons Attribution 3.0 License.

**Figure 11.4** From Figure 3 in Anagnostou, E., John, E.H., Edgar, K.M., (2016) 'Changing atmospheric $CO_2$ concentration was the primary driver of early Cenozoic climate'. Nature 533(7603), pp. 380–384. With permission from Springer Nature.

**Figure 11.5** From Figure 8.14 in Francis, J.E., Marenssi, S., Levy, R., Hambrey, M., et al (2009) 'From Greenhouse to Icehouse – The Eocene/Oligocene in Antarctica'. In F. Florindo and M. Siegert (eds.) Antarctic Climate Evolution, Developments in Earth and Environmental Sciences 8, Elsevier. With permission from Elsevier.

**Figure 11.6** From Figure 2 in Katz, M.E., Miller, K.G., Wright, J.D., et al. (2008) 'Stepwise transition from the Eocene greenhouse to the Oligocene icehouse' Nature Geoscience 1, 329–334. With permission from Springer Nature.

**Figure 11.7** From Figure 4 in Lauretano, V., Littler, K. Polling, M. Zachos, J.C. & Lourens, L.J. (2015) 'Frequency, magnitude and character of hyperthermal events at the onset of the Early Eocene Climatic Optimum'. Climate of the Past 11, 1313–1324. Published with the permission of Climate of the Past under the Creative Commons Attribution 3.0 License.

**Figure 11.8** From Figure 6.2 in Jansen, E., Overpeck, J., Briffa, K., et al. (2007) 'Paleoclimate'. Chapter 6 in the 4th Assessment Report of the IPCC. Published with the permission of the IPCC Secretariat.

**Figure 11.9** From Figure 2(d) in Frieling, J., Peterse, F., Lunt, D. J., et al. (2019) 'Widespread warming before and elevated barium burial during the Paleocene-Eocene Thermal Maximum: Evidence for methane hydrate release?' Paleoceanography and Paleoclimatology 34. With permission from AGU-WILEY.

**Figure 11.10** From Figure 1 in Holbourn, A., Kuhnt, W., Kochhann, K.G.D., et al., (2015) 'Global perturbation of the carbon cycle at the onset of the Miocene Climatic Optimum'. Geology 43 (2) 123–126. Published with the permission of the Geological Society of America.

**Figure 11.11** From Figure 7 in Holbourn, A.E., Kuhnt, W., Clemens, S.C., et al., (2013) 'Middle to late Miocene stepwise climate cooling: Evidence from a high resolution deep-water isotope curve spanning 8 million years'. Paleoceanography 28, 688–699. With the permission of AGU-WILEY.

**Figure 11.12** From Figure 10.25 Denton, G.H., Prentice, M.L., and Burckle, L.H., (1991) 'Cainozoic history of the Antarctic ice sheet'. In Tingey, R.J., (ed.) Geology of Antarctica. Oxford University Press, 365–433. Published with the permission of Oxford University Press.

**Figure 11.13** From Figure 14 in Sarnthein, M., Bartoli, G., Prange, M., et al. (2009) 'Mid-Pliocene shifts in ocean overturning circulation and the onset of Quaternary-style climates'. Climate of the Past **5**, 269–283. Published with the permission of Climate of the Past under the Creative Commons Attribution 3.0 License.

**Figure 12.1** Reproduced with the permission of André Berger.

**Figure 12.2** Reproduced with the permission of André Berger.

**Figure 12.3** Published with the permission of John Imbrie.

**Figure 12.4** From Figure 4 in Lisiecki, L.E., and Raymo, M.E. (2005) A Pliocene-Pleistocene stack of 57 globally distributed benthic $\partial^{18}O$ records. Paleoceanography **20**, PA1003. Published with the permission of AGU/Wiley.

**Figure 12.5** From Box 1 in Marshall, J., and Speer, K., (2012) 'Closure of the meridional overturning circulation through Southern Ocean upwelling'. Nature Geoscience 5, 171–180. With permission of Springer Nature.

**Figure 12.6** From Figure 1 in Marshall, J., and Speer, K., (2012) 'Closure of the meridional overturning circulation through Southern Ocean upwelling'. Nature Geoscience 5, 171–180. With permission of Springer Nature.

**Figure 12.7** From Figure 3 in Zeilhofer et al (2019) 'Western Mediterranean hydro-climatic consequences of Holocene ice-rafted debris (Bond) events'. Climate of the Past 15, 463–475. Climate of the Past Creative Commons 4.0, open access.

**Figure 13.1** From Figure 1 in Carlson, W.E., and Winsor, K. (2012) 'Northern Hemisphere ice-sheet responses to past climate warming'. Nature Geoscience **5** (9), 607–613. Reprinted with permission of Springer Nature.

**Figure 13.2** From Figure 2 in Zhang, Z., Yan,Q., Farmer, E.J et al. (2018) 'Instability of Northeast Siberian ice sheet during glacials'. Climate of the Past Discussions, https://doi.org/10.5194/cp-2018-79. Licensed under the Creative Commons Attribution 3.0 License.

**Figure 13.3** From Figure 1 in Patton, H., Hubbard, A., Andreassen, K., et al., (2017) Deglaciation of the Eurasian ice sheet complex. Quaternary Science Reviews 169,148–172. With permission from Elsevier.

**Figure 13.4** From Figure 13 in Patton, H., Hubbard, A., Andreassen, K., et al., (2017) 'Deglaciation of the Eurasian ice sheet complex'. Quaternary Science Reviews 169, 148–172. With permission from Elsevier.

**Figure 13.5** Reproduced with the permission of the Tyler Prize for Environmental Achievement.

**Figure 13.6** From Figure 1 in Ahn, J., and Brook, E.J. (2007) 'Atmospheric $CO_2$ and climate from 65 to 30 ka B.P'. Geophysical Research Letters 34 L10703, GL029551. Published with the permission of AGU/Wiley.

**Figure 13.7** Reproduced with the permission of the Tyler Prize for Environmental Achievement.

**Figure 13.8** From Figure 3 in Petit, J.R., Jouzel, J., Raynaud, D., et al (1999) 'Climate and atmospheric history of the past 420,000 years from the Vostok ice core, Antarctica'. Nature **399**, 429–436. Reprinted with the permission of Springer Nature.

**Figure 13.9**. From Figure 2 in Lüthi, D., Le Floch, M., Bereiter, B., et al (2008) 'High resolution carbon dioxide concentration record 650,000-800,000 years before present'. Nature **453**, 379–382. Reprinted with the permission of Springer Nature.

**Figure 13.10** Photographs supplied by Peter Clarkson.

**Figure 13.11** From Figure 1 in Buizert, C., Sigl, M., Severi, M., et al., (2018) 'Abrupt ice-age shifts in southern westerly winds and Antarctic climate forced from the north'. Nature 563, 681–685. Reprinted with the permission of Springer Nature.

**Figure 13.12** From Figure 3 in Brook, E., and Buizert, C., (2018) Antarctic and global climate history viewed from ice cores. Nature 558, 200–208. Reprinted with the permission of Springer Nature.

**Figure 13.13** From National Academies of Sciences, Engineering, and Medicine (NASEM) (2015). 'A Strategic Vision for NSF Investment in Antarctic and Southern Ocean Research'. With the permission of the US National Academy of Sciences.

**Figure 13.14** From Figure 7 in Saltzman, B., and Verbitsky, M. (1994) 'Late Pleistocene climate trajectory in the phase space of global ice, ocean state, and $CO_2$: observations and theory'. Paleoceanography 9, 767–779. With permission from AGU/WILEY.

**Figure 13.15** From Figure 5 in Beeman, J C., Gest, L., Parrenin, F., et al. (2019) 'Antarctic temperature and $CO_2$: near-synchrony yet variable phasing during the last deglaciation'. Climate of the Past 15, 913–926. Published with the permission of Climate of the Past under the Creative Commons Attribution 3.0 License.

**Figure 13.16** From Figure 2 in Stenni, B., Buiron, D., Frezzotti, M., et al., (2010) 'Expression of the bipolar see-saw in Antarctic climate records during the last deglaciation'. Nature Geoscience 4, 46–49. With the permission of Springer Nature.

**Figure 13.7** From Figure 3 in Pedro, J.B., Rasmussen, S.O., and Van Ommen, T.D (2012) 'Tightened constraints on the time-lag between Antarctic temperature and $CO_2$ during the last deglaciation'. Climate of the Past **8**, 1213–1221. Published with permission from Climate of the Past under the Creative Commons Attribution 3.0 License.

**Figure 13.18** From Figure 4.6.1 in Rintoul, S.R., Hughes, C.W., and Olbers, D., (2001) 'The Antarctic Circumpolar Current System'. In Siedler, G., Church, J., and Gould, J., (eds.) Ocean Circulation and climate; observing and modelling the global ocean. International Geophysics Series 77, 271–302, Academic Press. With the permission of Elsevier.

**Figure 13.19** From Figure 8 in Huybers, H. (2007) Glacial variability over the last two million years: an extended depth-derived age model, continuous obliquity pacing, and the Pleistocene progression, Quaternary Science Reviews 26, 37–55. With the permission of Elsevier.

**Figure 13.20** From Figure 1 in Huybers and Denton (2008) Antarctic temperature at orbital timescales controlled by local summer duration. Nature Geoscience 1, 787–792. With the permission of Springer Nature.

**Figure 13.21** From Figure 2 in Brook, E., and Buizert, C., (2018) Antarctic and global climate history viewed from ice cores. Nature 558, 200–208. With the permission of Springer Nature.

**Figure 13.22** From Figure 2 in WAIS Divide Project Members (2015) 'Precise interpolar phasing of abrupt climate change during the last ice age'. Nature 520, 661–665. With the permission of Springer Nature.

**Figure 13.23** From Box 2 in Brook, E., and Buizert, C., (2018) 'Antarctic and global climate history viewed from ice cores'. Nature 558, 200–208. With the permission of Springer Nature.

**Figure 14.1** From Figure 4.4 in Crucifix, M. (2009) 'Modeling the climate of the Holocene'. In Natural Climate Variability and Global Warming: A Holocene Perspective (eds R.W. Battarbee and H.A. Binney), Wiley-Blackwell, Chichester and Oxford, 98–122. With permission from WILEY.

**Figure 14.2** From Figure 7 in Johnsen, S. J., Dahl-Jensen, D., Gundestrup, N., et al. (2001) 'Oxygen isotope and palaeotemperature records from six Greenland ice-core stations: Camp Century, Dye-3, GRIP, GISP2, Renland and NorthGRIP'. Journal of Quaternary Science 16, 299–307. With permission from WILEY.

**Figure 14.3** From Figure 2 in Vinther, B.M., Buchardt, S.L., Clausen, H.B., et al (2009) Holocene thinning of the Greenland Ice Sheet. Nature **461**, 385–388. Reprinted with the permission of Springer Nature.

**Figure 14.4** From Figure 6 in Beer, J., and Wanner, H. (2012) 'Corrigendum to "Mid- to late Holocene climate change - an overview" [Quaternary Sci. Rev. 27 (2008) 1791–1828]'. Quaternary Science Reviews 51, 93–94. Published with the permission of Elsevier.

**Figure 14.5** From Figure 3 in Mulvaney, R., Abram, N.J., Hindmarsh, R.C.A., et al (2012) 'Recent Antarctic Peninsula warming relative to Holocene climate and ice-shelf history'. Nature 489, 141–145. Reprinted with the permission of Springer Nature.

**Figure 14.6** From Figure 3(a) in Renssen et al. (2012) 'Global characterization of the Holocene Thermal Maximum'. Quaternary Science Reviews 48, 7–19. With the permission of Elsevier.

**Figure 14.7** From Figure 4 in Crosta, X., Debret, M., Denis, D. et al (2007) 'Holocene long- and short-term climate change off Adélie Land, East Antarctic'. Geochemistry, Geophysics, Geosystems **8**(11), doi:10.1029/2007GC001718. Published with permission from AGU/Wiley.

**Figure 14.8** From Figure 5 in Waters, C.N., Zalasiewicz, J., Summerhayes C., et al., (2016) 'The Anthropocene is functionally and stratigraphically distinct from the Holocene'. Science 351 (6269), DOI:10.1126/science.aad2622 10pp. With the permission of the American Association for the Advancement of Scence.

**Figure 14.9** From Figure 2 in Studer, A.S., Sigman, D.M., Martínez-García, A., et al., (2018) 'Increased nutrient supply to the Southern Ocean during the Holocene and its implications for the pre-industrial atmospheric $CO_2$ rise'. Nature Geoscience. doi.org/10.1038/s41561-018-0191-. With the permission of Springer Nature.

**Figure 14.10** From Figure 10 in Broecker, W.S., Clark, E., McCorkle, D.C., et al., (1999) 'Evidence for a reduction in the carbonate ion content of the deep sea during the course of the Holocene'. Paleoceanography 14(6), 744–752. Published with the permission of AGU/Wiley.

**Figure 14.11** From Figure 12 in Broecker, W.S., Clark, E., McCorkle, D.C., et al., (1999) 'Evidence for a reduction in the carbonate ion content of the deep sea during the course of the Holocene'. Paleoceanography 14(6), 744–752. Published with the permission of AGU/Wiley.

**Figure 14.12** From Figure 3 in Ruddiman, W.F., Fuller, D.Q., Kutzbach, J.E., et al. (2016) 'Late Holocene climate: Natural or anthropogenic'? Reviews of Geophysics 54, doi:10.1002/2015RG000503. With the permission of WILEY.

**Figure 14.13** From Figure 2 in Ruddiman, W.F., Fuller, D.Q., Kutzbach, J.E., et al. (2016) 'Late Holocene climate: Natural or anthropogenic'? Reviews of Geophysics 54, doi:10.1002/2015RG000503. With the permission of WILEY.

**Figure 14.14** From Figure 1 in Zalasiewicz, J., Waters, C., Head, M.J., Poirier, C., Summerhayes, C.P., et al., (2019) 'A formal Anthropocene is compatible with but distinct from its diachronous anthropogenic counterparts: a response to W.F. Ruddiman's "three-flaws in defining a formal Anthropocene"'. Progress in Physical Geography. PPG-18–130. With permission from WILEY.

**Figure 14.15** From Figure 2a in Clark, P.U., Shakun, J.D., Marcott, S.A., et al., (2016) 'Consequences of twenty-first-century policy for multi-millennial climate and sea-level change'. Nature Climate Change 6, 360–369. With permission from Springer Nature.

**Figure 14.16** From Figure 4.5 in Crucifix, M. (2009) 'Modeling the climate of the Holocene'. In Natural Climate Variability and Global Warming: A Holocene Perspective (eds. R.W. Battarbee and H.A. Binney), Wiley-Blackwell, Chichester and Oxford, 98–122. With permission from WILEY.

**Figure 14.17** From Figure 4.6 in Crucifix, M. (2009) 'Modeling the climate of the Holocene'. In Natural Climate Variability and Global Warming: A Holocene Perspective (eds. R.W. Battarbee and H.A. Binney), Wiley-Blackwell, Chichester and Oxford, 98–122. With permission from WILEY.

**Figure 15.1** From Figure 1 in Kobashi, T., Kawamura, K., Severinghaus, J.P., et al. (2011) 'High variability of Greenland surface temperature over the past 4000 years estimated from trapped air in an ice core'. Geophysical Research Letters 38, L21501. doi:10.1029/2011GL049444. With permission of AGU/WILEY.

**Figure 15.2** From Figure 1 in Knudsen, M.F., Seidenkrantz, M.-S., Jacobsen, B.H., and Kuijpers, A., (2011) 'Tracking the Atlantic Multidecadal Oscillation through the last

8,000 years'. Nature Communications 2(1), 178. With permission of Springer Nature.

**Figure 15.3** From Figure 10, in Lamb, H.H. (1966) 'Britain's climate in the past'. In: The Changing Climate: Selected Papers by H.H. Lamb. Methuen, London, pp. 170–195. Published with the permission of Taylor and Francis.

**Figure 15.4** From Figure 7.1(c) in Folland, C.K., Karl, T.R., and Vinnikov, K.Y. (1990) 'Observed climate variations and change'. In: Report of Working Group 1, Climate change (eds. J.T. Houghton, G.J. Jenkins and J.J. Ephraums). The IPCC First Assessment Report. Cambridge University Press, Cambridge, pp 195–238. Published with the permission of the IPCC Secretariat.

**Figure 15.5** From Figure 4, in Ljungqvist, F.C., Krusic, P.J., Brattström, G., and Sundqvist, H.S. (2012) Northern Hemisphere temperature patterns in the last 12 centuries. Climate of the Past **8**, 227–249. Published with the permission of Climate of the Past under the Creative Commons Attribution 3.0 License.

**Figure 15.6** From Figure 5 in Wanner, H., Mercolli, L., Grosjean, M., and Ritzl, S.P. (2015) 'Holocene climate variability and change: a data-based review'. Journal of the Geological Society of London **172**, 251–253. Published with the permission of the Geological Society of London.

**Figure 15.7** From Figure 4 in Mayewski, P.A., and Maasch, K.A., (2006) 'Recent warming inconsistent with natural association between temperature and atmospheric circulation over the last 2000 years'. Climate of the Past Discussions **2**, 1–29. Published with permission from Climate of the Past under the Creative Commons Attribution 3.0 License.

**Figure 15.8** From Figure 8 in Stenni, B., Curran, M.A.J., Abram, N.J., et al., (2017) Antarctic climate variability on regional and continental scales over the last 2000 years. Climate of the Past 13, 1609–1634. Published with permission from Climate of the Past under the Creative Commons Attribution 3.0 License.

**Figure 15.9** From Figure 4 in Mulvaney, R., Abram, N.J., Hindmarsh, R.C.A., et al (2012) Recent Antarctic Peninsula warming relative to Holocene climate and ice-shelf history. Nature 489, 141–145. Reprinted with the permission of Springer Nature.

**Figure 15.10** From Figure 4 in Graf, W., Oerter, H., Reinwarth, O., et al,. (2002) 'Stable-isotope records from Dronning Maud Land, Antarctica.' Annals of Glaciology 35, 195–201. With the permission of the International Glaciological Society.

**Figure 15.11** From Figure 2 in Steinhilber, F., Abreu, J.A., Beer, J., et al (2012) '9,400 years of cosmic radiation and solar activity from ice cores and tree rings'. Proceedings of the National Academy of Sciences **109** (16), 5967–5971. Published with the permission of the National Academy of Sciences of the U.S.A.

**Figure 15.12** From Figure 3 in Steinhilber, F., Abreu, J.A., Beer, J., et al (2012) '9,400 years of cosmic radiation and solar activity from ice cores and tree rings'. Proceedings of the National Academy of Sciences **109** (16), 5967–5971. Published with the permission of the National Academy of Sciences of the U.S.A.

**Figure 15.13** From Figure 28 in Clette, F., Svalgaard, L., Vaquero, J.M., and Cliver, E.W. (2014) 'Revisiting the Sunspot number'. Space Science Review, DOI 10.1007/s11214-014-0074-2. With the permission of Springer.

**Figure 15.14** From Figure 5 in Cliver, E.W., (2017) 'Sunspot number recalibration: The ~1840–1920 anomaly in the observer normalization factors of the group sunspot number'. J. Space Weather and Space Climate 7 (A12). Open access journal for free download.

**Figure 15.15** From Figure 6, in Magny, M. (2007) 'West-Central Europe'. In: Lake Level Studies, Encyclopedia of Quaternary Science. Elsevier, Berlin, pp 1389–1399. Published with the permission of Elsevier and M Magny (author).

**Figure 15.16** From Figure 3 in Luterbacher, J., Werner, J.P., Smerdon, J.E., et al., (2016) 'European summer temperatures since Roman times'. Environmental Research Letters doi:10.1088/1748-9326/11/2/024001. With permission from IPO Publishing under the terms of the Creative Commons Attribution 3.0 licence.

**Figure 15.17** From Figure 4 in Kouwenberg, L., Wagner, R., Kurschner, W., and Visscher, H., (2005) 'Atmospheric $CO_2$ fluctuations during the last millennium reconstructed by stomatal frequency analysis of *Tsuga heterophylla* needles'. Geology 33 (1), 33–36. With permission from Geological Society of America.

**Figure 15.18** From Figure 6 in Ahn, J., Brook, E.J., Mitchell, L., et al., (2012) 'Atmospheric $CO_2$ over the last 1000 years: A high-resolution record from the West Antarctic Ice Sheet (WAIS) Divide ice core'. Global Biogeochemical Cycles 26, doi:10.1029/2011GB004247. With permission from WILEY.

**Figure 15.19** From Figure 3 in Sigl, M., Winstrup, M., McConnell, J.R., et al., (2015) 'Timing and climate forcing of volcanic eruptions for the past 2,500 years'. Nature 523, 543–549. With permission from Springer Nature.

**Figure 15.20** From Figure 9 in Sigl, M., Abram, N.J., Gabrieli, J., et al., (2018) '19th century glacier retreat in the Alps preceded the emergence of industrial black carbon deposition on high-alpine glaciers'. The Cryosphere 12, 3311–3331. With permission from the European Geophysical Union under creative commons attribution 4.0.

**Figure 15.21** From Figure 6 in Jevrejeva, S., Moor, J.C., Grinsted, A., et al., (2014) 'Trends and acceleration in global and regional sea levels since 1807'. Global and Planetary Change 113, 11–22. With permission from Elsevier.

**Figure 15.22** From Figure 6 in Leclercq, P.W., Oerlemans, J., and Cogley, J.G., (2011) 'Estimating the Glacier Contribution to Sea-Level Rise for the Period 1800–2005'. Surveys in Geophysics 32, 519–535. With permission of Springer.

**Figure 15.23** From Figure 3 in Bard, E., and Delague, G. (2008) 'Comment on "Are there connections between the Earth's magnetic field and climate?" by V Courtillot, Y. Gallet, J.-L. Le Mouël, F. Fluteau, A. Genevey, EPSL 253, 328, 2007'. Earth and Planetary Science Letters **265**, 302–307. Published with permission of Elsevier.

**Figure 15.24** Figure 16.14 in ACIA, 2005, Arctic Climate Impact Assessment. Cambridge University Press, Cambridge, 1042 pp. (also From Figure 6.1 in Wadhams, P. (2016) A Farewell To Ice: A Report from the Arctic. Allen Lane, Penguin Random House, UK, 240pp). With permission from Cambridge University Press via PLSClear.

**Figure 15.25** Photograph provided by Dorthe Dahl-Jensen.

**Figure 15.26** Photograph provided by Valérie Masson-Delmotte.

**Figure 15.27** From Figure 2 in Esper, J., Frank, D.C., Timonen, M., et al (2012) 'Orbital forcing of tree-ring data'. Nature Climate Change **2**, 862–866. Reprinted by permission of Springer Nature.

**Figure 15.28** From Figure 6a in Zemp, M., Frey, H., Gärtner-Roer, I., et al. (2015) 'Historically unprecedented global glacier decline in the early 21st century'. Journal of Glaciology 61 (228), 745–762, doi: 10.3189/2015JoG15J017. With the permission of the International Glaciological Society and M Zemp.

**Figure 15.29** From Figure 1 in Figure 1 in Lovejoy, S. (2015) 'Climate Closure'. EOS 96, doi:10.1029/2015EO037499. With permission from AGU/WILEY.

**Figure 15.30** From Figure 3 in Chen, X., and Tung, K.-K. (2018) 'Global surface warming enhanced by weak Atlantic overturning circulation'. Nature 559, 387–391. With permission from Springer Nature.

# Index

*Palaeoclimatology: From Snowball Earth to the Anthropocene*, First Edition. Colin P. Summerhayes.
© 2020 John Wiley & Sons Ltd. Published 2020 by John Wiley & Sons Ltd.